FOR REFERENCE

Do Not Take From This Room

THE OXFORD ENCYCLOPEDIA OF
THE HISTORY OF AMERICAN SCIENCE, MEDICINE, AND TECHNOLOGY

THE OXFORD ENCYCLOPEDIA OF
THE HISTORY OF AMERICAN SCIENCE, MEDICINE, AND TECHNOLOGY

Hugh Richard Slotten

EDITOR IN CHIEF

VOLUME 1

ABORTION DEBATES AND SCIENCE—LUMBERING

OXFORD

UNIVERSITY PRESS

OXFORD
UNIVERSITY PRESS

Oxford University Press is a department of the
University of Oxford. It furthers the University's objective
of excellence in research, scholarship, and education
by publishing worldwide.

Oxford   New York

Auckland   Cape Town   Dar es Salaam   Hong Kong   Karachi
Kuala Lumpur   Madrid   Melbourne   Mexico City   Nairobi
New Delhi   Shanghai   Taipei   Toronto

With offices in

Argentina   Austria   Brazil   Chile   Czech Republic   France   Greece
Guatemala   Hungary   Italy   Japan   Poland   Portugal   Singapore
South Korea   Switzerland   Thailand   Turkey   Ukraine   Vietnam

Oxford is a registered trade mark of Oxford University Press
in the UK and certain other countries.

Published in the United States of America by
Oxford University Press
198 Madison Avenue, New York, NY 10016
www.oup.com

The Library of Congress Cataloging-in-Publication Data
The Oxford encyclopedia of the history of American science,
medicine, and technology/Hugh Richard Slotten, editor in chief.
volumes   cm
ISBN 978-0-19-976666-6 (set: alk. paper)—
ISBN 978-0-19-020332-0 (vol. 1: alk. paper)—
ISBN 978-0-19-020333-7 (vol. 2: alk. paper)
1. Science—United States—History—Encyclopedias.   2. Medicine—United States—
History—Encyclopedias.   3. Technology—United States—History—Encyclopedias.
I. Slotten, Hugh Richard.   II. Title: Encyclopedia of the
history of American science, medicine, and technology.
Q127.U6O95   2014
509.7303—dc23   2014009198

9 8 7 6 5 4 3 2 1
Printed in the United States of America
on acid-free paper

EDITORIAL AND PRODUCTION STAFF

CONTENTS

LIST OF ENTRIES

INTRODUCTION

The Oxford Encyclopedia of American History carries on a tradition of scholarly publishing by Oxford University Press that began in 1478, 14 years before Christopher Columbus set out from Palos de la Frontera in Spain and sailed westward across the Atlantic. As for American history, Oxford's first contribution was John Smith's annotated *Map of Virginia*, published in 1612, five years after 104 English men and boys had founded the Jamestown settlement.

This tradition has continued over the centuries, including *The Oxford Companion to United States History* (2001), affectionately called "OCUSH," which I had the pleasure and privilege of editing along with a distinguished group of associates. To this long and noteworthy record, Oxford now proudly adds this new work. Building on the one-volume *Companion*, this twelve-volume set of six encyclopedias offers a far more capacious format, making it possible to include a greater number of entries and to provide more detailed, in-depth coverage than was possible in the briefer compass of the *Companion*.

The rationale for such a work is obvious. Awareness of, and interest in, our collective past is part of what makes us human, and the recording of history has characterized every great civilization. The case for this specific set of encyclopedias is equally compelling. From the time when the first human beings made their way from present-day Siberia across a now-vanished land bridge to what we call North America, the history of the peoples of this great geographic region has been rich in its diversity and interest. From the late sixteenth century onward, the various settlements in what Europeans called first the New World and then, after 1776, the United States have loomed large on the world stage—first in the realm of the imagination and then, increasingly, as unavoidable reality. The nation's political, social, economic, and cultural lives have always roused interest, sometimes as the target of criticism and ridicule, sometimes as an ideal to be envied and emulated. As Johann Wolfgang von Goethe wrote in an 1827 poem (in Daniel Platt's translation):

> *America, you've got it better*
> *Than our old continent. Exult!*
> *You have no decaying castles*
> *And no basalt.*
> *Your heart is not troubled,*
> *In lively pursuits,*
> *By useless old remembrance*
> *And empty disputes.*

The actual America of 1827, distracted by controversy over the tariff and by angry feuds within John Quincy Adams's unpopular and dysfunctional presidential administration, was, of course, irrelevant to Goethe. As others have done before and since, he was conjuring up a finer America of his dreams, a fresh, new society uncorrupted by the all-too-human flaws of tired, old Europe. (That strange reference to "basalt," some scholars believe, may be a metaphorical allusion to volcanoes as symbols of social unrest and revolutionary turmoil.) Of more substantive, less abstract interest has been America's global military, diplomatic, economic, and strategic role, which has grown steadily more significant and inescapable over the years. In short, for a variety of reasons, the case for a comprehensive reference work covering all of American history is clear.

But is such a work really needed in the twenty-first century, given the tsunami of "information" endlessly churning through the media, accessible at a keystroke? The answer is a decisive "Yes." This set of encyclopedias arose from a core assumption shared by the editors at Oxford University Press, by the gifted scholars constituting the editorial teams, and certainly by me. In an era in which the production and dissemination of knowledge have become wholly diffuse, often anonymous and unsourced, and frequently idiosyncratic and unreliable, we are convinced that there is still a place—indeed, a vital necessity—for a historical reference work that is authoritative, carefully planned, analytically sophisticated, written by scholars familiar with the latest findings, and carefully edited for accuracy, clarity, and readability.

Grounded in these basic principles, *The Oxford Encyclopedias of American History* are chronologically comprehensive, from the earliest human settlements to the twenty-first century, from the Stone Age to the Computer Age. The works are topically comprehensive as well. This is a daunting task, particularly because the field of American history has expanded dramatically in recent decades: the field has come to include much more extensive coverage of social, cultural, scientific, medical, and demographic topics than was true in the past. Recent scholarship has vastly broadened our knowledge of the historical experience of all Americans of the past, not just the elite figures who once dominated the story. This set of encyclopedias fully reflects this broader understanding of the American past.

HISTORY: AN ANCIENT DISCIPLINE, YET EVER NEW

History is among the oldest of the scholarly disciplines, tracing its origins to West African griots, ancient Chinese chroniclers, and Mesopotamians who combined the skills of historian, elocutionist, scribe, and stonemason, celebrating the exploits of their rulers in viva voce orations and on steles and tablets for future generations to transcribe and wonder at. The Greek historian Herodotus, who lived in the fifth century BCE, wrote an account of successive rulers of Persia, with whom Greece fought several wars. Filling his narrative with stories about the peoples and places he had heard about, he was an early practitioner of what is now called social and cultural history. Herodotus's younger contemporary Thucydides devoted his great talents to a history of the Peloponnesian War, the long conflict between Athens and Sparta. Focusing on battles and diplomacy and avoiding Herodotus's charming penchant for digression, Thucydides nevertheless recognized the importance of personality traits and social factors, such as the terrible plague that struck Athens in the midst of the war. In an authorial decision that would place him at odds with some participants in modern America's culture wars, Thucydides strictly excluded divine intervention from his causal forces, and he avoided suggesting that the gods favored one side or another in the conflict he was describing.

But history writing, this ancient pursuit, is also ever new, which helps account for its appeal to so many bright young scholars. It continually reinvents itself, not only because more things keep happening, but also because historians' understanding of their field, and of what merits their attention, is continually shifting and enlarging. In former times, the historian typically focused his attention (and it almost invariably was a "he") on politics, war, law, and diplomacy. Together with industrialization, finance, and other economic matters, these were considered history's driving forces, and thus the historian's proper subject matter.

But gradually a broader understanding emerged. George Bancroft, the prolific nineteenth-century historian, famously included a chapter on social history in his monumental 10-volume history of the United States (1834–1874). Henry Adams began his multivolume *History of the United States during the Administrations of Thomas Jefferson and James Madison* (1889–1891) with a chapter each on the "popular characteristics" of the American people, on "American ideals," and on the intellectual life of New England, the mid-Atlantic states, and the South. By offering new approaches to the history of ideas, Vernon L. Parrington's *Main Currents in American Thought* (1927) and Merle Curti's *The Growth of American Thought* (1943) helped create a new field, combining history and literature, that came to be known as American studies.

Later scholars have vastly expanded these initiatives. The twenty-first century's historians recognize that the history of social groups, class conflict, sexuality, popular culture, ideas and ideology, religious belief, literature, music, and the arts, as well as developments in science, medicine, technology, and the economy, all lie within their purview and merit careful attention and study. The traditional areas of politics, the law, military history, and foreign relations remain central, of course, but these fields, too, have been enriched by attention to the broader sociocultural context within which elections, legislative battles, legal decisions, wars, and diplomatic maneuverings unfold.

Indeed, all these subfields are interconnected. One cannot understand the social and military history of colonial America, for example, without understanding the devastating effects of the bacteria and viruses that the European newcomers unwittingly brought with them: lacking the immunity to these bacteria and viruses that comes with long exposure, Native peoples were decimated by disease. The history of slavery remains incomplete without attention to the changing technology of cotton production. One cannot grasp the full impact on America of World War I without also grasping the impact of the virulent influenza virus (the precise point of origin remains uncertain) that spread across the battlefields of Europe and home-front military bases, causing a global pandemic that took at least 50 million lives. The nation's military and political history is enriched by attention to the ways in which perceptions of wars, battles, elections, and political careers are shaped by the popular culture, from ballads and broadsides to cartoons, films, and television programs. Our picture of working-class America is enriched if we examine workers' families, leisure activities, religious lives, and consumption patterns, as well as their voting habits and participation in labor unions—important as voting patterns and unions obviously are. On and on the list could go, illustrating that American history, like all history, is really a seamless web, however we divide it into subfields for analytic and pedagogical purposes.

Along with the subject matter of history, the demographic profile of the profession of American history itself has changed. Once the domain of white males of British or northern European ancestry, in the twenty-first century the profession is far more diverse in terms of gender, race, and ethnicity. This has made for a livelier and more stimulating intellectual environment, as well as for a greater sensitivity to the diverse historical experience of the many subgroups within American society.

Historians' understanding of America's place in the world has evolved as well. In the past, U.S. historians tended to stress the uniqueness of the American experience. Underlying this view was the notion that America was not only different from, but also superior to, other nations, and that it was somehow immune to the historical processes that could be seen in the rise and decline of other once-powerful civilizations and nation-states, from Babylon, Greece, and Rome to Spain, Portugal,

Sweden, the Netherlands, and Great Britain. Sometimes, as in the writings of the New England Puritans, this view of America's uniqueness took on explicitly religious overtones. Even George Bancroft, a liberal Unitarian with a European education including a PhD from Germany's University of Göttingen, saw America as enjoying God's special favor. Such explicitly supernaturalist understandings of American history gradually faded, but notions of American exceptionalism survived, shaping historians' approach to their subject. This, too, has changed since the later twentieth century. U.S. historians now recognize that the forces shaping American history are hardly unique to the United States, but replicate processes also observable in other societies. This broader comparative dimension has both enriched American historical scholarship and added a chastening note of humility.

The heightened transnational awareness has not been only at the abstract level. American historians have vastly benefited from the work of foreign scholars, including the British historian E. P. Thompson, whose book on the making of the English working class was pathbreaking, and the French historians Lucien Febvre, Marc Bloch, Fernand Braudel, and Emmanuel Le Roy Ladurie, who did pioneering work in social history and founded the so-called *Annales* school, with its shift from providing a narrative of passing events to focusing on the *longue durée*, or long-term, processes that shape human societies over centuries.

Influential, too, has been the Italian historian Carlo Ginzburg, with his attention to the illuminating insights to be gained from "microhistory," the close analysis of specific events or small communities. Ginzburg's *Il formaggio e i vermi: Il cosmo di un mugnaio del Cinquecento* (1976; English trans., *The Cheese and the Worms: The Cosmos of a Sixteenth-Century Miller*, 1980), based on detailed Inquisition records that Ginzburg discovered in long-neglected archives of the Catholic Church, told of an obscure miller, Domenico Scandella, better known as Menocchio, burned at the stake in 1600 in the northeastern Italian town of Montereale for beliefs that he summed up this way:

I have said that in my opinion, all was chaos … and out of that bulk a mass

formed—just as cheese is made out of milk—and worms appeared in it, and these were the angels. There was also God, he too having been created out of that mass at the same time.

Such heresy shocked church authorities, and Menocchio went to his fiery death. In writing about this forgotten event in an obscure town, Ginzburg offered a wealth of fresh insights into the procedures and processes of the Inquisition, late medieval folk belief and oral culture, and the ramifications of the Counter-Reformation in Italy. Such scholarship proved appealing to American historians, who explored microhistories of their own to see what they would reveal. Laurel Thatcher Ulrich's Pulitzer Prize–winning *A Midwife's Tale* (1991) offered a biography of Martha Ballard, an obscure Maine midwife of the turn of the nineteenth century, based on Ballard's diary recording the unremarkable routines of her life. In the diary's "exhaustive, repetitive dailiness," Ulrich argued, lay its value as a historical document.

As this suggests, American historians' methodologies and sources have expanded as well. While retaining a strong sense of the value of their traditional text-based approaches, historians have learned from sociologists and political scientists the value of quantification in illuminating the past. Census records, voting data, tax lists, and many other forms of quantified information have yielded rich results. Such pathbreaking books using quantified data as Lee Benson's *The Concept of Jacksonian Democracy: New York as a Test Case* (1961) and Stephan Thernstrom's *Poverty and Progress: Social Mobility in a Nineteenth Century City* (1964) provided a means of testing cherished American myths about the "era of the common man," "rags to riches," and the United States as a "land of opportunity"—myths reinforced by Horatio Alger stories and the oft-told biographies of exceptional figures like Andrew Carnegie. (That Thernstrom's book was pathbreaking is indicated by its being relegated by the *American Historical Review* to a brief notice in a section called "Other Recent Publications" and by its being overlooked entirely by the *Journal of American History*, whose editor assumed that it was a work of sociology.)

Fully aware that statistically based findings must be interpreted for and presented in prose accessible to people who are not mathematicians, historians are in little danger of becoming computing automatons, spewing out numbers, statistics, and formulas. And the profession has learned that statistical data, like any other form of evidence, can be fudged and misrepresented and thus must be approached with due skepticism. Nevertheless, American historians have grown more conscious of the rich resources buried in the nation's vast repositories of quantified data. So, too, have they discovered the research value of material-culture products, visual materials, industrial designs, fashion shows, television commercials, and mundane everyday objects from toys, comic books, tombstones, and toilets to domestic structures. Approached with insight and imagination, such materials are invaluable historical sources.

To be sure, all this ferment and change within the profession of American history has raised troublesome questions. If historians focus too exclusively on subgroups within society, how can they formulate a larger conception of *American* history? Put another way, what meaning does the familiar phrase "one nation indivisible" from the original Pledge of Allegiance to the flag ("under God" was inserted later) retain in the twenty-first century, apart from the banal point that a second secession movement like that of 1861 would be unlikely to succeed? Or, again, if historians adopt a snapshot approach, documenting the experience of specific groups during one brief period of their history, how can they explain change over time? Other questions have been raised as well. As historians, from wholly laudable motives, turned their attention to the poor, disenfranchised, and disinherited—doing "history from the bottom up," as it was called—did they risk losing sight of the important role of powerful elites and interest groups in society? In abandoning the notion of American exceptionalism as self-serving and parochial, did they risk missing what might, in fact, be distinctive and unique about the American experience? At the practical level, could historians successfully integrate quantified statistical data and other new forms of evidence with the written sources upon which they have traditionally relied? Such questions have for decades energized the profession of American history, and echoes of these questions will be found in the pages of this set of encyclopedias.

THE OXFORD ENCYCLOPEDIAS OF AMERICAN HISTORY: A REFERENCE WORK FOR THE TWENTY-FIRST CENTURY

All these trends—the greater diversity of the profession of American history, historians' broader understanding of the discipline's scope and available range of sources and methodologies, the greater openness to multicultural and transnational perspectives—have informed the planning and execution of *The Oxford Encyclopedias of American History*. This becomes immediately evident as one looks at the basic structure of the works. The effort to bring organizational and intellectual coherence to such a vast and unwieldy topic has been formidable indeed. We have been reminded more than once of Immanuel Kant's much-quoted observation: "Out of the crooked timber of humanity, no straight thing was ever made." Yet we have given it a try—with, we hope, some success. To make this set of encyclopedias of maximum benefit to its users, we have divided the sprawling, almost boundless field of American history into six topical subcategories, each comprising two volumes and each with its own editor or editors in chief. Here are these six subcategories, with the name of the editor or editors in chief and his or her academic affiliation:

The Oxford Encyclopedia of American Business, Labor, and Economic History
Editor in Chief
Melvyn Dubofsky
Binghamton University, State University of New York

The Oxford Encyclopedia of American Cultural and Intellectual History
Editors in Chief
Joan Shelley Rubin
University of Rochester

Scott E. Casper
University of Nevada, Reno

The Oxford Encyclopedia of American Military and Diplomatic History
Editor in Chief
Timothy J. Lynch
University of Melbourne

The Oxford Encyclopedia of American Political and Legal History
Editors in Chief
Donald T. Critchlow
Arizona State University

Philip R. VanderMeer
Arizona State University

The Oxford Encyclopedia of the History of American Science, Medicine, and Technology
Editor in Chief
Hugh Richard Slotten
University of Otago

The Oxford Encyclopedia of American Social History
Editor in Chief
Lynn Dumenil
Occidental College

These editors, distinguished scholars all, have been assisted by teams of associate editors, ranging from prominent senior scholars to gifted younger scholars. Through extended discussions, these teams carefully formulated their entry lists to ensure comprehensive coverage of the core topics in their field, supplemented by mid-length and shorter entries on topics of interest. The editors then solicited entries from historians who had relevant expertise. This great project would be impossible without the participation of hundreds of scholars who welcomed the opportunity to offer an account of their work to a broader public, beyond the circle of col-

leagues who routinely review their books, evaluate their research proposals, and share panels at scholarly meetings. The specialization and compartmentalization of the profession of history can be rewarding, as one exchanges ideas with others who share one's specific interests, but it can also seem confining and a bit claustrophobic. Participation in a large-scale collaborative project like *The Oxford Encyclopedias of American History* provides an opportunity to move beyond one's professional comfort zone and to share one's research with an interested worldwide readership.

Incorporating the most recent research and interpretive approaches, the entries are designed to be of value to scholars, advanced students, and all those seeking high-level analytic treatment of key topics in American history. The entries include essential facts, of course—names, dates, and so on—but they go further, contextualizing the topic, conveying its human dimensions, suggesting its larger historical significance, and addressing matters of controversy or interpretive disagreement.

Along with matters of content, methodology, and scope, we have given high priority to the entries' organization and style. Historians have long hovered ambivalently between the social sciences and the humanities, unwilling to abandon the social scientists' commitment to methodological rigor, but also determined to retain the humanities' concern for clarity and persuasiveness of presentation. For all the changes since the days of Herodotus and Thucydides—and Edward Gibbon, Voltaire, Francis Parkman, George Bancroft, Henry Adams, and scores of other historians admired for their writing style—historians still take pride in writing lucid, accessible, even elegant prose. They pay attention not only to the information being conveyed, but also to the manner in which it is conveyed.

In choosing our contributors, describing their assignments, and editing their entries, therefore, we have emphasized style as well as content. Lewis Mumford once rather cattily observed that reading John Dewey was like riding on the Boston subway: one eventually got to one's destination, but considerably the worse for wear. We have made every effort to ensure that such a criticism will not be leveled against the contributors to this set of encyclopedias. Acquiring historical knowledge should not be the intellectual equivalent of a visit to the dentist.

USING THE ENCYCLOPEDIA

Taken together, the 12 volumes of *The Oxford Encyclopedias of American History* include more than 3,500 entries, arranged in alphabetical order within each two-volume topical set, for ease of research. Composite entries gather together discussions of similar or related topics under one entry title. For example, under the entry "Internal Migration" in *The Oxford Encyclopedia of American Social History*, the reader will find four subentries: "Colonial Era," "Nineteenth-Century Westward," "Twentieth Century and Beyond," and "African Americans." A headnote listing the various subentries introduces each composite entry.

A selective bibliography at the end of each entry directs the reader who wishes to pursue a topic in greater detail to the most important recent scholarly work. Most entries include references at the end that guide interested readers to related entries. Blind entries direct the user from an alternate form of an entry title to the entry itself. For example, the blind entry "Mormon Rebellion" tells the reader to look under "Utah War of 1857."

Throughout the planning and execution of this work we have been guided by a single goal: to make the full panorama of American history accessible and understandable to the maximum number of users. We hope that we have succeeded in this purpose.

Finally, the editors would like to extend sincere thanks to the professionals at Oxford University Press who aided immeasurably in the preparation of this work: Grace Labatt, who patiently and cheerfully oversaw the process of selecting the editors in chief; Stephen Wagley, the then-executive editor, who shared his wisdom and experience and saw the project successfully launched; Damon Zucca, the publisher, and Alixandra Gould, the executive editor, who aided us in thinking about the emerging world of online reference; and Eric Stannard and Andrew Jung, the developmental editors, who with unfailing good humor moderated early discussions among the editors over where specific entries should go, dealt with problems as they arose, and gently prodded us to maintain the pace. Our thanks to one and all.

Paul S. Boyer
Merle Curti Professor of History Emeritus,
University of Wisconsin-Madison
July 2011

PREFACE

Science, medicine, and technology have become increasingly important to all Americans. The importance of these three fields is in many ways one of the defining characteristics of modernity. Understanding their history is essential for educated individuals. Science, medicine, and technology are not static endeavors but processes, bodies of knowledge, tools, and techniques that are constantly growing and changing and must be understood in their proper historical context. The essays in this encyclopedia explore the changing character of science, medicine, and technology in the United States; the key individuals, institutions, and organizations responsible for major developments; and the concepts, practices, and processes underlying these changes. Especially since the early decades of the twentieth century, U.S. science, medicine, and technology have played dominant roles internationally. Essays in this two-volume work explore characteristics of American institutions and culture that help explain this development. At the same time, the encyclopedia situates specific events, theories, practices, and institutions in their historical context and explores their impact on American society and culture.

STRATEGY

This project was a mammoth undertaking and presented a number of major challenges to the Board of Editors. An important consideration in the planning for this encyclopedia was to try to maintain an approximate balance among science, medicine, and technology, especially in terms of the total number of words devoted to each subject. But we also encouraged authors to take into account important connections among all three historical subjects. For example, especially since biomedical research increasingly bridges science, medicine, and technology (including biotechnology and genetic engineering), the long, in-depth essay on "Research and Development (R&D)" necessarily deals with all three subjects.

The *Oxford Encyclopedia of the History of American Science, Medicine, and Technology* includes different categories of essays. A large number of the shorter essays are biographical entries and were originally published in the *Oxford Companion to United States History* (edited by Paul Boyer, 2001). All essays from the *Companion*, however, including the shorter biographical ones, were updated and in almost all cases extended, either by the author or by another expert, often a member of the editorial board. In particular, the editorial board recognized the importance of including the most recent scholarship in revised bibliographies. The vast majority of the material in this encyclopedia is new. The many essays specifically commissioned for this work from top scholars in their respective fields provide comprehensive coverage of the important themes in the history of American science, medicine, and technology. Although some readers will undoubtedly disagree with decisions made about the content of the encyclopedia, it is important to recognize that these decisions were judgments made by members of the Board of Editors who are experts in their respective areas of research.

Medium-length essays provide more in-depth analysis of central topics. Most of the essays on scientific and medical disciplines fall into this category. Disciplines are, of course, central to science, medicine, and technology, but they also are relatively modern creations. We instructed authors to avoid "presentism" by placing developments in their proper context and by paying attention to the use of terms has changed over time.

Essays selected for longer in-depth analysis (some over 10,000 words) include topics examining complex relationships such as "Gender and Technology," "Foundations and Health," "Law and Science," "War and Medicine," and "Literature and Science." Although essays have retained the accessible style of the original *Oxford Companion to United States History*, the longer essays in particular will especially appeal to advanced students and other scholars interested in an extended treatment of major themes, including an exploration of historiographic disagreements. The encyclopedia offers thorough coverage in core and emerging areas of research to provide students, other researchers, and general readers with an authoritative source of information and analysis by experts that are not covered by existing reference works.

USING THE ENCYCLOPEDIA

The *Oxford Encyclopedia of the History of American Science, Medicine, and Technology* includes nearly five hundred entries arranged alphabetically in two volumes. A headnote listing subentries is included with each broad, composite entry. The encyclopedia is designed to be easy to use. Authors have been encouraged to minimize the use of technical language and to explain complex issues clearly.

Readers will find a selective bibliography at the end of each essay that lists the most up-to-date scholarly sources on the subject. In many cases, annotations are included in bibliographies, especially for the important sources on a given topic. Cross-references can be found at the end of each essay. These "see also" lists direct readers from one entry to other related essays in the encyclopedia. Blind entries direct the reader from an alternate version of an entry title to the entry itself. The beginning of volume one contains the table of contents, the list of entries, the introduction, and the preface. At the end of volume two, the user can find a topical outline of entries providing an overview of the general subject categories and where entries fit within these broad categories; a directory of all contributors; and a comprehensive index, including not only all the entry titles covered in the encyclopedia but also many other related topics.

Acknowledgments

Many people assisted with this *Encyclopedia*, and I am very grateful for all the help I received. I was especially fortunate to have the generous support of Ron Numbers, who also served as a senior editor with the *Oxford Companion to United States History*. His broad expertise and deep knowledge of the profession were indispensable. I was also very honored to work with Charles Rosenberg and Sue Lederer who assisted especially with history of medicine entries; Steve Usselman (history of technology); Gregory Good (history of the physical sciences); Karen Rader, Marga Vicedo, and Constance Clark (history of the biological sciences); and Mark Solovey (history of the social sciences). A grant from the Humanities Division at the University of Otago supported the work of assistant editor Elspeth Knewstubb. Numerous professional colleagues gave advice and counsel. Although there are too many people to list here, I would like to take this opportunity to especially thank John Servos for his valuable assistance.

I enjoyed working with all the members of the Board of Editors and very much appreciated their patience and good humor as we all struggled to finish this huge project. At Oxford University Press, I would like to thank Damon Zucca, the publisher, and Alixandra Gould, the executive editor, for supporting and pushing forward the American History series; and the two development editors, first Eric Stannard and then Andrew Jung, who supervised and facilitated the work for this encyclopedia. Their organizational skills were truly amazing, and they handled their responsibilities with professional courtesy and patience.

Finally, I would like to acknowledge the excellent assistance of the late Paul Boyer, the series editor for the *Oxford Encyclopedias of American History* series (of which, this project was one of the six titles), and pay tribute to his work in the field. He was extremely helpful, especially during the early stages of this project when I was working on both this encyclopedia and another major research project. It was a tremendous honor to have had the opportunity to work with a master historian who made so many major contributions to the profession. Everyone involved with this project was very sad that he did not live to see the publication of these volumes.

Hugh Richard Slotten

COMMON ABBREVIATIONS USED IN THIS WORK

AD   *anno Domini,* in the year of the Lord

AH   *anno Hegirae,* in the year of the Hajj

b.   born

BCE   before the common era (= BC)

ca.   *circa,* about, approximately

CE   common era (= AD)

cf.   *confer,* compare

d.   died

diss.   dissertation

ed.   editor (pl., eds.), edition

f.   and following (pl., ff.)

fl.   *floruit,* flourished

l.   line (pl., ll.)

n.   note

n.d.   no date

no.   number

n.p.   no place

n.s.   new series

p.   page (pl., pp.)

pt.   part

rev.   revised

ser.   series

supp.   supplement

USSR   Union of Soviet Socialist Republics

Vol.   Volume (pl., Vols.)

THE OXFORD ENCYCLOPEDIA OF
THE HISTORY OF AMERICAN SCIENCE, MEDICINE, AND TECHNOLOGY

A

ABORTION DEBATES AND SCIENCE

The longstanding abortion conflict in the United States (U.S.) is generally understood as a debate between religious-based morality claims about the fetus (pro-life) and a combination of rights-based claims about a woman's bodily integrity (pro-choice) and public-health claims about the benefits of safe and legal abortions. Indeed much scholarship has addressed these perspectives. However, competing scientific claims have been central to legal and political struggles over abortion in the United States. Since the Supreme Court recognized a right to abortion in its 1973 *Roe v. Wade* decision (*Roe*) [410 U.S. 113 (1973)], abortion rights opponents have increasingly used scientific arguments—in particular about the alleged harm to women receiving abortions—to advocate for increased regulation. This essay examines and places into historical context these science-based arguments and their counterclaims by abortion rights supporters in the United States. The analysis focuses especially on the shifting role of medical and scientific authority in the political and legal debates involving abortion.

Public Health and the Need for Legal Abortion. Abortion was criminalized in the late 1800s in response to an aggressive campaign by the recently organized American Medical Association (AMA) to control both women's behavior and the practices of midwives and other health-care providers who were the routine providers of abortion services (Petchesky, 1990; Mohr, 1978; Luker, 1984; Smith-Rosenberg, 1985). The AMA's abortion campaign can be usefully understood as a key component of a larger battle underway by the organization at that time: the attempt by "regular" (university-trained and allopathic) physicians to attain professional dominance over the wide range of "irregular" health practitioners then in practice. The AMA's chief argument was that abortion was both immoral and unsafe in the hands of any other than appropriately trained regular physicians. As the historian Carroll Smith-Rosenberg (1985) has argued, the campaign to criminalize abortion in the late nineteenth century—which until that time had been unregulated in American society— was led by physicians, but concerns over the declining birth rates of Anglo-Saxon women, and the corresponding rise in births by immigrants from southern and eastern Europe, drew support from other sectors of society as well, including the press. In sharp contrast to what was to occur after the *Roe* decision, religious groups, including the Catholic Church, played merely a supporting role in this campaign and not a leadership one. Although scientific expertise featured heavily in physicians' arguments distinguishing themselves from other providers of health services, except for some very primitive discussions of the fetus, scientific explanations played a minor role in the anti-abortion campaign of that era. By the early 1870s, every state had passed statutes criminalizing abortion—except under narrow conditions to be determined by regular physicians.

However, for the next century, despite criminalization, millions of women had unsafe abortions either at the hands of an illegal provider or from having induced abortion themselves (Hall, 1970). A relatively small number of women with serious or life-threatening health conditions received physician-approved in-hospital abortions, and some women, to be sure, received illegal abortions from competent physicians who were acting as a matter of conscience (Joffe, 1995). Health consequences were severe during this "century of criminalization," especially for unmarried and low-income women (Reagan, 1997), with thousands of women dying and many more experiencing serious injuries (Joffe, 1995). Efforts to control this illegal behavior by implementing hospital abortion committees and publicly shaming the practice of abortion led to even higher rates of death and serious injury necessitating organ removal or lengthy hospital stays. Concerned for their patients as well as their ability to practice medicine in accordance with their desire to reduce harm by offering safe services, physicians began to advocate for abortion law reform in the mid-1960s (Joffe, 1995; Garrow, 1998).

Central to physicians' arguments was the need to legalize abortion to promote public health by reducing death and serious injury. As a result of their advocacy, several states reformed their abortion laws, allowing more women access to legal abortion services as provided upon physician approval. The AMA, reversing its century-old opposition to abortion, voted for legalization at its 1970 annual meeting (Garrow, 1998). Rates of hospitalization for women suffering the medical consequences of an unsafe abortion began to drop and, after the full legalization of abortion under *Roe*, the number of abortion-related deaths dropped to single digits almost overnight (Tietze, 1975). Studies conducted by the Centers for Disease Control and Prevention found the medical risks from legal abortion to be significantly less than those of a full-term delivery (Cates et al., 1977). According to abortion rights advocates, the public-health implications of illegal abortions remain the strongest argument for why abortion should be legal. In the early twenty-first century, across the globe where abortion was illegal, it was overwhelmingly unsafe and almost 70,000 women died each year as a consequence of unsafe abortion (Grimes et al., 2006).

Legal Protection for Abortion Rights in the United States. The *Roe* decision recognized abortion as a fundamental right for women upon which the government could not intrude until the point when the fetus could potentially live outside the woman's body (called viability), at approximately 28 weeks of pregnancy or the beginning of the third trimester. After that point, states could choose to limit abortion rights but they were obligated to include an exception to protect the health and life of the pregnant woman.

Immediately after the *Roe* announcement, a movement mobilized in opposition to abortion rights (called the pro-life movement), this time not led by physicians but initially by forces within the Catholic Church. Conservative political groups, some religiously affiliated (particularly groups of Protestant evangelicals) and some not, joined the opponents of abortion rights later. Those who supported legal abortion (called the pro-choice movement) grew complacent after the Supreme Court decision and have essentially been playing on defensive territory ever since the *Roe* victory (Staggenborg, 1991).

The first public fights over abortion after *Roe* revolved around efforts to pass a human life amendment to the U.S. Constitution that would grant the fetus the right to life from the moment of conception and thus reverse the *Roe* decision. In part because of the strength of the pro-choice movement's lobby in Congress in the mid-1970s—a strength that was to diminish markedly in the years ahead, with the election of Ronald Reagan in 1980—and because of the expansive potential of a human life amendment to outlaw not only abortion but also many forms of contraception, efforts to pass a human life amendment were ultimately unsuccessful. However, with the growth of a Religious Right, which had opposition to abortion as its centerpiece, and the formation of organizations opposed to abortion such as the Moral Majority, the Eagle Forum (led by Phyllis Schlafly), and the National Right to Life Committee, which established a national network of state chapters, abortion opposition shifted to a state-based incrementalist strategy to pass laws to restrict some aspects of the abortion right.

Prior to 1992, U.S. courts struck down restrictions such as waiting periods and mandatory information as violating the tenets of *Roe*, which set out abortion as a fundamental right. After several decades of conservative restructuring, the Supreme Court eventually found in favor of these types of laws, defining a new "undue burden" standard for abortion regulation [*Planned Parenthood of SE. Pa. v. Casey*, 505 U.S. 833 (1992) (*Casey*)]. Under *Casey*, through 2013, the courts continued to uphold the constitutionality of waiting periods and mandatory counseling laws that sought to persuade women against the abortion decision.

Mobilizing Scientific Arguments in the Name of Protecting Women.

Prior to *Roe* and in the period immediately following the decision, the major arguments of abortion opponents were couched primarily in moral terms, with, as mentioned, a greatly increased role of the Catholic Church (and later, Protestant denominations)

after legalization. However, by the 1990s and beyond, encouraged by the *Casey* provision allowing the regulation of abortion, opponents crafted a new body of ideologically produced research aimed at supporting two general claims against abortion that would justify further efforts to restrict access and allow the Supreme Court to overturn *Roe*. The first focused on the physical harms to women from abortions and the second on the psychological harms. In both efforts, abortion rights opponents sought to gain the public-health mantle for their position, a domain that had long been associated with abortion rights supporters.

Abortion and women's physical health. Regardless of the consensus within the public health and medical communities that legal abortion contributes to improvements in women's health, beginning after the *Roe* decision, opponents of abortion rights began to argue that abortions could actually harm women's physical health. The most successful of these claims involved the purported link between abortion and breast cancer. Although most medical and scientific experts have rejected these arguments, as late as 2013 five states required women be told of the risks of breast cancer before obtaining an abortion (Guttmacher Institute, 2014). During the second decade of the twenty-first century, this argument remained a leading significant scientific justification for making abortion illegal.

The abortion and breast cancer story began almost immediately after the *Roe* decision. Although early studies in the 1950s of the contribution of reproductive factors to breast cancer had noted a potential effect of abortion, the idea was given little scientific attention until the 1970s. Two scientists with religious-based opposition to abortion, Jose and Irma Russo, sought to research the potential risk of breast cancer resulting from abortion. Using rat models, they found that abortion disrupts the breast change occurring in pregnancy and thus contributes to breast cancer risk (Russo and Russo, 1980). Throughout the 1980s, numerous other studies furthered this hypothesis. Most studies finding a positive relationship did so by comparing women who had breast cancer with women who did not have breast cancer. Although these studies found higher reported rates of prior abortion among women who had breast cancer, other studies found no relationship between legal abortion and current rates of breast cancer using population-level analytic models.

In 1994, the debate entered the public arena as a result of a study published in the *Journal of the National Cancer Institute*. This study, like most of the previous ones on the topic, involved interviewing women with and without breast cancer about their life histories. Using this method, Daling et al. (1994) found that women with a history of abortion had a 50 percent higher risk of developing breast cancer than those without a history of abortion. The accompanying editorial pointed out the significant limitations of epidemiological studies using this method, in particular highlighting the limits of "recall bias" in which women with breast cancer are more likely to disclose having had an abortion than women not suffering from a cancer diagnosis because women with the diagnosis are seeking an explanation for their disease and are likely to recall many more events in the past. Despite this critique, abortion opponents accepted this study as "scientific fact." The most vocal advocate for the abortion-breast cancer link was Joel Brind, PhD, a professor of biochemistry at Baruch College in New York City. Attributing his new research interest to his conversion to Catholicism, Dr. Brind sought to use his scientific expertise to help the anti-abortion cause (Mooney, 2005). In 1996 he published a summary of the existing studies in which he concluded that abortion contributed to at least a 30 percent chance of increased breast cancer (Brind et al., 1996). Not long after, in 1997, the *New England Journal of Medicine*, one of the most prestigious medical journals in the United States, published the first study on the topic in which data were collected from women over time specifically to answer the question of what causes breast cancer. This study found no relationship between abortion and breast cancer. The authors explicitly criticized Brind for his prior work, noting his manipulation of poor-quality studies to create the appearance of scientific certainty (Melbye et al., 1997).

Regardless of this new evidence and the rejection of the position by the majority of the scientific community, the idea that abortion caused breast cancer gained ground in the public arena (Jasen, 2005). Brind and his supporters founded the Breast Cancer Prevention Institute to train spokespersons on the relationship between abortion and breast cancer or, as they referred to it, the "ABC link"; their work came to be supported by other organizations such as the Coalition on Abortion/Breast Cancer, founded with support from Concerned Women for America, a socially conservative organization dedicated to overturning *Roe*. Brind and his supporters, with backing from then-Representative Tom Coburn, pressured the National Cancer Institute to modify its web page in 1999 to say that the evidence on the subject was "inconsistent." Then under further pressure, this time from the George W. Bush presidential administration, the National Cancer Institute again revised its website to suggest an association between abortion and breast cancer (Mooney, 2005).

Significant controversy arose from this revision; under pressure from scientists and abortion rights supporters, in 2003, the National Cancer Institute removed the web page and scheduled a scientific conference to review the issue. The conference, which was attended by many of the researchers who previously had found a relationship in retrospective studies, concluded that there was no credible evidence of an association between breast cancer and abortion (National Cancer Institute, 2003). The following year, in 2004, the Collaborative Group on Hormonal Factors in Breast Cancer published in the *Lancet*, another premier medical journal, the results of a massive reanalysis of available data that again failed to find a link between abortion and breast cancer (Beral et al., 2004). Yet although this issue had essentially been resolved in the scientific community, it remained vibrant in the abortion wars of the early twenty-first century, as evidenced by those states requiring patients to be told of this link and the number of anti-abortion websites that included such information.

Abortion and women's mental health. The claim that abortion harms women's mental health dates to the early 1980s when abortion opponents began promoting a concept of a posttraumatic stress disorder from abortion. The concept was first introduced in 1981 by Vincent Rue, PhD, a psychologist opposed to abortion rights, in testimony before Congress in which he stated that he had diagnosed posttraumatic stress disorder in abortion recipients. In 1988 the president, Ronald Reagan, who had been elected in part because of massive support by abortion rights opponents, asked Surgeon General C. Everett Koop, MD, to investigate the effects of abortion on women's mental health.

Dr. Koop had come to the attention of the Reagan administration in large part because of his opposition to abortion rights. In 1975, he worked with the Reverend Billy Graham to form the first evangelical Christian anti-abortion group, the Christian Action Council. Despite his personal stance in opposition to abortion rights, after hearing testimony and reviewing available published literature, Dr. Koop eventually concluded that there were insufficient data showing that abortion had negative consequences on women's psychological health. He would openly chastise abortion opponents for promoting this line of thinking, stating, "The pro-life movement always focused—rightly, I thought—on the impact of abortion on the fetus.... They lost their bearings when they approached the issue on the grounds of the health effect on the mother" (Koop, 1991, pp. 274–275). Foreshadowing the politicization of the scientific research on abortion that was to become so prominent in subsequent decades, Koop also wrote, "In the minds of some [of Reagan's advisors], it was a foregone conclusion that the negative health effects of abortion on women were so overwhelming that the evidence would force the reversal of *Roe vs. Wade*" (Koop Says Abortion Report Couldn't Survive Challenge, 1989, p. A12).

The professional mental health community soon entered the debate on this growing controversy. After preparing a literature summary and recommendations for Dr. Koop, a panel from the American Psychological Association (APA) published their findings in the journal *Science*, concluding that "[s]evere negative reactions after abortions are rare and can best be understood in the framework of coping with a normal life stress" (Adler et al., 1990, p. 43). As we shall discuss, an APA task force updated its findings on this subject in 2008, evaluating the new research that had accumulated, and again reaffirmed that there was no evidence of widespread harm from abortion (Major et al., 2009). A future president of the American Psychiatric Association and a specialist in women's mental health, Nada Stotland, MD, weighed into the debate in an editorial in the *Journal of the American Medical Association* entitled, "The myth of the abortion trauma syndrome" (Stotland, 1992).

Despite these counterclaims from the professional mental-health community, the debate raged on. In 1987, David Reardon, a leading architect of the "abortion-hurts-women" argument, sought to expand the issue beyond a psychiatric definition of abortion trauma. To this end he published the first of many books, *Aborted Women: Silent No More*, in which he documented the stories of women who suffered emotionally after having an abortion (Reardon, 1987). Reardon also helped launch a new organization, known as Women Exploited by Abortion, whose chief activity consisted of public presentations by women who regretted their decision to have an abortion.

Frustrated that no one would take the claims of harm seriously, in 1988 Reardon founded the Elliot Institute specifically to conduct research to prove that abortion is bad for women. Reardon had obtained his PhD in biomedical ethics from Pacific Western University, an unaccredited correspondence school offering no classroom instruction. Although his research remains controversial, beginning in the 1990s he and colleagues became increasingly savvy about publishing their findings in reputable scientific journals and created a body of literature that alleges abortion's harmful effects on women. Their studies found higher rates of depression, anxiety, suicide, substance abuse, and psychiatric admissions among women who have abortions (Coleman, 2011). But experts significantly challenged these studies for their poor study design and inappropriate statistical methods (Charles et al., 2008).

In 2005, the South Dakota Task Force on Abortion, charged by the state legislature to assess the scientific facts of abortion, wholly adopted the findings of Reardon and his colleagues, concluding that there was a scientific, as well as a moral, reason to ban abortion. The task force specifically rejected an opposing body of scientific literature showing no evidence of harm and pointing out the significant methodological flaws in the studies that did show such evidence. Instead, it concluded, "Although the Task Force aggressively sought such contributions from all perspectives, this evidence was overwhelmingly in support of protecting life and preventing harm to women caused by abortion" (South Dakota Task Force to Study Abortion, 2005, p. 7).

Despite a near-professional consensus that abortions do not cause mental-health problems for the majority of abortion recipients (APA Task Force on Mental Health and Abortion, 2008), state legislatures like that in South Dakota implemented laws during the early twenty-first century requiring women to be told of the potential psychological consequences of abortion. Of the 19 states that included information on possible psychological responses to abortion, seven describe only the negative emotional responses (Guttmacher Institute, 2011). For example, in Texas, the state-authored booklet "A Woman's Right to Know," which all women obtaining abortion must be offered, included the following warning in 2012: "Some women have reported serious psychological effects after their abortion, including depression, grief, anxiety, lowered self-esteem, regret, suicidal thoughts and behavior, sexual dysfunction, avoidance of emotional attachment, flashbacks, and substance abuse. These emotions may appear immediately after an abortion, or gradually over a longer period of time" (http://www.dshs.state.tx.us/wrtk/default.shtm).

The Supreme Court has also demonstrated its acceptance of these arguments. In the *Gonzales v. Carhart* decision upholding the Partial Birth Abortion Ban Act of 2003, the Court gave a significant boost to the argument that women are psychologically harmed by abortion [550 U.S. 124 (2007) (*Carhart II*)]. Writing for the majority, Justice Kennedy argued, "While we find no reli-

able data to measure the phenomenon, it seems unexceptionable to conclude some women come to regret their choice to abort the infant life they once created and sustained...Severe depression and loss of esteem can follow" (*Carhart II* at 59).

A Change in Regulating Medical Practice.

The significance of *Carhart II* went well beyond its endorsement of discredited theories of the negative mental-health consequences of abortion. By upholding the ban on an infrequently used technique of later abortion, intact dilation and extraction (intact D&E; named "partial birth abortion" by abortion rights opponents although the name appears nowhere in medical literature), the Court signaled a shift in the role of medical authority in legal understandings of abortion. Initially, it overtly ignored the expert testimony of physicians as well as affidavits filed by the American College of Obstetricians and Gynecologists and other medical organizations. These experts had previously successfully argued that under some circumstances intact D&E was the safest method available to them [*Planned Parenthood Fed'n v. Ashcroft*, 320 F.Supp. 2d 957 (N.D. Cal. 2004), overruled by *Carhart II*].

More striking was the shift in the way in which physicians were treated in the decisions. Writing the majority *Roe* decision, Justice Harry Blackmun had clearly deferred to medical authority: "Up to those points [fetal viability], the abortion decision in all its aspects is inherently, and primarily, a medical decision, and basic responsibility for it must rest with the physician" (*Roe*, 410 U.S. at 166). By comparison, in *Carhart II*, Justice Anthony Kennedy, writing for the majority, stated, "The law need not give abortion doctors unfettered choice in the course of their medical practice, nor should it elevate their status above other physicians in the medical community" (at 163). In short, in direct contrast to *Roe*, physicians who performed abortions were suspect in the eyes of the 2007 Court, rather than experts upon which the Court should rely.

Yet another front in the early twenty-first century campaign to restrict the abortion provision and directly limit medical discretion and authority is evidenced in the disputes over the meanings of "health." *Roe* created health justifications for both the performance of abortion and the limitations upon it. Health as a concept was defined in the companion Supreme Court case *Doe v. Bolton* [410 U.S. 179 (1973)]. As two observers summarized, "There the Court explained that a physician could make the health determination in light of many factors—physical, emotional, psychological, familial, and the woman's age…The Court also pointed out that this broad definition benefited the woman and provided the physician with the room he needs to make his best medical judgment" (Lucas and Miller, 1981, p. 77). Health is further explicitly defined as including both physical and mental health.

More recently, to help courts find a justification to reverse this broad health claim, abortion rights opponents sought to narrow the definition of health within state laws regulating abortion. These new laws limit "health" to conditions that have a high likelihood of resulting in serious risk of substantial and irreversible physical impairment of a major bodily function. Explicitly targeted for elimination are abortion justifications based on the mental health of the pregnant woman. The goal was to narrow the discretion of physicians who perform abortions. In 2012, the six enacted state laws (in Alabama, Idaho, Indiana, Kansas, Nebraska, and Oklahoma) limiting abortion after 20 weeks following fertilization all contained similar limiting language: "Medical emergency means a condition which, in reasonable medical judgment, so complicates the medical condition of the pregnant woman as to necessitate the immediate abortion of her pregnancy to avert her death or for which a delay will create a serious risk of substantial and irreversible physical impairment of a major bodily function" [See Ala. Code §§ 26-23B-1–B-9 (2011), Idaho Code Ann. §§ 18-501–510 (2011), Ind. Code §§ 16-34-2-1–7 (2011), Kan. Stat. Ann. §§ 65-6722 (2011), Okla. Stat. tit. 63, §§ 1-745.1–5.11 (2011), and Neb. Rev. Stat. §§ 28-3,102–111 (2011)].

The desire to limit health exceptions to otherwise unpermitted abortions was also evident in the debate over partial-birth abortion. Prior to the successful implementation of a federal law limiting the use of intact D&E technique, between 1995 and 2000 28 states passed partial-birth abortion bans. When Nebraska's law was challenged in the first Supreme Court decision on partial-birth abortion, *Stenberg v. Carhart*, the Court struck down the law for its failure to include a health exception [530 U.S. 914 (2000) *(Carhart I)*]. However, seven years later, a reconfigured Supreme Court lacking Justice O'Connor, who authored the *Carhart I* decision, found in favor of the Federal Partial Birth Abortion Ban Act of 2003, despite a comparable lack of a health exception (Jones and Weitz, 2009).

One of the most pronounced debates over a health exception occurred in the state of Kansas, where George Tiller, MD, the prominent abortion-providing physician, was assassinated in May 2009. Dr. Tiller's practice was controversial because he provided post–24-week abortions to women whose fetuses had lethal or serious anomalies, often justified to protect the mental health of the pregnant woman. In addition, a limited number of women with serious health conditions of their own, including mental-health conditions, could receive abortions from Dr. Tiller (Joffe, 2011). Until 2011, under Kansas law women seeking such later abortions were required to be approved by a second physician, who had no financial relationship with the abortion provider. Dr. Tiller was subject to repeated legal actions by various political opponents, including the attorney general of Kansas and the abortion rights opposition group Operation Rescue, which took advantage of an 1887 law, unique to Kansas, allowing grand juries to be impaneled on the basis of citizen petitions. Dr. Tiller won all these court battles, although at significant personal and financial cost. In his last trial, several months before his death, the jury acquitted him in fewer than 45 minutes on the charge of technical violations of the requirement that a second physician approve all post–24-week abortions (Joffe, 2011). In 2011, more than two years after Dr. Tiller's murder and the closing of his practice, Dr. Ann Kristin Neuhaus, who had served as the second consulting physician for Tiller's later patients, faced charges before the state's Board of Healing Arts that she had improperly documented and approved abortions for those with mental-health conditions, and ultimately her medical license was revoked by the board (Hanna, 2012). As these cases indicate, the legal and policy dialog about the importance of a health exception in abortion regulations has shifted dramatically since *Roe*'s deference to the authority of the abortion-providing physician.

Mobilizing Scientific Arguments to Protect the Fetus. For abortion rights opponents, the declining role of medical authority in Supreme Court decisions suggests an openness on the part of the justices to rethink the unresolved question of when life begins. In *Roe*, Justice Blackman explained why the Court was not obligated to make a determination: "We need not resolve the difficult question of when life begins. When those trained in the respective disciplines of medicine, philosophy, and theology are unable to arrive at any consensus, the judiciary, at this point in the development of man's knowledge, is not in a position to speculate as to the answer" (*Roe*, 410 U.S. at 160). During the early twenty-first century, several strategies were employed by abortion rights opponents to demonstrate to the court that science had, in fact, come to consensus. These efforts included attempting to prove that the fetus is a separate person and arguing that the fetus is capable of feeling pain at gestational ages when abortions are permitted.

The fetus as a separate person. Although the belief that life begins at conception has been the basis for the pro-life movement's opposition to many forms of contraception as well as abortion, the movement's efforts during the early twenty-first century to enforce this position focused on "new" scientific discoveries about human biological development. For example, in 2005 the South Dakota Task Force on Abortion found that "new recombinant DNA technologies indisputably proved that the unborn child is a whole human being from the moment of fertilization" (South Dakota Task Force to Study Abortion, 2005, p. 31). Based on these findings, the state mandated that pregnant women obtaining an abortion be informed "[t]hat the abortion will terminate the life of a whole, separate, unique, living

human being." [S.D. Codified Laws § 34-23A-10.1 (2011)]. After considering a legal challenge to this requirement, the Court of Appeals found the mandatory statements of separation and uniqueness to have no bearing on a woman's right to have an abortion [*Planned Parenthood Minn., N.D., S.D.,* 653 F.3d at 669.].

Other efforts on the part of abortion rights opponents to expand on the idea that life begins at conception targeted the passage of state constitutional personhood amendments through the means of state ballot initiatives. Personhood as a strategy emerged in 2005 in Mississippi, followed by Michigan in 2006 and Georgia in 2007. All three states failed to gather enough signatures to put the initiative on the ballot, but in 2008 Colorado became the first state to actually vote on the issue. The initiative lost by an overwhelming majority. In 2011, although finally making it to the ballot in Mississippi, a similar measure was soundly rejected by voters. Of interest to readers of this essay is that although the national organization leading the effort to secure the passage of personhood amendments has a stated goal "to serve Jesus," the justification for the initiatives is put in scientific language. As *Personhood USA's* website explained in 2012, "The science of fetology in 1973 was not able to prove, as it can now, that a fully human and unique individual exists at the moment of fertilization and continues to grow through various stages of development in a continuum (barring tragedy) until natural death from old age" (http://www.personhoodusa.com/what-is-personhood).

The fetus and pain. In 2010, Nebraska enacted L.B. 1103 (2010), the Pain-Capable Unborn Child Protection Act, which bans abortions after 20 weeks postfertilization [codified at Neb. Rev. Stat. §§ 28-3,102–111 (2011)]. According to the proponents of the bill, at 20 weeks from fertilization the fetus has the physical structures necessary to experience pain. This ability to feel pain reflects the human status of the fetus that requires that it be protected from abortion. The justification for the law is based predominately on the work of K. J. S. Anand, MBBS, DPhil. Dr. Anand and colleagues have contended that the fetal structure for experiencing pain differs from that in adults and that these unique fetal mechanisms allow the fetus to experience pain. Dr. Anand has made a particularly compelling expert because he is well respected in the medical arena as a result of his contributions to the larger field of pain management in the care of babies born very prematurely (Anand, Stevens, and McGrath, 2007).

Opponents of the Nebraska ban contested the sidelined position of Dr. Anand and his colleagues, contrasting it with the consensus among most scientists in the field that the first pathways associated with pain perception are not complete before approximately 29 weeks of pregnancy (Lee et al., 2005) or 24 weeks (Royal College of Obstetricians and Gynaecologists, 2010). Those who find a later period for the onset of pain also focus on increasing evidence that the fetus never experiences a state of true wakefulness. Thus the fetus in the uterus does not feel pain, although that same fetus might feel pain after it is born. The newest scientific evidence bearing on the debate suggested that the fetus is unable to distinguish between sensations that are painful and simply being touched until much later in development (Fabrizi et al., 2011).

The Nebraska legislature was hoping to challenge the underpinnings of *Roe* by providing a scientific argument the Supreme Court could use to overturn *Roe* without having to reverse precedent. By accepting the notion of fetal pain, the Court could find that it needed to take into account new scientific evidence replacing the understanding of fetal development originally outlined in *Roe,* which determined when a fetus deserved protection by the state. The legal standard would then shift from a standard of potential fetal viability (approximately 24 weeks with current technologies) to a standard based on the potential of the fetus to experience pain at 20 weeks of gestation or earlier.

Following the success of the Nebraska statute, nine additional states have passed similar laws limiting abortion after or before 20 weeks. Mainstream reproductive rights organizations initially avoided filing a formal legal challenge to these laws, citing a reluctance to be baited by opponents into a fight over laws that limit only a small number of procedures that took place in the affected states (Kliff, 2011). Specifically, during the early twenty-first century abortions taking place after 20 weeks of pregnancy constituted only about 1.4 percent of all abortions performed yearly in the United States (Pazol et al., 2009), and very few abortions at that gestational age took place in the states that were first to pass such bans. The first legal challenge was made to the Arizona law. After it received a very hostile ruling from a Federal Court judge it was promptly overturned by the Ninth Circuit Court of Appeals and stayed by the U.S. Supreme Court that opted not to take the case. A similar Georgia law was under consideration in the state supreme court with no ruling as of 2013. However, a state law in Texas limiting abortion after 20 weeks went into effect in late 2013, severely limiting services in that state.

Conclusion. The story we have told in this essay is primarily one of how the competing sides in the abortion conflict in the United States have deployed competing scientific findings in legal and legislative battles since *Roe v. Wade's* 1973 landmark decision. Opponents of abortion rights have used scientific arguments about the relationship between abortion and breast cancer and the effect of abortion on women's mental health to counter earlier arguments by abortion rights supporters in the medical community about the public-health benefits of legal abortion. Debates over the existence and timing of a fetus' ability to feel pain break down the trimester system the courts have relied upon for decades. It is fair to say that the scientific arguments used by abortion rights proponents have had greater support in the mainstream scientific community, but the arguments used by abortion rights opponents, especially during the late twentieth and early twenty-first centuries, have held considerable sway in state legislatures and judicial circles, including the Supreme Court. The issues we relate in this essay also point to another significant development in the abortion conflict because the focus on science by both sides has corresponded with a decline of the medical authority that was once such a central actor in the abortion debate.

[*See also* Ethics and Medicine; Law and Science; Medicine: Overview; *and* Public Health.]

BIBLIOGRAPHY

Adler, N. E., H. P. David, B. N. Major, S. H. Roth, N. F. Russo, and G. E. Wyatt. "Psychological Responses after Abortion." *Science* 248, no. 4951 (1990): 41–44.

Anand, K. J. S., B. J. Stevens, and P. J. McGrath. *Pain in Neonates and Infants*, 3d ed. Edinburgh, U.K., and New York: Elsevier, 2007.

APA Task Force on Mental Health and Abortion. *Report of the APA Task Force on Mental Health and Abortion*. Washington, D.C.: The American Psychological Association, 2008.

Beral, V., D. Bull, R. Doll, R. Peto, and G. Reeves. "Breast Cancer and Abortion: Collaborative Reanalysis of Data from 53 Epidemiological Studies, Including 83,000 Women with Breast Cancer from 16 Countries." *Lancet* 363 (2004): 1007–1016.

Brind, J., V. M. Chinchilli, W. B. Severs, and J. Summy-Long. "Induced Abortion as an Independent Risk Factor for Breast Cancer: A Comprehensive Review and Meta-analysis." *Journal of Epidemiology and Community Health* 50, no. 5 (1996): 481–496.

Cates, W., Jr., D. A. Grimes, J. C. Smith, and C. W. Tyler Jr. "Legal Abortion Mortality in the United States. Epidemiologic Surveillance, 1972–1974." *Journal of the American Medical Association* 237, no. 5 (1977): 452–455.

Charles, V. E., C. B. Polis, S. K. Sridhara, and R. W. Blum. "Abortion and Long-term Mental Health Outcomes: A Systematic Review of the Evidence." *Contraception* 78, no. 6 (2008): 436–450.

Coleman, P. K. "Abortion and Mental Health: Quantitative Synthesis and Analysis of Research Published 1995–2009." *The British Journal of Psychiatry* 199, no. 3 (2011): 180–186.

Daling, J. R., K. E. Malone, L. F. Voigt, E. White, and N. S. Weiss. "Risk of Breast Cancer Among Young Women: Relationship to Induced Abortion." *Journal of the National Cancer Institute* 86, no. 21 (1994): 1584.

Eckholm, E. "New Laws in 6 States Ban Abortions after 20 Weeks." *New York Times*, June 26, 2011, p. A10.

Fabrizi, L., R. Slater, A. J. Meek, S. Boyd, S. Olhede, and M. Fitzgerald. "A Shift in Sensory Processing That Enables the Developing Human Brain to Discriminate Touch from Pain." *Current Biology* 21, no. 18 (2011): 1552–1558.

Garrow, D. J. *Liberty and Sexuality: The Right to Privacy and the Making of Roe v. Wade*. Berkeley: University of California Press, 1998.

Grimes, D. A., J. Benson, S. Singh, M. Romero, B. Ganatra, F. E. Okonofua, and I. H. Shah. "Unsafe Abortion: The Preventable Pandemic." *Lancet* 368, no. 9550 (2006): 1908–1919.

Guttmacher Institute. *State Policies in Brief. Counseling and Waiting Periods for Abortion*. New York: Guttmacher Institute, 2014.

Hall, R. E., ed. *Abortion in a Changing World: Volume II*. New York: Columbia University Press, 1970.

Hanna, John. "Kan. Doctor Loses License over Abortion Referrals." http://bigstory.ap.org/article/kansas-revokes-doctors-license-abortion-case.

Jasen, P. "Breast Cancer and the Politics of Abortion in the United States." *Medical History* 49, no. 4 (2005): 423–444.

Joffe, C. *Doctors of Conscience: The Struggle to Provide Abortion before and after Roe v. Wade*. Boston: Beacon Press, 1995.

Joffe, C. "Working with Dr. Tiller: Staff Recollections of Women's Health Care Services of Wichita." *Perspectives on Sexual and Reproductive Health* 43, no. 3 (2011): 199–204.

Jones, B. S., and T. A. Weitz. "Legal Barriers to Second-Trimester Abortion Provision and Public Health Consequences." *American Journal of Public Health* 99, no. 4 (2009): 623–630.

Kliff, S. "The Big Abortion Fight That May Not Happen." *Politico*, April 20, 2011. http://www.politico.com/news/stories/0411/53432.html.

Koop, C. E. *Koop: The Memoirs of America's Family Doctor*. New York: Random House, 1991.

"Koop Says Abortion Report Couldn't Survive Challenge." *New York Times*, March 17, 1989.

Lee, S. J., H. J. Ralston, E. A. Drey, J. C. Partridge, and M. A. Rosen. "Fetal Pain: A Systematic Multidisciplinary Review of the Evidence." *Journal of the American Medical Association* 294, no. 9 (2005): 947–954.

Lucas, R., and L. I. Miller. "Evolution of Abortion Law in North America." In *Abortion and Sterilization: Medical and Social Aspects*, edited by J. E. Hodgson. London and New York: Academic Press and Grune & Stratton, 1981.

Luker, K. *Abortion and the Politics of Motherhood. California Series on Social Choice and Political*

Economy. Berkeley: University of California Press, 1984.

Major, B., M. Appelbaum, L. Beckman, M. A. Dutton, N. F. Russo, and C. West. "Abortion and Mental Health: Evaluating the Evidence." *American Psychologist* 64, no. 9 (2009): 863–890.

Melbye, M., J. Wohlfahrt, J. H. Olsen, M. Frisch, T. Westergaard, K. Helweg-Larsen, and P. K. Andersen. "Induced Abortion and the Risk of Breast Cancer." *The New England Journal of Medicine* 336, no. 2 (1997): 81–85.

Mohr, J. C. *Abortion in America: The Origins and Evolution of National Policy, 1800–1900*. New York: Oxford University Press, 1978.

Mooney, C. *The Republican War on Science*. New York: Basic Books, 2005.

National Cancer Institute. *National Cancer Institute Boards Accept Scientific Workshop Findings on Early Reproductive Events and Breast Cancer*. National Institutes of Health, March 3, 2003. http://www.cancer.gov/newscenter/pressreleases/2003/bsacceptere.

Pazol, K., S. B. Gamble, W. Y. Parker, D. A. Cook, S. B. Zane, and S. Hamdan. "Abortion Surveillance—United States, 2006." *MMWR Surveillance Summary* 58, no. 8 (2009): 1–35.

Petchesky, R. P. *Abortion and Woman's Choice: The State, Sexuality, and Reproductive Freedom*, revised ed. Boston: Northeastern University Press, 1990.

Reagan, L. J. *When Abortion Was a Crime: Women, Medicine, and Law in the United States, 1867–1973*. Berkeley: University of California Press, 1997.

Reardon, D. C. *Aborted Women: Silent No More*. Chicago: Loyola University Press, 1987.

Royal College of Obstetricians and Gynaecologists. *Fetal Awareness Review of Research and Recommendations for Practice*. London: Royal College of Obstetricians and Gynaecologists, 2010.

Russo, J., and I. H. Russo. "Susceptibility of the Mammary Gland to Carcinogenesis. II. Pregnancy Interruption as a Risk Factor in Tumor Incidence." *The American Journal of Pathology* 100, no. 2 (1980): 497.

Smith-Rosenberg, C. "The Abortion Movement and the AMA, 1850–1880." In *Disorderly Conduct: Visions of Gender in Victorian America*, edited by C. Smith-Rosenberg. New York: Knopf, 1985.

South Dakota Task Force to Study Abortion. *Report of the South Dakota Task Force to Study Abortion: Submitted to the Governor and Legislature of South Dakota, December 2005*. Pierre, S.Dak.: State Government of South Dakota, 2005.

Staggenborg, S. *The Pro-choice Movement: Organization and Activism in the Abortion Conflict*. New York: Oxford University Press, 1991.

Stotland, N. L. "The Myth of the Abortion Trauma Syndrome." *Journal of the American Medical Association* 268, no. 15 (1992): 2078–2079.

Tietze, C. "The Effect of Legalization of Abortion on Population Growth and Public Health." *Family Planning Perspectives* 7, no. 3 (1975): 123–127.

Tracy A. Weitz and Carole Joffe

ACADEMY OF NATURAL SCIENCES OF PHILADELPHIA

Founded in 1812 in Philadelphia—then the commercial and scientific hub of the United States—the Academy of Natural Sciences is America's oldest natural history museum. An aspiring naturalist, Thomas Say, was one of seven founders whose primary goal was to use science as the basis for an intellectually stimulating social club for its elite membership.

Nevertheless, the academy fostered the growth of local science by providing a meeting place for serious naturalists, a reference library, and an ever-growing collection of natural objects. Lean, early years gave way to a period of relative prosperity after 1817 when William Maclure, a wealthy, philanthropic-minded geologist, was elected president. His 23-year tenure was marked by steady growth and a move toward popular education. By 1828, collections were opened to the public on a limited basis.

Scientific professionalism gained ground in the 1840s, when the academy became America's premiere center for natural science. Joseph Leidy, its most reputable scientist, ascended to the presidency in 1847. John James Audubon, Charles Darwin, Edward Drinker Cope, and other science luminaries joined the membership. By midcentury, academy members recognized the need to cater to popular interests to gain public support for a new building to accommodate the growing collections. Strict entrance requirements were relaxed and overcrowding became a serious concern. In 1865, the academy began paying its curators to care for the collections and manage visitors. The academy pioneered the display of mounted dinosaurs—now a museum staple—when Leidy's *Hadrosaurus* went on exhibit to wild acclaim, in 1868.

The academy's present building, completed in 1876, featured dedicated exhibit spaces and a public lecture hall. The separation of exhibit and study collections, beginning in the 1890s, ushered in an era of elaborate habitat displays and ambitious expeditions that ventured farther afield. Many dioramas constructed in the 1920s and 1930s were still on display in the early twenty-first century. The academy moved into applied research in 1948 with its new Environmental Research Division.

In 2011, the academy houses over 17 million specimens. Its library and archives comprise one of America's largest and finest specialty libraries. Current research and education programs focus on environmental themes.

[*See also* Audubon, John James; Biological Sciences; Botany; Cope, Edward Drinker; Dinosaurs; Leidy, Joseph; Museums of Science and Natural History; Popularization of Science; Say, Thomas; Science: Overview; *and* Zoology.]

BIBLIOGRAPHY

Orosz, Joel J. *Curators and Culture: The Museum Movement in America, 1740–1870.* Tuscaloosa: University of Alabama Press, 1990. Argues that curators reached a workable compromise between popularization and professionalization in early American museums.
Peck, Robert McCracken, and Patricia Tyson Stroud. *A Glorious Enterprise: The Academy of Natural Sciences of Philadelphia and the Making of American Science.* Philadelphia: University of Pennsylvania Press, 2012.
Rainger, Ronald. "The Rise and Decline of a Science: Vertebrate Paleontology at Philadelphia's Academy of Natural Sciences, 1820–1900." *Proceedings of the American Philosophical Society* 136 (1992): 1–32.

Paul D. Brinkman

ACQUIRED IMMUNODEFICIENCY SYNDROME

See HIV/AIDS.

ADVERTISING, MEDICAL

Medical advertising, or the commercial marketing of medical goods and services, has a long history in America despite the health profession's carefully nurtured reputation as a therapeutic rather than consumerist enterprise. Until the late nineteenth century physicians advertised their services to compete for clients, but as medicine professionalized after the Civil War, direct competition and advertising were officially banned. The bulk of medical advertising since then has been for drugs, devices, and other health-related goods and services.

In the late nineteenth and early twentieth centuries, the most prominently advertised medical goods were "patent medicines," whose manufacturers, facing intense competition, pioneered aggressive national advertising to establish famed brand names such as Carter's Little Liver Pills and Lydia Pinkham's Vegetable Compound. Entirely unregulated, the ads promised miraculous cures for virtually any ailment and touted invented statistics and fictional testimonials from customers and men of distinction.

This vibrant commercial market was not restricted to drugs. Companies also advertised an endless profusion of birth control devices, often described as "married women's friends" or "preventatives" to avoid running afoul of the law. Sickroom supplies for common diseases such as tuberculosis were also heavily marketed, including atomizers, chest expanders, thermometers, sputum cups, recliner chairs, and invalid beds. One very important category of goods included those sold to combat the menace of "germs," including sewer traps, ventilation systems, disinfectants, porcelain toilets, and home water filtration systems. These advertisements spread awareness of the new germ theory of disease and encouraged changes in private behaviors that complemented public-health initiatives and in some cases became models for public-health campaigns.

Patent medicine advertising provoked America's first federal advertising regulations. After Progressive-Era muckrakers exposed false claims and toxic or addictive ingredients in popular medicines, the 1906 Pure Food and Drug Act required drug companies to list their ingredients truthfully on the label, whereas the 1912 Shirley Amendment to that act outlawed false therapeutic claims on labels or in advertising. The patent medicine industry faded, replaced by a more staid but still well-advertised set of health-related "household items," such as Phillip's Milk of Magnesia, that would be categorized as "over-the-counter" medicines in the early twenty-first century.

These new regulations did not cover so-called "ethical" drugs, whose manufacturers sold only pure drugs of known provenance and who advertised only to physicians. This advertising took a range of creative forms such as sponsored medical research, advertisements in medical journals, direct mail to physicians, and office- and hospital-based "detailing" of physicians by sales representatives, all of which were exempted from federal oversight out of deference to medical expertise. To run in a respectable medical journal, however, the ads did have to receive a "Seal of Approval" from the American Medical Association's Council on Pharmacy and Chemistry.

Advertising to physicians became much more intense after World War II, when new discoveries and new prescription-only regulations vastly expanded the range and profitability of ethical drugs. Medical journals devoted more space to increasingly elaborate advertisements, whereas physicians also began to encounter a torrent of free gifts, intensified "detailing," sponsored conferences, and the like. Heeding the advice of a marketing consultant, even the American Medical Association ended its Seal of Approval program in 1955 and opened its journals to more advertising. Such moves did not go unnoticed; medical reformers decried the idea of "hawking medicines like soap," spurring congressional investigations by the end of the decade. In the 1962 Amendments to the Food and Drug Act, the Food and Drug Administration (FDA) finally received authority over prescription-drug advertising and imposed a new standard of "fair balance," that is, a thorough and fair summary of the drug's proven uses and a complete and accurate listing of potential problems and side effects.

Prescription-drug advertising targeted more than just physicians. "Institutional" ads for particular drug houses (rather than for specific drugs) had been common throughout the twentieth century. In the 1950s, companies promoted brand-name drugs to the public through a range of public-relations gimmicks and industry-produced "educational" material in popular magazines, cinema newsreels, and radio. Such informal marketing gave rise to the "celebrity drug" phenomenon. The FDA began to permit formal direct-to-consumer advertising in the 1980s, imposing the same fair-balance requirements as medical journal ads. Television and radio ads had no room for this lengthy material, however, so advertisers devised a workaround: they divided their campaigns into "health-seeking" spots that raised awareness of an illness without mentioning a specific drug and "reminder" spots that promoted the name of a drug without making any therapeutic claims. Faced with an absurd and counterproductive situation, in the 1990s the FDA allowed companies to list a toll-free number or Web site instead of providing the full fair balance on broadcast ads. With spending on direct-to-consumer advertisements surpassing $4 billion at the turn of the twenty-first century, debates continue about its impact on the practice of medicine in America.

[*See also* **American Medical Association; Birth Control and Family Planning; Ethics and Medicine; Germ Theory of Disease; Journals in Science, Medicine, and Engineering; Medicine: Overview; Pharmacology and Drug Therapy; Psychopharmaceutical Drugs; Public Health; Pure Food and Drug Act;** *and* **Radio.**]

BIBLIOGRAPHY

Greene, Jeremy, and David Herzberg. "Hidden in Plain Sight: The Popular Promotion of Prescription Drugs in the 20th Century." *American*

Journal of Public Health (May 2010): 793–803. Brief survey of direct-to-consumer marketing, before and after the FDA permitted it.

Lears, T. J. Jackson. *Fables of Abundance: A Cultural History of Advertising in America.* New York: Basic Books, 1994. Patent medicine companies as advertising pioneers.

Ott, Katherine. *Fevered Lives: Tuberculosis in American Culture since 1870.* Cambridge, Mass.: Harvard University Press, 1999. See Chapter 5 for sickroom commerce.

Tomes, Nancy. *The Gospel of Germs: Men, Women and the Microbe in American Life.* Cambridge, Mass.: Harvard University Press, 1999. Tracks how the public learned about "germs" through product advertising.

Tomes, Nancy. "The Great American Medicine Show Revisited." *Bulletin of the History of Medicine* 79 (2005): 627–663. Discusses reformers' efforts to combat doctors' susceptibility to drug advertising in the first two-thirds of the twentieth century.

Tone, Andrea. *Devices and Desires: A History of Contraceptives in America.* New York: Hill and Wang, 2001. Discusses birth-control advertising before and after it was decriminalized.

Young, James Harvey. *The Toadstool Millionaires: A Social History of Patent Medicines in the United States before Federal Regulation.* Princeton, N.J.: Princeton University Press, 1961. The classic account.

David Herzberg

AGASSIZ, LOUIS

(1807–1873), zoologist, geologist. Born in Switzerland, Agassiz was educated at the universities of Zurich, Heidelberg, and Munich. He earned degrees from the University of Munich in zoology (1829) and medicine (1830). Encouraged by Georges Cuvier and Alexander von Humboldt, Agassiz studied fossil fish at the Muséum d'histoire naturelle in Paris for eight months. In 1832, he became a professor of natural history at the University of Neuchâtel, Switzerland. During his time there he helped develop a museum of natural history, becoming its director. Between 1833 and 1843, Agassiz published a five-volume work, *Recherches sur les poissons fossiles*. In this work Agassiz analyzed over 1,700 ancient fishes from collections around Europe. *Recherches* was highly praised by European and American naturalists. In 1837 Agassiz announced his Ice Age theory, which held that northern Europe had once been covered by massive glaciers. His work on glaciers made him famous.

In 1846, burdened by debt, Agassiz accepted an invitation to lecture in the United States at Boston's Lowell Institute, where he was lionized. Harvard gave him a professorship in 1850, and he taught there until his death. His claim that human races are distinct species was welcomed by supporters of slavery, but his belief that science rather than the Bible is the authority on creation displeased religious conservatives. In his 1857 "Essay on Classification" Agassiz argued that taxonomic categories, classes as well as species, embody the thoughts of God, who also created the similarities that link embryos, fossils, and biogeography. When Charles Darwin's *Origin of Species* appeared in 1859, Agassiz contested it vigorously and never accepted evolution.

In the United States, Agassiz opened Harvard's Museum of Comparative Zoology in 1859, his greatest legacy. With other leading men of science, he worked to make American science more professional by founding the National Academy of Sciences in 1863. His trick of leaving a student alone with a single fish, telling him, "Study nature, not books!," became legendary. He steamed up the Amazon River in 1865–1866 collecting specimens for his museum, and in 1872 he sailed to San Francisco. His 1873 summer school on Penikese Island inspired others to build marine laboratories in Woods Hole. Agassiz's first wife, Cécile Braun, bore him three children and died in 1848. In 1850, he married Elizabeth Cary, who later founded Radcliffe College, where women could study under Harvard professors. His son, Alexander Agassiz, directed his museum after his death.

[*See also* **American Museum of Natural History; Creationism; Evolution, Theory of; Geology; Medicine: Overview; Museums of Science and Natural History; National Academy of Sciences; Paleontology; Religion and Science; Science: Overview;** *and* **Zoology.**]

BIBLIOGRAPHY

Lurie, Edward. "Agassiz, Jean Louis Rodolphe (1807–1873)." In *The History of Science in the United States: An Encyclopedia,* edited by Marc Rothenberg, pp. 12–13. New York and London: Garland Publishing, 2001.

Lurie, Edward. *Louis Agassiz: A Life in Science.* Chicago: University of Chicago Press, 1960.

Roberts, Jon H. "Louis Agassiz on Scientific Method, Polygenism, and Transmutation: A Reassessment." *Almagest 2* (May 2011): 76–99.

Winsor, Mary P. *Reading the Shape of Nature: Comparative Zoology at the Agassiz Museum.* Chicago: University of Chicago Press, 1991.

Mary Pickard Winsor;
updated by Elspeth Knewstubb

AGRICULTURAL EDUCATION AND EXTENSION

The system of agricultural education and extension consists of the land-grant colleges with their associated agricultural experiment stations and cooperative extension services and secondary schools that offer vocational instruction in agriculture. The 69 land-grant colleges that existed by the end of the twentieth century were established under the provisions of the Morrill Land Grant Act of 1862, which required that they offer residential instruction in agriculture.

The Hatch Act of 1887 appropriated federal funds for the establishment in each state of one or more experiment stations to undertake systematic study of agricultural problems and to formulate scientific knowledge that could be presented in college classrooms. The stations were usually located at the land-grant colleges and commonly shared faculty with them.

The experiment stations were required to disseminate their findings among farmers, but the printed word proved to be an ineffective form of communication, as were farmers' institutes. In 1903, Seaman A. Knapp introduced in Texas the demonstration method, by which farmers learned improved agricultural practices under the direction of a skilled adviser, later to be known as a county agent. Success with this teaching innovation led to boys' and girls' corn and tomato clubs, which developed into the 4-H Club movement for farm youth, and to home-demonstration work with rural women. The Smith–Lever Act of 1914 provided additional federal support for a nationwide educational program for all members of the farm family.

Vocational instruction in agriculture began around 1897 with nature study in the public schools of New York State and elsewhere. In the first decade of the twentieth century, some states authorized the establishment of agricultural high schools. These institutions disappeared when public high schools began to employ graduates of the land-grant colleges to offer courses in vocational agriculture and home economics. The Smith–Hughes Act of 1917 funded such educational programs. This system of agricultural education contributed greatly to the development of the United States in the twentieth century. By dramatically increasing agricultural productivity, it permitted a sharp reduction in farm population while providing abundant and low-cost food and fiber.

[*See also* **Agricultural Experiment Stations; Agricultural Technology; Agriculture, U.S. Department of; 4-H Club Movement; Higher Education and Science; High Schools, Science Education in; Home Economics Movement;** *and* **Morrill Land Grant Act.**]

BIBLIOGRAPHY

Eddy, Edward D. *Colleges for Our Land and Time: The Land-Grant Idea in American Education.* New York: Harper, 1957.

Marcus, Alan I. *Agricultural Science and the Quest for Legitimacy: Farmers, Agricultural Colleges, and Experiment Stations.* Ames: Iowa State University Press, 1985.

Scott, Roy V. *The Reluctant Farmer; the Rise of Agricultural Extension to 1914.* Urbana: University of Illinois Press, 1970.

Roy V. Scott

AGRICULTURAL EXPERIMENT STATIONS

By the 1880s, industrialization and urbanization in America were generating increasing demands for farm products, whereas the plains and western states were producing surpluses for a developing international market. At the same time, farmers were targeting the nation's railroads, banks, and capitalists as sources of economic destabilization and proposing a range of moderate to radical actions. The Hatch Act (1887) was part of the government's response to that challenge. This measure allotted funds, initially $15,000 annually, to establish agricultural research stations in every state addressing the needs of growers. An Office of Experiment Stations within the U.S. Department of Agriculture coordinated this decentralized system.

Expectations concerning the role of the agricultural stations generated tensions because their constituency included such diverse groups as farmers, agricultural businesses, and state and federal legislators. Moreover, the close geographical and structural relationship many of them had with the land-grant colleges placed strong educational demands on station personnel. The most effective late nineteenth-century agricultural experiment station administrators, such as William A. Henry of Wisconsin, Eugene Davenport of Illinois, and Liberty Hyde Bailey of New York, balanced these potentially conflicting demands while strengthening their states' agricultural economies and allocating some resources for scientific work. The 1906 Adams Act, for which these leading agricultural administrators strongly lobbied, provided each state with additional funds exclusively to support fundamental agricultural research. The Smith–Lever Extension Act (1914) provided funds to each state to pay county extension agents who would bring farmers the fruits of station work, while simultaneously strengthening the stations as sites of basic research by freeing station researchers from time-consuming extension work.

Most station research was directed toward increasing U.S. agricultural productivity. Plant pathology and economic entomology focused on reducing production losses to diseases and pests, while research on culture methods, fertilizers, and breeding sought to improve production directly by increasing yield. Chemistry, nutrition, genetics, and agricultural technologies were prominent research areas as well.

Over the years, station research revealed close links between basic and applied research and between science and industry. For example, concern for the dairy industry in the 1880s inspired chemist Stephen M. Babcock's research into the butterfat content of milk at the Wisconsin experiment station, resulting in a butterfat test that enabled producers to provide a richer, more standardized product. The discovery of vitamin A by Elmer V. McCollum at the Wisconsin station and by T. B. Osborne and L. B. Mendel at the Connecticut station originated in research on livestock nutrition. The genetics of corn was studied as both a basic and an applied science at various stations beginning in the 1910s. At the New York station in the 1920s, geneticist Rollins A. Emerson trained future Nobelists George W. Beadle and Barbara McClintock. At the Illinois station the development of hybrid corn, perhaps the single most important contribution of the experiment stations to American agriculture, bettered the yield and uniformity of the corn crop while increasing growers' dependence on the seed companies that had collaborated closely with the experiment station on the development of hybrids. Later technological innovations—such as the mechanical tomato picker and cotton picker, developed cooperatively among experiment stations and manufacturers—fostered more efficient cultivation and harvesting by the larger producers while throwing many farm laborers out of work.

From the mid-twentieth century on, critics of the agricultural research system, pointing to such developments, argued that experiment-station research had come to serve corporate needs more than those of American farmers. Such criticism, in turn, provoked many station scientists and government administrators to defend the station system's contributions to world agricultural production.

[*See also* **Agricultural Education and Extension; Agricultural Technology; Agriculture,**

U.S. Department of; Chemistry; Disease; Entomology; Food and Diet; 4-H Club Movement; Genetics and Genetic Engineering; Hybrid Seeds; McClintock, Barbara; Morrill Land Grant Act; Nobel Prize in Biomedical Research; *and* Pesticides.]

BIBLIOGRAPHY

Bush, Lawrence, and William B. Lacy. *Science, Agriculture, and the Politics of Research*. Boulder, Colo.: Westview Press, 1983.
Rosenberg, Charles E. *No Other Gods: On Science and American Social Thought*, revised ed. Baltimore: Johns Hopkins University Press, 1997.

Barbara A. Kimmelman

AGRICULTURAL TECHNOLOGY

In 1790, roughly 90 percent of the U.S. population resided in rural areas and most made their living as farmers. The next two centuries witnessed a continual decline in the share of the U.S. workforce engaged in agriculture. By 2010, only 1.6 percent of the civilian labor force worked on farms (Haines, Table Aa36-92; *Economic Report of the President*, p. 361). How an ever-shrinking share of the labor force provided the raw materials to feed and clothe the nation and a bounty of exports is largely a story of how scientific advances and a stream of new technologies transformed the agricultural landscape. Technological change is central to the story of American development, accounting for the growing farm output, the declining agricultural terms of trade, and the rising capitalization and size of farms. New technologies dramatically altered the day-to-day work patterns of farmers and led to enormous changes in land use. Even more, increased farm efficiency lowered food prices for American consumers, improved international competitiveness, and released labor to fuel the industrial and commercial sectors.

Most accounts of technological change in American agriculture stress the primacy of mech-

anization before the 1930s. Eli Whitney's saw gin, John Deere's steel plow, Cyrus McCormick's mechanical reaper, John Holt's combined harvester, and petroleum-fueled tractors all represent landmark innovations. The emphasis on mechanization fits comfortably within the induced innovation (or directed technical change) hypothesis, a popular model explaining innovation in American agriculture (Cochrane, 1979, p. 200; Hayami and Ruttan, 1985, p. 209). This model treats the rate and direction of technological change as endogenous responses to the forces of factor supply and product demand. The model's advocates suggest that rational farmers should have invested in saving labor because labor was the scarce factor in the United States. They further claim (incorrectly) that labor was becoming scarcer over the past two centuries. The standard account often begins with the decomposition of changes in output per unit of labor into changes in output per unit of land and changes in land available per unit of labor.

$$>\%\Delta \text{ Output/Labor} \cong \%\Delta \text{ Output/Land} + \%\Delta \text{ Land/Labor}$$

The changes in the two right-hand ratios are viewed as relatively independent. The $\%\Delta$ Output/Land ratio is associated with yield-enhancing "biological" innovations, and the $\%\Delta$ Land/Labor ratio is associated with labor-saving machinery. The two sources are viewed as products of different scientific/technological learning paths—chemistry, biology, and genetics as opposed to mechanics and engineering.

By this reasoning in the United States, the dominant trend before 1930 was the rising land-to-labor ratio associated with the mechanical path of development (Hayami and Ruttan, 1985, p. 171). Biological innovations such as hybrid corn became significant only after the 1930s when, according to the standard account, an increasing scarcity of land and a rapid decline in commercial fertilizer prices made such innovations profitable. For decades, this account has been standard fare for scholars. Many have even argued that most American farmers cared little about biological innovations (Cochrane, 1979, p. 201–202).

Problems with Standard Account. At the very core of the induced innovation story is the repeated assertion that over the nineteenth century, the price of land fell relative to the wage rate. But this was not so. The ratio of the price of an acre of farmland to the agricultural wage more than doubled between 1790 and 1850 and then tripled between 1850 and 1910. Trends after 1910 are also unkind to the notion that the price movements for land (or fertilizer) induced the biological innovations after 1930 (Hayami and Ruttan, 1985, p. 174–178). In actuality, the price of land relative to farm wages fell between 1910 and 1950. Thus, not only were farm wages (relative to land values) falling in the early period when, according to the model, they should have been rising, but also in the later period they were rising when they should have been falling. Movements in the relative price of fertilizer are similarly problematic. Over the entire span for which we have data, the two key relative price series that represent the empirical foundation for the induced innovation model were typically moving in the wrong direction. If one adheres strictly to the model's predictions, American farmers should have invested significantly in biological innovations before 1930. The historical record confirms this prediction.

Setting the record straight requires that we revise much of what has been written about American agricultural development. Specifically, we argue that biological innovations were highly important in the age before hybrid corn. In addition, there is no sound basis for systematically associating mechanical changes with only saving labor and biological changes with only saving land. We also emphasize that agriculturalists were keenly interested in biological innovation (for more detail see Olmstead and Rhode, 2008).

Agricultural production is an inherently biological process that uses organisms to transform sunlight and soil nutrients into food and fiber for humans. The efficiency of the transformation process is constantly threatened by evolving pests and diseases. Machines may become obsolete and

they can wear out, but a new machine of the outdated design will work more or less like the equipment it replaced. This is not true for biological technologies that underlie medical and agricultural sciences. Farmers and scientists have long understood the Red Queen effect: one has to run fast just to stay in one spot. A perfectly good plant variety can become worthless as pathogens evolve to overcome its defenses. To some extent, the challenges to the system are endogenous because production decisions can affect the rate of decay. Planting vast areas in a single crop intensified pest problems. Exogenous shocks, such as the introduction of invasive species from foreign lands, also dramatically reduced yields. In addition, yields often suffered as farmers pushed agriculture westward. Introducing production to new environments was inherently a biological learning process that required adapting organisms to meet new, often hostile, conditions.

For these reasons, trends in output per acre under cultivation understate the importance of biological innovations. Our account portrays a far more dynamic and creative story of American agricultural development than that found elsewhere—how could it be otherwise given the widespread neglect of biological innovation? We deal with how ecology affected agricultural development and in particular how technological change focused on overcoming the challenges imposed by climate, geography, and pests. An examination of the development of America's three nineteenth-century staple crops illustrates the need for a balanced approach that embraces both mechanical and biological innovations. Extending the analysis to study the livestock economy reinforces these points.

Cotton. By 1850, cotton had become the primary U.S. export crop and the nation dominated the global trade. The standard account relates how cotton took off after Eli Whitney invented the saw-tooth cotton gin (a machine that removed seeds from the lint) in 1793. Early models of the wondrous innovation allowed one worker to replace fifty to one hundred laborers picking the seeds by hand (Lakwete, 2003, pp. 22–23). With improvements, larger gins were many times more efficient. What better example of the substitution of machines for labor? However, there is more to the story. Southern historians have long credited the gin with facilitating the spread of the cotton industry over a vast domain. At the time of the gin's invention, the commercial production was confined to the coastal areas in the southeast that were suitable for growing Sea Island varieties and to patches in the Gulf region that could grow a Siamese variety. The seeds in these types of cotton did not stick so tenaciously to the lint, and they could be ginned with a more primitive roller gin. By breaking the technological constraint on growing fuzzy seeded upland cotton, Whitney's gin made it possible for landowners to convert *en masse* from less labor-intensive activities to cotton. Thus, in the longer run, the gin increased southern labor requirements. In the terminology of economics, the output effect was large and worked counter to the substitution effect. Clearly, the usefulness of the strict association of biological (or mechanical) innovations with crop yields (or labor productivity) warrants reconsideration (Olmstead and Rhode, 2008, pp. 98–113).

In addition to the saw gin, a succession of improved hoes and plows, mechanical seeders, and, after the 1920s, tractors all incrementally advanced productivity in cotton farming. But the next truly great mechanical innovation to revolutionize cotton production did not appear until roughly 150 years after Whitney's invention. J. D. Rust perfected a primitive spindle machine in 1928, but the surplus labor during the Great Depression limited demand for this innovation. This changed during and after World War II, but the machine needed to be perfected and cotton breeders needed to create a plant suitable for machine picking—a less woody plant that ripened more uniformly. The number of cotton-picking machines increased to roughly 4,000 in 1950 and to 38,000 by 1961. By 1970, virtually the entire U.S. cotton crop was mechanically harvested. The mechanization of the cotton harvest, coupled with the decline in cotton acreage, completed the restructuring of Southern rural labor relations that had begun in the 1930s with the spread of tractors. The number of Southern sharecroppers dwindled from 776,000 in 1930

to 121,000 in 1959 (Street, 1957; Olmstead and Rhode, 2006).

Both the saw gin and the mechanical picker changed the landscape of the cotton economy. The gin strengthened the slave economy of the antebellum South, and the mechanical picker contributed mightily to the depopulation of the rural South and to the northern migration of poor southerners, black and white. The story to this point appears to fit nicely with the standard emphasis on mechanization, but these mechanical innovations were accompanied by a stream of biological innovations that were no less revolutionary.

We mentioned how in the 1940s and 1950s cotton breeders redesigned the cotton plant to make it more compatible for mechanical picking. This represents one of many instances of breeders changing the cotton plant to adapt it to new environments, to increase its yield, to make it more resistant to pests, and to make it easier to pick. This story starts with the domestication of cotton in Asia, Africa, and the Americas before recorded history and continues with the early European settlers of North America.

Short-staple (upland) cotton may have been introduced at Jamestown as early as 1607. Over the pre-Revolutionary period, farmers across the South grew patches of low-quality and low-yielding cotton, mostly for home consumption. Planters experimented with seeds imported from around the world. In the late eighteenth century, a major advance was the introduction of long-staple Sea Island cotton varieties. Experiments continued with lower-quality upland cottons, and by 1800 an improved variety, Tennessee Green Seed, appeared in the Cumberland Valley. It rapidly gained favor elsewhere because it could be picked about 25 percent faster. Cotton production soared in response to the booming demand in England. However, in the 1810s and 1820s, a bacterial disease ("the rot") spread in Mississippi and neighboring states, causing devastating losses (Ware, 1936, p. 659; Olmstead and Rhode, 2008, pp. 100–112).

The first of a long chain of events that would transform Southern cotton production occurred in 1806 when a Natchez area planter, Walter Burling, returned from Mexico City with seeds of a high-quality cotton variety. Burling passed the seeds on to a fellow planter and local scientist, William Dunbar, who began the tedious experimentation process to adapt and increase the seed. By the 1810s, the Mexican cotton was prospering in the Natchez area, where it was mingled with the local varieties. The hybrid that emerged represented a vast technological improvement. It had long staple, matured early, and was relatively resistant to the rot. The diffusion of Mexican hybrids represents one of many cases of biological innovation helping overcome the Red Queen effect, because both the yields and the labor productivity of the older varieties were plummeting. The Mexican hybrids' most amazing attribute was that the large bolls opened widely, making them much easier to pick (Olmstead and Rhode, 2008, pp. 100–112).

Over the antebellum years, an enhanced understanding of breeding techniques contributed to further successes. By the early 1830s, Mississippi breeders created an even better variety, Petit Gulf, which also was relatively easy to pick and rot resistant. Cross-pollination in the field and the mixing of seeds at gins tended to reduce the average quality unless the grower selected seed with care. This led to the growth of specialized seed producers in Mississippi who shipped throughout the South. Many other advances in the 1840s and 1850s created the foundation stock for highly productive varieties later developed (Moore, 1988, pp. 12–16; Ware, 1936, p. 658).

The cumulative effect of the stream of biological innovations begun by Burling's 1806 introduction of Mexican cotton had worldwide implications. In the South, it led to huge advances in land and labor productivity. The quantity of cotton picked per day by a worker quadrupled between 1810 and 1860. This growth in picking productivity helps explain the South's increased specialization in cotton as slave owners transferred their assets to higher-productivity activities. Furthermore, the new American varieties proved difficult to grow elsewhere. The British tried. They hired American plantation overseers and attempted to grow the high-quality cotton in India, but without much success. The American South by the 1850s supplied over 75 percent of the cotton used in British mills. This dominance of the world cotton trade was based on the superiority of its product, which in turn was the result of conscious research endeavors of innovative plant breeders and seed companies in the South (Olmstead and Rhode, 2008, pp. 98–100; Ware, 1936, p. 658).

The work to improve cotton varieties did not stop after 1860. Through a long process of selection, breeders successfully adapted the plant to prosper on the more arid and windswept high plains of Texas, and scientists from the U.S. Department of Agriculture (USDA) developed the long-staple Pima in the first decades of the twentieth century from seeds imported from Egypt. In 1906, USDA scientists who were scouring Guatemala and Mexico in search of varieties that might better resist the boll weevil "discovered" Acala cotton. In 1920, the USDA introduced Acala to California, where it became the only variety planted on any scale for over 40 years (Olmstead and Rhode, 2008, pp. 166–168). This breeding process was repeated over and over to adapt cottons to new growing areas and to combat pests. The boll weevil represented the most serious challenge.

The boll weevil entered Texas in the early 1890s, and by 1920 it reached the eastern seaboard. Local communities were often devastated by the infestation. Everywhere, growers significantly changed cultural practices, opting for earlier planting, earlier-ripening and more resistant varieties, more fertilizer and insecticides, and giving more attention to cleaning their fields after the harvest to reduce the weevil's habitat. These modifications in farming practices allowed agriculturalists to continue cotton production. One consequence was the wholesale abandonment of most late-ripening and high-quality varieties, thus leading to a decline in the overall quality of U.S. cotton (Ware, 1936, p. 661; Lange et al., 2009, pp. 685–718).

Our account of technological change in the cotton industry illustrates several general points. Both mechanical and biological changes were important. The cotton story indicates how concentrating on ex post crude yield and quality indices can mask the importance of biological innovations. The Red Queen effect was ever at work. Because of the boll weevil, cotton yields and quality declined significantly. But, without the substitution to early varieties, it is doubtful that cotton production could have survived in large areas. Technological changes thus prevented yields and quality from declining even more. The large increase in hand-picking efficiency resulting from the diffusion of Mexican hybrids also calls into question the popular notion that machines only saved labor and biological innovations only saved land. As we shall demonstrate, there are just too many counterexamples to make this paradigm very useful.

Wheat. Mechanization has denominated the standard account of technological change in the production of wheat (and other small grains such as rye, oats, and barley). The achievements were indeed spectacular. Small grains had long been harvested with sickles—short curved knives. Men and women had to stoop over, grab a clump of stalks with one hand, and then cut it with the sickle. The grain was then gently laid on the ground so that seeds did not shatter. Scythes (a tool with a long handle and a curved blade two to three feet long) were used to mow grass and harvest grains. With a scythe, the harvester could stand erect and work much faster, but grains were apt to shatter. In the late Colonial Era, American farmers began adding a wooden basket to the rear of the scythe to catch the cut grain. This improved scythe was called a cradle. With the turn of the arm, the cradler could gently deliver the cut grain to the ground, thereby reducing shattering. One cradler could do the work of several workers with sickles. But cradling was heavy work, so harvesting evolved from a job done by both men and women to largely a man's task. There was still plenty of work for women and children. The grain (still attached to the stalk) needed to be gathered into small bundles and stacked in a shock to dry. The grain would then be threshed and winnowed by hand.

The threshing machine, which began diffusing in the 1820s, represented the first truly monumental change in grain mechanization. By the 1850s these machines were in general use in major

American wheat-growing areas. Farmers fed the grain still attached to the stalks into the machine, which separated the wheat from the straw and chaff. These were postharvest operations and there was no great urgency to complete the task. In the months following the harvest, commercial threshers moved from farm to farm with large horse-powered machines. The neighboring farmers supplied much of the labor and the horses.

The second fundamental mechanical change in wheat farming centered on the invention and diffusion of the mechanical reaping machine. The early models replaced cradlers, and after decades of refinement, later models could perform the tasks of those who gathered and shocked the cut grain. The most famous of the early inventors was Cyrus McCormick of Virginia, who patented his first machine in 1834. These early machines were pulled by four horses. McCormick moved his operation to Chicago in 1847–1848 to be closer to the growing Midwestern market, and in 1858 he made over 4,500 machines. Many other producers entered the market and total output soared. The reaper had significant effects on farm structure, vastly reducing the need for seasonal labor during the small-grain harvest. In addition, the reaper technology was applied to developing specialized mowers to cut grasses for livestock. This greatly reduced the labor required to maintain milk cows and led to an increase in dairy production (Hutchinson, 1930, pp. 250–275, 366; Olmstead, 1976, pp. 36–50).

These breakthrough inventions unleashed a wave of subsequent improvements that increased the machines' usefulness and reliability. By 1860, many reapers were equipped with automatic rakes that neatly deposited bundles of wheat on the ground to be gathered and shocked. This saved the labor of an additional worker. In the 1870s, an innovative twine binder was added to new reapers so that the bundles were tightly bound and easier to shock.

The crowning achievement in grain harvesting was the commercial development in the 1880s of the combined harvester, or, as it became known, the combine. The combine merged a reaper and a thresher into one machine. Early combines were huge, cumbersome machines suited only for the large-scale ranches of the arid West. Some of these harvesters had forty-foot cutting bars (as opposed to six to eight feet for the standard reaper) and were pulled by teams of forty or more draft animals. The evolution of the combine involved making the machines smaller and perfecting its components for use in harvesting crops other than small grains. Gradually, tractors replaced the horses. This allowed the combine to be profitably employed in the grain-growing regions east of the Rockies. Combines were rare in Kansas in 1918. By 1926, combines harvested about 30 percent of its wheat crop and by 1938, over 80 percent. The number of combines in the United States grew from 4,000 in 1920 to 61,000 in 1930 and then to 190,000 in 1940. By the late 1930s, combines harvested about one half of all U.S. wheat acreage.

The next important development was the spread of the self-propelled combine, which integrated the tractor into the machine. This required only one operator. In the 1940s, there was a reversal in the trend toward smaller machines as specialized custom harvesting services began to thrive. The share of U.S. wheat acreage harvested by combines rose to more than 75 percent by 1945 and to almost 95 percent by 1950 (Olmstead and Rhode, 2006). The combine spread to other crops. Combines had become the dominant harvesting technology for virtually every grain and field legume by 1980.

The standard account of technological change in wheat stops here—labor-saving machines (aided by the extension of railroads) facilitated both the increase in production and the westward shift in cultivation. But, as with cotton, biological innovations were crucial to the wheat industry's success. Between 1839 and 2009, U.S. wheat output increased 26 times and the center of production shifted about one thousand miles west from eastern Ohio into Nebraska. Almost all of this movement occurred before 1929 (Olmstead and Rhode, 2011, pp. 480–485). The change in location entailed moving to lands with far drier and harsher climates. In 1900, Mark Alfred Carleton, the USDA's most prominent wheat agronomist, noted that the regions of North America producing wheat were as "different from each other as though they lay in different continents"

and required entirely different varieties of wheat (Carleton, 1900, p. 9).

The march of wheat cultivation across North America was first and foremost a story of creating new crop varieties to meet new agroclimatic challenges. In the mid-seventeenth century, New England farmers experimented with numerous varieties in search of cold-tolerant and pest-resistant wheats, but without much success. Problems continued as settlers inched westward. In the 1840s, attempts to grow winter wheat in Wisconsin failed. Farmers only succeeded after they switched to a new variety of spring wheat (Olmstead and Rhode, 2008, p. 42).

The successful movement onto the even harsher conditions of the Great Plains was dependent on the introduction and breeding of hard red wheats that were entirely new to North America. Over the late nineteenth century, the premier hard spring wheat cultivated in North America was Red Fife. According to a popular account, David and Jane Fife of Otonabee, Ontario, selected and increased the grain stock from a single plant grown on their farm in 1842. Red Fife became the basis for the spread of the wheat frontier into Wisconsin, Minnesota, the Dakotas, and the Canadian Prairies.

Another key breakthrough came with the introduction of "Turkey" wheat, a hard red winter variety suited to Kansas, Nebraska, Oklahoma, and the surrounding region. The standard account credits German Mennonites, who migrated from southern Russia, with the introduction of this variety in 1873. Early settlers in Kansas had experimented with scores of soft winter varieties common to the eastern states, but these varieties could not survive the cold winters and hot, dry summers. In 1919, Turkey-type wheat made up over 80 percent of the wheat acreage in Nebraska and Kansas and nearly 70 percent in Colorado and Oklahoma. It accounted for 30 percent of total U.S. wheat acreage during that same year (Quisenberry and Reitz, 1974).

These innovations were often the direct result of government investments. In 1886, the Canadian Parliament created the federal experiment station system. The most acclaimed Canadian breeder, William Saunders, began systematically hybridizing high-quality local varieties with early-maturing stocks imported from around the world. In 1903, William's son, Charles Saunders, took over the work. The most valuable result of their combined research was Marquis, a cross between Red Fife and Red Calcutta, an early variety from India. Released in 1909, Marquis matured about 10 days earlier than Red Fife and was more resistant to disease. By 1916, it was the leading variety in the northern grain belt (Morrison, 1960; Olmstead and Rhode, 2008, pp. 30–33).

The rapid spread of Marquis was not an isolated case. Following extensive expeditions to the Russian Empire, Carleton introduced several durum varieties around 1900. These winter-hardy spring wheats proved relatively drought and rust resistant. By 1906, American durum production increased to 50 million bushels. In 1900, Carleton also introduced Kharkof, a Turkey-type hard-winter wheat that was well suited to the cold, dry climate in western and northern Kansas. By 1914 it accounted for one half of the state's crop. Without Turkey varieties, the wheat crop of Kansas circa 1920 probably would have been only half what it actually was, and farmers in Nebraska, Montana, and Iowa would have had to grow spring wheat (Olmstead and Rhode, 2008, pp. 30–37).

Wherever it is feasible, farmers generally prefer to grow winter wheat instead of spring wheat. Winter wheat is planted in the fall and the root system gets a head start before the plant goes dormant during the cold months. Spring wheat is planted after the ground thaws in the spring. Winter wheat generally offers higher yields and is less vulnerable to insects and diseases because it could be harvested earlier. However, in colder climates winter wheat suffers high losses to winterkill. The development of more hearty-winter varieties was a great achievement. Between 1869 and 1929, most of Kansas and Iowa, southern Nebraska, and parts of other states shifted from spring to winter wheat. This area that shifted to winter wheat accounted for almost 30 percent of U.S. output in 1929 (Olmstead and Rhode, 2008, pp. 37–39).

One of the crucial changes over the twentieth century was the shift to shorter wheat varieties that produced less straw and more grain, thereby increasing the "harvest index." The most celebrated example of this shift was the spread of the semidwarf varieties after the 1940s. This was one of a long list of innovations in wheat culture that depended on the introduction of foreign varieties into the United States. But the timing of the adoption of dwarf varieties also reflected the decline in the demand for straw caused by the spread of tractors and the change in harvest technologies. Farmers had previously rejected dwarf varieties because they were harder to harvest by hand or with a mechanical reaping machine. This all changed with the spread of the combine because now the extra straw slowed the rate of threshing.

Adapting wheat to new geoclimatic conditions and making it more compatible with new machines was only part of the story. As with cotton, myriad pests and diseases evolved to feed on the plant. Continued cultivation of wheat in a region invited infestations of the Hessian fly and scores of other insects. The Hessian fly had a devastating impact, leading many farmers to abandon growing wheat. Stem rust, blight, and smut also caused serious damage, and in many areas winterkill was always a threat.

Well before the American Civil War scientists and farmers, through a process of careful observation, developed an understanding of the life cycle of the major pests. With this knowledge, farmers changed varieties and cultural practices on a regular basis to limit losses to insects and diseases. Farmers learned that they could reduce the damage caused by the Hessian fly by sowing late in the fall after the first frost (or for spring wheat, early in the season) and by better cleaning their fields. Efforts to combat the fly also led to the search for new varieties. The most important was a bearded red winter wheat, Mediterranean, that was introduced in 1819 and became the dominant variety in the country by the 1850s. The wide diffusion of Red Fife, Turkey, Marquis, and scores of other varieties was in large part the result of their resistance to wheat rusts when first introduced. Eventually, diseases evolved to overcome the plants' defenses, and farmers (aided by plant breeders) had to switch to new varieties (Olmstead and Rhode, 2008, pp. 40–57).

The net result was a fundamental change in the character of the wheat varieties grown in North America. There were no hard red spring wheats in the United States before the mid-1850s and no hard winter wheats before 1873. In 1929, a decade before the presumed onset of the biological revolution, over 80 percent of U.S. wheat acreage was planted with varieties that did not exist in North America until 1873. In the newer wheat-growing areas of the Great Plains and the Pacific Northwest, new varieties were even more important (Olmstead and Rhode, 2008, pp. 30–36).

What would have happened if biological technologies had not changed? In the early 1950s the descendants of the durum wheat varieties that Carleton had introduced 50 years earlier succumbed to a new race of black stem rust. As a result, yields plummeted from 14.5 to 3 bushels an acre between 1951 and 1954 (Johnson and Gustafson, 1962, p. 120). Farmers scurried to adopt new varieties. The evidence for earlier epidemics is less well documented, but it is very likely that without replacing "worn-out" wheat varieties and adjusting cultural methods to combat pests, the yields would also have fallen precipitously.

Corn.   There were no landmark mechanical innovations in the corn sector that rivaled the importance of the reaper for wheat and the saw gin for cotton. This does not mean technology stood still. On the contrary, a succession of mechanical technologies, including a multibottom riding plow and improved planters, advanced farm productivity. The spread of check row planters following the Civil War made it possible to drop seeds at uniform distances in straight lines. This allowed horse-drawn cultivators to operate in perpendicular directions when weeding fields. This system both saved labor and increased yields. Mechanical corn pickers were commercialized relatively late and began to diffuse only in the late 1920s when coupled with tractors.

Corn holds a special place in the traditional history of American agricultural technology. The diffusion of hybrid corn in the 1930s has been portrayed as a watershed event marking the advent of biological innovation. Not only is this view inappropriate for the cotton and wheat

sector, but also it misrepresents the changes that transformed corn farming. Northeastern farmers typically grew flint corns. These early-maturing, slender-stalked varieties produced several ears with eight to ten rows of smooth, hard kernels. Southern farmers typically grew later-maturing, heavier-stalk gourdseed or dent corns that produced rounded, many-rowed ears with softer, dimpled kernels. The Corn-Belt Dents that transformed Midwestern agriculture were created in the nineteenth century by crossing southern dents and northern flints (Anderson and Brown, 1952, pp. 2–8).

The classic "founding story" of the Corn-Belt Dents highlights the development of Reid's Yellow Dent. As with many early advances, this depended on serendipity, keen observation, and close attention to detail. In 1845–1846, Robert Reid moved from southern Ohio to central Illinois. In 1847, he planted a field of Gordon Hopkins corn, a late red strain from the Shenandoah Valley. The variety performed poorly. He replanted the "missing hills" with Little Yellow, a local flint corn. Accidental crosses between the flint and dent possessed the best qualities of both types. Subsequent improvements resulted in an early high-yielding strain with 18 to 24 rows of relatively smooth kernels. Reid's Yellow Dent gained wide popularly in the Midwest. It served as the basis for many other improved varieties of open-pollinated corn and provided much of the germ stock used in the hybrid seed.

E. G. Montgomery, a Cornell University agronomist, observed in 1913: "Since 1840 there has been a rapid expansion of corn culture and great interest has been shown in the development of varieties adapted to various conditions and uses. It may be safely estimated that perhaps three-fourths of the present varieties of corn have been developed since 1840" (Montgomery, 1916, pp. 79 and 181). Varieties were matched to local conditions; yields dropped off sharply if the variety was moved far from its specific niche (especially to a different latitude). As with cotton and wheat, breeders also succeeded in adapting maize to harsher climates. Several breakthroughs, especially those of Andrew Boss at the University of Minnesota, facilitated the shift of the Corn Belt several hundred miles to the north (Olmstead and Rhode, 2008, pp. 80–84).

The pace of biological innovations did indeed accelerate with the more formal application of biological sciences to farming. A major advance occurred in 1908 when George H. Shull applied genetic theory to develop pure inbred lines of corn and produced a superior hybrid through single crossing. Farmers had long known that inbreeding produced runty-looking corn with low yields, which had little immediate appeal. By crossing known inbred lines, Shull created a means of reliably producing higher-yielding F1 seed. But because of stunting, the output of F1 seed was too low to be commercially viable. In 1917, Edward M. East and Donald Jones of the Connecticut Experiment Station made a key second advance. They multiplied the available seed by crossing the F1 stock to produce F2 seed, which was generally less productive than the original cross but sufficiently numerous for commercial uses. By the early 1920s, commercial seed firms such as DeKalb Corn and Pioneer Seed Co. commenced selling hybrid seed. Farmers adopted the seed in close accord with its economic advantages. Farmers in Iowa were the leaders, with initial adoption in the early 1930s. One half of Iowa corn was hybrid by 1938, and the diffusion process in that state was virtually completed by 1941. For the country as a whole, the spread of hybrid corn was somewhat slower; hybrid seed accounted for one half of the corn planted in 1943 and over 95 percent in 1959 (Manglesdorf, 1951; Griliches, 1957).

Hybrid corn initially offered yields 15 to 20 percent higher than open-pollinating varieties. Even after hybrid seed had fully diffused, corn yields continued to increase rapidly, primarily as a huge increase in the use of nitrogen fertilizer. There was a high degree of complementarity between improved varieties and the greater use of fertilizer, and probably more than half of the increase in corn yields was caused by the increased application of nitrogen (Sundquist et al., 1982, p. IV–4).

Fertilizer. Many technological changes transcended any one crop. Plows, seeding machines, and combines could be used for many crops. The ideas and techniques developed by the breeders of one crop stimulated advances in other crops. Technological changes associated with the development of improved fertilizers stand out for their wide application across the agricultural sector. There was an upsurge in yields coinciding with the introduction of hybrid corn, but we must be careful in deducing the cause of the higher yields. After all, the use of hybrid seed is specific to maize, sorghum, and a handful of other crops, but the yields of many additional crops also shot up. The increasing use of synthetic nitrogen fertilizers explains some of the general pattern. In 1909, after years of research, the German chemist Fritz Haber made a major breakthrough in synthesizing ammonia. Led by Carl Bosch, Badische Anilin–und Soda–Fabrik (BASF) overcame numerous technical obstacles and commercialized Haber's discoveries in 1913. Abrogation of the German monopoly and subsequent improvements in the Haber–Borsch process dramatically increased the supply of nitrogen while lowering its price. These changes indeed revolutionized American agriculture as cheaper nitrogen fertilizers became available after World War II (Smil, 2001, pp. xv, 60–107). Farm purchases of primary plant nutrients, which had doubled between 1910 and 1940, increased eightfold over the next 30 years. But other forces besides fertilizers were increasing agricultural yields.

Livestock Improvement. Livestock productivity for everything from chickens to cows also increased after World War II. These changes were not directly associated with lower fertilizer prices and suggest the broader impact of breeding and other scientific and applied advances across the agricultural sector. Improvements in the livestock sector date to the colonial period. The animals of 1940 bore little resemblance to those of 1700. Foreign introductions and domestic breeding led to new specialized breeds that gained weight faster and provided more and better services to their owners than the animals they replaced. In 1800, bovines were multiple-purpose animals, providing draft power, meat, and milk. By 1940, single-purpose cattle breeds specialized for either beef or dairy production dominated (and oxen used for draft power largely disappeared). Sheep experienced a similar evolution as different breeds suitable for producing wool or mutton emerged.

Production per animal soared. The output of milk per cow tripled between 1800 and 1940. The amount of wool clipped per sheep quadrupled over the same period. The age at which hogs, steers, and sheep were slaughtered declined. By 1940 hogs and steers generally reached market age in less than half the time they did in 1800. The quality of many animal products improved significantly. In addition, to breeding this reflected the deeper biological transformation that was built upon more intensive systems. Soybeans, corn, alfalfa, cottonseed products, and new grasses, along with processing and storage facilities, replaced or supplemented low-density grazing operations (Olmstead and Rhode, 2008, pp. 284–329).

Selective animal breeding dates back to ancient times, but the advent of modern genetic and veterinary sciences, the development of improved breed registries, and the spread of artificial insemination greatly accelerated productivity increases. Institutional innovations, such as dairy herd–improvement associations, first organized in 1906, and the national poultry improvement plan, which dates to the 1930s, stimulated genetic advances. The first known use of artificial insemination on U.S. dairy farms occurred in the mid-1930s, and by the mid-1970s, about one half of all dairy cows were bred artificially. With these changes came vast improvements in feed, including the use of concentrates and hormones, the control of diseases, and, in some cases, a wholesale restructuring of climatic and environmental conditions (Olmstead and Rhode, 2008, pp. 344–347).

Draft Power. The "first mechanical revolution" associated with the spread of reapers and threshers also entailed a biological revolution because there was a wholesale movement from oxen to equines. Horses and mules could maintain a faster pace more suitable to the new machines. Farmers bred much larger and more powerful horses. Many farmers also learned to manage

larger teams. The 40-horse teams pulling western combines in the 1890s were a novelty that amazed farmers in the East, who seldom employed more than four horses at a time.

The gasoline tractor was a revolutionary technology that transformed rural America. The early gasoline tractors were behemoths, patterned after the giant steam plows that preceded them (Olmstead and Rhode, 2001). They were useful for plowing, harrowing, and belt work but not for cultivating in fields of growing crops. Diffusion of tractors required a number of important refinements and the development of a variety of sizes suitable for a number of different tasks. At the same time, progress in the mass-production industries, coupled with competition among firms, helped drive down the tractor's price and improve its overall quality.

Several advances highlight the improvement in tractor design. The Bull (1913) was the first truly small and agile tractor, the Fordson (1917) was the first mass-produced entry, and the McCormick–Deering Farmall (1924) was the first general-purpose (GP) tractor capable of cultivating among growing row crops. Before the GP tractor, farmers of many crops needed to keep horses for cultivating. The Farmall also incorporated a power takeoff, enabling it to transfer power directly in implements under tow. Allied innovations, such as improved air filters, stronger implements, pneumatic tires, and the Ferguson three-point hitch, increased the tractor's life span and usefulness. After World War II, technological development involved refining designs, increasing tractor size, and adding driver amenities.

The number of tractors on American farms grew from about 1,000 machines in 1910 to 246,000 in 1920 and 920,000 in 1930. With the exception of the 1929–1933 period, the stock of tractors continued to increase, first surpassing the 4 million mark in 1953. Thereafter, the stock has remained relatively constant, but the horsepower per machine increased dramatically. Large farmers in the Great Plains and California were the earliest adopters. The development of the GP tractor in the mid-1920s made the machine suitable for more tasks and quickened the pace of diffusion in the Midwest. The South was the last region to achieve widespread adoption. Diffusion accelerated during and immediately after World War II. By 1960, tractors had largely replaced horses on commercial farms. The stock of farm horses and mules declined from 26.5 million in 1915 to 3.1 million in 1960 (Olmstead and Rhode, 2001, pp. 663–698).

The tractor greatly increased the amount of power available to farmers and saved the time that had been devoted to caring for draft animals. By 1960 tractors had reduced annual labor needs in agriculture by roughly 1.7 million workers. But the tractor was also one of the greatest land-saving innovations in the history of agriculture, allowing farmers to convert about 80 million acres of cropland and another 80 million acres of pastureland from supporting draft animals to providing products for human consumption. The cropland saved was roughly equivalent to adding acreage equal to two-thirds of the cropland harvested in the territory of the Louisiana Purchase to the nation's effective land base. But there's more. By allowing farmers to perform work better, the tractor had a direct impact on yields. Thus, it is very difficult to conclude whether one of the seminal mechanical inventions of all time was, on balance, land saving or labor saving (Olmstead and Rhode, 2001, pp. 663–698).

Conclusion. Scientific advances and technological changes continued to transform American agriculture in the early twenty-first century. The application of ground positioning systems to all sorts of equipment increased the efficiency of tillage operations and allowed farmers to more economically target irrigation water. Zero-tillage techniques are environmentally friendly and save labor, machines, and fuel. In 2012, machines performed many complex operations such as picking some fruits and vegetables. Where hand labor was still used, workers were more apt to be aided by machines. Although highly controversial (as were many technological innovations in the past), genetically modified crops offered the promise of further efficiency gains. Plants were being bred and modified to resist pests, thrive in arid or saline con-

ditions, and more efficiently absorb nutrients. The northern movement of winter wheat that started in the 1870s continued through the early twenty-first century. Changes after 2012 could never rival the impact of past technological innovations in releasing labor from farming because so few workers remained in the farm sector. However, changes were expected to be crucial in providing the food for a growing world population. To do this, farmers needed to counter new threats from pests and confront the environmental shocks expected with global warming. The historical record offers a sense of how farmers met past challenges.

During the nineteenth and twentieth centuries, scientists and agriculturalists created new biological technologies that allowed U.S. farmers to repeatedly push cultivation of the three major staple crops into environments previously thought too arid, too variable, and too harsh to farm. The climatic challenges that these farmers overcame rivaled the magnitude of the climatic changes predicted for the next hundred years in the United States. Public and private research was expected to be crucial in meeting the challenges associated with impending pests and climate changes. Because research has shown that there are longtime lags between investment in research and payoffs, a high priority should be given to reinvigorating stagnating public investment in farm productivity–oriented agricultural research (Alston et al., 2010; Olmstead and Rhode, 2011).

[*See also* **Agricultural Education and Extension; Agricultural Experiment Stations; Agriculture, U.S. Department of; Automation and Computerization; Barbed Wire; Biological Sciences; Biotechnology; Chemistry; Cotton Gin; Disease; Ecology; Engineering; Entomology; Food and Diet; Food Processing; Genetics and Genetic Engineering; Geography; Germ Theory of Disease; Global Warming; Hybrid Seeds; Machinery and Manufacturing; McCormick, Cyrus Hall; Meteorology and Climatology; Morrill Land Grant Act; Pesticides; Petroleum and Petrochemicals; Pure Food and Drug Act; Railroads; Research and Development (R&D); Technology;** *and* **Whitney, Eli.**]

BIBLIOGRAPHY

Alston, J. M., M. A. Andersen, J. S. James, and P. G. Pardey. *Persistence Pays: U.S. Agricultural Productivity Growth and the Benefits from Public R&D Spending*. New York: Springer, 2010. Prominent agricultural economists estimate the dynamic benefits from agricultural research. They note a reduction in agricultural production research results in a slower agricultural productivity growth.

Anderson, Edgar, and William L. Brown. "The History of the Common Maize Varieties of the United States Corn Belt." *Agricultural History* 26, no.1 (1952): 2–8. Corn scientists highlight the many crossings of southern dent corn varieties with northern flint varieties to produce the distinctive yellow maize that became the hallmark of the Corn Belt in the late nineteenth century.

Carleton, Mark Alfred. "The Basis for the Improvement of American Wheats." *USDA Division of Vegetable Physiology and Pathology Bulletin*, no. 24 (1900). America's leading wheat breeder in the early twentieth century gives a synopsis of important research and results.

Cochrane, Willard W. *The Development of American Agriculture: A Historical Analysis*. Minneapolis: University of Minnesota Press, 1979. This highly cited interpretation of American agricultural development emphasizes the importance of mechanization before 1940.

Griliches, Zvi. "Hybrid Com: An Explanation of the Economics of Technological Change." *Econometrica* 25 (1957): 301–322. This classic treatment of technological diffusion offers lessons that apply beyond the examination of the adoption of hybrid corn.

Haines, Michael R. "Population, by Region and Urban–Rural Residence: 1790–1990." Table Aa36-92. In *Historical Statistics of the United States, Earliest Times to the Present*, millennial ed., edited by Susan B. Carter, Scott Sigmund Gartner, Michael R. Haines, Alan L. Olmstead, Richard Sutch, and Gavin Wright. New York: Cambridge University Press, 2006. http://dx.doi.org/10.1017/ISBN-9780511132971.Aa1-109 (accessed 16 October 2012).

Hayami, Yujiro, and Vernon Ruttan. *Agricultural Development: An International Perspective*, revised and expanded ed. Baltimore: Johns Hopkins University Press, 1985. This is the most important treatment of the induced innovation model in explaining the creation and diffusion of agricultural innovations.

Hutchinson, William T. *Cyrus Hall McCormick: Seedtime, 1809–1856*. New York: Century Co., 1930. This is the single most influential history of one of the world's great inventors and inventions.

Johnson, D. Gale, and Robert L. Gustafson. *Grain Yields and the American Food Supply: An Analysis of Yield Changes and Possibilities*. Chicago: University of Chicago Press, 1962. An early attempt to measure the sources of increases in grain yields that clearly illustrates the devastating impact that stem rust can have on yields.

Lakwete, Angele. *Inventing the Cotton Gin: Machine and Myth in Antebellum America*. Baltimore: Johns Hopkins University Press, 2003. Lakwete argues that Eli Whitney's invention was not as critical as generally argued.

Lange, Fabian, Alan L. Olmstead, and Paul W. Rhode. "The Impact of the Boll Weevil, 1892–1932." *Journal of Economic History* 69, no. 3 (2009): 685–718. This article quantifies the destruction caused by the boll weevil as well as the adjustments farmers made to limit the damage.

Manglesdorf, Paul. "Hybrid Corn." *Scientific American* 183 (1951): 39–47. This is an analysis of the development of hybridization techniques.

Montgomery, E. G. *The Corn Crops: A Discussion of Maize, Kafirs, and Sorghums as Grown in the United States and Canada*. New York: MacMillan Co., 1916. Montgomery reviews the status of corn development on the eve of hybridization.

Moore, John Hebron. *The Emergence of the Cotton Kingdom in the Old Southwest, Mississippi, 1770–1860*. Baton Rouge: Louisiana State University Press, 1988. Moore offers a brief but insightful treatment of some of the biological innovations that revolutionized cotton production in the slave South.

Morrison, J. W. "Marquis Wheat—A Triumph of Scientific Endeavour." *Agricultural History* 34, no. 4 (1960): 182–188. Morrison details the careful and meticulous research efforts of William and Charles Saunders (and others) in creating the wheat variety that dominated production for decades in parts of the northern wheat belt.

Olmstead, Alan L. "The Civil War as a Catalyst of Technological Change in Agriculture." *Business and Economic History* 2d ser., 5 (March 1976): 36–50. Although reaper sales soared during the Civil War, institutional and technical forces already at work drove down reaper prices.

Olmstead, Alan L., and Paul W. Rhode. *Creating Abundance: Biological Innovation and American Agricultural Development.* New York: Cambridge University Press, 2008. This book upsets the long-held belief that mechanical innovations were the dominant source of American agricultural productivity advance from colonial times to the late 1930s. It examines the changes in crop technology that made possible the settlement of new geoclimatic zones and the need to combat agricultural pests.

Olmstead, Alan L., and Paul W. Rhode. "Adapting North American Wheat Production to Climatic Challenges, 1839–2009." *Proceedings of the National Academy of Sciences* 108, no. 2 (11 January 2011): 480–485. North American farmers overcame climatic challenges in moving wheat production into western areas with harsher climates that rivaled the magnitude of the challenges predicted over the next hundred years caused by global warming.

Olmstead, Alan L., and Paul W. Rhode. "Farms and Farm Structure." In *Historical Statistics of the United States, Earliest Times to the Present,* millennial ed., edited by Susan B. Carter, Scott Sigmund Gartner, Michael R. Haines, Alan L. Olmstead, Richard Sutch, and Gavin Wright, vol. 1, pp. 10–15, 39–88. New York: Cambridge University Press, 2006.

Olmstead, Alan L., and Paul W. Rhode. "Reshaping the Landscape: The Impact and Diffusion of the Tractor in American Agriculture, 1910–60." *Journal of Economic History* 61, no. 3 (September 2001): 663–698. This paper analyzes the impact of the tractor on land use, on labor demand, on the demand for draft animals, and on aspects of farm productivity.

Quisenberry, Karl S., and L. P. Reitz. "Turkey Wheat: The Cornerstone of an Empire." *Agricultural History* 48, no. 1 (1974): 98–114. Two USDA scientists analyze how the introduction of Turkey wheat made wheat culture commercially feasible in the southern Great Plains.

Smil, Vaclav. *Enriching the Earth: Fritz Haber, Carl Bosch, and the Transformation of World Food Production.* Cambridge, Mass.: MIT Press, 2001. Smil offers an original analysis of the development and the importance of the Haber–Bosch process for producing nitrogen. This process vastly increased the world's agricultural production capacity and was an important complement to the subsequent breeding advances of the Green Revolution.

Street, James H. *The New Revolution in the Cotton Economy: Mechanization and Its Consequences.* Chapel Hill: University of North Carolina Press, 1957. This is the classic historical treatment of the impact of cotton mechanization in the first half of the twentieth century.

Sundquist, W. Burt, et al. "A Technology Assessment of Commercial Corn Production in the United States." *Minnesota Agricultural Experiment Station Bulletin* no. 546 (1982): IV–4. This investigation surveys productivity advances in corn cultivation between 1949 and 1979.

U.S. President. *Economic Report of the President.* Appendix B, Table B-35, p. 361. Washington, D.C.: U.S. Government Printing Office, 2012.

Ware, J. O. "Plant Breeding and the Cotton Industry." In *Yearbook of Agriculture, 1936,* pp. 657–744. Washington, D.C.: U.S. Government Printing Office, 1936.

Alan L. Olmstead
and Paul W. Rhode

AGRICULTURE, U.S. DEPARTMENT OF

The U.S. Department of Agriculture (USDA) was established in 1862. The first commissioner of agriculture, Isaac Newton, organized the department by scientific fields such as entomology and botany, each represented initially by one person and later by a division. Early contact with farmers was limited to the distribution of seeds for trial and the publication of advisory pamphlets. The USDA began with a small budget and an uncertain relationship to the land-grant universities in the development of agricultural science. In 1887 the Hatch Act established state agricultural experimental stations as institutions for basic research in agriculture-related fields.

In the late nineteenth century the USDA began reorganizing its research activities to focus on specific problems, such as livestock diseases. The first new unit was the Bureau of Animal Industry, created in 1884. With permanent status and funding, an approving interest group (cattle ranchers in this case), and regulatory power, this new bureau made rapid progress against livestock diseases and became the model for other problem-oriented bureaus. Harvey W. Wiley, head of the Bureau of Chemistry from 1883 to 1912, led a long campaign against food adulteration, culminating in the 1906 Pure Food and Drug Act. The high-visibility activist Gifford Pinchot ran the Bureau of Forestry (later Forest Service) from 1898 to 1910, pursuing a policy of sustained use that has remained largely unchanged.

Working in the Bureau of Plant Industry during the early twentieth-century period when the boll-weevil infestation began to devastate the southern cotton industry, Seaman A. Knapp started "demonstration farms" to persuade farmers to try new methods developed by agricultural scientists, particularly those in the Bureau of Entomology. Knapp's success led to the Smith–Lever Act of 1914, which formed the Extension Service as a cooperative program of the USDA and land-grant colleges.

The USDA became a cabinet-level department in 1889. Secretary James Wilson (1897–1913) oversaw the expansion of scientific work in the bureaus, the establishment of federal experiment stations in new territories such as Hawai'i and Puerto Rico, and the introduction of new crops and varieties by the "plant explorer" David Fairchild. Henry C. Wallace (1921–1923) increased the department's work in agricultural economics, recognizing that merely improving production did not guarantee prosperity for farmers. His son, Henry A. Wallace (1933–1940), implemented such New Deal programs as the Agricultural Adjustment Acts, which introduced crop reduction and agricultural subsidies to boost prices, policies that in some form would long remain in place.

Direct contact between department scientists and farmers declined over time. In 1954 the bureaus were abolished and their scientific work reorganized into the Agricultural Research Service. The Food and Drug Administration was removed from the department in 1940. By World War II, the department was emphasizing pesticides to the near exclusion of other methods of insect control, opening it to a storm of criticism by environmentalists after the publication of Rachel Carson's *Silent Spring* in 1962. The controversy reflected a long-standing ideological divide between advocates of maximum resource utilization in the USDA and preservationists in the Department of the Interior. During the late twentieth century the USDA came under criticism for favoring large agribusinesses and big, rich, white farmers in its scientific research, economic policies, and regulatory actions.

[*See also* **Agricultural Education and Extension; Agricultural Experiment Stations; Agricultural Technology; Biological Sciences; Botany; Carson, Rachel; Disease; Entomology; Environmentalism; Food and Diet; Food Processing; Forest Service, U.S.; 4-H Club Movement; Home Economics Movement; Hybrid Seeds; Morrill Land Grant Act; Pesticides;** *and* **Pure Food and Drug Act.**]

BIBLIOGRAPHY

Dupree, A. Hunter. *Science in the Federal Government*. Baltimore: Johns Hopkins University Press, 1957.
Marcus, Alan I., and Richard Lowitt, eds. *The United States Department of Agriculture in Historical Perspective, Agricultural History 642*, special issue. Mississippi State: Agricultural History Society, 1990.

Richard C. Sawyer

AIDS

See HIV/AIDS.

AIRPLANES AND AIR TRANSPORT

The invention of the airplane by Wilbur and Orville Wright in 1903 fulfilled the ancient dream of human flight. However, after inventing the airplane, the brothers both failed to produce safe and practical production aircraft that pilots and mechanics could easily fly and maintain; they also failed to anticipate the future direction of aircraft design. Thus, when Louis Blériot flew across the Dover straits on 25 June 1909, his flight signaled French dominance in global aviation, a position it held throughout the "Great War." In response, the Wright brothers attempted to maintain their dominance through a series of international patent suits. But they succeeded only in severely constraining American aeronautics, which, already insufficiently supported by governmental, industry, and academic institutions, now entered a steady decline. America's aircraft industry increasingly looked to Europe for design inspiration, even for such basic aircraft as the Curtiss JN "Jenny" trainer, an iconic design. After entry into World War I, it proved incapable of meeting the need for thousands of combat aircraft, so American aviators obtained their aircraft from the Allies. One of these, the British de Havilland D.H. 4, became America's first great postwar commercial aircraft, a mainstay of the postal airmail service.

Developing Air Transport, 1918–1941. Various experiments took place in America and Europe before and during World War I with the carriage of mail, parcels, and passengers, but it was only after the war that practical, scheduled airmail and air transport services began. World War I had rapidly accelerated aviation's advancement, encouraging development of larger, more reliable, and more powerful piston engines, and, thereby, larger and more capable aircraft. By the middle of the war, Russia, Germany, Italy, Britain, and France all operated large multiengine biplanes. Three crossings of the North Atlantic in 1919 (by a flying boat, a bomber, and an airship) and flights (in stages) from Britain to India, Australia, and South Africa highlighted wartime advances in aviation technology.

The international foundation. Starting that year, British, Dutch, French, and German aviators established air services across Europe, into North Africa, and (via exported aircraft) into the Americas and Asia. America's only civil air venture in this period was the establishment in 1918 of a postal airmail service, although some individual designers and entrepreneurs (most notably Alfred Lawson) developed and attempted (unsuccessfully) to market rudimentary airliners.

Until World War I, practical craftsmanship rather than scientifically rooted engineering dominated aircraft design. The growing speed, power, and structural loadings of aircraft brought university-level mechanical engineering, fluid dynamics, and applied mathematics increasingly to the forefront of aircraft design. German researchers had created quantitative aerodynamics and aeronautical engineering. Now, fluid-dynamicists Ludwig Prandtl, Max Munk, and Theodore von Kármán, and engineers Hugo Junkers, Claude Dornier, and Adolf Rohrbach virtually "reinvented" the airplane, exemplified by Junkers' F-13 low-wing, all-metal enclosed-cabin monoplane of 1919, the most influential and successful air transport placed in service prior to the American air transport revolution of the early 1930s.

The first major international airlines—France's l'Aéropostale, Britain's Imperial Airways, Germany's Luft Hansa (later Lufthansa), Holland's KLM, Belgium's Sabena, and Australia's Qantas—largely depended upon direct government subvention to adequately finance their operations. Together these lines rapidly pushed airmail and transport routes across Europe and into Russia, to North Africa, the Middle East, across the South Atlantic and South America, throughout Africa, and through Egypt to India and the Far East.

The United States lingered behind all of these, being last to develop both a governmental infrastructure supporting civil air transport and a suitably networked airline system. Even so, by 1934, its air transport system (in route mileage, technical support, equipment, and sophistication of service) surpassed that of the rest of the world. The United States had not suffered extensive material, manpower, and economic losses in the war, and at war's end, it was the world's strongest economic and industrial power. Conscious of its wartime failure, the American aeronautical community worked assiduously afterward to redress weaknesses and shortfalls.

Steps on the road to recovery. It took approximately a decade for the United States to recover from its relative international aeronautical inferiority. A variety of organizational, legislative, and philanthropic activities highlighted key milestones that directly influenced the nation's subsequent rise to air transport dominance.

Advances in aeronautical research and development dramatically reshaped American design practice, leading to more efficient, higher-performance, and safer airplanes. The National Advisory Committee for Aeronautics (predecessor of today's National Aeronautics and Space Administration) adopted German aerodynamic theory, contracting in 1920 with Prandtl and hiring his former assistant Munk to bring science-rooted airfoil design to America. Then, designer John Northrop introduced the European-rooted streamline monocoque (shell) fuselage and cantilever (internally braced) wing to America with his wooden Lockheed Vega. Two years later, after having left Lockheed, he introduced the all-metal Northrop Alpha of 1929, with a "multicellular" (more properly multichannel) wing design adopted by many subsequent aircraft.

Important legislative initiatives transformed American air transport. Dissatisfied with the government's postal airmail system, Congress passed the Kelly Act in 1925, replacing it with a contract airmail system stimulating the rise of carriers that subsequently hauled passengers as well as mail. The next year, legislators passed the Air Commerce Act of 1926, establishing an Aeronautics Branch in the Department of Commerce (now the Federal Aviation Administration of the Department of Transportation), requiring licensing and certification of pilots, mechanics, airplanes, power plants, and routine air carrier inspections. It also passed army and navy five-year plans, rationalizing military aircraft acquisition, thereby furnishing manufacturers with badly needed financial and planning stability.

Inspired by the commercial possibilities of aviation, mining magnates Daniel and Harry Guggenheim established a philanthropic fund to promote aeronautics in 1926. Over the next four years it issued grants creating or supporting programs in aeronautical engineering and meteorology at leading American universities, funded faculty hiring and research facilities development, and imported leading foreign scientists (most notably the Hungarian von Kármán and Sweden's Carl Gustaf Rossby). It promoted safety research, most notably sponsoring Elmer Sperry's gyro-stabilized "artificial horizon" and Paul Kollsman's precision altimeter. These two instruments, combined with radio navigation, made practical "blind flying" a reality. The fund sponsored national "air-mindedness" and established an experimental airway with radio communication and weather forecasting and analysis, anticipating the subsequent federal national airways system of the early twenty-first century.

The burgeoning strengths of American aviation were revealed on a global stage in the spring of 1927 when Charles Lindbergh flew from New York to Paris. His flight demonstrated the reliability of new and more thermodynamically efficient air-cooled engines. It also stimulated the use of airmail, the establishment of aircraft companies, and aviation investment. So, too, were the weaknesses exposed of relying upon increasingly obsolescent and obsolete designs. In 1931, famed football coach Knute Rockne perished when a wooden-winged Fokker airliner disintegrated in flight. His accident stimulated introduction of streamlined all-metal transports, exemplified by the twenty-one-passenger Douglas DC-3 of 1935, the most successful and influential airliner of its time and the most widely produced airliner of all time. The DC-3 became an iconic symbol of air transport, the streamlined aesthetic, and modernism in general, as well as a great commercial and (as a troop and cargo transport in World War II) military success.

Maturation of transport and industry. In the years immediately after World War I, many air advocates anticipated that the rigid-frame dirigible airship would dominate long-range air transport. But airships were prone to destruction or damage from violent storms and (for those using hydrogen) susceptible to catastrophic fires, such as claimed Britain's R 101 in 1930 and Germany's Hindenburg in 1937. Pending development of reliable long-range landplanes, the flying boat thus constituted the only means of efficient long-range transoceanic air commerce. Flying boats offered the advantage of being able to safely alight at sea and featured prominently in interwar air transport, particularly that of Great Britain and France.

In America, airline entrepreneur Juan Trippe employed ever-larger and more efficient flying boats by Sikorsky and Boeing, with his Pan American Airways swiftly becoming the Western Hemisphere's dominant international air carrier. Trippe's rivalry with Britain's Imperial Airways over the Atlantic and Pacific skyways attracted the highest level of governmental and industry interest in both Britain and America, and although the two lines maintained an official cordiality, their rivalry was no less real for its civility. Trippe eventually proved dominant. In 1936, he extended flying boat service across the Pacific (the world's first regularly scheduled passenger service by airplane across any ocean) and in 1939, shortly before the outbreak of World War II, opened the first commercial flying boat service across the North Atlantic between Europe and America.

The surprising success of American commercial aviation in the interwar years generated foreign emulation, encouraging exports of American aircraft, particularly the soon-ubiquitous DC-3, which already carried 95 percent of American air commerce. By the end of the 1930s, the United States was exporting nearly 40 percent of its annual aircraft production overseas. The measure of America's civil aviation comeback is evident in its growth between the time of the Air Commerce Act and the attack upon Pearl Harbor. In 1926, the industry's 2,700 production workers produced 654 civilian aircraft and 532 military ones; air carriers flew 5,782 domestic passengers. In 1941, the industry's 423,000 production workers produced 6,844 civil aircraft and 26,277 military ones; air carriers flew 3.84 million domestic passengers.

Transitioning from Piston to Jet. More significantly, America possessed a specialized "dual-use" long-range aircraft design expertise unmatched by other nations, enabling the development of long-range bombers, transports, and flying boats. Accordingly, during World War II the United States manufactured a total of 299,293 aircraft. (By comparison, the Soviet Union produced 142,775, Great Britain produced 117,479, Nazi Germany produced 111,787, Imperial Japan

produced 68,057, and Fascist Italy produced 11,508).

Wartime development = postwar exploitation. Among these were three transport aircraft types that swiftly dominated postwar civil air commerce, the Douglas DC-4, Lockheed Constellation, and Boeing 377 Stratocruiser. These each had four engines and tricycle landing gear and were capable of flying across the Atlantic and in stages across the Pacific. Of the three, the Douglas DC-4 proved most successful: by 1951, 80 percent of the world's airliners were American built, with 56 percent of these being Douglas aircraft. By 1945, 175 flights crossed the Atlantic each day, and within five months of the end of the European war, on 23 October 1945, the United States began postwar commercial airline service across the North Atlantic. In February 1946, the United States and Great Britain signed the so-called "Bermuda agreement," which established the pattern of postwar bilateral civil aviation agreements.

Ironically, for all its undoubted aeroindustrial strength, in some respects the United States ended World War II as it had World War I: in a position of relative technological inferiority. If it dominated air commerce, with its landplanes so reliable that in the postwar world the flying boat had no further future, its civil aviation did not represent the leading edge of aviation technology. Instead, it was Britain that seemed poised to move into the global air transport lead because it had swiftly applied the gas turbine (jet) engine to commercial aircraft design. In July 1948, Britain flew the world's first turbopropeller (turboprop) airliner, the Vickers Viscount; a year later, in July 1949, it followed with the world's first pure-jet airliner, the de Havilland Comet, and Canada followed with a pure-jet experimental airliner of its own, the Avro C-102, a month later. Subsequently, the Viscount and Comet entered airline service in 1950 and 1952, respectively, and the C-102 was abandoned.

Air transport enters the jet age. Although the United States had moved swiftly to incorporate the jet engine and the advanced aerodynamics of the sweptwing on its military aircraft (to its benefit in the Korean War, where American swep-twing jet fighters clashed with Soviet equivalents), it was far slower to do so with commercial airplanes. Airline executives preferred sticking with the proven combination of a straight-wing airplane with radial piston engines driving large propellers. Thus, instead of the airlines, it was the U.S. Air Force's need for a jet-powered tanker to refuel its strategic bombers that led to the first American jet transport, the Boeing 707, an outgrowth of the military's KC-135 tanker-transport program, which had first flown in 1952 (and which itself benefited from the firm's experience with the B-47 and B-52 strategic bomber programs).

Britain won some American orders for the Comet (and so did Vickers with the Viscount), but a series of accidents caused by poor design grounded the Comet and forced extensive redesign and strengthening of its fuselage. Arguably, only these tragedies denied Britain the market success it otherwise might have enjoyed. The accidents bought the United States the time to catch up with a later and more advanced airplane—the 707—thus preserving American commercial air supremacy into the jet age.

Although the honor of making the first commercial transatlantic flight fell to Britain's revised Comet IV, flown by the British Overseas Airways Corporation (BOAC, successor to the prewar Imperial Airways and predecessor of British Airways) on 4 October 1958, it was the introduction of the faster sweptwing 707 later that month that truly transformed international air commerce. That year, for the first time, more passengers crossed the Atlantic by air than by sea. Within a decade, transatlantic passenger totals would exceed 6 million annually, carried primarily by 707 and Douglas DC-8 jetliners.

Technological choices and challenges. The fast speed of jet aircraft forced extensive re-equipping of the global air route traffic control system and, particularly, the provision for broad-area radar coverage. Since the early 1950s, commercial airliners had employed weather radar systems, and by the 1960s–1970s, the increasing power of avionics (aviation electronics), enhanced by the computer revolution, made flight sched-

uling, operations, and navigation both more precise and more reliable. The advent of space access, with the introduction of weather, communications, and ultimately navigation satellite systems such as the late-1980s Global Positioning System (GPS), greatly enhanced flight safety. As well, significant research began on assisting the pilot with comprehensive autopilot systems enabling automatic landings and with giving the pilot synthetic vision capabilities to generate "highway in the sky" flight path representation and cuing. All these efforts, coupled with advances in materials and optics technology, ultimately generated the "glass cockpit" instrumentation systems featured on modern commercial aircraft, whether intercontinental jet airliners or small feeder-line transports.

Various technologies transformed commercial aircraft from the era of the post–Korean War jetliner to the so-called "jumbo" or wide-body aircraft (first exemplified by the 747 of 1969). These included the invention of "fanjet" gas turbine engines that generated high thrust but with great fuel efficiency. The so-called "supercritical" wing delayed the onset of performance-robbing drag encountered as a plane flew closer to the speed of sound. With a distinctive "flat-top" airfoil profile, it enabled fuel-efficient flight at very high velocities just below the speed of sound. Accompanying these propulsion and design changes were new structural design concepts. So-called "composite" materials (such as carbon fiber) enabled engineers to design structures with high strength and great rigidity, but with far less weight and material than conventional metal structures. Finally, advances in the design of computer-based electronic engine and flight controls enabled aircraft that flew with more precise aircraft control and engine management.

However, some technological approaches, considered attractive in the 1950s, either failed upon adoption or never entered service. Chief among these was the supersonic transport (SST), which both the United States and Europe pursued assiduously from the mid-1950s. Whereas Britain and France partnered on the Anglo–French Concorde and the USSR developed the Tu-144 (both Mach 2 aluminum SSTs), the United States chose a more demanding Mach 2.7 titanium design, which it then abandoned in 1971 in the face of rising criticism and lukewarm support within the airline community. The Concorde continued in service until 2003, the Tu-144 was quickly retired from service. Although SST advocates have persisted in urging support for such aircraft, airline interest remains at best ambivalent, although more interest exists within the smaller business jet community.

Air Transport Adjusts to New Global Realities.
The progressive improvement and refinement of gas turbine engine technology resulted in the rapid replacement of piston-engine transports of all categories. By the mid-1970s, therefore, the jet engine was no longer restricted simply to long-range and wide-body jetliners. Instead, an increasing range of smaller jetliners and regional commuter aircraft manufactured in Europe and the Americas featured the benefits of turbine propulsion. As a consequence, the fleet mix of aircraft globally gradually changed.

Foreign resurgence, domestic decline. Although in the mid-1960s American-manufactured aircraft dominated all categories of commercial aviation, by the mid-1970s smaller foreign-made aircraft such as the Dutch Fokker F-28, the British BAC 1-11, and the Canadian DHC-7 were competing successfully for fleet orders issued by American carriers.

In 1972 the European Airbus consortium first flew the A-300 twin-engine wide body, progenitor of the Airbus family of jet transports. Largely ignored by American air transport authorities at the time, Airbus subsequently went on to enjoy great commercial success: in 2003, the one hundredth anniversary of Kitty Hawk, Boeing delivered fewer jetliners to customers than did Airbus. In the early twenty-first century, the Boeing–Airbus rivalry forced each to higher standards of excellence and smaller manufacturers offered new competition. As of 2012, the top four airliner manufacturers were Boeing and Airbus (essentially tied for first), Canada's Bombardier, and Brazil's Embraer. New competition also seemed likely to come from China

and the revitalized post-Soviet Russian aircraft industry.

As of 2012, the American aerospace industry had contracted from its World War II high of forty-seven companies to just three major manufacturers, only one of which—Boeing—produced commercial jetliners. America's regional aircraft market was dominated by aircraft of foreign origin and, globally, its air transport market share had fallen to less than 50 percent. General aviation—civil aviation not involving scheduled air operations—likewise had undergone a transformation, declining from an annual production rate in excess of 10,000 aircraft per year in the 1970s to a low of just 941 aircraft produced in 1992 and then rebounding sluggishly following Congressional passage of the General Aviation Recovery Act of 1994. But if the industry had contracted, its creativity—exemplified by new environmentally friendly and economically efficient designs such as the Boeing 787 Dreamliner—remained vibrantly alive.

Flight had always had profound social and cultural implications. The jetliner revolution ended the exclusivity of air transport, spawning deregulation, cheap tickets, and people roaming the globe. The terrorists of 11 September 2001 subverted commercial aviation, turning tools of cultural interchange, global awareness, communication, and commerce into vengeful and horrific weapons of mass destruction. But even they could not strip air transport of its most basic and enduring value: airline passengers in the early twenty-first century, if not perhaps enjoying the carefree comfort, inflight conveniences, and luxury of their predecessors, enjoyed an ease of routine global access unknown in previous times.

[*See also* **Computer Science; Engineering; Kármán, Theodore von; Lindbergh, Charles; Machinery and Manufacturing; Meteorology and Climatology; Military, Science and Technology and the; National Aeronautics and Space Administration; Postal Service, U.S.; Radio; Research and Development (R&D); Satellites, Communications; Space Program; Space Science; Technology;** *and* **Wright, Wilbur and Orville.**]

BIBLIOGRAPHY

Bilstein, Roger E. *Flight Patterns: Trends of Aeronautical Development in the United States, 1918–1929*. Athens, Ga.: University of Georgia Press, 1983. A classic and unsurpassed survey by American aviation's finest historian.

Bright, Charles. *The Jet Makers: The Aerospace Industry from 1945 to 1972*. Lawrence: Regents Press of Kansas, 1978. An incisive examination of how the American aircraft industry adjusted to the beginning of the jet age to the era of the 747.

Brooks, Peter W. *The Modern Airliner: Its Origins and Development*. London: Putnam, 1961. An outstanding and highly influential analytical study on the "generational" development of air transport.

Crouch, Tom D. *Wings: A History of Aviation from Kites to the Space Age*. New York: Smithsonian National Air and Space Museum in association with W. W. Norton & Company, 2003. The best and most comprehensive survey of flight to the modern era.

Daley, Robert. *An American Saga: Juan Trippe and His American Empire*. New York: Random House, 1980. The best and most insightful biography of Trippe and his leadership of Pan American Airways.

Davies, R. E. G. *A History of the World's Airlines*. London: Oxford University Press, 1967. The standard and indispensable reference on global airline development.

Davies, R. E. G. *Rebels and Reformers of the Airways*. Washington, D.C.: Smithsonian Institution Press, 1987. Interesting case studies of unappreciated airline leaders and executives by the doyen of airline historians.

Dierikx, Marc. *Clipping the Clouds: How Air Travel Changed the World*. Westport, Conn.: Praeger, 2008. A thoughtful examination on the meaning and impact of global air transport.

Hayward, Keith. *The World Aerospace Industry: Collaboration and Competition*. London: Duckworth and the Royal United Services Institute for Defence Studies, 1994. An extremely valuable survey of global aerospace in the post–Cold War Airbus era.

Higham, Robin. *Britain's Imperial Air Routes 1918 to 1939: The Story of Britain's Overseas Airlines*. Hamden, Conn.: G. T. Foulis, 1960. A detailed examination of Imperial Airways that makes for good comparison with the history of Pan American.

Leary, William M., Jr., ed. *The Airline Industry*. In *Encyclopedia of American Business History and Biography*. New York: Bruccoli Clark Layman/Facts on File, 1992. Excellent reference to key personages in air transport.

Leary, William M., Jr., ed. *Aviation's Golden Age: Portraits from the 1920s and 1930s*. Iowa City: University of Iowa Press, 1989. An extremely fine and nuanced collection of essays by leading historians about significant figures in aeronautics.

Linden, F. Robert van der. *Airlines and Air Mail: The Post Office and the Birth of the Commercial Aviation Industry*. Lexington: University Press of Kentucky, 2002. An extremely valuable reference on the sometimes tumultuous relationship between the Post Office and the advent of America's commercial airline industry.

Lynn, Matthew. *Birds of Prey: Boeing vs. Airbus—A Battle for the Skies*. New York: Four Walls Eight Windows, 1997. A comprehensive survey of the early years of the Boeing–Airbus rivalry, which continues to the present day.

Miller, Ronald, and David Sawers. *The Technical Development of Modern Aviation*. New York: Praeger, 1970. An outstanding survey of the interplay of technology, market need, and aircraft design and its influence upon commercial aviation.

Newhouse, John. *The Sporty Game*. New York: Alfred A. Knopf, 1982. An excellent accounting of the commercial jet aircraft business from the perspective of the transition point from first- to second-generation wide-body aircraft.

Robinson, Douglas H. *Giants in the Sky: A History of the Rigid Airship*. Seattle: University of Washington Press, 1973. A classic and still-definitive history of global airship development and use in air commerce.

Sabbagh, Karl. *21st Century Jet: The Making and Marketing of the Boeing 777*. New York: Charles Scribner's Sons, 1996. A revealing and informative look at the development of the world's first computer-designed jetliner and its subsequent marketing and introduction into commercial service at the height of the Boeing–Airbus rivalry.

Schlaifer, Robert, and S. D. Heron. *Development of Aircraft Engines and Fuels*. Boston: Graduate School of Business Administration, Harvard University, 1950. A superb study of the interrelationship of engines and aircraft design, covering the maturation of the piston engine and the early history of jet propulsion.

Smith, Henry Ladd. *Airways: The History of Commercial Aviation in the United States.* New York: Alfred A. Knopf, 1944. A classic survey history that has enduring value.

Sutter, Joe, and Jay Spenser. *747: Creating the World's First Jumbo Jet and Other Adventures from a Life in Aviation.* New York: HarperCollins, 2006. An excellent memoir by the designer of the 747, highlighting the complexities of technological and economic choice in aircraft design.

Whitnah, Donald R. *Safer Skyways: Federal Control of Aviation, 1926–1966.* Ames: Iowa State University Press, 1966. An excellent examination of the rise of federal air regulation, its impact on commercial aviation, and the interrelationships of government policy, technology, and airways evolution.

Richard P. Hallion

ALCOHOL AND ALCOHOL ABUSE

Alcohol looms large in American history, and attitudes toward it have been linked to myriad reformist causes; reflected many social concerns; and mirrored the prevailing cultural, political, and economic climate of successive eras.

Alcoholic beverages, whether rum distilled from West Indian sugar, home-brewed beer, or imported wines, were widely consumed in colonial America, and the physician-statesman Benjamin Rush (1745–1813) targeted them in his widely reprinted *Inquiry into the Effects of Ardent Spirits upon the Human Body and Mind* (1784). Rush was the first to consider intemperance an addictive disease. Fearful for the new republic, Rush recoiled at the prospect of intoxicated voters shaping its destiny—no small concern at a time when elections often featured heavy drinking. Annual per capita consumption of absolute alcohol when Rush wrote ranged between four and six gallons (more than twice the rate in 2007), and evidence suggests a further sharp rise between 1800 and 1830 (Hyman et al., 1980; LaVallee and Yi, 2011) The profitability of corn whiskey, heavy frontier drinking, the spread of saloons in cities, and the immigration of beer-drinking Germans and whiskey-swilling Irish all encouraged the na-

tion's bibulous tendencies. The years of the early republic also witnessed the rise of "firewater myths" about American Indian drinking, myths that posited a biological susceptibility of Native Americans to heavy and destructive drinking. Historians and anthropologists in the early twenty-first century have had a much greater appreciation for the diversity of drinking practices among native tribes and regard problem drinking among American Indians as primarily a response to the policies of cultural genocide and land annexation imposed by white Europeans.

The new nation's heavy drinking habits elicited a reaction within the Protestant churches, however, which linked salvation with temperance and other reforms. The American Society for the Promotion of Temperance, founded by evangelical clergymen in 1826, also gained support from farmers, industrialists, and homemakers. Indeed, the temperance campaign—really a series of reform drives—comprised the nineteenth century's longest and largest social-reform movement. Alcohol was seen as imperiling capitalist enterprise, domestic tranquility, and national virtue. By 1836 the American Society for the Promotion of Temperance, renamed the American Temperance Society, advocated total abstinence. In the early 1840s, Americans thronged to temperance rallies, "took the pledge" for sobriety, and in record numbers lobbied to end the licensing of saloons. The Washingtonian movement, a grassroots total-abstinence campaign, sponsored parades and speeches; offered recruits financial and moral assistance; and established institutions for inebriates—Washingtonian Homes—that relied on palliative care and moral suasion to restore residents to sobriety. The Washingtonian enthusiasm soon gave way to better-organized temperance fellowships, such as the Good Templars and the Blue Ribbon societies. The late Antebellum Era also saw renewed middle-class drives for local and state prohibition. In the 1850s, 11 states passed prohibitory legislation, although most were soon repealed.

Meanwhile, at midcentury, alcohol enjoyed a renaissance within regular medical practice, its utility as a "stimulant" (misidentified) rising with the ascent of supportive and stimulant therapies generally—although the trend was by no means universal. The use of alcohol at Massachusetts General Hospital, for example, showed a consistent rise through the 1840s, 1850s, and 1860s (Warner, 1997). Still other health practitioners, especially health reformers such as Sylvester Graham and John Harvey Kellogg, eschewed the use of alcohol for both moral and physiologic reasons (Nissenbaum, 1980; Markel, 2011). The American Medical Temperance Association, a group of reform-minded physicians, led by American Medical Association founder Nathan Smith Davis, organized in 1891 to promote abstinence among their own ranks and in the clinic.

The brewing and distilling industries expanded after the Civil War, and alcohol consumption, especially in the immigrant-rich cities, remained high. But the temperance movement revived as well, linking "Demon Rum" to concerns about immigration, workplace efficiency, social welfare, eugenics, and urban political corruption. Frances Willard's Woman's Christian Temperance Union redefined temperance, along with other reforms, as a women's issue involving home protection. At the union's prompting, after a long battle with American physiologists, Congress mandated the inclusion of "scientific" temperance instruction in high-school physiology texts across the United States (Pauly, 1990).

This era of social reorganization and professionalization also brought the first widespread attempt to medicalize drunkenness. The American Association for the Cure of Inebriates, founded in 1870 by physicians and reformers, promoted the concept of inebriety as a hereditary disease exacerbated by chronic debauchery. As their drinking progressed, contended the association, inebriates lost control of their actions and required restorative medical and moral treatment. Envisioning a new medical specialty to address this ailment, physician-reformers built a network of private institutions to treat habitual drunkards. California, Iowa, Massachusetts, New York, and other states followed suit. In this age of industrial capitalism, the goal was to restore inebriates' economic productivity as well as their willpower. A minority of state and private institutions treated women, whose heavy drinking was regarded as more deviant and harder to cure than that of men. Merging with the American Medical Temperance Association in 1904, the American Association for the Cure of Inebriates faded as the prohibition movement grew. By 1920, most of the inebriate institutions had closed, and habitual drunkenness was again viewed as primarily a moral, political, and legal issue.

The church-based Anti-Saloon League, meanwhile, founded in 1895 and supported by industrialists such as Henry Ford and Pierre du Pont, spearheaded the prohibition drive. Under the superintendent Wayne Wheeler, the league's innovative bipartisan lobbying approach secured prohibitory state legislation and, in 1919, ratification of the Eighteenth Amendment, establishing nationwide prohibition. A World War I reaction against German American–owned breweries and fears that alcohol would undermine the nation's military contributed to this success.

But many Americans, especially in the cities, rejected prohibition; speakeasies flourished and bootleg liquor flowed freely in many municipalities. With the right to vote in hand, by 1920 more women drank in public with men. The year 1929 witnessed the founding of the Women's Organization for National Prohibition Reform. Four years later, in the midst of the Depression, Repeal came—as remedy, some historians have argued, for both tax dollar shortages and labor unrest. The year 1933 ushered the nation into an era often referred to as an "Age of Ambivalence" about alcohol. Even historians remain divided on the relative "success" of Prohibition, with some citing reductions in alcohol-related diseases such as cirrhosis and Korsakoff's syndrome as real public-health benefits associated with the Eighteenth Amendment (Blocker, 2006; Burnham, 1993). Post Repeal, however, the reopened breweries and distilleries advertised heavily to win new customers. As old taboos faded, alcohol consumption spread widely. In the later twentieth century, wine connoisseurship expanded and U.S. wine production flourished in California. Vino-tourism became a staple of many states' tourism industries. Although the major breweries dominated the beer market, imported brands and local microbreweries also flourished. Single-malt scotch and microdistilled and specialty vodkas acquired new followings.

Simultaneously, however, antialcohol sentiment remained powerful in evangelical Protestantism; in such organizations as Alcoholics Anonymous (AA), established in 1935, and Mothers against Drunk Driving, founded in 1980; and in heightened concern about college-based binge drinking, alcohol-related domestic abuse, and fetal alcohol syndrome. Beginning in the early 1980s, these efforts, coupled with health and fitness concerns, spurred a slow decline in per capita alcohol consumption, a trend that began to slowly reverse at the dawn of the new millennium (La-Vallee and Yi, 2011, p. 6). Beginning in the early 2000s, red wine has been touted within medical and public-health circles for its potential cardiovascular and antiaging benefits, with "resveratrol" receiving much attention (Mayo Clinic, 2014). Recently, the rise of the Mediterranean diet as a dietary "gold standard," and red wine a staple among alcohol-permissive Mediterranean nations, has attracted additional attention under the banner of good health.

The second half of the twentieth century also saw renewed debate over the nature of alcoholism. AA founders businessman William Wilson and physician Robert Smith, along with the National Committee for Education on Alcoholism, led by Mrs. Marty Mann, orchestrated a mid-twentieth century crusade to treat alcoholism as a disease. In the 1950s, the biostatistician E. M. Jellinek of the Yale Center of Alcohol Studies promoted a multistage model of alcohol addiction based on his research on AA members. But the disease concept met criticism as well. The American Medical Association in the 1960s and 1970s encouraged physicians to treat alcoholism's "medical aspects" but argued that labeling alcoholism a disease did not relieve individuals of responsibility for their intoxicated behavior. The Supreme Court concurred, declining to exonerate persons for actions committed while drunk. Although the National Institute for Alcohol Abuse and Alcoholism was established in 1971, lending federal support and funding to alcoholism studies, some social scientists, including ones funded by the institute, mustered evidence discrediting the disease model.

The 1970s and 1980s witnessed the emergence of a broad-based public-health harm-reduction approach oriented toward preventing "alcohol problems," that is, problems such as drunk driving, domestic violence, and job impairment, as well as alcoholism. Twelve-step treatment modalities, modeled after AA but addressing specific drinker demographics, have proliferated as well: for example, the Wellbriety movement (American Indians), Rational Recovery (those who wish a faith-free rehabilitation), and Women for Sobriety (women).

As the twentieth century ended, what some called a neo-temperance movement gained momentum, linked to the anti-tobacco and anti-drug campaigns. In 2014, biomedical researchers, policy makers, treatment providers, and historians increasingly viewed alcohol dependence within a larger "alcohol, tobacco, and other drug" framework that appreciates the ways in which different chemical dependencies, as well as their demographics and histories, may affect one another and/or share an underlying neurotransmitter (dopamine) basis. As of 2010, the nation still had over 45,000 beer, wine, and liquor stores, and more than half of adult Americans regularly drank alcoholic beverages. Alcohol's central role in American culture, if somewhat diminished, seemed firmly entrenched.

[See also Food and Diet; Health and Fitness; Medicine: Overview; Psychiatry; Psychology; Psychopharmaceutical Drugs; and Public Health.]

BIBLIOGRAPHY

Blocker, Jack. "Did Prohibition Really Work?" American Journal of Public Health 96, no. 2 (2006): 233–243.

Burnham, John C. Bad Habits: Drinking, Smoking, Taking Drugs, Gambling, Sexual Misbehavior, and Swearing in American History. New York: New York University Press, 1993.

Burns, Eric. Spirits of America: A Social History of Alcohol. Philadelphia: Temple University Press, 2004.

Courtwright, David. Forces of Habit: Drugs and the Making of the Modern World. Cambridge, Mass.: Harvard University Press, 2001.

Golden, Janet. *Message in a Bottle: The Making of Fetal Alcohol Syndrome*. Cambridge, Mass.: Harvard University Press, 2005.

Hyman, Merton, Marilyn Zimmerman, Carol Gurioli, and Alice Helrich. *Drinkers, Drinking, and Alcohol-Related Mortality and Hospitalizations: A Statistical Compendium*. New Brunswick, N.J.: Center for Alcohol Studies, Rutgers University, 1980.

LaVallee, Robin, and Hsiao-ye Yi. *Surveillance Report #92: Apparent per Capita Alcohol Consumption: National, State, and Regional Trends, 1977–2009*. Bethesda, Md.: National Institute on Alcohol Abuse and Alcoholism Division of Epidemiology and Prevention Research Alcohol Epidemiologic Data System, August 2011.

Lerner, Barron. *One for the Road: Drunk Driving since 1900*. Baltimore: Johns Hopkins University Press, 2011.

Mancall, Peter. "'I Was Addicted to Drinking Rum': Four Centuries of Alcohol Consumption in Indian Country." In *Altering American Consciousness: The History of Alcohol and Drug Use in the United States, 1800–2000*, edited by Sarah W. Tracy and Caroline Jean Acker. Amherst: University of Massachusetts Press, 2004.

Markel, Howard. "John Harvey Kellogg and the Pursuit of Wellness." *Journal of the American Medical Association* 305, no. 17 (2011): 1814–1815.

Mayo Clinic Staff. "Red Wine and Resveratrol: Good for your Heart?" http://www.mayoclinic.org/diseases-conditions/heart-disease/in-depth/red-wine/art-20048281(Accessed 11 March 2014)

Nissenbaum, Stephen. *Sex, Diet, and Debility in Jacksonian America: Sylvester Graham and Health Reform*. Westport, Conn.: Greenwood Press, 1980.

Pauly, Philip J. "The Struggle for Ignorance about Alcohol: American Physiologists, Wilbur Olin Atwater, and the Women's Christian Temperance Union." *Bulletin of the History of Medicine* 64 (1990): 366–392.

Tracy, Sarah W. *Alcoholism in America from Reconstruction to Prohibition*. Baltimore: Johns Hopkins University Press, 2007.

Tracy, Sarah W., and Caroline Jean Acker, eds. *Altering American Consciousness: The History of Alcohol and Drug Use in the United States, 1800–2000*. Amherst: University of Massachusetts Press, 2004.

Warner, John, *The Therapeutic Perspective: Medical Practice, Knowledge, and Identity in America,* *1820–1885*. Cambridge, Mass.: Harvard University Press, 1986; with new preface, Princeton, NJ: Princeton University Press, 1997.

White, William L. *Slaying the Dragon: The History of Addiction Treatment and Recovery in America*. Ann Arbor: University of Michigan Press, 1998.

Sarah W. Tracy

ALVAREZ, LUIS WALTER

(1911–1988), Nobel Laureate in physics. Alvarez, whose family had roots in Spain as well as Ireland, was born in San Francisco, California. Thanks to the urging of his high school teachers—and his interest in science—he went on to study chemistry at the University of Chicago, where he switched to physics and earned a PhD in 1936. His sister Gladys, then secretary to Ernest Lawrence, got him a job at the Berkeley Radiation Laboratory where Lawrence and M. Stanley Livingston had developed the cyclotron a few years earlier. Alvarez fit well with the intense pace and can-do spirit of Lawrence's lab, quickly demonstrating what would become his hallmark traits: out-of-the-box problem solving, creativity, and inventiveness. He rapidly accumulated accomplishments. He conducted a number of noteworthy experiments at Berkeley; particularly important were his measurements of helium isotopes that opened inquiries into the feasibility of a hydrogen bomb. In late 1940, with Europe already at war, Alvarez left Berkeley to join the radar project at the MIT Radiation Laboratory. While there he contributed to a number of projects; he is best known for developing a radar system, the Ground Controlled Approach (GCA), which successfully allowed ground-based radar operators to guide pilots to a safe landing with verbal commands. Later in the war Alvarez also made contributions to the development of the first atomic bombs; particularly crucial was an invention that allowed precise detonation of the symmetrical implosion needed to deploy the plutonium weapon.

In the postwar Alvarez continued his illustrative career. One of his first projects was a proton linear accelerator that became a model for such devices for decades. Then, in 1953, Alvarez improved and enlarged a particle detection device, the hydrogen bubble chamber, which allowed the discovery of an unprecedented number of subatomic particles. This accomplishment led to his Nobel Prize in 1968. By this time he had already won the highly prestigious aviation prize, the Collier Trophy, as well as the Presidential Medal for Merit for his wartime work.

Alvarez also undertook numerous projects outside of physics. He analyzed film of President John F. Kennedy's assassination to pin down the number of shots fired and, as part of a Berkeley team, employed cosmic rays to search for hidden chambers in the Egyptian pyramid, Chephren. His most famous later work was performed with his son Walter, a geologist. Walter was conducting research in a gorge in Italy on a mysterious stratum of clay that marked the time when dinosaurs became extinct. Drawing on specialized techniques at the Berkeley laboratory and his own distinctive brand of problem solving, Luis Alvarez was able to discover enormous levels of iridium, and then, after carefully eliminating alternatives with his son, to infer that the iridium came from an asteroid. In 1979 the father-and-son team posited the theory, which came to be widely accepted, that dinosaurs died because an asteroid crashed into the earth, spewing smoke that blocked the sun (Alvarez, 1987; Heilbron and Seidel, 1989).

[See also Chemistry; Dinosaurs; Geology; High Schools, Science Education in; Lawrence, Ernest O.; Manhattan Project; Nobel Prize in Biomedical Research; Physics; and Science: Overview.]

BIBLIOGRAPHY

Alvarez, Luis. *Adventures of a Physicist.* New York: Basic Books, 1987. Readable and revealing autobiography.
Alvarez, Luis. *Nobel Lectures, Physics 1963–1970.* Amsterdam: Elsevier, 1972.
Heilbron, J. L., and Robert W. Seidel. *Lawrence and His Laboratory: A History of the Lawrence Berkeley Laboratory,* Vol. I. Berkeley: University of California Press, 1989. This includes the best schol-

arly account of Alvarez's career through World War II.

Catherine Westfall

ALZHEIMER'S DISEASE AND DEMENTIA

In contemporary medicine, dementia denotes a clinical syndrome involving the significant deterioration of cognitive abilities such as memory, language, and problem solving. Dementia may be caused by many different underlying conditions, including reversible conditions, brain injuries, and progressive neurodegenerative conditions. Although dementia can occur at relatively young ages, in the early twenty-first century it has been associated with aging and was much more common among the population aged 65 years old and older. Alzheimer's disease was the most common cause of dementia in 2011, accounting for an estimated 60 to 80 percent of all cases. The most widely accepted estimates were that about one in eight Americans 65 and older and nearly half of those 85 and older had Alzheimer's disease (Alzheimer's Association, pp. 5, 12).

In addition to their increasing prevalence and prominence as the United States became an aging society, dementing conditions like Alzheimer's became increasingly freighted with fear and dread through the nineteenth and twentieth centuries as American culture became more oriented around ideas of selfhood and self-fulfillment. Throughout this period, the market revolution and the erosion of traditional social hierarchies transformed American society and culture. By the end of the nineteenth century, selfhood was no longer an ascribed status bestowed at birth, but the product of individual choice and lifelong effort. In this context, the loss of cognitive abilities essential to the project of self-creation came to be widely seen as the most dreadful of all losses.

Origin of Modern Concepts of Dementia and Alzheimer's.   The modern concept of

dementia began to take shape in the mid-nineteenth century, when psychiatrists restricted the meaning of the term to include only intellectual impairments, with noncognitive symptoms moved into separate disease categories or regarded as secondary to the primary diagnosis of dementia. By the early twentieth century, physicians regarded dementia as a neuropsychiatric syndrome, the causes of which were to be found in distinct brain pathologies. Senile dementia was perhaps the most well-established form of dementia, with the clinical features of dementia correlated with microscopic neuritic plaques and neurofibrillary tangles found at autopsy in the brains of elderly patients. But the relationship between dementia and aging remained confused. Because most people at advanced ages exhibited at least some plaques and tangles, even if they had not shown signs of dementia in life, doctors could regard dementia either as a disease or as an extreme point on a continuum of brain aging. Although the issue remained controversial, it was clear by 1900 that aging was the greatest single risk factor for dementia and that the prevalence of dementia increased steadily with age, hence the continued use in the medical literature of "senility" and related terms through the 1970s.

Alzheimer's disease (AD) was established as a diagnostic category in 1910 by German psychiatrist Emil Kraepelin in the eighth edition of his influential textbook *Psychiatrie* to distinguish it from senile dementia. Kraepelin named the condition for his protégé Alois Alzheimer, who a few years earlier had published an article describing the case of a woman who showed the clinical features and pathological hallmarks of senile dementia, but was only 51 years old. The relatively young age of the patient seemed to Kraepelin to warrant a separate disease category, and he thus called AD a presenile dementia that occurred in people under the age of 65, in contrast to the much more common senile dementia. Alzheimer himself questioned the significance of the age distinction, and although AD was recognized in the neuropsychiatric literature in subsequent decades, it was widely recognized that both the clinical and the pathological features of AD and senile dementia were essentially identical.

American Psychiatry and the Psychodynamic Approach to Dementia. Through the 1920s, only a handful of publications on dementia and Alzheimer's disease appeared in the American medical literature, most of them written by psychiatrists like Solomon C. Fuller, the first African American psychiatrist in the United States, who had trained with Alzheimer in Germany and followed the Kraepelinian approach to these conditions. But beginning in the 1930s dementia began to attract significantly more attention from psychiatrists as they grappled with the fact that patients with dementia were becoming an increasingly large part of the population they encountered. Until the late nineteenth century, when elderly people with dementia could not be cared for at home, they were typically regarded as "merely senile" and cared for in poor houses. But as state governments assumed the cost of caring for the mentally ill around the turn of the twentieth century, elderly patients with dementia were increasingly likely to be classified as insane and admitted to state hospitals. Psychiatry viewed the proliferation of aged patients with dementia in the state hospital system with alarm, but could do little to stop the trend until the 1960s, when the deinstitutionalization movement and provisions of the 1965 Medicare and Medicaid legislation that made federal money available for nursing-home care led to a massive shift of older patients with dementia out of public psychiatric hospitals.

But from the 1930s through the 1950s, American psychiatrists led by David Rothschild, who worked in the Massachusetts state hospital system, argued that both AD and senile dementia were better thought of as psychodynamic disorders rather than brain diseases. Because correlations between the amount of brain pathology found at autopsy and the degree of dementia in life were far from perfect, they argued that social and psychological factors in the life history and personality of patients could best explain any particular case of dementia. This psychodynamic model seemed to bring dementia within the purview of mainstream psychiatric theory, which was dominated by psychoanalysis through the 1950s. It also resonated with the emerging field of social gerontology after World War II, and a more generalized psychodynamic approach applied to the broader problem of social adjustment to aging was taken up by many researchers in the aging field through the 1970s. The psychodynamic approach never had much currency in European psychiatry and faded quietly from view even in American psychiatry after the development of chlorpromazine and other drugs led to a resurgence of biological psychiatry. The rigorous clinical–pathological correlations done by the British team of Martin Roth, Gary Blessed, and Bernard Tomlinson in a series of studies from the late 1960s through the 1970s were widely seen as establishing the pathological basis of these disorders and ushering in a new era of research into the pathophysiology of dementia.

The AD Movement and the Reconceptualization of AD.

AD emerged as a major public issue in the late 1970s through the efforts of a coalition of caregivers and family members, biomedical researchers, and government officials who worked to increase public awareness of the condition and to get the federal government to commit money for research aimed at understanding the cause of Alzheimer's and related disorders and to finding an effective treatment, if not a prevention or cure. Central to their strategy was advancing the claim that chronic dementia in the elderly should be viewed as a disease process rather than a normal part of aging, part of a more general claim advanced within gerontology and geriatrics during this period that aging itself should not normally be accompanied by disease and disability. Perhaps the most prominent exponent of this claim was neurologist Robert Katzman, who argued in an influential 1976 article that AD and senile dementia ought to be regarded as a unified entity and that this entity should be called AD rather than senile dementia. Combining the categories meant that the problem was large—Katzman's editorial claimed AD to be the fourth or fifth leading cause of death in the United States—and with the aging of the baby-boom generation, it would soon become enormous. Calling the unified category AD made it a disease entity with a well-established pathological basis in the brain (i.e., a disease that was "real" and

thus worthy of a massive research effort into its cause and cure).

Two other developments were crucial to the rising prominence of AD. First was the creation in 1980 of the Alzheimer's Disease and Related Disorders Association (later shortened to the Alzheimer's Association) to provide support for family caregivers and to lobby for federal money for research. Within a decade the Alzheimer's Association grew into one of the largest and most influential health voluntary associations in the country. Second was the focus on AD by the National Institute on Aging (NIA), which was created in 1974. Gerontologist Robert Butler, the NIA's founding director, felt that the NIA needed a disease-specific focus to successfully compete for funding with the other National Institutes of Health, and he felt that AD as reconceptualized by Katzman was a perfect fit for the fledgling institute. With the combined efforts of the Alzheimer's Association, leading biomedical researchers, and the NIA, funding for AD research grew dramatically. In 1976 federal funding for AD research was less than $1 million; by 2005 it had risen to about $700 million (Ballenger, 2006, p. 114) and stood at an estimated $449 million for fiscal year 2013 (National Institutes of Health, 2012).

The Cholinergic Hypothesis and Drug Treatments.

The massive investment in research appeared poised to bring results by the mid-1980s when researchers established that the brains of AD patients had a deficit of the neurotransmitter acetylcholine, which was known to play an important role in memory, and researchers hypothesized that restoring or slowing the depletion of acetylcholine could be the basis of an effective treatment. After strategies aimed at boosting cholinergic levels in the brain failed to produce results, interest centered on drugs that inhibited the action of cholinesterase, an enzyme that metabolizes acetylcholine in the brain after synaptic transmission, thus keeping the overall level of acetylcholine in the brain higher.

The first cholinesterase inhibitor, tacrine, was mired in controversy over allegations of research misconduct in the clinical trials of safety and efficacy, but, in the wake of public outcry for its approval, it was nonetheless licensed in 1993 for treatment of mild to moderate AD. Three additional cholinesterase inhibitors passed more smoothly to approval by the U.S. Food and Drug Administration (FDA) in the following decade, as well as a drug that operates on a different neurotransmitter. But despite initial enthusiasm, none of these drugs proved to provide dramatic improvements for most patients with AD. Although disappointment and skepticism about the efficacy of the cholinesterase inhibitors grew in the first decades of the twenty-first century, enthusiasm began building around a new strategy for developing drugs aimed at clearing excess levels of beta-amyloid from the brain in Alzheimer's and thus, proponents claimed, intervening directly into the basic pathophysiology of the disease. Whether a newer generation of drugs would prove to be more effective than the cholinesterase inhibitors, prevention or cure remained, in 2012, on the far distant horizon.

Challenges to the AD Concept. At the turn of the twenty-first century, the concept of AD and dementia as created by Kraepelin and Alzheimer and modified by researchers in the 1970s remained firmly ensconced as the dominant approach to chronic, age-associated progressive dementia. Nonetheless, many critics have pointed out that the insistence that AD be regarded as a disease distinct from aging was always somewhat arbitrary. Because the prevalence of AD rises steadily with age, and because all of the putative biomarkers identified for AD could be found to some degree in the brains of all older people, it was as plausible to view AD as an extreme point on a spectrum of normal cognitive aging as to view it as a disease. Others pointed out that scientific progress was undermining the concept of a unified disease. For example, linking different genes to early- versus late-onset AD seemed to argue for a return to Kraepelin's original age-based distinction between senile dementia and Alzheimer's.

Other critics were more concerned with the impact of the AD concept than its scientific validity, arguing, for example, that returning to a concept of dementia as an extreme point on a spectrum of normal brain aging would better capture the diversity of the actual experience people have with dementia and avoid the paralyzing fear and stigma associated with the diagnosis of Alzheimer's. Perhaps the most interesting critics emerged from people with dementia themselves. As the diagnosis of dementia was made at earlier stages, people with early-stage dementia increasingly began to speak publicly for themselves, playing prominent roles in advocacy efforts and trying to establish a meaningful framework for living with the losses associated with the disease. By 2010, at least a dozen book-length memoirs had been published in English by people diagnosed with various forms of dementia.

[*See also* **Death and Dying; Disease; Gerontology; Hospitals; Medicare and Medicaid; Medicine: Overview; Mental Health Institutions; National Institutes of Health; Nursing; Pharmacology and Drug Therapy; Psychiatry; Psychology; Research and Development (R&D);** *and* **Societies and Associations, Science.**]

BIBLIOGRAPHY

Alzheimer's Association, "Alzheimer's Facts and Figures." http://www.alz.org/alzheimers_disease_facts_and_figures.asp (accessed 18 October 2012). Standard source for statistics about the prevalence and epidemiology of Alzheimer's and dementia.

Ballenger, Jesse F. "Disappearing in Plain Sight: Public Roles of People with Dementia in the Meaning and Politics of Alzheimer's Disease." In *The Neurological Patient in History*, edited by Stephen Jacyna and Stephen Casper. Rochester, N.Y.: University of Rochester Press, 2012.

Ballenger, Jesse F. *Self, Senility, and Alzheimer's Disease in Modern America: A History.* Baltimore: Johns Hopkins University Press, 2006. The only comprehensive cultural history of Alzheimer's and dementia in the United States.

Ballenger, Jesse F., et al., eds. *Treating Dementia: Do We Have a Pill for It?* Baltimore: Johns Hopkins University Press, 2009. This volume provides a variety of perspectives, including some from leading researchers in the AD field, on the development of drugs for dementia and the controversies that have surrounded them.

Bick, Katherine, Luigi Amaducci, and Giancarlo Pepeu, eds. *The Early Story of Alzheimer's Disease: Translation of the Historical Papers by Alois Alzheimer, Oskar Fischer, Francesco Bonfiglio, Emil Kraepelin, Gaetano Perusini*. Padova, Italy: Liviana Press, 1987. Valuable collection of English translations of the classic papers by Kraepelin and Alzheimer.

Fox, P. "From Senility to Alzheimer's Disease: The Rise of the Alzheimer's Disease Movement." *Milbank Quarterly* 67, no. 1 (1989): 58–102. Remains the most thorough account of the rise and success of the AD movement.

Katzman, R. "Editorial: The Prevalence and Malignancy of Alzheimer Disease. A Major Killer." *Archives of Neurology* 33, no. 4 (1976): 217–218.

Katzman, Robert, and Katherine L. Bick, eds. *Alzheimer Disease: The Changing View*. San Diego, Calif.: Academic Press, 2000. This book is a valuable collection of interviews of participants looking back at their involvement in the Alzheimer's disease movement and groundbreaking biomedical research of the 1970s.

National Institutes of Health. "Estimates of Funding for Various Research, Condition, and Disease Categories (RCDC)." 2012. http://report.nih.gov/categorical_spending.aspx (accessed 18 October 2012). The NIH's estimate of annual funding for various research, condition, and disease categories based on grants, contracts, and other funding mechanisms across the NIH.

Taylor, Richard. *Alzheimer's from the Inside Out*. Baltimore: Health Professions Press, 2007. Taylor is a retired psychologist, and his book is one of the most articulate and provocative critical memoirs of the AD experience.

Whitehouse, Peter J., and Daniel George. *The Myth of Alzheimer's Disease: What You Aren't Being Told about Today's Most Dreaded Diagnosis*. New York: St. Martin's Press, 2008. This book is a particularly interesting and comprehensive critique of the AD concept and the biomedical approach to dementia because Whitehouse was among the most prominent neurologists involved in establishing the cholinergic hypothesis and working with the pharmaceutical industry on drug treatments for AD.

Whitehouse, Peter J., Konrad Maurer, and Jesse F. Ballenger, eds. *Concepts of Alzheimer Disease: Biological, Clinical, and Cultural Perspectives*. Baltimore: Johns Hopkins University Press, 2000.

Jesse F. Ballenger

AMERICAN ASSOCIATION FOR THE ADVANCEMENT OF SCIENCE

The American Association for the Advancement of Science (AAAS) was founded in 1848 to facilitate communication among scientists and to establish an authoritative public presence for science in the larger community.

Growing out of the Association of American Geologists and Naturalists (1840), the early AAAS drew inspiration from the British Association for the Advancement of Science (1832). Its peripatetic annual meetings and published *Proceedings* soon made it the most distinguished national scientific organization. Leadership came from such eminent scientists as Joseph Henry of the Smithsonian Institution, Alexander Dallas Bache of the U.S. Coast Survey, William Barton Rogers of the Massachusetts Institute of Technology, and Louis Agassiz of Harvard University, all of whom shared a commitment to positioning scientific research in the foreground of public culture.

The growing sectional crisis and other factors led to a decline in membership and the cancelation of the 1861 meeting. The creation of the National Academy of Sciences (1863) and the growth of specialized scientific organizations, such as the American Chemical Society (1874), made postwar recovery difficult. Restructured and incorporated in the 1870s, the AAAS increased its membership and created a research fund. The psychologist James McKeen Cattell merged his journal *Science* into the AAAS in 1900 and in other aspects revitalized the organization. In 1907 the Smithsonian Institution provided office space and in 1945 the AAAS acquired its own headquarters in Washington, D.C. In Cattell's vision, the elitist National Academy of Sciences would be the upper house of American science, whereas the AAAS would serve as the lower house, accessible to all scientists and responsive to the public. In the 1920s, to create a united scientific front on matters of mutual concern, the AAAS encouraged the affiliation of specialized societies. During the late twentieth and early twenty-first centuries the AAAS advocated scientific education, encouraged racial and gender diversity in the scientific community, and addressed issues of public policy and scientific ethics.

[*See also* **Agassiz, Louis; Bache, Alexander Dallas; Cattell, James McKeen; Ethics and Medicine; Gender and Science; Gender and Technology; Henry, Joseph; Journals in Science, Medicine, and Engineering; National Academy of Sciences; Race and Medicine; Rogers, William Barton; Science: Overview; Smithsonian Institution;** *and* **Societies and Associations, Science.**]

BIBLIOGRAPHY

Kargon, Robert H. *The Maturing of American Science: A Portrait of Science in Public Life Drawn from the Presidential Addresses of the American Association for the Advancement of Science.* Washington, D.C.: American Association for the Advancement of Science, 1974.

Kohlstedt, Sally Gregory, Michael A. Sokal, and Bruce V. Lewenstein. *The Establishment of Science in America: 150 Years of the American Association for the Advancement of Science.* New Brunswick, N.J.: Rutgers University Press, 1999.

Sally Gregory Kohlstedt

AMERICAN ASSOCIATION FOR THE HISTORY OF MEDICINE

The American Association for the History of Medicine (AAHM) is the principal academic society for the history of medicine in North America. Its organizational development is simple enough for brief review and its history illustrates general features of this field of study. Established by a small group of academic physicians in 1924, the Association acquired the conventional elements of a scholarly society gradually. Originally founded as the American Section of the International Society of the History of Medicine through the initiative of Dr. E. B. Krumbhaar, the Associa-

tion developed independently but still maintains a close affiliation with the international organization. In 1938, Henry Sigerist, the director of the Institute for the History of Medicine at the Johns Hopkins University, led a revision of the Association's by-laws that opened the way for affiliation with the *Bulletin of the History of Medicine*. The *Bulletin*, which had previously separated itself from the *Bulletin of the Johns Hopkins Hospital*, established a permanent home at Sigerist's Institute and continued to serve as the official journal of the AAHM under an agreement refined in 1953 and 2004. In 1958, the Association became incorporated and in 1990 it created the History of Medicine Foundation to aid in developing its endowment.

The AAHM's early history paralleled that of an academic organization that also met for the first time in 1924, the History of Science Society, created by George Sarton, who established the study of the history of science at Harvard University. Sarton was quick to strike up a friendly rivalry with Sigerist, launching an attack on the hybrid nature of the history of medicine in the United States. Historians of medicine like Sigerist, Sarton suggested, were too closely allied with physicians to pursue first-flight historiography. Sarton, however, also endorsed the value of expert technical knowledge for historians and made no secret of his professional jealousies, stating an ambition to have for the history of science an institute just as fine as the one run by Sigerist at Hopkins.

Sigerist was, in fact, marking out a middle course for the AAHM, between the History of Science Society in the groves of academe and the professional territory of the AAHM's parent organization, the International Society for the History of Medicine, run almost exclusively by and for physicians. Trained as both a physician and an intellectual historian, Sigerist worked, often successfully, with the medical profession and pursued historical inquiry that was fiercely engaged with contemporary social and political analysis. He helped the Association to become the hub of a distinctive field of history and met with serious obstacles only late in his career, when his strong advocacy for state-sponsored medicine drew animosity in the 1940s.

The original hybrid nature of the AAHM continued to strengthen the organization, which in addition to historians and doctors included archivists and librarians as well as book collectors and vendors and professionals in public health, nursing, and dentistry. This diversity in membership might be traced to Sigerist's foundational alliance with academic physicians, who were themselves often bibliophiles and collectors, promoters of public health, and advocates for the archival preservation of professional papers.

The same hybridity may also create tensions. Tremors in academic history in the 1970s were felt as a kind of aftershock in the AAHM, spreading along the fault lines exposed by Sarton and Sigerist. A generation of historians trained in social history found the AAHM to be a marvelous partner in the development of graduate programs and departments. Papers from students of the new social history arrived in growing numbers before the Association's reviewers. But a mode of scholarship that felt vital and progressive to some had a deflationary feel to others, replacing a familiar historiography that seemed more supportive of conventional medical aspirations. Such tensions found expression in 2001 in a boycott by a portion of the Association's membership at the annual meeting in Charleston, South Carolina, in support of a call from the National Association for the Advancement of Colored People (NAACP). Over a decade later, however, the AAHM seemed stronger for its diverse membership as it faced the challenge of a digital realm that altered the value of journal subscriptions while holding out the allure of inexpensive international communication and influence.

[*See also* **Dentistry; History of Science Society; Journals in Science, Medicine, and Engineering; Medicine: Overview; Nursing; Public Health;** *and* **Societies and Associations, Science: Overview.**]

BIBLIOGRAPHY

Fee, Elizabeth, and Theodore M. Brown, eds. *Making Medical History: The Life and Times of Henry E. Sigerist*. Baltimore: Johns Hopkins University Press, 1997.

Miller, Genevieve. "The Missing Seal or Highlights of the First Half Century of the AAHM." *Bulletin of the History of Medicine* 50 (1976): 93–121.

Reverby, Susan M., and David Rosner. "'Beyond the Great Doctors' Revisited: A Generation of the 'New' Social History of Medicine." In *Locating Medical History: The Stories and Their Meanings*, edited by Frank Huisman and John Harley Warner, pp. 167–193. Baltimore: Johns Hopkins University Press, 2004.

Sarton, George. "The History of Science versus the History of Medicine." *Isis* 23 (1935): 313–320.

Christopher Crenner

AMERICAN INSTITUTE OF PHYSICS

The American Institute of Physics (AIP) is a not-for-profit umbrella organization comprising 10 Member Societies—leading professional societies in physics and allied sciences, which represent about 140,000 individuals—and 24 Affiliated Societies. AIP is a major publisher of scientific journals and provides diverse services to the Member Societies and to the physics community, including news and media coverage, government relations, demographic research, education and student services, and history programs. AIP publishes 13 journals and copublishes journals for some of its Member Societies and other publishing partners. AIP's headquarters is in College Park, Maryland; it also has offices in Melville, New York, and Beijing, China.

AIP was founded by the American Physical Society, the Optical Society of America, the Acoustical Society of America, and the Society of Rheology. The first governing board meeting was held in Washington, D.C., in May 1931. The Chemical Foundation helped defray expenses for the first year, and AIP opened an office in the Foundation's headquarters in New York City. The founding societies were joined by the newly formed American Association of Physics Teachers in January 1932, and AIP was incorporated on 1 June 1932 in the state of New York.

AIP was created to provide economies of scale in publishing for the Member Societies during the Great Depression, to respond to the "stop science" movement that blamed science for the decline in jobs, and to address the fragmentation of physics into subfields. AIP's constitution states that its objectives are the "advancement and diffusion of knowledge of the science of physics and its application to human welfare" through the publication of scientific journals, explaining physics to the public, cooperation with other organizations devoted to physics, promotion of unity of physicists, and fostering relations to other sciences, arts, and industries (Barton, 1956).

During its first year, the Institute assumed publishing operations for the Member Societies and took ownership of *The Review of Scientific Instruments* from the Optical Society. Early milestones include the founding of the *Journal of Chemical Physics* in 1933 and the launching of a corporate associates program in 1934. In 1936 the Institute assumed control of the journal *Physics* from the American Physical Society, renaming it *Journal of Applied Physics*.

The dramatic increase in the production of new physics PhDs that began after World War II and continued for the next two decades allowed AIP to grow steadily. It began publishing its flagship magazine, *Physics Today,* in 1948 and introduced the first of a number of English translations of Soviet physics journals in 1955. In 1956 AIP established an Office of Information and Public Relations, the first in a series of service divisions, and in 1958 it began launching new scientific journals. During the 1960s AIP added additional service divisions and began accepting new Member Societies: the American Crystallographic Society and the American Astronomical Society (1966), the American Association of Physicists in Medicine (1973), the American Vacuum Society (now AVS, 1976), and the American Geophysical Union (1986) (Hutchisson, n.d.).

In 1993 the Institute's headquarters was moved from Manhattan to a new facility in College Park, Maryland. A pioneer in electronic publishing, AIP put all of its own journals online by the late 1990s (Brodsky, 2008). To support a growing market for scholarly publishing in China, AIP opened an office in Beijing in 2010.

[*See also* **Chemistry; Journals in Science, Medicine, and Engineering; Physics; Printing and Publishing; Science: Overview;** *and* **Societies and Associations, Science.**]

BIBLIOGRAPHY

Barton, Henry A. "The Story of the American Institute of Physics." *Physics Today* 9 (January 1956): 56–60.

Brodsky, Marc H. "Oral History Interview." College Park, Md.: American Institute of Physics, 2008. http://www.aip.org/history/ohilist/31783.html (accessed 24 February 2012).

Hutchisson, Elmer. "The Story of the American Institute of Physics." Unpublished draft. College Park, Md.: Niels Bohr Library & Archives, American Institute of Physics, n.d.

H. Frederick Dylla
and R. Joseph Anderson

AMERICAN MEDICAL ASSOCIATION

The American Medical Association (AMA) was formed in 1847 by elite physicians hoping to improve the stature of their profession. This was no small task in the crowded and chaotic nineteenth-century medical marketplace, where homeopaths and other practitioners of alternative medicine challenged orthodox physicians' claims to authority and patients. Although AMA leaders aimed to improve standards of medical education and practice, regulate relations among physicians, represent the profession to the public, and promote "the usefulness, honor and interests of the medical profession," in its first 50 years the AMA won only lackluster support from physicians and possessed little public visibility or political power.

In 1901, the association reorganized itself, instituting a proportional representation system that gave state and local associations a voice in a newly created House of Delegates, which approved the association's policies and positions. Led by J. N. McCormack, national AMA representatives helped state and local medical societies reorganize themselves as constituent units in the national organization. As a result, membership increased dramatically, from fewer than 10,000 in 1900 to around 70,000 in 1910.

A more active and public profile accompanied this growth. The AMA's Council on Medical Education, for instance, monitored medical schools and their facilities, making the association a de facto arbiter of national standards governing medical education. The *Journal of the American Medical Association (JAMA)* became one of the world's preeminent medical journals, and the association began publishing specialized journals as well. AMA bureaus endeavored to safeguard the public's health by debunking quackery and so-called miracle cures, permitting approved food and drug advertisers to use the AMA seal, and sponsoring educational programs for the general public.

Historians of medicine have focused much attention on the association's objections to what its leaders termed "socialistic tendencies" in the organization of medical care. An AMA committee endorsed compulsory health insurance at the state level in the late 1910s, but the association soon recanted. Over the next three decades, the AMA (often represented by *JAMA*'s vocal editor Morris Fishbein) condemned advocates of federal health programs, government-sponsored health insurance, and other initiatives that threatened to disrupt the traditional physician–patient relationship or provide alternatives to fee-for-service medical practice. These battles peaked in the late 1940s when the AMA helped mobilize opposition to President Harry S. Truman's national health insurance plan and lent its support to private plans instead. This stance held through the 1960s and after, as the AMA leadership fought—and then reluctantly accepted—Medicare, Medicaid, and other programs considered government intrusions into medical practice.

The AMA grew slightly less conservative in the last quarter of the twentieth century and the beginning of the twenty-first century, as the balance of power shifted among the medical profession, the federal government, commercial insurers, and advocates of health-care reform. Nevertheless, with its substantial membership and tradition of vigorous advocacy, the association remained an important player in American health-care politics, widely regarded as "the voice of the American medical profession."

[*See also* Food and Diet; Food Processing; Health Insurance; Health Maintenance Organizations; Higher Education and Science; Journals in Science, Medicine, and Engineering; Medical Education; Medical Specialization; Medicare and Medicaid; Medicine: Overview; Pharmacology and Drug Therapy; Public Health; Science: Overview; *and* Societies and Associations, Science.]

BIBLIOGRAPHY

Burrow, James G. *AMA: Voice of American Medicine.* Baltimore: Johns Hopkins University Press, 1963.
Starr, Paul. *The Social Transformation of American Medicine.* New York: Basic Books, 1982.

Elizabeth Toon

AMERICAN MUSEUM OF NATURAL HISTORY

Founded in New York City in 1869 at the initiative of Albert Smith Bickmore, a Harvard-trained zoologist, the American Museum of Natural History (AMNH) was one of many late nineteenth-century institutions launched by a capitalist elite eager to assert its cultural dominance at a time of rapid immigration and urban growth. The AMNH founders included such notables as William E. Dodge Jr. and J. P. Morgan. Natural history museums, displaying specimens of New World flora, fauna, minerals, and fossils, had earlier been founded in Philadelphia (1812), Cincinnati (1820), Boston (1830), and Albany (1843). These institutions typically either catered to an elite membership or displayed entertaining "curiosities" with scant regard for educational objectives or organizational coherence. The AMNH pursued a more ambitious agenda. The curators organized the exhibits on scientific principles, arranged them in display cases with explanatory captions, and welcomed the public. The exhibits typically progressed from simpler to more complex forms, thereby helping popularize Charles Darwin's theory of evolution propounded in *The Origin of Species* (1859).

Other cities, following the scientifically oriented lead of the AMNH, the Smithsonian Institution (1846), and Harvard's Museum of Comparative Zoology (1859), opened natural history museums as well. The Milwaukee Public Museum (1882) displayed the collections of the Natural History Society of Wisconsin. Chicago's Field Museum of Natural History opened in 1893 as part of the World's Columbian Exhibition. Pittsburgh's Carnegie Museum of Natural History followed in 1896.

Initially, the AMNH and similar institutions elsewhere functioned as centers of scientific culture in their communities, sponsoring scientific meetings, hosting public gatherings to hear scientific papers, and even funding research. The AMNH's North Pacific Expedition (1897–1902), organized by anthropologist Franz Boas, produced important ethnographic findings. An AMNH expedition led by Roy Chapman Andrews in the 1920s made major fossil discoveries in Mongolia. As the natural sciences grew more specialized and as research funding shifted to universities and the federal government, this aspect of the role of natural history museums declined. But with time, the public education component revived and expanded. Again the AMNH led the way. Its exhibits, including the Hayden Planetarium (1935) and elaborate dioramas of birds and mammals in their natural habitats, became very popular. With continuing support from wealthy benefactors, the museum mounted exhibits ranging from "the Human Genome" and a "Grand Tour of the Universe" to a butterfly conservatory and a Hall of Biodiversity. In 2007, after premiering the Hollywood film *A Night at the Museum*, the AMNH introduced sleepovers for children. A 2007 exhibit in the Spitzer Hall of Human Origins offered, the museum claimed, "the most comprehensive evidence of human evolution ever assembled." As its sesquicentennial approached, the AMNH remained both a sought-after tourist destination and a leader in the popular diffusion of scientific knowledge.

[*See also* Boas, Franz; Evolution, Theory of; Museums of Science and Natural History; Science: Overview; *and* Smithsonian Institution.]

BIBLIOGRAPHY

"American Museum of Natural History." http://www.amnh.org/about-us/history. A detailed timeline of key events in the museum's history.

Conn, Steven. *Museums and American Intellectual Life, 1876–1926.* Chicago: University of Chicago Press, 1998. Chapter 2 (pp. 32–75) of this broad-ranging interpretive study deals with natural history museums, including the AMNH.

Paul S. Boyer

AMERICAN PHILOSOPHICAL SOCIETY

Benjamin Franklin organized the American Philosophical Society (APS) in 1743, but failure to attract wider support led to its collapse in 1745. Throughout the 1750s and 1760s, a small group of Philadelphians met intermittently to discuss science, and in 1766 they organized themselves into the American Society for Promoting and Propagating Useful Knowledge. Led by Charles Thomson, the membership consisted principally of liberal members of the Society of Friends (Quakers) who supported the Assembly Party in Pennsylvania politics. In 1767 a rival group, mostly Anglicans and Presbyterians aligned with Pennsylvania's Proprietary Party, organized an American Philosophical Society. Franklin, elected president of the latter body in 1768, oversaw the merger of the two groups and on 2 January 1769 presided over the first meeting of the American Philosophical Society, held at Philadelphia, for Promoting Useful Knowledge. Later that year the

APS, with the financial support of the Provincial Assembly, made 11 sets of observations of the Transit of Venus. These were published in 1771 in the society's *Transactions*, winning international recognition for American science. From the society's inception, it usually recruited its officers from the University of Pennsylvania faculty and its active members from the city of Philadelphia; yet until the 1840s, the APS served as a national scientific society, acting as a resource for the federal government and disseminating the research of American scientists through its *Transactions*.

With the emergence of specialized scientific societies and the creation of official national organizations for American science, the APS lost its national role and transformed itself into a general learned society. In 1917 the society helped organize the American Council of Learned Societies. Since the 1950s the APS has been a major archival repository of collections in early American history and the history of science; it awards approximately $700,000 each year in research grants. In 2001, the society opened a museum at its headquarters in Philosophical Hall, next to Independence Hall in Philadelphia. The museum's exhibitions, which feature material relating to science, art, and history, use the society's own collections as well as artifacts, specimens, instruments, books, manuscripts, and works of art from other institutions.

[*See also* **Franklin, Benjamin; Journals in Science, Medicine, and Engineering; Museums of Science and Natural History; Science: Overview;** *and* **Societies and Associations, Science.**]

BIBLIOGRAPHY

Carter, Edward C., II. *"One Grand Pursuit": A Brief History of the American Philosophical Society's First 250 Years, 1743–1993.* Philadelphia: American Philosophical Society, 1993.
Hindle, Brooke. *The Pursuit of Science in Revolutionary America, 1735–1789.* Chapel Hill: University of North Carolina Press, 1956.

Simon Baatz;
updated by Hugh Richard Slotten

AMERICAN SYSTEM OF MANUFACTURES

American manufacturers stole the show at London's 1851 Crystal Palace Exposition. Several American products especially impressed the British, among them Cyrus McCormick's reaper, Alfred C. Hobbs's unpickable lock, and most of all the guns: Samuel Colt's revolver and the Robbins and Lawrence rifle. These items impressed exposition visitors not only because of their excellence, but also because they were produced in large quantities and, in the case of the guns, with interchangeable parts.

Although historians debate the precise meaning of the American system of manufactures, most define it as the system of production that originated in the arms industry to manufacture guns with interchangeable parts. In nineteenth-century America, this system was often called the "armory system." A British commission was sent to the United States in 1853 to learn about the system that produced the articles displayed at the 1851 exposition. The commission called this pioneering mass-production process the "American system" of manufacture.

The idea of parts' interchangeability, which originated in France, was novel at the beginning of the nineteenth century and not important to most manufacturers. The American government, however, recognized the potential advantages of guns with interchangeable parts that could be easily repaired on the battlefield. At the urging of several presidents and secretaries of war, the Ordnance Department and Congress supported efforts by inventors and entrepreneurs to manufacture such arms. A number of armsmakers claimed to achieve interchangeability in the early nineteenth century, but as historian Merritt Roe Smith has shown (Smith, 1977), the first truly interchangeable arms were the rifles made by John Hall at his shop at the federal arsenal at Harpers Ferry, Virginia.

The American system of manufactures is vital to an understanding of the development of U.S. industry. Precision manufacture of interchangeable parts led to greater and greater division of labor and the invention of self-acting machines that could be operated by workers with less training and experience than traditional craftsmen. The system of modern mass production first introduced by Henry Ford at the Ford Motor Company can be traced directly to the armory system. American mass production manufactured consumer goods in unprecedented quantities and achieved a commercial monopoly of many products.

[*See also* **Agricultural Technology; Colt, Samuel; Ford, Henry; Machinery and Manufacturing; McCormick, Cyrus Hall; Military, Science and Technology and the; Springfield Armory;** *and* **Technology.**]

BIBLIOGRAPHY

Hounshell, David A. *From the American System to Mass Production, 1800–1932: The Development of Manufacturing Technology in the United States.* Baltimore: Johns Hopkins University Press, 1984.

Mayr, Otto, and Robert C. Post, eds. *Yankee Enterprise: The Rise of the American System of Manufactures.* Washington, D.C.: Smithsonian Institution Press, 1981.

Rosenberg, Nathan, ed. *The American System of Manufactures: The Report of the Committee on the Machinery of the United States 1855 and the Special Reports of George Wallis and Joseph Whitworth.* Edinburgh: Edinburgh University Press, 1969.

Smith, Merritt Roe. *Harpers Ferry Armory and the New Technology: The Challenge of Change.* Ithaca, N.Y.: Cornell University Press, 1977.

Lindy Biggs

AMNIOCENTESIS

Amniocentesis literally means "to prick the amnion." As a medical procedure, it was first developed in the 1930s to ease polyhydramnios, too much amniotic fluid, a serious, but rare condition in pregnancy. A sharp, hollow needle is introduced through a pregnant woman's abdominal and uterine walls, into the amniotic sac, to tap off some of the fluid it contains.

By the late 1960s, amniocentesis had become the first of several techniques for what later came to be called prenatal diagnosis. Because amniocentesis for prenatal diagnosis is a system, rather than a single technique, its development cannot be attributed to just one researcher or team of researchers in just one country at just one time.

The procedure begins with an amniotic tap, following which both the fluid and the fetal cells that are floating in it can be subjected to a variety of diagnostic tests. The first of these tests, developed in the mid-1960s—was a test for the sex of the fetus: obstetricians used microscopic examination of stained fetal cells to locate sex chromatin, a newly discovered cellular body that can be found in the cells of female, but not male, mammals. These obstetricians had patients who were known carriers of hemophilia, a very debilitating sex-linked disease usually limited to males; these doctors wanted to give their carrier patients the opportunity to terminate male pregnancies.

In the next decade other more common genetic or congenital conditions and diseases became diagnosable prenatally either because of improvements in the culturing of human fetal cells or in the development of very specific biochemical assays of amniotic fluid. Obstetric ultrasound had also developed to the point where it could be used to make the initial tap safer for the mother and the fetus. By the mid-1970s, obstetricians and medical geneticists (a specialty that experienced a period of rapid expansion as fetal diagnostic testing developed) could offer pregnant women prenatal testing for trisomy 21, or Down syndrome, a chromosomal anomaly that could be identified by karyotyping, a process that allowed photomicrographic imaging of chromosomes; similarly, spina bifida (when the developing spinal cord does not close completely) could be diagnosed by the presence of α-fetoprotein in amniotic fluid. Direct genomic analysis of the DNA in the fetal cells became possible early in the 1980s, permitting the diagnosis of such single-gene recessive diseases as sickle-cell anemia. When genomic diagnostic tests were later perfected for the genes that were known to increase the risk of early-onset breast and colorectal cancer (in the last decade of the twentieth century and the first decade of the twenty-first), these tests could also be done on the fetal cells obtained from amniocentesis.

Clinical trials assessing the safety of amniocentesis were successfully completed in several countries by the mid-1970s and since that time the procedure has been part of standard prenatal care in much of the developed world. Other prenatal diagnostic testing systems were subsequently developed, some of them less invasive (ultrasound, serum screening) and some of them useful earlier in pregnancy (chorionic villus sampling). None has been without social controversy, however, largely because the most common consequence of a positive diagnosis for a seriously disabling or life-threatening fetal condition is termination of the pregnancy.

[*See also* **Childbirth; Ethics and Medicine; Medicine: Since 1945;** *and* **Medicine and Technology.**]

BIBLIOGRAPHY

Cowan, Ruth Schwartz. *Heredity and Hope: The Case for Genetic Screening.* Cambridge, Mass.: Harvard University Press, 2008. See Chapter 3 and the conclusion.
Riis, Povl, and Fritz Fuchs. "Antenatal Determination of Foetal Sex in Prevention of Hereditary Diseases." *Lancet* (23 July 1960): 181. This article is the first published account of amniocentesis done to diagnose a genetic disease in the fetus. However, several knowledgeable people believe that other physicians, who happened not to be practicing in countries where abortion for what was then called "fetal indications" was legal, were also doing this at the same time. In Denmark, where Riis and Fuchs practiced, such abortions had been legal since the late 1930s.

Ruth Schwartz Cowan

ANATOMY AND HUMAN DISSECTION

Anatomy is the discipline that studies the normal structures of once-living beings; dissection is the process through which those structures are uncovered and identified as coherent parts of the once-whole creature. Human anatomy and, briefly, dissection emerged within Western medicine in the fourth century BCE, whereas the study of structures of other living macroscopic things developed as ancient philosophers questioned the regularities of natural forms.

From the Sixteenth to the Eighteenth Centuries. By the late sixteenth century, the formal study of human anatomy was well enshrined in the medical faculties of European universities as an erudite, text-based branch of learning, with dissections carried out on the corpses of executed felons to demonstrate the parts of the body to crowds of students. By the mid-eighteenth century, when references to organized teaching on human anatomy appeared in the British colonies, students in Europe had increasing opportunities to learn by performing their own hands-on dissections, largely in privately run anatomy courses, in addition to gaining anatomical knowledge through the more gentlemanly means of texts and lectures. The tensions between anatomy as the foundational descriptive science of the body and dissection as the messy, transgressive method by which it is best learned underlie the history of anatomical study in the United States.

The discipline of anatomy most likely arrived in the British colonies in books imported from the United Kingdom and continental Europe. Such illustrated texts would have served apprentices, informal students, and the self-taught until the establishment of home-grown medical schools. At the same time, colonists brought British laws that allowed for the occasional dissection of the bodies of executed criminals for educational purposes. Thus, Massachusetts granted the privilege "to anatomize…some malefactor" to students of medicine in 1647, and sporadic references to such experiences appeared throughout the colonial period (Sappol, 2002, p. 100). Such laws, and similar ones passed in the early Republic, maintained the associations among criminality, dissection, and posthumous punishment well established in the Old World. Where these laws did not exist and when no executed bodies were available, practitioners and their students dissected cadavers obtained illegally from graveyards. This practice became especially critical with the start of private medical lectures, such as those given on anatomy by Dr. Thomas Cadwalader in Philadelphia in 1750, and with the emergence of full-fledged medical schools in Philadelphia (1765), New York City (1767), and Boston (1783). These schools' founders trained in London, Edinburgh, and other European centers where human dissection had become the most favored way for students to learn anatomy.

From the Nineteenth to the Early Twentieth Centuries.

As the number of medical schools exploded in the early nineteenth century, from five in 1810 to sixty-five in 1860, so too did the demand for bodies intensify. As Michael Sappol argues, personally dissecting a human body became the *sine qua non* for claims to a scientific medical education and hence to professional legitimacy. At a time when there were no laws regulating the practice of medicine, competitors to the "regulars" flourished. Mastery of anatomical language and dissecting skills brought an aura of medical expertise beyond what sectarian practitioners and the self-taught could profess. The intense interest in human dissection as a means to learn normal human anatomy also stemmed from the rise in prestige of Parisian medicine in the 1820s to 1850s. In Paris, clinicians focused on extending knowledge of the clinical course of diseases seen at the bedside with the physical signs of disease seen at autopsy. This clinical-pathological revolution required familiarity with the normal appearances of organs and tissues and hence with normal anatomy learned from dissection.

The demand for bodies spurred much social tension. Grave-robbing and body-snatching scandals appeared in newspaper accounts throughout the nineteenth century. The middle classes became particularly outraged when one of their own ended up missing; the poor and dispossessed sometimes violently protested the disproportionate use of bodies from pauper graveyards. The political solution came with anatomy acts, passed state by state from the 1850s to the 1930s, as medical schools opened up and local populations discovered empty graves. Anatomy acts basically required that the bodies of those who died in pov-

erty, with no relatives or friends to claim the remains and pay for burial, be turned over to medical schools. Those who died in state institutions, such as mental hospitals and prisons, as well as those in city and county facilities, such as poorhouses, became the pool from which anatomists procured their materials. Although judged harshly by late twentieth and early twenty-first century commentators for this invidious discrimination against the abandoned poor, anatomy acts had their foundations in long-held legal principles. In English common law, it was the responsibility of those in whose possession a dead body was held to dispose of it, customarily through burial. In the vast majority of cases, this was the family or, at the very least, the executors of those who had property. The "state," be that city, county, or state governance, thus had the right to dispose of the dead bodies in its care. Because all anatomy acts specified that the remains were to be properly buried after dissection, states technically carried out their timeworn duties. Since the abandoned poor literally had no one to care for them (that is, no one with enough money for burial), moreover, authorities believed that they minimized the suffering of those whose loved ones had been targeted by grave robbers (Lawrence, 1998).

Not all body parts were properly buried, however. Indeed, the disposal of cadaveric parts before the 1960s requires much further research, although fragmentary evidence suggests that medical schools barely complied (Blakely and Harrington, 1997). Anatomists certainly kept interesting and illustrative parts from their dissections in anatomical collections in medical museums. Although several anatomists created private collections in the late eighteenth and early nineteenth centuries, such as John Warren, who donated his specimens to Harvard in 1847, institutional collections became the norm after the 1850s (McLeary, 2001, pp. 28–29). One of the most extensive collections of specimens, albeit largely pathological ones, for example, was started in 1862 by Surgeon General William Hammond as the Army Medical Museum to contain body parts sent in by Union surgeons during the Civil War that, it was hoped, would advance knowledge in military medicine (Barbian et al., 2012). The availability of industrially produced formaldehyde after the 1870s meant that both whole bodies and body parts could be kept for much longer periods of time, and this compound replaced alcohol as the key component of formulas for preserving anatomical and pathological specimens. Most nineteenth-century medical schools of any note established museums to house these physical samples of human anatomy, along with casts, models, and charts, for educational study and research. Such specimens allowed students to examine nicely dissected dried or bottled anatomical parts outside of the dissecting room, in relative comfort, and to compare them—ideally—with diseased versions of the same structures (McLeary, 2001).

Popular anatomy also flourished in the nineteenth century. Reformers argued that education on the proper names and functions of the human body would reduce the appeal of medical quackery and therapeutic superstitions, while encouraging healthful attention to hygiene, public health, and legitimate medical expertise. Lectures, museum displays open to the public, popular medical texts, and schoolbooks all brought knowledge of the human body to the masses (Sappol, 2002, pp. 168–211). At the same time, books designed to teach Americans about their bodies also helped to instruct them that "the" body was a male body (Sappol, 2002, p. 185). The lecturers and authors who introduced large audiences to anatomies that included the sexual organs claimed to provide information for morally responsible behavior—reproduction—but the underlying allusions to sexuality no doubt helped to draw consumers (Sappol, 2002, pp. 200–208).

By the late nineteenth century, anatomy's preeminent role as the foundational science for medical knowledge and practice became diluted by the rise of scientific competitors: experimental physiology, bacteriology, and chemistry. Anatomy courses began to cover histology and embryology along with gross (macroscopic) anatomy. As the supply of legal cadavers stabilized, moreover, anatomical teaching through lectures and dissection in the first year of medical school became a traditional staple, an often tedious rite of passage for each new class of aspiring doctors. Medical museums, long a crucial adjunct to the study of anatomy

and pathology, started to close in the 1930s and 1940s, losing their spaces to needs seen more critical to the modern medical enterprise such as research laboratories. In the 1970s and 1980s, pressure on the preclinical medical curriculum from other new sciences, notably genetics, along with other changes (problem-based learning, vertical integration learning) meant that the number of hours that medical students spent learning gross (macroscopic) anatomy started to decline. In the early twenty-first century, commenters noted that the proportion of hours that students spent dissecting had also fallen. Students, in turn, spent more time learning from computer programs, video recordings of dissections, scans from imaging technologies (X-ray, magnetic resonance imaging), and models (Bergmann et al., 2011).

From the Mid-Twentieth to the Early Twenty First Centuries. The social context of the late 1950s to the mid-1970s had quite a different impact on the study of human anatomy. The rise of social safety nets (e.g., Social Security death benefits), the push to close large mental institutions, and prison reforms led to a decline in the number of unclaimed bodies sent to medical schools for dissection. In response, anatomy programs began to encourage citizens to donate their bodies to science to further research and teaching. The Uniform Anatomical Gift Act, drafted in 1968, provided states with model legislation to allow organ donations and made the legal process of donating entire bodies for teaching and research explicit. Riding on the coattails of criticisms of the funeral industry, the rise in popularity of cremation, and the increasing possibilities for transplants as therapeutic interventions for failing organs, the pleas for body donation resulted in a relatively rapid shift in the source of cadavers. From the early 1980s, the bodies of respectable middle-class people, people with families and friends, replaced the bodies of the unclaimed poor in anatomy laboratories, although anatomy acts remained on the books in most states into the early twenty-first century (Warner and Rizzolo, 2006).

Although human dissection during the first years of medical school has been a well-known part of medical education since the mid-nineteenth century, using cadavers to study anatomy was never limited to junior medical students. More advanced students practiced surgical procedures on cadavers in the nineteenth century, and in the early twenty-first century body parts were still in high demand for use in residency training and continuing-education seminars (Roach, 2003). Undergraduate programs at universities with connections to medical schools now offered cadaver-based courses, although usually only advanced undergraduates were allowed to do the actual dissection. At the end of the twentieth century, moreover, anatomical displays reappeared in popular culture with the first exhibit of *Body Worlds* in 1995 in Tokyo and shows throughout the United States shortly thereafter. *Body Worlds* put plastinated human specimens, including whole bodies, on public display. Developed by Gunther von Hagens in the 1970s, plastination replaces water and fat with plastics and, in the process, fixes human tissues for nearly permanent preservation. Specimens can be touched, and they have no odor. Like popular anatomy lecturers in the nineteenth century, von Hagens promoted *Body Worlds* exhibits as educational and uplifting experiences for lay people. Displays of smokers' lungs and cross-sections of an obese body, for example, have been designed to offer quite visual lessons about the harms of smoking and obesity (Whalley, 2009). Given the history of human dissection, it is not surprising that controversies immediately arose over the extent to which the humans whose bodies, or body parts, were on display had given properly informed consent, and most exhibits in the early twenty-first century included documentation from donors. Critics have continued to abound, however, because *Body Worlds* and similar shows have been quite successful commercial enterprises. In 2014, it remained to be seen whether medical museums themselves would have a revival, with plastinated specimens of normal and pathological anatomy provided for medical students, and others, to study in conjunction with, or in place of, their own hands-on dissection of a cadaver.

[*See also* **Animal and Human Experimentation; Death and Dying; Ethics and Medicine; Medicine: Overview; Organ Transplantation; Physiology;** *and* **Public Health.**]

BIBLIOGRAPHY

Barbian, Lenore, Paul S. Sledzik, and Jeffery S. Resnick. "Remains of War: Walt Whitman, Civil War Soldiers and the Legacy of Medical Collections," *Museum History Journal* 5 (2012): 7–28.

Bergmann, E. M., Cees P. M. Van Der Vlueten, and Albert J. J. A Scherpbier. "Why Don't They Know Enough about Anatomy? A Narrative Review." *Medical Teacher* 33 (2011): 403–409. These Dutch authors review recent literature about the decline in cadaver dissection in medical schools, including those in the United States.

Blake, John B. "Anatomy." In *The Education of American Physicians: Historical Essays*, edited by Ronald L. Numbers, pp. 29–47. Berkeley, Calif., and Los Angeles: University of California Press, 1980. A useful overview.

Blakely, Robert, and Judith M. Harrington. *Bones in the Basement: Postmortem Racism in Nineteenth-Century Medical Training*. Washington, D.C.: Smithsonian Institution Press, 1997.

Lawrence, Susan C. "Beyond the Grave—The Use and Meaning of Body Parts: A Historical Introduction." In *Stored Tissue Samples: Ethical, Legal and Public Policy Implications*, edited by Robert F. Weir. Iowa City: University of Iowa Press, 1998.

McLeary, Erin. *Science in a Bottle: The Medical Museum in North America, 1860–1940*. PhD Dissertation, University of Pennsylvania, 2001.

Rizzolo, Lawrence J., and William B. Steward. "Should we Continue Teaching Anatomy by Dissection When...?" *The Anatomical Record (Part B: New Anatomist)* 289B (2006): 215–218. Reviews the ongoing debate over how much medical students actually learn from dissection compared to using computer programs and other educational media.

Roach, Mary. *Stiff: The Curious Lives of Human Cadavers*. New York: W. W. Norton, 2003. Roach documents many uses of cadavers in the twentieth century.

Sappol, Michael. *A Traffic of Dead Bodies: Anatomy and Embodied Social Identity in Nineteenth-Century America*. Princeton, N.J.: Princeton University

Press, 2002. A key text for understanding the history of anatomy and dissection in the United States before World War I.

Warner, John Harley, and James M. Edmonson. *Dissection: Photographs of a Rite of Passage in American Medicine, 1880–1930*. New York: Blast Books, 2009. A collection of striking images.

Warner, John Harley, and Lawrence J. Rizzolo. "Anatomical Instruction and Training for Professionalism from the 19th to the 21st Centuries." *Clinical Anatomy* 19 (2006): 403–414. Discusses attitudes toward dissection in medical education.

Whalley, Angelina. *Body Worlds—The Original Exhibition of Real Human Bodies*. Arts & Sciences, 2009.

Susan C. Lawrence

ANDREWS, ROY CHAPMAN

(1884–1960), naturalist and explorer. Roy Chapman Andrews was among the most prominent of an early twentieth-century generation of American "museum men." His scientific work was so thoroughly intertwined with the American Museum of Natural History (AMNH) from start to finish that he literally rose from janitor to director over the course of his tenure. Yet Andrews' public persona as an explorer belied what one biographer called his "boundless energy" and enthusiasm for natural history (Colbert, 1960, pp. 21–22).

Born in Wisconsin, Andrews developed excellent marksmanship skills as a boy and paid his way through college by practicing taxidermy. He graduated from Beloit College in 1906 and headed to New York City to pursue further scientific studies with the American Museum's ornithology curator Frank Chapman. Andrews infamously took the only job then available at the institution: washing floors and mixing clay in the taxidermy room. At the same time, he studied for and obtained a master's degree in biology at nearby Columbia University. In 1908, he accompanied AMNH mammalogy curators on a field expedition to British Columbia to collect and study whales. From field observations and specimens he compiled two published monographs and several short papers on Pacific Cetaceans.

By the late 1910s, Andrews had gained greater fieldwork experience and his scientific interests expanded (under the influence of museum director Henry Fairfield Osborn) to more far-flung locales. Combining "ability plus showmanship" (in the words of the *New York Times*), Andrews raised large sums of money to undertake multiple museum expeditions (in 1922, 1923, 1925, and 1928) to central Asia. The AMNH Central Asiatic expeditions resulted in important zoological and botanical collections, as well as a series of maps of Mongolia. Andrews later parlayed his swashbuckling, pistol-packing media image (developed during newspaper reporting of his travels) into popular lecture tours and documentary natural history films, in one of which he is shown reenacting the discovery of dinosaur eggs on a 1925 trip to Mongolia's Flaming Cliffs.

Andrews was elected president of New York's Explorer's Club (1931 to 1934) and, ultimately, director of the American Museum of American History (1935 to 1942). He retired from the museum to write popular natural history and travel books that would inspire subsequent generations of American paleontologists and zoologists. He is reputed to be the inspiration for Steven Spielberg's *Indiana Jones* film character.

[*See also* **American Museum of Natural History; Museums of Science and Natural History; Popularization of Science;** *and* **Science: Overview.**]

BIBLIOGRAPHY

Andrews, Roy Chapman. *Under a Lucky Star*. New York: Viking Press, 1943.

Colbert, Edwin. "Roy Chapman Andrews, Explorer." *Science* 132, no. 3418 (1960): 21–22.

Gallenkamp, Charles. *Dragon Hunter: Roy Chapman Andrews and the Central Asiatic Expeditions*. New York: Viking Press, 2001.

Rossi, Michael. "Fabricating Authenticity: Modeling a Whale at the American Museum of Natural History, 1906–1974." *Isis* 101, no. 2 (2010): 338–361.

Karen A. Rader

ANESTHESIOLOGY

Anesthesiology is a uniquely American contribution to the history of medicine. First publicly demonstrated at the Massachusetts General Hospital in Boston on 16 October 1846, by William Thomas Greene Morton for the removal of a jaw tumor from the patient Gilbert Abbott by the professor of surgery at Harvard University, John Collins Warren, surgical anesthesia has saved countless patients from feeling the agony created by an operation. Morton, however, was not the first to reveal that surgical pain could be obliterated nor was he the most deserving to receive credit for the discovery. Perhaps most interesting, surgical anesthesia evolved from public demonstrations by medical and dental students and itinerant chemists of the intoxicating effects of both diethyl ether and nitrous oxide.

In January 1842, William E. Clarke, a student at Berkshire Medical College, gave ether to the sister of one of his classmates for the removal of a molar. Although the anesthetic was a success, Clarke was dissuaded from continuing research into the phenomenon by his professor of midwifery (Stetson and William, 1992). Two months later, Crawford Long, a Pennsylvania-trained physician practicing in rural Jefferson, Georgia, successfully removed two tumors from the back of the neck of James Venable. Venable and Long were friends and had shared the effect of ether during frolics. Venable remembered nothing of the surgery. Long was trying to study the effect of ether in his practice, but did not feel he had sufficient cases to prove the effect and thus did not publish his results until after Morton's demonstration (Long, 1849). Two years later, Horace Wells, a dentist in Hartford, Connecticut, attended a public demonstration of the effects of nitrous oxide by Gardner Quincy Colton. After watching a participant cut his leg while under the influence of the gas and not feel pain, Wells believed that he could use nitrous oxide to remove teeth painlessly. The very next day, with nitrous oxide provided by Colton, a dental colleague removed Wells' own molar. Shortly thereafter, Wells attempted to demonstrate nitrous anesthesia at Massachusetts

General Hospital but was soundly jeered when the patient cried out during the extraction.

News of Morton's demonstration of surgical anesthesia traveled the world through private letters and by newspaper accounts. Carried by steamship, the first correspondence with versions of Morton's exploits reached England in December 1846, and within days ether anesthesia was being used in dental and more traditional surgical operations (Ellis, 1989). By June of 1847 operations under ether anesthesia were being performed in Australia, and the Reverend Doctor Peter Parker performed the first operation under ether in China in October of that year (Wilson, 1995; Sim et al., 2000). Over the next 40 years, as antisepsis became an accepted practice, the need for surgical anesthesia greatly increased.

At the turn of the twentieth century, the first American professional society devoted to the specialty of anesthesiology was created. In 1905, the Long Island Society of Anesthetists met for the first time at the Long Island College of Medicine in Brooklyn, New York. That group is important for several reasons. Meeting to advance the "art and science" of anesthesia, practitioners often demonstrated new technics and equipment and shared studies of the outcomes of anesthetics for certain operations. The need to find the optimum anesthetic for the patient and procedure has remained a driving force in the specialty. The fact that a group of physicians met together also demonstrated the need for professionalizing the practice of anesthetizing patients and the further need of a like-minded community of physicians who performed anesthetics in their daily practice. Over the years, the Long Island Society would transform itself from seven locally based members into the American Society of Anesthesiologists, which by the early twenty-first century had grown to over fifty thousand members in the United States and correspondents from around the world (Erickson, 2005).

At the turn of the twentieth century, it was apparent that there were not enough physicians interested in administering anesthetics to create a physician-only practice, as was the case in the United Kingdom. Thus surgeons trained nurses to administer anesthetics, and after the Great War schools for nurse anesthetists began appearing. By the middle 1930s, nurse anesthetists had created a national organization; shortly after World War II they developed a certifying exam.

In the late 1930s, it became apparent that there was a need to determine which physicians were qualified to practice as anesthesiologists. In response the American Board of Anesthesiology came into existence, in 1938, to certify physicians qualified to practice the specialty. At first the examination process consisted of three parts: a written essay exam, an oral exam, and a practice exam (Bacon and Lema, 1992). Over time, the essay exam was dropped in favor of a multiple-choice exam, which was later computer administered. By the second decade of the twenty-first century, the practical exam had been completely eliminated, but the oral exam remained as a final assurance of the ability to understand the interactions among the patient's medical condition, the surgical procedure, and the physician's ability to prescribe the anesthetic and modify it during the operation to assure the best possible outcome for the patient.

Several factors have helped create the modern specialty of anesthesiology. A tremendous need for anesthesiologists emerged during World War II. Different institutions across the country taught the principles of anesthesia in 90-day courses sponsored and supervised by the Subcommittee on Anesthesia of the National Research Council; many young physicians were exposed to the specialty for the first time and returned home to join the specialty (Waisel, 2001). In the 1960s, critical-care medicine evolved, often from anesthesiologists caring for postsurgical patients during prolonged periods in the recovery room. Eventually this developed into an intensive-care unit, which was often directed by anesthesiologists. In the mid-1980s, Ellison "Jeep" Pierce created the Anesthesia Patient Safety Foundation in an effort to decrease the mortality rates from anesthesia. The foundation has helped establish basic intraoperative monitoring criteria and has spent considerable effort upgrading anesthesia equipment and practice (Smith, 2005). The success of these efforts in the early twenty-first century can be assessed in a twofold manner. First, medical malpractice rates for anesthesiologists either decreased or remained level, whereas other specialties saw a steady rise. Second, the American Medical Association and other societies have imitated the foundation in an effort to decrease errors in all of medicine.

[*See also* **Dentistry; Medical Specialization; Medicine: From the 1870s to 1945; Medicine: Since 1945; Medicine and Technology; Nursing; Surgery;** *and* **War and Medicine.**]

BIBLIOGRAPHY

Bacon, D. R., and Lema, M. J. "To Define a Specialty: A Brief History of the American Board of Anesthesiology's First Written Examination." *Journal of Clinical Anesthesia* 4 (1992): 489–497.

Ellis, R. H. "Early Ether Anaesthesia: The News of Anaesthesia Spreads to the United Kingdom." In *The History of Anaesthesia*, edited by R. S. Atkinson and T. B. Boulton, pp. 69–76. New York: Parthenon, 1989.

Erickson, J. C., III. "In the Beginning: Adolph Frederick Erdmann and the Long Island Society of Anesthetists." In *The American Society of Anesthesiologists: A Century of Challenges and Progress*, edited by D. R. Bacon, M. J. Lema, and K. E. McGoldrick, pp. 1–8. Park Ridge, Ill.: Wood Library–Museum of Anesthesiology Press, 2005.

Fenster, J. *Ether Day*. New York: HarperCollins, 2001.

Long, C. W. "An Account of the First Use of Sulphuric Ether by Inhalation as an Anaesthetic in Surgical Operations." *Southern Medical and Surgical Journal*, 5 (1849): 705–713.

Sim, P., B. Du, and D. R. Bacon. "Pioneer Chinese Anesthesiologists." *Anesthesiology* 93 (2000): 256–264.

Smith, B. E. "The 1980s: A Decade of Change." In *The American Society of Anesthesiologists: A Century of Challenges and Progress*, edited by D. R. Bacon, M. J. Lema, and K. E. McGoldrick, pp. 173–191. Park Ridge, Ill.: Wood Library–Museum of Anesthesiology, 2005.

Stetson, J. B., and E. William. "Clark and His 1842 Use of Ether." In *The History of Anesthesia: Third International Symposium Proceedings*, edited by B. R. Fink, L. E. Morris, and C. R. Stephen, pp. 400–406. Park Ridge, Ill.: Wood Library–Museum of Anesthesiology, 1992.

Waisel, D. B. "The Role of World War II and the European Theater of Operations in the Development of Anesthesiology as a Physician Specialty in the USA." *Anesthesiology* 94, no. 5 (2001): 907–914.

Wilson, G. *One Grand Chain*, pp. 41–43. Melbourne: Australian and New Zealand College of Anaesthetists, 1995.

Wolfe, R. J. *Tarnished Idol: William Thomas Green Morton and the Introduction of Surgical Anesthesia: A Chronicle of the Ether Controversy.* San Anselmo, Calif.: Norman Publishing, 2001.

Wolfe, R. J., and Leonard F. Menczer, eds. *I Awaken to Glory: Essays Celebrating the Sesquicentennial of the Discovery of Anesthesia by Horace Wells, December 11, 1844–December 11, 1994.* Hartford, Conn.: Boston Medical Library in the Francis A. Countway Library of Medicine, in association with the Historical Museum of Medicine and Dentistry, 1994.

Wolfe, R. J., and Richard Patterson. *Charles Thomas Jackson: "The Head behind the Hands": Applying Science to Implement Discovery and Invention in Early Nineteenth Century America.* Novato, Calif.: Norman Publishing; November 1, 2007.

Douglas Bacon

ANIMAL AND HUMAN EXPERIMENTATION

Experimentation involving both animal and human subjects has played a vital, and as some would argue, an essential role in the expansion of the biomedical sciences and in the development of new vaccines, drugs, and procedures. The rise of the laboratory sciences, especially physiology, microbiology, and pharmacology, in the second half of the nineteenth century transformed American medical education and, eventually, American medical practice. Not all Americans welcomed the advent of the laboratory sciences and the increasing importance of animals in research. Following the lead of English animal welfarists, American animal protectionists accused physicians and researchers of privileging science over suffering, prompting explicit discussions about the ends and means of medical experimentation.

"The limits of justifiable experimentation upon our fellow creatures," observed the Canadian Anglo physician William Osler in 1907, "are well and clearly defined. The final test of every new procedure, medical or surgical, must be made on man, but never before it has been tried on animals" (Lederer, 1995, p. 1). Despite Osler's assumption of a robust consensus about appropriate animal and human experimentation, the use of both animals and human beings in research continued to cause controversy over the course of the twentieth century.

Before the 1860s most medical experimentation involved human beings rather than animals. In part, this reflected the assumption that animal and human bodies were sufficiently different as to render experiments on animals of little utility. Moreover, physicians had recourse to the bodies of enslaved people and the poor when they wanted to test new drugs or vaccines. When the Boston physician Benjamin Waterhouse (1754–1846) acquired samples of Edward Jenner's cowpox lymph, he first tested the vaccine on his own children. He subsequently performed a public trial at the Noddles Island Asylum in which he inoculated smallpox pus into 19 vaccinated children and two unvaccinated children. Only the unvaccinated orphans developed the disease (Hopkins, 1983). In another celebrated antebellum case, the army surgeon William Beaumont (1785–1853) entered into a contractual arrangement with a working-class French Canadian voyageur who had been shot at close range in the abdomen. When he failed to close the man's stomach wound, Beaumont seized upon this unprecedented opportunity to study digestion in a living, if not always willing, human subject. In perhaps the most notorious case of human experimentation, southern physician James Marion Sims (1813–1883) developed a technique for the successful treatment of vesicovaginal fistula by performing as many as 30 surgeries on the bodies of enslaved women entrusted by their owners to Sims. Once celebrated as the "father of American gynecology," Sims since the 1960s has become a much more problematic figure. Although some physicians have noted how Sims praised the heroism and "cooperation" of his patient-subjects in his surgeries, such critics as the journalist Harriet Washington have characterized the Southern physician as a "savior and sadist," condemning his "nightmarishly painful and degrading experiments" on women of color (Washington, 2006).

In the mid-nineteenth century, ambitious American physicians sought experiences in the great hospitals in Paris, Berlin, Vienna, and other European cities. There they witnessed experiments and demonstrations on living animals. When they returned to the United States, they brought such pedagogical innovations to American medical students. At Harvard Medical School, for example, students in the 1870s learned surgical techniques, observational skills, and instrumental approaches to such complex phenomena as digestion, respiration, and reproduction in live animals. Such European innovations alarmed such Americans as the New Yorker Henry Bergh (1813–1888) and Philadelphian Caroline Earle White (1833–1916), who organized societies to protest animal abuse. The American Society for the Prevention of Cruelty to Animals (founded in 1866) sought legislation to outlaw the use of living animals in research. The American Anti-Vivisection Society (founded in 1883) similarly sought legislation bans on experiments on live animals. Despite considerable agitation and the claim that animal experimentation led directly to human experimentation on the poor and unfortunate, medical researchers successfully avoided legislative restrictions on their methods until the mid-twentieth century.

In the face of strident criticism of their research activities, such investigators as Walter B. Cannon (1871–1945), a Harvard physiologist and chair of the American Medical Association's Committee on the Protection of Medical Research, encouraged experimentalists to adopt a uniform set of rules for defensible medical research. After sending the regulations to the deans of 79 American medical schools, Cannon reported by 1910 that 37 schools had adopted and were enforcing these rules. The physiologist also organized a systematic campaign to broadcast the benefits accruing to American society from research involving animals. He recruited eminent

physicians to describe how experiments involving dogs, cats, guinea pigs, mice, and rats had improved the safety of vaccines; brought improvements in the diagnosis and treatment of tuberculosis, cancer, plague, meningitis, rabies, dysentery, cholera, and typhoid fever; and saved lives through advances in surgery and obstetrics. Defenders of animal experimentation also pointed to the benefits that accrued to animals themselves through laboratory work that influenced control of such diseases as hog cholera, anthrax, foot and mouth disease, and Texas cattle fever.

High-profile successes were especially important in helping American researchers to avoid governmental supervision and restrictions on the use of living animals. "The Panama Canal," wrote Ernest Charles Schroeder, superintendent of the U.S. Bureau of Animal Industry in 1920, "would not have been built if animal experimentation had not revealed the etiology of yellow fever." Two years later, the use of dogs in the laboratory enabled two University of Toronto researchers— Frederick Banting and Charles Best—to isolate insulin, which materially transformed the experience and treatment of diabetes and earned Banting the Nobel Prize in 1923. (He shared the award with John MacLeod.)

In the 1960s the use of both animals and human beings in research aroused considerable public controversy. In the case of animals, public agitation and concerns, especially the use of dogs in research, brought unprecedented Congressional attention to the use of living animals in laboratories. Spurred by the 4 February 1966 publication of a photo-essay in *Life* magazine on "Concentration Camps for Dogs," documenting organized pet theft rings that procured dogs for research laboratories and medical schools, Congress enacted the Animal Welfare Act to regulate the transportation, sale, and handling of dogs, cats, and certain other animals intended to be used for purposes of research or experimentation. This initial legislation regulated the delivery of animals to the laboratory, not in the laboratory. Congress subsequently amended the act in 1976, 1985, 1990, 2002, 2007, and 2008 to include regulatory standards for the care, treatment, and experiences of laboratory animals.

Experimentation involving human subjects similarly aroused concern. Reports that American researchers were endangering the lives and welfare of such vulnerable populations as mentally impaired children (the hepatitis studies at the Willowbrook State School in New York) and the elderly (the injection of live cancer cells into patients at the Jewish Chronic Disease Hospital) appeared in the popular and professional press. In 1966 Henry K. Beecher (1904–1976), a professor at Harvard Medical School, received national attention when he described 22 examples of clinical research in which investigators compromised the safety and welfare of their research subjects without their knowledge or consent. These reports prompted congressional hearings and legislative proposals.

In 1972, Americans learned from a national wire service report that the U.S. Public Health Service had followed the progress of untreated syphilis in some four hundred African American men living in rural Alabama. More than any other single study, the so-called Tuskegee Syphilis Study (1932–1972) spurred the passage of the National Research Act. Signed into law in 1974, the act called for the creation of the National Commission for the Protection of Human Subjects of Biomedical and Behavioral Research to develop guidelines for experiments involving human subjects and for the establishment of regulations and supervision of human-subjects research. One outcome of the act was the introduction of institutional review boards to evaluate prospectively all research involving human subjects.

Since the 1970s, more revelations about government-sponsored human experimentation have emerged. In 1995, following reports about plutonium injection experiments and other research involving ionizing radiation, President William J. Clinton appointed an Advisory Committee on Human Radiation Experiments to consider some four thousand radiation experiments conducted under federal auspices in the years 1944 to 1974. In the wake of this attention, the federal government, universities, and such corporations as Quaker Oats reached financial settlements in lawsuits brought by research participants or their surviving family members. In 2010 Americans learned that in the years 1946 to 1948 the U.S. Public Health Service had conducted research involving the deliberate infection of Guatemalan prisoners, soldiers, and patients with syphilis, gonorrhea, and chancroid. In September 2011 the Presidential Commission for the Study of Bioethical Issues released its report of its historical Investigation of the 1940s U.S. Public Health Service STD Studies in Guatemala. In the view of the Commission, the Guatemala experiments "involved unconscionable basic violations of ethics, even as judged against the researchers' own recognition of the requirements of the medical ethics of the day."

The use of animals in laboratories remains contested. In 2013 the National Institutes of Health (NIH) announced plans to significantly reduce the use of chimpanzees in NIH-funded biomedical research and to provide retirement facilities for most of the chimpanzees it currently owns or supports. In accepting most of the recommendations offered by an Institute of Medicine committee, the NIH director, Francis S. Collins, MD, PhD, explained that the likeness of chimpanzees to humans "has made them uniquely valuable for certain types of research, but also demands greater justification for their use. After extensive consideration with the expert guidance of many, I am confident that greatly reducing their use in biomedical research is scientifically sound and the right thing to do."

[*See also* **Ethics and Medicine; Medical Education; Medicine: Overview; National Institutes of Health; Pharmacology and Drug Therapy; Physiology; Public Health; Public Health Service, U.S.; Race and Medicine;** *and* **Tuskegee Syphilis Study.**]

BIBLIOGRAPHY

Advisory Committee on Human Radiation Experiments. *The Human Radiation Experiments: Final Report of the Advisory Committee on Human Radiation Experiments.* New York: Oxford University Press, 1996.

Guerrini, Anita. *Experimenting with Humans and Animals: From Galen to Animal Rights.* Baltimore: Johns Hopkins University Press, 2003.

Hopkins, Donald R. *Princes and Peasants: Smallpox in History*. London: University of Chicago Press, 1983.

Jones, James H. *Bad Blood: The Tuskegee Syphilis Experiment*. New York: Simon & Schuster, 1993.

Lederer, Susan E. "Political Animals: The Shaping of Biomedical Research Literature in Twentieth-Century America." *Isis* 83 (1992): 61–79.

Lederer, Susan E. *Subjected to Science: Human Experimentation in America before the Second World War*. Baltimore: Johns Hopkins University Press, 1995.

National Institutes of Health. "NIH to Reduce Significantly the Use of Chimpanzees in Research." 26 June 2013. Retrieved 21 Jaunary 2014 from http://www.nih.gov/news/health/jun2013/od-26.htm. See this source for Collins quotation.

Numbers, Ronald L. "William Beaumont and the Ethics of Human Experimentation." *Journal of the History of Biology* 12 (1979): 113–135.

Presidential Commission for the Study of Bioethical Issues. *"Ethically Impossible:" STD Research in Guatemala from 1946–1948*. September 2012. http://www.bioethics.gov. This source includes a quotation in text about the U.S. Public Health Service Guatemala experiments.

Rader, Karen. *Making Mice: Standardizing Animals for American Biomedical Research, 1900–1955*. Princeton, NJ: Princeton University Press, 2004.

Reverby, Susan M. *Examining Tuskegee: The Infamous Syphilis Study and Its Legacy*. Chapel Hill: University of North Carolina Press, 2009.

Stark, Laura. *Behind Closed Doors: IRBs and the Making of Ethical Research*. Chicago: University of Chicago Press, 2011.

Washington, Harriet A. *Medical Apartheid: The Dark History of Medical Experimentation on Black Americans from Colonial Times to the Present*. New York: Doubleday, 2006.

Susan E. Lederer

ANIMATION TECHNOLOGY AND COMPUTER GRAPHICS

As a cinematic process or technology, animation refers to techniques for bringing objects or still images to life by creating the illusion of movement. This is accomplished by rapidly displaying a series of prepared or generated images in sequence. Most animation methods involve the creation and capture (e.g., filming) of still images. Single-frame recording techniques have long dominated animation production, although techniques such as multiple-frame, interpolation, live action, real-time, and algorithmic processes provide alternative methods.

The development of technologies to capture the motion of objects and present them as a series of images began in the nineteenth century. Experimentation with new cinematic technologies in the early twentieth century produced animated movies based on still images such as drawings. The first animated film was probably J. Stuart Blackton's "Humorous Phases of Funny Faces" (Vitagraph, 1906). Earl Hurd and John Bray's invention of celluloid ("cel") animation (1914), Winsor Mckay's introduction of the keyframe technique in works such as "Gertie the Dinosaur" (Vitagraph, 1914), Max Fleischer's Rotoscope technology (1917), and methods of animation production introduced by the Walt Disney studio through the 1930s improved the quality of film-based animation.

Computer Graphics and Computer-Generated Imagery (CGI).

Two qualities encouraged the application of computers to animation. First, animation is based on image creation. Computer applications such as drawing software for producing images were immediately relevant to the production process. Second, the creation of animated movies requires processes for organizing the work of a team of artists, editors, audio engineers, directors, and other creative talents. Hardware and software systems could reorganize animation production in ways that reduce resource requirements (staff, budget, etc.). Computer animation attracted attention both as a creative process and as an efficient system for making animated movies.

The term "computer graphics" was apparently coined around 1960 by William Fetter, an artist employed by Boeing who later created an iconic computer simulation of the human body known as the "Boeing Man." Drawing and computer-aided design (CAD) programs revealed the potential of computer graphics during the early

1960s, led by Design Augmented by Computers (DAC-1) and developed by International Business Machines, Inc. (IBM) and General Motors Research Laboratories, and Sketchpad, created by MIT graduate student Ivan Sutherland.

During the 1960s and into the 1970s, industrial and university laboratories introduced computer technology that demonstrated fundamental animation capabilities. In 1961, Edward E. Zajac at the Bell Telephone Laboratories produced "Simulation of a Two-Gyro Gravity-Gradient Attitude Control System," a computer-animated visualization of satellite motion. Zajac's software generated imagery from instructions based on algorithms or rules. Artists collaborating with Bell engineers developed the Beflix (Bell Flicks) animation system for the IBM 7094 computer in 1963, which led to animated films by Ken Knowlton, Leon Harmon, Lillian Schwartz, and Stan VanDerBeek during the 1960s. John Whitney, a former Lockheed employee, adapted computer-assisted anti–aircraft targeting technology to sequence images. His company, Motion Graphics, began to produce animated shorts by the late 1950s. Whitney's innovative slit-scan effects resonated in Douglas Turnbull's "Stargate" sequence for Stanley Kubrick's *2001: A Space Odyssey* (1968).

Artistic experimentation with computer graphics was also connected to academic work. In 1961, Stanford University students Larry Breed and Earl Boebert demonstrated MACS, a language for programming bit-mapped computer animation applied to the generation of card stunts at football games. That same year, MIT programmers developed the first interactive computer game, *Spacewar*. After Ivan Sutherland joined Dale Evans in the computer science department at the University of Utah, this department became a center for research in computer graphics. Contributors to the development of graphics, CGI techniques, and computer animation such as Ed Catmull, Henri Gauraud, and Jim Blinn emerged from this program, as did Evans & Sutherland, a company that produced hardware for graphics, CAD, and simulation applications. At Ohio State University, the artist Charles Csuri began to experiment with computers in the early 1960s, using them to trace and transform art works into frames for animated films.

The path from early experimentation with computer-based animation to animated movies led through CGI applications such as special effects and advertising during the 1970s and 1980s. Advertising as a short format was ideal for demonstrating new techniques while providing a source of revenue because of the high cost of producing CGI, which required massive computational resources and custom-developed software. Robert Abel & Associates, founded in 1971, provides an example of the mix of CGI projects undertaken during this period. Abel's company applied computer animation made with Evans & Sutherland vector graphics equipment to production of special effects for movies (including Disney's *Tron* and *The Black Hole*), advertising shorts, and even video games. In 1984, Abel's Image Research studio produced a stunning Super Bowl commercial called "Brilliance." It featured a model known as "The Sexy Robot," whose smooth movements were generated by motion-capture technology and graphics systems that represented the state of the art for commercial computer animation.

Mainstream Animation Production. CGI and mainstream movie-making converged during the 1980s. Disney's live-action film *Tron* (1982) used CGI extensively, with work parceled to computer graphics companies MAGI, Digital Effects, Information International, Inc. (III), and Abel. III had used digital image capture to represent an android's vision in the film *Westworld* (1974), the first significant use of computer graphics in a feature film. Two years later, III introduced three-dimensional (3D) CGI in the film *Futureworld* (1976). The *Tron* collaboration introduced CGI as a key component of mainstream filmmaking.

The Computer Graphics Lab (CGL) was founded in 1974 at the New York Institute of Technology (NYIT). Edwin Catmull and Alvy Ray Smith oriented the lab's researchers toward technology necessary for the production of entirely computer-animated movies. The group's accomplishments ranged from computer-based video editing to specialized techniques such as hidden-surface algorithms and mipmapping. A CGL team under the direction of Lance Williams worked on production of "The Works," which would have been the first 3D animated movie had the project been completed. By the mid-1980s, most of the CGL left NYIT to join the "Computer Division" of film director George Lucas' Industrial Light & Magic (ILM), founded in 1975 as a division of LucasFilm to create special effects for the original *Star Wars*. ILM advanced the production of visual effects through new technologies and creative techniques involving computer graphics and digital imaging. Catmull's group created the Computer Animation Production System (CAPS) for producing film-resolution images from 3D models, worked on digital compositing, and created methods for replacing effects based on physical models with computer animation. It produced innovative digital effects used in films such as *Star Trek II: The Wrath of Khan* (1982) and *Young Sherlock Holmes* (1985). One of its products was called Pixar-3D image rendering hardware. When the group was sold to computer entrepreneur Steve Jobs in 1986, this system gave the resulting company its name, Pixar Animation Studios.

At Pixar, ILM's technology was improved and became the 3D rendering system known as RenderMan. Under the direction of John Lasseter, Pixar utilized RenderMan to create its iconic short, "Luxo Jr.," in 1986. RenderMan became an industry standard. By the mid-1990s, computer-generated 3D animation was primed to become the primary method for producing feature-length movies. Two movies released by Walt Disney Pictures during this period, *The Lion King* (1994) and Pixar's *Toy Story*, exemplified the transition away from hand-drawn animation. Produced in-house by Walt Disney Features Animation, *The Lion King* upheld the stylistic conventions of traditional 2D animation, yet computational techniques produced a crucial scene (the wildebeest stampede) as well as numerous effects and simulated camera movements, such as zooms and pans. *Toy Story*, released only 18 months later, was an entirely digital production. Under Lasseter's direction, it was created with Pixar's 3D animation technology. Every frame was rendered by racks of SUN SPARCstation computers in a facility dubbed the "SUN farm." Hand-drawn animation was replaced entirely by 3D models and lighting techniques that produced a convincing illusion of depth and movement. Just as significantly, budget and staffing required for completing *Toy Story* was significantly less than that for *The Lion King*.

Over the decade following the release and phenomenal box office success of *Toy Story*, computer animation generally replaced traditional methods. New production techniques such as digital rotoscoping (based on live action) and machinima (real-time animation based on game technology) continued to emerge during this period, but frame-based, 3D digital animation in the early twenty-first century dominates production of commercial animation, especially for feature-length movies.

[*See also* **Automation and Computerization; Bell Laboratories; Computer Science; Computers, Mainframe, Mini, and Micro; Film Technology; Technology;** *and* **Television.**]

BIBLIOGRAPHY

Auzenne, Valliere Richard. *The Visualization Quest: A History of Computer Animation*. Rutherford, N.J.: Fairleigh Dickinson University Press, 1994.

Baker, Robin. "Computer Technology and Special Effects in Contemporary Cinema." In *Future Visions: New Technologies of the Screen*, edited by Philip Hayward and Tana Wollen, pp. 31–45. London: BFI, 1993.

Paik, Karen. *To Infinity and Beyond! The Story of Pixar Animation Studios*. San Francisco: Chronicle Books, 2007.

Price, David. *The Pixar Touch: The Making of a Company*. New York: Alfred A. Knopf, 2008.

Henry E. Lowood

ANOREXIA NERVOSA

Anorexia nervosa is a psychophysiological disorder characterized by prolonged refusal to eat or maintain normal body weight, an intense fear of becoming fat, a disturbed body image in which the emaciated patient feels overweight, and the absence of any physical illness that would account for extreme weight loss. The term anorexia (loss of appetite) is actually a misnomer, because genuine loss of appetite is rare and usually does not occur until late in the illness. In reality, most patients with anorexia nervosa are obsessed with food and constantly struggle to ignore natural hunger (Bruch, 1978).

As is true in the early twenty-first century, the majority of anorexic patients in the past were adolescent or young adult females from middle-class white families. Joan Jacobs Brumberg shows that the emergence of anorexia nervosa as a disease category in the late nineteenth century was closely linked to changes in the relationship between white middle-class parents and children during this time period. According to Brumberg, the bourgeois family protected girls from early marriage and the need to work outside the home but also intensified conflict between generations. This family dynamic set the state for anorexia nervosa. Brumberg argues that because food abundance was a sign of class status and parental affection, food refusal became one of the ways in which adolescent girls expressed their psychic distress with the restrictiveness of Victorian middle-class family life (Brumberg, 1988).

There are important differences between patients with anorexia nervosa today and those in the past, however. Although fear of fatness is a symptom common to both historic periods, severe body-image distortion is a characteristic that only appeared during the late twentieth century. The prevalence of anorexia nervosa was also very low until the 1970s (Parry-Jones and Parry-Jones, 1994).

Disease Classification. The modern disease classification of anorexia nervosa emerged during the mid-nineteenth century. In 1859, the American asylum physician William Stout Chipley published the first American description of what he called sitomania, defined as a type of insanity characterized by an intense dread or loathing of food (Chipley, 1859). Although Chipley found sitomania in patients from a broad

range of social, economic, and age groups, he identified a specific form of the disease that afflicted adolescent girls. Despite widespread acclaim for Chipley's work with patients with anorexia nervosa, most clinicians at this time generally rejected the idea that anorexia nervosa was a specific disease. Instead, they conceptualized it as either as a variant of hysteria that affected the gastrointestinal system or as a juvenile form of neurasthenia (Brumberg, 1988).

In the twentieth century, the treatment of anorexia nervosa changed to incorporate new developments within medical and psychiatric practice. Before World War II, two models for the cause and treatment of the disease emerged. The first came from the new field of endocrinology. In 1914, Morris Simmonds published a study of an extreme wasting syndrome called cachexia, which he attributed to destruction of the anterior lobe of the pituitary gland. Because patients with Simmonds disease and those with anorexia nervosa shared a common set of symptoms, endocrinologists suggested that a deficiency or absence of pituitary hormone was the cause of both conditions. Researchers at the Mayo Clinic argued that anorexia nervosa was a metabolic disorder caused by thyroid deficiency. Throughout the 1920s and 1930s, endocrinologists proposed several other hormones, including insulin, antuitrin, and estrogen, as treatments for anorexia nervosa (Brumberg, 1988).

The second major approach to the treatment of anorexia nervosa grew out of the work of Sigmund Freud, who suggested a link between anorexia nervosa and psychosexual development. According to Freud, all appetites were expressions of the libido. Thus, refusal to eat represented the patient's repression of her sex drive (Freud, 1918/1959). By the 1940s, psychoanalysis became the most common method for treating patients with anorexia nervosa (Brumberg, 1988).

During the 1970s, Hilde Bruch developed a more complex interpretation of the psychological roots of anorexia nervosa. According to Bruch, the families of anorexic patients often were characterized by extreme parental overprotectiveness, lack of privacy, and reluctance or inability to confront intrafamilial conflicts. Bruch claimed that anorexia nervosa represented a young woman's attempt to exert control in a family environment in which she felt powerless (Bruch, 1973). Bruch also raised public awareness of the disease with her book, *The Golden Cage: The Enigma of Anorexia Nervosa* (1978), written for a lay audience, and numerous articles in popular magazines.

Increasing Prevalence. In the wake of Bruch's work, the number of reported cases of anorexia nervosa increased tremendously. This phenomenon has led some experts to suggest that the popularization process may create a "sympathetic host environment" for the disorder (Striegel-Moore et al., 1986). As Bruch observed, anorexia nervosa was "once the discovery of isolated tormented women" but by the 1980s had "acquired a fashionable reputation of being something to be competitive about…This is a far cry from the twenty-years ago anorexic whose goal was to be unique and suggests that social factors may impact the prevalence of the disorder" (Bruch et al., 1988, pp. 3–4).

The dramatic increase in the incidence of anorexia nervosa since the early 1970s corresponded with increased cultural preoccupations with thinness, which suggests that anorexia nervosa is as much a product of culture as it is of biology and individual psychopathology (Brumberg, 1988). In addition, clinicians have recognized the race, gender, and class biases in earlier work on anorexia nervosa and found that the disease is more common in women of color, women from lower socioeconomic groups, lesbians, and young men than previously realized (Thompson, 1994).

[*See also* **Disease; Gender and Science; Medicine: Overview; Obesity; Psychiatry; Psychology; Public Health;** *and* **Race and Medicine.**]

BIBLIOGRAPHY

Bruch, Hilde. *Eating Disorders: Obesity, Anorexia Nervosa, and the Person Within.* New York: Basic Books, 1973.
Bruch, Hilde. *The Golden Cage: The Enigma of Anorexia Nervosa.* Cambridge, Mass.: Harvard University Press, 1978.
Bruch, Hilde, Danita Czyzweski, and Melanie A. Suhr. *Conversations with Anorexics.* New York: Basic Books, 1988.
Brumberg, Joan Jacobs. *Fasting Girls: The Emergence of Anorexia Nervosa as a Modern Disease.* Cambridge, Mass.: Harvard University Press, 1988. Examines the history of anorexia nervosa and its varying cultural contexts over time.
Chipley, William S. "Sitomania: Its Causes and Treatment." *American Journal of Insanity* 26 (1859): 1–42.
Freud, Sigmund. "From the History of an Infantile Neurosis." In *Collected Papers,* Vol. 3. Authorized translation under the supervision of Joan Riviere. New York: W. W. Norton, 1959. First published in 1918.
Parry-Jones, Brenda, and William L. Parry-Jones. "Implications of Historical Evidence for the Classification of Eating Disorders." *British Journal of Psychiatry* 165 (1994): 287–292.
Striegel-Moore, Ruth, Lisa R. Silberstein, and Judith Rodin. "Toward an Understanding of Risk Factors for Bulimia." *American Psychologist* 41 (1986): 246–263.
Thompson, Becky W. *A Hunger So Wide and So Deep: American Women Speak Out on Eating Problems.* Minneapolis: University of Minnesota Press, 1994. Reports on the experiences of 18 women with eating disorders.

Heather Munro Prescott

ANTHROPOLOGY

American anthropology has been integral to the emergence of American national identity, particularly through its study of the American Indian. This essay traces its emergence in colonial times through government support to academic anthropology under the leadership of Franz Boas and his students. It concludes with the postwar diversification of the discipline.

Anthropology and American Identity. American anthropology came to have its peculiar character as a result of its inextricable link to the nationalist aspirations and territorial expansion of the United States. The themes of manifest destiny and virgin land weave through and dominate the strands of American anthropology, focusing primarily on the American Indian up through World War II and thereafter on international expansion of the nation's political, economic, and cultural influence. Anthropology in America emerged as a profession during the nineteenth century alongside the American nation in an intimate and reciprocal relationship. Indeed, historian John Gilkeson (2010) argues that anthropologists throughout this period were among the foremost public intellectuals in the formation of American identity. Anthropologists valorized the uniqueness of America in contrast to Europe, particularly in its environment, including its Aboriginal population; the government and general public depended on anthropologists for accurate information about this diverse population and its changing and sometimes perilous relationship with American expansion and settlement. This essay focuses primarily on cultural anthropology and its ties to the larger American political and social context.

Preprofessional Studies of the American Indian. The earliest anthropologists during the eighteenth and early nineteenth centuries were not trained in this discipline because academic programs did not yet exist. Educated intellectuals who could afford to indulge their curiosity about the Aboriginal inhabitants of the New World shared the results of their research through learned societies and international correspondence. Catherine the Great of Russia spearheaded an international movement to collect vocabularies from around the world. American scholars were well situated to provide valuable and otherwise inaccessible firsthand data from their own experience and contact with missionaries, military personnel, and explorers. Such scholars identified Indians, understood as "American" in the same sense as flora and fauna, with the simplicity and natural quality of the New World (Deloria, 1998).

The American Philosophical Society, founded in Philadelphia in 1743 by Benjamin Franklin "for the promotion of useful knowledge," stood out for its constellation of members seeking to understand the diversity and historical relationship of the languages of the various American Indian tribes. Thomas Jefferson collected Indian vocabularies, now lost, and excavated an Indian mound on his Virginia estate. Albert Gallatin, the secretary of the Treasury during Jefferson's presidency, used questionnaires to obtain vocabularies and produced the first Indian linguistic classification. John Pickering, a noted colonial jurist, designed an alphabet in which to write these languages. Peter Stephen Du Ponceau, linguist and philosophe, focused on grammar and deemed all Indian languages "polysynthetic" (including multiple grammatical categories within a single word). Michigan governor Lewis Cass used his office to collect vocabularies. Indian agent and aspiring politician Henry Rowe Schoolcraft collected stories and linguistic materials from the relatives of his Ojibwe wife. These studies often addressed practical questions. Both success of settlement in unfamiliar land and European wars on North American shores depended critically on building alliances with local Indians. Effective alliance was based on effective communication and thus on knowledge of the languages, customs, and mutual relations of the Indian nations.

Professionalization. By the mid-nineteenth century, anthropology was emerging as a distinct discipline. Three major figures dominated this period: Daniel Brinton of Philadelphia continued the American Philosophical Society tradition of linguistic research. He collected and published eight volumes of what he recognized as "Aboriginal American literature" and held the first American professorship of anthropology at the University of Pennsylvania in 1887 (although he had no students and his aspirations for academic anthropology were abortive). Frederic Ward Putnam established a research program at Harvard's Peabody Museum of American Archaeology and Ethnology, taking up a professorship there a year after Brinton did at Penn. Putnam was an academic entrepreneur who also influenced anthropological developments in California, Chicago, New York, and Mexico.

The third founding figure was John Wesley Powell, a naturalist and self-taught geologist who turned to ethnology, or the study of cultures, in response to changing political constraints on exploration of the American West in the years following the Civil War (i.e., after 1865). In 1877, Powell established the Bureau of (American) Ethnology (BAE) under the auspices of the Smithsonian Institution ("American" was added to the name in 1892). Powell hired a small staff of talented amateurs who devoted themselves to cataloging and mapping the cultures and languages of the various Indian tribes. Government publication of their research produced a monumental classification of Native American languages into 55 separate linguistic families in 1892. The BAE staff also produced a synonymy of geographical terms and proper names in a comprehensive handbook of ethnological information (1906, 1912) and a handbook illustrating the diversity of American Indian grammatical structures (edited by Franz Boas, 1911, 1922). Powell accepted the theoretical framework of cultural evolution and cultural progress as presented in the works of amateur anthropologist Lewis Henry Morgan. Morgan's hierarchy of races, assumed to be equivalent to their cultures, justified land acquisition and the displacement of its original owners. Theoretical work was, however, peripheral to the work of the BAE.

This early professional anthropology arose largely out of the practical engagements between the American government and Native Americans. The federal government was motivated to support this work for its utility to political agendas of Western development. Thus the fascination of early eighteenth-century intellectuals with the Indian as noble savage rapidly gave way to a more pragmatic and cynical emphasis on the acquisition of land and resources, with increasingly frequent representations of the brutal and savage character of the Indian as an obstacle to the advance of American civilization.

Boasian Anthropology. Historians of anthropology have generally traced the beginning of American professional anthropology to the revolutionary work of Franz Boas (1858–1942), a German immigrant originally trained in psychophysics and geography (Stocking, 1968). Darnell (1998), however, has demonstrated considerable

continuity from the work of the BAE to what has been seen in retrospect as a Boasian rupture in American anthropological inquiry.

After spending a year among the Eskimo of Baffinland in 1883–1884, Boas began his fieldwork in British Columbia, Canada, on what came to be known in anthropology as the Northwest Coast culture area. This work occupied him for the rest of his life. Boas's early years in America were plagued by a lack of research funds, underemployment, and lack of a stable base for his visions of anthropology. He settled in New York City, with a joint appointment between Columbia University and the American Museum of Natural History in 1897. Thereafter he began to develop an academic program and a power base to challenge the anthropological establishment around Powell in Washington and Putnam in Cambridge. Boas, who had been a protégé and subordinate to Putnam and worked intermittently for the BAE, established organizational dominance through cementing alliances between museums and universities and producing a generation of students who would take important positions around the continent (Darnell, 1998).

The Boasian Theoretical Paradigm. Boas has been dismissed as a theorist by many of his post–World War II successors because anthropology at that time took a strong turn toward scientific objectivity and rejected its more psychologically oriented Boasian roots. Boas's insistence on cultural description based on qualitative ethnographic methods designed to capture what he called "the native point of view" infuriated those who envisioned an anthropological science based on deductive generalizing models (e.g., Leslie White or Marvin Harris).

Darnell (2001) contends, however, that the "Americanist tradition" that grew up around Boas and his first generation of students persists through "invisible genealogies." As in the case of the transition from the BAE to Boas, continuity was maintained across purported scientific revolutions. The persistent distinctive features of Americanist anthropology are its reliance on a symbolic or mentalist definition of culture, its sensitivity to the lack of correspondence of race, language, and culture in classifying ethnographic or cultural diversity, and its commitment to exploring what cultural forms mean for the members of a culture.

Boas's theoretical position has been so thoroughly absorbed by later Americanist anthropologists that it is now taken for granted. His critique of the premature generalization of cultural evolutionist theory demonstrated by counterexample that race, language, and culture cannot be fitted together in a way that conforms to a single classificatory schemed consisting of progressive stages of cultural development, as Morgan, Powell, and Brinton, for example, had assumed. In *The Mind of Primitive Man* (New York, Macmillan, 1911), Boas deconstructed these categories and argued for the essential racial or biological sameness of human capacity everywhere; environment or culture trumped heredity in their power to produce the distinctive features of particular cultures.

Boas's historical particularism sought the unique history of each American Indian tribe. In the absence of written records and reliable archaeological dating methods, Boas relied on geography and the distribution of culture traits to reconstruct past movements and interactions of tribes who borrowed from one another and adapted the borrowings to their own needs. Folklore elements proved most amenable to tracing particular historical trajectories.

History was, however, only half of the story: anthropology also needed psychology, which Boas defined in terms of the viewpoint of the members of a culture. Collection of texts (i.e., sustained records of the speech of fluent native speakers of Indian languages) would provide evidence of the mental world of the Indian in addition to purely linguistic information. Psychological questions of wider scope, however, could not be attacked until the basic mapping of the history of cultures and languages provided the data for further generalization. Both the historical and the psychological sides of Boas's theoretical position have been criticized by later scholars as mere salvage ethnography, based not on direct study of functioning cultures but on memories of the last generation to experience an authentic, uncontaminated cultural world. From Boas's standpoint,

however, history in the sense he meant it was inaccessible to an observer within the culture, and the psychology that interested him often belonged to the past. Thus elders' memories needed to be recorded while this was still possible.

Boas's theoretical position was elaborated by his first generation of students. Alfred Kroeber, who established anthropology at the University of California, argued that culture was "superorganic" and that culture in this sense was the distinctive concept defining anthropology as a discipline. Edward Sapir, the primary linguist among this generation, countered with an emphasis on the individual and culture as sides of the same coin and explored "the impact of culture on personality." With his protégé Benjamin Whorf, he argued that the categories of a language strongly influence (but do not fully determine) an individual's habitual patterns of thought. Paul Radin, the maverick among this cohort, sought Native American "primitive philosophers" and argued for the universality of philosophical questions (although not of formal systems of philosophy).

The agenda of Boasian anthropology shifted focus after the first decade of the twentieth century, moving from the mapping of Native American languages and cultures to studying the relationship between culture and the individual—in Boas's terms, from questions of history to those of psychology. In *Patterns of Culture* (Boston, Houghton Mifflin, 1934), Ruth Benedict, another of Boas's students, took the case for cultural relativism or tolerance for cultural diversity to the American public and hastened the breakdown of American isolationism during the interwar years. Benedict characterized whole cultures using terms adapted from abnormal psychology (the megalomaniac Kwakiutl of the Northwest Coast, the paranoid schizophrenic Dobuans in Melanesia, and the Apollonian self-control of the Zuni). Benedict's student and protégé Margaret Mead began the exodus from Americanist fieldwork on Native American reservations with *Coming of Age in Samoa* (New York, New American Library, 1928). The exoticism of Mead's challenge to the complacent American assumption of universal adolescent trauma established her as a public intellectual and well-known critic of American society based on her firsthand knowledge of diverse cultures. Her 17 years of monthly columns for the ladies' magazine *Redbook* brought her ideas to a wide audience beyond the academy. The so-called culture and personality school of American anthropology persisted robustly until the 1960s and fit well with the more individualistic and less collective tenor of American society. In contrast, other national traditions of anthropology did not take up the study of individual psychology or personality.

Benedict and Boas were also well known as outspoken critics of Nazi racist policies and racism in American society. Lee Baker (2010) argues persuasively that Boasian studies of race in America grew out of Boasian cultural studies of the American Indian.

Branching out after World War II. Two world wars during the twentieth century broke down much of the isolationism of the United States and catapulted it into international prominence. The American empire functioned somewhat differently from earlier formal European imperialism in that its sphere of influence, albeit still dominated by the implicit motif of manifest destiny, was indirect; colonies became protectorates. Even after nascent states gained nominal independence—in the American-dominated Pacific as in Africa, India, and Southeast Asia—postcolonial forces of global capitalism and emerging neoliberalism maintained much of the practical politics. At the same time, returning veterans on the G.I. bill flooded American universities, leading to tremendous expansion of higher education and government support for research, including in the social sciences with their promise of practical help in administering new territories with unfamiliar cultural patterns.

Under these new conditions, area studies, based in part on national character studies developed by Benedict, Mead, and others, proliferated. Anthropologists were key members of the interdisciplinary teams that pursued them. The new generation of anthropologists sought fieldwork sites outside North America, especially in the Pacific where many had established ties during their military service. As a result, research with Native

Americans was much diminished in disciplinary salience, a process accentuated during and after the intellectual and political upheavals of the 1960s. Parallels were rarely drawn between the internal colonialism of the American Indian case and the colonial regimes now studied around the world.

The postwar climate also challenged anthropologists to apply their traditional methods to new types of societies. Native American research had already turned increasingly from studies of traditional culture often remembered but no longer practiced to greater emphasis on acculturation or culture change and adaptation, especially after the introduction of the commissioner of Indian affairs John Collier's American Indian New Deal and the Indian Reorganization Act of 1934. Peasant societies did not fit the earlier model of isolated so-called "primitive" societies, yet they remained at least partly isolated from the economic and political centers of the postwar world. World systems theory as developed by the sociologist I. Wallerstein and other scholars turned, from the 1950s on, to the study of peasants who were neither proletariat nor primitive. Eric Wolf's *The People without History* (Berkeley, University of California Press, 1982) epitomizes the turn of anthropology toward the study of complex societies and the interdependency of even the most isolated cultures.

Meanwhile, some American anthropologists remained at home and studied their own society based on qualitative ethnographic methods, first developed in distant fieldwork sites in small-scale societies and then later transposed to study small communities, neighborhoods, and subcultures. Especially beginning in the 1960s, contemporary American society encouraged the study of race and ethnicity, with the traditional immigrant melting pot supplemented by increasingly large Afro-American and Hispanic sectors of the population.

Perhaps because of its long tradition of taking "the native point of view," anthropology, especially beginning in the late twentieth century, has been in the forefront of the critique of globalization for its effects on societies outside the mainstream and trapped by forces beyond their control.

From within the mainstream, some anthropologists have learned to "study up" (a term introduced by Laura Nader) through ethnographic studies of bureaucracy and administrative control (Price, 2004). Political activism, especially on behalf of peoples studied, has been common throughout anthropology's history (although there is a persistent conflicting aspiration to scientific objectivity, often associated with an apolitical professional stance and a value-neutral investigative model).

[*See also* Archaeology; Race Theories, Scientific; *and* Social Sciences.]

BIBLIOGRAPHY

Baker, Lee D. *Anthropology and the Racial Politics of Culture.* Durham, N.C.: Duke University Press, 2010. A meticulous and elegant comparison of anthropological approaches to Afro-American issues of race and Native American issues of culture.

Boas, Franz. *Race, Language and Culture.* New York: Free Press, 1940. Boas' own selection from his lifetime work two years before his death.

Darnell, Regna. *And Along Came Boas: Continuity and Revolution in Americanist Anthropology.* Amsterdam and Philadelphia: John Benjamins, 1998. Explores the transition from the Bureau of American Ethnology to Boasian anthropology.

Darnell, Regna. *Invisible Genealogies: A History of Americanist Anthropology.* Lincoln: University of Nebraska Press, 2001. Follows the major ideas of Boas and his early students and argues that their Americanist paradigm persists in largely unacknowledged forms in American anthropology today,

Deloria, Philip. *Playing Indian.* New Haven, Conn.: Yale University Press, 1998. This Native American historian explores the non–Native American fascination with the American Indian as integral to American identity.

Gilkeson, John. *Anthropologists and the Rediscovery of America, 1886–1965.* Cambridge, U.K., and New York: Cambridge University Press, 2010. An intellectual historian gives the context in which America experienced anthropology.

Price, David. *Threatening Anthropology: McCarthyism and FBI Surveillance of Activist Anthropologists.* Durham, N.C., and London: Duke University Press, 2004. The key source for exposing contemporary government and corporate suppression of anthropological critique of continuing colonialism.

Stocking, George W., Jr. *Race, Culture and Evolution: Essays in the History of Anthropology.* New York: Free Press, 1968.

Regna Darnell

ARCHAEOLOGY

The development of archaeological fields as disciplines had different trajectories influenced by the variable intellectual contexts of their origin. Thus, for example, what is often called "classical archaeology" has been heavily influenced by the development of antiquarianism in Europe and by access to classical Greek and Latin texts. In contrast, in this article we briefly discuss the development of "Americanist archaeology." This latter short-hand term refers to the disciplinary research in North America, where the indigenous peoples did not have written histories and were initially part of the subject matter for scholars who studied natural history rather than philology. The term Americanist archaeology is employed strictly to delimit the topical area of the discipline; certainly American researchers have in the past and continue in the present to study classical archaeology as well as Americanist archaeology.

American Revolution through the Civil War. The early history of Americanist archaeology focused on exploratory and speculative concerns relating to the origins of the First Nations. The question of the racial identification of American Indians engendered late eighteenth- and early nineteenth-century interest in the field that became Americanist archaeology. The American Academy of Arts and Sciences, founded during the American Revolutionary War (in 1780) by John Adams, John Hancock, and others, aimed, as one of its goals, "to promote and encourage the knowledge of antiquities of America."

During the same period, the king of France, who was interested in the development of the interior of North America as it related to France's colonies, sent a questionnaire to the governors of the various American colonies who sought French military aid during the Revolutionary War, inquiring about geography, zoology, botany, geology, and history. The governor of Virginia, Thomas Jefferson, responded with a 1784 report, "Notes on the State of Virginia," in which he included a section detailing the original inhabitants, describing a prehistoric mound that his slaves had excavated. He produced the first post facto stratigraphic interpretation in the Americas, indicating from his observations that the mound had several different levels, relating to discrete time periods, thus showing considerable antiquity for the ancient inhabitants.

Shortly after the end of the Revolutionary War, a group of American explorers established a settlement at Marietta, Ohio, on a large archaeological mound complex. Among the individuals who visited Marietta that first season in 1788 was Reverend Manasseh Cutler. Cutler felled a number of the largest trees on the mounds to count the tree rings to make an estimate of minimum age for the mounds. He reported that the trees had between 300 and 400 rings, with one tree having minimally 463 tree rings. Hence he concluded that the mounds were at least this old (with one tree ring equivalent to one solar year), thus providing the first published attempt in the United States to try to define mound antiquity in a scientific manner based on empirical data, in contrast to previous age estimates that were based on wild guesses with no factual basis.

News of the mounds "out West," as well as general curiosity about the inhabitants of the continent, convinced scholars that a more orderly means of data collection was needed. Consequently, learned associations such as the American Philosophical Society turned their attention to this problem. Thomas Jefferson, then president of that group, and his colleagues developed a circular that they sent to the leaders of all exploring expeditions going west, asking them

to obtain accurate plans, drawings and descriptions of whatever is interesting (where the originals cannot be had), and

especially of ancient fortifications, tumuli, and other Indian works of art; ascertaining the materials composing them, their contents, the purposes for which they were probably designed, etc.

Regarding the above,

the committee are desirous that cuts in various directions may be made into many of the tumuli, to ascertain their contents; while the diameter of the largest tree growing thereon, the number of its annulars and the species of the tree, may tend to give some idea of their antiquity. If the works should be found to be of masonry; the length, breadth, and height of the walls ought to be carefully measured, the form and nature of the stones described, and specimens of both the cement and stones sent to the committee.

—Jefferson et al., 1799, pp. xxxvii–xxxviii

These instructions set the parameters of much of the archaeological research in the United States for the next half century.

Another important early group with archaeological interests was the American Antiquarian Society. In its inaugural journal in 1812, the society proposed in its forward that research should focus upon

the collection and preservation of the antiquities of our country, and of curious and valuable productions in art and nature [which] have a tendency to enlarge the sphere of human knowledge, aid in the progress of science, to perpetuate the history of moral and political events, and to improve and instruct posterity.

The society was the venue of early Americanist archaeology publications, such as Caleb Atwater's seminal work in 1820 on the relics of Ohio (Atwater, 1820). Samuel Foster Haven, who became the librarian of the society, used his post to provide several summaries of American archaeology; the best known and most influential was his 1856

Archaeology of the United States, or, Sketches, Historical and Bibliographical, of the Progress of Information and Opinion Respecting Vestiges of Antiquity in the United States (Smithsonian Institution Contributions to Knowledge, Publication 71).

The first half of the nineteenth century saw a rapid territorial expansion beyond the Appalachian Mountains to the Pacific Ocean, with a concomitant broader understanding of the variations of North American Indian cultures as reflected in growing sophistication of the commentary on the continent's first peoples. The midcentury marks the shift from the period of speculative study that preceded it, where the collection of archaeological data was incidental to other pursuits, to the beginning of a "classificatory-descriptive" period (Willey and Sabloff, 1993). This new emphasis on the description of archaeological sites and materials became prominent roughly from the Civil War through World War I.

Post–Civil War through World War I.

The involvement of the federal government in archaeology after the Civil War through the U.S. National Museum and the Smithsonian Institution, and later the Bureau of American Ethnography, marked one of the major steps toward the professionalization of Americanist archaeology. Joseph Henry, then head of the Smithsonian (which had been established in 1846), saw the potential relevance of method and theory developed by researchers in Old World archaeology to the study of American antiquities and used the Smithsonian publication series as a means to introduce these ideas to U.S. researchers. Hence, beginning in 1860, he initiated a series of translated papers, summarizing important European advances. Based on this European work, Smithsonian staff member George Gibbs prepared two circulars in 1862 and 1863 with instructions on excavation methodology for collectors working to secure materials for government museums. Gibbs's instructions were quite sophisticated for the time. For example, Gibbs suggested that an excavator of a burial mound or shell midden should note if it "exhibited any marks of stratification," with any artifacts recovered noted as to "the depth at which they were discovered," and should

count the annual rings of any tree growing on a mound to get an estimate of age (Gibbs, 1862, pp. 395–396).

For much of the rest of the nineteenth century, J. Wesley Powell and other personnel from the Smithsonian Institution were in the forefront of investigations. This marked a time of "salvage ethnography," wherein the emphasis of many projects was to "salvage" information on the way of life of the First Nations, peoples whom they saw as rapidly disappearing. Concurrently archaeological sites were identified and archaeological collections were made and then curated at the Smithsonian.

An increasing contentious issue was the conflict between supporters of a "short history" and those researchers who argued for a "long history" of Indian settlement in the New World. Supporters of the short-history paradigm believed that the First Americans had been on the continent only for a short period, a millennium to a few millennia at most. Thus they believed that the archaeological remains they saw represented only the then-known distribution of First American tribes and nations. Supporters of the long history believed that the origins of the First Nations dated to at least the end of the Ice Age and that thus some of the patterns seen at archaeological sites represented cultural remains from groups inhabiting the landscape from several different time periods, rather than different cultures all contemporaneous at the same time period. Although an oversimplified generalization, one could view the government scientists as primarily short-history proponents, represented by scholars such as William H. Holmes and Alês Hrdlička.

Colleges and universities at this time also developed an institutional infrastructure for research trajectory, although initially exclusively through museums associated with these institutions. During the last decade of the nineteenth century, archaeological research programs were established at museums on three university campuses: Harvard University under Frederic W. Putnam, the University of Pennsylvania under Daniel G. Brinton, and the University of Chicago under Frederick Starr. Unfortunately, the programs at Pennsylvania and Chicago were stillborn and later had to be reinvented. But the Harvard program thrived and during the early part of the twentieth century provided the trained archaeologists who went on to found many of the departments around the country. Although the first course in Americanist archaeology may have been taught at Cornell University (in a one-time special lecture series), the first PhDs in the field were produced at Harvard University. The Harvard program expanded the training for the PhD degree from only two years post-BA study, such as was done at places like Columbia University at the turn of the century, to a program requiring minimally three years of class work, an additional time in independent field research, and a major dissertation manuscript. This model, for better or worse, became the "norm" among Americanist archaeology programs.

Putnam at Harvard's Peabody Museum was a proponent of the long history or Late Pleistocene antiquity of humans in the New World. Unfortunately the sites he chose to support his argument all failed subsequent rigorous evaluation. Researchers only established the antiquity of humans of at least 10,000 to 12,000 years ago in the Americas when Jesse D. Figgins recovered undisputed associations of humans with extinct Pleistocene fauna in New Mexico in 1926. Exactly how long ago humans colonized the New World has been a hotly debated issue. By the first decade of the twenty-first century, well-supported claims suggested an initial colonization sometime between 15,000 and 18,000 years ago.

As the new discipline grew, better methodological tools evolved. As more archaeology was done, researchers began looking for better ways to establish the chronological ordering of artifacts. During World War I, prehistorical ceramic analysis by the ordering technique known as "similary seriation" was first widely practiced in the American southwest and rapidly spread among other archaeological practitioners. And at this time, the recognition that one ought to actually excavate utilizing stratigraphic principles, rather than simply employing post facto interpretations from surmised stratigraphic associations, also became the norm among the new PhD practitioners in

the field. As it became apparent that there was a longer history of the First Americans, it was evident that better procedures needed to be devised. The former view that one only had to understand the ethnography of the surviving American Indian groups to interpret the archaeological remains no longer worked. Clearly, other cultures predated the contemporary groups. To unravel their history, the clues would have to come from the sites and their associations. The previously popular "direct-historical approach" could not answer all the questions. Tribal folklore equally was not up to the task. Increasingly researchers realized that the search for significant museum specimens, which had been the primary focus of the early archaeologists (and which still informed the general public's perception in the late twentieth and early twenty-first centuries, if one takes the *Indiana Jones* series as a marker), simply was not good science. Without context, remains often were mute, even meaningless. This resulted in a paradigm shift in data recovery.

Post World War I to the Early Twenty-First Century. With the Great Depression, archaeologists initiated a number of massive salvage projects under the aegis of the Works Progress Administration and its predecessors, particularly for the Tennessee Valley Authority projects. The needs for common typologies and classifications became clarion calls. Early archaeological artifact and feature typologies and classifications were idiosyncratic, rendering it nearly impossible to perform comparative studies. Thus the archaeologists decided that a critical need was to create typologies and classifications with generally accepted and agreed-upon definitions and terminology. Not only was there a need for an agreed-upon classification of diagnostic attributes of archaeological data but also several archaeologists began hoping to find a cultural trait-list mechanism that would allow them to integrate the discoveries from individual sites into larger cultural classificatory schemata. In the southwest, this gave rise to the Pecos Conference Classification of ceramics; in the Midwest it gave rise to McKern's Midwestern Taxonomic Method. Archaeologists were being almost literally buried by data, idiosyncratically described. Thus they hoped to identify data-reduction mechanisms to codify their finds in a systematic procedure meaningful to all.

During and after World War II, with more rigor applied to description and classification, it was perhaps only natural that American archaeology turned to "functionalism," which held sway in the larger arena of anthropology. Archaeologists attempted to define functionalist methodologies and tools suited to their particular studies. Furthermore, some scholars argued that it seemed intellectually sterile to do no more than collect, order, and describe materials in a systematic fashion—what did it all mean? One of the pioneering attempts to address this issue was proposed by Walter W. Taylor in a 1948 published version of his 1940 doctoral dissertation, *A Study of Archeology,* wherein he championed the "conjunctive approach," where one drew together in conjunction all possible lines of evidence to investigate cultural function and evolution. In hindsight, it is now popular to laud this book as paradigm breaking, but at the time its polemical and harsh language alienated the mainstream of Americanist archaeology.

In their typological definitions, archaeologists also became increasingly interested in style to help them understand how prehistoric cultural themes evolved over time and utilized style as a tool to track trade, diffusion, or contact between disparate groups. Style remained a much-discussed concept in archaeological monographs in the early twenty-first century.

Contextualization of these arguments was soon assisted by the development of new analytical methods in chemistry and physics relevant to analyses of archaeological materials. Although dating techniques such as radiocarbon assays, obsidian-hydration analysis, or potassium-argon dating had been discussed in technical journals in the 1930s, they had not been used and were essentially unknown to archaeology. When these new technologies became available in the 1950s, they revolutionized archeological practice and theory. Traditional time-space systematics became much more straightforward, allowing archaeologists to devote more time to other sorts of analyses. In

addition, rejecting sociocultural anthropologists' tendency to deemphasize environment as a relevant factor in the functioning of culture, archaeologists began (and continued into the twenty-first century) to be involved in issues relating to environmental archaeology, which focused upon the influence of environment on human culture. As technology has evolved, work in zooarchaeology, the study of zoological factors in archaeological interpretations—for example, the first domestication and uses of dogs or cattle, and paleoethnobotany, which deals with such issues as the first domestication or uses of wheat or corn, has grown to be of particular significance. In addition, during the early twenty-first century, geoarchaeological studies, which combine geological analyses with archaeological research, have increasingly become regular components of site projects.

Freed from time-space systematic issues and from typologies and classification arguments, archaeologists in the 1960s rejected the earlier belief that artifacts could not yield information about the way humans used material objects in the past and how they lived and began focusing on "cultural processual" studies, often included under the rubric of "New Archaeology." (A cautionary note: the archaeologists of the World War I period, with their new seriation and stratigraphic techniques, also termed what they did New Archaeology, in contrast to the older direct historical studies.) Lewis Binford, mainly because of his 1968 co-edited book *New Perspectives in Archaeology,* became the spokesperson for the paradigm shift, arguing that archaeologists should be particularly interested in elaborating cultural process. Claiming that the existing research approaches to date had been too simplistic, archaeologists began investigating "middle range theories" to link their time-space systematics with their more elucidatory arguments.

As with any new direction, some archaeologists were accused of being too zealous in following the cultural processual arguments. By the 1980s and 1990s, there was a backlash, with a group of scholars calling themselves "postprocessualists" or "contextual archaeologists." They argued that the human as actor or agent had become lost in the process focus, that the individual was a "wild card," and hence that the archaeological past could not be fully understood using "objective" procedures alone. Since that point, archaeologists have proposed a number of new archaeologies, all of which identify some aspect of prehistoric cultural studies that has been slighted or overlooked, and draw attention to other areas that, in their opinion, have been overemphasized. This has no doubt produced a very healthy situation because it allows for the exploration and growth of a number of fruitful techniques that enhanced the general archaeological toolkit.

[*See also* Anthropology; Biological Sciences; Geology; Higher Education and Science; Museums of Science and Natural History; Race Theories, Scientific; Science: Overview; Smithsonian Institution; *and* Social Sciences.]

BIBLIOGRAPHY

Atwater, Caleb. "Descriptions of the Antiquities Discovered in the State of Ohio and Other Western States." *Archaeologia Americana—Transactions and Collections of the American Antiquarian Society* 1 (1820): 105–267.

Binford, Sally R., and Lewis R. Binford, eds. *New Perspectives in Archaeology*. Chicago: Aldine, 1968.

Browman, David L. *Cultural Negotiations: The Role of Women in the Founding of Americanist Archaeology*. Lincoln: University of Nebraska Press, 2013.

Browman, David L., and Stephen Williams. *Anthropology at Harvard: A Biographical History, 1790–1940*. Peabody Museum Monographs 11. Cambridge, Mass.: Peabody Museum Press of Harvard University, 2013.

Browman, David L., and Stephen Williams, eds. *New Perspectives on the Origins of Americanist Archaeology*. Tuscaloosa: University of Alabama Press, 2002.

Gibbs, George. "Instructions for Archaeological Investigations in the United States." In *Annual Report of the Smithsonian Institution for 1861*, pp. 292–296. Washington, D.C.: Smithsonian Institution, 1862.

Hinsley, Curtis M., Jr. *Savages and Scientists: The Smithsonian Institution and the Development of American Anthropology 1846–1910*. Washington, D.C.: Smithsonian Institution Press, 1981.

Jefferson, Thomas, James Wilkinson, George Turner, Caspar Wistar, Adam Seybert, Charles Wilson Peale, and Jonathan Williams. *Transactions of the American Philosophical Society* 4 (1799): xxxvii–xxxix.

Melzer, David J. *Search for the First Americans*. Washington, D.C.: Smithsonian Books, 1993.

Melzer, David J., D. D. Fowler, and J. A. Sabloff, eds. *American Archaeology: Past and Future: A Celebration of the Society for American Archaeology*, 2d ed. Washington, D.C.: Smithsonian Institution Press, 1993.

Willey, Gordon R., and Jeremy A. Sabloff. *A History of American Archaeology*, 3d ed. London: Thames & Hudson, 1993.

Williams, Stephen. *Fantastic Archaeology: The Wild Side of North American Prehistory*. Philadelphia: University of Pennsylvania Press, 1991.

David L. Browman

ARMSTRONG, EDWIN HOWARD

(1890–1954), inventor of five important radio circuits, was born and grew up in the New York City area. His first invention, of the regenerative or "feedback" circuit that made long-distance radio communication possible, came in 1912, a year before his graduation from Columbia University with a degree in electrical engineering. This discovery led to a two-decade legal battle with Lee de Forest, who eventually won their patent dispute before the U.S. Supreme Court in 1934, although most engineers agreed Armstrong had the stronger case.

He served with the U.S. Army Signal Corps during World War I (1917–1919), rising to the rank of major, a title that stayed with him for life. In 1917–1918 he developed the superheterodyne circuit, which greatly improved radio tuning selectivity and the amplification of weak signals. In 1922 his superregenerative circuit helped to pave the way for two-way radio signaling.

Armstrong is perhaps best known for his fourth radio innovation—frequency modulation (FM) radio broadcasting, the basic patents for which were granted late in 1933. His wide-band (200-KHz) channel was largely static free and allowed for high-fidelity sound. He constructed the world's first FM station (W2XMN) in Alpine, New Jersey, near New York City, in 1938. Commercial FM began in the United States on 1 January 1941, although in a lower-frequency band (42–50 MHz) than used today (the change came four years later). Armstrong's final radio innovation, developed with John Bose in 1952, was the principle of multiplex FM transmission allowing one radio transmitter to carry two or more signals.

Armstrong's final years were dominated by his patent battle with the Radio Corporation of America (RCA) that began in 1948. Whereas other radio manufacturers had taken out licenses to manufacture FM receivers and transmitters, RCA sought to develop its own patent position in FM, although generally more committed to the commercial development of television. Armstrong brought suit for patent violations by RCA and for the next six years the case slowly worked through complex legal steps. But he was running short of funds and, in despair over the dismal state of FM broadcasting, he took his own life on the night of 31 January/1 February 1954.

[*See also* De Forest, Lee; Engineering; *and* Radio.]

BIBLIOGRAPHY

Frost, Gary L. *Early FM Radio: Incremental Technology in Twentieth-Century America*. Baltimore: Johns Hopkins University Press, 2010.

Lessing, Lawrence. *Man of High Fidelity: Edwin Howard Armstrong*. Philadelphia: Lippincott, 1956.

Morrisey, John W., ed. *The Legacies of Edwin Howard Armstrong*. New York: Radio Club of America, 1990.

Christopher H. Sterling

ARMY CORPS OF ENGINEERS, U.S.

The Corps of Engineers, founded in 1802, is the nation's oldest and largest construction agency: a builder of dams, canals, ports, roads, forts, airfields, missile emplacements, and space-exploration facilities as well as a diverse organization with a military command and a vast responsibility for water resources.

Combat engineers emerged from a mix of European traditions. During the Revolutionary War, British-trained surveyors staffed a small geographers department in General George Washington's army. French fort engineers planned the siege at the Battle of Yorktown and commanded the short-lived U.S. Corps of Artillerists and Engineers, founded in 1794. With the influence of Thomas Jefferson, the Congress endorsed a French-style force of scientific technicians in the 1802 legislation that created the Corps and its engineering school, the U.S. Military Academy at West Point.

The Corps' authority to cover river works began with the fortification of New Orleans after the War of 1812. The Corps also planned coastal defenses through the Board of Engineers for Fortifications (1816–1831). U.S. Topographical Engineers, a bureau of the Corps started in 1818 but made an independent command from 1831 to 1863, mapped roads, waterways, and rail routes into the West. With the General Survey Act of 1824, Congress loaned Corps engineers to canal companies and launched an ambitious program of army-directed navigation improvement.

Dam building for flood control, a twentieth-century mission, evolved from the levee work of the Mississippi River Commission, a Corps-led organization founded in 1879. Army engineers also pioneered iron framing, flood modeling, underwater explosives, floating bridges, atomic bomb facilities for the Manhattan Project, and enormous lock-and-dam projects such as the Panama Canal; the Saint Lawrence Seaway project, completed in 1959; the 434-mile McClellan–Kerr waterway along the Arkansas River; and the 10-state Pick–Sloan project of the Missouri River basin. The Clean Water Act of 1972 made the Corps a protector of wetlands, a role frequently at odds with its development mission. In recent decades, the Corps weathered increasing criticism for the negative environmental and economic impact of some of its civil projects. In 2005,

Hurricane Katrina breached the levees designed by the Corps, flooding New Orleans and causing over $100 billion in damage. After experts traced the flooding during Hurricane Katrina to design flaws in the levee system, the Corps faced charges of graft and corruption. Congress attempted to address these charges and concerns with the 2007 Water Resources Development Act, increasing Congressional oversight over the Corps' funding and projects.

[*See also* **Airplanes and Air Transport; Canals and Waterways; Dams and Hydraulic Engineering; Engineering; Environmentalism; Geological Surveys; Jefferson, Thomas; Manhattan Project; Military, Science and Technology and the; Missiles and Rockets; Panama Canal; Rivers as Technological Systems; Roads and Turnpikes, Early;** *and* **Space Program.**]

BIBLIOGRAPHY

Grathwol, Robert P., and Donita M. Moorhus. *Bricks, Sand, and Marble: U.S. Army Corps of Engineers Construction in the Mediterranean and Middle East, 1947–1991.* Washington, D.C.: U.S. Government Printing Office, 2010.
History, Office of, and U.S. Army Corps of Engineers. *The History of the U.S. Army Corps of Engineers.* Honolulu, Hawai'i: University Press of the Pacific, 2003.
Shallat, Todd A. *Structures in the Stream: Water, Science, and the Rise of the U.S. Army Corps of Engineers.* Austin: University of Texas Press, 1994.
Traas, Adrian G. *Engineers at War.* Washington, D.C.: U.S. Government Printing Office, 2011.
Todd A. Shallat

ARTHRITIS

A disease or diseases of the joints, generally with swelling, redness, inflammation, and pain—the term arthritis is postclassical Latin, derived from the medical Greek term for joint, arthro-, and the suffix for disease, -itis. The etymologists at the Oxford English Dictionary date the word to the third century of the common era. Although joint diseases were certainly known to the ancients, distinctions between gout and various forms of arthritis were not consistently made until the modern period.

Historians Roy Porter and G. S. Rousseau (1998, pp. 8–9) argue that the great medical systematizers of the sixteenth and seventeenth centuries worked to draw distinctions among rheumatism, gout, and rheumatoid arthritis. Physicians such as Thomas Sydenham (1624–1689) believed that rheumatism was found in individuals with a predisposition to it and similar maladies because of a combination of heredity and habits described at the time as "constitution." Sydenham linked rheumatism and related joint diseases to scurvy and other conditions found in the thin, weak, and malnourished. Physicians in early nineteenth-century Paris, famed for their correlation of symptoms to lesions at autopsy, sought to distinguish gout, rheumatoid arthritis (also known as rheumatic gout), and other conditions. By 1888, John Kent Spender, working in the English spa community of Bath, described osteoarthritis in its modern understanding, as a degenerative condition of the joints generally found in the elderly. By 1890, Archibald Garrod had written an entire manuscript laying out the microscopic pathology of the joints in rheumatoid arthritis, osteoarthritis, and a variety of other manifestations of painful disease in the joints (*A Treatise on Rheumatism and Rheumatoid Arthritis,* London: Charles Griffin, 1890).

That a physician from Bath would have made noted observations on joint disease is hardly surprising because arthritis in its various forms has been and remains an important source of business for spa towns all over the world. In the United States, Hot Springs, Arkansas; Saratoga Springs, New York; Colorado Springs, Colorado; and thousands of other communities blessed by seacoasts, natural springs, mountains, and desert competed for the trade of the rich and lame. Warm weather, hydrotherapy, physical manipulation, and various types of canes, crutches, and splints were of value in managing the pain and mobility compromise of arthritis, whatever its cause. Given a relatively younger population than in the early twenty-first century, the heavy physical labor done by most people, and the nearly universal exposure to tuberculosis, arthritis in nineteenth-century Americans was likely commonly the result of traumatic injury, infection that did not heal or that seated itself in the joints in the preantibiotic era, and other acute conditions such as rheumatic fever. The frequency of tuberculosis of the bones likely accounts in part for the suspicion of nineteenth-century physicians that arthritis and tuberculosis were caused by a shared defect of a person's constitution. Their response to treatment in similar climates also supported this linkage. The development of antibiotics in the 1940s, particularly those which treated tuberculosis and gonorrhea, dramatically altered the populations affected by crippling joint diseases.

In modern usage, gout is caused by deposition of uric acid crystals in the joints, typically in patients who consume too much meat or whose kidneys are unable to secrete enough of the substance into the urine. By the last decades of the nineteenth century, osteoarthritis was being distinguished from rheumatoid arthritis. Osteoarthritis is classically seen in elderly patients, caused by wear and tear on the joint cartilage, which responds by wearing thin, developing cysts and other signs of damage, and forming calcium deposits at sites where it attempts to repair itself. In contemporary usage, rheumatoid arthritis is seen across the age spectrum, from young to old, is more common in women, and causes rapid and crippling damage through repeated episodes of inflammation.

Chronic debility of the joints caused by wear and tear over a lifetime was certainly familiar to doctors in nineteenth-century America, but for the academically inclined practitioner, by 1900 degenerative osteoarthritis became a diagnosis of exclusion, to be distinguished from gout, inflammatory conditions such as rheumatoid arthritis, and acute cause of joint damage such as infections and trauma. More clinically oriented practitioners might have questioned the value of making such distinctions at a time when all of these joint diseases were treated essentially the same, with rest, pain relief, mobility aids, and an expectation of continued decline.

The twentieth century saw the development of new fields of knowledge and practice, pharmaceuticals for treatment of arthritic pain, and surgical procedures to fuse and replace damaged joints. Antibiotics reduced the frequency of direct infections of the joints and indirect damage from processes like rheumatic fever. The growth of the field of orthopedics, the development of physiatry or rehabilitation medicine, in part descended from the spa doctors, and the emergence of the medical specialty of rheumatology all meant new specialists working on new treatments for joint diseases.

The first decades of the twentieth century also saw the emergence of a theory that most arthritis might be caused by infection of the joints. In an era when rheumatic fever was still a common cause of joint damage and given that white blood cells and often streptococci or staphylococci could be isolated from painful, swollen joints, this theory of focal infections made some sense. It was also known at the time that the teeth, tonsils, and intestines often harbored bacteria, and these were considered prime candidates for the source of bacteria that were subsequently found in the joints. From 1910 to 1940, this theory led many patients to have tooth extractions, tonsillectomies, colectomies, and other procedures in the hope of curing arthritis, ulcerative colitis, and a wide variety of other ailments.

By the 1930s and 1940s, infectious arthritis could be treated with antibiotics; inflammatory arthritis, including gout, could be managed with colchicine, gold salts, and later steroids; and osteoarthritis of the aged could be palliated with walkers, canes, spa treatments, and pain medication. Gold salts, steroids, and by the 1990s monoclonal antibodies such as infliximab, which reduced tumor necrosis factor activity at the joint, were all designed to stop joint damage from autoimmune reactions (inflammation). Other medications, such as salicylic acid from willow bark and acetylsalicylic acid (aspirin) and later ibuprofen and naproxen, came to be called nonsteroidal antiinflammatory medications; they reduced local inflammatory reactions in the body and treated pain, but did not stop ongoing damage to the joints. Bayer started selling purified acetylsalicylic acid under the trade name Aspirin in 1899, but the beginning of World War I in 1914 led to nullification of the German-held patents in the United States, Canada, Britain, France, and other countries at war with Germany, which led to widespread production of the drug by many companies and continuing lawsuits over the patents and profits well into the mid-1920s. The year 1929 saw the introduction of gold salts to reduce inflammatory damage to joints in rheumatoid arthritis (Smyth et al., 1985, p. 125). This was followed by the use of antimalarial agents and D-penicillamine, introduced in the 1950s and 1960s largely based on empiric trials without a clear explanation or understood mechanism in the treatment of rheumatoid arthritis.

Cortisone and other glucocorticoid agents were released to great fanfare in 1949, after extensive research on the therapeutic properties of products of the human adrenal gland, and analysis and synthesis work by many groups, notably those working at the Mayo Clinic under Philip S. Hench (1896–1965). Steroids, used throughout the body both as pills and as injections, and injected directly into joints to treat local inflammation quickly became the standard treatment for many kinds of arthritis and dozens of other diseases. Almost as quickly, they were found to have serious side effects and were linked to many patient deaths. So although they remained valuable in stopping noninfectious acute damage to joints, the search for alternatives continued. From the 1950s onward, as steroid therapy was being refined with the development of more potent agents and physicians gained experience in when and how they were most beneficial and how to minimize side effects, orthopedic surgeons worked to develop surgical alternatives. By the 1960s, the first total hip replacements were being implanted in Britain and the United States, and the 1970s and 1980s saw refinements of their design and rapid growth in their use. The 1990s saw the introduction of the monoclonal antibodies infliximab and etanercept (tumor necrosis factor-α inhibitors), products of molecular biology and the biotechnology industry as the newest—but also most expensive—therapies to stop inflammation from damaging joints and other organs.

Steroids and new monoclonal antibodies are the potent modifiers of immune system function, working to protect patients from damage caused by their own immune cells. This concept of autoimmunity developed in the beginning of the twentieth century with the recognition that the immune cells, designed to protect an individual against infectious diseases, could sometimes attack the body's own tissues. This concept of an overactive or misdirected immune response became the explanation of choice for a number of diseases that had patterns of worsening and improving in unpredictable fashions and that could leave lasting damage to the body's tissues after an "attack." Other diseases grouped with arthritis under this autoimmune label included systemic lupus erythematosus, ulcerative colitis, multiple sclerosis, and other rarer conditions. All of these diseases had been seen as psychosomatic from the 1870s to the 1940s and became widely accepted as immune only after the availability of direct tests of immune function allowed them to be characterized more precisely and the psychological and psychogenic explanations lost favor.

The treatment of arthritis, transitioning as it has since the twentieth century from hydrotherapy, pain control, and fusion surgeries to antiinflammatory drugs, monoclonal antibodies, and joint replacement, reflects the power of modern science, the growth of the pharmaceutical and medical-device industries, and the splitting of the medical profession into ever more specialized practices.

[See also Disease; Gerontology; Health and Fitness; Medical Specialization; Medicine: Overview; Pharmacology and Drug Therapy; Public Health; Surgery; and Tuberculosis.]

BIBLIOGRAPHY

Anderson, Julie, Francis Neary, and John V. Pickstone, in collaboration with James Raftery. *Surgeons, Manufacturers, and Patients: A Transatlantic History of Total Hip Replacement.* Basingstoke, U.K., and New York: Palgrave Macmillan, 2007.
Beeson, P. B. "Fashions in Pathogenetic Concepts during the Present Century: Autointoxication,

Focal Infection, Psychosomatic Disease, and Autoimmunity." *Perspectives in Biology and Medicine* 36, no. 1 (1992): 13–23.

Mann, Charles C. *The Aspirin Wars: Money, Medicine, and 100 Years of Rampant Competition.* New York: Knopf, 1991.

Porter, Roy, ed. *The Medical History of Waters and Spas.* London: Wellcome Institute for the History of Medicine, 1990. Medical history. Supplement 0025–7273; no. 10.

Porter, Roy, and G. S. Rousseau, *Gout: The Patrician Malady.* New Haven, Conn.: Yale University Press, 1998.

Rooke, Thom W. *The Quest for Cortisone.* East Lansing: Michigan State University Press, 2012.

Silverstein, Arthur M. *A History of Immunology*, 2d ed. Amsterdam and Boston: Academic Press/ Elsevier, 2009.

Smyth, Charley J., Richard H. Freyberg, and Currier McEwen. *History of Rheumatology in the United States.* Atlanta: Arthritis Foundation, 1985.

Carla C. Keirns

ARTIFICIAL INTELLIGENCE

Artificial intelligence, a field seeking to express human intelligence through machinery, predates the modern computing age. As a field of computer science, it includes two broad approaches to the study of human intelligence. The first approach attempts to identify laws of reasoning and express those laws in symbols that may be manipulated by a computer. The second approach attempts to use computer hardware and software to model the functions of the human brain. Both approaches have experienced fluctuating fortunes because failures to achieve goals undermined early successes in the field.

From the start of the computer era, researchers sought to develop a computer system that would be indistinguishable from a human being. This standard, articulated by the mathematician Alan Turing (1912–1954) in 1951, would identify a system as possessing artificial intelligence if it could produce responses to questions that were indistinguishable from those produced by a human being.

From the start, researchers pursued both the symbolic approach to artificial intelligence and the efforts to build an artificial brain. In 1951, MIT professors created an artificial neuron and workers at Carnegie Mellon University began developing a program that would prove theorems within a limited field of mathematics. Researchers quickly identified goals that would remain central to the field, including the preparation of a program that would translate text from one language to another or one that could play chess. The term "artificial intelligence" was coined in 1955 by MIT professor John McCarthy (1927–2012).

The field of artificial intelligence quickly fell into a pattern of early successes for simple problems, followed by the inability to solve more complicated problems that more closely modeled human intelligence. An MIT program called SHRDLU, completed in 1969, was typical of this pattern. SHRDLU was designed to process English commands and use them to manipulate objects with a robot arm. It worked well for a simple environment filled with toy blocks but it was unable to be adapted for more realistic settings.

In the 1970s and 1980s, the pattern of successes on simple problems but failure with more complicated tasks caused the major American and European funding agencies to reduce their support for artificial intelligence research. For almost a decade, the field suffered what researchers called "the AI winter."

The symbolic approach to artificial intelligence recovered briefly in the early 1980s. One of the early successes was the rule-based systems used to capture human knowledge about some task or activity. As had happened to the field as a whole, the work on rule-based systems also suffered from the inability to meet inflated expectations. Although expert systems became commercial products by the middle of the decade, they were never able to accomplish all that had been claimed for them. The Japanese government invested heavily in them for their 5th Generation Computing project and had little to show in return. Early hopes of building a system that could diagnose disease were also frustrated.

Artificial intelligence resurged in the early 1990s as researchers looked at new approaches to modeling the operation of the brain. These efforts, identified in this period as neural nets, proved capable of recognizing patterns in data, such as objects in photographs and suspicious actions in financial transactions. Perhaps because earlier forms of this work had failed to meet the expectations set for it, researchers tended to identify this work as "pattern recognition" or "machine intelligence" rather than artificial intelligence.

By limiting the goals of their work, researchers were able to achieve substantial success in the 1990s and 2000s with programs that played chess, recognized images in photographs, diagnosed difficult mechanical problems, identified songs, translated text, and even played the television game show *Jeopardy*. However, the field seemed little closer to the goal of building a system that would pass Turing's test and be indistinguishable from a human being.

[*See also* **Computer Science;** *and* **Computers, Mainframe, Mini, and Micro.**]

BIBLIOGRAPHY

Buchanan, Bruce G. "A (Very) Brief History of Artificial Intelligence." *AI Magazine* 26, no. 4 (Winter 2005): 53–60.
Crevier, Daniel. *AI: The Tumultuous History of the Search for Artificial Intelligence*. New York: Basic Books, 1993.
Mackenzie, D. "The Automation of Proof: A Historical and Sociological Exploration." *Annals of the History of Computing, IEEE* 17, no. 3 (1995): 7–29.
McCorduck, Pamela. *Machines Who Think*. San Francisco: W. H. Freeman, 1979.
Nilsson, Nils J. *The Quest for Artificial Intelligence: A History of Ideas and Achievements*. Cambridge, U.K.: Cambridge University Press, 2010.

David Alan Grier

ASTHMA AND ALLERGY

A gasping child carrying an inhaler in one hand and a ball in the other has become the modern face of asthma in the early twenty-first century. The game is abandoned because he can't breathe. The house behind him might be in a decaying urban neighborhood, where dust and roaches worsen symptoms, or a "tight" house in the suburbs, where the insulation and efficient windows keep in heat but also the humidity that grows dust mites and pollutants from cooking and household furnishings. The same child is diagnosed and monitored with breathing measurements from a handheld spirometer and treated with safe bronchodilators through a home nebulizer. The conditions that increase the child's vulnerability and the technologies that treat him are nearly all new, introduced since the 1970s. What is also new is the number of people influenced by asthma in America, with Centers for Disease Control and Prevention estimates in 2012 at 8.3 percent of the population diagnosed with a disease that a century earlier was considered a rare "disease of civilization" found only in the economic elite.

At the beginning of the twentieth century, the most famous person in the United States with asthma was the vice president, Theodore Roosevelt (1858–1919), soon to be elevated to the presidency by an assassin's bullet. In 1863, as a young Theodore Roosevelt wheezed, coughed, and struggled for breath, his parents frantically searched for help, consulting doctors, trying home remedies, and eventually "fleeing" to Saratoga Springs in the hopes that a change of the air would help their five-year-old son breathe. Internationally, it was probably Marcel Proust (1871–1922), son of a physician and one of France's literary heroes, who spent most of his life in his bedroom indisposed from his many ailments, chief among them asthma. At the start of the twenty-first century it was probably Jackie Joyner-Kersee (1962–), the runner who won six Olympic medals despite her asthma.

The history of asthma starts with the ancient Greeks, who gave us the word. Asthma is mentioned in the *Iliad* of Homer, where it describes a winded hero after a battle, and in the Hippocratic treatises where it describes shortness of breath in pregnancy, old age, and associated with a variety of other symptoms. In the medical language of the day, then, asthma was synonymous with shortness of breath, but was not linked with a specific understanding of what caused that breathing difficulty or how it might be treated. This broader understanding of "asthma" as gasping, wheezing, or coughing ("noisy breath" in the Chinese characters) was maintained in Europe until the nineteenth century.

Asthma has frustrated modern scientists and doctors, who have looked for diseases to manifest in the same way in most people who have the condition. Traditional medical ideas allowed for contributions to disease from a person's inherited traits, but as much or more from that person's lifetime of exercise, diet, and habits, which all combined to build a body that resisted illness or succumbed to it.

From the seventeenth through the nineteenth centuries, a series of physicians who suffered from asthma themselves wrote about asthma as a specific disease, rather than a broad symptom. From Sir John Floyer's (1649–1734) *Treatise on Asthma* (1698) to Henry Hyde Salter's (1823–1871), *Asthma: Its Pathology and Treatment,* first (1860) and second editions (1868), these physicians offered testimony to the nature of their suffering, describing vividly the feelings of suffocation and fear and their own and their patients' responses to various treatments. Floyer's *Treatise* went through four London editions and one in French. It does not seem to have appeared in an American edition, but was held by early U.S. medical libraries in Boston, New York, Philadelphia, Baltimore, and Washington, D.C. By the 1860s, however, Americans had much greater access to European medical knowledge, with Henry Hyde Salter's 1860 edition appearing in a Philadelphia edition by 1864. Theodore Roosevelt Sr. (1831–1878) read Salter's book on asthma to guide treatment of his sickly, asthmatic son Theodore (1858–1919), who went on to be the 26 president of the United States (1901–1908).

Henry Hyde Salter agreed with ancient authorities on the value of moderation in diet and exercise, but had at his disposal remedies from Asia and the Americas that ancient European medicine had not known. From the Americas, stimulants such as tobacco and cocaine to open the breathing passages combined with lobelia and ipecac, used to induce vomiting and give the lungs more space to expand. Stramonium from India and teas and acupuncture from China, all used to thin the mucus and ease clearing of the lungs by coughing, rounded out the medicines for asthma available to Americans and Europeans in the late nineteenth century, derived from a worldwide colonial trade in medicinal plants. Beyond these medicinal plants, the nineteenth century saw an efflorescence of "patent" medicines—an area in which U.S. entrepreneurs were both successful and notorious—combinations of usually common ingredients sold as powders, pills, syrups, and sprays. Those for asthma typically contained lobelia, black tea, stramonium, cocaine, opium, or other well-recognized treatments for the condition (Oleson, 1891, pp. 9, 92–93).

Although travel had long been the most potent prescription for breathing problems, until the nineteenth century it was only available to a small economic elite. New transportation technologies such as faster sailing ships, steamships, and railroads made it possible for many more people to travel long distances, including those who traveled to mountains, seasides, and other regions for their health. Patients with asthma, allergic rhinitis (hay fever), and many other conditions sought out the clean air of mountain regions and the sea air of coasts and beaches, and railroads, land speculators, and others looking for settlers in the American West frequently made claims for the health-giving properties of a place. By some estimates, a quarter of the settlers in Colorado from 1850 to 1900 had traveled there for their own health or that of a family member. Hay fever and "hay-asthma" sufferers were lured to the White Mountains of New Hampshire, Michigan's Mackinac Island, and the Rockies, all of which were low in pollen during the critical autumn pollen season when American ragweed caused sneezing and wheezing across the country. From the 1930s to the 1970s, hundreds of parents sent their children to Colorado, Switzerland, and other salubrious climates to improve their breathing, whereas those who stayed at home were treated with allergy shots, adrenaline, theophylline, and steroids (after 1949).

Since the term "allergy" was coined in 1906 by Clemens von Pirquet (1874–1929), an Austrian pediatrician, allergic diseases have also grown in frequency—or at least recognition of their manifestations. The classic allergic diseases, including asthma, allergic rhinitis (hay fever), eczema (allergic dermatitis), urticaria (hives), and food allergies, were grouped together by allergists and immunologists in the first decades of the twentieth century, who sought to understand them with new laboratory methods and treat them with vaccines and other medications designed to alter the body's immune reactions.

The pharmaceutical industry produced a steady stream of new treatments for asthma in the twentieth century, most of which were still in use during the early twenty-first century. Adrenaline was isolated from the adrenal gland and then synthesized in the last decade of the nineteenth century, becoming commercially available as Epinephrin (without the final E for the brand name) in 1900. The German synthetic pharmaceutical industry brought out theophylline in 1922, isoprenaline (isoproterenol) in 1941, and ipratropium in 1975. British companies produced sodium cromoglycate in 1967 and salbutamol (albuterol) in 1969. Inhaled steroids were marketed starting in the 1970s, leukotriene inhibitors in the 1990s, and anti-immunoglobulin E in the 2000s.

These new pharmaceuticals were used alongside other kinds of treatments, most notably psychotherapy. Helen Flanders Dunbar (1902–1959) and Franz Alexander (1891–1964) developed the concept of the "smothering" mother and an approach to psychoanalytic psychotherapy for asthma that dominated writing, treatment, and the public understanding of the condition for much of the twentieth century. New York–based allergist M. Murray Peshkin (1892–1980) believed so strongly in this theory that in the 1920s he began advocating for "parentectomy" for selected severely asthmatic children and by the 1940s had partnered with an institution in Denver to become a long-term treatment facility for them. Isolation of the children in the Colorado Rockies and psychotherapy for them and their parents were designed to alter their family dynamics and allow for their return home. This rationale for treatment began to break down in the 1960s as the psychoanalytic perspective was challenged and

new pharmaceuticals were able to manage symptoms with far fewer risks and side effects.

According to a 2011 National Health Interview survey, almost 19 million Americans had asthma (18.9 million), including 7.1 million children. In the early twenty-first century, about four thousand people died of asthma every year in the United States. The disease has remained the most common cause of childhood hospitalization and a frequent cause of lost days from work and school absence. It has shown a social gradient in its impact, with greater severity, hospitalization rates, and mortality rates in the poor and socially disadvantaged groups and with concentration of suffering and deaths among the urban poor. Asthma has inspired debates about the quality of air and housing in cities, about the differences good health care can make for individuals and communities, and about issues of social role and personal identity when illness interferes with school and work.

Despite a series of putative mechanisms—based on many of the most important changes in medical theory and practice, such as nerves, germs, immunity, psychoanalysis, and genetics—asthma has remained fundamentally a clinical category most useful in diagnosis and treatment of the individual patient. Medications such as adrenaline, allergy shots, and steroids have served to alter treatment and sometimes became de facto diagnostic tools and inspired and justified new theories about the fundamental causes of the disease. This bidirectional feedback between clinical practice and basic science highlights the inductive and empirical processes of scientific discovery, as well as links between laboratory and clinic, that are fundamental to biomedical sciences.

[*See also* **Centers for Disease Control and Prevention; Disease; Food and Diet; Health and Fitness; Medicine: Overview; Occupational Diseases; Pediatrics; Pharmacology and Drug Therapy; Psychotherapy; Public Health;** *and* **Public Health Service, U.S.**]

BIBLIOGRAPHY

Brewis, R. A. L. *Classic Papers in Asthma*. London: Science Press, 1990.

Jackson, Mark. *Allergy: The History of a Modern Malady*. London: Reaktion, 2006.

Jackson, Mark, *Asthma: The Biography*. Oxford and New York: Oxford University Press, 2009.

Keirns, Carla. *Measured Breath: A Short History of Asthma*. Baltimore: Johns Hopkins University Press, forthcoming.

Mitman, Gregg. *Breathing Space: How Allergies Shape Our Lives and Landscapes*. New Haven, Conn.: Yale University Press, 2007.

Oleson, Charles W. *Secret Nostrums and Systems of Medicine, a Book of Formulas*, pp. 9, 92–93. Chicago: Oleson & Co., 1891.

Peumery, Jean-Jacques. *Histoire Illustrée de l'Asthme: de l'Antiquité á Nos Jours*. Paris: R. Dacosta, 1984.

Strobel, Martine. *Asthma bronchiale: die Geschichte seiner medikamentosen Therapie bis zum Beginn des 20. Jahrhunderts*. Stuttgart: Wissenschaftliche Verlagsgesellschaft mbH, 1994.

Carla C. Keirns

ASTRONOMY AND ASTROPHYSICS

Astronomy encompasses the observation and theory of celestial objects, but traditionally also includes the study of meteors ("shooting stars") and the aurora (the "northern lights")—phenomena that, strictly speaking, occur in Earth's upper atmosphere but are of celestial origin.

Colonial Origins. Science was valued in colonial America from the early seventeenth century. Because astronomy was one of the seven classical liberal arts, it was part of the curriculum of Harvard College (founded 1636). For the first two decades, Harvard students were still taught Ptolemy's geocentric system of the universe and Aristotelian physics, but the ideas of Copernicus and Isaac Newton encountered little resistance when introduced into the English-speaking New World.

The colonies included active observers and careful theorists. The first exclusively astronomical colonial treatise was *An Astronomical Description of the Late Comet of Blazing Star*, written in 1665 by Reverend Samuel Danforth of Roxbury, Massachusetts, recounting his own observations of Hevelius's comet in December 1664. Danforth noted that the tail always seemed to point away from the sun and correctly calculated that the comet was nearest Earth on 18 December 1664. He concluded that the comet was "a Celestial Luminary, moving in the starry Heavens," not simply an "exhalation" from Earth as taught by Aristotelian theory. Twelve years later, recent Harvard graduate Thomas Brattle observed the bright comet of 1680 with a telescope that had been donated to Harvard in 1672 by John Winthrop Jr., first governor of Connecticut and a founding member of the Royal Society of London. Brattle's observations were among those cited by Isaac Newton in his *Philosophia naturalis principia mathematica* (known as his *Principia*, 1687) in his determination of the comet's orbit (Yeomans, 1977).

Three eighteenth-century colonial astronomers deserve special note. John Winthrop IV (a descendent of John Winthrop Jr.), Hollis professor at Harvard, observed sunspots, comets, eclipses, and several transits of Mercury—plus both the 1761 and the 1769 transits of Venus—across the face of the sun. He applied Newton's laws of motion and gravitation to deduce the density of a comet and contended that the enormous observed speeds of meteors implied that they were solid bodies that must originate in outer space and Earth intercepted them in its orbit. David Rittenhouse of Norriton, Pennsylvania (20 miles from Philadelphia), constructed what may have been the first telescope built in the United States, as well as clocks, orreries (models of the solar system that moved through clockwork), and scientific instruments. Rittenhouse was one of at least 19 colonial observers who timed the 1769 transit of Venus—a rare event of keen interest worldwide because of its potential to yield an exact measure of the distance from Earth to the Sun (Hindle, 1964). Samuel Williams, Winthrop's successor as Hollis professor, during the total solar eclipse of 27 October 1780, described and drew the phenomenon now called Baily's beads (where the very edge of the sun glows in valleys between mountains on the edge of the Moon) 56 years before British astronomer Francis Baily saw, described, and drew them himself.

Colonial almanacs—calendars and handbooks of miscellaneous useful information essential to daily life—were compiled exclusively by recent Harvard graduates during the seventeenth century and informed the public about current astronomy. Indeed, Zechariah Brigden's almanac for 1659 printed the first colonial astronomical treatise on the Copernican system. A best-selling almanac was lucrative, so after about 1700 almanac publication became competitive. By the mid-eighteenth century, Benjamin Franklin's *Poor Richard's Almanack* (Philadelphia, Pennsylvania) and Nathanael Ames's *Astronomical Diary* (Boston, Massachusetts) each sold 50,000 to 100,000 copies annually to a collective colonial population of 2.5 million. Astronomical essays in almanacs recounted European telescopic discoveries, the nature of planets and comets, the elliptical shape of planetary orbits, the rotation of the sun as shown by sunspots, and rudiments of Newton's laws.

Evening scientific lectures became popular entertainment. Dramatic speakers such as Isaac Greenwood, Theophilus Grew, and Ebenezer Kinnersley included demonstrations with orreries or views through small telescopes. Even Sunday sermons could include "natural philosophy" (the term for science until the nineteenth century); indeed, famous Puritan leader and Newtonian scientist Cotton Mather wrote books for ministers to "demonstrate, that *Philosophy* is no *Enemy,* but a mighty and wondrous *Incentive* to *Religion*" so that alongside the Bible as the Word of God, the universe should be studied as the Work of God (Mather, 1721).

Nineteenth-Century Independence. After the American Revolution (1775–1783) resulted in colonial independence from Great Britain, American astronomers looked less to the Royal Society of London and began creating their own institutions and publications. Before the end of the nineteenth century, American astronomy had developed into a community of researchers whose work rivaled that in Europe (Gingerich, 1984; Lankford and Slavings, 1997).

Positional astronomy and nation-building. Until the mid-nineteenth century, astronomical research primarily consisted of observing and measuring changes in the positions of visible celestial objects. Positional astronomy for the terrestrial purposes of timekeeping, navigation by sea, or surveying on land was called practical astronomy (Waff, 1999). After the Louisiana Purchase (1803) and other major western land acquisitions extended the new United States across the North American continent to the Pacific Ocean, it became a national priority to establish borders, map resources, navigate shorelines and rivers, record land grants and transfers, and locate cross-continent routes for military commercial purposes.

Americans made fundamental improvements to practical astronomy. In 1836, Captain Andrew Talcott, a topographic engineer for the Army Corps of Engineers (a bureau charged with mapping military routes and conducting surveys of strategic interest), improved latitude determination by measuring zenith distances (star angles *down* the meridian from the zenith) instead of altitudes (star angles *up* the meridian from the horizon). Zenith distances eliminated uncertainties in the star's position from atmospheric refraction (bending of starlight caused by the greater thickness of the atmosphere near the horizon). Within a decade, the "Talcott method" for determining latitude became standard worldwide (Chauvenet, 1960).

Finding longitude amounts to a difference in time between an unknown location and a reference location to the east or west. Boston mathematician Nathaniel Bowditch edited and corrected a British handbook for ship navigators; the third edition was retitled *The New American Practical Navigator* and came out under Bowditch's own name in 1802. Bowditch also simplified a method called lunar distances for finding longitude at sea, which became widely used on American ships. Although the marine chronometer of eighteenth-century British clockmaker John Harrison solved the longitude problem reliably enough for open-ocean mariners, it was not precise enough for navigating the treacherous shoals along the eastern seaboard of the United States, which annually wrecked countless ships. Staunching such drastic losses required precise maps, so in 1807, President Thomas Jefferson chartered the U.S. Coast Survey, appointing Ferdinand Rudolph Hassler to chart all coastal waterways (Theberge, 2007). Within half a century, the Coast Survey grew to be the most powerful science and technology funding body in the nation, employing most of the country's practical astronomers; it evolved into today's National Oceanic and Atmospheric Administration (NOAA).

In the 1840s, Sears Cook Walker and other Coast Survey astronomers used the newly invented telegraph to revolutionize longitude measurements on land. Because the telegraph could transmit signals hundreds of miles nearly instantaneously, precise clocks reading local solar or sidereal time at each of two widely separated locations could be compared simultaneously; the difference in local times represented their difference in terrestrial longitude. This telegraphic "American method of longitudes" became the standard geodetic method worldwide until it was supplanted by radio techniques in 1922 (Bartky, 2000; Stachurski, 2009). As a spin-off, the "American method of transits" used telegraph technology within the walls of an individual observatory to increase the precision and speed of astronomical data collection by creating a paper record of astronomers' timings of astronomical events directly alongside standard time signals.

Refracting telescopes and observatories.

One or two eighteenth-century observatories could be considered the first established in the United States (Milham, 1938; Shy, 2002). The oldest extant U.S. observatories were founded in 1838: the Hopkins Observatory at Williams College in Williamstown, Massachusetts, and the Hudson Observatory (later renamed the Loomis Observatory) at Western Reserve Academy in Hudson, Ohio. In 1845, the Cincinnati Observatory in Cincinnati, Ohio, installed a 12-inch refracting (lens) telescope from the Munich firm of Merz and Mahler, the largest refractor in the United States. In 1847, Harvard College Observatory in Cambridge, Massachusetts, installed a 15-inch Merz and Mahler refractor, tied in

aperture with the 15-inch Merz and Mahler refractor installed in 1839 at the Central Astronomical Observatory of the Russian Academy of Sciences in Pulkovo near St. Petersburg (Jones and Boyd, 1971). Thus, by 1847 the young nation possessed one of the world's two largest refractors.

The Harvard 15-inch was the last world's largest refractor in America to be built by European artisans. All later ones were built by Cambridgeport, Massachusetts, portrait painter-turned-optician Alvan Clark and his sons, George Bassett Clark and Alvan Graham Clark. Between 1864 and 1897, the Clarks figured lenses for the world's largest refractors five times, culminating with the 40-inch lens for the Yerkes Observatory at Williams Bay, Wisconsin, in 1897 (Warner and Ariail, 1995). For precision equatorial mounts (a telescope mount with a polar axis tilted to be parallel to Earth's rotation axis at that latitude), the telescope engineers of choice became Worcester Reed Warner and Ambrose Swasey; Warner & Swasey also designed and built mechanical parts for several large twentieth-century reflecting (mirror) telescopes.

From the 1840s into the 1890s, the United States and Europe raced to build and retain title to the world's largest astronomical refractor because national pride in astronomical achievements was seen as important for nation-building and international scientific and industrial stature. The magnitude of the observatory-building movement in nineteenth-century America was unique in the world; modern counts reveal that by 1900, more than two hundred observatories, large and small, were built, many funded and named after private philanthropists (Aubin et al., 2010; Bell, 2002; Hutchins, 2008). Lick Observatory atop 4,200-foot Mount Hamilton, California, became the first major mountaintop observatory; results from its 36-inch Clark/Warner & Swasey refractor installed in 1888 demonstrated the research value of siting a large astronomical instrument at high altitude.

Many nineteenth-century U.S. observatories also possessed a specialized instrument of precision for positional astronomy, usually a telescope fixed on a horizontal east–west axis so its movement was restricted to its north–south meridian. The best nineteenth-century meridian instruments allowed skilled naked-eye observers to measure celestial positions to several hundredths of a second of arc for runs of observations. With such extraordinary precision for visual observations—not superseded by photography until the early twentieth century—positional astronomy became a source of valuable new data about the dynamic/gravitational interactions among astronomical objects. Results informed key developments in lunar theory by Simon Newcomb, George W. Hill, and Ernest W. Brown (Dick, 2003; Wilson, 2010); understanding of the motions of stars in our "sidereal system" (what we now know as our galaxy) from the measurements of proper motions—motions of stars across the line of sight—by Lewis Boss and others; and insight into the internal physical makeup of Earth itself from the discovery by Seth Carlo Chandler of Boston in 1891 of the wandering variation of Earth's rotational axis, known as the Chandler wobble (Carter and Carter, 2003).

Solar system discoveries. Nineteenth-century American astronomers made fundamental contributions to knowledge of objects in the solar system. Chemical analysis of meteorites that fell near Weston, Connecticut, in 1807 by Benjamin Silliman of Yale College revealed they were of a nickel–iron composition unlike terrestrial rocks. By collating observations of the Leonid meteor shower of 1833, Denison Olmsted of Yale deduced that meteor showers occur when Earth, in its annual orbit around the sun, passes through a stream of meteoroids in space. In 1847, Maria Mitchell on Nantucket Island, Massachusetts, became the first American to discover a comet, rocketing the young woman into national prominence as the nation's first female astronomer; in 1865, she became the first professor of astronomy at Vassar Female College. In the 1850s, Yale professor Elias Loomis found that a magnetic compass needle responded to displays of the aurora; by analyzing observations from many observers, he calculated that auroral streamers occurred 50 to 500 miles above Earth, in an oval belt that circled the northern (and southern) magnetic poles. In 1870, Loomis demonstrated that auroral displays were more intense when the sun showed many sunspots. In 1866, Daniel Kirkwood at Indiana University demonstrated that gaps in the distribution of orbits of asteroids between Mars and Jupiter occur at orbital periods that are at resonances (1/2, 2/5, 1/3) with the orbital period of Jupiter. Later, he showed that similar Kirkwood gaps occur in the rings of Saturn.

The rise of astrophysics. In 1859, Gustav Kirchhoff in Germany produced colored flames by tossing salt or flakes of metal into the flame of the new Bunsen burner invented by Robert Bunsen and directed the flames' light through prisms. Kirchhoff and Bunsen discovered that each burning chemical element emitted its own unique pattern of bright spectral lines at specific wavelengths (colors). They also found that if brilliant light is shown through a cool gas of that element, the gas absorbs light at those identical wavelengths. This important discovery led to the invention of astronomical spectroscopy: dispersing (spreading out into a rainbow) the light of celestial objects through prisms to determine their chemical composition, the presence of magnetic fields, the ionization (removal of electrons from atoms) of gases, and the object's radial velocity (its speed and direction of movement along the line of sight). After the advent of faster, more sensitive dry-plate photography in the 1880s, photography could record spectra even of faint objects. Ultimately, spectroscopy united astronomy, physics, and chemistry into the "new astronomy" of astrophysics, named after the title of a popular book, *The New Astronomy*, by Allegheny Observatory director Samuel Pierpont Langley in 1888.

Many early American developers of astrophysics were solar physicists. While watching a total solar eclipse in 1870 through a spectroscope, Charles A. Young of Dartmouth discovered the reversing layer of the sun (a cooler layer of gases above the sun's bright visible surface that produces the dark Fraunhofer lines in the sun's spectrum). By the 1870s, astrophysicists had found that finely ruled gratings were superior to prisms for high dispersion of spectra; two premier makers of precise gratings were astrophotography pioneer Lewis

M. Rutherfurd of New York City and physicist Henry A. Rowland of Johns Hopkins University. In the early 1890s, George Ellery Hale invented a spectroheliograph to photograph spectra of the sun's chromosphere (ruby-colored irregular upper layer and its dramatic flame-like prominences); by 1908, the spectroheliograph had become adopted worldwide (Meadows, 1984). With the backing of the new Carnegie Institution of Washington, Hale in 1904 founded the Mount Wilson Solar Observatory, which in 1920 (after the institution became home to large nighttime telescopes) was renamed the Mount Wilson Observatory.

Meanwhile, in the 1880s, Harvard College Observatory under director Edward C. Pickering began classifying stars based on their colors, spectral lines, brightness, and other properties. He built astrographs (special photographic telescopes with a wide field of view); instead of mounting a high-dispersion spectroscope at the eye end, he mounted a thin, low-dispersion objective prism over the front aperture to disperse the light from all celestial objects in the field of view into small, bright spectra that showed only principal light and dark lines, to compare many stars at once. Pickering and his team—many of whom were women—created an alphabetical classification of stellar types, subsequently modified by Annie Jump Cannon into the letter sequence standard today that groups stars in spectral classes from blue-white to red (O, B, A, F, G, K, M). In the early twentieth century, Henry Norris Russell of Princeton University and Ejnar Hertzsprung of Denmark independently analyzed what relationships might exist among the stellar types and whether some evolved into others as stars aged. They devised a scatter graph now known as the Hertzsprung–Russell or H–R diagram that plotted stars by spectral type and absolute magnitude (actual brightness independent of distance), whose different regions are now known to correspond to specific nuclear-fusion processes inside a star (DeVorkin, 2000).

Twentieth-Century Broadening. In the twentieth century, astronomy broadened to include the study of celestial objects at electromagnetic wavelengths other than visible light. In the

space age with the launch of interplanetary spacecraft, astronomy also came to encompass planetary science, which includes aspects of geology and meteorology. Cosmology, the study of the origin of the universe, became an observational science.

Reflecting telescopes and observatories. With the completion of the 40-inch Yerkes refractor, large, massive lenses suspended only by their circumference reached their physical limits (Osterbrock et al., 1988). In the 1850s, European opticians began experimenting with casting disks of crown (plate) glass and covering them with a thin reflective coating of silver. In 1862, New York doctor and wealthy amateur astronomer Henry Draper reported on observations with his 15.5-inch reflecting telescope, introducing silver-on-glass mirrors to American astronomers (Draper, 1864). The chemical process for silvering a mirror, however, involved heat and was risky. In 1880, a cheap, simple, reliable silver-deposition process invented by Pittsburgh telescope maker John A. Brashear made silver-on-glass telescope mirrors practical; Brashear's technique remained standard into the twentieth century.

Glass mirrors opened the way for giant telescopes. America's first large silver-on-glass reflector—a 36-inch fashioned by British optician A. A. Common for Canadian amateur Edward Crossley in 1885—was acquired by Lick Observatory in 1895 (Osterbrock et al., 1988). The Crossley reflector's success for astronomical photography under James E. Keeler led others to try even bigger mirrors. Yerkes optician George W. Ritchey fashioned the mirror for a 60-inch reflector installed at Mount Wilson Solar Observatory in 1908, followed by a one hundred–inch reflector in 1918. Until 1948, the one hundred–inch Hooker reflector on Mount Wilson was the world's largest telescope.

Three fundamental engineering advances in the 1930s allowed even larger mirrors. First, Corning Glass Works invented the low-expansion borosilicate glass Pyrex, less subject than crown glass to changing its figure with changes in day–night temperatures. Second, Corning cast Pyrex mirrors as a thin disk with ribbing on the back for strength, reducing the mass of the resulting mirror. Third, John D. Strong at the California Institute of Technology invented a method for vaporizing a thin coating of aluminum onto the glass. Almost as reflective as silver, aluminum is far more durable and does not tarnish. Telescope mirrors in the early twenty-first century are still aluminized using Strong's process. These three advances enabled the next world's largest telescope: the two hundred–inch (five-meter) Hale reflector at Palomar Observatory on Palomar Mountain, California, completed in 1948 (Preston, 1996).

The Palomar two hundred–inch telescope itself represented the practical limit of a massive monolithic (one-piece) telescope mirror and its precision equatorial mount. Three major advances—all in electronic computer control technology—in the late twentieth century enabled truly mammoth telescopes. Actuators and software were developed to control the optical figure of multiple thin mirrors so they functioned as a single large mirror within optical tolerances. Altazimuth mounts (which move only horizontally and vertically, regardless of latitude, so as to minimize flexure) were improved to be capable of tracking celestial objects under ultrafine computer control precisely enough for long-exposure images. The first successful altazimuth-mounted segmented-mirror reflector was the Keck I ten-meter (four hundred–inch) telescope at the W. M. Keck Observatory in Hawaii (completed in 1993). Its design principles have been adopted worldwide for even larger reflectors, including current twenty-first-century designs for telescopes with segmented mirrors up to 30 meters (1200 inches) in diameter. Last, adaptive optics—real-time feedback methods for instantaneously "detwinkling" star images to compensate for ever-changing distortions and blurring from fluctuations in the atmosphere's temperature and density—were first implemented for astronomical purposes in 1999 at the Keck II telescope and adopted in observatories worldwide by the early twenty-first century (Duffner, 2009).

The cosmic distance ladder. Around 1880, Pickering at Harvard devised a classification of variable stars (stars that vary in brightness), which he modified in 1911. Still used today, it classified variables by their light curves (pattern of variation over time): whether they brightened and dimmed regularly or irregularly over short times (hours to days) or long times (a year or more) or even exploded cataclysmically. Brightness measurements from photometers combined with spectroscopic observations revealed that some variables were eclipsing binaries (pairs of stars orbiting each other that block each other's light as seen from Earth), whereas others pulsated (expanded and contracted); still others throw off outer layers of gas. One class of variable stars became exceptionally important to cosmology.

Around 1912, Henrietta Swan Leavitt, a "computer" or computational assistant at Harvard, discovered that many variables had light curves similar to that of Delta Cephei in the constellation Cepheus. Leavitt discovered 25 such Cepheid variables in the Small Magellanic Cloud (one of two companion galaxies to our Milky Way), with periods ranging from 1.25 to 129 days; the longer-period Cepheids were significantly brighter than the shorter-period Cepheids. Because their location in the same relatively small galaxy meant the stars were all at about the same distance, their brightness differences meant actual differences in the stars' absolute luminosities. By 1917, Harvard College Observatory director Harlow Shapley (Pickering's successor) realized that Leavitt's discovery of a relationship between Cepheid period and luminosity could allow vast distances to other star systems to be measured, simply by timing the periods of their constituent Cepheids. After 1956, when Walter Baade of Mount Wilson Observatory showed that there are two distinct types of Cepheids, the Shapley–Leavitt period–luminosity relationship became a powerful yardstick for determining distances to other galaxies.

Astronomers worldwide had long observed that some faint fuzzy nebulae appeared to have a spiral structure. Wondering whether spiral nebulae were other solar systems in the process of forming, early twentieth-century wealthy amateur astronomer Percival Lowell hired Vesto M. Slipher to photograph spectra of spiral nebulae from Lowell Observatory in Flagstaff, Arizona, to

92 · ASTRONOMY AND ASTROPHYSICS

see whether they were rotating. Slipher discovered that the spectral lines of most spiral nebulae were shifted toward the red end of the spectrum, implying they were unexpectedly receding from Earth. Slipher speculated that the spiral nebulae might be distant "island universes" outside the Milky Way system of stars. In April 1920, Harlow Shapley of Harvard and Heber D. Curtis of Lick Observatory debated the nature of the spiral nebulae before the National Academy of Sciences in a famous "Great Debate" (Shapley argued they were part of the Milky Way, whereas Curtis argued they were outside it; the prevailing view at the time was that Shapley won, although later Curtis was proven correct).

In 1923, Edwin P. Hubble's long-exposure photographs through the Mount Wilson one hundred–inch reflector revealed that the Great Spiral Nebula in Andromeda was actually a vast system of individual stars. Hubble also discovered that the galaxy (as we now recognize the Andromeda Nebula to be) possessed more than a dozen Cepheid variables. Using the Shapley–Leavitt period–luminosity relationship, Hubble calculated that the Andromeda Nebula was more than 1 million light-years distant. In 1929, Hubble announced that the red shifts in spectral lines revealed that the more distant a galaxy, the faster it was receding, a relationship now known as Hubble's law. In the early 1930s, Hubble's Mount Wilson colleague Milton L. Humason demonstrated that one could tell a galaxy's distance just by knowing its red shift. These discoveries implied the universe was expanding (Berendzen et al., 1976).

Building on early work from Belgian cosmologist Georges Lemaître proposing that the universe started very small as a "primeval atom," in the 1940s, nuclear physicist George Gamow at George Washington University advocated that the universe originated in a colossal explosion (which detractors nicknamed the Big Bang) that produced chemical elements such as hydrogen and helium. Other astronomers later showed that the rest of the elements were made inside stars.

Beginning in the 1930s, observations pointed to the fact that far more mass was needed for galaxies and other large structures in the universe to remain gravitationally bound systems than was observed in stars and clouds of gas and dust. Observations included high velocities of galaxies in galaxy clusters and later of stars revolving around the centers of spiral galaxies as measured by Vera C. Rubin and W. Kent Ford of the Carnegie Institution of Washington. In 1933, Fritz Zwicky at the California Institute of Technology presented strong evidence for some kind of invisible dark matter ("dunkle materie" in German) whose presence is known only by its significant gravitational influence. The "cold dark matter" (CDM) cosmological theory, proposed in 1984 by three astronomers at the University of California, Santa Cruz (George R. Blumenthal, Sandra M. Faber, and Joel R. Primack), plus Cambridge University astronomer Martin J. Rees, suggested that the observations were well explained by the predicted existence of dark matter that was slow moving in the early universe (Blumenthal et al., 1984).

In the 1990s, other observations suggested that the expansion of the universe was accelerating; one hypothesis for the unknown repulsive force, given the name "dark energy," is the cosmological constant Lambda (Λ) proposed by Albert Einstein in 1916. The existence of dark energy—known to comprise about three quarters of the density of the universe—was confirmed in 1998 by the observations of supernovae by two teams of observers, one led by Saul Perlmutter of Lawrence Berkeley National Laboratory and the other by Adam G. Riess of Johns Hopkins University and Brian P. Schmidt of Australian National University.

Radio astronomy. In 1931, Bell Telephone Laboratories physicist and radio engineer Karl Jansky in New Jersey discovered hissing static at a shortwave radio frequency. Because it followed the 23-hour 56-minute sidereal day instead of the 24-hour solar day, Jansky concluded it was of celestial origin, identifying it as the center of the Milky Way in the constellation Sagittarius. In 1937, radio engineer Grote Reber built the first dedicated dish radio telescope and made the first systematic survey of celestial radio sources around the entire sky (Sullivan, 2009).

In 1951, Harvard physicists Harold I. Ewen and E. M. Purcell mapped the structure of the Milky Way's spiral arms by tracing signals emitted from clouds of neutral hydrogen gas at the radio wavelength of 21 centimeters. In 1965, Bell Labs physicists and radio engineers Arno Penzias and Robert Wilson discovered the cosmic microwave background radiation. In 2013, the biggest radio telescopes consisted of multiple dishes linked together as interferometers to simulate the high resolution of a much larger aperture, such as the Karl Jansky Very Large Array (VLA) near Socorro, New Mexico. After the invention of radar in the 1940s, large radio telescopes have also been used to bounce radio signals off the Moon, Venus, asteroids, and other solar-system bodies to ascertain their surface structures and precise distances (Butrica, 1996).

Infrared astronomy. Celestial objects radiating at near-infrared (heat) wavelengths somewhat longer than red visible light can be observed using cryogenically cooled detectors on telescopes at mountaintop observatories. Even at the highest and driest earthbound sites, such as in the Peruvian Atacama or at the South Pole, however, residual atmospheric water vapor blocks observations in the far infrared out to millimeter wavelengths (one thousand micrometers at the edge of short radio waves), especially between thirty and three hundred micrometers. By the mid-twentieth century, astronomers were lofting cryogenically cooled instruments into the upper atmosphere to altitudes of 25 miles (40 kilometers) via balloons and sounding rockets to observe at wavelengths longer than 3 micrometers.

The U.S. National Aeronautics and Space Administration (NASA) modified several aircraft to create high-altitude airborne infrared observatories. The Gerard P. Kuiper Airborne Observatory (KAO)—a Lockheed C-141A Starlifter cargo plane modified to carry a 36-inch (0.91-meter) reflector peering out of an open cavity near the nose—enabled a wealth of discoveries from 1974 to 1995 about the structure of Pluto's atmosphere, the nature of the interstellar medium (gas and dust between stars), and the birth of stars. In 2011, NASA (in collaboration with the German Aerospace Center DLR) began science missions on the Stratospheric Observatory for Infrared Astronomy (SOFIA), a short-bodied commercial Boeing 747SP airliner modified to carry a telescope in the rear of the plane, having an effective aperture of one hundred inches (2.5 meters).

Spacecraft astronomy. NASA, founded in 1958, revolutionized astronomy for the solar system and beyond (Beatty et al., 1999; Kraemer, 2000); although not noted below, some missions were partnered with international space agencies. The 12 Apollo astronauts who landed on the Moon between 1969 and 1972 returned to Earth some eight hundred pounds of rocks from both relatively smooth dark lunar maria (lava "seas") and rugged highlands; glassy orange soils confirmed the Moon had been volcanically active.

The first U.S. satellite, *Explorer 1* (launched in 1958), discovered the existence of the vast Van Allen radiation belt of charged particles around Earth. Sun-orbiting spacecraft *Pioneer 6, 7, 8,* and *9,* launched in the 1960s, first explored the nature of the solar wind and interplanetary medium (subatomic particles and electromagnetic fields between the planets). A host of solar observatories launched between 1962 and 2010 revolutionized understanding about the Sun, discovering coronal holes as a source of the solar wind, mapping the interior and far side of the Sun, and observing the Sun at various wavelengths through several 11-year sunspot cycles.

NASA spacecraft have flown by and/or orbited all the planets in the solar system (Beatty et al., 1999; Kraemer, 2000). For the rocky planets, spacecraft revealed Mercury to have a surface heavily cratered with unusual volcanic structures and a density unexpectedly high and Venus to be a hothouse with a crushing atmosphere of carbon dioxide, with 85 percent of its surface shaped by volcanic activity. Spacecraft showed Mars has gullies and flows of debris suggestive of current sources of liquid water, and the spectral signature of hydrogen suggested that water still exists a meter below the surface. NASA's robotic rovers Sojourner (landed in 1997), Spirit and

Opportunity (both landed in 2004), and Curiosity (landed in 2012) crawled over the Martian surface for close-up photographs and spectroscopic analysis of soils.

The outer planets Jupiter, Saturn, Uranus, and Neptune were first photographed from close up by one or more of four spacecraft (*Pioneer 10* and *11, Voyager 1* and 2) that flew by 1979–1989. Orbiting spacecraft in the 1990s studied both Jupiter and Saturn in greater detail. Collectively, the NASA spacecraft revealed that Jupiter's innermost satellite, Io, has over four hundred active volcanoes spewing sulfur and sulfur dioxide; Saturn's largest moon, Titan, has a smoggy atmosphere; Saturn's rings display complex structures that resemble spokes and braids; and Uranus and Neptune have magnetic poles skewed at a large angle from their axes of rotation. NASA spacecraft have also investigated asteroids and comets, and *New Horizons* (launched 2006) is due to fly by Pluto— no longer classified as a planet—in 2015 and to travel through the Kuiper Belt (a distant ring of mysterious Pluto-like objects beyond the orbit of Neptune).

To explore the larger cosmos, since the 1960s NASA has lofted more than a dozen astronomical telescopes, culminating with the four Great Observatories for Space Astrophysics, all of which have made substantial contributions to astronomy. The Hubble Space Telescope (launched 1990), with its 2.4-meter mirror, observes primarily at visible wavelengths (Smith, 1989). The Compton Gamma Ray Observatory (launched 1991) observed gamma rays and hard X-rays (the most energetic X-rays). The Chandra X-ray Observatory (launched 1999) observes soft X-rays. The Spitzer Space Telescope (launched 2003) observes objects across infrared wavelengths.

Meanwhile, the Cosmic Background Explorer (COBE, launched 1989) discovered that the temperature of the cosmic microwave background varies slightly in different directions around the sky, revealing an anisotropy (unevenness) in the distribution of matter and energy throughout the universe that is important for the formation of large-scale structures in the universe, as predicted by the Lambda CDM cosmology (the CDM cosmology was modified in 1996 to include dark energy). In 2001, the Wilkinson Microwave Anisotropy Probe (WMAP) began long-term, systematic, high-resolution measurements of temperature variations and other cosmological parameters (Boslough and Mather, 2008; Impey, 2012).

Twenty-First Century and Perspective. Telescopes and other astronomical instruments are so large, specialized, and such a major investment that astronomy in the early twenty-first century has become "big science"—indeed, aspects of high-energy astrophysics (such as the search for dark matter particles) overlap with particle physics. Meanwhile, several instruments have been built to try to detect gravitational waves predicted by Einstein's general theory of relativity to be radiated by supernovae, colliding black holes, and other violent astronomical events. The biggest "telescopes" were the twin Laser Interferometer Gravitational-wave Observatories (LIGO) in Livingston, Louisiana, and Hanford, Washington, the first of which began initial operation in 2001 (Bartusiak, 2000). Upgraded detectors, known as Advanced LIGO, are expected to go into operation around 2014.

Digital techniques are transforming astronomy in observation, analysis, and theory. Since 2000, the Sloan Digital Sky Survey by an automated 2.5-meter telescope at Apache Point Observatory in New Mexico has detected and cataloged more than 930,000 distant galaxies and 120,000 quasars in a quarter of the sky (away from the plane of the Milky Way). Intensive computation is key in real-time astronomical discovery, such as with the Palomar Transient Survey (of the California Institute of Technology and University of California) to identify supernovae exploding in distant galaxies (Law et al., 2006). Supercomputing is enabling detailed simulations of astrophysical processes, yielding predictions that can be tested against observations of the actual universe, effectively turning astronomy—including cosmology—from a purely observational science into an experimental science (Klypin et al., 2011). Computational techniques have also been key to definitively confirming the detections of extrasolar planets (planets in orbit around other stars) discovered by both ground-based telescopes and a precision photometer aboard NASA's Kepler spacecraft, launched in 2009, as well as characterizing their masses and orbits.

Role of popularization in astronomy. Since the nineteenth century, astronomy has experienced an ongoing tension between amateurs and professionals. Although some professionals sought to distance themselves from amateurs, others have recognized the value of the contributions of skilled amateur observers and welcomed them into specialized organizations to coordinate observations (Percy and Wilson, 2000; Rothenberg and Williams, 1999). Examples include the American Association of Variable Star Observers (AAVSO) and the American Meteor Society (AMS), both founded in 1911, and the International Occultation Timing Association (IOTA, founded in 1974 by David W. Dunham). In the early twenty-first century, amateurs participated in classifying galaxies from the Sloan Digital Sky Survey through the international "citizen science" online collaboration of Galaxy Zoo (founded in Great Britain around 2007) and through lending the idle CPU cycles of their home personal computers to Einstein@Home to help perform computations on observations from the LIGO searches for gravitational waves. Annually since 1976, adventurous amateurs have helped collect meteorites from the Antarctic ice sheet through the U.S. Antarctic Search for Meteorites (ANSMET) program headed first by William Cassidy of the University of Pittsburgh and subsequently by Ralph Harvey of Case Western Reserve University (Cassidy, 2003).

Astronomy is a beautiful science that has always attracted members of the lay public, some of whom have become donors of equipment, telescopes, or even entire observatories, or underwriters of major astronomical research projects or expeditions to observe total solar eclipses, transits of Venus, or the southern heavens. As astronomy has relied more on taxpayer funding through the National Science Foundation, NASA, and public universities, education and public outreach to disseminate astronomical findings at a level accessible to lay audiences has grown in importance.

Public astronomical societies around the nation have also been important in sponsoring popular talks and starparties, where members of the general public can look through impressively large telescopes (24 inches and larger) constructed by amateurs, notably Dobsonian reflectors of a revolutionary inexpensive, portable altazimuth design pioneered by John Dobson, cofounder of the San Francisco Sidewalk Astronomers in 1967 (Dobson and Sperling, 1991).

The first U.S. periodical specifically dedicated to astronomy was the short-lived monthly *Sidereal Messenger,* published from 1846 to 1848, by Ormsby MacKnight Mitchel, founder and first director of the Cincinnati Observatory. In 1849, Benjamin Apthorp Gould, first director of the Dudley Observatory, established the *Astronomical Journal,* the first U.S. research journal of astronomy, still published in the early twenty-first century. In 1882, two decades after Mitchel's death, William Wallace Payne of Carleton College Observatory established a periodical also named *The Sidereal Messenger* both for popular articles and for astronomical research, published 10 times per year. In 1892, the periodical changed its name to *Astronomy and Astro-Physics;* three years later, it split into *Popular Astronomy* (published until 1951) and *The Astrophysical Journal,* still published in 2013.

In the 1930s, two popular magazines were founded: *The Sky* (founded 1935), covering astronomical discoveries, and *The Telescope* (founded 1931), reporting on amateur telescope making. In 1941, they merged to become the monthly *Sky and Telescope,* covering both astronomical research and technology. In 1973, *Astronomy* magazine began monthly publication, aimed more toward a complete lay audience. Other U.S. magazines of astronomy still publishing in the early twenty-first century in the United States included *Mercury* (published by the Astronomical Society of the Pacific, online only in 2012), *The Griffith Observer* (published by the Griffith Observatory and Planetarium since 1937), and *Journal of the Antique Telescope Society* (published at least annually since 1993). Additional nineteenth- and twentieth-century U.S. astronomy magazines were shorter lived.

The invention of the planetarium projector in the 1920s by Zeiss in Germany brought the stars indoors and allowed dramatic shows to be produced for astronomical education and entertainment. The first planetarium in the United States was the Adler Planetarium in Chicago (1930); within five years, three other major planetariums—Fels in Philadelphia, Griffith in Los Angeles, and Hayden in New York City—had also opened. All continue to operate into the twenty-first century. Because of the difficulty and expense of obtaining planetarium projectors from Zeiss in the mid-twentieth century, in 1936 Philadelphia journalist and astronomy lecturer Armand Spitz designed a simple projector for schools and smaller planetarium domes that relied primarily on pinhole projection rather than lenses (King and Millburn, 1978). For decades, the stars, the planets and their motions, and other astronomical objects were optically projected onto a darkened large white dome from an elaborate machine in the center that was controlled manually. By the twenty-first century, planetariums started converting to computer-controlled digital projectors, which allow such special effects as flying through galaxies in space. Many planetariums have been directed by professional astronomers; in addition to public shows, they may also offer popular lectures or short courses in popular astronomy, observing, or telescope making.

[*See also* **Army Corps of Engineers, U.S.; Bowditch, Nathaniel; Cartography; Chemistry; Clocks and Clockmaking; Einstein, Albert; Franklin, Benjamin; Geological Surveys; Geology; Hale, George Ellery; Hubble, Edwin Powell; Hubble Space Telescope; Instruments of Science; Jefferson, Thomas; Maritime Transport; Meteorology and Climatology; Military, Science and Technology and the; Mitchell, Maria; National Academy of Sciences; National Aeronautics and Space Administration; Photography; Physics; Pickering, Edward Charles; Religion and Science; Rowland, Henry A.; Silliman, Benjamin; Space Program; Space Science;** *and* **Telegraph.**]

BIBLIOGRAPHY

Aubin, David, Charlotte Bigg, and H. Otto Sibum, eds. *The Heavens on Earth: Observatories and Astronomy in Nineteenth-Century Science and Culture*. Durham, N.C., and London: Duke University Press, 2010. Collection of articles focuses exclusively on European observatories and culture, but its exploration of the broader context of astronomy in nineteenth-century culture worldwide is also relevant to the United States.

Bartky, Ian R. *Selling the True Time: Nineteenth-Century Timekeeping in America*. Stanford, Calif.: Stanford University Press, 2000.

Bartusiak, Marcia. *Einstein's Unfinished Symphony: Listening to the Sounds of Space-Time*. New York: Berkley Books, 2000.

Beatty, J. Kelly, Carolyn Collins Peterson, and Andrew Chaikin. *The New Solar System*. Cambridge, U.K.: Cambridge University Press, 1999. Authoritative account of spacecraft contributions to knowledge of planetary science in the solar system.

Bell, Trudy E. "The Roles of Lesser-Known American Telescope Makers in 19th-Century American Observatories." *Journal of the Antique Telescope Society* 23 (2002): 9–18. Includes detailed tables of instruments in U.S. observatories and of American telescope makers.

Berendzen, Richard, Richard Hart, and Daniel Seeley. *Man Discovers the Galaxies*. New York: Science History Publications/Neale Watson, 1976.

Blumenthal, George R., Sandra M. Faber, Joel R. Primack, and Martin J. Rees. "Formation of Galaxies and Large-Scale Structure with Cold Dark Matter." *Nature* 311 (1984): 517–525.

Boslough, John, and John Mather. *The Very First Light: The True Inside Story of the Scientific Journey Back to the Dawn of the Universe*. New York: Basic Books, 2008. A collaboration between a Nobel laureate and a science journalist recounts the development and discoveries of COBE.

Butrica, Andrew J. *To See the Unseen: A History of Planetary Radar Astronomy* (NASA SP-4218). Washington, D.C.: National Aeronautics and Space Administration, 1996.

Carter, Bill, and Merri Sue Carter. *Latitude: How American Astronomers Solved the Mystery of Variation*. Annapolis, Md.: Naval Institute Press, 2003.

Cassidy, William A. *Meteorites, Ice, and Antarctica: A Personal Account*. Cambridge, U.K., and New York: Cambridge University Press, 2003.

Chauvenet, William. *A Manual of Spherical and Practical Astronomy.* New York: Dover Publications, 1960. Complete and unabridged reprint of the two-volume fifth edition (1891) of a textbook of enormous influence in American astronomy, widely credited as being the first to introduce German methods to U.S. astronomers. The second volume is on instruments, including history and theory of their use.

DeVorkin, David H. *Henry Norris Russell, Dean of American Astronomers.* Princeton, N.J., and Oxford: Princeton University Press, 2000.

Dick, Steven J. *Sky and Ocean Joined: The U.S. Naval Observatory, 1830–2000.* Cambridge, U.K., and New York: Cambridge University Press, 2003.

Dobson, John L., and Norman Sperling. *How and Why to Make a User-Friendly Sidewalk Telescope.* Berkeley, Calif.: Everything in the Universe, 1991.

Draper, Henry. *On the Construction of a Silvered Glass Telescope, Fifteen and a Half Inches in Aperture, and Its Use in Celestial Photography.* Washington, D.C.: Smithsonian Institution, 1864. Also reprinted by the Smithsonian in 1904 (no. 1459).

Duffner, Robert W. *The Adaptive Optics Revolution: A History.* Albuquerque: University of New Mexico Press, 2009.

Gingerich, O., ed. *The General History of Astronomy.* Vol. 4, Part A, *Astrophysics and Twentieth-Century Astronomy to 1950,* edited by Michael Hoskin. Cambridge, U.K.: Cambridge University Press, 1984. Half a dozen articles in this volume on key subjects were important in the preparation of this article, including "Variable Stars" (Helen Sawyer Hogg); "The Impact of Photography on Astronomy," (John Lankford); "The New Astronomy," (A. J. Meadows); "Telescope Building, 1850–1900," and "Building Large Telescopes, 1900–1950" (Albert Van Helden).

Hindle, Brooke. *David Rittenhouse.* Princeton, N.J.: Princeton University Press, 1964.

Hutchins, Roger. *British University Observatories 1722–1939.* Aldershot and Burlington, U.K.: Ashgate, 2008. Hutchins devotes more than 50 pages comparing British observatories to their competitors in other nations, including the United States.

Impey, Chris. *How It Began: A Time-Traveler's Guide to the Universe.* New York: W. W. Norton, 2012.

An up-to-date account of the findings of modern cosmology by a well-known observational astronomer.

Jones, Bessie Zaban, and Lloyd G. Boyd. *The Harvard College Observatory: The First Four Directorships.* Cambridge, Mass.: Belknap Press of Harvard University Press, 1971.

King, Henry C., with John R. Millburn. *Geared to the Stars: The Evolution of Planetariums, Orreries, and Astronomical Clocks.* Toronto: University of Toronto Press, 1978.

Klypin, Anatoly A., Sebastian Trujillo-Gomez, and Joel Primack. "Dark Matter Halos in the Standard Cosmological Model: Results from the Bolshoi Simulation." *Astrophysical Journal* 740 (2011): 102–119. Description of the Bolshoi cosmological simulation and comparison with observational results.

Kraemer, Robert S. *Beyond the Moon: A Golden Age of Planetary Exploration, 1971–1978.* Washington, D.C., and London: Smithsonian Institution Press, 2000.

Langley, Samuel Pierpont. *The New Astronomy.* Boston: Ticknor and Co., 1888. This popular book collected a series of articles Langley had previously written for *Century* magazine; the book was reprinted in 1891 by Houghton Mifflin Co.

Lankford, John, with Ricky L. Slavings. *American Astronomy: Community, Careers, and Power, 1859–1940.* Chicago and London: University of Chicago Press, 1997. Primarily a sociological analysis of influence among professional and amateur astronomers.

Law, Nicholas M., et al. "The Palomar Transient Factory: System Overview, Performance, and First Results." *Publications of the Astronomical Society of the Pacific* 121 (2006): 1395–1408. First paper describing how supercomputers assist nightly telescopic detection of supernovae in other galaxies.

Mather, Cotton. *The Christian Philosopher: A Collection of the Best Discoveries in Nature, with Religious Improvements.* London: Eman Matthews, 1721. Edited, with an introduction and notes, by Winton U. Solberg. Champaign: University of Illinois Press, 2000. Solberg's one hundred–page introduction examines the origins, sources, contents, reception, and influence of Mather's book.

Milham, Willis I. *Early American Observatories: Which Was the First Astronomical Observatory in America?* Williamstown, Mass.: Williams College,

1938. This 58-page booklet extensively explores 11 observatories erected and equipped between 1769 and 1840, including a temporary one by David Rittenhouse for the transit of Venus in 1769 and a permanent one at Rittenhouse's Philadelphia residence in 1786.

Osterbrock, Donald E., John R. Gustafson, and W. J. Shiloh Unruh. *Eye on the Sky: Lick Observatory's First Century.* Berkeley: University of California Press, 1988.

Percy, John R., and Joseph B. Wilson, eds. *Amateur–Professional Partnerships in Astronomy: Proceedings of a meeting held at University of Toronto, Toronto, Canada. 1–7 July 1999.* San Francisco: Astronomical Society of the Pacific Conference Series, 2000.

Preston, Richard. *First Light: The Search for the Edge of the Universe.* New York: Atlantic Monthly Press, 1987; reprinted New York: Random House, 1996. A history of the Palomar Observatory.

Rothenberg, Marc, and Thomas R. Williams. "Amateurs and the Society during the Formative Years." In *The American Astronomical Society's First Century.* Washington, D.C.: American Astronomical Society/American Institute of Physics, 1999.

Shy, Jeffrey R. "Early Astronomy in America: The Role of the College of William and Mary." *Journal of Astronomical History and Heritage* 5 (2002): 41–64. Shy presents strong circumstantial evidence from primary documents that a permanent observatory was constructed at the College of William and Mary in the spring of 1788, but destroyed before 5 November 1789 during the French occupation of the college—contending that it was possibly the first observatory in the nation, certainly the first at an educational institution.

Smith, Robert W. *The Space Telescope: A Study of NASA, Science, Technology, and Politics.* Cambridge, U.K., and New York: Cambridge University Press, 1989.

Stachurski, Richard. *Longitude by Wire: Finding North America.* Columbia: University of South Carolina, 2009. History of the American telegraphic method of determining longitude; some research was originally published in *Professional Surveyor* magazine.

Sullivan, Woodruff T. *Cosmic Noise: A History of Early Radio Astronomy.* Cambridge, U.K.: Cambridge University Press, 2009. A 574-page definitive history of the formative years of radio astronomy.

Theberge, Captain Albert E. *The Coast Survey 1807–1867.* Vol. I of *The History of the Commissioned Corps of the National Oceanic and Atmospheric Administration.* Last update 30 November 2007. Major historical work published online. http://www.lib.noaa.gov/noaainfo/heritage/coastsurveyvol1/CONTENTS.html (accessed 20 December 2012).

Waff, Craig B. "Astronomy and Geography vs. Navigation: Defining a Role for an American Nautical Almanac, 1844–1850." In *Proceedings, Nautical Almanac Office Sesquicentennial Symposium, U.S. Naval Observatory, March 3–4, 1999,* edited by Alan D. Fiala and Steven J. Dick, pp. 83–128. Washington, D.C.: U.S. Naval Observatory, 1999.

Warner, Deborah Jean, and Robert B. Ariail. *Alvan Clark & Sons: Artists in Optics.* Richmond, Va.: Willmann–Bell, in association with the National Museum of American History, Smithsonian Institution, 1995. Revised and expanded second edition of a classic 1968 work, with a forty-five-page biography of the Clarks and a two hundred–page directory of their telescopes and owners (plus a catalog of Alvan Clark's paintings). Widely regarded as a major reference.

Wilson, Curtis. *The Hill–Brown Theory of the Moon's Motion: Its Coming-to-Be and Short-Lived Ascendancy (1877–1984).* New York: Springer, 2010. Part biography, part mathematical analysis.

Yeomans, Donald K. "The Origin of North American Astronomy—Seventeenth Century." *Isis* 68 (1977): 414–425.

Trudy E. Bell

ATLANTIC CABLE

A cable telegraph link between the United States and Great Britain was conceived soon after the first telegraph systems were developed in those countries in the 1840s. Little progress was made until 1854, however, when the New York City paper merchant Cyrus W. Field (1819–1892) became the principal promoter of the project. With a group of his wealthy New York friends,

Field formed an American company to undertake the project. After completing construction of the Newfoundland–New York overland route and organizing the American Telegraph Company to control telegraph lines along the eastern seaboard, Field was unable to convince American investors to back the Atlantic cable. Accompanied by Samuel F. B. Morse, inventor of the first American telegraph, he traveled to London to raise the necessary funds by forming the Atlantic Telegraph Company (1856) and winning the support of the British government. Work began in August 1857, but the cable broke while being laid. Raising additional capital, Field completed the cable in August 1858 between Newfoundland and Ireland. (Transatlantic messages were then relayed to England via an England–Ireland cable.) People on both sides of the Atlantic celebrated the cable's success, and Queen Victoria sent a celebratory message to President James Buchanan, but after operating for only about a month, the cable failed. Unable to raise sufficient capital in America for a third attempt, especially after the outbreak of the Civil War, Field once again turned to British investors. The first permanent Atlantic cable, completed in 1866, was largely a British project, and British investors, cable companies, and engineers would continue to dominate the world's cable industry.

For countries on both sides of the Atlantic the undersea cable opened up a new era in global communications. Before the introduction of the Atlantic cable, telegraph messages between Europe and North America were largely dependent on shipping. Instead of waiting many days for news or information to arrive by ship from across the Atlantic, the Atlantic cable allowed for near-instantaneous communication. Ocean-spanning telegraph cables effectively broke down the traditional constraints of time and space on a global scale. The Atlantic cable was thus especially significant as a technology crucial in the development of the modern process of globalization.

[*See also* Internet and World Wide Web; Morse, Samuel F. B.; Technology; Telegraph; *and* Telephone.]

BIBLIOGRAPHY

Carter, Samuel. *Cyrus Field: A Man of Two Worlds.* New York: G. P. Putnam's Sons, 1968.

Coates, Vary T., and Bernard Finn. *A Retrospective Technology Assessment: Submarine Telegraphy: The Transatlantic Cable of 1866.* San Francisco, Calif.: San Francisco Press, 1979.

Hearn, Chester G. *Circuits in the Sea: The Men, the Ships, and the Atlantic Cable.* Westport, Conn.: Praeger, 2004.

Paul Israel; updated by Hugh Richard. Slotten

ATOMIC ENERGY COMMISSION

The Atomic Energy Commission (AEC), made up of five members appointed by the president of the United States, was created by Congress in 1946. Establishing the principle of civilian control of atomic energy, Congress assigned the new agency responsibility for developing and testing nuclear weapons and for encouraging peaceful uses of the new technology. David E. Lilienthal (1899–1981), a former director of the Tennessee Valley Authority, became its first chairman. As the Cold War progressed, the agency focused its resources on weapons, expanding the U.S. stockpile, and, after January 1950, undertaking a crash program to build a hydrogen bomb. The agency's military emphasis proved a source of frustration and disappointment for Lilienthal, whose interests lay in the nonmilitary uses of atomic energy. But Cold War tensions and the still-rudimentary state of the technology prevented major strides in civilian applications.

Although the AEC stood at the center of numerous controversies in its early years, the most divisive occurred in 1954 when it stripped J. Robert Oppenheimer, leader of the scientific effort to build the atomic bomb during World War II, of his security clearance for alleged Communist associations. The chairman of the AEC, Lewis L. Strauss, took the lead, for both personal and political reasons, in ending Oppenheimer's career as a scientific adviser to the AEC. At about the same time, the AEC's atmospheric testing of nuclear weapons set off a major public debate over the

health effects of the radioactive fallout they produced. The AEC's claims that fallout posed no
significant health hazard aroused much dissent,
permanently undermining the agency's credibility.

By the mid-1960s, the AEC was devoting increasing attention to promoting the peaceful uses
of atomic energy. Its encouragement of the nuclear power industry helped stimulate a boom in
reactor construction. But public faith in the AEC
was further shaken by its dual mandate to promote and regulate nuclear power. By the early
1970s, controversies over reactor safety, radiation
standards, environmental protection, waste disposal, and nuclear weapons proliferation had
severely damaged confidence in the AEC. In response to widespread criticism, Congress abolished the AEC in 1974, dividing its functions
between the Energy Research and Development
Administration (later a part of the Department of
Energy) and the Nuclear Regulatory Commission.

[*See also* **Environmentalism; Environmental
Protection Agency; Manhattan Project; Mili-
tary, Science and Technology and the; Nuclear
Power; Nuclear Regulatory Commission; Nu-
clear Weapons; Oppenheimer, J. Robert;** *and*
Public Health.]

BIBLIOGRAPHY

Hewlett, Richard G., and Francis Duncan. *Atomic
Shield, 1947–1952*. University Park: Pennsylva-
nia State University Press, 1969.
Hewlett, Richard G., and Jack M. Holl. *Atoms for
Peace and War, 1953–1961*. Berkeley: University
of California Press, 1989.
Walker, J. Samuel. *Containing the Atom: Nuclear Reg-
ulation in a Changing Environment, 1963–1971*.
Berkeley: University of California Press, 1992.

J. Samuel Walker

ATOMS FOR PEACE

Atoms for Peace was a program in which the U.S.
government sought to domesticate nuclear weapons and nuclear power for both the United States
and the international community. On 8 December 1953, in a speech before the United Nations
(UN) General Assembly, U.S. President Dwight
D. Eisenhower articulated a profound shift in U.S.
nuclear policy. Rather than restrict nuclear knowledge and technology to the few nations that had
so far developed atomic and thermonuclear weapons, Eisenhower proposed a framework in which
any nation seeking such knowledge might enter
into a set of agreements for the peaceful development of nuclear technology. In language that
matched both the perceived urgency and the gathering threat of the atomic armaments race, Eisenhower (1953) refused "to confirm the hopeless
finality of a belief that two atomic colossi are
doomed malevolently to eye each other indefinitely across a trembling world." In the remainder
of the speech, Eisenhower proposed a new UN
agency, which became the International Atomic
Energy Agency, to which the nuclear powers
would donate fissionable materials and nuclear
technology. In turn, nations agreeing to an array of
safeguards could apply to the new agency for permission to build a reactor using the agency's
supply of fissionable material. Often seen as part
of American propaganda, Atoms for Peace was a
wonderful example of the unexpected consequences of seemingly benevolent action.

Atoms for Peace was a mechanism for creating
a market for nuclear power, but it was also part of
the administration's transformation of American
strategy, the New Look. In this paradigmatic shift,
Eisenhower countered the Soviet advantage in
conventional weapons and troops with nuclear
weapons. Atoms for Peace would use only a tiny
fraction of the vast amount of fissionable materials
that the United States was then producing for its
thermonuclear arsenal. Despite claims that atomic
energy would make electricity vastly cheaper,
there was no demand among U.S. power companies for nuclear reactors. Instead, the U.S. government had to produce a standard reactor design
and provide utilities with vast subsidies to even
consider building a commercial reactor. Establishing a market for nuclear power required more
than subsidies; it required a massive change in
the laws related to nuclear weapons and the de-
classification of much research on the design and

engineering of nuclear reactors. An August 1955 UN Conference on the Peaceful Uses of Atomic Energy provided the United States with a platform to display its newly declassified knowledge and technology. Fifty thousand visitors took in the conference's highlight—the U.S. display of a working swimming pool reactor identical to those at Oak Ridge and operated by Union Carbide and Carbon for the U.S. Atomic Energy Commission. By August 1955, the United States had negotiated 24 bilateral agreements with nations seeking nuclear reactors for either research or power; by 1965 the number had increased to 39.

Among the countries acquiring nuclear materials and technology within the Atoms for Peace framework were India and Pakistan, both of which used the technology as foundations for their own development of nuclear weapons. Although designed as a program to halt the proliferation of nuclear weapons, nations could easily subvert the programs' safeguards and monitoring programs. In the early twenty-first century it is no small irony that we still live with the legacy of Atoms for Peace.

[See also **Atomic Energy Commission; Military, Science and Technology and the; Nuclear Power; Nuclear Regulatory Commission;** *and* **Nuclear Weapons.**]

BIBLIOGRAPHY

Balogh, Brian. *Chain Reaction: Expert Debate and Public Participation in American Commercial Nuclear Power, 1945–1975.* New York: Cambridge University Press, 1991. The best available account of how the U.S. nuclear power industry came into existence.

Eisenhower, Dwight D. "Atoms for Peace." Address by Mr. Dwight D. Eisenhower, President of the United States of America, to the 470th Plenary Meeting of the United Nations General Assembly, 8 December 1953. http://www.iaea.org/About/atomsforpeace_speech.html (accessed 9 April 2012).

Krige, John. "Atoms for Peace, Scientific Internationalism, and Scientific Intelligence." *Osiris* 21 (2006): 161–182. An excellent starting place for understanding the complex legacies of Atoms for Peace.

Pilat, Joseph E., ed. *Atoms for Peace: A Future after Fifty Years?* Washington, D.C.: Woodrow Wilson Center Press, 2006. This edited collection contains a copy of Eisenhower's Atoms for Peace speech.

Michael Aaron Dennis

AUDUBON, JOHN JAMES

(1785–1851), artist and naturalist famed for his striking color portraits of North American birds and mammals. Born in Les Cayes, Santa Domingo (now Haiti), the illegitimate son of a French naval officer, Jean Audubon, and one of his mistresses, Jeanne Rabin, Audubon grew up in France. With little formal schooling, the youthful Audubon remained free to pursue his budding interest in the outdoors. Fearful of conscription during the Napoleonic Wars, he fled to the United States in 1803 and took up residence at Mill Grove, a 284-acre farm on the outskirts of Philadelphia, Pennsylvania, that his father had purchased several years earlier. Although an attempt to profitably mine lead on the property failed, Audubon became increasingly adept at identifying local birds and regularly enjoyed the company of one of his neighbors, Lucy Bakewell. In 1807 he established a general store in frontier Kentucky, and when that business seemed on secure footing, he married Lucy. Five years later Audubon became an American citizen during the patriotic fervor ignited by the War of 1812. Economic crisis in the trans-Appalachian West, the result of wartime embargoes of European goods, led to a string of business failures, and Audubon found himself bankrupt and in debtor's prison in 1819 (Rhodes, 2004).

For the next several years Audubon held a variety of positions, including a stint as a taxidermist at the Western Museum in Cincinnati, while pursuing his dream of producing portraits depicting all the birds of North America. Traveling widely, he filled his folio with vivid watercolors of the avifauna he encountered, including more than two dozen species that proved new to science. In 1824, he ventured back to Philadelphia and two years later to England, to seek a publisher for his growing body of work. He found success with the London engraver Robert Havell Jr., with whom he would collaborate for more than a decade to produce the elegant four-volume, double-elephant folio edition of *Birds of America* (1827–1838), printed on 39 by 26 inch paper.

Breaking with the conventions of natural history illustration of his day, Audubon's masterpiece featured vivid, life-size color engravings of 435 species in their natural habitats (Blum, 1993). Although critics charged that he plagiarized from the Scottish American artist-naturalist Alexander Wilson and produced overly anthropomorphic depictions of birds, the response to his work proved overwhelmingly positive. In 1842 he published a popular octavo version of *Birds of America* and later, with his son John Wodehouse Audubon and John Bachman, *The Viviparous Quadrupeds of North America* (1845–1848), although this publication never enjoyed the same critical acclaim or popular success as his work on birds.

Audubon's widely reproduced bird portraits not only garnered him much fame in America and abroad but also helped foster a more sympathetic attitude toward their subjects. Not surprisingly, late nineteenth-century promoters of an emerging wildlife protection movement adopted his name for a series of local, state, and national organizations devoted to rescuing native birds from commercial exploitation, widespread population decline, and possible extinction. The National Audubon Society, first organized in 1905 as the National Association of Audubon Societies, remains a potent force for wildlife protection and appreciation to this day.

[*See also* **Biological Sciences; Environmentalism; Fish and Wildlife Service, U.S.;** *and* **Zoology.**]

BIBLIOGRAPHY

Blum, Ann Shelby. *Picturing Nature: American Nineteenth-Century Zoological Illustration.* Princeton, N.J.: Princeton University Press, 1993. A masterful study that places Audubon's artistic and scientific achievement in broader context.

Rhodes, Richard. *John James Audubon: The Making of an American.* New York: Alfred A. Knopf, 2004.

Mark V. Barrow Jr.

AUTISM

Autism has never been free from controversy. Throughout its relatively short history, it has attracted seemingly endless public and professional interest, yet in the early twenty-first century there was still no consensus about even the most fundamental questions of etiology, diagnosis, prognosis, and treatment. In 2012, experts considered it a spectrum of neurological developmental disorders with multiple, complex causes involving interactions between genes and environments. But even that statement is contentious. Indeed, autism is perhaps one of the few psychiatric classifications to have repeatedly resisted processes of medicalization. Although physicians may diagnose children or prescribe pharmaceuticals to treat peripheral symptoms, therapies are largely carried out by paraprofessionals and parents and overseen by educators and psychologists. Further, parents and people with autism themselves have shaped treatment and research in unusual ways, serving not only as political lobbying outfits, but also as recognized experts. The history of autism thus reflects less the forward march of medical knowledge than shifting institutional arrangements and ethical commitments to identifying and treating human difference and disability (Eyal et al., 2010).

The term autism, based on the Greek *autos* ("self"), was first introduced by Swiss psychiatrist Eugen Bleuler in 1908 to describe states of social withdrawal among adult schizophrenic patients. Only in 1943 did autism become a syndrome when American psychiatrist Leo Kanner published a landmark paper reporting 11 cases of children with "autistic disturbances of affective contact" (Grinker, 2007). In 1944, Austrian pediatrician Hans Asperger independently employed Bleuler's term in his thesis using four cases to describe what he called "autistic psychopathy." The very fact that the syndromes they proposed were considered distinct entities for half a century is testament to autism's tumultuous history.

Although autism remained an extremely rare disorder for the next four decades, Kanner speculated that it was widespread but hidden behind a diagnosis of "mental retardation." As children considered "retarded" began to be institutionalized in the postwar years with increasing frequency, Kanner and others thought a large contingent of them had undiagnosed autism. During this period, however, the salient medicolegal distinction was between mental retardation and mental illness, with autism assimilated to the childhood psychoses and its causes sought in the early family environment. Regnant theories considered autism the result of pathological mother–child attachments during infancy. Correspondingly, treatment approaches in the 1950s and 1960s modeled therapy after an ideal of "good parenting" and situated the therapist as a kind of substitute parent (Nadesan, 2005).

A sea change in the history of autism began in the mid-1960s with the rise of parent activism. Bernard Rimland, an experimental psychologist and father of an autistic boy, published the monograph *Infantile Autism* in 1964. The book had a major impact for several reasons. First, reviewing a range of extant research, Rimland argued persuasively for neurological and genetic—not psychological—origins of autism. Second, the book contained a checklist that parents filled out and returned to Rimland. Using these contacts, he cofounded the American National Society for Autistic Children (NSAC) in 1965, which became a crucial vehicle for promoting research, exchanging experiences, and forging durable alliances among parents and sympathetic experts. For instance, two NSAC parents founded the *Journal of Autism and Childhood Schizophrenia* (now the *Journal of Autism and Developmental Disorders*), a leading venue for publishing scientific research about autism.

Also in the mid-1960s, Rimland and other NSAC parents began working with psychologists such as O. Ivar Lovaas and Eric Schopler to develop and disseminate behavioral treatments for autism. These methods included parents as "co-therapists" and turned homes and schools into experimental therapeutic environments. Together, parents and professionals ultimately inverted the prevailing hierarchy of blame and expertise. Initially stigmatized and silenced by the psychiatrists who considered them pathological agents, parents came to be considered "experts on their own children" (Eyal et al., 2010; Silverman, 2011).

These methods and the parents and professionals who practiced them were well placed by the 1970s, when the deinstitutionalization of mental retardation began in earnest. Outside the institution, in early-intervention and special-education facilities, the distinction between mental illness and mental retardation decreased in importance. The focus had shifted to the specific delays and needs of individual children who were all treated with the same arsenal of behavioral and educational therapies. No longer hidden behind the appearance of mental retardation, autism could now be considered comorbid with it. These shifts were codified in successive editions of the *Diagnostic and Statistical Manual* (DSM) of the American Psychiatric Association. DSM-III (1980) reclassified autism as a "pervasive developmental disorder" (PDD) rather than a form of "childhood schizophrenia." DSM-III-R (1987) linked autism and mental retardation as combinable diagnoses. DSM-IV (1994) added Asperger's disorder to form the autism spectrum. In the rewriting of these criteria, a major role was played by the American National Society for Autistic Children, as well as by Lorna Wing, British psychiatrist and mother of an autistic girl.

The last chapter in the contemporary history of autism belongs to autistic "self-advocacy." Although autism was initially a rare diagnosis reserved for very severely affected children, some individuals so diagnosed learned to communicate, reflect on their experience, and advocate for others. Temple Grandin, Donna Williams, Daniel Tammet, and a number of others have written autobiographical accounts that have created new vocabularies and images for thinking about autism, thereby making it more intelligible for parents and experts, while also expanding the image of autism to include previously undiagnosed and high-functioning individuals (Hacking, 2009).

People with autism have been at the forefront of the neurodiversity movement, which seeks to resituate autism as a natural form of human neurological diversity.

The Centers for Disease Control estimate that the prevalence of autism in the United States increased from four per ten thousand children in 1989 to about one hundred per ten thousand in 2009. Similar increases have been estimated for the United Kingdom, Canada, Japan, Sweden, Denmark, and China. Some cast these increases as an "epidemic" caused by environmental toxicity or vaccination. Indeed, the vaccine controversy garnered massive media coverage. Groups of parents, physicians, and researchers believed that measles, mumps, and rubella (MMR) inoculations or mercury in a vaccine preservative called thimerosal were causing autism. Many parents, especially those associated with the Defeat Autism Now! movement, experimented with alternative medical treatments to repair vaccine injuries or to rid the body of environmental toxins (Silverman, 2011). Although a series of epidemiological studies found no link between MMR or thimerosal and autism, many parents opted not to vaccinate their children, and outbreaks of measles have been reported (Offit, 2008). Most social scientists and epidemiologists explain the increase in autism diagnoses as largely the result of changes in diagnostic criteria, greater availability of services, and increased awareness of autism among experts and parents. Although the possibility cannot be ruled out that *de novo* genetic mutations and gene–environment interactions had increased the total number of developmentally disabled children, it seems hard to dispute that these children were now being given the diagnosis of autism and not a different diagnosis because of the changes to diagnostic criteria and awareness and ultimately owing to the rise of parent activism and the institutional changes described above.

[*See also* **Centers for Disease Control and Prevention; Disease; Genetics and Genetic Engineering; Journals in Science, Medicine, and Engineering; Medicine; Mental Health Institutions; Mental Illness; Pharmacology and Drug Therapy; Psychiatry; Psychological and Intelligence Testing; Psychology; Psychopharmaceutical Drugs; Research and Development (R&D);** *and* **Social Sciences.**]

BIBLIOGRAPHY

Centers for Disease Control and Prevention. http://www.cdc.gov/mmwr/preview/mmwrhtml/ss5810a1.htm (accessed 17 October 2011.

Eyal, Gil, Brendan Hart, Emine Onculer, Neta Oren, and Natahsa Rossi. *The Autism Matrix: The Social Origins of the Autism Epidemic.* Cambridge, U.K.: Polity Press, 2010. Argues that the recent rise in autism should be understood an "aftershock" of the deinstitutionalization of mental retardation in the mid-1970s. This entailed a radical transformation not only of the institutional matrix for dealing with developmental disorders of childhood, but also of the cultural lens through which we view them. It opened up a space for viewing and treating childhood disorders as neither mental illness nor mental retardation, neither curable nor incurable, but somewhere in between. It also opened up a space for parent activism to forge a network of expertise that bypassed the psychiatric establishment at a time when it was still wedded to psychogenic theories.

Feinstein, Adam. *A History of Autism: Conversations with the Pioneers.* Malden, Mass.: Wiley–Blackwell, 2010. The author interviewed major actors in the history of autism, experts as well as parent activists. The book thus reconstructs the history of the discovery of autism by Kanner and Asperger, the emergence of parental activism, and major turning points in research and therapy from the point of view of the actors involved.

Grinker, Roy Richard. *Unstrange Minds: Remapping the World of Autism.* New York: Basic Books, 2007. Counters the notion that there is an autism epidemic through an account of broader shifts in American psychiatry and special education. The author interviewed parents and professionals in several countries (India and South Africa, among others) and provides several suggestive cross-cultural examples and makes illuminating comparisons with diagnostic practices in South Korea and France.

Hacking, Ian. "Autistic Autobiographies." *Philosophical Transactions of the British Royal Society, B* 364 (2009): 1467–1473. Analyzes several memoirs written by individuals on the autistic spectrum. Makes the argument that this new genre of "autistic autobiographies" represents a turning point in the history of autism because it provides others, especially parents and experts, with a language to represent autistic subjectivity as "thick" and meaningful and thus to facilitate communication between autistics and others.

Nadesan, Majia Holmer. *Constructing Autism: Unravelling the "Truth" and Understanding the Social.* London: Routledge, 2005. A historical account of the discovery of autism and the subsequent epidemic. The discovery of autism by Kanner and Asperger is related to broad changes in the surveillance and classification of childhood disorders. The autism epidemic is explained by reference to the rise of biogenetic and cognitive models and theories in psychology and medicine.

Offit, Paul A. *Autism's False Prophets: Bad Science, Risky Medicine, and the Search for a Cure.* New York: Columbia University Press, 2008. A critical account of the major controversies surrounding autism (MMR, thimerosal, facilitated communication, and secretin). Written in an accessible style by a major vaccine expert. Presents a scathing critique of the medical professionals, activists, lawyers, and journalists that propelled the antivaccine movement (for a partly contrasting view, see Silverman, 2011).

Silverman, Chloe. *Understanding Autism: Parents, Doctors, and the History of a Disorder.* Princeton, N.J.: Princeton University Press, 2011. The first to analyze two key archives: the Bruno Bettelheim papers and the Amy L. Lettick papers. Provides a meticulously detailed and thoughtful analysis of parents' historical and contemporary involvement in scientific research and medical and behavioral treatments. Offers an important counterweight to popular and scholarly portrayals of parents as desperate and irrational.

Gil Eyal and Brendan Hart

AUTOMATION AND COMPUTERIZATION

Automation and computerization are related but distinct concepts that have their origins in production engineering but are applied to many forms of economic, social, and political activity. Both are employed in efforts to reduce the costs of labor and to make institutions more responsive to market forces. Both have substantially altered the labor force and changed the way in which organizations make decisions.

Automation. Traditionally, automation is a three-stage activity that strives to create a process to produce goods with as little human intervention as possible. The first step analyzes the human labor in a production process and breaks the entire process down into a collection of individual steps. The second stage replaces the human labor in some or all of those steps with machine labor. The third stage of the activity employs another machine to coordinate or control all steps in the process.

The first step of automation was identified by Adam Smith in the first chapter of *The Wealth of Nations* (1776). Smith described the process of making pins and divided it into the steps of straightening wire, cutting wire into short segments, sharpening one end of the segment, and so forth. The "important business of making a pin is, in this manner, divided into about 18 distinct operations," he explained.

Smith also noted the second step of automation, the extent to which "labour is facilitated and abridged by the application of proper machinery." In Smith's time, most of the "proper machinery" was hand controlled and hence did not automate a production process. Perhaps the best example of an automatic machine was Jesse Rasmussen's dividing engine (1773), which engraved regular markings on a circular piece of metal. It was followed by Joseph Marie Jacquard's controlled loom (1801) and Thomas Blanchard's duplicating lathe (1822).

The third step of automation, which moved from automatic machinery to automatic processes, was described by Charles Babbage in his book *On the Economy of Machines and Manufactures* (1832). After reviewing Smith's analysis of pin-making, he described how each step of the process could be controlled by a single mechanism. This mechanism is "highly ingenious in point of contrivance," he wrote, "and, in respect to its economical principles, will furnish a strong and interesting contrast with the manufacture of pins by the human hand."

Few products are as simple as pins and hence few were produced by truly automated processes in the nineteenth century. However, the new products of that era were produced by increasingly sophisticated divisions of labor, more complicated machines, and a few devices that could control certain aspects of production. The bicycle, which was first mass produced in the 1880s, marked an important step in the process of automation, according to historian David Hounshell. The workers were able to use a variety of automatic machines to manufacture bicycle parts but they were unable to solve "a fundamental problem" of automation: the final assembly of complex goods.

The problem of final assembly was solved by the automobile manufacturer Henry Ford. Ford developed an assembly process that relied heavily on human labor but coordinated the work with an automated conveyor system or assembly line. This system carried partially assembled automobiles past workers, who would add one part to the final product. "When we call the new system 'automatic' or 'mechanical,'" explained the twentieth-century management consultant Peter Drucker, "we do not mean that the machines have become automatic or mechanical. What has become automatic and mechanical is the worker."

Scientific Management.

At the start of the twentieth century, Ford was one of many engineers thinking about how to incorporate human labor into a tightly controlled manufacturing process. The most prominent of these engineers was a group led by Frederick Winslow Taylor. Taylor and his followers promoted a technique that they called "scientific management." This technique focused on the first two steps of automation: the division of labor and the application of proper tools. Taylor claimed that this technique identified the "one right way" for doing any kind of job.

More than any other engineer, Taylor laid the foundation for the third step of automation by restricting the freedom and judgment of the worker. In scientific management, the workers did the labor but the engineers controlled the manufacturing process. Taylor argued that such an approach would create "an almost equal division of work and responsibility between management and workmen."

Critics of Taylor have argued that scientific management has never created an equal partnership but has undermined the position of the worker. Writing 80 years after Taylor, Harry Braverman argued that such techniques created deskilled jobs, "labor from which all conceptual elements have been removed along with them most of the skill, knowledge and understanding of the production process."

Computerization.

Strictly defined, computerization is a specialized form of automation that developed after the introduction of the stored program computer in 1946. In this specialized form, computers can operate either as automatic tools that perform one step of a production process or as control devices that manage the entire production process.

Because of their ability to manipulate symbols, computers found applications in jobs that had previously been considered the domain of skilled human workers. Beginning in 1952, computers replaced employees who managed information. They handled tasks that included bookkeeping, inventory tracking, scheduling, and typesetting. The introduction of the microprocessor in the early 1970s allowed computers to be placed in small, portable objects that could move beyond the offices and plants that had traditionally held computing equipment. The expansion of digital communications in the 1980s and 1990s allowed computers to coordinate the activities of organizations that were dispersed across a large geographic area.

Because of the flexible nature of computers, these machines could be simultaneously an automatic tool and an element of a large system that controlled production. A single machine could do bookkeeping tasks, analyze production figures, and prepare a production plan for the organization

Computerization has tended to restrict the actions of certain kinds of workers and strengthened the position of those who had designed and implemented the production system. However, computerization also tended to elevate a third class of workers who understand the production system and know how to interpret the information that comes from computers. Management consultant Peter Drucker has identified them as "knowledge workers."

The effects of computerization can be seen in as simple an organization as a restaurant wait staff. Traditionally, restaurants had a uniform and undifferentiated staff of waiters who dealt with customers. This staff would consist of a single manager and a group of people, equal in status and pay, who handled all the interactions with the customers: greeting them at the table, taking the order, getting the order properly filled by the kitchen, delivering food, reconciling the bill, and clearing the dishes.

Wait staffs were computerized by the introduction of order management software. This software kept track of the food ordered by each table and the bill incurred by each table. It reduced a substantial fraction of waiters' work to a series of simple tasks that included delivering the proper food to a table and clearing the plates when the diners were done with their meals. These tasks could be done by a staff with far less skill than the traditional waiter; hence, they were assigned to a group of low-wage workers who did not even get the title "waiter" but were called something like "food runner."

However, the system still needed a staff of knowledge workers, employees who would interact with customers, operate the system, manage the food runners, and make judgments about specific events. This group retained the title of waiter and had more compensation than the food runners because they were responsible for a larger number of customers and had greater authority. In a pattern repeated in organization after organization, this new division of labor created a few positions that required sophisticated skills and a much larger number of jobs that were directed by others through computer technology and hence required fewer skills.

Computerization has not only changed production but also restructured organizational management. As computers and computer software moved into every part of company operations, they turned some managerial tasks into routine activities that could be handled by less-skilled employees. At the same time, they also created highly sophisticated tasks that favored individuals who could make refined decisions from large amounts of data.

The use of both computerization and automation expanded widely during the second half of the twentieth century. It tended to increase the productivity of workers but also had the effect of regimenting labor, disciplining management, and favoring individuals who have the skill and understanding to make judgments about large, complicated systems. Both computerization and automation have been blamed for the decline of manufacturing jobs as well as some white-collar jobs in the United States.

Impact of Automation and Computerization. Many writers have made the claim that automation and computerization would end unskilled labor. Human "labor is being systematically eliminated from the production process," wrote the journalist Jeremy Rifkin in 1995. "Within less than a century, 'mass' work in the market sector is likely to be phased out in virtually all of the industrialized nations of the world."

Although both automation and computerization have certainly altered the labor market by increasing the demand for certain labors and decreasing demand for others, they have not entirely eliminated manual or unskilled labor. Instead, they have identified four factors that influence how labor can be replaced by capital in the form of machines, computers, software, and disciplined processes. In 1776, Adam Smith noted that this replacement is controlled by the demand for the good or service being produced. Goods or services with large markets are more likely to have more sophisticated divisions of labor and hence are easier to computerize or automate.

However, the extent of automation and computerization is also dependent upon the cost of the three steps of automation, the cost of analyzing a production process, developing tools, and creating a procedure that automatically controls the process. In some cases, the cost of producing machines and control proves to be more than simply moving the work to a lower-cost labor market. In other cases, the analysis of the labor process reveals that the work involves tasks that are difficult to adapt to a computer or a machine. The most common forms of such work are tasks that require human judgment, compromises based on broad knowledge of the process, and an understanding of all the interests involved.

Rather than bringing an end to mass labor, automation and computerization have produced a dynamic and unstable labor market. Automation has created large demands for certain skills in one year and made those same skills obsolete a decade later. Computerization has spread that same phenomenon to workers who process information and make decisions about organizations.

[*See also* **Bicycles and Bicycling; Computer Science; Computers, Mainframe, Mini, and Micro; Engineering; Ford, Henry; Motor Vehicles;** *and* **Technology**]

BIBLIOGRAPHY

Babbage, Charles. *On the Economy of Machinery and Manufactures*. London: C. Knight, 1832.

Braverman, Harry. *Labor and Monopoly Capital*. New York: Monthly Review Press, 1974.

Cortada, James, *The Digital Hand*, Vols. 1–3. Oxford: Oxford University Press, 2004, 2006, 2008.

Drucker, Peter F. *New Society*. New York: Harper, 1950.

Hounshell, David. *From the American System of Mass Production 1800–1932*, p. 190. Baltimore: Johns Hopkins University Press, 1984.

Kanigel, Robert. *The One Best Way: Frederick Winslow Taylor and the Enigma of Efficiency*. New York: Viking, 1997.

Levy, Frank, and Murnane, Richard. *The New Division of Labor: How Computers Are Creating the Next Job Market*. New York: Russell Sage Foundation, 2004.

Rifkin, Jeremy. *The End of Work: The Decline of the Global Labor Force and the Dawn of the Post-Market Era*. New York: G. P. Putnam's Sons, 1995.

David Alan Grier

B

BABY AND CHILD CARE

The most widely read and influential child-rearing manual of the second half of the twentieth century. Written during World War II by the pediatrician Dr. Benjamin Spock, with the assistance of his first wife, Jane Cheney Spock, *The Common Sense Book of Baby and Child Care* by the mid-1990s had more than 46 million copies in six editions (1946, 1957, 1968, 1977, 1985, and 1992). The book's success owed much to Spock's readable prose; his precise and accessible advice; and his effort, not always successful, to convince anxious parents that good child rearing was a matter of "common sense."

Despite its popularity, *Baby and Child Care* proved controversial after 1968, when the Reverend Norman Vincent Peale and others charged Spock (then a leader of the anti–Vietnam War peace movement) and his "permissive" approach to child rearing with having produced a generation of spoiled, radical youths. In the 1970s, Gloria Steinem led feminists in casting the Freudian Spock as an oppressor of women. With good reason, Spock disputed the charge of permissiveness, insisting that *Baby and Child Care* championed a flexible approach encompassing either "moderate strictness" or "moderate permissiveness." Spock met some of the feminist objections in 1977 and later editions by adopting nonsexist language, advocating an expanded parenting role for fathers, and acknowledging the existence of two-career families and their need for day care.

Scholars generally rejected the view that *Baby and Child Care* was responsible for the upheavals of the 1960s. The historian William Graebner presented Spock as a social engineer whose "democratic" approach to child rearing, based on Progressive-Era educational theory, reflected interwar anxieties about aggression and totalitarianism. Whereas Michael Zuckerman argued that Spock's advocacy of a confident parent presiding over frictionless parent–child relationships prepared children to be cooperative and amicable adults in a postwar corporate order, Nancy

Pottishman Weiss contended that Spock's advocacy of unfailing maternal confidence was burdensome and counterproductive for many women readers.

[*See also* Childbirth; Pediatrics; *and* Social Sciences.]

BIBLIOGRAPHY

Apple, Rima D. *Perfect Motherhood: Science and Childrearing in America.* New Brunswick, N.J.: Rutgers University Press, 2006.

Graebner, William. "The Unstable World of Benjamin Spock: Social Engineering in a Democratic Culture, 1917–1950." *Journal of American History* 67 (December 1980): 612–629.

Weiss, Nancy Pottishman. "Mother, the Invention of Necessity: Dr. Benjamin Spock's *Baby and Child Care.*" *American Quarterly* 29 (Winter 1977): 519–546.

Zuckerman, Michael. "Dr. Spock: The Confidence Man." In *The Family in History*, edited by Charles E. Rosenberg, pp. 179–207. Philadelphia: University of Pennsylvania Press, 1975.

William Graebner

BACHE, ALEXANDER DALLAS

(1806–1867), geophysicist, educator, and science administrator. A great-grandson of Benjamin Franklin, Bache received specialized scientific training from the U.S. Military Academy at West Point, where he graduated first in his class in 1825. After serving in the Corps of Engineers for two years, he was appointed professor of natural philosophy and chemistry at the University of Pennsylvania. He conducted not only basic physical and geophysical research but also practical research, including, most importantly, a federally supported investigation of steam-boiler explosions on steamships.

During the period from 1836 to 1842, Bache continued to pursue scientific research but also became heavily involved in educational reform. In 1836, he resigned his post at the University of Pennsylvania to accept a position as president of Girard College, a new institution in Philadelphia. Because the school's opening was delayed, Bache assisted in the reorganization of Philadelphia public schools and, from 1839 to 1842, served as superintendent of the Central High School in Philadelphia, where he stressed a curriculum that included advanced scientific and technical training.

In 1843, Bache was appointed superintendent of the U.S. Coast Survey. Under Bache's command, the Coast Survey not only supported the country's dramatic economic and commercial growth but also became its preeminent scientific institution. To perform the practical task of mapping with great exactness and precision, Bache's institution completed extensive scientific research, from astronomical and geophysical studies to observations of the Gulf Stream and studies of microscopic animals from the ocean bottom. He successfully mobilized popular support for the survey and convinced politicians that the scientific research was compatible with commercial interests.

With support from elite scientists including Joseph Henry, Benjamin Peirce, and Louis Agassiz, Bache became the major leader of the antebellum American scientific community. Bache and his supporters sought to raise the standards of science in the United States by cultivating government support and by emulating European patterns. Bache had met many leading scientists in Europe during a tour he completed in connection with his work organizing Girard College. Bache utilized the resources of the Coast Survey to participate in "world-class" scientific research and to influence or dominate other prominent institutions in the United States, including the American Association for the Advancement of Science and the Smithsonian Institution.

The Coast Survey under Bache supported more scientists—either directly or indirectly, through his practice of employing consultants—than any other American institution. As the most prominent scientific institution in this period, the Coast Survey helped shape a geographical style for nineteenth-century American science. One of the first scientist-entrepreneurs in the United

States, Bache helped to recast modern science from an individualistic enterprise to a highly structured organized activity dependent on government support.

During the Civil War, Bache served as vice president of the U.S. Sanitary Commission and provided technical and scientific advice to the army and navy. He also played the major role in persuading Northern members of Congress to pass new legislation in 1863 establishing the National Academy of Sciences. Bache became the first president of the new institution.

[*See also* **Agassiz, Louis; American Association for the Advancement of Science; Army Corps of Engineers, U.S.; Astronomy and Astrophysics; Cartography; Chemistry; Engineering; Franklin, Benjamin; Geography; Geophysics; Henry, Joseph; Higher Education and Science; High Schools, Science Education in; Military, Science and Technology and the; National Academy of Sciences; Oceanography; Science: Overview; Shipbuilding; Smithsonian Institution;** *and* **Steam Power.**]

BIBLIOGRAPHY

Slotten, Hugh Richard. "The Dilemmas of Science in the United States: Alexander Dallas Bache and the U.S. Coast Survey." *Isis* 84 (1993): 26–49.
Slotten, Hugh Richard. *Patronage, Practice, and the Culture of American Science: Alexander Dallas Bache and the U.S. Coast Survey*. New York: Cambridge University Press, 1994.

Hugh Richard Slotten

BAIRD, SPENCER FULLERTON

(1823–1887) naturalist, museum director, and science administrator. Born in Reading, Pennsylvania, Baird was the son of Samuel Baird, a lawyer, and Lydia McFunn Biddle. With his brother Will, he developed an early interest in natural history, corresponding with John James Audubon and other naturalists. After receiving a BA in 1840 and an MA in 1843 from Dickinson College, Baird was appointed professor of natural history at Dickinson in 1846. That same year, he married Mary Helen Churchill; they had one daughter, Lucy Hunter Baird. As he published research and amassed a large natural history collection, Baird became a noted ornithologist and ichthyologist.

In 1850 Baird was appointed assistant secretary and first curator of the U.S. National Museum at the newly founded Smithsonian Institution. He served as assistant to the first secretary, physicist Joseph Henry, who focused on creating research laboratories rather than the development of a large museum. Baird initially managed the International Exchange Service, which disseminated scientific publications to scholarly institutions both in the United States and abroad. During his 37-year career at the Smithsonian, however, Baird primarily devoted himself to establishing a great national museum in the nation's capital, amassing over 2.5 million objects during his tenure. Baird set new standards for modern systematic ornithology, notably in *The Birds of North America* (1858), and quickly accepted Charles Darwin's theory of evolution. He established an international network of natural history collectors, ensured that naturalist/collectors accompanied the western exploring expeditions, and served as mentor to a generation of American naturalists. During the 1870s and 1880s, he oversaw preparation of Smithsonian and government exhibits at international expositions, setting a new standard for displays at world's fairs and museums. His protégé, George Brown Goode, became the leading American museum theorist of the nineteenth century. A highlight of Baird's career was the opening of the first U.S. National Museum building in 1881 on the National Mall.

Baird served as the second secretary of the Smithsonian from 1878 to 1887. He also oversaw the creation of the Bureau of American Ethnology and the National Zoological Park as part of the Smithsonian. In addition to his Smithsonian duties, Baird was concerned about the decline in

American fisheries and served concurrently as the first Commissioner of Fish and Fisheries from 1871 to 1887, establishing the U.S. Fish Commission Laboratory at Woods Hole, Massachusetts, as a center for marine research.

[*See also* **Academy of Natural Sciences of Philadelphia; American Museum of Natural History; Audubon, John James; Environmentalism; Evolution, Theory of; Fish and Wildlife Service, U.S.; Fisheries and Fishing; Henry, Joseph; Museums of Science and Natural History; Smithsonian Institution;** *and* **Zoos.**]

BIBLIOGRAPHY

Allard, Dean C. *Spencer Fullerton Baird and the U.S. Fish Commission. A Study in the History of American Science.* New York: Arno Press, 1978. Focuses on Baird's work in ichthyology and contains a detailed list of archival collections, based on Allard's 1967 dissertation.

Deiss, William A. "Spencer F. Baird and his Collectors." *Journal of the Society for the Bibliography of Natural History* 9, no. 4 (1980): 635–645.

Goode, George Brown. "The Published Writings of Spencer Fullerton Baird, 1843–1882." *U.S. National Museum Bulletin* 20 (1883): i–xvi, 1–377. Contains a detailed bibliography.

Rivinus, Edward F., and Youssef, Elizabeth M. *Spencer Baird of the Smithsonian.* Washington, D.C.: Smithsonian Institution Press, 1992.

Spencer Baird and Ichthyology at the Smithsonian 1850–1900. National Museum of Natural History. http://vertebrates.si.edu/fishes/ichthyology_history/index.html (accessed 9 April 2012).

Spencer F. Baird's Vision for a National Museum. Smithsonian Institution Archives. http://siarchives.si.edu/history/exhibits/spencer-f-bairds-vision-national-museum (accessed 9 April 2012).

Spencer Fullerton Baird, 1823–1887, Smithsonian Secretary. Smithsonian Institution Archives. http://siarchives.si.edu/history/spencer-fullerton-baird (accessed 9 April 2012).

Pamela M. Henson

BARBED WIRE

One of the most important technological innovations of the nineteenth century, the introduction of barbed wire offered farmers a cheap, durable way to fence their property. For centuries, fences had been made of stone, wood, and other natural materials. As Americans migrated west, they continued to erect fences with the native materials available to them. Timber was expensive and in many places hard to secure. Smooth, galvanized wire fences were in use, but they failed with changes in temperature and proved an ineffective deterrent to livestock. The era of cheap fencing started with Michael Kelly's 1868 patent for a "thorny fence," but ultimately developed based on Joseph Glidden's patent for "Improvement in Wire Fences" in 1874.

In 1873, Glidden, Jacob Haish, and Isaac Ellwood saw Henry Rose's "wooden strip with metallic points," spikes nailed to a wooden strip, on display at the DeKalb (Illinois) County Fair. Within two years, each of the three men had a separate patent for barbed wire, although it was Glidden's innovation, the twisting of two strands of wire together to keep the barbs in place and a process for mass production, that ultimately proved most successful. Within the year, Glidden and Isaac Ellwood formed the Barb Fence Company on the merits of Glidden's work. Production of 10,000 pounds of wire in 1874, most of which was made by hand, rose to 600,000 pounds in 1875 with the erection of a steam-powered factory. In 1876, they sold half of the business for $60,000 plus royalties to Washburn and Moen, the largest plain-wire manufacturer in the United States.

The mass production of cheap, rust-proof steel in the mid-1870s made barbed wire more durable and drove its price down further, and by the turn of the century nearly all new fences were barbed wire, with production reaching 200,000 tons annually. But although the new fencing material allowed settlers to secure their property, it also forever altered Native American living patterns and cut herds off from what cattleman considered shared water supplies and grazing land. Conflict in Texas, Wyoming, New Mexico, and other states led to "fence-cutting" wars, sweeping changes to American land laws, and ultimately the end of free-range herding.

In the twentieth century, barbed wire was adopted as a military weapon during land campaigns and as an escape deterrent in prisons and military camps. Today, barbed wire in many forms continues as an inexpensive form of property protection.

[*See also* **Agricultural Technology; Iron and Steel Production and Products;** *and* **Military, Science and Technology and the.**]

BIBLIOGRAPHY

Glidden's Patent Application for Barbed Wire. National Archives and Records Administration. http://www.archives.gov/education/lessons/barbed-wire/index.html (accessed 24 February 2012).

Hornbeck, Richard. "Barbed Wire: Property Rights and Agricultural Development." *The Quarterly Journal of Economics* 125 (2010): 767–810.

McCallum, Henry D., and Frances T. McCallum. *The Wire That Fenced the West.* Norman: University of Oklahoma Press, 1965.

Neil Dahlstrom

BARDEEN, JOHN

(1908–1991), theoretical physicist, Nobel laureate. Bardeen was born in Madison, Wisconsin, where his father, George Russell Bardeen, was dean of the University of Wisconsin Medical School. After earning an M.A. in electrical engineering at Wisconsin, Bardeen worked for three years at the Gulf Research Laboratory in Pittsburgh and then entered Princeton University's graduate program in mathematics, where he embarked on a study of electron interactions in solids. As a junior fellow at Harvard (1935–1938) he consolidated his characteristic experimentally grounded pragmatic approach to physics. While an assistant professor of physics at the University of Minnesota (1938–1941) he began work on superconductivity, the phenomenon in which certain metals

and alloys abruptly lose all electrical resistance below a certain temperature—a phenomenon that had baffled physicists since its discovery in 1911.

After directing an engineering group at the Naval Ordnance Laboratory in Washington, D.C., during World War II, Bardeen joined William Shockley's new semiconductor research group at Bell Telephone Laboratories in New Jersey. Here he launched the group on a program of basic research that in December 1947 led to his invention, with the experimental physicist Walter Brattain, of the first transistor, a key component in the electronics revolution. For this achievement he shared the 1956 Nobel Prize for Physics.

Conflicts with Shockley over further transistor research at Bell Labs led Bardeen in 1951 to move to the University of Illinois where, with Leon Cooper and J. Robert Schrieffer, he solved the puzzle of superconductivity by 1957. Central to their theory was the attractive interaction between electrons inside superconductors resulting from the coupling of electrons to quantized sound waves in the solid. For this he shared a second Nobel Prize in 1972, making him the first person ever awarded two Nobel Prizes in the same field.

Bardeen continued to consult for businesses after his return to academia, including Haloid (which became Xerox) and General Electric. He served on numerous high-level government scientific-advisory bodies under presidents Eisenhower, Kennedy, and Reagan. As a member of Reagan's White House Science Council, he strongly opposed the Strategic Defense Initiative. Bardeen remained active in cutting-edge research until his death.

[*See also* **Bell Laboratories; Mathematics and Statistics; Military, Science and Technology and the; Nobel Prize in Biomedical Research; Physics; Science: Overview; Shockley, William; Solid-State Electronics;** *and* **Strategic Defense Initiative.**]

BIBLIOGRAPHY

Hoddeson, Lillian, et al., eds. *Out of the Crystal Maze: A History of Solid State Physics, 1900–1960.* New York: Oxford University Press, 1992.

Hoddeson, Lillian, and Vicki Daitch. *True Genius: The Life and Science of John Bardeen: The Only Winner of Two Nobel Prizes in Physics.* Washington, D.C.: Joseph Henry Press, 2002.

Riordan, Michael, and Lillian Hoddeson. *Crystal Fire: The Birth of the Information Age.* New York: W. W. Norton, 1997.

Lillian Hoddeson

BARTRAM, JOHN AND WILLIAM

(1699–1777) and (1739–1823), botanist, father of William; and naturalist, respectively. Born on a farm near Darby, Pennsylvania, John Bartram spent most of his adult life on his farm at Kingsessing, four miles from Philadelphia. In 1728, John established the first botanic garden on the continent at Kingsessing, filling it over the years with both local specimens and those collected on botanical explorations. By 1730 his interest in botany had attracted the attention of Philadelphia's scientific community. Around 1734, John was introduced to Peter Collinson, a London merchant and naturalist who became his mentor and patron. Collinson and other British plant enthusiasts purchased seeds and plants from Bartram, thereby helping finance his botanical explorations. In 1736 he explored to the sources of the Schuylkill River in Pennsylvania. He followed this exploration up in 1738 with one to Virginia and Blue Ridge, an expedition of 1,100 miles. He also made shorter expeditions, including one to the New Jersey coast and another to southern Delaware's cedar swamps. During the early 1740s, John traveled over the Catskills and into the Indian country of New York. He returned from each expedition with specimens and observations valued by the European scientific community. Bartram's expeditions and observations made him famous by midcentury. Other American naturalists sought his acquaintance and advice, his manuscripts were circulated among London botanists, and Linnaeus's student Peter Kalm spent a great deal of time at the botanical garden at Kingsessing during his trip to North America from 1748 to 1751. Named king's botanist (George III) in

1765, Bartram used his stipend to finance an expedition to South Carolina, Georgia, and Florida.

William Bartram, the son of John and his second wife, Ann Mendenhall, while still a teenager accompanied his father on botanical collecting trips and sketched specimens for John's correspondents in Britain and Europe. William attempted careers as a merchant and then as an indigo planter. After failing at both, William followed his father's footsteps as a naturalist. In 1768, John Bartram's patron, Collinson, arranged for William to draw natural history specimens for London physician John Fothergill. Fothergill soon became an important patron to the Bartrams. With Fothergill's financial support, William undertook a four-year expedition (1773–1777) exploring the Carolinas, Georgia, and Florida. His account of his travels, *Travels through North and South Carolina, Georgia, East and West Florida* (1791), contained an extensive exposition of the southern flora and fauna. For this it was praised, although some condemned William's flowery style of writing. This style, with its rich descriptions of nature, influenced English romantic writers such as William Wordsworth and Samuel Taylor Coleridge. In 1778 William returned to the family farm at Kingsessing, now owned by his brother John Jr., William refused offers to become professor of botany at the University of Pennsylvania and to join the Freeman Red River Expedition in 1806. He remained at his family's garden, where he provided encouragement and training to the succeeding generation of American naturalists, including Thomas Say and Alexander Wilson. The careers of the Bartrams, father and son, demonstrate the importance of European patronage for American science during the Colonial Era.

[*See also* **Botanical Gardens; Botany; Say, Thomas;** *and* **Science: Overview.**]

BIBLIOGRAPHY

Barrow, Mark V., Jr. "Bartram, William (1739–1823)." In *The History of Science in the United States: An Encyclopedia,* edited by Marc Rothenberg, pp. 72–73. New York and London: Garland Publishing, 2001.

Bell, Whitfield J., Jr. "Bartram, John." In *Complete Dictionary of Scientific Biography*. Vol. 1, pp. 486–488. Detroit, Mich.: Charles Scribner's Sons, 2008.

Berkeley, Edmund, and Dorothy Smith Berkeley. *The Life and Travels of John Bartram: From Lake Ontario to the River St John*. Tallahassee: University Presses of Florida, 1982.

Slaughter, Thomas P. *The Natures of John and William Bartram*. 2d ed. Philadelphia: University of Philadelphia Press, 2005.

Marc Rothenberg;
updated by Elspeth Knewstubb

BEADLE, GEORGE WELLS

See Lederberg, Joshua.

BEAUMONT, WILLIAM

(1785–1853), physician and scientist, the first American physiologist to achieve international renown. A native of Lebanon, Connecticut, Beaumont earned a license to practice medicine in 1812 after a brief apprenticeship with physician Benjamin Chandler in northern Vermont. From 1812 to 1839, Beaumont mostly served in the medical department of the U.S. Army: on the Canadian border during the War of 1812 and at various frontier posts in the Great Lakes region during the 1820s and 1830s. During this time he educated himself in physiology and performed the important research into human digestion that defined his career. After his retirement from the army, Beaumont established a successful private practice in St. Louis, Missouri, where he lived until his death.

Beaumont achieved permanent fame for his research on human digestion. In June 1822, while at Fort Mackinac in Michigan Territory, Beaumont treated a French-Canadian voyageur, Alexis St. Martin, who had been shot in the stomach. The wound healed in such a manner as to leave a permanent gastric fistula, or opening. This enabled Beaumont to insert food into St. Martin's stomach, observe the digestive process, and remove gastric juice for analysis. Beaumont pursued his experiments at different times over the next decade, working under extremely difficult conditions in rough facilities at army hospitals. Beaumont was encouraged and supported in his research by Joseph Lovell, the Surgeon General of the U.S. Army. With Lovell's help, Beaumont published his first article and sought to arrange a tour of Europe to exhibit St. Martin's gastronomy. Much of Beaumont's most valuable work was carried out in Fort Crawford at Prairie du Chien, Wisconsin.

In 1832 Beaumont drew up a contract to safeguard his rights to his study subject, St. Martin, in preparation for his proposed tour of European universities. This contract is significant in medical history because it was the first to deal with the use of human subjects in medical research. St. Martin was to be paid a surgeon's wage for the tour and assigned as an aide to Beaumont. The tour never took place. St. Martin escaped across the border from Washington, D.C., to Canada. Beaumont's book on his experiments with St. Martin, *Experiments and Observations on the Gastric Juice and the Physiology of Digestion* (1833), was published despite the canceled tour and established that digestion was a chemical process. This finding was quickly accepted in both the United States and Europe.

[*See also* Ethics and Medicine; Medicine: Overview; Physiology; Research and Development (R&D); Science: Overview; *and* War and Medicine.]

BIBLIOGRAPHY

Horsman, Reginald. *Frontier Doctor: William Beaumont, America's First Great Medical Scientist*. Columbia: University of Missouri Press, 1996.

Myer, Jesse S. *Life and Letters of Dr. William Beaumont*. St. Louis: C. V. Mosby Co., 1912.

Pitcock, Cynthia DeHaven. "Beaumont, William (1785–1853)." In *The History of Science in the United States: An Encyclopedia*, edited by Marc Rothenberg, pp. 73–75. New York and London: Garland Publishing, 2001.

Reginald Horsman;
updated by Elspeth Knewstubb

BEHAVIORISM

The school of experimental psychology known as behaviorism was first articulated by John B. Watson (1878–1958) in a 1913 article in *Psychological Review* entitled "Psychology as a Behaviorist Views It." Rejecting the notion of consciousness, Watson viewed all behavior as conditioned by external experience that could be observed, measured, and controlled. As espoused by its founder, however, behaviorism had more to do with defining psychology's purpose and function than with any specific methodology beyond a seemingly rigorous empiricism. Watson belonged to the first generation of American-trained psychologists—a cohort determined to make psychology no longer a stepchild of philosophy but an empirical science.

Watson's formulation of behaviorism characterized psychology as a science whose primary objective was to predict and control human behavior. Accordingly, he promoted behaviorism as a management tool, a pedagogical method, an advertising technique, and a child-rearing method. (Watson's own child-rearing manual, *Psychological Care of Infant and Child* [1928], enjoyed considerable influence.) In this sense, Watson's behaviorism, which characterized *Homo sapiens* as "organic machines," belonged on the spectrum of Progressive-Era social thought along with scientific management and technocracy linked to the work of Thorstein Veblen.

Behaviorists promised to unlock the mechanism that governed human action. Their notion of a malleable human nature inspired the new professional managerial classes that saw the intractability of the "human factor" as the last obstacle to a rationally managed society. The 1920s witnessed the transformation of large foundations from charitable trusts to professionally managed enterprises. As foundation support became more problem oriented, it increasingly focused on scientific research related to social conduct and behavior. Within this context, human behavior became the focus of a new synthesis in an increasingly integrated research community. The "social and behavioral sciences" became invariably linked.

Behaviorism, emphasizing the primacy of environment over instincts, held special appeal for reformers. Watson, however, never contended that all human beings were equal, only that they were the same. By denying the existence of consciousness, behaviorism desanctified both the outer world of nature and the realm of inner experience, reducing both to manipulable objects. If behaviorism represented the freedom to remake the individual, it also raised the specter of conditioning human behavior into predetermined channels. Behaviorism, wrote the social philosopher Horace Kallen, made human beings "as equal as Fords."

As a popularizer of self-help psychology, Watson made behaviorism a household word in the 1920s. Behaviorism's claims to demystify psychology and to simplify the complexities of modern life appealed strongly to middle-class Americans. By the 1930s, it had become the dominant paradigm in American experimental psychology. Although few psychologists accepted the more radical aspects of Watson's extreme materialism, behaviorism's objective methodology powerfully influenced the direction of American psychology. Influenced by Watson's popular writings, B. F. Skinner, of Harvard University, became the preeminent behaviorist of the post–World War II generation. Skinner's major achievement, the development of operant conditioning, was based on the notion of designing an environment so controlled that the subject conditioned itself to behave in ways predetermined by the experimenter. Although his experimental subjects were pigeons, Skinner promoted a vision of human social engineering and scientific management in his utopian novel *Walden Two* (1948).

Behaviorism helped legitimize experimental psychology among the natural sciences and provided scientific underpinning to the belief in American exceptionalism that characterized American social science in the twentieth century. Behaviorism's ascendancy reflected a preoccupation with order and efficiency among the shapers of modernist America.

[*See also* **Psychology; Skinner, B. F.;** *and* **Social Sciences.**]

BIBLIOGRAPHY

Buckley, Kerry W. *Mechanical Man: John Broadus Watson and the Beginnings of Behaviorism.* New York: Guilford Press, 1989.

Mills, John A. *Control: A History of Behavioral Psychology.* New York: New York University Press, 1998.

O'Donnell, John M. *The Origins of Behaviorism: American Psychology, 1870–1920.* New York: New York University Press, 1985.

Rutherford, Alexandra. *Beyond the Box: B. F. Skinner's Technology of Behavior from Laboratory to Life, 1950s–1970s.* Toronto: University of Toronto Press, 2007.

Kerry W. Buckley

BELL, ALEXANDER GRAHAM

(1847–1922), inventor and scientist best known for his invention of the telephone. Born in Scotland, his family immigrated to Canada in 1870. Two years later Bell moved to Boston as a teacher of the deaf. At that time, he began researching methods to transmit several telegraph messages simultaneously over a single wire. This line of research ultimately led to his invention of a version of the telephone in 1876.

In the mid-1870s Thomas Edison invented the quadruplex, a system for sending four simultaneous telegraph messages over a single wire. Afterward, a major focus of invention in the telegraph industry was the development of methods that could send more than four messages simultaneously. Several electricians, including Bell and Elisha Gray, developed designs capable of subdividing a telegraph line into 10 or more channels. These harmonic telegraphs used several reeds that responded to specific acoustic frequencies. They worked well in the laboratory, but proved unreliable on real-world telegraph lines.

One of Bell's deaf pupils was Mabel Hubbard, daughter of Boston businessman Gardiner Greene Hubbard. Hubbard hoped to establish a federally chartered telegraph company to compete with the Western Union monopoly by contracting with the Post Office to send low-cost telegrams. Hubbard saw great promise in the harmonic telegraph and backed Bell's experiments. Bell, however, was more interested in transmitting the human voice and wanted to devote his time to that project. Hubbard pushed Bell to focus on the harmonic telegraph, even threatening to withhold Mabel's hand after Bell began courting her. Bell and Hubbard worked out an agreement that Bell would devote most of his time to the harmonic telegraph but continue developing his telephone concept.

The telephone thus arose out of Elisha Gray's and Alexander Graham Bell's work on harmonic telegraphs. It was a short conceptual step for both men from transmitting musical tones to transmitting the human voice. Bell filed a patent describing his method of transmitting sounds on 14 February 1876, just several hours before Gray filed a caveat, a statement of concept, on a similar method. Thus, the question of priority of invention has been controversial ever since. It is most likely that both men independently invented the telephone as an outgrowth of their work on harmonic telegraphy. The Patent Office issued a patent to Bell on 7 March 1876, one of the most valuable patents in history.

At this time, Bell did not have a fully functioning instrument. He first produced intelligible speech on 10 March 1876 when he directed his assistant, Thomas Watson, to "come here, I want to see you." Over the next few months, Bell continued to refine his instrument to make it suitable for public exhibition. In June he demonstrated his system to the judges of the Centennial Exposition in Philadelphia, a test witnessed by Brazil's emperor Dom Pedro and celebrated British physicist Sir William Thomson.

Gardiner Hubbard organized the Bell Telephone Company in July 1877 to commercialize Bell's telephone. A few months later, Western Union organized a rival telephone company using the patents of Thomas Edison and Elisha Gray. Both companies competed fiercely for about two years. In November 1879, however, Western Union agreed to exit the telephone business in exchange for the Bell Company's commitment not to enter the telegraph market. In one of the biggest failures of vision in American business history, Western Union agreed to hand over its telephone business to Bell for a mere 20 percent of telephone rentals until Bell's patents expired in 1894.

Bell served as the company's technical advisor until he lost interest in telephony in the early 1880s. Although his telephone invention rendered him independently wealthy, he sold off most of his stock holdings early and did not profit as much as he might have had he retained his shares. Thus, by the mid-1880s, his role in the telephone industry was peripheral at best. He devoted his time to other pursuits, including an improved phonograph, photophone (a telephone that operated by means of light), hydrofoil watercraft, and heavier-than-air flight. He also succeeded his father-in-law Gardiner Hubbard as president of the National Geographic Society between 1898 and 1903. He died at his Nova Scotia estate on 2 August 1922.

[*See also* **Airplanes and Air Transport; Bell Laboratories; Deafness; Edison, Thomas; Science: Overview; Sound Technology, Recorded; Technology; Telegraph;** *and* **Telephone.**]

BIBLIOGRAPHY

Alexander Graham Bell Family Papers. Library of Congress Manuscript Division and Library of Congress's American Memory. http://memory.loc.gov/ammem/bellhtml/bellhome.html (accessed 25 October 2012).

Bruce, Robert V. *Bell: Alexander Graham Bell and the Conquest of Solitude*. Ithaca, N.Y.: Cornell University Press, 1990.

Hochfelder, David. *The Telegraph in America: 1832–1920*. Baltimore: Johns Hopkins University Press, 2012.

John, Richard R. *Network Nation: Inventing American Telecommunications*. Cambridge, Mass.: Belknap Press of Harvard University Press, 2010.

Tosiello, Rosario Joseph. *The Birth and Early Years of the Bell Telephone System, 1876–1880*. New York: Arno Press, 1979.

David Hochfelder

BELL LABORATORIES

Bell Laboratories, or more formally Bell Telephone Laboratories, was the research and development subsidiary of the American Telephone and Telegraph Company, which until 1984 was the privately owned national telephone monopoly of the United States. Bell Labs is generally considered the preeminent industrial research laboratory of the twentieth century. One measure of its prominence is that 13 scientists have shared in seven Nobel Prizes for work done at Bell Labs. Bell Labs was responsible for thousands of incremental and major improvements in the telephone system such as the crossbar, an improved electromechanical telephone switch (1938), customer long-distance dialing (1951), electronic telephone switching (1965), and touch-tone telephones (1963), but it is best known for a long list of fundamental scientific discoveries and technical contributions that had implications far beyond AT&T's telephone business. Initially headquartered in New York City, Bell Labs opened a research facility in Murray Hill, New Jersey, in 1941. By 1959, Murray Hill was not only the headquarters for Bell Labs but also the largest industrial research lab in the country, with over 4,200 employees. There were additional facilities elsewhere.

AT&T established Bell Labs in 1925 from what had been the engineering department of AT&T subsidiary Western Electric. Bell Labs researchers almost immediately began making fundamental contributions. In its first decade alone, Clinton Davisson provided experimental verification of the wave nature of the electron, Harold Black discovered negative feedback, and Lloyd Espenschied and Herman Affel invented broadband coaxial cable. In 1936, Bell Labs undertook fundamental research in solid-state physics, a program that led in 1947 to John Bardeen, Walter Brattain, and William Shockley's invention of the transistor, the first solid-state amplifier and switch. The transistor became the underlying fundamental device of modern electronics. Bell Labs freely shared transistor technology with others, hosting seminars in 1951 and 1952. Solid-state physics yielded many subsequent innovations including, at Bell Labs, the solar cell by Gerald Pearson, Daryl Chapin, and Calvin Fuller in 1954 and the charged-coupled device by William Smith and George Boyle in 1969. Bell Labs researchers applied transistors, solar cells, and other technologies in the development of

Telstar, the first active communications satellite, launched with NASA in 1962. Much of the fundamental research led back to improvements in the telephone system. In 1965, AT&T installed the first electronic telephone switch, the 1ESS. It used tens of thousands of transistors in place of electromechanical relays. Over the following years, electronic switches supplanted electromechanical ones.

Bell Labs' contributions went far beyond solid-state physics. Karl Jansky invented the science of radio astronomy in 1930. Claude Shannon conceived a new discipline, information theory, in 1948. Information theory provides the theoretical underpinning for all modern communication media. Dennis Ritchie and Ken Thomson invented Unix, the first hardware-independent computer language, in 1971. Later versions provided the underlying language of the Internet. Bell Labs also undertook government research. At the height of the Cold War, the Department of Defense funded over one-third of all Bell Labs' work.

Bell Labs succeeded in part because, with AT&T a monopoly, it could take a long view. It had a secure, steady source of income and could innovate without competitive pressure. In 1984, the AT&T monopoly ended when a government antitrust suit concluded with AT&T's agreement to divest its local telephone operations. Bell Labs continued as a part of AT&T but slowly narrowed its focus and decreased its size in response to changed conditions. In 1996, AT&T broke up again, spinning off its manufacturing businesses as Lucent Technologies. The Labs itself split, with approximately three-quarters of the staff, the Bell Labs name, and Murray Hill going to Lucent. In 2012, Bell Labs was the research and development arm of the Alcatel–Lucent Corporation.

[See also **Bardeen, John; Bell, Alexander Graham; Computer Science; Electronic Communication Devices, Mobile; Engineering; Internet and World Wide Web; Military, Science and Technology and the; National Aeronautics and Space Administration; Nobel Prize in Biomedical Research; Physics; Radio; Research and Development (R&D); Satellites, Communications; Shockley, William; Solid-State Electronics; Telegraph;** and **Telephone.**]

BIBLIOGRAPHY

Fagan, M. D., et al., eds. *A History of Engineering and Science in the Bell System*, 7 vols. Murray Hill, N.J.: AT&T, 1975–1985.
Gertner, Jon. *The Idea Factory: Bell Labs and the Great Age of American Innovation.* New York: Penguin Press, 2012.
Noll, A. Michael, and Michael Gelelowitz, eds. *Bell Labs Memoirs.* New Brunswick, N.J.: IEEE History Center, 2011.

Sheldon Hochheiser

BERKNER, LLOYD

(1905–1967), physicist, engineer, and science administrator. Lloyd Viel Berkner grew up in Wisconsin, North Dakota, and Minnesota. He studied electrical engineering at the University of Minnesota and after graduating in 1927 worked for federal government agencies. During the 1930s, Berkner conducted ionospheric research with the U.S. Bureau of Standards and the Carnegie Institution of Washington. His main contributions focused on developing methods and instrumentation for investigating the structure of the ionosphere and its significance for radio propagation. He learned during this period the importance of international cooperation for geophysical research. Berkner traveled internationally and met other researchers in Germany, England, Australia, and New Zealand. While in Australia for an extended period, he worked on the installation of an automatic ionospheric sounder.

During World War II, Berkner, a naval reserve officer, supervised aviation electronics engineering with the Radio and Engineering Group within the Engineering division of the U.S. Navy's

Bureau of Aeronautics, including the procurement of newly developed radar systems for naval aircraft. This national security involvement continued after the war. Berkner played a major role during the Cold War using the resources of the national security state to advance science and engineering. Despite not having an advanced science degree, scientists accepted him as a leading member of their community. And as a government researcher, he was accepted as a leading statesman and policy maker.

Berkner worked to maintain the involvement of the country's scientists and engineers in military research and development after the war. He was instrumental in the establishment by the War and Navy Departments of the Joint Research and Development Board and served as the first chair. He also was involved in convincing the military services to operate "summer studies" that used the country's elite technical experts to help define and solve military problems. One of these summer study programs at MIT resulted in the creation in the arctic regions of the Distant Early Warning line, a system of radar stations in Canada, Alaska, and other countries that would warn of a Soviet bomber attack. Recognizing that national security during the Cold War also involved soft power, Berkner was concerned about the relationship between foreign policy and science and technology. A government committee he helped organize made recommendations to the State Department about the potential contributions of science and scientists to the country's foreign policy.

Berkner's other major contributions to science during the Cold War mainly involved problems in geophysics. His proposal in 1950 for a global study of geophysical processes led five years later to the International Geophysical Year. Berkner played a leading role in organizing and promoting this international cooperative effort. And when the implications for space exploration became clear, he also worked to link the International Geophysical Year to the American space program. The International Geophysical Year provided a crucial justification for establishing the space program. Berkner then served as the first chair of the Space Science Board set up before the creation of

the National Aeronautics and Space Administration in 1958.

[*See also* Engineering; Geophysics; International Geophysical Year; Military, Science and Technology and the; National Aeronautics and Space Administration; Physics; Science: Overview; Space Program; *and* Technology.]

BIBLIOGRAPHY

Hales, Anton L. "Lloyd Viel Berkner, 1905–1967." *Biographical Memoirs of the National Academy of Sciences.* Washington, D.C.: National Academy of Sciences, 1992.
Needell, Allan A. *Science, Cold War, and the American State: Lloyd V. Berkner and the Balance of Professional Ideals.* Amsterdam: Harwood Academic, 2000.

Hugh Richard Slotten

BETHE, HANS

(1906–2005), theoretical physicist and nuclear weapons expert, Nobel laureate, was born in Strassburg, which was then part of Germany. His death marked the end of an era. He was the last of the young physicists who, from 1925 to the early 1930s, established quantum mechanics, the theory that made possible the intellectual mastery of the atomic and nuclear world. Bethe's impressive abilities were first demonstrated in the two encyclopedic *Handbuch der Physik* articles he published in 1933. In them, he gave a masterly exposition of the application of quantum mechanics to atomic, molecular, and solid-state physics.

In 1933, after Hitler came to power, Bethe lost his position as an assistant professor in Tübigen because his mother had been Jewish before she converted to Protestantism. In February 1935 Bethe immigrated to the United States after accepting a position at Cornell University, where he stayed for the rest of his life.

When in the 1930s the frontier of physics shifted to nuclear physics, Bethe became an acknowledged leader in this field, coauthoring with Stanley Livingston and Robert Bacher three lengthy articles in the *Reviews of Modern Physics* that became known as the *Bethe Bible* of nuclear physics. His mastery of nuclear physics made it possible for him to put forward in 1938 his explanation of energy generation in stars, for which he won the Nobel Prize in 1967.

Bethe was at the center of nuclear weapons development: first as head of the Theoretical Division of the Los Alamos Laboratory during World War II; then in helping design H-bombs; and, subsequently, trying to restrain the further expansion of the atomic arsenal and preventing the proliferation of atomic weapons.

Bethe was held in high esteem as an outstanding scientist, as an expert in weaponry and nuclear matters, and as a citizen. He was invited to join the committee that advised President Eisenhower on scientific matters. This committee would become the President's Science Advisory Committee (PSAC) after the Soviet satellite *Sputnik* was launched in October 1957. Bethe served on the panels of PSAC that dealt with disarmament and with strategic military problems. He played a key role in the signing by the United States and the Soviet Union of the Test Ban Treaty of 1963, which forbade atmospheric and underwater nuclear tests and limited underground testing to low-yield weapons.

[*See also* **Atomic Energy Commission; Military, Science and Technology and the; Nobel Prize in Biomedical Research; Nuclear Power; Nuclear Regulatory Commission; Nuclear Weapons; Physics; President's Science Advisory Committee;** *and* **Quantum Theory.**]

BIBLIOGRAPHY

Schweber, Silvan S. *In the Shadow of the Bomb: Bethe, Oppenheimer, and the Moral Responsibility of the Scientist.* Princeton, N.J.: Princeton University Press, 2000.

Silvan S. Schweber

BICYCLES AND BICYCLING

The history of bicycling in the United States is one of peaks and valleys, inventions and reinventions. Although the attention devoted to bicycling in the early twenty-first century pales in comparison to past heydays, its future appears bright because of both the bicycle's simplicity and its ability to address a host of complex problems.

The two-wheeled "dandy horse" or "swift-walker," powered by its rider pushing against the ground, gained popularity in East Coast cities around 1819 but was soon banned from sidewalks as dangerous to pedestrians. Pedal power caught on in the United States at the 1868 New York Athletic Games in the form of the "velocipede," with a metal frame, solid rubber tires, and front-wheel pedals, an 1866 innovation by the Frenchman Pierre Lallement. Improving upon this in the first two months of 1869 alone, U.S. firms flooded the patent office with more successful applications than had previously been granted to human-powered vehicles in 80 years. Riding rinks and academies popped up in cities and small towns alike, yet after this brief vogue the velocipede's popularity faded as it was banned from many roads.

By the late 1870s, the "high wheeler" emerged, with its steel frame, brakes, and, by the 1880s, ball bearings and a front wheel up to five feet high. The Bostonian Albert Pope converted several Connecticut factories into bicycle production and by 1882 was the world's largest manufacturer, rolling out 1,000 bikes per month. Faster than their predecessors, high wheelers were also more dependent on roads, and in 1880 Pope and others founded the League of American Wheelmen (LAW, now LAB, or League of American Bicyclists) to advocate for good roads with bike access. Cycling was further popularized through local clubs with posh clubhouses and group rides. Approximately 100,000 high wheelers were on the road in the United States by 1887. Young, athletic males were the primary market, but even among them the high wheeler's appeal was limited by the risk of dramatic headers (i.e., head-first crashes).

The "safety" bicycle addressed this with a profile similar to earlier boneshakers, improved by a chain drive, and was an instant and even bigger hit after being introduced to U.S. audiences in the late 1880s, not least because of its popularity with women. The adoption of pneumatic tires in the early 1890s made it even more attractive. Able to move twice as fast as a horse and carriage, these bikes were relatively light and available in a wide array of models manufactured by some three hundred firms by the mid-1890s, bicycling's golden age. As prices dropped, more than 4 million Americans owned safeties by 1896, a craze so pervasive that the government opened a separate patent office just for bicycle-related innovations. Promoted by feminists such as Frances Willard, the bicycle allowed women independent mobility and gave them reason to wear bloomers rather than sweeping skirts. Although bike manufacturers, repairers, and accessory-makers thrived; clothing fashions transformed; bike-related advertisements proliferated in beautiful Art Nouveau designs; and artists, playwrights, and songwriters celebrated the bicycle, everyone from pastors to barkeepers blamed cycling for declining business. Despite warnings of moral depravity for those choosing a ride over church on Sundays and such maladies as "bicycle twitch," cyclists of all ages and classes took to the roads.

Responding to the bicycle craze, New York City passed the nation's first comprehensive traffic code in 1897. At the same time, nearly as dramatically as the boom had come about, riding began a significant decline for reasons including changing tastes and the failure of the industry to improve technology most useful for recreational riders, such as gearing. Overproduction resulting from techniques pioneered by bike manufacturers—the assembly line, planned obsolescence, and marketing incentives—was a factor as well. By 1909, cyclists had almost disappeared from roadways as the automobile became affordable and entered its own golden age. As the century wore on, most highways were built without provision for nonautomotive transportation, and bicycling ceased to be perceived as a serious endeavor.

On the competitive side, after the first recorded bicycle races in East Coast cities in the 1860s and 1870s, amateur and professional road and especially track racing quickly became favorite spectator sports. In 1899, drafting behind a train, Charles M. Murphy biked the world's first subminute mile. The athletic color line was broken that year by the African American Marshall "Major" Taylor, who won the world championship, although blacks had been banned from the League of American Wheelmen starting in 1894. In the 1920s, six-day velodrome races—where men rode continuously, stopping only when absolutely necessary—drew tens of thousands of fans to major venues including Madison Square Garden. The American Bicycle League's national championships first included women in 1937, but by then public interest in competitive cycling had waned. U.S. rider Greg LeMond's 1983 world championship and 1984 victory in the grueling Tour de France (est. 1903) brought competitive cycling back into the public imagination. From 1999 to 2005, Lance Armstrong's first-place finishes in the Tour de France broadened cycling's popularity, but his record and that of his U.S. team was marred by subsequent findings of widespread doping.

Innovations in gearing, shifting, braking, frame materials, and manufacturing, among other areas, continued both overseas (primarily in Europe early on and later in Asia) and in the United States. Modern mountain bikes, invented in California in the 1970s and mass-marketed beginning in the 1980s, broadened cycling's appeal not only on mountain trails but also in urban environments replete with potholes, curbs, and narrow passages. Organizations such as Rails to Trails and Adventure Cycling sought to expand opportunities for recreational cycling nationwide, and ad hoc groups such as Critical Mass riders in San Francisco called attention to cyclists' needs.

In the first decade of the twenty-first century, utility biking was on the rise, with the number of bike commuters increasing by 57 percent, although other modes still dwarfed this figure. To improve personal mobility and health in the face of increasing congestion, parking limitations, pollution, and sedentary lifestyles, U.S. municipalities began introducing sophisticated bike-share systems following European examples in the late

2000s, often funded by advertising revenue and thus spearheaded by media concerns rather than bike companies. Bike share is just one example of the increasing public recognition of bicycling's health and environmental benefits and the outgrowth of efforts by grassroots organizations and some governmental agencies to make roads and attitudes friendlier toward cyclists.

[*See also* **Environmentalism; Gender and Technology; Health and Fitness;** *and* **Motor Vehicles.**]

BIBLIOGRAPHY

Alliance for Biking and Walking. http://www.peoplepoweredmovement.org/site/ (accessed 24 February 2012).
Herlihy, David V. *Bicycle: The History*. New Haven and London: Yale University Press, 2004.
Mapes, Jeff. *Pedaling Revolution: How Cyclists Are Changing American Cities*. Corvallis: Oregon State University Press, 2009.
Nye, Peter. *Hearts of Lions: The Story of American Bicycle Racing*. New York: W. W. Norton, 1988.
Smith, Robert A. *A Social History of the Bicycle: Its Early Life and Times in America*. 1972; revised ed., *Merry Wheels and Spokes of Steel: A Social History of the Bicycle*. San Bernardino, Calif.: Borgo Press, 1995.

Hannah Borgeson

BIOCHEMISTRY

As a biological discipline, biochemistry entails applying chemistry to biological problems to explain the chemical processes that occur within living cells. The development of biochemistry in the United States occurred during the closing decades of the nineteenth century. During this time, biochemistry had a strong tradition in Britain and Germany and many American scientists pursued training in European physiology and chemistry laboratories. As a result, American biochemistry's experimental and intellectual origins defined how the first generation of American biochemists would study the "chemistry of life."

The first generation of American biochemists conducted an array of research pursuits on the chemistry of life in a variety of institutional settings: agricultural chemistry in agricultural research stations, medical chemistry in hospitals and medical schools, and physiological chemistry in colleges and universities. Biochemistry's development as a major biological discipline was particularly tied to the rise of agricultural chemistry in the United States, but most significant was its close association with medical reform and applied clinical research. Biochemistry, or biological chemistry, served as an umbrella term that united these diverse research traditions and programs dedicated to the study of chemical processes in living organisms.

Samuel Johnson, professor of analytical and agricultural chemistry at the Sheffield Scientific School of Yale University, introduced the nation's first course in physiological chemistry in 1874. This first course in biochemistry was held in a modest laboratory under the direction of Russell Chittenden. Trained in agricultural chemistry, Chittenden left New Haven in 1878 to pursue further training under Professor Willy Kühne at Heidelberg. Chittenden returned to Yale and was appointed as the first chair of physiological chemistry in the United States in 1882. When he later became director of the Sheffield School, Lafayette Mendel succeeded him as head of the department of physiological chemistry.

Biochemistry solidified as a biological discipline at the turn of the twentieth century as professional biochemists taught specialized courses and carried out research programs. In 1905, Harvard assigned Carl Alsberg and Lawrence Henderson as its first full-time biochemistry instructors. Researchers adapted and integrated the new discipline into existing institutional structures. At Yale, Chittenden and Mendel developed a research and training program based on chemical and nutritional physiology; at Johns Hopkins, John J. Abel linked physiological chemistry with pharmacology; and at Michigan, Victor Vaughan did so with hygiene. By 1905, biochemists could publish in the *Journal of Biological Chemistry*, and one year later, biochemists had their own professional society when Abel and William Gies founded the American Society of Biological Chemists.

Biochemistry and Clinical Medicine.
The development of American biochemistry was closely tied to efforts to reform American medical education. American medical schools began to teach biochemistry in response to the introduction of rigorous entrance requirements and preclinical education standards. Between 1910 and 1920, most American medical schools established departments of biochemistry, medical chemists were replaced by biochemists, and courses in medical chemistry were replaced by those in biochemistry. Clinical laboratories in teaching hospitals also served as active sites for biochemical diagnosis. Stanley Benedict at Cornell, Otto Folin at Harvard, and Donald Van Slyke at the Rockefeller Institute developed and introduced new analytical methods to diagnose specific diseases.

In similar fashion, biochemists affiliated with medical schools during the 1920s and 1930s introduced new research techniques and laboratory technologies that enabled researchers to monitor and measure molecular changes occurring in metabolic processes. The use of X-ray crystallography to investigate protein structure and the use of isotopes to trace chemical changes occurring in physiological processes were among two such developments. The first half of the twentieth century witnessed the elucidation of several major metabolic pathways, including glycolysis, cholesterol biosynthesis, the Krebs cycle, and the urea cycle, among others. Biochemists studied the physiology of digestion, metabolism, and respiration; characterized enzymes, hormones, and proteins; and documented the transformation of molecules within organisms.

In the 1930s, a number of American medical schools and universities established basic and clinical biochemistry departments of international renown by extending positions to scientists immigrating to the United States from war-torn Europe. Hans Clarke, chair of the biochemistry department at Columbia, brought on several such immigrant biochemists, including

Rudolf Schoenheimer, whose introduction of isotopic tracers to study metabolic changes transformed basic biochemical research. American biochemists were involved in war-related projects during World War II, contributing to the industrial production of penicillin and the isolation of blood fractionation products for transfusion. The institutionalization of basic biochemical research and applied clinical biochemistry reflected the character of midcentury American biochemistry.

Biochemistry and Molecular Biology.
During the second half of the twentieth century, biochemistry became closely associated with molecular biology. In the 1930s, Warren Weaver, director of the Rockefeller Foundation's division of natural sciences, promoted the application of chemical and physical techniques to the study of biological problems. Institutional support for biochemistry from the Rockefeller Foundation was followed by increased funding from government agencies like the National Institutes of Health and the National Science Foundation as well as private agencies and foundations.

This shift in patronage for biochemical research coincided with a series of discoveries in biochemical genetics that would inform the development of another biological discipline, molecular biology. During the 1940s and 1950s, researchers studying the interplay between biochemistry and genetics demonstrated the importance of proteins and nucleic acids in inheritance. In 1941, George Beadle and Edward Tatum showed that genes controlled enzyme synthesis. In 1952, Alfred Hershey and Martha Chase demonstrated the genetic role of DNA, which confirmed Oswald Avery, Colin MacLeod, and Maclyn McCarty's 1944 discovery that DNA was the carrier of genetic information. One year later, James Watson and Francis Crick discovered the chemical structure of DNA. Although biochemists had been traditionally concerned with the structure and function of enzymes, nucleic acids, and proteins, the first generation of molecular biologists focused on how these molecules were involved in patterns of genetic inheritance. Because of the close overlap between the two fields, the advent of molecular biology proved to be promising and challenging for biochemists.

Legacy. The development of American biochemistry was a primarily twentieth-century phenomenon. The field emerged out of efforts to discredit vitalism, or the belief that a vital force was the underlying cause of the chemical processes in living organisms, and solidified as a biological discipline as biochemists pursued parallel programs of basic biochemical and applied clinical research. During the first half of the twentieth century, biochemists contributed to the identification, purification, and production of antibiotics, blood products, hormones, vaccines, vitamins, and other synthetic products. The development of molecular biology during the second half of the twentieth century extended the scope of biochemical research as biochemists engaged in biotechnological and genetic engineering pursuits beginning in the 1970s and 1980s. In the early twenty-first century, biochemical research has influenced multiple fields of biomedical research.

[*See also* **Agricultural Experiment Stations; Biological Sciences; Biotechnology; Chemistry; Genetics and Genetic Engineering; Higher Education and Science; Medical Education; Medicine: From the 1870s to 1945; Medicine: Since 1945; Molecular Biology; National Institutes of Health; National Science Foundation; Penicillin; Pharmacology and Drug Therapy; Physiology; Science: Revolutionary War to 1914; Science: 1914 to 1945;** *and* **Science: Since 1945.**]

BIBLIOGRAPHY

de Chadarevian, Soraya, and Harmke Kamminga. *Molecularizing Biology and Medicine: New Practices and Alliances, 1910–1970s.* Amsterdam: Harwood Academic Publishers, 1998. An edited collection of essays documenting the relationship among biochemistry, molecular biology, and clinical medicine.

Fruton, Joseph. *Proteins, Enzymes, Genes: The Interplay of Chemistry and Biology.* New Haven, Conn.: Yale University Press, 1999. A broad history of biochemistry and molecular biology with an extensive bibliography, written by a biochemist turned historian of biochemistry.

Kohler, Robert E. *From Medical Chemistry to Biochemistry: The Making of a Biomedical Discipline.* Cambridge, U.K.: Cambridge University Press, 1982. The standard institutional and disciplinary history of American biochemistry. Reprint edition available.

Morange, Michel. *A History of Molecular Biology.* Translated by Matthew Cobb. Cambridge, Mass.: Harvard University Press, 1998.

Todd M. Olszewski

BIOLOGICAL SCIENCES

The biological sciences encompass a wide range of objects and problems—from cells to whales, from eating to evolution. The degree to which these studies have been perceived as a cohesive science, or even as a federation of sciences, has varied greatly over time. Americans participated significantly in the shaping of many biological sciences from the eighteenth century to the beginning of the twenty-first. Some areas of biology were among the first regions of science in which Americans became world leaders. Knowledge imported and developed by American biologists had substantial impact, both through its application in medical and agricultural technologies and through its influence on ways that Americans thought about themselves and their relations with the natural world.

Colonial and Antebellum Eras. Prior to 1800, no coherent conception of "biological science" existed in either England or its North American colonies. If we set aside modern categories, however, we can find relevant knowledge and discussion among and between professional physicians and a loosely defined group of collectors and writers known as naturalists. American colonial physicians, for the most part, sought to absorb the so-called "humoral theories" about the human body and its ills that were prominent in European medical circles—particularly at the University of Edinburgh, where leading American doctors such as Benjamin Rush completed their education. Although some colonial naturalists were university trained, others simply collected, described, and named specimens according to a loose familiarity with Linneaus's classification system, and most participated in natural history as a means to develop social networks. Benjamin Smith Barton, a nephew of the American Philosophical Society's second president David Rittenhouse, studied medicine at Edinburgh before returning to the States without a degree and taking a position lecturing on botany and natural history at the College of Philadelphia; Barton edited one of the nation's major medical journals (*Philadelphia Medical and Physical Journal*, 1802–1805) and wrote the first American textbook on botany (*Elements of Botany*, 1803) as well as treatises on medicinal plants and American Indian archaeology. Jane Colden, daughter of the Edinburgh-trained physician-turned-politician Cadwallader Colden, learned botanical nomenclature and illustration from her father; she corresponded with an international community of plant naturalists and developed a well-known (but unpublished) botanical journal that other naturalists relied on for understanding the flora of the New York area. With European science casting aspersions on their country, colonial naturalists were "practically devoted to a method of natural history study that privileged observation and testimony over theory and experiment" (Lewis, 2010). Thomas Jefferson, a lawyer by training, devoted nearly one-third of his *Notes on the State of Virginia* to the vast climate and organism differences between the United States and Europe; he also admonished the French naturalist Buffon's "new theory of the tendency of nature to belittle her productions on this side of the Atlantic" without having seen them himself.

Indeed, the real biological novelties of North America were its climate and its plants and animals. Many early European travelers reported on the organisms of the New World, and in the early 1700s European naturalists such as Mark Catesby and Peter Kalm traversed the Eastern seaboard and described uniquely American organisms according to the new standards of scientific taxonomy. Colonials such as the Pennsylvania nurseryman John Bartram collected plants and seeds for the growing network of botanists and plant enthusiasts in Europe. Some "invasive species" succeeded well by naturalists standards, although there was little formal scientific understanding of why. For instance, rice accompanied enslaved West Africans across the Middle Passage throughout the New World to Brazil, the Caribbean, and the southern United States, along with the informal agricultural knowledge and practices that ensured its successful farming on rice plantations. Jefferson introduced grapevines, olive trees, and oranges at his Monticello estate in the first half of the 1770s; although privately he blamed the failure of these crops on the American environment, he continued to argue publicly that colonial territories had great potential for the "cultivation" and betterment of domesticated animals, plants, and people (including Native American Indians and enslaved Africans).

After the Revolutionary War, natural history drew strength from and significantly enhanced American nationalism. Government explorers, beginning with the Lewis and Clark Expedition, collected plants as part of their efforts to reinforce U.S. claims of sovereignty over western territories; at the same time, they sought to assess the agricultural potential of these areas and to determine the abundance of animal species that could be successfully raised and exploited for profit. Historians have demonstrated that, although Lewis and Clark did not shy away from Jefferson's instruction to develop a comprehensive Linnaean catalog of the Columbia River basin's flora and fauna, they tended "toward a holistic viewpoint that mixed specific notations on plant and animal species with comments on patterns of living among Indians" (Lang, 2004). At the same time, Eastern naturalists showcased distinctively American plants and animals to heighten national self-consciousness and to demonstrate that Americans could participate in the cosmopolitan world of science. Charles Wilson Peale's museum, established on the second floor of Philadelphia's Independence Hall in 1802, linked a mounted bald eagle and the skeleton of the "American mastodon," excavated and reconstructed by Peale, with his paintings of George Washington, Thomas Jefferson, and other national heroes. Peale's vision of natural history as a tool for democratic education relied on wealthy collectors, but also prefigured a strategy of popular science outreach that would later become a hallmark of some American biological sciences like ornithology and ecology.

By the mid-nineteenth century, while American medicine struggled to understand and apply the new germ theory of disease, U.S. naturalists became increasingly professionalized and they largely controlled the scientific study of the biological organisms of North America. Beginning in the late 1830s, the Harvard botanist Asa Gray coordinated efforts to describe and classify the continent's plants. The U.S. government showed increasing interest in such projects, but toward more practical ends: for instance, the Patent Office acted (from 1839 to 1845) as a de facto clearinghouse for seed sharing of common garden varieties of vegetables and corn. The zoologist Louis Agassiz, who emigrated from Switzerland to teach at Harvard in 1846, sought a similar classificatory achievement with animals, but was supplanted after about 1860 by a network of collectors led by the Smithsonian Institution's Spencer Baird (1823–1887). Natural history museums, displaying specimens of New World flora, fauna, minerals, and fossils, were founded in Philadelphia (1812), Cincinnati (1820), Boston (1830), and Albany (1843). These institutions catered primarily to local elites who pursued science as a form of recreation, but ultimately, through their collections and libraries, museums became national centers of natural history research as well as popular education.

Civil War to World War II. After the Civil War, a network of scientists associated with the Smithsonian's National Museum, the U.S. Commission of Fish and Fisheries, and the Department of Agriculture sought to manage the organisms of North America. Although too late to save the passenger pigeon, which became extinct in 1914, those scientists involved in the so-called "conservation movement" helped to preserve the bison. They also introduced a vast number of new species and varieties into American agriculture, including important pasture grasses, but they experienced mixed success in keeping out insects and weeds. The Ecological Society of America, founded in 1915, promoted the further development and application of biological knowledge to wildlife management decisions. From the 1890s through the 1930s, the growth of field work (accomplished both through the proliferation of field stations in the West and the expansion of research at botanical gardens) and the introduction of new quantitative research methods ensured that American ecology remained influential in the world as well as central to the ongoing economic and geographical expansion of the country.

Throughout the reform-intensive Progressive Era, American naturalists increasingly thought about the implications of their science for education and social change, as well as for religion, ethics, and public health. A number of scientists delivering the prestigious Lowell Lectures in Boston in the mid-nineteenth century, for example, emphasized the existence and wisdom of the Creator and the prevalence of progressive change. These commonplaces took a new turn with the publication of Charles Darwin's *Origin of Species* in 1859 and with the new prominence of materialistic ideas among such British scientific intellectuals as Thomas H. Huxley (1825–1895) and Herbert Spencer (1820–1903). Starting in the 1890s, an object-based biological curriculum known as nature study was introduced in American elementary schools; nature study flourished during a period when cultural institutions such as zoos, botanical gardens, natural history museums, and national parks promoted the idea that direct knowledge of nature would enrich the citizens of an increasingly diverse, urban, and industrial nation. Epidemics such as malaria and yellow fever that had swept through the South during the Civil War generated intense medical interest but the continued separation of clinical and public-health practices from laboratory investigations ensured that they remained little understood after 1900.

American scientists of this period agreed that evolution was a fact, but few saw it as random or amorally competitive. Some, such as Edward Drinker Cope (1840–1897) and Lester Ward, argued explicitly for the "neo-Lamarckian" view of evolution as the progressive emergence of intelligence in the living world. Most American biologists had a looser expectation—that some progressive, or "orthogenetic," force guided life in certain directions, most notably toward humanity and Anglo-American civilization. This evolutionary message was conveyed to mass audiences in the decades around 1900 through popular magazines, world's fairs, and the new secondary-school subject of biology. Protestant fundamentalists responded by pressing for laws restricting the teaching of evolution. The resulting 1925 Scopes trial in Tennessee publicized the tensions between fundamentalist Christians and scientists but had relatively little impact on the teaching of biology; science educators may have eliminated the word "evolution" from their texts, but they continued to teach "civic biology" and the principles of progressive development. The emergence of research and land-grant universities after 1870 produced a significant increase in the number of academically trained American biological scientists and reinforced the idea that the biological sciences formed a single major scientific unit, even while new tensions emerged between pure and applied research. The new Johns Hopkins University in Baltimore established the first academic biology department in 1876, and similar programs arose during the next two decades at the University of Pennsylvania, Clark University, Columbia University, and the University of Chicago. From coast to coast, many states also established public universities with Colleges of Agriculture where laboratory and field research on crop plants and domesticated animals flourished. In 1871, even Harvard University embraced this trend, founding its Bussey Institution on newly donated land in nearby Jamaica Plains to foster teaching and research on agricultural and applied sciences. But in 1908—after Bussey re-

searchers' repeated clashes with zoologists on the main campus—it was reorganized as a graduate school of applied biology.

In 1883, biological scientists took the lead in creating the American Society of Naturalists. One of the first American organizations limited to professional scientific specialists, it had become by the 1890s a federation of yet more specialized societies created by, among others, morphologists, anatomists, physiologists, and psychologists. The most important institutional center for academic biology was the Marine Biological Laboratory established in Woods Hole, Massachusetts, in 1888. Directed by Charles Otis Whitman (1842–1910), this scientific summer colony became a locus for both basic biological research and the informal interactions of a vibrant scientific community. In this same period, the number of American women and other minorities participating in the biological sciences at colleges, universities, and research stations started to grow; this trend was enhanced by broader educational and social reforms as well as by the presence of some prominent mentors in certain fields like embryology (Ross Harrison at Yale) and cytology (Edmund Beecher Wilson at Bryn Mawr).

The first generation of laboratory-based academic biologists trained during this period focused primarily on the study of cells and embryonic development. They hoped to gain a unified understanding of metabolism, development, reproduction, heredity, and evolution. Among the notable discoveries of this period were Jacques Loeb's 1899 demonstration of artificial parthenogenesis and E. B. Wilson and Nettie Stevens's recognition in 1904 that chromosomes play a major role in determining sex. The most important American advance in biology, however, came from a project begun at Columbia University (1910–1928) and continued at the California Institute of Technology (1928–1942) by Thomas Hunt Morgan to map the locations of genes on chromosomes, using the fruit fly (*Drosophila*); this work was awarded the Nobel Prize in Physiology and Medicine in 1933.

By the 1920s a confederation of zoologists, botanists, and agricultural researchers were developing genetics as both a fundamental science and the basis for problems as varied as improving corn crops and understanding cancerous tumors. Genetic researchers initially retained close ties to agricultural breeders and fanciers, which gave them access to long-inbred homogeneous strains of animals—from show dogs and cattle to pet mice—and plants for their controlled experimental studies. Such collaborations were formative for the development of new laboratory-based methods and applied knowledge in these fields. Harvard graduate students Clarence Cook Little, Leslie Dunn, and George Snell developed and studied the genetics of mouse tumors and immunology in strains of inbred mice obtained from Granby, Massachusetts, fancier Abbie Lathrop; Snell's research on histocompatibility complexes in animal models later informed development of successful tissue and organ transplant work. Barbara McClintock first observed cross-shaped interaction of homologous chromosomes during meiosis while working with a group of plant breeders and cytologists in the plant breeding department at Cornell's College of Agriculture; her development of a new chromosome staining technique also enabled the development of cytogenetics as a subfield.

Spurred on by investments made by the Rockefeller Foundation in laboratory technologies, such as the ultracentrifuge and the electron microscope, and in the standardization of a variety of research materials, "molecular biology" (a term coined in 1938 by Warren Weaver) became an approach favored by American interwar researchers at the intersection of biological research fields. After isolating genetic mutants of the bread mold *Neurospora*, biochemist Edward Tatum collaborated with geneticist George Beadle to use this system to discover the genetic control of cellular enzymatic production. The experimental techniques and instruments that Wendell Stanley and his colleagues developed for studying the biochemistry of tobacco mosaic virus were generalized to research on other diseases such as poliomyelitis and influenza and to studies of genes and cell organelles.

These developments were paralleled by activities in America's rapidly modernizing medical schools. In the 1880s and 1890s, Harvard and Johns Hopkins made laboratory research integral to their medical programs. The Hopkins anatomist Franklin P. Mall fostered the work of Florence Sabin (1871–1953), Herbert Evans (1882–1971), and George Corner (1889–1981) on human embryology, sex hormones, and reproduction. The new science of bacteriology developed at both medical schools and research centers such as the Rockefeller Institute for Medical Research (later Rockefeller University) in New York; in 1944, while working at the Rockefeller Institute for Medical Research, bacteriologist Oswald Avery would discover that genes and chromosomes were made of deoxyribonucleic acid (or DNA). Influential medical scientists such as the Johns Hopkins pathologist William Welch (1850–1934) and the Harvard physiologist Walter Cannon (1871–1945) fended off the attacks of anti-vivisectionists by publicizing the laboratory origins of new therapies against infectious diseases, most notably diphtheria.

In the early twentieth century, many leading biologists and some medical researchers extended their interest in improving human life to include "race betterment," or eugenics. The most prominent of these, Charles B. Davenport (1866–1944), established the Eugenics Record Office (ERO) alongside the Station for the Experimental Study of Evolution at Cold Spring Harbor, New York, in 1910. Researchers at the Station for the Experimental Study of Evolution like Edwin Carleton MacDowell studied animal models of alcoholism and other conditions they believed to be genetically and socially damaging "traits." The ERO collected data on human heredity to demonstrate what many geneticists believed to be the ill effects of miscegenation, the inferiority of recent immigrant groups, and the value of laws to sterilize the "unfit"; ERO fieldworkers' pedigree and observational data about Carrie Buck and her family led to the U.S. Supreme Court upholding Virginia's sterilization laws in1927. By the end of the 1930s, declining immigration, advances in genetics, the expansion of the social sciences, and the rise of Nazism had discredited the eugenics movement but not the more general ambition to explain and improve on human social phenomena in biological terms. Biologists turned to birth control and family planning, population policy, and circumscribed problems in medical genetics. In their place, social scientists and psychiatrists came to the fore as interpreters of "human nature," although research fields such as "psychobiology," which emphasized the experimental study of animal models to understand the hereditary basis of intelligence, continued to flourish into the 1940s and beyond.

Post–World War II Developments. The scientific boom triggered by World War II initially proved less important for biology than for a number of other disciplines. Biological scientists did important war work, notably in the development of antibiotics and the prevention of malaria and other tropical diseases through insect control. But the war did not lead to "big science" in biology and it did not give biologists the same public visibility that the physicists acquired through their role in the development of the atomic bomb. Nevertheless, postwar federal funding for the biological sciences came from the Atomic Energy Commission (which supported major initiatives in genetics and ecology), the National Institutes of Health (whose support for biomedical research grew dramatically beginning around 1957), the Department of Agriculture, and the newly founded National Science Foundation. Private organizations, ranging from the Rockefeller Foundation to the March of Dimes, which raised money for poliomyelitis research, also provided significant funding. In the long run, this pattern of gradual growth and diversified patronage persisted and proved advantageous for biologists.

The development of molecular biology, spurred by the determination of the structure of DNA in 1953 by the American James D. Watson and the Englishman Francis Crick, was the most important early postwar event in the biological sciences. The antipolio vaccines developed by Jonas Salk and Albert Sabin (1906–1993) seemed the culmination of biomedical researchers' efforts to eliminate the major infectious diseases in the United States. The oral contraceptive, based on a drug synthesized by the chemist Carl Djerassi (1923–) in 1951, and on the research of the endocrinologist Gregory Pincus (1903–1967) and others in the 1950s, was marketed in 1960 by the G. D. Searle Company. It was the most visible of the many important drugs introduced by an alliance of biologists, chemists, and corporations.

Postwar biologists of all political and disciplinary inclinations played important economic and social roles by contributing agricultural and medical ideas and applications, as well as assessing problems resulting from scientific-industrial developments. Plant geneticists responded to concerns about overpopulation and famine by developing more productive varieties of tropical and domestic crops, thereby making possible the so-called "green revolution." Radiobiologists, supported at Oak Ridge and Argonne National Laboratory after the war, produced radioisotopes and other technologies that became important for medical research and clinical treatments. In the 1950s, the biochemist Linus Pauling and the geneticist Hermann J. Muller (1890–1967) participated prominently in scientific-political campaigns against fallout from atmospheric nuclear testing. Rachel Carson's *Silent Spring* (1962) indicted the indiscriminate use of pesticides. The microbiologist René Dubos (1901–1982) emerged as the philosopher of the environmental movement. Within this activist context, biologists became sufficiently confident to make evolution an explicit part of the secondary-school curricula they developed in the 1960s.

Late twentieth-century changes in the biological sciences were so rapid and substantial that a brief sketch must suffice. Numbers tell an important story: although the annual production of PhDs in the physical sciences remained about the same in 1995 as in 1970, recruitment of biological scientists increased by two-thirds during that period. Much of this growth occurred in molecular biology. The development of recombinant-DNA technology in the mid-1970s led, after an initial controversy about safety and ethics, to the creation of a major new science-based biotechnology industry; this industry inspired and was itself spurred on by new biological technologies like engineered bacteria (Ananda M. Chakrabarty was awarded a U.S. Patent for his "oil eating bacteria" in 1981), automated DNA and protein sequencing and synthesizer machines, and the polymerase chain reaction (for which Kary Mullis won the Nobel Prize in 1993). Geneticists working through the Department of Energy and the National Institutes of Health first obtained funding in 1990 for a 15-year big science project to map the human genome; the project's first director was James Watson, then-director of the revamped Cold Spring Harbor Laboratories, which had (since 1968) embraced a program of basic research on human cancers. A looser network of scientists introduced a series of technologies, ranging from amniocentesis to in vitro fertilization, which transformed the possibilities for human reproduction. The emergence of acquired immunodeficiency syndrome in the 1980s stimulated research in immunology and cell biology while simultaneously ending the widespread belief that the battle against infectious diseases had been won.

During this same period, issues involving organisms and their evolution once again became prominent in the American context. Population geneticists, ethologists, and organismal biologists in both university and museum contexts had continued to investigate evolutionary problems like sexual selection and animal language throughout the 1950s and 1960s. But in the 1970s, the new discipline of sociobiology, led by the Harvard entomologist Edward O. Wilson, directly challenged the primacy of social scientists as interpreters of human behavior. Wilson also addressed the problem of declining global biodiversity. New arguments that dinosaurs were warm-blooded social animals, as well as the claim that they had been wiped out by a meteorite, increased public attention to evolutionary history. In 1987, the U.S. Supreme Court upheld science educators' claim that evolution, but not creationism, should be taught in public schools. More recently some small local schools have challenged this federal biology education policy—most prominently, when the Dover, Pennsylvania, school board ruled in 2004 that intelligent design must be taught as an alternative to evolution; ultimately, these approaches were ruled unconstitutional.

By the 1990s, biology had replaced physics as the most important and visible of the natural sciences in America. Through their activities in medicine and agriculture, molecular biology, and the field sciences, American biologists were transforming the conditions of people's lives.

In 2003, the U.S. National Institutes of Health and the private corporation Celera Genomics co-announced they had completed the Human Genome Project; for the next decade, American biomedical researchers successfully investigated human genetic diseases with markers like bRAC1 (a recessive gene for inherited breast cancer). Applied ecologists, working in remote field stations as well as college and university laboratories, made advances in understanding biodiversity-ecosystem function, which further illuminated the effects of global climate change. Building on recombinant DNA technology, industrial biologists devised a new generation of genetically modified crops. Through their investigations into genetics, ecology, and evolutionary history, these American scientists were now redefining the meaning of life itself.

[*See also* **Agricultural Education and Extension; Agricultural Experiment Stations; Agriculture, U.S. Department of; American Museum of Natural History; Biochemistry; Biotechnology; Botanical Gardens; Botany; Chemistry; Columbian Exchange; Conservation Movement; Creationism; DNA Sequencing; Dust Bowl; Ecology; Entomology; Environmentalism; Environmental Protection Agency; Ethics and Medicine; Evolution, Theory of; Fish and Wildlife Service, U.S.; Fisheries and Fishing; Forest Service, U.S.; Gender and Science; Genetics and Genetic Engineering; Geography; Geological Surveys; Geology; Higher Education and Science; Human Genome Project; Hybrid Seeds; Instruments of Science; Lewis and Clark Expedition; Medical Education; Medicine; Missionaries and Science and Medicine; Molecular Biology; Museums of Science and Natural History; National Institutes of Health; National Park System; National Science Foundation; Nobel Prize in Biomedical Research; Oceanography; Paleontology; Physics; Physiology; Psychology;**

Religion and Science; Science; Scopes Trial; Scripps Institution of Oceanography; Sex and Sexuality; Smithsonian Institution; Sociobiology and Evolutionary Psychology; Stem-Cell Research; Zoology; and Zoos.]

BIBLIOGRAPHY

Allen, Garland. "The Eugenics Record Office at Cold Spring Harbor, 1910–1940: An Essay in Institutional History." *Osiris* 2 (1986): 225–264. An archivally rich account of early genetics and "race betterment" research at Cold Spring Harbor, with a focus on the Progressive values of efficiency and rational control of society.

Appel, Toby A. *Shaping Biology: The National Science Foundation and American Biological Research, 1945–1975.* Baltimore: Johns Hopkins University Press, 2000. Describes the federal government's increased postwar investment in basic research, with attention to tensions over the merits of funding basic versus applied research and funding individuals versus institutions.

Barrow, Mark. *Nature's Ghosts: Confronting Extinction from the Age of Jefferson to the Age of Ecology.* Chicago: University of Chicago Press, 2009. An overview of research on endangered species, as well as a thoughtful examination of the relations between American naturalists and conservationists.

Benson, Keith, Jane Maienschein, and Ronald Rainger, eds. *The Expansion of American Biology.* New Brunswick, N.J.: Rutgers University Press, 1991. The second edited collection (a follow-up to Rainger, Benson, and Maienschein, 1988) of case-study examinations of important themes and episodes in the history of twentieth-century biology.

Bocking, Stephen. *Ecologists and Environmental Politics: A History of Contemporary Ecology.* New Haven, Conn.: Yale University Press, 1997. A comparative study of the participation of ecologists in environmental politics in the United States, England, and Canada.

Buhs, Joshua. *The Fire Ant Wars: Nature, Science, and Public Policy in Twentieth-Century America.* Chicago: University of Chicago, 2004. Uses the 1918 fire ant invasion and entomologists' ongoing responses to it as a window on the development of American concepts of nature and stewardship.

Carney, Judith. *Black Rice: The African Origins of Rice Cultivation in America.* Cambridge, Mass.: Harvard University Press, 2001. Explores crops, landscapes, and agricultural practices in Africa and America and demonstrates the critical role West African slaves played in the creation of the system of rice production that provided the foundation of the Carolinas' wealth.

Clark, Constance. *God and Gorillas: Images of Evolution in the Jazz Age.* Baltimore: Johns Hopkins University Press, 2008. Details the efforts of biologists to popularize evolution in museums and popular magazines in the 1920s.

Clarke, Adele E. *Disciplining Reproduction: Modernity, American Life Sciences, and "Problems of Sex."* Berkeley: University of California Press, 1998. An overview of the development of reproductive life sciences in the United States from 1910 to 1963.

Comfort, Nathaniel. *The Science of Human Perfection: How Genes Became the Heart of American Medicine.* New Haven, Conn.: Yale University Press, 2012. A history of medical genetics emphasizing what the author calls a "hybridization of science and medicine" that has consistently defined American health care.

Comfort, Nathaniel. *The Tangled Field: Barbara McClintock's Search for the Patterns of Genetic Control.* Cambridge, Mass.: Harvard University Press, 2003. A biographical study of McClintock, a geneticist who integrated classical genetics with microscopic observations of the behavior of chromosomes.

Creager, Angela N. H. *Life Atomic: A History of Radioisotopes in Science and Medicine.* Chicago: University of Chicago Press, 2013. Evaluates the histories of American Cold War biology and medicine through the lens of radioisotope tracing as a research and therapeutic technology.

Creager, Angela N. H. *The Life of a Virus: Tobacco Mosaic Virus as an Experimental Model, 1930–1965.* Chicago: University of Chicago Press, 2001. Focusing on research conducted by Wendell Stanley's lab, describes how tobacco mosaic virus served as a model system for virology and molecular biology.

Creager, Angela, and María Jesús Santesmases. "Radiobiology in the Atomic Age: Changing Research Practices and Policies in Comparative Perspective." *Journal of the History of Biology* 39, no. 4 (Winter 2006): 637–647. Excellent thematic discussion that introduces a special issue of the *Journal of the History of Biology* dedicated to the history of radiobiological research.

Dietrich, Michael R., and Brandi H. Tambasco. "Beyond 'the Boss and the Boys:' Women and the Division of Labor in Drosophila Genetics in the United States, 1934–1970." *Journal of the History of Biology* 40, no. 3 (September 2007): 509–528. Establishes and discusses the gendered division of labor within the American Drosophila research community in the mid-twentieth century.

Ewan, Joseph, and Nesta Dunn Ewan. *Benjamin Smith Barton, Naturalist and Physician in Jeffersonian America.* St. Louis: Missouri Botanical Garden Press, 2007. The definitive biography on Barton and his work.

Fitzgerald, Deborah K. *The Business of Breeding: Hybrid Corn in Illinois, 1890–1940.* Ithaca, N.Y.: Cornell University Press, 1990. An account of how corn genetics research at land-grant universities contributed to the commercialization of seed selling to American farmers in the early twentieth century.

Gronim, Sara Stidstone. "What Jane Knew: A Woman Botanist in the Eighteenth Century." *Journal of Women's History* 19 (2007): 33–59. Places the work of botanist Jane Colden in the context of colonial science and women's history.

Haraway, Donna J. *Primate Visions: Gender, Race, and Nature in the World of Modern Science.* New York: Routledge, 1989. A postmodern history of American primatology research and popularization from the 1890s to the 1970s.

Henson, Pamela. 'What Holds the Earth Together': Agnes Chase and American Agrostology." *Journal of the History of Biology* 36, no. 3 (Autumn 2003): 437–460. Discusses Chase's scientific and mentorship contributions to field biology and grass research during the early twentieth century.

Humphreys, Margaret. *Yellow Fever in the South.* Baltimore: Johns Hopkins University Press, 1999. Chronicles the devastating impact of disease on the Civil War South, as well as researchers' early attempts to investigate the cause of illness.

Kay, Lily E. *The Molecular Vision of Life: Caltech, the Rockefeller Foundation, and the Rise of the New Biology.* New York: Oxford University Press, 1993. A social and scientific history of early Rockefeller Foundation patronage for American molecular biology and the "science of man."

Kevles, Daniel J., and Leroy Hood, eds. *The Code of Codes: Scientific and Social Issues in the Human Genome Project.* Cambridge, Mass.: Harvard University Press, 1992. Provides historical reflections of those involved in the Human Genome Project, as well as the founding of the American biotechnology industry.

Kingsland, Sharon. *The Evolution of American Ecology, 1890–2000.* Baltimore: Johns Hopkins University Press, 2005. An examination of the roots of American ecology across various institutional contexts, with attention to the consistent evolutionary frame in which research was pursued.

Kohler, Robert E. *Lords of the Fly: Drosophila Genetics and the Experimental Life.* Chicago: University of Chicago Press, 1994. A "follow the fly" history of the opportunities and pitfalls of bench top work on *Drosophila* starting with Morgan's "fly room" at Columbia.

Kohlstedt, Sally G. *Teaching Children Science: Hands-On Nature Study, 1890–1930.* Chicago: University of Chicago Press, 2010. A history of nature study and its role in American science education, centering on its Progressive Era development and expansion.

Lang, William. "Describing a New Environment: Lewis and Clark and Enlightenment Science in the Columbia River Basin. *Oregon Historical Quarterly* 105, no. 3 (Fall 2004): 353, 360–389. A consideration of the scientific approach of the Lewis and Clark expeditions and the consequences of that approach for naturalists and American expansion.

Larson, Edward. *Summer for the Gods: The Scopes Trial and America's Continuing Debate Over Science and Religion.* New York: Basic Books, 1997, 2006 (with new afterword). The Pulitzer Prize–winning legal and scientific history of America's 1925 Scopes "Monkey" Trial.

Lewis, Andrew J. *A Democracy of Facts: Natural History in the Early Republic.* Philadelphia: University of Pennsylvania Press, 2010. An overview of the methods and accomplishment of colonial American naturalists.

Lindee, M. Susan. *Suffering Made Real: American Science and Survivors at Hiroshima.* Chicago: University of Chicago Press, 1994. Explores radiation genetics and clinical research done with atomic bomb survivors, with an emphasis on the political and social consequences of this research for American science.

Maienschein, Jane. *Transforming Traditions in American Biology, 1880–1915*. Baltimore: Johns Hopkins University Press, 1991. Examines the origins of American academic biology through the careers of Johns Hopkins University's first four biology PhDs: E. B. Wilson, T. H. Morgan, R. G. Harrison, and E. G. Conklin.

Milam, Erica. *Looking for a Few Good Males: Female Choice in Evolutionary Biology*. Baltimore: Johns Hopkins University Press, 2010. Utilizes the problem of sexual selection as a window on the history of American evolutionary biology research ranging from population genetics to theories of the animal mind.

Mitman, Gregg. *Reel Nature: America's Romance with Wildlife on Film*. Cambridge, Mass.: Harvard University Press, 1999. Discusses the development of natural history film and its impact on science, the entertainment industry, and popular culture.

Mitman, Gregg. *The State of Nature: Ecology, Community, and American Social Thought, 1900–1950*. Chicago: University of Chicago Press, 1992. Examines the relationship between political issues in early-twentieth-century American society and the sciences of evolution and ecology, with a focus on work at the University of Chicago.

Pauly, Philip J. *Biologists and the Promise of American Life: From Meriwether Lewis to Alfred Kinsey*. Princeton, N.J.: Princeton University Press, 2000. A provocative but thoughtful overview of how American life scientists have linked their studies of nature with what the author calls a "desire to culture"—animals, land, and people.

Pauly, Philip J. *Fruits and Plains: The Horticultural Transformation of America*. Cambridge, Mass.: Harvard University Press, 2007. Looks at how horticultural pursuits shaped the environmental and scientific history of America.

Porter, Charlotte M. *The Eagle's Nest: Natural History and American Ideas, 1812–1842*. University: University of Alabama Press, 1986. Focuses on natural history and nationalism in Philadelphia after the War of 1812.

Rader, Karen A. *Making Mice: Standardizing Animals for American Biomedical Research, 1900–1955*. Princeton, N.J.: Princeton University Press, 2005. A history of the development of genetically inbred mice and their uses in biological and medical research in the early to mid-twentieth century.

Radick, Gregory. *The Simian Tongue: The Long Debate about Animal Language*. Chicago: University of Chicago Press, 2007. Chronicles how the invention of the Edison phonograph revitalized research on the evolution of primate language in the early decades of the twentieth century.

Rainger, Ronald, Keith R. Benson, and Jane Maienschein, eds. *The American Development of Biology*. New Brunswick, N.J.: Rutgers University Press, 1988. A collection of historical case studies on American biology in the nineteen and twentieth centuries.

Rasmussen, Nicholas. *Picture Control: The Electron Microscope and the Transformation of Biology in America*. Stanford, Calif.: Stanford University Press, 1997. Discusses the development of one of the key visualization technologies in modern biomedical research, with an emphasis on collaborations among the Rockefeller Foundation, academic biologists, and industry.

Rogers, Naomi. *Dirt and Disease: Polio before FDR*. New Brunswick, N.J.: Rutgers University Press, 1992. A social, cultural, and scientific history of the polio epidemic in the United States during the early twentieth century.

Rosenberg, Charles E. *No Other Gods: On Science and American Social Thought*. Baltimore: Johns Hopkins University Press, 1976; reissued in 1997. A series of essays that provides a broad and thought-provoking examination of the ways in which social institutions and values have shaped American biological and medical research.

Rumore, Gina. "Preservation for Science: The Ecological Society of America and the Campaign for Glacier Bay National Monument." *Journal of the History of Biology* 45, no. 4 (2012): 613–650. A case study of the founding of the Preservation Committee of the ESA, with emphasis on how it revealed the methodological ambitions American ecologists had for their science in the 1920s and 1930s and how they understood the role of place in biological field studies.

Schloegel, Judith Johns, and Karen A. Rader. *Ecology, Environment, and 'Big Science': An Annotated Bibliography of Sources on Environmental Research at Argonne National Laboratory, 1955–1985*. Published as ANL/HIST-4. Lemont, Ill.: Argonne National Laboratory, 2005. An overview of themes and sources on postwar environmental research and "big science" at one national laboratory.

Slack, Nancy. *G. Evelyn Hutchinson and the Invention of Modern Ecology*. New Haven, Conn.: Yale

University Press, 2011. An examination of the life and work of the leading American ecologist of the twentieth century.

Smocovitis, Vassiliki B. *Unifying Biology: The Evolutionary Synthesis and Evolutionary Biology*. Princeton, N.J.: Princeton University Press, 1996. The definitive history of the evolutionary synthesis, in America and beyond, in which the author argues that this historical episode was a part of a larger process of unifying the biological sciences during the twentieth century.

Vicedo, Marga. *The Nature and Nurture of Love: From Imprinting to Attachment in Cold War America*. Chicago: University of Chicago Press, 2013.

Weidman, Nadine. *Constructing Scientific Psychology: Karl Lashley's Mind-Brain Debates*. New York: Cambridge University Press, 1999. Details the early history of psychobiology in relation to genetics and Progressive-Era social aspirations.

Philip J. Pauly and Karen A. Rader

BIOLOGY

See Biological Sciences.

BIOMEDICAL RESEARCH

See Nobel Prize in Biomedical Research.

BIOTECHNOLOGY

The word "biotechnology" became suddenly familiar in the media in the United States in the year 1980. The term described the activities of a new type of small company with a distinctively new business plan: it would produce therapeutic human proteins by altering the DNA of bacteria. The stockbroker Nelson Schneider had launched this interpretation the previous September. He had held a seminar for Wall Street analysts in which he discussed the prospects of what he introduced as biotechnology and evoked huge enthu-

siasm. In June 1980, the Supreme Court permitted the patenting of an organism. When the Genentech Company was floated four months later, its stock rose faster than any other had ever done in the history of the New York Stock Exchange. A range of new scientific techniques were now permitting the organized engineering of new bacteria whose genes could be combined with human genes to produce useful chemicals. At this moment, when an oil crisis suggested that old energy-intensive industries had had their day, the success of electronics demonstrated the possibility of an industrial revolution, and biotechnology was touted as the route to both a cure for cancer and industrial renewal.

Before the 1980s. To many Americans in the 1980s, biotechnology was brand new and exclusively concerned with the application of the techniques of molecular biology. This was, however, too narrow a view. The intensity of commercial and indeed political excitement at that high point obscured a long historical trajectory. Indeed, the frequent incantation of the "Frankenstein" story as a warning of unanticipated consequences pointed to an ancestry reaching back to Mary Shelley's 1817 novel. That work and its enduring popularity testify to the feelings of power and danger the combination of biology and technology evoked, even at the beginning of the nineteenth century. In Shelley's day both words were newly minted; they were only brought together at the beginning of the twentieth century. When that innovation was made, it was initially in far-off Hungary, and its attraction was felt first primarily in Europe. However, its common-sense appeal was seen as so obvious that the word was almost immediately adopted by a Chicago fermentation consultancy seeking avenues beyond beer in Prohibition-Era America. Although a few small companies such as the Pfizer Corporation successfully made organic acids by fermentation, they were isolated. With continuing access to cheap oil, ingenious routes to the manufacture of most needed chemicals from hydrocarbons, and the end of federally mandated Prohibition in 1932, the idea did not catch on widely in the United States. In Europe, where oil was imported and countries feared war, the word biotechnology did come into limited currency. Both in Germany and in the United Kingdom, the word was used to describe the manufacture of chemicals using natural organisms. More sinister was the connotation of engineering entire human populations through eugenics.

During World War II, the standing of biotechnology as a concept was moved from the periphery of industrial and scientific cultures to their very center. Penicillin, which British researchers first identified as a drug at the beginning of the war, was derived from the liquid exuded by a mold. Because its production from conventional chemical techniques using oil or coal proved entirely unsatisfactory, the mass manufacture of cheap penicillin required the development of a new technology for growing micro-organisms and for extracting the by-products of their growth. U.S. pharmaceutical companies drew on their own experience with related products and laboratory-scale experiments, on European developments in the Netherlands and Germany, and on the determination to manufacture a life-saving wonder drug. The result was the technology of the stirred tank fermenter and associated separation techniques for purifying penicillin from perplexingly weak solutions by switching between two different solvents.

The production of other antibiotics and even hormones also used the technology developed in the United States for penicillin manufacture. Such drugs as tetracycline and its relatives ("the golden horde") and streptomycin used to cure tuberculosis were developed within a few years. On the back of such new products and the exploitation of a combination of biology and engineering, previously small, ethical pharmaceutical companies such as Merck and Pfizer became global giants within a decade. The technologies they deployed prospered accordingly. In 1958 the antibiotics specialist and chemical engineer Elmer Gaden founded the *Journal of Microbiological and Biochemical Engineering* to forward and promote this dynamic combination. But the name was clumsy and he shortly adopted the long-established European word when he changed the name to *Biotech-*

nology and Bioengineering. In launching his new title he explained, "Biotechnology embraces all aspects of the exploitation and control of biological systems and their activities. Some of these form the bases for old and well established industries. Industrial 'fermentations' and the isolation and purification of chemical products from natural products are examples." Gaden also looked forward to new products. Gasoline made from alcohol was one, but others included food made from fungi grown on crude oil. Such "single cell protein" foods (to use a term coined at MIT in 1966) used as human or animal feed were seen as potentially solving the problem of world hunger. The manufacture of such products drew directly on the expertise developed in antibiotic manufacture.

But the experts on antibiotic manufacture could not keep biotechnology to themselves. The 1960s saw a growth explosion in the health industries, in the proportion of the U.S. gross national product devoted to medicine, and in the science of molecular biology. The former was significant because it would offer the prospect of huge rewards for the discovery of new drugs, especially for treating diseases resistant to the approaches of the past, such as cancer. Molecular biology was significant because scientific developments promised a revolution in capabilities.

Molecular Biology. Above all, perhaps, the 1960s and early 1970s saw a revolution in the perceived prospects for molecular biology. In retrospect, the critical moment in its history was the 1953 description of the structure of DNA by Crick and Watson. At the time, however, this was not widely seen as being of revolutionary significance. That meaning only became widely shared after subsequent developments. In 1961, Nirenberg and Matthei at the National Institutes of Health showed how RNA codes for proteins, and the sequence of DNA, RNA, and protein was complete. This was followed in 1967 by Arthur Kornberg's synthesis of short stretches of DNA, making up a viral gene. Thus, by the late 1960s, the sequence of developments since 1953 was seen as earth-shattering, even by laypeople. President

Johnson interrupted a speech celebrating the two hundredth anniversary of the *Encyclopaedia Britannica* in December 1967 to say that Kornberg's achievement "was the most important story you ever read, your daddy ever read or your granddaddy every read."

In 1973 Stanley Cohen and Herbert Boyer succeeded in transferring a section of DNA from one organism to another. As in the early nineteenth century, the development of the new biological capabilities aroused alarm as well as enthusiasm. There was concern about genetically engineered organisms escaping into the environment and causing incurable diseases (a prospect popularized by Michael Crichton's novel *The Andromeda Strain*). At the same time, the influential and articulate Nobel Prize–winning molecular biologist and physician Joshua Lederberg warned against excessive regulation. Against remotely possible dangers, there were both possible benefits and even remotely possible transformative benefits.

This rhetorical conflict weighing transformative benefits, albeit speculative, against risks, some of which were also speculative, characterized the 1970s. During the second half of the decade the courts and Congress proved the battleground on which these conflicts were fought. The new biological techniques offered the prospect of adding human genes to bacteria or fungi to allow them to express human proteins on order. These in turn could be used, it was hoped, to treat the large number of illnesses caused by the body's own genes failing to code for the right proteins. Of course, if the ambitious hopes defined in these public arenas were even partly justified, substantial profits would ensure. Private funding flowed into a few companies that explored the limits.

In practice, only a few proteins were obviously therapeutic. For 40 years diabetes had been treated by injecting insulin, albeit from animals and slightly different from the natural human variety. Now, two small, new, scientist-led companies, Genentech and Biogen, sought ways of producing human insulin microbially to replace the animal insulin normally used by diabetics. The success of Genentech paved the way for further investor enthusiasm. Losses now, it seemed, would

be more than compensated by huge profits later. When the American Cancer Society, on the prompting of its lay president Mary Lasker, urged research on the antiviral protein interferon, a premium was put on its economical production. Biogen would succeed in its microbial production in 1980. Another company, Genex, managed to produce a third protein, the human growth hormone, which physicians could use to treat patients whose growth had been stunted by their own body's failure to produce the hormone. A fourth company, Cetus, succeeded in producing an instrument that would make such developments easier. The polymerase chain reaction (PCR) machine enabled rapid reproduction of a given DNA sequence.

Each of these companies had a novel structure compared with the traditional pharmaceutical corporations. Rather than being large businesses with substantial marketing and distribution departments and professional business leadership, they were small, led and largely manned by scientists with continuing academic associations, and their major product was intellectual property rather than drugs. Thus, Genentech licensed its discovery to Eli Lilly, the traditional market leader in insulin production, and Biogen licensed its beta-interferon to Schering Plough and other companies. Because they rewarded their staff with stock options, scientists working for these companies were rewarded with both the satisfaction of working on pathbreaking products and the prospects of becoming wealthy. Rather than practicing the risk aversion of large companies, these organizations were risk and prospect hungry. The close relationship with universities also changed academe. Some of the most distinguished scientists at major universities would found new companies; thus, Harvard's Walter Gilbert was both a founder and a chief executive officer of Biogen in 1978 and was awarded the Nobel Prize for Chemistry in 1980.

The new developments were turned into a narrative by a congressional body established in the 1970s, the Office of Technology Assessment. The only federal body to review the area publically, it published a 1984 report entitled "Commercial

Biotechnology: An International Analysis." It distinguished between the "old biotechnology associated with traditional fermentation industries and the "new biotechnology" associated with the exploitation of molecular biology, particularly recombinant DNA and cell fusion. Although the report also included novel bioprocess technologies, it soon dropped these from the category. In the characterization of new biotechnology, great emphasis was given to the academic base. The index gave separate entries to 14 different universities. This academic linkage attracted even large companies. Thus, in 1981 the German Hoechst organization offered Harvard Medical School's Massachusetts General Hospital $67 million in exchange for first rights of commercial exploitation of its intellectual developments for a whole decade.

Industrial Development. Biotechnology was closely associated with recovery from the "stagflation" of the late 1970s. Its academic base was supported by the encouragement given to academic enterprise by changes in the taxation of capital gains (1978) and changes in the rules governing the rights to patents based on federally funded research (the Bayh–Dole Act of 1980). States and regions viewed these new developments as potential ways to move them from old industries to high technology. San Diego and the Research Triangle region in North Carolina proved particularly successful. Worldwide, the United States proved the most successful and innovative home for biotechnology and its commercial prosperity.

The range of products produced through biotechnology grew rapidly from the early focus on proteins. Entrepreneurs developed a range of related but different products using the expertise of molecular biologists and the same funding patterns. These have included hybridomas, cloned mammals, and, most recently, "personalized medicine," products tailored for patients based on knowledge of their distinct genomic makeup. Diagnostics have long been an important application of biotechnology. One of the first uses of hybridomas was in rapid pregnancy testing. At the

beginning of the twenty-first century, biotechnology is making possible personalized medicine in which patients' genomes are examined to identify which drugs would be most appropriate for them. Above all, crops, typically genetically modified to build resistance to herbicides, have been a major product. In this sense the concept of biotechnology has kept the broad and unbounded range of meanings implied by the word's structure.

During the 30-year period beginning in 1980, the industry came to be characterized by episodically overinflated hopes. Venture capital–funded scientists occasionally succeeded in selling individual small organizations for hundreds of millions of dollars. By the end of the period, about one-third of all the venture capital available in the United States for the support of early-stage development was directed toward biotechnology. Intellectual property had been a major product of most companies. According to the Organisation for Economic Co-operation and Development (OECD), the total number of alliances between companies for research or technology transfer in biotechnology increased from 28 in 1990 to 360 in 2006. Whereas biotechnology had been a new category in 1980, according to the OECD, in 2006 the United States had 3,301 biotech research and development (R&D) firms, providing jobs in total to 1.36 million people, of whom 150,000 were employed in R&D.

[*See also* **Alzheimer's Disease and Dementia; Biological Sciences; Cancer; Chemistry; Diabetes; Disease; DNA Sequencing; Eugenics; Food Processing; Genetics and Genetic Engineering; Germ Theory of Disease; Higher Education and Science; Human Genome Project; Instruments of Science; Journals in Science, Medicine, and Engineering; Lederberg, Joshua; Medicine; Medicine and Technology; Military, Science and Technology and the; Molecular Biology; National Institutes of Health; Nobel Prize in Biomedical Research; Office of Technology Assessment, Congressional; Penicillin; Pharmacology and Drug Therapy; Science;**

Technology; Tuberculosis; War and Medicine; *and* Watson, James D.]

BIBLIOGRAPHY

Bud, Robert. *The Uses of Life: A History of Biotechnology.* Cambridge, U.K.: Cambridge University Press, 1993. This author's interpretation. Links developments in industrial microbiology before the commercialization of molecular biology with the subsequent commercial enthusiasm for biotechnology.

Cambrosio, Alberto, and Peter Keating. *Exquisite Specificity: The Monoclonal Antibody Revolution.* New York: Oxford University Press, 1995. Authoritative, sociologically sophisticated interpretation of the rise of hybridomas.

Dutfield, Graham. *Intellectual Property Right and the Life Science Industries: Past, Present and Future.* Singapore: World Scientific, 2009.

Hughes, Sally Smith. *Genentech: The Beginnings of Biotech.* Chicago: University of Chicago Press, 2011. An unusual in-depth study of a biotech company.

Krimsky, Sheldon. *Biotechnics and Society: the Rise of Industrial Genetics.* New York: Praeger, 1991. The best view of the impact on academe of the enthusiasm for industrial genetics.

U.S. Office of Technology Assessment. *Commercial Biotechnology: An International Analysis.* Washington D.C.: U.S. Government Printing Office, 1984. http://www.princeton.edu/~ota/ns20/alpha_f.html (accessed 9 April 2012). This report was a key document in making a narrative out of sometimes apparently chaotic diversity.

Van Beuzecom, Brigitte, and Anthony Arundel. *OECD Biotechnology Statistics* 2009. http://www.oecd.org/dataoecd/4/23/42833898.pdf (accessed 9 April 2012). This series provides key authoritative data.

Vettel, Eric James. *Biotech: The Countercultural Origins of an Industry.* Philadelphia: University of Pennsylvania Press, 2006. An excellent evocation of the changing attitudes to application and commercialization in the life science faculties of key West Coast universities during the postwar era and how this experience framed the reception of innovations in molecular biology.

Robert Bud

BIRTH CONTROL AND FAMILY PLANNING

Margaret Higgins Sanger (1879–1966) helped coin the term "birth control" in 1914 as she began a lifelong mission to secure women's access to safe, effective, and legal methods of fertility limitation. However, Americans have long employed fertility-control practices that have ancient origins. Attitudes toward birth control are mediated by class and culture, and they have changed over time according to shifting definitions of gender roles, childbearing, sexuality, morality, legality, and population management. Contraception raises ongoing debates over who controls reproduction: a woman, her husband or partner, the government, religious leaders, or medical/scientific experts.

Early America. In early America, European women brought with them knowledge of herbal contraceptives and abortifacients, including savin, rue, tansy, and pennyroyal. Euro-American women sought new abortifacients like snakeroot from American Indians and cotton root from enslaved Africans and in turn shared medicinal knowledge. Women also recognized that breastfeeding increased child spacing. Although Europeans used barrier methods like condoms and pessaries (diaphragms or intrauterine devices), they are rarely mentioned in American sources. Women and midwives managed childbirth and fertility within the domestic sphere. Until the early nineteenth century, a woman could legally terminate a pregnancy prior to "quickening," the perception of fetal movements in the second trimester of pregnancy.

In the seventeenth and early eighteenth centuries, most Euro-Americans viewed large families positively because children provided an economic safety net and laborers in an agrarian and artisanal economy. However, fertility rates began to decline in free populations in the late eighteenth century. Starting with the urban middle classes, women redefined the American Revolutionary rhetoric of equality, rationality, and self-control and invented a new small-family ideal by intentionally limiting fertility. With more time and resources, middle-class women focused on rearing and educating fewer children and pursued interests outside the home. White women of all classes averaged 7 children in 1800, 5 by 1850, and 3.5 by 1900 (Klepp, 2009, p. 8). In the new market-oriented republic, children became a responsibility rather than a labor source.

For enslaved African Americans, reproductive autonomy was restricted by slave owners' legal control over women's procreative bodies. Colonial laws mandated that slavery was inherited based upon the mother's free or enslaved status. Enslaved women faced a dilemma: unfree children perpetuated the system of slavery, but childbirth also ensured the continuity of African American communities. Some slave owners increased their labor force by raping enslaved women and by practicing enforced "breeding." Childbearing became a site of slave resistance, and slaveholders decried black women and midwives' contraceptive and abortion practices. The legacy of slaveholders' abusive regulation of slaves' reproduction informed African Americans' twentieth-century protests against coercive state-sponsored sterilization and birth control.

Contraceptive Proliferation and Restrictive Legislation.

In the 1830s, a popular health movement, an expanding consumer marketplace, and increasing literacy facilitated the dissemination of contraceptive information and technologies. Health reformers promoted sex education and contraceptive methods like postcoital douching. They debated the health benefits and hazards of popular methods like periodic abstinence and *coitus interruptus* (withdrawal prior to ejaculation). Consumers learned about contraceptives and abortion providers through books, newspapers, magazines, public lectures, and word of mouth. Charles Goodyear's discovery of vulcanization in 1839 expedited the production of more pliant and durable rubber products, including condoms, douching syringes, intrauterine devices (IUDs), and diaphragms. By the mid-nineteenth century, consumers could purchase contraceptives from reputable pharmacies, hardware stores, mail-order vendors, and door-to-

door salespeople. However, there were no regulations regarding product safety and efficacy, and some religious leaders and physicians increasingly linked contraception and abortion with immorality.

In the last decades of the nineteenth century, middle-class suffragists, social reformers, and Free Love advocates found common ground in a voluntary motherhood ideology that asserted a woman's right to refuse sexual intercourse. Radical activists promoted reproductive self-sovereignty and female sexual desire in a culture that considered women passionless. Conservative reformers argued that voluntary motherhood strengthened women's traditional maternal role. All affirmed "natural" forms of birth control like periodic abstinence and rejected abortion and barrier methods. Some voluntary motherhood advocates joined purity reformers who sought to limit a flourishing sexualized popular culture. In 1873, under purity reformer Anthony Comstock's sponsorship, Congress passed the Comstock Act, which prohibited using the federal postal service to disseminate materials deemed obscene, including contraceptive and abortion information and devices. Twenty-four states responded with similar legislation supported by purity reformers and the American Medical Association (AMA). Many physicians took this opportunity to claim the role of scientific and moral arbiters. Although Comstock and his allies pursued contraceptive purveyors, apart from several high-profile cases, enforcement was difficult because devices like vaginal syringes could be used for "hygienic" purposes. Veiled in euphemisms, an illicit contraceptive market thrived.

The Birth Control Movement.

Margaret Sanger founded the American birth control movement while working in the 1910s as a visiting nurse among New York City's impoverished immigrant women. She witnessed the physical, emotional, and economic consequences of continuous childbirth and bungled illegal abortions. Sanger's participation in radical socialist and feminist causes also informed her birth control campaign. She reprised earlier calls for women's right to sexual expression and argued that women's social and political freedom hinged upon their reproductive self-determination. Sanger was indicted in 1914 for publishing material considered obscene under the Comstock Law. She issued an influential pamphlet on contraception entitled *Family Limitation* before fleeing to Europe, where she observed innovative physician-run contraceptive clinics employing diaphragms. When Sanger returned to New York in 1916, her charges were dropped. However, when Sanger and her sister opened America's first birth control clinic, police promptly raided and closed it. When the women were convicted and jailed, they used the press coverage to advocate the repeal of the increasingly unpopular Comstock Laws. In 1918, Judge Frederick Crane ruled on Sanger's case that physicians could prescribe contraceptives to prevent disease and promote health. When Sanger opened her first legal contraceptive clinic in New York City in 1923, she extended this definition to include fertility limitation as necessary for women's health. Condom manufacturers interpreted Crane's ruling as legalizing their product to prevent venereal disease, which coincided with World War I veterans' popularization of condoms as inexpensive prophylactics and contraceptives.

The birth control movement emerged alongside eugenics and anti-immigration movements. Whereas eugenics encompassed a variety of ideologies, proponents generally advocated selectively breeding humans to increase the number of allegedly "fit" people and to decrease hereditarily "unfit" populations. Some Americans of northern European descent feared their numbers were diminishing, and African Americans and immigrants from southern and Eastern Europe were apparently increasing. Leaders like Theodore Roosevelt maintained that middle- and upper-class "old stock" Americans were committing "race suicide" by using birth control, whereas the working classes and the unfit were incapable of voluntary contraception. Some eugenicists promoted restrictive immigration laws and compulsory surgical sterilization to restrict the growth of the unfit, including criminals, epileptics, the mentally ill, and the poorly defined "feeble-minded." Prominent scientists legitimized eugenics and its rhetoric became widespread.

As Sanger and her colleagues organized a nationwide network of birth control leagues and clinics in the 1920s, they distanced the movement from Sanger's earlier radicalism by arguing that contraception improved maternal/child welfare and solved societal problems like poverty through population control. Because they were even more controversial than birth control, the clinics did not provide abortions. Sanger championed physician-run clinics that promoted woman-controlled, safe, and effective diaphragms and spermicides. Sanger's rival Mary Ware Dennett, who founded an early birth control league, contended that Sanger's strategy forced women to rely on physicians for contraception access. Nevertheless, birth control activists, often socially prominent women, established over 650 contraceptive clinics by 1939 (Hajo, 2010, p. 21). Women without access to these predominantly urban clinics continued to use popular nonprescription methods like suppositories, withdrawal, condoms, and vaginal douches, often prepared with caustic substances like Lysol. Although birth control activists anticipated that state and federal governments would fund contraceptive services as part of broad-based public-health programs accessible to impoverished women, most clinics were privately funded.

Racial Implications. Recognizing African American women's interest in birth control, in 1930 the Urban League and Sanger's Birth Control Clinical Research Bureau opened a Harlem clinic staffed by black physicians and nurses. Birth control clinics sparked debates in the African American community that continued throughout the century. Whereas prominent activists including sociologist W. E. B. Du Bois supported birth control, Black Nationalists like Marcus Garvey condemned it as "race suicide," arguing that an increasing black population amplified political clout. African Americans and the poor had reason to be concerned. In the 1927 *Buck v. Bell* case, the Supreme Court upheld a Virginia eugenics statute and ordered the compulsory surgical sterilization of allegedly feeble-minded 21-year-old Carrie Buck. Thirty states followed with compulsory sterilization laws that were often implemented against African Americans and the poor. Black activists questioned whether birth control initiatives, like the North Carolina program funded by philanthropist–activist Clarence Gamble in 1937, targeted African Americans. Birth control's links to population control blurred boundaries between humanitarianism and social control.

Contraceptive Legitimacy and the Pill.

Sanger's commitment to physician-prescribed contraceptives allowed her to mobilize physicians' increasing scientific and political power. Still, until the late 1930s, the AMA's official position was anti-contraception, despite longstanding pro–birth control lobbying by prominent physicians like obstetrician/gynecologist Robert Dickenson. Sanger employed the rhetoric of eugenics and recruited prominent pro-contraception eugenicists as board members to counter anti-contraception physicians. However, she rejected any eugenic programs that discouraged women from using birth control. Sanger again used the courts to forward her cause when customs agents confiscated a clinic physician's contraceptive order from Japan. In *U.S. v. One Package* in 1936, the Circuit Court judge exempted physicians from the Comstock Law. After the ruling, the AMA endorsed physician-prescribed birth control in 1937. In a 1938 *Ladies Home Journal* survey, 79 percent of readers approved of birth control. Sanger declared a victory, but the Catholic Church responded by successfully lobbying to close clinics in Massachusetts and Connecticut. Catholic leaders affirmed longstanding beliefs that sexual intercourse should be confined to marriage for the divinely ordained purpose of unimpeded procreation.

In 1942, Sanger's organizations were restructured to form Planned Parenthood Federation of America. Its new agenda linked family health with population control and national strength, but Sanger objected that this stifled her goal of women's reproductive self-determination. Following World War II as journalists publicized Nazi atrocities, eugenics was discredited and eugenicists also refocused on population control. Cold War tensions exacerbated fears that a "population explosion" would increase poverty and famine in developing countries, causing them to turn to Communism. In 1952, new organizations like John D. Rockefeller III's Population Council and Sanger's International Planned Parenthood Federation addressed these concerns. By the 1960s, the U.S. government funded international family-planning programs, giving federal birth control policy wide-reaching implications.

Sanger continued her quest for improved contraceptives. In 1953, with funding and advice from philanthropist–activist Katharine McCormick, Sanger tapped biologist Gregory Pincus, who later collaborated with fertility specialist Dr. John Rock to develop an estrogen/progestin pill. The Food and Drug Administration approved "the Pill" in 1960, and it quickly became American women's favored contraceptive. However, the high estrogen doses caused side effects, including blood clots and strokes. After several deaths from the Pill and life-threatening uterine infections from the 1970 Dalkon Shield IUD, women's health activists successfully lobbied for safer low-dose contraceptives, improved IUDs, and side-effect disclosures with prescriptions.

Oral contraceptive mass-marketing occurred in tandem with the rise of second-wave feminism and relaxed sexual norms. Feminists asserted that effective contraception facilitated gender equality and women's full participation in economic and political life. The Pill was nearly 100 percent effective, increasing women's ability to separate sex from reproduction. In the landmark decision *Griswold v. Connecticut* (1965), the Supreme Court invalidated a Connecticut law prohibiting the dissemination of contraceptives and confirmed married people's right to privacy. *Eisenstadt v. Baird* (1972) extended this right to unmarried people. In 1968, Pope Paul VI responded to cultural changes by reconfirming the Catholic Church's stance against abortion and artificial contraception. Catholic leaders allowed only periodic abstinence and the rhythm method, but many American Catholic women embraced the Pill. Despite Catholic opposition, in 1968 President Lyndon Johnson funded contraception services as part of his war on poverty. President Richard Nixon followed suit, funding family planning under the Public Health Service Act.

Reproductive Rights. In the 1970s, white feminists made abortion access central to their reproductive rights agenda, and they were energized by the 1973 *Roe v. Wade* Supreme Court decision legalizing abortion. Activist women of color challenged white feminists to expand their reproductive rights agenda to encompass minority women's concerns, including abolishing coercive birth control and sterilization and assuring federally funded maternal/child health care and child care. Activists protested the 1973 cases of ten Mexican American women in Los Angeles and two young African American girls in Alabama who were surgically sterilized without appropriate consent. They also publicized coerced sterilizations of American Indian women in federal hospitals and questions regarding Puerto Rican contraceptive trials. Puerto Rican and Black Nationalist organizations argued that abortion and contraception constituted genocide and called for women of color to reproduce to enhance the groups' political and economic power. Feminist women of color argued instead for reproductive autonomy, including the right to have children or to choose abortion without government coercion. White feminists and feminists of color joined forces to protest the 1976 Hyde Amendment that barred most Medicaid-funded abortions. In the 1980s and 1990s, state and federal legislators debated linking birth control—including long-term injectable and implantable forms—to welfare benefits. Minority women protested and continued to battle for reproductive justice in the context of broader civil rights agendas.

Birth control and abortion were increasingly politicized in the last decades of the twentieth century. The pro-life movement gained momentum following *Roe v. Wade,* along with an antifeminist backlash. President Ronald Reagan promoted an anti-abortion agenda, including a "global gag rule" in 1984 that defunded international family-planning agencies offering abortion counseling, followed by a similar domestic legislation. Although most anti-abortion organizations pursued peaceful protests, others promoted intimidation and violence, which impeded women's access to all contraceptive services. Planned Parenthood joined feminist groups like the National Organization for Women and the National Association for the Repeal of Abortion Laws in mass protests and intensive lobbying, resulting in the passage of the Freedom of Access to Clinic Entrances Act in 1994. The debate continued: President Bill Clinton reversed both gag rules in 1993, President George W. Bush reinstated them, and President Barack Obama reversed them again in 2009.

In the early twenty-first century, there are a variety of choices in birth control methods, including improved barrier methods, hormonal patches, vaginal rings, injections, and three-year implants. Amid pro-life organizations' protests, the Food and Drug Administration approved over-the-counter emergency contraceptives in 2006 and a prescription-only five-day pill in 2010. Emergency contraception reinvigorated debates over access to hormonal contraception. Reproductive rights activists applauded President Obama's 2010 Affordable Care Act provisions mandating insurance coverage for contraceptives without co-payments. However, in the 2011 Congressional budget battles, conservatives opposed this measure as well as federal funding for Planned Parenthood clinics. Despite Americans' wide acceptance of contraception, reproductive rights remains controversial a century after Margaret Sanger initiated the birth control movement.

[*See also* **Abortion, Debates and Science; American Medical Association;** *Baby and Child Care;* **Childbirth; Ethics and Medicine; Eugenics; Gender and Science; Gender and Technology; Goodyear, Charles; Health and Fitness; Health Insurance; Health Maintenance Organizations; Hygiene, Personal; Medicare and Medicaid; Medicine; Midwifery; Military, Science and Technology and the; Pharmacology and Drug Therapy; Pincus, Gregory Goodwin; Postal Service, U.S.; Public Health; Public Health Service, U.S.; Pure Food and Drug Act; Race and Medicine; Religion and Science; Sanger, Margaret; Sex and Sexuality; Sex Education;** *and* **War and Medicine.**]

BIBLIOGRAPHY

Chesler, Ellen. *Woman of Valor: Margaret Sanger and the Birth Control Movement in America*. New York: Simon and Schuster, 2007.

Critchlow, Donald T. *Intended Consequences: Birth Control, Abortion, and the Federal Government in Modern America*. New York: Oxford University Press, 1999.

Engelman, Peter. *A History of the Birth Control Movement in America*. Santa Barbara, Calif.: Praeger, 2011.

Gordon, Linda. *The Moral Property of Women: A History of Birth Control Politics in America*, 3d ed. Urbana: University of Illinois Press, 2002.

Hajo, Cathy Moran. *Birth Control on Main Street: Organizing Clinics in the United States, 1916–1939*. Urbana: University of Illinois Press, 2010.

Klepp, Susan E. *Revolutionary Conceptions: Women, Fertility, and Family Limitation in America, 1760–1820*. Chapel Hill: University of North Carolina Press, 2009.

Marsh, Margaret, and Wanda Ronner. *The Fertility Doctor: John Rock and the Reproductive Revolution*. Baltimore: Johns Hopkins University Press, 2008.

Roberts, Dorothy. *Killing the Black Body: Race, Reproduction and the Meaning of Liberty*. New York: Pantheon Books, 1997.

Silliman, Jael M., Marlene Fried, Loretta Ross, and Elena Gutierrez, eds. *Undivided Rights: Women of Color Organize for Reproductive Justice*. Cambridge, Mass.: South End Press, 2004.

Tone, Andrea. *Devices and Desires: A History of Contraceptives in America*. New York: Hill and Wang, 2001.

Susan Hanket Brandt

BLACKWELL, ELIZABETH

(1821–1910), first woman medical school graduate in the United States, key figure in opening the medical profession to women. Born in Bristol, England, Blackwell came to the United States in 1832. She was the third of nine children and part of a family that, through marriage, came to include such feminist pioneers as her sister-in-law Lucy Stone and Antoinette Brown Blackwell (1825–1921), the first ordained woman minister in the United States. Blackwell's father had moved the family to the United States because of business difficulties. When he died in 1838, he left the family with little, so Blackwell and her two older sisters supported the family by operating a school.

Blackwell's decision to study medicine was prompted in part by a friend who died of uterine cancer, claiming that she would have sought medical advice earlier if a woman doctor had been available. Following considerable effort and difficulty, including opposition from the medical establishment, Elizabeth Blackwell was admitted to Geneva Medical College in New York, from which she graduated in 1849. After postgraduate study in London and Paris, she returned to New York City, setting up a practice in 1851. She and her sister Emily (1826–1910) established the New York Infirmary for Women and Children in 1857. The Infirmary was run by women, and many early woman doctors gained experience there. The sisters also established the Women's Medical College of the New York Infirmary in 1868.

In 1858, Blackwell visited Britain for a year to lecture and practice medicine. On her return to the United States in 1859, she concentrated on the New York Infirmary and the establishment of the Medical School. Eventually, tension with her physician sister over the management of the infirmary led to her 1869 decision to return to England. Although she initially remained active in the women's medical movement, she soon became attracted to other causes. Indeed, her interest in such topics as moral reform, spiritualism, hygiene, and animal welfare soon overshadowed her medical interests. She became a prolific author of works on reform topics, writing such books as *The Human Element in Sex* (1880) and *Counsel to Parents on the Moral Education of Their Children* (1880). Her memoirs, *Pioneer Work in Opening the Medical Profession to Women* (1895), focus on her early life and medical career in the United States. Blackwell's historical significance derives less from her contributions to medical practice than from the example she set for women interested in the medical profession.

[*See also* **Animal and Human Experimentation; Gender and Science; Hygiene, Personal; Medical Education;** *and* **Medicine.**]

BIBLIOGRAPHY

Dupree, Marguerite. "Blackwell, Elizabeth." In *Dictionary of Medical Biography,* edited by W. F. Bynum and Helen Bynum, Vol. 1, pp. 224–225. Westport, Conn., and London: Greenwood Press, 2007.

Sahli, Nancy Ann. *Elizabeth Blackwell, M.D. (1821–1910): A Biography.* Stratford, N.H.: Ayer Company, 1981.

Nancy A. Sahli;
updated by Elspeth Knewstubb

BLALOCK, ALFRED

(1899–1964), surgeon, first to perform the so-called "blue baby" operation. A native of Culloden, Georgia, Blalock graduated from the University of Georgia and earned an MD at Johns Hopkins Medical School in 1922. He served as an intern and assistant resident at the Johns Hopkins Hospital until 1925, when he became a resident surgeon and eventually head of surgery at Vanderbilt University Hospital. At Vanderbilt, Blalock pursued what would become a lifelong research interest: the physiological effects of diminished blood volume and insufficient oxygenation of the blood by the lungs. His canine research on the role of severe blood loss in triggering traumatic shock, as well as ways of preventing this through plasma and whole-blood transfusions, played an important role in the treatment of wounded soldiers during World War II. In a 1938 experiment on hypertension, he joined a dog's left subclavian artery to the left pulmonary artery. He also conducted the first successful canine kidney transplant, connecting the organ's vascular system to a cervical artery. These experiments served as a model for the later surgical breakthrough for which he is best known.

Returning to Johns Hopkins Medical School in 1941, Blalock occupied the posts of surgeon in chief and director of the surgery department. Here he continued research on problems of the pulmonary–vascular system. In collaboration with Dr. Helen B. Taussig (1898–1986), director of the cardiac clinic of Hopkins' pediatric depart-

ment, he devised a procedure for the surgical treatment of pulmonary stenosis, a syndrome of congenital heart malformations, including a narrowing of the pulmonary artery, that impede blood oxygenation and circulation, causing a bluish discoloration of the skin, particularly in infants. (The condition's formal name, Tetralogy of Fallot, derives from a nineteenth-century French physician, Étienne-Louis Arthur Fallot.) On 29 November 1944, Blalock performed the first "blue baby" operation on a 15-month-old girl by connecting one end of the left subclavian artery to the left pulmonary artery, increasing blood flow to the lungs. The operating team included Taussig and surgical technician Vivien T. Thomas (1910–1985), an African American who lacked a college degree but who had achieved great proficiency in surgical technology and techniques as Blalock's laboratory assistant, first at Vanderbilt and then at Hopkins.

Blalock and Taussig described the procedure, now known as the Blalock–Taussig shunt, in a 19 May 1945 article in *The Journal of the American Medical Association.* As word spread, surgeons from other institutions arrived to learn the technique, and Blalock and Taussig offered clinical demonstrations at medical schools throughout the United States and abroad. Before his death in 1964, Blalock received numerous honors, including election to the National Academy of Sciences. Now performed worldwide, the surgery he and Helen Taussig pioneered, with the significant contribution of Vivien Thomas, has saved the lives of many thousands of infants.

[*See also* **Medical Specialization; Medicine; Military, Science and Technology and the; National Academy of Sciences; Pediatrics; Physiology; Surgery;** *and* **War and Medicine.**]

BIBLIOGRAPHY

Alan Mason Chesney Medical Archives of the Johns Hopkins Medical Institutions. "The Blue Baby Operation." http://www.medicalarchives.jhmi.edu/page1.htm (accessed 25 October 2012).

Harvey, A. McGehee, et al. *A Model of Its Kind: A Centennial History of Medicine at Johns Hopkins.*

2 vols. Baltimore: Johns Hopkins University Press, 1989.

Ravitch, Mark M., ed. *The Papers of Alfred Blalock.* 2 vols. Baltimore: Johns Hopkins University Press, 1966.

David Cortes;
updated by Paul S. Boyer

BLINDNESS, ASSISTIVE TECHNOLOGIES AND

Assistive technologies provide the blind with alternative means of literacy and mobility. Blind people sometimes engage the sighted for help, and proper use of a sighted assistant or guide dog is a technical blindness skill.

Literacy. Valentin Haüy, founder of the school for the blind where Louis Braille lived from 1821 until his death in 1852, was one of several Europeans who created tactile alphabets. Not satisfied with Haüy's raised letters, Braille devised the dotted code that bears his name as well as a special code for music. Others have revised Braille's music code and created dotted systems for mathematics and computer science. In the United States, until about 1920, braille competed with other systems of tactile writing (including alphabets and another dotted code). Dotted codes are both easier to read with the fingers and readily written using a pointed tool and a template.

Tactile books were produced using stereotype plates. In 1893, Frank H. Hall of the Illinois school for the blind introduced the braille stereotype machine. Readable braille embossed on both sides of a page became common after 1900.

In 1892, Hall introduced a braillewriter with a stationary head that could emboss any combination of six dots in a two-by-three rectangle. Later braillewriters used a stationary platen and movable embossing head. In 1951, the Perkins School for the Blind introduced the Perkins Brailler, which became the most common mechanical tool for producing single copies of braille. Because standard braille takes

up considerably more space than 10- or 12-point type, blind people frequently corresponded by typing.

A small proportion of America's blind read Braille; the rest depend on sound recordings. Since 1931, the National Library Service for Blind and Physically Handicapped Individuals has been the central node of a network that distributes reading material in braille or talking-book format. With braille and talking books, printed materials in 14-point type qualify for free U.S. mail delivery. The development of digital sound now allows the blind to download talking books and magazines.

Electronic Technologies for Reading and Writing. In the 1960s, Stanford engineer John Linvill introduced the Optacon, a mechanico-electronic device that forms tactile images of letters. A blind person familiar with the alphabet could thus read print. A decade later, Ray Kurzweil introduced the Kurzweil Reader, a tabletop machine that combined multifont optical scanning with text-to-speech translation. Successively smaller models led, in 2008, to Kurzweil reading software that can be installed on a smartphone.

More significant for many blind people have been the personal computer and the Internet. With screen-reading software, the blind can read aurally what they had just written. This technology thus allows for electronic correspondence between the blind and the sighted on equal terms. By 2011, standards for nonvisual accessibility were making increasing numbers of digital applications available for use by the blind.

Refreshable braille came into its own in the 1990s. Braille displays—devices that form a row of braille characters by raising dots through holes in a flat surface—can be attached to computers and driven by screen-reading software. They are also built into braille note-takers—portable computers without monitors and with a configuration of keys similar to that of mechanical braillewriters.

In the nineteenth century, American schools for the blind began making use of handcrafted tangible aids for teaching geography and other subjects. During the Great Depression, the Works Progress Administration used hand-embossed plates for mass production of maps for the blind. By the end of the twentieth century, semiautomated means of producing tactile graphics were becoming common.

Orientation and Mobility. Until World War II and for some time afterward, blind Americans used standard wooden crook-handled canes painted white and red. During the war, a counselor at the Veterans Administration facility in Valley Forge, Pennsylvania, introduced the "long white cane." Swung in a wide arc rather than tapped immediately in front of the blind traveler, the long cane not only identifies objects in the blind person's path, but also detects changes in the surface, indicates elevation changes, reveals the nature of objects by their sound when tapped, and creates potentially helpful echoes. By 1960, the long white cane had become basic to the method of teaching orientation and mobility known as "structured discovery."

Fiberglass, carbon fiber, and aluminum have since replaced wood. In addition to straight one-piece canes, telescoping and folding canes have become common. Attempts to augment the capabilities of canes through electronic enhancements have not roused much interest among the blind.

Other technologies intended to facilitate travel by the blind include braille signage, truncated domes, and audible traffic signals. Many blind people regard the last two as unnecessary because proper use of the cane should indicate where a sidewalk ends, and audible signals may confuse a blind pedestrian who is listening for the noise made by cars. Because the noise of traffic has traditionally been a blind person's guide to safe street crossing, the organized blind have pushed manufacturers to have hybrid and electric cars mimic the sound of a gasoline engine.

Engineers at the Smith–Kettlewell Eye Institute have created a transmitter for mounting on street signs. The transmitter beams an infrared signal with information on location and direction to blind pedestrians who carry receivers. As of 2011, these had been installed on a few corners in San Francisco.

Dennis Hong and his students at Virginia Tech have developed a prototype automobile that provides tactile signals to a blind driver from sensors mounted on the exterior of the car. Hong and his team look forward to applying their technology to other areas of interest to blind people.

Universal Design. Disability activists advocate "universal design," or consideration of requirements for use by the disabled, in the design of new products (or their subsequent redesign). In the case of touch-screen devices, pressure from the organized blind has led several manufacturers to modify their products so that they can be used either visually or aurally.

Digital technology is inadvertently responsible for the most successful example of universal access. Text in digital format is not necessarily associated with any particular sensory modality. As electronic books came of age at the start of the twenty-first century, there was no technical reason why they should be any less accessible to the blind than to the sighted.

[*See also* **Computer Science; Computers, Mainframe, Mini, and Micro; Internet and World Wide Web; Keller, Helen; Medicine and Technology; Ophthalmology; Optometry; Printing and Publishing;** *and* **Technology.**]

BIBLIOGRAPHY

Braille into the Next Millennium. Washington, D.C.: National Library Service for the Blind and Physically Handicapped, 2000. Twenty-three essays on the history and other aspects of braille and tactile graphics.

Eriksson, Yvonne. *Tactile Pictures: Pictorial Representations for the Blind 1784–1949*. Gothenburg, Sweden: Acta Universitas Gothoburgensis, 1998.

Ferguson, Ronald F. *The Blind Need Not Apply: A History of Overcoming Prejudice in the Orientation and Mobility Profession*. Charlotte, N.C.: Information Age Publishing, 2007. Describes the history of the adoption of the long white cane, the development of "structured discovery," and the struggle of blind people to be recognized as capable travel instructors.

NLS: *That All May Read*. http://www.loc.gov/nls/ aboutnls.html (accessed February 27, 2012). This site has links to pages on the history, legal basis, organization, and functions of the National Library Service for Blind and Physically Handicapped Individuals.

Edward T. Morman

BOAS, FRANZ

(1858–1942), founder of modern American cultural anthropology. Born in Minden, Westphalia, Germany, into a freethinking Jewish household, Boas attended Heidelberg, Bonn, and Kiel universities, receiving his doctorate in physics from Kiel in 1881. After a year in the German army and another two years of reading, Boas went to Baffin Island on an expedition to study the cultural geography of the Eskimos. His experiences among the Eskimos created in him a desire to understand the laws of human nature and prompted him to make a gradual transition from cultural geography to ethnology.

In 1887, hoping for career opportunities denied him in the conservative, anti-Semitic climate of Bismarck's Germany, Boas immigrated to the United States. Yet for almost a decade after his arrival, he confronted serious obstacles in securing and holding professional positions, primarily because of the virulent anti-Semitism that pervaded the nation. In 1896, Frederic W. Putnam appointed him assistant curator at the American Museum of Natural History in New York City; he advanced to the curatorship in 1901. He was simultaneously serving as a lecturer at Columbia University, where he became professor of anthropology in 1899, a position he held until his death. Before 1920, Boas had trained such prominent anthropologists as Alfred L. Kroeber, Robert H. Lowie, Edward Sapir, and Alexander A. Goldenweiser. Students trained after 1920 included Melville J. Herskovits, Ruth Benedict, Margaret

Mead, and Otto Klineberg. So great was Boas' impact on American anthropology that by the 1950s virtually all the anthropologists in America had studied under him or one of his students. Boas helped establish the American Anthropological Association in 1902 and served as its president from 1907 to 1909. He also served as president of the American Association for the Advancement of Science in 1931.

Boas's scholarship—especially his pathbreaking *The Mind of Primitive Man* (1911)— profoundly influenced the concepts of "race" and "culture." Debunking the concepts of cultural hierarchies and of "race" as a supraindividual organic entity, Boas almost single-handedly ushered in the modern conceptions of race and cultural relativism. In his research studying the Indians in northwestern America, Boas focused on their folklore and art; he introduced the innovation of relying on trained native speakers to document unwritten languages. In addition, Boas had a profound influence on the Harlem Renaissance. He worked with both the great novelist and folklorist Zora Neale Hurston and the cultural critic Alain L. Locke, as well as opening the pages of the *Journal of American Folklore*, one of the major journals that he served as editor, to numerous African American folklorists.

All too aware of the insidious implications of racism during World War II, Boas suffered a fatal heart attack while denouncing Nazi propaganda in 1942.

[*See also* American Association for the Advancement of Science; American Museum of Natural History; Anthropology; Journals in Science, Medicine, and Engineering; Mead, Margaret; Race Theories, Scientific; Social Sciences; *and* Societies and Associations, Science.]

BIBLIOGRAPHY

Baker, Lee D. *From Savage to Negro*. Berkeley: University of California Press, 1998.

Stocking, George W., Jr. *Race, Culture, and Evolution*. New York: Free Press, 1968.

Williams, Vernon J., Jr. *Rethinking Race*. Lexington: University Press of Kentucky, 1996.

Vernon J. Williams Jr.

BOTANICAL GARDENS

Modern botanical gardens arose at sixteenth-century universities with medical faculties (the first was at the University of Pisa dating from 1543) and at guild "physic gardens" (the Apothecaries' Guild in London founded their Chelsea Physic Garden in 1673). They received further impetus from the explorations by the European empires and became showcases of national pride. The Royal Botanical Gardens at Kew is the epitome of this impulse because it took on a scientific character under the leadership of Sir Joseph Banks in the aftermath of the 1768–1771 voyage around the world of HMS *Endeavour*, led by Captain James Cook. Banks sailed with Cook and made botanical collections with Daniel Solander that reflected the reach of the British Empire.

These trends, particularly nationalism, had their expression in the British North American colonies and later the United States, first through Bartram's Garden in Philadelphia. Founded in 1728, John Bartram and, later, his sons John Jr. and William, who collected in the south in 1773–1777, created the most distinctive and thorough collection of North American plants and helped send those plants to England and Europe, collaborating with English merchant Peter Collinson. The Bartrams helped make American plants known in the Old World. Given the importance of Philadelphia as a cultural center, the Bartrams were connected to important figures such as Benjamin Franklin and the botanist/physician Benjamin Smith Barton.

The Missouri Botanical Garden opened in St. Louis in 1859. Its founder, Henry Shaw, was inspired to build such a garden after visiting the Duke of Devonshire's grounds at Chatsworth during an 1851 visit to England. At the suggestion of William Hooker, director of the Royal Botanical Gardens at Kew, and Asa Gray of Harvard, Shaw enlisted the help of St. Louis physician George Engelmann, who was already playing a key role in cataloging and identifying the new plants discovered in the American West. Upon Shaw's death, Missouri flourished as a scientific garden, balancing—as all botanical gardens did—research, display, and education. Its graduate studies began with the founding in 1885 of the Henry Shaw School of Botany at Washington University in St. Louis. Systematic botany has been a consistent cornerstone of science at Missouri, with its Flora of Panama project begun in 1926, with its final publication in 1987 serving as a model for many such projects including those in Madagascar, China, the Venezuelan Guyana, North America, and Missouri.

The New York Botanical Garden was founded in 1888 building on the efforts of physician David Hosack, who also privately founded the Elgin Botanical Garden in Manhattan in the first decade of the nineteenth century, and by members of the Torrey Botanical Club, which was founded in 1867. This reflected the growing interest in the study of plants. Columbia University botanist Nathaniel Lord Britton and his wife Elizabeth Knight, herself active in the Torrey Botanical Club, reinitiated the idea of a botanical garden in New York City inspired by a visit to Kew. In 1891, the New York State Legislature set aside land for the Garden in the Bronx. Britton mobilized further support from the civic and social leadership of the city, including J. Pierpoint Morgan, Cornelius Vanderbilt, and Andrew Carnegie, to ensure its success. Appointed director in 1896, Britton led the Garden until 1929, initiating research projects on the *Cactaceae* and into the flora of Puerto Rico.

In the twentieth century, botanical gardens such as Rancho Santa Ana (1927) in Claremont, California, showcasing that state's unique flora, and Fairchild Tropical (1936) in Miami, based on economically important tropical plants, were formed around these important regional collections.

Botanical gardens in the United States have reflected the contributions of such physicians as Benjamin Smith Barton, David Hosack, Asa Gray, and George Englemann. They also have been an expression of the nation's extending reach and influence both throughout the continent and into Panama (Missouri), Puerto Rico (New York), and the tropics (Fairchild).

[*See also* **Bartram, John and William; Biological Sciences; Botany; Franklin, Benjamin; Gray, Asa;** *and* **Medicine.**]

BIBLIOGRAPHY

Ewan, Joseph, and Nesta Dunn Ewan. *Benjamin Smith Barton: Physician and Naturalist in Jeffersonian America.* St. Louis: Missouri Botanical Garden Press, 2007. An exposition of early American science through a thorough examination of Barton as a pivotal figure.

Kleinman, Kim. *The Museum in the Garden: Research, Display, and Education at the Missouri Botanical Garden since 1859.* PhD Diss. Union Institute, 1997. A historical study of the Missouri Botanical Garden as a natural history museum.

Mickulas, Peter. *Britton's Botanical Empire: The New York Botanical Garden and American Botany, 1888–1929.* New York: New York Botanical Garden Press, 2007. A study of the impact of the founding director on the New York Botanical Garden.

Kim Kleinman

BOTANY

Interest in the plant world was common in early modern Europe, although the terms *botany* and *botantiste* only developed in the late seventeenth century to denote those natural historians learned in the uses, characters, classes, orders, genera, and species of plants (Schiebinger, 2004, p. 6). The reasons for widespread interest in the plant world were both economic and medical and intricately tied for many years first with European colonialism and then with American nationalism. Professionalization and laboratory-based study eventually led to specialization within the discipline, and the twentieth and twenty-first centuries have been marked by further technological and disciplinary change.

The discovery of the New World led to great interest among European naturalists in collecting and cataloging the new flora (as well as other aspects of American natural history). This interest was not only scientific, but also driven by economic and medical considerations. Part of the impetus behind European voyages of discovery and colonialism was the potential profits of acquiring costly spices and medicinal plants. Christopher Columbus in 1492 sought a sea route to the lucrative spice, silk, and dye markets of China and India. Plants figured in colonial expansion into the Americas as an exercise in both discovery and profit making. Columbus's discovery began a frenzy of plant movements between old world and new. In 1494, Columbus returned to Hispaniola with sugar cuttings, citrus fruits, grapevines, olives, melons, onions, and radishes. The valuable sugar crop was acclimatized in the Americas by 1512 and in full production in Hispaniola by 1525. Plants from the Americas were taken back to Europe for classification and cultivation. Some of these proved extremely valuable as food, luxury items, or medicines. Peruvian bark (also known as *Cinchona officinalis*), for example, the source of alkaloid quinine, proved vital for European colonizing efforts in tropical climates (Schiebinger, 2004, pp. 3–4).

Quinine was perhaps the most obvious example of botany serving medical purposes, but much early botany was intricately tied to the search for new drugs and other *materia medica* as well as to economic schemes. Many natural historians who studied plants and their uses in the eighteenth century were also medical men. Carl Linnaeus, whose new binomial classification system for plants revolutionized eighteenth-century botany, was also a practicing physician. He held the chair of botany and medicine at Uppsala University, Sweden. Linnaeus's aim was to cultivate valuable plants within Sweden's borders and thus stem the flow of Swedish money to Asia. Linnaeus, like many other European collectors of the eighteenth century, was dependent for his plant specimens on his own collector Peter Kalm and American colonists.

Colonist Botanists. Colonists collecting and cataloging plants specimens in North America in the eighteenth century were part of a wider network of colonial knowledge. They adopted and implemented Linnaeus' system of classification and corresponded with and received financial support from European and English naturalists. A small circle of American naturalists emerged in the eighteenth century along the Eastern seaboard. Many were in contact with each other as well as with Europeans (Keeney, 2001, p. 90). In Philadelphia, John Bartram established the first botanical garden in the United States, filling it with locally collected specimens and with specimens collected on expeditions funded by a London mentor and patron, Peter Collinson, who was an important member of a network of natural history correspondence.

Other American botanists, such as John Clayton and Cadwallader Colden and his daughter Jane, also collected for British and European audiences. Linnaeus's decision to send Peter Kalm to collect specimens of North American plants in 1749–1750 was likely influenced by his correspondence with Collinson, John Bartram, and Cadwallader Colden (Müller-Wille, 2005, p. 39). These early American botanists usually undertook their own fieldwork, with the exception of Jane Colden. Bartram traveled as far afield as Florida from his Philadelphia base. Through cataloging American flora, they laid the groundwork for the great synthetic works of Asa Gray and other American botanists in the nineteenth century.

The decades following the American Revolution saw the severance of much correspondence with Britain and the beginning of a tradition of botanizing more tied to American national identity. Naturalists in the early republic imagined themselves as the center of a network of American plant collectors. They enlisted the assistance of domestic naturalists, physicians, travelers, gardeners, and others as collectors for their projects. American inclusion in the natural history circle of Peter Collinson was a casualty of the Revolution. American naturalists, particularly the well-educated, Eastern elite, sought to champion a more democratic natural historical practice (Lewis, 2005, p. 67). The late eighteenth and early nineteenth centuries were marked by the establishment of learned and scientific societies in the new republic. Journals (both those focused on scientific matters and more general journals with some coverage of science) grew in number too. There were twelve in 1800, forty in 1810, slightly less than one hundred in 1825, and nearly six hundred by the middle of the century (Keeney, 1992, pp. 26–27).

Nineteenth-Century Botanizing. Botany became increasingly popular as a pursuit in the nineteenth century, both among college-trained natural historians, who worked from specimens others collected and sought to advance science, and among interested amateurs, who collected, grew, and painted plants for a variety of reasons including interest, health and exercise, and, in the case of many young women botanizers, the respectability of the pursuit. New networks grew up around societies, journals, and institutions, such as the Harvard network centered around Asa Gray, professor in natural history there from 1842 (Keeney, 1992, p. 34). Asa Gray and his mentor, John Torrey, both produced vast synthetic works on the flora of North America, working from specimens and observations of collectors spread throughout America. The sheer scale of these works showed that the classification system was not up to the task, leading them to introduce de Jussieu's natural system of classification in place of the Linnaean system. From this more comprehensive base, Gray began to draw conclusions about the geographical distribution of species, a great contribution to botany.

These nineteenth-century knowledge networks and the popularity of botany included women. When Jane Colden had learned the Linnaean classification system and examined and classified over three hundred plants in the 1750s, she had been a rarity. Few women undertook such rigorous study (Gronim, 2007, p. 33). In the nineteenth century, however, botany increased in accessibility and desirability as a pursuit for women. Indeed, by midcentury, botanizing was so popular among young women that some critics sought to refute charges that it was too effeminate an activity for boys and young men (Keeney, 1992, p. 70). Botany was acceptable for women in part because of its combination of gentility with exercise and fresh air. At first women usually refrained from fieldwork, but by the later nineteenth century, as the women's sphere expanded in other areas, more engaged in the fieldwork aspect of botany. Gentility was less of a hindrance to the active fieldwork side of botanizing, but it also ceased to be a draw card to this branch of science (Keeney, 1992, p. 82). Moreover, as botany became increasingly professionalized in the last decades of the nineteenth century, many women and other "amateur" botanists without university educations and access to laboratories were shut out from the tightening professional networks.

Professionalization. After the Civil War and the dissemination of Charles Darwin's *The Origin of Species*, botany's intellectual networks gradually changed to include only interactions among professional botanists and among amateurs, rather than between the two groups. The Morrill Act of 1862 created the land-grant institution system. This fueled teaching and research in the agricultural sciences at the college level. This expansion of teaching also led to the foundation of gardens focused on botanical research in New York and Missouri and the growth of government agencies like the U.S. Department of Agriculture. The number of people with a serious scientific interest in plants, who wished to distance themselves from their amateur counterparts, was increasing (Smocovitis, 2006, pp. 942–943). By the early twentieth century, professional journals and professional societies, including the Botanical Society of America, had been established, making the old exchange networks no longer viable (Keeney, 1992, pp. 123–124; Smocovitis, 2006). The focus of botany as a science changed from a branch of natural history to one of biology. Laboratory-based experimentation as well as observation became part of the discipline, requiring specialized college training and equipment.

The Desert Botanical Laboratory near Tucson, Arizona, was established in 1903 with support from the Carnegie Institution, the first pure research laboratory in botanical science. Laboratory science changed methods of observation and encouraged the evolution of new areas of specialization particularly ecology, which related botanical life to its environment (Kingsland, 2005). Early twentieth-century botanical advances in crop breeding and growing also changed the science of agriculture. For botany, the move to a biological focus and the alliance of the Botanical Society of America with the American Institute of Biological Sciences from 1947 to 2000 led to the dissolution of botany departments around the country in the second half of the twentieth century. Increasingly botany and zoology departments were merged into single biology departments. Technological and scientific developments throughout the twentieth century dramatically altered botany, making it dependent on ecological databases and molecular genetics. By the beginning of the twenty-first century botany had come under the aegis of several academic disciplines including biology, molecular biology, agriculture, genetics, and ecology (Keeney, 2001, pp. 90–91). The adoption of electronic communication technology and the promotion of botany online moved the focus of American botanists to a global stage, and plants scientists and societies such as the Botanical Society of America have developed a voice in the discussion of policy issues such as biodiversity loss, climate change, and even the teaching of evolution in high schools (Smocovitis, 2006, p. 951).

[*See also* **Agricultural Education and Extension; Agricultural Technology; Agriculture, U.S. Department of; Biochemistry; Biological Sciences; Botanical Gardens; Ecology; Genetics and Genetic Engineering; Medicine; Molecular Biology;** *and* **Science.**]

BIBLIOGRAPHY

Dupree, A. Hunter. *Asa Gray: 1810–1888.* Cambridge, MA: Harvard University Press, 1959.

Fitzgerald, Deborah. *The Business of Breeding: Hybrid Corn in Illinois, 1890–1920.* Ithaca, NY: Cornell University Press, 1990.

Gronim, Sara Stidstone. "What Jane Knew: A Woman Botanist in the Eighteenth Century." *Journal of Women's History* 19, no. 3 (2007): 33–59.

Hindle, Brooke. *The Pursuit of Science in Revolutionary America, 1735–1789*. Chapel Hill: University of North Carolina Press, 1956.

Keeney, Elizabeth. *The Botanizers: Amateur Scientists in Nineteenth-Century America*. Chapel Hill: University of North Carolina Press, 1992.

Keeney, Elizabeth. "Botany." In *The History of Science in the United States: An Encyclopedia*, edited by Marc Rothenberg, pp. 89–91. New York and London: Garland Publishing, 2001.

Kingsland, Sharon E. *The Evolution of American Ecology, 1890–2000*. Baltimore: Johns Hopkins University Press, 2005.

Lewis, Andrew J. "Gathering for the Republic: Botany in Early Republic America." In *Colonial Botany: Science, Commerce, and Politics in the Early Modern World*, edited by Londa L. Schiebinger and Claudia Swan, pp. 66–80. Philadelphia: University of Pennsylvania Press, 2005.

Müller-Wille, Staffan. "Walnuts at Hudson Bay, Coral Reefs in Gotland: The Colonialims of Linnaean Botany." In *Colonial Botany: Science, Commerce, and Politics in the Early Modern World*, edited by Londa L. Schiebinger and Claudia Swan, pp. 34–48. Philadelphia: University of Pennsylvania Press, 2005.

Schiebinger, Londa L. *Plants and Empire: Colonial Bioprospecting in the Atlantic World*. Cambridge, MA: Harvard University Press, 2004.

Smocovitis, Vassiliki Betty. "One Hundred Years of American Botany: A Short History of the Botanical Society of America." *American Journal of Botany* 93, no. 7 (2006): 942–952.

Elspeth Knewstubb

BOWDITCH, NATHANIEL

(1773–1838), astronomer and navigator. Born in Salem, Massachusetts, Bowditch ended his formal education at age 10. After working in his father's workshop, he was indentured, at the age of 12, as a bookkeeping apprentice to a retailer specializing in the sale of ship supplies and equipment. Between 1795 and 1803 he made five voyages on merchant ships, serving on the last voyage as master and part owner. He became head of the Essex Fire and Marine Insurance Company in 1804. In 1823, he became actuary of the Massachusetts Hospital Life Insurance Company of Boston, a position he held until his death.

Self-educated, Bowditch mastered mathematics, celestial mechanics, French, and German and familiarized himself with the scientific research underway in Europe. He was initially known for his *New American Practical Navigator* (1802), a manual for mariners often referred to as *Bowditch's Navigator*, which contained detailed tables and instructions. Initially an attempt to correct errors in a British publication for navigators by John Hamilton Moore, Bowditch's new work became the standard publication on navigation for the shipping industry in the Western Hemisphere. The U.S. Hydrographic Office purchased the copyright in 1866 and has continued to publish revised editions of the work. The U.S. Navy was still using Bowditch's *Navigator* during the second decade of the early twenty-first century.

Bowditch also published extensively on celestial mechanics, mathematics, and physics, establishing himself as the leading American celestial mechanician and mathematician of the first quarter of the nineteenth century. He was elected a member of the American Academy of Arts and Sciences in 1799 and the American Philosophical Society nine years later. Bowditch became one of the few American men of science with an international reputation. He was elected to the Royal Society of London, the Royal Society of Edinburgh, and the Royal Irish Academy. Bowditch was not interested in a teaching position in higher education. He declined an offer of the Hollis Professorship at Harvard, and he also rejected an offer from Thomas Jefferson of a professorship at the University of Virginia.

Bowditch's greatest contribution to American science was his translation and annotation of Pierre Simon Laplace's *Traite de mécanique céleste* (1799–1825), a five-volume summary of the progress of physical astronomy since Isaac Newton. Between 1814 and 1817 Bowditch translated into English and annotated the four volumes published to that point. He began publishing the volumes beginning in 1829; the final publication was completed posthumously in 1839. Adding the missing steps in Laplace's demonstrations, reporting recent improvements and discoveries in mathematics, and providing the sources from which Laplace had drawn, Bowditch's commentary proved crucial in the education of the next generation of American astronomers and mathematicians.

[*See also* **Astronomy and Astrophysics; Maritime Transport; Mathematics and Statistics; Physics; Science;** *and* **Technology.**]

BIBLIOGRAPHY

Berry, Robert Elton. *Yankee Stargazer: The Life of Nathaniel Bowditch.* New York: McGraw-Hill, 1941.
Greene, John C. *American Science in the Age of Jefferson.* Ames: Iowa State University Press, 1984.
Ingersoll, Henry. *Memoir of Nathaniel Bowditch.* Boston: J. Munroe and Company, 1841.

    Marc Rothenberg and Hugh Richard Slotten

BP GULF OIL SPILL

See Deepwater Horizon **Explosion and Oil Spill.**

BRIDGMAN, PERCY

(1882–1961), Harvard physicist, philosopher of science, and Nobel laureate. Educated at Harvard, AB *summa cum laude* (1904), AM (1905), PhD (1908), Percy Williams Bridgman stayed on, becoming Professor Emeritus in 1954. In 1946 Bridgman was awarded the Nobel Prize in Physics for his studies of the properties of matter under very high pressure made possible by his invention, while a graduate student, of a seal that became tighter as pressure was increased. Leakage thus prevented, there was no upper limit to attainable pressures except for that imposed by the strength of the apparatus (the containing vessel, the connecting tubes, and the pressure-transmitting piston). As advances in contemporary industrial

metallurgy produced stronger steels and alloys, he improved his apparatus, soon surpassing the 3,000-atmosphere limit reached by nineteenth-century French investigators, and by the 1940s he was regularly reporting pressures as high as 100,000 atmospheres.

Bridgman exemplified the traditional American experimentalist. He believed in the unimpeachable moral worth of pure science, that is, science for its own sake, and an empiricist methodology. Science grows by the accumulation of facts. He took for granted the Newtonian principles of the uniformity of nature and the existence of absolute time and space. However, when relativity and quantum mechanics revolutionized physics, disaffirming the empirical validity of these classical fundamentals, he was shocked. In response, he set out to cleanse physics of all contamination by unrealizable ideals and ensure that no such revolutions would further be necessary because physics would forever after remain firmly grounded in experience. To this end, he enunciated his influential operational definition that the meaning of a concept is synonymous with its determining operations, primarily referring to physical measurement. Bridgman believed that his operational method was an extension of the epistemology of special relativity. Einstein denied this.

At the same time, the operational method was quickly appropriated by the positivists and behavioral psychologists and became "operationism" or "operationalism," a usage Bridgman abhorred as an aberration of his simple precept. Moreover, it was during his disagreements with the positivists and behaviorists that he first concluded that scientific knowledge exists only in the understanding mind of the individual—the locus of a radical cognitive and moral freedom. Subsequent application of his operational razor to social institutions led him to discover the artificiality of social truths and the primacy of individual subjectivity.

Although Bridgman freely contributed to military research in both World War I and World War II, he deplored the transition he witnessed during his lifetime from small-scale pure science to government-sponsored big science. He saw this transition as a degradation of democracy and a betrayal of scientific freedom and the intellectual integrity of the scientist. In 1961, suffering from incurable bone cancer, he took his own life.

Bridgman is remembered as the founder of high-pressure physics. His legacy lives on today because researchers in fields ranging from physics and chemistry to geophysics, cosmology, and biology refer to his work as the starting point for their investigations.

[*See also* Behaviorism; Biological Sciences; Chemistry; Einstein, Albert; Geophysics; Instruments of Science; Iron and Steel Production and Products; Military, Science and Technology and the; Nobel Prize in Biomedical Research; Physics; Quantum Theory; *and* Science.]

BIBLIOGRAPHY

Bridgman, Percy Williams. *The Logic of Modern Physics*. New York: Macmillan, 1927. First articulation of the operational definition. Launches Bridgman into a lifelong excursion into the field of interpretational scientific philosophy. Illustrates his simple, straightforward style and earnestness of expression.

Bridgman, Percy Williams. *The Intelligent Individual and Society*. New York: Macmillan, 1938. Contains Bridgman's operational analysis of social institutions. Exemplifies his highly principled operational attitude.

Hemley, Russell J. "Percy W. Bridgman's Second Century." *High Pressure Research* 30, no. 4 (December 2010): 581–619. A comprehensive overview of contemporary high-pressure research, acknowledging Bridgman's contributions.

Walter, Maila L. *Science and Cultural Crisis: An Intellectual Biography of Percy Williams Bridgman (1882–1961)*. Stanford, Calif.: Stanford University Press, 1990. The only full-length biography of Bridgman. Based on original sources. Describes his often contentious entanglement in a surprisingly wide range of scientific, philosophical, and political arenas. Reveals him as an important, if underappreciated, thinker.

Maila L. Walter

BRIN, SERGEY

See Software.

BROOKLYN BRIDGE

When it opened on 24 May 1883, the Brooklyn Bridge was the longest suspension bridge in the world and among the most celebrated creations of the nineteenth century. In 1867, the New York State legislature chartered a company to build a bridge across the East River, between Brooklyn and lower Manhattan, to provide more reliable transportation than the existing ferries for a growing stream of commuters. John Augustus Roebling (1806–1869), a German-born wire manufacturer, designed a span that incorporated many innovations, including extensive use of steel. When Roebling died of injuries in 1869, as construction was about to begin, his son, Washington Roebling (1837–1926), took over as chief engineer. Although crippled by the bends while supervising excavation for the bridge towers, the younger Roebling conquered innumerable technical problems during the long course of construction.

The Brooklyn Bridge helped pave the way for the consolidation of the cities of Brooklyn and New York in 1898. Its success spurred a national wave of bridge building. Although the bridge carried massive vehicular traffic, generations of New Yorkers found its elevated walkway a haven from the city's bustle.

From its opening, the Brooklyn Bridge served as a symbol of the new industrial America. Although the novelist Henry James deemed it a "monster" obliterating the city of his youth, the cultural critic Lewis Mumford judged it an artistic success as "a fulfillment and a prophecy" of the Machine Age. The painters John Marin and Joseph Stella saw the bridge as embodying modernity, whereas the poet Hart Crane in *The Bridge* (1930) viewed it as an affirmation of faith in the possibilities of America. In the twenty-first century, the Brooklyn Bridge continued to move commuters into and out of New York City's downtown business district while conjuring up hope and harmony.

[*See also* **Engineering; Iron and Steel Production and Products;** *and* **Technology.**]

BIBLIOGRAPHY

McCullough, David. *The Great Bridge.* New York: Simon and Schuster, 1972.
Trachtenberg, Alan. *Brooklyn Bridge: Fact and Symbol.* New York: Oxford University Press, 1965.
                                        Joshua B. Freeman

BUILDING TECHNOLOGY

Construction in North America has been heavily influenced by unique geographical, economic, and cultural pressures. While related to parallel developments in Europe and Asia, American building has often been a proving ground for new technologies and—by necessity—an important arena for experimentation in structure, construction, and building systems.

Eighteenth and Nineteenth Centuries. European settlers in North America found rich sources of building material, including stone, clay, and timber, which allowed them to build using familiar methods from their home countries. But as settlement expanded beyond the eastern seaboard, building technologies in America developed to match the country's unique range of climates and natural resources, and builders keenly adopted techniques from indigenous builders as well as from immigrant cultures whose homelands matched the new territories' weather patterns. Heavy timber, stone foundations, and fired brick were supplemented in the south by stucco and tile that built on Spanish traditions and light timber construction that reflected lessons learned from native populations. More technically advanced materials such as glass and iron were often imported from Europe, but these trades also sprang up in larger settlements, forming the basis for increasingly sophisticated material industries that soon provided domestic sources.

Brick and iron both saw important advances in the early nineteenth century, which began a long process of industrialization for the building trades. Horse and steam power were used to consolidate clay into strong pressed brick in Philadelphia by 1835, setting off a slow but thorough transformation of masonry from a local craft to a regional industry. Masonry buildings were heavy, but slow to build and thus not always adequate to the commercial need for quick, agile construction. In New York, the pioneering work of James Bogardus and Daniel Badger produced cast-iron facades and towers constructed entirely of prefabricated elements, offering a system of construction that was lightweight and transportable, separating production from the job site. A similar process industrialized lumber in the Midwest, allowing balloon frame, and later platform frame, houses to be quickly and economically constructed with standardized light timber elements that could be easily shipped by barge or, later, rail. Elevators began to prove their commercial feasibility around 1870, allowing office buildings to rise beyond their traditional, stair-based limits of five to six stories.

Railroads provided access to new fields of resources—the primeval forests of the West, for instance, and the iron range of the upper Midwest. But they also provided models for new methods of construction. Commercial construction, which had been limited by its reliance on brick for its structure, adopted methods of bridge-building, first relying on cast-iron columns as a cost- and space-effective substitute for masonry and later as a method of staying iron frames against wind. Surplus rails were used in foundations, particularly in Chicago, to supplement traditional stone footings, allowing the large, open basements that were increasingly required for new types of mechanical equipment such as elevator and water pumps. America developed its own glass industry as well, allowing large windows to fill the more open spaces of iron-framed buildings. These developments allowed high-quality office space to populate the upper floors of buildings that reached 10, 12, or even 16 stories, and the term "skyscraper" was soon applied to this new building type. Skyscraper efficiency was enhanced by the strength and workability of steel, which dominated tall building construction by the mid-1890s.

Advances at the Turn of the Century. Although structural advances were more visible, improvements in construction also made buildings more fire resistant and efficient. Terra cotta fireproofing employed local materials to jacket vulnerable metal frames, whereas improvements in building codes after a series of devastating fires in the late 1800s demanded more conscientious construction and planning for fire stairs and exits. Meanwhile, electricity went from being a novelty in the 1880s to a luxury in the 1890s and a commonplace in commercial and urban residential buildings by 1910. The tungsten filament light bulb, on the market by 1912, completed the electrical revolution, offering a cost-effective alternative to large, poorly insulated windows. Electricity also powered faster, safer, and more efficient elevators, making skyscrapers as tall as New York City's 792-foot-tall Woolworth Building (1913) economically feasible.

By World War I concrete offered a legitimate competitor to steel in building and industrial construction. Ernest Ransome, an English expatriate, constructed a handful of buildings around San Francisco in the 1880s. Acceptance of the new material was relatively slow, but by 1910 it was extensively used for factory construction, in particular by Albert Kahn for Ford's plants in Detroit, and in 1903 its viability for tall building construction was proven by the 15-story Ingalls Building in Cincinnati. Kahn's twisted-bar system was gradually replaced by more sophisticated forms of reinforcement, whereas new road construction spurred on by the automobile energized the growth of the concrete industry. Its fireproof qualities made it an attractive material in the wake of several fires in the early 1900s, in particular the Triangle Shirtwaist Factory fire of 1911. These influenced more thorough building codes and the industry-led development of automatic fire sprinklers, which had been used in factories and mills during the nineteenth century but now saw more widespread use in commercial and institutional buildings and saved thousands of lives.

The 1920s saw explosive economic growth and with it a tremendous expansion of America's building industry. Standardization of steel products, a broad industrialization of the glass industry, and the growth of scientific management in building construction all contributed to more efficient and economical buildings. Electric arc welding added speed and strength to metal building construction, and a growing body of experimental work on steel's strength allowed more ambitious and more efficient structures. Residential construction saw its own lumber-based version of prefabrication in catalog houses, which arrived in railcars and could be assembled by any modestly skilled carpenter. Although efforts to combine developments in metal construction with the explosive residential market largely failed, future generations would find inspiration in the visionary work of Buckminster Fuller, whose Dymaxion House projects sought to combine automotive and aeronautical engineering with domestic programs. New mechanical technologies, in particular ducted ventilation, better insulation, and air conditioning, all contributed to improvements in interior environmental control. Willis Carrier's pioneering use of condensers and evaporative cooling in 1902 led to packaged air conditioning systems by the 1920s. Initially used to lure patrons into movie theaters and restaurants, air conditioning quickly became a standard commercial and higher-end residential amenity in the southern United States, and by the 1950s its miniaturization made it ubiquitous throughout the country. As with steel, a body of experimental work provided the nascent air conditioning industry with reliable data, and human comfort became as scientific a field as structural engineering.

World War II and the Postwar Era. The quasi-public project to construct the 102-story Empire State Building, which set a record for skyscraper height in 1931, illustrated the ability of the American construction and engineering industries, but this ability lay mostly fallow during the Great Depression. Government-funded projects such as dams, bridges, and sports stadiums supplanted commercial construction, employing laborers while providing opportunities for research and development in new materials and techniques. The war effort in the 1940s accelerated the pace of experimentation. New factory buildings were erected using prefabricated steel trusses and advanced management methods; Henry Ford and architect Albert Kahn collaborated on bomber and tank factories in Detroit, for example, that were constructed and operational within months of their commissioning. Military technology, in turn, provided both capacity and new expertise to the building industry after the war. Advanced production techniques for aircraft glass and aluminum, for instance, were adapted by manufacturers for building cladding systems, making the ubiquitous glass curtain wall of the 1950s one of the most visible legacies of wartime production. Spurred by both reconstruction abroad and unprecedented economic growth at home, commercial and residential construction adopted prefabrication, standardization, and scientific management methods on unprecedented levels. The automobile and the airliner both shaped new building types, and suburbs, shopping malls, airport terminals, and sports stadiums all turned demographic pressures into innovative developments in construction.

Such economic growth also challenged conventional notions of scale, and building projects on unprecedented scales changed American cities in the 1950s and 1960s. Large-scale urban redevelopment formed the basis for large, mixed-use complexes in Boston, New York, and elsewhere, and although the complex integration of highways, rail lines, and commercial space that these projects represented showed the abilities of engineers and builders, they also cast the social and cultural implications of such technological faith in stark relief. New systems of elevatoring and wind bracing allowed much taller skyscrapers such as Chicago's John Hancock and Sears Tower, both by Fazlur Khan and Bruce Graham of Skidmore, Owings, and Merrill (SOM), and more reliable cladding systems began to address issues of water infiltration, daylighting, and insulation that had

plagued the first generation of postwar skyscrapers. Contemporaneous developments in concrete reinforcement, mixtures, and placement also enabled larger concrete spans and towers, including the Louisiana Superdome, which set the record for uninterrupted span, and Toronto's CN Tower, which remains the tallest structure in North America.

Ecology and Energy—New Imperatives. Given the United States' profligate use of energy, it is notable that developments in energy efficiency—in particular passive solar design, lightweight construction, and environmentally controllable building skins—occurred alongside such energy-intensive projects. The "drop-out" movement of the 1960s, taking the visions of Buckminster Fuller's technocentric approach as part of its mantra, pioneered a renewed interest in living self-sufficiently and in transferring technology developed for the military into peaceful uses. The *Whole Earth Catalog* produced a marketplace for new technologies on a backyard scale and it also presented a cobbled-together ethic that introduced a more personal, even radical view of technology and society to the developing counterculture. Passive solar houses had been built experimentally since the 1930s, but the application of basic principles of climatic design and solar orientation became the basis for increasingly mainstream residential designs, even as solar energy struggled to prove itself in the marketplace. On a commercial level, the Hooker (later Occidental) Chemical headquarters building, designed by Cannon for a prominent site in Niagara Falls, New York, deployed an early double skin, a technique that allowed the façade to respond to changes in exterior temperature and sunlight to either heat or cool its interior passively.

Under the somewhat misleading title of "sustainability," more materials, systems, and techniques have been deployed in the early twenty-first century that seek to reduce both the embodied and the life cycle energy of buildings. However, the Americas have lagged far behind Europe and Asia in bringing successful products to market and in legislating for buildings that reduce their considerable carbon penalty. Commercial efforts to provide leadership in these areas have often been compromised by financial affiliations with manufacturers, yet a renewed interest in ecological techniques found a keen audience at the end of the twentieth century, with buildings such as California's State Office Building in Sacramento by Sim van der Ryn and Phoenix's Public Library by Will Bruder offering very public examples of a new awareness. Parallel to these efforts at downsizing carbon footprints in building, there has been a paradoxical emphasis on building larger and larger structures, often overseas. The export of American expertise in projects such as Dubai's Burj Kaliffa (2010) demonstrates that the country's knowledge now typically exceeds its own economic ability, and this neatly reverses the need in early America to import both building materials and expertise. The rise of digitally aided design and construction has, meanwhile, overturned the industry's long-standing emphasis on mass production as a means of increasing efficiency, with new methods of production and fabrication enabling complex construction such as Los Angeles' Walt Disney Concert Hall, whose willfully curving skin exemplifies both the stylistic emphasis of its designer and, perhaps more importantly, the ability of its fabricators to produce high-tolerance cladding systems in a huge variety of arbitrary and unique shapes. As this ability is matched by engineering software that can accurately model environmental and structural performance in buildings, the potential for more economic, more efficient, and more evocative construction remains largely untapped.

[*See also* **Airplanes and Air Transport; Dams and Hydraulic Engineering; Electricity and Electrification; Elevator; Empire State Building; Engineering; Environmentalism; Ford, Henry; Forestry Technology and Lumbering; Illumination; Iron and Steel Production and Products; Military, Science and Technology and the; Motor Vehicles; Railroads; Refrigeration and Air Conditioning; Skyscrapers; Steam Power;** *and* **Technology.**]

BIBLIOGRAPHY

Ali, Mir. *Art of the Skyscraper: The Genius of Fazlur Khan.* New York: Rizzoli, 2001.
Banham, Reyner. *The Architecture of the Well-Tempered Environment.* Chicago: University of Chicago Press, 1969.
Bergdoll, Barry, and Peter Christensen. *Home Delivery: Fabricating the Modern Dwelling.* New York: Museum of Modern Art, 2008.
Carrier, Willis H. "The Economics of Man-Made Weather." *Scientific American* (April 1933): 199–202.
Condit, Carl. *American Building Art.* Oxford: Oxford University Press, 1961.
Elliott, Cecil. *Technics and Architecture: The Development of Materials and Systems for Buildings.* Cambridge, Mass.: MIT Press, 1992.
Elzner, A. O. "The First Concrete Skyscraper." *The Architectural Record* 15 (June 1904): 531–544.
Fenske, Gail. *The Skyscraper and the City: The Woolworth Building and the Making of Modern New York.* Chicago: University of Chicago Press, 2008.
Friedman, Donald. *Historic Building Construction,* 2d ed. New York: W. W. Norton, 2010.
Gayle, Carol, and Margot Gayle. *Cast-Iron Architecture in America: The Significance of James Bogardus.* New York: W. W. Norton, 1998.
Gray, Lee E. *From Ascending Rooms to Express Elevators: A History of the Passenger Elevator in the 19th Century.* Mobile, Ala.; Elevator World, 2002.
Hunt, William Dudley Jr. *The Contemporary Curtain Wall: Its Design, Fabrication, and Erection.* New York: F. W. Dodge, 1958.
Kolarevic, Branko, ed. *Architecture in the Digital Age: Design and Manufacturing.* New York: Spon Press, 2003.
Marks, Robert W. *The Dymaxion World of Buckminster Fuller.* New York: Reinhold, 1960.
Misa, Thomas J. *A Nation of Steel: The Making of Modern America, 1865–1925,* paperback ed. Baltimore: Johns Hopkins University Press, 1999.
Peters, Tom F. *Building the Nineteenth Century.* Cambridge, Mass.: MIT Press, 1996.
Platt, Harold L. *The Electric City: Energy and the Growth of the Chicago Area, 1880–1930.* Chicago: University of Chicago Press, 1991.
Purdy, Corydon T. "The Evolution of High Building Construction." *Journal of the Western Society of Engineers* 37 (August 1932): 201–211.

Sadler, Simon. "Drop City Revisited." *Journal of Architectural Education* 58 (2006): 5–14.

van der Ryn, Sim. *Design for Life*. Salt Lake City: Gibbs Smith, 2005.

Wermiel, Sara. *The Fireproof Building*. Baltimore: Johns Hopkins University Press, 2000.

Thomas Leslie

BUREAU OF STANDARDS, U.S.

See Research and Development (R&D).

BUSH, VANNEVAR

(1890–1974), inventor, engineer, wartime administrator. Armed with a dual Harvard–Massachusetts Institute of Technology (MIT) doctorate in mathematics and electrical engineering, Bush, a native of Everett, Massachusetts, became a pioneering designer of analog computers as a professor of engineering (and later dean) at MIT from 1919 to 1938. The differential analyzer (1928), a mechanical array of precisely machined gears, cams, and shafts capable of solving previously unsolvable differential equations of interest to the electrical industry and researchers in many specialized fields, brought Bush public attention as the inventor of a "mechanical brain." While at MIT, he played an important role updating the engineering curriculum and coordinating research activities. He also developed an interest in the role of the professional engineer in society. Beginning in 1932, he served as MIT's first vice president and dean of engineering.

Arriving in Washington, D.C., in January 1939 as president of the Carnegie Institution (1939–1953) and chair of the National Advisory Committee for Aeronautics, Bush soon began the work of organizing American science for the coming war. In June 1940, with President Franklin Delano Roosevelt's approval, he established the National Defense Research Committee (NDRC) to mobilize researchers. In June 1941, the NDRC became part of a new and larger organization, the Office of Scientific Research and Development (OSRD), also led by Bush. Carving out a space between the military, which would use the new weapons developed under OSRD contract in academic and industrial research laboratories, and the corporations building these weapons, Bush helped facilitate a profound transformation in military attitudes. When World War II began, the U.S. armed services were bastions of technological conservatism; by war's end they were technological enthusiasts. Among the weapons Bush championed were radar and the atomic bomb.

In *Science—The Endless Frontier* (1945), an OSRD report to Roosevelt articulating the federal government's duty to fund basic research in American universities, Bush called for a new federal agency, a National Research Foundation. In 1950, after years of debate, President Harry S. Truman established the National Science Foundation (NSF). The NSF, however, was not the organization envisioned in Bush's report. Indeed, Bush had by then concluded that the armed services had taken over American science; his *Modern Arms and Free Men* (1949) called on the nation's political leaders to reassert control of the military. By the time of his death, Bush had become a severe critic of the very world of government–science relations he had helped bring into existence.

[*See also* Atomic Energy Commission; Computer Science; Computers, Mainframe, Mini, and Micro; Engineering; Mathematics and Statistics; Military, Science and Technology and the; National Science Foundation; Nuclear Weapons; Office of Scientific Research and Development; *and* Science.]

BIBLIOGRAPHY

Owens, Larry. "Vannevar Bush and the Differential Analyzer: The Text and Context of an Early Computer." *Technology and Culture* 27 (1986): 63–95.

Zachary, G. Pascal. *Endless Frontier: Vannevar Bush, Engineer of the American Century*. New York: Free Press, 1997.

Michael Aaron Dennis

C

CANALS AND WATERWAYS

The first waterway engineers in the Americas were the prehistoric builders who constructed fish weirs and ditches. In the ninth century AD, on the Salt and Gila Rivers of the Tucson Basin, the Hohokam built more than one hundred miles of gravity irrigation canals. The Pueblo people at Mesa Verde maintained a sophisticated canal-fed reservoir system for more than four hundred years.

Europeans exploring the Atlantic seaboard encountered dangerous reefs and sandbars. Flatboat commerce on the Ohio–Mississippi braided through rapids, log-jams, and snags. In 1785, George Washington organized a canal company that cleared rocks from the Potomac River. Eighteenth-century builders also excavated rapids on the Santee, James, Delaware, and Susquehanna. The 1804 Middlesex Canal connected Boston Harbor to the Merrimack River. In 1830, on the Merrimack at Pawtucket Falls, a masonry

dam impounded an 18-mile mill pond. In 1859, in protest against fish-killing river impoundments, New Englanders with crowbars and axes destroyed a low wooden dam.

Boston's Middlesex Canal Company became the model for semipublic stock corporations supported by state or municipal bonds. In 1808, Secretary of the Treasury Albert Gallatin asked Congress to supplement local investments with $23 million of federal money for roads and canals. Gallatin's plan fell to President James Madison's veto in 1817. New Yorkers, twice denied federal assistance, raised more than $7 million in state revenues and bonds for the 364-mile Erie Canal, completed in 1825. The success of the Erie Canal sparked rivalries among port cities competing for inland trade. Encouraged by the U.S. Supreme Court's outspoken nationalism in *Gibbons v. Ogden* (1824), Congress purchased $300,000 in canal stock to finance a cut through the Delaware Peninsula. Virginians chartered the Chesapeake and Ohio Canal. By 1830, New Jersey, Ohio, Indiana, and Illinois had also launched ambitious projects, and Pennsylvania spanned the Alleghenies with the Main Line Canal from Philadelphia to Pittsburgh, opened in 1834.

The U.S. Army Corps of Engineers, trained at West Point, planned most of the early republic's biggest waterway projects. Army engineers opened Philadelphia to winter commerce with ice-resistance harbors and a breakwater in Delaware Bay. On the Mississippi–Ohio system, where stumps and logs blocked steamboat commerce, the army's topographical bureau ran a fleet of snag-pulling machine boats. By 1836, government snag crews had opened the Red River to Texas. Army engineers also supervised lighthouses, navigation dams, harbor dredging, and a gated ship canal that connected Lake Superior to the lower Great Lakes.

Lavish appropriations for rivers and harbors consolidated the power of Congress in the wake of the Civil War. Under the spendthrift Grant administration, 1869 to 1877, the Corps of Engineers became a powerful link in the political alliance between Congress and shipping interests. Midwestern grain farmers wanted navigation channels as an alternative to shipping by rail. Below New Orleans, where shallow deposits of mud had closed deepwater shipping, Congress financed a funnel of navigation jetties, completed in 1879. On the Ohio at Davis Island, a Corps-built "movable dam" opened barge navigation. In 1894, Congress authorized the first two large navigation locks on the 29-step "aquatic staircase" that opened a nine-foot channel from St. Louis to St. Paul. The Refuge Act of 1899 further expanded federal jurisdiction by making the Corps responsible for the removal of shipwrecks and the dumping of hazardous waste. Here began the Corps's far-reaching permit authority to regulate construction in wetlands.

New agencies with rival constituencies fragmented the federal management of water resources during the Progressive Era. The U.S. Reclamation Service, created in 1902, assumed control of more than one hundred western projects that irrigated millions of acres. Renamed the Bureau of Reclamation, the agency shattered construction records with the 280-foot-high, rock-masonry Theodore Roosevelt Dam on Arizona's Salt River, completed in 1911. Increasingly massive dams of reinforced concrete plugged canyons for irrigation on Wyoming's Shoshone River (1910), at Arrowrock on the Boise River (1915), at Elephant Butte on the Rio Grande (1915), and in Oregon's Owyhee Canyon (1932).

The Corps of Engineers, meanwhile, had acquired a vast federal mandate to protect valleys from murderous floods. The mission evolved from the Corps's scientific work on the dynamics of seasonal flooding. In 1879, the Corps-led Mississippi River Commission began setting construction standards as a levee-oversight bureau. In 1917, a year of terrible flooding in Louisiana and California, Congress granted the Corps a limited authority to redesign dangerous floodways. Soon after 1927 a flood of record topped levees in the cotton lowlands, Congress extended the Corps's authority to flood-works nationwide.

Whereas the Corps maintained locks and levees and the Bureau of Reclamation built signature dams, no single state or federal bureau coordinated the chaotic rush for hydropower. Always controversial, dam-building for hydropower tested the limits of federal authority, pitting the property rights of corporations against the conservationists who advocated public control. The Woodrow Wilson administration fought for public regulation via the Federal Power Commission, created in 1920. The Corps of Engineers, in 1925, realized the kilowatt potential of Muscle Shoals on the Tennessee River with Wilson Dam. During the Great Depression, from 1929 to 1941, Congress added powerhouses to towering dams at Boulder Canyon on the Colorado (future Hoover Dam), Shasta in California, Grand Coulee and Bonneville dams on the Columbia River, and Fort Peck Dam on the Missouri. Beginning at Norris Dam on the Clinch River, the New Deal's Tennessee Valley Authority (TVA) reengineered Appalachia with more than 50 powerhouse projects. Publicly controlled, the TVA was the nation's leading producer of hydropower at the close of World War II.

Bigger was better in the postwar 1950s and 1960s when monumental construction epitomized

America's supremacy as an industrial power. With floodways, dredging, tributary dams, and more than 3,000 miles of levees, the Mississippi River and Tributaries Project (MR&T) became one of the world's largest networks of river and harbor projects, extending flood protection for 36,000 square miles. The Mississippi River Gulf Outlet brought container ships to New Orleans. The 445-mile McClellan–Kerr Waterway moved oil shipments from Tulsa. Five large dams and 1,500 miles of levees channelized the Missouri. Great Lakes cargo reached world markets via the U.S.–Canadian St. Lawrence Seaway, opened in 1959. Although the Florida segment was never fully completed, the Intracoastal Waterway from New York to Texas provided a navigation network of more than 3,000 miles.

There was never a time, not even when the country was at war, when the large dams and canals were not controversial. Presidents Franklin Roosevelt, Harry Truman, and Dwight Eisenhower all denounced the Congressional system of omnibus legislation that bankrolled wasteful projects. Indian tribes sued in federal court over back-flooded reservations. Private utilities fought the public control of hydropower on the Snake River in Idaho–Oregon and Kings Canyon in the California Sierras. At Colorado's Dinosaur National Monument, where the Bureau of Reclamation hoped to impound the Green River at Echo Canyon, the Sierra Club galvanized environmental protest. Chemical pollution in slackwater freightways opened another path of protest. On 22 June 1969, the nation reacted in horror when the black oily surface of Cleveland's Cuyohoga River burst into flames. Shocking as it was, the Cuyohoga—so polluted that boats near the Sherwin–Williams paint factory changed colors after four days in the river—was a small problem compared with the tons of industrial waste being dumped into larger streams. In 1975 the State of Virginia banned fishing after Allied Chemicals confessed to the routine dumping of roach killer into the James River. General Electric, meanwhile, was using the Hudson to dissolve cancer-causing polychlorinated biphenyls (PCBs). Chicago protected Lake Michigan by reversing the flow of its poisonous river, polluting the Ohio system. Niagara Falls flushed more than a ton of toxins each day.

The National Environmental Policy Act of 1970 and subsequent clean water legislation challenged engineers to rethink the benefit–cost equations that justified river construction. The Corps of Engineers responded with greener nonstructural solutions to flooding—notably in Florida and Louisiana where flood works had aggravating the fraying and erosion of wetlands. In 1997, the Federal Energy Regulatory Commission (FERC) responded to environmental protest by calling for the removal of Edwards Dam in Maine. In the Pacific Northwest, meanwhile, a Corps study concluded that the most effective way to restore seagoing salmon and steelhead was to breach Snake River dams. Yet the dam-it, ditch-it tradition of monumental construction survived in some of the Corps's most extravagant projects: in Alabama, for example, where funding for the $2 billion Tennessee–Tombigbee Waterway survived the threat of a presidential veto. Completed in 1984, the 232-mile waterway rerouted coal and heavy shipping to the Gulf of Mexico via the Port of Mobile.

In all, about 75,000 large dams have been built in the nation's rivers. More the 200,000 river miles have been channelized for navigation. The Corps of Engineers maintains more than 200 barge locks and an estimated 14,000 miles of flood levees. The Bureau of Reclamation in its 17 western states provides irrigation to about 10 million acres of farmland. A federal network of scenic rivers guards more than 12,000 river-miles in 38 states.

[*See also* **Agricultural Technology; Army Corps of Engineers, U.S.; Cancer; Conservation Movement; Dams and Hydraulic Engineering; Environmentalism; Environmental Protection Agency; Erie Canal; Fish and Wildlife Service, U.S.; Fisheries and Fishing; Hoover Dam; Hydroelectric Power; Maritime Transport; Panama Canal; Rivers as Technological Systems; Sierra Club; Steam Power; Technology;** *and* **Tennessee Valley Authority.**]

BIBLIOGRAPHY

Armstrong, Ellis L., ed. *History of Public Works in the United States, 1776–1976.* Chicago: American Public Works Association, 1976.

Bartlett, Richard A. *Rolling Rivers: An Encyclopedia of America's Rivers.* New York: McGraw–Hill, 1984.

Maass, Arthur. *Muddy Waters: The Army Engineers and the Nation's Rivers.* Cambridge, Mass.: Harvard University Press, 1951.

Palmer, Tim. *America by Rivers.* Washington, D.C.: Island Press, 1996.

Shallat, Todd A. *Structures in the Stream: Water, Science, and the Rise of the U. S. Army Corps of Engineers.* Austin: University of Texas Press, 1994.

Todd A. Shallat

CANCER

Although cancer is an ancient disease, it only gained notice as a significant public-health issue in the past 150–200 years. From the late eighteenth and early nineteenth centuries, physicians reported rising numbers of deaths from this group of diseases among their patients, observations that were reinforced from the 1840s by mortality statistics that indicated an increase in cancer deaths in Europe and later in the United States. At first commentators were not sure whether the rise was real or an artifact of better diagnosis, the systematic collection of mortality statistics, or greater cancer awareness among physicians and the public. Nor was it clear why, if the rise was real, it was happening at all. Maybe it was a product of changes in the diet, morals, or living conditions associated with urbanization or the working conditions and pollution that came with industrialization, or maybe it was the result of an aging population—a penalty, perhaps, of public-health progress. Maybe climate, heredity, or race played a part. There were a multitude of apparent causes, but little consensus as to which explained the rise.

By the twentieth century, uncertainty about the reality of the rise began to dissipate. Debate continued as to its cause, the reliability of the statistics, the contributions of particular types of cancer to overall mortality, and how mortality and incidence might best be monitored. But the general consensus was that cancer was emerging as a major public-health issue and that cancer mortality displayed a disturbing upward trend. In the early 1920s, this group of diseases surpassed tuberculosis as a cause of mortality, and by the early 1930s it was the nation's number two killer disease after heart disease. Cancer mortality continued to grow for much of the rest of the century. The question was what to do about it.

Before the 1940s. The dominant answer in the twentieth century was early detection and treatment. From this perspective the goal was to catch the disease as early in its life as possible, even before it turned malignant, and to treat it the moment cancer or its possibility was discovered. Physicians had long urged people to see their doctors as soon as they spotted something wrong, but the message before the twentieth century was not promoted consistently, nor was there agreement as to the best therapeutic interventions. Some advocated surgery. Others saw surgery as risky and to be avoided, especially in advanced and internal cancers. So physicians also used a mix of other interventions—including herbal remedies, caustics (which ate away surface tumors), palliation for the incurable, and advice on prevention, including recommendations on diet, regimen (lifestyle), and exposure to environmental and occupational carcinogens. The development of anesthesia and antiseptic and aseptic techniques boosted enthusiasm for cancer surgery, and later X-rays (discovered in 1895) and radium (1898) joined the therapeutic armamentarium. Other therapeutic interventions began to be sidelined, as did preventive approaches. Indeed, surgery and radiotherapy were sometimes defined as preventive procedures: *preventing* the possibility of cancer turning malignant and the further development of malignancies already established in the body.

The work of the Johns Hopkins surgeon William Stewart Halsted, a late nineteenth-century pioneer of the radical mastectomy, illuminates the growing centrality of surgery to cancer therapy. Halsted based his operation on the belief that breast cancer spread in a centrifugal manner from the primary tumor to nearby structures. Halsted was not the first to argue that breast cancer began as a local condition that later affected other body parts. His argument was that earlier surgeons did not remove enough, often failing to remove all cancer cells, which led to future recurrences. To catch these cells, surgeons had to cut more aggressively, he claimed. They had to remove not only the breast but also surrounding body tissues including the skin, neighboring lymph nodes, muscles, and parts of the rib cage or shoulder.

Halsted's methods rapidly became central to breast cancer therapy in the United States. He trained his students in his methods, and they colonized leading American hospitals and medical schools, trained their own students in his approach, and applied it to other cancers. It was Halsted's students who helped to found the first national campaign against cancer in 1913, the American Society for the Control of Cancer (ASCC). Adapting the local, centrifugal model of breast cancer to broader anticancer programs, the ASCC launched efforts to promote early detection and treatment of the disease or its possibility. The window of opportunity for successful treatment was short, the ASCC claimed. Patients had to go to their physicians as soon as the disease or its possibility was identified, before it spread out from the local site, and physicians had to act with urgency.

The problem, the ASCC argued, was that most patients saw their physicians after the best chance for successful treatment had gone. The early signs of cancer were not obvious, and pain or debility often arrived too late to encourage people to see their physicians before the disease spread and became incurable. But even if cancer was suspected, physicians claimed that patients still delayed. Many were frightened by the disease or its treatments. Others were unduly pessimistic about the possibility of a cure or unaware of the warning signs of the disease. Still others were swayed by quacks, purveyors of patent medicines, or friends and family, all of whom could dissuade people from seeking appropriate help. Even regular physicians were a problem. Too often, the ASCC complained, they were ignorant of the disease and its treatment, pessimistic about the possibility of suc-

cessful treatment, or hesitated too long when confronted with cancer.

In its efforts to combat such problems, the ASCC launched vast educational programs to teach people about the "early warning signs" of the disease and to persuade them to go to a regular physician at the first indication of what might be cancer and, from the late 1910s, to go for regular medical checkups even if they felt well. They also warned the public of the dangers of quackery, of ineffective and hazardous home remedies, and of the bad advice of friends, family, the media, and ignorant physicians. Yet, as its educational program took off in the 1920s and 1930s, the ASCC became alarmed that its own efforts were actually exacerbating existing fears of the disease and its treatment and consequently encouraging delay by frightening people into inaction. The organization sometimes seemed unsure which way to turn. It believed that public education was essential to efforts to promote early detection and treatment; the problem was to stop it promoting delay.

Despite these difficulties, the ASCC's model of early detection and treatment dominated anticancer programs in the 1920s and 1930s. A growing number of state health departments adopted this approach, as did the Veterans Administration, and state and private agencies created networks of cancer clinics and hospitals into which patients might be channeled. The economic depression of the 1930s frustrated many of these efforts. Then in 1936 the appointment of an activist Surgeon General, Thomas Parran, revolutionized a hitherto conservative Public Health Service and led to the creation of the federally funded National Cancer Institute (NCI) in 1937. The hope was that this would help the struggling state cancer programs, and it did so in part through radium.

Radium, along with X-rays, had been employed since the beginning of the twentieth century as an alternative and supplement to surgery, the mainstay of cancer therapy. Many physicians doubted that they could ever truly replace surgery; indeed radium therapy often required a surgery of access to implant radium or radon (the gas of radium) containers into the body, and both radium and X-rays were used to reduce inoperable tumors to operable size. Nevertheless, radiotherapy gained a reputation for offering the possibility of cancer treatment without the fearful mutilation and pain associated with surgery, and demand for X-rays and radium increased, especially in the 1920s and 1930s. In the midst of economic depression, however, many physicians, hospitals, and health agencies could not afford to purchase radiotherapeutic equipment, a particular problem in the case of radium, which was extremely expensive and in short supply. The creation of the NCI was the federal government's response to the radium shortage. In 1938 the Cancer Institute allocated half its budget to purchase radium, which it loaned out to hospitals across the country for the *routine* treatment of poor patients, and state cancer agencies used these loans to bolster their faltering anticancer efforts. The plan was to purchase similar quantities of radium in future years. However, fears that this would be the beginning of socialized medicine prompted the NCI to abandon future purchases, to focus increasingly on medical training and research, and to find other ways of supporting state cancer programs.

The 1940s to the 1970s. Between the 1940s and 1970s campaigns against cancer were transformed. Programs of early detection and treatment expanded. Efforts to encourage people (predominantly white women and men) to learn the early warning signs of the disease and to go for regular checkups from a physician were joined by programs that sought to encourage women to undertake regular breast self-examinations (late 1940s) and to have routine mammograms (1960s) and pap smears (1960s) for the detection of breast and cervical cancer, respectively. Cancer organizations urged children to monitor their parents; women to monitor their mothers, their daughters, and their men; and men to listen to both and to learn the danger signs of the disease. From the 1960s, educational programs also targeted African Americans and other minorities, previously marginal to cancer education efforts. A major revolution in health practices had occurred in which millions of men and women practiced some form of self-surveillance and subjected themselves to routine surveillance by physicians and others.

If programs of surveillance expanded, so too did other anticancer efforts. Attacks on "quackery" became increasingly aggressive in the 1950s and 1960s. Prewar training programs expanded to ensure that family physicians were not the liability they had been in the 1930s and to build specialist cadres of cancer physicians (oncologists), nurses, and pathologists—the last had long been the final arbiter of whether a growth was cancerous or precancerous. Cancer services grew dramatically as mortality rates continued to rise, the disease gained growing media attention, federal funding for hospital construction increased, health insurance coverage grew, and Medicare and Medicaid were created in the 1960s. Family physicians hoped to be the port of first call for patients, referring suspicious cases on to the growing numbers of specialist cancer centers. But increasingly Americans went directly to specialists and bypassed family physicians.

Within cancer centers, surgery and radiotherapy remained important therapeutic interventions. In surgery, physicians adopted ever more radical procedures in the 1950s and 1960s, and women were often asked to submit to a one-step procedure in which the diagnostic biopsy and mastectomy were undertaken in a single operation. A few American surgeons criticized such methods and called for a reconsideration of Halsted's local centrifugal model of cancer. But such critics were vilified by colleagues, who blocked efforts to test surgical procedures until well into the 1960s. Radiotherapy also changed with the emergence of new technologies in the 1940s and 1950s— cobalt came to displace radium therapy, and more powerful X-ray machines, linear accelerators, cyclotrons, and other radiation equipment were developed, sometimes financed with Cold War monies. New combinations of surgery, radiotherapy, and other modalities emerged. A medical specialty of radiotherapy separated out from diagnostic radiology, following an earlier European trend.

The major new therapeutic innovation was chemotherapy, which joined surgery and radiotherapy as a means of treating cancer, especially in the 1960s. Its origins can be traced, in part, to the discovery of the anticancer properties of the nitrogen mustards during research for the U.S. Army in World War II. There followed a search for analogous compounds that might work on the leukemias and lymphomas, but the nitrogen mustards failed to produce lasting remissions, and enthusiasm for them began to evaporate. Then in the summer of 1947 Sidney Farber, a specialist in children's diseases in Boston, injected a two-year-old boy with an experimental drug named aminopterin, one of a class of pharmaceuticals called antifolates that impair the function of folic acid. The boy had acute lymphoblastic leukemia, a disease that almost always killed. But astonishingly Farber's patient went into remission, albeit temporary. Farber's work with aminopterin suggested that antifolates might work on certain cancers of the blood.

Building on such work, in 1955 the NCI established a vast national screening effort to develop and test anticancer drugs, promising new chemical compounds being investigated in animal models and through (randomized double-blind) clinical trials on patients. Specialist hospitals and centers established to provide cancer care became centers of research, their large populations of patients and teams of experts being invaluable for therapeutic trials. But chemotherapy raised as many problems as it solved. Although growing numbers of children survived leukemia, chemotherapy did not work as well in other cancers and not always as well in cases of childhood leukemia. Physicians began to mix "cocktails" of pharmaceuticals, to increase doses of drugs to near the limits of human tolerance, and to continue treatment long after remission had occurred. Patients often found these procedures unbearable and many died, sometimes of cancers caused by the chemicals. Chemotherapy also came to be used as an adjuvant to surgery: (combination) chemotherapy used to destroy any remaining malignant cells after surgery, perhaps using doses that would not have been enough to destroy the original tumor.

The period from the 1940s to the 1970s thus began with the expansion of programs of early detection and treatment to include new therapies, services, and means of surveillance. It also saw significant developments in other areas. The first of these areas was research, funding for which dramatically increased from the 1940s. A key figure was Mary Lasker, the wife of advertising magnate Albert Lasker. In 1944, Lasker and her allies seized control of the American Society for the Control of Cancer, renamed it the American Cancer Society, and turned it into a formidable fund-raising machine. But her ambitions did not end there. After her husband's death from colon cancer in 1952, Lasker became convinced that only the federal government had the resources to tackle cancer. Using her extensive network of political, business, medical, and social connections and the substantial assets of her late husband's estate, she began to campaign for more funding for cancer research. Sidney Farber was a key part of this campaign, and together they helped to ensure a vast expansion of federal support for cancer research in the 1950s and 1960s and to initiate President Richard M. Nixon's 1971 "war on cancer."

But Farber's vision of cancer research was controversial. In testimony to Congress, he argued that the cure of cancer could be achieved with little further basic research: many, if not most cures, he claimed, were developed before their mechanism of action was unraveled. For many scientists in universities and medical schools, such claims set off alarm bells. They argued that Farber's dismissal of basic research was tantamount to running before one could walk, and they claimed that most breakthroughs in science came from promoting independent research in universities and medical schools, not from what was being proposed—organized centrally directed research such as the cancer chemotherapy program or a program to find viral causes of cancer. Such criticisms gained momentum in the mid-1970s when the American economy dipped after the oil crisis, and budgets tightened. Increasingly opponents argued that the "war" had proved unable to deliver on its promise of a cure, that the Special Virus Cancer Program had failed to find viral causes except in a few rare cancers, that hopes of a vaccine based on such research had faded, and that the war's focus on finding a cure was mistaken. Critics argued that preventive measures were more likely to reduce incidence and mortality from cancer and that more needed to be done to help people through the recovery process and face the prospect of death.

A second area of significant development was prevention, which gained growing attention following epidemiologic research in the late 1940s that identified cigarette smoking as a cause of lung cancer, the incidence of which was rising rapidly. For much of the 1950s it remained unsettled whether the statistical association of smoking and lung cancer truly indicated a cause. This changed with a 1964 report of U.S. Surgeon General, a shift in official attitudes toward the acceptance of epidemiologic proof that smoking "caused" cancer, and the triumph of multicausal explanations of the onset of cancer. In some ways, this was not entirely new. Nineteenth- and early twentieth-century physicians had seen cancer as the outcome of many different causes, dependent on both constitution and environment, neither of which was sufficient to promote cancer. Diet, nervous tension, environment, heredity, occupational exposure, and individual susceptibility might all contribute to the onset of the disease, but it was rarely the case that any one factor was sufficient in itself. Rather, cancer was the outcome of an often-unknowable combination of factors. What was different about the new interest in the 1960s and 1970s in multifactorial causation was that it was built upon statistical calculations of risk and a new acceptance that statistical association could be deemed a cause under certain conditions.

In the 1960s and 1970s interest in smoking was joined by a growing concern about the impact of occupation, environmental pollution, and diet on cancer causation. But advocates of prevention argued that the focus on finding a cure (combined with a continuing emphasis on therapeutic prevention) trumped other efforts to prevent the onset of cancer by changing behaviors such as smoking or diet or by reducing exposure to occupational and environmental carcinogens. The cancer establishment represented by Lasker, Farber, and their allies, they claimed, was dominated by physicians more interested in treatment than prevention and timorous in the face of opposition from those industries that caused cancer and whose products, processes, or profits would be the target of preventive measures.

A final area of significant development was postoperative and terminal/palliative care, both of which had roots in the criticism that anticancer campaigns paid more attention to getting patients to their physicians than to what happened to them after treatment. Postoperative recovery gained significant public attention in the 1940s and 1950s with the establishment of patients' groups such as Reach to Recovery (a mastectomy support group formed in 1953) that were critical of the quality of postoperative care, the growth of commercial organizations (that provided supplies and services for postoperative patients such as bust forms, surgical bras, and colostomy equipment), and the development of new professional groups of rehabilitationists, psychologists, and nurses who specialized in cancer. Palliative care of the sick and dying gained growing attention also at this time. A new generation of hospices was created from 1974, inspired in part by the English nurse Cicely Saunders, who had established hospices and palliative care in the United Kingdom in the 1950s.

The 1980s to the 2010s. By the 1980s, the field of cancer looked very different from that in the 1940s. Cancer research and services had expanded dramatically, and new emphases on cancer prevention, postoperative recovery, and end-of-life care had emerged. At the same time, the antiquackery programs of the 1950s and 1960s had been joined by an interest in incorporating what came to be known as complementary and alternative medicine into cancer care. Radical surgery had declined in favor of more conservative procedures, and the one-step procedure in breast cancer faced growing criticism from feminists, ethicists, and surgeons themselves. The enthusiasm of the 1960s for mammography had been joined by concerns about whether its benefits outweighed its costs, especially for older women. New screening techniques and recommendations emerged, including the PSA test (1980s/1990s) for prostate cancer and the first recommendation for routine screening for colorectal cancer (1990s).

Critics continued to suggest that a focus on finding cancer cures trumped prevention; that the Reagan and Bush administrations emphasized personal responsibility and lifestyles over preventing industrially produced carcinogens; that the American Cancer Society was too close to the very industries that created these carcinogens; and that the NCI was too cautious toward industry and its congressional supporters. Anticancer campaigns trumpeted breakthrough cures, but critics questioned such claims. Echoing critics of the 1970s, the epidemiologist and biostatistician John C. Bailar III argued in 1986 that mortality rates had not improved since the 1950s. In his view, postwar efforts against cancer had largely failed.

The major new therapeutic development of the late twentieth century was targeted therapy—the use of drugs or other substances to block the growth and spread of cancer by interfering with specific molecules involved in tumor growth and progression. Targeted therapy had roots in a series of amazing discoveries made in the 1990s, in particular of proto-oncogenes and tumor suppressor genes (the genes that drive and suppress cellular growth) and of the complex regulation of signaling systems used by cells—both normal and cancerous—to communicate with each other and their environments. These discoveries and others opened up the possibility of interfering with cancer cell division and spread in a variety of new ways, for example, by targeting proteins involved in the cell-signaling pathways that govern basic cellular functions and activities such as cell division, movement, responses to specific external stimuli, and death. They also allowed scientists to identify people who might be at risk from common cancers such as those of the breast and colon. Paradoxically, they also destabilized these disease categories. For example, in 2012 one study noted that the category "breast cancer" actually comprised 10 completely different diseases, each with a specific abnormality that might be targeted.

But targeted therapy had its problems. Pharmaceutical companies were cautious about the expense of trials, especially when the market for each individual drug was limited to the few with a particular mutation. Insurance companies, governments, health services, and patients worried about the high cost of those drugs that came on the market. And critics argued that targeted therapy was unlikely to significantly reduce cancer

mortality without the development of many more drugs. It seemed that for everyone who benefited from a drug there were many others who gained nothing, and those who did respond could develop resistance to the medication. Critics argued that targeted therapy should not develop at the expense of prevention, much as earlier critics had argued a similar case against surgery, radiotherapy, and chemotherapy. The causes of cancer were well known, it was claimed. It just needed the political will to tackle them.

In prevention, the use of chemicals to prevent cancer (chemoprevention) gained growing attention in the 1990s—drugs included tamoxifen (for breast cancer), BCG (for bladder cancer), finasteride (prostate cancer), and aspirin (colon cancer). Vaccines also gained new medical and commercial attention, despite the disappointments of the NCI's virus program of the 1970s. In the early twenty-first century, the U.S. Food and Drug Administration approved vaccines against the hepatitis B virus to prevent liver cancer and against human papillomavirus (HPV) types 16 and 18 to prevent cervical cancer. Cervical cancer had effectively been turned into a sexually transmitted disease because HPV infections were the most common sexually transmitted infections in the United States and now recognized as a cause of cervical and other cancers. It was no surprise then that the executive order issued in 2007 by Texas's governor Rick Perry requiring all girls entering sixth grade to be vaccinated against HPV to prevent cervical cancer generated controversy. It mixed questions of cancer prevention with concerns about child sexual activity and the roles of government, parents, pharmaceutical companies, and advocacy groups in these issues.

Mortality Decline. Then in the 1990s age-adjusted cancer mortality rates began to decline, falling 3 percent between 1990 and 1995, the first such decline since records began. This was not true for all cancers; death rates from liver cancer and female lung cancer continued to increase. However, after years in which cancer mortality rates had either increased or leveled out, death rates for cancers of the lung (in males), prostate, female breast, colon-rectum, pancreas, leukemia, and ovary began to decrease, as did death rates for stomach and uterine cancers. The reasons for this were not always clear, and commentators urged caution in interpreting the statistics and the need for continued monitoring. Hundreds of thousands still died of the disease each year. Yet the decline persisted. From 1999 to 2008 cancer death rates fell by more than 1 percent per year in men and women of every racial/ethnic group with the exception of Native Americans/Alaskans, among whom rates remained stable.

Such statistics created a sense of optimism that something had happened in the long campaign against this group of diseases. Cancer remained a major object of public-health concern, but physicians began to hope that it was turning from a killer into a chronic disease. Cancer would never disappear, but a future seemed possible in which people died with the disease but not of it. Experts disagreed as to the best route to this future, when it might come about, or what it meant to live with a potential killer disease (or the risk of it) within. Critics charged that different stakeholders were exploiting the fall in cancer mortality to promote their favored anticancer approaches. If earlier physicians had been divided about the best way to stem rising mortality, physicians in the early twenty-first century were divided about how to maintain the fall.

[*See also* Disease; Medicine; National Institutes of Health; Pharmacology and Drug Therapy; Public Health; Public Health Service, U.S.; Radiology; *and* Surgery.]

BIBLIOGRAPHY

Aronowitz, Robert A. *Unnatural History: Breast Cancer and American Society*. Cambridge, U.K., and New York: Cambridge University Press, 2007. With Lerner (later in list), the standard history of breast cancer in the United States.

Breslow, Lester, et al. *A History of Cancer Control in the United States, with Emphasis on the Period 1946–1971*, 4 vols. Bethesda, Md.: Division of Cancer Control and Rehabilitation, National Cancer Institute, Department of Health, Education, and Welfare, Public Health Service, National Institutes of Health, National Cancer Institute, Division of Cancer Control and Rehabilitation, 1977. A valuable historical account of approaches to cancer control and the roles of various governmental, philanthropic, medical, and scientific institutions in its development.

Cantor, David, ed. *Cancer in the Twentieth Century*. Baltimore and London: Johns Hopkins University Press, 2008. An edited collection of essays documenting efforts to promote cancer education, prevention, treatment, and research in the United States and Europe.

Gardner, Kirsten E. *Early Detection. Women, Cancer, and Awareness Campaigns in the Twentieth-Century United States*. Chapel Hill: University of North Carolina Press, 2006. The standard history of women's involvement in American cancer campaigns.

Goodman, Jordan, and Vivien Walsh. *The Story of Taxol. Nature and Politics in the Pursuit of an Anti-Cancer Drug*. Cambridge, U.K.: Cambridge University Press, 2001. The story of an anticancer drug discovered by publicly funded scientists in the bark of the Pacific yew and of how private industry subsequently developed this drug.

Keating, Peter, and Alberto Cambrosio. *Cancer on Trial: Oncology as a New Style of Practice*. Chicago and London: University of Chicago Press, 2012. The best account of the development of cancer clinical trials and of how this practice emerged after World War II.

Kutcher, Gerald. *Contested Medicine: Cancer Research and the Military*. Chicago: University of Chicago Press, 2009. A history of cancer research, particularly valuable for its account of the connections between research and Cold War military concerns.

Lerner, Barron H. *The Breast Cancer Wars: Fear, Hope, and the Pursuit of a Cure in Twentieth-Century America*. New York: Oxford University Press, 2003. With Aronowitz (earlier in list), the standard history of breast cancer in the United States.

Löwy, Ilana. *Between Bench and Bedside. Science, Healing, and Interleukin-2 in a Cancer Ward*. Cambridge, Mass.: Harvard University Press, 1996. The standard history of this drug and a valuable account of immunotherapy more generally.

Löwy, Ilana. *Preventive Strikes: Women, Precancer, and Prophylactic Surgery*. Baltimore: Johns Hopkins University Press, 2009. A valuable history of breast cancer and especially of precancers (lesions that might turn cancerous) and prophylactic breast surgery to prevent the disease in women.

Löwy, Ilana. *A Woman's Disease: The History of Cervical Cancer.* Oxford: Oxford University Press, 2011. The standard history of cervical cancer in the United States and elsewhere.

Patterson, James T. *The Dread Disease: Cancer and Modern American Culture.* Cambridge, Mass.: Harvard University Press, 1987. The standard survey of American cultural attitudes toward cancer and campaigns against this group of diseases.

Proctor, Robert N. *Cancer Wars. How Politics Shapes What We Know and What We Don't Know about Cancer.* New York: BasicBooks, 1995. One of the best accounts of the politics around the knowledge of cancer and how it shapes efforts to combat this group of diseases. Author argues that the causes of cancer are known, albeit often obscured by stakeholders, and that they can be tackled.

Rettig, Richard A. *Cancer Crusade. The Story of the National Cancer Act of 1971.* Princeton, N.J.: Princeton University Press, 1977. The standard account of President Nixon's "war on cancer."

Siegel, Rebecca, Deepa Naishadham, and Ahmedin Jemal. "Cancer Statistics, 2012." *CA: A Cancer Journal for Clinicians* 62 (2012): 10–29. Along with Wingo et al. (later in list), a valuable account of mortality trends in the late twentieth and early twenty-first centuries.

Wailoo, Keith. *How Cancer Crossed the Color Line.* New York: Oxford University Press, 2011. An account of how anticancer efforts started with a focus on white women and later came to include African Americans and others.

Wailoo, Keith, Julie Livingston, Steven Epstein, and Robert Aronowitz, eds. *Three Shots at Prevention: The HPV Vaccine and the Politics of Medicine's Simple Solutions.* Baltimore: Johns Hopkins University Press, 2010. An edited collection that explores the complex issues raised by the HPV vaccine from a variety of historical perspectives.

Wingo, Phyllis A., Cheryll J. Cardinez, Sarah H. Landis, Robert T. Greenlee, Lynn A. G. Ries, Robert N. Anderson, and Michael J. Thun. "Long-Term Trends in Cancer Mortality in the United States, 1930–1998." *Cancer* 97 (2003): 3133–3275. Along with Siegel et al. (later in list), a valuable account of mortality trends in the late twentieth century.

David Cantor

CANNON, WALTER BRADFORD

(1871–1945), physiologist. Walter Bradford Cannon was born in 1871 in Wisconsin. His high school teachers in Saint Paul, Minnesota, recognized his scholastic abilities and encouraged him to attend Harvard College, even helping him to obtain a much-needed scholarship. Although punctuated by loneliness, Cannon's time at Harvard broadened his interests considerably. Zoology proved especially compelling for Cannon, and his high marks afforded him the opportunity to conduct biological research with Charles B. Davenport. This propensity toward research continued after he graduated summa cum laude in 1896 and enrolled at Harvard Medical School. Cannon quickly sought out the guidance of the physiologist Henry Pickering Bowditch, who suggested using the new technology of X-rays to help visualize the mechanisms of digestion. These early studies foreshadowed Cannon's future contributions to gastric research.

After receiving his MD in 1900, Cannon quickly climbed the ranks at Harvard from instructor in zoology to assistant professor in 1902, succeeding Bowditch as chair of the Department of Physiology in 1906. He held this position until his retirement in 1942. Beyond Harvard, Cannon assumed leadership of the American Medical Association's Council on the Defense of Medical Research in 1908 and dedicated much of his career to championing the value of animal experimentation against the outcries from anti-vivisectionist groups. A belief in justice and social responsibility also pervaded his political activities, which included efforts in the 1930s to provide medical aid to the embattled Spanish republic and to advocate for improved American–Soviet relations.

The young professor extended his studies of digestion in the 1910s, based on the observation that movements of the digestive tract were inhibited when an animal was startled or provoked. A series of experiments on the relationship of visceral function and emotional states, paying special attention to the remarkable effects of adrenaline, led Cannon to propose his notion of the sympathetic nervous system's function in emergency situations—the "fight or flight" response. His 1915 book on the subject, *Bodily Changes in Pain, Hunger, Fear, and Rage*, received widespread attention in scientific and public arenas. In the 1920s, Cannon generalized his research to answer the larger question of how the body maintains a constant internal physiological state in the face of a fluctuating external environment. The result, Cannon's theory of homeostasis, was built on the physiologist Claude Bernard's notion of the *milieu intérieur* and described the myriad physiological processes that ensure the body's internal stability. In 1932, Cannon published his theory as another book for a general audience, entitled *The Wisdom of the Body*, which addressed the possibility of applying homeostatic principles to stabilizing society.

Cannon's research in the 1930s continued to investigate the chemical mediation of the sympathetic nervous system. These years were also marred by illness, as Cannon received one cancer diagnosis after another, all stemming from his early research with X-rays. Walter Cannon died in 1945, having made foundational contributions to research on digestion, nerve transmission, and hormones. His theory of homeostasis provided an influential framework for understanding the regulation of the body's physiology.

[*See also* **American Medical Association; Animal and Human Experimentation; Biological Sciences; Davenport, Charles Benedict; Physiology; Science;** *and* **Zoology.**]

BIBLIOGRAPHY

Benison, Saul, A. Clifford Barger, and Elin L. Wolfe. *Walter B. Cannon: The Life and Times of a Young Scientist.* Cambridge, Mass., and London: Belknap Press of Harvard University Press, 1987.

Benison, Saul, A. Clifford Barger, and Elin L. Wolfe. *Walter B. Cannon: Science and Society.* Cambridge, Mass.: Boston Medical Library in the Francis A. Countway Library of Medicine, 2001.

Brooks, Chandler McC., Kiyomi Koizumi, and James O. Pinkston, eds. *The Life and Contributions of Walter Bradford Cannon, 1871–1945: His Influence on the Development of Physiology in the Twentieth Century.* Brooklyn: State University of New York Downstate Medical Center, 1975.

Tulley Long

CARDIOLOGY

Cardiology, the medical specialty devoted to the diagnosis and treatment of patients with diseases of the heart and blood vessels, owes much of its success to a long series of technological and procedural innovations. Beginning in the second half of the nineteenth century, technology transformed the way doctors evaluated patients and the way patients viewed their doctors. Increasingly, specialists were distinguished by the tools they used and the tests and procedures they performed. Technology has played a major role in defining cardiology as a specialty since Dutch physiologist Willem Einthoven invented the electrocardiograph (ECG) in 1902. Within a decade, his complex instrument migrated from the laboratory to the hospital and from Europe to North America.

The Invention of a Medical Specialty. A combination of scientific, technological, intellectual, and social factors contributed to the creation of American cardiology a century ago. Physicians had cared for patients with complaints attributed to heart disease for generations, but these caregivers were general practitioners. In the 1920s, some doctors in New York, Boston, and a few other big cities perceived and portrayed themselves as heart specialists. Their diverse interests (including science, practice, and public health) were reflected in the original goals of the American Heart Association (AHA), founded in 1924, and the *American Heart Journal*, launched the following year. The association influenced patient care and practice by encouraging the creation of cardiac clinics and the diffusion of the electrocardiograph as a diagnostic tool. These developments, together with the public's growing awareness about the high prevalence and economic consequences of heart disease, drove demand for cardiologists. The earliest heart specialists used the stethoscope, chest X-ray, and fluoroscope to help them detect and classify heart disease. Doctors who owned an electrocardiograph and interpreted the tracings it produced were often considered cardiologists by patients and general practitioners.

Between the world wars, doctors found the electrocardiograph helpful in detecting and characterizing heart disease, but they still had few effective treatments to offer cardiac patients. Boston cardiologist Paul Dudley White published the first edition of his classic book *Heart Disease* in 1931. He divided cardiac disease into three main groups in order of frequency: coronary, hypertensive, and rheumatic heart disease were common; syphilitic and thyroid heart disease were fairly common; and bacterial endocarditis and congenital heart disease were uncommon. The sections on therapy demonstrate that doctors prescribed an assortment of drugs and nonpharmacologic treatments, but the text also reveals the limitations of contemporary therapeutics. Digitalis, morphine, nitroglycerin, and quinidine were the main cardiac drugs available to practitioners in the 1930s that are still used today. Dozens of drugs have disappeared, and most present-day medicines had not been discovered. There were no antibiotics to cure infections, potent diuretics to treat congestive heart failure and fluid retention, or antihypertensives to lower blood pressure. Heart failure was treated with digitalis and diuretics (mercury compounds or purine derivatives like theophylline), but these drugs had significant side effects or had to be administered by injection. Rest was a mainstay of therapy for heart failure and angina pectoris.

Dividends from America's Investment in Cardiovascular Research.

The pace of medical invention and innovation accelerated dramatically in America after World War II when the federal government began to spend millions (and eventually billions) of dollars on research. The National Heart Act, signed into law in 1948, created the National Heart Institute and its program of extramural grants for research. Another notable event took place that year: the AHA was transformed from a professional society into a voluntary health organization that sponsored very successful public fund-raising drives generating millions of dollars for research. The decade after World War II was marked by the introduction of several new drugs that dramatically changed the way doctors cared for patients with cardiovascular disease. Antibiotics, oral diuretics, and effective antihypertensive medications were all introduced at this time. Penicillin, available at the end of the war, was effective in treating and preventing recurrent streptococcal infections that could cause acute rheumatic fever. As the incidence of rheumatic fever declined, so did its dreaded cardiac complications. Two anticoagulants, intravenous heparin and oral dicumarol, were useful in preventing and treating blood clots that could cause heart attacks, pulmonary emboli, and strokes.

Doctors and patients were optimistic at midcentury about promising developments in the diagnosis and treatment of cardiovascular disease. The most sensational therapeutic advances would be in the emerging field of cardiovascular surgery. Johns Hopkins surgeon Alfred Blalock performed an operation in 1944 that signaled a new era in the treatment of heart disease. Pediatric cardiologist Helen Taussig and Blalock's African American laboratory technician Vivien Thomas helped him develop the "blue baby operation" for a complex congenital heart defect that was invariably fatal. The surgery was not curative, but the babies and children who underwent it had marked improvement in their exercise tolerance and lived significantly longer. News of the operation electrified the medical community and the public. Four years later, Horace Smithy of Charleston, South Carolina, Charles Bailey of Philadelphia, and Dwight Harken of Boston performed successful operations for mitral stenosis, a complication of rheumatic fever that mainly affected young adults. They performed these procedures by putting a finger or an instrument inside a patient's closed, blood-filled, beating heart and splitting open the tight valve leaflets. Soon, several surgeons were collaborating with scientists and engineers in an attempt to invent a heart–lung machine that could take over the functions of these organs for an hour or more. This would make it possible to stop the flow of blood through the heart so a surgeon could see and operate inside it. Open-heart surgery, introduced as a routine clinical procedure at the University of Minnesota and the Mayo Clinic in 1955, had profound implications for patients with specific types of heart disease, for cardiologists who were responsible for diagnosis, and for thoracic surgeons whose field was evolving into cardiothoracic surgery.

The possibility of operating on patients with heart disease encouraged the invention of new diagnostic methods. A generation of cardiologists tried to make an accurate diagnosis by integrating information derived from a patient's history, physical examination (with emphasis on abnormal heart sounds and murmurs heard through a stethoscope), X-ray and fluoroscopic findings, and electrocardiograph patterns. The shortcomings of this strategy in some situations became more evident as surgeons began to operate inside the heart. Cardiac catheterization provided new and unique information. This technique involved inserting a thin, flexible tube into a peripheral vein or artery and pushing it into the heart. It was first employed as a diagnostic aid in 1945, four years after André Cournand's group at Bellevue Hospital in New York City reported how catheterization could be used to study cardiopulmonary physiology in humans. The technology was transformed from a research tool into a clinical test by the end of the decade. Innovations in cardiac surgery stimulated the development and diffusion of catheterization, a technology that helped cardiologists and surgeons classify heart defects, quantitate their severity, and plan treatment. The scope of cardiac surgery expanded significantly in 1960 with the introduction of artificial valves to replace diseased ones that caused symptoms such as shortness of breath and premature death.

New Technologies Designed to Combat Coronary Artery Disease.
Coronary artery disease, far more common that congenital heart defects and acquired valve disorders, affected millions of Americans. It could cause chronic angina pectoris and acute myocardial infarction (AMI) with its risk of sudden death. During the 1960s, three innovations had major implications for the care of patients with coronary heart disease: the

coronary care unit (CCU), coronary angiography, and bypass surgery. Technologies introduced the previous decade, such as continuous electrocardiograph monitoring, pacemakers, and transthoracic (external) defibrillators, enabled doctors to treat some life-threatening heart rhythm disorders. In 1960, electrical engineer William Kouwenhoven and his colleagues at Johns Hopkins combined defibrillation with closed-chest cardiac massage and mouth-to-mouth respiration to inaugurate the modern era of cardiopulmonary resuscitation (CPR).

The CCU concept was implemented in 1962 as a care model that targeted a specific group of patients—those at risk of sudden death in the context of a heart attack. Vulnerable patients were admitted to a special hospital area staffed by nurses who were trained to use new electronic technologies for the rapid diagnosis and treatment of life-threatening arrhythmias and to perform CPR. The successful treatment of ventricular fibrillation, an arrhythmia that was invariably fatal prior to the advent of the defibrillator, provided compelling evidence that the CCU model saved lives. It transformed the care of patients, the careers of cardiologists, and the boundaries of nursing practice in less than a decade. Nurses were given the authority to institute immediate treatment because cardiac arrest leads to irreversible brain damage in less than four minutes. There was no time to wait for a doctor to run to a patient's bedside and discharge a defibrillator that delivered a shock that could be lifesaving. This led physicians to train a special corps of nurses to use the device without personal supervision. Hundreds of hospitals opened CCUs in the mid-1960s, which, in turn, created job opportunities for cardiologists who were hired to direct them.

The CCU model was designed to treat patients with a heart attack, an acute complication of coronary artery disease. During the 1960s, there was increasing interest in developing more effective ways to treat angina pectoris, a chronic condition characterized by chest discomfort that limited a patient's activity level and could lead to unemployment. Progressive angina could culminate in a heart attack and sudden death. Doctors had used catheters to diagnose congenital heart defects and valve disease for a decade when Cleveland Clinic cardiologist Mason Sones Jr. invented selective coronary angiography in 1958. This catheter-based imaging technology documented the location and severity of blockages in the coronary arteries. Few heart specialists adopted the technique until Sones's surgical colleague René Favaloro developed an operation for angina a decade later. Coronary artery bypass surgery (CABG) involved using a vein segment to bypass a blocked coronary artery. The rapid diffusion of CABG in the early 1970s stimulated demand for cardiologists who could perform coronary angiography.

Catheter-based cardiac diagnosis involved inserting tubes into the blood stream and advancing them into the heart. These procedures were termed "invasive" because the operator penetrated a patient's body with a tool. During the final third of the century, several new "noninvasive" diagnostic technologies were introduced. Echocardiography, ultrasound imaging applied to the heart, provided unique information about the heart's structure and function without inserting a catheter. Several other technologies moved from the experimental stage to standard practice at rates that reflected their perceived value as a diagnostic aid, their cost, reimbursement policies, and market forces. The advent of coronary angiography and CABG encouraged the use of treadmill exercise testing to determine whether a patient's chest pain was likely to be a manifestation of coronary disease. Nuclear medicine techniques were applied to the heart to measure its pumping function and as an adjunct to exercise testing to evaluate chest pain. Holter monitors were wearable devices that recorded an electrocardiograph continuously for several hours to help identify and characterize abnormal heart rhythms. Technological innovations were not limited to diagnosis. Implantable electronic cardiac pacemakers, invented in Sweden and the United States in the late 1950s, made it possible to treat patients with symptomatic bradycardia and heart block resulting from defects in the organ's electrical conduction system. Prior to permanent pacemakers, such patients were subject to recurrent fainting, heart failure, and sudden death.

Medicare Revolutionizes American Medicine. Scientific discoveries and technological innovations changed cardiology practice dramatically during the final third of the twentieth century. Meanwhile, political and economic forces had a profound effect on the pace of discovery and the diffusion of innovations into patient care. Between 1948 and 1965, the National Heart Institute spent more than three-quarters of a billion dollars supporting its mission, and its leaders could point with pride to many practical discoveries made by researchers it had helped to support. But as government-sponsored research led directly and indirectly to discoveries and innovations, concern grew that medical advances were diffusing too slowly into practice. Congress passed laws in the mid-1960s designed to accelerate the clinical application of new knowledge and to make health care available to more Americans. The Social Security Amendments of 1965 Act established Medicare, a program that entitled all U.S. citizens 65 years of age and older to government-funded hospital insurance benefits. They could also pay a premium that would extend their coverage to physician services. Medicare, combined with the fact that more Americans had health insurance as a benefit of employment, stimulated demand for specialized heart care. Reimbursement policies and consumer demand catalyzed cardiology's evolution from a technology-oriented into a technology-dominated specialty. The field grew steadily as new equipment and techniques were invented, indications for their use were liberalized, and access to them was enhanced.

The advent of sophisticated new technologies and growing demand for specialized heart care services contributed to a surge in the number of cardiology trainees beginning in the early 1970s. Government grants had helped create an academic infrastructure capable of rapid expansion, and fellowship programs met the demand for cardiologists trained to carry new diagnostic tools and therapeutic techniques into community hospitals across the country. Between 1961 and 1976, the number of cardiology fellows increased tenfold from 142 to 1,409 (Fye, 1996). By the mid-1970s, the nation's cardiology and cardiothoracic surgery training programs were producing a record number of practitioners to meet the demand.

Cardiac Catheters Transformed into Treatment Tools.
Cardiology entered a new era in the late 1970s when medical heart specialists began using unique catheters designed to perform therapeutic procedures. German cardiologist Andreas Grüntzig did the first percutaneous transluminal coronary angioplasty (PTCA) for the treatment of angina in 1977. The procedure involved threading a catheter with an inflatable balloon near its tip into a narrowed coronary artery. Once positioned in the blocked segment, the balloon was blown up to expand the vessel's lumen and increase the flow of blood through it. PTCA migrated to the United States a few months later. Many patients with angina chose angioplasty over coronary artery bypass surgery because it often resulted in symptom relief and meant they could avoid the discomfort, risks, and recovery time of an operation. Postprocedure angiograms revealed immediate, dramatic results in successful cases. Before PTCA, cardiologists used the catheter as a diagnostic tool to help decide which patients might benefit from surgery. Now, a new type of heart specialist, an interventional cardiologist, used a catheter to perform therapeutic procedures on patients with angina. Cardiologists and cardiac surgeons had to negotiate new roles and rules in hospitals where PTCA was introduced in the 1980s. The tradition whereby cardiologists were diagnosticians who treated heart patients with pills and referred some for surgery was suddenly disrupted. Medical and surgical heart specialists who had been collaborators in referral centers were now competitors. By 1990, the number of angioplasties surpassed the number of bypass operations performed annually in America.

Catheters were originally used to withdraw blood samples from the heart or to measure pressures in its chambers. During the last third of the century, catheters designed to record the heart's electrical activity were introduced into clinical practice. They were used to help understand the mechanism of rhythm disorders ranging from bradycardia resulting from heart block to tachycardia originating in the atria or ventricles. Initially, the data gained from so-called electrophysiology studies were used to guide therapy that might involve inserting a permanent pacemaker for bradycardia or prescribing drugs for tachycardia. Soon, clinical investigators collaborated with electrical engineers and others to develop catheters that could be used to treat abnormal heart rhythms. One arrhythmia—ventricular fibrillation—was invariably fatal and accounted for a high percentage of sudden deaths. A frequent and much-feared complication of acute myocardial infarction, sudden death also occurred unexpectedly in outpatients without chest pain, especially those with reduced left ventricular function. The CCU provided a safety net for men and women during the first days after a heart attack, but the majority of cardiac arrests occurred outside a hospital. The first defibrillators were so big that they were hospital-bound and had to be pushed around on a wheeled cart. Progressive miniaturization of batteries and electronic components contributed to the development of defibrillators so small that they could be implanted under the skin, much like a pacemaker. By the turn of the century, automatic implantable defibrillators were poised to become part of the cardiologist's armamentarium. Automatic external defibrillators would become part of the landscape of airports and other public buildings.

Two catheter-based treatment strategies were introduced in the early 1980s to interrupt a heart attack. The goal was to open a coronary artery blocked by a fresh blood clot—the cause of the life-threatening event in most cases. One strategy was to inject a thrombolytic (clot-dissolving) drug through a catheter placed in the blocked artery. Meanwhile, some cardiologists advocated performing an emergency PTCA to open an occluded artery in the context of an AMI. Most patients did not have access to these aggressive approaches, however, because the catheter techniques were limited to hospitals equipped and staffed to perform them around the clock. By the middle of the decade, it had been shown that injecting a thrombolytic agent into a peripheral vein was also effective. This meant that patients with

AMI admitted to community hospitals without a catheterization laboratory could be treated within the four-hour window required to prevent irreversible heart muscle damage. Biotechnology, a term introduced around World War I, meant little to cardiologists or their patients until the mid-1980s when a California genetic engineering company produced tissue plasminogen activator (t-PA), a thrombolytic agent that was shown to have an advantage over streptokinase, an older and much cheaper drug, in randomized clinical trials.

Clinical Trials Validate Therapies and Prevention Strategies.

The modern randomized clinical trial, invented shortly after World War II, became the gold standard for evaluating new drugs and devices used to treat or prevent a range of cardiovascular disorders. Large multicenter clinical trials, made possible by more powerful computers and industry support, were designed to detect small but statistically significant differences between treatment or prevention strategies. Cardiologists could not have imagined how the number of randomized trials would increase from a trickle to a torrent by the end of the century. They were central to evidence-based medicine, a model pioneered in Canada and the United Kingdom that was designed to rationalize decisions relating to patient care. Initially controversial, evidence-based medicine gained momentum rapidly during the 1990s. It became a critical component of the trial–guideline–education process. By the beginning of the twenty-first century, the parallel clinical trial, practice guideline, and continuing education movements combined to create one of the greatest paradigm shifts in the history of medicine. Evidence was the new currency of medical decision making that had long depended on a combination of clinical experience and common wisdom based mainly on expert opinion. The recent phenomenon of evidence-based medicine continues to have profound implications for cardiovascular research, patient care, cardiology practice, and corporate profits. Guidelines are now woven into the fabric of modern medicine—especially in cardiology and in America.

During the second half of the twentieth century, cardiologists focused their efforts on using new technologies and drugs to diagnose and treat cardiovascular disease. Coronary artery disease and its consequences in terms of angina pectoris, myocardial infarction, rhythm disturbances, and heart failure attracted increasing attention with the advent of the CCU, coronary angiography, bypass surgery, and angioplasty. The notion that certain "risk factors" were associated with atherosclerosis and coronary disease gained momentum as epidemiologists and others published data demonstrating a relationship between these common problems and high blood pressure, an elevated serum cholesterol level, and cigarette smoking. The prevalence of cardiovascular disease gave pharmaceutical companies a powerful incentive to invest in research to discover drugs that could be used to lower blood pressure and cholesterol. Once a novel drug was shown to be effective, its manufacturer was motivated to support a randomized clinical trial hoping to prove that their product was better than a competitor's. Statins, a class of cholesterol-lowering drugs, produced a seachange in the practice of cardiology in the 1990s, as evidence accumulated that their use was associated with a reduced risk of adverse cardiovascular events in selected patients, especially those with other risk factors.

Progressive Subspecialization and Concern about Health-Care Costs.

Innovations in diagnosis (e.g., echocardiography, coronary angiography, and invasive electrophysiology testing) and therapy (e.g., pacemakers, bypass surgery, and angioplasty) greatly stimulated the growth of cardiology during the 1980s. Heart specialists had a plethora of powerful tools to help them care for patients with a wide range of cardiac problems. The introduction of these and many technologies into practice contributed to changes in the content and duration of cardiology training. Prior to World War II, most American cardiologists were self-identified specialists whose formal training ended with an internship and a year of residency. They gained experience caring for patients with cardiovascular disease by working in a cardiac clinic or functioning as an assistant to a doctor with a reputation as a cardiologist. In the 1960s, most academic medical centers began offering formal fellowships in cardiology to doctors who had completed three years of internal medicine training. New technologies contributed to progressive subspecialization which, in turn, was accompanied by the founding of journals and organizations that focused on new clinical fields such as interventional cardiology, cardiac electrophysiology, and echocardiography.

Technology will remain central to the care of patients with cardiovascular disease for the foreseeable future. Some innovations have helped doctors diagnose and treat disorders of the heart and blood vessels, whereas others have facilitated basic research aimed at preventing them. There is growing recognition that physicians have a responsibility to use technology wisely. Many things contribute to the complex equation of health-care costs in the United States, but technology is an important factor. Technology often helps doctors make a correct diagnosis and initiate optimal therapy. Although many current clinical technologies are complementary, some provide redundant information that adds nothing but cost when used to evaluate the same problem in a single patient. America and its cardiologists confront three issues: new treatment technologies will continue to be invented and marketed; the population is aging and more individuals will develop cardiovascular disease; and the trajectory of health-care costs associated with innumerable innovations in patient care is unsustainable. Future historians will evaluate how the public, medical professionals, lawmakers, entrepreneurs, and other interested agents addressed these challenges and how cardiology and the care of patients with cardiovascular disease changed as a result.

[See also Automation and Computerization; Biotechnology; Blalock, Alfred; Disease; Health and Fitness; Health Insurance; Hospitals; Instruments of Science; Journals in Science, Medicine, and Engineering; Mayo Clinic; Medical Specialization; Medicare and Medicaid; Medicine; Medicine and Technology; National Institutes of Health; Nursing; Penicillin; Pharmacology and Drug Therapy; Public Health; Race and Medicine; Research and Development (R&D); Surgery; and Technology.]

BIBLIOGRAPHY

Acierno, Louis J. *The History of Cardiology*. Pearl River, N.Y.: Parthenon, 1994.

Bertrand, Michel E., ed. *The Evolution of Cardiac Catheterization and Interventional Cardiology*. St. Albans, U.K.: European Society of Cardiology, 2006.

Burch, George E., and Nicholas P. DePasquale. *A History of Electrocardiography*, with a new introduction by Joel D. Howell, 2d ed. San Francisco: Jeremy Norman, 1990.

Comroe, Julius H., Jr., and Robert D. Dripps. *The Top Ten Clinical Advances in Cardiovascular-Pulmonary Medicine and Surgery 1945–1975*. Washington, D.C.: U.S. Government Printing Office, 1978.

English, Peter C. *Rheumatic Fever in America and Britain*. New Brunswick, N.J.: Rutgers University Press, 1999.

Fye, W. Bruce. *American Cardiology: The History of a Specialty and Its College*. Baltimore: Johns Hopkins University Press, 1996.

Fye, W. Bruce. "The Power of Clinical Trials and Guidelines, and the Challenge of Conflicts of Interest." *Journal of the American College of Cardiology* 41 (2003): 1237–1242.

Hurst, J. Willis, C. Richard Conti, and W. Bruce Fye, eds. *Profiles in Cardiology*. Mahwah, N.J.: Foundation for Advances in Medicine and Science, 2003.

Jeffrey, Kirk. *Machines in Our Hearts: The Cardiac Pacemaker, the Implantable Defibrillator, and American Health Care*. Baltimore: Johns Hopkins University Press, 2001.

Lüderitz, Berndt. *History of the Disorders of Cardiac Rhythm*, 3d ed. Armonk, N.Y.: Futura Publishing, 2002.

Postel-Vinay, Nicolas, ed. *A Century of Arterial Hypertension, 1896–1996*. Translated by Richard Edelstein and Christopher Coffin. New York: John Wiley & Sons, 1996.

Reiser, Stanley J. *Medicine and the Reign of Technology*. New York: Cambridge University Press, 1978.

Rothstein, William G. *Public Health and the Risk Factor: A History of an Uneven Medical Revolution*. Rochester, N.Y.: University of Rochester Press, 2003.

Westaby, Stephen, and Cecil Bosher. *Landmarks in Cardiac Surgery*. Oxford: ISIS Medical Media, 1997.

W. Bruce Fye

CAROTHERS, WALLACE HUME

(1896–1937), chemist, inventor. Carothers, as a research chemist for the DuPont Company from 1928 to 1937, made seminal contributions to the emerging discipline of polymer science; at the same time his laboratory produced two major technological inventions: neoprene synthetic rubber (1930) and nylon (1934). Carothers demonstrated that industrial researchers could make important scientific contributions as well as technological innovations.

Carothers came from a family of modest circumstances; his father taught at a business school in Des Moines, Iowa. He was able to attend Tarkio College in Missouri, graduating with a degree in chemistry in 1920. Carothers then enrolled in the graduate program at the University of Illinois, where he worked with the eminent organic chemist Roger Adams. After earning his PhD in 1924, Carothers taught at Illinois for two years before becoming an assistant professor at Harvard University.

In 1927 he was recruited by Charles M. A. Stine, director of the central research laboratory of the DuPont Company in Wilmington, Delaware. Stine's new fundamental research initiative was intended to explore the fundamental science underlying the company's products. For example, at this time DuPont's most important raw material was cellulose, which was made into plastic, paint, fibers, and films. Yet, the chemical structure of cellulose remained uncertain. Large molecules, such as cellulose, were the subject of scientific debate regarding whether they were just larger versions of ordinary organic molecules or aggregates of smaller molecules held together by what were called colloidal forces. Stine suggested to Carothers that large molecules or polymers would be a research topic of interest to DuPont. Intrigued by this project, Carothers accepted DuPont's generous salary offer and joined the company in 1928.

Over the next two years Carothers and his team of PhD assistants pursued a research program that demonstrated unambiguously that polymers were just large organic molecules. Carothers accomplished this by building large-chain molecules step-by-step using standard chemical reactions. Using a novel experimental apparatus, his team was able to make very long molecules that produced strong fibers. After this 1930 breakthrough, four years passed before the discovery of a synthetic fiber, nylon, that had commercial potential.

Two weeks before the first synthetic fiber was produced, another significant discovery was made by Carothers' team. Work on polymers made from acetylene yielded a rubber-like substance that, on analysis, turned out to be chemically analogous to natural rubber.

During the 1930s DuPont developed nylon synthetic fibers and neoprene synthetic rubber into very successful products while Carothers completed his classic studies of polymers. At the same time, Carothers' mental health, which had always been an issue, began to deteriorate, eventually leading to his suicide in 1937.

[*See also* Chemistry; Nylon; Research and Development (R&D); Science; *and* Technology.]

BIBLIOGRAPHY

Furukawa, Yasu. *Inventing Polymer Science: Staudinger, Carothers, and the Emergence of Macromolecular Chemistry*. Philadelphia: University of Pennsylvania Press, 1998.

Hermes, Matt. *Enough for One Lifetime: Wallace Carothers, Inventor of Nylon*. Washington, D.C.: American Chemical Society and the Chemical Heritage Foundation, 1996.

Hounshell, David A., and John Kenly Smith Jr. *Science and Corporate Strategy: DuPont R&D, 1902–1980*. New York: Cambridge University Press, 1988.

McGrayne, Sharon Bertsch. *Prometheans in the Lab: Chemistry and the Making of the Modern World*. New York: McGraw–Hill, 2001.

John Kenly Smith Jr.

CARSON, RACHEL

(1907–1964), nature writer, environmentalist. Raised on a farm in western Pennsylvania, Carson attended the Pennsylvania College for Women, gaining a magna cum laude in English and biology in 1929. Encouraged by zoology professor Mary Scott Skinker, Carson pursued graduate study at Johns Hopkins University. She earned a master's degree in marine zoology in 1932. Her thesis focused on the embryological development of the kidney system in a catfish species. In 1935, she went to work at the U.S. Fish and Wildlife Service, where she embraced the New Deal's commitment to environmental conservation. Carson's responsibilities with the Fish and Wildlife Service mainly involved work as a science writer, synthesizing the research of others, rather than working independently on research of her own choosing. She produced numerous publications for the agency until she left in 1953 to write independently.

Carson's *The Sea around Us* (1951), a hugely successful book about the oceans, was at once scientifically knowledgeable and compellingly lyrical. In this book she explored the origins and history of the oceans and the relationship between marine life (including humans) and the wider physical environment. The book became a bestseller, earning Carson enough money to resign from the Fish and Wildlife Service to work independently. Carson's writings alerted the public to the emerging field of ecology. She is best known for *Silent Spring* (1962), an impassioned defense of unspoiled nature for its own sake and a powerful warning that the chemical despoilation of the environment, especially by the insecticide DDT, threatened human health. An international best seller, the book eloquently popularized the idea that human beings must recognize themselves as part of nature rather than purely as masters of it.

In *Silent Spring*, which first appeared in the *New Yorker* magazine, Carson lucidly explained the intricate interconnectedness of nature and how chemical herbicides or insecticides applied by spraying the earth or the air could diffuse through the local soil and then be carried through ground and surface water to distant areas and accumulate in the wild food chain. Her book roused a barrage of ridicule and denunciation from the chemical industry, parts of the food industry, academic scientists allied with both, and some powerful sectors of the media. However, *Silent Spring*'s carefully documented analyses were reviewed and endorsed by the President's Science Advisory Committee, and Carson's views helped inspire the environmental protection movement that began in the 1960s.

[*See also* **Conservation Movement; Ecology; Environmentalism; Environmental Protection Agency; Fish and Wildlife Service, U.S.; Pesticides; President's Science Advisory Committee;** *and* **Zoology.**]

BIBLIOGRAPHY

Freeman, Martha, ed. *Always, Rachel: The Letters of Rachel Carson and Dorothy Freeman, 1952–1964.* Boston: Beacon Press, 1995.

Kimler, William C. "Carson, Rachel Louise." In *Complete Dictionary of Scientific Biography*, Vol. 17, pp. 142–143. Detroit: Charles Scribner's Sons, 2008.

Lear, Linda. *Rachel Carson: Witness for Nature.* New York: Henry Holt, 1997.

Daniel J. Kevles;
updated by Elspeth Knewstubb

CARTOGRAPHY

Although mapmaking long predated the achievement of American independence, the revolution stimulated the demand for cartography, as did westward expansion. The Northwest Ordinance of 1785 mandated extensive surveys of the new lands of the interior and established a method of surveying and mapping that would extend to the Pacific Ocean by 1900. The General Land Office (1812) instituted the rectangular survey, an ongoing source of land maps for the young nation.

Nineteenth-Century Origins. The early commercial map industry centered around firms in Philadelphia. Perhaps the single most important map of the new nation was the "Map of the United States with the contiguous British and Spanish Possessions," created by John Melish just as the war against Britain was ending and advertised as including the latest intelligence from western explorers. The map oddly extends to the Pacific Ocean, almost anticipating national expansion, and this may account for its contemporary popularity and historical significance. Five presidents in the early republic owned copies of the map, and it was also used to negotiate the nation's western border with Spain in 1819.

Though the Melish map was symbolically important, most topographical mapping in the nineteenth century was conducted or sponsored by the federal government. The expedition of Meriwether Lewis and William Clark began this tradition, which provided the intelligence for Samuel Lewis' "A Map of Lewis and Clark's Track across the Western Portion of North America" (1811). In 1818 the War Department established a topographic bureau devoted to maps for internal improvement, followed by the Corps of Topographic Engineers (1838), which was folded into the U.S. Geological Survey in 1879. These federal efforts produced several detailed maps of the west, notably the seven-sheet traverse map created by John Fremont and Charles Preuss in the 1840s, "Topographical Map of the Road from Missouri to Oregon." In the 1850s, the Department of War launched the Office of Explorations and Surveys in the pursuit of a transcontinental railroad route. Among the most impressive products of this effort was Gouverneur Kemble Warren's "Map of the Territory of the United States from the Mississippi to the Pacific Ocean" (1855).

The federal government also invested heavily in coastal and hydrographic mapping throughout the century. In 1842 the Navy created the Depot of Charts and Instruments, and at the same time the U.S. Coast Survey became the nation's leading source of maps and charts. Under the direction of Alexander Dallas Bache the Survey attracted the nation's best mapmakers and many skilled German immigrants. This infusion of talent had enormous consequences. For instance, the introduction of lithography broadened the speed and ease of map production, whereby maps could be drawn rather than engraved into copperplate. Similarly, the Coast Survey aggressively applied photography to lithography on the eve of the Civil War, which widened the availability of maps and charts for both strategic and civilian purposes. The circumstances of the war also spawned other techniques, such as solar (salt) printing that was used by both Union and Confederate forces. On the homefront, the hunger for news kept several commercial map firms in business throughout the war.

For much of the first half of the century, these federal agencies sought to create increasingly precise representations of the landscape, but gradually another purpose for maps emerged, one that emphasized the organization of information rather than topography. For instance, the Office of the U.S. Army Surgeon General and the Smithsonian Institution broke new ground in meteorology by mapping rainfall and charting patterns of climate in the 1850s. This paved the way for the explosion of weather mapping in the late nineteenth century, primarily by the U.S. Signal Service (later the U.S. Weather Service). This "thematic" use of cartography was also championed by the Superintendent of the Ninth Census, Francis Amasa Walker, who compiled a pathbreaking *Statistical Atlas of the United States* (1874) to profile the nation's resources as well as the characteristics of its population. The *Atlas* maps were the work of German immigrant Julius Bien, the most capable and admired commercial lithographer of the late nineteenth century who set new standards for cartographic detail and representation. Together, Walker and Bien engineered the first national statistical atlas, one that mapped everything from geology and agriculture to the nation's ethnic groups and its race of wealth and literacy.

The concurrent explosion of commercially made maps after the Civil War was made possible by the introduction of wax engraving, which facilitated the insertion of type, and which made it easier to update and revise maps of westward growth, especially the railroads. The result was a flood of maps from new commercial firms such as Rand McNally, and this bolstered Chicago's role as a hub of map production. These new firms also produced atlases, bird's-eye views, and a uniquely American invention, the county atlas. This same urbanization prompted the founding of the Sanborn Insurance Company (1867), which was responsible for detailed maps of real estate in American cities, maps that remain useful for historical researchers today.

Cartography prior to World War II. In the twentieth century, the federal government continued to sponsor some of the most important cartographic work in the United States, as in the Department of Agriculture's soil-mapping enterprises in the early twentieth century. In 1918, President Woodrow Wilson prioritized cartographic knowledge by establishing a committee to prepare for the postwar peace. Led by geographer Isaiah Bowman, "The Inquiry" committee oversaw the creation and organization of an enormous number of maps designed to address the anticipated problems of the postwar world, such as the distribution of ethnicities and resources. This ambitious undertaking represents the fruition of decades of experimentation in thematic cartography during the nineteenth century.

Commercial cartography continued to grow in the early twentieth century, driven especially by the nation's growing dependence upon the automobile. By the 1920s oil companies, state governments, and motor clubs distributed road maps across the country, a practice that would continue into the 1970s. Road maps emphasized locations and routes and, like the railroad maps that preceded them, were schematic more than detailed. These maps poured forth from H. M. Gousha and Rand McNally in the mid-twentieth century, frequently as promotions for particular businesses, oil companies, and tourist bureaus, and in the process became icons of modern American culture.

The nation's first academically trained cartographer was John Paul Goode, who taught for years at the University of Chicago and authored *Goode's School Atlas,* which remains in print. Goode also designed a map projection in the 1920s, important for its emphasis on equal area representation (the "homolosine" projection). Similarly influential was Erwin Raisz, who authored the authoritative text in the field (*General Cartography,* 1938). Raisz was particularly adept at representation, and his evocative techniques spread through his courses and textbooks.

The rise of aviation fostered a sense of perspective and play in mid-twentieth-century mapping. This artistic sensibility was captured by Charles Owen and Richard Edes Harrison, who captured a sense of global interdependence through oblique and other pictorial representations of World War II (as in Harrison's *Look at the World,* 1944). Harrison and Raisz also challenged the concept of a map as a representation of absolute place; the latter, for instance, popularized "cartograms," or graphic images that emphasized proportional, rather than absolute, space and distance. These methods were popularized later by Michael Kidron and Ronald Segal in their *State of the World Atlas* of 1981 and the "red" and "blue" scaled maps created after the 2004 and 2008 elections.

The rise of aviation also transformed the practice and science of mapmaking. Aerial photography allowed mapmakers to view the earth from above on a systematic scale. This adoption of photogrammetry—deriving geographical properties of a region from a photograph—was widely adopted by the Army Map Service. The practice is as old as photography itself, but was only widely adopted in twentieth-century mapmaking.

Cartography since World War II. The postwar years were a tremendously fertile time for map design and execution. Military demands during World War II and the ensuing Cold War created several new purposes for cartography, and many who later became influential academic cartographers were first employed by the Office of Strategic Services during the war, such as Arthur Robinson and Richard Hartshorne. Robinson would launch one of the leading postwar departments of cartography at the University of Wisconsin and create the projection that bears his name, made famous by its adoption by the National Geographic Society in 1976.

Perhaps the most consequential area of cartography in the late twentieth century has been in the advent of geographic information systems (alternatively geographic information science, GIS). GIS has a complex history rooted in the military intelligence of World War II, and intellectually its origins lie in the thematic mapping enterprises of the nineteenth century (as in Walker, above). But GIS was made possible by the advent of digital technology, whereby data are manipulated and organized in spatial terms to achieve particular ends, such as analyzing problems of marketing, crime, and urban planning. This approach has dramatically widened the practice and interest in cartography and geography in recent decades. The attention to maps has been extended further by the user-driven capabilities of online tools such as Google Maps. Ironically, this has also led to a decline in the teaching of cartography as drafting in favor of geovisualization. The dominance of digital methods is now nearly complete.

[*See also* **Agriculture, U.S. Department of; Army Corps of Engineers, U.S.; Bache, Alexander Dallas; Highway System; Internet and World Wide Web; Lewis and Clark Expedition; Geography; Geological Surveys; Geology; Military, Science and Technology and the; Motor Vehicles; Photography; Railroads; Roads and Turnpikes, Early;** *and* **Smithsonian Institution.**]

BIBLIOGRAPHY

Dodge, Martin, Rob Kitchin, and Chris Perkins, eds. *The Map Reader: Theories of Mapping Practice and Cartographic Representation.* Chichester, West Sussex: John Wiley & Sons, 2011.

Ehrenberg, Ralph E., ed. *Library of Congress Geography and Maps: An Illustrated Guide.* http://www.loc.gov/rr/geogmap/guide/ (accessed 27 February 2012).

McElfresh, Earl B. *Maps and Mapmakers of the Civil War.* New York: Harry N. Abrams, 1999.

Ristow, Walter W. *American Maps and Mapmakers: Commercial Cartography in the Nineteenth Century.* Detroit, Mich.: Wayne State University Press, 1985.

Schulten, Susan. *Mapping the Nation: History and Cartography in Nineteenth-Century America.* Chicago: University of Chicago Press, 2012.

Schwartz, Seymour I., and Ralph E. Ehrenberg. *The Mapping of America.* New York: Harry N. Abrams, 1980.

Susan Schulten

CARVER, GEORGE WASHINGTON

(1864 [?]–1943), botanist, agricultural chemist, was born in the southwestern Missouri farm community of Diamond Grove (now Diamond), near Joplin. His mother was a slave owned by Moses Carver, a German immigrant farmer; his father is thought to have been a slave on a nearby farm. After Emancipation, George and a brother were reared as the foster children of Moses Carver and his wife. He early showed botanical aptitude, earning the nickname "the plant doctor." After graduating from a high school in Kansas and briefly attending Simpson College in Indianola, Iowa, he enrolled at Iowa State College of Agricultural and Mechanic Arts (now Iowa State University), where he earned a BS degree in 1894 and an MS degree in 1896. At Iowa State he supervised the college's greenhouse and conducted botanical experiments. In 1896, Booker T. Washington invited him to join the faculty of Tuskegee Institute in Alabama, where, despite occasional clashes with the strong-willed Washington, he would remain for the rest of his life.

As director of Tuskegee's agricultural research program, Carver developed an extension program bringing up-to-date crop information, as well as home economics training, to impoverished African American farmers of the South. To counter the soil exhaustion resulting from the monoculture cultivation of tobacco and cotton, Carver emphasized the importance of crop rotation to restore nitrogen to the soil. He encouraged the cultivation of peanuts, soybeans, and sweet potatoes to increase farmers' income while also benefitting the soil. A 1916 pamphlet for farmers, "Help for the Hard Times," offering practical advice for profitable vegetable gardening, was typical of his down-to-earth approach. As a researcher, he directed his laboratory experimentation toward results that would directly benefit farmers seeking to diversify their crops. Carver developed myriad plant-based commercial products, from lubricants, dyes, and wood stains to facial creams and medicinal products. He won fame for promoting peanuts as a cash crop; as a food, including peanut butter; and as the basis for numerous useful products. He reported his experimental findings in a series of Tuskegee publications from 1898 until 1942, a year before his death. As a matter of principle, Carver secured only three patents (in 1925–1927, on soybean-based cosmetics and wood stains), believing that his findings should be freely available.

Although teaching at an all-black institution in the Jim Crow era and pursuing his research outside the established, white-dominated scientific world of major universities and research institutes, Carver nevertheless won acclaim for his achievements. He also became an iconic cultural figure because popular magazines like *American* and *Reader's Digest* cited his career, often patronizingly, as proof that even those born in the lowliest circumstances could achieve great things through persistence and hard work. In the process, his genuine achievements were sometimes exaggerated and mythologized. Recent scholarship has highlighted the evolution of his environmental consciousness. The Moses Carver farm and reconstructed homestead, designated a National Monument by Congress in 1943, are now maintained by the National Park Service.

[*See also* **Agricultural Education and Extension; Agricultural Technology; Botany; Chemistry; Food and Diet; Food Processing; Home Economics Movement;** *and* **National Park System.**]

BIBLIOGRAPHY

Hersey, Mark D. *My Work Is That of Conservation: An Environmental Biography of George Washington Carver.* Athens: University of Georgia Press, 2011. Traces the evolution of Carver's view of man's relationship to the natural environment.

McMurry, Linda O. *George Washington Carver: Scientist and Symbol.* New York: Oxford University Press, 1982. Well-researched biography, sorting out facts from legend and myth.

Paul S. Boyer

CATTELL, JAMES MCKEEN

(1860–1944), psychologist and editor. After graduating in 1880 from Lafayette College, Cattell spent two years attending lectures at German universities before becoming a Fellow in Philosophy at Johns Hopkins University. There he was introduced to experimental psychology—then emerging from philosophy—and to laboratory science. In 1884, he moved on to the University of Leipzig, where he first worked in Wilhelm Wundt's laboratory (established in 1879), measuring the duration of several mental processes. In 1886 he became the first American to earn a German PhD in experimental psychology (Sokal, 2010).

After two years in England, Cattell assumed a University of Pennsylvania professorship, where he carried out major experimental programs that reinforced his reputation as an experimentalist. He moved to Columbia University in 1891, where he established a leading academic program and developed an influential set of "mental tests." With collaborators he used standard laboratory procedures—which measured (among other traits) reaction times, short-term memory, and the sensitivity of the senses—to gather quantitative data on psychological differences. But they lacked a functional view of how these traits helped people live, their tests produced trivial results, and psychologists soon abandoned them (Sokal, 1987). Cattell then left the laboratory, but in 1901 his experimental reputation led to his election as the first psychologist in the National Academy of Sciences.

From 1894 (when he founded *The Psychological Review* with Princeton colleague James Mark Baldwin), Cattell owned, edited, and published many major scientific journals. Most significantly, in 1894 he took control of the weekly *Science*, and in 1900 it became (even while privately owned) the official journal of the American Association for the Advancement of Science (AAAS, Sokal, 1980). In 1900, Cattell also took over *The Popular Science Monthly* (*The Scientific Monthly* after 1915) and his reputation attracted prominent contributors. In 1906, he issued the first edition of the directory *American Men of Science*. (He used its data in his studies of the psychology of scientific eminence.)

American scientists respected Cattell's achievements and appreciated his defense of academic freedom—at Columbia and elsewhere—through articles in his journals. But this defense usually exhibited an unpleasant self-righteousness that often included ad hominem attacks on others that put off many (Sokal, 2009).

Through the 1930s Cattell continued his editorships, chaired the AAAS Executive Committee, and acted as psychology's grand old man. But his last years proved disappointing. Younger scientists found *Science* out of touch with their interests and discounted *American Men of Science*'s continuing studies of eminence. Cattell continued to alienate others: through the 1930s the AAAS under his leadership hired and fired four Permanent Secretaries and his behavior at public meetings scandalized American psychologists.

[*See also* **American Association for the Advancement of Science; Journals in Science, Medicine, and Engineering; National Academy of Sciences; Printing and Publishing; Psychological and Intelligence Testing;** *and* **Psychology.**]

BIBLIOGRAPHY

Sokal, Michael M. "James McKeen Cattell, Columbia University, and Academic Freedom at Columbia University, 1902–1923." *History of Psychology* 12 (2009): 87–122.
Sokal, Michael M., ed. *Psychological Testing and American Society, 1890–1930.* New Brunswick, N.J.: Rutgers University Press, 1987.
Sokal, Michael M. "*Science* and James McKeen Cattell." *Science* 209 (4 July 1980): 43–52.
Sokal, Michael M. "Scientific Biography, Cognitive Deficits, and Laboratory Practice: James McKeen Cattell and Early American Experimental Psychology, 1880–1904." *Isis* 101 (2010): 531–554.

Michael M. Sokal

CELLULAR PHONES

See **Electronic Communication Devices, Mobile.**

CENTERS FOR DISEASE CONTROL AND PREVENTION

The Centers for Disease Control and Prevention, founded in Atlanta, Georgia, in 1946, is the world's premier public-health institution. It evolved from a World War II unit of the U.S. Public Health Service—Malaria Control in War Areas—which was charged with keeping the South, where many troops were trained, malaria free. Originally called the Communicable Disease Center, its purpose was to control communicable diseases throughout the United States. Several name changes reflected its expanding mission, but the acronym CDC continued.

Dr. Joseph W. Mountin, founder of the CDC, envisioned an institution that would assist state health departments in disease control through laboratory work and epidemiology. The "disease detectives" of its Epidemic Intelligence Service, organized in 1951, quickly achieved fame. This unit, conceived by Dr. Alexander Langmuir, introduced the concept of disease surveillance, first used effectively against poliomyelitis and influenza in the 1950s. Routine disease surveillance would become the cornerstone of public-health practice.

The CDC played a vital role in the global elimination of smallpox and discovered the Legionnaires' disease bacterium, the hantavirus (agent of a serious respiratory disease outbreak in the American Southwest), and the causes of toxic shock syndrome and Lassa and Ebola fevers. In 1981, the CDC first identified a new fatal disease, subsequently named acquired immunodeficiency syndrome (AIDS). Among its successes in environmental health were the removal of lead from gasoline and the development of a serum test for dioxin, used to detect exposure to Agent Orange among Vietnam War veterans. The CDC's massive swine flu inoculation campaign of 1976 elicited bitter criticism when the expected epidemic never materialized and serious side effects from the vaccine came to light. In the 1980s, the CDC added accidents, violence, and lifestyle issues to its concerns. Disease prevention, a requisite to achieving the goal of a healthy people in a healthy world, became the CDC's emphasis beginning in the 1990s.

[*See also* **Disease; Environmentalism; HIV/ AIDS; Influenza; Malaria; Medicine; Poliomyelitis; Public Health; Public Health Service, U.S.; Science;** *and* **Smallpox.**]

BIBLIOGRAPHY

Centers for Disease Control and Prevention. "Notable Milestones in NIOSH History." http://www.cdc.gov/niosh/timeline.html (accessed 25 October 2012).
Etheridge, Elizabeth W. *Sentinel for Health: A History of the Centers for Disease Control.* Berkeley: University of California Press, 1992.

Elizabeth W. Etheridge

CGI (COMPUTER-GENERATED IMAGES)

See Animation Technology.

CHALLENGER DISASTER

On 28 January 1986, in the skies over Cape Canaveral, Florida, the space shuttle *Challenger* exploded seventy-three seconds after liftoff, killing all seven astronauts on board: Greg Jarvis, Christa McAuliffe, Ron McNair, Ellison Onizuka, Judy Resnick, Dick Scobee, and Mike Smith. Although shuttle launches had become almost routine in the five years and twenty-four trips since the first liftoff, *Challenger* flight STS 51-L stood out for carrying the high school teacher McAuliffe. As a result, millions of Americans watched the launch and explosion live on television. The Teacher in Space program was intended to boost the National Aeronautics and Space Administration's (NASA's) declining budget and public interest. Instead, the disaster highlighted budget and management troubles at NASA.

President Ronald Reagan appointed former secretary of state William Rogers to head a scientific commission to investigate the explosion. The commission's report, issued two months later, traced the cause to the rubber O-rings used to seal the joints in the shuttle's solid rocket boosters. The unusually cold weather on the morning of the launch stiffened the O-rings, allowing fuel to leak from the boosters, igniting a deadly fireball. The Rogers Commission also discovered that NASA managers had not heeded warnings by engineers at both Morton–Thiokol and the Rockwell Corporation on the evening of 27 January 1986 that a launch in temperatures of less than 50 degrees was likely to be hazardous. Nearly three years and $2.4 billion later, NASA relaunched the U.S. space program with its next shuttle, *Discovery.*

[*See also* **Missiles and Rockets; National Aeronautics and Space Administration; Space Program; Space Science;** *and* **Technology.**]

BIBLIOGRAPHY

McDonald, Allan J. *Truth, Lies, and O-Rings: Inside the Space Shuttle Challenger Disaster.* Gainesville: University Press of Florida, 2009.
U.S. Senate. *Rogers Commission Report.* Washington, D.C.: U.S. Government Printing Office, 1986.
Vaughan, Diane. *The Challenger Launch Decision: Risky Technology, Culture, and Deviance at NASA.* Chicago: University of Chicago Press, 1996.

Sarah K. A. Pfatteicher

CHEMISTRY

Chemistry has its origins in alchemy, pharmacy, and natural philosophy. It emerged as a separate discipline beginning in the 1780s. Chemistry in colonial and early republic America, like all the sciences, remained essentially a branch of European and British science until the 1820s. During these early years American leaders envisioned chemistry and science as a way to promote the power, prestige, and interests of the United States.

Chemistry also had advanced from a qualitative descriptive science to a specialized, quantitative, experimental science, and by the late 1800s from an empirical science to an increasingly theory-based or theoretical science. The specialized disciplines that emerged in the nineteenth and twentieth centuries included analytical chemistry, inorganic chemistry, organic chemistry, biochemistry, industrial chemistry, physical chemistry, and quantum chemistry. The article examines the development of chemistry and its impact on American society from colonial times to the twenty-first century.

Colonial and Early Republic Chemistry: The Quantitative Study of Chemical Identity and Composition.

The history of American chemistry begins with John Winthrop Jr. (1606–1676), son of the first governor of the Massachusetts Bay Colony. Within a year of his arrival in Boston in 1631, Winthrop brought chemicals, apparatus, and books from England and established the first chemical laboratory and scientific library within the present boundaries of the United States. His reading "Of the Manner of Making Tar and Pitch in New England" before the Royal Society in London in 1662 made him the first American colonial to present a scholarly paper to a scientific organization. Most colonists, however, paid little attention to the science of chemistry. Those interested studied with apothecaries or in medical schools and adopted a mainly empirical and descriptive methodology. John Morgan (1735–1789) at the University of Pennsylvania Medical School in Philadelphia was the first to present a complete course of chemical lectures, and in 1767 James Smith (1740–1812) at Columbia Medical School (King's College) in New York City became the first professor to have the word *chemistry* in his title. At the College of Philadelphia, Benjamin Rush (1745–1813), one of the colonies' leading physicians and signer of the Declaration of Independence, held the first colonial chair in chemistry. His chemistry course, based on his textbook *A Syllabus of a Course of Lectures on Chemistry* (1770), consisted of seven units: salts, earths, inflammables, metals, waters, vegetables, and animal substances.

The beginning of the American political revolution coincided with the chemical revolution, which emphasized the study of composition and thereby clarified the meaning of a chemical element and a compound and of combustion. Its chief spokesman, Antoine Lavoisier (1743–1794) in Paris, defined a chemical element and compound, demonstrated the law of conservation of mass, introduced the system of nomenclature that formed the basis of the chemical nomenclature still in use at the beginning of the twenty-first century, and, after learning that Joseph Priestley (1733–1804) discovered oxygen gas in 1774, correctly interpreted the results of Priestley's combustion experiments. Oxygen, and not phlogiston, produced combustion. Following Lavoisier's lead, chemistry moved from a speculative qualitative science to a quantitative analytical one. In the United States, chemists such as Samuel Mitchill (1764–1831) at Columbia and James Woodhouse (1770–1809) at Pennsylvania overwhelmingly adopted Lavoisier's new quantitative chemistry. Woodhouse, who in 1792 founded the Chemical Society of Philadelphia, the first such society in the United States, for 10 years (1794–1804) debated Priestley on combustion, mainly in Mitchill's widely read *Medical Repository* (published 1797–1824). Priestley came to Philadelphia in 1794 to escape political persecution in England.

Antebellum Chemistry: Chemical Identity and Composition.

The quantitative study of chemical identity and composition continued throughout the nineteenth century. In the United States, with its vast untouched mineral resources, the nineteenth century initiated a long period of descriptive and analytical chemistry. The first published paper addressing chemical identity and composition, John De Normandie's "An Analysis of the Chalybeate Waters of Bristol in Pennsylvania," had already appeared in 1769 in Volume 1 of the *Transactions of the American Philosophical Society*. But the lack of apparatus, manpower, and public interest hindered progress until President Thomas Jefferson appealed to chemists to apply themselves to the useful arts and sciences and encouraged the American public to support their study. The useful arts and sciences were important in promoting the country's prestige, power, and interests. Robert Hare (1781–1858) at the University of Pennsylvania Medical School and Benjamin Silliman (1779–1864) at Yale best exemplify the American outlook. In 1801 Hare invented the oxyhydrogen blowpipe, the first laboratory apparatus capable of fusing mineral samples. Later he developed two types of electric batteries, the calorimotor in 1816 and the deflagrator in 1820. Visiting scientists considered Hare's personally funded laboratory unsurpassed in the world. His *Compendium of Chemistry* (1827), a textbook of plant and animal chemistry, inorganic chemistry, and physics, contained more than two hundred illustrations of his apparatus. Apparatus was in such short supply in the United States that when Silliman went to Europe to study in the early 1800s, he purchased $9,000 worth of apparatus and books. American chemists continued to have most apparatus made to order or imported from Europe until the 1840s, and only late in the century did demand support domestic production of laboratory apparatus and chemicals.

At the dawn of the nineteenth century few American universities (Pennsylvania, William and Mary, Harvard Medical School, Dartmouth, Columbia, and Princeton) offered separate courses of instruction in chemistry. During this period chemistry advanced from being one component of the college natural philosophy course, usually taught by a clergyman, to a separate academic subject offered in the third or fourth years. Silliman was instrumental in the transition. He regarded the importing of chemically based products from Europe an affront to the nation's independence and maintained that the United States would achieve status as a world power only when it produced its own chemists and became chemically self-sufficient. As professor of chemistry and natural history at Yale from 1802 to 1853, Silliman firmly established chemistry in the undergraduate curriculum through a combination of popular lectures filled with pyrotechnic demonstrations, pronouncements on chemistry's utility, and acknowledgment of God's omnipotence and the compatibility of science and religion. Eschewing European textbooks, he published *Elements of Chemistry* (1830), the most widely used American text of the antebellum period. His students, and Hare's, established chemistry departments in colleges from New England to Kentucky and Tennessee.

Although chemistry before the U.S. Civil War concentrated on analytical data gathering and accurate measurement, the period produced a notable theoretician in Josiah Cooke (1827–1894). Appointed chemistry professor at Harvard in 1850, Cooke established the college's chemistry laboratory with his own funds and used its facilities to determine precise atomic weights. His 1854 article, "Numerical Relations between the Atomic Weights and Some Thoughts on the Classification of the Elements," introduced a type of periodic table based on experimentally determined atomic weights. Foreshadowing Dmitri Mendeleev's work, Cooke believed he could predict the properties of any undiscovered element and its compounds in any given series of elements.

The study of chemistry went beyond the college. Popular lecturers strove to inform the public. Amos Eaton (1776–1842) at Rennselaer Institute in Troy, New York, presented lectures and simple experiments to New England's and New York's farmers, mechanics, and housewives for more than 30 years. In 1819, a public speaker identified only as Dr. Russell offered lectures and experiments to the citizens of New Orleans. The following year the English medical doctor John Cullen charged participants in Richmond, Virginia, $10 each for a series of lectures and demonstrations.

Chemistry in the Gilded Age: The Maturing of Chemistry. In the antebellum period John Pitkin Norton (1822–52) at Yale was among the first of the new agricultural chemists to apply the principles of chemistry to improve farming in the United States. American scientists continued the applied research approach in the years after the Civil War. Land-grant colleges established by the Morrill Land Grant Act of 1862 stressed the practical aspects of chemistry and trained both agricultural and analytical chemists. Connecticut used Norton's ideas to open the country's first agricultural experiment station in 1875. Other prominent agricultural chemists included John Lawrence Smith (1818–1883), who taught at the University of Virginia (1852–1854) and the University of Louisville (1854–1866), and John William Mallet (1832–1912), who served as chemistry professor at the universities of Alabama, Louisiana (now Tulane), and Virginia and in 1883 organized the chemistry department at the University of Texas. Smith carried out extensive analyses of soils and minerals; Mallet conducted an exhaustive chemical study of cotton, including the nutrients essential for its growth.

To pursue a graduate degree in the post–Civil War years American chemists had to go abroad, especially to Germany. Led by Harvard, Yale, and then Johns Hopkins, American institutions began imitating the German universities in offering laboratory instruction and graduate degrees in science. Yale granted its first PhD in science (engineering) to Josiah Willard Gibbs (1839–1903) in 1863; in 1877 Harvard awarded Frank Gooch (1852–1929) its first PhD in chemistry. Universities excluded women and minorities from graduate study until much later. Ellen Swallow Richards (1842–1911), the first woman to enroll as a full-time science student at an American university, received a BS in chemistry from the Massachusetts Institute of Technology (MIT) in 1873. Although she later became the most prominent nineteenth-century American female chemist, no American university allowed her to pursue a graduate degree. Two decades later, in 1894, Fanny Ryan Mulford Hitchcock at the University of Pennsylvania and Charlotte Fitch Roberts (1859–1917) at Yale became the first of only 13 nineteenth-century women to receive doctorates in chemistry from American universities. African Americans waited even longer. In 1916, the University of Illinois granted Saint Elmo Brady a PhD in chemistry, and Columbia in 1947 awarded Marie M. Daly (1921–2003) her PhD in chemistry. Daly was the first African American woman to receive the degree.

By the 1880s, chemistry in the United States was moving toward parity with Europe. Credit for this achievement resulted from (1) professionalizing of the chemical community with the organization of the American Chemical Society in New York City in 1876; (2) the proliferation of journals such as *American Chemist* (1870), *American Chemical Journal* (1879), and *Journal of the American Chemical Society* (1879); (3) the great expansion of research into areas previously neglected, especially structural organic chemistry and physiological chemistry; and (4) the spread of scientific education including the founding of MIT (1861) and Johns Hopkins (1876). Gibbs best epitomized the stature American chemistry had achieved. His lengthy two-part paper, "On the Equilibrium of Heterogeneous Substances," published in 1875 and 1878 in the *Transactions of the Connecticut Academy of Arts and Sciences*, introduced the phase rule and free energy to the study of chemical equilibrium and inaugurated physical chemistry in the United States.

Twentieth-Century Developments: Electron Theory of Bonding and Molecular Structure. Whereas descriptive and empirical analytical, organic, and inorganic chemistry dominated the nineteenth century, organic and physical chemistry dominated the first half of the twentieth century. With the discovery of subatomic particles, especially the electron in 1897 and the proton in 1911, chemists showed a new interest in electron theories of bonding (valence) and molecular structure. In 1902, Gilbert N. Lewis (1875–1946), then at MIT, proposed that an atom had its electrons arranged singly at the corners of a cube and that the sharing of an electron pair, one electron from each of two cubic atoms, accounted for the chemical bond holding the atoms in a molecule. The sharing of two electron pairs accounted for a double bond. Lewis published his theory of the shared electron pair bond in 1916 after moving to the University of California at Berkeley. He abandoned the cubic atom that same year because it failed to account for a triple bond in a molecule, but his electron pair, represented by the familiar pair of dots in a chemical formula, became the basis of modern bonding theory. Lewis summarized his ideas on valence in a 1923 monograph *Valence and the Structure of Atoms and Molecules* and at that time introduced his well-known electron theory of acids and bases. An acid was an electron pair acceptor, and a base was an electron pair donor. Longtime rival Irving Langmuir (1881–1957), a General Electric research chemist, applied Lewis's electron pair theory to the elements following neon $Z = 10$ (where Z = atomic number, or number of protons), and in 1919 he renamed Lewis's shared electron pair, or nonpolar bond, the covalent bond.

Lewis also made Gibbs's thermodynamics intelligible to chemists. Gibbs's publications were highly mathematical and lacked concrete examples to illustrate his thermodynamic principles, nor had Gibbs established a following of graduate students to continue his program. For over 20 years Lewis and his Berkeley colleagues, particularly William Giauque (1896–1971), a future Nobel Prize winner, provided the experimental evidence supporting thermodynamic functions such as Rudolf Clausius's entropy and Gibbs's free energy while introducing important new relations such as Lewis's fugacity. This work appeared in *Thermodynamics and the Free Energy of Chemical Substances,* another classic Lewis published in 1923.

With the continuing maturing of chemistry in the United States in the early twentieth century, international recognition followed. Harvard's Theodore Richards (1868–1928) in 1914 became the first American chemist and only the second American scientist to receive a Nobel Prize in science. He received the prize for his precise atomic weight determinations of 25 elements. Langmuir received the Nobel Prize in 1932 for his research on monomolecular films and surface chemistry. He was the first American industrial scientist to win the prize. At Columbia, Lewis'

former student Harold Urey (1893–1981) earned the 1934 prize for his spectroscopic discovery of deuterium (H-2, heavy hydrogen) by evaporation of liquid hydrogen.

Linus Pauling (1901–1994) at the California Institute of Technology emerged as the most influential American chemist during this period. In a series of publications beginning in 1929, Pauling transformed Lewis's intuitive and qualitative electron pair bond into a mathematical quantitative theory called the valence bond method, which introduced the novel idea of hybrid atomic orbitals such as sp^2 and sp^3 in chemical bond formation. Pauling's theory of the chemical bond was completely compatible with the new quantum mechanical theory of matter that European physicists introduced in 1925–1926. Pauling further developed the valence bond method in his highly influential *The Nature of the Chemical Bond and the Structure of Molecules and Crystals* (1939), which remains required reading for chemists in the early twenty-first century. His combination of hybrid orbitals in the carbon atom and crystallographic techniques enabled him in the 1940s and 1950s to determine the structure of crystals and complex molecules such as the proteins. For his elucidation of the chemical bond and the structure of molecules Pauling received the 1954 Nobel Prize.

Although the valence bond method dominated into the early 1960s, an alternative theory of chemical bonding had emerged during the period 1926–1932. The molecular orbital theory that Robert Mulliken (1896–1986) at Chicago and Friedrich Hund (1896–1997) at Leipzig developed independently offered less graphic descriptions although more satisfying explanations of molecular polarity, oxygen's paramagnetism, and molecular spectra. Molecular orbital theory remains crucial to the study of chemical bonding and molecular structure. Mulliken received a belated Nobel Prize in 1966.

In organic chemistry, including biochemistry, research centered on discovering reaction pathways and on structural determinations of naturally occurring macromolecules to synthesize them. Melvin Calvin (1911–1997) at the Lawrence Radiation Laboratory, Berkeley, used radioactive carbon (C-14) to trace the reactions occurring in photosynthesis and received the 1961 Nobel Prize for his contributions. At the Glidden Company of Chicago, Percy Julian (1899–1975), the first African American to hold the title of chief research chemist for a major corporation, received more than 130 patents for synthetic hormones, steroids, and drugs. Harvard's Robert Woodward (1917–1979), the 1965 Nobel recipient, succeeded in synthesizing numerous biologically important compounds including quinine (1944), cholesterol (1951), cortisone (1951), chlorophyll a (1960), vitamin B-12 (1976), and numerous antibiotics such as tetracycline (1954).

In the 1940s the field of nuclear chemistry emerged. Enrico Fermi (1901–1954) at the University of Rome had tried in the mid-1930s to produce new elements with atomic numbers greater than 92 (uranium) by bombarding uranium with the newly discovered neutron. His results were inconclusive, but in 1940 Edwin McMillan (1907–1991) and Philip Abelson (1913–2004) at Berkeley succeeded in their uranium-neutron bombardment experiments to produce neptunium, an element with 93 protons ($Z = 93$). Glenn Seaborg (1912–1999) and his colleagues at Berkeley continued the search for other transuranium elements, and in December 1940 they identified and named plutonium, element 94 ($Z = 94$). The Manhattan Project's secret wartime research on plutonium showed in 1941 that its isotope Pu-239, like U-235, was fissionable, but the announcement of plutonium's discovery came five years later with the dropping of the second atomic bomb on Nagasaki, Japan, on 9 August 1945. After the war, Seaborg identified and named eight new elements: americium, curium, berkelium, californium, einsteinium, fermium, mendelevium, and nobelium ($Z = 95$–102). He and McMillan shared the 1951 Nobel Prize. Crucial to the wartime fission research were the theoretical contributions of Yale chemist Lars Onsager (1903–1976), which provided the basis for the gaseous diffusion method of separating uranium's isotopes. Onsager received the Nobel Prize in 1968.

After World War II, chemistry and the chemistry profession in the United States changed dramatically. A blurring of the dividing lines between chemistry and the other sciences had occurred. Chemists bemoaned the loss of job opportunities as physics, geology, genetics, pharmacology, environmental science, and chemical engineering usurped various areas previously belonging to chemistry. Biochemistry emerged as a separate entity, chemical physics rivaled physical chemistry, and physicists dominated environmental science. Despite the disruption, American chemists from 1955 to 2011 won outright or shared forty-two Nobel Prizes in chemistry for studies ranging from the structure, synthesis, and physical chemistry of macromolecules to carbon-14 dating and inorganic and organic reaction mechanisms.

Most of the awards since the early 1990s were for significant achievements in biochemistry or molecular biology, although Ahmed Zewail (b. 1946), an Egyptian-born chemist at Caltech, won in 1999 for his invention of a new research area he named femtochemistry. Femtochemistry is the study of fast reactions (1 femtosecond = 10^{-15} seconds). With femtosecond spectroscopy chemists have reached the time scale at which chemical reactions occur. None occurs faster than this. Chemists can study atoms and molecules in slow motion, see what happens when chemical bonds break and form and thereby understand why some reactions occur and others do not, and explain why the speed and yield of reactions depend on temperature. A better view of catalytic behavior, the mechanisms of life processes, and the synthesis of future medicines has resulted.

The Stanford chemist Roger Kornberg (b. 1947) is a recent recipient in molecular biology. He received the 2006 Nobel Prize for his detailed crystallographic pictures showing the copying of genetic information from DNA into RNA. Messenger RNA carries the information out of the cell nucleus where it begins the construction of proteins essential for life. Nearly 50 years earlier Kornberg's father, Arthur Kornberg (1918–2007) at Stanford, won the 1959 prize in medicine for his work on the synthesis of DNA.

Other important research areas in which American chemists have made significant contributions include environmental chemistry; nanocatalysis; metal-organic framework materials;

catalysts for the efficient production of oxygen and hydrogen gases from the solar-powered electrolysis of water, the two gases then stored for use in fuel cells; and bioorganic metallic chemistry.

In environmental chemistry David Keeling (1928–2005), beginning in 1960 at the Scripps Institution of Oceanography in San Diego, provided irrefutable scientific evidence that directly connected the increasing concentration of atmospheric carbon dioxide, a known greenhouse gas, and the increasing industrial and domestic consumption of fossil fuels. Keeling proved that society has in fact contributed significantly to global warming. A decade later the atmospheric studies of F. Sherwood Rowland (1927–2012) at the University of California, Irvine, and his former research associate, Mario Molina (b. 1943), showed that chlorofluorocarbon molecules escaping into the atmosphere depleted the stratospheric ozone layer and permitted increasing levels of dangerous ultraviolet light to strike the earth. Rowland and Molina shared the 1995 Nobel Prize for their work.

In nanocatalysis John Fackler and Wayne Goodman at Texas A&M demonstrated the enhanced catalytic activity of nanogold catalysis in oxidation reactions. Such studies of metal-organic or organometallic gold complexes since the 1990s have established their importance in the nanocatalytic oxidation of carbon monoxide. Carbon monoxide's oxidation is essential for the production of cleaner automobile exhaust. At the University of California at Los Angeles, Omar Yaghi's 1998–1999 pioneering studies on the synthesis and application of metal-organic frameworks demonstrated the ability of these crystalline porous materials to absorb, capture, or store molecules such as hydrogen, methane, and carbon dioxide. Hydrogen storage is important for automobile fueling, methane for automobile fueling and for transporting natural gas, and carbon dioxide capture for reducing power plant and automobile emissions. Yaghi named this new branch of chemistry reticular chemistry and defined it as chemistry that investigates the linking of building blocks by strong bonds into predetermined structures.

At MIT Daniel Nocera and his group in 2008 described how their preparation of a new catalyst led to the development of a simple, efficient, and inexpensive process for storing solar energy. Their solar storage process provides a solution to a longstanding and formidable problem that has hindered solar energy's development and application. The catalyst (cobalt metal, a phosphate, and an electrode) greatly increased the yield of oxygen from the electrolysis of water. Another inexpensive catalyst they developed in 2011 produces hydrogen gas at the other electrode. Solar energy striking a photovoltaic cell generates an electric current that, when passed through the electrolytic cell containing the water and electrodes, produces oxygen and hydrogen gas. Combining the two gases in a fuel cell, or storing them separately for later combining in a fuel cell, produces water as the only product and generates carbon-free electricity for home or building consumption. Nocera envisioned facilities equipped with solar panels, photovoltaic cells, electrolytic cells, and fuel cells, enabling them to use any excess solar-generated electricity to produce oxygen and hydrogen in an electrolytic cell, store the gases, and then later combine them in a fuel cell to produce electricity when photovoltaic electricity is unavailable. His solar storage process would eliminate the need for a central electrical generating system that requires the consumption of fossil fuels.

The accomplishments of American chemists in universities and industry discussed above, particularly in the post–World War II years, have demonstrated that they consistently rank among the world's best. University chemistry departments, led over the years by such notable chemists as William A. Noyes and Roger Adams at Illinois, Farrington Daniels at Wisconsin, Joel Hildebrand at Berkeley, William Lipscomb at Harvard, Paul Flory at Cornell and Stanford, Henry Talbot at MIT, Henry Taube at Stanford, and Roald Hoffman at Cornell, have remained in the forefront of chemical research. Since the early 1980s, the University of California at Berkeley has consistently granted the most chemistry PhDs, while Iowa State and the universities of Illinois, Texas, Michigan, North Carolina, Wisconsin, and Washington have excelled in both undergraduate and graduate education. No other country can match that record.

State of the History of American Chemistry.
Historians of science have not given adequate attention to American chemists and chemistry. Despite increasing membership in the American Chemical Society's History of Chemistry Division and financial support from the Chemical Heritage Foundation's Beckman Center for the History of Chemistry at the University of Pennsylvania, historical research and writing on chemistry in the United States continues in its infancy. No comprehensive history and few books on the history of American chemistry and chemists exist, although the bibliography contains several entries that deal with specialized periods and branches of American chemistry.

Two different methodologies have defined the history of science and have divided somewhat loosely historians of science into two groups, those with scientific backgrounds and those from the humanities and social sciences. The first group, often called internalists, tends to emphasize the historical development of scientific ideas, or theories, and their impact on society with less emphasis on the influence of social-cultural-economic factors and the institutions of science. The second group, or externalists, deemphasizes the development of scientific ideas and assigns a greater role to social-cultural-economic factors and the institutions of science. Both approaches are necessary to give a balanced, fuller, and richer history of science. The history of chemistry has followed this pattern. Edgar F. Smith (1854–1928), at Pennsylvania, the first historian of chemistry in the United States, and Aaron Ihde (1909–2000), the dean of history of chemistry in the United States, began their careers as chemists. Their books reflect this background although Ihde's *Development of Modern Chemistry* (1964), written 50 years after Smith's *Chemistry in America* (1914), skillfully interweaves the development of chemical ideas and their impact on society and social and institutional developments. More recent histories have either been topical or downplayed many of the specific and general developments in chemistry in favor of anecdotal and biographical information. The trend during the second decade of the early twenty-first century was biographies and collective biographies of chemists, including compilations of Nobel Prize winners and women, topical and periodic histories, and histories of chemical industries and institutions.

Writing the history of chemistry in the United States somewhat parallels the position of chemistry within the sciences. Books on colonial and early republic science included chemistry with the natural sciences or medicine. Smith's book treated chemistry as a full-fledged, independent discipline. When chemistry splintered into several branches, so did its historical scholarship. The history of chemistry moved from straightforward, chronologically organized surveys to more specialized interpretive studies. The interconnectedness of chemistry with most scientific disciplines has hindered current historians of science from pinpointing successes in chemistry and has made the writing of comprehensive histories an increasingly formidable undertaking.

[*See also* **Biochemistry; Molecular Biology; Petroleum and Petrochemicals; Pharmacology and Drug Therapy;** *and* **Physics.**]

BIBLIOGRAPHY

"American Chemical Society." http://www.acs.org.
Annals of Science. http://www.tandfonline.com/loi/tasc20.
Brock, William H. *The Norton History of Chemistry.* New York: W. W. Norton, 1992. A good, comprehensive history that includes the contributions of American chemists.
Browne, Charles Albert, and Mary Elvira Weeks. *A History of the American Chemical Society.* Washington, D.C.: American Chemical Society, 1952.
Bruce, Robert V. *The Launching of Modern American Science 1846–1876.* Ithaca, N.Y.: Cornell University Press, 1987. Discusses developments in chemistry for that time period in which American science was empirical.
Bulletin for the History of Chemistry. http://www.scs.illinois.edu/~mainzv/HIST/bulletin/.
Chandler, Alfred D., Jr. *Shaping the Industrial Century: The Remarkable Story of the Evolution of the Modern Chemical and Pharmaceutical Industries.* Cambridge, Mass.: Harvard University Press,

2005. Emphasis is on the business and economics of these industries.

"Chemical Heritage Foundation." http://www.chemheritage.org.

Chemical Heritage Magazine. http://www.chemheritage.org/magazine/.

Coffey, Patrick. *Cathedrals of Science: The Personalities and Rivalries That Made Modern Chemistry.* New York: Oxford University Press, 2008. Discusses the rivalries and personalities of several significant twentieth-century physical chemists including Lewis, Langmuir, and Pauling.

Farber, Eduard, ed. *Great Chemists.* New York: Interscience Publishers, 1961. Biographical collection of chemists.

Gillispie, Charles, ed. *Dictionary of Scientific Biography.* 16 vols. and 2-vol. Supplement II. New York: Charles Scribner's Sons, 1970–1980, 1990. Biographical collection of scientists, including American chemists.

Giunta, Carmen. "Classic Chemistry." Le Moyne College. http://web.lemoyne.edu/~giunta/.

Greene, John C. *American Science in the Age of Jefferson.* Ames: Iowa State University Press, 1984. Includes discussions on colonial and early republic chemists.

ICIS. http://icis.com.

Ihde, Aaron J. *The Development of Modern Chemistry.* New York: Harper & Row, 1964. Still the most comprehensive book on the history of chemistry but ends with the 1960s.

James, Laylin K., ed. *Nobel Laureates in Chemistry, 1901–1992.* Washington, D.C.: American Chemical Society, 1993.

Journal of Chemical Education. http://pubs.acs.org/journal/jceda8.

Miles, Wyndham D., ed. *American Chemists and Chemical Engineers.* Washington, D.C.: American Chemical Society, 1976. A biographical dictionary.

"Nobel Foundation." http://www.nobelprize.org/nobel_organizations/nobelfoundation/.

Rossiter, Margaret W. *Women Scientists in America.* 3 vols. Baltimore: Johns Hopkins University Press, 1984, 1998, 2012. A pioneering study that includes women chemists.

Servos, John W. *Physical Chemistry from Ostwald to Pauling: The Making of a Science in America.* Princeton, N.J.: Princeton University Press, 1990.

Skolnik, Herman, and Kenneth M. Reese. *A Century of Chemistry: The Role of Chemists and the American Chemical Society.* Washington, D.C.: American Chemical Society, 1976.

Smith, Edgar F. *Chemistry in America: Chapters from the History of the Science in the United States.* New York: D. Appleton, 1914. Good but dated.

Stranges, Anthony N. *Electrons and Valence: Development of the Theory, 1900–1925.* College Station: Texas A&M University Press, 1982. Gives significant attention to American chemists, particularly Lewis and Langmuir.

Tarbell, D. Stanley, and Ann T. Tarbell. *Essays on the History of Organic Chemistry in the United States.* Nashville, Tenn.: Folio Press, 1986.

Technology and Culture. https://www.press.jhu.edu/journals/technology_and_culture/.

Thackray, Arnold, Jeffrey L. Sturchio, P. Thomas Carroll, and Robert Bud. *Chemistry in America. 1876–1976: Historical Indicators.* Boston: Reidel, 1985. Examines research concerns in American chemistry and chemical engineering.

Woodward, Walter. *Prospero's America: John Winthrop Jr., Alchemy, and the Creation of New England Culture 1606–1676.* Chapel Hill: University of North Carolina Press, 2010. Discusses Winthrop's alchemical knowledge, connections among religion, metallurgy, and healing, and the attempt to establish scientific research in New England.

Anthony N. Stranges

CHILDBIRTH

Prior to the rise of an organized medical specialty in obstetrics in the late eighteenth century, childbirth was a largely single-sex domestic experience in which a woman was attended at home by female friends, relatives, and often a midwife. These female companions offered moral support to the mother and assisted the midwife during labor and delivery. Although it was acknowledged that labor and delivery could pose dangers to the lives of both mother and child, birthing was viewed primarily as a natural process, one that required little intervention. In her own home, attended by other women, a mother was free to move about to alleviate her discomfort during contractions. This often meant that mothers remained standing up, walking, or seated in specially designed birthing chairs. After the birth of the baby, friends and relatives assisted the new mother during her convalescence, which was typically short if the birth was normal and free of complications.

Scientific Training and Male Physicians.

The first major change in childbirth practices for white women began after 1750 when affluent colonists in urban areas began to call upon scientifically trained male doctors to attend them in their homes during delivery. Physicians assured these women that knowledge of anatomy and access to drugs like ergot (a medicinal substance that promotes uterine contractions) and opiates (for pain relief), as well as their use of obstetrical forceps, could ensure faster and less painful labors and safer deliveries. The interventions used by these early physicians, however, posed greater risks for normal births. For example, the well-known diary of an eighteenth-century midwife, Martha Ballard, reveals that the use of opiates by a physician during a local woman's labor resulted in the temporary alleviation of the mother's pains, but also had the apparent effect of slowing her uterine contractions, thus prolonging the labor and placing the lives of mother and child at greater risk. As the record of Martha Ballard illustrates, the transition from midwives to physicians was gradual and involved a great deal of overlap. It would not have been uncommon for a woman to call both a midwife and a physician to her lying-in because they were initially believed to fulfill two distinct roles.

The availability of pain medication and women's interest in drug interventions grew over time and by 1850 both ether and chloroform were being demanded by mothers and used by physicians to reduce labor pains. Despite the scientific education that eighteenth- and nineteenth-century physicians brought with them to their obstetrical practices, many left medical school having never participated in, or even witnessed, an actual labor and delivery. Inexperienced physicians were more likely to intervene unnecessarily during childbirth, interfering in normal labor and delivery processes, and placing the health of the mother and baby at risk. Prior to the widespread acceptance of a germ theory of disease, the insertion of unsanitized hands and forceps into the woman's cervix during labor to extract the infant increased the likelihood that an infection known as puerperal (childbed) fever would arise. Greater medical oversight increased the degree to which labor and delivery were seen as pathological events that required expert care and medical intervention.

Economic status, geography, ethnicity, and race significantly influenced the way women experienced childbirth. The obstetrical services available to poor women depended on where they lived. Those who resided in small towns, rural areas, or urban ethnic neighborhoods continued to use midwives and to depend on their female support network for services well into the twentieth century.

From the Home to the Hospital.

In the early twentieth century, childbirth began to move from the home to the hospital. Doctors encouraged this development because hospital births centralized obstetric care, enabled them to control the birthing environment, and provided a regular supply of patients for the clinical education of medical students. Women also actively participated in the shift from home to hospital births. Seeking the modern experience of a scientific childbirth and the alleviation of pain, women increasingly ceded control over their births in exchange for medical oversight, a perceived reduction in uncertainty, and access to the most advanced medical technologies of the day. By the early twentieth century, the widespread acceptance of the dangers posed by germs coupled with a growing faith in the power and authority of scientific medicine helped bolster the belief that the hospital was the safest place for a woman to go through labor and delivery. Extended hospital stays of up to several weeks also offered women a prolonged respite from household responsibilities and the care of other children.

Despite the fact that both doctors and mothers helped drive the increase in hospital births, this change resulted in a redistribution of power between doctors and mothers. Doctors increasingly looked for pathology in their obstetrics cases and intervention in the childbirth process became routine as procedures such as episiotomies and the use of forceps became commonplace. Mothers, particularly of the upper and middle classes, continued to demand more effective and safer anesthesia during childbirth, prompting the early-twentieth-century development of drugs such as scopolamine, a narcotic and amnesiac famous for inducing the experience of "twilight sleep." Under the influence of scopolamine a woman's mind went to sleep while her body continued to go through the labor and delivery experience. The screaming and thrashing that the mother could experience during a scopolamine birth prompted physicians to restrain laboring women to prevent them from injuring themselves while unconscious.

Continuing concern about the issues of maternal and infant mortality, a highly popular cause among the newly enfranchised female reform community, prompted Congress to pass the Sheppard–Towner Act in 1921 to provide programs for prenatal, obstetric, and postnatal care for poor women. The act was short-lived, however, and was defunct by 1929. By the 1930s, childbirth was widely accepted as an event that required medical oversight whenever possible. By the 1940s, women expected not only to survive their parturition but also to enjoy the experience, as confidence in the safety of modern medicine and the belief in painless childbirth became the norm.

Opposition to the Medicalization of Childbirth. The rising expectations surrounding childbirth left more and more women feeling dissatisfied with their medical birthing experiences. Women began opposing the medicalization of childbirth and demanding more control over the process. Following the 1944 publication of the book *Childbirth without Fear*, by British obstetrician Grantly Dick-Read, women began to demand more access to education and preparation for childbirth. Groups like the Milwaukee Natural Childbirth Association (1950), the Boston Association for Childbirth Education (1953), and the International Childbirth Education Association (1960) formed to provide women with resources for educating themselves about pregnancy, labor and delivery, and postnatal care. As a result, women increasingly came to oppose the routine use of anesthesia, called for less medical intervention, and even advocated a return to what they called "natural childbirth." By the 1960s, birth reformers were promoting such innovations as Lamaze breathing techniques, birthing at home, the presence of fathers during the birth, and the utilization of nurse-midwives instead of, or in addition to, doctors. Nevertheless, most American women continued to experience childbirth in a medicalized context over which they had only limited control.

In the 1970s, a generation of women went to the hospital to have their babies, much like their mothers had in the 1950s. However, educated by decades of childbirth reform, enlightened by feminist theory, and emboldened by the information available through sources like the health education book, *Our Bodies, Ourselves*, mothers in the final decades of the twentieth century struggled to redefine their childbirth experiences on their own terms. Women increasingly sought greater access to natural childbirth practices, rooming-in (in which mother and baby are kept together in the same room), breastfeeding support, and the ability to have friends and family accompany them into the delivery room. Simultaneously, throughout this era the rates of cesarean section, a surgical procedure to remove the infant through the mother's abdomen, have steadily climbed. Despite a twenty-first-century resurgence in the utilization of certified midwives for home or hospital births, particularly among the more affluent segments of society, the Centers for Disease Control reported in 2010 that nearly one-third of all U.S. births were the result of a cesarean section.

[*See also* **Anatomy and Human Dissection; Anesthesiology;** *Baby and Child Care;* **Biological Sciences; Birth Control and Family Planning; Gender and Science; Germ Theory of Disease; Hospitals; Medical Education; Medicine; Medicine and Technology; Midwifery; Pharmacology and Drug Therapy; Physiology; Race and Medicine;** *and* **Surgery.**]

BIBLIOGRAPHY

Banks, Amanda Carson. *Birth Chairs, Midwives, and Medicine.* Jackson: University Press of Mississippi, 1999.

Hoffert, Sylvia D. *Private Matters: American Attitudes toward Childbearing and Infant Nurture in the Urban North, 1800–1860.* Urbana: University of Illinois Press, 1989.

Kline, Wendy. *Bodies of Knowledge: Sexuality, Reproduction, and Women's Health in the Second Wave.* Chicago: University of Chicago Press, 2010.

Leavitt, Judith Walzer. *Brought to Bed: Childbearing in America, 1750–1950.* New York: Oxford University Press, 1986.

Leavitt, Judith Walzer. *Make Room for Daddy: The Journey from Waiting Room to Birthing Room.* Chapel Hill: University of North Carolina Press, 2010.

McMillen, Sally G. *Motherhood in the Old South: Pregnancy, Childbirth, and Infant Rearing.* Baton Rouge: Louisiana State University Press, 1990.

Sandelowski, Margarete. *Pain, Pleasure, and American Childbirth: From Twilight Sleep to the Read Method, 1914–1960.* Westport, Conn.: Greenwood Press, 1984.

Ulrich, Laurel Thatcher. *A Midwife's Tale: The Life of Martha Ballard, Based on Her Diary, 1785–1812.* New York: Alfred A. Knopf, 1990.

Wertz, Richard W., and Dorothy C. Wertz. *Lying-in: A History of Childbirth in America.* New York: Schocken Books, 1977.

Wolf, Jacqueline. *Deliver Me from Pain: Anesthesia and Birth in America.* Baltimore: Johns Hopkins University Press, 2009.

Jessica Martucci and Sylvia D. Hoffert

CHOLERA

For most physicians of the early nineteenth century, both in the United States and abroad, "cholera morbus" was a pathological discharge of bile through diarrhea and vomiting. However, as more physicians began to notice a new and particularly deadly epidemic coming out of South Asia, this diagnosis changed, and as the century wore on, American physicians, like their counterparts elsewhere, began to use the term for the often deadly epidemic disease. Then, as during more recent periods of history, physicians and lay professionals relied on epidemic presence to distinguish cholera from other types of diarrhea. Specifically in the case of the United States, preexisting environmental conditions and endemic pathogens made cases of diarrhea common, further complicating diagnosis.

The United States suffered three major epidemics of cholera during the nineteenth century. The first occurred between 1832 and 1834. Successive epidemics, from 1849 to 1855 and from 1865 to 1868, spread first to cities on the Atlantic and Gulf coasts and reached the West Coast both by overland and by maritime routes. In all cases, cholera reached the United States as part of global pandemics thought to have originated in South Asia. Although cholera appeared sporadically during subsequent pandemics, its impact was minimal and rarely reached beyond urban ports.

1832 to 1834. Cholera first appeared in the United States in waves, utilizing busy ports of entry along the Eastern seaboard, including those in the St. Lawrence Valley, New York, and New Orleans. The best-documented early entry point, in the summer of 1832, is Quebec, where the disease arrived in the early part of June. From there, symptoms of cholera began to appear further along the Great Lakes and border areas, such as Lake Champlain (Peters et al., 1885). Already reports of the international panic over the spread of the disease in Asia and Europe had reached the interior of the United States; those reports influenced reactions as well as diagnoses. The disease spread by other routes, too, and broke out simultaneously in the Hudson River Valley, as far west as Detroit and at points along the Ohio River. Although it is difficult to arrive at accurate death tolls, local physicians and others concerned with the public health often reported deaths to city councils and sometimes to newspapers. In Albany, significant as a port city from the confluence of the Hudson River and the Erie Canal, 311 died in a population of 24,000. New York City deaths were about 2,000 of a population of about 200,000 in 1832; the city continued to suffer high death rates over the next three years (Whitney, 1833, p. 333).

Cholera spread into the interior of the United States down the Ohio and Mississippi rivers. The final major site of importation was New Orleans.

Although lasting only three weeks—from the end of October through the middle of November—the epidemic there resulted in an estimated 4,350 to 5,000 deaths, about 15 percent of a city already depleted by recent yellow fever. The sheer magnitude of mortality, combined with the abandonment of the city by many residents, created shortages of burial staff and equipment (Duffy, 1971, pp. 1156–1157). Reports of outbreaks in New England (reaching as far northeast as Rhode Island and southeastern Massachusetts), the mid-Atlantic states, and the Southeast may have spread along local ferry routes or been carried by trade from the larger Atlantic ports.

1849 to 1855. The second major cholera epidemic hit the United States in at least two separate locations at the end of 1848—the first on 2 December, in New York, and the second on 11 December, in New Orleans, where newspapers reported more than 800 deaths within two weeks, following introduction of the disease on an immigrant ship. Some reporters, such as Dr. Joseph Jones, estimated more than double that number. During the next year more than 3,000 cholera deaths were recorded, with mortality reaching between 450 and 1,448 for the following five years. There were scattered reports of cholera in New Orleans between 1855 and 1866. It is not clear whether these represent diagnostic looseness or persistence of the same infection (Duffy, 1962, pp. 143–144).

The 1848 cholera outbreak used many of the same water routes as had the 1832 epidemic, although the dominant movement was upstream. It reached Louisville on 22 December, Cincinnati on the 25th, and St. Louis on the 27th. Mortalities in all these locations were high: in St. Louis, 4,557 cholera deaths occurred in a population of 63,471 in 1849 (Roth, 1993, pp. 242–243). Over the next seven years, this epidemic crossed the continent, moving with migrants heading toward Oregon or the gold fields of California. Cholera graves along the trail are a significant part of the mythology of that great trek (Chambers, 1938). Cholera in California, however, probably came more commonly via shipping than overland. Cholera killed 500 to 600 in San Francisco and was even deadlier in Sacramento, killing 750 in a population of 6,000 (Roth, 1993, pp. 331–333, 391–392).

In New York, cholera's spread was initially slowed by quarantine and other public measures, with 61 cases and 32 deaths being reported in 1848. However, in the spring of 1849 the disease broke out in the notorious Five Points (although this was denied by city authorities). By autumn the death toll had reached 5,071. Here and elsewhere, however, it is best to be skeptical of such figures. Authorities often underrepresented deaths to calm the public, and during an epidemic doctors and officials used the prevailing diagnosis without carefully investigating each case (Duffy, 1971, p. 1159).

1865 to 1868. By 1866 the multiplicity of routes of travel made it nearly impossible to identify the path of the epidemic. Railroads, bridges, canals, and improved roads facilitated transportation across the country. One can still see a broad pattern of spread, however. Again, cholera entered through Atlantic ports. As cholera swept Europe in 1865, ships arriving in New York began reporting cholera victims on board. The first, the *Atalanta*, arrived in early November, reporting 60 cases of cholera and 15 deaths (Rosenberg, 1962, p. 185). Use of hospital ships and a lazaretto on War Island successfully kept the disease within the city (Peters et al., 1885). Cholera worries were also significant in fostering creation of the Metropolitan Board of Health, the first permanent municipal board of health in the United States. Pointing to the inability of patronage-based institutions to stem cholera, reformers urged a new state agency to keep the city clean and identify and isolate cholera victims (Duffy, 1971). Despite the ambition of modern medical techniques, no evidence suggests the actions of this new board led to the reduction of infection or death rates. Nevertheless, New Yorkers credited medical professionals with protecting them from the epidemic. Still, by the summer of 1866 cholera had come ashore. Deaths totaled 1,435 by the end of the year (Peters et al., 1885, p. 38).

Only beginning to recover from the Civil War, New Orleans, too, was hit in October, although cases began to be diagnosed as early as August.

There were 1,294 deaths during 1866 and another 681 in 1867. In 1868, only 129 deaths were attributed to cholera in the city (Duffy, 1962, p. 444).

As before, cholera spread from these ports to other urban centers and to tertiary population centers in the interior and along the coasts. It infected some remaining Union troop encampments, such as that in Newport, Kentucky, where it was introduced from nearby Cincinnati. Greater familiarity with sanitation limited its impact, and it ran its course without further spread (Macnamara, 1876).

Later Nineteenth-Century Outbreaks. This postwar epidemic was the last. Minor recurrences of cholera surfaced between 1873 and 1875, mainly in the Mississippi Valley, from New Orleans as far north as Minnesota, Indiana, and Kentucky. Control at ports, European and American, rather than American sanitary reform, seems to account for its absence. In the last two decades of the nineteenth century, strengthened quarantine measures, given greater authority by the new germ theory and the discoveries of Robert Koch, contributed. However, the administrations of these controls were still maintained by health officials who often distrusted the new science of bacteriology and were limited in their actions by political infighting and cronyism (Markel, 1999).

The Twentieth and the Twenty-First Centuries. For most of the twentieth century, cholera's absence reinforced belief that it no longer threatened the United States. For scientists and practitioners alike, it would become effectively a tropical disease. Only in the Pacific theater of World War II did American physicians encounter cholera.

In the early 1990s, the appearance of pandemic cholera in South and Central America made it again a significant topic of consideration in the United States. Between 1990 and 1994, 195 cases were reported in the United States, of which 134 were associated with travel to Latin America—75 of which were traced to a single flight from Lima, Peru. Interestingly, 4 cases were also attributed to the consumption of seafood from the Gulf of Mexico, advancing the argument that the *Vibrio*

cholerae is endemic in at least some areas in the Americas (Steinberg et al., 2001, p. 799).

In October 2010, cholera reappeared in Haiti nine months after a major earthquake damaged the country's infrastructure. By the end of May 2011, the U.S. Centers for Disease Control reported 321,066 cases and 5,337 deaths, refocusing American attention on cholera (CDC, 2011).

[*See also* **Biological Sciences; Centers for Disease Control and Prevention; Columbian Exchange; Disease; Germ Theory of Disease; Life Expectancy; Maritime Transport; Medicine; Public Health; Public Health Service, U.S.; Race and Medicine;** *and* **War and Medicine.**]

BIBLIOGRAPHY

CDC—*Haiti Cholera Outbreak*. Centers for Disease Control and Prevention. 2011. http://www.cdc.gov/haiticholera/ (accessed 18 October 2012).

Chambers, John S. *The Conquest of Cholera: America's Greatest Scourge*. New York: Macmillan Co., 1938. A history of three major cholera outbreaks in America in 1833, 1849, and 1873.

Duffy, John. "The History of Asiatic Cholera in the United States." *Bulletin of the New York Medical Association* 47, no. 10 (October 1971): 1152–1168.

Duffy, John. *The Rudolph Matas History of Medicine in Louisiana*. Baton Rouge: Louisiana State University Press, 1962.

Macnamara, N. Charles. *A History of Asiatic Cholera*. London: Macmillan Co., 1876.

Markel, Howard. *Quarantine!: East European Jewish Immigrants and the New York City Epidemics of 1892*. Baltimore: John Hopkins University Press, 1999. Focuses on the racial politics associated with public-health measures in the cholera and typhus epidemics in 1892 New York.

Peters, John Charles, Ely McClellan, John Brown Hamilton, and George Miller Sternberg. *A Treatise of Asiatic Cholera*. New York: William Wood and Company, 1885.

Rosenberg, Charles E. *The Cholera Years: The United States in 1832, 1849, and 1866*. Chicago: University of Chicago Press, 1962. This book remains widely influential in the history of medicine and demonstrates the impact of epidemic disease on American society.

Roth, Mitchell. "The Western Cholera Trail: Studies in the Urban Response to Epidemic Disease in the Trans-Mississippi West, 1848–1850." PhD thesis. University of California, Santa Barbara, 1993.

Steinberg, Ellen B., Katherine D. Greene, Cheryl A. Bopp, Daniel N. Cameron, Joy G. Wells, and Eric D. Mintz. "Cholera in the United States, 1995–2000: Trends at the End of the Twentieth Century." *Journal of Infectious Diseases* 184, no. 6 (2001): 799–802.

Whitney, Daniel. *The Family Physician, and Guide to Health: In Three Parts…; Together with the History, Causes, Symptoms and Treatment of the Asiatic Cholera: Glossary Explaining the Most Difficult Words That Occur in Medical Science, and a Copious Index; To Which Is Added an Appendix.* New York: H. Gilbert, 1833.

Nicholas E. Bonneau
and Christopher Hamlin

CHRISTIAN SCIENCE

See Medicine: Alternative Medicine.

CLOCKS AND CLOCKMAKING

Since the invention of the mechanical clock somewhere in Europe before 1300, making and using timepieces has become a global enterprise. Significant American contributions to that enterprise fall into three main categories: changing clockmaking from a craft to an industry in the nineteenth century, inventing new types of highly accurate clocks for the electronic age, and structuring synchronized systems—from distant clocks linked by telegraph in the nineteenth century to today's global positioning systems (GPS).

Mechanical Clocks: From Craft to Industry. Colonial Americans initially adopted European timekeeping practices and instruments. To provide communal time, they installed public tower clocks and bells throughout the eighteenth and nineteenth centuries. Making tower clocks remained a craft until after the Civil War, when E. Howard and Company and Seth Thomas Clock Company began to make them in quantity for customers worldwide.

Colonial domestic clocks were uncommon, costly indicators of a person's wealth and social standing. Imported English clocks and watches, along with other English goods, dominated the material life of the English colonies. The craft of clockmaking grew steadily in the eighteenth century as men trained in making, finishing, and repairing timepieces immigrated to the United States. Relatively few clocks were made in the colonies before the Revolution, but by 1800 clocks were becoming cheaper and more widely available to a growing American middle class.

Contributing to that availability were new ways of making clocks to meet growing demand. At the end of the eighteenth century, brothers Simon and Aaron Willard set up workshops close to Boston, assembled a critical mass of skilled artisans to produce clocks, and introduced an American style of quantity production before industrialization. Simultaneously, a few Connecticut clockmakers abandoned the English tradition of making high-style clocks with movements of scarce and expensive brass. Instead, they built movements almost entirely from native wood.

Connecticut's head start in manufacturing wooden clocks enabled entrepreneurs in the Naugatuck River Valley to transform clockmaking from a craft to an industry as the nineteenth century began. Eli Terry, Seth Thomas, and Silas Hoadley pioneered the use of water-powered factories with machines for mass-producing uniform, interchangeable clock parts. This manufacturing style, applied to other products, became known as "the American system" of manufacturing. Wooden clocks from Connecticut were among the earliest mass-produced consumer goods in the United States. Sold largely to rural buyers by itinerant merchants, these clocks played an early and significant role in transforming the rural North from overwhelmingly agricultural to a modern market society based on industrial capital and consumerism.

The same region of Connecticut gave birth to a successful brass industry. After 1840, the brass-movement shelf clock, with a pendulum and a steel spring for power, became the standard mechanical clock produced in the United States until World War II. By the 1870s, just seven companies, all born in western Connecticut, were producing millions of timepieces. Although never a large industry compared with others in the U.S. economy, clockmaking was nevertheless one of the proving grounds of the American Industrial Revolution.

Electronic Time: Quartz and Atomic Clocks. Americans contributed to a major shift in clocks and clockmaking in the twentieth century. The science underlying timekeeping, long focused upon improving the mechanical pendulum clock, unexpectedly headed in a new direction in the 1920s with research on quartz crystals and radio frequencies. Aiming to monitor and maintain precise electromagnetic wave frequencies carrying telephone messages, engineers at Bell Telephone Laboratories—Warren Marrison and Joseph W. Horton—developed a clock in 1927 that kept stable time using a quartz crystal that vibrated 100,000 times a second. Only a small number of experimental and commercial quartz clocks were made in the 1930s, but World War II brought dramatic improvements in quartz technology. Wartime research and development focused on measuring and controlling the relatively low frequencies in military radios and the higher frequencies in the microwave region for radar, sonar, and loran.

In the postwar world, physicists and electronic engineers were the new clockmakers. Based on the fundamental research of Isidor Rabi, they exploited quantum mechanics to extract and count the "beats" of electromagnetic radiation from atoms. In doing so they devised the world's most accurate clocks. The world's first laboratory atomic clock, based on ammonia, was completed in 1949 at the National Bureau of Standards (now National Institute of Standards and Technology). Commercial clocks based on the signature frequency of cesium had widespread success. In 1956, the National Company, in Malden, Massachusetts, sold the first commercially available cesium clocks, mostly to military customers. The user base expanded after Hewlett Packard introduced

a more compact version of the cesium clock in 1964. The cesium clock proved so stable that it caused a fundamental shift in timekeeping. Since 1967, by international agreement, the length of the second is no longer a fraction of an astronomical day. It is instead the duration of 9,191,631,770 cycles of energy released from an isotope of cesium—cesium 133. Clocks based on hydrogen and rubidium are also among the best contemporary clocks, with commercial rubidium clocks most common for controlling frequency for broadcasting, cell phone systems, and satellite navigation.

Synchronized Clock Systems. For connecting large expanses of the nation, providing accurate time at a distance has been an American preoccupation since the nineteenth century. With the help of the telegraph and innovative electromechanical distribution systems, astronomical observatories in the United States determined time for their own purposes and distributed it to jewelers, port authorities, and the railroads. The most enduring electrical time service, lasting well into the twentieth century, was a partnership between the U.S. Naval Observatory and Western Union.

Beginning in the 1970s, an artificial constellation developed by the U.S. military and made up of the satellites of the NAVSTAR GPS became a one-way radio source for the world's most accurate time and location information. In the early twenty-first century, GPS time, based in on-board atomic clocks, was available continuously and globally to just about everyone with a receiver, military or civilian. GPS clocks set the tempo globally for transportation, communications, computers, finance, military operations, and even space exploration.

[*See also* **American System of Manufactures; Astronomy and Astrophysics; Bell Laboratories; Computer Science; Computers, Mainframe, Mini, and Micro; Electronic Communication Devices, Mobile; Engineering; Machinery and Manufacturing; Military, Science and Technology and the; Physics; Quantum Theory; Rabi, Isidor I.; Railroads; Research and Development (R&D); Satellites,**

Communications; Space Program; Telegraph; *and* Telephone.]

BIBLIOGRAPHY

Forman, Paul. "Atomichron: The Atomic Clock from Concept to Commercial Product." *Proceedings of the Institute of Electrical and Electronic Engineers* 73 (1985): 1181–1204.

Rip, Michael, and James Hasik. *The Precision Revolution: GPS and the Future of Aerial Warfare.* Annapolis, Md.: Naval Institute Press, 2002.

Zea, Philip, and Robert C. Cheney. *Clock Making in New England. 1725–1825.* Sturbridge, Mass.: Old Sturbridge Village, 1992.

Carlene E. Stephens

CLONING

In the biological sciences, the word *cloning* has been applied to a variety of phenomena. In the early twentieth century, botanists began using the word cloning to describe the asexual copying, or budding, of a plant. Soon after, microbiologists adopted the term "clone" to describe populations of bacteria that had descended from a common ancestor. The discovery of DNA in the 1950s and the advent of the molecular era in the 1970s spurred molecular biologists to begin referring to the replicates of recombinant DNA strands as clones and the process of creating them as cloning. Yet despite these multiple meanings, for many people cloning calls to mind an idea of copying whole animals, particularly humans.

This latter meaning has been intertwined with the success of a particular experimental technique, nuclear transplantation. Nuclear transplantation is a microsurgical technique in which the nucleus of one cell is removed and placed into an early egg cell, or oocyte, of that species that has had its own nucleus previously extracted or inactivated. The resulting combination yields an embryo that will develop into a genetic duplicate of the donor nucleus. Almost any cell, no matter how differentiated, can be used as the donor because nearly all nuclei retain the necessary components required

of any cell (there are exceptions, of course, such as red blood cells, which do not have nuclei). If a nucleus from a specialized cell is used, the procedure is referred to as somatic cell nuclear transfer. One implication of this experimental work is that theoretically the technique could be used to create a genetic duplicate, or clone, of a human being.

The first successful nuclear transplantation experiment was carried out by Robert Briggs, Thomas King, and Marie DiBerardino in 1952 at the Institute for Cancer Research at the Lankenau Hospital Research Institute in Philadelphia, Pennsylvania (which became the Fox Chase Cancer Center in 1974). Briggs and his team of researchers carried out nuclear transplantation experiments in frogs (*Rana pipiens*) to investigate larger questions about development and to solve problems in cancer research. In the following decades, John Gurdon, Robert McKinnell, John Moore, and several others used the technique to explore problems in cellular differentiation, cancer development, and speciation. By the 1970s, researchers had successfully carried out nuclear transfer experiments in several species, including other amphibians, fish, and insects.

Nuclear transfer first became connected to human cloning in the 1960s, when public intellectuals began to imagine how biology was going to revolutionize science and society. Taking cues from J. B. S. Haldane and Aldous Huxley, the Nobel laureate Joshua Lederberg publicly predicted that nuclear transfer research would lead to human cloning. Lederberg, who wrote approvingly of the possibility, desired to instigate more discussions about the potential benefits and pitfalls of contemporary scientific research to anticipate any associated hazards. Lederberg's work stirred a negative response from the burgeoning bioethics community, particularly Paul Ramsey and Leon Kass.

In 1978, journalist Peter Rorvik published *In His Image: The Cloning of a Man*, claiming that he had personally witnessed the cloning of a wealthy businessman. The book prompted a significant response from a worried public and a disbelieving scientific establishment, eventually leading to investigation by the U.S. Congress. In the early 1980s, controversy again engulfed nuclear transfer research when Karl Illmensee and Peter Hoppe claimed to have used nuclear transplantation techniques to successfully clone mice for the first time. Charges of fraud followed and their work could not be replicated.

By the mid-1980s, however, the first mammal had indisputably been cloned, and in 1997 biologists at the Roswell Institute in Scotland succeeded in using a nucleus from a fully specialized mammary cell to create a sheep named Dolly. Dolly's birth—the first mammal achieved through a somatic cell nuclear transfer—ignited another round of bioethical discourse on the topic of human cloning.

Today, somatic cell nuclear transplantation is used to create stem cells and is associated with therapeutic cloning, thus situating it squarely within bioethical and cultural controversies surrounding stem-cell research.

[*See also* **Biological Sciences; Botany; Cancer; DNA Sequencing; Ethics and Medicine; Lederberg, Joshua; Medicine; Molecular Biology; Nobel Prize in Biomedical Research; Research and Development (R&D);** *and* **Stem-Cell Research.**]

BIBLIOGRAPHY

DiBerardino, M. A. *Genomic Potential of Differentiated Cells.* New York: Columbia University Press, 1997.
Gurdon, John, and J. A. Byrne. "The First Half-Century of Nuclear Transplantation." *Proceedings of the National Academy of Sciences of the USA* 100, no. 14 (July 2003): 8048–8052.
Maienschein, Jane. *Whose View of Life? Embryos, Cloning, and Stem Cells.* Cambridge, Mass.: Harvard University Press, 2003.

Nathan Crowe

COLDEN, CADWALLADER AND JANE

(1688–1776) and (1724–1760), respectively. Cadwallader Colden and his daughter, Jane, were colonial New Yorkers with strong interests in the natural world. In the eighteenth century, science did not yet have the formal boundaries between fields that it would come to have in the nineteenth century. With interests in medicine, botany, meteorology, and astronomy, Cadwallader was like many men of his period who pursued science. Jane worked solely in botany and so was doubly unusual because few women in this period engaged in formal science. For them both, however, their scientific work was as much about sociability as it was about knowledge itself. Through their scientific interests they developed networks of friendship with people with similar interests in South Carolina, Pennsylvania, Britain, the Netherlands, and Sweden.

Cadwallader Colden, born in Scotland and educated at the University of Edinburgh, moved to New York in 1719, where he was appointed to political office. In the 1720s Cadwallader wrote reports on the climate and the economy of New York for authorities in London and a book on the Iroquois that promoted New York's strategic importance to Britain. He assisted with the astronomical observations that determined New York City's longitude, which were published in the *Philosophical Transactions* of the Royal Society of London. However, by the end of the decade he found himself politically marginalized.

Eventually Cadwallader would find friendship and esteem by turning to the like-minded outside New York. In 1729 Cadwallader and his wife moved to a farm up the Hudson River, where they would rear their eight children. There Cadwallader overcame his isolation through developing his scientific interests. He examined local plants, applying a new system for categorizing plants developed by the Swedish naturalist Carl Linneaus. His work was published in a Swedish learned journal, and Linneaus named a newly discovered plant genus for him, a mark of real honor in the scientific world. Cadwallader then turned to Isaac Newton's explanation of planetary motion and attempted to extend Newton's insights in a book published in London in 1752. Although experienced mathematicians saw the flaws in Cadwallader's ideas, his book was nonetheless greeted respectfully. He also published articles in the London press on an earthquake in New York and on local medical treatments. Through this work, Cadwallader developed a network of admiring correspondents on both sides of the Atlantic.

Of all of his endeavors, however, his correspondents most valued his botany. By the 1750s Cadwallader felt too old to look for new plants and so he taught the Linnean system to his daughter, Jane. Jane Colden's mastery of botany soon surpassed her father's. She analyzed over three hundred fifty plants, including two not previously described by European botanists. Her description of one new plant was published in a Scottish learned journal, the first Linnean description of a plant by a woman to appear anywhere. Although her botanical journal was not published, botanists in New York during and after the American Revolution relied on it to develop their floras of the region.

[See also Astronomy and Astrophysics; Botany; Gender and Science; Medicine; Meteorology and Climatology; and Science.]

BIBLIOGRAPHY

Gronim, Sara Stidstone. "What Jane Knew: A Woman Botanist in the Eighteenth Century." *Journal of Women's History* 19 (2007): 33–59. An examination of Jane Colden's work in the context of eighteenth-century practices of gender and sociability; winner of the Margaret W. Rossiter Prize for the History of Women in Science.

Hoermann, Alfred R. *Cadwallader Colden: A Figure of the American Enlightenment.* Westport, Conn.: Greenwood Press, 2002. A thorough and reliable account of all facets of Cadwallader Colden's intellectual life, although without an awareness of recent work that has significantly reconceptualized both the Enlightenment and early modern science.

Sara Stidstone Gronim

COLT, SAMUEL

(1814–1862), firearms manufacturer. Born in Hartford, Colt's early years included work in his father's dyeing and bleaching business, a year at sea, and a stint as "Dr. Samuel Coult" delivering popular chemistry lectures. In 1835–1836 he secured patents on a revolving breech pistol, allowing six shots to be fired without reloading, an advance over single-shot pistols. His first manufacturing venture, in Paterson, New Jersey, failed in 1842, but during the Mexican War he fulfilled a military order for one thousand pistols in collaboration with the Connecticut armory headed by Eli Whitney Jr. Back in Hartford, Colt in 1855 began manufacturing firearms on a sprawling site along the Connecticut River. Colt's Patent Firearms Manufacturing Company prospered as his revolvers, with their distinctive ornamental scrollwork, gained popularity in the West. Indeed, "Colt" became a generic synonym for a revolver. Colt's advanced manufacturing techniques, including interchangeable parts, precision machine tools, and assembly-line methods, offered a prototype of U.S. industrial production. In 1853–1856, he briefly operated a factory in England as well. An early business tycoon, Colt marked his 1856 marriage by building an imposing Italiante mansion, Armsmear, overlooking his factory. The Civil War brought boom times as the Colt factory produced pistols and muskets for the Union Army, but Colt himself died of rheumatic fever in 1862. After an 1864 fire, a new main factory—now an iconic Hartford landmark—featured a blue onion-shaped dome spangled with gold stars and surmounted by a rearing white stallion, an early corporate logo. Production for the civilian, military, and private-security markets eventually included a wide range of weapons, as well as watches, typewriters, sewing machines, bicycles, and other products marketed under various brand names. Colt's original factory and mansion are now national historic landmarks. Hartford's Museum of Connecticut History displays the company's firearms collection. An important figure in the history of technology, Samuel Colt also epitomizes the free-for-all entrepreneurial capitalism of the antebellum era and hovers over the mythic Old West, his legendary six-shooter featured in countless dime novels and Western films. Colt himself summed up his career succinctly: "The good people of this world [sic] are very far from being satisfied with each other, & my arms are the best peacemakers."

[*See also* **American System of Manufactures; Bicycles and Bicycling; Clocks and Clockmaking; Machinery and Manufacturing; Military, Science and Technology and the; Research and Development (R&D);** *and* **Technology.**]

BIBLIOGRAPHY

Hosley, William N. *Colt: The Making of an American Legend.* Amherst: University of Massachusetts Press, 1996.
Tucker, Barbara M. *Industrializing Antebellum America: The Rise of Manufacturing Entrepreneurs of the Early Republic.* New York: Palgrave Macmillan, 2008.

Paul S. Boyer

COLUMBIAN EXCHANGE

Prior to 1492, the Americas and Eurasia-Africa, except for occasional connections via the Bering land bridge between present-day Alaska and Siberia, had been separated for millions of years. During the Great Ice Age, in which vast ice sheets covered large parts of the North American and Eurasian continents, water levels were significantly lower, which allowed people, animals, and plants to traverse the land bridge. The end of the Great Ice Age about 11,000 years ago, however, raised water levels and covered the land bridge. As a result, organisms diverged in their evolution, and human beings on either side of the Atlantic Ocean developed their own ways of life, including their own crops, domesticated animals, and diseases. When the voyages of Christopher Columbus and other European sailors established contacts between these great land masses, they triggered an unprecedented exchange of people, diseases, plants, and animals between the Old and New Worlds that unleashed massive consequences for the human populations and for the entire biosphere. Although this was a multifaceted, two-way exchange, some of the facets were more momentous and fortuitous than others. The

colonies established by England, France, Portugal, and Spain in the New World after 1492 were vast arenas of agricultural, cultural, and social experimentation.

In 1492, the native crops of the Old and New Worlds were entirely different, with the exception of cotton, which was cultivated on both sides of the Atlantic. The most important Native American crops were maize (corn), white potatoes, sweet potatoes, manioc (also known as cassava), and tomatoes. Among the more important Eurasian-African crops (a longer list, as would be expected of a much larger area) were wheat, barley, rice, sugarcane, oats, rye, soybeans, bananas, carrots, coffee, okra, cabbage, and oranges. The initial encounter between European and Native American methods of food production occurred on the island of Hispaniola. Because wheat did not grow well in the tropical Caribbean, the Spaniards began to eat bread made out of ground manioc. Sugarcane and bananas, however, grew exceptionally well in the Caribbean and became an important part of the diet of the Spaniards.

In addition to plants that provided sustenance for human populations, the Columbian Exchange also facilitated the transfer of weeds. The Old World had more species of weeds because it was much larger in area and because it had more species of grazing animals to whose teeth and hooves Eurasian and African grasses and herbs had been obliged to adapt. Thus, many of America's most aggressive weeds, such as dandelions, crabgrass, wild oats, thistle, kudzu, and tumbleweed, originated in Europe. Only a few weeds from the Americas, such as amaranth, established themselves in the Old World.

The Eastern and Western hemispheres also exchanged many kinds of wild animals. For instance, black and brown rats were brought to America and gray squirrels and muskrats to the Old World. The exchange of domesticated animals, however, was more important. Native Americans were much less effective at animal domestication, possibly because they had fewer animals with which to work. Their livestock—llamas, guinea pigs, and turkeys—included none that were ridden, pulled heavy loads, provided hides or fertilizer in significant amounts beyond the local level, or supplied large quantities of nourishment (i.e., meat and milk) for human consumption. The domesticated animals of Eurasia-Africa, which included horses, cattle, sheep, goats, pigs, donkeys, and chickens, were a major source of nourishment, leather, fiber, power, and fertilizer. The downside of introducing these Old World grazing animals onto a landscape formerly lacking a significant grazing animal population was the possibility of exceeding the carrying capacity of the land.

The exchange of cultivated plants and livestock increased food production on both sides of the Atlantic, changed the eating habits of many people in both the Old and the New Worlds, and, according to many historians, helped to fuel a population explosion in the Old World. American white potatoes, for instance, enabled farmers in cool, rainy northern Europe to extract more nourishment from the soil than ever before. The introduction of the white potato in Ireland helped increase that island's population from 3 million people in 1700 to 8 million people in 1840 (O'Hara, 2002, p. 6). Hundreds of millions of Chinese are dependent on the New World's maize and sweet potatoes. Multitudes of Africans depend on maize, peanuts (which were first grown in the valleys of present-day Peru), and manioc. It is difficult to imagine how the Americas could support their hundreds of millions without wheat, rice, beef, chicken, and other foodstuffs of Old World origin.

Most of the world's most troublesome human diseases originated in the Old World. America had its own, like Chagas disease and syphilis, but nothing as devastating and influential as the Old World's smallpox, measles, whooping cough, chicken pox, bubonic plague, malaria, yellow fever, diphtheria, amoebic dysentery, and influenza. With its greater area and bigger animal and human populations, Eurasia-Africa was also home to more kinds of germs. In addition, human beings had lived in dense populations, fertile ground for germs, for much longer in Eurasia-Africa than in the Americas. They had practiced irrigation agriculture for much longer, creating conditions for the propagation of waterborne infections, and had traded longer and more intensively, ensuring the wide diffusion of infections. Above all,

they had for thousands of years lived among domesticated animals and vermin-carrying creatures like rats, with which they shared and mutually cultivated a great number of pathogens. No American disease figured importantly in the Old World, with the possible exception of syphilis, a venereal disease that many claim Europeans acquired from the Native Americans of the Caribbean during the early sixteenth century. Skeletal evidence indicates that syphilis, or some infection like it, was indeed present in the pre-Columbian Americas. Evidence for it before 1492 in the Old World is not so clear, but the question remains unresolved (Aufderheide, 1998, pp. 166–171).

The effects of the Columbian Exchange on the size and distribution of human populations are clear and spectacular. The impact of Old World diseases on Native Americans radically reduced their number, conceivably by as much as 90 percent, opening their two continents and nearby islands to massive shifts of Europeans and Africans across the Atlantic. The racial and ethnic blending between Africans and Europeans, Native Americans and Europeans, and Africans and Native Americans resulted in mulattos, mestizos, and zambos (respectively), which is most clearly evident in Latin America. The racial and ethnic composition of Latin American populations is directly related to the size of pre-Columbian populations, the position of the area within the Atlantic slave trade, and the willingness of Europeans (especially Spaniards and Portuguese) to procreate with Africans and Native Americans. The full significance of the Columbian Exchange cannot be fully measured, because it reversed more than 100 million years of divergent evolution.

[*See also* **Botany; Diphtheria; Disease; Food and Diet; Influenza; Malaria; Sexually Transmitted Diseases; Smallpox;** *and* **Yellow Fever.**]

BIBLIOGRAPHY

Aufderheide, Arthur C., and Conrado Rodriguez-Martin. *The Cambridge Encyclopedia of Human Paleopathology.* Cambridge, U.K.: Cambridge University Press, 1998.

Cook, Noble David. *Born to Die: Disease and New World Conquest, 1492–1650.* Cambridge, U.K.: Cambridge University Press, 1998.

Crosby, Alfred W. *The Columbian Exchange: Biological and Cultural Consequences of 1492.* 30th anniversary ed. Westport, Conn.: Praeger, 2003.

Diamond, Jared. *Guns, Germs, and Steel: The Fates of Human Societies.* New York: W. W. Norton, 2005.

Dunmire, William W. *Gardens of New Spain: How Mediterranean Plants and Foods Changed America.* Austin: University of Texas Press, 2004.

O'Hara, Megan. *Irish Immigrants, 1840–1920.* Mankato, Minn.: Capstone Press, 2002.

Salaman, Redcliffe. *The History and Social Influence of the Potato.* 2d ed. Cambridge, U.K.: Cambridge University Press, 1985.

Alfred W. Crosby;
updated by Michael R. Hall

COMPTON, ARTHUR H.

(1892–1962), physicist and Nobel laureate. Born in Wooster, Ohio, Compton received his BS from the College of Wooster and his PhD from Princeton University (1916), where he studied the properties of matter and radiation, focusing specifically on the properties of thermionic and photoelectric emission. Compton began his career by teaching for a year at the University of Minnesota (1916–1917). He then worked as a research engineer in the Westinghouse Research Laboratories in Pittsburgh, Pennsylvania (1917–1919). After World War I, he was awarded a fellowship by the National Research Council and traveled to study X-ray scattering at the Cavendish Laboratory in Cambridge, England, from 1919 to 1920. While there he worked with J. J. Thomson, Ernest Rutherford, and Sir William Bragg. On his return to the United States, he accepted a position as professor of physics at Washington University in St. Louis, Missouri (1920–1923). In 1916, he married Betty Charity McCloskey; they had two sons.

In 1922, while doing research on X-rays, Compton discovered what came to be called the Compton effect. X-rays were long known to behave like ordinary electromagnetic waves under various experimental conditions. Compton now showed that when X-rays collide with electrons in a substance like carbon, they behave like particles. This discovery at long last established the validity of Albert Einstein's light-quantum hypothesis of 1905 and became a milestone in the creation of quantum mechanics. For his discovery, Compton received the Nobel Prize in Physics in 1927.

Compton moved to the University of Chicago in 1923, where in the 1930s he carried out important cosmic-ray research using sensitive electrometers. Physicists used these electrometers in a dozen expeditions worldwide to map cosmic radiation intensity. Based on his discovery that cosmic rays were not as intense near the magnetic equator, he proved that they were not gamma-ray photons, as Robert Millikan had argued, but rather charged particles that could be deflected by the magnetic field of the earth.

During World War II, Compton directed the Manhattan Project's Metallurgical Laboratory in Chicago, where Enrico Fermi achieved the first nuclear chain reaction on 2 December 1942. His work in Chicago became the basis for the production reactors at Hanford, Washington. The plutonium used in the first atomic device came from Hanford. In 1945, Compton became chancellor of Washington University, St. Louis, and taught there until his retirement in 1961. The deeply religious son of a Presbyterian minister, Compton in his public lectures and writings always insisted upon the compatibility of science and religious faith.

[*See also* **Einstein, Albert; Fermi, Enrico; Manhattan Project; Millikan, Robert A.; Nobel Prize in Biomedical Research; Nuclear Power; Physics; Quantum Theory; Religion and Science;** *and* **Science.**]

BIBLIOGRAPHY

Johnson, Marjorie, ed. *The Cosmos of Arthur Holly Compton.* New York: Alfred A. Knopf, 1967.

Seidel, Robert W. "Compton, Arthur Holly (1892–1962)." In *The History of Science in the United States: An Encyclopedia,* edited by Marc Rothenberg, pp. 132–133. New York and London: Garland Publishing, 2001.

Stuewer, Roger H. *The Compton Effect: Turning Point in Physics.* New York: Science History Publications, 1975.

Roger H. Stuewer;
updated by Elspeth Knewstubb

COMPTON, KARL TAYLOR

(1887–1954), American scientist administrator and statesman. Compton was born in Wooster, Ohio, the eldest of four children. His younger brother Arthur also became a prominent scientist. Karl Compton excelled early: he skipped a grade and spent his last two years in high school at Wooster College, where he earned a bachelor of philosophy in 1908. After earning a master's degree and serving as an instructor in Wooster's chemistry department, he entered graduate school at Princeton. Working with his advisor, Owen W. Richardson, he made the initial confirmation of Einstein's theory of the photoelectron effect. Compton went on to earn a PhD in physics in 1912.

After a stint as a physics instructor at Reed College in Portland, Oregon, Compton returned to Princeton in 1915. As a young Princeton professor he established a research reputation in electronics and spectroscopy and became a consultant with General Electric. When World War I started he drew on his industrial connections to link industrial and academic physics; he also served as a scientific advisor to the U.S. Embassy in Paris. When World War I ended Compton returned to Princeton. In addition to establishing an illustrious research and teaching career, he also assumed a number of important administrative duties. In 1927 he was named Director of Research at Princeton's Palmer Laboratory and became president of the American Physical Society. In 1930, after the onset of the Great Depression, he became president of the Massachusetts Institute of Technology (MIT), spearheading reforms that had an impact far beyond MIT. Driven by the conviction that science could be an important spur for industrial progress, he pressed for a tighter connection between science and engi-neering as well as more rigorous academic engineering standards. He also helped found and became in 1931 the first chair of the board of the American Institute of Physics. Then, in 1933, he accepted the invitation of President Franklin Roosevelt to chair the Scientific Advisory Board, meant to provide increased funding for science as well as continuing scientific advice to the highest levels of government.

Although the board was short-lived, Compton had achieved a high profile in Washington. Thus, with World War II brewing, he was asked in 1940 to be a member of the National Defense Research Committee (NDRC) and was put in charge of the division that gathered experts to study radar. When the NDRC became a part of the Office of Scientific Research and Development, Compton chaired the U.S. Radar Commission to the United Kingdom. In 1945 he also served on the Interim Committee that advised President Harry Truman on the use of the first atomic bombs.

In the postwar, Compton continued to serve in both government and academe. In 1948 Truman appointed him to head the Joint Research and Development Board to oversee scientific preparedness. In that year he ended his long tenure as MIT president to become chairman of the MIT Corporation, a post he held until his death (Stratton, 1992; Kevles, 1978).

[*See also* **American Institute of Physics; Chemistry; Einstein, Albert; Engineering; Higher Education and Science; Military, Science and Technology and the; Office of Scientific Research and Development; Physics;** *and* **Science.**]

BIBLIOGRAPHY

Compton, Karl T. *A Scientist Speaks: Excerpts from Addresses by Karl Taylor Compton During the Years 1930–1949 When He Was President of the Massachusetts Institute of Technology.* Cambridge, Mass.: MIT Press, 1955.

Hewlett, Richard G., and Oscar E. Anderson Jr. *The New World: A History of the United States Atomic Energy Commission,* Berkeley: University of California Press, 1990.

Kevles, Daniel J. *The Physicists: The History of a Scientific Community in America*. New York: Alfred A. Knopf, 1978.

Stratton, Julius A. *National Academy of Sciences, Biographical Memoirs*, Washington, D.C.: National Academies Press, 1992.

Catherine Westfall

COMPUTER SCIENCE

Computer science, the study of computational activities, emerged in the mid-twentieth century as a combination of mathematics, philosophy, and electrical engineering. As it developed, it retained elements of all three, as it embraced more and more subjects, including cognitive psychology, graphic design, and systems engineering. Although its practitioners have often debated whether computer science truly has the attributes of a true science, the field has shown itself to be remarkably powerful and flexible in its ability to abstract and systematize most forms of human activity.

The origins of computer science predate the development of the electronic computer. In 1937, the English mathematician Alan Turing laid the foundation for the field in a paper that attempted to identity the basic properties of computational processes. In writing this paper, Turing described a hypothetical automatic computing machine that was unlike any computing device of the time and would have been difficult to construct with the technology of the time. Still, this abstraction established the qualities of computation, qualities that are identified as "Turing computable."

Many of the early computers were built by radio engineers, who were primarily interested in the design of electronic computing circuits and less concerned about the nature of computation. Some researchers thought that the task of programming these machines would be relatively straightforward and would be handled by mathematicians or even mathematics students. Nonetheless, a few early researchers started to consider the broader nature of computation and began to develop ideas that were not wholly part of mathematics or engineering but drew from both. The

mathematician John von Neumann described computing machines in his 1943 paper, the First Draft Report on the electronic discrete variable automatic computer (EDVAC). In three subsequent papers, he and two coauthors began to develop the ideas that quickly became common in the new field, the ideas that defined and analyzed complexity. In these papers, the authors began exploring computational processes and analyzing how much time they would require and how much memory they would need for their calculations.

The ideas of computer science were first disseminated in conferences. The first were held in the fall of 1945 and the winter of 1946 under the auspices of the Massachusetts Institute of Technology (MIT) and Harvard. At first, they considered the design of computing circuits and the calculation of numerical qualities, such as the solution of differential questions or linear systems. By the end of the decade, these conferences had started to consider other kinds of computing problems, notably those of sorting data. In 1950, the Institute for Radio Engineers (IRE) began publishing a scholarly journal for problems of computer science, the *Transactions on Computing*. This was followed in 1952 by the *Journal of the Association for Computing Machinery* (ACM). The two organizations, the IRE and the ACM, jointly sponsored a semiannual conference, which would serve as the fundamental forum for computer science for the next 35 years.

During the 1950s, the new computer scientists worked primarily on the problems of designing and programming the new machines. A key problem was described as "automatic coding," a term that referred to the ability of describing a computation in a sophisticated language that would be translated into the instructions that would drive the computer. A major accomplishment in this work was Chomsky's Hierarchy, a description of languages that could be processed by a machine. Identified by Noam Chomsky in 1956, it opened the way for the development of computer languages such as Fortran (1957), LISP (1958), COBOL (1959), and ALGOL (1960).

During the last years of the 1950s, computer scientists also focused on the problems of managing computing resources so that the machines could perform the greatest amount of work. This research led to the development of operating systems, including the Compatible Time Sharing System (1961) and the Multics (1964) project at MIT.

The term "computer science" was an invention of the late 1950s. The *Communications of the ACM* began using the phrase in 1959. Throughout the 1960s and 1970s, it referred equally to the study of computing circuits, machine design independent of the underlying circuitry, and the problems of programming. By the late 1960s, it was increasingly used to refer to problems of software—programs—or problems that could be expressed in a symbolism like those used to create software. In 1968, the North Atlantic Treaty Organization (NATO) sponsored a conference on software engineering that focused attention on the problems involved in creating large programs.

The year 1968 saw two other events that marked the movement of interest toward software problems. The IBM Corporation, which produced the computer with the largest installed base, the System/360, announced that it would no longer give software away for free. This action opened the way for the independent software industry. That same year, Stanford researcher Douglas Engelbart demonstrated a software system that he claimed would enhance, rather than replace, human ability. This system, which included elements similar to those used by modern systems, encouraged computer scientists to work on problems that included human beings as well as machines and programs.

The 1970s saw a rapid expansion of computer science as researchers developed fundamental concepts in a variety of areas including computer graphics, data structures, database design, computer simulation, artificial intelligence, and parallel processing. As computation began to expand into more areas of human endeavor, the two major professional organizations in the field, the Association for Computing Machinery and the Institute of Electrical and Electronic Engineers (IEEE, the successor of the IRE) began to define computer science in more and more universal terms. Although they resisted claiming that computer science was a means of studying the natural world in the same way that physics or biology considered physical phenomena, they still conceived it as providing universal, abstract laws and principles that would guide computing practice.

The two organizations identified four areas that constituted the core of computer science. The first was the theory of computation, a branch that worked to understand the tasks that could and could not be computed. Next was the study of algorithms and data structures, the means for manipulating and storing data. Third was the study of programs and programming languages, the work that had begun in the 1950s and built upon the achievements of linguistics and logic. Only the last, the study of computer elements and architecture, dealt directly with computing machines themselves.

[*See also* **Artificial Intelligence; Automation and Computerization; Computers, Mainframe, Mini, and Micro; Engineering; Journals in Science, Medicine, and Engineering; Mathematics and Statistics; Psychology; Software; Technology;** *and* **Von Neumann, John.**]

BIBLIOGRAPHY

Akera, Atsushi. "Edmund Berkeley and the origins of ACM." *Communications of the ACM* 50, no. 5 (May 2007): 30–35.

Aspray, William. "John von Neumann's Contributions to Computing and Computer Science." *Annals of the History of Computing* 11, no. 3 (1989): 189–195.

Bergin, Tim. "A History of the History of Programming Languages." *Communications of the ACM* 50, no. 5 (May 2007): 69–74.

Ensmenger, Nathan. *The Computer Boys Take Over: Computers, Programmers, and the Politics of Technical Expertise.* Cambridge, Mass.: MIT Press, 2010.

Gupta, G. K. "Computer Science Curriculum Developments in the 1960s." *Annals of the History of Computing* 29, no. 2 (2007): 40–54.

Hodges, Andrew. *Alan Turing: The Enigma.* New York: Simon and Schuster, 1983.

King, W. K., and Land, S. K. "A Historical Perspective of the IEEE Computer Society: Six Decades of Growth with the Technology It Represents." In *Proceedings of the 2009 IEEE Conference on the History of Technical Societies*, pp. 1–6. Philadelphia: University of Pennsylvania, 2009.

David Alan Grier

COMPUTERIZED AXIAL TOMOGRAPHY

See Radiology.

COMPUTERS, MAINFRAME, MINI, AND MICRO

The terms mainframe, minicomputer, and microcomputer refer to three classes of computers that were generally defined by the markets that they reached. Mainframe computers were large systems that were sold or leased to large corporate customers. Minicomputers were smaller machines that were generally acquired by small and midsize businesses, as well as the smaller departments of large customers. Microcomputers were desktop machines that were sold to individuals or employed by individuals in the workplace. Taken as a whole, the story of these three classes of machines illustrates the growing sophistication of computer systems, the shift from hardware to software as the point of innovation, and the relentless drive to standardize computing technology.

Before the Mainframe: 1946 to 1952. The basic ideas for computing machinery were disseminated at the Moore School Lectures, a series of presentations that were held at the University of Pennsylvania in the summer of 1946. The talks were given four months after the announcement of the ENIAC computer, which had been constructed at the university's Moore School of Engineering. The speakers included almost every American researcher who had been involved in computing projects over the course of World War II and covered topics as diverse as

numerical analysis, computing circuits, and promising ideas for memory technology.

At the center of the lectures was a key paper on computing machinery, John von Neumann's (1903–1957) "Draft Report on the EDVAC." This paper, written the prior summer as part of the ENIAC project, described the basic form of the stored program computer. This idea divided the computer into three fundamental units: the control unit, the memory unit, and the processor that could manipulate data. The program was coded as a string of numbers, which was stored in the memory. The memory fed this string to the control unit, which decoded the instructions and then directed the operation of the entire machine.

After the end of the Moore School Lectures, the participants returned to their own institutions and began to build their own computers based on the von Neumann paper. Von Neumann built a machine at the Institute for Advanced Study at Princeton and freely offered his ideas and plans to others. Some of the projects that borrowed extensively from von Neumann's ideas and plans include machines at Cambridge University, Los Alamos Laboratory, the University of Illinois, the National Bureau of Standards, MIT, and the University of Pennsylvania itself.

These early machines shared the scale and complexity of the mainframes but they really predate the mainframe era. Although many did valuable computations for the military or for commercial clients, all were closer to engineering prototypes than to production machines. They were sensitive devices and had to be managed by skilled engineers. Furthermore, each of these projects modified von Neumann's original idea in some way. None of them was capable of running a program that was written to another machine.

Mainframes in the Ascendency: 1952 to 1968. The first proper mainframes were the machines that moved beyond the laboratory and were installed in customer offices. The first of these machines included the LEO in the United Kingdom (1951), the UNIVAC I (1951), the IBM 701 (1952), and the BESM (1952) in the Soviet Union. The design of the IBM 701 was directly based on von Neumann's machine. It was marketed to government laboratories and large engineering firms. In all, IBM installed 19 copies of the machine.

Although all 19 of the IBM 701 machines could run the same programs, they could not execute the programs for other computers, such as the UNIVAC I. The UNIVAC I was designed for government agencies and corporate offices. It could compute with the decimal arithmetic that was found in accounting offices but could not easily handle the scientific notation required by technical problems.

During the 1950s, computer engineers created some of the fundamental ideas of computer technology and deployed them on mainframe computers. These ideas include magnetic core memory, magnetic disk storage, transistor switching circuits, programming languages, and operating systems.

The announcement of the IBM System/360 series in 1964 marked the start of the dominant age of the mainframe. The System/360 offered five different models, each running at different speeds and having different capacities. All five models were capable of running the same program without any changes. Because of this common design, the System/360 quickly became the best-selling computer of the time. It shared the market with machines produced by Burroughs, General Electric, RCA, Honeywell, UNIVAC, and NCR, but none of these machines offered the range of performance provided by the different versions of the System/360. Because of its popularity and its range of performance, the System/360 became the initial market for the early commercial software industry.

Minicomputers Opening a New Market: 1968 to 1984. As the name suggests, minicomputers were smaller, less expensive, and less powerful than mainframe computers. However, they were not miniaturized versions of mainframe computers but had an independent path of development. Minicomputers trace their origins to the Whirlwind computer of MIT (1947–1951). Although this computer's construction borrowed heavily from von Neumann's ideas, it was designed to be a machine that interacted directly

with humans. It was originally intended to operate a flight simulator to train bomber pilots.

One of the junior designers on the Whirlwind project, Ken Olsen (1926–2011), took the basic ideas of Whirlwind and used them to design a relatively small computer that he called the Programmable Data Processor-1, or PDP-1 (1960). This machine was not commercially successful, but its successors, notably the PDP-10 (1965), found large markets. Because they were inexpensive, they were commonly purchased by universities and used to train students. They supported an interactive programming environment, which allowed many individuals to work with the computer simultaneously. A common programming tool on the minicomputers was the BASIC language, which was invented at Dartmouth College.

The market for minicomputers grew rapidly in the 1970s. Ken Olson's PDP computers, which were manufactured by Digital Equipment Corporation (DEC), were the most popular models of minicomputers but they shared the market with machines from Data General, Prime, Honeywell, Burroughs, and NCR.

Because they were inexpensive, minicomputers were commonly used in the development of networks. In the early 1970s, the designers of the ARPAnet, the precursor to the Internet, used minicomputers to construct their network. Rather than find a way to connect every kind of computer to a common network, they built a network with a single model of minicomputer and then connected other machines to those computers. By the early 1980s, minicomputers were often connected in networks. The original vendor of such machines, DEC, eventually started marketing their computers with the phrase "The Network Is the Computer." At the end of the decade, minicomputers formed the largest market of computing equipment.

Microcomputers, A Common Basis for Computing: 1984 to 2000. Microcomputers changed both the scale and the environment for computation. When microcomputers began appearing on the market in the mid-1970s, most computer manufacturers retained control of the fundamental design of their machines. The only exceptions were the companies, such as Amdahl Corporation, that built machines that copied the underlying design of the IBM System/360. However, microcomputers were based on computer processors that were found on a single chip or a small group of chips. Using these chips, the computer manufacturers would no longer have control over the fundamental technology of computation, the design of the instructions that could be used to control the computer.

Microcomputers were introduced in the mid-1970s, although the processor chips began to appear in the late 1960s. The first of these chips was Intel's 4004 processor, which was designed to be the basis of a handheld electronic calculator. This processor was soon followed by the 8008 chip, which was turned into a computer product by the Canadian firm Micro Computer Machines in 1973. The successor to this chip became the basis of the Altair 8800, which was introduced in 1975. This machine quickly drew the attention of hobbyists, who began to improve the design or develop software for it. Bill Gates (1955–) developed a BASIC language compiler for the Altair and eventually founded Microsoft. Steve Jobs (1955–2011) and Steve Wozniak (1950–) created their own computer, which they dubbed the Apple.

Although the early microcomputers were based on processor chips, they were not identical machines nor could they yet run common software. Two firms, Intel and Motorola, sold competing and incompatible processors. Furthermore, computer vendors could use these processors in incompatible ways. As had happened for both mainframes and minicomputers, the market quickly focused on families of computers that had a range of capabilities. The most successful of the early families was the Apple II.

The success of the Apple II was partially the result of the spreadsheet program VisiCalc, which was released in 1979. Spreadsheets were an important early product. They allowed users to manipulate numbers in ways that were not easy to duplicate on larger machines.

IBM entered the microcomputer market in 1983 and quickly began to drive the market of its own making. IBM not only had a large customer base, but also had a well-recognized brand and a strong financial base that would survive economic challenges. Many early microcomputer manufacturers failed because of a lack of funding. The IBM PC quickly became popular with corporate customers.

Apple followed the announcement of the IBM PC with its Macintosh computer. This machine, which was based on research that Steve Jobs had seen at Xerox Parc Laboratories in Palo Alto, was radically different from the IBM product or even the Apple II. It had a graphics screen, used a windowed interface, and employed a mouse to direct the machine. Although such elements would eventually become standards for microcomputers, they failed to secure a dominant place in the market for Apple as IBM quickly became the largest supplier of microcomputers.

The success of IBM encouraged other companies to offer identical machines, called IBM clones, which were based on the same chips. In 1988, IBM tried to reclaim the microcomputer market by redesigning its machines, but the new models failed to find a market. IBM eventually retreated and offered machines that were compatible with its original design. It would never be able to regain the dominance it held in the mid-1980s. Less than six years after introducing its new models, it would be surpassed by two other companies in the microcomputer market.

As the standard IBM design began to dominate the market, the distinction among different brands of microcomputers started to fade. By the early 1990s, microcomputers were distinguished by their speed, the amount of memory they contained, and the extra features they offered, such as graphics processors. Only Apple maintained a unique design in its Macintosh. Increasingly, people interacted with their machines through software. Increasingly that software was a form of the Windows operating system by Microsoft. Microsoft released version 3.0, the version that would be the basis for subsequent designs, in 1990.

Expansion, Not Substitution. In general, microcomputers did not replace minicomputers nor had minicomputers replaced mainframes. Each form had expanded the market for computer technology and found new applications for computing technology. The mainframe actually experienced a resurgence after 1992, when the Internet was opened to commercial development. Now called servers, these machines provided centralized services and common repositories of data. However, these new versions of the mainframe had as much in common with microcomputers and minicomputers as they did with their larger progenitors of the 1960s. They were often built with the same processor chips used in microcomputers and they utilized software that had been originally developed for minicomputers.

The distinctions of mainframe, minicomputer, and microcomputer belong to certain eras and markets rather than to the size and scale of computing equipment. To some extent, those market distinctions still exist. Customers still need large, centralized servers, midsize departmental computers, and individual machines. Although each of those market segments can claim specific products that fulfill certain needs, all use common technologies that were developed over the three periods that produced mainframes, minicomputers, and microcomputers.

[*See also* **Automation and Computerization; Computer Science; Engineering; ENIAC; Internet and World Wide Web; Mathematics and Statistics; Military, Science and Technology and the; Software; Technology;** *and* **Von Neumann, John.**]

BIBLIOGRAPHY

Ceruzzi, Paul. *A History of Modern Computing*, 2d ed. Cambridge, Mass.: MIT Press, 2003.
Cortada, James. *The Digital Hand*, Vols. 1–3. Oxford: Oxford University Press, 2004, 2006, 2008.
Kidder, Tracy. *Soul of a New Machine*. Boston: Little Brown, 1981.
Manes, Stephen, and Paul Andrews. *Gates: How Microsoft's Mogul Reinvented an Industry*. New York: Touchstone, 1994.
Pugh, Emerson. *Building IBM*. Cambridge, Mass.: MIT Press, 1995.
Waldrop, Mitch. *The Dream Machine*. New York: Viking Press, 2001.

David Alan Grier

CONANT, JAMES B.

(1893–1978), president of Harvard University, science administrator, diplomat. Born in Dorchester, Massachusetts, Conant received his BA and PhD from Harvard, where he taught organic chemistry from 1916 until he became Harvard's president in 1933. Active in poison-gas research in World War I, Conant in World War II oversaw the military application of science as head of the National Defense Research Committee and deputy director of the Office of Scientific Research and Development (OSRD). As chairman of OSRD's Manhattan Project executive committee and a member of President Franklin Delano Roosevelt's top policy group and President Harry S. Truman's interim committee, Conant played a vital role in developing cooperative atomic research with Great Britain and participating in critical decisions regarding the development and use of atomic weapons. He made a crucial suggestion in June 1945 to use one of the new atomic bombs on a Japanese war plant in a populated area. President Truman adopted Conant's idea for shortening the war by attacking Hiroshima, a major city but also an important industrial center supporting the Japanese war effort. In 1946, troubled by criticism of the U.S. atomic bombings of Japan, Conant played a central role, with McGeorge Bundy, in preparing secretary of war Henry Stimson's influential essay, *The Decision to Use the Atomic Bomb* (*Harper's*, February 1947).

Subsequently, fearing the nuclear destruction of civilization, Conant sought international control of atomic weapons, opposed development of the hydrogen bomb, and questioned the benefits of nuclear power. A typical Cold War liberal, he championed expanded defense spending and military preparedness while nevertheless opposing the militarization of academic research and trying to shield academia from McCarthyism. Although he defended civil liberties, he believed that Communist teachers should be banned from America's schools. From 1946 to 1962, Conant served as an adviser to the Atomic Energy Commission.

Leaving Harvard in 1953, Conant served as high commissioner for Germany (1953–1955) and as the first U.S. ambassador to the German Federal Republic (1955–1957). He devoted his final years to educational reform. From 1957 to 1962, he directed a Carnegie Foundation study of secondary schools in the United States. His influential book *The American High School Today* (1959), a response to the Soviet Union's 1957 launch of its *Sputnik* space satellite, called for more rigorous instruction in science, mathematics, and foreign languages.

[*See also* **Atomic Energy Commission; Chemistry; Diplomacy (post-1945), Science and Technology and; High Schools, Science Education in; Manhattan Project; Mathematics and Statistics; Military, Science and Technology and the; Nuclear Power; Nuclear Weapons; Office of Scientific Research and Development; Physics;** *and* **Science.**]

BIBLIOGRAPHY

Conant, James B. *My Several Lives: Memoirs of a Social Inventor.* New York: Harper & Row, 1970.
Hershberg, James G. *James B. Conant: Harvard to Hiroshima and the Making of the Nuclear Age.* New York: Alfred A. Knopf, 1993.

Peter J. Kuznick;
updated by Hugh Richard Slotten

CONDON, EDWARD

(1902–1974), theoretical physicist and research director. Born in New Mexico, Condon played a leading role in the transfer of quantum mechanics from Europe to the United States in the 1920s. Unlike many physicists of his generation who had studied abroad and established academic research schools in theoretical physics, Condon ended a productive career at Princeton University in 1937 to become associate director of research at the Westinghouse Electric and Manufacturing Company.

Westinghouse had just begun to support research that had no direct connection to the company's existing product lines. This research and development (R&D) strategy, which still anticipated the accrual of commercial benefits over the long term, gradually began to emerge among science-based firms in the late 1920s, and Westinghouse institutionalized it to a far greater extent than any of its rivals in the electrical industry. The company's research personnel and managers, however, possessed little or no expertise in theoretical physics. This lack of knowledge gave Condon the freedom to choose topics of investigation that matched his own academic interests. Research in nuclear physics and mass spectrometry, for example, produced good science, but yielded no viable commercial products. Only in the field of vacuum tube electronics did Condon's interests match those of the company, culminating in a successful wartime program to develop microwave radar technology. The notion of fundamental research as the driver of technological innovation, misguided as it was, became ingrained in corporate strategic planning after 1945.

Condon's liberal proclivities, which, to the consternation of his superiors at Westinghouse, became more pronounced at the end of the war, caught the attention of Henry Wallace, a New Dealer who had served as secretary of agriculture in the 1930s and Franklin Roosevelt's vice president during the war. A critic of big business and now secretary of commerce, Wallace handpicked Condon in 1945 to transform the department's National Bureau of Standards from an institution known for the maintenance of physical standards and routine testing of industrial materials into a national center for scientific research tuned to the needs of small businesses. Competition for resources from other federal research agencies and rigid civil service requirements, however, frustrated Condon's plans, forcing him to rely on the military departments for institutional support.

The permanent military preparedness and national security policies that accompanied the onset of the Cold War allowed Condon to access resources otherwise not available to the Bureau of Standards, but he did so at considerable professional and personal cost. Condon's association with Wallace bolstered critics on the House Un-American Activities Committee (HUAC), who had first publicly questioned his loyalty in 1947. Facing pressure to resign from Wallace's successors in the Department of Commerce, Condon left the Bureau in 1951 to become director of research at the Corning Glass Works. He departed in 1954 when the company tried unsuccessfully to have his recently revoked security clearance reinstated. Scarred by his security problems and recurring confrontations with HUAC, Condon quietly returned to academia, holding faculty appointments at Washington University and the University of Colorado.

[*See also* **Military, Science and Technology in the; Physics; Quantum Theory; Research and Development (R&D);** *and* **Science.**]

BIBLIOGRAPHY

Kevles, Daniel J. *The Physicists: The History of a Scientific Community in Modern America.* Cambridge, Mass.: Harvard University Press, 1987.

Lassman, Thomas C. "Government Science in Postwar America: Henry A. Wallace, Edward U. Condon, and the Transformation of the National Bureau of Standards, 1945–1951." *Isis* 96 (March 2005): 25–51.

Lassman, Thomas C. "Industrial Research Transformed: Edward Condon at the Westinghouse Electric and Manufacturing Company, 1935–1942." *Technology and Culture* 44 (April 2003): 306–339.

Wang, Jessica. "Science, Security, and the Cold War: The Case of E. U. Condon." *Isis* 83 (June 1992): 238–269.

                                    Thomas C. Lassman

CONSERVATION MOVEMENT

Launched in 1908 as a national crusade, the conservation movement involved the wide range of concerns later embraced by the environmental movement. Its intellectual origins date to the western land surveys of the nineteenth century, but it belongs to the realm of politics as much as to science. In the Progressive Era, two main branches, utilitarian and preservationist, emerged. Gifford Pinchot (1865–1946), a wealthy Pennsylvania forester who in 1898 became head of the federal government's small division of forestry (renamed the U.S. Forest Service in 1905), led the utilitarian wing. He advocated multiple-purpose use of the national forests. An astute strategist, he won the support of industries and interest groups eager to exploit the forests for profit by proposing a system of government regulation that eliminated wasteful competition and conflict. Close to President Theodore Roosevelt, Pinchot spearheaded an expanding program focused on "wise use" of natural resources, coordinated with other departments and agencies concerned with federal lands. The National Reclamation Act of 1902, establishing a federal agency to oversee irrigation projects in the Southwest, exemplified this objective.

Opposition to these policies arose in the western states most affected by them, and congressional opposition soon followed. The conservation movement was, in effect, Pinchot's public-relations crusade to create broad popular support for policies that until then had been promoted by narrow interest groups and bureau chiefs like himself—policies that western opponents identified with eastern corporations and elitist eastern bureaucrats. Through magazine articles and a 1908 White House conference, Pinchot crafted a public constituency for conservation.

The forest service's timber doctrine—of continual yield management (cutting no more timber than annual growth replaced)—became the foundation for a wildlife-preservation policy and the central doctrine of the U.S. Fish and Wildlife Service. A former forest service ranger, Aldo Leopold, carried over from forestry the notion that game populations were an agricultural crop to be harvested periodically to prevent overpopulation and preserve their range and food supply. Leopold also learned from Pinchot to cultivate an interlocking coalition of support groups constituting an effective wildlife lobby.

The preservationist wing of the movement, originally a part of Pinchot's grand concert of interests, split off after the Hetch Hetchy controversy (1913). This conflict focused on whether the Hetch Hetchy Valley, a part of Yosemite National Park, should be used as a water reservoir for San Francisco—the position Pinchot supported—or preserved for its natural beauty, as advocated by John Muir, a nature writer and activist well known to readers of mass-circulation magazines. Although the Hetch Hetchy Valley became a reservoir, disappointed preservationists helped in 1916 to establish the National Park Service, a federal bureau that rivaled the utilitarian forest service. The first director, Stephen Mather, proved as adept as Pinchot at buttressing his agency with the support of friendly industries and interest groups whose managers understood the commercial benefits awaiting those who helped meet the leisure needs of a rapidly growing urban middle class.

[*See also* Environmentalism; Fish and Wildlife Service, U.S.; Forestry Technology and Lumbering; Forest Service, U.S.; Geological Surveys; Leopold, Aldo; Muir, John; National Park System; *and* Sierra Club.]

BIBLIOGRAPHY

Dupree, A. Hunter. *Science in the Federal Government: A History of Policies and Activities to 1940.* Cambridge, Mass.: Belknap Press of Harvard University Press, 1957.

Hays, Samuel P. *Conservation and the Gospel of Efficiency: The Progressive Conservation Movement, 1890–1920.* Pittsburgh, Pa.: University of Pittsburgh Press, 1999.

Penick, James Lal, Jr. "The Progressives and the Environment: Three Themes from the First Conservation Movement." In *The Progressive Era,* edited by Louis L. Gould, pp. 115–131. Syracuse, N.Y.: Syracuse University Press, 1974.

James Lal Penick Jr.

CONTRACEPTIVES

See Birth Control and Family Planning.

COPE, EDWARD DRINKER

(1840–1897), paleontologist, zoologist, and neo-Lamarckian evolutionary theorist. Born to a prosperous Quaker family in Philadelphia, Cope showed a precocious interest in natural history, publishing his first scientific paper at 19. He traveled in Europe during the American Civil War, visiting collections and meeting prominent naturalists. Returning home, he married a distant cousin, Annie Pim, in 1865. They had a daughter, Julia, in 1866. Cope's father, who wanted his son to be a farmer, eventually provided the financing to enable him to become a gentleman naturalist.

Although largely self-taught, Cope benefitted from the sponsorship and training of Joseph Leidy—then America's foremost paleontologist—at Philadelphia's Academy of Natural Sciences. Membership in the Academy and access to its collections provided early opportunities for research. Cope made many fossil-collecting trips to the American West, often with federally funded geological surveys. He also employed talented fossil hunters.

In 1868, Cope proposed a theory whereby embryological development sometimes accelerates through preordained stages of growth, yielding a higher level of organization. Later individuals inherit this organization, creating, in effect, a new species. Contemporary naturalists took Cope's neo-Lamarckism seriously, yet his ideas are difficult to decipher—even Charles Darwin could make little sense of them.

Cope is infamous for his feud with paleontologist Othniel Charles Marsh. Exactly when or why their feud began is unclear, but by the 1870s they were bitter rivals fighting over access to fossils and priority of publication. In 1890, their private feud became a public scandal when they aired their grievances on the front page of the *New York Herald.* Both of their reputations were damaged by the scandal.

Cope was one of the most productive and influential American biologists of the nineteenth century and his scientific accomplishments are manifold. He published more than 1400 articles. He named and described more than one thousand vertebrate species. In 1895, he was elected president of the American Association for the Advancement of Science. He authored "Cope's rule," which states that mammalian species tend to increase in size over time. The zoology journal *Copeia* is named for him. Finally, Cope—along with Alpheus Hyatt—was America's leading neo-Lamarckian evolutionary theorist.

In the 1880s, disastrous mining investments drained Cope's finances, forcing him to accept a teaching position and to sell his fossil collections. He suffered for much of his adult life from chronic and mysterious infections. He worked through his illnesses, however, sometimes resorting to self-medication with formalin and belladonna. He died at his Philadelphia home on 12 April 1897.

[*See also* **Academy of National Sciences of Philadelphia; American Association for the Advancement of Science; Biological Sciences; Dinosaurs; Evolution, Theory of; Geological Surveys; Journals in Science, Medicine, and Engineering; Leidy, Joseph; Marsh, Othniel Charles; Paleontology; Science;** *and* **Zoology.**]

BIBLIOGRAPHY

Bowler, Peter J. "Edward Drinker Cope and the Changing Structure of Evolutionary Theory." *Isis* 68 (1977): 249–265.

Osborn, Henry Fairfield. *Cope: Master Naturalist.* Princeton, N.J.: Princeton University Press, 1931. This hagiographic biography remains the single best source on Cope's life and career.

Paul D. Brinkman

CORI, GERTY AND CARL

(1896–1957) and (1896–1984), respectively, a husband-and-wife team of biochemists whose work on carbohydrate metabolism helped to elucidate how an organism regulates its blood glucose concentrations. Both Carl and Gerty Cori were born in Prague. Carl spent his childhood in Trieste, where his father directed the Marine Biological Station. This setting afforded Carl an early appreciation of biology. The daughter of a successful businessman, Gerty was educated privately. In 1914, she passed the final examination required to study medicine at Carl Ferdinand University in Prague, where she met Carl. The two married in 1920, shortly after receiving their MD degrees. After struggling through poor nutrition, growing anti-Semitism, and unpaid postdoctoral work in postwar Austria, the Coris seized the opportunity to immigrate to Buffalo, New York, in 1922, where Carl had obtained a position at the New York State Institute for Study of Malignant Diseases. Their relocation enabled them to become U.S. citizens, pursue research without clinical duties, and begin their lasting and fruitful scientific collaboration.

At Buffalo, the Coris' early physiological experiments in animals attempted to quantify all processes involved in increasing or decreasing blood sugar. The Coris gradually developed sensitive new biochemical techniques for measuring glucose, glycogen, and other molecules. Such careful, quantitative analysis suggested a cycling of carbohydrates: muscle glycogen gets broken down to produce energy, leading to a build-up of lactic acid. The lactic acid diffuses into the bloodstream and to the liver, which converts it to glucose to be used or recycled back into glycogen by the muscle.

With the "Cori cycle" outlined by 1930, the couple investigated its underlying mechanism by isolating and characterizing the other intermediate molecules and enzymes involved. This shift in focus toward biochemistry and enzymology coincided with a move to St. Louis in 1931, where Carl assumed the chair of Washington University's pharmacology department. Gerty served as a research assistant, becoming professor of biochemistry after Carl opted to take the chair of that department in 1946. Throughout this period, the Coris' style of meticulous analysis, as well as the use of in vitro tissue preparations instead of whole animals, facilitated their biochemical investigations. They characterized many enzymes and molecules in the glycogenolytic pathway, including the glycogen-cleaving phosphorylase and the resulting intermediate glucose 1-phosphate. This work ultimately earned them the 1947 Nobel

Prize for Physiology or Medicine, making Gerty the first American woman to receive the award.

In the 1950s, the Coris continued their work in enzymology and studied its implications for human health and disease. Students and postdoctoral fellows flocked to their laboratory, including Severo Ochoa and Arthur Kornberg. Gerty Cori died in 1957 after a long struggle with an incurable anemia. After retiring in 1966, Carl moved to Boston, where he maintained a laboratory at Massachusetts General Hospital and studied how gene expression influences enzyme synthesis until his death in 1984.

Carl and Gerty Cori's collaborative research on carbohydrate metabolism reflects the promise and power that biochemistry held in the mid-twentieth century for understanding broader physiological and biological processes.

[*See also* **Biochemistry; Biological Sciences; Disease; Gender and Science; Nobel Prize in Biomedical Research;** *and* **Physiology.**]

BIBLIOGRAPHY

Cohen, Mildred. "Carl Ferdinand Cori, 1896–1984." *National Academy of Sciences Biographical Memoirs* 61 (1992): 79–109.

Cori, Carl. "The Call of Science." *Annual Review of Biochemistry* 38 (1969): 1–20.

Larner, Joseph. "Gerty Theresa Cori, 1896–1957." *National Academy of Sciences Biographical Memoirs* 61 (1992): 111–135.

Tulley Long

COTTON GIN

The cotton gin (short for engine), a mechanical device that separated cotton seed from cotton fiber, had been in existence in various forms since at least the fifth century when American inventor Eli Whitney introduced his version in Georgia in 1792. A Yale graduate who started his first business making nails at the age of 15, Eli Whitney responded to the complaints of local planters about the need for a machine to effectively separate green-seed, or short-staple cotton, which then had to be separated by hand before baling. Within 10 days, Whitney built a prototype, a hand-cranked drum cylinder that used teeth to pull cotton fiber through a metal grate to separate the seeds, passing it to a rotating brush that removed the lint. Whitney received a patent for the machine in 1794.

Whitney's invention was both revolutionary and timely. In the United States, long-staple cotton was being grown in the Deep South, but climate limited its expansion. Recently introduced short-staple cotton, or green-seed cotton, offered tremendous prospects, but the traditional method of separating seed from fiber, based on a device created in India centuries before called a churka, was ineffective with green-seed cotton. Whitney's cotton gin marked the beginning of a transition from high-quality, long-staple cotton to lower-quality short-staple cotton that could be produced in staggering quantities. A single machine could produce fifty pounds of cotton a day, compared with one pound per day produced by hand.

Low on funds, Whitney formed a partnership with Phineas Miller. Instead of selling gins, they built their own ginning operations, charging for services instead of selling licenses for gins. The plan was ineffective and, combined with technical flaws in the original design, a series of improvements by competitors led to a number of court battles, all unsuccessful until 1802. In 1807, the patent expired, with both Whitney and Miller failing to make any substantial financial gains.

The gins that were developed from Whitney's initial design helped transform the antebellum South, facilitating the rapid increase in cotton production and subsequent demands for African slaves and textile mills. By the outbreak of the Civil War in 1861, the United States was producing three quarters of the world's cotton, and approximately one of every three southerners was a slave.

Post–Civil War, large-scale factory ginning operations controlled larger portions of total cotton output. In the mid-1880s and early 1890s, Robert Munger of Texas built a pneumatic ginning system, which, followed by other enhancements, ultimately ended small-scale ginning.

[*See also* **Agricultural Technology;** *and* **Whitney, Eli.**]

BIBLIOGRAPHY

Eli Whitney's Patent for the Cotton Gin. National Archives & Records Administration. http://www.archives.gov/education/lessons/cotton-gin-patent/ (accessed 28 February 2012).
Lakwete, Angela. *Inventing the Cotton Gin: Machine and Myth in Antebellum America.* Baltimore: Johns Hopkins University Press, 2003.

Neil Dahlstrom

CREATIONISM

Although impossible to document quantitatively, the overwhelming majority of Americans rejected evolution until well into the twentieth century. They especially damned its Darwinian version, not only for its scientific shortcomings but also for its materialistic implications, its incompatibility with traditional readings of the Bible, and its apparent advocacy of ethics at odds with the teachings of Jesus. Representative of such creationist views was the Calvinist cleric-geologist Edward Hitchcock (1793–1864). Although he saw no theological barriers to admitting the antiquity of life on earth, a local flood, or even an Edenic creation limited to particular species, he believed that evolution made God unnecessary and promoted materialism. "After all," he insisted, "the real question is, not whether these hypotheses accord with our religious views, but whether they are true." It seemed preposterous to him—and to many other scientists and laymen—that anyone would claim that humans were "merely the product of transformation of the radiate monad through the mollusk, the lobster, the bird, the quadruped, and the monkey." Even more pointed was Charles Hodge (1797–1878), the leading Calvinist theologian in America, who devoted a small book to answering the question *What Is Darwinism?* (1874). The answer: "It is Atheism."

The issue of human evolution understandably provoked the greatest outrage. One conservative Protestant, H. L. Hastings, wrote a tract called *Was Moses Mistaken? or, Creation and Evolution* (1896), in which he addressed this "delicate" topic:

I do not wish to meddle with any man's family matters, or quarrel with any one about his relatives. If a man prefers to look for his kindred in the zoological gardens, it is no concern of mine; if he wants to believe that the founder of his family was an ape, a gorilla, a mud-turtle, or a monar [moner], he may do so; but when he insists that *I* shall trace *my* lineage in that direction, I say No *sir!*

The Great War in Europe (1914–1918) and the apparent breakdown of social norms at home focused increasing attention on the evils of evolution. No one contributed more to this backlash than the Presbyterian layman and thrice-defeated Democratic candidate for the presidency of the United States William Jennings Bryan (1860–1925). Early in 1922 Bryan helped to launch a crusade aimed at driving evolution out of the churches and schools of America. Since early in the century he had occasionally alluded to the silliness of believing in monkeys as human ancestors and to the ethical dangers of thinking that might makes right, but until the outbreak of the war he saw little reason to quarrel with those who disagreed with him. The European conflict, however, exposed the darkest side of human nature and shattered his illusions about the future of Christian society. Obviously something had gone awry, and Bryan soon traced the source of the trouble to the paralyzing influence of Darwinism on the human conscience. As he explained to one young correspondent, "The same science that manufactured poisonous gases to suffocate soldiers is preaching that man has a brute ancestry and eliminating the miraculous and the supernatural from the Bible." By substituting the law of the jungle for the teaching of Christ, evolution threatened the principles he valued most: democracy and Christianity.

Anti-evolutionists such as Bryan typically went out of their way not to appear antiscience. The Northern Baptist minister William Bell Riley, founder of the World's Christian Fundamentals

Association, outlined the reasons why fundamentalists opposed the teaching of evolution. "The first and most important reason for its elimination," he explained, "is the unquestioned fact that evolution is not a science; it is a hypothesis only, a speculation." Bryan often made the same point, defining true science as "classified knowledge… the explanation of facts." This view came straight from the dictionary. "What is science?" asked one Holiness preacher who opposed evolution. "To the dictionaries!" he answered. There one learned that science was "certified and classified knowledge."

Unfortunately for the anti-evolutionists, no one active in their movement possessed so much as a master's degree in biology. S. James Bole (1875–1956), professor of biology at fundamentalist Wheaton College, Illinois, stood virtually alone as a creationist with advanced training in biology. After earning an AM degree in education from the University of Illinois, he had gone on to study pomology (fruit culture) at the Illinois school of agriculture—and abandoned evolution for creation and biblical inerrancy. Two other prominent anti-evolutionists had exposed themselves to science while studying medicine. Arthur I. Brown (1875–1947), an American-born surgeon who had settled in Vancouver, British Columbia, became, in his own estimation, "one of the leading surgeons of the Pacific Coast." His handbills touted him as "one of the best informed scientists on the American continent," and hosts routinely introduced him as a famous scientist, perhaps the "greatest scientist in all the world." By the early 1920s he was publishing anti-evolution pamphlets on such subjects as *Evolution and the Bible* (1922) and *Men, Monkeys, and Missing Links* (1923). The other anti-evolutionist with medical training was Harry Rimmer (1890–1952), a young Presbyterian minister and self-styled "research scientist" who had attended San Francisco's Hahnemann Medical College, a small homeopathic institution that required only a high-school diploma or its "equivalent" for admission. A lively debater, Rimmer insisted that the Bible contained no scientific errors.

According to the editor of *Science*, "the principal scientific authority of the Fundamentalists" was the Canadian George McCready Price. As an adolescent he and his widowed mother had joined the Seventh-Day Adventist church, founded by the prophetess Ellen G. White. During one of her trance-like visions God had allowed her personally to witness the creation, which she saw and related how it had occurred in six literal days of twenty-four hours. On other occasions God revealed the role played by Noah's flood in laying down the fossil-bearing rocks—in a single year, not millions. Collapsing the geological column into the work of one year deprived evolutionists of the time they needed and focused attention on the deluge rather than on the creation. Price repeatedly insisted that the idea of successive geological ages represented the crux of the evolution problem. He thus devoted comparatively little attention to biological questions. "What is the use of talking about the origin of species," he asked as early as 1902, "if geology can not prove that there has actually been a succession and general progress in the life upon the globe?" Besides, the development of Mendelian genetics, which seemingly allowed for only "definite and predictable" variations, had so eroded confidence in the efficacy of natural selection that Price felt he needed only to write the "funeral oration" for Darwinism. As he observed in *The Phantom of Organic Evolution* (1924), his most extensive critique of the *biological* arguments for evolution, "A dead lion needs no bullets."

During the period between the 1920s and the 1960s the most pressing biological question among creationists centered on the fixity of species since Eden. Most flood geologists, as Price's followers came to be known, eagerly welcomed "every bit of *modification during descent* that can reasonably be asserted." The underlying problem was the shortage of space on Noah's ark. "If we insist upon fixity of species we make the Ark more crowded than a sardine can," argued one flood geologist. "If we agree to all [variation] that can be demanded, we simplify the Ark problem greatly," because all present-day species could have descended from relatively few passengers on the ark. Harold W. Clark, a student of Price's who had gone on to earn a master's degree in biology from the University of California, Berkeley, and who in the late 1920s began promoting his mentor's views as "creationism" rather than anti-evolutionism, defended limited Darwinian natural selection—within genera, families, and even orders—against the "extreme creationism" of those who insisted that God had created every species. In 1941 Frank C. Marsh, another former student of Price's and the holder of a PhD in biology from the University of Nebraska, suggested naming the originally created kinds *baramins*.

Despite Price's prominence as an anti-evolutionist, few creationists outside his Adventist communion saw the necessity of abandoning their traditional old-earth views. That began to change with the publication in 1961 of *The Genesis Flood*, by John C. Whitcomb Jr., an Old Testament scholar, and Henry M. Morris, a hydraulic engineer. Their sensational book carried Price's hitherto marginal flood geology to masses of Bible-believing Christians. In attacking evolution, the authors followed Price in focusing on its geological and paleontological foundations rather than on biology. The success of *The Genesis Flood* led to the formation in 1963 of the Creation Research Society (CRS), dedicated to promoting a "young-earth" version of creationism, which limited the history of life on earth to no more than 10,000 years. The phrase "creation research" in the society's name bordered on the oxymoronic. Even Walter E. Lammerts (1904–1996), the principal founder of the society and the person who coined the term, admitted the impossibility of researching creation because "we were not there to watch God do it!" As he explained to one perplexed questioner, the CRS investigated extant evidence of God's creative activity, not the act of creation itself. For years creationists tended to expend their limited resources on library research (to find fallacies or inconsistencies in the writings of evolutionists) or on low-cost field studies (to find evidence that might support flood geology).

In the early 1970s the partisans of flood geology, hoping to gain entrance into public-school classrooms, began referring to their Bible-based views as "scientific creationism" or "creation science." In his popular textbook *Scientific Creationism* (1974), Morris argued that creationism could "be taught without reference to the book of Genesis or to other religious literature or to religious doctrines." Although Morris and his creationist colleagues occasionally appealed to scientific and philosophical authorities, for the most part they worked outside the context of established science and philosophy. References to the names of scientists who questioned evolutionary orthodoxy served largely as literary ornaments.

In *Epperson v. Arkansas* in 1968 the U.S. Supreme Court ruled that banning the teaching of human evolution violated the constitutional requirement of separating church and state. In response, two states, Arkansas and Louisiana, passed laws requiring the balanced treatment of creation science and evolution science in public schools. At first glance the ensuing legal battles between creationists and evolutionists seemed to provide one more example of the so-called warfare between science and religion, but appearances were deceiving. For example, in the Arkansas trial in 1981, the plaintiffs, who opposed creation science, overwhelmingly represented religious organizations; virtually all the experts testifying in support of creationism possessed graduate degrees in science.

Scientific creationism continued to flourish, but by the mid-1990s the focus of attention among creationists and evolutionists alike was shifting to a new form of anti-evolutionism called "intelligent design" (ID). From a small but well-funded base at the Discovery Institute in Seattle, these anti-evolutionists sought to foment a veritable "intellectual revolution." Unlike creation scientists, the advocates of ID largely ignored geological arguments associated with the age of the earth and the extent of Noah's flood and focused instead on finding biological evidence of a god-like designer. Most controversial of all, they tried "to reclaim science in the name of God" by rewriting the rules of science to allow the inclusion of supernatural explanations of phenomena, insisting that "*The ground rules of science have to be changed.*"

Leading this effort was a Presbyterian lawyer on the faculty of the University of California, Berkeley, Phillip E. Johnson (b. 1940). Although he targeted evolution, he identified the real enemy as scientific naturalism, which ruled all God talk out of science. Most troubling of all was the widespread acceptance of so-called methodological naturalism among even evangelical Christians, who, despite their beliefs, left God at the laboratory door and refrained from invoking the supernatural when trying to explain the workings of nature. In contrast to "metaphysical naturalism," which denied the existence of a transcendent God, methodological naturalism implied nothing about God's existence and activities. But to Johnson, it smacked of thinly veiled atheism.

In one of the canonical ID texts, *Darwin's Black Box: The Biochemical Challenge to Evolution* (1996), the biochemist Michael J. Behe argued that biochemistry had "pushed Darwin's theory to the limit…by opening the ultimate black box, the cell, thereby making possible our understanding of how life works." The "astonishing complexity of subcellular organic structure" led him to conclude—on the basis of scientific data, he asserted, "not from sacred books or sectarian beliefs"—that intelligent design had been at work. "The result is so unambiguous and so significant that it must be ranked as one of the greatest achievements in the history of science," he proudly declared. "The discovery [of intelligent design] rivals those of Newton and Einstein, Lavoisier and Schroedinger, Pasteur and Darwin." As newspapers and magazines spread the news of Behe's discovery of what he called "irreducibly complex" organic structures—such as the bacterial flagellum, which propels microscopic organisms—he won recognition as a modern-day William Paley (1743–1805), the famous natural theologian of the early nineteenth century.

The first major legal test of intelligent design began with the Dover (Pennsylvania) Area School District Board's decision to make students "aware of gaps/problems in Darwin's theory and of other theories of evolution including, but not limited to, intelligent design." To implement this action, the board instructed ninth-grade biology teachers to read a statement to their classes describing Darwinism as "a theory…not a fact." At a subsequent trial to determine the constitutionality of this directive, in 2005, Behe, the star witness for the defense, acknowledged that intelligent design lacked a mechanism for explaining how designed structures arose. Shortly before Christmas Judge John E. Jones III (b. 1955) handed down his verdict, excoriating the Dover school board for its actions, which he described as a "breathtaking inanity." Although a conservative Republican and a practicing Lutheran, Jones ruled that ID was "not science" because it invoked "supernatural causation" and failed "to meet the essential ground rules that limit science to testable, natural explanations." Thus, even this nonbiblical version of creationism violated the establishment clause of the Constitution. Judge Jones rejected as "utterly false" the assumption "that evolutionary theory is antithetical to a belief in the existence of a supreme being and to religion in general." His conclusion: "It is unconstitutional to teach ID as an alternative to evolution in a public school science classroom."

Despite the turn-of-the-millennium efflorescence of intelligent design, young-earth creationism continued to flourish. Since its founding in 1972, Henry M. Morris's Institute for Creation Research (ICR) had led the way. But after a quarter century it was eclipsed by Answers in Genesis (AiG), a Kentucky-based operation (located just south of Cincinnati) begun in 1994 by the Australian Ken Ham, an alumnus of the ICR. As creationism's newest star, he packed in audiences almost everywhere he went, speaking to well over 100,000 people a year. In less than a decade he and his AiG colleagues had created a network of AiG organizations around the world. In 2007 AiG opened a $27-million, state-of-the-art creation museum. Buoyed by its success, AiG rolled out plans to build a Noah's Ark theme park (including a Tower of Babel), costing more than $150 million. The governor of Kentucky, a Democrat, announced his support for this jobs-creating project; in his 2012 budget he included a $43-million tax break for the creationist park while slashing the budget for higher education by 6.4 percent. The state legislature voted an additional $2 million to improve the roads leading to the park.

In 2010 the Gallup organization, which had been tracking belief in creation and evolution since 1982, found that 40 percent of Americans affirmed that "God created humans in their present form about 10,000 years ago." Some

38 percent believed that God had guided evolution to create humans; 16 percent thought that God had nothing to do with the evolution of humans.

[*See also* Biological Sciences; Evolution, Theory of; Genetics and Genetic Engineering; Geology; Higher Education and Science; High Schools, Science Education in; Hitchcock, Edward; Paleontology; Religion and Science; *and* Science.]

BIBLIOGRAPHY

Clark, Constance A. *God—Or Gorilla: Images of Evolution in the Jazz Age.* Baltimore: Johns Hopkins University Press, 2008.

Davis, Edward B. "Science and Religious Fundamentalism in the 1920s." *American Scientist* 93 (2005): 253–260.

Forrest, Barbara, and Paul Gross. *Creationism's Trojan Horse: The Wedge of Intelligent Design.* New York: Oxford University Press, 2004.

Gallup. "Four in 10 Americans Believe in Strict Creationism." http://www.gallup.com/poll/145286/four-americans-believe-strict-creationism.aspx (accessed 18 October 2012).

Israel, Charles A. *Before Scopes: Evangelicalism, Education, and Evolution in Tennessee, 1870–1925.* Athens: University of Georgia Press, 2003.

"Kentucky's Noah's Ark Theme Park: When a Recession, Culture, and Political Pandering Collide." http://www.dailykos.com/story/2012/05/15/1091214/-Kentucky-s-Noah-s-Ark-Theme-Park-When-a-Recession-Culture-and-Political-Pandering-Collide (accessed 30 January 2013).

Laats, Adam. *Fundamentalism and Education in the Scopes Era: God, Darwin, and the Roots of America's Culture Wars.* New York: Palgrave Macmillan, 2010.

Larson, Edward J. *Summer for the Gods: The Scopes Trial and America's Continuing Debate over Science and Religion.* New York: Basic Books, 1997.

Larson, Edward J. *Trial and Error: The American Controversy over Creation and Evolution,* 3d ed. New York: Oxford University Press, 2003.

Lienesch, Michael. *In the Beginning: Fundamentalism, the Scopes Trial, and the Making of the Antievolution Movement.* Chapel Hill: University of North Carolina Press, 2007.

Moran, Jeffrey P. *American Genesis: The Evolution Controversies from Scopes to Creation Science.* New York: Oxford University Press, 2012.

Moore, James. "The Creationist Cosmos of Protestant Fundamentalism." In *Fundamentalisms and Society: Reclaiming the Sciences, the Family, and Education,* Vol. 2 of *The Fundamentalism Project,* edited by Martin E. Marty and R. Scott Appleby, pp. 42–72. Chicago: University of Chicago Press, 1993.

Numbers, Ronald L. *The Creationists: From Scientific Creationism to Intelligent Design,* expanded ed. Cambridge, Mass.: Harvard University Press, 2006.

Numbers, Ronald L. "The Scopes Trial: History and Legend." In *Darwinism Comes to America,* edited by Ronald L. Numbers, pp. 76–91, 187–192. Cambridge, Mass.: Harvard University Press, 1998.

Pennock, Robert T. *Tower of Babel: The Evidence against the New Creationism.* Cambridge, Mass.: MIT Press, 1999.

Ruse, Michael. *The Evolution–Creation Struggle.* Cambridge, Mass.: Harvard University Press, 2005.

Shapiro, Adam. *Trying Biology: Scopes, Textbook Reform and School Antievolution in America.* Chicago: University of Chicago Press, 2013.

Talbot, Margaret. "Darwin in the Dock: Intelligent Design Has Its Day in Court." *New Yorker,* December 5, 2005, pp. 66–77.

Toumey, Christopher P. *God's Own Scientists: Creationists in a Secular World.* New Brunswick, N.J.: Rutgers University Press, 1994.

Ronald L. Numbers

D

DALTON, JOHN CALL, JR.

(1825–1889), pioneer physiologist, was born in Chelmsford, Massachusetts. After receiving his undergraduate and medical degrees from Harvard, he traveled to Paris for postgraduate training. Dalton's exposure to French physiologist Claude Bernard led him to seek a career in teaching and research rather than practice. In 1855, he was chosen as chair of physiology at the College of Physicians and Surgeons of New York. Unlike other medical school professors, Dalton did not combine practice with his academic duties. For this reason he is considered America's first professional physiologist.

Dalton spent his career at the elite New York institution, where he taught thousands of medical students. Unlike other American physiology professors, Dalton supplemented his lectures with live animal demonstrations and experiments (vivisection). This practice was controversial, especially

after the founder of the American Society for the Prevention of Cruelty to Animals, Henry Bergh, focused the public's attention on it. Dalton became his generation's most persistent defender of vivisection. He spoke on the subject and published two books that contained examples of how animal experiments led to discoveries that changed medical practice.

Dalton's main research interests related to the physiology of the digestive and nervous systems. His most important contribution to the literature was his textbook *A Treatise on Human Physiology*, which first appeared in 1859 and went through seven editions. Dalton's most imposing publication was his *Topographical Anatomy of the Brain* (1885), a huge atlas illustrated with actual photographs mounted in the book.

Dalton was one of a small but influential group of medical scientists and scientifically oriented physicians who considered the endowment of laboratories and research to be a vital step in improving American medical education. He served as president of the College of Physicians and Surgeons of New York from 1883 until his death on 12 February 1889.

[*See also* **Animal and Human Experimentation; Medical Education; Medicine;** *and* **Physiology.**]

BIBLIOGRAPHY

Fine, Edward J., Tara Manteghi, Sidney H. Sobel, and Linda Lohr. "John Call Dalton, Jr., M.D., America's First Neurophysiologist." *Neurology* 55 (2000): 859–864.

Fye, W. Bruce. "John Call Dalton, Jr., Pioneer Vivisector and America's 'First Professional Physiologist.'" In *The Development of American Physiology: Scientific Medicine in the Nineteenth Century*, pp. 15–53, 237–244. Baltimore: Johns Hopkins University Press, 1987.

W. Bruce Fye

DAMS AND HYDRAULIC ENGINEERING

In colonial America, small streams powered rural grist mills and sawmills. In the early nineteenth century, entrepreneurs built factories powered by the flow of large rivers; the Lowell mills in Massachusetts, where capitalists diverted the Merrimack River to turn large waterwheels and energize textile looms, became a particularly famous example of this. Builders also erected dams to inundate rocky stretches of rivers and facilitate the operation of transportation canals. By midcentury, the Army Corps of Engineers was actively removing "snags" and otherwise improving the navigability of major waterways such as the Ohio River.

Urban growth prompted large-scale dam construction. In 1887 San Francisco began receiving water from a dam 146 feet high; in the 1890s New York City began building a 297-foot dam. In the West, privately financed canal and reservoir projects led the way in "reclaiming" and cultivating desert land through irrigation. In 1902 the U.S.

Reclamation Service (later renamed the Bureau of Reclamation) started building large western dams. As hydroelectric power systems grew in the 1890s, corporations began transmitting electricity from remote waterpower sites to distant, urban "load centers." Niagara Falls became the site of a renowned early hydroelectric power plant. By 1910 such systems were operating throughout America.

During Theodore Roosevelt's presidency, plans for "conserving" water resources became a prominent component of Progressivism. Despite conflicts over the proper roles of government and private enterprise, Congress by the mid-1920s had authorized the Corps of Engineers to devise multiple-use strategies for developing America's rivers. "Multipurpose" planning garnered additional attention because of southern California's desire to dam the Colorado River for irrigation, power generation, and municipal water supply. In 1928 Congress authorized $177 million to build the 726-foot-high Boulder (later Hoover) Dam on the Colorado and to fund other improvements desired by California legislators. Catastrophic floods along the Mississippi River in 1927 also fostered federal support for water-control projects.

Large dams constituted an important part of President Franklin Delano Roosevelt's New Deal; starting in 1933 with the Tennessee Valley Authority (a government agency that supplanted private electric companies in parts of the South), the Roosevelt administration championed water projects nationwide. These included the Bureau of Reclamation's Grand Coulee Dam in Washington State and the Corps of Engineers' Fort Peck Dam in Montana. During the 1940s and 1950s, dam building emerged as a staple of the national economy. Although sometimes decried as "pork-barrel" waste, water projects proved politically effective in infusing federal funds into local economies.

By the 1960s, thousands of dams impounded rivers throughout America, prompting fears that too many wild rivers and fragile wetlands had been destroyed by reservoirs of limited social value. As early as 1909, naturalist John Muir had (unsuccessfully) opposed construction of the Hetch Hetchy Dam in Yosemite National Park. By the time environmentalism became a political force in the 1970s, dams were no longer considered an unalloyed public good. By the turn of the century, efforts to reduce water consumption were superseding interest in new water projects, and the demolition of dams to aid spawning fish was gaining support. Dams remained an important part of the American landscape, but little new construction was planned.

[*See also* **Army Corps of Engineers, U.S.; Canals and Waterways; Conservation Movement; Electricity and Electrification; Environmentalism; Hoover Dam; Hydroelectric Power; Lowell Textile Mills; Muir, John; National Park System; Rivers as Technological Systems; Sierra Club;** *and* **Tennessee Valley Authority.**]

BIBLIOGRAPHY

Hunter, Louis C. *A History of Industrial Power in the United States, 1780–1930: Water Power.* Charlottesville: University Press of Virginia, 1979.
Jackson, Donald C. *Building the Ultimate Dam: John S. Eastwood and the Control of Water in the West.* Lawrence: University Press of Kansas, 1995.
Malone, Patrick M. *Waterpower in Lowell: Engineering and Industry in Nineteenth-Century America.* Baltimore: Johns Hopkins University Press, 2009.

Donald C. Jackson

DANA, JAMES DWIGHT

(1813–1895), geologist, mineralogist, zoologist. Born in Utica, New York, in 1813, James Dwight Dana was the eldest of 10 children of Harriet and James Dana, a hardware merchant. He grew up in a religious household and showed an early interest in mathematics, natural history, music (piano and guitar), and art. In 1830, Dana went to study science with the renowned teacher Benjamin Silliman Sr. at Yale College. He graduated in three years and took an appointment as mathematics instructor to midshipmen in the U.S. Navy. During a 16-month cruise in the Mediterranean, Dana visited France and Italy and climbed Mt. Ve-

suvius, the account of which became his first scientific paper. In 1836, he returned to Yale as Silliman's teaching assistant, a coveted position that provided time and resources for him to complete *A System of Mineralogy* (1837), an ambitious attempt to classify all minerals by their physical characteristics. He revised his *Mineralogy* in 1844 and included an original Latin binomial nomenclature, but in 1850 Dana abandoned that method in favor of the dominant chemical classification system. Dana's responsiveness to scientific change revealed a flexible and synthetic mind and ensured his *Mineralogy* would remain the standard reference.

But Dana also had a dogmatic side. In 1838, he underwent a religious conversion, and in search for the Supreme Architect's plan in nature, he signed on as the mineralogist and geologist of the U.S. Exploring Expedition to the South Seas under the imperious Captain Charles Wilkes. For four years, he sailed the Pacific; explored coral reefs, atolls, and volcanoes; and took on the duties of Expedition's marine zoologist. The round-the-world voyage brought back vast collections for Dana to describe and classify and in 1844 he returned to Yale to write. Over the next 12 years he produced three massive volumes: *Zoophytes* (1846), *Geology* (1849), and *Crustacea* (1854). He also became Silliman's coeditor of the *American Journal of Science* and his son-in-law when he married Henrietta Silliman. In 1855, he became the Silliman Professor of Natural History at Yale.

The Wilkes Expedition made Dana's reputation; he became president of the American Association for the Advancement of Science in 1854–1855. It also gave him a global perspective on geological processes, which underlay his *Manual of Geology* (1862). America, for Dana, was the exemplar of geological history because its form and structure were simple (meaning unitary, not unsophisticated). The interior plains, periodically flooded by rising sea levels, were bordered by mountains thrust up when a cooling and contracting earth pushed the ocean basins against the permanent continent. Dana was a chauvinist and a catastrophist. He thought the American fossil record clearly revealed mass exterminations and divine creations of species as well as the overall progress of life leading to humans. For Dana, there was no conflict between the truths of evangelical Protestantism and science. Not surprisingly, he initially rejected Darwin's theory of evolution. But in 1859, Dana suffered the first of many breakdowns in health. He was unable to write and teach or read Darwin. Dana would not accept evolution until the 1870s, by which time his leadership in American science had been eclipsed.

[*See also* **American Association for the Advancement of Science; Creationism; Evolution, Theory of; Geology; Journals in Science, Medicine, and Engineering; Mathematics and Statistics; Paleontology; Religion and Science; Silliman, Benjamin; Wilkes Expedition;** *and* **Zoology.**]

BIBLIOGRAPHY

Lucier, Paul. *Scientists and Swindlers: Consulting on Coal and Oil in America, 1820–1890.* Baltimore: Johns Hopkins University Press, 2008.

Prendergast, Michael L. "James Dwight Dana: The Life and Thought of an American Scientist." PhD diss. University of California, Los Angeles, 1978.

Rossiter, Margaret. "A Portrait of James Dwight Dana." In *Benjamin Silliman and His Circle,* edited by Leonard G. Wilson, pp. 105–127. New York: Science History Publications, 1979.

Paul Lucier

DAVENPORT, CHARLES BENEDICT

(1866–1944), zoologist, geneticist, eugenicist, and science administrator. Born near Stamford, Connecticut, Davenport was first tutored at home and in 1879 attended Brooklyn Collegiate and Polytechnic Institute. After a brief period as an engineer, he obtained an AB (1889) and a PhD in zoology (1892) from Harvard. In 1894 Davenport married Gertrude Crotty, a biologist studying at the forerunner of Radcliffe College.

Until 1899 Davenport held an instructorship at Harvard and then a post at the University of Chicago. From 1904 to 1934 he directed the Station for Experimental Evolution at Cold Spring Harbor, New York, which he helped establish with support from the Carnegie Institution of Washington. In 1910, Davenport founded the Eugenics Record Office (ERO) with funds first from Mary Harriman, the widow of the founder of the Union Pacific Railroad, and starting in 1918 until its closing in 1939 from the Carnegie Institution. Between 1910 and 1924, ERO trained about two hundred women with biology degrees as eugenic field workers and employed them to collect information about the inheritance of diverse traits in individuals and families in several communities.

Best known as a skilled scientific administrator, during his prolific career Davenport's main scientific concerns centered on genetics and eugenics. In the area of genetics, Davenport was one of the first American biologists to support both statistical methods to study the distribution of traits in populations and experimental methods to study the patterns obtained in breeding crosses. Influenced by British statistician Karl Pearson's biometrical approach to the study of variations in populations, Davenport published *Statistical Methods with Special Reference to Biological Variation* (1899), a book that introduced biometrical methods in the United States. He realized the significance of the newly rediscovered work of Gregor Mendel, whose plant breeding crosses showed the existence of patterns in the inheritance of certain traits of organisms. Contrary to other authors who viewed the biometrician and the Mendelian approaches as incompatible, Davenport saw the value of combining both to address the problems of variation and inheritance in a research program he called experimental evolution.

Like many students of inheritance in his time, Davenport was also eager to apply knowledge of heredity to improve the human race. In partnership with his wife, Davenport studied the inheritance of human traits such as eye color and hair forms. He also argued that broader characteristics such as temperament, intelligence, feeblemindedness, and criminality were inherited. Convinced of the influence of heredity on social issues, he became a prominent supporter of eugenics, the program to research human inheritance and implement policies to improve the human race by controlling breeding. His many writings in this area included *Eugenics* (1910) and *Heredity in Relation to Eugenics* (1911).

A tireless worker, Davenport promoted genetics and eugenics as an editor and organizer of science. He edited several journals, including *Genetics* and the *Journal of Physical Anthropology*. He became the chairman of the eugenics section of the American Breeders Association, the president of the Third International Congress on Eugenics held in New York City in 1932, the founder of six diverse associations, and the president or vice president of ten societies. He published over four hundred titles, including eighteen books. He was elected to the American Philosophical Society in 1907 and the National Academy of Science in 1912.

[*See also* **American Philosophical Society; Biological Sciences; Eugenics; Gender and Science; Genetics and Genetic Engineering; Journals in Science, Medicine, and Engineering; National Academy of Sciences; Science;** *and* **Zoology.**]

BIBLIOGRAPHY

Allen, G. E. "The Eugenics Record Office at Cold Spring Harbor, 1910–1940: An Essay on Institutional History." *Osiris* 2d ser., 2 (1986): 225–264.

MacDowell, E. Carleton. "Charles Benedict Davenport, 1866–1944: A Study of Conflicting Influences." *Bios* 17 (1946): 2–50.

Riddle, Oscar. "Charles Benedict Davenport (1866–1944)." *Biographical Memoirs* 25 (1948): 75–110.

Rosenberg, Charles E. "Charles B. Davenport and the Beginnings of Human Genetics." *Bulletin of the History of Medicine* 35 (1961): 266–276.

Marga Vicedo

DDT

See Pesticides.

DEAFNESS

Medical treatments and technological interventions to cure deafness are premised on the notion that, as an editor put it in the *National Magazine* in July 1927, "deafened humanity" "has…hoped for hearing throughout the centuries" (p. 489). That basic idea—that the deaf want nothing more than to hear—can be found guiding otologists (hearing specialists) and others interested in treating deafness throughout American history. Many Deaf—with a capital "D"—people, however, reject the central premise of treatment. They argue that the Deaf are a linguistic minority, and from that perspective, a focus on medical and technological treatments of individuals is inappropriate. The Deaf call for less attention to individual medical or technological intervention and greater effort to break down the barriers to Deaf participation in American life.

Treatments and Technologies. Deafness has many causes, complicating any simple approach to treatment, whether medical or technological. Causes include heredity, genetic mutation, injury, disease, exposure to loud noise, and aging. Deafness can be congenital (occurring from birth) or adventitious (arising after birth). Otologists further distinguish between conductive deafness (mainly problems with the ear drum or with the three tiny bones of the middle ear, the malleus or hammer, the incus or anvil, and the stapes or stirrup), versus nerve deafness (problems in the inner ear where the cochlear nerve passes sound as electrical impulses to the brain or problems with the brain itself understanding the impulses received from the cochlea). Deafness is usually rated along a scale from mild to profound, can be temporary or permanent, and can occur in one ear (unilateral) or both ears (bilateral).

Given the complexity of causes, a variety of medical approaches to treating deafness have developed. For instance, a mastoidectomy, or the removal of infected air-containing spaces in the skull behind the ear, was first successfully performed in the 1770s. Although antibiotics have rendered mastoidectomies largely unnecessary, the stapedectomy operation pioneered by Dr. John J. Shea Jr. in the 1950s is still commonly used to treat deafness resulting from loss of mobility of the stapes bone. Twenty-first-century stapedectomies commonly involve the use of lasers that were unavailable to Shea. These are just two of the many surgical treatments for deafness.

Perhaps the most common treatment for deafness or hearing loss, however, is a technological one, the hearing aid. Hearing aids have a history dating back to at least the seventeenth century. Prior to the late nineteenth century, these devices were crude—a wad of cotton with a string attached, inserted into the ear canal to function as an artificial ear drum, for example—or unwieldy—the familiar nineteenth-century trumpet- or horn-style hearing aid. Truly effective hearing aids that people wanted to use were the product of late-nineteenth- and early-twentieth-century experimentation in areas unrelated to medicine: Alexander Graham Bell's experiments with the electrical transmission of speech, Thomas Edison's invention of the carbon transmitter that converted sounds to electrical signals, and Guglielmo Marconi's work on wireless transmission of sound. The first electric hearing aids, produced in the 1890s, were so large they had to sit on a table. Because few people wanted everyone to know they were deaf or hard of hearing, the devices often were designed to be built into chairs, hidden under floral arrangements, or otherwise concealed. The breakthrough to miniaturization came with the development of the transistor after World War II. Hearing aids then could be contained in small boxes, which a man could slip in his jacket pocket or a woman could hook over her brassiere. Soon smaller models that fit in the frame of a pair of eyeglasses or over the ear came out. Since the 1990s, digitization has further reduced the size of hearing aids and improved their function.

The next step in the technology was the development of the cochlear implant. Cochlear implants send sound as electrical signals directly to the cochlear nerves, bypassing the middle ear. The device consists of two parts. One part is a removable microphone and speech processor that (usually) fits behind the ear like a more conventional hearing aid. The other part is a transmitting coil permanently affixed to the side of the patient's head with a wire extended inside the skull into the cochlea (hence cochlear *implant*). First approved for use on adult otology patients in the United States in 1984, the age for implantation has been gradually lowered; in the year 2000, the age for implantation was lowered to 12 months.

Neither hearing aids nor cochlear implants are a panacea, however. Neither is appropriate for every deaf or hard-of-hearing individual, and both require education and training to maximize their effectiveness for individual patients. Some patients get no benefit at all. Hearing aids and implant technology work most successfully for patients adventitiously deafened (i.e., who had some prior experience of hearing).

Deaf Culture. Beyond questions of effectiveness, for many in the Deaf community the push for implant technology raises the specter of eugenics. In the late nineteenth and early twentieth centuries, eugenicists—including Alexander Graham Bell, whose inventions contributed to the technology necessary for hearing aids and cochlear implants—targeted (congenital) deafness as one of the "defects" they wanted to eliminate from the human race. Bell stated the case most clearly in an address he gave to the National Academy of Sciences in 1883, the same year Francis Galton coined the term eugenics, titled "Memoir upon the Formation of a Deaf Variety of the Human Race." Believing the problem to be intermarriage and reproduction among the congenitally deaf, Bell and others sought ways to break down deaf communities that particularly developed around schools for the deaf. A teacher by profession, Bell promoted oralism (lip-reading and audible speech) over manualism (sign language) as the best approach to deaf education. He believed giving the deaf language skills to communicate with mainstream Americans would enable them to participate more fully in American life. Although oralism dominated deaf education in the United States by the end of the nineteenth century, Deaf people continued to use sign language among themselves, an important point in the emergence of a more activist Deaf culture, particularly in the latter part of the twentieth century.

Bell and the eugenicists were wrong; a marriage between two congenitally deaf individuals is no more likely to yield congenitally deaf offspring than any other marriage. For many adherents of Deaf culture, however, the late-twentieth- and early-twenty-first-century emphasis on cochlear implant technology represents a new eugenic push. Most controversially, they object to allowing hearing parents to make the decision to implant their congenitally deaf children before those children can make their own decisions about whether to become part of Deaf culture. Likewise, although the emerging technology of genetic testing of embryos for congenital defects will enable Deaf couples to choose to have nonhearing children, the Deaf fear that the real impact will be to reduce the number of Deaf children being born. The result will be the elimination of Deaf culture, amounting to cultural genocide.

This necessarily brief history of medical and technological interventions in the area of deafness reminds us that people with impairments are people first and foremost. The definition of deafness as a disability is less the product of a hearing impairment and more the product of social constraints and the tendency to medicalize anything that differs from the social norm. Despite impressive advances and the seemingly benign nature of technology, the Deaf and other people with disabilities have their own ideas about how they can best participate in mainstream society, and these ideas should be respected.

[*See also* **Bell, Alexander Graham; Blindness, Assistive Technologies and; Disease; Edison, Thomas; Electricity and Electrification; Eugenics; Genetics and Genetic Engineering; Keller, Helen; Medicine; Medicine and Technology;** *and* **Surgery.**]

BIBLIOGRAPHY

Baynton, Douglas C. *Forbidden Signs: American Culture and the Campaign against Sign Language*. Chicago: University of Chicago Press, 1996. The best discussion of oralism versus manualism in deaf education, including the role of Alexander Graham Bell.

Chu, Eugene, and Robert K. Jackler. "The Artificial Tympanic Membrane (1840–1910): From Brilliant Innovation to Quack Device." *Otology & Neurology* 24, no. 3 (2003): 507–518. Discusses some of the earliest technological interventions in treating deafness, suggesting (through implication) why the Deaf have reason to question medicalization.

Deafness in Disguise: Concealed Hearing Devices of the 19th and 20th Centuries. Washington University School of Medicine. http://beckerexhibits.wustl.edu/did/index.htm (accessed 9 April 2012). A fascinating and lavishly illustrated look at the evolution of hearing aids.

Padden, Carol, and Tom Humphries. *Inside Deaf Culture*. Cambridge, Mass.: Harvard University Press, 2005. A very readable broad introduction to Deaf culture (including the reaction to cochlear implants) for those unfamiliar with it.

Russell L. Johnson

DEATH AND DYING

Throughout time humans have struggled to answer basic questions regarding the meaning of death and when death occurs. Scientific and technical advances coupled with the increasing secularization of American society have forced a redefinition of death and dying from a spiritual tradition based on natural processes to the medicalization of the body negotiated by physicians and the state. The modernization of death practices evolved hand in hand with concepts of hygiene, secularization, and rational methods of governance. Religious leaders abdicated their role over the death process only to be replaced by bioethicists who directed the physicians' role. The medicalization of death and the dying process has distanced humans from experiencing the natural cycle of life and death. The evolution of death and dying can be best viewed by the practices related to the care of the body: the changing definition of death, euthanasia, legal ownership and disposal of the body, organ transplantation, end-of-life care, and near-death experiences.

Burial Practices. For monotheistic religions, being formed in the image of God meant that any intrusion into the body was seen as a violation against God. Dead bodies had to be maintained intact to rise up again on Judgment Day. This resulted in religious objections to autopsy and dissection until the early nineteenth century. Beginning in the Middle Ages, an officer of the church, the sexton, emerged and assumed from the family the duties of assisting the dead. During the early American period, the position of the sexton merged with that of woodworking craftsman and the tasks of display, coffining, and transportation of the body organized under the specialized occupation of the undertaker were added. By the 1840s, the process of embalming had been imported from Europe and incorporated into the duties of the American undertaker. Burial practices remained essentially unchanged until the late nineteenth century, when under increasing pressure from population growth, diminishing urban land, and modernization, cremation became a popular alternative. Supporters of cremation assuaged religious concerns and encouraged the practice on the basis of improvements in hygiene. In the late twentieth century, environmentalists, concerned about the potentially harmful effects of decomposing bodies on the environment, including chemical contamination, designated cremation as a type of environmentally friendly "green funeral" and advocated for its widespread use.

Death in Numbers. In 1900 the U.S. government established the Division of Vital Statistics, which created a system for the collection, tabulation, and analysis of death statistics that resulted in the mandated reporting and certification of all deaths. Prior to standardized reporting, coroner's records and church ledgers provided the only demographic information for death certification. Death certification became the single most important measure of the nation's health, which informed political and health officials on growing medical problems, salient epidemics, and measures to prevent unnecessary deaths. The collection of death statistics assisted late twentieth-century public-health officials in their political efforts to enact legislation to effect societal issues such as the regulation of drunken driving, motor vehicle safety design, abolition of cigarette smoking, and gun control.

Defining Death. Traditionally, physicians equated the lack of a heartbeat and respiration with death. The popular belief in colonial times of cruenation (the bleeding of the dead body in the presence of the murderer) and the fear of being mistakenly buried alive illustrated the blurred line between life and death that existed in early America. The legal changes in the definition of death from a spiritual and religious meaning to one strictly medical mainly occurred beginning in the 1950s and early 1960s in response to scientific advances, first renal dialysis and later organ transplantation and artificial respiration. In 1958 Pope Pius XII removed religious obstacles to ending artificial life support when he acknowledged that it was not "in the competence of the Church to determine" the moment of death. Physicians no longer had the moral obligation to initiate or prolong artificial respiration if the patient's vital functions had ceased (Pius XII, 1958, pp. 393–398).

In 1968 an ad hoc committee at Harvard University under the direction of Henry Beecher, an anesthesiologist, rejected the notion that death resulted with the permanent cessation of heart activity. Instead, the committee argued that irreversible coma was to be the new criterion for death. The characteristics of death (unresponsiveness, lack of reflexes, absence of muscular movements or spontaneous respirations, and a "flat" electroencephalogram) could only be determined by a physician. In 1970, Kansas was the first state to include brain death in the legal determination of death. Later developments called for the use of brain angiography as a measure of the absence of intracranial blood flow as a final criteria confirming brain death. The Harvard Committee's new definition of brain death was modified 13 years later, which resulted in a more careful definition of death as cessation of all neurological activity including the brainstem.

Who Owns the Body? The postmortem use of individual body organs led to legal questions over the property rights of the family. Bodies, body parts, and even individual cells eventually attained legal status in U.S. courts. The uniquely American concept of a "quasi-property" status of a corpse was an ingenious invention to allow families the sepulchral rights to make funeral arrangements and provide a "decent burial." In 1905 in *Winkler v. Hawkes and Ackley*, the courts upheld the first American case regarding postmortem organ retention. Numerous subsequent court cases in the late twentieth century have supported the right of medicolegal authorities to retain organs in criminal investigations. However, the family's emotional attachment to whole organs has in most cases led to attempts to comply with the family's wishes for deposition of the remains.

Organ Transplantation.   Concomitant with the introduction of the brain-death criteria, the federal government codified organ transplantation in the Uniform Anatomical Gift Act of 1968. The act set regulations and standards for donation of organs or other body parts after death and permitted families to authorize donation. The act also granted a patient the ability to designate an anatomical gift before death. The code emphasized that human organs were gifts and not for sale. The National Organ Transplant Act (1974) later created a network under the direction of the United Network for Organ Sharing to assist in the equitable distribution of organs by organ-procurement organizations. With improved methods to determine compatibility, the use of human tissues (bone, tendon, skin) also began in earnest, although the practice remained unregulated. The unsuccessful attempts to significantly increase cadaver organ donation through the 1980s resulted in federal legislation that required hospital personnel to seek permission from the next of kin for donation. In addition, some states enacted "presumed consent" laws that permitted procurement of organs without the decedent's known prior objection.

The guidelines for organ procurement were expanded still further in 1992 with the inclusion of heart-beating donors when patients were removed from mechanical support, pronounced dead, and subsequently reattached to life support for the purpose of procuring organs. In the eyes of some critics, this confirmed Ruth Richardson's prediction that using the term "brain death," proponents of organ transplantation would create a new category of organ donors who were not completely dead at the initial time of donation. The procedure has continued to meet resistance in the transplant community.

End-of-Life Care.   The American Hospital Association in 1973 adopted a "Patient's Bill of Rights," which recognized the right of patients to refuse medical treatment. The first widely publicized case involving end-of-life care in the United States occurred in 1976 when Karen Ann Quinlan, a 21-year-old woman, arrived in an emergency room in a persistent vegetative state. Her father eventually went to court to obtain authority to discontinue all "extraordinary procedures" keeping his daughter alive. The court determined that Quinlan's parents had the authority to withdraw life support and to appoint a health-care proxy. The case instantly altered the "right to die" from a personal matter to a legal, ethical, and social issue. By 1977, eight states (California, New Mexico, Arkansas, Nevada, Idaho, Oregon, North Carolina, and Texas) had signed right-to-die bills into law. In another case, *Cruzan v. Director* (1990), the U.S. Supreme Court affirmed the rights of patients and their families to refuse life support and establish binding medical directives and living wills.

Physician-Assisted Suicide.   America followed the English common law tradition that disapproved of both suicide and efforts by physicians and others to assist someone to commit suicide, although euthanasia had been practiced publicly and privately for centuries. Highlighting the issue of doctor-assisted suicide was Dr. Jack Kevorkian, a retired Michigan pathologist, who from 1990 to 1995 assisted in 131 suicides. Many of Kevorkian's patients were later proven not to have had a fatal organic disease. Kevorkian was eventually convicted and sentenced to prison but not before he focused world attention on the issue of euthanasia and physician-assisted suicide. In 1997 Oregon enacted the Death with Dignity Act, the first law in American history permitting physician-assisted suicide. The Hemlock Society, a national right-to-die organization founded in California in 1980, published a self-help guidebook *Final Exit* (1991) for those contemplating suicide.

Hospice Care.

Beginning in the late eighteenth century, doctors replaced the clergy in managing the pain and fear of the final moments of life for the dying patient. The final duty of the physician was to request the family's permission for an autopsy. The practice of specialized care for the dying was not introduced to the United States until the early 1960s. In 1969, Elisabeth Kübler-Ross, a Swiss American psychiatrist, published the book *On Death and Dying* (1969), which proposed five stages of death: denial, anger, bargaining, depression, and acceptance. She encouraged the development of a hospice movement under the belief that euthanasia robbed people from completing and finalizing the death process. By 1982 the U.S. government had created legislation to include hospice care in Medicare benefits and established a system of accreditation.

The Near-Death Experience.

The development of new medical techniques for cardiac resuscitation in the 1960s allowed physicians to resuscitate patients but also created the condition referred to as the near-death experience. Patients who formally would have died lingered near death and later returned to health to retell their experiences. The 1975 publication of *Life after Death*, which described a patient's near-death experience, and other works such as *On Life after Death* by Kübler-Ross unleashed a flood of literature recounting similar events. The near-death experience challenged existing medical theories of death and dying, forcing many physicians and medical theorists to reevaluate metaphysical explanations in medicine. Thus, for some Americans, death and dying have once again become demedicalized and returned to a spiritually awakening experience.

[*See also* Animal and Human Experimentation; Ethics and Medicine; Forensic Pathology and Death Investigation; Law and Science; Medical Malpractice; Medicine; *and* Public Health.]

BIBLIOGRAPHY

Abel, Emily K. *The Inevitable Hour: A History of Caring for Dying Patients in America*. Baltimore: Johns Hopkins University Press, 2013. A historical review of the institutional practice of end-of-life care confirming the transition of modern medical care from the relief of pain and suffering in the final stages of life to an emphasis on cure.

Alexander, Eben. *Proof of Heaven: A Neurosurgeon's Journey into the Afterlife*. New York: Simon & Schuster, 2012.

Baker, Robert, and Laurence McCullough. *The Cambridge World History of Medical Ethics*. Cambridge, U.K.: Cambridge University Press, 2009. An exhaustive review of the multicultural, ethical, and religious practices influencing medical ethics and the practice of medicine with chronological timeline of the development of the field.

Beecher, Henry. "A Definition of Irreversible Coma: Report of the ad hoc Committee of the Harvard Medical School to Examine the Definition of Brain Death." *Journal of the American Medical Association* 205, no. 6 (1968): 337–340.

Dowbiggin, Ian. *A Merciful End: The Euthanasia Movement in Modern America*. New York: Oxford University Press, 2003. An examination of the ethical and social complexities of the euthanasia movement at the turn of the century.

Faust, Drew Gilpin. *This Republic of Suffering: Death and the American Civil War*. New York: Alfred A. Knopf, 2008. A detailed portrayal of nineteenth-century concepts of death and dying and the modernization of burial practices in the wake of the Civil War.

Kübler-Ross, Elizabeth. *Death and Dying*. New Brunswick: Routledge, 1969.

Laderman, Gary. *Rest in Peace: A Cultural History of Death and the Funeral Home in Twentieth-Century America*. New York: Oxford University Press, 2003.

Moore v. Regents of University of California. California 2nd Court of Appeals, 1998.

Pius XII. "The Prolongation of Life." *Pope Speaks* 4, no. 4 (1958): 393–398.

Richardson, Ruth. "Fearful Symmetry: Corpses for Anatomy, Organs for Transplantation?" In *Organ Transplantation Meanings and Realities*, edited by Stuart Younger, Renée Fox, and Laurence O'Connell, pp. 66–100. Madison: University of Wisconsin Press, 2001. A collection of essays by well-known scholars that frame and debate issues of organ transplantation in the late twentieth century.

Jeffrey M. Jentzen

DeBAKEY, MICHAEL

(1908–2008), pioneer cardiovascular surgeon, was born in Lake Charles, Louisiana, the son of Lebanese immigrants. He received his undergraduate and medical degrees from Tulane University and trained in surgery in New Orleans and Europe. DeBakey returned to Tulane after World War II; in 1948 he was appointed chief of surgery at Baylor University College of Medicine in Houston. He remained at the institution for the next six decades, serving as president (1969–1979) and chancellor (1979–1996) in addition to his responsibilities to the surgery program.

DeBakey was a pioneer in the development of modern cardiovascular surgery, inventing or improving major operations on the aorta and other blood vessels. His aggressive approach to aortic aneurysms or dissections revolutionized the care of patients with those life-threatening problems. DeBakey's interest in cardiovascular surgery was matched by his passion for developing an artificial heart and left ventricular assist devices to treat patients with severe heart failure. Clinical problems surrounding diseased blood vessels and failing hearts stimulated his research interests. DeBakey's vital role in launching research initiatives and mentoring teams to study specific problems was acknowledged in 1963 when he won the prestigious Albert Lasker Clinical Research Award.

During an active surgical career that spanned more than a half century, DeBakey operated on more than 60,000 patients, including many international celebrities. He helped educate thousands of medical students and residents. His surgical residents and associates were inspired and intimidated by DeBakey, described as a perfectionist who was demanding, driven, and supremely self-confident.

In addition to his central role in developing the Baylor College of Medicine and Houston's Methodist Hospital, DeBakey was very active nationally as an advocate for research. He was a prominent member of an influential group of physicians, scientists, politicians, and concerned citizens that lobbied effectively on behalf of the federal funding of research. He also contributed significantly to the creation of the V.A. Hospital system and the National Library of Medicine. DeBakey supported the efforts of presidents Kennedy and Johnson to establish Medicare when organized medicine and most doctors opposed the plan. His professional accomplishments and public service were acknowledged by the receipt of dozens of awards culminating in the Congressional Gold Medal in 2008. He died in Houston on 11 July 2008.

[*See also* Cardiology; Medicare and Medicaid; Medicine; Research and Development (R&D); *and* Surgery.]

BIBLIOGRAPHY

Mattox, Kenneth L. "Michael E. DeBakey: The Consummate Leader." *Methodist DeBakey Cardiovascular Journal* 5 (2009): 32–36.
"Michael Ellis DeBakey: A Conversation with the Editor." *American Journal of Cardiology* 79 (1997): 929–950.

W. Bruce Fye

DEEPWATER HORIZON EXPLOSION AND OIL SPILL

At about 9:45 P.M. on 20 April 2010, methane gas escaping from an oil well being drilled from a massive drilling rig on the ocean surface into the floor of the Gulf of Mexico five thousand feet below ignited and exploded, starting an inextinguishable fire that caused the drilling rig to collapse into the sea after burning for 36 hours. Although 115 workers on the rig escaped by lifeboat, 11 were killed.

The rig, *Deepwater Horizon*, a structure that could be dynamically positioned, had drilled the last phase of an exploration well on the Macondo oil prospect, 41 miles south of Louisiana, within the U.S. economic exclusion zone, and the crew had just completed cementing the well closed to hold it for production later. A failure occurred at the wellhead on the ocean floor, allowing water, and then drilling mud, cement, and methane, to shoot up through the marine riser (the connecting pipe) and onto the drilling platform, where the methane ignited. The crew's attempts to stop the upsurge by engaging the blowout preventer at the wellhead failed.

On 22 April, the day the rig sank, observers detected oil in the ocean. Leaking from the wellhead, the Macondo well ultimately discharged an estimated 4.9 million barrels of oil into the Gulf, making it the largest marine oil spill ever in American waters, far exceeding the *Exxon Valdez* spill in Alaska's Prince William Sound in 1989. After numerous attempts, the well finally was capped on 15 July 2010. The oil spread on the sea floor, on the surface, and through the water column, bringing considerable harm to marine and wildlife along the Gulf Coast and harming the fishing and tourist industries. Almost five hundred miles of the coastlines of Texas, Louisiana, Mississippi, Alabama, and Florida were affected by the drifting oil. Boom material, skimmer boats, and other equipment were used to screen off bays and estuaries, including chemical dispersants and burning in limited amounts. But oil appeared in marsh grasses and other shore vegetation and in shoreline rocks and sand. Tar balls in nets resulted in officials closing shrimp fishing in 2010. Shellfish showed oil or its effects through the summer of 2010 and later. Dolphins and whales were reportedly dying at twice the normal rate. Experts disagree on the amount of oil remaining in the environment; some estimates suggest 75 percent. Evaporation, consumption by microbes, and dispersal in the Gulf's Loop Current apparently did not eliminate most of the spilled oil.

President Obama established the National Commission on the BP *Deepwater Horizon* Oil Spill and Offshore Drilling in May 2010. Released in January 2011, the Commission's report cited shortcuts in safety and testing, taken to save money, that might have prevented the blowout. The report further cited an industry culture that mitigates against adequate risk management and safety assessment and planning.

In 2012, experts predicted that litigation regarding responsibility for the blowout and subsequent explosion and fire, for the deaths of the 11 crew members, and for environmental and economic damage would be long and complicated.

[*See also* Environmentalism; Environmental Protection Agency; Ethics and Professionalism in Engineering; *Exxon Valdez* Oil Spill; Fish and Wildlife Service, U.S.; *and* Fisheries and Fishing.]

BIBLIOGRAPHY

Deep Water, The Gulf Oil Disaster and the Future of Offshore Drilling: Report to the President. Washington, D.C.: National Commission on the BP *Deepwater Horizon* Oil Spill and Offshore Drilling, January 2011.

Report of Investigation into the Circumstances Surrounding the Explosion, Fire, Sinking, and Loss of Eleven Crew Members Aboard the Mobile Offshore Drilling Unit Deepwater Horizon in the Gulf of Mexico, April 20–22, 2010. Washington, D.C.: U.S. Coast Guard, September 2011.

Stephen Haycox

DEFENSE ADVANCED RESEARCH PROJECTS AGENCY

Established in *Sputnik's* wake in 1958, the Advanced Research Projects Agency (ARPA) became the U.S. military's leading patron of high-risk and high-reward research in areas ranging from computing to materials science to ballistic missile development and ballistic missile defense. The agency's primary role has been to prevent the United States from being the victim of a technological surprise. Given this mission, the agency is able to fund research in a variety of intellectual domains and institutions that might otherwise seem foreign to the military. Although most famous for the developments that led to the Internet, much of the agency's work remains classified and historians have had access to only a small quantity of archival materials for serious research. Although originally known as ARPA, the word "defense" was added to the agency's name in 1972; hence the acronym DARPA.

Sputnik's October 1957 success created a panic in Washington, D.C.; among the many organizational responses was the creation of

DARPA within the Office of Secretary of Defense Neil McElroy so that a single agency within the defense establishment would have supervision over advanced weapons program research. The new agency would do research and development until such time as the technology became operational and was turned over to one of the military services or was judged of little value to the armed services. Initially, work focused upon ballistic missile development, missile defense, and detecting the testing and use of nuclear weapons tests through both seismographic and satellite-based detection methods. As the ballistic missile programs reached maturity, DARPA began to investigate problems relating to the command and control of military forces at all levels, ranging from the battlefield to presidential control of the U.S. nuclear arsenal. Research also focused on the development of an array of technologies, including satellite-based navigation systems (Transit, a forerunner of the Global Positioning System), remote sensing, computer networks, computer chip manufacturing, and materials science. Also important has been DARPA's ability to fail, that is, to investigate promising technologies and realize that they do not meet the military's needs. Such research is terminated, but lessons are learned.

Among DARPA's most famous technological innovations is the ARPANET, a predecessor to the Internet. Although many claim that the Internet emerged from the desire to develop a system to command U.S. nuclear forces, the reality is far more mundane. In the early 1960s, the U.S. military found itself with an array of computer systems, almost none of which could communicate with each other. Beginning in 1962, DARPA's Information Technology Processing Office (IPTO) began funding research to develop the technology allowing multiple users to use the same mainframe computer, what became known as time-sharing. In turn, DARPA sought to develop technologies to allow researchers at remote locations to share the time-sharing resources at geographically disperse computing centers. Research on computer networks produced the software that made the ARPANET and, later, the Internet possible. Such work also allowed the military's various computer systems to communicate with each other, greatly reducing costs because computer resources could now be shared over great distances. Although the Internet remains the most famous of the IPTO projects, other research funding led to the computer graphics and software innovations that are the basis of contemporary computing.

What remains striking about DARPA is its organization. DARPA does not have its own laboratories or research facilities. Program managers are brought in from academia and industry to run research and development efforts based upon their areas of expertise. In turn, the program managers provide interested and able universities and corporations with contracts for specific research and development projects. Because DARPA managers have often been experts in their fields, such as J. C. R. Licklider and Ivan Sutherland in the IPTO, they have had the ability to contract with the most able available researchers for specific projects. Regular turnover and constant movement among the program managers means that DARPA is constantly reinvigorated with new ideas and practices. More recently, DARPA has used competitions among research groups in academia and industry to speed the development process, as in the case of developing vehicles that might navigate difficult terrain autonomously.

Academic fields, such as artificial intelligence (AI), and even entire disciplines, such as computer science, have benefitted from program managers whose interests lie not only in specific technologies, but also in the training of a new generation of researchers. Grants at universities always contained funds for graduate students and postdoctoral fellows. In the early twenty-first century, military needs continued to drive DARPA funding as well as its orientation. In 2012, after 10 years of constant warfare, DARPA increasingly viewed its mission as finding technological solutions for the warfighter. This included not only research on medical advances to keep soldiers alive after suffering serious wounds from improvised explosive devices, but also the development of computer-based translation hardware and software that rendered English into various Arabic dialects and vice versa. DARPA's success at funding transformation research and development led to the establishment of DARPA-like organizations within the intelligence community (IARPA) and the Department of Energy (ARPA-E).

[*See also* **Artificial Intelligence; Computer Science; Computers, Mainframe, Mini, and Micro; Internet and World Wide Web; Military, Science and Technology and the; Missiles and Rockets; Nuclear Weapons; Research and Development (R&D); Satellites, Communications; Software; Space Program; Space Science;** *and* **Technology.**]

BIBLIOGRAPHY

Abbate, Janet. *Inventing the Internet.* Cambridge, Mass.: MIT Press, 1999.
Norberg, Arthur L., and Judy E. O'Neill. *Transforming Computer Technology: Information Processing for the Pentagon, 1962–1986.* Baltimore: Johns Hopkins University Press, 1996. Based upon extensive archival work and oral history interviews, this work remains the best available study of DARPA practices and their effects upon the production of knowledge.
Watson, Robert J. *Into the Missile Age, 1956–1960.* Volume IV of *The History of the Office of the Secretary of Defense.* Washington, D.C.: U.S. Government Printing Office, 1997.

Michael Aaron Dennis

DE FOREST, LEE

(1873–1961), engineer and inventor, was born in Iowa and grew up in Alabama. Famous for his development of the three-element vacuum tube he called the "Audion," de Forest was an innovator in radio and later in film and television. He eventually held more than 180 patents. He studied at the Shefield Scientific School, then a part of Yale University, where he earned one of the first PhD degrees (1899) based on research on radio waves.

During the twentieth century's first decade, de Forest was a principal in more than a dozen short-lived wireless companies. His chief technical innovation was to add a third element, a tiny screen-like grid, between the existing elements of a diode vacuum tube. This triode or Audion of 1906 turned out, after further research by de Forest and others, to have the ability to amplify weak wireless signals. He conducted occasional experimental broadcasts from the stage of New York's Metropolitan Opera in 1908 and 1910. Desperate for funds, he sold part of his Audion patent to AT&T, which used Audion-type amplifiers to build and operate the first transcontinental telephone service beginning in 1915. His experimental station 2XG broadcast news of the 1916 presidential election. After World War I, de Forest moved to California and operated experimental radio transmitters in the San Francisco area into the 1920s.

De Forest sought to develop sound motion pictures in the 1920s. He applied for his first Phonofilm patent in 1919 for a system that recorded what was, in effect, an image of the sound track right on the motion picture film, resolving the problem of synchronizing image and sound. Nearly two hundred short films were made using Phonofilm, but disagreements with inventor Theodore Case and others, and the lack of major studio interest, led to the dissolution of de Forest's film company in 1926. Hollywood studios eventually adopted a modified version of the sound-on-film system to which de Forest had contributed some ideas. He was honored with an Oscar for this work in 1959.

His work with television in the late 1930s was more limited and centered on potential applications of the technology for military and other uses. The final decades of his life were devoted to promoting his historical role, especially as the "father" of radio.

[*See also* **Engineering; Film Technology; Military, Science and Technology and the; Radio; Sound Technology, Recorded; Telephone;** *and* **Television.**]

BIBLIOGRAPHY

Carneal, Georgette. *Conqueror of Space: An Authorized Biography of the Life and Work of Lee de Forest.* New York: Horace Liveright, 1930.

Forest, Lee de. *Father of Radio.* Chicago: Wilcox & Follett, 1950. Autobiography.

Forest, Lee de. *Television Today and Tomorrow.* New York: Dial Press, 1942.

Hijiya, James A. *Lee de Forest and the Fatherhood of Radio.* Bethlehem, Pa.: Lehigh University Press, 1992.

Christopher H. Sterling

DELBRÜCK, MAX

(1906–1981), pioneering figure in the emerging field of molecular biology and Nobel laureate, corecipient of the Nobel Prize for Physiology or Medicine in 1969 with Alfred Hershey and Salvador Luria for work on the mechanisms of viral replication in bacterial hosts. Delbrück was born in Germany. His training at Göttingen in the late 1920s and research at Bristol, Copenhagen, and Zurich brought him into contact with leading architects of the quantum revolution, including Niels Bohr. Bohr's speculation that mechanistic and purposive descriptions of life might stand in a complementary relation inspired Delbrück's transition from physics to biology and guided his subsequent research. From 1932 to 1937 Delbrück worked under Lise Meitner at the Kaiser Wilhelm Institute for Chemistry in Berlin. In a collaborative paper with Nikolai Timoféeff-Ressovsky and Karl Zimmer in 1935, he explored gene structure and mutation by investigating the effects of ionizing radiation. Erwin Schrödinger drew attention to this work in *What Is Life?* (1944).

Amid increasingly turbulent political circumstances prior to World War II, Delbrück left Berlin for the California Institute of Technology in 1937 on a Rockefeller Fellowship. After a brief interaction with T. H. Morgan's group in fruit fly genetics, Delbrück teamed up with Emory Ellis to focus on bacteriophage and established himself as a leading intellectual among phage researchers. Following a period at Vanderbilt (1940–1947), he returned to Caltech as professor of biology (1947–1976) and from the 1940s through the 1970s organized annual summer courses on phage genetics and, after 1953, the sensory physiology of the photo-tropic fungus *Phycomyces* at the Cold Spring Harbor Laboratory in New York. Known to colleagues as a sharp critic and respected leader, Delbrück helped to make these institutions important centers for research in molecular biology, to develop and refine quantitative methods for the investigation of foundational questions about simple living systems, and to draw hundreds of young scientists into the field.

[*See also* **Genetics and Genetic Engineering; Germ Theory of Disease; Molecular Biology; Morgan, Thomas Hunt; Nobel Prize in Biomedical Research; Physics;** *and* **Quantum Theory.**]

BIBLIOGRAPHY

Fischer, Ernst Peter, and Carol Lipson. *Thinking about Science: Max Delbrück and the Origins of Molecular Biology.* New York: W. W. Norton, 1988. Biography coauthored by one of Delbrück's last graduate students at Delbrück's request.

Hayes, William. "Max Ludwig Henning Delbrück, 4 September 1906–10 March 1981." *Biographical Memoirs of Fellows of the Royal Society* 28 (1982): 58–90. Reprinted in *Biographical Memoirs of the National Academy of Sciences* 62 (1993): 67–117.

Kay, Lily E. "Conceptual Models and Analytical Tools: The Biology of Physicist Max Delbrück." *Journal of the History of Biology* 18, no. 2 (1985): 207–246.

Daniel J. McKaughan

DEMOGRAPHY

Demography examines the sources and consequences of population increases or decreases. This, in turn, involves assessing the rates of birth, death, and geographic movement—fertility, mortality, and migration. Historically, American demography fits into a three-stage progression characteristic of societies that now have low birth and death rates. Convenient labels for these three stages are Malthusian frontier, neo-Malthusian, and post-Malthusian. (The term "Malthusian" comes from Thomas Robert Malthus, a pioneering English theorist of demography). Originating in the theory of demographic transition, this periodization scheme locates a turbulent transitional era, between two periods of relative stability, when a demographic pattern of roughly balanced *high* birth and death rates gives way to one of relatively equal *low* birth and death rates.

Whereas "demographic transition" theorists once portrayed the decline in fertility as a response to a reduction in mortality, this theory is today usually seen as a useful description, or first approximation, of historical experience rather than a guide to cause and effect or a detailed blueprint that adequately captures all of the features of the demographic history of the United States or any other society. In comparative terms, the most distinctive feature of American demographic history is the extremely rapid growth of people of European and African origins, from 250,000 in 1700 was projected to be 275 million by the year 2000—a thousandfold increase. Over these same three hundred years, the population of Europe multiplied only fivefold.

The human impact of this demographic transition is profound. Before 1800, in the traditional or pretransitional era of high fertility, the average woman who survived to age 50 had 7 children; in the 1990s, the average was 2 children. The high fertility of the traditional era produced a very young aggregate population. About half of the pre-1800 population was under age 16 by the 1990s, the median age was 33 years. Only one in 40 Americans in 1800 was over 65 compared with 1 in 8 at the end of the twentieth century.

Mortality rates have changed dramatically as well. Until about the 1870s, the average life expectancy in the United States was around 45 years, compared with 76 years in the 1990s. The impact of mortality decline is somewhat misleading if portrayed in terms of averages, however, because relatively few individuals actually die around the average age of death. Instead, infants and young children once died at radi-

cally higher rates than today, pulling the average down. In the pre-1870s period, about one in six infants died before his or her first birthday and one in four died before age five. In the 1990s, fewer than 2 percent of American infants died before their fifth birthday. Instead of infectious diseases, which took a heavy toll on the young, the major killers at the end of the twentieth century were chronic diseases related to aging, especially cardiovascular disease and cancer, which together accounted for nearly two-thirds of all deaths.

Three elements characterize the demography of the earliest stage of American population history: (1) an extremely high rate of overall growth that, despite a substantial contribution of immigration, was mostly the result of natural increase—the difference between birth and death rates; (2) high fertility caused by markedly younger marriage ages for women than in western Europe; and (3) mortality that was high compared with that in the late twentieth century, but moderate in comparison to contemporary death rates in Europe.

The term "Malthusian frontier" summarizes the larger economic and cultural context of this demographic regime of rapid natural increase in early America. Writing in 1798, Malthus linked mortality rates and marriage age to the tenuous but ultimately equilibrating relationship between population size and food supply. Because Malthus lived in an era when sustained growth of economic productivity was nearly inconceivable, long-term population growth seemed impossible. Demographic expansion, he theorized, would ultimately be halted by what he called the "positive check" of higher mortality caused by famine and malnutrition.

Malthus, however, viewed eighteenth-century America as an exception to his general rule that resource constraints would limit population growth. America's seemingly boundless frontier and low population density meant that land was cheap and labor expensive. Because couples could acquire land relatively easily, they could, Malthus reasoned (as had Benjamin Franklin a half century earlier), marry earlier than their counterparts in Europe. This relaxation of what Malthus called the "preventive check" of late marriage spurred American population growth.

Over the course of the nineteenth century, couples married later and, more significantly, began to practice family limitation. Fertility fell by 50 percent, although few couples apparently made use of contraceptive devices. These nineteenth-century trends can be attributed in part to the declining availability of agricultural land as the frontier moved toward closure. Urbanization also played a role in this large-scale demographic transition. Before 1800, during the Malthusian-frontier era, only one in twenty Americans lived in a town or city. Since 1970, with the process of urbanization nearly complete, about three fourths of Americans have resided in places with populations over 2,500. Although large numbers of foreigners arrived in the United States from 1840 to 1920, in no decade did immigration account for more than one-third of the total population increase. High (although declining) rates of natural increase continued as a distinctive feature of American demographic history during this transitional stage.

The third phase of American demographic history, the closing decades of the twentieth century, was characterized by the increasing irrelevance of marriage, demographically speaking. The wide gap between fertility rates of married and unmarried women shrank dramatically in this period. By the 1990s, further, Americans were marrying later than at any time in the nation's history.

The three-period framework fits the historical experience of African Americans as well. As with the population as a whole, rapid rates of aggregate growth characterized the early historical demography of the American black population, especially in the era of slavery. Of the 10 to 11 million Africans brought to the New World in the slave trade, some 600,000 to 650,000 were imported into the area that became the United States. During the nineteenth century, the annual natural increase among American slaves was over 2 percent, only slightly under the rate for the white population. Compared with slaves in other regions of the Americas, the enslaved population in the United States had both higher fertility and lower mortality. Since emancipation, blacks and whites have experienced the same trends in mortality and fertility, although both rates have been

consistently higher for blacks. The black–white difference in life expectancy at birth was six years in 1990 compared with eight years in 1900.

Prior to the twentieth century, the indigenous Indian population experienced a radically different demographic trajectory from that of European and African Americans. Instead of rapidly increasing, the numbers of Indians declined precipitously, a demographic catastrophe owing largely to extremely high death rates from diseases of European origin—most importantly smallpox, typhus, and measles. Having no previous experience with these diseases, whole populations were nearly wiped out by epidemics. Estimates of the numbers of Indians in North America before European contact are varied and disputed. One conjecture places the figure at more than 5 million in the conterminous U.S. area in 1492. By 1800, the Indian population was about 600,000. It reached its nadir of 250,000 in the 1890s. The Indian population rebounded after 1900, however, reaching nearly 1.9 million in the 1990 Census.

The public discussion of population-related issues over the course of American history has typically reflected these broad trends. During the Colonial Era, Americans exulted in, and British officials worried about, the colonies' exploding population. The relationships among land availability, population, and the social order were central to nineteenth-century demographic thinking. Thomas Jefferson and James Madison believed that territorial expansion would sustain the egalitarian economic basis of republican political institutions. During the Antebellum Era, both southern and northern writers tied the eventual extinction of slavery to a limitation of its territorial expansion into cheap lands. At century's end, historian Frederick Jackson Turner saw the closing of the frontier as the end of an epoch in American history.

Differential fertility among various groups attracted considerable comment during the period of transition to lower birth rates. Native-born New Englanders, the vanguard group in the control of fertility within marriage, were thought to be on a path toward "race suicide." By the close of the twentieth century and the beginning of the twenty-first century, population questions often intertwined with value debates over such issues as abortion, women's status, and intergenerational equity.

Even in the third stage of relative demographic stability, important changes still occurred. The baby boom of the post–World War II era, peaking in 1957, nearly doubled total fertility rates. A "baby bust" of equal magnitude then surprised the experts. These fluctuations gave rise to the prospect of the nation's having too few people of working age to fund the retirement of the baby boomers beginning in the second decade of the twenty-first century. A new wave of immigration, principally from Latin America and Asia, was another important development of the late twentieth century with demographic implications.

Despite these changes, by the late twentieth century fertility fluctuated narrowly around replacement levels, mortality was declining at a markedly lower pace than in the three quarters of a century after 1880, urbanization had nearly ended, and Americans had become less residentially and geographically mobile. With only 3 percent of the workforce engaged in agriculture, it was obvious that the long-term shift from farm to city was over. At the beginning of the new century, demographic phenomena were much more stable than they had been during the previous century and a half.

[*See also* **Birth Control and Family Planning; Columbian Exchange; Disease; Food and Diet; Franklin, Benjamin; Jefferson, Thomas; Life Expectancy; Smallpox;** *and* **Typhus.**]

BIBLIOGRAPHY

Anderton, Douglas L., Richard E. Barrett, and Donald L. Bogue. *The Population of the United States.* 3d ed. New York: Free Press, 1997.

Haines, Michael R. "The Population of the United States, 1790–1920." In *The Cambridge Economic History of the United States.* Vol. II. *The Long Nineteenth Century,* edited by Stanley Engerman and Robert Gallman. Cambridge, U.K.: Cambridge University Press, 2001.

Livi-Bacci, Massimo. *A Concise History of World Population.* Cambridge, Mass.: Blackwell, 1992.

Preston, Samuel H., and Michael R. Haines. *Fatal Years: Child Mortality in Late Nineteenth-Century America.* Princeton, N.J.: Princeton University Press, 1991.

Wells, Robert V. *Revolutions in Americans' Lives: A Demographic Perspective on the History of Americans, Their Families, and Their Society.* Westport, Conn.: Greenwood Press, 1982.

Daniel Scott Smith

DENTISTRY

Persons in North America have experienced pain, disability, and sometimes death as the result of dental disease since the establishment of the earliest colonies. Until the late eighteenth century most persons in North America relied on self-help to treat dental disease; for example, many patients used "tooth forceps" to pull teeth that caused pain as a result of decay, accident, or violence. Understanding of the pathology and epidemiology of dental disease increased begining in the second half of the nineteenth century. By the mid-twentieth century caries and periodontal disease had been established as the major causes of dental pain, disfigurement, and inability to achieve appropriate nutrition in the population.

The formal practice of dentistry in what became the United States began in the late eighteenth century when immigrants from Europe began to practice a branch of surgery that called its adherents "operators for the teeth" or "surgeon-dentists." These practitioners also trained local artisans to practice their trade. Dental practice in the first four decades of the nineteenth century consisted mainly of extractions and the construction and fitting of dentures. A substantial transformation in dentistry and, almost immediately, of general surgery, occurred in the mid-1840s when Horace Wells, a Connecticut surgeon-dentist, demonstrated the effective use of nitrous oxide ("laughing gas") as a general anesthetic. Surgeons, including dentists, soon expanded anesthetic agents to include ether.

Dentistry emerged as an organized profession in the last four decades of the nineteenth century. The American Dental Association (ADA) emerged in 1866, an American Academy of Dental Science in 1876, and a National Association of [state] Dental Examiners in 1883. Schools of dentistry, like schools of medicine, proliferated during these years. By 1900, 50 dental schools, 21 of them proprietary, had been established across the country. By 1924, as a result of pressure from the Dental Education Council of America, only four of these schools were privately owned. In the early 1920s, the Carnegie Endowment for the Advancement of Teaching commissioned an evaluation of dental schools, modeled on its earlier surveys of education for medicine and social work. William J. Geiss, a dentist in New York City who conducted this survey, reported in 1926 on gaps in dental education. He recommended that dentistry become either a specialty of medicine or a university-based discipline that prioritized the hiring of full-time faculty in the basic sciences and clinical disciplines. The leaders of the ADA and the most prominent schools of dentistry chose the second alternative.

Dental practice and, as a result, postgraduate education of specialists licensed as general dentists became increasingly specialized during the first half of the twentieth century. Orthodontics became the first formally organized specialty in 1900, when Edward H. Angle (1855–1930) founded a school of orthodontics in St. Louis, Missouri. Angle subsequently standardized orthodontic treatment and published books and pamphlets on appliances to correct what he termed malocclusions of the teeth. Other specialties soon emerged: periodontics (treatment of disease of the gums) in 1918; oral surgery, also in 1918; oral pathology in 1936; endodontics (treatment of the roots of teeth) in 1919; and public health (1937). By the end of the twentieth century dentistry recognized nine distinct specialties, each with formal associations and systems of training and qualifying for practice. However, most dentists continued to focus on the restoration of diseased or broken teeth with fillings or the insertion of dentures and bridges.

Providing access to preventive and restorative dental care became an activity of government as a result of the hiring of dentists by state health departments early in the twentieth century and the establishment in 1919 of a dental division of the U.S. Public Health Service (PHS). The scope and influence of public-health dentistry expanded considerably after the establishment and validation by the PHS in the mid-1940s of a pilot program in Grand Rapids, Michigan, to fluoridate the public water supply to reduce the incidence of dental caries. The use of fluoride in water supplies and by direct application to children's teeth gradually spread worldwide, despite widespread and continuing claims that it causes harm. A systematic review published in a peer-reviewed journal in 2007, which met prevailing international standards for conducting such studies, concluded that "fluoridation of drinking water remains the most effective and socially equitable means of achieving community-wide exposure" to effective prevention of caries.

Research in oral biology and on methods of preventing and treating caries, periodontal disease, and other dental disorders proliferated after the establishment by the U.S. Congress in 1948 of the National Institute for Dental Research (now the National Institute for Dental and Craniofacial Research). In 2012, the priorities of this institute included the "development of molecular-based oral health care," strengthening the "pipeline of researchers," and "eliminat[ing] disparities in oral, dental and craniofacial health."

These disparities have persisted despite considerable effort to improve access to dental care for Americans, especially seniors receiving long-term care and persons with low incomes. A landmark article in 1975 concluded that "despite the knowledge and ability to radically improve the dental health of the population of the United States…the current system has been unable to do so." The author identified the major impediments to improved dental health as "problems of organization and performance" of dentists and their work. A report by the U.S. Surgeon General in 2003 concluded that the "oral health of Americans has improved in recent years yet considerable gaps in the provision of dental care remain." Since the late 1990s many state Medicaid programs have reorganized dental services for persons with low incomes and increased reimbursement of dentists' fees. Moreover, the ADA embraced expansion of the number and scope of practice of members of allied health professions providing and coordinating dental services, particularly in underserved communities. Nevertheless, an article published early in 2012, although acknowledging the "extraordinary growth of scientific knowledge related to oral health and the development of evidence-based dentistry," concluded that "large segments of the [United States] population are disproportionately burdened by oral diseases." The authors identified "challenges remaining in completing the decades-long paradigm shift to prevention and the important role to be played by the dental education community" in eliminating these disparities.

[*See also* Anesthesiology; Biological Sciences; Disease; Hygiene, Personal; Medical Education; Medical Specialization; Medicare and Medicaid; Medicine; National Institutes of Health; Public Health; Public Health Service, U.S.; *and* Surgery.]

BIBLIOGRAPHY

Farmicola, A. J., H. L. Bailitt, T. J. Beazoglou, and L. A. Tedesco. "The Inter-relationship of Accreditation and Dental Education: History and Current Environment." *Journal of Dental Education* 72, suppl. (February 2008): 53–60. Replaces previous review articles on the history of dental education.

Garcia, R. L., and S. Woosung. "The Paradigm Shift to Prevention and Its Relationship to Dental Education." *Journal of Dental Education* 76 (January 2012): 36–45. Replaces previous scholarship on the relationship between education and practice in dentistry.

Harris, R. Roy. *Dental Science in a New Age: A History of the National Institute for Dental Research*, 2d ed. Iowa City: Iowa State University Press, 1992. A competent, but truncated, survey that has no competitors.

Schoen, M. H. "Dental Care and the Health Maintenance Organization Concept." *The Milbank Memorial Fund Quarterly: Health and Society* 53 (Spring 1975): 173–193. A seminal article by an author who had a strong reputation and unparalleled expertise in the organization of dental services and how to remedy its deficiencies.

Stanton, M. W. "Dental Care: Improving Access and Quality." Agency for Healthcare Research and Quality. *Research in Action*, Issue 13 (July 2003). http://www.ahrq.gov/research/dentalcare/dentria.htm (accessed 18 October 2012). An able summary of the findings of the 2003 Surgeon General's report on dental services that cites relevant literature.

Yeung, C. A. "A Systematic Review of the Efficacy and Safety of Fluoridation." *Evidence-Based Dentistry* 9 (2007): 39–48. Reports the methods and findings of a systematic review that meets international standards.

Daniel M. Fox

DIABETES

Diabetes is a disease marked by high blood glucose levels. It results when certain cells in the pancreas, known as beta cells, are unable to produce enough insulin or when the body's tissues have developed an insensitivity to the hormone. The body needs insulin to move glucose, which is its main fuel source, from the blood stream into the tissues. When left unchecked, high blood glucose levels can lead to retinopathy, kidney disease, neuropathy, and an increased risk for cardiovascular disease and hypertension.

Aretaeus of Cappadocia, a second-century Greek physician, coined the term diabetes (which is Greek for siphon) and provided the first detailed clinical description of the disease. He painted a disturbing picture of individuals with emaciated bodies, tortured by insatiable thirst and unable to stop drinking or urinating. Aretaeus had no effective treatment to recommend, but he could take some comfort in knowing that diabetes was "not very frequent among mankind." Two thousand years later that was no longer the case. The Centers for Disease Control and Prevention (CDC) predicted that one of every three people born in the United States in 2000 would develop the disease. Small wonder public-health officials at the beginning of the twenty-first century referred to a diabetes "epidemic," a term historically linked to infectious diseases. The financial burden of this disease—in 2007 it ran to $174 billion in direct and indirect costs in the United States—was surpassed only by the human tragedy in lives lost and in the suffering caused by such sequelae as blindness, renal failure, and amputations (CDC, 2011). In the early twenty-first century, diabetes emerged as one of the most challenging chronic diseases because it proved difficult to control and could not be cured.

Diabetes first began attracting attention in the United States at the end of the nineteenth century. Before then, articles appeared occasionally in scholarly journals, but medical practitioners encountered the disease so rarely that it seldom piqued their interest. To William Osler, one of the most highly regarded medical professionals of his day, diabetes was still "a rare disease" in 1892. But even as he wrote, changes were underway. U.S. Census reports revealed a 150 percent increase in the diabetes mortality rate between 1850 and the end of the century (Purdy, 1890, p. 18). Diabetes may not yet have claimed many lives, but no other disease was showing such exponential growth.

Searching for a Cause. Physicians' explanations for this disturbing trend revealed deep-seated anxieties about how the forces of industrialization, urbanization, and immigration were transforming American life in the decades following the Civil War. For many, diabetes symbolized the nation's abandonment of a life of "frugality" in favor of "lazy comforts." Most striking to the modern reader is the widespread belief among physicians that diabetes primarily afflicted the affluent (in contrast to the early twenty-first century, when, at least in the United States, it was more often considered a disease of poverty). To some extent medical reasoning proved sound. Not only did the monied classes have a better chance of surviving to an age when they might develop diabetes, but also they were more likely to receive medical care (and thus a diagnosis) and to eat enough to get fat (and obesity was considered a risk factor). Race and class stereotypes buttressed the association between diabetes and wealth: diabetes was held to flourish where "civilized humanity" abounded, afflicting refined individuals with more advanced nervous systems. Thus, what appeared to be a relatively low rate among African Americans was explained by recourse to racist beliefs about their "primitive"

nature and lack of "nervous strain." In contrast, Jews, stereotypically cast as "neurotic" city dwellers, overweight, and highly ambitious, were widely believed to experience rates up to six times greater than any other "race."

Modernity and race could not, however, explain why a particular individual developed the disease. Here physicians turned to an entire grab bag of possibilities. In addition to obesity and nervous strain, hereditary explanations proved popular, but almost anything seemed capable of triggering the release of sugar into the urine. Among the many causes considered were infections, trauma, sexual excesses, syphilis, and tuberculosis.

As physicians spun theories about the causes of diabetes, they also faced the challenge of attending to an increasing number of diabetics. They learned quickly, as Aretaeus realized centuries earlier, that polydipsia (excessive thirst) and polyuria (excessive urination) made diagnosis relatively easy. They recognized as well that among the young the disease usually began abruptly, often ending in death within a few years, whereas among the elderly the onset tended to be gradual and the symptoms—at least initially—were milder. Despite this distinction, the consensus was that all diabetic symptoms hailed from the same fundamental pathology, differing in degree rather than kind.

Treatment, in contrast to diagnosis, proved daunting. Only restricted diets seemed capable—and then just barely—of extending a patient's life, yet even here few agreed whether it was best to restrict carbohydrates, fats, protein, or simply calories. In the early twentieth century, the American physician Frederick M. Allen advanced a diet that was widely employed, but it proved so restrictive that it became known as "starvation therapy." Although its advocates insisted that they were extending the lives of their patients, its critics recoiled at the image of some of the so-called success stories. In one particularly famous case, three years of Allen's diet had left 15-year-old Elizabeth Hughes—daughter of Charles Evans Hughes, U.S. secretary of state and later chief justice of the U.S. Supreme Court—weighing under 50 pounds and barely able to walk.

Insulin. The discovery of insulin in 1921/1922 changed forever the lives of diabetics. Working at the University of Toronto, John J. R. Macleod, Frederick G. Banting, Charles H. Best, and James B. Collip carried out the scientific work leading to the isolation of this pancreatic hormone. Medical researchers had only known about the importance of the pancreas in the pathophysiology of the disease for about 30 years, with earlier research tending to focus on the kidneys and the liver. In 1889, Oskar Minkowski and Joseph von Mering at the University of Strassbourg had removed the pancreas from a dog while studying this organ's role in digestion, only to learn that they had unintentionally created a diabetic condition. Subsequent research by them and others drew attention to clusters of cells scattered throughout the pancreas, known as the islets of Langerhans and believed to be the site of an "internal secretion." Macleod and his team were the first to isolate this secretion, which they subsequently named insulin.

Preliminary tests of insulin produced remarkable results, reducing blood sugar levels dramatically. Within five weeks of taking her first insulin injections, Elizabeth Hughes had gained 10 pounds and expressed her joy that she could now expect "a normal, healthy existence." (She lived another 50 years.) Newspapers hailed the new breakthrough, which they sometimes labeled a "cure," sharing stories about individuals who had been "rescued from death."

Early scientific work on diabetes often occurred outside the United States, largely because of the lack of financial and institutional support for research that marked much of the nineteenth century. By the 1920s, however, this was no longer the case. For example, the American drug firm Eli Lilly played a critical role in the early development and production of insulin and underwrote large clinical trials supervised by Elliott P. Joslin in Boston, Frederick M. Allen in Morristown, New Jersey, and Rollin T. Woodyatt in Chicago. These insulin trials eventually revealed a more complex story than the initial triumphalist accounts presented. Insulin was not a cure, but a new, imperfect tool for lowering blood glucose levels. It saved many lives, but it left diabetics in an ongoing battle to stave off renal failure, cardiac arrest, blindness, neuropathy, and gangrene. Additionally, insulin-dependent diabetics faced the new risk of taking too much of the hormone and triggering a hypoglycemic attack that could lead to diabetic coma and death.

Insulin made no impact on the increasing diabetes rates. Fifteen years after the discovery of insulin, the U.S. Public Health Service (PHS) conducted a National Health Survey, which revealed steady increases in diabetes morbidity and mortality rates. After World War II, with time and money for research more readily available, the PHS undertook a community study to estimate the number of undiagnosed diabetics in the nation. It chose Oxford, Massachusetts (Elliott P. Joslin's hometown), which it considered representative of small towns throughout the country, and in 1947 tested 70 percent of its almost five thousand inhabitants. The results proved disturbing: for every four individuals known to have diabetes, investigators uncovered three undiagnosed cases. As an article in *Time* magazine pointed out, this finding suggested that there were "some 2,800,000 U.S. diabetics…and half of them don't know it" ("Diabetes Up?" *Time*, 50, no. 14 (6 October 1947), p. 52).

Increased Detection. The government proposed additional detection drives, followed by "prompt treatment by the family physician." This program received the full support of the American Diabetes Association, a group founded by specialists in 1940 to raise awareness of the disease, advocate on behalf of patients, and explore better ways of treating those struggling to manage their blood sugar levels. With hindsight, it is easy to see some of the flaws of the 1947 investigation: the PHS's assumptions about who was most at risk of developing diabetes led it to choose a town whose population lacked the racial, ethnic, and economic diversity of much of the rest of the country. By framing the solution, moreover, around private medical practice—public-health departments were directed to refer rather than treat anyone they found to have high blood sugar—this approach further excluded anyone who lacked the resources (or the inclination) to seek private care.

Yet evidence began mounting in the second half of the twentieth century that the populations experiencing the greatest rate increases in diabetes incidents were not the middle-class whites common in Oxford, but Native Americans and African Americans. Whether this finding revealed an actual increase or a more accurate assessment of the disease burden in these communities remained unclear. Nor was there consensus about what might be responsible for the higher rate, although the search began immediately for a possible genetic predisposition among the newly designated "at-risk" populations. However, the government's long-held policy of collecting information about race and ethnicity, but not about class, meant that explanations grounded in biological race may have reflected—as they continue to do today—the kind of data collected rather than the actual distribution of the disease.

In 1974, as diabetes became the fifth leading cause of death in the United States, Congress passed the National Diabetes Mellitus Research and Education Act. Funding increased immediately for basic research into the causes, cures, and means of preventing diabetes; to encourage the translation of new knowledge into clinical practice; and to promote the dissemination of information to diabetics and health-care professionals. The legislation also authorized the establishment of a network of Diabetes Research and Training Centers (DRTCs) to coordinate the efforts among researchers, government agencies, and health-care providers in combating the disease. These centers continue to play a vital role in the nation's response to diabetes.

In the wake of increased attention and funding, diabetes specialists sought to standardize the terminology used to differentiate different types of diabetes. This signaled a move away from the earlier belief that all diabetic symptoms stemmed from the same fundamental pathology. In 1979, insulin-dependent diabetes mellitus (IDDM) and non–insulin dependent diabetes mellitus (NIDDM) were officially distinguished from one another. The former, known today as type 1, is considered an autoimmune disease that attacks the beta cells of the pancreas, rendering the body incapable of producing insulin. The latter, known

as type 2, has a more obscure pathology but is marked by the tissues' resistance to the insulin produced by the pancreas. A third common form of the disease, gestational diabetes, occurs only during pregnancy and then disappears. Notably, the proliferation of other designations, such as latent autoimmune diabetes of adults (LADA or type 1.5) and maturity onset diabetes of the young (MODY), suggests that the definitions established in 1979 did not fully capture this condition's complex clinical picture.

Pharmaceutical Industry. As the government's interest in diabetes grew, so too did the interests of the pharmaceutical and medical technology industries. They developed oral diabetic medicines, first marketed in the 1950s, as well as new insulins, lancets, syringes, glucose monitors, test strips, and insulin pumps, enabling diabetics to better and more effectively manage their blood glucose levels. The new technologies not only turned huge profits for these industries but also altered radically both the definition and the experience of the disease. To name only one striking change, an increasing number of people learned that they had diabetes before they ever experienced the symptoms described so graphically by Aretaeus in the second century. Their "symptom," rather, was a number recorded on a device that registered the concentration of blood glucose. In this way, individuals who felt healthy were redefined as at risk and prescribed drugs that added to the industry's profit margin while treating conditions that were rendered "visible" only through clinical tests.

Investments in research and the production of new drugs and medical devices reflect the focus on acute care that has defined the American health-care system since the beginning of the twentieth century. This system draws from laboratory knowledge while overlooking social and cultural understandings of disease. It also favors research scientists and specialists rather than the clinicians and public-health experts who encounter the disease most often either by treating patients or by looking at community health risks. Guided by a disease rather than a health focus, the American system makes substantial investments in costly medical treatments and gives far less attention and investments to costly changes in the built environment that might facilitate healthier habits and ultimately disease prevention. By all measures, a medical rather than a public-health approach proved ineffective in reversing the sharp rise in diabetes morbidity and mortality rates. At the beginning of the twenty-first century, estimates were that almost 26 million people in the United States were living with the disease (CDC, 2011).

Current Assessment. In 2012, most agreed that whatever role heredity might play in predisposing someone to diabetes, such an exponential increase in rates over the previous century could only be explained by changes in how and where Americans work, eat, and live. Decreased levels of physical activity, the increased consumption of high-calorie processed foods, exposure to high stress levels, and soaring obesity rates all drove up the diabetes rate. But there the consensus ended, largely because few agreed about who or what is responsible for the changes in how Americans live and the environments they inhabit. As in the late nineteenth century, when diabetes came to symbolize a nation's abandonment of "frugality" and embrace of "luxury," diabetes in the early twenty-first century is often blamed on a culture that has chosen "excess" and "overconsumption" over moderation. Critics contend that this perspective focuses too much attention on individual behaviors and too little on the social determinants of health, which can either facilitate or obstruct an individual's decision to make healthy choices: the safety of neighborhoods, opportunities to exercise, the accessibility of healthy foods, levels of stress, and access to medical care. In short, it fails to address the fact that diabetes, once most prevalent among the middle class, has become as great a threat (if not greater) to those living in poverty.

Global statistics indicate that the United States is far from alone in its experience with diabetes. According to the World Health Organization, 171 million people lived with diabetes in 2000, and experts estimated that this number would increase to 366 million by 2030 (Wild et al., 2004, p. 1047).

Such high diabetes rates are a problem not only in most economically developed countries but also increasingly in some of the poorest countries of the world. There, urbanization, changes in the way people work, and the globalization of processed foods increasingly are producing what epidemiologists call a "double burden": countries, which are still battling malnutrition, infectious disease, and high infant and maternal mortality rates, are also struggling to reverse increasing rates of chronic disease. Although we have much still to learn about the global history of diabetes, the history of diabetes in the United States suggests that reductions in the global burden of this disease in the twenty-first century will depend upon efforts not only to increase access to quality medical care but also to address the many social determinants of health.

[*See also* **Cardiology; Centers for Disease Control and Prevention; Disease; Food and Diet; Food Processing; Health and Fitness; Medicine; Medicine and Technology; Obesity; Osler, William; Pharmacology and Drug Therapy; Public Health; Public Health Service, U.S.; Race and Medicine; Research and Development (R&D);** *and* **World Health Organization.**]

BIBLIOGRAPHY

American Diabetes Association. *The Journey and the Dream: A History of the American Diabetes Association.* New York: American Diabetes Association, 1990. ADA's own account of the first 50 years of its existence.

Barnett, Donald, and Leo Krall. "The History of Diabetes." In *Joslin's Diabetes Mellitus,* edited by C. Ronald Kahn et al., 14th ed., pp. 1–20. New York: Lippincott Williams & Wilkins, 2005. Concise overview of major moments in the history of diabetes, from the early twentieth century to the present.

Bliss, Michael. *The Discovery of Insulin,* Chicago: University of Chicago Press, 1982. Classic story about insulin's discovery, with special attention to the scientists' personalities and the controversy over who deserved the Nobel Prize.

Centers for Disease Control and Prevention (CDC). *National Diabetes Fact Sheet, 2011.* http://www.cdc

.gov/diabetes/pubs/pdf/ndfs_2011.pdf (accessed 16 October 2012).

Feudtner, Chris. *Bittersweet: Diabetes, Insulin, and the Transformation of Illness*. Chapel Hill: University of North Carolina Press, 2003. Historical study of insulin's transformation from an acute disease to a chronic illness. Based on letters between Elliott Joslin and his patients.

Greene, Jeremy. *Prescribing by Numbers: Drugs and the Definition of Disease*. Baltimore: Johns Hopkins University Press, 2007. Examines the way new diagnostic tests and pharmaceuticals have radically altered contemporary understandings of chronic disease.

Humphreys, Margaret, et al. "Racial Disparities in Diabetes a Century Ago: Evidence from the Pension Files of U.S. Civil War Veterans." *Social Science and Medicine* 64 (2006): 1766–1775. Demonstrates low rates of diabetes among black Civil War veterans and concludes that high rates today among black males reflect historical developments.

Mauck, Aaron Pascal. "Managing Care: The History of Diabetes Management in Twentieth-Century America." PhD diss. Harvard University, 2010. Examines links between the rise in diabetes rates and changes in the management of chronic disease over the course of the twentieth century.

Purdy, Charles. *Diabetes: Its Causes, Symptoms, and Treatment*. Philadelphia: F. A. Davis, 1890.

Tattersall, Robert. *Diabetes: The Biography*. New York: Oxford University Press, 2009. Concise and lively history of diabetes from ancient times to the present.

Tuchman, Arleen Marcia. "Diabetes and Race. A Historical Perspective." *American Journal of Public Health* 101, no. 1 (2011): 24–33. Critical look at scholarship, both past and current, that purports a link between diabetes and biological race.

Tuchman, Arleen Marcia. "Diabetes and the Public's Health." *The Lancet* 374, no. 9696 (2009): 1140–1141. Brief exploration of why twentieth-century public health efforts to reduce diabetes rates have not worked.

Wild, Sarah, Gojka Roglic, Hilary King, Anders Green, and Richard Sicree. "Global Prevalence of Diabetes." *Diabetes Care* 27, no. 5 (2004): 1047–1053. http://www.who.int/diabetes/facts/en/diabcare0504.pdf (accessed 16 October 2012). Uses data on diabetes prevalence to estimate rates for 2000 and to project rates for 2030.

Arleen Marcia Tuchman

DINOSAURS

Dinosaur paleontology is an international endeavor. Dinosaurs were first recognized as a distinct group of extinct animals by Richard Owen, an English comparative anatomist, in 1842. Since then, many important contributions have been made by American scientists on American fossils.

Strange fossils were known from America even before Owen's time. In 1787, Caspar Wistar reported a large thigh bone to the American Philosophical Society. William Clark, co-leader of the Lewis and Clark expedition, recorded large fossil ribs in present-day Montana in 1806. Edward Hitchcock, an Amherst professor, made and described an enormous collection of fossil footprints in New England. All of these fossils are now thought to have been dinosaurs.

In 1855, Ferdinand Vandiveer Hayden collected Cretaceous fossil teeth in present-day Montana and gave them to comparative anatomist Joseph Leidy. Recognizing their similarity to certain English fossils, Leidy designated *Palaeoscincus Troodon, Trachodon,* and *Deinodon* as the first American dinosaurs. In 1858, Leidy described another Cretaceous dinosaur, *Hadrosaurus foulkii,* from a relatively complete fossil from Haddonfield, New Jersey. With its greatly reduced forelimbs, Leidy concluded that its posture was kangaroo-like. *Hadrosaurus,* the world's first mounted dinosaur, created a sensation when it was exhibited in 1868 at Philadelphia's Academy of Natural Sciences.

In 1877, local amateurs discovered a superabundance of gigantic Jurassic dinosaurs in multiple localities in Colorado and Wyoming. Yale's Othniel Charles Marsh and Philadelphia's Edward Drinker Cope raced to describe and name these new forms first, including *Apatosaurus, Camarasaurus, Allosaurus, Stegosaurus,* and many other iconic dinosaurs. They resorted to underhanded tactics to acquire specimens and penned hasty, poorly illustrated descriptions to establish priority. In 1890, their unseemly competition exploded into public scandal when Cope and Marsh exchanged accusations of dishonesty, incompetence, and misuse of government funds on the front page of the *New York Herald*. Both of their reputations were tarnished by the scandal.

A new generation of museum-based paleontologists succeeded Cope and Marsh in the 1890s, led by Henry Fairfield Osborn of New York's American Museum of Natural History. Osborn and his rivals in Pittsburgh and Chicago developed meticulous and efficient techniques for collecting, cleaning, and mounting fossils that yielded better, more lifelike results. The sensational museum exhibits they created, including gigantic sauropods and Osborn's *Tyrannosaurus rex*, as well as the media attention these exhibits attracted, helped make dinosaur a household word. Meanwhile, casts of Pittsburgh's *Diplodocus* mounted in European and Latin American museums disseminated *dinomania* internationally.

The heyday of American museum paleontology was the 1920s, when ambitious, expensive expeditions scoured the globe for fossils. Walter Granger, for example, found dinosaur eggs and many new kinds of Cretaceous dinosaurs during a high-profile expedition to central Asia. Dinosaur paleontology diminished from the Depression through World War II.

A dinosaur renaissance began in the postwar period. A worldwide expansion of collecting produced thousands of new specimens. New techniques, like histology, were applied to dinosaurs, raising new questions regarding physiology, behavior, and evolution. John Ostrom, Robert Bakker, and others worked on dinosaur metabolism, for example, in the 1960s and 1970s. John Horner used an abundance of fossil and trace evidence to draw new conclusions about dinosaur nesting behavior and ontogeny among *Maisaura* in the 1980s. Abundant feathered dinosaurs, discovered in China since the 1990s, support the commonly held hypothesis that birds are descended directly from dinosaurs. There are more dinosaur researchers now than ever. Dinosaur paleontology is pursued at many American universities and is a staple in America's natural science museums.

Today, dinosaurs are pop-culture icons. Dinosaurs like *Tyrannosaurus rex* are commonly known. Books and movies about dinosaurs, including *Jurassic Park*, are best sellers and blockbusters. Prehistoric-themed television shows like *Dinosaur Train* are immensely popular with children. Museums are often judged by the quality and quantity of their dinosaur exhibits.

[*See also* **Academy of Natural Sciences of Philadelphia; American Museum of Natural History; Cope, Edward Drinker; Evolution, Theory of; Hitchcock, Edward; Leidy, Joseph; Lewis and Clark Expedition; Marsh, Othniel Charles; Museums of Science and Natural History; Paleontology;** *and* **Science.**]

BIBLIOGRAPHY

Brinkman, Paul D. *The Second Jurassic Dinosaur Rush: Museums & Paleontology in America at the Turn of the Twentieth Century.* Chicago and London: University of Chicago Press, 2010. Draws heavily on archival sources and tells the lesser-known story of paleontology in America in the post Cope–Marsh era.

Colbert, Edwin H. *Men and Dinosaurs: The Search in Field and Laboratory.* New York: E. P. Dutton and Company, 1968. A thorough and authoritative account of dinosaur paleontologists and their achievements.

Paul D. Brinkman

DIPHTHERIA

Diphtheria is a serious acute disease that usually strikes young children. It is most frequently spread by person-to-person contact or inhalation. Its first symptoms resemble the common cold; within two or three days a pseudo-membrane may form in the upper respiratory organs. Case mortality for the youngest and oldest victims may be as high as 20 percent. Overall, between 5 and 10 percent of the cases lead to death, which is generally caused by suffocation. The diphtheria bacillus secretes a toxin that can cause other symptoms, including inflammation of the heart, skin lesions, and peripheral neuropathy. Treatment consists of antitoxin, antibiotics, oxygen, bed rest, and airway maintenance.

Diphtheria has probably existed since antiquity, but it was not identified as a specific disease until the nineteenth century. Mid-seventeenth-century epidemics in New England and Virginia were probably the earliest incidences of diphtheria in English-speaking North America. The particularly deadly New England "throat distemper" epidemic of 1735–1740 was almost certainly diphtheria. In his 44-page pamphlet on "angina suffocativa or sore throat distemper," published in 1771, Samuel Bard of New York urged immediate isolation of patients to prevent spread of the disease.

The work of Pierre Bretonneau of Tours, France, underlies much of the modern understanding and treatment of diphtheria. In addition to giving diphtheria its name and establishing criteria for its clinical diagnosis, in 1825 Bretonneau performed the first successful tracheostomy for diphtheria. Tracheostomy (or tracheotomy) is a surgical procedure that creates an opening through the neck into the windpipe so that a patient may breathe despite the presence of the pseudo-membrane.

Diphtheria established itself as an endemic disease in American cities during the 1850s, and in the 1880s it emerged as the leading killer of children in the United States. Fear of spread of diphtheria justified measures ranging from placarding the homes of diphtheria cases to forced removal to contagious disease hospitals of the most impoverished patients.

Intubation through the mouth, introduced in 1885 by Joseph O'Dwyer of the New York Foundling Hospital, soon largely superseded tracheotomy. O'Dwyer's technique and new instruments for opening access to the windpipe quickly became standard in the management of diphtheria cases.

The etiology of diphtheria was established, in 1883 and 1884, when bacteriologists Edwin Klebs in Switzerland and Friedrich Loeffler in Germany, respectively, identified and cultivated the Klebs-Loeffler bacillus (now known as *Corynebacterium diphtheriae*). Over the next few years the practice of swabbing the throat of suspected diphtheritics become routine in American cities. The swabs were cultured for laboratory confirmation of the clinical diagnosis.

In 1888, Emile Roux and Alexandre Yersin of the Pasteur Institute in Paris filtered the bacillus out of its culture fluid and found that the fluid itself could sicken laboratory animals, thus demonstrating the existence of an exotoxin that acted apart from the bacillus itself. Two years later, Emil Behring of Berlin established that patients' bodies produce an antibody capable of neutralizing the toxin. He further showed that this antitoxin could be harvested in the blood serum of animals that had been injected with the toxin.

In December 1891 Behring electrified the medical world with his dramatic cure of a dying child using antitoxin taken from a horse. Diphtheria is therefore especially significant in modern medical history because antitoxin was the first effective therapeutic developed through bacteriological research. Anti-toxin quickly became available on a commercial scale.

At its 1895 meeting, the American Pediatric Society resolved that "the evidence thus far produced regarding the effects of diphtheria antitoxin, justifies its further and extensive trial." In 1897 a committee of the society concluded that treatment with antitoxin greatly reduced the likelihood of death and the necessity for intubation.

Despite the committee's report and similar findings from others, skepticism about the efficacy of antitoxin remained and recent retrospective analysis has shown that the effect of these early trials of antitoxin may not have been as great as proponents claimed. Nonetheless, American doctors continued using antitoxin, American pharmaceutical companies and public-health agencies continued producing it, and American medical scientists continued research into antitoxin standardization and development. Since 1997, antitoxin has been available only from the Centers for Disease Control and Prevention. Physicians and health authorities maintain their interest in laboratory confirmation of diagnosis, although it is clear that the likelihood of curing a case is greatest if antitoxin is administered well before the throat culture is ready to be conclusively read.

In 1892, bacteriologist Hermann Biggs established the New York City Health Department's laboratory of pathology and bacteriology, the first such facility in the world. Under the direction of William H. Park, diphtheria diagnosis and antitoxin production and distribution soon became the laboratory's principal activities, setting an example for public-health agencies around the country.

At the Massachusetts Department of Health in 1909, Theobald Smith demonstrated that a mixture of toxin and antitoxin can confer immunity to diphtheria. In 1916 Alfred Hess of New York introduced active immunization with toxin-antitoxin. Despite reservations about it—because of its potential toxicity and possible allergic reactions to the horse serum—public-health authorities encouraged vaccination with toxin-antitoxin.

In 1913 Bela Schick, then at the University of Vienna and later to move to New York, announced a simple test to determine susceptibility to diphtheria. In contrast to smallpox vaccination, which had been mandatory in many American jurisdictions for decades, the campaign for vaccination against diphtheria relied on education of the public through the mass media. An example is the extensive press coverage of the relatively small 1925 outbreak in Nome, Alaska. Headline stories around the United States described the great effort taken to get antitoxin to Nome in a 5-day, 674-mile sled relay involving no fewer than 24 mushers and 150 dogs.

Although contemporaries had reason to question the efficacy of antitoxin in its early years, product standardization and further scientific developments were probably responsible for dramatic declines, first in case mortality and later in incidence. During the 1920s, the French veterinarian Gaston Ramon showed that diphtheria toxin could be rendered harmless although still able to stimulate the body's production of antitoxin. Called toxoid or anatoxine, it was soon was adopted as an immunizing agent.

During the 1920s, the United States annually experienced about 150,000 diphtheria cases and 13,000 deaths. By 1945, the number of cases had dropped to fewer than 20,000. In the postwar period, immunization has led to the virtual eradication of diphtheria in the United States. Fifty-two cases were reported between 1980 and 2004 and a total of only 5 since the year 2000.

[*See also* Disease; Germ Theory of Disease; Medicine; *and* Public Health.]

BIBLIOGRAPHY

Colgrove, James. "The Power of Persuasion: Diphtheria Immunization, Advertising, and the Rise of Health Education." *Public Health Reports* 119, no. 5 (2004): 506–509.

Condran, Gretchen A. "The Role of Medicine in Mortality Decline: The Case of Diphtheria in the Late Nineteenth Century." *Journal of the History of Medicine and Allied Sciences* 63, no. 4 (2008): 484–522.

"Diphtheria." In *Epidemiology and Prevention of Vaccine-Preventable Diseases. The Pink Book: Course Textbook.* 12th ed. Atlanta: Centers for Disease Control and Prevention, 2012. http://www.cdc.gov/vaccines/pubs/pinkbook/dip.html.

Hammonds, Evelynn M. *Childhood's Deadly Scourge: The Campaign to Control Diphtheria in New York City, 1880–1930.* Baltimore: Johns Hopkins University Press, 1999.

Liebenau, Jonathan. "Public Health and the Production and Use of Diphtheria Antitoxin in Philadelphia." *Bulletin of the History of Medicine* 61 (1987): 216–236.

Ziporyn, Terra. *Disease in the Popular American Press: The Case of Diphtheria, Typhoid Fever, and Syphilis, 1870–1920.* New York: Praeger, 1988.

Edward T. Morman

DIPLOMACY (POST-1945), SCIENCE AND TECHNOLOGY AND

The role of science and technology as instruments of American foreign policy is often overlooked, in part because it calls for integrating insights from academic disciplines that barely, if ever, intersect. The field of diplomatic history and American foreign policy has long since moved beyond the narrow study of the relations between high-level state officials to incorporate the place of culture, ideology, and gender into its analyses of interstate relationships—but it usually stops short of science and technology. Of course there is an extensive literature on the role of the atomic and hydrogen bombs in diplomacy beginning in 1945, but even here the weapons are taken as an already available negotiating tool and reduced to their performance characteristics. The dynamics of the vast research and development complex that produced them is not considered, although it was a constitutive component of American power in the postwar period and shaped the strategic options that were available to negotiators in interstate relations. Historians of American science and technology, for their part, have enriched our understanding of the many ways in which research and development in academia, in think tanks, in industry, and in government laboratories were harnessed to the chariot of the national security state that emerged after 1945. They have shed light on the functioning of the national innovation system but, although they recognize that the system provides invaluable resources for the expression of American power, they usually do not study how those resources were mobilized as instruments of statecraft. Put rather crudely, if diplomatic historians have tended to focus on use and to ignore the innovation that lies behind it, historians of science and technology have tended to focus on innovation, pushing use to the background. The former black-box the knowledge (in all its forms) that bubbles up from the national research system and that is constitutive of American power. The latter probe into the black box and describe the intricacies of the processes therein that produce knowledge, but they rarely follow science and technology beyond the laboratory and the national frame as it circulates into the wider world.

In recent years there has been a welcome move by scholars on both sides of the divide to dissolve the disciplinary boundaries between the histories of science and technology and that of American diplomacy and foreign policy (Adas, 2006; Engerman, 2007; Hecht, 2012; Krige, 2006; Krige and Barth, 2006; LaFeber, 2000; Westad, 2000). A narrow conception of the Cold War as a bipolar confrontation between the superpowers has yielded to a far more nuanced understanding of the postwar period. Collaboration and cooption coexisted with competition, nation building with war mongering, spectacular demonstrations of scientific and technological prowess with wide-ranging restrictions on the flow of knowledge (Connelly, 2008; Cullather, 2010; Krige, 2008; Manela, 2010; Van Vleck, 2009; Wang, 2010). This essay is situated within this emerging framework. It takes a thematic and conceptual approach, rather than a predominantly chronological one, highlighting the many different ways in which the natural and social sciences and technologies were mobilized as instruments of foreign policy after World War II.

Leadership is the capacity to move proactively, and ahead of others, into uncertain territory, taking bold initiatives and investing resources to meet a difficult challenge, inspiring other stakeholders, and even demanding that they 'follow' for fear of being marginalized. Here the ideology of leadership serves as a focal point that gives meaning *both* to the U.S. government's sense of its role in the world *and* to the federal government's massive investment in research and development after 1945. It was as leader of the "free world" that successive American administrations devoted vast resources to science and technology to confront and fight their enemies, to collaborate with friend and foe alike, to modernize the third world, to "Americanize" the globe, and to symbolically express national power and prestige. Disaggregating the relationship among science, technology, and foreign policy along these axes has its limitations—some topics are handled more than once under different themes, there is a greater emphasis on the first two or three decades of the Cold War than on the later period (as is to be expected with archival-based scholarship), and transitions are sometimes abrupt. At the same time, by their very sweep, these fragments collectively hint at the immense richness of a field of research that is still fluid and in formation.

Periodization. The geopolitical situation was not static during the Cold War. Containment and confrontation gave way to "peaceful coexistence" and détente, followed by another period of intense confrontation in the 1980s that ended with the implosion of the Soviet Union. Nor was the course of history in the 1950s and 1960s simply defined by relationships between

Washington and Moscow (and their allies). Its physiognomy was also shaped by the emergence of new and influential actors on the global stage: the United Nations and its agencies and the tens of decolonized states in Africa and Asia whose historical trajectories moved at a different rhythm, intersecting with those of the dominant powers in complex ways. In addition, the world order began to be reshaped in the 1970s by technologically supported global interconnections that spawned interdependence, diluted state sovereignty, and enabled a power shift in favor of nonstate actors (multinational corporations, giant foundations, and well-organized issue-driven pressure groups, for example). This essay cannot hope to do justice to this rich tapestry; it seeks only not to lose sight of it.

Leading. The United States emerged from World War II as the major economic, industrial, and military power on the globe with broad international support for its democratic ideals. We should not forget, however, that there was also a major asymmetry in scientific and technological capability between America and its allies and enemies. The debate on how best to maintain that preeminence got under way before the war was over. It was focused by the consolidation of Communist regimes in Eastern Europe and in China in the late 1940s, followed by the outbreak of the war in Korea in June 1950, and it was given urgency by the first Soviet atomic test in August 1949. National Security Council Document 68 (NSC68), an official doctrine approved in April 1950, described the United States as being faced with an existential threat from an implacable enemy that espoused values that were completely irreconcilable with those that the country held dear. The United States could not meet "hordes with hordes," however. For President Eisenhower, the security of the West, a budget that balanced military and civilian needs, and the protection of domestic liberties and pluralistic institutions could only be achieved by substituting firepower for manpower, quality for quantity. The permanent military preparedness needed to meet the Communist threat called for ceaseless innovation to maintain a comparative technological advantage over friend and foe

alike. This quest for scientific and technological preeminence, or "leadership," was called for by Vannevar Bush in 1945, was enthusiastically embraced by the scientific elite, and was fueled by Soviet achievements in the 1950s. It provided—and still provides—the dominant rationale for the sponsorship of research and development (R&D) both for and beyond the needs of the national security state. The federal government's R&D budget, initially modest, increased dramatically after the Korean War broke out and was given another enormous boost by the launch of *Sputnik* in 1957. By 1967, at the height of the Apollo lunar program, federal spending on defense and nondefense R&D had reached a staggering $82 billion (in constant FY2008 dollars), more than quadrupling in a decade. Private industry added half as much again (American Association for the Advancement of Science [AAAS], 2008, 1953).

The pursuit of scientific and technological "leadership" resonated with, and gave substance to, a national security ideology premised on the assumption that peace and freedom were indivisible and that a threat to democracy anywhere was a threat to democracy everywhere, so also to the United States. The "internationalists" in Congress who promoted this view gradually asserted their authority over "isolationists" who balked at foreign entanglements. As Michael Hogan puts it, it became widely accepted by 1950 that the "leadership of the free world was a sacred mission thrust upon the American people by divine Providence and the laws of both nature and history" (Hogan, 1998, p. 15). Assuming the mantle of global leadership—some would say dominance—and proclaiming its rejection of imperial modes of governance, the United States sought to make the world safe for democracy.

The quest for leadership is the node that fuses the pursuit of scientific and technological superiority with a foreign policy that aspires to political, economic, ideological, and military dominance. It binds the innovation system to America's global ambitions, embedding the national in the international. Securing scientific and technological "leadership" became the single most important argument for the ongoing support for R&D by government and industry alike, a spontaneous ideology that was so widespread as to be invisible. Similarly, falling behind in any significant domain of science and technology ignited declinist fears and encouraged the view that its competitors were overtaking the United States. After 1945, constant innovation and the rapid transformation of new knowledge into useful devices were widely understood to be necessary for maintaining national vitality and for the successful projection of American power abroad.

Confronting.   From the dawn of the atomic age, nuclear weapons and their delivery systems were at the heart of international diplomacy. Their capacity to change the dynamics of foreign policy was evident immediately after the Trinity test in July 1945 that confirmed the functioning of the implosion device on the plutonium bomb. President Truman's former enthusiasm to have Stalin enter the war against Japan turned to chagrin: here was another way to end the war quickly without offering the Soviets a foothold in the Pacific theater. The atomic bomb was the winning weapon, and the United States did all that it could to retain its monopoly after 1945. The first successful Soviet test in 1949 created a window of opportunity for the development of the hydrogen bomb, opposed strongly by some physicists as genocidal, but supported equally volubly by others as essential to security and to stop the United States from becoming a second-class world power. The arms race was under way, carried on a wave of scientific and technological enthusiasm that transformed the practice of the sciences and the training of physicists, in particular, and that satisfied the open-ended requests of all the armed services for the most up-to-date delivery systems. In parallel, first Britain (1952), then France (1960), and then China (1964) successfully tested atomic weapons as much to enhance security as to affirm their status as great powers: one was nuclear or one was negligible, as a French minister of defense put it in 1963.

By 1955 the four powers (Britain, France, the United States, and the USSR), meeting in Geneva, recognized that nuclear war with the weapons already available would be suicidal for the combatants. Nuclear stockpiles continued to grow

nevertheless, driven by mutual suspicion and distrust. By 1960 the U.S. nuclear weapons stockpile counted almost 20,500 warheads; the Soviet Union's numbered some 1,600 (National Resources Defense Council [NRDC], 2002). During the 1960s the United States deployed over a thousand intercontinental ballistic missiles (ICBM) and hundreds of submarine-launched missiles in the Polaris (and later Poseidon) systems and developed accurately guided, multiple independently targetable reentry vehicles (MIRVs) for both. The Soviet ICBM arsenal matched this in size by the end of the decade. During the 1970s the United States' advantage was gradually offset by the formidable SS-17s and SS-18s and then the SS-20s that were protected in hardened silos, were highly accurate, could carry up to as many as 10 MIRVed warheads, and benefited from improved command and control systems.

Each incremental increase in capability rendered the use of nuclear weapons less likely. The Cuban missile crisis in 1962, which brought the world to the brink of nuclear war, persuaded President Kennedy and President Johnson after him that nuclear weapons were effective in deterrence but unusable in offense. Nuclear strategy was devised not to win wars but to avert them, as one theorist put it. Although thousands of warheads were stationed in the European theater, new policies like "flexible response" were devised to fight wars using conventional means behind the nuclear shield.

Several international efforts were made to contain the threat of nuclear war and to limit the escalation of nuclear capabilities. In 1963 the superpowers signed a partial-test-ban treaty that forbade all nuclear tests except those conducted underground. A treaty on the nonproliferation of nuclear weapons (the NPT) was signed in London, Moscow, and Washington in 1968. Agreement between its proponents was only possible because it did not require any of the existing nuclear powers to renounce their weapons, while denying them to any signatory that did not already have them, most notably Germany. The NPT was followed in 1972 by a treaty limiting the use of anti–ballistic missile systems to defend countries from an attack by nuclear-tipped strategic missiles (the ABM treaty).

These treaties were the high-water mark of nuclear diplomacy between the superpowers. The United States unilaterally withdrew from the ABM treaty in 2002. A comprehensive test-ban treaty was adopted in 1996 but in 2012 still waited on signature and/or ratification by numerous nuclear weapons states, including China, India, North Korea, Pakistan, and the United States.

The massive quantitative increase in the nuclear arsenals of both superpowers in the 1960s and 1970s was followed by a qualitative change in the destructive power of America's arsenal in the 1980s. Weapons designers at Livermore and Los Alamos, the two major national laboratories, worked on so-called third-generation nuclear weapons (succeeding the atomic and hydrogen bombs). These devices focused the explosive power of a bomb, increasing the energy it delivered on a target a thousandfold. They could also be used to produce intense beams of directed energy, as in an X-ray laser "pumped" by a concentrated nuclear explosion. Complementing these innovations, President Ronald Reagan's Strategic Defense Initiative (SDI) planned for the deployment of a variety of high-power directed beam weapons to destroy incoming enemy ballistic missiles. Although SDI was presented as purely defensive, it also offered offensive possibilities. Placed in outer space, a variety of powerful beams could do considerable damage in nanoseconds to installations on Earth. American policy makers insisted that this option was of no strategic interest, even if it was technically possible. The Soviets were not convinced and refused to accept on trust that a space-based offensive capability would not be used against them simply because the United States' existing ballistic missile system rendered it strategically redundant.

The deployment of dual-use technologies in space increased after the Reagan years, stretching to the limit the constraints imposed by the treaty on the "peaceful uses" of outer space negotiated in the early 1960s. In 2012, space-based communications systems were strategically important as force multipliers in war. They also served as information highways that were central to the functioning of the global economy. The development of antisatellite weapons by several countries, including China, posed a grave threat to these assets.

The credibility of a strategic posture based on ballistic missiles topped with nuclear weapons depended on intelligence of many kinds that was derived from many different sources. The Corona series of spy satellites provided invaluable images of Soviet missile complexes, identified ABM sites and defensive missile batteries, and provided inventories of Soviet bombers and fighters and data on surface and submarine ocean fleets, in addition to verifying arms control agreements. Other modes of information gathering relied directly on human insight and learning. The Central Intelligence Agency (CIA) and other government agencies mobilized countless scientists and engineers as informal intelligence gatherers to gain access to security-related information, much of which was readily available in an open society. The agencies used these intelligence gatherers to glean what knowledge they could of the state of military programs abroad by chatting with colleagues from the Soviet Union and other sensitive countries at international conferences and workshops, debriefing them when they returned home.

The humanities and social sciences played their part too. The Department of State placed great store in training diplomats in foreign languages for new international responsibilities. The military needed people who were competent in foreign languages for intelligence purposes. They also invested over $20 million in machine translation between 1945 and the mid-1960s, one of the first uses of computers for nonnumerical tasks. Organizations like the American Institute of Physics supported the translation of leading Soviet science periodicals into English. The Eisenhower administration was so convinced of the strategic importance of learning foreign languages that it was identified as second only to improved training in mathematics and science in the National Defense Education Act passed in 1958, in the wake of the *Sputnik* shock. Area studies also blossomed with support from both the federal government and foundations. Columbia University's Russian Institute and Harvard's Russian Research Center led the way in the late 1940s; by the mid-1960s Russian studies were offered in many major universities across the nation. "Sovietology" sought to combine academic excellence with political pluralism to produce useful knowledge for those who made American foreign policy. The Cold War channeled billions of dollars into R&D that satisfied the exigencies of the national security state, but it also provided unprecedented resources that secured America's preeminence at the cutting edge of knowledge production in many domains of natural and social science and engineering.

Fighting. A long peace reigned over much of the industrialized north since 1945 despite—or thanks to—the massive concentration of lethal weaponry in the hemisphere. The suicidal consequences of unleashing their arsenals on each other, which kept their confrontation "cold," did not stop the United States and the Soviet Union (and not only them) from becoming entangled in "hot" wars, sometimes far beyond their shores in Asia, Latin America, and Africa. In 1992 Francis Fukuyama saw the collapse of the Soviet Union as signaling the "end of history"; his celebration of the triumph of liberal democracy was short-lived. The People's Republic of China emerged as a major global player, the number of nuclear powers proliferated, and radical Muslim fundamentalism gained in strength. During the late twentieth and early twenty-first centuries, the expansionist ambitions of both Iraq and Iran, an ongoing crisis of governance in Afghanistan, and the terrorist attacks of 11 September 2001 dragged the United States into a series of wars in the Middle East whose costs were enormous and whose end was difficult to define and to discern. To meet these new and different threats, the federal defense budget increased dramatically (in 2012 dollars) from just under $300 billion in 2000 to almost $700 billion in 2010 (Office of Management and Budget [OMB], 2012). In 2012, federal spending on defense and nondefense R&D (in constant 2008 dollars) was around $140 billion, an all-time high, and considerably above the peak it reached in the mid-1980s, about $95 billion (AAAS, 2008).

These dollars translate into the development and deployment of increasingly sophisticated weapons systems that hold the promise of achieving victory at terrible cost to the enemy with lim-

ited loss of American life. In the Vietnam War the then-new technology of the armed helicopter, widely known as the Huey, significantly enhanced battlefield mobility. Nearly 20 million gallons of Agent Orange and other novel herbicides and defoliants denied cover and food to enemy forces and permitted the more efficient use of aerial power, destroying 25 million acres of agricultural land in central and southern Vietnam. A group of brilliant physicists and chemists gathered together in the Jason group devised an innovative "electronic battlefield" in which hidden sensors detected the movements of trucks and guerillas through dense vegetation. The output from the sensor network was distinguished from false alarms using two IBM 360 computers, displayed on a large screen, and transmitted to dedicated attack aircraft, ideally on call at all times. Although the actual value of the system is contested, it did form the basis for what is today called Network-Centric Warfare, or a "sensor-to-shooter" link in military parlance. So, too, the origins of precision-guided air-to-ground weaponry as well as the now ubiquitous infrared search and weapon-aiming equipment can be traced back to the Vietnam War. Each conflict served as a laboratory and a proving ground for the next.

Eisenhower's ambition to replace personnel by capital-intensive nuclear weapons, to overwhelm quantitative enemy superiority with qualitative technological advantage, has become the general leitmotif of U.S. military thinking. In 2012 all three branches of the armed services had weapons that were unmatched by those of any other country. The U.S. Air Force, for example, was in the position of acquiring five different types of stealth aircraft; no other nation had any. Global positioning system (GPS) satellites combined with joint surveillance and target radars guide "drones" (unmanned aerial vehicles) that strike with ruthless precision; no one else could do this on the same scale. Military supremacy derived from scientific and technological preeminence, spectacular displays of technical prowess, embedded journalists who send optimistic reports back to the home front, and relatively few U.S. casualties (about 4,500 in nine years of Operation Iraqi Freedom) make high-tech war an attractive instrument of diplomacy for American presidents, politicians, and the public alike.

Collaborating. Collaboration is usefully distinguished from cooperation. Cooperation typically involves the circulation of knowledge through conferences, workshops, and informal face-to-face encounters, and it is usually limited to data sharing. Collaboration requires a greater investment than cooperation. In collaboration data are not simply pooled; knowledge is coproduced. This knowledge is a key resource for the stakeholders involved, whose capacity to exploit collaboration to their advantage is determined by what they bring to the table and what others want from them. No partner gives away a competitive edge in a collaborative engagement unless there is something to gain from it. Collaboration requires reciprocity for it to succeed; cooperation is less demanding.

The confrontational rhetoric of the early Cold War can easily blind one to the parallel cooperative and collaborative efforts undertaken by multiple actors in the United States to defuse tension and to build alliances. In his Annual State of the Union address in 1958, in the wake of the Soviet launch of *Sputnik*, President Dwight D. Eisenhower warned that Moscow was waging "total Cold War" in which "trade, economic development, military power, arts, science, education, the whole world of ideas—all are harnessed to the same chariot of expansion." The International Geophysical Year (IGY) was in full swing even as he spoke. The IGY ran from July 1957 to December 1958. It is usually remembered as inaugurating the space race, which animated Eisenhower's address. But it was also one of the first major cooperative events between scientists and their national societies in the United States and the Soviet Union. It engaged tens of thousands of practitioners from more than three score countries. And it provided a vast amount of data on the properties of the earth, the oceans, and the upper atmosphere. The IGY not only paved the way for closer American–Soviet cooperation in fields like meteorology and oceanography that called for global data collection, but also stimulated the creation of international scientific networks that transcended national boundaries and provided an opportunity for elites to maintain lines of communication and of solidarity, even when tensions persisted between their governments. The physicist-led Pugwash conferences that have brought together influential scientists, policy makers, and state officials since 1955 to reduce the threat of nuclear war represent just one of the most visible transnational networks that have sought to dissolve national borders and to promote peace as a dominant foreign-policy objective in a divided world.

Collaboration burrows to the core of national scientific and technological assets and can forge close international alliances on the basis of reciprocity in knowledge flows. In 1958 the United States formally suspended the draconian restrictions on nuclear sharing demanded by the 1946 McMahon Act. It now allowed for military cooperation with countries that had made "substantial progress" in the nuclear field—a form of words intended to single out the United Kingdom and to exclude France. The move was made only after weapons scientists had assured the U.S. Congressional Joint Committee on Atomic Energy that the British were well advanced with their hydrogen bomb and the United States had something to learn from them. Thereafter, there was a continuous exchange of sensitive knowledge across the Atlantic that consolidated the Anglo-American "special relationship." Britain remained a junior partner all the same. The country had to tie itself to American priorities in research to benefit from what the United States had to offer. It was obliged to sacrifice a degree of control over its "independent" nuclear deterrent. And it was also led to believe that its privileged nuclear links could be jeopardized if it did not support American foreign-policy objectives in the United Nations, Europe, and the Middle East.

The U.S. State Department and the Atomic Energy Commission also tried to exploit American scientific and technological leadership to promote the construction of a "United States of Europe" around a variety of technological platforms. In the late 1950s, with the active support of President Eisenhower, it enthusiastically supported the formation of the European Atomic Energy Community, Euratom. Euratom was formally established, along with the European Economic Community, by the Treaties of Rome signed in March 1957. In supporting a joint research effort under its auspices, Washington hoped to lever the United States' putative lead in nuclear power research, development, and production to consolidate a united Europe, to create markets for American reactor firms, and to force European governments (particularly France) to funnel limited amounts of money, brainpower, and industrial capacity into a collaborative civilian nuclear effort at the expense of national and military weapons programs.

France was again the target of American concern in the mid-1960s when it complemented its nuclear arsenal with the successful launch of its own satellite using an adapted ballistic missile developed domestically. Now it was the National Aeronautics and Space Administration (NASA) that sought to lever America's lead in the space sector by offering to share advanced rocket technology with a floundering civil, European-wide, launcher development program. Again, the hope was that the development of a robust, supranational civil satellite launcher would divert scarce resources away from national missile programs. Europe's relative penury, and its dependence on American science and technology to modernize efficiently, provided Washington with an opportunity to lever its leadership in strategic sectors like the nuclear and space programs to build a united Europe and to contain "runaway modernity" on the continent.

The promise of broad technological collaboration in civil nuclear energy was also the incentive offered to non–nuclear weapons states if they agreed to renounce acquiring weapons of mass destruction. This was the grand bargain struck in the treaty on the nonproliferation of nuclear weapons in 1968. It offered the "fullest possible exchange" of equipment, materials, and scientific and technological information pertinent to the peaceful use of nuclear energy if the then-non-weapons states formally accepted their existing status. Brokered after long and difficult negotiations between the United States and the Soviet Union, the nonproliferation treaty established an international regime that is variously described as

entrenching "nuclear apartheid," as impotent in the face of the determination of countries like Israel, India, and Pakistan to develop nuclear weapons, as unable to reign in the actions of particularly dangerous proliferators like Iran and North Korea, and as having played a major role in stabilizing world order for 40 years and beyond Cold War rivalries.

Technological denial has the inverse effect to technological collaboration. In the late 1960s NASA strongly encouraged extensive European technological participation in the development of the space shuttle, only to back off under pressure from White House staffers in the new Nixon administration. They were hostile to sharing advanced American aerospace technology when, in their view, the United States got so little in return. The results were predictable: Gaullist forces in France, determined to secure the political and commercial benefits of an autonomous access to space, successfully championed the development of a European heavy launcher, Ariane. Technological collaboration locks allies into one's orbit; technological denial reinforces aspirations to national autonomy, strains alliances, and breeds resentment.

The close relationship among science, technology, and national security after 1945 imposed tight restrictions on the free circulation of sensitive knowledge. The Atomic Energy Act of 1946 broke all precedent by declaring that all knowledge that was pertinent to nuclear weapons, including the production of nuclear material and its use for the production of energy, was restricted data and so was exempt from the sacred principles enshrined in the Freedom of Information Act. No government action was needed to classify it: it was "born secret." These tight controls were somewhat relaxed in 1954 under pressure from American industries that wanted to sell nuclear reactors as part of the Atoms for Peace program. Special arrangements were also made with some European allies, led by Britain, to collaborate on weapons development and uranium enrichment behind security walls. The cumulative effect of the born-secret policy was daunting. In 1995 the Department of Energy estimated that it held at least 280 million classified pages of material that would take nine thousand person-years of effort to review, an impossible task. This mass exploded after new restrictions on the free circulation of information were imposed after the terrorist attacks of 2001 and the fear that biological weapons might be used against the United States. In the early twenty-first century, international scientific and technological collaboration was increasingly regulated by a range of restrictions that penetrated into the laboratory, even when unclassified work was being done, frustrating American scientists and corporations and alienating traditional allies.

Modernization. In 1945 there were 51 sovereign states in the world; 20 years later, in 1965, the number had more than doubled to 117. The Third World emerged with its own identity—desperately poor, increasingly overcrowded, and trapped in a cycle of misery. No new nation engaged in the postcolonial process of state formation could remain indifferent to the competing models of society that were being promoted by Washington and Moscow to propel them into modernity. For the United States, the teeming masses posed a threat to world order, a breeding ground for Communism that had to be eradicated by transforming traditional habits and social structures.

In January 1949 President Harry Truman committed the United States to intervene on behalf of the wretched of the earth. In "Point IV" of his inaugural address he pledged to use America's preeminence in science, technology, and industry to help people in "underdeveloped areas" learn to help themselves. In one stroke he defined the Third World in terms of a lack, he undertook to help steer it on the path to development and modernization, and he singled out "our imponderable resources in technological knowledge" as the single most important instrument that would be brought to bear "to help them realize their aspirations for a better life."

The language of "development" provided a policy goal that all could share. It fused the local aspirations of national elites with the global ambitions of the industrialized powers. It was not tarnished by the presumption of superiority that had inspired the "civilizing mission" embarked on by the European imperialist powers. It was denuded of political overtones by defining progress as the mobilization of science and technology for human improvement. And it opened space for the U.S. government and foundations to merge humanitarian sentiments with expansionist ambitions, benevolently making the world over in America's image.

The superpowers rushed to build dams, hydroelectric schemes, and model farms to curry favor with potential client states, but with little regard for local conditions. Rivers silted up, good earth turned saline, generators lay idle, and tractors bogged down in the mud or were pillaged for metal.

Eisenhower's Atoms for Peace program, announced to rapturous applause in the United Nations in December 1953, raised utopian spirits. Its aim, the president said, was to exploit the peaceful atom for agriculture and medicine and to provide energy for the "power-starved areas of the world." In August 1955, 10 years after the atomic bombing of Japan, U.S. authorities launched a major campaign to distribute nuclear reactors for research and power throughout the developing world. A huge international conference in Geneva was presided over by Homi J. Bhabha, secretary to the Department of Energy in India. "For the full industrialization of the underdeveloped areas, for the continuation of our civilization and its further development, atomic energy is not merely an aid: it is an absolute necessity," Bhabha enthused. It would enable the teeming millions in his country to *"reach a standard of living equivalent to the US level"* (Krige, 2010, p. 152, Bhabha's emphasis). Bhabha thus sanctified the work of social scientists who identified modernization with convergence on the American social model. In the late 1950s Walt Whitman Rostow, later to be a national security adviser to Lyndon B. Johnson, charted the "stages of economic growth" that every country needed to pass through to arrive at that goal. His "non-Communist manifesto" is remembered as emblematic of the ideology that welded together a diverse group of social scientists dedicated to identifying the economic, political, and cultural obstacles to the American transformative agenda so as better to eliminate them.

Demography, not a lack of atomic energy, was the greatest threat to development. It was given added urgency after the first Chinese nuclear test in 1964—in merely 15 years a backward peasant society had taken giant strides into modernity with Soviet, not American, help. The Ford and Rockefeller Foundations, strongly encouraged by the administration, pumped millions of dollars into training local officials on the ground in India to familiarize villagers with the benefits of contraception, sometimes forcibly sterilizing men and inserting intrauterine devices (IUDs) into women. An international population establishment flourished under the umbrella of the United Nations, who dispatched highly paid, jet-setting consultants to expensive hotels all over the Third World, where they blithely dispensed advice that was indifferent to the idiosyncrasies of native culture and tradition. In parallel, concerted efforts were made to engineer genetically modified, highly productive strains of wheat and rice that flourished in a wide variety of climates and soils—when nourished by fertilizer and irrigation. Impressive increases in yields of "Mexi-Pak" wheat and IR-8 rice were announced with much fanfare in 1968 as signaling the onset of an incremental, democratically supported "green revolution." Brought about by American researchers and their indigenous collaborators, it would pull the carpet from under the feet of the Red Menace.

The often brutal proxy wars in developing countries in Africa, Asia, and Latin America, inflamed by superpower rivalry and a global arms trade, contrast with the peace that reigned in Europe after 1945. The social transformations undertaken in the name of modernization in the Third World were often imposed from above and backed by force if necessary. National elites colluded in a process that selectively benefited them at the expense of a destitute and growing peasant population. Existing hierarchies in Third World countries were reinforced by the access that local elites had to resources made available by the industrialized "north." Education and training, science and technology, economic aid, and military supplies were available to the privileged few and denied to the disempowered many. Efforts to accelerate modernization from without became, unwittingly or not, enrolled in a repressive struggle for power and authority within.

Modernization, as a theory of stepwise progress toward the American social model, died with détente. Population control gradually morphed into family planning with the aim of empowering the poor, especially women, to manage their own fertility. The green revolution had patchy results. Notwithstanding its uneven record, the dream of finding a universal technological fix that can eliminate poverty throughout the world continues to fire the imaginations of presidents, pop stars, and philanthropists, spurred on by multinational corporations. Their genuine engagement is oblivious to history and will doubtless founder for ignorance of its lessons.

"Americanization." The preponderance of power in America's favor at the end of World War II, as well as the growing conviction that the nation had been called on to defend democracy wherever it was threatened, provided successive presidents with the historic and ideological legitimacy for engaging in "foreign entanglements" that were otherwise anathema to those of an "isolationist" bent. Inspired by a conception of American exceptionalism that exalted their own social and political system to the pinnacle of what was possible and genuinely convinced that the United States was the greatest force for good in history, policy makers sought to export the American way abroad. The construction of durable alliances, the modernization of systems of production, the penetration of new markets, the creation of a consumer culture, and the refashioning of social values were bundled together in a package that was lived, with varying degrees of enthusiasm, as "Americanization" by those on the receiving end. Communication technologies, including film, radio, and television, were essential to this ambitious agenda, offering the alluring prospect of a life of freedom and plenty and raising the hope that "you too can be like us." America's capacity, except at moments of intense paranoia, to show the darker sides of poverty and race relations in the country only enhanced its status as an open society and sharpened the difference with its immensely secretive and oppressive Communist rival.

The Marshall Plan, or European Recovery Program, announced by President Truman's secretary of state George C. Marshall at a Harvard commencement ceremony in June 1947, helped build what has famously been called an "empire by invitation" in a debilitated and vulnerable postwar Western Europe. It was underpinned by visits to the United States lasting from one to three months by managers and workers, engineers, and economists who were instructed in the technological and managerial demands of the "politics of productivity"—the key to peaceful labor relations, agricultural and industrial growth, and political stability. These productivity missions were later supplemented by a vast effort at public diplomacy in which every imaginable medium was used to advertise the benefits of American generosity to the participating nations. Resisted by some as a form of colonization and welcomed by others as a liberation from the cramping boredom of bourgeois class culture, the "irresistible empire," along with security guarantees offered by the United States through the North Atlantic Treaty Organization (NATO), helped to lock continental Western Europe into the American sphere of influence in one of the most volatile phases of the Cold War. As one left-wing Italian militant admitted ruefully many years later, "The American myths kept their promises and won through."

Concerns to integrate European science and technology more productively into the American research system, to mutual advantage, inspired foundations, scientific statesmen, and government agencies to raise the level and change the orientation of research and education in postwar Europe in many fields, including genetics, physics, and operations research. The transatlantic circulation of people and ideas, but also of equipment, experimental practices, and management techniques, transformed European science along American lines while also providing the conditions needed for a collaborative enterprise. By the late 1950s the United States was determined that Europe establish a solid base in science and technology to share the burden of defense. The failure of a NATO-inspired committee to establish an institution like the Massachusetts Institute of Technology (MIT) in Europe in the early 1960s, linking basic research with the needs of industry and the military, was indicative of the resistance of

the intellectual and political elite to American models that ran against the grain of indigenous academic traditions—in addition to being seen in France as a Trojan horse that would exploit European ideas to the United States' advantage.

The development of the first French scientific satellite launched in 1965 with NASA's help sheds light on the multiple sites in which American practices were adopted. A young team of French researchers spent extended periods of time at the Goddard Space Flight Center in Greenbelt, Maryland, learning how to organize a large project along American lines, with a single project manager, milestones, design reviews, testing procedures, and so on. The major French aerospace company Matra abandoned its traditional hierarchical management structure for a more fluid American matrix structure that broke up work by projects and accorded considerable autonomy to each project leader. Left-wing scientists and engineers who admired Soviet achievements in space found it far more congenial to work with counterparts in the American space agency, where the openness and lack of hierarchy contrasted sharply with Soviet secrecy and the sense that one was always speaking to an intermediary, not to a responsible official. From the United States' point of view, a collaborative effort under its tutelage was an opportunity to build a research and business community that spoke the same technical language and that was organized along the same lines as their American homologues, facilitating knowledge circulation, creating new markets, and consolidating the Atlantic community. From the European point of view the standardization that went with Americanization enabled them to jump-start their exploration of space by creatively applying American practices to local needs.

The concept of Americanization is deemed by many historians to be too general and totalizing to be of much analytical value. It has the advantage, however, of alerting us to the hold that "America" had on the imagination of people all over the globe. It alerts us to the depth to which the postwar American project sought to refashion everyday social practices and values in line with those that prevailed in the United States. Navigating between the attractions of the American way and its threat to deeply ingrained customs, traditions, and values, those on the receiving end selectively appropriated, adapted, reconfigured, or simply rejected the model on offer to satisfy the specificities of particular situations. The project to refashion the world in America's image not only had to take account of the agency of those at whom it was directed. Above all, it was necessarily constrained by the stamina of local cultures.

Symbolizing. Former president George H. W. Bush characterized the Cold War as a struggle for the soul of mankind, a struggle for a way of life. Spectacular technological displays were one of the most important symbols used to demonstrate the superiority of that way of life to a public entranced by an idiom of progress founded on the mastery of nature. This faith in the transformative power of science and technology was not new to the period after 1945, of course, nor was the fusing of national prestige with awe-inspiring demonstrations of technological achievement. The giant, forty-five-foot-tall Corliss engine that powered most of the exhibits at the Centennial Exhibition in Philadelphia in 1876 was an extraordinary engineering feat. It stirred national pride and reassured an American people reconstituting its identity after a divisive civil war that they were a match for their rivals Britain and France, if not streaking ahead of them. What is different in the Cold War is the escalation of a traditional scientific and technological rivalry between nations into a rivalry between two world systems that were fundamentally incompatible with each other. The United States was not simply competing against the Soviet Union: freedom was pitted against tyranny. The success of American laissez-faire capitalism, Washington's aspirations to global leadership, and the security guarantees it gave to allies hung on the proper functioning of a complex rocket engine. The ability of Soviet state socialism to provide for the masses and to liberate them from drudgery hung on filling stores with consumer goods and cramped apartments with modern kitchens. The promise of the green revolution in agriculture to hold back the red wave that threatened to engulf the Third World by producing high-yielding versions of wheat and rice was enhanced by the sight

of robust stalks in well-organized fields sprouting alongside the tangled disorder that was the hallmark of traditional agricultural production. Spectacular visual demonstrations and displays of scientific and technological ingenuity moved to center stage in the Cold War battle for hearts and minds.

Hands-on experience, along with massive media coverage and organized training, was used to market Atoms for Peace to a broad and receptive audience. At the international meeting in Geneva in 1955, a working swimming-pool-type reactor was flown in from Oak Ridge and installed in the grounds of the United Nations building, a mere mile from the city center. Public visits were interrupted every day to allow specially selected official delegates from over 70 countries to see a reactor go critical. The distinguished guests gradually removed the control rods from the pool to bring the reactor up to power by turning a handle themselves; as the device went critical the darkened room was bathed in the eerie glow of Cerenkov radiation. Posters lining the wall of the exhibit celebrated the benefits of radioisotopes produced in research reactors for agriculture, medicine, and industry. Special training programs for foreign students were arranged at Oak Ridge and at Argonne National Laboratory. Traveling exhibits, working models, films, and lectures introduced hundreds of thousands of people throughout the world to the benefits of the nuclear age, distracting attention away from the massive stockpile of lethal weapons being assembled on Eisenhower's watch.

The launch of *Sputnik* by the Soviet Union in October 1957, and the orbiting of the dog Laika in a giant capsule a few weeks later, did not pose a significant threat to American security. They were, however, dramatic displays of Soviet technological achievement and they struck a blow to American prestige and self-confidence. They raised doubts in the country's mind about the security of the mainland to surprise attack, they unsettled allies, they undermined the United States' credibility as the world's scientific and technological leader, and they enhanced Soviet claims to the superiority of Communism. The failed launch of the first American satellite in December 1957, in the presence of over a hundred newspaper and television reporters, amplified fears that the United States had fallen badly behind its Communist rival. The conquest of space was transformed from a technological and industrial challenge into a domestic political issue and a struggle for global influence that helped propel John F. Kennedy into the White House in 1961. Recognizing the impact of space spectaculars on "the minds of men everywhere" who were being called upon to choose between "freedom and tyranny," within months—and shortly after Yuri Gagarin became the first human in space—the new President asked Congress to commit the nation to a manned lunar landing before the decade was out.

More modest forms of technology were also enrolled in the psychological propaganda war between the superpowers. The "economic miracle" that got under way in Western Europe in the late 1950s, filling stores with glittering consumer goods unavailable in the Communist bloc, provided the Eisenhower administration with an opportunity to regain the ground lost in perceptions of Soviet missile and space superiority. In July 1959 an American exhibition in Moscow displayed a range of American consumer goods, including three fully equipped kitchens. One of them, General Electric's all-electric lemon-yellow kitchen, was packed with modern appliances and installed in a full-scale, ranch-style American house through which thousands of visitors circulated every day. There Vice President Richard Nixon exalted the benefits of American consumer culture in an exchange with the Soviet premier Khrushchev in an attempt both to undermine popular support for the Soviet regime and to pressure its planners to divert resources away from arms into domestic goods. The kitchen as symbol of American freedom brought Cold War technological rivalry into the gendered domestic space of the nuclear family. It presented the individual consumer, home ownership, private property, and the market economy as defining features of the superiority of capitalism over Communism.

The enthusiasm for technological prowess in the early Cold War had turned to skepticism by the 1970s. The defeat in Vietnam, urban decay and unrest, environmental damage, and the failure of modernization projects in the developing world generated a counterculture that decried technological fixes and sought more holistic approaches to social problems. Human spaceflight was criticized as a waste of scarce resources that were better spent on improving conditions on earth. The failed promise of "energy too cheap to meter," the risks of nuclear accidents, and the difficulty of disposing of nuclear waste broke the hegemonic discourse that had uncritically promoted nuclear power's benefits for two decades. Although scientific and technological preeminence was still pursued, along with constant military preparedness, it lost its role as a bearer of national pride and as a barometer of progress.

This is not to imply that the propaganda value of scientific and technological excellence lost its pertinence; on the contrary, in the early twenty-first century it was reborn as an essential asset of what is sometimes called soft power. The confrontation with the Soviet Union in the early Cold War provided an immense stimulus to sectors of American industry, notably in the domains of aerospace, computers, and miniaturization. These advances fueled the technological revolution in communications and transportation in the 1970s that were associated with the globalization of trade and the dilution of the independence of the nation state. Issues demanding state intervention—climate change, epidemics like severe acute respiratory syndrome (SARS), the international traffic in drugs, terrorism—became increasingly global in scope and required collaboration with other states, including those that were relatively small and weak. To lead from a position of interdependence requires a more nuanced exercise of power, a willingness to achieve one's foreign policy goals by cooption that secures legitimacy rather than by coercion that generates opposition. This thinking inspired policies that emphasize the possibility of using the attractions of American science and technology abroad as one of a repertoire of instruments of soft power available to secure desirable foreign-policy objectives.

The concept of soft power was given an important boost by the disastrous foreign-policy fallout of Operation Iraqi Freedom. It was alluded to by President Obama's first Secretary of State, Hillary

Clinton in the Senate hearings that preceded her nomination and institutionalized as an option in U.S. foreign policy. It starts from the premise that the attraction of American science and technology, along with its culture, ideology, and institutions, can be deployed to achieve diplomatic goals without recourse to "hard" coercive power. Indeed, a 2002 Pew Global Attitudes Project revealed that 80 percent of people in much of the world admired American science and technology compared with 60 percent who were attracted by its cultural exports and just 30 percent who favored the dissemination of U.S. ideas and customs abroad (Nye, 2004, pp. 69–72). The marketing of U.S. achievements in science and technology could still serve as an important instrument of diplomacy, but only by advertising the virtues of the American way of life using techniques quite different from those that were deemed effective in the early Cold War. This at least was the concept driving Clinton's "twenty-first-century statecraft." A section in the State Department's Office of Policy Planning sought to complement traditional foreign-policy tools by levering the networks, technologies, and demographics of an interconnected world to win hearts and minds. In 2012, the Internet and social media platforms, not nuclear reactors, space spectaculars, and modern kitchens, symbolized American freedom.

[*See also* **American Association for the Advancement of Science; American Institute of Physics; Atomic Energy Commission; Atoms for Peace; Automation and Computerization; Bush, Vannevar; International Geophysical Year; Internet and World Wide Web; Military, Science and Technology and the; National Aeronautics and Space Administration; Nuclear Weapons; Radio; Research and Development (R&D); Rockefeller Institute, The; Science; Social Sciences; Space Program; Space Science; Technology;** *and* **Television.**]

BIBLIOGRAPHY

Adas, Michael. *Dominance by Design. Technological Imperatives and America's Civilizing Mission,* Cambridge, Mass.: Belknap Press of Harvard University Press, 2006. A study of the American way of war over the past century.

American Association for the Advancement of Science. *Federal Spending on Defense and Nondefense R&D. Outlays for the Conduct of R&D, FY1952–2015, Billions of Constant FY2014 Dollars.* 2014. http://www.aaas.org/sites/default/files/Function_0.jpg (accessed 23 June 2014).

Balancing Scientific Openness and National Security Controls at the Nuclear Weapons Laboratories. Washington, D.C.: National Academies Press, 1999. Argues that American scientific and technological leadership requires the minimum regulation of knowledge flows.

Bischof, Günter, and Dieter Stiefel, eds. *Images of the Marshall Plan in Europe. Films, Photographs, Exhibits, Posters.* Innsbruck, Austria: StudienVerlag, 2009. The use of cultural artifacts to encourage Western Europeans to adopt the American way of life.

Connelly, Matthew. *Fatal Misconception. The Struggle to Control World Population.* Cambridge, Mass.: Belknap Press of Harvard University Press, 2008. Highlights the misguided and harmful campaigns led by U.S. foundations to reduce fertility in the Third World.

Cullather, Nick. *The Hungry World. America's Cold War Battle against Poverty in Asia.* Cambridge, Mass.: Harvard University Press, 2010. A critique of the rhetoric surrounding the "green revolution" and the flawed definition of development that engenders it.

Cullather, Nick. "Miracles of Modernization. The Green Revolution and the Apotheosis of Technology." *Diplomatic History* 28, no. 2 (2004): 227–254. Emphasizes the importance of spectacle and demonstration to winning hearts and minds.

Day, Dwayne A., John M. Logsdon, and Brian Latell, eds. *Eye in the Sky. The Story of the Corona Spy Satellites.* Washington, D.C.: Smithsonian Institution Press, 1998. A description of the multiple roles of a key intelligence-gathering tool.

De Grazia, Victoria. *Irresistible Empire. America's Advance Through 20th-Century Europe.* Cambridge, Mass.: Belknap Press of Harvard University Press, 2005. A detailed study of the dissemination of American consumer culture in Europe.

Deitchman, Seymour J. "The 'Electronic Battlefield' in the Vietnam War." *The Journal of Military History* 72, no. 3 (2008): 869–887. Benefits from access to classified material.

Doel, Ronald, and Allan Needell. "Science, Scientists, and the CIA: Balancing International Ideals,

National Needs, and Professional Opportunities." *Intelligence and National Security* 12, no. 1 (1997): 59–81. Discusses the subterranean network of informal intelligence gathering by scientists.

Edgerton, David. *The Shock of the Old. Technology and Global History since 1900*. Oxford: Oxford University Press, 2007. A stirring plea to direct attention away from innovation and onto use.

Engerman, David C. "American Knowledge and Global Power." *Diplomatic History* 31, no. 4 (2007): 599–622. Distinguishes between U.S. knowledge for, of, and as global power and has an excellent bibliography.

Engerman, David C. *Know Your Enemy. The Rise and Fall of America's Soviet Experts*. New York: Oxford University Press, 2009. Analyzes the work done by eminent scholars in Sovietology in the American academy.

Engerman, David C., Nils Gilman, Mark H. Haefele, and Michel J. Latham. *Staging Growth. Modernization Development, and the Global Cold War*. Amherst: University of Massachusetts Press, 2003. A collection of readings that describes the genesis of modernization theory and its implementation in Africa and Asia.

Forman, Paul. "Behind Quantum Electronics: National Security as Basis for Physical Research in the United States, 1940–1960." *Historical Studies in the Natural Sciences* 18, no. 1 (1987): 149–229. Describes and deplores the transformation in the practice of physics in response to the demands of the military–industrial complex.

Friedberg, Aaron L. *In the Shadow of the Garrison State. America's Anti-statism and its Cold War Grand Strategy*, Princeton, N.J.: Princeton University Press, 2000. Describes the place of technological innovation in the construction of the national security state.

Galison, Peter. "Removing Knowledge." *Critical Inquiry* 31, no. 1 (2004): 229–243. A critical analysis of the scale of the classified realm of knowledge.

Hecht, Gabrielle, *Being Nuclear. Africans and the Global Uranium Trade*. Cambridge, Mass.: MIT Press, 2012. Uses the production of uranium as a node to link nuclear proliferation with colonial and postcolonial power relations.

Hogan, Michael J. *A Cross of Iron. Harry S. Truman and the Origins of the National Security State, 1945–1954*. Cambridge, U.K.: Cambridge University Press, 1998. A study of the place of ideology in the construction of the national security state.

Kaiser, David. "Cold War Requisitions, Scientific Manpower, and the Production of American Physicists after World War II." *Historical Studies in the Physical Sciences* 33, no. 1 (2002): 131–159. On training a new generation to meet the explosive demand for science graduates in the mid-1950s.

Krige John. *American Hegemony and the Postwar Reconstruction of Science in Europe*. Cambridge, Mass.: MIT Press, 2006. How U.S. authorities, scientific statesmen, and foundations tried to re-shape scientific practices in postwar Europe.

Krige, John. "The Peaceful Atom as Political Weapon: Euratom and American Foreign Policy in the Late 1950s." *Historical Studies in the Natural Sciences* 38, no. 1 (2008): 5–44. How the United States' presumed leadership in all things nuclear was mobilized to divert resources into a supranational civil program.

Krige, John. "Techno-utopian Dreams, Techno-political Realities." In *Utopia/Dystopia. Conditions of Historical Possibility*, edited by Michael Gordin, Helen Tilley, and Gyan Prakash, pp. 151–175. Princeton, N.J.: Princeton University Press, 2010. An analysis of the utopian promotion of the peaceful atom, with particular reference to developing countries.

Krige, John, and Kai-Henrik Barth, eds. "Global Power Knowledge." In *Science, Technology and International Affairs*, Osiris Vol. 21. Chicago: University of Chicago Press, 2006.

Krige, John, Angelina Long, and Ashok Maharaj. *NASA in the World. 50 Years of International Collaboration in Space*. New York: Palgrave Macmillan, 2013. Treats Western Europe, the Soviet Union/Russia, India, and Japan.

Lafeber, Walter. "Technology and U.S. Foreign Relations." *Diplomatic History* 24, no. 1 (2000): 1–19. A leading diplomatic historian calls for greater attention to technology.

Leffler, Melvyn P. *For the Soul of Mankind. The United States, the Soviet Union, and the Cold War*. New York: Hill and Wang, 2007. An analysis organized around the choices made by political leaders.

Leffler, Melvyn P., and Odd Arne Westad, eds. *The Cambridge History of the Cold War*, 3 vols. Cambridge, U.K.: Cambridge University Press, 2010. An invaluable resource that covers the topic in depth and breadth.

Lundestad, Geir. "'Empire by Invitation' in the American Century." *Diplomatic History* 23, no. 2

(1991): 189–217. A contradiction in terms that highlights the ambiguities surrounding the American presence in postwar Europe.

Maddock, Shane J. *Nuclear Apartheid. The Quest for American Atomic Supremacy from World War II to the Present*. Chapel Hill: University of North Carolina Press, 2011. A scathing analysis of the unintended consequences of America's ceaseless quest for nuclear superiority.

Maier, Charles S. "The Politics of Productivity. Foundations of American Economic Policy after World War II." *International Organization* 31 (1977): 607–633. On the significance of the notion of "productivity" as an organizing principle for U.S. postwar economic policy.

Manela, Erez. "A Pox on Your Narrative: Writing Disease Control into Cold War History." *Diplomatic History* 34, no. 2 (2010): 299–323.

Martin-Nielsen, Janet. "'This War for Men's Minds': Birth of a Human Science in Cold War America." *History of the Human Sciences* 23, no. 5 (2010): 131–155. A fine analysis of linguistics as a Cold War science.

National Resources Defense Council. National Resources Defense Council Environmental Issues; Nuclear Energy, Nonproliferation, and Disarmament Main Page; All Nuclear Energy, Nonproliferation, and Disarmament Documents. Archive of Nuclear Data from NRDC's Nuclear Program; Table of Global Nuclear Weapons Stockpiles. 1945–2002. http://www.nrdc.org/nuclear/nudb/ datab19.asp (accessed 18 October 2012).

Nye, Joseph, S., Jr. "Soft Power." *Foreign Policy* 80 (1990): 153–172. One of the earliest statements of the concept by an eminent scholar of international affairs who was also an adviser to the Carter and Clinton administrations.

Nye, Joseph, S., Jr. *Soft Power: The Means to Success in World Politics*. Cambridge, Mass: PublicAffairs, 2004. Stresses soft power as a form of 'attraction', and includes useful global data on what others like about America.

Office of Management and Budget. *Fiscal Year 2012. Historical Tables. Budget of the U.S. Government*. Office of Management and Budget, Table 3.2. http://www.gpo.gov/fdsys/pkg/budget-2012-tab/pdf/budget-2012-tab.pdf (accessed 18 October 2012).

Oldenziel, Ruth, and Karin Zachmann, eds. *Cold War Kitchen. Americanization, Technology, and European Users*. Cambridge, Mass.: MIT Press, 2009. Describes the reception of the American kitchen in the Soviet Union and across Europe.

Osgood, Kenneth. *Total Cold War. Eisenhower's Secret Propaganda Battle at Home and Abroad*. Lawrence: University Press of Kansas, 2006. The first wide-ranging study of the topic, with important chapters on Atoms for Peace and Sputnik.

Paarlberg, Robert L. "Knowledge as Power. Science, Military Dominance, and U.S. Security." *International Security* 29, no. 1 (2004): 122–151. Describes and analyses the current state of U.S. military superiority.

R&D Budget and Policy Program. *Guide to R&D Funding Data—Total U.S. R&D*. Washington, D.C.: American Association for the Advancement of Science, 1953–.

Sokolski, Henry D. *Best of Intentions. America's Campaign against Strategic Weapons Proliferation*. Westport, Conn.: Praeger, 2001. A sympathetic assessment of the United States' nuclear nonproliferation initiatives since the late 1940s.

Van Vleck, Jennifer. "An Airline at the Crossroads: Ariana Afghan Airlines, Modernization and the Cold War." *History and Technology* 25, no. 1 (2009): 3–24.

Wang, Zuoyue. "Transnational Science during the Cold War: The Case of Chinese/American Scientists." *Isis* 101, no. 2 (2010): 367–377.

Westad, Odd Arne. *The Global Cold War. Third World Interventions and the Making of Our Times*. Cambridge, Mass.: Cambridge University Press, 2007. The first major history to define the Cold War in global terms.

Westad, Odd Arne. "The New International History of the Cold War: Three (Possible) Paradigms." *Diplomatic History* 24, no. 4 (2000): 5551–5565. Elevates ideology, technology, and the Third World to core variables for studying the dynamics of the Cold War.

Westwick, Peter J. "'Space-Strike Weapons' and the Soviet Response to SDI." *Diplomatic History* 32, no. 5 (2008): 955–979. On Reagan's Strategic Defense Initiative.

Zierler, David. *The Invention of Ecocide. Agent Orange, Vietnam, and the Scientists who Changed the Way We Think about the Environment*. Athens: University of Georgia Press, 2011. Merges diplomatic, environmental, and technological history, concentrating on the scientists who opposed the use of chemical warfare in Vietnam.

John Krige

DISABILITIES, INTELLECTUAL AND DEVELOPMENTAL

Instructing Idiots. In 1845, Samuel G. Howe, the superintendent of the Perkins Institution for the Blind in Boston, read an article by the Englishman John Conolly in the *British and Foreign Medical Review*. In the article, Conolly described his visits to the Bicêtre Hospital, where he had observed the "idiot school" of Edouard Séguin. Opened in 1842, the French school had shown that pupils with limited abilities could learn. Encouraged by Conolly's article, Howe persuaded the Massachusetts legislature to appropriate funds for a school for "idiots" that opened in 1848.

In 1847 Hervey Wilbur, a physician in Barre, Massachusetts, also read reports about Séguin's work with idiots. The next year he opened a private school for "idiotic children." Three years later, he moved to New York to become the superintendent of that state's newly opened school for weak minds. Added to these events, Séguin immigrated to the United States in 1850. During the 1850s Howe, Wilbur, and Séguin published reports about their successes in educating idiots. Interest in this education spread throughout the nation, and by 1865, Pennsylvania, Ohio, Connecticut, Kentucky, and Illinois also opened residential schools.

Burden of the Feeble-Minded. The two decades after the opening of "idiot schools" was a period of educational optimism, but the seeds of a shift to pessimism after the Civil War were just below the surface. In 1858, Isaac Kerlin, the superintendent of the Pennsylvania Training School for Feeble-Minded Children, published *The Mind Unveiled; or, A Brief History of Twenty-Two Imbecile Children*. In it he introduced the "moral imbecile." Devoid of moral judgment but slightly intelligent, moral imbeciles were a threat to the good order of every community. Complicating this growing pessimism were Census counts in 1870 and 1880 that suggested there were more feeble minds than had been thought by prewar estimates. Also, economic downturns in the 1870s made it difficult for superintendents to find community employment for their educated inmates. In this context, by the mid-1880s superintendents were increasingly using the rhetoric of the "burden of the feeble minded."

Between 1880 and 1910 this rhetoric became a rationale for institutional expansion. Soon superintendents worried less about education and more about methods and costs of operating large institutions. To meet these costs, they turned to the inmates themselves. Now likely never to leave the facility, inmates would pass their time working on the institution's farm or in caregiving for less capable fellow inmates.

Menace of the Feeble-Minded. In the first decades of the twentieth century, the image of intellectually disabled people changed from social burdens to social menaces. Two events precipitated this change: the development of intelligence testing and the emergence of the eugenics movement. These developments reached their apogee in the psychologist Henry Goddard's 1912 *The Kallikak Family: A Study in the Heredity of Feeble-Mindedness*.

According to Goddard, families like the Kallikaks were numerous, but their inferiority was not readily recognized. To classify this newly identified group of feeble minds, Goddard constructed a new label, "morons." Unlike "low-grade idiots" and "middle-grade imbeciles," "high-grade morons" usually remained outside the protection of the institution. Among the public, inferior in intelligence, breeding frequently, and prone to vice, morons were especially menacing because they looked so ordinary. To control the propagation of this menace, superintendents and their eugenics supporters turned to a familiar solution, institutional incarceration; but they also set their sights on a new means of control, sterilization.

By 1920 it had become clear to local authorities that not all feeble minds, especially feeble-minded children, could be institutionalized. Given this awareness, public schools developed special classrooms for "mentally deficient" children. Such children, these authorities claimed, could learn basic work skills for their eventual placement in residential institutions.

Institutions and the Community. The expansion of residential institutions on the one hand and public school classes for "special children" on the other continued through the 1920s. The years of the Great Depression followed by World War II, however, saw little institutional construction. State revenues were down, but the demand to take more residents from economically distressed families and communities had never been greater. To accommodate the demand, institutions admitted more and more new clientele. Even before the nation entered the war in 1941, residential institutions for the mentally deficient were crowded. Exacerbating this situation was the reduction of available staff, many of whom had been drafted into the armed services.

By the end of the war, public institutions had become "snake pits." In the 1950s the first organization of parents, the National Association of Retarded Children, incorporated. Despite the association's advocacy for services, it was not until the Kennedy administration took office in 1961 that "mental retardation" gained widespread national attention. To this end, the states used new federal resources to launch institutional construction in the late 1950s and 1960s.

As these new facilities appeared, the beginnings of their destruction also emerged. By the mid-1960s, new realities developed that changed demands for improving public residential institutions to calls for their closure. The first source was economic. The states had used federal funds to construct and upgrade their institutions. This funding, however, did not include support for hospital maintenance and staffing costs. By the early 1970s many state officials had concluded that the costs of their residential facilities were stressing their state's budget. Accompanying this economic reality was a second source of pressure. Parents, academics, and media representatives began to argue that residential institutions were inhumane places. In the place of the institution, these critics advocated for the "normalization" of services in ordinary communities. Several federal judicial rulings of the period supported the new perspective. By the mid-1970s the community had become the principal locus of services for intellectually disabled people.

Federal legislation in the form of Medicaid provided a way to shift service costs from the states to the federal government, while also transferring the place of services from institutions to communities. By the year 2000 states had reduced the populations of their residential facilities and, in many cases, had completely closed their institutions. Across the nation, residents of public institutions moved to community apartments, group homes, and other residential facilities. In public schools, too, students with disabilities were increasingly "mainstreamed" in ordinary classrooms, where "least restrictive alternative" specified the new benchmark of their education.

For intellectually disabled citizens who have never lived in an institution, the community has been the only source of their education, jobs, and daily life. For disabled people who moved from residential institutions to communities, life has often improved. Some have acquired regular jobs, gotten married, and become parts of the social mix of ordinary American communities.

But others have fared less well. Progressive federal mandates of the 1970s have been followed by federal stinginess since the 1980s. Also, in many states the principal benefactors of the Medicaid assistance are for-profit operators of nursing facilities. In these "homes," disabled adults live with a dozen or more residents and their days are routine and predictable. Likewise, in many public schools mainstreaming has become an empty goal because many disabled students spend much of their day in "special classes," even if schools no longer call them special.

[*See also* **Medicine;** *and* **Mental Illness.**]

BIBLIOGRAPHY

Ferguson, Philip M. *Abandoned to Their Fate: Social Policy and Practice toward Severely Retarded People in America, 1820–1920.* Philadelphia: Temple University Press, 1994. Focuses especially on the Rome (NY) State Custodial Asylum.

Noll, Steven. *Feeble-Minded in Our Midst: Institutions for the Mentally Retarded in the South, 1900–1940.* Chapel Hill: University of North Carolina Press, 1995. Overview of the institutional experience in the American South.

Noll, Steven, and James W. Trent Jr., eds. *Mental Retardation in America: A Historical Reader*. New York: New York University Press, 2004. A collection of articles detailing the American and Canadian histories of institutions, as well as community advocacy and reform.

Trent, James W., Jr. *Inventing the Feeble Mind: A History of Mental Retardation in the United States*. Berkeley: University of California Press, 1994. Follows the changing constructions of intellectual disabilities from 1840 to 1990.

Trent, James W., Jr. *The Manliest Man: Samuel G. Howe and the Contours of Nineteenth-Century American Reform*. Amherst: University of Massachusetts Press, 2012. Traces the earliest proponent of education for "idiots" in the United States.

Wright, David. *Downs: The History of a Disability*. Oxford: Oxford University Press, 2011. Surveys Down's syndrome from its nineteenth-century identification to current ethical issues associated with it.

James W. Trent Jr.

DISEASE

Disease has played a profound but ever-changing role throughout American history. In the early twenty-first century the faith that the conquest of disease is a realistic possibility remains pervasive. Some believe that medical science can develop curative therapies for virtually all diseases; some believe that environmental and behavioral changes can eliminate many diseases; and still others turn to alternative medicine as a panacea. However dissimilar, all assume that disease is "unnatural" and that its virtual elimination is a distinct possibility.

The faith that disease is unnatural and can be conquered rests on a fundamental misunderstanding of the biological world. There are, for example, millions of micro-organisms. Some are harmless, whereas some are parasitic and have the potential to cause infection. Others play vital symbiotic roles that nourish and maintain organic life. Nor can efforts to destroy pathogenic micro-organisms through a variety of drugs meet with permanent success, if only because such micro-organisms have the potential to develop resistant properties. Similarly, environmental changes can enhance the virulence of micro-organisms or magnify other health risks. Even malignant cells—which are regarded as the "enemy"—are hardly aliens that invade our bodies; they grow from our own normal cells. And efforts to arrest the aging process and thus avoid the diseases that accompany old age represent a utopian dream of avoiding death and perpetuating life indefinitely. Nowhere are these generalizations better illustrated than in the history of disease in America from its beginnings to the present.

The Columbian Encounter and the Early Colonial Era. When Europeans first reached the Americas, they encountered a hitherto unknown indigenous population as well as a novel natural and biological environment. Amerindians probably had migrated from Asia to Alaska across a land bridge produced by a lowering of the oceans during the last Ice Age. Many pathogens responsible for infectious diseases that took a heavy toll in Asia, Europe, and Africa probably did not survive the migration through the harsh climate of Siberia and Alaska. Peoples of the Americas were thus isolated from many of the epidemic and endemic diseases that had profoundly shaped population structures elsewhere. The absence of contact with diverse populations also gave them a far more homogeneous genetic inheritance.

These and other factors gave pre-Columbian America a unique disease environment. Many of the diseases characteristic of Europe, Asia, and Africa—malaria, smallpox, bubonic plague, and infectious diseases associated with childhood (measles, mumps, whooping cough)—were unknown. The greatest health risks to the Amerindian population included accidents, wildlife diseases associated with hunting and food gathering, warfare, and sporadic famines and food shortages. The relative absence of domesticated livestock minimized zoonotic (animal-transmitted) diseases, and low population density and the absence of commercial contacts among tribes reduced the dangers of epidemic and endemic infectious disease. Nevertheless, life expectancy at birth for Amerindians was generally in the low thirties on the eve of colonization, although the causes of morbidity and mortality differed from the rest of the world.

The migration of Europeans to the Americas that commenced at the end of the fifteenth century had a catastrophic impact on the indigenous population. The introduction of new infectious diseases imported from other continents into a population often lacking immunological defenses led to extraordinarily high mortality rates. Genetic homogeneity may have also enhanced vulnerability. Moreover, children as well as adults did not receive the kind of care that might have mitigated the impact of these new diseases; neither custom, tradition, nor religion provided any guide. Whatever the reasons, the Amerindian population suffered a precipitous decline in the period following the first contacts with Europeans. Diseases such as smallpox, measles, whooping cough, chicken pox, and malaria—to cite only a few—exacted a heavy toll. On the eve of colonization the population of the future contiguous United States was between 2 and 12 million. When the nadir was reached in the early twentieth century, the number of Native Americans had fallen to about 250,000 (Stuart, 1987, pp. 51–54). Disease and the ensuing social demoralization—not military conquest—played the major role in this demographic disaster.

Those who migrated from England to America faced their own health problems. The Atlantic crossing, which could last three or four months, posed its own risks. Within their new environment the settlers faced harsh conditions, partly because they did not understand that eastern North America had a continental climate compared with Western Europe's oceanic climate. The former resulted in far greater extremes of cold, heat, and humidity, which compounded the burdens of adjustment. The construction of housing, securing an uncontaminated water supply, and the development of an adequate and varied food supply took time. During the period of adjustment (often aptly described as a process of "seasoning"), many new settlements experienced extraordinarily high mortality rates that, if unchecked, threatened their very existence.

The period of seasoning varied from place to place. In New England the process of adjustment was brief; within a short time, mortality rates dropped and inhabitants enjoyed unprecedented levels of health. In seventeenth-century Andover, Massachusetts, the average age of death among the first generation was nearly 71, and infant and child mortality was correspondingly low (Greven, 1970, pp. 26). Nor was Andover unique. Low mortality rates resulted in rapid population growth in seventeenth-century New England. By the end of the century the region's population was about 90,000 (Wells, 1975).

The Chesapeake areas of Virginia and Maryland and the southern colonies, by contrast, remained dangerous places. The importation of such tropical diseases as malaria and yellow fever into a region with a warm and moist climate proved devastating. Mortality rates in the South exceeded those of New England, the Middle Atlantic colonies, and even Great Britain. On average, seventeenth-century white male New Englanders who survived to the age of twenty outlived their Maryland counterparts who had also survived to the age of twenty by more than two decades. Much the same was true in the Carolinas and Georgia. In 1699 the population of Virginia had reached about 63,000. Yet between 1607 and 1699 somewhere between 50,000 and 100,000 persons had migrated to that colony (Morgan, 1975, pp. 405–410). The greater resistance of African Americans to the ravages of such tropical diseases as malaria undoubtedly contributed to the growth of slavery in the South. The pattern of regional variation in mortality and morbidity would persist until well into the twentieth century.

The Era of Infectious Diseases and Epidemics.

During the eighteenth century there were major changes in the disease ecology of the American colonies. Natural population growth, high rates of immigration, and the geographic mobility that accompanied the increase in trade and commerce enhanced the movement of infectious pathogens. Smallpox and yellow fever epidemics appeared in the port cities. Because many infectious diseases had not gained a foot-hold in the American colonies, the population included a disproportionately high number of susceptible persons. The result was a partial repli-cation of the harsh disease environment charac-teristic of England and Europe. Although the colonial population continued to grow, its curve resembled a saw-tooth shape on an upward gra-dient because of the impact of epidemic diseases with their high mortality rates in specific years. However hard hit by infectious diseases, colonial America nevertheless had lower mortality rates than those of England and Europe.

During the eighteenth century mortality rates among the young from such diseases as measles, mumps, whooping cough, and a variety of respira-tory and intestinal disorders rose dramatically, especially in more densely populated towns. Al-though the spectacular epidemics of smallpox and yellow fever were the most feared, the major causes of death were intestinal disorders, including typhoid fever and various forms of dysentery. Seasonal patterns as well as population density shaped morbidity and mortality patterns. Intes-tinal diseases were more frequent in warmer months because of stagnant water, contaminated food, and large insect populations that could transmit malaria and yellow fever. Respiratory and pulmonary diseases peaked in cold weather. Because infectious diseases that killed the young were by far the dominant cause of morbidity and mortality, the proportion of aged persons in the population remained low; chronic and degenera-tive diseases accounted for only a small propor-tion of total deaths.

The morbidity and mortality patterns in place by the late eighteenth century persisted in one form or another for much of the nineteenth cen-tury. Nevertheless, a changing social and physical environment as well as population movements both to and within the United States contributed to a significant modification of the earlier disease environment, especially in urban areas. The immi-gration of destitute groups such as the Irish into densely populated neighborhoods where squalor and unhygienic conditions prevailed dramatically increased risks to health. Infants and young chil-dren were particularly susceptible to infectious

diseases. Intestinal disorders continued to take the highest toll, but other diseases associated with population and/or unsanitary conditions—typhus, typhoid, smallpox, and respiratory disor-ders—loomed large. Population growth exceeded the ability of municipal governments to provide a safe water supply or a sanitation system to remove organic waste and to ensure clean streets (which were usually covered with heaps of animal wastes). Housing codes were all but absent; inadequate ventilation and crowding permitted the rapid transmission of infectious diseases. Tuberculosis and other pulmonary diseases emerged as major causes of mortality. Occupations that posed a threat to health went largely unregulated. Urban areas also continued to experience periodic epi-demics related to the quickened pace of trade and commerce. Cholera became an international dis-ease during the nineteenth century as more rapid ocean transportation magnified the ability to move pathogens. Major cholera epidemics swept across the United States in 1832, 1848, and 1866, result-ing in 200,000 to 300,000 deaths (Chambers, 1938). In addition to cholera epidemics, South-ern cities experienced virulent outbreaks of yellow fever. New Orleans in particular was hard hit. During the epidemic of 1853, for example, be-tween a quarter and a half of the city's population of 150,000 fled. Of the remainder, 40,000 devel-oped yellow fever. In a five-month period during the epidemic, 9,000 of the 11,000 deaths were at-tributed to the disease, and the disposal of corpses became one of the most serious problems (Duffy, 1966, p. 167).

Rural areas and small towns, by contrast, often escaped the infectious diseases that plagued urban areas although health indicators declined during the first half of the nineteenth century. In 1830, for example, urban death rates were between two and three times higher than those in rural areas; small towns tended to fall midway between the two. In certain circumstances, however, the advantages conferred upon rural inhabitants proved a liability. During the Civil War, young men recruited from rural areas, lacking the immunity of their urban counterparts who had survived the infectious dis-eases of infancy and childhood, died in large num-bers when they encountered unhygienic conditions and dangerous pathogens in crowded military camps. Indeed, the overwhelming number of Civil War deaths occurred not from battlefield wounds, but from respiratory and enteric disorders as well as smallpox, measles, malaria, and other infectious diseases. In the Union army 67,000 died on the field of battle, and an additional 43,000 died of wounds and injuries. By contrast, 224,000 died from disease (*The Medical Department*, n.d., p. 27). Confederate data, although less complete because of the destruction of records from a fire, suggest a similar pattern.

By the early nineteenth century a torrent of migrants began a process of westward expansion that reshaped the nation in profound ways. This massive migration was marked by high morbidity and mortality. About 6 percent of those traveling on the Oregon–California trail perished. Moreover, the initial stages of settlement often elevated health risks. The initial settlement of the Ohio Valley was along rivers. Land-clearing practices and periodic overflows of rivers created stagnant pools of water. This in turn led to fertile breeding grounds for the Anopheles mosquito, which transmitted malaria from infected to susceptible persons. By midcentury malaria was endemic in this region. Subsequent movement of the population away from waterways and other changes contributed to the disappearance of malaria after 1870. The harsh conditions of life in frontier regions also led to high mortality rates, particularly among infants and children.

Although infectious diseases remained by far the major causes of mortality, their distribution varied by region and class. Malaria, yellow fever, and hookworm, for example, were largely confined to the South. Social class and race were important elements in morbidity and mortality patterns. Lower class and minority ethnic and racial groups tended to have higher mortality rates. Nutritional levels and sanitary conditions undoubtedly exacerbated the impact of infectious diseases on these groups. But more prosperous groups did not escape the threat of infectious disease; infant mortality remained high at all levels of the population. The institution of slavery in particular elevated health risks. Slave infants weighed only 5.5 pounds and mortality was twice that of the white antebellum population. More than half perished before their fifth birthday. The conditions of life led to a decline in life expectancy among Americans generally for much of the nineteenth century despite the rising standard of living.

The Era of Chronic Disease. Beginning in the late nineteenth century the United States, as well as England and many European nations, experienced what has become known as the second "epidemiological (or health) transition." The first, which occurred perhaps 10,000 or more years ago, involved the development of agriculture, which created a more stable food supply. The result was a more sedentary population that increased in both size and density. Population growth in turn heightened the potential for epidemic and endemic infectious diseases, which require a host that permitted the survival of invading pathogens. During the second epidemiological transition, infectious diseases began to decline as the major cause of mortality, to be replaced by chronic and degenerative diseases. This unparalleled transformation had a profound impact on all human beings.

By 1940 most of the infectious diseases associated with childhood (viral diseases such as measles, mumps, whooping cough, and chicken pox and bacterial diseases that included scarlet and rheumatic fever), although still prevalent, no longer posed a threat to life. The dramatic decline in infant and child mortality meant that far more people would survive to old age. In 1850 only 4 percent of people were aged 60 or older; by 2008 the figure was 17 percent (or 51.7 million of a total population of 304 million). In these circumstances long-term diseases, notably cardiovascular–renal diseases and malignancies, slowly became the major causes of mortality (Grob, 2002).

What was the cause or causes of this massive shift in morbidity and mortality patterns? Most scholars agree that medical interventions were relatively unimportant. Prior to World War II, the primary function of medicine was to diagnose disease. With the exception of surgical procedures, antitoxins (e.g., the diphtheria antitoxin), smallpox immunization, and drugs such as digitalis and insulin, physicians had few effective therapies to deal with infections. Antibiotic therapy did not come into widespread use until the 1940s, and the development of vaccines for most viral infectious diseases lay in the future. Of the 15 leading causes of death in 1900, infectious diseases accounted for 56 percent of the total. By 1940 infectious diseases were minor causes of mortality (Erhardt and Berlin, 1974, pp. 21–23). Indeed, after World War II the widespread use of antibiotic therapy fostered a belief that infectious diseases— historically the major elements in mortality— would no longer play decisive roles in either morbidity or mortality.

It is far easier to describe than to explain the decline in mortality from infectious diseases. Many scholars have attributed it to economic growth and a rising standard of living. The difficulty with such global explanations is that they are not based on empirical data that shed light on the precise mechanisms responsible for the mortality decline for specific diseases. Some have pinpointed dietary improvements as the most important factor. Yet the relationship between diet—excluding severe malnutrition, which rarely existed in the United States—and most infectious diseases is tenuous at best. Moreover, economic growth involves more than living standards; it includes rising levels of literacy, education, and a variety of other complex social changes. Some of these changes and their interactions—including housing arrangements, population density, water and food purity, personal hygiene, individual behavioral patterns, public-health interventions— may have had a more direct influence on mortality levels. Although the importance of economic development in the reduction of mortality is generally recognized, no consensus exists on the precise role of specific factors.

In some cases the reduction in mortality from infectious diseases followed specific public-health interventions. Typhoid fever, for example, was generally transmitted by contaminated water. The construction of central sewer systems did not seem to have a major impact, but reduced mortality did follow the introduction of water filtration. The reduction in mortality from tuberculosis, on the other hand, presents far greater complexi-

ties. Mortality began to fall well before overt efforts were made to contain the disease. Improved diets and a reduction in exposure thanks to better housing and the building of sanitariums may account for growing resistance to the disease, but the evidence for these explanations remains inconclusive. The fall in infant and childhood mortality from diarrheal diseases probably followed changes in baby-feeding practices, improvement in the milk supply (pasteurization), and public-health authorities' efforts to sensitize parents to more effective means of care and prevention. Greater attention of sick children may explain why measles (which often led to respiratory complications) declined as a cause of mortality even while incidence remained high. Scarlet and rheumatic fever, by contrast, probably declined because of changes in bacterial virulence and a decline in household population density.

To point to the importance of the decline of infectious diseases as the major element in mortality is not to imply that before 1900 long-duration or chronic diseases were unimportant. The fact of the matter was that high death rates among the very young tended to mask the presence of chronic illnesses and disabilities among both young and older adults. Indeed, pension records of Union Army Civil War veterans revealed high prevalence rates of a variety of chronic conditions. Moreover, the industrial changes that transformed American society magnified older and created novel occupational risks. Dusty environments, unsafe machinery, the use of toxic substances, high rates of industrial accidents, and crowded workplaces magnified occupational diseases that often led to severe disabilities.

The mortality decline that began in the late nineteenth century reflected a dramatic increase in survival rates among infants and children (which was the primary factor in the increase in longevity in the twentieth century). Longevity among the elderly increased as well, but not as spectacularly. As infant and child mortality fell, more Americans survived to old age, and the median age and the proportion of elderly increased commensurately. The change in the age distribution of the population mirrored a shift in the causes of mortality. In the nineteenth century death was associated with infancy and childhood. In the twentieth century death was increasingly associated with old age. Indeed, the longer individuals survived, the more likely were they to die from cancer, cardiovascular, cerebrovascular, or pulmonary diseases. By the close of the twentieth century these diseases accounted for more than two-thirds of total mortality.

The prevailing presumption is that long-duration diseases involve a complex blend of genetic, environmental, and behavioral factors. Yet despite claims to the contrary, their etiology remains unclear. In recent decades the emphasis has been on the role of risk factors in disease etiology, which in turn leads to the claim that a major role for medicine is to prevent disease by modifying the risks responsible for the disease. Thus managing risk, in addition to the treatment of disease, has become a major theme in medical practice.

The efforts to demonstrate the importance of risk factors in disease etiology, however, has been fraught with pitfalls. To be sure, the pioneering work of Austin Bradford Hill and Richard Doll in demonstrating the relationship between smoking and lung cancer is an exception. But many of the subsequent epidemiological studies attempting to isolate risk factors for a variety of other diseases suffered from major methodological shortcomings. The examples of coronary heart disease (CHD) and cancer are illustrative.

Between 1900 and 1920 the death rate from CHD remained more or less stable. About 1920 an upward trend began and continued unabated for 30. The rise was particularly concentrated among young males in their 30s and 40s. Interestingly enough, the increase in CHD mortality was international in scope and occurred in more than two dozen countries among people with quite different lifestyles. Somewhere in the 1950s mortality rates began to fall. Between 1950 and 1998 age-specific death rates for all heart diseases fell from 307 to 127 per 100,000 and heart disease became largely a disease of the very old (Grob, 2002, p. 249). The decline has persisted until the present.

What explains this pattern? In 1944 Paul Dudley White—perhaps the nation's most famous cardiologist—expressed puzzlement. "Why should the robust and apparently most masculine young males be particularly prone to this disease?," he wrote in his authoritative *Heart Disease* (p. 482). In the 1950s and 1960s Ancel Keys developed the thesis that dietary fat raised serum cholesterol and led to atherosclerosis. He insisted that three critical variables explained CHD, namely, age, blood pressure, and serum cholesterol. At the about the same time, the Framingham Heart Study—a community study that began in 1948 and has tracked three generations of residents—seemed to indicate that hypertension, obesity, smoking, and a family history of heart disease played crucial roles in CHD. A variety of other studies came to similar conclusions.

The emphasis on risk factors as major elements in the etiology of CHD, however, is not entirely persuasive. Why, for example, did CHD mortality rise among younger men in their thirties and forties between the 1920 and 1950s, particularly because risk factors did not begin to be a factor until the late twentieth century? Why was mortality rising during the depression of the 1930s and World War II when rich diets and sedentary lifestyles were uncommon? To be sure, cigarette smoking became prominent in the 1920, but its health effects did not appear until several decades later. Moreover, mortality from CHD began to decline during the 1960s, well before the emphasis on risk factors to promote behavioral change and the introduction of statins to lower cholesterol had any effect. Recent claims that medical interventions and behavioral modifications were responsible for the decline in CHD mortality after 1980 ignore the fact that the decline began two decades earlier, well before either factor was operative. Comparative data from other countries also failed to substantiate the claim that risk factors such as high-fat diets explained CHD mortality.

Like CHD, cancer presents similar enigmas. Speculation about its etiology has involved a myriad of competing theories. Heredity, microbes, viruses, irritations, occupation, diet, environment, and such psychological factors as stress were all advanced at one time or another as causal elements in cancer genesis. Unlike CHD, however, cancer mortality rates have not fluctuated radically in the past half century, although the mix of cancers has changed. The striking relationship between smoking and lung cancer, as well as evidence that exposure to a relatively small number of chemicals and radiation, fostered the emergence of an explanatory model that emphasized an environmental and behavioral etiology for most cancers. Indeed, a Harvard Report on Cancer Prevention attributed 60 percent of total cancer deaths to smoking and adult diet/ obesity; the remainder included other behavioral factors ("Harvard Report on Cancer Prevention," 1996, S3).

Such claims rely on epidemiological studies employing cohort analysis and observational studies. In employing this methodology, investigators monitor disease rates and lifestyle factors (e.g., diet and physical activity) and then infer conclusions about the relation between them. Thus they identify risk factors, such as high-fat diets, that presumably cause cardiovascular disease. The problem is that risk factors are associations. At best, cohort analysis and observational studies can generate hypotheses, but say nothing about causation. Indeed, the etiology of most common cancers (breast, prostate, colon) is unknown although research has shown that genetic changes (e.g., impairment of tumor suppression genes) play a major role.

Although cancer, cardiovascular–renal disease, diabetes, mental illnesses, and other long-duration diseases account for the bulk of morbidity and mortality, infectious diseases remain significant elements. HIV/AIDS, influenza, sexually transmitted diseases, and the appearance of resistant strains of a variety of pathogens all suggest that their eradication is an illusion. Social, economic, and environmental changes have the capacity both to diminish and to enhance their virulence. "Each type of society creates its own diseases," noted the distinguished biologist René Dubos a half century ago. Conceding that it was possible to develop new effective medical interventions, he nevertheless argued that society would always confront diseases, although "the diseases will only be different from those of the past (Dubos, 1961, p. 71)." So long as human beings are mortal, the prevention and conquest of disease will remain nothing more than an illusion.

[*See also* **Alzheimer's Disease and Dementia; Arthritis; Asthma and Allergy; Autism; Cancer; Cardiology; Cholera; Columbian Exchange; Death and Dying; Diabetes; Diphtheria; Epilepsy; Food and Diet; Genetics and Genetic Engineering; Germ Theory of Disease; HIV/ AIDS; Hospitals; Influenza; Life Expectancy; Malaria; Medicine; Medicine and Technology; Obesity; Occupational Diseases; Pediatrics; Penicillin; Pharmacology and Drug Therapy; Poliomyelitis; Public Health; Public Health Service, U.S.; Rabies; Race and Medicine; Science; Sexually Transmitted Diseases; Sickle-Cell Disease; Smallpox; Surgery; Technology; Tuberculosis; Typhoid Fever; Typhus;** *and* **Yellow Fever.**]

BIBLIOGRAPHY

Aronowitz, Robert A. *Making Sense of Illness: Science, Society, and Disease.* New York: Cambridge University Press, 1998.

Brandt, Allan M. *No Magic Bullet: A Social History of Venereal Disease in the United States since 1880,* expanded ed. New York: Oxford University Press, 1987.

Chambers, John Sharpe. *The Conquest of Cholera: America's Greatest Scourge.* New York: Macmillan, 1938.

Condran, Gretchen A. "The Role of Medicine in Mortality Decline: The Case of Diphtheria in the Late Nineteenth Century." *Journal of the History of Medicine and Allied Sciences* 63, no. 4 (October 2008): 484–522.

Condran, Gretchen A., Henry Williams, and Rose Cheney. "The Decline in Mortality in Philadelphia from 1870 to 1930: The Role of Municipal Services." *Pennsylvania Magazine of History and Biography* 108, no. 2 (1984): 153–178. Reprinted in *Sickness and Health in America,* edited by J. Leavitt and R. Numbers, pp. 422–436. Madison: University of Wisconsin Press, 1986.

Crosby, Alfred W., Jr. *The Columbian Exchange: Biological and Cultural Consequences of 1492.* Westport, Conn.: Greenwood, 1972.

Dubos, René. *The Dreams of Reason: Science and Utopias.* New York: Columbia University Press, 1961.

Duffy, John. *Epidemics in Colonial America.* Baton Rouge: Louisiana State University Press, 1953.

Duffy, John. *Sword of Pestilence: The New Orleans Yellow Fever Epidemic of 1853.* Baton Rouge: Louisiana State University Press, 1966.

English, Peter C. *Rheumatic Fever. A Biological, Epidemiological, and Medical History of Rheumatic Fever in America and Britain from the 18th to the 20th Century.* New Brunswick, N.J.: Rutgers University Press, 1999.

Erhardt, Carl L., and Joyce E. Berlin, eds. *Mortality and Morbidity in the United States.* Cambridge, Mass.: Harvard University Press, 1974.

Greven, Philip J., Jr. *Four Generations: Population, Land, and Family in Colonial Andover, Massachusetts.* Ithaca, N.Y.: Cornell University Press, 1970.

Grob, Gerald N. *The Deadly Truth: A History of Disease in America.* Cambridge, Mass.: Harvard University Press, 2002.

"Harvard Report on Cancer Prevention." *Cancer Causes and Control* 7 (Supplement 1996): S3–S59; 8 (Supplement 1997): S1–S50.

The Historical Statistics of the United States, 5 vols. Cambridge, U.K.: Cambridge University Press, 2006.

Humphreys, Margaret. *Malaria, Poverty, Race, and Public Health in the United States.* Baltimore: Johns Hopkins University Press, 2001.

Kunitz, Stephen J. "Mortality Changes in America, 1620–1920." *Human Biology* 56 (1984): 559–582.

Maulitz, Russell C., ed. *Unnatural Causes: The Three Leading Killer Diseases in America.* New Brunswick, N.J.: Rutgers University Press, 1988.

McKinlay, John B., and Sonja M. McKinlay. "The Questionable Contribution of Medical Measures to the Decline of Mortality in the United States in the Twentieth Century." *Milbank Memorial Fund Quarterly* 55 (1977): 405–428.

Morgan, Edmund S. *American Slavery American Freedom: The Ordeal of Colonial Virginia.* New York: W. W. Norton, 1975.

Mukherjee, Siddhartha. *The Emperor of All Maladies: A Biography of Cancer.* New York: Charles Scribner's Sons, 2010.

Preston, Samuel H. *Fatal Years: Child Mortality in Late Nineteenth-Century America.* Princeton, N.J.: Princeton University Press, 1991.

Rosenberg, Charles E. *The Cholera Years: The United States in 1832, 1849, and 1866.* Chicago: University of Chicago Press, 1962.

Rosenberg, Charles E. *Our Present Complaint: American Medicine, Then and Now.* Baltimore: Johns Hopkins University Press, 2007.

Rosner, David, ed. *Hives of Sickness: Public Health and Epidemics in New York City.* New Brunswick, N.J.: Rutgers University Press, 1995.

Stallones, Reuel A. "The Rise and Fall of Ischemic Heart Disease." *Scientific American* 243 (1980): 53–59.

Stuart, Paul. *Nations within a Nation: Historical Statistics of American Indians.* New York: Greenwood Press, 1987.

Weatherall, David J. *Science and the Quiet Art: The Role of Medical Research in Health Care.* New York: W. W. Norton, 1995.

Wells, Robert V. *The Population of the British Colonies in America before 1776.* Princeton, N.J.: Princeton University Press, 1975.

White, Paul Dudley. *Heart Disease,* 3d ed. New York: Macmillan, 1944.

Gerald N. Grob

DISNEY, WALT

See **Animation Technology and Computer Graphics.**

DNA SEQUENCING

DNA (deoxyribonucleic acid) sequencing is a technology that determines the precise linear order of the chemical units along a DNA strand, which consists of four nucleobases—adenine (abbreviated A), cytosine (C), guanine (G), and thymine (T). The sequence of DNA emerged as a key subject of scientific interest among molecular biologists when it became widely understood in the 1950s as the most important component of the cell, the "master plan" that would direct the production of a wide variety of molecular structures. By the late-1950s, not only was the double-helical structure of DNA elucidated, but also its biological significance as a genetic entity was firmly established.

In 1958, Francis Crick, one of the codiscoverers of the structure of DNA, boldly articulated the idea that the linear, one-dimensional sequences of DNA would contain all the genetic instructions necessary for protein synthesis. His "sequence hypothesis" proposed that "the specificity of a piece of nucleic acid is expressed solely by the sequence of its bases, and that this sequence is a (simple) code for the amino acid sequence of a particular protein" (Crick, 1958, p. 152). In this hypothesis, Crick redefined the problem of gene action and protein synthesis in terms of information flow and coding between nucleic-acid sequences and amino-acid sequences, referring biological information to the precise linear order of the units along a nucleic acid. He further insisted that protein "folding is simply a function of the order of the amino acids" (Crick, 1958, p. 144), proclaiming that the diversity of biological macromolecules and its organization is specified in a linear, one-dimensional sequence of simple units. The DNA sequences of a human being would then provide "a recipe for constructing a person."

The presumed significance of the DNA sequence became central to molecular biologists as the problem of protein synthesis was redefined as one of information flow and coding. Biologists theorized that RNA (ribonucleic acid), a template for protein synthesis, carries the genetic instructions from the DNA, as it was synthesized from the DNA sequences through complementary base-pairing. On this interpretation, the specificity of the gene was linear, one-dimensional. Many prominent molecular biologists and mathematicians, such as Crick, Sidney Brenner, and George Gamow, worked to elucidate the biochemical or mathematical relationship among DNA, RNA, and amino-acid sequences. In the end, it was Marshall Nirenberg and Heinrich Matthaei, two young biochemists at the National Institutes of Health, who first showed that a triplet of bases in a particular order along the nucleic-acid strand (the genetic code) specifies an amino acid. Their biochemical solution of the coding problem in the 1960s came before the experimental determination of any actual DNA sequences.

DNA sequencing remained a technological challenge for several reasons until the early 1970s. DNA molecules were difficult to separate because they were so similar in their chemical properties, and the length of naturally occurring DNA molecules was much greater for proteins, with no base-specific enzymes that could cleave DNA sequences

into manageable pieces. Fred Sanger at Cambridge and Walter Gilbert at Harvard were two principle pioneers of DNA sequencing in the 1970s. Sanger had determined the exact amino-acid sequence of a small protein, insulin, in the late 1940s and turned to sequencing RNA and DNA. In 1975, Sanger invented his "plus and minus" method for DNA sequencing that laid the subsequent foundation for the later generations of DNA-sequencing techniques. He further developed dideoxy, sometimes called Sanger sequencing, a method still used to rapidly sequence genes. In 1977, Sanger's team produced the first complete genome sequence of an organism, the bacteriophage phi-X 174, one of the tiniest viruses with few genes and single-stranded DNAs. Walter Gilbert and Allan M. Maxam in turn developed a DNA-sequencing method that used chemicals rather than enzymes. In the 1980s and 1990s the two methods developed by Sanger and Gilbert were standardized, speeded up, and, in large part, automated. Leroy Hood at the California Institute of Technology and Applied Biosystems, a start-up company, developed the first commercial gene-sequencer machine.

Rapid developments in DNA sequencing and the need to analyze an immense amount of DNA sequence data led to the emergence of genomics and bioinformatics as new disciplines. DNA sequencing became a large-scale, international research program through the Human Genome Project in the 1990s, which aimed to produce a "book of man." The Human Genome Project was an ambitious, technologically driven program that systematically applied the tools of DNA sequencing and mapping to decipher the entire human genome. The automation of DNA sequencing, first achieved by Hood's team, along with other technical developments such as the invention of polymerase chain reaction (PCR) by Kary Mullis and pulsed-field gel electrophoresis developed by David Schwartz, played a critical role in the experimental implementation of the Human Genome Project. The first draft of the human genome sequence was published in 2001.

[*See also* **Biochemistry; Biological Sciences; Genetics and Genetic Engineering;** *and* **Molecular Biology.**]

BIBLIOGRAPHY

Crick, Francis H. C. "On Protein Synthesis." *Symposia of the Society for Experimental Biology* 12 (1958): 138–163.

De Chadarevian, Soraya. "Sequence, Conformation, Information: Biochemists and Molecular Biologists in the 1950s." *Journal of the History of Biology* 29 (1996): 361–386.

Kevles, Daniel J., and Hood, Leroy, eds. *The Code of Codes: Scientific and Social Issues in the Human Genome Project.* Cambridge, Mass.: Harvard University Press, 1992.

"Special Issue: The Human Genome." *Science* 291 (2001): 1145–1434.

Strasser, Bruno J. "World in One Dimension: Linus Pauling, Francis Crick, and the Central Dogma of Molecular Biology." *History and Philosophy of the Life Sciences* 28 (2006): 491–512.

Doogab Yi

DOBZHANSKY, THEODOSIUS

(1900–1975), geneticist and evolutionary biologist, one of the most influential biological scientists of the twentieth century. Born in a small town in the Ukraine, Dobzhansky graduated with a biology degree from the University of Kiev in 1921 and became an instructor at Kiev's Polytechnic Institute. He went to the University of Leningrad in 1924 to assist Yuri Filipchenko, head of the Department of Genetics, and immigrated to the United States in 1927 to work in Thomas Hunt Morgan's laboratory at Columbia University. When Morgan moved to the California Institute of Technology in 1928, Dobzhansky also relocated and spent more than 15 years at Caltech as an assistant professor of genetics (1929–1936) and full professor (1936–1940). From 1940 to 1962 Dobzhansky was Professor of Zoology at Columbia University. In 1962 he went to The Rockefeller Institute, becoming emeritus in 1970, at which point he moved to University of California, Davis, as an adjunct professor of biology.

Primarily a laboratory geneticist, Dobzhansky trained as a naturalist and made fundamental contributions to the study of evolution in natural populations. He developed a new experimental system that blurred the field–lab boundary by collecting many wild fruit flies, *Drosophila pseudoobscura,* from many regions in North and South America and then analyzing them in the laboratory. In 1937 he published *Genetics and the Origin of Species,* which fused Mendelian genetics with Darwinian natural selection. This book influenced many scientists, including other major contributors of the evolutionary synthesis, Ernst Mayr, G. G. Simpson, and G. Ledyard Stebbins. Many biological scientists have hailed it as the twentieth century's most influential book on evolution.

Dobzhansky stressed the importance of genetic variation to support evolutionary change, which not only was pivotal to the evolutionary synthesis, but also served as a basis for his views on human evolution and the social sciences. He counteracted eugenics and racism through discussions of biological and cultural evolution in *Heredity, Race and Society* coauthored with L. C. Dunn (1946, rev. 1952) and *Mankind Evolving* (1962). Dobzhansky joined other biologists in campaigns to undermine Trofim Lysenko's influence in the Soviet Union and counteract antievolutionists in the United States. In his support of teaching evolution in American public schools, Dobzhansky penned his most well-known statement, "Nothing in biology makes sense except in the light of evolution" (*The American Biology Teacher* 35, no. 3 [March 1973]: 125–129).

[*See also* **Biological Sciences; Eugenics; Evolution, Theory of; Genetics and Genetic Engineering; Mayr, Ernst; Morgan, Thomas Hunt; Race and Medicine; Rockefeller Institute, The; Simpson, George Gaylord;** *and* **Zoology.**]

BIBLIOGRAPHY

Adams, Mark B., ed. *The Evolution of Theodosius Dobzhansky: Essays on His Life and Thought in Russia and America.* Princeton, N.J.: Princeton University Press, 1994.

Ayala, Francisco J. "Theodosius Dobzhansky, January 25, 1900–December 18, 1975." *Biographical Memoirs of the National Academy of Sciences* 55 (1985): 163–213.

Smocovitis, Vassiliki Betty. *Unifying Biology: The Evolutionary Synthesis and Evolutionary Biology.* Princeton, N.J., Princeton University Press, 1996.

Melinda Gormley

DREW, CHARLES RICHARD

(1904–1950), pioneering blood plasma scientist, surgeon, teacher. Born in Washington, D.C., Charles Drew graduated from McGill University Medical School in Montreal in 1933, ranking second in a class of 137. During a two-year fellowship at Columbia University's medical school (1938–1940), he did research on blood banking, setting up Presbyterian Hospital's first blood bank, and became the first African American to receive the doctor of science degree. Drew served as medical director of the Blood for Britain Project in 1940 and also of a 1941 American Red Cross pilot project involving the mass production of dried plasma. Drew's work proved pivotal to the success of the Red Cross's blood-collection program, a major lifesaving agent during World War II. In 1941 Drew became chairman of Howard University's Department of Surgery and chief surgeon at Freedmen's Hospital, where he worked tirelessly to build Howard's surgical residency program. Between 1941 and 1950, he trained more than half of the black surgeons certified by the American Board of Surgery. During the war years, Drew had spoken out against the Red Cross's blood segregation policy. When he died at the age of 45 after an auto accident in North Carolina, a legend sprang up that he had bled to death after being turned away from a whites-only hospital. Although the legend was false, persisting medical discrimination against African Americans perpetuated it. Throughout his career, Drew was committed to making medical care and training available to citizens of all races and economic levels.

[*See also* **Biological Sciences; Medical Education; Medicine; Military, Science and Technology and the; Race and Medicine; Science; Surgery;** *and* **War and Medicine.**]

BIBLIOGRAPHY

Love, Spencie. *One Blood: The Death and Resurrection of Charles R. Drew.* Chapel Hill: University of North Carolina Press, 1996.
Wynes, Charles E. *Charles Richard Drew: The Man and the Myth.* Urbana: University of Illinois Press, 1988.

Spencie Love

DUBOS, RENÉ JULES

(1901–1982), French-born American microbiologist, medical scientist, and environmentalist. A prominent ecological thinker of the twentieth century, Dubos espoused the philosophy that the health or disease of a living organism—microbe, man, society, or the earth itself—can be understood only in the context of relationships formed with its environment. His bench research informed and influenced treatment of infectious diseases, development of antibiotics, and studies of both the human microbiome and environmental health. Later, his ecological views influenced conservation, urban planning, and architecture (Moberg, 2005).

Following studies of agronomy in France, Dubos earned a PhD at Rutgers University under soil microbiologist Selman Waksman with a thesis proving that local soil properties determine which bacteria decompose cellulose and which chemicals they produce. This led Oswald Avery of The Rockefeller Institute (now The Rockefeller University), who was trying to destroy cellulose capsules of pneumococci, to hire him. Applying his thesis model, Dubos succeeded in 1930. In 1939, he published "Studies on a Bactericidal Agent Extracted from a Soil Bacillus," describing his discovery of the antibiotic tyrothricin. Within months, he published its bacterial, chemical, and clinical properties. Learning of this work, Oxford scientists broadened studies of penicillin and Waksman began the search that led to streptomycin. In 1942, even before antibiotics were in general use, Dubos warned that bacterial resistance to antibiotics should be expected.

The death of his first wife from tuberculosis (TB) in 1942 spurred him toward TB research, then a field almost ignored. He created a culture medium that enabled the first accurate, quantitative studies of the bacilli, their disease-causing properties, and methods to standardize TB vaccines. Acting on his beliefs that infection is the rule, disease the exception, and prevention better than cure, he published *Mirage of Health* (1959), a seminal book still in print.

In further research Dubos showed how subtle environmental stresses of malnutrition, toxins, and crowding increase susceptibility to disease. This led him to revise the germ theory, conceived by Louis Pasteur, stating that a microbe is necessary but not sufficient to cause disease. When the same stresses were applied to newborn animals, they induced lifelong deleterious effects; a phenomenon Dubos labeled "biological Freudianism." Other studies revealed that the digestive tract is an ecosystem that cannot develop fully or function normally unless it harbors a robust, diverse population of microbes.

Dubos became a public figure in the environmental movement after winning a 1969 Pulitzer Prize for *So Human an Animal*. Calling himself a "despairing optimist," he served as a wise provocateur and championed the view that the quality of our daily lives is intricately interwoven with the quality of the earth itself. He promoted a humanistic biology in numerous books, essays, interviews, and lectures, and coined memorable aphorisms to simplify profound insights, among them "think globally, act locally," "trend is not destiny," "improving on nature," and "creative adaptations." His constructive ideas stressed the human obligation to cultivate a healthy earth.

Dubos received many international honors, including membership in the National Academy of Sciences, 41 honorary degrees, the Lasker Award, and the Tyler Ecology Award.

[*See also* Biological Sciences; Conservation Movement; Disease; Ecology; Environmentalism; Germ Theory of Disease; Medicine; National Academy of Sciences; Penicillin; Rockefeller Institute, The; Science; *and* Tuberculosis.]

BIBLIOGRAPHY

Dubos, René. *Mirage of Health: Utopias, Progress, and Biological Change*. New York: Harper & Brothers, 1959. Reprint ed. New Brunswick, N.J.: Rutgers University Press, 1987.

Dubos, René. "Studies on a Bactericidal Agent Extracted from a Soil Bacillus." *The Journal of Experimental Medicine* 70 (1939): 1, 10, 11, 17, 249, 256.

Moberg, Carol L. *René Dubos, Friend of the Good Earth*. Washington, D.C.: ASM Press, 2005.

Carol L. Moberg

DU PONT, PIERRE S.

(1870–1954), chemist, business executive. Pierre du Pont made the DuPont Company a model of the modern science-based, professionally managed, industrial corporation. Along with two cousins, he oversaw the transformation of DuPont from a family firm that made dynamite and black powder in 1902 into a large, diversified chemical corporation by the 1920s. To manage its diverse product line, DuPont in 1921 devised a new corporate structure that consisted of semiautonomous product-based divisions. In subsequent decades this organizational structure was adopted by many other companies.

Pierre du Pont was born in 1870, the eldest son of Lammot du Pont, an inventive chemist who became interested in the manufacture of Alfred Nobel's powerful new explosive, dynamite. In 1884 he was killed in an explosion at the plant. Despite its dangers, dynamite soon replaced black powder for blasting. Pierre joined the DuPont Company after earning a degree in chemistry from the Massachusetts Institute of Technology (MIT) in 1890. In 1899 his concerns about the future of DuPont under the leadership of the older generation led Pierre to join his cousin, T. Coleman, in steel and railway enterprises in Cleveland. In 1902, the two cousins, along with another cousin, Alfred I., bought the company and then proceeded to buy out many of their competitors until they had control of about 60 percent of the American dynamite business. To improve its products and processes, DuPont established two research laboratories, the Eastern Laboratory (1902) and the Experimental Station (1903). One major research initiative was to develop nitrocellulose or smokeless powder to replace black powder as a propellant in guns. The military was the primary customer for this product, so when government orders declined in 1909, the company began its first tentative attempt to diversify by buying the Fabrikoid Company that made artificial leather from nitrocellulose.

In 1913 the DuPont Company was broken up as a result of an antitrust conviction. Although the dynamite business was split with two new spin-off companies, DuPont, at the request of the War Department, retained all of its smokeless powder capacity. When World War I broke out in Europe a year later, DuPont became a major supplier of smokeless powder for the Allies. Because Pierre feared that the war might end suddenly, leaving DuPont with large plants and canceled orders, DuPont initially charged about twice the prewar price for powder. Although DuPont eventually lowered its prices, company sales and profits soared. Between 1914 and 1916 company annual sales increased from $25 to $318 million and net earnings from $5.6 million to $82 million. With this enormous influx of money, DuPont began to invest in chemical industries.

Managing a number of different businesses such as plastics, dyestuffs, and paints proved so difficult that in 1921 DuPont decided to create semiautonomous divisions for each product line. This new structure allowed the company to focus management attention on each product and to add new product divisions as DuPont continued to diversify in the 1920s.

Although Pierre had served as company president only from 1915 to 1919, he was the driving force in the company's management from 1902 to 1921. In the 1920s Pierre retired from active management, serving as board chair until 1940, except for a stint as president of General Motors (GM) from 1920 to 1923. During World War I, DuPont had invested heavily in GM, which suffered a severe financial crisis during the postwar economic downturn. After stabilizing the situation at GM, Pierre turned the company over to Alfred P. Sloan, who would reorganize GM into an organization similar to the one Pierre had instituted at DuPont.

[*See also* **Chemistry; Iron and Steel Production and Products; Military, Science and Technology and the; Plastics; Railroads; Research and Development (R&D);** *and* **Science.**]

BIBLIOGRAPHY

Chandler, Alfred D., and Steven Salsbury. *Pierre S. du Pont and the Making of the Modern Corporation.* New York: Harper & Row, 1971.

John Kenly Smith Jr.

DUST BOWL

The name Dust Bowl applied to the high plains of Texas, Oklahoma, New Mexico, Colorado, and Kansas during the later 1930s as immense dust storms blew across the region, darkening the sky and depositing soil hundreds of miles to the east. At its peak the Dust Bowl covered nearly 100 million acres, with similar conditions extending northward into Canada. In 1938, the worst year for erosion, farmers lost an estimated 850 million tons of topsoil.

Severe wind erosion led to a precipitous drop in farm income, impaired health, and caused widespread damage to houses and machinery. Those conditions, combined with national economic depression, turned many people into refugees; in the worst-hit counties, one-third to one-half of the population left, many migrating to California. For those who stayed, bankruptcies were common in both town and country.

The causes of this environmental catastrophe are disputed. Some historians see the farmers as innocent victims of drought; others argue that agricultural practices were heavily to blame. During World War I and the 1920s, wheat farming expanded rapidly into the windy, drought-prone plains. Native grasses that had evolved a high degree of climatic resilience abruptly disappeared under the plow. For a while crops were abundant and profits high, but then began a record-breaking drought that withered the fields and left them bare.

Severe but short-lived droughts recurred in the decades after the 1930s, but none had the impact of the Dust Bowl years, leading many observers to conclude that farmer ingenuity and improved technology had made another disaster impossible. In truth, although a constant flow of federal dollars along with irrigation from deep aquifers managed to stave off a repeat catastrophe, the future of the region remains volatile and uncertain.

[*See also* **Agricultural Technology;** *and* **Environmentalism.**]

BIBLIOGRAPHY

Hurt, R. Douglas. *The Dust Bowl: An Agricultural and Social History.* Chicago: Nelson–Hall, 1981.

Worster, Donald. *Dust Bowl: The Southern Plains in the 1930s.* New York: Oxford University Press, 2004.

Donald Worster

E

EASTMAN, GEORGE

(1854–1932), inventor. Born on a farm in upstate New York, George Eastman moved with his family in 1860 to the city of Rochester. His father died two years later, and his mother supported the family by operating a boarding-house. Eastman attended a private school in Rochester during the 1860s but had a limited education. Partly to help support his mother, he took a job as an office boy with an insurance agent when he was 13. He later worked as a bookkeeper for a Rochester bank.

Eastman became interested in photography in 1877. Frustrated with the complicated and messy wet plate technology in use at the time, he decided to experiment with newly introduced gelatin dry plates—glass plates coated with a gelatin emulsion that would retain photosensitivity even after the emulsion had dried. From reading photographic journals and experimenting in his mother's kitchen, he developed his own dry plates. He

then invented machinery to produce the dry plates and received financial backing from a family friend to establish a new company in 1881, the Eastman Dry Plate Company, to manufacture and sell the plates. Driven by an interest in selling to a wider market of nonexpert photographers, Eastman made a major innovation by replacing the dry plate with a flexible and transparent film.

In 1888, Eastman began to sell a new camera he had developed based on his innovations. He coined the term "Kodak" for the name of the camera. The Kodak camera contained a roll of film mounted on a holder. After each new photograph, the user could unwind the roll for a new exposure. When finished with the entire one-hundred-exposure roll of film included with the camera, the consumer-photographer would simply return the entire package to the company, which would then develop the film, print the photographs, reload new film, and return the camera to the customer. The Kodak camera played a crucial role in democratizing photography. Expertise was no longer a requirement. Eastman sold the camera using the slogan "You press the button, we do the rest."

Other innovations pioneered by Eastman and his company continued this process of creating a mass market for photography. In 1889, he replaced the paper film with celluloid, an innovation also important in the development of motion picture projection. His cheap and easy-to-use Brownie camera, introduced in 1900 and designed especially for children, expanded the practice of photography to millions of Americans. The company he established, known as the Eastman Kodak Company after 1892, grew to become a major multinational corporation responsible for significant innovations through the acquisition of patents and support of research and development. Later in his life, after accumulating immense wealth, he became a major philanthropist, contributing tens of millions of dollars to universities and charities, especially in Rochester.

[*See also* **Engineering; Film Technology; Photography; Research and Development (R&D); Technological Enthusiasm;** *and* **Technology.**]

BIBLIOGRAPHY

Brayer, Elizabeth. *George Eastman: A Biography.* Rochester, NY: University of Rochester Press, 2006.
Jenkins, Reese V. *Images and Enterprise, Technology and the American Photographic Industry, 1839 to 1925.* Baltimore: Johns Hopkins University Press, 1975.

Hugh Richard Slotten

ECOLOGY

The development of modern ecology was strongly shaped by the earlier tradition of plant geography rooted in Alexander von Humbolt's explorations in South America and continuing throughout the nineteenth century as European botanists attempted to systematically map and classify vegetation. This interest in mapping and classifying flora and fauna later served as the central focus of a nas-

cent ecology in America, in part because of the wide variety of habitats and the realization that nature was quickly being altered by human activities. In 1889 the U.S. Department of Agriculture supported a biological survey by the naturalist C. Hart Merriam in the San Francisco Mountains in Arizona. Based on his study, Merriam developed his system of life zones, claiming that lines of equal temperature determined three broad belts, each with a characteristic flora and fauna. Merriam further divided the Boreal, Austral, and Tropical life zones into subzones. Although widely criticized for being too simplistic, Merriam's life zones continued to be modified and attracted adherents throughout the first half of the twentieth century. More broadly influential were slightly later attempts to classify vegetation by the first generation of American plant ecologists, notably Frederic Clements and Henry Chandler Cowles. These two ecologists shaped the broad contours of American ecology up until World War II, although their work rested heavily upon earlier developments in European ecology.

The Danish botanist Eugenius Warming and German plant physiologist Andreas Schimper had a particularly strong influence on Clements, Cowles, and other American plant ecologists. Both Warming and Schimper emphasized the relationships between plants and the environment, and they continued the earlier European tradition of mapping vegetation. Impressed by Warming's work, the British botanist Arthur Tansley organized the first International Phytogeographical Excursion (IPE, 1911) to show a distinguished group of European and American botanists various plant communities in the British Isles. Cowles and Clements, who participated in this event, later organized a second IPE throughout North America in 1913. Tansley's efforts to map vegetation gave rise to the British Ecological Society in 1913, which served as a model for the Ecological Society of America formed two years later, largely as a result of the efforts of Cowles and his students.

Ecology during the First Half of the Twentieth Century. Research in ecology during the first half of the twentieth century can be conveniently characterized by two broad categories: autecology and synecology. Autecology focused upon the relationships between individual organisms and the physical environment, whereas synecology was the study of communities of organisms. Both of these broad ecological domains were heavily influenced by physiology, which served as a model science for early ecologists. The rigorous quantitative and experimental methods perfected in physiological laboratories were emulated both directly and indirectly by ecologists. For example, in 1903 the Carnegie Institution of Washington established an ecological laboratory in Tucson, Arizona, that quickly became a major center for studying the physiological adaptations of plants to extreme environments. In a more general way, synecologists invoked physiology as an intellectual guide for studying the processes that governed plant and animal communities.

This physiological perspective shaped Cowles' early approaches to plant ecology, although he was also strongly influenced by recent developments in geography and geology. As a graduate student at the University of Chicago, Cowles studied the changes in vegetation on the sand dunes bordering Lake Michigan. He traced the developmental history (succession) of plant communities starting with a few hardy pioneer species growing on the beach and ending in a mature hardwood forest that stabilized the dune on which it grew. Cowles's dissertation (1899) on plant succession became a classic paper. It encouraged other plant ecologists to study succession, but it also inspired zoologists, such as Cowles's student Victor Shelford, to study the distribution of insects in the various dune communities. Through his charismatic teaching, Cowles attracted a large number of students who carried on his legacy and helped make the University of Chicago a major center of ecological research during the first half of the twentieth century.

Frederic Clements made the analogy between organismal development and succession explicit. For Clements, a plant community actually was a "complex organism" that underwent an orderly and predictable developmental process culminating in a mature *climax community*. In contrast to Cowles's emphasis on local geological and geo-

graphical factors, Clements claimed that vegetation was determined by climate. According to Clements, vegetation in each geographic region developed into a unique climax community that was in harmony with the climate. Following Warming and Schimper, Clements also emphasized the physiological basis of ecology by claiming that ecologists could study the development of a community with the same precision attained in the physiological laboratory. To this end he perfected a number of analytical and experimental techniques, most notably the use of quadrats and transects for precisely quantifying the distribution and abundance of various species. Thus, despite criticisms that his physiological perspective was overly simplistic, Clements made major contributions to quantitative and experimental methods in ecology. An early critic, Henry Allen Gleason, challenged Clements' physiological perspective by denying that plant communities were comparable to organisms. Instead, Gleason claimed that a community was simply a collection of plants sharing characteristics allowing them to grow in the same general environment.

Despite these criticisms that would later undermine his influence, Clements's voluminous writings, particularly *Plant Succession* (1916), did much to set the research agenda for pre–World War II ecology. Not only was this work influential among plant ecologists, but also animal ecologists such as Victor Shelford adopted many of Clements's ideas about communities. In their book, *Bio-ecology* (1939), Clements and Shelford attempted a synthesis of plant and animal ecology. Combining plant and animal ecology became a major focus of ecology after World War II, and *Bio-ecology* established the idea of *biomes* as large biogeographic regions characterized both by the plants and the animals that lived there. Clementsian ecology also played an influential role in agronomy, forestry, and land management during the 1930s, particularly through federal programs aimed to alleviate the destructive effects of the Dust Bowl.

Although the idea of succession was a central part of ecology during the first half of the twentieth century, animal ecologists in particular were also interested in interactions among species making up communities. Early animal ecologists such as Stephen Forbes and Charles C. Adams studied competition and predation among animal species. In his famous essay "The Lake as a Microcosm" (1887), Forbes presented competition and predation not only as warfare among individuals, but also as mechanisms for stabilizing the community as a whole. Both Forbes and Adams were also interested in how energy was transferred through a community by the process of predation. Adams in particular attempted a broad theoretical foundation for ecology based on ideas borrowed from physical chemistry such as equilibrium, energy flow, and system. This approach presaged later developments in both community and ecosystem ecology after World War II.

The tension between the physiological perspective that shaped much of early ecology and the more Darwinian emphasis on competition and predation sparked some attempts at synthesis. Shelford's student Warder Clyde Allee worked within the physiological tradition to study animal behavior, emphasizing the importance of cooperation rather than competition. During the 1930s and 1940s he joined his University of Chicago colleague Alfred Emerson in developing an evolutionary population ecology based upon the idea of group selection. According to this view, natural selection acted primarily on populations rather than individuals, favoring those populations that had the greatest cooperation among their members. Group selection remained influential among ecologists until the 1960s, when ecologists rejected it in favor of theories based on individual fitness.

Post–World War II. As with many biological disciplines, ecology was transformed by the events of World War II. Prior to the war, there were fewer than one thousand ecologists in the United States, but beginning in the 1950s that number increased rapidly so that by the late 1970s membership in the Ecological Society of America reached six thousand. Fueled largely by increased government support for science—and later by environmental concerns—the post–World War II growth of ecology led to increasing specialization and important intellectual changes.

The influence of Clementsian ecology dwindled after World War II. During the 1950s, detailed quantitative studies of the vegetation of Wisconsin by John Curtis and his colleagues and of plant communities in the Great Smoky Mountains by Robert Whittaker undercut Clements's idea that communities are comparable to organisms and vindicated Gleason's alternative individualistic hypothesis. Although classifying communities and studying succession remained important, the emphasis in community ecology shifted much more to the interactions among populations within the community. Both theoretical and experimental studies on predation and competition became the central focus of community ecology. Furthermore, the study of communities was transformed by the emergence of two newer ecological specialties: population ecology and ecosystem ecology. Although built upon earlier work, both specialities flourished during the decades following World War II.

The newer population-based approach to studying communities is illustrated by the work of the Yale University ecologist G. Evelyn Hutchinson and his students during the 1950s and 1960s. Responding to the earlier work of the British ecologist Charles Elton, the Soviet biologist Georgii Gause, and mathematical theoreticians such as Alfred Lotka and Vito Volterra, Hutchinson and his students pursued theoretical and field studies demonstrating the importance of competition and predation as regulatory processes in communities. The concept of the ecological niche was a prominent feature of this work. For example, Hutchinson's student Robert McArthur demonstrated how competition was reduced among related species of warblers because each species fed in a slightly different location in the pine forests they inhabited. Ambitious experimental studies were also a major feature of the new population-based approach to community ecology. For example, Joseph Connell selectively removed competitors and predators from enclosures containing different species of barnacles in intertidal marine habitats to demonstrate how the interaction of competition, predation, and the physical environment regulated the zonation of barnacles in the tidal column.

Advances in population genetics and evolutionary biology also influenced population ecology.

The trend toward a more Darwinian approach was exemplified by George C. Williams's influential *Adaptation and Natural Selection* (1966), which harshly criticized the group selection theories used by Warder Clyde Allee and other ecologists. Instead, Williams emphasized natural selection acting on individuals. Many younger biologists were also strongly attracted to broader theoretical approaches that bridged population ecology and population genetics. Robert MacArthur, Edward O. Wilson, and Richard Levins joined population geneticists in the development of a more interdisciplinary population biology. These approaches were later reflected in a number of new textbooks that took an explicitly Darwinian perspective on ecology, notably Eric Pianka's *Evolutionary Ecology* (1974), which was a major departure from earlier ecology textbooks.

Occurring during the same time period in which population ecology flourished, the growth of ecosystem studies constituted another important shift in ecology. The British ecologist Arthur Tansley coined the term *ecosystem* in 1935 as an alternative to Clements' idea of the community as a "superorganism." Ecosystems were composed of functional groups or *trophic levels*: producers, consumers, and decomposers. Along with the physical environment, these trophic levels were viewed as machine-like components of an interrelated functioning entity. G. Evelyn Hutchinson and his postdoctoral student Raymond Lindeman explored this idea during the early 1940s, but it quickly caught on after World War II through the work of another Hutchinson student, Howard Odum and his older brother Eugene. The ecosystem as a central ecological concept was popularized by Eugene Odum's highly successful textbook, *Fundamentals of Ecology*, first published in 1953. By the 1960s, ecologists were actively engaged in studying a wide variety of terrestrial and aquatic ecosystems. These studies centered on quantifying the flow of energy through the various trophic levels and the biogeochemical cycling of important nutrients, radioactive elements, and toxic chemicals. The success of ecosystem ecology during this period was aided by financial support from the Atomic Energy Commission (AEC) and other government agencies. For example, AEC contracts sustained Eugene Odum's early research, and later the Oak Ridge National Laboratory funded an important ecological research unit where other ecosystem ecologists studied radiation ecology.

Ecosystem ecologists also pioneered a trend toward long-term ecological studies using large teams of scientists. This was partly a response to increased government spending on scientific research after World War II, particularly at the national laboratories where ecosystem ecology flourished. It was also a response to "big science" initiatives in the physical sciences typified by activities during the International Geophysical Year (1957–1958). American participation in the International Biological Program (IBP, 1968–1974) fostered large-scale ecosystem studies of biomes employing computer modeling and massive data collection by large teams of ecologists. Independent of the IBP, another important long-term project supported by the U.S. Forest Service and organized by Herbert Bormann and Gene Likens studied a watershed at Hubbard Brook in New Hampshire. Beginning in the 1960s and continuing today, Hubbard Brook exemplified a trend toward long-term cooperative studies by interdisciplinary teams of scientists. As a consequence of the IBP and projects such as Hubbard Brook, the National Science Foundation began a network of Long-Term Ecological Research (LTER) projects beginning in 1980, which eventually included more than 25 sites.

[*See also* **Agriculture, U.S. Department of; Atomic Energy Commission; Biological Sciences; Botany; Conservation Movement; Dust Bowl; Environmentalism; Forest Service, U.S.; Geography; Geology; Hutchinson, G. Evelyn; International Geophysical Year; National Science Foundation; Odum, Eugene and Howard; Physiology;** *and* **Zoology.**]

BIBLIOGRAPHY

Bocking, Stephen. *Ecologists and Environmental Politics: A History of Contemporary Ecology.* New Haven, Conn.: Yale University Press, 1997.

Golley, Frank Benjamin. *A History of the Ecosystem Concept in Ecology: More Than the Sum of the*

Parts. New Haven, Conn.: Yale University Press, 1993.

Hagen, Joel B. *An Entangled Bank: The Origins of Ecosystem Ecology*. New Brunswick, N.J.: Rutgers University Press, 1992.

Hagen, Joel B. "Teaching Ecology during the Environmental Age, 1965–1980." *Environmental History* 13 (2008): 675–694.

Kingsland Sharon E. *The Evolution of American Ecology, 1890–2000*. Baltimore: Johns Hopkins University Press, 2005.

Kingsland, Sharon E. *Modeling Nature: Episodes in the History of Population Ecology*. Chicago: University of Chicago Press, 1985.

Kohler, Robert E. *Landscapes and Labscapes: Exploring the Lab–Field Border in Biology*. Chicago: University of Chicago Press, 2002.

McIntosh, Robert P. *The Background of Ecology: Concept and Theory*. Cambridge, U.K.: Cambridge University Press, 1985.

Mittman, Gregg. *The State of Nature: Ecology, Community, and American Social Thought, 1900–1950*. Chicago: University of Chicago Press, 1992.

Nicolson, Malcolm, and Robert P. McIntosh. "H. A. Gleason and the Individualistic Hypothesis Revisited." *Bulletin of the Ecological Society of America* 83 (2002): 133–142.

Slack, Nancy G. *G. Evelyn Hutchinson and the Invention of Modern Ecology*. New Haven, Conn.: Yale University Press, 2010.

Tobey, Ronald C. *Saving the Prairies: The Life Cycle of the Founding School of American Plant Ecology, 1895–1955*. Berkeley: University of California Press, 1981.

Worster, Donald. *Nature's Economy: A History of Ecological Ideas*. 2d ed. Cambridge, U.K.: Cambridge University Press, 1994.

Joel B. Hagen

EDISON, THOMAS

(1847–1931), best known today as America's greatest inventor. Crucial to Edison's success were his development of the research and development laboratory and his transformation of invention into a broader process of innovation. His most famous invention is the electric light bulb, but more significant was the system of incandescent electric lighting that launched the electric light and power industry. In addition, he invented the phonograph and helped to found the sound recording industry, produced the first successful motion picture apparatus, devised the commercial telephone transmitter, and made significant contributions to telecommunications technology, battery technology, ore milling, and Portland cement production.

Edison was born in 1847 in the canal town of Milan, Ohio, and grew up in Port Huron, Michigan. He attended school briefly but was principally educated at home by his mother Nancy, with books from his father Samuel's library. Beginning in 1863, Edison worked as a telegraph operator in the Midwest before moving to Boston, where he became a full-time inventor in 1868. Moving to New York the next year, he soon acquired a reputation as a first-rank telegraph inventor and established a series of manufacturing shops in Newark, New Jersey, where he employed experimental machinists to assist in his inventive work. By the end of 1873, Edison had set up his first laboratory in his Newark shop and then separated it from manufacturing operations in 1875. The following year Edison built his famous laboratory in Menlo Park, New Jersey. Over the next five years he turned this "invention factory" into the first research and development laboratory, thus laying the cornerstone of modern industrial research.

When Edison arrived at Menlo Park he had a reputation as an ingenious inventor of telegraph technology. Especially important were his quadruplex telegraph, which enabled four messages to be transmitted simultaneously over one wire, and his earlier improvements in stock tickers. At his new laboratory Edison developed the carbon-button transmitter that proved crucial to the commercial success of the telephone. His work in telephony led him to conceive and invent a machine for recording and playing back sound that he called the phonograph, which helped to make Edison's international reputation as the "inventor of the age" and gave him his famous nickname, "The Wizard of Menlo Park." Edison's reputation helped him to secure support for his effort, beginning in the fall of 1878, to develop a commercial electric light system.

In his earlier inventive work Edison had relied on assistants who were skilled craftsmen and, like himself, were self-taught in electricity. The electric light research called for a new type of researcher represented by college-educated Francis Upton, a graduate of Bowdoin College in Maine, who received the first master's degree in science at Princeton and then studied under Hermann von Helmholtz, a leading German physicist. Although he had advanced education in physics, Upton considered the Menlo Park laboratory another school and felt that he learned more about electricity from Edison than he had in his scientific coursework. At this time, inventors like Edison were often working ahead of scientific understanding of the apparatus, and things like telephones, electric lights, and generators were the subject of considerable scientific investigation. Although most of the staff continued to be self-taught, Edison also hired PhD chemists and graduates of engineering schools.

With the electric light Edison broadened the notion of invention to include far more than simply embodying an idea in a working artifact. His vision encompassed what the twentieth century would call innovation—invention, research, development, and commercialization. At his Menlo Park laboratory Edison combined an existing tradition of invention centered in the machine shop with sophisticated laboratory research into basic and applied science and technology. In developing new inventions at the laboratory Edison relied increasingly on teams of researchers to improve all aspects of his inventions and move them rapidly from research to development and commercialization. With the electric light Edison also recognized that as the new technology generated in the laboratory reached a commercial stage, continued research and development at the manufacturing shops would play a crucial role in commercial success. Edison famously described this process with his well-known statement that "invention is one percent inspiration and 99 percent perspiration."

Edison's success at Menlo Park helped to make his laboratory a model for others in the electrical industry, including Alexander Graham Bell, Edward Weston, Bell Telephone, and Western Electric. Menlo Park also provided the model for the even larger laboratory Edison himself built in West Orange, New Jersey, in 1887. Having developed a process of research and development and innovation at Menlo Park, Edison applied it at West Orange to a wide variety of technologies— most notably sound recording, motion pictures, ore milling, Portland cement manufacture, and storage batteries. The West Orange Laboratory would come to be surrounded by the factories and company offices Edison established to manufacture and market his inventions. Edison's process of research and development at Menlo Park and later West Orange helped lay the groundwork for modern industrial research, and his success at moving inventions from laboratory to market provided a model process of innovation.

[*See also* **Bell, Alexander Graham; Bell Laboratories; Electricity and Electrification; Electronic Communication Devices, Mobile; Engineering; Film Technology; Illumination; Instruments of Science; Research and Development (R&D); Sound Technology, Recorded; Technology; Telegraph;** *and* **Telephone.**]

BIBLIOGRAPHY

Friedel, Robert, and Paul Israel with Bernard S. Finn. *Edison's Electric Light: The Art of Invention.* Baltimore and London: Johns Hopkins University Press, 2010.

Israel, Paul. *Edison: A Life of Invention.* New York: John Wiley & Sons, 1998.

Millard, Andre J. *Edison and the Business of Innovation.* Baltimore: Johns Hopkins University Press, 1990.

Pretzer, William S., ed. *Working at Inventing: Thomas Edison and the Menlo Park Experience.* Dearborn, Mich.: Henry Ford Museum & Greenfield Village, 1989. Reprint ed. Baltimore: Johns Hopkins University Press, 2002.

Paul Israel

EINSTEIN, ALBERT

(1879–1955), physicist and Nobel Prize laureate, is considered one of the most significant and revolutionary scientists of all time. Born in Ulm,

Germany, Einstein spent his early years in Munich, where his father and uncle owned an electrotechnical company. He dropped out of the Luitpold-Gymnasium at age 16 and, after spending a year at the progressive Aargau cantonal school in Switzerland, entered the Polytechnic in Zurich as the youngest student ever to be admitted, studying physics and mathematics from 1896 to 1900.

Unable to find academic employment, he worked as substitute teacher and private tutor and, in 1902, became a technical expert at the Federal Office for Intellectual Property in Bern. He spent seven productive years in this "secular cloister," publishing more than sixty scientific papers before becoming an extraordinary professor of physics at the University of Zurich in 1909. It was in the "miracle year" 1905 that Einstein published a remarkable series of papers that secured his recognition in the physics community. He completed work on the special theory of relativity, on the determination of molecular dimensions, and on the photoelectric effect, and published for the first time the calculations on the equivalence of mass and energy, $E = mc^2$, that would forever become linked to his name. After spending a year at the German University in Prague and two years as a professor at the Federal Institute of Technology in Zurich, Einstein moved to Berlin in 1914 when he was appointed a member of the Prussian Academy of Sciences, professor at the University of Berlin, and director of the Kaiser-Wilhelm-Institute of Physics.

For almost a decade, he was engaged in an intensive effort to unify relativity and gravitation. He published his theory of general relativity in early 1916. It was during World War I that Einstein began formulating his pacifist position that ran counter to the nationalist spirit embraced by most of his academic colleagues. When in the fall of 1919 two British eclipse expeditions led by Arthur S. Eddington confirmed gravitational light bending, one of the significant predictions of general relativity, Einstein's name became known beyond the confines of his profession.

In spring 1921 Einstein traveled for the first time to the United States. He accompanied a Zionist delegation to New York, Chicago, and Boston, raising funds for the establishment of a Hebrew University in Palestine. He also lectured widely on relativity, including at Princeton University.

In the early 1930s, Einstein returned for three consecutive visits to the United States and spent three academic winter terms at the California Institute of Technology in Pasadena. He there established close relations to a number of scientists, as well as to Upton Sinclair and Charlie Chaplin. Upon Hitler's ascent to power in early 1933, Einstein publicly announced that he would not return to Germany "where civil liberty, tolerance and equality of all citizens before the law" no longer prevailed. After leaving Pasadena, he resigned from the Prussian Academy, which was about to expel him, and spent the summer of 1933 in Belgium. He traveled to England, where he delivered the Herbert Spencer Lecture at Oxford University and also began efforts on behalf of aid to refugees from Nazi Germany.

Einstein returned to the United States in mid-October 1933. Upon arrival, he was invited by President F. D. Roosevelt to the White House. He joined the newly established Institute for Advanced Study in Princeton, where he spent the next two decades pursuing the goal of unifying quantum mechanics, gravitation, and relativity. He collaborated with younger physicists on many scientific articles, among which the 1935 Einstein-Podolsky-Rosen paper is best remembered for pointing to the inadequacies of quantum mechanics. Together with Leopold Infeld, he published a book on *The Evolution of Physics*. He devoted much of his energy and resources to assisting refugees seeking entry and employment in the United States.

Einstein became a U.S. citizen on 1 October 1940. Unlike many of his colleagues, Einstein did not participate in the development of the first atomic bomb. Nevertheless, his name and work continued to be associated in the public mind with the unleashing of atomic power, primarily because in the summer of 1939 Einstein signed a letter to President Roosevelt alerting him to the possibility that Nazi Germany would develop the capability of nuclear power and urging him to authorize atomic research in the United States. He worked briefly as a consultant for the Research and Development Division of the U.S. Navy

Bureau of Ordnance. He also raised U.S. war bonds by recopying his original 1905 paper on the special theory of relativity and donating an as yet unpublished scientific manuscript to the Fourth War Loan drive. In early 1944, the two manuscripts were auctioned for a total of $11.5 million in Kansas City and were afterward deposited at the Library of Congress in Washington, D.C.

In 1946, Einstein became the chairman of the Emergency Committee of Atomic Scientists, whose aim was to encourage and further the peaceful uses of atomic energy and on whose behalf he maintained a large correspondence, seeking support for civilian control of nuclear weapons. In 1949 he published his primarily scientific *Autobiographical Notes*, followed a year later by a collection of essays entitled *Out of My Later Years*.

In 1952, Einstein was offered, but declined, the presidency of the State of Israel, whose creation he had wholeheartedly supported. He continued to speak out for civil and human rights and openly criticized the House Committee on Un-American Activities. His political and humanitarian statements and activities and his association with progressive movements and personalities earned Einstein a two-thousand-page file at the Federal Bureau of Investigation.

During the last days of his life, Einstein signed the Russell-Einstein Manifesto, released 9 July 1955 in London, which urged "governments of the world to realize, and to acknowledge publicly, that their purpose cannot be furthered by a world war" and "to find peaceful means for the settlement of all matters of dispute between them."

Einstein died on 18 April 1955 in Princeton after the rupture of an aortic aneurysm. His ashes were scattered at an undisclosed location.

[*See also* **Manhattan Project; Military, Science and Technology and the; Physics; Quantum Theory;** *and* **Science.**]

BIBLIOGRAPHY

Isaacson, Walter. *Einstein: His Life and Universe.* New York and London: Simon & Schuster, 2007.

Sayen, Jamie. *Einstein in America.* New York: Crown Publishers, 1985.

Schulmann, Robert, and David Rowe, eds. *Einstein on Politics: His Private Thoughts and Public Stands on Nationalism, Zionism, War, Peace, and the Bomb.* Princeton, N.J.: Princeton University Press, 2007.

Stachel, John, Diana Kormos Buchwald, et al., eds. *The Collected Papers of Albert Einstein*, Volumes 1–13. Princeton, N.J.: Princeton University Press, 1987–2012.

Diana Kormos Buchwald

ELECTRICITY AND ELECTRIFICATION

Electricity—so named by the Englishman William Gilbert around 1600—was known since ancient times in the form of static electricity. From the seventeenth century onward, such scientists as Robert Boyle, Henry Cavendish, Alessandro Volta, G. S. Ohm, and the American Benjamin Franklin added to electrical knowledge. Franklin, whose *Experiments and Observations on Electricity* (1751–1753) won international attention, is best remembered for his 1752 experiment with a kite and a key in a thunderstorm, which demonstrated that lightning is an electrical discharge. By the early nineteenth century, Michael Faraday and other scientists were developing techniques of generating electricity. Working independently of Faraday, the American Joseph Henry began research on electromagnetism in 1827. Henry constructed an electromagnetic motor in 1829 and later discovered electrical induction, crucial to generating power, and demonstrated the oscillatory nature of electrical discharges.

Practical applications came slowly and piecemeal, long before anyone conceived of electrification as a universalizing process. Most early electrical technologies, including fire-alarm systems, railway signaling, burglar alarms, doorbells, servant-calling systems, and the telephone, were modifications of the telegraph, first demonstrated in 1838. These devices relied on batteries to supply a modest direct current. A much more powerful current was needed for practical lighting, heating, electroplating, and electric motors. Such applications developed only after about 1875 when improved generators and dynamos became available. After 1878, arc lighting, a powerful but crude form of illumination, drew crowds to demonstrations in city centers and expositions. Large cities quickly adopted lights for streets and public places such as theaters and department stores. Once Thomas Edison's firm installed incandescent lighting systems across the country, beginning in New York City in 1881, however, most indoor sites and street-lighting companies chose his technology. Edison and his assistants developed not only a practical incandescent lightbulb (1879), but also the now familiar system of wiring, wall switches, sockets, meters, insulated transmission lines, and central power plants. Edison designed this distribution system to compete on price with gaslight, while offering brighter and safer illumination. Rapidly adopted by the wealthy for fashionable indoor venues, including theaters, clubs, ocean liners, expensive homes, and the New York Stock Exchange, electric lighting became a prestigious and sought-after form of illumination.

Initially, each electrical technology had a separate energy source, as well as different financial backers. Lighting utilities, factories, and streetcar lines maintained their own power plants and delivery systems, with no uniform standards for wiring or current. The private systems installed by hotels, skyscrapers, and large private homes in the 1880s were incompatible with one another, but they did have the advantage of not requiring overhead wires—which soon became so numerous in the major cities as to constitute a public nuisance—or costly underground conduits. This pattern of development merely continued the earlier piecemeal commercialization of electricity.

Put another way, electrification was not a centrally coordinated process that spread out from central stations like ripples in a pond. Rather, there were many different generating locations, owned not only by utilities, but also by streetcar systems, department stores, hotels, railroads, telegraph systems, and wealthy individuals, any of whom could purchase stand-alone generators. Likewise, until the 1930s many factories also had their own power stations. The degree of such fragmentation varied from one nation to another, being considerably greater in Britain than in the United States, for example. In the early twentieth century, London had 65 electrical utilities and 49 different types of supply systems. Anyone moving house within London risked entering a new grid because there were 10 different frequencies, 32 different voltage levels for transmission, and 24 for distribution.

Commercial Development and Standardization. The electrical industry was one of the most dynamic sectors of the American and European economies between 1875 and 1900, growing into a $200-million-a-year industry in the United States with the backing of investors like J. P. Morgan, who financed Edison's work. Once commercial development began, a flurry of mergers reduced the field from 15 competitors in 1885 to only two large corporations, General Electric and Westinghouse in 1892. Railroads, once America's largest corporations, were now a mature industry, in contrast to the rapidly expanding electric traction companies, local utilities, and equipment manufacturers that collectively exemplified the spread of managerial capitalism (as opposed to partnerships and family firms). From its inception, the electrical industry also relied heavily on scientific research and development, a fact formalized when General Electric founded the first corporate research laboratory in 1900.

At first electricity was largely confined to cities. In both Europe and the United States, public electric lighting appeared in the late 1870s as a spectacular form of display in city centers and at expositions and fairs and then gradually replaced gas lighting. Many inventors developed streetcar systems, which became competitive and practical in the middle to late 1880s. Burgeoning cities eagerly sought electric trams to replace dirty, slow horsecars. In the United States streetcars became practical after 1887 when Frank Sprague's new motor proved itself in hilly Richmond, Virginia. By 1890, two hundred cities had ordered similar systems. By 1902, $2 billion had been invested in electric railways, and a typical urban family of four spent about $50 a year on fares. Electric trams were faster and cleaner than horsecars. They greatly expanded urban housing districts, and the lines soon reached into the rural hinterland. They improved mobility for women and older children, who no longer needed a horse-drawn vehicle to visit town or the new (and heavily electrified) amusement parks built at the ends of the urban lines, starting in the early 1890s.

Electricity spread into factories with equal speed, starting with lighting in textile and flour mills. From a worker's point of view, incandescent lighting improved visibility and reduced pollution and the danger of fire, but it also made possible round-the-clock shifts. Furthermore, as electric motors and cranes provided more horsepower for production, they brought radical changes in the construction and layout of factories, most strikingly in Henry Ford's assembly line (1913), an innovation partly anticipated by Edison's experiments with automating iron mining in the 1890s. The assembly line was literally impossible in any complex industry before electricity freed machines from fixed, steam-driven overhead drive shafts.

As electrical systems reached more customers, utilities improved technologies for generating power and achieved economies of scale. They began to sell current and service so cheaply that the myriad small plants could no longer compete. Samuel Insull, Edison's former private secretary, early grasped the importance of consolidating power production and maximizing consumption. Insull left Edison to head Chicago's Commonwealth Edison in 1892, and he remained a leading figure in utility development for a generation. Insull convinced traction companies and factories to abandon their power plants and to purchase electricity from him. Through astute marketing he created one of the world's largest electrical utilities. As others copied his methods, holding companies created regional power companies and linked the many local systems into a national power grid. Private companies proved agile in the consolidation process. They possessed readier access to capital and had fewer jurisdictional problems than government-run utilities and by the 1920s owned all but a fraction of the national generating capacity.

Between the 1880s and the 1940s the spread of electrification, first in cities and towns and then in rural areas, provided a major economic stimulus and transformed everyday life. An array of electric appliances—from fans, irons, and mixers to vacuum cleaners, refrigerators, and washing machines—eased labor for middle-class housewives. Domestic servants largely disappeared, men did less housework, and women were saddled with more responsibilities. For example, men and boys had typically been responsible for beating rugs and helping with spring cleaning, but after the vacuum cleaner became common, this work fell almost exclusively to mothers and daughters. A different form of electric power, the storage battery, enhanced the manufacture and use of new technologies, including both the automotive and the aviation industries. (As batteries shrank in size after 1950, they had many more applicants, not least in the medical field.) Electricity was also crucial to the new mass media—radio, films, and recordings—as well as to night baseball, introduced in Cincinnati, Ohio, in 1935.

During the Depression of the 1930s, the federal government promoted public utilities, in part to create a yardstick to measure the price and performance of private power companies. The government built a system of dams on the Tennessee River—administered by the Tennessee Valley Authority (TVA), which sold power to rural cooperatives—as well as systems of dams on the Colorado and Columbia Rivers. Because private power had generally ignored farmers, only 10 percent of whom had electricity as late as 1935, President Franklin D. Roosevelt in 1935 established the Rural Electrification Administration (REA) to bring power to this neglected sector of the nation. Rural electrification spread comparatively slowly in the South and Middle West, where customers were widely dispersed, but more rapidly in the arid West, where farmers wanted electric pumps for irrigation, and in areas served by interurban trolleys. The REA and the TVA organized cooperatives and made available loans and technical expertise. By 1945, thanks to the New Deal, most of America was electrified. Electricity had important military applications as well, playing a crucial role in World War II and the Cold War era, for example, in the development of radar, rocketry, and the mainframe computers essential to ballistic missiles and space technology.

Changes beyond the Functional. Electric lighting dominated public spaces and changed the culture in ways that went far beyond the functional. American cities became the most intensively lighted in the world, not least because of the spread of electric advertising. Spurred by the marketing campaigns of Westinghouse, General Electric, and the utilities, the illuminated skyline became a source of civic pride. Even small cities aspired to emulate New York City's "Great White Way," where millions of flashing bulbs in Times Square and the theater district created a scintillating artificial environment. Nightlife expanded as hundreds of brightly lit amusement parks emerged as early as the 1890s, followed by stadiums and other outdoor venues.

As early as 1903, American cities were far more brightly lit than their European counterparts: Chicago, New York, and Boston had three to five times as many electric lights per inhabitant as Paris, London, and Berlin. This indicated more than prosperity and wealth. Levels and methods of lighting varied from culture to culture, and what was considered dramatic and necessary in the United States often seemed a violation of tradition elsewhere. Many European communities continued throughout the twentieth century to resist electric signs and spectacular advertising displays. At the 1994 Winter Olympics in Lillehammer, Norway, for example, the city council refused corporate sponsors the right to erect illuminated signs.

Once American families acquired electrical lighting, they had less reason to cluster at night around the hearth, giving rise to a pattern of dispersed privacy. With power available at the flick of a switch, consumers ceased to associate lighting with physical work such as hauling wood and ashes or cleaning lamps. Electricity also extended the range of usable space. Domestic activity after sunset was no longer confined to the hearth and the range of the kerosene lamp. In commerce, immense department stores, office buildings, and eventually malls could be built with adequate illumination far from any natural light source.

In industry, the expansion of the electrical grid made it possible to locate a factory virtually anywhere, without regard for proximity to coal supplies or water power. Because not only factories but also shops and other businesses could spring up wherever the grid reached, electrification facilitated urban deconcentration. The rapidity of electricity adoption in industry and commerce varied considerably by nation. The United States took the lead, closely followed by Germany. In 1928 one British expert estimated that 73 percent of manufacturing power was electrical in the United States compared with 67 percent in Germany, 48 percent in Britain, and less for other European countries. In the same years air conditioning and climate control began to have important industrial applications, notably in food processing and brewing. Later still, industries pioneered the use of computers and the electrical transmission of information.

But if electrification homogenized space, delivering light, power, climate control, and information to any site, it also facilitated the concentration of people in cities. Indeed, night satellite photographs reveal the location of thousands of cities as intense blobs of light. Electricity, a scientific curiosity in 1800 and still a novelty for the rich in 1880, had become indispensable by the late twentieth century. Domestic consumption increased faster than that of industries, which tended to become more efficient energy users. The typical family of 2010 used as much electricity in a month as their grandparents had during an entire year in 1940. Families acquired more electrical devices, notably televisions, computers, and clothes driers. As houses grew larger and air conditioning became a new norm, peak electrical demand shifted from night to day and from winter to summer. After the 1980s, power outages, once largely caused by storms and malfunctions, became common on hot afternoons because demand exceeded supply. For all its advantages, electricity also proved inseparable from blackouts, air pollution, and global warming.

As consumers grew more aware of these linkages, reducing electrical use became a political issue and governments began to rate appliances and mandate greater use of alternatives to oil, coal, and gas. In the European Union in 2012, for example, all stoves, refrigerators, freezers, and other large appliances had to be sold with a clearly posted efficiency rating. Between the early 1970s and the late 1990s people used more appliances, but their new devices were more efficient. Refrigerator manufacturers tripled their efficiency and washing machines halved their use of electricity. Nevertheless, coal- and oil-fired power plants cause acid rain and global warming, and voters increasingly pressed governments to adopt windmills and solar power.

Two models of future energy production and consumption had emerged by about 2010. The first model relied on centralized service using increasing amounts of nuclear power instead of burning coal. France exemplified such a system, generating more than 80 percent of its electricity in nuclear plants. This model anticipated that windmills and solar arrays would supply no more than a quarter of the total energy. The second model, developed in various forms, deemphasized centralized power stations and the construction of nuclear plants. Instead, it relied on radical decentralization, based on a smart electrical grid akin to the Internet. It would minimize large power plants in favor of microinstallations in every home, automobile, business, and office building.

Yet even as such plans were debated during the early twenty-first century, more consumers in all parts of the world were becoming high-energy users, and they were plugging in portable computers and mobile telephones. Electricity had become an integral part of daily life from the electrical alarm, electric toothbrush, and electric toaster at dawn until one sleeps again at night, while protected by electric burglar alarms and smoke detectors. Millions in the emerging middle classes of China, India, Indonesia, Brazil, and elsewhere around the globe were rapidly adopting the full range of electrical devices. Reconciling their growing electrical dependency with global warming was one of the major challenges of the new century.

In the early twenty-first century, there were signs of a shift in consumer awareness. Voluntary curbs on electrical demand took place once a year during the worldwide Earth Hour. This movement began in 2007 in Sydney, Australia, and spread to London and then to eighty-eight countries during the following two years. Held on the last Saturday in March, individuals extinguished their lights and most of their appliances for one hour in a voluntary blackout that rolled around the globe. With this symbolic gesture, citizens could tell politicians they wanted to make reductions in electricity demand. In 2012 the enthusiasm for this movement seemed to be weakening, but the demand for more power increased.

[*See also* **Agricultural Technology; Airplanes and Air Transport; Automation and Computerization; Computer Science; Computers, Mainframe, Mini, and Micro; Edison, Thomas; Electronic Communication Devices, Mobile; Film Technology; Ford, Henry; Franklin, Benjamin; Global Warming; Henry, Joseph; Hydroelectric Power; Illumination; Machinery and Manufacturing; Military, Science and Technology and the; Nuclear Power; Radio; Refrigeration and Air Conditioning; Research and Development (R&D); Rural Electrification Administration; Solid-State Electronics; Space Science; Telephone; Television; Tennessee Valley Authority; Urban Mass Transit; *and* Westinghouse, George.**]

BIBLIOGRAPHY

Hughes, Thomas P. *Networks of Power: Electrification in Western Society, 1880–1930.* Baltimore: Johns Hopkins University Press, 1983.

Nye, David E. *Electrifying America: Social Meanings of a New Technology, 1880–1940.* Cambridge, Mass.: MIT Press, 1990.

Nye, David E. *When the Lights Went Out: A History of Blackouts in America.* Cambridge, Mass.: MIT Press, 2010.

Platt, Harold L. *The Electric City: Energy and the Growth of the Chicago Area, 1880–1930.* Chicago: University of Chicago Press, 1991.

Rose, Mark H. *Cities of Light and Heat: Domesticating Gas and Electricity in Urban America.* University Park: Pennsylvania State University Press, 1995.

Tobey, Ronald C. *Technology as Freedom: The New Deal and the Electrical Modernization of the American Home.* Berkeley: University of California Press, 1996.

David E. Nye

ELECTRONIC COMMUNICATION DEVICES, MOBILE

Although mobile electronic communication dates back to the early twentieth century, consumer use of now ubiquitous mobile digital devices is a post-1990 phenomenon. Development of viable land mobile services required technical breakthroughs to manufacture efficient transmitters, end-user devices, and the availability of sufficient spectrum channels to allow multiple users at any one time.

Origins. The use of cumbersome, complex, and expensive radio equipment while on the move originated aboard merchant vessels and naval ships early in the twentieth century. Such "point-to-point" applications allowed ships beyond sight of land to maintain contact or call for assistance. Radio-assisted rescues, such as that of survivors of the *Titanic* disaster in 1912, captured the public imagination. Following limited use of wireless before and during World War I, all military forces depended on effective mobile radio systems in World War II. The army's "walkie talkie" portable two-way radio was one widely publicized device, and the use of radio to coordinate thousand-aircraft bombing missions was another.

Civilian (although still institutional) application of mobile two-way radio first appeared in police vehicles in the 1930s. The first urban Mobile Telephone Service (MTS) system was opened by AT&T in St. Louis in mid-1946 and expanded to 25 other cities within a year. But because too few channels were allocated in the 150-MHz band, these early automobile MTS systems could serve no more than 23 users at a time in any city. A market with more than 250 subscribers made waiting for an open channel almost impossible. MTS required operators and one could talk or listen, but not both at the same time, and any subscriber could easily listen in to calls by others. MTS required the use of expensive equipment and usage was billed at high rates, limiting use to business and government. Despite these drawbacks, waiting lists to obtain service kept growing over the next three decades. At its peak by 1980, MTS only served about 120,000 customers nationwide.

Analog Mobility. The first popular mobile consumer communications device was the portable radio, the first complex and heavy examples of which appeared in the 1920s. Although only providing an ability to *receive* signals, ever-smaller vacuum tube radios in the 1930s and 1940s encouraged people to take radio with them. Development of automobile radios (again in the 1930s, although widely available only decades later)—and especially the innovation of tiny transistor radio receivers in the mid-1950s (and integrated circuits a decade later)—greatly encouraged the trend to portable radio listening.

Portable devices to listen to recordings took more time to develop. The 1950s saw the development of small portable record players as well, although available technology did not encourage widespread adoption beyond a teenage audience. Development of the cassette audio tape allowed for devices that could be carried while playing. High-powered audio "boom boxes" that played both radio stations and cassette tapes became popular in the 1970s, with their very size and loudness becoming the basis of personal pride more than their portability.

The product often credited as the first modern consumer mobile communication device, the Sony "Walkman," was introduced in 1979–1980. Designed to play compact cassette audio tapes, some 80 million were sold by the early 1990s. By then, "discman" players allowed portability for compact disc (CD) players along similar lines. The addition of the ability to tune television broadcasts in the mid-1980s created the analog "Watchman," but its tiny screen size, although contributing to portability, did not encourage viewing. They were manufactured until 2000. However, these devices were one-way in that they could only play recorded or broadcast material.

The first portable consumer audio tape recorders appeared in the late 1950s, and video tape recorders (camcorders) came on the market in the mid-1980s, rapidly replacing movie cameras. The battery-powered devices were initially heavy, clumsy, and difficult to use, but shrank in size and cost by the late 1990s to become widely popular.

They placed control over the material recorded (and presumably viewed) in consumer hands.

The first mobile means of *two-way* consumer communication was Citizen's Band (CB) radio in the 1970s. On frequencies originally set aside in the 1950s and intended for what the Federal Communications Commission (FCC) called "land mobile" users (such as long-haul truckers), the ease of CB radio use quickly expanded as equipment costs plummeted and the FCC stopped trying to closely regulate users. Being able to send as well as receive radio communications while in an automobile or truck was a novel experience for most consumers. The CB boom burned out by the early 1980s, to be replaced by electronic pagers and mobile telephones.

Consumer paging devices appeared in 1974 with the first Motorola Pageboy. Some 3 million pagers were in use by 1980 and 22 million a decade later. But the limited paging device (it could only receive a message, either text or audio) soon gave way to the more capable mobile telephone.

Years of cellular telephone development and prototype experimentation, chiefly by Motorola and AT&T, culminated with the first cell[ular] phone service aboard the Metroliner trains between New York and Washington in 1969 and experimental cell phone services in Chicago and Baltimore by 1977. Pushed by growing demand to act, the FCC approved national Advanced Mobile Phone Service (AMPS) analog cell phone service in 1982, allocating more frequencies to make the system viable. Cellular transmission technology had originated in AT&T's Bell Labs in 1947 and was slowly improved over the next three decades. It allowed reuse of spectrum frequencies in multiple areas (or cells) that, connected together, could blanket an urban region, thus allowing vastly more users to be accommodated at a time than the existing MTS. Lack of sufficient computer power for switching calls among cells was the primary factor that delayed its introduction.

Initial cell phones were the size (and nearly the weight) of bricks and were intended for use in automobiles rather than as personal carry-about portables. Chicago saw the first operating AMPS system in late 1983. Soon made available in 90 cities, AMPS service reached 1 million customers within 4 years, rapidly outstripping most predictions of demand. Mobile telephones began to shrink in size, gain in capability, and drop in cost, which increased that demand. The transition to more efficient digital service began by the late 1980s.

Digital Devices. Pushed by manufacturers and the growing success of analog mobile phones, the FCC approved an expanded Personal Communication Service (PCS) or digital cell phone service on the 1.9-Ghz band in 1991. But the commission did not mandate a single system (as happened in Europe) and three different standards developed (CDMA, TDMA, and GSM). Cell phones dropped in price and rose in capability and what was once seen as a luxury quickly became a necessity. The number of phones in service soon skyrocketed—to nearly 34 million by 1995, 109 million five years later, and nearly twice that by 2005. By 2007, cell phones were in use by more than 2 *billion* people around the world, and mere voice service was no longer the driving application.

Increasing functionality of the cell phone (as well as longer-lasting batteries) has pushed this growth, beginning with an ability to take and send photos (and eventually video), then e-mail, and finally full Internet access. The Blackberry personal digital assistant (PDA), for example, first appeared with a black-and-white screen in 1999, melding these functions plus a capacity for scheduling. "Generations" of cell phones, billed as 3G or 4G, each with more features and often greater speed, appeared. Apple's feature-laden iPhone was introduced in early 2007, although tied to a contract with AT&T (until 2011). Millions were sold in the next several years. A competitor, the Motorola Android device does as much, although in different ways. But as portable digital devices became more central in American life, so did their cost—for the devices themselves and the typical two-year service contracts and some user charges. Indeed, with their networks pushed by rising data demand, cellular carriers instituted new pricing policies in 2011 that limited heavy users by charging sharply higher rates for more monthly time and/or faster connection speeds.

Portable entertainment systems progressed as well. Indeed, the distinction between "work" and "play" blurred, with many devices providing both. The first compressed digital audio MP3 players appeared in 1999, based on a German invention of two decades before. The most famous MP3 player, Apple's iPod, came out in late 2001, its small size achieved using a smaller hard drive than competing devices. Its easy-to-use navigation and clean design helped to set new standards for portable music devices. Millions of subsequent models were sold over the next decade. The iPad, introduced in 2010, combined elements of laptop computers and game boards and soon thousands of applications were available for it.

[*See also* **Bell Laboratories; Computers, Mainframe, Mini, and Micro; Internet and World Wide Web; Military, Science and Technology and the; Photography; Radio; Sound Technology, Recorded; Telephone;** *and* **Television.**]

BIBLIOGRAPHY

Agar, John. *Constant Touch: A Global History of the Mobile Phone.* Duxford, England: Icon Books, 2003.
Bowers, Raymond, Alfred M. Lee, and Cary Hershey. *Communications for a Mobile Society: An Assessment of New Technology.* Beverly Hills, Calif.: Sage, 1978. Precellular technology and industry.
Bringing Information to People: Celebrating the Wireless Decade. Washington, D.C.: Cellular Telecommunications Industry Association, 1994.
Du Gay, Paul, et al. *Doing Cultural Studies: The Story of the Sony Walkman.* Thousand Oaks, Calif.: Sage, 1997.
Levinson, Paul. *Cellphone: The Story of the World's Most Mobile Medium and How It Has Transformed Everything!* New York: Palgrave Macmillan, 2004. Informal but also insightful.
Meurling, John, and Richard Jeans. *The Mobile Phone Book: The Invention of the Mobile Phone Industry.* London: Communications Week International, 1994.
"Mobile Radio." In *A History of Engineering and Science in the Bell System: Transmission Technology 1925–1975*, edited by E. F. O'Neill, chap. 14. Murray Hill, N.J.: AT&T Bell Laboratories, 1985.
Murray, James B., Jr. *Wireless Nation: The Frenzied Launch of the Cellular Revolution in America.* Cambridge, Mass.: Perseus, 2001. Key protagonists and the vital role of spectrum.
Schiller, Michael Brian. *The Portable Radio in American Life.* Tucson: University of Arizona Press, 1991.

Christopher H. Sterling

ELEVATOR

The now ubiquitous modern elevator can trace its roots to the textile mills in England where freight and people were moved through the mill buildings using a "teagle" or tackle system with hoisting ropes and belts powered by the mill. Moving people and freight vertically in a safe and efficient manner led inventive individuals to develop new devices, systems, and technologies to answer the increasing demand for elevators, especially from architects and owners who wanted to build taller buildings to maximize land. Most commonly associated with the improvement and ultimate widespread use of the elevator was Elisha Graves Otis (1811–1861), who is credited with making a major safety advance that he demonstrated at the New York Crystal Palace Exhibition in 1854. His invention allowed the steam-powered car to be held in place in the case of rope failure. His "Improved Hoisting Apparatus" received a patent in 1861 and permitted a safe and smooth ride and exact braking. The first commercial application of this new technology that traveled forty feet per minute was in the five-story E. V. Haughwout Company in New York in 1856–1857.

The Equitable Life Assurance Society Building (1868–1870) in New York received two 130-foot-high steam-powered elevators installed by Otis Tufts, which allowed the building to be almost twice as tall as previous buildings. However, steam-powered elevators could only go so high and travel so fast and thus were replaced by the hydraulic elevator. The Baldwin–Hale elevator utilized buckets of water filled to serve as a counterweight and then emptied. The elevator ran smoothly and fast, but it was also dangerous because it was only controlled by its brake.

Rapidly increasing building heights stimulated the improvement of the hydraulic elevator, which by 1880 was being widely manufactured. Locating the cylinder, piston, and valves within the shaft saved space, along with placing the pulley sheaves at the top and bottom of the shaft. It still required an operator, but this person had better hand controls and an automatic brake. The elevator could travel higher and higher by increasing the amount of hoisting cables and the sheaves, which, unlike earlier systems, did not require vastly larger cylinder and pistons that supplied the power.

Between 1880 and 1886, the U.S. Patent Office issued 328 patents related to elevators or related devices. Utilizing the Sprague electric motor and based on a screw-and-nut system, the Sprague–Pratt elevator received a patent in 1888 and was installed in a number of buildings. The Otis Brothers' drum-type elevators remained the better alternative.

The Otis Brothers experimented with the electric motor and installed two in 1889 in the Demarest Carriage Company in New York. In 1903, the Otis Brothers manufactured an electric elevator without any drums or gears. Its cables were run over the sheaves and attached directly to the counterweights. Its advantage was the falling counterweight that produced kinetic energy and reduced the load on the motor when lifting the car. In addition, the Otis Brothers developed the motor controls so that the car's starting, stopping, and speed could be smoothly and accurately regulated.

Over the course of the twentieth century, innovations to the elevator continued so that they were even safer, faster, smoother, and more powerful. The cars would also become more luxurious, such as the glass elevators in John Portman's Atlanta Hyatt Regency Hotel (1967). Containing 104 elevators installed by Westinghouse and improved by the Schindler Group, the Willis Tower in Chicago, still commonly referred to as the Sears Tower (1974) in 2012, was the tallest building in the United States and ninth tallest in the world, at 1,451 feet with 110 stories. This building and dozens of others around the world demonstrate how the modern elevator allows for almost unlimited vertical access.

[*See also* **Building Technology; Dams and Hydraulic Engineering; Skyscrapers;** *and* **Steam Power.**]

BIBLIOGRAPHY

Goetz, Alisa, ed. *Up, Down, Across: Elevators, Escalators, and Moving Sidewalks*. New York: Merrell Publishers, 2003.

Goodwin, Jason. *Otis: Giving Rise to the Modern City*. Chicago: Ivan R. Dee, 2001.

Gray, Lee E. *A History of the Passenger Elevator in the 19th Century*. Mobile, Ala.: Elevator World, 2002.

Leslie N. Sharp

EMPIRE STATE BUILDING

Thanks to its familiar silhouette and status as the world's tallest building from 1931 until 1973, the Empire State Building, on New York City's Fifth Avenue between 33rd and 34th Streets, ranks as the twentieth century's most famous skyscraper. Designed by the architectural firm Shreve, Lamb, and Harmon and erected on the site of the Waldorf–Astoria Hotel, it stands 102 stories (1,250 feet) high, with 85 floors of office space. A tourist promenade on the 86th floor offers a 360-degree view. The 200-foot tower, originally planned as a dirigible mooring mast, provides a grand crown. The idea for the building originated with John J. Raskob, chairman of the Democratic National Committee in 1928, when Alfred E. Smith was the party's presidential nominee. After Smith's loss to Herbert Hoover, Raskob decided to erect the world's tallest building, with Smith as president. He chose the name to emphasize the relationship between the building and Smith, the former governor of the Empire State. It was the first twentieth-century building to hold the "world's tallest" title that did not bear a corporate name such as Metropolitan Life, Woolworth, or Chrysler. Planned in the booming 1920s but completed during the Great Depression, the building attracted few tenants and for a time was dubbed the "Empty State Building." Further bad luck came in 1945 when a B-25 bomber crashed into the 78th floor, killing 14. Although the building is said to be Art Deco in design, the phrase did not exist at the time; the architects themselves simply described it as "modern." Seeking a practical, utilitarian structure, they little dreamed that it would become a tourist mecca, the symbol of New York, a national icon, and the setting for many movies, from *King Kong* (1933) to *Sleepless in Seattle* (1993).

[*See also* **Building Technology; Elevator; Engineering;** *and* **Skyscrapers.**]

BIBLIOGRAPHY

Tauranac, John. *The Empire State Building: The Making of a Landmark.* New York: Charles Scribner's Sons, 1995.

Willis, Carol, ed. *Building the Empire State.* New York: W. W. Norton, 1998.

John Tauranac

ENGINEERING

Engineering has never been an easily defined or a static pursuit. As a result, the activity of engineering, which is as old as human history, does not map perfectly onto the profession of engineers, which is a relatively modern development. In the case of the United States the chronological coherence between the professionalization of engineering and the development of American identity and the nation created opportunities for engineers to play unusually important roles in the development of the nation and its ethos. Engineering professionalization began in the seventeenth and eighteenth centuries with the formalization of engineering education and the creation of state corps of military and civil engineers. Professionalization came to a head with the formation of professional societies in the nineteenth century, and trained engineers became essential to industry at the turn of the twentieth century. Engineers who developed key military and consumer technologies that in turn shaped culture, politics, and the economy

heavily shaped World War II and the Cold War. Engineers have been as central to building the nation as any professional group, including lawmakers, because engineers have designed the infrastructure and the technologies that have defined the nation. Engineers have also played iconic and culturally suggestive roles, from the surveyors of the eighteenth century to post–Civil War inventors and the nerds of the 1980s. Engineers are both representative of the nation's citizens (and their rags-to-riches stories) and notable for the difficulties that women and people of color have encountered in entering the profession. They have simultaneously represented the ideal American worker and the threat of immigration. In many ways, understanding the changing roles of American engineers and engineering since the eighteenth century is a way to understand the United States itself.

For over two hundred years, the United States has been a technological nation and its eras have often been defined and punctuated by technological developments—for example, the era of railroads, the machine age, the information age, and the space age. Many debates have ensued about the factors necessary for America's technological growth. Some historians point to the flow of capital, whereas others emphasize the recognition of intellectual property in the Constitution. Others follow Alexis de Tocqueville and his recognition of the practical bent of Americans. Still more consider the role of widespread public education, particularly at the postsecondary level, through the Morrill Acts of 1862 and 1890. These factors are all important in explaining the development of engineering in America. With the exception of de Tocqueville's cultural observations, all of these factors emphasize decisions made by the state. The state itself, through institutions like the Army and the Bureau of Reclamation, has been the most important of a number of large institutions, including corporations and universities, which have shaped American engineering.

European Background and Colonial Engineering. Both Great Britain and France had important engineering traditions in the eighteenth century that influenced the development of engineering in colonial America. These two nations had evolved distinct cultures of engineering by this time. The French had a highly organized, formalized culture of engineering that was dominated by the state. The French military began the formalization of engineering organization and training with the creation of a Corps des Ingénieurs du Genie Militaire in 1675. A civilian version, focusing on road and bridge engineering, the Corps des Ponts et Chaussées, was founded in 1716. Schools for training engineers and standardizing engineering practices followed in the eighteenth century with the civil engineering school (L'Ecole Royale des Ponts et Chaussées), founded in 1747, and its military analogue (L'Ecole Royale du Génie de Mezières) a year later. These institutions would be important examples for American politicians intent on creating engineering organizations in the early Republic. The French culture of engineering was shaped by the kind of formal, rigorous, mathematical training these schools offered as a way of opening the military to social groups in France beyond the aristocracy (Alder, 1997). The state, even through its upheaval in the French Revolution and Napoleonic wars, remained the focus of French engineering. This tight coupling of engineers, rational methods, and the state would serve as both a positive and a negative model for American engineering in the nineteenth century, especially in a military context. Americans first looked to the French for organizational models during the Revolution, forming the first Corps of Engineers on March 11, 1779.

Although there was also organized military engineering in England, it did not adopt rational mathematical training with the same vigor as the French. Nor did the British create an array of different state organizations for engineering. Part of the difference between the French and British traditions of engineering can be explained by the different status of the army in the two nations. France's military vulnerabilities were largely land based (especially after the loss of colonies in the Seven Years' War), whereas Britain's were oriented toward the sea. Whereas the construction of land defenses was central for French national security, in England the naval defenses were key. The British Navy was the more powerful military branch and ship design was the more important application of the state's attention to engineering. Therefore, the military influence on the construction of British infrastructure was much less than in France. Private citizens played a much larger role in building Great Britain's roads, canals, and bridges. The most famous of these projects was the Bridgewater Canal, commissioned in 1759 by Francis Egerton, the third Duke of Bridgewater, to transport coal from mines in Worsley to the factories of Manchester, about six miles away. This canal, and its privately financed construction, would be exemplary for the engineering of American canals just a few years later. Colonial engineers had often apprenticed in the British Isles, and they brought their approaches and experience to the colonies. Their experiences came from a much more informal and artisanal culture of engineering.

Surveying constituted the most important engineering activity of the colonial period, and here practices were usually borrowed from the British. However, the American landscape offered a number of challenges the English countryside did not, and American surveyors drew on the experience of British surveyors in diverse locations from India to Egypt. Practices were transferred tacitly through apprenticeships, but also through publications. The Penns and Calverts hired Charles Mason and Jeremiah Dixon to survey the boundary between their colonies (present-day Maryland, Pennsylvania, and Delaware), and Mason and Dixon proceeded to publish a long description of their methods in the *Philosophical Transactions of the Royal Society* in 1769. Accurate surveying required sophisticated instruments, like transits and theodolites, which almost always came from London, although some Americans, like David Rittenhouse, did build their own instruments. Surveying also provided an opportunity to some engineers to earn, and occasionally lose, large sums of money on land speculation; therefore, in the colonial as well as the early national period, it was common for engineers to work on surveying projects interspersed with infrastructure, design, and construction projects. Surveying offered many young men the opportunity to see many different parts of the colonies and then nation, and as a result engineers and surveyors were often some of the most extensively traveled citizens of the British Empire and then American republic. Having seen such a variety of landscapes, they often influenced decisions concerning national development, especially in transportation.

Engineering from the Early Republic to the Civil War.
Engineering in the early Republic was often divided between approaches drawn from French and British traditions. On one hand, the creation of the U.S. Military Academy at West Point had clearly French roots. On the other hand, projects like the Middlesex Canal in Massachusetts replicated English practices. The American landscape and its scale, however, were like neither France's nor Britain's. As a result, distinct engineering practices began forming. One of these unique elements was the connection between surveying and engineering, which was further bolstered by the creation of the public lands survey (sometimes called the rectangular survey) to measure lands west of the 13 original colonies so they could be sold and settled. The first two surveyors general had also worked on engineering projects—Rufus Putnam, who had been an engineer with George Washington in the Revolution, and Jared Mansfield, a Yale graduate who had taught mathematics at West Point. Furthermore as the nation grew physically, demographically, and economically, new traditions emerged. By the Civil War, American engineering had developed a variety of distinct engineering cultures and practices.

Engineering in the Early Republic.
The engineering challenges facing Americans in the early Republic were considerable. The nation did not have a well-developed group of people to draw on to provide possible solutions to the challenges of communication, transportation, manufacturing, and resource extraction. As a number of European countries had discovered in the sixteenth through eighteenth centuries, the state depended on such experts. Although in many cases such expertise could be developed through apprenticeships and hands-on experience, in other cases politicians preferred to rely on expertise that emerged from formal education. This was a central tension in the early United States—whether to rely on average citizens to develop needed skills and methods or to create a government-sponsored system of education to produce experts. In the nineteenth century, the United States chose a middle route and did both. In still other cases, experts could be imported, but often with political as well as economic costs and in some cases without producing better results than locals might achieve.

West Point and the Army Corps of Engineers.
President Thomas Jefferson established the U.S. Military Academy (USMA) in 1802 and also reorganized a new Corps of Engineers, which was based at the USMA in West Point, New York. From its inception, West Point has been principally an engineering school, modeled on the French engineering schools of the eighteenth century. Whereas the Grand Écoles for military and civil engineering were established well before the French Revolution broke out in 1789, the École Polytechnique was a creation of the Revolutionary period. West Point was to be America's École Polytechnique, but the École Polytechnique was less a military academy than an elite scientific academy for the training of powerful civil servants and bureaucrats. Some politicians opposed the creation and continued funding of West Point because it seemed to valorize an aristocratic bearing that was inconsistent with American democracy. The Academy also aimed to train career military officers, which many Americans saw as fundamentally antidemocratic. In addition, the French Revolution and its aftermath did not do much to popularize French political institutions in the time of such unrest. Some Americans worried that political unrest might be imported with French institutions. Yet the Enlightenment rationality and rigor that the French engineering traditions represented had its supporters in the United States as well—not least Jefferson himself. For more than its first decade, West Pointers learned engineering from French textbooks and in some cases from French polytechniciens, like Claudius Crozet. Through several different superintendents between 1802 and 1817, the emphasis on French methods shifted, but the Academy gained both stability and a commitment to keep apace of

French engineering methods with the appointment of Sylvanus Thayer in 1817. Thayer, a graduate of both Dartmouth and West Point, as well as the engineer who directed the defenses of Norfolk, Virginia, in the War of 1812, left the United States in 1815 to study at the École Polytechnique in Paris. He arrived back in the United States to become the superintendent of West Point with an extensive collection of French mathematics, natural philosophy, and engineering textbooks, which formed the basis for a new curriculum and a rededication of the USMA to rationalist French engineering. Under Thayer the USMA trained engineers, the best of whom were then commissioned to the Corps of Engineers.

These book-learned engineers often struggled on their first projects, which primarily involved designing and building American coastal fortifications. When Joseph Totten found that young graduates were not ready to take on practical tasks in engineering projects, he instituted an early kind of an experiment station at Fort Adams in Rhode Island to offer them hands-on training after West Point. Here they could test materials and learn engineering in the field (Johnson, 2009). After that they could be assigned to sites all over the nation to build its coastal defenses, harbors, and ports and to work on hydrological projects. These West Point engineers played crucial roles in the development of the nation before the Civil War, often overseeing railroad route surveys and becoming chief engineers for the railroad industry, which was growing rapidly by the 1830s. Dozens of the most sought-after railroad engineers, like Herman Haupt and John Gunnison, were West Point graduates. In addition, some of the nation's leading scientists were also West Pointers with experience in the Corps—men like Alexander Dallas Bache, who headed both the Franklin Institute and the U.S. Coast Survey over his distinguished career (Slotten, 1994). Although occasionally under attack for its elitist attitudes, West Point and the Corps of Engineers were the central institutions for building the nation before the railroad boom in the 1830s. Before the Civil War, a majority of formally educated engineers came from West Point, making it the most important generator of engineering experts in the United States.

In this period one of the main jobs of officers in the Corps of Engineers was to create surveys that would inform national development. During the War of 1812 topographical engineers were explicitly authorized to undertake military surveys and create maps for strategic purposes. Between 1813 and 1831 10 topographical engineers operated as a branch of the Corps of Engineers and generally produced surveys for both the military and the commerce-oriented projects of the federal government. The best known of these projects was Major Stephen H. Long's western expedition, between 1816 and 1823, to survey the northern plains from Illinois to the Rocky Mountains. In 1831 the Corps of Topographical Engineers was separated from the Corps of Engineers, but remained part of the War Department and was still overseen by the chief engineer of the United States. In 1838 under Secretary of War Joel Poinsett, a further refinement led to all civil engineering projects being assigned to the Corps of Topographical Engineers under the supervision of Colonel John James Abert, West Pointer and the chief of the Topographical Bureau. All military projects, including surveys, were then handed off to the Corps of Engineers under newly appointed chief engineer Joseph Totten. Totten's command would last into the Civil War. With clearer roles being afforded civil and military engineers in the employ of the federal government, the Corps of Engineers could focus their attention on fortifications and the settlement of the West, whereas the Topographical Corps could undertake expeditions to determine the resources that would fuel western expansion (Goetzmann, 1959). By the mid-nineteenth century there were several federal organizations devoted to surveying and mapping; however, by 1850 surveying was also becoming an activity distinct from engineering. Distinctions between surveyors and engineers would become more hierarchical during the period of professionalization after the Civil War, but late in the antebellum period the two professions were simply on parallel tracks.

Canal building. The federal government was not the only employer of engineers during the early Republic. Canals, and later railroads, created great demand for engineers. In the late eighteenth century, assessing the expertise of the engineers remained a complicated task. The 26-mile-long Middlesex Canal, which was to connect the Charles and Merrimack rivers from Boston to Lowell, was first commissioned by a group of private citizens with the approval of the Massachusetts General Court. The Middlesex Canal Company appointed local "man of learning" Loammi Baldwin to plan and build the canal in 1793. Baldwin tried to hire locals to survey the course, but found that their judgment and methods were questionable and produced sizeable errors (Morison, 1974). Baldwin and the locals also discovered that they lacked the skills to excavate through rock, to seal the canal so that the water did not drain out, and to build the locks that would allow barges to move through level changes. The canal company contacted William Weston, a young English engineer who claimed to have worked with James Brindley, engineer of the Bridgewater Canal in Manchester. Weston had been hired to design and build a canal to connect the Schuylkill and Susquehanna rivers in Pennsylvania. Weston offered advice on the problems Baldwin had encountered and showed the canal builders how to run accurate levels. The builders still struggled to cut the course of the canal through rock, to build functional locks, and to stop the canal from leaking. In the end, the canal was functional, if imperfect. Weston clearly knew more than Baldwin and company, but Weston was not familiar with local conditions, local materials, and local labor. The history of the Middlesex canal illustrates the difficulty of how to blend knowledge of the local with knowledge of how engineering problems were solved elsewhere in the world.

The Erie Canal, proposed in 1808, started in 1817, and completed in 1825, constitutes a much more successful engineering achievement. Again, it is impossible to find a trained engineer in the mixture of men working on the canal. Unlike the Middlesex canal, however, the Erie Canal produced a large number of engineers who would go on to engineer and build other canals and roads and eventually design railroad routes and bridges, earning it the name "the Erie School of Engineering." The Erie Canal physically dwarfed all

other American canal projects of the period. Its 363-mile-long route included a rise of nearly 600 feet, requiring 83 locks to connect the Hudson River and the Atlantic Ocean to Lake Erie and the Great Lakes. The chief engineer of the canal was Benjamin Wright, whose previous work had been as a surveyor and lawyer who settled land disputes. He was originally in charge of the middle and eastern sections of the canal prior to his promotion to chief. Wright subsequently designed the Delaware and Hudson canal and Chesapeake and Ohio canal before moving onto railroad work in 1832 with the New York and Erie Railroad Company. Canvass White, another Erie Canal engineer, traveled to England in 1817 to look at canals and brought back both lock designs and ways of making hydraulic cement, which can cure underwater, for lining the canal and to prevent leaking. After the canal, White not only worked on more canals, but also set up a company to manufacture hydraulic cement. John Jervis started work on the canal as an axe man in 1817. By 1819 he had risen to become the lead engineer of the center section. After the canal, he had a distinguished career as one of the nation's best known engineers, working on railroads and rising to president of the Chicago and Rock Island Railroad in the 1850s. He also designed both New York's and Boston's water supplies, including the Croton Aqueduct, which brings water from Westchester to Manhattan, and Boston's Lake Cochituate reservoir and the aqueduct connecting it to the Brookline reservoir just outside of Boston. Wright, White, Jervis, and dozens of others were all graduates of the Erie School of Engineering, which was in fact no school at all, just a grand project that sustained apprenticeships for a large number of skilled men.

The Railroad era. By 1830 the era of the great canals was giving way to the beginnings of railroad construction. Railroads demanded an increasing number of engineers, and both practically trained engineers and West Pointers plied their trade on the railroads. Although American railroads were largely developed by private interests, the Corps of Engineers remained an important state resource. Many veterans of the Topographical Engineers also became railroad route surveyors, including Stephen H. Long, who worked on the Baltimore and Ohio route survey. The Pacific Railroad survey in 1853 involved no fewer than four West Point graduates under the supervision of West Pointer John Gunnison. In addition to the challenges of planning routes, railroad engineering required bridges that would stand up to very large dynamic loads.

There was often a shortage of engineers with credentials, references, or experience. Private engineering schools starting springing up to fill these demands—places like the Rensselaer Polytechnic Institute in Troy, New York, and Union College in Schenectady, New York, started new and rigorous programs in the 1820s. During the Jacksonian era, informal education through mechanics institutes was also popular, and these institutes often offered programs to train entry-level railroad surveyors and engineers. By the 1840s, older colleges, like Harvard and Yale, also developed new schools to train engineers. As a result, at the height of the railroad boom, there were several paths to becoming an engineer.

Although the railroad was in some sense an import from Britain, American inventor and engineer Oliver Evans was experimenting with high-pressure steam engines in the 1780s, earning America's third patent for his high-pressure engine in 1790. Evans had the idea for making the engine mobile, but found no investors for his design. Instead he concentrated on industrial engines. Colonel John Stevens of New Jersey also worked on the mobile engine and built steamboats; he received the first U.S. charter for a railroad, the New Jersey Railroad, in 1815. Like Oliver Evans, Stevens's success came from steam engines, but not railroads. By the 1820s using steam for land transportation had become an attractive option, and American entrepreneurs turned to Britain for the first locomotives. American engineers quickly moved in to provide locally designed and manufactured engines. Matthias Baldwin of Philadelphia was probably the most famous and successful (Brown, 1995). Baldwin's first locomotive rolled into action in 1832 and his firm survived into the 1950s producing locomotives.

The period between the founding of the new nation and the Civil War was a formative one for American engineers. There was much change— politically, economically, demographically, and organizationally. American engineers were developing both the social structures and the tools needed by the federal government and private industry. American engineering had multiple cultures, which were often in tension with each other. Engineers were essential to westward expansion through surveys and railroads, but also centrally involved in building the eastern cities and their water supplies, factories, and local transportation infrastructure. Some engineers swore by theory-driven methods at West Point, whereas others never opened a book and learned their trade through the hands-on informal apprenticeships of the canals and railroads. Engineers were in high demand at times, but were also vulnerable to the severe economic panics that characterized the nineteenth century. They designed and built the machines that supported the growth of agriculture, like the McCormick reaper and the cotton gin. There was no geographical center for American engineering in the nineteenth century. Engineers tended to be highly peripatetic, chasing interesting and lucrative work or government assignments, and constituted some of America's most well-traveled, least rooted citizens.

Engineering in the Civil War. The long association of engineers with the military meant that engineers would be central participants in the Civil War. The fact that nearly all military engineers and a large percentage of government-employed engineers had been educated at West Point (with a few coming from the southern military academies such as the Citadel and Virginia Military Institute) made the division of engineers into Union and Confederate supporters especially poignant. West Point–trained engineers were also well represented in the command structure of both Union and Confederate armies, with Robert E. Lee being the most famous and highest ranking former member of the Corps of Engineers.

The Union maintained control of the Corps of Engineers and the Topographical Engineers and, to address wartime needs, combined them. Some engineers resigned their positions in the Corps to join the Confederacy. Joseph Totten commanded the Corps until his death in 1864, when Richard Delafield took over. The Corps were responsible for the Union's defenses as well as advising and creating routes for logistical support. They also destroyed infrastructure that would be critical to the Confederacy's progress. The Confederacy also created a Corps of Engineers, although its 10 officers were dwarfed by the Union's 43. Still, the Confederacy's Corps of Engineers faced bigger engineering challenges and was in some sense more active than the Union's in planning and constructing roads, railroads, and waterways to move materiel and support the war effort. Filling gaps in the Confederacy's infrastructure was of great importance because Civil War battles were fought primarily in the South and prior to the war the transportation technologies were sparser.

Engineering Education and the Federal Government.

Demand for trained engineers and agriculturalists attracted the attention of the federal government in the 1850s. In 1857, Vermont representative Justin Smith Morrill introduced a bill to offer each state 30,000 acres of federally owned land for each congressman, as determined by the 1860 census, in exchange for the creation of an educational institution to teach "such branches of learning as are related to agriculture and the mechanic arts, in such manner as the legislatures of the States may respectively prescribe, in order to promote the liberal and practical education of the industrial classes in the several pursuits and professions in life." Morrill's bill passed through Congress in 1859, but was vetoed by James Buchanan on the eve of the Civil War under pressure from oppositional southern states. Morrill brought the bill up in 1861 and Lincoln signed it into law on July 2, 1862. The passage of the bill led to the creation of the "land-grant colleges"; Iowa created the first by designating Ames College its agricultural and mechanical college; Ames became Iowa State University. Kansas created the first new institution with the founding of Kansas State University in 1864. States had the discretion to direct the new resources at an existing institution or to found a new agricultural and mechanical

school. The land-grant colleges were required to offer programs in agriculture, the mechanical arts, and military tactics, but they were not restricted from offering anything else. These institutions educated large numbers of engineers who were essential to the industrialization of postbellum America. Engineering disciplines were also formed under pressure to differentiate courses of study relevant for civil, mechanical, electrical, geological, and eventually chemical engineering. By the end of the nineteenth century the Society for the Promotion of Engineering Education was formed to standardize and approve engineering educational programs (Seely, 2005).

A second Morrill Agricultural College Act in 1890 fostered the creation of land-grant colleges in states that had been part of the Confederacy in 1862. The 1890 act required that each state show that race was not a criterion for admission or or else designate a separate institution for people of color. All states chose the latter option, and this led to a dual structure of land-grant colleges in the southern states with legal segregation—one college for white students and another for students of color. Although there was a demand for black doctors and lawyers to serve black clients, the employment and economic structure of engineering did not create such opportunities. There were black engineers to be sure; they were a very small minority of working engineers and an even smaller minority of professional society members. Black inventors were more common than university-trained engineers, but making distinctions between inventors and engineers is notoriously difficult (Fouché, 2003). Women were also admitted to most of the land-grant colleges but were often restricted from taking engineering courses. The land-grant colleges served to solidify a social structure already in place. The exclusion of certain classes and groups of people was further strengthened by the formation of professional associations.

The Formation of Professional Societies, 1830–1910.

The growth of engineering activity and the increasing options for formal education in engineering led to a concerted effort, beginning in the 1830s and culminating in the 1850s, to form a national professional association.

Put another way, engineers began to create the formal trappings of a profession at this time. Other professions, like medicine and the law, were also generating professional societies and pursuing professional structures. Professions looked to each other for organizational models. All professions aimed to make clear distinctions between practitioners with credentials and those without. Engineers and doctors alike believed their livelihoods were potentially threatened by charlatans who called themselves engineers or physicians, but who did not uphold their values or do competent work. There was recognition that professions required the public trust. If engineering work were needed, then only a certified engineer would be a reliable provider of services. Every engineering failure affected all engineers and made them potentially less trustworthy and deserving of the public's respect. The need for public trust and respect was particularly important for high-profile jobs where lives were at stake, like the design of urban water systems or railroad bridges.

If the rationale for creating professional organizations was clear, the path to creating them was anything but smooth, and the credentials professional societies would validate remained unclear. Formal education could not be a requirement at this point because too many competent, and even famous, engineers lacked it. Societies could not limit professional membership to those with a credential that was too rare since they recognized the need to meet the demand for expert engineers. In contrast, many of the engineers involved in the founding of professional societies were well educated, so they were highly sympathetic to the idea of an engineering elite. Professionalization is a process of exclusion; some individuals who perform many of the functions of a job will be excluded from membership in the professional society or from recognition as engineers because of their credentials, social status, or race and gender. The problem is where to draw that line so that the criteria are clear and rational to potential members, those excluded, and the public. Criteria cannot appear either capricious or irrelevant to the jobs at hand.

There were some precursors to the movement to create engineering professional societies in the middle of the nineteenth century. The Franklin Institute, founded in 1824, had been an intellectual home for engineers and the work the Institute undertook in the 1840s testing materials to address the threat of exploding steam boilers was the sort of high-profile, socially relevant activity that professional societies wanted to promote (Burke, 1966). The Franklin Institute was also involved in the training of engineers and in publishing new ideas in engineering through its journal (Sinclair, 1974). And the Institute's journal provided a forum for the translation of the most important engineering methods from Europe. Still, the Franklin Institute never entered the business of excluding some less competent or educated engineers from the ranks of the nascent profession.

Other groups tried to create professional civil engineering associations in the 1830s and 1840s, but none survived more than a year or so without folding. The first recorded call for a professional society came in 1836, in an address to the Charleston and Cincinnati Railroad. This call to arms received little attention beyond the immediate vicinity and the railroad was still only a pipe dream. In 1839 40 engineers gathered in Baltimore to attempt another society, this time electing a president, Benjamin Latrobe Jr. Forty engineers probably constituted a tenth of the full-time, practicing engineers in the country at the time (Wisely, 1974). The attendees at the Baltimore meeting included many of the country's most important civil engineers—including Benjamin Wright, Claudius Crozet, John Jervis, and Latrobe. The Franklin Institute volunteered to host the society and merge its considerable resources with whatever the new society might develop. A committee of five engineers drew up a plan for creating a constitution and bylaws for the new society. In doing so, they recognized the challenges that American engineers faced were much greater than those in, say, England, who might reasonably attend an annual meeting in London. The scale of the United States was such that travel distances would prohibit an annual meeting of a substantial number of members. The committee did begin the work of dividing engineers hierarchically into "members" and lower-status "associates." However, there was dissent over such requirements, and by 1840 this attempt at professional formation collapsed.

The effort to create a national civil engineering society was renewed in the late 1840s when local professional engineering societies in New York and Boston formed and began to publish journals. Boston's society was particularly elite, with rigid admission requirements. One of the founders of the Boston society, James Laurie, orchestrated another effort at a national society with a meeting of civil engineering practitioners in New York. The 12 men who attended this meeting on 5 November 1852 were named Founders, the highest membership class, of a new American Society for Civil Engineers and Architects (the forerunner of the American Society of Civil Engineers [ASCE]). Ten more men accepted early invitations to be involved and were named charter members. Membership was open to all types of engineers—civil, mechanical, and mining—as well as architects and men of science. The aim of the society was to exchange knowledge and, through that process, to elevate the engineering profession to a learned one. Within a year the society had 55 members. By 1875 there were 408 members (Wisely, 1974). The ASCE had staying power, although the creation of the American Institute of Architects in 1857 led the society to vote to remove the term "architects" from its name. Disputes over membership requirements continued well into the twentieth century, but the society provided a venue for discussions about how to distinguish an engineering hierarchy.

After the ASCE the next professional association to form was the American Institute for Mining Engineers in 1873. The American Society for Mechanical Engineers convened in New York in 1880 under the leadership of steel manufacturing innovator Alexander Lyman Holley. Electrical engineers formed the American Institute of Electrical Engineers in 1884 at the Franklin Institute, with such illustrious founding members as Thomas Edison, Nikola Tesla, Elihu Thomson, and Edwin Houston. Although there were efforts at consolidating societies or at least coordinating them in the early twentieth century, all of these groups have persisted as professional societies.

The final step in professionalization was licensure. Wyoming became the first state to license engineers to discredit disreputable water specula-

tors who called themselves engineers to give credibility to their services. Adopting such a title threatened the respectable profession of engineering, and in 1907 Wyoming law made it unlawful to claim the title without proper qualifications. The ASCE membership was initially split on the question of whether to allow states to determine credentials or whether only the ASCE should vet the qualifications of engineers. Following the examples of law and medicine, the ASCE began increasingly to support state licensure in the early twentieth century, writing its own model licensing statute in 1910. The qualifications and credentials needed to gain a license changed considerably over the twentieth century, but licensing was accepted as a tool of professionalization early in the twentieth century, and professional societies became partners of the state in vetting the qualifications of engineers.

Industrial America. Many of the structures that shaped modern, professional, industrial engineering were in place by Reconstruction. These institutional structures, including universities and technical institutes, professional associations and publications, government organizations, and a well-known operational definition of what engineers did and why they did it, were socially important. However, the development of large corporations in the late nineteenth century provided a new resource and challenge for engineers. By 1925, 75 percent of engineers were employed by corporations; by 1965 the number of self-employed engineers had fallen to 5 percent (Reynolds, 1991b). Firms clearly provided many job opportunities for engineers, but they also obscured the face of the engineer. Engineering college graduates could usually find work, but corporate positions often meant subverting one's professional identity to that of the firm's. Unlike doctors, whose professional identity demanded autonomous self-employment, company employment compromised the professional autonomy of the engineer. Compared with the railroad chief engineers who commonly moved from position to position every few years, engineers in industrial America could spend their careers with a single firm. The culture of engineering was changing;

engineers were less peripatetic but more invested in their communities and shaping the long-term trajectory of the nation's economy. They saw opportunities in staying with the same firm and working their way up the corporate ladder. David Noble has argued that this relationship inevitably co-opted engineers and prevented them from countering any of the excesses of industrial capitalism (Noble, 1977).

Large corporations and research and development (R&D) laboratories.

Industrialization meant not only more corporations but also larger ones. High-technology industries, like electrical utilities, internal combustion engines, and petroleum and chemicals, led the way. Their products required considerable capital investments so large firms had competitive advantages. Markets for their products became increasingly national, instead of local or regional, and competition between firms stiffened, leading to firm consolidation. Vertical and horizontal integration provided further advantages, although these strategies also drew concern from antitrust lawyers.

The pressures industrialization created on engineers and their professional organizations came to a head in the 1910s. Sociologist Peter Meiksins and historian Edwin Layton have analyzed these developments (Layton, 1986; Meiksins, 1988). Layton's book *Revolt of the Engineers* describes the efforts of progressive reformers like Morris Cooke and Frederick Haynes Newell to reshape engineers' collective relationships with American business. Cooke and Newell saw engineers as being in a unique position to humanize business and to force social responsibility onto corporations. Ultimately these reforms failed, but for a generation engineers grappled again with their position in an ever-changing world. Meiksins paints a nuanced picture of the failure of these reforms. He breaks early twentieth century engineering into three cultures. An elite of academics, consultants, and entrepreneurs generally denied conflict between engineering and business. These men, mostly autonomous engineers and those least involved in the day-to-day corporate world, often controlled the national engineering societies. Cooke and Newell belonged to the group of

"patrician reformers"—activists who sought to unify the engineering professions in light of the need for engineers to earn the trust and respect of the public as their highest goal. A third group was composed of the rank-and-file engineers, who sought higher status and better employment options for themselves. In the decade of the 1910s the interests of the patrician reformers and rank-and-file engineers aligned. They challenged the elite's conception of the relationship between capitalism and engineering. Buoyed by wartime service and sacrifice, reformers pursued a social agenda. They wanted engineers to cultivate social responsibility and provide a check on out-of-control industrialists. Rank-and-file engineers wanted to agitate for better wages and work conditions. They believed the sheer number of engineers in the corporation blocked paths of advancement. The post–World War I economic recession and accompanying Red Scare of 1919–1920 forced many of the rank-and-file engineers to retreat from their quasi-unionist positions, and the alliance between the two nonelite groups was broken. The prosperity of the 1920s led to a return to a probusiness position. David Noble sees the 1910s as an anomaly and the return to uncritical acceptance of American business practices the normal and inevitable state of engineering-business relations (Noble, 1977).

The evolution of large corporations went hand in hand with the increasing science dependence of new technologies. Science-based industries like electrotechnology and chemicals created industrial research laboratories that employed both scientists and engineers, and distinctions between the two types of professionals were often blurred. Research and development became one of the functions of the firm. Large laboratory facilities were also part of the R&D missions of chemical firms like DuPont, Eastman Kodak, and Standard Oil, electrical firms like General Electric and American Telephone and Telegraph, and eventually automobile and aviation firms like Ford, General Motors, and Boeing. Thomas Edison was particularly proud of his "Invention Factory," first in Menlo Park and then in West Orange, New Jersey. By the mid-twentieth century, a research department was as much a part of corporate culture as a mailroom. For some corporations, like Bell Labs, research *was* their corporate identity (Gertner, 2012). Firms invested heavily in R&D by investing in scientific and engineering personnel.

Scientific engineering. Developing complex new technologies required increasingly scientific engineering training, which drove major curricular changes at the land grant schools and technical institutes (Seely, 1993). Colleges like the Massachusetts Institute of Technology (MIT) emerged at the vanguard of scientific engineering, and faculty fought over the role of corporate influence on MIT's curriculum (Lecuyer, 2009). Between the end of the nineteenth century and World War II, the curriculum of engineering schools began to move out of the machine shops and drafting rooms and into the lecture hall and physics and chemistry laboratories. School culture had won out over shop culture, but not without a cost. Engineers had made a clean break from trades and paraprofessional occupations like drafting, but doing so also meant knowing less about the work done on the shop floor and sacrificing another element of the engineer's tenuous claim to autonomy. The engineering media often published frustrated accounts from both senior engineers and managers who complained that new graduates knew very little about the real work of engineering. Management methods, like scientific management, often exacerbated tensions by shifting the power into the hands of supervisors who might have no experience of shop floor work but who controlled the minute details of the workplace practices (Meyer, 1981; Slaton, 2001). On the other hand, the new, nascent high-tech industries, including vacuum tube manufacturing and aviation, also offered many high-paying opportunities to engineering graduates, even during the Great Depression.

The scientization of engineering was more than just a response to corporate demand. It was also a professional strategy to raise the status of the engineer, especially in corporate settings. Engineers used the imprimatur of science to make clear distinctions between themselves and tradespeople. Engineers had two claims to "white-collar"

status—engineers could and did rise into corporate management, and they were scientific in their pursuits. Although numerous historians, particularly Edwin Layton, have written that engineering is more than the mere application of existing science to real-world problems, engineers themselves have been loath to reject the epistemological model of engineering as applied science (Layton, 1971; Kline, 1995).

The relationships between the engineering disciplines were also redefined in the industrial settings of the turn of the twentieth century. Mechanical and electrical engineers were joined in the early twentieth century by chemical engineers, who formed their professional society in 1908. These three "new" professions dominated the corporate world of engineering. Civil engineers could also be corporate employees, at large construction firms or railroads, for example, but their social structure was still dominated by government work. Municipal services had to expand rapidly to accommodate larger populations and deal with sanitary services. Engineers planned extensive urban engineering projects, such as the reversal of the Chicago River to avoid the continuous contamination of Lake Michigan (Cronon, 1992). By World War I, civil engineers were also planning out new interconnected paved road systems to facilitate the use of automobiles. States formed departments of transportation and hired engineers to oversee them. In addition, the Army Corps of Engineers had taken on the new mission of flood control. In the twentieth century the Corps' work had much less of a military character outside of wartime, but remained an important institution for hydrological engineering.

From World War I to World War II. From the European perspective World War I was a major engineering effort. However, given the late U.S. entry into the conflict, the roles for American engineers were more limited. New technologies were used extensively for the first time in the war, and they substantially changed the nature of the conflict. These technologies included smokeless gunpowder, long-range artillery, machine guns, poison gas, mines, tanks and other vehicles, airplanes and other aviation technologies, radio, and submarines. Older technologies, like railroads, also played essential roles in shaping the conflict.

In the United States the National Advisory Committee for Aeronautics and the National Research Council were both formed in 1915 to ensure that, even in isolationism, the United States would keep apace with scientific and engineering developments during the war. However, the war also revealed institutional weaknesses in enlisting science and engineering for war. The Naval Consulting Board, headed by Edison, argued for a naval research lab, but failed to produce enough of a vision to garner support to build the lab and assemble a staff. The Army Signal Corps was a success during the war in developing new communications technologies, but after the war no personnel were left in place. American engineers were called upon to advise on the reconstruction of Europe. Herbert Hoover, transitioning from his career as a mining engineer into politics, led the World Engineering Federation and found like-minded European engineers who argued that they should use engineering expertise to rebuild and stabilize the new political and physical landscapes of Europe (van Meer, 2012).

Many engineers experienced the layoffs and difficulties of the Great Depression just as other Americans did. Engineers were often forced often to turn to new lines of work. When industry began picking up again in the late 1930s engineers were rehired. New Deal programs like the Civil Works Administration hired engineers (as many as two per county in some states) to perform hydrological and road surveys. Electrification and dam building also called for engineering expertise. The Tennessee Valley Authority was created in 1933 to generate hydroelectric power for the largely rural region of Tennessee, Kentucky, Alabama, Mississippi, and parts of Virginia, Georgia, and North Carolina, while also providing flood protection and improving the navigability of rivers. Befitting a Depression-era federal project, the Tennessee Valley Authority hired a number of unemployed, unskilled workers, including African Americans and women, but engineers were hired as well to plan hydroelectric facilities, power distribution, and industrial plants. The planning for the Hoover Dam in Nevada was undertaken by

the Bureau of Reclamation during the 1920s, while the $48 million dam was constructed by a conglomeration of six large construction companies, primarily from California, led by superintendent and chief engineer Frank Crowe. It was the signature engineering accomplishment of the Great Depression.

Demands for engineers shot up with the U.S. entry into World War II in 1941. Both the military and the industrial war mobilization created countless opportunities for engineers. Furthermore, there were opportunities for men lacking formal engineering education to learn skills, which they could sometimes also parlay in engineering positions after the war. Short-term courses were created to train women to replace male engineers who had gone off to war. Some nine hundred women received two and a half years of aeronautical engineering courses over ten months through the Curtiss-Wright Cadettes at seven different universities. They replaced a largely male workforce at Curtiss-Wright designing aircraft throughout wartime. Efforts to develop new military technologies, such as radar and atomic weapons, relied on highly skilled elite engineers, in addition to the better known contributions from scientists. However, much more mundane engineering was also key to the war in both Europe and Asia. In the Pacific, island conquests meant work for Navy Construction Battalions, or Seabees, to create new airstrips, bases, and fuel storage. Many of these men returned home and rejoined the workforce as engineers.

Engineering since World War II. Industrial demobilization after World War II was limited by the threats of the Cold War. Particularly in the industries relevant to national defense (the so-called military industrial complex), employment stayed high. These industries employed large numbers of engineers, but so did consumer industries, like automobiles, which experienced surprising growth in demand after the war.

After the war, about 50 percent of veterans took advantage of the GI bill to pursue higher education, many of these in technical and engineering subject areas. Since the GI bill benefits went predominantly to men, the ratio of male to female college students rose. Women's participation in engineering, low even before the war, dropped in the 1950s. The 1950 census showed 6,475 female and 518,781 male engineers, making women a 1.2 percent minority (Mack, 2001). Even women who had played important roles in industrial engineering during the war were typically relieved of their positions. Social pressure to open engineering education and jobs to women did not take hold until the 1960s, when MIT built its first female dorm and Caltech and Georgia Tech first admitted women to their undergraduate degree programs.

Space. The *Sputnik* launch in 1957 stung American engineers, who had imagined themselves far in advance of their Soviet counterparts. Given the threat of nuclear war, *Sputnik* was received by the American public as a huge threat; the government had to respond. The National Defense Education Act of 1958 provided low-cost loans to students. It also aimed to expand the college-going population and to direct those students into particularly strategic subjects like science, engineering, mathematics, and foreign languages. Behind the act was a nativist sentiment, following McCarthyism, that Americans needed better incentives to study mathematics to become the teachers of the next generation. Foreign mathematicians might be fine for programming the nation's small number of computers, but American students needed qualified American teachers. Secondary educators needed to encourage more Americans to aspire to the science and engineering fields.

Between the *Sputnik* launch and the mid-1970s the Space Race played an important public role in generating and maintaining interest in engineering in general and in aerospace engineering in particular. After World War II, the United States had captured and recruited German aeronautical engineer Wernher von Braun to head a missile program in the United States, initially at White Sands, New Mexico, and then in 1950 at Redstone Arsenal in Huntsville, Alabama. In 1958 von Braun's three-stage Jupiter-C rocket launched the first U.S. satellite, *Explorer 1*, into space—providing the counterpoint to *Sputnik*. In another

response to *Sputnik* in 1958, the National Advisory Council of Aeronautics was renamed the National Aeronautics and Space Administration (NASA). Von Braun and his team were also transferred to NASA, whose Marshall Space Flight Center was located at Redstone. Project Mercury, a manned space flight program, began in 1958 too. NASA was one of the few public faces of Cold War engineering. Much of the high-tech engineering during the Cold War was either classified, esoteric, or both. The transistor, despite its obvious importance, was hard to understand—what did it do? Computers were still nearly room-size machines that used punched cards. Nuclear power, let alone nuclear weapons, seemed more about physics and chemistry than engineering, although this was an erroneous perception. Automobiles and radios were not so mysterious anymore. The space program was critical public relations for the engineering profession; it captured the public's imagination, at least for the decade of the 1960s.

Engineering in the 1960s and 1970s. Technology and engineering in the 1960s was Janus-faced, although both faces looked to the future. If space flight was an exciting and limitless possibility, then war and environmental degradation were catastrophic and apocalyptic possibilities. In 1972 an influential report titled *The Limits to Growth* predicted the catastrophic collapse of civilization because of the exhaustion of planetary resources. Many engineers took this prediction as a challenge to design better technologies to elude such limits to growth (McCray, 2013). Engineers knew they had a role to play in both sides of technology's relationship to society and some struggled with this reality. Responding to social critics like Theodore Roszak and Herbert Marcuse, some engineers saw that they had unleashed technological capacities and society lacked the tools and discipline to manage them (Wisnioski, 2012). Like the engineers of the World War I era, they wondered how engineers might lead the way to better social responsibility. Engineering educators both called for and resisted external pressures to make engineers better trained to understand social realities and be less technocratic and short sighted. On occasion, engineers of the 1960s seemed genuinely surprised that they were the targets of antiestablishment protest. They had failed to understand their role as part of the establishment, as employees of the large corporations who supplied the war effort, resisted civil rights laws and labor unions, and polluted the air and water. Engineers were not just observers of the military-industrial-academic complex; they were at the heart of it. Some conservative engineers believed in the mission of the complex, but others simply failed to understand their central role in it.

Engineers in the 1970s experienced the country's worst economic downturn since the Great Depression. This recession was particularly hard on engineers because it came on the heels of a downsized military budget. The oil crisis hit the petroleum, automobile, and aviation industries particularly hard, and these industries were all critical for engineering employment. Engineers peaked at 1.6 percent of the U.S. workforce in 1970; by 1980 they had dropped to 1.2 percent (Panel on Engineering Employment, 1985). This is the only decadal drop in the percentage of engineers in the workforce of the twentieth century. Still, the absolute number of engineers grew over the decade, primarily because of the growth of computer specialists, who doubled in number over the 1970s to 750,000. However, labor statistics do not break down any distinctions among programmers, engineers, and technicians, in part because such distinctions were only beginning to solidify. New types of programs were also established to educate computer specialists, with the first degree program in computer engineering established in 1972 at Case Western Reserve University.

Personal computers and consumer electronics grew a new industry dependent on the skills of engineers. These industries relied on a diverse array of different engineering specialties, from electrical to chemical to manufacturing. Some of these engineers also rose to be public figures and the public face of engineering, as well as leading entrepreneurs. Men like Steve Jobs, Steve Wozniak, and Bill Gates represented both the founders of the personal computer industry and the "geek culture." Over the course of one generation from the mid-1970s to the turn of the twenty-first century, the personal computer industry and geek culture became mainstream, representing the dominant public perception of engineers. Geek culture eventually extended to technophiles, whether engineers or not. Consequently, engineers were becoming increasingly associated with so-called high technology, although many continued to work in traditional areas such as infrastructure design, transportation technologies, and industrial manufacturing.

Title IX and Underrepresented Groups in Engineering.   The Equal Employment Opportunity Act of 1972 was an important piece of legislation to promote equal access to employment opportunities without regard to worker's gender. It is best known for its Title IX, which outlawed gender discrimination in educational programs receiving federal funding. Although Title IX is commonly credited with opening high school and college athletics to women, it also opened engineering education. Programs could no longer received federal assistance and deny women's admission to degree programs. In the decade after the passage of Title IX, women went from receiving just 2 percent of engineering degrees to 10 percent (Mack, 2001). Despite its promising start, the effects of gender equity legislation have been very slow in engineering. There are many disputes regarding the reasons why (Rosser, 2012).

The Society of Women Engineers, founded in 1950, had over 21,000 members in 2012, and across all disciplines women earned approximately 17 percent of engineering degrees in the first decade of the twenty-first century. Still, aggregate statistics overlook important disparities in representation. Mechanical and aerospace engineering have traditionally had the lowest percentage of undergraduates and women in the workforce (<10 percent), whereas the newer disciplines of environmental and biomedical engineering have approached near gender parity for undergraduates. Furthermore, some universities have been much more successful at recruiting and mentoring female students. MIT's 2010 graduating class was 42 percent women but this included all majors, not only engineering. The National

Society of Professional Engineers, a professional society and licensing organization, reported that in 2004 there were around 200,000 female engineers in the workforce versus 1.5 million men, or just over 13 percent. Of course statistics are problematic because the definition of who is counted as an engineer is still more complicated than that of who is counted as a medical doctor. But the numbers tend to agree that in the early twenty-first century women constituted less than 20 percent of the engineering workforce and a slightly larger proportion of undergraduate engineering students. During this period, concerns tended to focus on areas where numbers of women in engineering were declining relative to just a generation earlier, most prominently in the computer fields (Misa, 2010).

The racial dynamics of the engineering profession also changed considerably over the course of the twentieth century. Although African Americans have always been underrepresented, other racial groups have moved to dominate certain subdisciplines in engineering. The long history of exclusion of African Americans has been an important factor in their underrepresentation (Slaton, 2010). Hispanic Americans and Native Americans have similarly low numbers and have historically also been excluded from the educational opportunities available to white men. In contrast, starting in the 1970s Asian immigrants and Asian Americans have been overrepresented relative to their percentage of the population, particularly in the electrical and computer engineering fields. Engineering, along with science and the health-care fields, has been the profession traditionally allowing entry to the American professional classes, especially among immigrant groups.

The social and educational structures that were shaped in the nineteenth century continued to have a legacy for engineers in the twenty-first century. Professionalization has brought access and respectability to engineering and has yielded a large population of engineers playing critical roles in shaping American technology. However, it also defined a profession that was fully integrated into corporate America and had less space for the iconic engineer-inventors of an earlier time.

[*See also* **Animation Technology and Computer Graphics; Army Corps of Engineers, U.S.; Biotechnology; Chemistry; Computer Science; Ethics and Professionalism in Engineering; Gender and Technology; Genetics and Genetic Engineering; Higher Education and Science; Journals in Science, Medicine, and Engineering; Machinery and Manufacturing; Military, Science and Technology and the; Physics; Research and Development (R&D); Rivers as Technological Systems; Science; Solid-State Electronics; Space Program; Technology;** *and* **Tennessee Valley Authority.**]

BIBLIOGRAPHY

Alder, Ken. *Engineering the Revolution: Arms and the Enlightenment in France, 1763–1815.* Princeton, N.J.: Princeton University Press, 1997. One of the best studies of the organization and education of engineers in eighteenth-century France.

Bix, Amy Sue. "From Engineeresses to Girl Engineers to Good Engineers: A History of Women's American Engineering Education." *National Women's Studies Association Journal* 16, no. 1 (Spring 2004): 27–49.

Brown, John K. *The Baldwin Locomotive Works: A Study in Industrial Practice, 1831–1915.* Baltimore: Johns Hopkins University Press, 1995.

Burke, John G. "Bursting Boilers and the Federal Power." *Technology and Culture* 7, no. 1 (January 1966): 1–23. A seminal article on the first intervention of the U.S. government into technology regulation in response to exploding steamship boilers in the 1830s and 1840s.

Calhoun, Daniel Hovey. *The American Civil Engineer: Origins and Conflict.* Cambridge, Mass.: Harvard University Press, 1960. An older study of the profession of civil engineering and internal improvements, covering the period from 1770 to 1850.

Calvert, Monte A. *The Mechanical Engineer in America, 1830–1910: Professional Cultures in Conflict.* Baltimore: Johns Hopkins Press, 1967. A historical study that focuses on the tensions between shop and school cultures in the development of mechanical engineering.

Cronon, William. *Nature's Metropolis: Chicago and the Great West.* New York: Norton, 1992. An environmental history of Chicago from the mid-nineteenth century to the early twentieth century.

Dilts, James D. *The Great Road: The Building of the Baltimore and Ohio, the Nation's First Railroad, 1828–1853.* Stanford, Calif.: Stanford University Press, 1993.

Fouché, Rayvon. *Black Inventors in the Age of Segregation: Granville T. Woods, Lewis H. Latimer, & Shelby J. Davidson.* Baltimore: Johns Hopkins University Press, 2003.

Gertner, Jon. *The Idea Factory: Bell Labs and the Great Age of American Innovation.* New York: Penguin, 2012. An examination of the culture of Bell Labs, particularly the interactions among science, engineering, and business.

Goetzmann, William. *Army Exploration in the American West, 1803–1863.* New Haven, Conn.: Yale University Press, 1959. A study of the Corps of Topographical Engineers.

Johnson, Ann. "Material Experiments: Environment and Engineering Institutions in the Early American Republic." In *National Identity: The Role of Science and Technology. Osiris 24,* edited by Carol Harrison and Ann Johnson. Chicago: University of Chicago Press, 2009. An article that details the role of the Corps of Engineers and the Franklin Institute in experimental strength of materials.

Kline, Ronald. "Construing 'Technology' as 'Applied Science': Public Rhetoric of Scientists and Engineers in the United States, 1880–1945." *Isis* 86, no. 2 (June 1995): 194–221. Kline examines engineers' and public perspectives on the relationship between technology and science.

Kranakis, Eda. *Constructing a Bridge: An Exploration of Engineering Culture, Design, and Research in Nineteenth-Century France and America.* Cambridge, Mass.: MIT Press, 1997. A comparative study of experiential American bridge design and formal, mathematical French bridge engineering.

Larson, John Lauritz. *Internal Improvement: National Public Works and the Promise of Popular Government in the Early United States.* Chapel Hill: University of North Carolina Press, 2001. A study of the politics and economics of building public works.

Layton, Edwin. "Mirror-Image Twins: The Communities of Science and Technology in 19th Century America." *Technology and Culture* 12, no. 4 (October 1971): 562–580. A seminal article that dispels the notion that engineering is merely the application of existing science to practical problems. Layton shows the knowledge-producing practices of engineers.

Layton, Edwin. *Revolt of the Engineers: Social Responsibility and the American Engineering Profession.* Baltimore: Johns Hopkins University Press, 1986. An examination of the movement for socially engaged engineering in the first decades of the twentieth century, led by Morris Cooke and Frederick Haynes Newell.

Lecuyer, Christophe. "Patrons and the Plan." In *Becoming MIT: Moments of Decision,* edited by David Kaiser. Cambridge, Mass.: MIT Press, 2010 A chapter in an excellent collection about the history of MIT that looks at the pressure to make MIT more responsive to business needs.

Mack, Pamela. "What Difference Has Feminism Made to Engineering in the Twentieth Century?" In *Feminism in Twentieth-century Science Technology and Medicine,* edited by Londa Schiebinger, Angela Creager, and Elizabeth Lunbeck, pp. 149–168. Chicago: University of Chicago Press, 2001.

McCray, W. Patrick. *The Visioneers: How a Group of Elite Scientists Pursued Space Colonies, Nanotechnologies and a Limitless Future.* Princeton, N.J.: Princeton University Press, 2013. An investigation of the development of visionary engineers' plans for space colonies and nanotechnologies as reactions to the 1972 *Limits to Growth* report (see below).

Meadows, Donella, Dennis Meadows, Jørgen Randers, and William H. Behrens. *The Limits to Growth.* New York: Universe Books, 1972. A study, based on cybernetic computer simulations, of the limited resources available for population, industrial, and economic growth in the 1970s.

Meiksins, Peter. "The 'Revolt of the Engineers' Reconsidered." *Technology and Culture* 29, no. 2 (April 1988): 219–246. A reflection on and further investigation of Layton's *Revolt of the Engineers* (above).

Merritt, Raymond H. *Engineering in American Society, 1850–1875.* Lexington: University Press of Kentucky, 1969.

Meyer, Stephen. *The Five Dollar Day: Labor Management and Social Control in the Ford Motor Company, 1908–1921.* Albany: State University of New York Press, 1981. A study of the social consequences of Fordist management.

Misa, Thomas, ed. *Gender Codes: Why Women Are Leaving Computing.* New York: Wiley-IEEE Computer Society Press, 2010.

Morison, Elting Elmore. *From Know-how to Nowhere: The Development of American Technology.*

New York: Basic Books, 1974. A classic study of the complex role of engineering knowledge in the development of technology.

Noble, David F. *America by Design: Science, Technology, and the Rise of Corporate Capitalism*. Oxford: Oxford University Press, 1977. A controversial, Marxist book about the failure of engineers to resist the excesses of capitalism and the reasons for their inability to prevent being co-opted by business.

Panel on Engineering Employment Characteristics, Committee on the Education and Utilization of the Engineer, National Research Council. "Front Matter." *Engineering Employment Characteristics*. Washington, D.C.: National Academies Press, 1985.

Reynolds, Terry S. "The Engineer in 19th Century America." In *The Engineer in America: A Historical Anthology from Technology and Culture*, pp. 7–26. Chicago: University of Chicago Press, 1991a. An introduction to a series of articles drawn from the journal *Technology and Culture* on engineering in America. An overview of engineering's development in the nineteenth century.

Reynolds, Terry S. "The Engineer in 20th Century America." In *The Engineer in America: A Historical Anthology from Technology and Culture*, pp. 169–190. Chicago: University of Chicago Press, 1991b. From the same volume as the previous Reynolds reference. An overview of engineering's development in the twentieth century.

Rosser, Sue. *Breaking into the Lab: Engineering Progress for Women in Science*. New York: New York University Press, 2012.

Seely, Bruce. "Patterns in the History of Engineering Education Reform: A Brief Essay." In *The Engineer of 2020: Visions of Engineering in the New Century*. Washington, D.C.: National Academies Press, 2005.

Seely, Bruce. "Research, Engineering, and Science in American Engineering Colleges, 1900–1960." *Technology and Culture* 34 (April 1993): 344–386. A study of the role of research in the development of American engineering education.

Shallat, Todd A. *Structures in the Stream: Water, Science, and the Rise of the U.S. Army Corps of Engineers*. Austin: University of Texas Press, 1994.

Sheriff, Carol. *The Artificial River: The Erie Canal and the Paradox of Progress, 1817–1862*. New York: Hill and Wang, 1996.

Sinclair, Bruce. *Philadelphia's Philosopher Mechanics: A History of the Franklin Institute, 1824–1865*. Baltimore: Johns Hopkins University Press, 1974.

Slaton, Amy. *Race, Rigor and Selectivity in U.S. Engineering*. Cambridge, Mass.: Harvard University Press, 2010. A study examining reasons for the underrepresentation of African Americans in engineering.

Slaton, Amy. *Reinforced Concrete and the Modernization of American Building*. Baltimore: Johns Hopkins University Press, 2001. An important study about engineers' roles in shaping hierarchical labor practices.

Slotten, Hugh. *Practice, Patronage and the Culture of American Science: Alexander Dallas Bache and the Coast Survey*. Cambridge, U.K.: Cambridge University Press, 1994. A biography of Bache, a West Point graduate, who wielded considerable influence over science and engineering in the nineteenth century.

Smith, Merritt Roe. *Harpers Ferry Armory and the New Technology: The Challenge of Change*. Ithaca, N.Y.: Cornell University Press, 1977. A study of the challenges faced by the Army in implementing interchangeable parts manufacture in the federal armories.

Van Meer, Elisabeth. "The Transatlantic Pursuit of a World Engineering Federation: For the Profession, the Nation, and International Peace." *Technology and Culture* 53, no. 1 (January 2012): 120–145. Details the development of "engineering internationalism," an effort by engineers, including Herbert Hoover, to form expert groups to reconstruct Europe after World War I.

White, Richard. *Railroaded: The Transcontinentals and the Making of Modern America*. New York: Norton, 2011. A study of the business and geography of railroad management.

Wisely, William H. *The American Civil Engineer, 1852–1974: The History, Traditions, and Development of the American Society of Civil Engineers, Founded 1852*. New York: American Society of Civil Engineers, 1974. An institutional history of the ASCE.

Wisnioski, Matthew. *Engineers for Change: Competing Visions of Technology in 1960s America*. Cambridge, Mass.: MIT Press, 2012. A study of engineers' reactions to and engagement with 1960s counterculture and engineering education reforms.

Wozniak, Steven, and Gina Smith. *iWoz: Computer Geek to Cult Icon: How I Invented the Personal Computer, Co-founded Apple and Had Fun Doing It*. New York: Norton, 2007.

Ann Johnson

ENIAC

The Electronic Numerical Integrator and Computer (ENIAC) was an early electronic computing machine that marked the start of the computer age. Although the machine had little in common with the design of modern computers and the claims of its inventors have been marked by controversy, it demonstrated the value of electronic computation and the power of the stored computer program.

The ENIAC was constructed by the University of Pennsylvania between 1943 and 1946 for the U.S. Army Ballistics Testing Laboratory at Aberdeen, Maryland. It was conceived by John Mauchly, a physics professor at nearby Ursinius College, and J. Presper Eckert Jr., a skilled engineer who had just finished his studies at the University of Pennsylvania, although a number of other people were involved in the design and construction. The ENIAC project was particularly notable for the prominent presence of women as the first programmers of the machine. Princeton mathematics professor John von Neumann became interested in the ENIAC project as it was undergoing construction and served as an advisor. In this role, von Neumann wrote a paper on the design of general-purpose computing machinery, the *First Draft Report on the EDVAC*, which was widely circulated and became the foundation for the common design of electronic computers.

The ENIAC was large and highly complex, costing over $500,000 at the time, containing over 18,000 vacuum tubes and occupying over 1,500 square feet of space. Known by its classification "Project PX" during the war, the ENIAC was funded by the army to have a more efficient method of calculating ballistics range tables, which until then had been calculated by a combination of teams of human computers and differential analyzers, electromechanical analog calculating machines that were slow and required frequent repair and recalibration. The ENIAC sped up the calculations considerably. Whereas it would take hours for a group of human computers to calculate an artillery trajectory, the ENIAC could solve the same problem in 30 seconds, faster than the actual travel time of the projectile. In its initial design, the ENIAC was programmed by rewiring several panels, and although the calculations were fast, the process of resetting the wiring was time-consuming. In 1948 the machine was retrofitted to allow for a program to be stored in its memory.

After the war, Mauchly and Eckert realized the potential of the new computer and attempted to commercialize their work on the ENIAC. They eventually founded the UNIVAC Computer Corporation, which became a division of Sperry Rand. They received a patent on the ENIAC design in 1964, but that patent was challenged by the Honeywell Corporation and was overturned in 1973 on the grounds that the key elements of the machine had been disclosed prior to the patent filing. Nonetheless, the ENIAC was an early test bed for many computational ideas and it was the centerpiece of a summer class, the 1946 Moore School Lectures, which introduced the concepts of electronic computation to the electrical engineering community.

[*See also* **Automation and Computerization; Computer Science; Computers, Mainframe, Mini, and Micro; Engineering; Military, Science and Technology and the; Software;** *and* **von Neumann, John.**]

BIBLIOGRAPHY

Goldstine, H. H., and Goldstine, A. "The Electronic Numerical Integrator and Computer (ENIAC)." *IEEE Annals of the History of Computing* 18 (Spring 1996): 10–16.

Marcus, Mitchell, and Akera, Atsushi. "Exploring the Architecture of an Early Machine: The Historical Relevance of the ENIAC Machine Architecture." *IEEE Annals of the History of Computing* 18 (1996): 17–24.

McCartney, Scott. *ENIAC: The Triumphs and Tragedies of the World's First Computer.* New York: Walker and Co., 1999.

Neukom, H. "The Second Life of ENIAC." *IEEE Annals of the History of Computing* 28, no. 2 (April–June 2006): 4–16.

Neumann, J. von. "First Draft of a Report on the EDVAC." *IEEE Annals of the History of Computing* 15, no. 4 (1993): 27–75.

Stern, Nancy. *From ENIAC to UNIVAC: An Appraisal of the Eckert–Mauchly Computers*. Bedford, Mass.: Digital Press, 1981.

David Alan Grier and Eric Rouge

ENTOMOLOGY

American entomology grew out of two traditions, the interplay of which—complementary and adversarial—drove the discipline's development. First, a cadre of nineteenth-century gentlemanly amateurs adapted collecting and naming conventions from European natural history and applied these to American insects. After the publication of Darwin's *Origin of Species* (1859), they also borrowed and developed evolutionary theories. But American entomologists refused to be handmaidens to their Old World peers: they were motivated by a fierce patriotism to establish the national identity of their science, as made clear by the title of Thomas Say's *American Entomology* (1824–1828).

Harvard was an early center of entomological work, a position that was strengthened by the founding of the Museum of Comparative Zoology. In the 1850s, the Smithsonian took over from amateur organizations and state surveys the coordination of expeditions, the maintenance of a national collection, and the publication of entomological research. At the end of the century, John Henry Comstock established a research program in evolutionary entomology at Cornell. Professional recognition of this form of entomology came in 1907, with the founding of the Entomological Society of America.

The second entomological tradition—and the one that was peculiarly American—developed from a body of technical knowledge. Since colonial times, European immigrants to North America had developed ways of working with and controlling insects—making possible sericulture, apiculture, and the protection of food crops. Economic entomology, as this mode of practice would come to be known, became formalized in the decades after the Civil War because of changes in agricultural practices and the creation of new institutional spaces. The planting of huge monocultures in the Midwest encouraged the irruption of insect pests on scales never before seen. Land grant universities and agricultural experiment stations employed entomologists to control these outbreaks. The federal government also became a major patron of economic entomology, especially after Leland O. Howard (1857–1950) assumed control of the Bureau of Entomology in 1894. Professional recognition for economic entomology came in 1889, with the founding of the American Association of Agricultural (later Economic) Entomology.

Through most of the nineteenth century, these two traditions complemented one another. Entomological naturalists received funding, facilities, and respect because the objects of their study were so obviously of public concern. In turn, their research provided techniques for economic entomologists. Economic entomologists sought to restore nature to a balance that humans had disrupted by controlling insects with natural predators, advocating changes in agricultural practices, and the selective application of arsenicals. This technical work required an intimate acquaintance with insect systematics and bionomics. The intertwining of practices meant that entomologists did both basic and applied research, and memberships in the two professional organizations overlapped significantly.

Around 1900, the traditions began to diverge, becoming adversarial by the second third of the twentieth century, in part because of the failure of biological and cultural control programs in high-profile cases and insect systematics becoming too baroque for practical utility, but mostly because of the introduction of chlorinated hydrocarbon insecticides in the years after World War II. So potent were the chemicals, and so seemingly safe, that some economic entomologists saw their discipline as no longer allied to biology but a part of chemistry, the main problem to solve being the most efficient way to apply the new insecticides. Especially on the federal level, there was a desire to completely eradicate pest species. Many entomological naturalists protested these changes, even as the turn to insecticides shaped their own practices. Theodosius Dobzhansky (1900–1975) based his theories of population genetics, in part, on evidence that insects evolved resistance to chemicals. Meanwhile, Edward F. Knipling (1909–2000) sought ways to eradicate insects by manipulating their chromosomes. Work on insect physiology—especially on insect hormones—took off in America during the 1950s as an alternative method of controlling insects. The polarization, however, could not stop the merging of the two professional organizations in 1953, although the unification was fractious.

The final decades of the twentieth century saw a return to the traditions complementing each other. During the 1970s, chlorinated hydrocarbons were outlawed for many uses because of the risk they posed to human health and the environment, thus forcing economic entomologists to return to the balancing of chemical, biological, and cultural control—a practice known as integrated pest management. As had been the case at the end of the nineteenth century, this approach required a thorough understanding of insect natural history, and there was a renaissance in insect systematics and ecological studies. Still, pest control defined entomology—and let it avoid the fate of other organism-based disciplines, which were cannibalized by functional disciplines in the second half of the twentieth century. Entomology remained a coherent field, nourished by the creative tension of its two traditions.

[See also Agricultural Education and Extension; Agricultural Experiment Stations; Agricultural Technology; Agriculture, U.S. Department of; Biological Sciences; Chemistry; Disease; Dobzhansky, Theodosius; Environmentalism; Evolution, Theory of; Food and Diet; Food Processing; Morrill Land Grant Act; Pesticides; Physiology; Say, Thomas; Smithsonian Institution; and Zoology.]

BIBLIOGRAPHY

Palladino, Paolo. *Entomology, Ecology, and Agriculture: The Making of Science Careers in North America, 1885–1985*. Amsterdam: Harwood Academic Publishers, 1996.

Perkins, John H. *Insects, Experts, and the Insecticide Crisis: The Quest for New Pest Management Strategies*. New York: Plenum Press, 1982.

Sorenson, Conner. *Brethren of the Net: American Entomology, 1840–1880*. Tuscaloosa: University of Alabama Press, 1995.

Joshua Buhs

ENVIRONMENTALISM

Environmentalism encompasses a range of perspectives connecting concern for the natural world with a social movement that emerged in the last third of the twentieth century. The origins of environmentalism are seen in many strands of human experience, including religion, philosophy, and mythology, as well as economic and political theory. Wherever people faced the realities of their dependence on nature alongside an often paradoxical degradation of it, they confronted the foundation of what is now considered environmentalism. The distinguishing feature of environmentalism in the 50 years up to 2012 has been the emerging role of modern science in defining problems. In the early twenty-first century, concerns were based on precise comparisons over multiple ecosystems, and scientific models can predict the impact of change into the future. The ways in which humans create that change can be established in detail. This role for science and its effect on economics, politics, and society has made environmentalism a high priority in shaping the modern world. The history of environmentalism, particularly in the United States, reveals the significance of science in establishing current priorities.

Early-Twentieth-Century Environmental Impulses. The rapid growth of industrial society during the 1800s in much of Europe and across the expanding United States put new pressure on the way humans extracted and transported resources to developing markets. This period of transformation, the Industrial Revolution, afforded large populations access to new products. At the same time, people could experience comfort and leisure in new ways. Industrial society did not free the working class from labor, nor did it relieve laborers from long hours, tedious tasks, and dangerous conditions in fields and factories. In the democratized nations of the world, however, industry itself became an engine for rapid redistribution of resources and centralizing the production of goods. With these changes and the accompanying technology came unprecedented extraction of minerals from the earth, food crops and livestock from rural areas, and timber from forests. The corresponding concentration of population in cities meant that former limits to growth were reset.

The response to these changing limits took various forms. Among those who observed change from a transcendentalist perspective, the concern became more about the loss of a previous version of the natural world. In that view, often labeled preservationism, nature had informed human experience and values. To be human was to commune with the world in ways that might disappear in a rapidly industrializing system of destructive practices. Preservation of nature emerged as a distinctive movement in the early twentieth century. It carried with it a sense that to preserve nature was an important element of humanity. At the same time, other observers recognized that continued exploitation of the natural world could lead to a depletion of resources, with equally dangerous implications for human well-being. That practical view, labeled conservationism, also saw nature as threatened by industrialization and sought to reconcile the recent exploitative past with an industrial future. The two views, preservationism and conservationism, were not articulated separately, although their strongest advocates have long held distinct roles in the history of environmental ideals. For example, John Muir championed the preservation of spectacular wilderness places, best represented by Yosemite Valley in California, whereas President Theodore Roosevelt's chief forester, Gifford Pinchot, managed vast swaths of the American West for the purpose of eventual timber harvests at a sustained rate. These perspectives sometimes clashed, as when Hetch Hetchy Valley, near Yosemite, was slated as the site of a dam that would destroy the scenic site but provide electric power to growing California cities. The clash,

however, belies the underlying union of idealists seeking accommodations in unrelenting industrial growth that would allow for more varied kinds of experiences in nature for future generations.

Balancing Environmental Impulses in an Embattled World.

As Europe and eventually the United States plunged into war, priorities shifted away from protecting wild places. Interest in managing resources for sustained use and providing for ever-increasing industrial markets, however, expanded vigorously. New opportunities for modern production spurred investment and helped accelerate modes of transport, especially across the North American continent. Some aspects of this economic growth proved frail, especially as years of drought devastated the Great Plains. Crop failure, erosion, and the ensuing "Dust Bowl" became a legacy of the period and carried a stern warning for land managers, agricultural scientists, and policy planners ever after. The reversal of growth and collapse of markets plunged the United States into economic depression. A major component of the federal government's New Deal envisioned by President Franklin D. Roosevelt brought conservation once again to the forefront. Across the country, jobs programs put people to work on projects that blended the priorities of preservation and conservation. Large parcels of land under government control, originally set aside with vague ideas about stewardship of resources and wildlife preservation, became testing grounds for new conservation management techniques. With expanded government oversight and the addition of bureaus that could command the labor of thousands of workers, remote parcels of land became interconnected forests, parks, byways, riverways, and preservation areas. Rustic lodges and interpretive centers claimed for the public what had in some locations become commercialized natural attractions surrounded by hotels and amusements. The American park ideal—established in Yellowstone and Yosemite during decades of relative prosperity—became a public institution through the efforts of the Civilian Conservation Corps in the 1930s. Infrastructure for recreation and wildlife reserves meant increased access for tourism, which eventually heightened concern for wildlife and natural areas even further.

In less visible but equally lasting ways, scientists began to play an active role in setting conservation priorities. The Ecological Society of America, founded in 1915 by botanists, zoologists, foresters, meteorologists, and other scientists, came to represent the potential for conservation of contributions from diverse fields. These scientists aimed to build the new scientific discipline they had begun to call ecology. Ecology had obvious implications for wildlife preservation and resource conservation, but most practitioners of the new science had academic positions and theoretical priorities. As they worked through issues related to predator control, erosion, wetland drainage, and other conservation efforts, however, their science took on decidedly practical elements. Particularly as a result of the needs of nations at war in the late 1930s and into the 1940s, ecologists saw their work applied to conservation for resource management.

After World War II, social and industrial processes that had supplied armies across two oceans were reshaped to convert raw materials into commercial goods at home. Ecologists benefited from this reshaping in at least two ways. First, they became consumers of goods and infrastructure that enabled them to travel, take measurements, record data, and publish findings more efficiently and broadly than previously imagined. Second, they found a new range of study subjects as global ecosystems became accessible and underwent startling changes before their eyes. Not only could they travel rapidly to remote areas with newly developed equipment, but also they arrived at those places to find unexpected impacts on natural systems underway.

Studies of pollution and human impacts once relied on dramatic and usually catastrophic changes that had to be observed only after significant damage was done. Ecological studies, beginning in the 1950s, were more subtle. The reach of human impacts was the subject of a symposium in 1956 entitled "Man's Role in Changing the Face of the Earth." There, over 50 ecologists, anthropologists, economists, historians, and other scholars presented evidence of systemic changes to nature

shaped by modern life. The event itself and the publication produced from it provided a deeper intellectual foundation for an emerging environmental critique of human activities. That critique and the role of ecology became stunningly clear with the publication of *Silent Spring* in 1962. Rachel Carson, a well-known nature writer and seasoned government scientist, wrote the book in response to her studies of the effects pesticides had on songbirds. The implication went far beyond birds, however, and Carson made the point with haunting clarity. The book signaled the need for more concerted attention to the problems wrought by humanity and revealed by science.

The Environmental Movement. Concerns over pollution and its effects on human health crystallized, especially after Rachel Carson's tour-de-force. The environmental conditions created by human activity were recognized as primarily negative. These conditions ranged from toxins and radioactivity in air and water to destruction of wildlife habitat and declines in distinctive plant and animal populations. The response to those conditions was based initially in the scientific awareness of changes—subtle at times, dramatic at others. Later, the environmental movement followed the trend of social organization and demonstration established by the civil rights movement of the 1960s and paralleling the equal rights for women and antiwar movements of the 1970s. These factors shaped environmentalism. Elements of preservation and conservation from an earlier period became historical footnotes, even as the foundational ideals they represented persist.

Ecologists had learned to trace radioactive particles through ecosystems. At first, they wanted to know where radiation from atomic weapons had spread. Ongoing atomic testing and the wide distribution of radiation became an environmental concern in its own right. As ecologists began to examine a wider range of radioactive trace elements, those that exist at low levels in all ecosystems, they realized that they could track the flow of energy and cycling of nutrients with great precision. Along with nutrients, ecologists could follow other molecules, toxins, and pollutants of all sorts. Finding the distribution of pollution in an ecosystem, and identifying the source of pollutants, opened a new chapter in government involvement in environmental regulation.

Shortly after *Silent Spring*, President John F. Kennedy refocused the attention of his scientific commission to look into issues of particular environmental importance. The President's Science Advisory Committee had previously addressed concerns surrounding national security, but shifted its emphasis to the effects of pesticides that Carson had identified. Only recently had efforts to regulate water and air quality been enacted in the Clean Water Act (1960) and the Clean Air Act (1963). A ban on the most pernicious pesticide, DDT, and the 1963 Limited Nuclear-Test-Ban Treaty added to a new wave of regulation based largely in scientific awareness of environmental risks. Beyond legislation, scientists and lawyers sought additional legal protection with the creation of the Environmental Defense Fund in 1967.

These growing governmental efforts reflected a surging interest in the environment. With a population of over 200 million, the United States had become highly urbanized, and yet across the country, Americans saw clear reminders of the majesty and legacy of a nation rich in resources and natural wonders. They visited wilderness areas, national parks and forests, wildlife refuges, and scenic coastlines. In 1964, a strong coalition of preservationists, scientists, and outdoor recreationists had succeeded in lobbying for a Wilderness Act that set a schedule for protection of millions of acres. Even with pollution and urbanization on the rise, such grassroots successes made environmental decline seem anything but inevitable. The social movement gained momentum, and each victory for nature became a corrective to the irresistible forces of urban and suburban expansion, economic development, industrial growth with its corresponding pollutants and toxic by-products, and growing consumerism with its own corresponding waste. Sometimes seen as a starting point for the environmental movement, the first Earth Day celebration in 1970 was really a culmination of at least eight years of mounting expectations. Tension over how environmental ideals would survive the relentless process of modernization simmered beneath Earth Day festivities.

Environmentalists had much to celebrate in 1970. The year before, the federal government had passed a sweeping National Environmental Policy Act and created a bureau to oversee environmental regulation. The Environmental Protection Agency (EPA) would review environmental impact statements for projects with federal funding that posed a potential risk to natural systems. Although this expanded regulatory oversight promised care of the environment, it also pointed to the greatest fear of many activists. Bureaucratization of environmental protection meant red tape, increased power to special interests, and delocalization of protection in the eyes of many skeptics.

Retrenching Environmentalism. With new environmental laws on the books and new regulatory agencies to enforce them, the environmental movement as a social effort took shape. In making a more visible statement, environmentalists also sought the roots of their movement. They turned to nature writers of the past century, reaching back to Henry David Thoreau and his reflections on a year of living simply near a Massachusetts pond, *Walden*. John Muir's descriptions of the Sierra Mountains and Yosemite served as an inspiration. The written work of the first professional game manager, Aldo Leopold, enjoyed emerging popularity in this period. Originally, his essays were not widely read outside of forestry and game management journals in the 1930s. Many of his later essays—written in a shack on an abandoned farm along the Wisconsin River north of his home at the University of Wisconsin—were collected in 1949, the year after his death. More than 20 years later still, Leopold's *A Sand County Almanac* began speaking to environmentalists in large numbers. He had written on wildlife, wilderness, and ethics, with his Midwestern upbringing and years of surveying for the U.S. Forest Service in New Mexico providing him with a hunter's and a resource manager's appreciation of nature. His experience as a wildlife scientist in states where game had become scarce in some areas and overabundant in others, combined with an awareness of emerging ecological principles of animal populations, led him to present a distinctive perspective on the shared responsibilities of individuals, communities, and government for conservation. Those essays became touchstones of environmentalism beginning in the 1970s. Like Rachel Carson, Leopold enabled readers to comprehend the scientific basis of environmental change and degradation, even as the words called readers to action on an ethical basis.

Environmentalists also found inspiration in the work of Paul and Anne Ehrlich, biologists who pointed to the damage done by a rapidly expanding human population in *The Population Bomb* (1968) and several other books on biology and society. Barry Commoner emphasized environmental damage at the ecosystem level, including a clear description of ecological principles for the readers of *The Closing Circle* (1971). Ecologists could point increasingly to air and water pollution, despite improving regulatory efforts. When toxins were found buried beneath a neighborhood known as Love Canal—part of Niagara Falls, New York—in 1978, lawsuits and sensational media coverage brought the issue of environmental health to new audiences around the world. Nuclear meltdown and the release of radiation at the Three Mile Island power plant in Pennsylvania in 1979 reawakened lingering Cold War fears of nuclear attack. In this case, however, the enemy was the citizens' own desire for electricity rather than the Soviet Union posing a threat. The Soviets themselves suffered an even more destructive meltdown in 1986 in Chernobyl, Ukraine. Additional environmental damage appeared in backyards somewhere every day. Only the largest toxic dumps or most harmful chemical releases attracted national or global attention. Many of the major sites were listed for costly cleanup in the 1980 Comprehensive Environmental Response, Compensation, and Liability Act in the United States. Also known as the Superfund Act, the EPA and other regulatory agencies used the authority in this new law to get the largest polluters with the deepest pockets to fund the most dangerous sites first. Priority listing was far from comprehensive and lawsuits proved inefficient in securing funding. The innumerable smaller dumps and discharge points received local attention, but rarely attracted funding to halt the leakage of toxins into soil, groundwater, rivers, lakes, and the air overhead.

In their turn, wildlife issues took center stage for environmentalists. New Zealand neuropsychologist Paul Spong began research in the mid-1960s at the Vancouver Public Aquarium in British Columba, Canada. His initial observations and experiments led to an awareness that the captive orca whale he was studying had more complex behavioral responses than anticipated. He soon developed a fascination with the animal that grew into a kind of friendship. He went on to call for the orca's release, which cost him his job. He then convinced a Canadian activist group to take up the cause to "save the whales." That group, primarily focused on protesting nuclear arms, shifted its mission to whales and other environmental concerns, taking on the name Greenpeace in 1975. The World Wildlife Fund (now known as the World Wide Fund for Nature in most countries, other than the United States and Canada) was established in the early 1960s, taking the giant panda as its logo. The scope of that organization's efforts focused on protection of habitat for particular endangered species. Those efforts also made the group symbolic of a broader environmental priority, especially in affluent countries where funding could be set aside to buy up habitat in developing countries.

The different levels of environmental damage, and the risks involved, led to a complex and rarely coordinated network of social awareness raising, local political response, legal action, state and federal regulation, media outcry, and nongovernmental organization. These diverse activities evoked the full spectrum of human aspirations for effecting change, from rational, scientific evidence to personal, emotional testimony. The effectiveness of such activities differed on local stages compared with regional and global situations.

Environmentalism Globalized. Even as environmental concerns took a local turn in neighborhoods and communities, the global causes and consequences of human impacts on the environment reshaped the movement on the international level. Government regulations in one country rarely matched those of neighboring nations, and the spread of toxins and pollutants knew no borders. The issues that best exemplified global concerns involved the atmosphere: acid rain, stratospheric ozone depletion, and, of course, global warming. The release of pollution into the atmosphere had long spurred concern for air quality, but those concerns, like most environmental issues, remained local or regional and tended to be dealt with inside the jurisdiction of a particular government. When pollution and changes in atmospheric chemistry could be tracked farther, as had emerged with studies and regulation of nuclear fallout, concern extended to keep up with the growing clouds of disturbance.

Acid rain became a familiar local concern as early as the nineteenth century, when industrial cities like Manchester, England, produced particles that could acidify moisture in the atmosphere in high enough concentrations to cause corrosion of metals when that rain fell nearby. By the 1960s in eastern Canada, acid rain that developed out of industrial pollution produced in the manufacturing cities of the United States was detected. Eville Gorham, the Canadian scientist responsible for this discovery, published his findings in numerous scientific journals, but they failed to attract much attention at first. Significant changes in surface-water quality eventually led to more public announcements of the phenomenon, and regulatory action in the United States began in earnest when scientists showed that lakes in the Adirondack Mountains of New York were affected.

Regulation to limit pollutants that led to acid rain proved costly to industry. Even as the damage became more clearly understood, political and philosophical opposition to regulation during the presidency of Ronald Reagan made the United States a reluctant partner in international negotiations to protect global air quality. The scientific evidence, however, became unassailable, and George H. W. Bush signed new legislation to protect air quality in 1991. At the same time, the United States finalized an agreement with Canada to limit emissions that contributed to acid rain, at last protecting Canadian lakes from U.S. pollution, although the damage had already been done.

A similar series of scientific discoveries, regulatory obstacles, industry innovations, and multinational agreements led to the ban of chlorofluorocarbons (CFCs) in products that had caused depletion of ozone in the upper atmosphere. This ozone depletion, although still cause for concern because of the increased exposure to harmful solar ultraviolet radiation, has reversed notably in the decades since the 1987 ban on CFCs. Organisms in the far southern latitudes, including human populations, continue to see increased disease rates related to ultraviolet exposure.

By far the most expansive environmental topic of concern over the past 25 years is known variously as climate change and global warming. Scientific evidence dating back over a century consistently and increasingly points to the role of humans in accelerating the accumulation of carbon dioxide in the earth's atmosphere. The subsequent "greenhouse effect" created by elevated carbon dioxide levels leads to higher average temperatures in most parts of the planet. The most pronounced effect of climate change, because local conditions vary around their averages, has been increased severe weather events in most areas and consistent warming at the poles. The northern polar region, a frozen ocean for much of the planet's history, has seen increasing areas of open water in recent decades. Virtually all climate models converge around predictions that the North Pole will be open ocean in summer by the turn of the twenty-second century. The scientific consensus emerging around evidence and models is unquestionable. Less clear is what to do about this global phenomenon and how to respond to claims that it represents a human-induced catastrophe for the environment that supports life as we know it.

Early Twenty-First Century. As much as science has shaped environmentalism in the late twentieth and early twenty-first centuries, expectations for the resolution of environmental problems depend increasingly on the coordination of economic, political, and social action. Longstanding ideals of stewardship and management remain the basis for much decision making, informed by scientific evidence and technological options. Changing human behavior and organizational systems, although complex, remain at the heart of environmental ideals.

[*See also* Agricultural Education and Extension; Agricultural Experiment Stations; Agricultural Technology; Agriculture, U.S. Department of; Anthropology; Audubon, John James; Biological Sciences; Botany; Canals and Waterways; Carson, Rachel; Conservation Movement; *Deepwater Horizon* Explosion and Oil Spill; Dust Bowl; Ecology; Environmental Protection Agency; *Exxon Valdez* Oil Spill; Fish and Wildlife Service, U.S.; Fisheries and Fishing; Forestry Technology and Lumbering; Forest Service, U.S.; Global Warming; Hydroelectric Power; Leopold, Aldo; Meteorology and Climatology; Muir, John; National Park System; Nuclear Power; President's Science Advisory Committee; Religion and Science; Sierra Club; Three Mile Island Accident; Wind Power; *and* Zoology.]

BIBLIOGRAPHY

Alm, Leslie R. *Crossing Borders, Crossing Boundaries: The Role of Scientists in the U.S. and Canada.* Westport, Conn.: Praeger, 2000. A detailed analysis of the political debates surrounding acid rain in the United States and Canada.

Armstrong, Philip. "Cetaceans and Sentiment." In *Considering Animals: Contemporary Studies in Human–Animal Relations.* Edited by Carol Freeman, Elizabeth Leane, and Yvette Watt, pp. 169–182. Burlington, Vt.: Ashgate, 2011. A compelling illustration of new scholarship in animal studies that informs the place of animals in human history, including a clear connection to how animal protection becomes part of environmentalism.

Bocking, Stephen. *Nature's Experts: Science, Politics, and the Environment.* New Brunswick, N.J.: Rutgers University Press, 2004. A careful examination of the relationship between science and policy making around ecological concerns in North America.

Hagen, Joel B. *An Entangled Bank: The Origins of Ecosystem Ecology.* New Brunswick, N.J.: Rutgers University Press, 1992. A history of science focusing on the rapid development of ecology in the middle of the twentieth century.

Hays, Samuel P. *Beauty, Health, and Permanence: Environmental Politics in the United States, 1955–1985.* New York: Cambridge University Press, 1987. A comprehensive overview of environmental policies in the post–World War II era.

Klingle, Matthew. *Emerald City: An Environmental History of Seattle.* New Haven: Yale University Press, 2009. A recent exemplar of environmental history that focuses on the plight of urban spaces and the response of city dwellers to dramatic changes in their connection to nature.

McNeill, J. R. *Something New under the Sun: An Environmental History of the Twentieth-Century World.* New York: W. W. Norton, 2000. A readable historical account of twentieth-century environmental history, containing additional social commentary.

Rome, Adam. *The Bulldozer in the Countryside: Suburban Sprawl and the Rise of American Environmentalism.* New York: Cambridge University Press, 2001. A scholarly and engaging introduction to the ways environmental historians have begun to look at suburban development and changing attitudes toward residential planning.

Soulé, Michael E., and Gary Lease, eds. *Reinventing Nature? Response to Postmodern Deconstructionism.* Washington, D.C.: Island Press, 1995. Scientists at the forefront of conservation biology consider the ways environmentalism has turned on definitions of nature.

Wang, Zuoyue. *In Sputnik's Shadow: The President's Science Advisory Committee and Cold War America.* New Brunswick, N.J.: Rutgers University Press, 2009. A detailed and insightful account of science advice at the highest levels during a period of rapid transition in U.S. science and technology policy.

Wilson, Edward O. *The Creation: An Appeal to Save Life on Earth.* New York: W. W. Norton, 2007. A recent volume from a leading environmental scientist who has for decades led scientists in advocating for a more coherent view of nature and a more human response to environmental concerns.

Worster, Donald. *Nature's Economy: A History of Ecological Ideas,* 2d ed. New York: Cambridge University Press, 1994. A classic discussion of this history of ecological ideas that blends environmental concerns with social sensibility about nature in diverse settings.

Young, Christian C. *The Environment and Science: Social Impact and Interaction.* Santa Barbara, Calif.: ABC-CLIO, 2005. A comprehensive examination of scientific developments set in the context of environmental, social, and political change.

Christian C. Young

ENVIRONMENTAL PROTECTION AGENCY

Established by the Richard M. Nixon administration in October 1970, in the wake of the first Earth Day that April, the Environmental Protection Agency (EPA) became the agent and symbol of a vast expansion of the federal government's role in addressing a widely perceived environmental crisis. Initially proposed by a Nixon advisory committee seeking ways to streamline the federal bureaucracy, the EPA brought together under one institutional umbrella several longstanding federal programs aimed mostly at pollution control. Through existing and new legislation that it implemented and defended in court, including Clean Water and Clean Air acts, the Toxic Substances Control Act, and laws regulating pesticides and hazardous wastes, the EPA spearheaded the most far-reaching federal intervention into the American economy since the New Deal.

During its first three decades, the EPA grew steadily in its size, responsibilities, and regulatory strategies, even as its fortunes fluctuated with the political climate. Although its founding laws and early administrators envisioned an "ecological" protection encompassing nonhuman as well as human life, the agency found its scientific footing by stressing environmental threats to public health. Initially quite decentralized, the EPA under President Jimmy Carter centralized and coordinated its programs, largely through quantitative risk assessments (starting with carcinogenic risks) and cost–benefit analyses. Early appointees in the Ronald Reagan administration assaulted the EPA's programs and damaged morale, but in the wake of the Love Canal exposé—a scandal involving long-term toxic pollution by the Hooker Chemical Company in Niagara Falls, New York—Congress mandated a new and ambitious EPA initiative to control industrial wastes.

Friendlier Republican administrations of the later 1980s and early 1990s restored money and muscle to the EPA's older programs while expanding its regulatory tools. Experiments with dispute mediation and market-based incentives carried over into the Bill Clinton administrations. During the 1990s, the EPA tightened pollution standards and began monitoring and regulating the effects of pollution on nonhuman species. In response to the environmental-justice movement, the agency also moved to shed its white, middle-class image by attending more closely to the disproportionate share of environmental risk borne by minorities and the poor.

Subsequent presidential administrations, in the way they treated the EPA and its work, reflected the stark divide consolidating between Republicans and Democrats over the role and worth of federal environmental policy. The George W. Bush administrations deprioritized the agency's work, often to the chagrin of appointed administrators. Although the administration of Barack Obama early on proved much friendlier to new regulatory initiatives from the EPA, Republican opposition to global warming and other environmental concerns combined with the "Great Recession" to weaken the agency's hand.

[*See also* Cancer; Carson, Rachel; Environmentalism; Global Warming; Pesticides; *and* Public Health.]

BIBLIOGRAPHY

Andrews, Richard. *Managing the Environment, Managing Ourselves; A History of American Environmental Policy.* New Haven, Conn.: Yale University Press, 1999.
Russell, Edmund. "Lost among the Parts per Billion: Ecological Protection at the United States Environmental Protection Agency, 1970–1993." *Environmental History* 2 (1997): 29–51.
U.S. Environmental Protection Agency. *The Guardian: Origins of the Environmental Protection Agency.* Washington, D.C.: U.S. Environmental Protection Agency, 1992.

Christopher Sellers

EPIDEMIOLOGY AND POPULATION HEALTH

See Public Health; Medicine.

EPILEPSY

The modern understanding of epilepsy as a brain disorder dates back to one of the many anonymous physicians whose writings are collectively attributed to the ancient Greek physician Hippocrates (c. 460 BCE–c. 370 BCE). Hippocratic physicians argued that the "falling sickness," also known as the "sacred disease," was the result of a somatic process originating in the brain, not of a visitation from the gods. Building on Hippocratic writings, the Greek physician Galen (c. 131 CE–c. 200 CE) and his followers attributed epilepsy to a thick humor accumulating in the cerebral ventricles. They advised those with epilepsy to avoid cold and wet foods and to seek out hot, dry climates, if possible. Yet, despite such medical writings, many continued to subscribe to the idea that epilepsy was the result of a demonic possession that literally seized the sufferer.

Common treatments for seizure disorders, from Galen to the early twentieth century, included the pharmacological, especially the use of drugs derived from plants; the dietary, including abstinence from meat; and the magical, such as the wearing of amulets and spells. In some cases, especially those involving traumatic epilepsy, surgeons turned to trephining, the surgical opening of the skull, often through the drilling of a hole. Initially, this was done to release vapors and humors but, by the eighteenth century, more often surgeons turned to trephining to remove a localized pathological condition affecting the brain and its membranes.

Changing Understandings. Medical discussions of epilepsy began to change radically in the nineteenth century, as neuroanatomists and neurophysiologists advanced scientific understandings of how the brain worked. Particularly influential were two British neurologists, John Hughlings Jackson (1835–1911) and William Gowers (1845–1915). In 1881, Gowers published *Epilepsy and Other Chronic Disorders*, an important synthetic work that linked seizures to the physiology of the human nervous system; he also offered an influential evolutionary theory of the anatomy

of the brain and spinal cord and attributed certain focal epilepsies to localized irritations of the cerebral cortex. Such work enabled epilepsy to be distinguished from other convulsive disorders, such as eclampsia, toxemia, and syncope. Jean-Martin Charcot (1825–1893), a pioneering French neurologist and professor of anatomical pathology, further differentiated epilepsy from hysteria. These ideas rapidly traveled across the Atlantic as a result of Americans studying in Paris with Charcot and in London with Hughlings Jackson and Gowers.

More traditional negative views of epilepsy resurfaced in the form of medical arguments that attributed epilepsy, particularly when accompanied by psychic distress, to a degenerate heredity and linked epilepsy to mental illness and mental deficiency. Policy makers, along with medical professionals, used the idea that idiopathic epilepsy easily passed from one generation to the next to justify the segregation of those with seizure disorders in specialized institutions, often called colonies, across Western Europe and the United States. Doctors used what they referred to as the "stigmata of degeneration," which included misshapen skulls and overly large ears, along with "fits," to identify those who should be isolated and, in some cases, sterilized.

In the twentieth century, neurological medicine moved out of large state-funded institutions into university hospitals, although researchers continued to rely on the institutionalized for their experimental work, especially with drug therapies. Although bromides had been used to control seizures since the late nineteenth century, new drugs with fewer side effects, such as phenobarbital (introduced in 1912), began to appear. The introduction of phenytoin in 1937 facilitated the control of petit mal seizures (short disturbances of brain function often involving staring episodes) for the first time. In addition, the introduction of electroencephalography in the mid-1930s enabled the recording of the electrical activity of the brain from the surface of the scalp and thus facilitated the diagnosis of epilepsy. In the 1940s, specialized institutions for epileptics began to shrink—not because new drugs allowed the quicker release of patients but because they enabled patients to avoid institutional care altogether. For the most part, only patients with multiple developmental problems, including seizure disorders, remained.

Twentieth Century and Beyond. The debate over how best to classify and talk about seizure disorders, which dates back to Hippocrates, continued during the twentieth century. In the 1930s, Harvard neurologist William G. Lennox led an unsuccessful campaign to refer to epilepsy as a cerebral dysrhythmia. In the early twenty-first century, the International League against Epilepsy (ILAE) referred to seizure disorders as epilepsies, not epilepsy, following a suggestion first made by Gowers more than a century ago. This suggestion, however, had not yet been widely accepted. In 2010, the ILAE updated its classification system to reflect modern neuroimaging, genomic technologies, and new concepts in molecular biology. However, the organization admitted that the new system continued to have substantial limitations. Among the many changes was the replacement of "convulsion," described as a lay term, with "seizure" and the revival of a late-nineteenth-century term "focal" (in place of "partial") for seizures produced by specific cerebral dysfunctions. Despite steady advances in scientific understandings of the etiology of seizures, many still were considered "idiopathic," with their origin unknown. Further, despite the steady development of new anticonvulsants, some seizures remained intractable. In the early twenty-first century, the most promising therapies for intractable seizures involved major surgical procedures, including temporal resections and hemispherectomies.

Therapeutic advances have diminished but not eliminated the social stigma attached to epilepsy. Since the early twentieth century, patient and family advocacy groups, such as the Epilepsy Foundation of America, have launched educational campaigns and lobbied for better research funding. Most discriminatory laws, including those that forbade the marriage of epileptics, have been revoked. Despite such improvements, many with epilepsy in the twenty-first century continued to hide their diagnosis; others continued to have trouble finding employment.

[*See also* **Anatomy and Human Dissection; Disease; Hospitals; Medicine; Mental Health Institutions; Mental Illness; Molecular Biology; Pharmacology and Drug Therapy;** *and* **Physiology.**]

BIBLIOGRAPHY

Dwyer, Ellen. "Stories of Epilepsy, 1880–1930." In *Framing Disease: Studies in Cultural History*, edited by Charles Rosenberg and Janet Golden, pp. 248–270. New Brunswick, N.J.: Rutgers University Press, 1992. Medical and familial efforts to explain epilepsy, especially in institutionalized populations.

International League against Epilepsy. "Revised Terminology and Concepts for Organization of the Epilepsies: Report of the Commission on Classification and Terminology." http://www.ilae-epilepsy.org/Visitors/Centre/ctf/ctfoverview.cfm (accessed 25 October 2012). Latest and most widely recognized statement of classification and terminological issues relating to the epilepsies.

Scott, D. F. *The History of Epileptic Therapy*. Frome, U.K. and London: Butler and Tanner, 1993. Useful history of drug therapies through the 1990s.

Temkin, Owsei. *The Falling Sickness: A History of Epilepsy from the Greeks to the Beginnings of Modern Neurology*. Baltimore: Johns Hopkins University Press, 1945, 1971. A classic text that remains invaluable.

Ellen Dwyer

ERIE CANAL

Begun in 1817 and completed in 1825, the Erie Canal stretched 363 miles across New York from Albany to Buffalo, linking the Hudson River and Lake Erie through the Mohawk River gap in the Appalachian Mountains. Its length surpassed that of all other existing canals in Europe or the Americas. The state of New York built and financed the project, thanks in part to the promotional efforts of DeWitt Clinton, the former mayor of New York City. Because of his crucial role, it was often referred to as "Clinton's Ditch."

This pioneering engineering achievement was 40 feet wide and four feet deep, with 83 locks and 18 aqueducts. (It was later enlarged to 70 feet wide and 7 feet deep; after 1905 it was rebuilt and renamed the Erie Barge Canal.) The self-taught engineers who designed the original canal, such as Benjamin Wright, Canvass White, Nathan S. Thomas, and John B. Jervis, drew largely from English technology. Engineering programs at nearby universities, including programs at Rensselaer Polytechnic Institute in Troy and at Union College in Schenectady, played an important role in enlarging the canal through the late nineteenth and early twentieth centuries. Union College alone contributed over 30 graduates to the enlargement of the canal.

Celebrated in American folklore, the Erie Canal had a transforming influence. It increased settlement in western New York and the Great Lakes region, carried immigrants and merchandise west, and brought western grain to the Hudson. It created cities such as Buffalo, Rochester, and Syracuse; increased wheat production in western New York; and stimulated lateral canals to the Finger Lakes, Lake Champlain, and Lake Ontario. By 1845 the canal carried over a million tons of goods annually. Spreading ideas as well as goods, the canal promoted such diverse social reforms as religious revivalism, the women's rights movements, and temperance. Into the 1880s the Erie Canal surpassed the competing New York Central Railroad in its freight traffic, but by the late twentieth century it served mainly as a corridor for recreational and commercial development.

[*See also* Canals and Waterways; Engineering; *and* Railroads.]

BIBLIOGRAPHY

Shaw, Ronald E. *Erie Water West: A History of the Erie Canal 1792–1854.* Lexington: University of Kentucky Press, 1966.
Sheriff, Carol. *The Artificial River: The Erie Canal and the Paradox of Progress 1817–1862.* New York: Hill and Wang, 1996.

Union College. *Erie Canal—175th Anniversary.* http://www.eriecanal.org/UnionCollege/175th.html (accessed 26 October 2012).

Ronald E. Shaw;
updated by Hugh Richard Slotten

ETHICS AND MEDICINE

In the ninth edition of his enormously successful guide to the things that concerned the "reputation and success" of the physician, *Book on the Physician Himself*, Baltimore physician D. W. Cathell (1889, p. 52) speculated about why America was the only nation where the medical profession had a written code of ethics. "Possibly old countries from long custom can dispense with them," Cathell explained, "But in this Young Land of Freedom the very nature of society requires that physicians shall have some general system of written ethics to define their duties, and, in case of doubt, regulate their conduct toward each other and the public in their intercourse and in competition" (p. 52). In a medical marketplace crowded with many different schools of medical theory and practice, Cathell exhorted orthodox physicians about their duty to familiarize themselves with the Code of Ethics of the American Medical Association (AMA). This code, adopted by the fledgling organization in 1847, Cathell characterized as a "great oracle" on the "moral aspects of various subjects" and also "the balance-wheel that regulates all professional action, and neither Professor Bigbee nor Dr. Littlefish can openly ignore it without overthrowing that which is vital to his standing among medical men" (p. 55). Cathell's explicit discussion about medical ethics stressed avoiding consultations with "quacks" and irregular physicians such as homeopaths and eclectics and exercising caution when called in to attend patients under the care of their regular colleagues. He was primarily concerned with the professional relationships of practitioners with each other and with their patients. He also advised physicians about activities long considered the purview of moral medical practice, including declining to give directions for avoiding conception, refusing to provide abortions unless necessary to save the life of the mother, preserving the secrets of the sick room and patient confidences, and judicious truth telling to patients with mortal illness.

Like Cathell, this essay considers ethics and medicine in American history, together with issues and controversies over morality in medicine. Although some philosophers use the terms ethics and morals interchangeably, for the purpose of this essay the term ethics refers to a system or set of moral principles. The *Online Oxford English Dictionary* defines ethics as the science of morals and the department of study concerned with the principles of human duty. In other words, morals refers to beliefs about good and evil, right and wrong, that influence how individuals conduct themselves in their personal and societal relationships. Ethics refers to ways in which rules for acting morally are codified and provided with a philosophy or theory that undergirds them.

Before the AMA's Code of Ethics.
Medical ethics was not invented in America. Cathell acknowledged that the AMA's Code of Ethics owed much, indeed everything, to the English physician Thomas Percival (1740–1804). In the late eighteenth century, Percival served as physician adviser to the Manchester Infirmary, then one of the largest hospitals in England. In 1789 in the wake of controversy over the plan to increase the number of physicians to care for patients sick with typhoid and typhus, the hospital trustees asked Percival to draft a set of guidelines in an effort to resolve the dispute. Completed in 1792, the code was revised and originally published for private use in 1794 as *Medical Jurisprudence*. In 1803 Percival published an expansively revised edition under the title *Medical Ethics; or, a Code of Institutes and Precepts Adapted to the Professional Conduct of Physicians and Surgeons*. As philosophers Robert Baker and Laurence McCullough (2009) have noted, Percival owed an enormous debt to the Scottish physician John Gregory (1725–1773), whose 1772 publication of *Lectures on the Duties and Qualifications of a Physician* reconceptualized medical morality. Drawing on the language and logic of such Enlightenment philosophers as

David Hume (1711–1776), Francis Hutcheson (1694–1746), and Adam Smith (1723–1790), Gregory created an ethics based on sympathy, which served to transform medicine from a trade into an art "most beneficial and important to mankind."

First published in the United States in 1803, Percival's Code of Ethics was embraced by an influential number of American physicians as they established the first formal medical organizations in the early Republic. In 1808 a group of physicians, members of the Boston Medical Association, drew extensively on Percival when they created the *Boston Medical Police*, a set of ethical rules for the governance of medical practice. Many of the precepts in this text concerned fees and the need for physicians to set fixed prices for the performance of such medical services as variolation (using smallpox) and vaccination (using cowpox to prevent infection with smallpox): "General rules are adopted by the faculty in every town, relative to the pecuniary acknowledgements of their patients; and it should be deemed a point of honor to adhere to them; and every deviation from, or evasion of these rules, should be considered as meriting the indignation and contempt of the fraternity" (*Boston Medical Police*, 1808, p. 6). Keeping the fees for medical services fixed served to regulate the medical marketplace and to forestall physicians from competition for patients. This set of rules was adopted by at least 13 other local and state medical societies in 11 states between the years 1810 and 1842.

In May 1846 New York physician Nathan Smith Davis, committed to the reform of medical education and to improving the cultural authority of the medical profession, convened a meeting that called for the creation of a national medical association. At the second convention of the nascent association held in Philadelphia in 1847, 250 delegates from 24 states and the District of Columbia adopted a constitution, bylaws, and the name, the American Medical Association. One of the first acts of the association was the creation of the Code of Ethics, modeled on Percival's ethical code and adapted to the local circumstances of the United States.

Why did these American physicians readily embrace Percival? For some American physicians, Percival's writings conformed to their own medical training at the University of Edinburgh. Two students of the Scottish physician John Gregory—Philadelphia physician Benjamin Rush (1745–1813) and New York physician Samuel Bard (1742–1821)—became prominent medical educators. When he returned from Edinburgh in 1769 after finishing his medical degree, Rush began practicing medicine in Philadelphia and was named professor of chemistry at the newly founded College of Philadelphia (1765), the first medical school founded in the British Colonies. In New York, Samuel Bard, who received his medical degree from Edinburgh in 1765, was one of six physicians and surgeons invited by the trustees of Kings College to establish a medical faculty. Although this college closed, Columbia College opened a medical faculty in 1787, whose faculty included Bard and other Edinburgh-trained physicians.

In Philadelphia, Benjamin Rush lectured medical students on various aspects of medical ethics. As the historian Chester Burns has noted, Rush stressed the values of humanity and patriotism. By participating in civic and forensic cases and providing evidence about matters relating to poison, abortion, and evaluations of mental soundness, physicians, Rush argued, contributed to the welfare of the nation. By extending medical knowledge and the provision of medical care to the poor, physicians, the doctor explained, served the cause of humanity. Rush also identified moral duties of patients, which included the selection of an educated physician, compliance with the physician's clinical judgments and prescriptions, and quick payment of the fees for services rendered.

Another reason for the ready embrace of Percival's medical ethics was the desire on the part of some physicians to distinguish themselves from their competitors in a tumultuous medical marketplace. By the 1830s, potential patients had many choices among practitioners. In addition to orthodox physicians, homeopaths and various botanic healers, among others, offered services to Americans. Moreover, the Jacksonian impulse in American politics, the intense repudiation of mo-

nopolies and privilege, had prompted repeals in virtually every American state of traditional medical licensing laws. Despite the regular medical profession's insistence on the need to protect the public from quacks and poorly trained healers, the belief that medical societies functioned as closed guilds rather than as safeguards for public welfare eroded public support for licensing (until the 1870s). The AMA remained throughout the nineteenth century a marginal enterprise with lofty goals, if little influence, on the practice of medicine and healing in the United States.

By the 1880s, the Code of Ethics adopted by the AMA to uphold medical tradition was viewed by some prominent American physicians as an impediment to the advance of medical science. As the historian John Harley Warner (1999) has argued, these physicians rejected the code as both antiquated and unnecessary in light of their growing confidence in the authority of science over medical practice. New ideals of science undergirded by the laboratory and the experimental sciences rendered the code's insistence on the importance of refusing to consult with nonregular practitioners, especially homeopaths and eclectics but also the emerging practitioners of osteopathy, not only obsolete but also a danger to the patient's best interests. Controversy over the consultation clause proved extremely bitter—in 1887 when the International Medical Congress met in Washington, D.C., the majority of the scientific leaders of American medicine boycotted the meeting. Among those absent were members of the Association of American Physicians, newly founded in 1885 by seven prominent physicians for "the advancement of scientific and practical medicine." One year later, the president of the Association of American Physicians, the New York pathologist Francis Delafield, pointedly informed the audience attending the group's inaugural meeting: "We want an organization in which there will be no medical politics, and no medical ethics" (Delafield, 1886, p. 1).

In 1903 the AMA eliminated the consultation clause and proposed a major revision of its code of ethics. Reconfigured as "Principles of Medical Ethics," the new statement included a more explicit and robust professional obligation to medical science, condemned the practices of fee splitting and rebates (or kickbacks) for prescriptions and appliances, and excluded the earlier sections that demanded obedience and loyalty from the laity. At the same time, the AMA in adopting these principles moved from a policy of ethical enforcement to ethical advice. Rather than expelling members for violations of the profession's norms and moral traditions, the new policy encouraged education in professional conduct by medical societies and by more experienced physicians. "If you hear that a young fellow just starting out has made mistakes or is a little 'off color,'" urged the Johns Hopkins medical professor William Osler (1905, p. 368), "go out of your way to say a good word to him, or for him. It is the only cure; any other treatment only aggravates the malady." Optimism about the gentle instruction of erring physicians proved short-lived. In 1910 a bitter dispute between the Chicago physician Frank Lydston and George H. Simmons, secretary of the AMA and editor of its journal, roiled the profession. Lydston publicly assailed Simmons for his arbitrary and unfair abuse of power, charging that Simmons had fraudulently obtained a degree from an Illinois homeopathic college. This dispute prompted the AMA's "New Principles of Medical Ethics" in 1912, which restored the organization's ability to discipline offending physicians. By this time the association, extensively reorganized in 1901 into a more effective representative body covering the entire country and the majority of the medical profession (excluding physicians of color and women doctors), was growing in numbers and influence.

Medical Morality versus Public Morality. One vehicle for professional uplift in the nineteenth century was the reform of medical education. Deemed essential to medical education and to the formation of medical identity, the study of human anatomy and dissection of the dead body required a much greater supply of cadaveric material than was legally available to physicians (chiefly bodies of executed prisoners and, in the antebellum period, of slaves). When American physicians participated—either directly or indirectly—in the clandestine and illicit "resurrection" of the dead bodies of their fellow Americans, public morality was outraged. As one American physician lamented in 1883, "We brand as infamous laws which on the one side subject surgeons to heavy damages for malpractice, and on the other withhold from students of surgery the only means by which they can acquire their knowledge" ("Shall Physicians Learn" 1883, p. 46). This physician insisted on a professional obligation to overcome "foolish sentimentality over the dead body" and to insist on their rights to sufficient anatomical material for education. In the wake of "doctor's riots" that erupted upon the discovery of a disturbed grave site or a loved one's body in the dissecting room, doctors and legislators produced a legal compromise: anatomy acts that made the bodies of the unclaimed dead available to medical schools or state anatomical boards, which then distributed materials to appropriate institutions. In his 1938 autobiography, the physician William Aughinbaugh (1938, p. 45) described how he had financed his medical education by supplying cadavers culled from the See Me No More Cemetery, a local Negro graveyard "on the outskirts of town, in a desolate section carefully avoided by the superstitious blacks." Together with four medical school classmates—co-founders of what the physician facetiously labeled the Hippocratic Exhumation Corporation, Aughinbaugh opened freshly dug graves, broke open the caskets, and removed the corpses, but carefully left all items of dress behind. "Our grave-robbing activities," the doctor conceded, "may seem heartless, but the lack of a dissecting material was a problem that confronted every medical student of my day, and most of them solved it as we did.…I had no qualms of conscience. Neither had any of my associates, for it aided us in acquiring medical knowledge needed in order to fight disease" (p. 45). By the mid-twentieth century, the once adequate anatomical material collected from prisons, almshouses, tubercular hospitals, and other public institutions became scarcer as welfare societies and government agencies, especially the Veterans Administration, increasingly provided some meager burial benefits. By this time, a growing number of Americans voluntarily bequeathed their bodies to medical schools, which had established "willed body programs" aided by changes in anatomical gift statutes.

The dissection of the dead body was not the only conflict between medical morality and public morality. In the second half of the nineteenth century, elective abortion became the subject of controversy as an influential cohort of physicians affiliated with the AMA aggressively campaigned to criminalize abortion. Prompted by fears that abortion rates had dramatically increased, especially among white, married, Protestant women, such physicians as the Harvard professor of obstetrics Horatio R. Storer successfully overcame the reluctance of some fellow physicians, who shared with the public the belief that abortion in the early months of pregnancy was not a crime. Regular physicians had professional reasons for their animus toward abortion, including concern about the dangers to women from criminal abortion and a growing scientific understanding of fetal development. They also had personal reasons; some physicians expressed acute uneasiness about dramatically altered gender norms. Storer, for example, not only opposed abortion, but also rejected liberalized divorce laws and bitterly fought the entry of women into medical schools and medical societies. As the historian James Mohr (1979) has argued, the AMA and the vigorous efforts of regular physicians would prove to be the single most important factor in the profound alteration of American society from one that tolerated the practice of elective abortion to one that officially prohibited the practice in explicit criminal statutes. By 1890 every American state had enacted a law criminalizing abortion.

Throughout the period of criminal abortion (the 1870s to 1973), American physicians consistently distinguished between abortion undertaken to end pregnancy for social reasons and that undertaken for therapeutic indications (some argued that improved maternal outcome was sufficient; others argued only to save the life of the mother). But the lines between these two were not always clear. As the historian Leslie Reagan (1998) has compellingly demonstrated, despite the organized medical profession's opposition to abortion, some American physicians continued to provide abortion services to some of their female patients. In the 1930s, for example, amid the economic privations of the Great Depression,

physicians proved more willing to provide pregnancy termination for married patients for social reasons. In periods of more aggressive enforcement of criminal abortion statutes, some physicians retreated from providing such services except for therapeutic indications. But what were the therapeutic indications for abortion? By the 1930s the medical indications for therapeutic abortion were receiving greater attention and wider discussion than before within the medical profession. As physicians debated the evidence for terminating pregnancy, they also began to develop a shared view that decisions about abortion should not be the province of individual medical decision making but instead should involve at least two other physicians. "The therapeutic abortion committee," observed the Chicago physician H. Close Hesseltine in 1940, "is a very good idea. It would give moral support and medico-legal protection to the institution and to the individuals responsible" (Hesseltine et al., 1942, p. 561). At the many American hospitals where therapeutic abortion committees were implemented, the number of abortions performed notably declined. By the 1960s, some doctors increasingly criticized such committees as not only chaotic but also discriminatory, forcing women into the hands of untrained abortionists. In 1973, the U.S. Supreme Court ruled that such committees were unconstitutional. In *Doe v. Bolton*, a parallel ruling to the better known *Roe v. Wade* decision, the Supreme Court found such committees violated a woman's right to health care and interfered with the physician's right to practice medicine. In the nearly four decades since the Roe and Doe decisions, the medical profession, like the American public, continued to be divided on the issue of abortion. In perhaps no other area of medical practice have legislators attempted to regulate medical encounters more than in reproductive medicine. In the first half of 2011, for example, more than 80 abortion-related restrictions were enacted across the United States, including new rules about waiting periods; required information regarding abortion procedures, fetal development, and alternatives; and prohibitions on medicalized (as opposed to surgical) abortion.

Hippocrates American Style. Medical prohibitions on abortion have an ancient lineage in Western medicine. In the traditional Hippocratic Oath, the most revered, misquoted, and versatile proclamation of Western medical morality, the physician promises, among other things, not to give a deadly drug to enable someone to end his life and similarly swears to refuse to provide a woman with an abortive remedy. In the modern era, the oath was adopted by medical faculties as both a public affirmation of medical ethics and a claim to ancient lineage. In the United States, professional investment in the Hippocratic Oath waxed and waned between the 1760s and the 1950s. When the College of Philadelphia conferred its first MD degrees in 1771, John Morgan, the professor of medicine, informed the graduates that the oath prescribed by Hippocrates, which had been generally adopted in universities and medical schools, would not be used. Although the Philadelphia school did not administer the oath, Morgan laid out the professional obligations of the physician, including commitments to patient welfare, that were the traditional purview of the oath. By the 1850s, an increasing number of American medical schools administered a version of the Hippocratic Oath as a "graduating pledge." As the historian Dale Smith (1996) notes, by the end of the century the use of the oath by medical schools had declined, citing a 1928 survey by Dean Eben Carey of Marquette University School of Medicine (Carey, 1928, p. 159). When he surveyed his fellow deans from 79 American and Canadian medical schools, Carey found that only 14 colleges reported the use of the Hippocratic Oath at graduation. At the University of Michigan, the swearing of the oath was reserved for those students inducted into the medical honor society, Alpha Omega Alpha.

In a small number of medical schools, the oath was read only in the context of courses in medical ethics and medical history. Although such courses were rare before World War II, the growing number of Catholic hospitals fostered more explicit attention to the moral problems in medical and nursing practice for Roman Catholic physicians and nurses. In 1897 the Jesuit Charles Coppens published the series of lectures he delivered to medical students at the John A. Creighton Medical College in Omaha, Nebraska. His lectures included arguments against craniotomy (the destruction of fetal life to save the life of the mother), abortion, and so-called venereal excesses (masturbation, lust, prevention of conception). He also lectured students about Catholic doctrine against euthanasia, the importance of infant baptism, and the legal aspects of insanity. At most schools, however, as the historian Chester Burns (1980) has noted, medical ethics received little attention in the medical curriculum. In 1909, for example, when the AMA Council on Medical Education asked a committee of more than one hundred medical school faculty members around the country to produce a comprehensive medical curriculum, they recommended that students receive some 305 hours in hygiene, medical jurisprudence, and medical ethics and medical economics. Asked to provide a more realistic recommendation, the Committee reduced the number of recommended hours to these subjects to 120 hours. But in 1910, when the American Association of Medical Colleges surveyed schools to see what was actually taught to medical students, they found that on average, schools devoted 37 hours to hygiene, 21 to medical jurisprudence, and 9 to medical economics. Medical ethics received no mention. This situation did not materially change until the 1960s (Burns, 1980).

The infrequent use of the Hippocratic Oath at medical graduations does not mean that the oath was unimportant to American physicians. In the medical autobiographies popular in the 1930s and 1940s, for example, physicians consistently invoked the oath as a ritual of medical initiation. Both physician William E. Aughinbaugh and surgeon Bertha van Hoosen reprinted the oath in their respective autobiographies (*I Swear by Apollo*, 1938; *Petticoat Surgeon*, 1947). Perhaps these physicians were themselves responding to the frequent appearances of the oath in American popular culture. In 1933, when the playwright Sidney Kingsley's play *Men in White* debuted on the Broadway stage, the theater program included the oath in its entirety. The oath also featured in the successful Metro-Goldwyn-Mayer Studio's film adaptation in 1934. In the film version, the young Dr. Ferguson played by the actor Clark Gable observes as his fellow interns recite the Hippocratic Oath in the hospital library surrounded by busts of Hippocrates and other great doctors. In 1935 the Nobel Prize–winning novelist Sinclair Lewis published a short story in the popular Hearst magazine *Cosmopolitan* entitled "The Hippocratic Oath." Although the doctor in Lewis's story initially dismisses the oath as a "solemn churchly formula," the physician confronted with the moral dilemma of informing the woman he loves about a secret he learned in confidence from a patient concedes that the commitments embodied in the oath are sacred to the physician: "In itself an oath is nothing as in itself a Cross is only a trick of ebony and silver, and a Flag a streaky swatch of cotton wool. It is the feeling behind the oath: that a doctor must never betray anyone who depends upon him, no matter how vile" (Lederer, 2002, p. 97) Many Americans seemed to share these sentiments. Even if they were unaware of the specific precepts of the Hippocratic Oath, these Americans assumed that physicians were bound to different moral rules and higher ethical obligations. In 1950 when Duke University Hospital refused admission to a young African American man traumatically injured in an automobile accident because the "Negro beds were full," students at a local black college wrote to the president of the university expressing shock and disbelief: "in light of this incident, we are forced to wonder if the [Duke] staff has ever heard of the Hippocratic oath and the teaching of the Nazarene" (Love, 1996, p. 222). As the historian Spencie Love (1996) has noted, the university's apparent breach of its Christian principles and the violation of Hippocratic commitment to patient welfare both disturbed and distressed many Americans.

In the second half of the twentieth century, American medical educators made a renewed investment in the Hippocratic Oath. This interest in the oath may have resulted from the rapid advance of medical specialization in the postwar era and a desire for some unifying principles or ritual in the light of the intellectual and social fragmentation of the profession. Some historians have pointed to World War II and the tragic participation of the medical profession in Nazi war crimes as an impetus for interest in greater emphasis on medical ethics in the medical profession. In 1948 the World Medical Association, formed in response to medical developments in National Socialist Germany, endorsed a new physician's oath, the Declaration of Geneva. In so doing, they explicitly modified the Hippocratic Oath, omitting all references to deities and such specific practices as abortion, euthanasia, and surgery. Instead, physicians pledged to privilege responsibility for patient welfare over any considerations of nationality, race, politics, and social status. As a condition of Germany's membership in the World Medical Association, all German physicians were required to promise to adhere to the Declaration of Geneva upon graduation from medical school.

In 1969 when the physician Ralph Crawshaw (1970) surveyed the deans of 85 American medical schools, he reported that 78 medical colleges administered a professional oath— a modified Hippocratic Oath, Declaration of Geneva, or other variations (Crawshaw, 1970, pp. 145–50). By 1989 nearly all American medical schools had adopted some form of professional oath at graduation. In 1993, medical educators introduced a new ritual of initiation into moral medical practice that has since spread to nearly all American medical schools, as well as to schools of nursing, pharmacy, and osteopathy. This ritual, the white coat ceremony, first took place in 1993 at the Columbia University College of Physicians & Surgeons where entering medical students received a short white coat from their professors and publicly swore a version of the Hippocratic Oath. Fostered by the Arnold Gold Foundation for Medical Humanism and the Robert Wood Johnson Foundation, the ceremony, as the Gold Foundation's website explains, "is a hands-on experience that underscores the bonding process. [The white coat—the mantle of the medical profession] is personally placed on each student's shoulders by individuals who believe in the students' ability to carry on the noble tradition of doctoring. It is a personally delivered gift of faith, confidence and compassion." In the early twenty-first century, more and more medical schools have encouraged students to create their own oath or pledge for public recitation at the white coat ceremony.

Becoming Bioethical. In the early twenty-first century, the explicit attention and concerns about medical professionalism and the moral obligations of physicians arrived in the wake of a major upheaval and reorganization of ethics in medicine. Beginning in the 1960s (some historians say earlier) a new ethics in medicine and the life sciences challenged earlier professional preoccupations with clinical medicine, fostering a broader public discourse about the meaning and implications of medical practices and medical decision making and giving rise to a new, methodologically sprawling academic discipline with its own cadre of practitioners, journals, and credentialing, a collection of disparate discourses, journals, theories, and practices.

The challenge to traditional medical ethics, the purview largely of physicians and of some Catholic and Protestant theologians, required a new language, at the very least a new word, to distinguish itself from these older, more conventional discussions. The word was *bioethics*. First introduced in print in 1970, bioethics was coined by the University of Wisconsin research oncologist Van Rensselaer Potter (1911–2001) to identify "a new discipline that combines biological knowledge with knowledge of human value systems." After his initial discussion of bioethics in the journal *Perspectives in Biology and Medicine*, Potter expanded his vision of the new field in a 1971 book *Bioethics: Bridge to the Future*, in which he called for "a new science of survival," one that would encompass humanity's long-range environmental risks and challenges. Here he sought not only a discipline to bridge to the future, but also a bridge to the two cultures—the science and the humanities—made famous in the 1959 lecture by the British scientist and novelist C. P. Snow. Like Snow, Potter believed that the inability of the sciences and the humanities to speak to each other imperiled humanity's future, but he found hope, perhaps unexpectedly, in the U.S. Congress. "The obligation to the future," observed Potter, "has been recognized and the need for combining science with the talents of the humanities is being met with new legislation calling for an 'Office of Technology Assessment'" (Potter, 1971, p. ix) Such an office, Potter believed, would enable informed social policy to guide humanity.

At about the same time, a small group of Catholic physicians and laypersons also began using the word bioethics to refer to a more narrow range of activities in biology and medicine. According to R. Sargent Shriver, the first director of the Peace Corps and the husband of President John F. Kennedy's sister Eunice Kennedy Shriver, in 1970 he came up with the word "bioethics" as he was discussing the possibility that the Joseph Kennedy Foundation would provide some funding into the study of the religious and ethical aspects of modern medicine and the life sciences. "Because of the need to bring biology and ethics together, I thought of 'bioethics,'" the Catholic layman Shriver explained. "And the people latched onto it as the name of the Institute. Our idea was that we were starting an ethics institute regarding this new science, with primary emphasis on biology with ethics.... I know full well I proposed the word. But I don't think it was a stroke of genius. It was as easy to come up with the word bioethics as falling off a log" (Reich, 1994, p. 325). Shriver's reference to the institute was to the Joseph and Rose Kennedy Center for the Study of Human Reproduction and Bioethics that opened in July 1971 at Georgetown University. Funded by a one-time gift from the Kennedy Foundation, the institute later changed its name to the Kennedy Institute of Ethics. As the bioethicist Albert R. Jonsen notes, the institute began an intensive summer course in bioethics in 1974, which continues in the early twenty-first century. In the late 1970s the group also established a graduate program in bioethics.

In 1971 when the Kennedy Institute of Ethics opened, there was only one other institution devoted to bioethics inquiry. In March 1969, Daniel Callahan, a philosophically trained writer and editor, joined the Columbia University psychiatrist Willard Gaylin to create the Institute of Society, Ethics, and the Life Sciences in Hastings-on-Hudson, New York. With financial support from such individuals as Elizabeth Dollard and John D. Rockefeller III and foundations such as the Rockefeller Foundation and the National Endowment for the Humanities, the institute brought together a wide range of philosophers, physicians, biologists, sociologists, and clergymen. In 1971 the organization launched a new journal, the *Hasting Center Report*, which became a major forum for discussing such issues as genetic counseling, heart transplantation and public policy, and the implications of a new definition of death.

There is no doubt that bioethics attracted increasing professional and public attention in the 1970s. There is less agreement about what gave rise to bioethics as both an intellectual and a cultural project. In a provocative and far-reaching analysis, anthropologist Atwood Gaines and ethicist Eric Juengst have analyzed what they identify as "origin myths in bioethics," stories about the beginning of the movement that have implications for the direction, scope, and future of bioethics. In so doing, they direct attention to the fact that bioethics is, in fact, not one thing but many. Although generally discussed in the singular (like "the blues"), it is nonetheless important to keep in mind that there are multiple, overlapping, and even conflicting aspects of bioethics. Gaines and Juengst identify three major types of origin stories in bioethics: bioethics as reactive to major changes in biomedical knowledge and technology or to the radical cultural pluralism of the 1960s; bioethics as a proactive effort to anticipate and guide social and medical futures; and bioethics as a continuation of medical ethics.

As bioethics enters its third or fourth decade (depending on one's reckoning of its birth date), the number and variety of such origin stories has grown. Many of the first-generation participants in bioethics have been asked to provide or have volunteered their own recollections and reminiscences about the development of bioethics. The physician Edmund Pellegrino (1920–2013), a major physician pioneer in bioethics, for example, has acknowledged both the irresistible pull of such invitations and the problematic character of such participant histories. Nevertheless, these participant-observer projects have offered a major source of interpretation and raw material for more formal analyses by historians. Pellegrino, like another major bioethics pioneer, Al Jonsen, locates the origins of bioethics in the profound changes in medical practices that fostered fears about the "dehumanization" of medicine in the 1960s. He

points to the claims of such physicians as the Public Health Service Surgeon General Luther Terry, who in 1964 warned, "The greatest single challenge to the physician will be to reconcile automation and humanism in treating the patient" (Pellegrino, 1999, p. 75). As the chair of medicine at the University of Kentucky, Pellegrino introduced ethics and medical humanities in his teaching in the 1960s. In 1972 as the founding director and dean of medicine at the State University of New York at Stony Brook, he created a division of humanities and social sciences, in which a philosopher, historian, and sociologist contributed to the education of medical students, nursing students, and students in the allied health professions. Pellegrino continued his formal association with the bioethics movement as the chancellor of the University of Tennessee, at Yale University, and at Georgetown University, where he served as the director of the Kennedy Institute of Ethics and the Center for Clinical Bioethics.

Pellegrino was one of the forty-six first-generation bioethicists who attended the conference convened by the bioethicist Albert R. Jonsen in 1992 to examine the beginning of bioethics. This conference became the basis of his 1998 book, *The Birth of Bioethics*. Like Pellegrino, Jonsen located the origins of bioethics in the changing medical interventions and technologies of the 1960s. The "impersonal machines that intervened between doctor and patient" (Jonsen, 1998, p. 11), Jonsen argues, brought new challenges to traditional questions of benefits and harms, as well as new questions of professional authority. Bioethics, Jonsen conceded, inherited some of the traditional ideas and rationales of an earlier medical ethics, even as it brought new participants and intellectual resources to the contemplation of the new biology and the new medicine. In addition to theologians, philosophers, and legal scholars, Jonsen pointed to the unprecedented investment and involvement of federal and state governments in questions of health and human values. He also acknowledged the "hot button" issues of the 1960s that received extensive attention and fostered intense controversy, including human experimentation, genetics, organ transplantation, death and dying, and reproduction.

In so doing, Jonsen drew extensively on the work of the historian David J. Rothman, whose 1991 book, *Strangers at the Bedside*, offered the first, and still most influential, historical analysis of the ways in which law and bioethics transformed medical decision making in the second half of the twentieth century. Rothman framed bioethics as responsive to scandal at the heart of the biomedical enterprise, namely the exposé of the rampant and pervasive abuses of human subjects. The critical period of change, Rothman argues, came in the decade between 1966 and 1976. The year 1966 marked the year that the Harvard anesthesia researcher Henry K. Beecher rocked the stolid world of the *New England Journal of Medicine* with the publication of 22 examples of routine experiments at major American research institutions and funded by major American foundations and the federal government that threatened the health and safety of vulnerable human subjects. The subjects of Beecher's 22 cases included "mentally defective" children, mentally retarded and delinquent children, the very elderly, soldiers in the armed forces, charity patients, the terminally ill, alcoholics, and newborn infants. The revelation of these abuses, together with the staggering announcement in July 1972 of the Public Health Service's 40-year study of untreated syphilis in African American men (the Tuskegee Syphilis Study), resulted in the National Research Act. Signed into law by President Richard Nixon (three weeks before his resignation from office) in July 1974, this act established the National Commission for the Protection of Human Subjects of Biomedical and Behavioral Research (Albert Jonsen served on this panel, along with three physicians, three lawyers, a public member, and another ethicist), which provided an unprecedented and public niche for the growing bioethics movement. The year 1976 marked a decade of profound legal and ethical challenges to death and dying in the United States, from the introduction of brain death in 1968 to foster the retrieval of usable organs for transplantation to the New Jersey State Supreme Court decision in 1976, which allowed the parents of Karen Ann Quinlan to authorize the hospital to wean her from the mechanical respirator that doctors believed essential to her survival. Transferred to a long-term care facility, Quinlan actually lived for another nine years after the "right to die" victory. The research abuses of the 1960s helped to trigger the contemporary bioethics movement. These prominent examples of the profound failures within biomedicine to respect basic human rights resonated with more fundamental social questions in the period about the nature and legitimacy of medical authority within clinical medicine. Expressed most succinctly in the attention to and diffusion of informed consent and patient self-determination, the broader context of rights-based movements in the 1950s and 1960s, including the civil rights movement, the second women's rights movement (and later the women's health movement), the consumer's rights movement, and the patient's rights movement, was also crucial to the rise of bioethics.

In contrast to these accounts, the historian Tina Stevens has argued that bioethics should be understood as the product of a longer American ambivalence about progress. Bioethics, rather than offering new tools for patient's rights, won legitimacy in the United States because it functioned, in Stevens's words, as a form of "cultural inoculation, an immunization that forestalled more virulent attacks of radical critics, who were mistrustful of biomedicine's undergirding role in a technological society" (Stevens, 2000, p. 2). She locates the taproot of bioethics in the postatomic responsible-science movement of the late 1940s and 1950s, when intellectuals responded to the invitations from geneticists and other biomedical scientists to contribute to a dialogue explicitly undertaken to avert the profound uneasiness of atomic scientists. Some commentators have endorsed Stevens's historical analysis of bioethics as the most palatable (and least threatening) alternative to more radical social critics and her characterization of bioethics as handmaiden or apologist for biomedicine, even as they challenge her controversial claims about the ways in which the transplantation community with its critical need for cadaveric material co-opted the bioethics movement to achieve public support for a new definition of death that expanded the supply of transplantable organs.

Sociologist John Evans (2002) similarly identified the genetics community as critical to the development of bioethics and what he labels a "thinning" of the public debate over the ends of medical science. Using the study of human genetic engineering, Evans analyzes the debate in the 1950s and 1960s between geneticists and theologians over the goals of science. In *Playing God?: Human Genetic Engineering and the Rationalization of Public Bioethical Debate*, he describes how the scientific community responded to the unwanted incursions of theologians, both Protestant and Catholic, and others who wanted a more robust discussion about the implications and meaning of new genetic technologies by creating a new profession, bioethics. The success of bioethics, Evans argues, dramatically altered the public debate, effectively marginalizing theologians and religious groups. Despite their initial professional legitimacy, bioethicists, he claims, have not resolved the enduring issues of how a society determines what scientific advances and technological interventions are ethical.

Whether bioethics was born in the 1950s or 1960s, whether it was triggered by the political milieu of civil rights activism and challenges to orthodoxy or by the technological advances of biomedicine, there is little doubt that it remains a highly visible, if heterogeneously amorphous, enterprise. In addition to its formal embodiment in federal advisory committees charged with directing our societal investment in such biomedical scientific developments as synthetic biology or human cloning, academic centers, independent think-tanks, journals, and graduate programs, bioethics has influenced everyday experience in the clinic and the laboratory in unexpectedly bureaucratic ways. If bioethics at its beginning represented a challenge to traditional medical authority, it has increasingly since the 1980s served to constitute and legitimate medical authority and to promise solutions to our profound uneasiness about medicine and society. But as the historian Charles Rosenberg poignantly reminds us in his reflections on the bioethical enterprise, many, if not all, of our most pressing problems remain unsolvable: "We are well aware that there is no ultimate solution for pain and death, no way to explain the brutal randomness with which suffering is distributed. These are aspects of the human condition" (Rosenberg, 1999, p. 44).

[*See also* Abortion Debates and Science; American Medical Association; Amniocentesis; Animal and Human Experimentation; Death and Dying; Ethics and Professionalism in Engineering; Law and Science; Medicine; Organ Transplantation; *and* Stem-Cell Research.]

BIBLIOGRAPHY

Aughinbaugh, William E. *I Swear By Apollo: A Life of Medical Adventure*. New York: Farrar & Rinehart, 1938.

Baker, Robert. "The History of Medical Ethics." In *Companion Encyclopedia of the History of Medicine*, edited by W. F. Bynum and Roy Porter, pp. 852–887. New York: Routledge, 1993.

Baker, Robert, and Laurence B. McCullough. *The Cambridge World History of Medical Ethics*. Cambridge, U.K.: Cambridge University Press, 2009.

Boston Medical Police. Boston: Snelling and Simons, 1808.

Burns, Chester R. "Medical Ethics and Jurisprudence." In *The Education of American Physicians: Historical Essays*, edited by Ronald L. Numbers, pp. 273–289. Berkeley: University of California Press, 1980.

Carey, Eben J. "The Formal Use of the Hippocratic Oath for Medical Students at Commencement Exercises." *Bulletin of the Association of American Medical Colleges* 3 (1928): 159–166, on p. 159.

Cathell, D. W. *Book on the Physician Himself and Things That Concern His Reputation and Success*, 9th ed. Philadelphia: Davis, 1890.

Crawshaw, Ralph. "The Contemporary Use of Medical Oaths." *Journal of Chronic Diseases* 23 (1970): 145–150.

Delafield, Francis. "Chronic Catarrhal Gastritis, with Opening Remarks, by the President." *Transactions of the Association of American Physicians* 1 (1886): 1–10.

Evans, John H. *Playing God? Human Genetic Engineering and the Rationalization of Public Bioethical Debate*. Chicago: University of Chicago Press, 2002.

Faden, Ruth R, Tom L. Beauchamp, and Nancy M. P. King. *A History and Theory of Informed Consent*. New York: Oxford University Press, 1986.

Fox, Renée C., and Judith W. Swazey. *Observing Bio-ethics*. New York: Oxford University Press, 2008.

Gaines, Atwood D., and Eric T. Juengst. "Origin Myths in Bioethics: Constructing Sources, Motives and Reason in Bioethic(s)." *Culture, Medicine, and Psychiatry* 32 (2008): 303–327.

Hesseltine, H. Close, F. L. Adair, and M. W. Boynton. "Limitation of Human Reproduction. Therapeutic Abortion." *American Journal of Obstetrics and Gynecology* 39 (1940): 549–562.

Jonsen, Albert R. *The Birth of Bioethics*. New York: Oxford University Press, 1998.

Jonsen, Albert R. *A Short History of Medical Ethics*. New York, Oxford University Press, 2000.

Jonsen, Albert R., Shana Alexander, Judith P. Swazey, Warren T. Reich, Robert M. Veatch, Daniel Callahan, Tom L. Beauchamp, Stanley Hauerwas, K. Danner Clouser, David J. Rothman, Daniel M. Fox, Stanley J. Reiser, and Arthur L. Caplan. "The Birth of Bioethics: Report of a Conference." *Hastings Center Report* 23 (1993): S1–S16.

Kleinman, Arthur, Renée C. Fox, and Allan M. Brandt. "Introduction to Bioethics & Beyond." *Daedalus* 128 (1999): vii–x.

Konold, Donald Enloe. *A History of American Medical Ethics, 1817–1942*. Madison: State Historical Society of Wisconsin for the Department of History, University of Wisconsin, 1962.

Lederer, Susan E. "Hippocrates American Style: Representing Professional Morality in Early Twentieth-Century America." In *Reinventing Hippocrates*, edited by David Cantor, pp. 239–256. Aldershot, England: Ashgate, 2002.

Lederer, Susan E. "Medical Ethics and the Media: Oaths, Codes and Popular Culture." In *The American Medical Ethics Revolution*, edited by Robert Baker, Arthur Caplan, Linda Emanuel, and Stephen Latham, pp. 91–103. Baltimore: Johns Hopkins University Press, 1999.

Lederer, Susan E. "Research without Borders: The Origins of the Declaration of Helsinki." In *Twentieth-Century Ethics of Human Subjects Research: Historical Perspectives on Values, Practices, and Regulations*, edited by Volker Roelcke and Giovanni Maio, pp. 199–218. Stuttgart: Steiner, 2004.

Lederer, Susan E. *Subjected to Science: Human Experimentation in America before the Second World War*. Baltimore: Johns Hopkins University Press, 1995.

Love, Spencie. *Of One Blood: The Death and Resurrection of Charles R. Drew*. Chapel Hill: University of North Carolina Press, 1996.

Mohr, James C. *Abortion in America: The Origins and Evolution of National Policy*. New York: Oxford University Press, 1979.

Numbers, Ronald L. "William Beaumont and the Ethics of Human Experimentation." *Journal of the History of Biology* 12 (1979): 113–135.

Osler, William. "Unity, Peace, and Concord: A Farewell Address to the Medical Profession of the United States." *Journal of the American Medical Association* 45 (1905): 365–369.

Pellegrino, Edmund D. "The Origins and Evolution of Bioethics: Some Personal Reflections." *Kennedy Institute of Ethics Journal* 9 (1999): 73–88.

Potter, Van R. *Bioethics: Bridge to the Future*. Englewood Cliffs, N.J.: Prentice Hall, 1971.

Reagan, Leslie J. *When Abortion Was a Crime: Women, Medicine, and Law in the United States, 1867–1973*. Berkeley: University of California Press, 1998.

Reich, Warren Thomas. "The Word 'Bioethics': Its Birth and the Legacies of Those Who Shaped It." *Kennedy Institute of Ethics Journal* 4 (1994): 319–335.

Reich, Warren Thomas. "The Word 'Bioethics': The Struggle over Its Earliest Meanings." *Kennedy Institute of Ethics Journal* 5 (1995): 19–34.

Rosenberg, Charles E. "Meanings, Policies, and Medicine: On the Bioethical Enterprise and History." *Daedalus* 128 (1999): 27–46.

Rothman, David J. *Strangers at the Bedside: A History of How Law and Bioethics Transformed Medical Decision Making*. New York: Basic Books, 1991.

"Shall Physicians Learn Anatomy or Not?" *Medical and Surgical Journal* 48 (1883): 46.

Smith, Dale C. "The Hippocratic Oath and Modern Medicine." *Journal of the History of Medicine* 51 (1996): 484–500.

Stevens, M. L. Tina. *Bioethics in America: Origins and Cultural Politics*. Baltimore: Johns Hopkins University Press, 2000.

Warner, John Harley. "The 1880s Rebellion against the AMA Code of Ethics: 'Scientific Democracy' and the Dissolution of Orthodoxy." In *The American Medical Ethics Revolution: How the AMA's Code of Ethics Has Transformed Physicians' Relationships to Patients, Professionals, and Society*, edited by Robert B. Baker et al., pp. 52–69. Baltimore: Johns Hopkins University Press, 1999.

Susan E. Lederer

ETHICS AND PROFESSIONALISM IN ENGINEERING

The history of ethics and professionalism in American engineering is at its heart the story of a search for identity. To ask what it means to be ethical or professional is to ask what it means to be an engineer. Who does and does not belong to the profession, and how are members expected to behave? How do membership and the obligations that go with it change over time?

Although it is possible to argue over precisely when engineering became a profession (or even whether it is one today), most scholars point to the formation of so-called professional societies in the latter half of the nineteenth century as marking the formal organization of American engineering into the American engineering profession. The founding of the American Society of Civil Engineers (ASCE) in 1852 marked the start of the trend that would continue for the next five decades. Engineers were not alone in establishing professional societies in these years; the American Medical Association dates to 1847 and the American Bar Association to 1878. From the beginning, each of these professional societies wrestled with the question of whom to admit and under what conditions; somewhat later they would also address the question of whom to expel and under what conditions.

In broadest terms, a professional is one with specialized knowledge who acts on behalf of others (typically the public) and who has an obligation to use that knowledge ethically, that is, for the benefit of those being served. But the nature of that specialized knowledge changes over time, as does the understanding of what it means to make appropriate use of that knowledge. Debates have tended to focus on two issues. First, who should be considered a member of (or be admitted to) the "club"? And second, who should control admission to and practice of the profession?

In the early days of the ASCE, for instance, only those with a personal connection to founding members could join; membership required recommendation by an existing member and a positive vote by the entire membership. In the early

twentieth century, new membership rules requiring applicants to have supervised engineering work effectively excluded women, as Ruth Oldenziel (1999) has argued, because the required supervisory work was itself only open to men. In 2012, many societies required that "members" (as opposed to "affiliates") have an engineering bachelor's degree or higher; and discussions had been underway for years about making the master's degree the first professional degree. If professional societies debated whom to admit to their ranks, they also worried about whom to exclude. Are software engineers, sanitary engineers, domestic engineers, and the like really engineers? Can one group of "engineers" prevent another group from utilizing the term? Implicit in each of these delineations is the fact that some group had established itself as the arbiter of membership (Pfatteicher, 1996; Wisely, 1974).

Who Is an Engineer? Decisions about who counts as an engineer and who does the counting have been shaped by developments in three broad areas: society as a whole, other professions, and engineering itself.

Societal trends shape a profession's sense of self. From scientific management and the Progressive Era of the early twentieth century to the environmental movement and technological skepticism of the 1960s and 1970s, broad social movements had the power to affect engineers' sense of self. Even demographic changes, such as women's increasing participation in higher education and the workforce, altered thinking about membership standards and professional practice (Bledstein, 1976; Layton, 1986; Oldenziel, 1999).

Pressure from other professions also affected the structure of engineering. In the late nineteenth century, engineers' desire to emulate doctors and lawyers affected the way they structured their membership requirements, and in the early twenty-first century a similar desire helped drive the push to increase educational requirements for engineering practice. At the turn of the twentieth century, the growth of architecture and engineering led to rifts between the two groups, with early licensing laws being used to carve out exclusive areas of practice for each (Haber, 1991; Perrucci and Gerstl, 1969; Pfatteicher, 1996).

Just as factors outside of engineering shaped what happened inside the profession, so too did developments within engineering itself change how engineers understood themselves. Large civil engineering projects of the late 1800s (including skyscrapers and the Brooklyn Bridge), electrical power projects of the 1930s (the Tennessee Valley Authority [TVA] and the Hoover Dam), nuclear weapons and power in midcentury, and even disasters such as the collapse of the World Trade Center all pushed engineers to explore the range of their responsibilities (Pfatteicher, 1996, 2010; Wisely, 1974).

The Nature of Engineering. As each of these broad factors has caused engineers to reenvision themselves and their duties, three characteristics of the engineers themselves have presented challenges to the creation of a radically new form of engineering organization.

The practice of engineering requires both progressivism and conservatism—the simultaneous boldness of vision to create new things and the caution to do those things safely. Edwin Layton's seminal work, *The Revolt of the Engineers* (1986), emphasized this duality, as have many other observers before and since. It should come as no surprise, then, that similar attitudes have shaped engineers' approach to designing their own profession. There is boldness evident, for example, in engineers' depiction of their profession as crucial to the future of humanity. The various works of Samuel Florman and Henry Petroski provide some of the most enduring examples. At the same time, there is caution evident in engineers' reluctance to dramatically alter the education that produces engineers for that very future (Brown et al., 2009; Pfatteicher, 2010).

Another dichotomy frequently used to describe engineering is the tension between the technological and economic components of the work, or what has been described as "design under constraint." In this view, engineers must dream up new technical solutions, but always under the constraint that those solutions be financially viable for the company or government sponsoring

them. Again, Layton's *Revolt of the Engineers* emphasizes the importance of this tension, describing how it shaped the organization and regulation of engineering in the Progressive Era. In the early twenty-first century, "Introduction to Engineering" textbooks tended to take this tension as an inherent and defining feature of engineering practice.

Finally, engineering is a diverse discipline, if not in its practitioners, at least in its areas of practice. Despite various attempts since the mid-nineteenth century, engineers have been unsuccessful at creating a single umbrella organization to which all engineers might belong. Unlike doctors, who have the American Medical Association, or attorneys, who have the American Bar Association as the ultimate overseer of the profession, engineers have no single comprehensive organization. Ron Kline (2008) notes the failed attempts of the American Association of Engineering, the Engineering Council, and the Federated Association of Engineering Societies—all short-lived experiments in unifying the diversity of engineers. Today, most engineers who belong to a professional society will first join their disciplinary society—the American Institute of Chemical Engineers (AIChE), American Society of Civil Engineers (ASCE), American Society of Mechanical Engineers (ASME), and Institute of Electrical and Electronics Engineers (IEEE,) for example—although broader groups do exist. The National Society of Professional Engineers, for instance, seeks to serve all Professional Engineers (that is, those who have obtained a PE license), but the vast majority of engineers are not required to be licensed to practice. ABET, Inc. (formerly the Accreditation Board for Engineering and Technology, now known simply by its acronym), serves as the formal accrediting body for engineering and engineering technology programs in the United States, but its focus is almost exclusively on higher education, not on the practice of engineering (except insofar as it helps to ensure graduates are capable of practicing).

In the broad sweep, it is instructive to divide the history of ethics and professionalism into three rough eras. From roughly 1850 to 1900, the desire for professional autonomy drove engineers to a laissez-faire approach to defining and regulating their practice. From the early 1900s to mid-century, the increasing diversity of engineering practice led to tensions that supported the development of ethics codes and licensing laws. The second half of the twentieth century brought a new focus on social responsibility as a reaction to increasingly critical views of technology. In each phase, educational patterns, membership demographics, and regulatory standards in engineering marked the guiding professional philosophy of the era (Pfatteicher, 1996).

The Age of Autonomy. For the first half century of American professional engineering societies (roughly 1850 to 1900), autonomy and individualism ruled the day. Engineering education was largely informal and idiosyncratic; society membership was small and elite; and regulation was nonexistent.

In late-nineteenth-century America, most engineers had no official credentials with which to impress prospective clients. In 1852, only half a dozen schools in the United States offered engineering degrees. A few engineers had trained in formal apprenticeships, but the vast majority had learned on the job. There were no licensing laws or other regulations to certify the competence of the few hundred engineers practicing in the United States. An engineer's reputation was his most important credential, and success in one job was necessary to obtain the next position. The dozen founding members of the ASCE came together to provide professional fellowship and to endorse one another's standing as elite members of the field (Pfatteicher, 1996; Reynolds, 1991, pp. 7–26).

This first professional engineering society in the United States was a loose confederation of old friends who valued their independence. These men understood that they relied not only on their individual reputations, but also on the character of the entire discipline. As a result, the behavior of engineers ranked as an important topic of discussion in society publications, meetings, and correspondence. But for the first half century, the members of the ASCE and other national engineering societies rejected explicit codes of behavior as a means of supervising one another,

choosing instead to rely on more informal methods.

The society directors maintained a hands-off policy with respect to members' actions. An ASCE member, for instance, was by definition a person knowledgeable in engineering, fair in business dealings, and upstanding as a citizen, and the founding members accepted only those applicants whose qualifications they could confirm through personal knowledge. By carefully controlling who gained admittance to the society, the directors felt they relieved themselves of the need to police members. To the extent that the directors wished to limit the use of the title "engineer" to ASCE members, anyone not fulfilling these select criteria was not a bad engineer; he was, by the ASCE definition, not an engineer at all.

In exchange for the privilege of being able to draw on the reputation of the profession, each engineer had responsibilities to that profession. Building safe structures was such a basic duty as to be mentioned only rarely, and engineering practice varied from project to project. As a result, reputations were built not only on engineering knowledge and the success of an engineer's projects, but also on business sense and personal morality.

These early society members did not want the ASCE to be a regulatory body, but they did feel a need to distinguish themselves from nonmembers. One effective means of accomplishing this was to investigate failures. Engineering disasters threatened the reputation of all engineers, ASCE members or not. But society members found that by conducting inquiries and publishing reports declaring the cause of disasters to be unqualified designers or poor materials, they could use failures to indicate the value of hiring a trained professional.

Autonomy and exclusivity were deeply valued by nineteenth-century engineers and shaped the professional societies that emerged in this period as members simultaneously chose to define their responsibilities individually and joined together to distance themselves from nonmembers.

A New Generation of Leaders. For the next 50 years (roughly 1900 to 1950), American engineering grew at a rapid pace, with society membership doubling every 10 years in the early part of this period. In the face of such growth, members could no longer personally vouch for every new applicant's qualifications, and new, more detailed membership requirements emerged that emphasized formal education. As the profession expanded, so too did the array of jobs in which engineers found themselves. This diversity presented special challenges to defining who could be called an engineer and how such professionals ought to behave. The profession in this half century entered a proscriptive phase in which certain behaviors were ruled out, but autonomy continued to be the ideal (Pfatteicher, 1996; Reynolds, 1991, pp. 169–190, 367–398).

Turbulent as the years between 1900 and 1950 were for the American engineering profession, the changes between 1900 and 1910 were especially dramatic. In 1905, for the first time, over half of all American engineers held bachelor's degrees. Society membership, always disproportionately weighted toward college-trained engineers, now became overwhelmingly so. Membership in the ASCE (still the predominant engineering society) doubled between 1902 and 1909. And more important, the society leaders' characteristics were changing. The ASCE, 50 years old in 1902, of necessity was led by a new generation of engineers as the founding members retired or died. When they took control, they held different expectations than their predecessors of the society's duties and functions.

Ironically, it was these middle-age members rather than the younger ones who were most troubled by the dramatic changes in the profession and who led the push for increased regulation of the field. They had been reared, as engineers, to expect certain things from their profession; they now found their experiences falling short of those expectations and began to view self-regulation as a way to restore engineering to an earlier state. Opposition to regulation often came from younger members of the profession, who found their experiences and dilemmas missing from the measures drafted by older, more traditional engineers and from the oldest engineers, who were established enough in their careers to have no incentive to want to be told how to conduct themselves.

Commonalities across Specialties. In addition to these demographic changes, the proliferation of specialization and the splinter groups that resulted reshaped the profession. The founders of the ASCE had broadly defined civil engineering as all nonmilitary (and nonrailroad) engineering. Architects split from the civil engineers in 1857, before the ASCE had been firmly established (Levy, 1980). In the late 1800s, four more major groups had defined themselves apart from civil engineering. Mining engineers were the first to go, founding their own society in 1871; mechanical engineers formed their own association in 1880; electrical engineers followed suit in 1884 (Sinclair, 1980; McMahon, 1984). The mining, mechanical, and electrical engineering groups joined with the ASCE to become the four Founder Societies, purportedly working in parallel, each in its own specialty. But the ASCE leaders continued to view themselves as the umbrella group for what they considered the subspecialties of mining, mechanical, and electrical engineering. Tensions rose in the upper ranks of the ASCE every time one of the other founder societies exerted its independence, and the ASCE continually felt the blows of competition for members and for control over the discipline of engineering.

The increasingly varied makeup of the ASCE and her sister societies could not be changed, because this was an age of specialties, in engineering as well as in other professions. Finding consensus among such a varied group would prove difficult. But engineering's heterogeneity made it all the more important that professional society members find a way to express what they all held in common. The conflicting interests of these different types of engineers at once encouraged the adoption of ethics codes and licensing laws as guidelines for how to interact with one another and instigated arguments over the content of those policies. Even deciding who was an engineer and what constituted engineering came up for frequent discussion (Hughes, 1960; Reynolds, 1991, pp. 343–365).

Corporate Engineering. In addition to the growing specialization, engineers increasingly found themselves entering corporate workplaces.

Engineers had quickly understood that within a corporation their ability to make decisions would be sharply curtailed if it were not protected. They soon realized that this lack of authority would affect the safety of engineering works. First, the corporation, they argued, introduced new financial constraints on engineering projects. Daniel Calhoun (1960) argued that such constraints already existed, but it is true that engineers perceived a change that stirred them to action. Design decisions would no longer be based solely on engineering considerations (if they had had ever been). But furthermore, if corporate managers now had the authority to oversee design, how could engineers ensure the safety of their products? In the nineteenth century, engineers decided how to balance design and financial considerations; in the early twentieth century, engineers made design suggestions, and managers figured out how to fit them in a budget. With decision making thus dispersed, engineers were less "responsible" or "liable" than they had been. Codes and laws attempted to retain some control for engineers over their work, but acknowledged that, ultimately, decisions might be made by nonengineers (Layton, 1986; Noble, 1977).

It was in the context of these new surroundings that engineers again faced the question of self-regulation. The arguments remained as heated as those in earlier years, but this time had quite different results. Local engineering societies, splinter groups of civil engineers, and nonengineers began to propose licensing laws in a variety of states. American engineers quickly realized they would have to act or be subjected to control by nonengineers.

The First Ethics Codes.   Between 1910 and 1915, all major U.S. engineering societies adopted codes of ethics, in part to thwart proposals from outside the profession that would force licensure on engineers. These early codes prohibited certain behaviors, ranging from advertising to competitive bidding, but otherwise left practice open to individual judgment. The arguments for and against adopting a code had not changed, but the atmosphere in which ASCE members made their decision was substantially different from that

faced by nineteenth-century society members who had opposed self-regulation. Licensing was no longer a mere suggestion or even a looming possibility, but a reality in two states. Nineteen more states would follow in the next ten years. ASCE members felt increasingly constrained and wanted to assert their continuing self-control. In the nineteenth century, ASCE members had chosen to emphasize and maintain their autonomy from one another as well as from nonengineers. But in the face of increasing legislation, twentieth-century members of the society chose instead to focus on establishing their authority as an indication of their professionalism (Pfatteicher, 1996).

Other professions were adopting codes in the early years of this century. The American Bar Association's code dates to 1908 and the American Institute of Architects' code to 1909; the American Medical Association had revised its nineteenth-century code in 1912. In a renewed burst of interest in the 1920s, according to code historian Hunter Hughes (1960), "every profession, semiprofession, and business that had not adopted a code of ethics before the War got one soon after." Indeed, at least 130 such groups, from clothing manufacturers to optometrists to stock breeders, had adopted or revised their codes in the 1910s or 1920s.

The efforts of doctors, lawyers, and engineers were all part of what Burton Bledstein (1976) has called "the professionalization of America." The emerging professions had much in common: education, clientele, problems. And many scholars have emphasized the similarities to illustrate just how widespread the trend was. But the differences between engineering and the other professions presented unique problems in regulating engineering.

Doctors, lawyers, and engineers all worked with and depended upon others. Doctors interacted with nurses, pharmacists, and orderlies in serving patients. Lawyers enlisted the aid of paralegals in working for a client. But in each case, a clear hierarchy existed, giving physicians and attorneys penultimate authority over their practices. Although doctors increasingly had to submit to the decisions of hospitals and insurers and lawyers could be overruled by judges and juries, engineers envisioned medicine and the law as fields with clearly defined authority for professional practitioners.

Engineers faced a more complex web of interactions. They relied upon and had to negotiate with contractors (like architects), fabricators, construction teams, and maintenance crews to serve their clients, the owners of the structures they built. On the one hand, if doctors and lawyers needed codes to help them determine their appropriate responsibilities and allegiances, engineers, according to regulation proponents, needed all the more help. But on the other hand, according to opponents, the unique relationships in engineering complicated the matter of drafting an acceptable code.

Despite the difficulties, engineering societies across the board opted to develop codes of ethics, even if they were imperfect. In May 1911, the American Institute of Consulting Engineers (AIConsE or AICE) had become the first national engineering society in America to adopt a code. The American Institute of Electrical Engineers (AIEE) and the American Institute of Chemical Engineers (AIChE) soon followed, in March and December 1912, respectively. The American Society of Mechanical Engineers (ASME) and ASCE followed suit in 1914 (Pfatteicher, 1996; Hughes, 1960).

The adoption of ethics codes did not prevent the spread of registration laws, however. By 1950, every state had adopted an engineering licensing statute. All of the state laws followed the basic pattern of a model law drafted by the ASCE in 1910. "All" engineers were required to register, but many were exempted, especially government and corporate engineers, who presumably were already being monitored and were not directly in contact with the public. Generally, state examining boards were made up of engineers approved by the state society. Most states required some sort of examination in general engineering knowledge. And all states provided a mechanism for revoking a license once issued (Pfatteicher, 1996).

The Point of No Return. Whereas in its early years the ASCE had been dominated by several dozen elite, established engineers with no need for a code to boost their reputations or guide them in the practice they had been at for so long,

by 1910, the society had gained younger, less established, but college-educated members specializing in particular areas of engineering and faced competition from at least half a dozen other national engineering societies. These younger engineers wanted laws and codes to help distinguish them from the vast numbers of new graduates being produced each year, to guide them through their dealings with one another and employers who were often no longer engineers, and to return them to the golden age in which they believed their predecessors had practiced. But even these new members would have been content to define their roles informally. Increasing infringements on civil engineers' territory by engineers from other specialties and by nonengineers (most notably, architects) convinced ASCE members to codify their duties before others did it for them (Pfatteicher, 1996).

All of the codes and laws adopted between 1907 and 1950 attempted to reconcile the uneasy alliance that was engineering in those years. Society leaders needed to find a way to bring together the increasingly disparate interests and concerns of the members and, to exclude undesirable practitioners, set limits on what constituted acceptable engineering practice. Early twentieth-century self-regulation, as a result, defined engineers' responsibilities to one another and to their profession, but as a rule did not explicitly address responsibility to the public the profession was meant to serve. The generations to follow would adapt the codes and laws to new circumstances, but the existence of self-regulation had by midcentury become a given in American engineering.

Engineers and Society. The second half of the twentieth century brought a shift to definitions of ethics and responsibility, moving away from the proscriptive measures adopted in the first part of the century and toward language that emphasized the aspirations of the profession. In educational patterns, society membership, and regulatory standards, engineers sought to persuade an increasingly skeptical public of their trustworthiness.

The number of professionally trained instructors in the country's engineering colleges had risen from perhaps 25 or 30 in 6 institutions in the mid-1850s to approximately 7,500 teachers in 148 schools in 1953. Undergraduate enrollment in these colleges had grown from around 250 to 128,000, and the total number of engineers in the country expanded from a few hundred to over 400,000 from the mid-nineteenth to the mid-twentieth century. The vast number of new engineers entering the field each year indicated that engineering had become a vigorous profession, but with this growth came an increasing concern over competition for jobs, especially among young engineers (Pfatteicher, 1996).

The issue that young members pushed most strongly was the need for the society to develop a social consciousness. As the newly formed Committee on Younger Member Publications explained, the public was often unaware of engineers' work and engineers, in turn, often failed to understand the public's reticence toward certain engineering projects. When the public objected to a new engineering project in favor of preserving a scenic landscape, for example, the Committee argued that engineers too often responded with derision.

Several reports in the 1960s came to the unsurprising conclusion that engineers were more interested in theoretical and economic issues than in political or religious issues and even less interested in esthetic or social issues. In 1969, for instance, Joel Gerstl and Robert Perrucci published a landmark sociological study of the engineering profession in which they presented in statistical terms what the young ASCE members had been arguing: engineers were overwhelmingly machine oriented. Whether this characteristic resulted from nature or training (in other words, whether antisocial types were drawn to the field or whether engineering education made engineers this way) Gerstl and Perrucci left up for debate. Whatever the origins of this tendency toward the technical, it would take more than the comments of a few dissatisfied, junior ASCE members to redirect the society's efforts toward the public interest.

Subordination.

Beginning with the influx of newly minted engineers after 1900, the type of employment held by ASCE members, and that of civil engineers in general, had shifted dramatically. By midcentury, the vast majority of American civil engineers were employees rather than employers or independent consultants. This subordination stood in stark contrast to the authority of the founding members of the ASCE, who had all been well advanced in their careers and had generally acted as independent consultants, making them essentially self-employed (Layton, 1986; Noble, 1977; Pfatteicher, 1996).

The subordinate position of the majority of American civil engineers raised ethical and legal concerns for ASCE members. When, if ever, should an engineer's judgment be overruled by a nonengineer manager? Should engineers' primary loyalties be with their profession or their employers or the public? Could an engineer be held responsible by peers or the courts for an action required by an employer? The term and issue of "whistleblowing" grew out of such jockeying between the engineer-employee and nonengineer-manager by the end of 1960s.

The nature of engineering projects themselves also presented engineers with challenges. Many projects in the nineteenth and even early twentieth century could be overseen by a single engineer. But business and government control of projects and the increasing complexity of the structures engineers were designing, combined with the vast expense of such ventures, meant that most projects were now controlled by groups of individuals, not all of whom were engineers. The problem with this new arrangement, in the eyes of many engineers, was that such broad diffusion of authority left little responsibility for any individual engineer.

The licensing laws and codes of ethics of the early twentieth century had assumed and advocated engineers to be in complete control of their work. But could a late twentieth-century engineer be held solely responsible for a decision jointly made by a group consisting not only of engineers, but also of businessmen, government employees, and others? Would he or she want to be? In the face of such questions, the intent and content of ethics codes and licensing laws came up for renewed consideration. In 1986, two engineers in Missouri became the first to have their licenses revoked for malpractice, in response to their role in the deadly collapse of two elevated pedestrian walkways at the Hyatt Regency hotel in Kansas City. There might be situations, ASCE members began to believe, in which authority was not as attractive a goal as earlier members had thought (Pfatteicher, 1996).

Professional Differences. Changes in engineers' relationship to other building professionals accompanied a shift within the engineering profession. The field, which had once been dominated by independent engineers, increasingly consisted of large, multimember firms employing dozens and even hundreds of engineers. Although doctors and lawyers also saw their workplaces changing, this new composition was especially problematic in the engineering and architectural professions, where not every member needed to be licensed. Physicians and attorneys, even in group practice, had to be licensed individually. But engineers and architects often worked in firms where only the top engineers or architects held licenses to certify and seal the work of everyone in the office. This, too, posed novel ethical questions about who was responsible for what in a design (Layton, 1986; Noble, 1977; Pfatteicher, 1996).

By the end of the 1950s, then, the engineering profession had become considerably younger and more subordinate than it had been in the early decades of the century. By the mid-1960s, the ASCE members recognized that they were in a clearly different position than the other engineering specialties and other professions with whom they had so often compared themselves. The most notable characteristic of the ASCE by the mid-1960s was its heterogeneity when compared with other specialties and professions. The ASCE makeup was roughly balanced among members in private practice, those in industry, and those working for the government. The other national engineering societies were all heavily dominated by engineers in industry. Societies for architects, attorneys, and physicians were largely populated by members in private practice (Pfatteicher, 1996).

Such heterogeneity caused special problems for the ASCE. The American Medical Association,

American Bar Association, and American Institute of Architects, with so many members in private practice, naturally emphasized their broad professional functions. The AIME, ASME, IEEE, and AIChE, weighted toward industry engineers, leaned toward the more technical aspects of their work. The ASCE, meanwhile, was forced to strike a balance between professional and technical activity—a balance that was constantly shifting as the membership of the ASCE leaned slightly more toward one type of employment or another. Furthermore, the content of any policy, such as a code of ethics, would have to be adjusted to keep up with the changing demographics of ASCE membership. The society's heterogeneity also stimulated a great deal of discussion about unity in the engineering profession and about the public image of the engineer (which easily became muddled when civil engineers' self-understanding was unclear).

Technological Criticism. As American engineers watched their legal protections erode, their public image suffered from growing criticism of the effects of technology. These criticisms came in sharp contrast to what historian Howard Segal (1985) has called the "technological utopianism" of pre–World War II America. Technology, which as Segal and colleague Alan Marcus have explained (1989, pp. 315–361) had been viewed by American engineers and nonengineers alike as a "social *solution*" prior to 1950, came to be seen instead as a "social *question*," neither all good nor all bad, but able to be shaped and used for either. Although some technology critics used this newly popular skepticism to argue against leaving professional engineers and scientists in control of technology and science, the ASCE leaders held that the dangerous potential of technology required that qualified, trained professionals be in charge of technological decision making.

This environment of criticism of technology encouraged Congress to pass a series of laws aimed at technology control and oversight. The 1965 Water Resource Planning Act, the 1968 Wild and Scenic Rivers Act, the 1969 National Environmental Policy Act, and the Clean Water, National Dam Safety, and Federal Water Pollution acts, all in 1972, each encroached on the authority of civil engineers involved in such work as dam building, levee construction, and bridge erection. Congress also established several regulatory and oversight agencies in the early 1970s, such as the Environmental Protection Agency (1970), the Office of Technology Assessment (1972), and the Occupational Safety and Health Administration (1973).

Civil engineers felt the effects of these new laws more than engineers in most other specialties because they were the ones building the structures being regulated. To protect their authority on such projects, ASCE members responded with a series of revised codes and model laws reaffirming civil engineers' commitment to building safe structures.

Discussions of what larger role civil engineers should and would play in this new age of technological criticism repeatedly appeared in the civil engineering press, particularly in the late 1960s and early 1970s, and many members of the engineering community called for greater social responsibility on the part of engineers. To ensure that engineers would continue to be in decision-making positions, all civil engineers had a duty to maintain engineering's reputation and authority by supporting the ASCE and its code of ethics and defending ASCE-supported licensing laws. The ways in which an engineer could aid the profession's reputation were many: Some suggested that civil engineers volunteer as scouting leaders or Big Brothers; others recommended that engineers involve themselves in public relations or politics (Pfatteicher, 1996).

By the 1970s, the ASCE members advocating increased social responsibility were a varied lot. Some were professors or students of civil engineering, some were members of or heads of private engineering firms, and still others practiced in industry. Across the board, the "typical" ASCE member had, by the early 1970s, come to see that American civil engineers needed to defend their authority actively if they were to maintain control of their profession and their work. Whether the changes faced by late twentieth-century engineers were more substantial than those faced by earlier generations is not so much the point as the fact

that some civil engineers of the second half of the twentieth century, believing their world to be newly complex, used such changes to justify their actions and policies.

Education for a New Era. One issue raised repeatedly in the ASCE was how to educate the engineers who would be playing such a substantial role in shaping this new world. No longer would it be sufficient (if it had ever been) to teach engineers only the science of engineering. Particularly because engineers increasingly had to deal with nonengineers, whether they be corporate managers, politicians, or the public (which felt a new right to offer opinions on public projects), engineers needed to learn how to communicate technical subjects and decisions to those untrained in engineering. This rising interest in teaching engineers nontechnical material came at a time when the engineering curriculum was beginning to suffer from an overload of even technical courses. Four-year curricula had become the norm for engineers in the first half of the century, but following World War II, five-year undergraduate degrees became increasingly common. In the face of already full semesters, administrators frequently chose to remove humanities courses in favor of more specialized and detailed engineering courses. Indeed, it appears likely that the squeeze put on nonengineering coursework may have stimulated the many articles supporting nontechnical courses. But ASCE members soon came to see that more than coursework would be required to convince the public (and lawmakers) that civil engineers ought to remain in control of their work (Brown et al., 2009; Pfatteicher, 2010).

Hold Paramount the Public Safety. Between 1950 and 1980, American engineering societies revised their codes. The most notable difference was that whereas the earlier codes had comprised a negative list of what was considered unprofessional, the new codes explained the duties of a professional engineer. Instead of excluding unacceptable behavior and leaving members free to do whatever else they chose, the revised codes now detailed the primary duties required of an engineer and made dereliction of

these duties grounds for discipline by the societies. The earlier codes assumed an engineer was
ethical until he was found otherwise; the replacement codes assumed an engineer was unethical
unless he fulfilled certain fundamental obligations
(Pfatteicher, 1996).

Engineering societies had come to feel that
their codes should reflect a new hierarchy of engineers' responsibilities. Public welfare should be
the primary concern of engineers. Second should
come engineers' duty to their clients, employer, or
employees. Third should be engineers' loyalty to
their profession. This ordering turned the earlier
code on its head. The drafters of the early codes—
and even their predecessors who had opposed
adopting a code—felt engineers' primary responsibility should be to their fellow engineers; preserving the reputation of the profession came first.
Once that obligation had been fulfilled, engineers
owed a duty to those with whom they worked:
subordinates and superiors. The public did not
warrant any mention at all.

Recent Activities. Developments between
1980 and 2010 served to cement the notion that
engineers must attend to the effects of their work
on the public. Spurred by National Science Foundation funding in the 1980s, scholars in engineering, philosophy, and history contributed to
an emerging literature on engineering ethics, with
special emphasis on authoring textbooks for use
in undergraduate engineering programs. In the
1990s, the engineering accreditation agency
(ABET, Inc.) and state licensing examiners began
to assume that engineering colleges and their
graduates should be able to demonstrate their familiarity with issues of ethics and professional responsibility and adapted accreditation criteria and
licensing exams accordingly. The National Academy of Engineering (NAE), a supporter of these
efforts, implemented its own programs aimed
at fostering further emphasis on ethics among
engineers, including launching an Online Ethics
Center in 2007 and announcing its Grand Challenges project in 2008. The 14 Grand Challenges
for Engineering in the Twenty-First Century
ranged from making solar energy economical
to engineering better medicines to preventing

nuclear terror and were intended as a "call to
action" for NAE members and other engineers,
particularly those engaged in higher education.

[*See also* **Brooklyn Bridge; Building Technol-
ogy; Dams and Hydraulic Engineering; Engi-
neering; Environmental Protection Agency;
Hoover Dam; Journals in Science, Medicine,
and Engineering; Nuclear Power; Nuclear
Weapons; Office of Technology Assessment,
Congressional; Skyscrapers;** *and* **Tennessee
Valley Authority.**]

BIBLIOGRAPHY

Bledstein, Burton J. *The Culture of Professionalism:
The Middle Class and the Development of Higher
Education in America*. New York: W. W. Norton,
1976. On the interconnectedness of developments in American universities and the concept
of professionalism in the nineteenth century.
Brown, John K., Gary Lee Downey, and Maria Paula
Diogo. "The Normativities of Engineers: Engineering Education and History of Technology."
Technology and Culture 50 (2009): 737–752. Introductory essay to a special issue of *Technology
& Culture* that highlights key themes in the history of technology, with many related to professionalization and ethics.
Calhoun, Daniel Hovey. *The American Civil Engi-
neer: Origins and Conflict*. Cambridge, Mass.:
MIT Press, 1960. The most recent comprehen-
sive, academic history of the ASCE.
Calvert, Monte A. *The Mechanical Engineer in America.
1830–1910*. Baltimore: Johns Hopkins University
Press, 1967. Precursor to Sinclair's *Centennial
History*, covering the earlier history of mechanical engineering in the United States.
Haber, Samuel. *The Quest for Authority and Honor in
the American Professions, 1750–1900*. Chicago:
University of Chicago Press, 1991. Classic work
on the sociology of professionalization in
America.
Hughes, Hunter. "The Search for a Single Code."
Consulting Engineer 15 (1960): 112–121. A good,
brief overview of the repeated attempts to unify
the American engineering profession.
Kline, Ronald R. "From Progressivism to Engineering Studies: Edwin T. Layton's *The Revolt of
the Engineers*." *Technology and Culture* 49 (2008):

1018–1024. This review of Layton's classic work provides a useful historiography of engineering professionalization since Layton's original edition in 1971 and highlights key themes and controversies in the historical record.

Layton, Edwin T., Jr. *The Revolt of the Engineers: Social Responsibility and the American Engineering Profession*, 2d ed. Baltimore: Johns Hopkins University Press, 1986. Classic discussion of the tensions between engineers and their employers that shaped concepts of social responsibility.

Levy, Richard Michael. "The Professionalization of American Architects and Civil Engineers." PhD diss. University of California, Berkeley, 1980. Analysis of the origins of architecture and engineering as professions, with particular emphasis on conflicts and overlaps between the two.

Marcus, Alan I., and Howard P. Segal. "Technology as a Social Question: The 1950s to the Present." In *Technology in America: A Brief History*. Chicago: Harcourt Brace Jovanovich, 1989. An overview of the increasing criticism of the engineering profession in the latter twentieth century and the shift from viewing technology as a "social solution" to perceiving technology as a "social question."

McMahon, A. Michal. *The Making of a Profession: A Century of Electrical Engineering*. New York: Institute of Electrical and Electronics Engineers, 1984. On the origins of the group that would become the Institute of Electrical and Electronics Engineers (IEEE).

Merritt, Raymond H. *Engineering in American Society, 1850–1875*. Lexington: University Press of Kentucky, 1969. A social history of the early years of professionalization of American engineers.

Noble, David. *America by Design: Science, Technology, and the Rise of Corporate Capitalism*. New York: Oxford University Press, 1979. Classic history of the intersections of technology and economic forces.

Oldenziel, Ruth. *Making Technology Masculine: Men, Women, and Modern Machines in America, 1870–1945*. Amsterdam: Amsterdam University Press, 1999. A useful introduction to the interplay between gender and the engineering profession in the United States.

Papke, David R. "The Legal Profession and Its Ethical Responsibilities: A History." In *Ethics and the Legal Profession*, edited by Michael Davis and Frederick A. Elliston. Buffalo: Prometheus Books, 1986. An overview of the development of ethics in the American legal profession, useful for comparison with engineering.

Perrucci, Robert, and Joel Gerstl. *Profession without Community: Engineers in American Society*. New York: Random House, 1969. Groundbreaking sociological study of the personality of engineers and their profession.

Pfatteicher, Sarah K. A. "Death by Design: Ethics, Responsibility, and Failure in the American Civil Engineering Community, 1852–1986." PhD diss. University of Wisconsin–Madison, 1996. Details the development of engineering ethics codes and the evolution of the concept of responsibility, with special emphasis on the connections to engineering disasters.

Pfatteicher, Sarah K. A. *Lessons amid the Rubble: An Introduction to Post-Disaster Engineering and Ethics*. Baltimore: Johns Hopkins University Press, 2010. Draws on the collapse of the Twin Towers and other major disasters to explore the meaning of professionalism in engineering, with implications for practice and education.

Reynolds, Terry, ed. *The Engineer in America: A Historical Anthology from Technology and Culture*. Chicago: University of Chicago Press, 1991. Useful collection of articles on the history of engineering, from the journal of the Society for the History of Technology.

Reynolds, Terry. *Seventy-Five Years of Progress: A History of the American Institute of Chemical Engineers, 1908–1983*. New York: American Institute of Chemical Engineers, 1983. The classic history of the AIChE by a noted historian of technology.

Segal, Howard P. *Technological Utopianism in American Culture*. Chicago: University of Chicago Press, 1985. A history of technological optimism in the United States.

Sinclair, Bruce. *A Centennial History of the American Society of Mechanical Engineers, 1871–1970*. Toronto: ASME and University of Toronto Press, 1980. Academic history of the professional society for mechanical engineers in the United States.

Sullivan, William M. *Work and Integrity: The Crisis and Promise of Professionalism in America*. San Francisco: Jossey–Bass, 2005. Based on a Carnegie Foundation study on preparation for the professions of medicine, nursing, law, engineering, and the clergy, this book provides a comprehensive and accessible introduction to the concept of professionalism, its origins, development, and future.

Wear, Delese, and Mark G. Kuczewski. "The Professionalism Movement: Can We Pause?" *The American Journal of Bioethics* 4 (Spring 2004): 1–10. Opening essay to a special issue on the state and meaning of professionalism in the American medical profession. Useful for comparison with engineering developments.

Wisely, William H. *The American Civil Engineer, 1852–1974: The History, Traditions, and Development of the American Society of Civil Engineers.* New York: American Society of Civil Engineers, 1974. Comprehensive, internal history of the ASCE.

Sarah K. A. Pfatteicher

EUGENICS

As the 2010 publication of the 586-page *Oxford Handbook of the History of Eugenics* attests, eugenic ideology captured a global following in the late nineteenth and early twentieth centuries. And since the 1960s, scholars have vigorously debated the impact, meaning, practice, and legacies of a movement grounded in evaluating human worth.

When Francis Galton, a British statistician, coined the term "eugenics" in 1883, he took the word from a Greek root meaning "good in birth." He defined eugenics as the science of improving human stock by giving "the more suitable races or strains of blood a better chance of prevailing speedily over the less suitable" (Paul, 1999, p. 4; Kevles, 1998, p. ix). Analyzing the lineage of prominent British families, he reasoned that most moral and mental traits, such as courage, intellect, and vigor, were passed on to offspring. Yet he also noted that the "worthiest" families produced the fewest children, believing that the results would be disastrous if something were not done. He proposed that "those highest in civic worth" should be encouraged to have more children (defined as "positive eugenics") and those unworthy should be encouraged to have fewer or none ("negative eugenics") (Paul, 1999, pp. 3–5).

Eugenics gained authority and legitimacy in the early twentieth century, when the rediscovery of Gregor Mendel's laws of segregation and independent assortment led to the establishment of genetics. Working with peas in 1865, Mendel had found that hereditary material is transferred from parent to child. His contemporaries, however, were not impressed, and it was not until 1900 that scientists appreciated the significance of his findings. Although eugenicists had been arguing for the importance of heredity in their quest for "race betterment" since the 1870s, they had lacked the scientific evidence to suggest how characteristics were transmitted to offspring. Mendel's laws established genetics as a serious science and lent legitimacy to the eugenic claim that social undesirables—including alcoholics, prostitutes, and even unwed mothers—would produce more of their kind by passing down their supposed genetic flaw to their children.

Although eugenics influenced social policy and practice across the globe, it elicited tremendous popular and professional support in the American context because it linked two issues of great concern to the white middle class in the early twentieth century: race and gender. Social and economic changes threatened to undermine established race and gender hierarchies. The fear that "race suicide" would result from the fact that the birthrate of the American-born white middle class was dropping well below that of immigrants and the working class heightened public alarm over this loss of authority.

In Progressive-Era America, eugenic ideology appealed to reformers representing a wide range of interests and politics, who applied their own varied definitions of eugenics. Social radicals such as Charlotte Perkins Gilman and Margaret Sanger embraced it as a civilizing force that would further the rights of women, as well as improve the race. Nativists such as Madison Grant (author of *The Passing of the Great Race*) viewed it as a justification for restraining the liberties of immigrants and the procreative powers of sexually promiscuous women. What they had in common was a vision of the future in which reproductive decisions were made in the name of building a better race, although they may have disagreed on how to go about achieving this goal.

Indeed, one of the strengths of the eugenics movement was its widespread popular appeal to a diverse audience, in large part because of its decidedly vague definition. Eugenic ideology represented a complex combination of popular and scientific beliefs and interests that has forced scholars to probe into particular contexts to understand its historical impact. From "Fitter Families for Future Firesides" contests at state fairs to eugenic sterilization practices and intelligence testing, eugenics became an established part of the American landscape in the early to mid-twentieth century.

There is less consensus over what happened to eugenics in the second half of the century. Like many controversial historical topics in U.S. history, American eugenics went through a scholarly transformation during the latter half of the century. Although the movement held enormous sway during the first half of the century, historians did not openly critique the movement until the 1960s. A new generation of scholars, influenced by the social upheavals of the decade, approached past events in American history with greater skepticism than their forebears. Just as biology textbooks were finally renouncing the legitimacy of eugenic principles, scholars began their attack on the insidious role of progressives bent on curbing the population of the so-called "unfit" in the early twentieth century. And for the first time, scholars linked the American eugenics movement to the Nazis and the Holocaust.

Although the "Nazi connection" drew greater attention to the abuse of power in the U.S. eugenics movement, it also, unfortunately, distorted the local history. Linking American eugenics to genocide in Germany provided fodder for sensational histories and the occasional journalistic frenzy, but prevented most from integrating the story into mainstream social history. In other words, rather than ask why so many Americans embraced eugenic and hereditarian ideals in the first half of the twentieth century, many scholars vilified a small number of individual racists as responsible for generating an embarrassing mistake.

Despite this somewhat limiting approach to the history of eugenics, intellectual debates over the role of nature versus nurture in human development (or heredity versus environment) kept eugenics in the spotlight and led to the production of a wider range of scholarship (focusing on the social and cultural implications of eugenics). If anything, such interest has only increased, as the Human Genome Project and other technological developments have raised the bar of genetic engineering and its implications for society's future. One thing is certain: this controversial movement in American history remains a subject worthy of attention and debate.

[*See also* **Biological Sciences; Ethics and Medicine; Gender and Science; Genetics and Genetic Engineering; Human Genome Project; Intelligence, Concepts of; Psychological and Intelligence Testing; Race and Medicine; Sanger, Margaret;** *and* **Social Sciences.**]

BIBLIOGRAPHY

Bashford, Alison, and Philippa Levine, eds. *Oxford Handbook of the History of Eugenics*. Oxford and New York: Oxford University Press, 2010.

Currell, Susan, and Christina Cogdell, eds. *Popular Eugenics: National Efficiency and American Mass Culture in the 1930s*. Athens: Ohio University Press, 2006.

Kevles, Dan. *In the Name of Eugenics: Genetics and the Uses of Human Heredity*. Cambridge, Mass.: Harvard University Press, 1998.

Kline, Wendy. *Building a Better Race: Gender, Sexuality, and Eugenics from the Turn of the Century to the Baby Boom*. Berkeley: University of California Press, 2001.

Larson, Edward J. *Sex, Race, and Science: Eugenics in the Deep South*. Baltimore: Johns Hopkins University Press, 1995.

Ordover, Nancy. *American Eugenics: Race, Queer Anatomy, and the Science of Nationalism*. Minneapolis: University of Minnesota Press, 2003.

Paul, Diane. *Controlling Human Heredity, 1865 to the Present*. Amherst, N.Y.: Humanity Books, 1999.

Wendy Kline

EVOLUTION, THEORY OF

Until the late 1850s, when the British philosopher Herbert Spencer introduced the term "evolution" to describe the history of the universe as a progression from the simple to the complex, writers discussing what came to be called evolution typically labeled it the "development" theory or, when referring to the organic world, the "transmutation" hypothesis. Few Americans paid any attention to such speculations until after 1844, when Robert Chambers, a Scot, anonymously published *Vestiges of the Natural History of Creation*, a synthesis of organic and inorganic development theories that elicited more ridicule than respect in the United States. Intense discussion of organic evolution did not begin until the appearance of the British naturalist Charles Darwin's *Origin of Species* in 1859. Although Darwin himself rarely used the term "evolution" to describe his views, they came to typify what others meant by evolution. Indeed, by the 1870s American commentators were commonly using evolution and "Darwinism" interchangeably.

Darwin's primary goal in writing the *Origin of Species* was to overthrow "the dogma of separate creations," which he regarded as "utterly useless" as a *scientific* explanation. By the mid-1870s the overwhelming majority of Darwin's fellow naturalists in America—botanists, zoologists, geologists, and anthropologists—had come to agree with him. Some were already calling evolution an "ascertained fact," although virtually none assigned as much importance as Darwin did to natural selection as the primary agent of organic change. Even Darwin's leading American champion, the Harvard botanist Asa Gray, broke with his English friend in attributing the appearance of human beings and complex organs (such as the eye) to special divine intervention. Until the second third of the twentieth century, few American biologists identified natural selection as *the* mechanism of evolution.

During the early scientific debates over evolutionary theory and the efficacy of natural selection, religious leaders typically sat on the sidelines, many of them doubting that evolution would ever

be accepted as serious science. By the mid-1870s, however, American naturalists were becoming evolutionists in such large numbers that the clergy could scarcely continue to ignore the issue. Some liberals simply baptized evolution as God's method of creation. But most religious leaders—whether Protestant, Catholic, or Jewish—rejected evolution, especially as it applied to human beings, or remained silent on the subject. Darwinism seemed not only to deny design in nature but also, more important, to undermine the historical and ethical teachings of the Bible. Americans who believed that God had created human beings in his image took umbrage at Darwin's assertion in *The Descent of Man* (1871) that "Man is descended from a hairy quadruped, furnished with a tail and pointed ears."

Despite widespread criticism of evolution in the late nineteenth and early twentieth centuries, no group mounted an organized crusade against it until after World War I. Concerned by the increasing exposure of the nation's youth to evolution in high schools and emboldened by erroneous rumors that Darwinism (that is, evolution) lay on its "death-bed," Protestant fundamentalists in the 1920s sought to outlaw the teaching of human evolution in the public schools of America, which they accomplished in Tennessee, Mississippi, and Arkansas. In a celebrated 1925 trial, a court in Dayton, Tennessee, found John Thomas Scopes, a high school science teacher, guilty of violating the state's anti-evolution law. Although the state supreme court later overturned Scopes conviction, the groundswell of public opposition to evolution convinced many science teachers and most textbook publishers to soft-pedal discussions of evolution.

After seven decades of debating possible mechanisms of evolution—the inheritance of characters acquired by use and disuse, the influence of climatic changes and environmental catastrophes, the role of dramatic mutations and elusive internal forces—American biologists in the 1930s and 1940s finally reached a consensus about the central role played by natural selection operating on minute variations. Together various geneticists, systematists, paleontologists, embryologists, and botanists forged what came to be called the modern or evolutionary synthesis. Above all, as William B. Provine has pointed out, the so-called synthesis squeezed out mechanisms that allowed for purpose and design in evolution.

In the 1960s evolution reappeared in American classrooms with a vengeance, as school districts across the land adopted the federally funded Biological Sciences Curriculum Study textbooks, which featured evolution as "the warp and woof of modern biology." Outraged conservative Christians launched a counterattack that continued for the rest of the century. For a hundred years following the publication of the *Origin of Species*, anti-evolutionists had been united by their antipathy to human evolution, not by agreement on the mode of creation. They typically experienced little difficulty accommodating the evidence of ancient life forms with their reading of Genesis, either by interpreting the "days" of the first chapter of Genesis as geological ages or by assuming a huge chronological gap between the creation "in the beginning" and the much later Edenic creation. Beginning in the 1960s, however, large numbers of them turned their backs on such accommodating schemes in favor of young-earth, or scientific, creationism, which collapsed virtually the entire geological column into the year of Noah's flood and shrank earth history to a mere six thousand to ten thousand years. In the early 1980s two states, Arkansas and Louisiana, passed legislation mandating the teaching of this "creation science" whenever "evolution science" was taught, but the U.S. Supreme Court in 1987 ruled that such laws violated the First Amendment to the Constitution, requiring the separation of church and state. This ruling prompted anti-evolutionists in the 1990s to push instead for the teaching of "intelligent design," based on the complexity of organic structures, and for treating evolution as a mere "theory."

By 2000 virtually all Americans, creationist and evolutionist alike, accepted the reality of "microevolution" (which for conservative Christians meant change within the originally created "kinds" of plants and animals), but the country remained bitterly divided over "macroevolution." A Gallup poll in 2010 revealed that 40 percent of Americans, including a quarter of college graduates, believed that "God created human beings pretty much in his present form at one time within the last 10,000 years." Another 38 percent thought that "Human beings have developed over millions of years from less advanced forms of life, but God guided this process." Only 16 percent supposed that "Human beings have developed over millions of years from less advanced forms of life" with God having no part in the process.

[*See also* **Anthropology; Biological Sciences; Botany; Creationism; Geology; Gray, Asa; High Schools, Science Education in; Paleontology; Religion and Science; Science; Scopes Trial;** *and* **Zoology.**]

BIBLIOGRAPHY

Larson, Edward J. *Evolution: The Remarkable History of a Scientific Theory*. New York: Modern Library, 2004.

Mitman, Gregg A., and Ronald L. Numbers. "Evolutionary Theory." In *Encyclopedia of the United States in the Twentieth Century*, edited by Stanley I. Kutler, Vol. 4, pp. 859–876. New York: Charles Scribner's Sons, 1996.

Moore, James R. *The Post-Darwinian Controversies: A Study of the Protestant Struggle to Come to Terms with Darwin in Great Britain and America, 1870–1900*. Cambridge, U.K.: Cambridge University Press, 1979.

Numbers, Ronald L. *The Creationists: From Scientific Creationism to Intelligent Design*. Expanded ed. Cambridge, Mass.: Harvard University Press, 2006.

Numbers, Ronald L. *Darwinism Comes to America*. Cambridge, Mass.: Harvard University Press, 1998.

Numbers, Ronald L., and John Stenhouse, eds. *Disseminating Darwinism: The Role of Place, Race, Religion, and Gender*. Cambridge, U.K.: Cambridge University Press, 1999.

Provine, William B. "Progress in Evolution and Meaning in Life." In *Julian Huxley: Biologist and Statesman of Science*, edited by C. Kenneth Waters and Albert Van Helden, pp. 165–180. Houston, Tex.: Rice University Press, 1992.

Roberts, Jon H. *Darwinism and the Divine in America: Protestant Intellectuals and Organic Evolution, 1859–1900*. Madison: University of Wisconsin Press, 1988.

Ronald L. Numbers

EXXON VALDEZ OIL SPILL

Shortly after midnight on 24 March 1989, the oil tanker *Exxon Valdez*, having left the port of Valdez, Alaska, and exiting Alaska's pristine Prince William Sound with 53 million U.S. gallons of crude oil, ran onto Bligh Reef, a submerged but lighted rock escarpment. The grounding shredded three of the ship's 11 cargo tanks and punctured 5 more, spilling 11 million gallons of oil. Containment equipment that should have been in place for such a contingency was not, and after four days, a storm with seventy-mile-per-hour winds drove the oil onto the sand and rock beaches of the Sound and eventually nearly five hundred miles from the spill site, fouling one thousand miles of irregular shoreline. Scientists estimate mass mortalities of sea otters, harbor seals, and unprecedented numbers of seabirds, as well as invertebrates on beaches and the shoreline, from both oil and toxic chemicals used in the attempted cleanup. The spill constituted an emotional, psychological, and economic tragedy for people living in communities on the Sound, including two Native villages whose residents depended on subsistence harvest of the water's resources. One of the greatest environmental disasters in American history, oil still lingers among the rocks and in the sand of the shore.

The ship had detoured from the established ship transit channel through the Sound because of floating icebergs, a maneuver permitted by the U.S. Coast Guard but a violation of its navigation plan. Having executed the course change, Captain Joseph Hazelwood, who, in violation of federal and state laws, had consumed alcohol prior to the ship's departure, went to his cabin, leaving the ship under the command of the third mate and an unqualified able seaman, another violation. In court, Hazelwood pled "no contest" to illegal alcohol consumption.

Analysts concluded that the accident was the result of a failure of the Exxon Corporation to assess and attend sufficiently to risk management and of federal and state agencies to effectively monitor and enforce policies and regulations intended to prevent accidents and to contain pollution. In addition to the Coast Guard's lax oversight of tankers navigating the Sound, the State of Alaska failed adequately to regulate the Alyeska Pipeline Service Company, which had primary responsibility for spill containment and cleanup. Additionally, years prior to the spill, federal courts invalidated a restrictive state risk mitigation plan that might have prevented the spill. Further, Exxon Corporation had automated many of the ship's navigation functions, allowing a smaller crew, but burdening them with fatiguing and debilitating responsibilities and working conditions.

Exxon paid a $150 million criminal fine, $125 million of which was forgiven in acknowledgement of its cleanup effort. The corporation paid $100 million for damages to wildlife and land and $900 million in civil fines, three fourths of which capitalized restoration work by a newly created, joint federal–state *Exxon Valdez* Oil Spill Trustee Council. Subsequent federal court action exacerbated the continuing impact of the spill because a class action suit to compensate affected fishermen and communities in the Sound was not settled for 20 years.

[*See also Deepwater Horizon* **Explosion and Oil Spill; Environmentalism; Environmental Protection Agency;** *and* **Ethics and Professionalism in Engineering.**]

BIBLIOGRAPHY

Exxon Valdez Oil Spill Trustee Council. http://www.evostc.state.ak.us/ (accessed 10 April 2012).
Spill: The Wreck of the Exxon Valdez, Implications for Safe Transport of Oil; Alaska Oil Spill Commission: Final Report. Anchorage: Alaska Oil Spill Commission, 1990.

Stephen Haycox

F

FARNSWORTH, PHILO TAYLOR

(1906–1971), one of the most important American inventors of electronic television. Lacking formal education or strong industrial backing, he succeeded in developing important elements of a working television system in the 1920s.

Farnsworth grew up on farms in Idaho and Utah. He showed an early proficiency for working with electrical equipment. By 1922 he was describing ideas for an electronic television system to his high school teacher. Four years later he met investors George Everson and Leslie Gorrell, who agreed to back his further experimentation. After his marriage in 1926, he moved to California to continue his television work. Early in 1927, he applied for his first television patents—five more applications followed in 1928 as experiments continued. The first public demonstrations of his system occurred in September 1928. By mid-1929, Farnsworth first transmitted an electronic television signal. In 1930–1931, he demonstrated

his work for potential investors, including RCA, first for Vladimir Zworykin and then for David Sarnoff.

During 1931, Farnsworth shifted his work to the Philco Company in Philadelphia, where he conducted experimental broadcasts. A 10-day public demonstration of his system was offered at Philadelphia's Franklin Institute in August 1934 (by this point, his television pictures had 220 lines of definition) and to British authorities later that year, after which Farnsworth and John Logie Baird signed a patent licensing contract. Similar agreements were signed with major European firms working in television. Farnsworth cameras were used in television coverage of the 1936 Olympics. By the late 1930s, he had applied for dozens more patents, but money was tight and tensions were rising with Philco, which was concerned about expenses and lack of revenues. After intermittent patent disagreements over several years, Farnsworth and RCA signed cross-licensing agreements in the fall of 1939.

Farnsworth moved to Indiana in 1938 and continued his television and other electronic research for nearly three decades. For a period in the late 1940s, television receivers bearing his name were sold. His company was absorbed by ITT in 1949. During the 1950s he explored 3D television, fusion energy, and military and industrial applications of television. He retired in the mid-1960s and returned to Utah, where he died in Salt Lake City.

[*See also* **Armstrong, Edwin Howard; Military, Science and Technology and the; Radio;** *and* **Television.**]

BIBLIOGRAPHY

Everson, George. *The Story of Television: The Life of Philo T. Farnsworth.* New York: W. W. Norton, 1949. First biography, written by a financial backer.

Farnsworth, Elma G. *Distant Vision: Romance and Discovery on an Invisible Frontier—Philo T. Farnsworth, Inventor of Television.* Salt Lake City, Utah: Pemberly Kent, 1990.

Godfrey, Donald G. *Philo T. Farnsworth: The Father of Television.* Salt Lake City: University of Utah Press, 2001.

Christopher H. Sterling

FERMI, ENRICO

(1901–1954), nuclear physicist, Nobel laureate, and inventor of nuclear power. Born in Rome, Fermi obtained his doctorate at the University of Pisa in 1922. After postdoctoral studies with Max Born at the University of Göttingen in Germany, he returned to the University of Florence, where he mathematically analyzed the statistical behavior of electrons using quantum mechanics creating Fermi–Dirac statistics, independently of Paul Dirac. Particles whose behavior obeys them are known as "fermions." Appointed full professor at Rome shortly thereafter, he formulated the theory of nuclear beta-decay to explain the range of energy in nuclear emission of electrons in terms of a "weak" nuclear force carried by a "neutrino," a particle that was experimentally detected only in 1956

by American physicists Fred Reines and Clyde Cowan. In a series of nuclear physics experiments, Fermi and his team of nuclear physicists in Rome found that neutrons slowed by collisions were efficient agents of nuclear transformations. He discovered a number of new artificially radioactive isotopes and won the Nobel Prize in Physics in 1938 "for his demonstrations of the existence of new radioactive elements produced by neutron irradiation, and for his related discovery of nuclear reactions brought about by slow neutrons." Fermi used the Nobel Prize money to escape to the United States from Mussolini's Italy and joined the faculty of Columbia University.

At Columbia, Fermi learned of the discovery of fission from Niels Bohr early in 1939. With another émigré physicist, the Hungarian-born Leo Szilard, he devised and built a series of prototype nuclear reactors, or "piles," which used purified graphite to slow neutrons emitted in uranium fission to create a controlled chain reaction as well as to transform natural uranium-238 into plutonium-239, which had only just been discovered by Glenn Seaborg and which was to become an important fissile alternative to uranium-235, a rare isotope that was used in the first atomic bomb designs. After the United States entered World War II, he continued this work at the University of Chicago as a member of the Manhattan Project. On 2 December 1942, Fermi achieved the first self-sustaining nuclear chain reaction. His piles were developed at the Oak Ridge Laboratory and scaled up by the DuPont Corporation at Hanford, Washington, where they converted uranium into plutonium on an industrial scale. This first successful nuclear device tested at Alamogordo, New Mexico, in July 1945, and "Fat Man," the bomb dropped on Nagasaki, Japan, on 9 August 1945, used plutonium made in these reactors.

After the war, Fermi launched the Institute for Nuclear Studies at Chicago, where he was appointed professor in 1946, and pioneered scientific computation at Los Alamos. A member of the Atomic Energy Commission's General Advisory Committee, he opposed the development of the hydrogen bomb, which he had conceived in 1942. Revered and much beloved in the scientific community, he died in 1954 of stomach cancer.

[*See also* American Institute of Physics;
Atomic Energy Commission; Einstein, Albert;
Manhattan Project; Mathematics and Statis-
tics; Nobel Prize in Biomedical Research;
Nuclear Power; Nuclear Regulatory Com-
mission; Nuclear Weapons; Oppenheimer,
J. Robert; Physics; Quantum Theory; Sci-
ence; *and* Seaborg, Glenn T.]

BIBLIOGRAPHY

Cooper, Dan. *Enrico Fermi and the Revolutions in
Modern Physics*. New York: Oxford University
Press, 2009. This is the most recent biography, although it is intended for nonspecialists.
Fermi, Laura. *Atoms in the Family*, reprint ed. Chicago: University of Chicago Press, 1995. Enrico
Fermi's wife wrote this personal biography of her
life with the physicist.
Segrè, Emilio. *Enrico Fermi, Physicist*. Chicago: University of Chicago Press, 1970. This is still the authoritative biography, written by Fermi's student
and collaborator.

Robert W. Seidel

FEYNMAN, RICHARD

(1918–1988), theoretical physicist, Nobel laureate. Raised in an assimilationist Jewish household in Far Rockaway, New York, Richard Phillips
Feynman attended the Massachusetts Institute of
Technology for his undergraduate studies before
beginning doctoral work in physics at Princeton
University. His graduate work was interrupted by
the outbreak of World War II. Feynman was recruited to work at the top-secret Los Alamos Laboratory in New Mexico, one of the central sites of
the sprawling wartime Manhattan Project, whose
aim was to design and build nuclear weapons.
Early on, Feynman emerged as a young leader at
the laboratory, working alongside senior physicists like Hans Bethe in the theoretical physics
(or T) division. In fact, Feynman was promoted
to group leader within T Division early in 1944,
making him the youngest group leader at Los

Alamos. His research during the war focused
on ways to calculate how neutrons—tiny, elec-
trically neutral constituents of atomic nuclei—
would behave inside a slab of fissionable material
like uranium.

Immediately after the war, Bethe recruited
Feynman to teach at Cornell University. While
there, Feynman returned to a challenge that
had stymied the world's theoretical physicists
for decades: how to rectify quantum theory
(which treated matter at the scale of atoms and
parts of atoms) with special relativity (which
treated matter at speeds comparable to the speed
of light). For years, every effort to unite the two
main pillars of modern physics had led to
mathematical nonsense: infinities spoiled every
calculation, rather than yielding finite numbers.
Feynman invented an idiosyncratic means of
breaking down the calculations and representing them, piecemeal, by simple line drawings
(now known as "Feynman diagrams"). With
the aid of the diagrams, Feynman demonstrated
how to isolate and ultimately remove the infinities. His solution was formally quite distinct
from other efforts, pursued independently by
Julian Schwinger and Sin-Itiro Tomonaga, although Freeman Dyson ultimately demonstrated
their mathematical equivalence. Feynman shared
the Nobel Prize in Physics with Schwinger and
Tomonaga for this work in 1965.

Feynman left Cornell for the California Institute of Technology in 1950, where he taught
for the remainder of his career. In addition to
pathbreaking research in quantum theory, nuclear physics, condensed-matter physics, and
gravitation, Feynman also became a celebrated
teacher and popular author. His three-volume
Feynman Lectures on Physics (1964) remains a
pedagogical classic, and his autobiographical
essays—*"Surely You're Joking, Mr. Feynman!"*
(1985) and *"What Do You Care What Other
People Think?"* (1988)—became best sellers.
Late in life, Feynman made headlines when he
served on the official investigating commission
into the 1986 *Challenger* space shuttle explosion. His most memorable contribution came
during a televised news conference in February
1986, when he dipped a piece of rubber O-ring

from a space shuttle booster rocket into a cup of ice water to demonstrate how quickly the rubber lost its elasticity. Failure of the rubber O-ring seals during the cold January launch eventually emerged as the leading cause of the shuttle disaster.

[*See also* **Bethe, Hans;** *Challenger* **Disaster; Manhattan Project; Nobel Prize in Biomedical Research; Nuclear Weapons; Physics; Quantum Theory; Space Program;** *and* **Space Science.**]

BIBLIOGRAPHY

Works by Richard Feynman
Brown, Laurie, ed. *Selected Papers of Richard Feynman, with Commentary*. River Edge, N.J.: World Scientific, 2000. A collection of many of Feynman's most significant publications in physics, with helpful discussion provided by Brown, a theoretical physicist and former student of Feynman's.
Feynman, Michelle, ed. *Perfectly Reasonable Deviations from the Beaten Track: The Letters of Richard P. Feynman*. New York: Basic Books, 2005. A selection of Feynman's previously unpublished correspondence, edited and with commentary from his daughter.

Secondary Works
Gleick, James. *Genius: The Life and Science of Richard Feynman*. New York: Pantheon, 1992. A lively biography of Feynman written by an award-winning science journalist.
Kaiser, David. *Drawing Theories Apart: The Dispersion of Feynman Diagrams in Postwar Physics*. Chicago: University of Chicago Press, 2005. A history of how Feynman's idiosyncratic approach to quantum theory, centered around his diagrammatic methods, became the dominant approach to high-energy physics.
Krauss, Lawrence. *Quantum Man: Richard Feynman's Life in Science*. New York: W. W. Norton, 2011. An accessible introduction to Feynman's main contributions to physics, written by an accomplished theoretical physicist and popular author.

David Kaiser

FILM TECHNOLOGY

Unlike the novel, poetry, the theater, painting, sculpture, and other preindustrial art forms, the motion picture is the product of technology. The "marks" of technology are clearly visible and audible on the films themselves, which range from silent to sound, from black and white to color, and from narrow-screen to widescreen. The development of motion picture technology was uneven and highly mediated. It did not, like Athena, spring, fully formed, from the head of Zeus, but evolved slowly within a horizon of other industrial technologies, ranging from photochemistry, optics, photography and cinematography, phonography, radiotelegraphy, telephony, and audio electronics to various digital technologies. Mediation comes in a variety of forms—often through the conflicting needs and demands of the marketplace into which a particular technology is introduced. In 1953, Twentieth Century–Fox, for example, designed its widescreen cinema format Cinema-Scope with four-track magnetic stereo sound for it to compete with Cinerama's seven-track magnetic stereo sound. But exhibitors balked at the additional cost of the installation of four-track sound in the theater, forcing Fox to alter the technology and to provide a traditional monaural soundtrack alongside the magnetic tracks, resulting in a diminution of the width of the overall image (and in monaural sound in most theaters).

Mediation plays a crucial role in what Rick Altman (2004) has called "crisis historiography." Altman argues that the identity of a new technology is both socially and historically contingent: it depends on the way users develop and understand it. In other words, it is subject to an identity crisis (thus the term "crisis historiography") whereby its initial identity is subject to redefinition. Thomas Edison's phonograph, for example, was designed for one purpose—the recording of business dictation—but transformed by its users to accomplish something else—the recording of music. This unintended use spawned the phonograph industry.

Although motion picture technology evolved slowly, its initial invention, as André Bazin (1967) has suggested, "imagined" its ultimate development. In the 1890s, films were made in sound, color, and widescreen (whereas British inventor William Friese-Greene took out patents for a motion picture three-dimensional [3D] process in 1893). The fact that it took decades for these technologies to be perfected provides further evidence of the sort of mediation that characterized the unevenness of the cinema's development. Sound awaited the invention of the audion tube by Lee de Forest in 1906, which enabled the amplification of sound, the advent of electrical recording (ca. 1924), which extended the frequency range that could be captured by the microphone, and improvements in sound-on-disc and sound-on-film recording equipment (ca. 1922–1926). Color, which was always available through various technologies, took the longest to become a universal norm. It is true that three-strip Technicolor (ca. 1932–1934) provided a technology that was finally capable of reproducing the full color spectrum, but Technicolor's reliance on color-sensitive black-and-white negatives and the addition of dyes in the printing process represented a stage of technological development that was essentially incomplete. It was only with the introduction of color-negative, -positive, and -intermediate film materials, associated with the advent of Eastman Color in 1950–1953, that the basic technological problems related to color film had been solved. But economics also played a crucial role in the history of motion picture color. It was not until 1965, when an ancillary market for color features opened up on network television, that Hollywood had an economic incentive to make all of its films in color. That year NBC offered its viewers an all-color, prime-time television schedule. CBS and ABC immediately followed suit. Big-budget Hollywood features in color were central to this new programming strategy; the value of color films sold to television escalated, whereas black-and-white films became less and less desirable for prime-time airing and commanded significantly smaller rental or license fees. Widescreen, which failed to catch on with Grandeur and other large-format processes in 1929 and 1930, did not become an industry norm until 1953–1954 when Cinerama,

CinemaScope, VistaVision, and Todd–AO proved to be popular box office attractions.

At each stage of its invention, the cinema renegotiated its identity as its basic technology, which was often developed for other purposes in other fields, was redefined through its users. Thus microphones, amplifiers, speakers, and other electronic equipment developed for the telephone and the radio (as well as the engineers who used them) gave early film sound telephonic and radiophonic characteristics before they were gradually "repurposed" for the medium of film. The space constructed by early sound films (ca. 1926–1927), for example, rarely matched the space depicted on screen. It was not until 1929 and 1930 that the codes of sound perspective had been established.

The following essay looks at major periods of technological development in the cinema in an effort to highlight moments of crisis in the cinema's identity. Those periods include the invention of the cinema and the major technological innovations that have taken place since that initial coming into being—the advent of sound, color, widescreen, and digital cinema.

The Invention of the Cinema. The invention of the cinema began with the phonograph. Edison's cylinder phonograph was invented in 1877 (and then languished on the inventor's shelf until 1888 when it was "perfected"). That same year (1888), Edison filed a caveat (an intention to develop an invention) with the Patents' Office for a motion picture device. In it, he announced: "I am experimenting upon an instrument which does for the Eye what the phonograph does for the Ear, which is the recording and reproduction of things in motion.... The illusion is complete and we may see & hear a whole Opera as perfectly as if actually present although the actual performance may have taken place years before."

Modeled on his earlier invention, the prototype of Edison's moving image machine looked and operated remarkably like the phonograph. It consisted of a rotating cylinder designed to record and play back a sequence of microphotographs arranged in a spiral (like grooves) around the circumference of a drum driven by a hand-cranked feed screw. A lens for filming and viewing occupied a position similar to that of the phonograph's recording stylus/playback horn.

Although Edison and his assistant, W. K. L. Dickson, soon abandoned this design (because it did not work), the concept of the cinema as an extension of the phonograph remained: Edison's Kinetoscope, when it was commercially exploited in April 1894, combined recorded images with recorded sound (in the form of music). The music was loosely synchronized to the images, played back on an Edison phonograph, and heard through stethoscope-like ear tubes. Even more important, the marketing of Edison's new invention borrowed a page from that of the phonograph. Edison's films were exhibited in Kinetoscope parlors on banks of peep-show machines, recalling the arcade-like phonograph parlors designed to exploit that device commercially in the late 1880s. But others reimagined Edison's invention, adapting it for projection on a large screen, transforming Edison's individual viewer into mass audiences. With projection, the cinema moved to new spaces—the legitimate theater, the lecture hall, vaudeville, and the fair ground—and took on features of other public amusements that occupied those spaces—the lantern slide show, the lecture, and the vaudeville or sideshow attraction, becoming the "cinema of attractions" that, as Tom Gunning (2000) observes, characterizes pre-1908 filmmaking practice. During the first few years of this period, the technology itself emerged as an attraction as spectators did not come to see individual films but the Cinematographe, the Vitascope, and the Biograph, the machines that projected them.

Sound. The coming of sound marked a major transformation of the essential nature of the cinema, imposing a new identity on it from without. Western Electric, RCA, General Electric, and other corporations that had invested heavily in the research and development of sound recording and transmission technologies, ranging from the telephone and public address systems to the record industry and radio broadcasting, sought to maximize the return on their investments by exploiting their patents in the highly profitable film industry. Eventually, they formed

partnerships with two studios at the bottom of Hollywood's economic pyramid, Warner Bros. and Fox. These two studios were eager to improve their positions within the film industry and willing to take a gamble on sound. One patent-holder, RCA, even formed its own studio, as well as a distribution organization and an exhibition circuit, to exploit its patents. The resultant company was RKO Radio Pictures, which quickly joined the ranks of the five major studios in Hollywood. Typically, all the other studios sat on the sidelines, waiting to see which system won out before investing any money in the conversion to sound.

Drawing on the preexisting association of the phonograph with music, Warner Bros. designed its sound-on-disc Vitaphone system for the reproduction of music in the theater, seeking to eliminate the cost of live accompaniment. Theaters thus initially placed loud speakers in the orchestra pit. But when Al Jolson's ad-libbing in *The Jazz Singer* (1927) gave speech greater novelty value than music, "talking" pictures became the norm and loud speakers were placed *behind* the screen to better match the source of on-screen human speech. By the time that sound on disc had given way to sound on film (ca. 1930), the priority of dialog over music and sound effects had established itself as the natural order of things in Hollywood, putting in place a hierarchy of film sound that persists to this day.

The advent of stereo magnetic sound in the postwar era threatened this order. Studios such as Twentieth Century–Fox, with their Cinema-Scope process, placed three speakers behind the screen and sought to "travel" dialog from speaker to speaker as the actors moved across the screen. Having internalized the codes of monaural sound, audiences refused to accept speech from any but the center speaker, forcing sound mixers to keep all significant dialog in the center channel (another practice that continues to this day).

Stereo sound took its time in becoming a norm. Associated with 70 mm processes in the 1950s, 1960s, and 1970s, six-track magnetic sound did not become a norm until the early 1990s when 5.1 sound became the basic configuration for the major digital sound systems, such as Dolby Digital and Digital Theater Systems (DTS) (5.1 sound features three channels behind the screen, two surround channels and one [0.1] low-frequency channel). Dolby Stereo brought high-fidelity, four-track stereo sound within the financial reach of the average neighborhood theater and rapidly developed trade-name recognition among general audiences. In 1975, Dolby's stereo optical system was introduced with the release of *Tommy*. For the first time since CinemaScope, four-track stereo accompanied a 35 mm film. But unlike Cinema-Scope, Dolby's stereo tracks could be optically printed at the same time that the film itself was printed, avoiding the costly process of magnetic striping and of transferring the sound to the various magnetic tracks.

The success of *Star Wars* in 1977, which featured an elaborate and spectacular sound design, made Dolby a household name. Dolby SVA (Stereo Variable Area) sound upgraded the low end of the exhibition marketplace. At the high end, Dolby perfected a six-track 70 mm system that incorporated an extended bass response that transformed 70 mm showings of *Star Wars* into multisensory events.

At the turn of the century, advocates of digital technology proclaimed that the age of Edison was over. The phonograph had given way to the compact disc. And the compact disc became a model for digital sound. The 5.1 sound for the DTS system was even played in theaters on a compact disc kept in synchronization with the projected film image.

Color. During the silent era, color was either "natural" or "applied." Natural color systems, such as Kinemacolor, Gaumont Chronochrome, Prizma Color, and two-color Technicolor, relied on filters to secure black-and-white records of the original scene and on filters or dyes to reproduce them. Applied color processes involved the application of color (after the fact, as it were) to a black-and-white image—as was done in the case of films that were hand painted, stencil colored, tinted, or toned. Applied color practices drew extensively on other, noncinematic technologies, such as lantern slide painting, the use of colored lights on the stage, and the techniques of painting itself. Color was something that was added on to an original black-and-white image, like paint to an empty canvas. The evolution of color from applied to natural can be seen on a technological level in the shift from hand painting, tinting, and toning (in which paint or dyes are arbitrarily applied to the image) to the Technicolor imbibition process (in which dyes are added to the image in proportion to color information contained within black-and-white color records) and finally to dye coupler color technology (in which dyes contained within the film negative itself are released in proportion to color information within the color negative). A color that was initially outside the image eventually comes from within it. The digital intermediate (DI) process effectively takes this development to the next level. Unlike colorization, which applied color to black-and-white images by electronic means and which clearly was a step backward in the evolution of color cinema, DI works *within* the picture elements of the image (the pixels) to control and manipulate color. This is actually depicted visually in the first major digital intermediate film, *Pleasantville* (1998), in which the characters' "true colors" are released from beneath the surface of their black-and-white appearances.

Ironically, the evolution of color from novelty to norm involved the suppression of distinctions between color and black and white and the adoption of the codes of black-and-white cinema, putting color through an "identity crisis." The use of color in the first Technicolor, live-action short *La Cucaracha*, for example, relied extensively on prior use of color in other representational forms. The film's producers enlisted the famous stage designer Robert Edmond Jones to sculpt the film's mise-en-scène with colored lights (an obtrusive technique that Technicolor subsequently banned for over 20 years). Jones imposed codes belonging to the theater onto the cinema. At the same time, the "meanings" of individual colors in the film did not arise from within the film but were borrowed from extra-cinematic cultural stereotypes—green represented jealousy; red conveyed anger. The successful incorporation of color into classical Hollywood narratives lay not in the arbitrary imposition of extra-filmic color codes, but in the careful application of certain codes already present in the cinema, specifically the prior conventionalization of black and white as a diegetic norm, as the signifier of dramatic verisimilitude.

Color supplanted black-and-white film as a norm by adopting its codes and conventions (i.e., through a process of codification whereby variation in the saturation of color empowered color as a narrative tool). *Snow White and the Seven Dwarfs* (1937), for example, attempts to create a clearly defined color system through the narrativization of color, that is, by identifying specific characters and situations with specific colors. Thus the Disney film uses earth tones and desaturated colors for the characters (Snow White and the Dwarfs) and settings (the forest, the cottage) associated with nature and natural cycles, and it uses artificial, saturated colors for the Queen, who is associated with the unnatural, with magic, with the reversal of the laws of nature. The color spectrum, which ranges from desaturated pastels to highly saturated primaries, becomes crucial to the creation of an illusion of reality within what is perhaps the least realistic of dramatic forms—the animated cartoon. Thus the film moves back and forth between the stylized naturalism of Snow White and the demonic expressionism of the Queen: yet even at this latter end of the spectrum, when color is spectacularized, as in the Queen's lab and in her subsequent transformation from beautiful queen to ugly witch, the color continues to work within the fairy tale's diegetic reality, expanding the film's range of expression without rupturing the coherence of its fantasized illusion of reality.

The most recent development in motion picture color has been the introduction of the DI process in postproduction. The DI process begins with the scanning of the original camera negative and its transformation into a data file. Once this is done, a digital workflow occurs in which various operations are folded into this master file, including conforming the negative, importing and integrating visual effects elements, incorporating an edit decision list, performing color correction and timing, and outputting the finished digital file to film. Although it includes virtually all postproduction operations, the DI process has, over the years, become primarily associated with color grading. DI colorists, for example, can isolate individual elements of the image and manipulate them without changing anything else, enabling filmmakers to change the saturation of one particular color of one particular object in a shot, to control the brightness of light coming through a window, to tinker with general atmosphere, or to alter skin tone. DI does for movies what Photoshop does for still images. With DI, filmmakers can quite literally paint with cinema.

Widescreen. Between 1948 and 1952, the average weekly attendance at motion pictures in the United States fell from 90 million to 51 million, largely as a result of new patterns in leisure-time entertainment. Consumers abandoned passive entertainment, such as film-going, in favor of active participation in new forms of recreation, such as gardening, hunting, fishing, boating, golfing, and travel. These activities filled greater and greater blocks of the leisure time available to postwar audiences, and television satisfied their need for short-term, passive entertainment.

The American motion picture industry responded to these new patterns by redefining the nature of the cinematic experience, providing more participatory forms of entertainment, which were modeled in part on concepts of presence associated with the legitimate theater. New motion picture technologies involved spectators with the on-screen action in ways that provided them with an enhanced illusion of participation. Thus Cinerama, introduced in September 1952, informed its audiences that "you won't be gazing at a movie screen—you'll find yourself swept right into the picture, surrounded by sight and sound." Ads for CinemaScope told potential spectators that it "puts YOU in the picture." Ads for Todd–AO's *Oklahoma!* (1955) declared that "you're in the show with Todd–AO."

Cinerama achieved its remarkable sense of participation by filling the spectator's field of peripheral vision, encompassing an angle of view that was 146 degrees wide and 55 degrees tall. This was accomplished by filming with three interlocked 35 mm cameras, which were equipped with wide-angle, 27 mm lenses set at angles of 48 degrees to one another. In the theater, three interlocked projectors in three different booths were used to project the three separate strips of film onto a huge, deeply curved screen. Seven-track stereo sound further enhanced the illusion of immersive participation.

CinemaScope, which was innovated by Twentieth Century–Fox, attempted to duplicate Cinerama, streamlining it for adoption by the film industry as a whole. The cornerstone of the CinemaScope system was an anamorphic lens, which compressed a wide angle of view onto 35 mm film; a similar anamorphic lens on the projector in the theater decompressed the image, producing it on a slightly curved, highly reflective screen.

Digital Cinema. "Digital cinema" properly encompasses the digitization of each aspect of the film chain, from production and postproduction (editing) to distribution and exhibition (projection). Digital technology was employed in the late 1970s with the development of computer-controlled cameras to film special-effects sequences in films such as *Star Wars* (1977). In 1982, Industrial Light & Magic developed technology for creating images on a computer for the Genesis sequence in *Star Trek II: The Wrath of Khan*. In 1984, George Lucas developed EditDroid and SoundDroid, computerized, electronic nonlinear editing systems. In the 1980s, James Cameron relied increasingly on computer-generated imagery (CGI) for special-effects work in *The Terminator* (1984) and *The Abyss* (1989), resulting in the three-dimensional "pseudopod" character in *The Abyss* and the "liquid-metal man" in *Terminator 2* (1991). The early 1990s also witnessed an industry-wide shift in postproduction from linear to nonlinear editing. At the same time, digital sound was introduced. In 1995, *Toy Story*, the first completely computer-generated film, was released. By the end of the decade, Lucas and others began filming live-action scenes with digital cameras. *The Phantom Menace* (1999) contained over two thousand digital effects shots and *Attack of the Clones* (2002) was shot entirely on twenty-four-frame, progressive high-definition digital video. The completion of the so-called "digital revolution" in the

cinema occurred in June 1999 when *Phantom Menace* was distributed and exhibited in electronic form. Digital projection marked the completion of the film chain, but it also exposed the weakest link in the chain. Exhibitors balked at the cost of digital projectors, which were, at that time, $100,000 per unit compared with $35,000 for a film projector. The revolution stalled at the exhibition phase.

Advocates for digital cinema celebrated the technology as revolutionary and compared it with the coming of sound, color, and widescreen. Although digital cinema may transform the way films are made, distributed, and exhibited, it does not transform the experience that spectators have of moving images and sound in the theater in the way sound, color, 3D, and widescreen did. It merely duplicates the experience spectators have always had with 35 mm film. The fact that digital cinema lacks any novelty value hampered attempts to convert movie theaters to digital projection. In an attempt to compensate for this problem, proponents of digital cinema have sought to give digital cinema a novelty value through the use of digital 3D. The gamble paid off. The stalled rollout of digital cinema gave way to dramatic expansion in the number of digital screens, both domestic and international. Following the phenomenal success of digital 3D films such as *Avatar* (2009), *Alice in Wonderland*, and *Toy Story 3*, the conversion of theaters has increased dramatically, doubling each year since 2006.

Digital cinema has succeeded as a new technology by borrowing its identity from other, analog technologies—from 35 mm film, which it has sought to emulate in terms of its "look," and from 3D whose novelty value it "borrowed" in an attempt to become a new norm. The history of technology in the cinema is a constantly shifting series of identities it has taken up from the different technologies that have informed its development.

[*See also* **Animation Technology and Computer Graphics; Chemistry; De Forest, Lee; Eastman, George; Edison, Thomas; Electricity and Electrification; Photography; Radio; Sound Technology, Recorded; Technology; Telephone;** *and* **Television.**]

BIBLIOGRAPHY

Altman, Rick. *Silent Film Sound.* New York: Columbia University Press, 2004.

Bazin, André. *What Is Cinema?* Vol. 1. Translated by Hugh Gray. Berkeley: University of California Press, 1967.

Belton, John. *Widescreen Cinema.* Cambridge, Mass.: Harvard University Press, 1992.

Coe, Brian. *The History of Movie Photography.* Westfield, N.J.: Eastview Editions, 1981.

Fielding, Raymond. *A Technological History of Motion Pictures and Television.* Berkeley: University of California, 1967.

Gunning, Tom. "The Cinema of Attraction: Early Film, Its Spectator, and the Avant-Garde." In *Film and Theory: An Anthology,* edited by Robert Stam and Toby Miller. New York: Blackwell, 2000.

Happe, L. Bernard. *Basic Motion Picture Technology.* New York: Hastings House, 1975.

Higgins, Scott. *Harnessing the Technicolor Rainbow: Color Design in the 1930s.* Austin: University of Texas Press, 2007.

Limbacher, James L. *Four Aspects of the Film.* New York: Brussel & Brussel, 1968.

Manovich, Lev. *The Language of New Media.* Cambridge, Mass.: MIT Press, 2001.

McKernan, Brian. *Digital Cinema: The Revolution in Cinematography, Postproduction, and Distribution.* New York: McGraw–Hill, 2005.

Neale, Steve. *Cinema and Technology: Image, Sound, Colour.* Bloomington: Indiana University Press, 1985.

Rodowick, D. N. *The Virtual Life of Film.* Cambridge, Mass.: Harvard University Press, 2007.

Rosen, Philip. *Change Mummified: Cinema, Historicity, Theory.* Minneapolis: University of Minnesota Press, 2001.

Ryan, Roderick T. *A History of Motion Picture Color Technology.* London and New York: Focal Press, 1977.

Salt, Barry. *Film Style & Technology: History & Analysis.* London: Starword, 1983.

Society of Motion Picture and Television Engineers. *Elements of Color in Professional Motion Pictures.* New York: SMPTE, 1957.

Spellerberg, James. "Technology and Ideology in the Cinema." *Quarterly Review of Film Studies* 2, no. 2 (August 1977): 288–301.

Weis, Elisabeth, and John Belton, eds. *Film Sound: Theory and Practice.* New York: Columbia University Press, 1985.

Winston, Brian. *Technologies of Seeing: Photography, Cinematography, and Television.* London: British Film Institute Publishing, 1996.

Yumibe, Joshua. *Moving Color: Early Film, Mass Culture, Modernism.* New Brunswick, N.J.: Rutgers University Press, 2012.

John Belton

FISH AND WILDLIFE SERVICE, U.S.

The U.S. Fish and Wildlife Service is a federal agency within the Department of Interior responsible for wildlife research, management, and law enforcement. It was created in 1940 from the merger of the Bureau of Fisheries and the Bureau of Biological Survey. The former agency originated in 1871, with the establishment of the U.S. Commission on Fish and Fisheries (better known as the U.S. Fish Commission), which was created at the prodding of the famed American naturalist Spencer Fullerton Baird. Its original mandate was to investigate the decline of native fish stocks in coastal and inland waterways and develop appropriate remedies to reverse those declines. As founding director, Baird hoped the U.S. Fish Commission would focus on pure ichthyological research, but a more practically minded Congress demanded an emphasis on artificial propagation and restocking. The Bureau of Biological Survey, in turn, began in 1885, when the newly founded American Ornithologists' Union successfully petitioned Congress to create an Office of Economic Ornithology and Mammalogy within the Department of Agriculture. The primary moving force behind this effort, and the founding director of the resulting office, was the physician/naturalist Clinton Hart Merriam, who aimed to create an authoritative inventory of North American fauna and delineate the patterns of biographical distribution of the continent's species. Following the passage of the Lacey Act (1901) and the Migratory Bird Act (1918), Merriam's agency was charged with enforcing federal wildlife law and, beginning in 1903, administering federal wildlife refuges.

The Bureau of Biological Survey, established under that name in 1906, remained modest in size and scope until after 1915, when Congress began providing funds for a federal predator control program (Dunlap, 1988). Over the next several decades, this new mandate not only greatly increased the budget and size of the bureau but also embroiled it in a bitter controversy over the appropriateness of a federal agency carrying out an eradication campaign aimed at several large mammalian predators, such as the wolf, coyote, and mountain lion. In the 1950s and 1960s, growing scientific evidence about the value of predators in maintaining healthy ecosystems, along with the emergence of the modern environmental movement, resulted in successful calls to abandon this campaign.

Ironically, one of the key sources of growing environmental concern in postwar America was the work of Rachel Carson, who joined the Bureau of Fisheries in 1936 as a junior aquatic biologist. Carson rose to the rank of editor in chief in the newly established Fish and Wildlife Service in 1949 before retiring to pursue a career as a writer. After authoring a series of best-selling books about the sea, Carson completed *Silent Spring* (1962), a blockbuster book that ignited concern about the dangers of synthetic pesticides and popularized notions from the science of ecology.

Increased public interest in environmental issues during the postwar period pushed the Fish and Wildlife Service to devote more attention to the problem of endangered species. Concern about the fate of the whooping crane, bald eagle, American alligator, and other declining wildlife led to the creation of the Committee on Rare and Endangered Wildlife Species (CREWS) in 1964 (Barrow, 2009). Composed of agency biologists, CREWS issued the first federal endangered species list, a document that fostered a series of Endangered Species Acts (1966, 1969, and 1973) providing increasingly stringent protection for the most at-risk species. Critics have decried the controversial law, administered by the Fish and Wildlife Service and the National Marine Fisheries Service, for unduly restricting the rights of property owners, whereas supporters have credited it with rescuing numerous species from the brink of extinction.

[*See also* **Agriculture, U.S. Department of; Baird, Spencer Fullerton; Biological Sciences; Botany; Canals and Waterways; Carson, Rachel; Environmentalism; Environmental Protection Agency; Fisheries and Fishing;** *and* **Zoology.**]

BIBLIOGRAPHY

Barrow, Mark V., Jr. *Nature's Ghosts: Confronting Extinction from the Age of Jefferson to the Age of Ecology.* Chicago: University of Chicago Press, 2009. Places the work of the U.S. Fish and Wildlife Service in the context of broader changes in attitudes toward, scientific research on, and policies related to species threatened with extinction.

Dunlap, Thomas R. *Saving America's Wildlife.* Princeton, N.J.: Princeton University Press, 1988. A concise and authoritative history of the agency's involvement in predator control.

Mark V. Barrow Jr.

FISHERIES AND FISHING

Long before European contact, Native peoples all along North American coastlines drew heavily upon local marine resources. In the Pacific Northwest, native peoples took salmon, whales, seals, sea otters, abalone—in short, a variety of all usable marine life living both inshore and offshore. Similar patterns marked East Coast and Gulf peoples as well, and in all areas, the use of marine resources played important roles in shaping settlement patterns and the social development of Americans before European contact (Bragdon, 1999).

Offshore Fishing. Offshore fish stocks drew Europeans to North America shortly after John Cabot's (Zuan Chabotto) voyages for the English crown in 1498 (and some allege even before Columbus as well). Basque, French, Portuguese, and later English adventurers mounted season-long voyages to Newfoundland as early as the first decade of the sixteenth century.

Because its lean flesh lent it to thorough air- and salt-curing, Europeans focused most intently upon codfish for their export production. Continental European fishermen, with access to cheap sources of salt needed to cure fish, rarely landed on the Newfoundland shore, preferring instead to remain at sea for their entire voyage. English fishermen, who had to pay more for salt, needed shore stations to air-dry their catch before using just enough salt to finish the cure. Consequently, English fishermen built seasonal shoreside fishing on the Newfoundland shore; these camps, occupied regularly during the fishing season throughout the sixteenth century, represent the earliest permanent, if not continuously occupied, European settlements in North America. Similar camps were soon dotted near shore islands and headlands along the Canadian and New England shores before 1620 (Vickers, 1994).

More permanent, planned colonies in New England also drew heavily upon regional fisheries resources. Lacking tropical products for export to Europe, New England coastal communities used fishing and fishing vessels as a means to enter the Atlantic trading systems of the seventeenth and eighteenth centuries. Merchants bought fish from local fishermen and shipped top-quality fish to Europe and substandard pieces of varying degrees of edibility to the West Indies as food for slave masters to feed slave populations. Fishermen also shipped their products: when not fishing in the spring, summer, and fall, they used their schooners in the coasting trade connecting ports all along the North American coastline. In addition to creating a product for trade and an infrastructure needed to earn freight charges, fishing also fostered shipbuilding, blacksmithing, provisioning trades, and other ancillary industries needed by the fisheries export sector. Consequently, fishing allowed merchants to expand operations and to accumulate capital enough to rival even Chesapeake and West Indian plantation owners. In time, these profits provided much of the investment funds needed to begin New England's industrial development in the early nineteenth century (Vickers, 1994; Bolster, 2008).

Inshore Fishing. Whereas codfish exports dominated coastal economies, inshore fishing provided key subsistence for European settlers. Early settlement accounts list many species that settlers used to meet their various needs. Although biological factors limited their exportability, inshore runs of river herring, shad, coastal schools of mackerel, and menhaden, along with striped bass and sturgeon, all provided key inputs in local subsistence. As early as the mid-seventeenth century, these species—essential for survival but not commercially important—fell under local regulation and management to ensure their long-term survival.

Commercial Fisheries. Between 1810 and 1860, the ecological footprint of American commercial fisheries expanded dramatically. Growing populations of factory workers in eastern seaboard industrial centers created new markets for cheap fish protein. More importantly, because those markets sat close to eastern seaboard fishing grounds, fishermen could sell their catch fresh and dispense with costs and limitations of curing fish for export. As a result, American fishermen began commercially targeting more species— previously avoided for their inability to take an effective cure—broadening the ecological impact of human fishing within the ecosystem. In addition to cod, fishermen now targeted mackerel, haddock, halibut, hake, and herrings (and in particular river herring) to supply fresh markets in New York, Boston, Providence, Fall River, and New Bedford. The traditional cod fishery similarly thrived, and cod fishermen turned to new bait-intensive gear (in particular tub-trawls, or long-lines) to meet it. As a result, demand for bait species—especially menhaden—grew as well. In short, the period before the American Civil War can be seen as one of technological, ecological, and commercial expansion whose reliance upon previously untargeted stocks gave the impression that fishing on the eastern seaboard was limitless (McFarland, 1911).

Following the Civil War, American fisheries expanded geographically and consolidated commercial footings in older fisheries. American fishing firms—some led by New Englanders looking for better supplies of halibut—established operations in the Pacific Northwest after 1865. Fishing in the Gulf of Mexico and the Great Lakes also expanded as rail networks allowed those products to reach growing urban centers throughout the United States. Back in the northeast, the fisheries underwent commercial expansion, consolidation, innovation, and growth. In cod, halibut, and mackerel fisheries, what had been a largely owner-operator fishing fleet (that is, the vessels were owned at least in part by the captains that commanded them) before the Civil War emerged through the 1870s as a fishery dominated by corporate-owned vessels (McFarland, 1911; Goode, 1884–1887).

Outside New England, fishing firms expanded to take advantage of cheaper labor and previously unexploited fish stocks. In the Chesapeake and southern Atlantic coasts, African American laborers soon dominated the menhaden, alewife, shad, and other commercial fisheries. On the West Coast, eastern firms hired Native Americans and Asian immigrants—often on unfavorable terms—to catch sardines, salmon, pilchards, halibut, and cod. Not only did these fish serve West Coast markets, but also the cheap wages paid to western fishermen resulted in increased firms' product streams of Pacific fish in eastern urban markets (Taylor, 1999; Garrity-Blake, 1994; Finley, 2011; Chiang, 2008).

Calls for Restraint in Fishing. As fishing intensified in the northeast and expanded to other coasts, calls for restraint also mounted. In the 1870s, conflicts within the commercial fishery over the health of inshore fish stocks in southern New England led to the creation of the U.S. Fish Commission in 1872. Tasked with the promotion of the American fisheries, the commission and its successors sponsored extensive studies of commercial fish stocks' natural histories, abundances, distributions, and prospects for artificial propagation. They also took sides in the recurring debates over the effects of new fishing gear on fish stocks, more typically siding with industry against conservation. Unlike European discussions, however, American fisheries science—beyond artificial propagation—failed to receive significant federal support until after World War I. Reliance upon European work and the commission's proindustry stances meant that much of American understandings of fish stock recruitment, migratory patterns, effects of fishing on stock size, and causes for annual catch fluctuations remained lightly explored until the 1920s (Rozwadowski, 2002; Finley, 2011).

The advent of mechanized fishing—in the form of both beam trawling and otter trawling—in North American waters pushed fishing and fisheries science to new levels between the World Wars. Whereas sailing vessels had been towing nets—called beam trawls—along southern New England's ocean floor since the 1880s, the construction in 1905 of a coal-fired steam trawler, the F/V *Spray*, by the Bay State Fishing Co., led to a fundamental reorganization of the fishery. Initially unprofitable, the *Spray's* owners soon found a way to make the vessel pay, and by 1912, more steam trawlers were either built or on the way.

The volumes of fish these vessels landed—and wasted as dead or mangled discards—alarmed hook-and-line fishermen throughout the northwest Atlantic. Despite these protests, however, federal investigations into the effects of otter trawling paved the way for the gear's expansion. Otter and beam trawling's ecological impacts soon became as evident as their economic effects. By the early 1930s, New England experienced its first ecological collapse—in the haddock fishery—brought about largely because of the destructive consequences of otter trawling. Similarly, West Coast fisheries also began to show signs of overfishing by the 1930s and 1940s. Declines in Pacific halibut led to joint U.S.–Canadian fishing agreements in the 1930s. Fears for Pacific salmon runs in the Pacific Northwest also grew, and in California, the sardine fishery began to waver shortly before World War II. Although most fishermen and scientists interpreted such signs not as collapse but as stock migrations, historical hindsight now suggests that these marked the beginning of the twentieth-century's decline in marine fish stocks (Dewar, 1983; Finley, 2011).

Post World War II. Following World War II, fishing resumed intensively. American trawlers targeted more species in more areas and for more markets than ever before, and U.S. foreign policy used fishing as a means to extend influence and bolster allies. In the Pacific, U.S. fishing policy relying upon maximum sustainable yield theory helped American fishermen enter foreign fishing grounds and exclude foreign fleets from American ones. In the Atlantic, Cold War politics led the U.S. State Department to invite allied European fleets to fish in American waters, whereas Soviet and Warsaw Bloc nations similarly looked to the Northwest Atlantic to feed domestic populations. Furthermore, fueled by optimistic forecasts for the potential yield of world ocean resources, American foreign policy also looked to global fisheries products as an important support for decolonized and developing nations (Finley, 2011; Weber, 2002).

By the 1970s, fisheries policies in Atlantic fishing grounds led to domestic and international conflict. Large factory fishing vessels from allied and nonallied nations alike took orders of magnitude for more fish than had ever before been harvested from northwest Atlantic fishing grounds. Unsustainable catches mobilized New England fishermen to push for the exclusion of non-U.S. fishers from U.S. grounds, culminating in a joint U.S.–Canadian declaration in 1976 of two hundred–mile exclusive economic zones (EEZs) that forced out all others. In the fisheries sciences, many also began to question the ecological foundations of U.S. projections of world fisheries productions from the 1960s. Realizing that the levels forecast would result in the removal of all fish—and a good proportion of plankton too—from the world's oceans, U.S. fisheries science began a sea change wherein supporting industrial expansion as a top priority was forced to make room for consideration of sustainable catch levels. At sea, however, the two hundred–mile EEZ, combined with the federal subsidies designed to grow the American fleet that accompanied the move, soon allowed American fishermen in small boats to wreak the havoc done by foreign fishers in factory ships. By the 1980s, fish stocks in California, Washington, Oregon, New England, and the Gulf states had declined dramatically (Weber, 2002).

[*See also* Environmentalism; Environmental Protection Agency; Fish and Wildlife Service, U.S.; *and* Oceanography.]

BIBLIOGRAPHY

Bolster, William Jeffrey. "Putting the Ocean in Atlantic History: Maritime Communities and Marine Ecology in the Northwest Atlantic, 1500–1800. *The American Historical Review* 113 (February 2008): 19–47.
Bragdon, Kathleen. *Native People of Southern New England, 1500–1650.* Norman: University of Oklahoma Press, 1999.
Chiang, Connie Y. *Shaping the Shoreline: Fisheries and Tourism on the Monterey Coast.* Seattle: University of Washington Press, 2008.
Dewar, Margaret. *Industry in Trouble: The Federal Government and the New England Fisheries.* Philadelphia: Temple University Press, 1983.
Finley, Carmel. *All the Fish in the Sea: Maximum Sustainable Yield and the Failure of Fisheries Management.* Chicago: University of Chicago Press, 2011.
Garrity-Blake, Barbara. *The Fish Factory: Work and Meaning for Black and White Fishermen of the American Menhaden Industry.* Knoxville: University of Tennessee Press, 1994.
Goode, George Brown. *The Fisheries and Fisheries Industries of the United States.* Section V, Vol. 1. Washington, D.C.: U.S. Government Printing Office, 1884–1887.
McEvoy, Arthur. *The Fisherman's Problem: Ecology and Law in the California Fisheries, 1850–1980.* New York: Cambridge University Press, 1986.
McFarland, Raymond. *A History of the New England Fisheries.* New York: D. Appleton & Co., 1911.
Rozwadowski, Helen. *The Sea Knows No Boundaries: A Century of Marine Science under ICES.* Seattle: University of Washington Press, 2002.
Taylor, Joseph E., III. *Making Salmon: An Environmental History of the Northwest Fisheries Crisis.* Seattle: University of Washington Press, 1999.
Vickers, Daniel F. *Farmers & Fishermen: Two Centuries of Work in Essex County, Massachusetts, 1630–1850.* Chapel Hill: University of North Carolina Press, 1994.
Weber, Michael. *From Abundance to Scarcity: A History of U.S. Marine Fisheries Policy.* Washington, D.C.: Island Press, 2002.

Matthew McKenzie

FLEXNER REPORT

The Flexner Report (1910), an evaluation of American medical education, has become the exemplar of critical investigations of higher education. Abraham Flexner (1866–1959), a non-physician with little knowledge of medicine or medical training, was employed by the Carnegie Foundation for the Advancement of Teaching, established in 1905 by Andrew Carnegie. The foundation shared the widespread concern that many medical schools were poorly equipped to teach the new discoveries in the medical sciences and clinical medicine. In 1909, under an agreement between the foundation and the American Medical Association (AMA), Flexner participated in the AMA's second inspection of medical schools. Flexner's independently written report won wide publicity because of its descriptions of each medical school in the United States and Canada and its harsh denunciations of the education provided by many of them. The report gained a reputation as having been responsible for the closing of many medical schools, but mergers and closings had been underway for a decade. (Between 1904 and 1910 the number of medical students decreased by one fourth.) Flexner also failed to observe that the weakest schools produced few graduates. Using as a model the renowned Johns Hopkins University medical school, which opened in Baltimore in 1893, Flexner advocated laboratory instruction as the principal form of preclinical education and hospital training as the core of clinical training. The former proved to be prohibitively expensive for most medical schools and the latter, which was widely adopted after midcentury, has been criticized for neglecting ambulatory patients and their social environment.

Late in the twentieth century, the Flexner Report became a source of inspiration for some critics of medical education because it advocated the close integration of patient care, research, and teaching. Patient care in university hospitals would lead clinical faculty members to perform patient-centered research. The faculty would then teach the research methods and findings to medical students. These critics claimed that current methods of research funding encouraged medical school researchers to engage in laboratory research, which prevented them from becoming skilled clinicians and clinical teachers. In addition, the critics believed the need for revenue by medical schools led clinical faculty members to provide large amounts of patient care to paying patients, which reduced their time for teaching and research. Flexner's model was perceived as integrating the major functions of medical schools and strengthening the education of medical students.

[*See also* American Medical Association; Foundations and Health; Higher Education and Science; Hospitals; Medical Education; *and* Medicine.]

BIBLIOGRAPHY

Cooke, Molly, David M. Irby, William Sullivan, and Kenneth M. Ludmerer. "American Medical Education 100 Years after the Flexner Report." *New England Journal of Medicine* 335 (2006): 1339–1344.

Flexner, Abraham. *Medical Education in the United States and Canada.* New York: Arno Press, 1910.

Rothstein, William G. *American Medical Schools and the Practice of Medicine: A History.* New York: Oxford University Press, 1987.

William G. Rothstein

FOOD AND DIET

If one had to sum up the history of Americans and their food in a word, it would likely be "abundance." Although the first English settlers suffered difficult times, most were soon much better fed than their counterparts across the Atlantic. Thanks mainly to better diets, George Washington's Revolutionary War troops were, on average, much taller than the British soldiers facing them. Citizens of the new republic prided themselves on what a Philadelphia physician called their "superabundance" of food. For most of the free population, this meant lots of meat, accompanied by breads made from corn, rye, and, increasingly, wheat. Fruits and vegetables were abundant in season, and wild animals inland and plentiful fish and seafood along the coasts provided additional sources of protein. The winter and early spring diet comprised preserved pork, bread, beans, and root vegetables—filling, if monotonous.

By the 1830s new roads, canals, and steamboats brought vast new areas of farmland into the market economy, making a wider variety of foodstuffs available for longer durations. Food reformers now cautioned against excessive indulgence. The minister and temperance advocate Sylvester Graham, warning that meat, alcohol, and spicy foods sapped the body's vital force, condemned such foods as processed white flour that had been altered from its God-given natural state.

America's slave population, totaling nearly 4 million by 1860, experienced a very different dietary environment. Slave families typically received a scant weekly ration of cornmeal and fatty pork. Some supplemented this unbalanced fare with fish, small game, eggs, and vegetables they provided for themselves.

After midcentury, the expanding railroads transported affordable supplies of wheat, pork, and beef to the growing cities; market gardening and dairy farms proliferated around them; and steamships brought exotic foods from abroad. By 1900, skilled chefs were turning out elaborate multicourse meals in the style of French haute cuisine for the wealthy. The growing middle and upper-middle classes could readily purchase the abundant foods but could not afford the servants to prepare and serve them in this fashion and were thus amenable to calls by a new generation of food reformers for dietary restraint.

Science and New Nutrition.

The scientific basis for the reformers' crusade was the so-called New Nutrition: the discovery by chemists of proteins, carbohydrates, and fats, each with its unique physiological function. Proper nutrition now meant consuming as much of these as necessary—any less was unhealthful; any more, wasteful. Urging immigrant workers to economize, the reformers insisted that the proteins in beans were fully as nutritious as those in beefsteak. The middle classes, heeding the call to choose foods on the basis of their "physiological economy" rather than taste, made culture heroes out of dietary faddists like John Harvey Kellogg, who amplified Graham's theories with purgative nostrums based on recent scientific discoveries that the colon harbored large amounts of bacteria. The "scientific cooking" advocate Fannie Farmer offered simple menus and exact recipes in her *Boston Cooking School Cook Book* (1896). Women in the new profession of home economics, teaching about food and health in the schools, similarly insisted that science rather than taste should guide one's food choices.

For the urban immigrant poor, meanwhile, providing even subsistence nutrition for their families proved difficult. In hard times, such as the Depression of the 1890s, it was more difficult still. Impure water, tainted milk, and spoiled meat contributed to illness, infant mortality, and periodic epidemics in the slums. Tougher public-health measures such as the 1906 Pure Food and Drug Act and milk pasteurization gradually ameliorated the worst of these dietary hazards, but their health ultimately improved mainly because of more ample and varied diets.

During World War I, the federal Food Administration used the New Nutrition to persuade Americans to substitute beans, whole grains, and fresh vegetables for the meat and wheat being shipped to Europe. Meanwhile, the discovery of vitamins in the early twentieth century gave rise to a new nutritional paradigm. Its dissemination was encouraged by the transformation of food production by mass-production industries characterized by large capital investments, mechanization, complex distribution networks, and large promotion and advertising budgets. With servants having practically disappeared from middle-class homes, housewives were encouraged to buy labor-saving processed foods such as canned goods, as well as vitamin-rich citrus fruits and milk, which were said to be essential for children's health. Although still little understood, vitamins proved to be a food promoter's dream. Citrus growers, dairymen, the grain-milling industry, pickle producers—almost anyone could and did make extravagant claims. When synthesized vitamin pills became available in the late 1930s, food producers insisted that such supplements were unnecessary: a "balanced diet" would provide more than enough nutrients.

Neither the Depression of the 1930s nor World War II undermined confidence in America's abundant food supply. Indeed, the Depression-Era agricultural crisis was defined as one of overproduction of food and maldistribution of income. And despite wartime rationing, many doubted that the shortages were real. Recurring rumors insisted that food supplies were more than adequate, but that government incompetence or crooked middlemen were keeping them off the market.

Mass Production and Prepared Food.
In the postwar "baby boom" years, 1946 to 1963, the long-term tendency of food preparation to move outside the home intensified as the food industries sold harried young mothers and homemakers on the "convenience" of their products. Frozen foods and other new kinds of processed, precooked, and packaged foods became popular. From 1949 to 1959, chemists developed more than four hundred additives to help food survive these new processes. Restaurants, especially the proliferating fast-food chains, welcomed this development: with food preparation reduced to defrosting, frying, or adding hot water, unskilled labor could replace expensive, often temperamental cooks.

Gastronomical considerations took a backseat to all of this, but few seemed to notice because haute cuisine had long since fallen out of favor. In the 1920s, Prohibition had deprived expensive restaurants of the income from alcohol that had padded their profit margins. During World War II, a preoccupation with fine food had seemed unpatriotic. By the 1950s, food tastes were no longer an important mark of social distinction. Most Americans seemed satisfied by beefsteak, pizzas, fried chicken, canned-food casseroles, Jell-O molds, and frozen TV dinners. Regional differences, already undermined in the 1920s and 1930s, practically disappeared under the onslaught of mass-produced foods aimed at supposedly homo-geneous Middle American tastes. Government officials, educators, journalists, and the food industries insisted that Americans were "The Best Fed People on Earth."

The self-satisfaction eroded in the 1960s with the realization that, amid massive agricultural surpluses, millions of poor citizens could not afford an adequate diet. Programs were instituted to distribute surplus commodities and food stamps to the poor. As middle-class concerns over the healthfulness of their own diet increased, a new dietary paradigm, which one might call Negative Nutrition, arose. Whereas earlier nutritional systems had emphasized consuming healthful foods, Negative Nutrition warned *against* eating certain foods, particularly those treated with potentially harmful pesticides and chemical fertilizers and those robbed of nutrients by overprocessing. Veterans of the New Left, meanwhile, redirected their critique of capitalism toward its effects on food and the environment. The giant corporations, they charged, used their immense advertising resources to brainwash Americans into eating overprocessed, denutrified, unhealthful, and environmentally hazardous products. They pointed out, for example, that the spread of cattle ranching in South America to meet U.S. demands for beef was contributing directly to the destruction of the rain forests. Both health and morality, they insisted, dictated a preference for "organic" and "natural" foods, preferably grown by small producers.

The food industry responded nimbly, reformulating and repackaging their products with labels such as "Natural" and "Nature's Own." However, new findings in nutritional science reinforced another aspect of Negative Nutrition, as specific foods came to be identified as dangerous. Rising rates of heart disease were now blamed on high levels of cholesterol in many of America's favorite foods. Sugar, long linked to diabetes and now thought by some to be a factor in other diseases and psychological disorders, was called an addictive substance manipulated by food processors to "hook" children on nutritionally deficient products. Themes from the Graham and Kellogg eras resurfaced, as vegetarianism, once the domain of cranks, became a serious option for many. Issues relating to obesity added a new twist to the nutrition debate. In the later nineteenth century a full figure had been a mark of beauty for woman and a sign of health, wealth, and substance for men. Since the 1920s, however, evidence had accumulated of a relationship between excessive weight and higher mortality rates.

Paradoxes of American Abundance. As in previous eras, food reformers mustered impressive scientific support. By the mid-1970s, the federal government was supporting research on diet and health and urging Americans to lose weight and reduce the animal fat, sugar, and sodium in their diets. Organizations such as the American Heart Association underscored the need for dietary change. "Low-fat," "lite," "no-cal," "cholesterol-free," and "sodium-free" products now lined supermarket shelves.

The result reflected the paradoxes of American abundance. Many took to frenetic diet-and-exercise regimens, yet the average weight of Americans continued to rise. Although consumption of full-fat dairy products and red meat fell, that of other fats soared, as did that of sugar and sodium. Dietary self-denial was undermined by the foreign travel boom, which encouraged indulgence and helped once again to make food tastes a sign of social distinction. Consumers now had an unprecedented choice of foods and ways to consume them. As more women entered the workplace, the trend for food production to move outside the home accelerated. "Take-home" foods boomed, as did eating out, particularly at fast-food and other chain restaurants.

By the early twenty-first century, as obesity and associated medical problems loomed ever larger as a public-health issue, diet, nutrition, and weight loss became national obsessions. Parents protested the introduction of fast foods and sweetened soft drinks in public schools. Lawyers planned lawsuits against fast-food chains, analogous to the lawsuits against tobacco companies, for knowingly endangering their customers' health. Weight-loss programs proliferated, including the "Atkins Diet," heavy on protein and low on carbohydrates, publicized by the cardiologist Dr. Robert C. Atkins in *Dr. Atkins' New Diet Revolution* (1972 and many later editions). An extreme manifestation of the preoccupation with food and diet was the growing popularity of expensive gastric bypass ("stomach stapling") surgery as a last-ditch means of cutting intake and losing weight.

Persisting moralism, in the form of guilty consciences, impeded indulging in the abundance of food choices, but the targets of the guilt constantly shifted, as experts regularly warned of new food dangers and absolved old ones. With the Negative Nutrition now superimposed on older nutritional ideas, many Americans simultaneously tried to eat more of the foods that were supposed to prevent or cure illness and promote general good health and less of those foods deemed unhealthy or likely to cause weight gain. Americans seemed doomed by their past to both celebrate their food abundance and avoid enjoying it too much.

[*See also* **Alcohol and Alcohol Abuse; Canals and Waterways; Carson, Rachel; Chemistry; Diabetes; Disease; Environmentalism; Food Processing; Graham, Sylvester; Health and Fitness; High Schools, Science Education in; Home Economics Movement; Obesity; Public Health; Pure Food and Drug Act; Railroads;** *and* **Roads and Turnpikes, Early.**]

BIBLIOGRAPHY

Belasco, Warren. *Appetite for Change.* New York: Pantheon, 1989.

Cummings, Richard. *The American and His Food.* Chicago: University of Chicago Press, 1940.

Hooker, Richard. *Food and Drink in America: A History.* Indianapolis, Ind.: Bobbs–Merrill, 1981.

Levenstein, Harvey. *Fear of Food: A History of Why We Worry about What We Eat.* Chicago: University of Chicago Press, 2012.

Levenstein, Harvey. *Paradox of Plenty: The Social History of Eating in Modern America.* New York: Oxford University Press, 1993.

Levenstein, Harvey. *Revolution at the Table: The Transformation of the American Diet.* New York: Oxford University Press, 1988.

Nissenbaum, Stephen. *Sex, Diet, and Debility in Jacksonian America.* Chicago: Dorsey Press, 1980.

Root, Waverly, and Richard de Rochemont. *Eating in America: A History.* New York: William Morrow, 1976.

Stearns, Peter. *Fat History: Bodies and Beauty in the Modern West.* New York: New York University Press, 1997.

Wharton, James. *Crusades for Fitness: The History of American Health Reformers.* Princeton, N.J.: Princeton University Press, 1982.

Harvey Levenstein

FOOD PROCESSING

Just as other sectors of the U.S. economy became industrial in the second half of the nineteenth century, so too did food. Meatpacking and canning are perhaps the most common examples, but grain production and milling, baking (bread, cookies, cakes, and pies), and breakfast cereals also shifted toward industrial production by the turn of the twentieth century. By the 1920s, fruits and vegetables would also begin to fit neatly into an industrial system. Yet, it is wrong to say that the American diet was industrial before late in the twentieth century. Although the industrial revolution reshaped American life, it was an evolution in food processing that transformed what Americans ate (Panschar, 1956, p. 46). As the twin forces of industrialization and urbanization became more powerful in the early twentieth century, fewer Americans had the means (or perhaps the desire) to grow and process their own foods. The movement of rural Americans into higher-paying jobs in cities, often with minimal cooking facilities, led both native-born Americans and new immigrants like Anzia Yezierska and her family to rely on delicatessens, cafeterias, saloons, tearooms, and street vendors for their daily meals (Yezierska, 1925, p. 27; Turner, 2009, pp. 217–232).

Grain. Because wheat was so important to the European diet, it was commonly referred to as "the King of Cereals." Its standing in the American context was no different. Early milling was dominated by the gristmill, featuring heavy millstones that crushed grain into flour. One of the first American innovators in milling was Oliver Evans, with his gravity-driven automated mill.

His mill, perfected by 1787, relied on conveyors to move grain and flour as it was ground, sieved, dried, and stored from the top to the bottom of his building. His elevator (an endless-bucket system inside a closed chute) and hopper-boy (two radial arms set with teeth that slowly raked the flour to cool it after grinding) were widely adopted by millers throughout the country. Although Evans's invention relieved much of the hard work and labor required in milling and incrementally improved quality (Pursell, 1995, pp. 27–28), it was the importation of Hungarian rolling technology in the 1870s and the use of "high milling" of the middling purifier, which separated the bran, germ, and endosperm and produced very-high-quality patent flour from hard wheat, that transformed Minneapolis's more traditional mills into the truly industrial factories that made "Mill City" into the center of American flour production (Vulté and Vanderbilt, 1916, pp. 61–63). By the 1880s, rolling and sifting technologies allowed steam-softened wheat to be milled into consistent blends of flour that gave millers, and ultimately bakers, control over their products. In combining protein-rich hard wheat with starch-rich soft wheat, millers could produce specialized flours for all sections of the baking industry as well as for home use. The three major classes of flour were bread (high protein/low starch), cake (high starch/low protein), and all purpose (intermediate mix of protein and starch), used for products that need some gluten development, but also require a light crumb, like biscuits, cookies, crackers, and pie dough (Panschar, 1956).

An additional advantage of flour produced in the industrial flour mills of Minneapolis and other northwestern cities was that it was relatively shelf stable. Unlike flour produced using traditional millstone technologies, which ground the oily germ into the flour, rolling-mill flour did not become rancid quickly (Vulté and Vanderbilt, 1916, p. 63). Flour's new stability allowed it to be shipped farther via expanding railroads and stored longer in urban warehouses, transforming a once perishable product into a commodity (Cronon, 1991). In scaling up and further refining flour production, companies like Washburn-Crosby,

General Mills, and Pillsbury made more flour available more cheaply to all Americans than ever before (Chandler, 1977). This flour was also more predictable in its protein and starch content, allowing bakers and housewives to make more consistent products. The flavor, color, and texture of industrial flour also met Americans' preference for fine white flour (Cummings, 1940; Levenstein, 1988, p. 22). The availability of large volumes of flour at any time of year encouraged growth in the baking industry.

With the development of turbomilling in the late 1950s, millers no longer needed to blend hard and soft wheat to create patent flours because the technology enabled millers to separate protein molecules from starch molecules. Using turbomilling, which combined air-separation, flow-dynamics, and centripetal-force technologies, millers produced an even wider array of protein/starch blends in flour (Larson, 1959, pp. 194–197). Industrial bakers could at last control the protein and starch content of their dough with exacting precision, giving them more consistency in their product as it flowed through the factory.

Meat. Although corn has a much longer history in America than wheat, its significance to food processing lies in meat rather than grain production. In fact, cornmeal is more closely linked to deprivation and malnutrition than any other cereal produced in the United States, despite its importance to the southern diet both during and after slavery. Even before the Civil War, corn fattened both hogs and cattle across the nation. With the closing of the West and the rise of Chicago as the center of railroading and meat processing, corn moved from the hinterlands into midwestern cities to finish pork and, increasingly, beef.

The shift from Cincinnati to Chicago as the heart of American meat production signaled the ascendance not only of railroads but also of beef (Horowitz, 2006, p. 29). With the opening of the Ohio River Valley in the early 1800s, Cincinnati became a center for pork slaughtering because of its location. River and canal systems brought country pigs to the city to be converted into meat or, more precisely, barrel pork. Industrial pork was not fresh, but rather cured and salted so that it could be shipped vast distances. It was also seasonal when farmers took advantage of the pigs' summer foraging and additional weight from corn to slaughter in the fall and winter. The same rivers that brought pigs into the city facilitated pork's dissemination to eastern cities and western farmers. By midcentury, it was uncommon for middling Americans not to have meat, in the form of cured pork, on a daily basis. Even the poorest American could afford pork regularly (Horowitz, 2006, pp. 12–13). The key to producing barrel pork was the mechanization of the slaughterhouse. Pigs were herded to the top of a four-story building, where they began their journey from animal to food. Once at the top, animals were struck in the head with a mallet and then had their throats cut and were hung from their hind feet to be bled out. The pig flowed from one operation to another until it was cut into pieces and layered with salt, saltpeter (potassium nitrate), and sugar and then filled with water. The barrels of pickled pork could be stored and shipped great distances to feed Americans from shore to shore.

As railroads began to traverse the Mississippi Valley, Chicago became the center of meatpacking because it connected farmers to cities (Cronon, 1991, pp. 73–74). Not only did the vast planes of the trans-Mississippi West produce ever more wheat and corn, but also its grasslands were ideal for grazing cattle. The destruction of the Plains Indians and the buffalo allowed ranchers to raise cattle in ever-larger numbers (Cronon, 1991, pp. 213). New railheads allowed live cattle to move quickly from field to slaughter in urban areas. As the rail hub linking the West to the East, Chicago became a logical place for slaughtering. Although Chicago meatpackers did not invent the disassembly line, they certainly perfected it, largely by adding refrigeration to both the slaughterhouse and the packing shed. Keeping meat cold rather than preserved transformed Americans from pickled pork eaters into fresh beef eaters (Horowitz, 2006, pp. 44). Gustavus Swift's refrigerated railcar solved the shipping problem. Industrial refrigeration in the plant, on the rails, in the branch house, and at the butchers provided fresh meat year-round. Mass beef was also cheap beef, and thus pork became the "other red meat" and beef became synonymous with American prosperity in the first decade of the twentieth century. By colluding with railroads and forcing many local slaughterhouses and distributors out of business, the Meat Trust controlled 90 percent of the chilled beef in the country at the turn of the twentieth century. It was not until the 1970s, when the packinghouse moved closer to feedlots and vacuum packing enabled processors to portion meat, that butchers were completely driven out of meat processing. Rather than shipping a side of beef to towns and cities, now beef simply came in a box and unskilled employees could wrap, weigh, and price it for consumers to pick up in the meat department (Horowitz, 2006, pp. 144).

When Herbert Hoover promised Americans a chicken in every pot, he was promising them an expensive and special food. Chicken did not become an everyday item until after World War II. Although chickens were ubiquitous on farms throughout the nineteenth century, they were largely eaten only after their laying days were over. Eggs rather than meat were a chicken's contribution to the American diet into the twentieth century. As eastern cities grew along the Atlantic coast, specialized chicken farmers brought their live broilers to markets. These birds varied dramatically in quality and consumers rarely knew whether there would be enough meat on the chicken to feed a family. Throughout the 1920s and 1930s, scientists at land-grant universities sought to develop heartier broilers that could be mass produced. During the early 1920s, poultry researchers solved one of the major limitations of industrial broiler production: leg weakness. By simply adding cod liver oil to poultry feed, agricultural scientists increased the vitamin D in the chicken's diet and prevented leg weakness (Boyd, 2001, p. 638). In solving the vitamin deficiency, boiler hens could now be raised indoors and throughout the year rather than just in the warmer months. Raising chickens in confinement gave farmers the controls they needed to intensify production. With virtually all farms having electrification by the 1950s, broiler production became truly industrial as heated and air-conditioned henhouses spread across the South and Midwest.

Expanding production quickly translated into cheaper chicken. By the 1960s, a plump chicken in every pot was no longer a political slogan, but a reality. In addition to expanding the quantity of birds available for market, the chicken industry also made the broiler meatier. Delaware's "Chicken of Tomorrow" contest sought to create broad-breasted meaty chickens to maintain the state's status as the poultry capital of the country. Between 1948 and 1951, contest winners provided the breeding stock for virtually the entire industry, thus redefining chicken as a single type of meaty bird. These improved chickens weighed approximately 50 percent more in 1965 than they did in 1935 and by the mid-1990s broilers were almost double the weight of Depression-Era birds (Boyd, 2001, p. 637). The use of subclinical antibiotics in poultry feed facilitated weight gain in birds and helped to suppress disease, which was a chronic problem, in new massive facilities (Boyd, 2001, p. 647). The transformation of chicken from a luxury item to a cheap dietary staple depended not only on more and better birds, but also on an industrial system that integrated feed, vaccines, antibiotics, waste management, automated disassembly lines, rural electrification, refrigerated trucks, and paved roads, among many other features of postwar life.

Dairy. Dairy farming has a very long history in the United States, dating back to the first settler in the Massachusetts Bay colony. In the Jeffersonian world of the yeoman farmer, the dairy cow played a critical role. Cows not only provided manure essential for the mixed farming of mid-Atlantic and northeastern farmers, but also converted cellulose into protein though milk (Stoll, 2002, p. 49). Although milk was highly perishable, it could be relatively easily processed into other products, including butter and cheese, that could be stored for much longer periods of time, sometimes even for many months in cool weather.

Throughout much of the nineteenth and early twentieth centuries, dairying was an unrelenting, labor-intensive way to make a living. The defining feature of dairying is that even in the early twenty-first century lactating cows must be milked in the morning and in the evening. Managing the roughly 120 pounds of urine and manure each cow excreted daily only added to a farmer's work. Even when cows were grazing in the field, farmers needed to continually move them from one spot to the next so that the manure would be evenly spread and not kill or overnourish the field, rendering it unusable for several years (Stoll, 2002, p. 52). Until the twentieth century, dairying was a local endeavor, with cows living very close to consumers. In New York and other cities, cows were very common until the sanitary movement of the 1880s sought to separate animals from humans to prevent a host of diseases (Goodwin, 1999), not to mention concerns over the quality of swill milk. Cows fed on the spent grains of local brewers produced milk that was low in fat and had a blue hue. Many women and reformers thought this milk was dangerous. Although the milk was low in fat, it was not in fact dangerous and spent grains continued to be used as a high-protein component of dairy feed in the early twentieth century. It was the growth of large cities, the speed of railroads, and the availability of ice for refrigeration that allowed country milk to move quickly into urban markets (DuPuis, 2002, p. 5). These systems supported the sanitary movement's efforts to divorce consumers from producers. Yet, given milk's susceptibility to infection and rancidity, speed, ice, and country air were not always enough. Louis Pasteur's work on bacteria helped transform milk from "white poison" to a nutritious and wholesome food (DuPuis, 2002, pp. 5 and 77). With the spread of pasteurization, dairies shifted from small family operations into large factories that converted dangerous raw milk into a safe commodity that could be delivered right into American homes on a daily basis. From the 1910s through the 1940s, milk became an idealized new staple food that nourished generations of Americans (DuPuis, 2002). Yet by the Depression, dairy farmers felt the strain of the overproduction of milk (Hamilton, 2008, pp. 164–165). The economic pressure on dairy farmers only increased with the spread of paper cartons and milk trucks in the 1950s. These technologies not only allowed large commercial dairies that pasteurized and distributed milk to urban areas, suburban districts, and towns to reduce costs, but also kept milk prices low, supporting the Americans' demand for cheap

and plentiful milk (Hamilton, 2008, pp. 165–175). Postwar technological systems and suburbanization conspired to force small dairy operations out of the market, leaving only the largest and most productive operations to supply ever-cheaper milk across the country.

The market for milk largely echoed the market for meat, fresh fruits and vegetables, baked goods, and many other food products. Bigger facilities that utilized the most advanced technological systems provided economies of scale and minimized cost to consumers. Whereas consumers saw their grocery bills fall and it was easier to feed growing families more nutritiously than ever before, small and medium-size farmers were squeezed between falling prices and rising input costs (Anderson, 2008). By the time John F. Kennedy brought Camelot to Washington, D.C., in 1960, an entirely new era in food and food processing began to radically transform the American diet. These new, highly processed foods, including canned goods, industrial pies and cakes, soda pop, and frozen orange juice, moved beyond staple products like grains, meat, and milk. The age of truly industrial food arrived with the Baby Boom. By the 1980s, boomers/yuppies sought out a wide variety of food options, many of which were hyperprocessed and needed no preparation at all. Microwavable "ethnic" foods and a wide array of snack foods reflected Americans' embrace of a multicultural palate while at the same time reducing the need to cook. The advent of hyperprocessed food coincided with larger structural changes in the United States, including deindustrialization, dual-income families, rising divorce rates, the farm crisis, latch-key kids, and the obesity "epidemic." From the second half of the twentieth century through the first decade of the twenty-first century, processed foods made women's lives easier while at the same time inciting fierce debates over the health of the nation. At the same time, technological systems and globalization allowed for more fresh food to flow at lower prices into American markets to satisfy the United States' hunger for authentic ethnic foods and fresh foods out of season. By the 1990s, what was and was not a processed food was almost impossible to determine, even for the organic food movement, which sought to reverse the previous 150 years of food industrialization while simultaneously embracing much of the century's technological advances (Guthman, 2004).

[*See also* **Agricultural Education and Extension; Agricultural Experiment Stations; Agricultural Technology; Canals and Waterways; Food and Diet; Morrill Land Grant Act; Pure Food and Drug Act; Railroads; Refrigeration and Air Conditioning; Rivers as Technological Systems; Rural Electrification Administration;** *and* **Technology.**]

BIBLIOGRAPHY

Anderson, J. L. *Industrializing the Corn Belt: Agriculture, Technology, and the Farm Belt.* DeKalb: Northern Illinois University Press, 2008.

Boyd, William. "Making Meat: Science, Technology, and American Poultry Production." *Technology and Culture* 42, no. 4 (October 2001): 631–664.

Chandler, Alfred D. *The Visible Hand: The Managerial Revolution in American Business.* Cambridge, Mass.: Belknap Press of Harvard University Press, 1977.

Cronon, William. *Nature's Metropolis: Chicago and the Great West.* New York: W. W. Norton and Company, 1991.

Cummings, Richard Osborn. *The American and His Food: A History of Food Habits in the United States.* Chicago: University of Chicago Press, 1940.

DuPuis, E. Melanie. *Nature's Perfect Food: How Milk Became America's Drink.* New York: New York University Press, 2002.

Goodwin, Lorine Swainston. *The Pure Food, Drink, and Drug Crusaders, 1879–1914.* Jefferson, N.C.: McFarland & Company, 1999.

Guthman, Julie. *Agrarian Dreams: The Paradox of Organic Farming in California.* Berkeley: University of California Press, 2004.

Hamilton, Shane. *Trucking Country: The Road to America's Wal-Mart Economy.* Princeton, N.J.: Princeton University Press, 2008.

Horowitz, Roger. *Putting Meat on the American Table: Taste, Technology, Transformation.* Baltimore: Johns Hopkins University Press, 2006.

Larson, Robert A. "Milling." In *The Chemistry and Technology of Cereals as Food and Feed,* edited by Samuel A. Matz, pp. 194–197. Westport, Conn.: AVI Publishing Company, 1959.

Levenstein, Harvey. *Revolution at the Table: The Transformation of the American Diet.* New York: Oxford University Press, 1988.

Panschar, William F. *Baking in America: Economic Development.* Vol. I. Evanston, Ill.: Northwestern University Press, 1956.

Pursell, Carroll. *The Machine in America: A Social History of Technology.* Baltimore: Johns Hopkins University Press, 1995.

Stoll, Steven. *Larding the Lean Earth: Soil and Society in Nineteenth-Century America.* New York: Hill and Wang, 2002.

Turner, Katherine Leonard, "Tools and Spaces: Food and Cooking in Working-Class Neighborhoods, 1880–1930." In *Food Chains: From Farmyard to Shopping Car,* Warren Belasco and Roger Horowits, eds., pp. 217–232. Philadelphia: University of Pennsylvania Press, 2009.

Vulté, Herman T., and Sadie V. Vanderbilt. *Food Industries: An Elementary Text-Book on the Production and Manufacture of Staple Foods.* Easton, Pa.: Chemical Publishing Company, 1916.

Yezierska, Anzia. *Bread Givers.* New York: Persea Books, 1925.

Gabriella M. Petrick

FORD, HENRY

(1863–1947) Born on a farm near Dearborn, Michigan, automobile manufacturer Henry Ford held various jobs as a young man in Detroit, including machine-shop apprentice, traction car operator, and engineer for the Edison Illuminating Company. He designed and built his first prototype automobile in 1896. In 1903 he formed the Ford Motor Company and began small-scale commercial production. In the turbulent world of early automobile manufacturing, Ford initially gained prominence for race cars and his long, ultimately successful, battle with George B. Selden, who tried to gain a monopoly over automobile manufacturing by taking out a series of patents in 1895.

1908 to 1914, Years of Genius. By 1914, three achievements—the 1908 Model T, the moving assembly line, and the Five-Dollar Day—

coalesced to create Ford's worldwide reputation and place in the pantheon of U.S. heroes. The Model T thrived in an America where, away from urban and long-haul rail systems, terrible roads punished flimsy cars and their passengers at every turn and nearly nonexistent repair facilities left them to their own mechanical wits. The T offered an exceptionally strong steel frame, high wheel clearance, and fix-it-yourself simplicity. Surging demand forced the Ford team to a series of production breakthroughs between 1908 and 1914 that came to be known collectively as the "Assembly Line," one key element in what also came to be known as "Fordism" ("Fordismus" in Europe). Fordism embedded "Mr. Ford" within a manufacturing concept understood to be revolutionary: mass production aimed at a low profit margin and high volume, specialized machine tools and single-skill workers, and just-in-time materials handling. Fordism was also understood to mean superb wages for immigrant factory workers together with an Americanization program including home inspections and English-language schools. The company's January 1914 announcement of the Five-Dollar Day, backed by technological achievement at the highest level, put Ford's image as a beneficent hero in newspapers all over the world.

Fordism and Mr. Ford.

Fordism was endlessly analyzed in business and engineering journals during the Model T's astonishing production run, when the ugly, indestructible little cars defined automobility for a generation (1908 to ca. 1925). In the public mind, however, Mr. Ford gave soul and identity to mass production's complexities. A dapper, articulate team leader marked by one innovative miracle after another, he also understood the working man's life and needs. Doubling the daily wage, cutting the work shift from nine to eight hours, and teaching unlettered immigrants the subtleties of American life, he remained until his death in 1947 the plain-spoken working man's friend for millions of Americans. Convincing evidence suggests that this image credibly reflects the agile leader of the innovation team that secretly designed the "T" at the Piquette Street plant and then responded to a whirlwind of

market demand by transforming the industrial world's manufacturing practice. Ford's feel for workers shows in the early years at Piquette Street; Henry walked across the street from his home, up the stairs, and through a machine shop into his second-floor office, one of the men. It can also be argued, however, that the company's exponential growth transformed Ford, from his former supple leadership to an increasingly rigid addiction to control and from plain-spoken kinship with workers to a reclusive obsession with privacy.

Obsession with Control.

As early as 1912, the run-up to a fully integrated assembly line led to a drastic increase of supervisors (from 2 to 14.5 percent) and a de-skilling of most workers. This, and the relentless pace of the line, resulted in extreme worker turnover (ca. 380 percent in 1913). From this perspective the Five-Dollar Day marks the first move in Ford's lifelong battle to control his workforce and his company. Doubling the daily wage came with house inspections aimed at separating workers from their cultures of origin. Soon, home inspectors were replaced with a network of in-factory spies and, by the mid 1920s, brutal enforcers. In 1919 Ford bought out all stockholders and forced out independent-thinking senior managers. During General Motors' rise to market dominance, with annual model updates, multiple models and colors, and attractive advertisements, Ford refused to consider changing the T's core concept until 1926's catastrophic market share loss forced him to shut down production for a half year before releasing the beautifully engineered Model A. But neither the A nor the innovative V-8 engine (1932) brought the company back to its earlier market dominance.

For Ford himself, fame took its toll, as his eccentricities and prejudices became increasingly evident. In an abortive effort to end World War I through arbitration, he chartered a "peace ship" in December 1915 and sailed to Europe. His awkward and unschooled responses during the 1919 *Chicago Tribune* libel trial invited ridicule and pity. Ford's newspaper *The Dearborn Independent*, distributed in the 1920s through Ford dealers, disseminated virulent anti-Semitism. To settle a libel suit, Ford issued a retraction and halted publication in 1927. His bitter anti-unionism led to outbreaks of bloody violence at Ford plants during the Great Depression, most notably in 1932 and 1937. Only in 1941 did Ford sign a contract with the United Automobile Workers union, forced, it is said, by his wife Clara. Adolf Hitler quoted Ford with approval in his 1924 manifesto *Mein Kampf* and in 1938 the Third Reich awarded Ford the Grand Cross of the German Eagle.

Henry and Edsel.   In later life Ford increasingly withdrew from day-to-day corporate operations and retreated to "Fair Lane," his ca. 1,300-acre estate and the nearby historical museum and Greenfield Village. Henry's stark contrast with his only son, Edsel, shows most vividly in the location of their homes. Henry's estate straddled the Rouge River in working-class Dearborn, one mile upstream from the big plant. Edsel's mansion, Gaukler Point, occupies 87 acres of prime shoreline in Grosse Pointe Shores, then the heartland of Detroit's elite. Henry shunned public events, whereas Edsel's civic commitments remain memorialized in Diego Rivera's portrayal of Edsel and Detroit Institute of Arts (DIA) director William Valentiner in the bottom right corner of the Rivera Murals' south panel. Edsel and his wife Eleanor's patronage of the arts helps explain why the DIA remains one of America's great encyclopedic museums.

The Ford Foundation, established by Henry and Edsel in 1936, ultimately received many millions in nonvoting Ford Motor Company stock, making it one of America's wealthiest foundations. It operated with a board and a worldview independent of its primary benefactor, and increasingly after Henry's death it funded individuals and programs that might have displeased its benefactor.

Oral history reminiscences suggest that Henry's grandchildren blamed Ford for Edsel's early death from stomach cancer, caused, it was said, by Henry's ceaseless and capricious interventions in corporate practice even as he retreated into the private world of the estate, the museum, and the Village.

The Museum, Greenfield Village, and Fairlane.   Beginning in the early 1920s Ford's

agents scoured New England, the Midwest, and Great Britain for artifacts relating to the history of technology and domestic life. In October of 1929 the unfinished complex opened to a who's who of America's industrial elite. Over a national radio hookup Thomas Edison switched on a replica of his first incandescent light for its fiftieth anniversary. Work on the museum and village continued for over a decade, with Henry continually issuing sometimes minuscule design change orders. The Village took shape as a nostalgic re-creation of eighteenth- and nineteenth-century America dotted with shrines to Henry's heroes (e.g., Thomas Edison, the Wright Brothers, the McGuffey Reader, a one-room schoolhouse). By contrast, the museum's eight-acre main floor exhibited some of the world's finest collections of machines—steam engines, machine tools, farm equipment, locomotives, automobiles, airplanes—each laid out in a triumphantly self-confident display of linear progress. Nowhere was this concept more elegantly achieved than in the design, completed by 1940, of the entrance as coordinated with the vast hall's central aisle. One entered a replica of Philadelphia's Independence Hall, home for the Declaration of Independence and the U.S. Constitution, and proceeded into the main hall's central aisle, which was framed by two facing lines of bulky nineteenth-century steam tractors, themselves framing an enormous dynamo taken from the 1910 Highland Park plant. Ford intended a three-century celebration of progress: from eighteenth-century civics to nineteenth-century steam to twentieth-century electricity. The complex also housed a fully accredited K–12 school, where students could study amid physical reminders of technological progress. An elusive figure, Ford shunned visitors but often roamed the grounds of the Village alone at night. Oral histories of house servants at Fair Lane also tell of a secluded life with few visitors. After dinner Clara ordinarily retired to the solarium and listened to the radio, whereas Henry spent time in his powerhouse and workshop where first-name familiarity with the small technical crew provided a semblance of the early Piquette Street innovation team. Few guests were welcome, most notably Edison and Charles Lindberg.

Henry, Diego Rivera, and Charles Sheeler. Improbably, Diego Rivera was invited to lunch soon after arriving in Detroit in 1932 to paint the now world famous industrial murals at the heart of the DIA (1933). In Dearborn, the historical museum was not yet open to the public but Ford invited Rivera to spend a day there alone, contemplating the collections. He spent a month in the Rouge sketching for the murals. Rivera, the flamboyant Marxist, was commissioned and later defended from public outrage by Edsel Ford, but his cordial relationship with Henry continues to surprise. The improbable warmth between these two men reveals another facet of Ford's complex character. Despite his increasingly eccentric and often brutal obsession with privacy and control, he never lost a refined sense of the well-designed machine as beautiful. In this he shared soul space not only with Rivera but also with Charles Sheeler, who was commissioned in 1927 to photograph the Rouge and whose industrial landscape portraits stand with the Rivera murals as contrasting contemplations of industrial beauty on a massive scale— Rivera's murals pulsing with humanity, Sheeler's nearly devoid of any signs of unengineered nature or of human beings.

Ambivalent Common Man. Ford's obsession with control coupled with an enduring sense of beauty in the machine reveal a man caught in the classic ambivalence of the modernist technological aesthetic. It has been argued that his near-psychotic obsessions were only allowed because of the PR protection of the Ford Motor Company. Another reading, favored here, suggests that Ford was an ordinary adult of his time, caught between exultant awe at transformative inventive forces and deep insecurity about one's place as a single, tiny individual set against forces so vast. Sheeler's tiny running man, barely visible against a vast factoryscape in "American Landscape" (1930), could be imagined as Mr. Ford, wandering alone in his private nocturnal world of nostalgic shrines and gleaming machines. In this reading, Ford's fortune did less to protect him from the consequences of near-psychotic tendencies than to supply him with resources to act out the neurotic impulses of an American society suspended between confidence and anxiety. Most of his fellow citizens could not afford the luxury of such excesses and had to go back to work the next day.

[*See also* **Automation and Computerization; Edison, Thomas; Internal Combustion Engine; Lindbergh, Charles; Machinery and Manufacturing; Motor Vehicles; Technology;** *and* **Wright, Wilbur and Orville.**]

BIBLIOGRAPHY

Davis, Donald Finlay. *Conspicuous Production: Automobiles and Elites in Detroit, 1899–1933*. Philadelphia: Temple University Press, 1988.
Jardim, Anne. *The First Henry Ford: A Study in Personality and Business Leadership*. Cambridge, Mass: MIT Press, 1970.
Lacey, Robert. *Ford: The Men and the Machine*. Boston: Little, Brown, 1986.
Lewis, David L. *The Public Image of Henry Ford: An American Folk Hero and His Company*. Detroit: Wayne State University Press, 1976.
Meyer, Stephen, III. *The Five Dollar Day: Labor, Management, and Social Control in the Ford Motor Company, 1908–1921*. Albany: State University of New York Press, 1981.
Nevins, Allan, with Frank E. Hill. *Ford*, 3 vols. New York: Charles Scribner's Sons, 1954–1963.
Staudenmaier, John M. "Henry Ford's Relationship to 'Fordism': Ambiguity as a Modality of Technological Resistance." In *Resistance to New Technology: Nuclear Power, Information Technology, and Biotechnology*, edited by Martin Bauer. Cambridge, U.K.: Cambridge University Press, 1995.
John M. Staudenmaier, S. J.

FOREIGN RELATIONS

See Diplomacy (post-1945), Science and Technology and.

FORENSIC PATHOLOGY AND DEATH INVESTIGATION

Four hundred years after the first colonists arrived in America, elected nonphysician coroners continued to investigate suspicious and unexplained deaths. For centuries, physicians have attempted to replace the coroner with forensically trained physician experts modeled after the centralized medicolegal institutes of continental Europe. Federal and state governments have worked to strengthen the office of the coroner by providing constitutional legitimacy. Both coroners and physician medical examiners point to democratic ideals in support of their offices in an ongoing feud over medical authority where medical examiners have vowed to abolish the coroner, without success. At the beginning of the twenty-first century, death investigation in the United States has continued to resemble a "patchwork quilt" involving variability in quality and practice and reflecting the ongoing contest to establish medical authority (Bonnie, 2003, p. 3).

English Law Comes to America. As early as 1192, King Henry II established the position of English coroner as a political instrument to retain chattels of the deceased for the crown. Initially called "keepers of the pleas of the crown," the name was later shortened to crowner and eventually coroner. The duties of the coroner included burying the dead, taking charge of treasure troves, and impaneling inquest juries to establish the cause and manner of death. Initially occupied by ordinary citizens and passed down through families, the position of coroner was soon claimed by educated men seeking higher positions in a system of government dominated by inherited titles.

The American colonies inherited their legal traditions from England. The English king gave local government officials in the colonies—including justices of the peace, sheriffs, and coroners—the authority to investigate deaths. American coroners pledged to act in accordance with the normal practices of English coroners. Following the Revolutionary War and the implementation of a new Constitution, the office of coroner evolved with the changing legal structure of the nation.

Unlike in England, the coroner in the American colonies remained a low-level appointee whose most visible duty was to hold inquests and recover property in cases of violent or untimely death. American coroners were common men,

farmers, craftsmen, and undertakers who possessed no specialized medical or legal knowledge; they were simply called to republican ideals of public service.

In the English common-law system the determination of a defendant's guilt or innocence rested within a system of traditional laws and the power of a lay jury composed of citizens of the locale. The U.S. judicial system, with its adherence to the lay jury, adversarial legal procedures, and representative system of government increasingly controlled by political parties, helped to develop the peculiar system of American death investigation. Medical experts were not deemed essential to the Anglo-American legal system and physicians enjoyed no special standing either in English or in American courts (Brock and Crawford, 1994, p. 27).

Unique to the Anglo-American system, the coroner's inquest jury had the statutory duty to determine not only the cause but also the manner of death. Legislative acts of the colonies, literally copied from the English law, instructed coroners to "declare of the death of this man, whether he died of felony, or by mischance; and if of felony, whether of his own, or of another's; and if by mischance, whether by the act of God or man" (Massachusetts Province Laws, 1770). In contrast to the continental system where judges established the manner of death and decided about guilt and innocence, the initial inquest rulings of American coroners determined whether a crime had been committed or a death had simply resulted from natural or accidental causes. This single process gave the coroner and his inquest jury the power to determine whether a specific case would advance to trial or be cut short in deference to the accused. The coroner's verdict served as a formal check and balance, restraining overreaching justices and frustrating aggressive prosecutors.

The continental system of death investigation, unlike the English system, depended on medical experts who relied on textual authorities and scholarly university research. Courts used these experts to establish medical facts in evidence. As a result, jurists frequently used scientific methods documented in writing to answer legal questions of guilt and innocence. Anglo-American physicians, by contrast, presented evidence to the courts through oral testimony as common witnesses and possessed only as much authority as the jury was willing to give them in each instance. The coroner's inquest, as a result, did not provide rational written medicolegal texts or inquiry. According to Catherine Crawford (1994, pp. 94–98), the continental system, with its reliance on medical experts, encouraged scientific research and the development of forensic science in contrast to the English common-law system that was based on ancient tradition and judicial precedent.

Autopsies were relatively common in colonial America and physicians regularly consulted to provide evidence in provincial courts, especially in cases of contested causes of death such as infant murder. The practice of performing postmortem examinations to determine the cause of death was well established when the early English colonists brought it to North America. The first recorded postmortem in colonial America was in Massachusetts in 1639; a physician determined that a servant boy died from head injuries inflicted by his master. During the colonial and early republic period, physicians engaged in matters of legal medicine with a republican enthusiasm for public service and rarely received remuneration for their testimony.

Legal Medicine in the New Republic. In the United States prior to the nineteenth century, medicine and law were distinct specialties that did not relate in practice, nor was medical expertise an important element of many trials. Populist attitudes of the early nineteenth century included a distrust of experts, intellectuals, and those individuals who held themselves apart from the common people. American physicians influenced by the Scottish Enlightenment, especially the relatively large number who studied medicine in Edinburgh and Glasgow, emphasized the use of rational scientific medicine to assist the law. Dr. James S. Stringham (1775–1817), a Columbia professor of chemistry, delivered the first lectures on medical jurisprudence in the United States in 1804. He was appointed professor of medical jurisprudence in 1813, one of only two chairs of medical jurisprudence established during the first

decades of the nineteenth century. Benjamin Rush (1746–1813) emphasized the relationship of law and medicine with the publication of "On the Study of Medical Jurisprudence" in 1811 and encouraged physicians to study medical jurisprudence. Rush urged physicians to share their medical knowledge in courts of law as part of their republican commitment to civic responsibility, patriotism, and public service.

In 1815, Theodric Romeyn Beck (1791–1855) became professor of the Institutes of Medicine and lecturer of medical jurisprudence at the College of the Western District of the State of New York at Fairfield. At the same time, Walter Channing, one of Rush's students, became the first professor of medical jurisprudence at Harvard, a position he held for 40 years. Harvard appointed Channing to the rank of professor of midwifery and medical jurisprudence, which was not an illogical combination considering the medicolegal issues related to complications of pregnancy and the high numbers of infanticide at the time.

Beck first gained an international reputation with the publication of his two-volume work, *Elements of Medical Jurisprudence* (1823). The *Elements* discussed legal aspects of rape, poisons, problems of sexuality and reproduction, mental illnesses, and wounds, as well as practical methods of death investigation for coroners. Beck defined the duties of physicians at the death scene and the need for thorough postmortem and toxicology examinations in all cases of suspected homicide. He encouraged the involvement of physicians in solving crimes. The *Elements* became a classic; it remained the primary textbook on medical jurisprudence in the United States and continental Europe for more than 50 years.

In Philadelphia medical jurisprudence was given a prominent place in medical instruction. Moreton Stillé (1822–1855) was a recognized expert in legal medicine, collaborating with a Philadelphia attorney, Frances Wharton, in a *Treatise on Medical Jurisprudence* (1855). The pathological anatomy described in the book was based on collaborative work Stillé conducted with Carl von Rokitansky (1804–1878), the famous Austrian pathologist. Another Philadelphian physician,

John J. Reese (1818–1892), taught medical jurisprudence at the University of Pennsylvania and published an authoritative volume on medical jurisprudence and toxicology (1874).

The growth of medical science in the early nineteenth century helped to propel the development of legal medicine in the United States. American medical students ambitiously pursued courses in medical jurisprudence that were taught in 37 regular medical schools. The growing field of chemistry had major medicolegal applications in the study of poisons and became a common topic in American trials. As noted by James Mohr (1993), by the mid-nineteenth century medical jurisprudence had become a special subject of study in most medical schools as well as a general area of interest for many American physicians.

The explosion of medical malpractice claims that began in America during the 1840s perpetuated a growing animosity and distrust between physicians and lawyers. Physicians had hoped that training in medical jurisprudence would prepare them to become better expert witnesses in the courtroom to assist the law. Physicians initially enthusiastic about assisting the courts confronted the reality of adversarial justice and marketplace professionalism. In the eyes of the court, physicians received no special standing or compensation as medical experts. Despite the initial interest, most physicians eventually lost their early enthusiasm for legal matters because they harbored contempt and distrust of legal and judicial inquiry and resented not being paid for their time. By the mid-nineteenth century, physicians had lost their initial optimistic view of medical jurisprudence and instead confronted the problems of corrupt coroners and disrespectful lawyers in American courts. The abusive treatment of physicians at the hands of lawyers, coupled with corrupt coroners and the lack of scientific medicine in the courtroom, was an important factor in the creation of the American Medical Association in 1847. As early as 1857, as Mohr (1993) has pointed out, the American Medical Association recommended mandatory medical training for coroners and paying of physicians for legal services when they were forced to testify in court. Physicians also called upon

their medical colleagues to abolish the system of coroners.

From its origins in England, the position of coroner grew in political strength in the United States during the early nineteenth century. Enacted into state constitutions, the coroner became a constitutionally elected office with steadily growing political influence. It also frequently served as a stepping stone to a higher political position. No longer appointed, the democratically elected coroner continued to pose a major obstacle to the use of medical experts. By law, the coroner determined whether inquests and autopsies would be necessary and, in the event they were, could steer the eventual outcome to his financial advantage by selecting the jury and controlling the investigation. Often influenced by bribes and political favoritism, coroners developed a reputation among physicians for political corruption and for being ignorant of medicine.

Coroners and Corruption. As the legal historian David Garland (2000) has observed, during the nineteenth century, the United States developed the most decentralized, fragmented, and least bureaucratized central state administrative apparatus of any modern democracy. In contrast to European nations, the United States did not have a culture of deference toward elites or established groups and institutions. As a consequence of the retention of local autonomy in criminal justice activities by the states, mandated by the federal Constitution, electoral politics affected criminal justice more directly and extensively in the United States than in any other liberal democracy. The growth of state institutions was always limited and shaped by democratic controls and elected officials, which resulted in highly politicized and partisan state institutions. Rules and procedures for criminal and civil procedures varied from state to state; in the great majority of states the local positions of county coroner, district attorney, state judges, sheriff, and police chief were all elective. Physicians argued that coroners were selected on the basis of political affiliation rather than scientific training.

Physicians, frustrated not only with the overt incompetence and corruption of coroners but also with the disparaging treatment they received in courts at the hands of lawyers, called for medical expertise in criminal investigations. In 1865, the state of Maryland became the first state to legislate that coroners must be physicians. In a lecture celebrating the American centennial in 1876, Stanford Chaillé (1830–1911), a New Orleans physician, gave a speech entitled "Origin and Progress of Medical Jurisprudence, 1776–1876." In it he condemned the legal system that rejected medical evidence and expertise and decried poorly trained physicians who performed medicolegal autopsies. Chaillé observed on the teaching of medical jurisprudence, "There are very few of the medical colleges in which it was taught, and still fewer in which it takes rank as a distinct and independent branch along with the other departments" (Burns, 1980, p. 279). A young Boston lawyer, Theodore Tyndale, spearheaded a reform movement critical of both coroners and physicians and called for the appointment of medical examiners who were "able and discrete men of science." Tyndale argued that the lay coroner controlled the expert testimony of untrained physicians by limiting their testimony and manipulating the legal process.

In 1877, Massachusetts became the first state to appoint physician medical examiners to the investigation of sudden death. Unfortunately, the statute limited the medical examiner's investigation to violent deaths, which restricted the investigative authority of the medical examiner. The increasing involvement of physicians in death investigation catalyzed the creation of medicolegal societies in both Massachusetts and New York, which provided a forum for physicians and lawyers to present and discuss medicolegal cases and served as a sounding board for legislators on improving medicolegal practice. These medicolegal societies published their transactions and papers in the first American journals dedicated to forensic medicine.

By the late nineteenth century, in response to urbanization, immigration, and a growing fear of urban crime, the Progressives ignited a nationwide call for improvements in death investigation and the removal of corrupt coroners. Leading the reform movement was Richard Spencer

Childs (1882–1978), a young businessman who believed the practice of electing lay coroners to a position that required medical expertise threatened American democracy. In 1915, in response to allegations of political corruption, New York City successfully replaced the coroner with a scientifically trained physician medical examiner. Charles Norris (1867–1935), a pathologist at Bellevue Hospital, assumed the position of chief medical examiner of New York City. But the process of removing coroners was slow. In 1928, Essex County New Jersey (Newark) became the second major city to establish a medical examiner and Maryland became the first state in 1938 to create a statewide medical examiner system.

The Creation of a Medical Specialty. In the early twentieth century elite physicians began to include forensic medicine in their pathology practices. The Philadelphia pathologist William Scott Wadsworth (1868–1955), Harrison Stanford Martland (1883–1954) in Newark, and George Burgess Magrath (1870–1938) and Timothy Leary (1870–1954) in Boston performed thousands of autopsies for coroners and became the first generation of full-time pathologists committed to the practice of forensic medicine. The leading American pathologists Ludvig Hektoen (1863–1951), a Chicago bacteriologist and immunologist, and Victor Vaughan (1851–1929), the renowned bacteriologist and chairman of pathology at the University of Michigan, sought to improve legal medicine. They established a committee in conjunction with the National Research Council to study the factors influencing death investigation in America. The pathologists convinced Abraham Flexner of the Rockefeller Institute to fund the study. The National Research Council publicized the work of the committee in the 1928 publication *The Medical Examiner and the Coroner,* which openly called for the removal of the lay coroner. Four years later, Oscar T. Schultz (1877–1947), a pathologist on the committee, published recommendations for educational reform. The 1928 and 1932 publications by the National Research Council strongly argued for the application of scientific medicine to the criminal justice system. They called for the establishment of medicolegal institutes similar to those in continental Europe affiliated with state universities and staffed by academic pathologists trained in legal medicine.

Pathologists centered in major urban centers helped to professionalize forensic pathology by appropriating Progressive-Era reforms in occupational medicine and public health. In Newark, Harrison Martland described the medical effects of chronic industrial exposure to radium in women employed as dial painters. Martland's discovery helped to support the growing industrial health movement and later the work of the Nuclear Regulatory Commission. Other pathologists utilized autopsy findings while working in medical examiner offices to recognize and describe growing urban public-health issues. Milton Helpern (1902–1977), the New York City medical examiner, identified the growing problems of intravenous drug abuse and coal gas poisoning. Physician medical examiners, as well as their lay coroner counterparts, began to study the public-health implications of alcoholism, chronic disease, suicides, and the growing numbers of automobile deaths. Coroners leveraged their affiliation with public-health officials, which allowed them to combat public criticism concerning their lack of medical expertise. The Progressive reform movement in local government proved incapable of removing the lay coroner, who continued to control death investigation in most large urban centers.

In 1932 Harvard Medical School established the first academic chair of legal medicine in the United States and appointed Dr. George B. Magrath (1870–1938) to the position. A survey of 77 U.S. medical schools that same year revealed that only 6.5 percent gave a course in legal medicine, 58 percent gave lectures on the subject, and 35 percent did nothing (Burns, 1980, p. 286). In 1938 with the financial assistance of the philanthropist Frances Glessner Lee, Harvard created the first department of legal medicine. Alan Richards Moritz (1899–1986), a Cleveland pathologist, was chosen to head the department. Moritz attempted to improve death investigation throughout the United States by training physicians in forensic medicine through countless lectures on the need for a uniform approach to death investigation.

In addition to Moritz, Richard Childs, a leader in the National Municipal League, a Progressive national civic organization, made the creation of model medicolegal systems staffed by forensically trained physician medical examiners a major reform goal. In promulgating his cause, Childs attacked lay coroners as corrupt, unintelligent, and incapable of making medical decisions to determine the cause and manner of death. Childs pressured state legislators to abolish coroners and enact medical examiner offices. He regularly published a list of the states with coroners and labeled them "The Best Places for Murder."

In 1951 the National Municipal League published its proposal entitled "A Model State Medico-Legal Investigative System." The report argued that hospital-based pathologists without forensic training were incapable of interpreting injuries and determining the cause and manner of death in medico-legal autopsies. The report reiterated the call to abolish the coroner system. The Model Death Investigative legislation had limited success because only 10 states enacted medical examiner statutes abolishing the office of coroner.

Childs' attempts at abolishing lay coroners in the United States, although initially successful, stalled mainly because of the political strength of constitutionally elected coroners, as well as the desire of physicians privately performing autopsies for the coroner to maintain fee-for-service reimbursement to supplement their income. To defend their claim of legitimacy, coroners argued that their position was supported by an American democracy where individual citizens spoke through their elected officials. They asserted that because politically appointed physician medical examiners were beholden to their appointing authorities, they were not independent like coroners. Coroners also pointed to the increased expense of medical examiner systems, although medical examiners successfully refuted this argument in almost every instance. By the end of the twentieth century, however, medical examiner reform had stalled, leaving only 60 percent of the American population under the jurisdiction of physician medical examiners, with the remaining 40 percent still controlled by coroners.

Forensic Education Takes Shape.

One claim that supporters of the medical examiner system could not refute was the paucity of trained forensic pathologists. By the mid-twentieth century, professional pathology groups assisted by the American Medical Association stepped in to assist in the professionalization of forensic pathology. Pathology had become a separate medical specialty in America in 1936, governed by the American Board of Pathology. In 1958, pathologists working within the College of American Pathologists, a national professional pathology society, successfully created the subspecialty of forensic pathology through the American Board of Pathology. Forensic pathology became officially defined as the subspecialty within the field of pathology that deals with the investigation of cause and manner of death and the performance of medicolegal autopsy and ancillary studies. Pathologists could qualify for board certification by either demonstrating previous experience in the field or completing a year of specialized training in forensic pathology.

Academic pathologists argued that the training of forensic pathologists should take place in academic medical centers where trainees could benefit from research and teaching. Medical school pathologists argued that training centered in municipal morgues was no more than technical training. In reality, forensic casework, sufficient to train pathologists in forensic medicine, was primarily located in big-city morgues. Elite research hospitals and medical schools eschewed any affiliation with government-run morgues, fearing the stain of corrupted bodies and politics. In reaching a final compromise, organizers allowed municipal medical examiners to continue to train and teach forensic pathologists, but insisted the candidates first fulfill the requirements for certification in anatomic pathology by the American Board of Pathology.

Training programs were led by forensic pathologists who developed broad "schools" of practice. Each program developed a unique theory of practice based on the procedures and philosophy of the chief medical examiner and influenced by the realities of the local politics and culture. The Massachusetts program, dominated by Harvard, was heavily weighted toward academics, whereas Miami emphasized the importance of autopsies and Minneapolis required hands-on participation in death scene investigations. These programs formed the foundation for teaching forensic pathology in America. Although forensic training centered on the medicolegal autopsy, trainees also received instruction in criminalistics, anthropology, firearms analysis, courtroom testimony, and toxicology methods and interpretation. At first primarily a male-dominated specialty, by the end of the century there were more females in forensic training programs than males.

The federal government also participated in the training of forensic pathologists to supply the needs of the military. In the 1950s the Armed Forces Institute of Pathology (AFIP), the successor to the Army Medical Museum and long concerned with the pathology of trauma, established a unit devoted to forensic pathology. By 1958, the unit had developed an active teaching program and a registry of interesting forensic pathology cases. In conjunction with the College of American Pathologists, in 1962 the AFIP offered the first fellowship-training program using the facilities of the Maryland State Medical Examiner Office in Baltimore. In 1988 Congress created the Office of the Armed Forces Medical Examiner as a component of the AFIP. Located in Washington, D.C., the office was responsible for investigating the death of military personnel, all individuals at military bases, and senior elected officials in the federal government in addition to war-related fatalities and has provided important training for active-duty forensic pathologists. Postmortem examinations have been performed at the military mortuary in Dover, Delaware.

Despite the creation of board certification, legislated standards and court procedures could not limit the performance of medicolegal autopsies to board-certified forensic pathologists. This eventually created a fractured specialty where non-board-certified pathologists continued to compete with board-certified forensic pathologists for medicolegal autopsies. Many pathologists failed to appreciate the benefit of board certification and declined to obtain extra training. Despite early advances, forensic pathology lacked funding, governmental regulation, and the professional stature of European counterparts. The diluted requirements for certification left many forensic pathologists questioning whether the specialty existed at all.

Professional Organizations. During the second half of the twentieth century, American death investigation continued to be a patchwork of jurisdictions located at the county or state level with varying practices and procedures. Unlike countries in continental Europe, where the police controlled the investigation and determined the need for autopsy, American coroners and medical examiners were mostly detached from law enforcement and delegated the scene investigation to lay death investigators. These investigators examined bodies, took photographs, and interviewed the family and witnesses regarding the circumstances of the death. Many forensic pathologists eschewed going to crime scenes and instead used the lay investigator as their eyes and ears at the death scene. Frequently, because these investigators were retired policemen in search of additional governmental pensions, police influence continued in many medical examiner and coroner offices.

To establish political support for medical examiners, forensic pathologists created a professional organization, the National Association of Medical Examiners (NAME), in 1966. NAME was dedicated to removing coroners and replacing them with physician medical examiners. The organization provided annual educational programs on topics of forensic pathology and legal medicine. In 1977, NAME created an Inspection and Accreditation Program to provide guidelines for the support and administration of medical examiner offices. Although voluntary, accreditation signified that the office had at least maintained minimum standards endorsed by a consensus of the profession. The accreditation process provided politicians with benchmark information to assess the quality of the death investigation system and helped legitimate medical examiners'

requests for needed resources. Although never able to remove coroners, NAME has been recognized as the professional voice of forensic pathology in the United States.

The Federal Government Steps In. Following the initial 1928 National Research Council report, the federal government convened a number of conferences to explore ways to improve death investigation. For years the federal government had been critical of state and local officials in investigating deaths. Earlier attempts to standardize death investigation, including the 1951 Model Death Investigation Legislation, ended in failure because of a lack of finances, organization, or authority. Professional pathology organizations such as the College of American Pathologists continued to provide educational programs in forensic pathology through its committee on forensic pathology, but because of the conflicting agendas of its pathologist membership, it did little to enforce standards of practice or push for governmental reform. Instead, the College of American Pathologists turned a blind eye to death investigation, preferring to support its noncertified members in the lucrative practice of performing autopsies for coroners.

In 2003 the National Institutes of Justice, the research and policy evaluation office of the U.S. Department of Justice, published the first cohesive American standards for death scene investigation. For the first time, procedures and practices to be performed at the death scene were formally delineated. The publication in 1996 of *Death Scene Investigation: A Guide for the Professional Death Investigator* helped to establish the profession of death investigator by outlining the essential skills and laying a foundation for professional certification. Ten years later the National Association of Medical Examiners published *The Forensic Autopsy Performance Standards*, which provided formal recognition of the differences between hospital and forensic autopsies. The *Standards* not only addressed the variability in practice of forensic pathologists in areas of autopsy performance, scene investigation, and procedures for identification, but also provided a consensus document on acceptable practice and emphasized the importance of including experts from other fields. The cross-examinations of defense attorneys in the courtroom also influenced death investigation in the United States. Defense attorneys hammered dilettante pathologists in court who deviated from the newly created standards for their lack of certification, accreditation, and experience.

In 2003 the Institute of Medicine, the health arm of the National Academy of Science responsible for providing independent advice to decision makers, responded to concerns about the adequacy of death investigation in the United States by sponsoring a workshop entitled *Medicolegal Death Investigation System*. The goal of the workshop was to "highlight not only the status and needs of the medicolegal death investigation system as currently administered by medical examiners and coroners but also its potential to meet emerging issues facing contemporary society in America" (Institute of Medicine, 2003, p. 1). The workshop was one more attempt by the federal government to identify and improve defects in death investigation. Recognizing the patchwork of laws governing death investigation and lack of standards for coroners, the chairmen of the workshop conceded that, similar to health care, the country lacked a unified system of death investigation. One key result of the workshop was to confirm the desire of stakeholders to eliminate the lay coroner and to enhance professionalism by supporting medical examiner systems.

In 2009 the National Academy of Sciences released the long-awaited two-year study on the state of death investigation in the United States, *Strengthening Forensic Science in America: A Path Forward*. The report called for certification of forensic pathology and investigators as well as the accreditation of all forensic facilities, including medical examiner offices. The report further supported the role of academic institutions in partnering with medical examiner offices to provide scientifically based death investigation and research in forensic science. It also mandated that only board-certified forensic pathologists should be allowed to perform autopsies on suspected homicide cases. For the first time the report indicated the federal government's intent to abolish lay coroners. The report established board-certified forensic pathologists as the gold standard and lead experts in death investigation. In doing so, it significantly enhanced the professional authority and status of the forensic pathologist, something the medical community had been unwilling or unable to accomplish.

At the beginning of the twenty-first century physicians and the public came to acknowledge the medical expertise of forensic pathologists. With the advent of stricter legal rules governing expert medical testimony in state and federal courts, the introduction of inspection and accreditation programs, the increasing expectations of the work of medical experts, and the complexity of medicolegal cases, untrained pathologists were less inclined to become involved in medicolegal cases. *Strengthening Forensic Science: A Path Forward* not only served as a call to arms for serious improvement of forensic science in America but also fostered the hope of the republican ideals of early physicians to create experts in legal medicine who might assist the legal system.

[*See also* **Death and Dying; Ethics and Medicine; Law and Science; Medical Malpractice; Medicine;** *and* **Public Health.**]

BIBLIOGRAPHY

Bonnie, Richard. *Medicolegal Death Investigation System: Workshop Summary*. Institute of Justice. Washington, D.C.: National Academies Press, 2003.

Brock, Helen, and Catherine Crawford. "Forensic Medicine in Early Colonial Maryland, 1660–1760." In *Legal Medicine in History*, edited by Michael Clark and Catherine Crawford. Cambridge, U.K.: Cambridge University Press, 1994.

Burns, Chester R. "Medical Ethics and Jurisprudence." In *The Education of American Physicians*, edited by Ronald L. Numbers, pp. 273–290. Berkeley: University of California Press, 1980. A valuable resource on the rise and fall of legal medicine education in the medical curriculum of the United States.

Crawford, Catherine. "Legalizing Medicine: Early Modern Legal Systems and the Growth of Medico-legal Knowledge." In *Legal Medicine in History*, edited by Michael Clark and Catherine

Crawford. Cambridge, U.K.: Cambridge University Press, 1994. A valuable edited history of Anglo-American legal medicine.

Garland, David. *Peculiar Justice*. Cambridge, Mass.: Harvard University Press, 2012. A history of the death penalty in the United States notable for a lucid description of the political influence on the police powers of local governments.

Institute of Medicine. *Medicolegal Death Investigation System: Workshop Summary*. Washington, D.C.: National Academies Press, 2003.

Jentzen, Jeffrey M. *Death Investigation in America: Coroners, Medical Examiners, and the Search for Medical Certainty*. Cambridge, Mass.: Harvard University Press, 2009. A detailed examination of the development of the death investigation profession and forensic pathology in the United States during the twentieth century.

Johnson, Julie. "Coroners, Corruption, and the Politics of Death." In *Legal Medicine in History*, edited by Michael Clark and Catherine Crawford, pp. 268–294. Cambridge, U.K.: Cambridge University Press, 1994. Along with Jentzen (see preceding), reviews the establishment of medical examiner systems and conflicts with coroners in the United States.

Long, Esmond R. *A History of American Pathology*, pp. 267–277. Springfield, Ill.: Charles Thomas, 1962. The standard history of the creation of pathology as a medical specialty.

Luke, James. "Law—Medicine Notes: Forensic Pathology." *New England Journal of Medicine* 295 (1976): 32–34.

Massachusetts Province Laws, 1770. Boston Public Library.

Mohr, James C. *Doctors and the Law: Medical Jurisprudence in Nineteenth-Century America*. New York: Oxford University Press, 1993. The best account of the medicalization of death investigation in the United States during the nineteenth century through the attempts of the American Medical Association and physician advocates.

National Research Council Committee on Identifying the Needs of the Forensic Sciences Community. *Strengthening Forensic Science in America: A Path Forward*. Washington, D.C.: National Academies Press, 2009.

Schultz, Oscar. *Possibilities and Need for Development of Legal Medicine in the United States*. Washington, D.C.: National Research Council, October 1932. Combined with the previous source, the work called for the creation of medical examiner systems centered in state universities similar to European medicolegal institutes.

Schultz, Oscar, and E. M. Morgan. *The Coroner and the Medical Examiner*. Bulletin of the National Research Council 64. Washington, D.C.: National Research Council, 1928. The findings of an early-twentieth-century review of medical examiners and coroners offices by an elite group of American physicians that called for a total overhaul of the death investigation system.

Timmermans, Stefan. *Postmortem: How Medical Examiners Explain Suspicious Deaths*. Chicago: University of Chicago Press, 2006. A sociologist's analysis of the professional expertise, authority, and scientific practices of modern medical examiners.

Tyndale, Theodore H. "The Law of Coroners." *Boston Medical and Surgical Journal* 154 (1877): 246–247.

Watson, Katherine D. *Forensic Medicine in Western Society: A History*. New York: Routledge, 2011. A standard history of the English coroner system.

Jeffrey M. Jentzen

FORESTRY TECHNOLOGY AND LUMBERING

America's lumbering includes four basic aspects: felling, transportation of the resulting logs, sawmilling, and moving the mills' cut to market. All four underwent significant technological change over the years.

Continental Europe provided much of the initial technology used in North America. Resistance to labor-saving innovations out of fear of job loss had kept English production simple and labor intensive; tools were inefficient, having been developed for agrarian use, not lumbering. Living under labor-short conditions, Americans drew upon Dutch, German, Polish, Danish, and Scandinavian sources for better means of production. Booms for collecting and holding logs, water-powered reciprocal saws, and the rafting of logs and lumber all had European roots; continental workmen were hired to build many of the earliest sawmills. Over time American innovations improved upon this foreign base.

As markets grew and lumbering expanded, production underwent a series of transitions. In sawmills, water and steam power increasingly replaced human and animal power, which enabled the dispersal of mills away from waterways. Subsequently, internal combustion engines and electric-powered equipment replaced steam power. Each of these changes, and the accompanying technological adaptations, was driven by a need for efficiency, increased output, and lower costs. As sawmill production rose, loggers struggled to keep up with demand, but increased output of logs came from more and larger logging camps rather than improved technology.

Change also came to the marketing of lumber. Rafting lumber to nearby outlets gave way to long-distance marketing, first via sailing vessels (some specifically adapted to the trade) and then in steamships. On land, marketing passed from wagons to railroads to trucks. Initially lumbermen put most of their investment capital into sawmills, but over time more and more capital went to purchase timberlands so as to insure stable supplies of raw material. This shift accelerated as logging railroads became important; lumbermen believed they had to own timberland tributary to their rail lines to insure the profitability of their investments therein as well as in associated yarding and loading equipment.

There were regional variations in the patterns of change. The first major lumber-producing region, northern New England, had relatively gentle terrain and numerous rivers well adapted to log drives. Sleighs and ice roads were used in winter to move logs to streamside for downstream drives on spring floods. Splash dams were erected on smaller streams to store water until a sufficient head was built up to float logs downstream. As logging grew on the Allegheny Plateau and in upstate New York, log chutes, flumes, and other adaptations to more mountainous conditions emerged. Gravity thus came to supplement animal power in getting out logs. Over time this technology was transferred southward down the Appalachians and to the Sierra Nevada and other Far Western mountains.

On the Delaware, Hudson, Susquehanna, and other rivers in the mid-Atlantic states, log rafts were used initially, but in time transplanted New Englanders introduced free-floating log drives as a cheaper alternative. Equipment such as the peavy (a sturdy pike pole with a movable hook) and huge booms to store driven logs were also introduced from New England. Lumbering moved into the upper Great Lake states in the 1840s and after; conditions and patterns of production there approximated those of New England. From the 1830s steam-powered circular saws replaced water-powered reciprocal saws in mills and by the 1870s dominated in all major centers of production. In time these too were superseded, this time by band saws that could handle large logs more efficiently than circular saws and took a smaller kerf, thus reducing waste.

By the 1880s crosscut saws had replaced axes as the main felling tool. Earlier they had been used only in cutting downed trees into log lengths. Axes had been modified into straight-handled, double-bitted tools superbly efficient in felling, but now they came to be used in conjunction with crosscuts. After World War II power saws superseded crosscut saws, but the felling ax continued as a major supplement.

In the Gulf South and Far West, winters brought rain and mud, rather than snow, and log-moving techniques adapted to New England and the Lake States had to be replaced. Heavy-duty carts and big wheels (devices with two huge wheels and a long, levered tongue that when pulled lifted one end of the logs off the ground, allowing horses to pull them to landings at streamside or railside with relative ease) became key means of moving logs, now largely done in summer and fall, whereas railroads increasingly replaced waterways in transporting logs to sawmills and their cut to market.

Skid roads (cleared paths with small, notched logs set crosswise at close intervals and greased to ease the pulling logs to landings by oxen) were a far Western staple until John Dolbeer developed a steam-powered winch for the task. When equipped with steel cables, these ground-lead "steam donkeys" could handle larger logs than horses or oxen and get them to landings with greater speed, but they devastated the land, destroying groundcover, increasing erosion, and starting many forest fires. Subsequent systems of high-lead yarding moved logs at least partially through the air, reducing environmental damage while enabling steeper slopes than ever before to be logged, but high-lead systems introduced new elements of danger into an industry already far more dangerous than most.

The technology of logging and milling was relatively simple, entry was relatively easy, the resource base was widely distributed, and economies of scale were limited. As a result, ownership was fragmented and often in the hands of men directly involved in production. This led to an industry marked by numerous small improvements and minor inventions, but slow in adapting major changes in technology and management.

Until the twentieth century, production processes were adapted to old-growth forests and a wood-intensive society. Circumstances changed as old-growth stands dwindled. Technologies adapted to smaller, second-growth trees and aimed at the use of by-products became increasingly important. Building on the work of Charles H. Herty, much southern pine was redirected to the manufacture of newsprint. Veneers and laminated products replaced solid wood, first in hardwoods and then in softwoods; in turn, plywood gave way to chipboard. Technological changes driven by environmental concerns followed. Rubber-wheeled feller-bunchers that cut and loaded trees in a single operation and other logging equipment that was relatively easy on the land became common; initially much of this technology was imported from Scandinavia and other places where old growth had become scarce earlier than in the United States.

As uncut timberland became less available and the price of logs escalated, plants became larger and more integrated. Older sawmills largely devoted to the production of lumber became less competitive. Technological innovation and centralization of production in large, multiproduct plants became common; plants became fewer, larger, and more technologically sophisticated. Leading companies turned to research at facilities such as that of the Weyerhaeuser Company in Federal Way, Washington. Meanwhile, with more logs coming from federally owned forests, controls of logging and of sawmill technology increased.

By the 1920s, the Forest Service was insisting that logs from its timber sales be milled in modern, band saw–equipped plants that were served by standard-gauge railroads. Controls on destructive logging technology, especially Caterpillar-type tractors and ground-lead yarding, followed. Although the resource base continued to be widely scattered, such requirements helped move the industry away from its roots as a resource-intensive semi-frontier industry. In the early twenty-first century, it remains relatively fragmented.

[*See also* **Agricultural Technology; Canals and Waterways; Electricity and Electrification; Environmentalism; Forest Service, U.S.; Hydroelectric Power; Internal Combustion Engine; Printing and Publishing; Railroads; Steam Power;** *and* **Technology.**]

BIBLIOGRAPHY

Brown, Nelson Courtland. *Logging: The Principles and Methods of Harvesting Timber in the United States and Canada,* revised ed. New York: John Wiley & Sons, 1949.

Cox, Thomas R. *The Lumberman's Frontier: Three Centuries of Land Use, Society, and Change in America's Forests.* Corvallis: Oregon State University Press, 2010.

Hindle, Brooke, ed. *America's Wooden Age: Aspects of Its Early Technology.* Tarrytown, N.Y.: Sleepy Hollow Press, 1975.

Van Ravenswaay, Charles. "America's Age of Wood." *Proceedings of the American Antiquarian Society* 80 (1970): 49–66.

Williams, Michael. *Americans and Their Forests: A Historical Geography.* Cambridge, U.K.: Cambridge University Press, 1989.

Thomas R. Cox

FOREST SERVICE, U.S.

Initially established in 1876 as a one-man research government bureau simply charged with compiling data and information about American forest conditions, the U.S. Forest Service's mission today is managing 193 million acres of diverse forest and grassland landscapes for multiple uses. In the words of the agency's founding chief, Gifford Pinchot, its mission is to provide "the greatest good for the greatest number in the long run." To do that, the Forest Service has evolved into one of the world's leading scientific research organizations. In 1905, Congress renamed the bureau the U.S. Forest Service and gave it direct managerial control over the national forests and grasslands.

The Service's work is carried out by three administrative branches. Two branches—the National Forest System and State & Private Forestry—largely concern themselves with land management efforts. The third, Research and Development, employs five hundred plus researchers (of a total thirty thousand agency employees) who conduct studies in biological, physical, and social science fields in all 50 states and U.S. territories. This branch is made up of the following major areas: Resource Use Sciences, Quantitative Sciences, Forest Management Sciences, and Environmental Sciences. Research is carried out at experimental forests and ranges, research stations, and the Forest Products Laboratory and with research partners like colleges and universities. The first research station was established in 1908 at Fort Valley, Arizona.

The creation of the Forest Service grew in part out of the late nineteenth-century fear of an impending "timber famine" and calls to protect watersheds for urban and agricultural water supplies. The agency's research touted the questionable link of "forest influences" to climate moderation and water flow. Congress nonetheless passed the 1891 Forest Reserve Act, which authorized the establishment of forest reservations (renamed "national forests" in 1907) from public lands. The use of selective data to shape policy raised the question of the independence of research from administrators' aims, an issue that remains today.

The 1897 Forest Management Act's declaration that the reserves necessary for "securing favorable conditions of water flows and to furnish a continuous supply of timber" made clear that the federal forests would be open for use under regulated management. The law remained the main statutory basis for the management of the national forests until 1960 and was ultimately replaced by the National Forest Management Act in 1976.

Congress passed that and several other laws in response to the general public's growing discontent over how the Forest Service was managing its lands. During most of the agency's first half century, timber harvest levels on national forest lands had remained relatively low. In the decades following World War II, an expanding economy and the exhaustion of private forest lands led to unbalanced management of national forests. The agency's declining federal budget in the 1960s led to rapidly increasing timber harvest levels—at the expense of other land uses—to maintain the agency's revenue base. Finding itself at odds with environmental groups and under increasing criticism from both outside and inside the agency, the Forest Service in the 1990s turned to using science to inform policy through its newly announced "ecosystem management" policy.

[*See also* **Conservation Movement; Environmentalism;** *and* **Forestry Technology and Lumbering.**]

BIBLIOGRAPHY

Lewis, James G. *The Forest Service and the Greatest Good: A Centennial History*. Durham, N.C.: Forest History Society, 2005.
Steen, Harold K. *Forest Service Research: Finding Answers to Conservation's Questions*. Durham, N.C.: Forest History Society, 1998.

James G. Lewis

FOUNDATIONS AND HEALTH

Foundations might best be defined by what they are not. They are not controlled by national, state, or local governments and they are not designed—or, per federal law, even allowed—to make a profit. In terms of the history of public health, their contributions must be distinguished from those of a larger universe of nonprofit organizations and government agencies, which include hospitals, universities, charities, benevolent societies, voluntary associations, and federal entities like the National Institutes of Health (NIH) and the National Science Foundation (NSF; a government agency and not a private foundation). All the same, foundation activities to a great extent overlap with the purposes and conduct of these and many other similar but formally distinct institutions. Since the nineteenth century, foundations evolved in tandem with the emergence of multinational corporations, the expansion and contraction of government spending, and the proliferation of what has come to be known as the nonprofit sector.

To give a more positive definition, foundations are nongovernmental, nonprofit institutions that manage endowed funds and distribute a periodic percentage of those funds for charitable purposes. One or more originating donors provide the capital for a foundation's endowment and then authorize a self-perpetuating board of trustees to govern in accordance with an explicit mission as stated in a founding charter or a similar binding legal document. In turn, most boards hire a professional staff to conduct day-to-day business, manage the disbursal of funds, and grow the foundation through additional private gifts, government support, or donations from the public.

Although individual foundations work toward a variety of goals, foundations tend to define their purpose as serving (depending on their preferred vocabulary) the "betterment of humanity," the "improvement of mankind," or the "promotion of public welfare"—in a word, philanthropy. For that reason, trustees and officers have often found the improvement of public health to be an ideal goal for their undertakings. The measurable outcomes and political palatability of health initiatives give the typical, entrepreneurial-minded leaders of foundations a sense of maximizing their ability to change the world. Still, although they often fall short of such lofty aims, the actions and ideologies of foundations have played an integral part in the production and organization of scientific knowledge in the United States since the end of the Civil War. That history places these American institu-

tions, more prolific and influential than in any other nation, at the center of a wide-ranging debate, not only about how to improve health outcomes around the world, but also about the proper balance of science, private wealth, and public health within a democratic society.

Foundations in Civil Society. Many American readers will know the names of the most prominent foundations from program credits on public radio and public television or from the lists of donors at charity events and museums. These names include Carnegie, Rockefeller, and Russell Sage, whose fortunes seeded foundations in the early 1900s; Ford, Robert Wood Johnson, and MacArthur, whose endowments came to prominence in the second half of the twentieth century; and Gates and Clinton, who led the creation of a new wave of high-profile foundations since the 1990s. Those who work in higher education or in the nonprofit sector might be familiar with a wider range of foundations thanks to the grant applications and funding requests that have become an inescapable part of such lines of work. Beyond the household names and billion-dollar trusts, smaller foundations have also proliferated since the end of the Civil War.

According to the Foundation Center, a New York City–based nonprofit that began in 1956 and acts as a clearinghouse for resources on philanthropic organizations, private foundations of all types and sizes numbered 81,777 in 2011 (the most recent year with available data). Collectively, those foundations held assets worth more than $662.3 billion and distributed nearly $49 billion in annual grants. Although the immense number of organizations and the sheer amount of money make the precise categorization of all philanthropic giving difficult to calculate, the Foundation Center's 2013 report, *Key Facts on U.S. Foundation*, lists "health" as the largest category of foundation expenditures, accounting for 28 percent of the grants made by the largest 1,122 foundations, with more than 850 of those foundations contributing to health-related initiatives.

Organized philanthropy has existed alongside other forms of charitable giving since the American Revolution, but an unprecedented explosion of nonprofit foundations emerged from a nineteenth-century debate between capitalists who had accumulated vast industrial wealth and reformers aiming to correct the social inequality exacerbated by that wealth. On one hand, the originators and supporters of the first large, general-purpose foundations were looking for ways to alleviate the negative consequences of unbridled capitalism by devoting private fortunes to the promotion of public good. On the other hand, progressives saw industrial money as tainted by the exploitation of wage workers. These critics maintained that the public good must be determined through democratic processes rather than by the whims of the powerful. The sides of this debate maintained their basic contours over time even as they shifted in ways that scrambled the political spectrum (Friedman and McGarvie, 2003, pp. 217–225; Zunz, 2012, pp. 17–22; Anheier and Hammack, 2013, pp. 19–20; Dowie, 2001, pp. xxxvi–xxxvii, 1–21).

The social critic Dwight Macdonald captured the underlying dynamic in these interactions when, writing in the *New Yorker* magazine in 1954, he referred to the Ford Foundation as "a large body of money completely surrounded by people who want some" (Macdonald, 1956, p. 3). Macdonald likely had in mind those clamoring for grants at the doorstep of foundations, but his barb applied just as well to those who would dissolve foundations to redistribute their wealth and those in politics and business who hoped to manipulate foundations to serve their purposes. All of this has left foundation employees to defend their intentions and assert their authority on politicized and, at least in the early years, precarious terrain.

Legal Precedents. Foundations emerged from a long evolution of centuries-old legal precedents. The notion of an independent institution designed for philanthropic purposes dates back to the Roman legal code when emperors first granted charters to corporations and encouraged them to engage in charitable activities. However, U.S. law on foundations descends more directly from an act of the English Parliament during the reign of Elizabeth I. The Statute on Charitable Uses of 1601 first defined the activities that could be listed

as charitable gifts. These included, among others, relief for the poor, sick, and aged, as well as medical care for soldiers and sailors (Friedman and McGarvie, 2003, pp. 20, 34; Macdonald, 1956, pp. 36–37).

Although modern statutes expand on those original uses, the Elizabethan list has inspired the codification of philanthropic activity throughout the English-speaking world. The U.S. government, of course, abolished all English laws after the American Revolution, but the general tenets of English common law have continued to guide the courts in legal disputes over the disbursal of estates to charitable causes. Adherence to these traditions at times hindered the formation of large foundations because the established principle of *cy pres* (which held that the original intention of the donor should be followed to the letter) allowed heirs of large fortunes to challenge many charitable gifts. Therefore, unless outside donations had been carefully defined, the funds of many estates often reverted to the relatives of the deceased. Although these restrictions enabled many small, typically religious foundations to form, only a handful of wealthy individuals managed to sustain large, secular foundations in perpetuity, either by creating trusts during their lifetime or by developing strong working relationships with their offspring. By the second half of the nineteenth century, reformers and industrialists alike began to press for new laws that would allow donors to leave open-ended bequests giving a single foundation the freedom to pursue a wider range of philanthropic activities (Lagemann, 1999, pp. 6–26; Zunz, 2012, pp. 8–17, 76–85).

But even prior to that reform movement, other legal interpretations had begun to pave the way for the creation of bigger, private foundations. In Massachusetts, for example, the capital accumulation of Boston merchants, combined with the evangelical spirit of the Protestant establishment, led to a groundswell of family trusts and religious charities. The State Supreme Court freed the hand of these organizations when it ruled in 1829 that trustees who acted within reasonable standards of prudence and discretion could not be held negligent if the market value of their assets declined. In a similar vein, an 1847 essay by the Yale professor

Lawrence Bacon laid down the first systematic ex- plication of the internal governance of nonprofit trusts.

These more favorable developments estab- lished the basic standards of professionalism and organization that evolved into the institutional structure of modern foundations. Then, in 1893, New York passed a law, known as the Tilden Act, allowing donors to create trusts with no stipula- tion on using funds. Other states followed with similar laws, creating momentum for more uni- form national policies on philanthropy. The move- ment culminated in 1913, when passage of the Sixteenth Amendment legalized federal income tax. In the act implementing the tax code, Con- gress included high margins on large incomes and granted exemptions to charitable donations. Both statutes proved to have a stimulating effect on foundation creation (Anheier and Hammack, 2013, pp. 38–39; Lagemann, 1999, pp. 6–26; Zunz, 2012, pp. 8–17).

Social Origins.   The new income tax also led to lasting skepticism that wealthy donors only cre- ated foundations to avoid paying taxes. Among the tens of thousands of trusts, and with the exten- sive legal resources of the rich, many foundations have indeed served such self-interested purposes. In the bigger picture, however, such an interpreta- tion of foundations discounts their deep roots in American traditions of community welfare and faith in scientific progress. As the United States emerged as a world power over the course of the nineteenth century and industrialists accumu- lated both money and influence as a result, the na- tion's elite gained confidence that they could solve many of society's ills. They need only work hard and apply the sound principles of rational man- agement. In addition, at the dawn of the twentieth century, economic expansion began to spread to the middle class. The new middle classes added political strength to movements that sought to end poverty and cure disease by investigating underlying causes. Foundations became a pivotal tool in the implementation of these confident, or perhaps hubristic, social schemes.

Alexis de Tocqueville, after his famous tour of the United States in the 1830s, noted the pro-

clivity of Americans to form associations. Like the New England religious charities and family trusts mentioned above, many of the groups observed by Tocqueville directed their energies to serving the poor and sick in their community, usually within the context of a Christian mission. These efforts became more national, and more secular, once the end of the Civil War drew northern at- tention to the cause of rebuilding the devastated Confederate states. The cause of southern recon- struction led to the creation of what the historian of philanthropy Olivier Zunz describes as the first modern foundation: the Fund for Southern Edu- cation founded by George Peabody, who had made a fortune financing Atlantic merchant ship- ping (Friedman and McGarvie, 2003, pp. 29–48; Zunz, 2012, p. 10).

Peabody limited his fund to the support of ex- isting schools and therefore only reached white southerners. But it would be the education of Af- rican Americans that became the central cause of large foundations throughout the era of Jim Crow. Although derided at the time and in some sub- sequent histories as paternalistic and failing to address the structural forces of segregation and racism, the potential to transform freed slaves into model citizens struck many philanthropists as a golden opportunity. The John F. Slater Fund for the Education of Freedmen, begun in 1882 by a Connecticut mill owner, dedicated all of its assets to the construction of schools for African Ameri- cans. Julius Rosenwald, part owner and future president of Sears & Roebuck, followed in 1895 with his own eponymous foundation. The Rosen- wald Fund helped create, among other schools, Booker T. Washington's Tuskegee Institute (An- heier and Hammack, 2013, pp. 36–37; Friedman and McGarvie, 2003, pp. 161–178; Zunz, 2012, pp. 30–43).

The targeted giving of Peabody, Slater, and Rosenwald presaged the more sweeping efforts soon to be devised by John D. Rockefeller and Andrew Carnegie. The fact that these early foun- dations chose education as their cause was nei- ther random nor coincidental but reflected the philosophy of charitable giving that informed the formation of general-purpose foundations. Andrew Carnegie most famously enumerated the principles of this philosophy in an essay published in the *North American Review* in 1889. Although titled simply "Wealth," Carnegie's doctrine soon became known as the "Gospel of Wealth." The steel magnate believed that he, and rich men like him, had an obligation to give back to the society that made them rich, but he was determined to do so in ways that applied the same management principles that earned him his wealth. That meant making strategic, efficient decisions to achieve measurable goals. In his view, small-scale solutions like feeding the poor or treating disease encouraged the needy to beg. True charity should instead teach them skills to support themselves. So rather than building libraries, Carnegie gave grants to communities that agreed to build and maintain their own libraries. Rather than buying supplies for teachers, the Carnegie Foundation for the Advancement of Teaching invested in teachers' pensions, thereby elevating the respectability of the teaching profession as a whole (Dowie, 2001, pp. 4–5; Friedman and McGarvie, 2003, pp. 222–225; Sealander, 1997, pp. 17–18; Zunz, 2012, pp. 1–2, 22–30).

Although foundations have continued to cite the "Gospel of Wealth" as an inspiration, Carnegie's creed has never been without its detractors. In fact, the popularity of Carnegie as a historical personality has led to exaggerations of the importance of his credo. For one thing, Carnegie's gospel butted heads with the Social Gospel, an actual Christian doctrine. Whereas the captains of industry, following the so-called social Darwinism of the sociologist Herbert Spencer, argued for the survival of the fittest, Protestant reformers espoused the biblical dictum that the meek shall inherit the earth. The Christian social worker and settlement house founder Jane Addams eschewed the large-scale giving of foundations in favor of personal engagement with the poor and the sick. Trustbusters in the administrations of Theodore Roosevelt and William Howard Taft disavowed conglomerations of money and power, no matter their end purpose. And, by the turn of the century, community chests and other local charities began to form national associations like United Way, soliciting small donations from the middle class and not the rich. These varying philosophies of philanthropy conflicted, at times with vituperation,

but they came to recognize that they grew out of a common American tradition. They all shared a belief that industrial growth placed unique burdens of inequality and disease on society and a faith that American progress could attack those ills at the source (Friedman and McGarvie, 2003, pp. 226–228; Sealander, 1997, pp. 17–18; Zunz, 2012, pp. 18–21).

From Educational Funds to General-Purpose Foundations. Just as the Christian tradition of charity and the devastation of the American South combined to motivate the emergence of the modern foundation, the legal reforms, social ferment, and unprecedented wealth of the late nineteenth century would foment a new creation: the general-purpose foundation. Rockefeller and Carnegie, the two wealthiest men in the United States at the turn of the century, had already initiated an era of what historian Judith Sealander has called scientific philanthropy. But their ambition, along with the practical difficulties of executing millions of dollars of donations, led both men to look for a more enduring structure to guide their foundations (Friedman and McGarvie, 2003, pp. 217–239; Sealander, 1997, pp. 9–25).

The solution, in the words of Frederick T. Gates, the Baptist minister hired by Rockefeller to be his philanthropic adviser, was "wholesale philanthropy." Instead of directly targeting individual causes or diseases, the general-purpose foundations would disburse grants to established organizations that supported a broad range of causes. Full-time professionals such as Gates took charge of building relationships and disbursing grants. Then, in the first decades of the new century, foundation officers developed this basic concept of wholesale philanthropy into elaborated working models to guide their work. The historian Stephen C. Wheatley (1988) identifies two trends within these somewhat ad hoc schemes. In the first, the venture capital theory, foundations funded basic research into large-scale problems often neglected by political and educational institutions occupied by routine administrative tasks. Second, foundations took institutional leadership, usually by funding university academic programs or building separate research institutes, to expand the infrastructure of scientific and academic institutions. When describing their own work, foundation officials tended to reflect more on the idealism of basic research than on the at times uncomfortable political implications tied to intervening in higher education. In truth, although the two tendencies complemented each other, the more politicized aspects of foundation tended to have the most lasting influence, for better or for worse (Friedman and McGarvie, 2003, pp. 221–222; Fosdick, 1989, pp. xiii–xv).

The partnership between Gates and the senior Rockefeller developed the systemization of the general-purpose foundation more than any other. Rockefeller admired Carnegie's "Gospel of Wealth," even writing his fellow industrialist a letter of appreciation, but whereas Carnegie turned later in life to a secular brand of philanthropy, the founder of Standard Oil had a long-standing reputation for religious donations and transformed that personal predilection into a lasting and influential vision of scientific giving. As the story goes, Rockefeller tithed 10 percent of his income, beginning with his first-ever paycheck, to his Baptist church. Later, as his income and his charity increased, he channeled his contributions through the American Baptist Education Society as a way to avoid the burden of evaluating multiple original donors. In 1890, a Rockefeller gift and the administrative guidance of Gates, who served as secretary of the society, led to the founding of the University of Chicago. Although both men intended it to be a religious institution, Gates sought out support across denominations to both avoid sectarian divisions and attract wider interest in donations to match the original bequest. As a result, the University of Chicago, much like the Carnegie teacher pensions, demonstrated how foundation initiatives acted as a pluralizing and secularizing force in American higher education (Anheier and Hammack, 2013, pp. 48–51; Berliner, 1985, pp. 26–27; Fosdick, 1989, pp. 1–4; Kevles, 1992, pp. 195–197; Zunz, 2012, pp. 26–30).

Not long after the chartering of the university, Gates became a full-time employee for Rockefeller and began to expand his boss's philanthropic endeavors beyond single gifts and initiatives. The first step in the process came in 1901 when the duo founded the independent Rockefeller Institute for Medical Research in New York City, later renamed Rockefeller University. Rather than aligning their new medical school with an existing college, Gates and Rockefeller made the institute independent. This gave them a freer hand to shape the school into their vision for a model of public-health education. Beginning the following year, the General Education Board distributed money to other medical schools across the country, in addition to its support for agricultural life and African American schools in the South. Then, in 1909, the Rockefeller Sanitary Commission joined the efforts of Wickliffe Rose, a Tennessee professor who worked with the Peabody Education Fund and who later became a leading executive in Rockefeller's philanthropies, to eradicate the hookworm parasite in the American south. All of these initiatives coalesced in 1913 when New York State chartered the general-purpose Rockefeller Foundation. The new organization absorbed several of Rockefeller's earlier efforts, operated as a grant-making body in its own right, and coordinated policy across his many endeavors (Berliner, 1985, pp. 35–75; Fosdick, 1989, pp. 1–43; Sealander, 1997, p. 12; Zunz, 2012, pp. 40–43).

Perhaps the biggest contribution of the Rockefeller Foundation to the reorganization of scientific knowledge grew out of a report originally funded by the Carnegie Foundation for the Advancement of Teaching. In 1910, Abraham Flexner published his *Medical Education in the United States and Canada*, which advocated a system for educating doctors modeled on the four-year curriculum adopted by the recently opened Johns Hopkins University School of Medicine. The report recommended a hierarchical academic structure that integrated medical schools into university bureaucracies. Such an arrangement replaced profit-seeking motives and informal apprenticeships with standardized pedagogical practices. Flexner soon moved on to work for the General Education Board and, with the help of money from the Rockefeller Foundation, began to implement his policies. His specific recommendations grew out of ideas that had long percolated in academic circles and coincided with a broader transformation in the culture of higher education in the early twentieth century. More than changing medical education in and of itself, the report demonstrated the newfound power of general-purpose foundations. By endorsing the report, by donating to schools that adopted Flexner's principles, and, just as important, by refusing to give to those that did not, the Rockefeller Foundation acted as a gatekeeper to the medical profession. In large part because of these funding decisions, scores of medical schools closed their doors and others turned over entire teaching staffs in the name of reform (Berliner, 1985, pp. 92–138; Sealander, 1997, p. 19; Wheatley, 1988, pp. 45–54; Zunz, 2012, pp. 25–26).

Around the same time, the Russell Sage Foundation, founded by Sage's widow, Olivia, in 1907, served a similar rationalizing function in the professionalization of social workers. But the new general-purpose foundations also sparked the professionalization of philanthropy itself. In addition to Frederick Gates, Wickliffe Rose, and Abraham Flexner, Henry Pritchett (the Carnegie advisor who hired Flexner) and Flexner's brother Simon (the first director of the Rockefeller Institute for Medical Research) joined a vanguard of experts who dedicated their careers to refining a science of large-scale giving. John D. Rockefeller Jr., emerging out of the shadow of his father as the first president of the foundation, would soon become the most prominent name among the new class. Although the work of these new philanthropists did lead to some direct advancements in public health—the eradication of hookworm in the United States most notable among them— the professionalization of philanthropy remains the most discernible legacy of foundation work (Berliner, 1985, pp. 92–138; Sealander, 1997, pp. 16–31; Zunz, 2012, pp. 22–30).

Foundations and the Federal Government. The golden age of general-purpose foundations, and of the rationalized philanthropy they spawned, did not arrive without opposition. In fact, the birth of the Rockefeller Foundation coincided with the climax of public controversy over Standard Oil, John D. Rockefeller's main commercial endeavor. The anti-trust lawsuit that would

soon dissolve the company went before the Supreme Court in 1910, so when the Rockefeller Foundation petitioned Congress for a national charter during that same year, the request met a legislative brick wall. Newspapers around the country accused the foundation of acting as a Trojan horse of corporate perfidy within the citadel of democracy. One Rockefeller staffer came to believe that senators would even rise to speak against the Ten Commandments if their boss happened to mention he liked them. The Senate voted not once, but twice to deny the charter. For the next few years, whatever goodwill had arisen between industrialists and reformers would fall apart under the pressure of the trust-busting and labor unrest of the era. The birth pangs of the Rockefeller Foundation marked the beginning of a century-long effort of philanthropists learning to accommodate the prerogatives of government and negotiate the denunciations of populist politicians (Berliner, 1985, pp. 132–133; Dowie, 2001, pp. 13–14; Fosdick, 1989, pp. 26–28; Kohler, 1991, pp. 49–52; Lagemann, 1999, p. 25; Sealander, 1997, pp. 224–234; Zunz, 2012, p. 21).

The Rockefellers managed to sidestep the problem of a national charter by securing a state charter from New York in 1913. The friendly terms allowed by the state that passed the Tilden Act had attracted the Carnegie Corporation two years earlier. But solving the immediate issue would not eliminate the larger political problems either for Rockefeller or for the other large foundations. In response to a wave of violent incidents involving organized labor, Congress created a Commission on Industrial Relations, often called the Walsh Commission after the Kansas City attorney who served as its chair, to investigate the causes of the turmoil. Over the course of the proceedings, the Rockefellers (junior and senior), Andrew Carnegie, John Glenn, director of the Russell Sage Foundation, and Julius Rosenwald all testified. Walsh seized the opportunity to grill the country's most prominent philanthropists over what he believed to be the inherent abuse of power by private foundations. Adding fuel to the fire, just as the Commission began its investigations, violence erupted at a Colorado mine in which Rockefeller Jr. owned a primary stake. In what became known as the Ludlow Massacre, the Colorado National Guard and mining camp guards opened fire on striking workers, killing as many as 25 people, including 2 women and 11 children. When the Commission completed its hearings, the final report issued legislative proposals that would have placed harsh limits on the operations of foundations. However, because the controversy had become so intense, Congress declined to act (Berliner, 1985, pp. 132–133; Dowie, 2001, pp. 13–14; Fosdick, 1989, pp. 26–28; Kohler, 1991, pp. 49–52; Lagemann, 1999, p. 25; Sealander, 1997, pp. 224–234; Zunz, 2012, p. 21).

The negative publicity surrounding the public role of foundations began to die down only when the United States entered World War I. Beginning while combat still raged and accelerating in the war's aftermath, foundations began to adapt their existing programs for application abroad. Starving populations received U.S. aid funded by American philanthropy, and European academics earned access to private grants. Under the direction of Wickliffe Rose, the Rockefeller Foundation founded international versions of its Sanitary Commission and Education Board. These bodies internationalized the effort to eradicate hookworm and started new initiatives to attack malaria and influenza. The Flexner-inspired and Rockefeller-funded medical reforms also extended overseas with the creation of the Peking Union Medical College and the expansion of the London School of Hygiene and Tropical Medicine. The postwar Republican administrations in Washington welcomed the expansion of organized—and private—philanthropy both at home and abroad. Herbert Hoover, first as secretary of commerce and then as president, spearheaded efforts to coordinate government agencies with foundation programs (Friedman and McGarvie, 2003, pp. 246–256; Zunz, 2012, pp. 104–110).

Despite the welcoming attitude of the executive branch, the expanded role of foundations once again faced some headwinds. For example, from 1925 to 1930, the Board of Regents of the University of Wisconsin, backed by the Progressive Party founded by Senator Robert La Follette, enacted a rule prohibiting the school from accepting any money from private foundations. Then, Hoover's public-private partnerships floundered in the midst of the Great Depression. The New Deal policies of Franklin Roosevelt moved away from the use of foundation support, choosing instead to empower federal agencies to address crises of agriculture, poverty, and public health. In the face of policy changes, and suffering from the stock market crash themselves, foundations scaled back and refocused their efforts. Those financial pressures, along with concerns that philanthropic funds had overinvested in established fields that promised fewer new breakthroughs, convinced foundation professionals to push more academic grants toward less developed disciplines. Exemplary of this approach was Warren Weaver, a University of Wisconsin mathematician who learned a more foundation-friendly attitude from his roots as a La Follette progressive. When he left Wisconsin in the 1930s to become the head of the natural sciences division at the Rockefeller Foundation, Weaver persuaded his new employers to explore how new technologies in physics and chemistry could be applied to biological sciences. His policy recommendations led to investments in nuclear physics and the chemistry of radiation that helped build Ernest O. Lawrence's cyclotron at the University of California (Friedman and McGarvie, 2003, pp. 263–280; Kevles, 1992, pp. 206–211; Kohler, 1991, pp. 265–302; Zunz, 2012, pp. 125–136).

Weaver's contributions to these new technologies continued the tradition of foundations reshaping scientific disciplines. Indeed, he came to be known as the father of molecular biology, although in truth, scientists themselves determined the precepts and content of the new field (Kevles, 1992, pp. 208–209). But Weaver's efforts existed within broader patterns of philanthropy. As foundations retooled their policies during years of economic downturn and shifted resources in an era of activist government, they became the leading advocates for science as a progressive, modernizing force in society and a decisive force in the development of scientific research and public-health institutions in the United States. Although New Deal programs brought a more activist approach to social policy and changed the relationship between the government and foundations, the Depression-Era government lacked both the tax base and the political clout to dedicate significant financial resources to science. Researchers only began tracking federal obligations to research and development after the war, but even in 1947 the government dedicated more than two-thirds of research and development resources to the military while giving only 1.7 percent ($10.6 million) to the Department of Education, Health, and Welfare (*Historical Statistics of the United States*, Table Cg 182-202). As indicated by Stanley L. Engerman and Gavin Wright, the economists who compiled the data, prewar spending on research and development would have been much smaller with the largest expenditures dedicated to agriculture and aeronautics. The entire annual budget for the NIH, for example, did not exceed $1 million until 1943 (*NIH Almanac*, Appropriations, Section 2).

Meanwhile, the one hundred largest private foundations together spent, on average, about $45 million per year in all areas between 1921 and 1940, with amounts nearly double that average from 1929 to 1931 during the early years of the Great Depression. Of that total, one-third, or a yearly average of around $15 million, went to health-related causes (*Historical Statistics of the United States*, Table Bg 28-40). If anything, these numbers understate the amounts of money donated to health and science by foundations in the second quarter of the twentieth century. The data exclude numerous foundations that might be smaller than the top one hundred but still donated significant, if difficult to pinpoint, amounts to health-related causes (Anheier and Hammack, 2013, pp. 43–48, 55–57, 168–169). With the federal government yet to enter the field, no other section of U.S. society contributed as much in ideas, personnel, and wealth as private foundations in the first four decades of the twentieth century.

Building the Postwar Establishment. The dominant role of foundations in scientific research forever changed when the United States joined the Allied effort in World War II. Policy makers turned to foundation personnel for their expertise in a wide variety of technical and bureaucratic areas and, in the process, pulled private foundations into the political mainstream once and for all. At the same time, the massive increase in government spending to support the war effort gave federal agencies a decided advantage in determining research funding and expenditures. In the decades that followed, the government would only increase its role as the major sponsor of science in the academy. With their free hand diminished, foundations at first found themselves searching to define their new space within civil society.

Historians now agree by and large that the warfare state of the 1940s played a majority role in pulling the United States out of the economic devastation of the 1930s. At the time, scientists and politicians alike trumpeted astonishing technological advancements, with the atomic bomb the most dramatic of all, as nothing less than the key to securing an Allied victory and sustaining postwar prosperity. This interpretation underestimates the contributions of full employment—both on the battlefield and in manufacturing—to the war effort and to economic growth. Nonetheless, the possibilities of scientific achievement captured the public imagination. Under the leadership of public officials such as Vannevar Bush, who headed the wartime Office of Scientific Research and Development and whose book, *Science: The Endless Frontier*, argued for long-term investment into basic research, the government leveraged public support into sustained government spending on science. Because of the war, the incipient postwar rivalry with the Soviet Union, and perhaps Bush's own fame as a rocket scientist, increased funding to other military technologies—and, in particular, the development of nuclear weapons and atomic energy—gained the most attention (Kevles, 1992, pp. 211–215).

However, the postwar wave of appropriations also extended to medicine and the health sciences. The Hill-Burton Hospital Construction Act of 1946 converted wartime support for hospitals into a peacetime domestic program. Then, in 1948, Congress dramatically increased funding to the NIH, a government agency created in 1937 and dedicated to research in medicine and other health-related disciplines, by quadrupling its budget from $8 million to $24 million (*NIH Almanac*, Appropriations, Section 2). The budget doubled again in 1950 when it exceeded $50 million. That same year, the advocacy of Vannevar Bush culminated with the creation of a new federal agency, the NSF, designed to support basic science across all fields, including the biological sciences, by issuing grants to academic scientists. The budgets for both agencies grew exponentially over the next few decades. In 2012, the NSF listed their annual budget as around $7 billion, which accounted for 20 percent of all federally funded research in U.S. higher education ("NSF at a Glance," http://www.nsf.gov/). In addition, the NIH invested nearly $31 billion dollars in medical research in 2012. Of that budget, 80 percent went to universities, medical schools, and research institutes in the form of grants ("NIH Budget," http://www.nih.gov/; Anheier and Hammack, 2013, pp. 75–78).

In the face of these ballooning public expenditures, even the collective donations of the 100 largest foundations had little hope of keeping pace. Their total grant giving by 1957 had grown to $248 million, up from $40 million in the last year before the war. But the amount donated to health, although higher in terms of raw dollars, had been cut in half as a percentage of all foundation giving—down to 14 percent from 30 percent in 1940 (*Historical Statistics of the United States*, Table Bg 28-40). This lower distribution of funds resulted in part from a recognition on the part of the big foundations that they could no longer compete with federal money. In 1957, the budget for the NIH alone, not to mention the NSF and other related agencies, amounted to $177 million dollars, more than four times the amount spent on health by the big foundations. The NIH budget doubled by 1960 and broke $1 billion six years later. In 2012, the $30 billion of NIH annual spending dwarfed the $6.8 billion in health grants awarded not just by the top 100 but by the 1,100 largest foundations ("Foundation Stats," http://data.foundationcenter.org/).

Beyond the growth of federal funding for scientific research, the growth of Social Security, Medicare, Medicaid, and various other government social programs revolutionized the way that U.S. society cared for the poor, sick, and elderly. That left foundations searching for new roles to play within a transformed bureaucratic system. The class of professional philanthropists added unique funding expertise to government roles in the Office of Scientific Research and Development and in military service. In turn, as the war wound down, a new generation of professionals, trained in the wartime bureaucracy, injected their experience into a burgeoning philanthropic sector. In the following decades, as government support revolutionized the funding mechanisms in higher education, the work and money of foundations became more and more entangled with the federal bureaucracy. This development engendered truculent critiques aimed at foundation contributions to what President Dwight Eisenhower warned had become an inexorable military-industrial complex (Friedman and McGarvie, 2003, pp. 319–325; Lagemann, 1999, pp. 196–201; Zunz, 2012, pp. 170–189).

Besides the organizational transformations initiated by increased government funding of science, the unprecedented economic expansion of the postwar years did fill the coffers of private foundations, albeit at a pace far below the increases in federal research and development. The new wealth expanded the endowments of existing philanthropies and led to exponential increases in the number of new foundations. In terms of both money and influence, the Ford Foundation marked the pinnacle of this postwar explosion, a fitting development considering the contribution of Henry Ford's streamlined manufacturing principles to the efficiency of U.S. war production. The pioneer of the automobile industry never had the same zeal for philanthropy as his fellow magnates in steel and oil. But in 1936 Ford did endow a foundation dedicated to charitable giving in and around Detroit, Michigan. Both Henry and his son, Edsel, then arranged their estates so that the bulk of their nonvoting shares in the Ford Motor Company devolved to the foundation, thus avoiding the large inheritance tax instituted during the New Deal and allowing their family to retain control of the corporation. After the two men died within four years of each other, Edsel's son, Henry

Ford II, became the president of the world's largest foundation in 1947. The grandson brought a renewed focus to a family empire that had begun to drift under the guidance of his elderly grandfather. On the philanthropic side, the younger Ford hired H. Rowan Gaither Jr., a former researcher at the Massachusetts Institute of Technology and chair of the RAND Corporation, a private think tank initially created by the military. Gaither led a full-scale examination of how the Ford Foundation's resources could best be used. The resulting Gaither Report pushed the foundation into philanthropy on an international scale, funneling money into the reconstruction of Europe and the developing world as well as into large-scale studies in behavioral science and civil rights. By 1960, the more than $3.5 billion assets of the Ford Foundation dwarfed the combined endowments of the Rockefeller, Carnegie, and Sage foundations (Friedman and McGarvie, 2003, pp. 319–320; Lagemann, 1999, pp. 195–196, 298–299; Zunz, 2012, pp. 173–181).

Many of the new Ford initiatives dovetailed so well with government programs that in 1950 Paul Hoffman, a former Studebaker executive and administrator of the Marshall Plan, transitioned directly from his government position into the presidency of the Ford Foundation. In fact, even the increased wealth of private foundations only managed to supplement the gargantuan spending of government institutions in the 1950s and 1960s. With the NSF and NIH pouring money into the medical and natural sciences, the Russell Sage Foundation, the Kellogg Foundation, and many others decided to shift their resources to the social sciences and the arts. By 1970, federal funding dominated university research to such an extent that the government owned almost 30 thousand patentable inventions developed by academic scientists. Even if foundation money had funded the majority of a research project, the government retained ownership of the intellectual property if even one dollar of federal money had been used (Anheier and Hammack, 2013, pp. 88–94; Friedman and McGarvie, 2003, pp. 363–383; Zunz, 2012, pp. 147–148).

For some smaller foundations these sorts of government interventions challenged the very nature of their philanthropic activities. As one example, the Wisconsin Alumni Research Foundation had, since 1925, filed patents on inventions made by scientists on the University of Wisconsin campus and then invested the resulting proceeds to build an endowment for the support of scientific research at the school. By the 1960s, with the government claiming exclusive rights to more and more university science and then neglecting to develop that science into useful inventions, the Wisconsin Alumni Research Foundation watched their patenting and licensing activity begin to dwindle. More important, across the country the creation of new technologies based on groundbreaking university research had stalled out. Less than 1 percent of the 30 thousand government-claimed inventions had been developed into an actual product in the marketplace. Recognizing the dilemma, Howard Bremer, patent counsel for the Wisconsin Alumni Research Foundation, began working with Norman Latker, who held the same position at the NIH, to negotiate institutional agreements between government agencies and university administrators that allowed schools to retain patent rights and commercialize their research. Over the course of 20 years, Bremer and Latker banded together with officials at other universities and research institutes hoping to patent and license their research without foregoing government funding. Their work culminated with passage of the Bayh-Dole Act in 1980, which for the first time adopted universal rules to govern the relationship between scientific research conducted at universities and the federal funding agencies that supported that research. The legislation marked just one of the many negotiations among private money, higher education, and public policy that turned universities into the economic engines of high-tech metropolitan centers.

In other venues, the integration of philanthropic funding and government agencies led to more immediate results. In particular, all of the major foundations embraced the liberal, anti-Communist consensus of the foreign-policy establishment. The internationalized initiatives of the Ford Foundation joined the Rockefeller philanthropies and the Carnegie Corporation in efforts to develop the economies of countries in Latin America, Africa, and Asia. Much as they did in previous decades, all three institutions proved more influential in training personnel and generating theoretical knowledge than in directly curing disease or saving lives. Paul Hoffman, the Marshall Plan administrator who then became president of the Ford Foundation, was just one example of a revolving door between foundation offices and government agencies. The two longest serving secretaries of state of the postwar era, the Republican John Foster Dulles and the Democrat Dean Rusk, both left leadership positions in the Rockefeller Foundation to join the cabinet. McGeorge Bundy, a foreign policy adviser for Lyndon Johnson, and John McCloy, U.S. High Commissioner of occupied Germany, did the reverse, leaving government posts to join the Ford Foundation. In addition, Ford was a major donor to the Massachusetts Institute of Technology's Center for International Studies where Walt W. Rostow and Max Millikan devised methods for using state-sponsored economic development programs to counter the appeal of Communism in the Third World. Their modernization theory served as a guiding principle for the Central Intelligence Agency's covert interventions around the world and eventually for the military operations in Vietnam (Friedman and McGarvie, 2003, pp. 320–325; Zunz, 2012, pp. 146–151).

New Politics Transform the Institutional Landscape. Not least because of the quagmire in Southeast Asia, foundations would not escape the cultural turmoil of the 1960s and 1970s. Antiwar protesters and leftist intellectuals charged that philanthropic funding of academic research implicated the major foundations in the excesses of industrial warfare and corporatized universities. From the other side, a new generation of conservatives accused the old-money foundations of facilitating runaway government spending and propping up the agenda of the liberal elite. By this point, however, foundations had become so enmeshed in the political establishment of the United States that their critics had far less opportunity to pose existential threats. Too many legal precedents now existed, and federal

and nonprofit research funding had become too intertwined on campuses, for opponents to challenge modern foundations in the way the Progressives had challenged the Rockefellers. Still, the new generation of critiques drew attention to the growing political implications of foundation work. In response, established philanthropies once again looked to shift resources away from controversy even as new foundations emerged with explicit ideological agendas (Anheier and Hammack, 2013, pp. 81–84; Lagemann, 1999, pp. 70–75; Zunz, 2012, pp. 220–230).

As the civil rights movement gathered momentum in the early 1960s, a number of philanthropic organizations, including the Ford Foundation, offered their support for the cause. In response, segregationists resurrected charges of undue political influence. Soon a coalition of southern Democrats and conservative Republicans pushed legislation designed to draw a clear delineation between political and philanthropic organizations. Since the 1930s, federal law had prevented private foundations from contributing to candidates' campaigns and, during the Red Scare of McCarthyism and the subsequent political backlash, Congress began to strengthen tax laws to reinforce such restrictions. But proposals in the 1960s expanded to target lobbying efforts and limit the ability of politicians to work at foundations. These efforts culminated with the Tax Reform Act of 1969. The act amended Section 501 (c) (3) of the tax code to set limits on lobbying and required government officials to resign before accepting foundation employment. It also established a minimum percentage of assets that they must spend on an annual basis. Therefore, unlike the Walsh Commission report of 1915, opponents succeeded in passing legislation, but the 1969 reforms still failed to staunch the tide of foundation creation. Their money, ideas, and personnel had become too ensconced in the policy-making process for one law to reverse the trend (Friedman and McGarvie, 2003, pp. 341–361; Lagemann, 1999, pp. 70–75; Zunz, 2012, pp. 220–230).

During the following decade another commission, this time organized by philanthropists including John D. Rockefeller III and titled the Commission on Private Philanthropy and Public

Needs, recommended a permanent body to oversee the operations of foundations and to define the boundaries of the nonprofit sector. After the Carter administration decided against forming a government agency for that purpose, several philanthropic councils took the initiative in 1980 to form the Independent Sector, an advocacy group that brought representatives from both foundations and their main benefactors into a single coalition. The timing proved fortuitous because Ronald Reagan's administration would soon pursue policies that championed charitable giving as an American ideal even as it defunded programs that provided federal support to charities and nonprofits. In response, and behind the leadership of the Independent Sector, foundations published journals and magazines and even funded university centers with the aim of defending their work but also studying the effectiveness of philanthropic policies. By the 1990s, foundations had completed their transition from an unrecognized and novel creation in the late nineteenth century to a leadership role in an expanding complex of nonprofit organizations (Friedman and McGarvie, 2003, pp. 375–377; Lagemann, 1999, pp. 33–37, 289–290; Zunz, 2012, pp. 232–263).

Throughout this period, as the rhetoric of the culture wars brought heightened attention to the legal and institutional structures of foundations, professional philanthropists once again adapted to changing circumstances. Already firmly planted within a complex system of interactions between public and private funding mechanisms and therefore limited in their ability to embark on the same large-scale enterprises that had marked the heyday of Rockefeller and Carnegie, foundations looked for ways to leverage their funds. In the health and medical fields, that meant moving away from the massive outlays of cash needed to eradicate a disease, found a new institute, or reform an entire field. Instead, some of the richest philanthropies sought to raise the standards of excellence at existing institutions either by rewarding the highest achievers or by investing in underserved regions and populations.

The Ford Foundation, for example, spent almost half a billion dollars in the 1950s supporting religious colleges, nonprofit hospitals, and private medical schools. Later in the 1970s, the foundation gave $100 million dollars to support historically black colleges and minority researchers (Anheier and Hammack, 2013, p. 85). In an example of even narrower targeted giving, the Howard Hughes Medical Institute, the eponymous foundation of the famous businessman, aviator, and film mogul, extended grants to individual scientists. Because the institute began as a single research center in 1953 and has continued to operate its own laboratories, it has retained a legal status as a medical research organization rather than a true private foundation. Nevertheless, especially after Hughes's death in 1976, it began giving grants to other institutions nationwide and soon grew to become the largest private funder of biomedical research. Its $16.9 billion endowment in 2013 placed the Hughes Institute second only to the Gates Foundation in size. Individual grant recipients, known as "Hughes Investigators," have received competitive grants based on demonstrated skills and potential and not with the expectation of accomplishing a defined goal or specific agenda ("How We Advance Science," Howard Hughes Medical Institute, http://www.hhmi.org/about/). In other words, rather than picking and choosing grantees as a way to modify the methods of scientific knowledge, the Hughes Institute has aimed to support and advance those who excel within the existing structure. Unlike Abraham Flexner, who sought to reform medical education, or Warren Weaver, who helped to invent a new scientific discipline, the more recent era of giving, typified by the institutional support of Ford and the individual grants of Hughes, has hoped to maximize the achievement of the existing system by increasing competitiveness and raising national standards across the board (Anheier and Hammack, 2013, pp. 84–90).

Foundations and Health into the Twenty-First Century. The economic expansion of the 1990s and early 2000s led to (yet another) expansion of the aggregate wealth and total number of private foundations. Although never wholly without controversy, the professionalization and bureaucratization of the nonprofit sector gave new foundations a host of resources to navigate the perils of the American political landscape. In the twenty-first century, even more than in earlier periods, initiatives in public health have given foundations an established way to avoid attracting negative attention as well as measurable ways to prove how much they benefit the interests of the public.

The Bill and Melinda Gates Foundation, funded by billions of dollars in Microsoft stock and by far the largest endowment in the world, exemplifies these twenty-first-century trends in philanthropy. In 2006, billionaire investor Warren Buffett announced that he would donate $30 billion of his personal fortune to the Gates Foundation. Once completed, Buffett's pledge will double the size of the foundation and make it almost four times larger than any other philanthropy. With that vast well, they choose to focus on a few select areas that supplement, rather than compete with, government programs and for-profit corporations ("Who We Are," http://www.gatesfoundation.org/). In the United States, the Gates Foundation has focused on educational programs such as offering college scholarships and providing computers to inner-city schools, but three-fourths of its money ($3.3 billion of the $4.4 billion in grants delivered in 2011, according to data from the Foundation Center) has gone to global health, including initiatives to combat HIV/AIDS, infectious diseases, and malnutrition.

In a similar vein, the Robert Wood Johnson Foundation, which grew out of the community philanthropy of the chair of the Johnson & Johnson Corporation to become a national foundation in 1972, has narrowed its funding activities to maximize its effects. The Johnson Foundation, however, in 2013 continues to devote all of its $9 billion endowment to public health and health care. In fact, by the 1980s, it had become the largest foundation dedicated solely to health. In place of the institution-building efforts of earlier foundations or efforts to raise national standards of research, Johnson focuses on improving the delivery and quality of health-care services. These efforts included campaigns to reduce the use of tobacco products in the 1990s and in the twenty-first century to lower childhood obesity. The Johnson Foundation has also worked to encourage the underprivileged to enroll in health insurance and, more recently, has turned to funding policy studies that analyze how to improve the performance of both private insurers and federal health programs (Anheier and Hammack, 2013, pp. 94, 138–140; Robert Wood Johnson Foundation, "About RWJF," http://www.rwjf.org/).

During and after the passage of the 2010 Affordable Care Act, the Robert Wood Johnson Foundation and similar philanthropies such as the Kaiser Family Foundation worked to disseminate up-to-date information on the latest policy proposals and best practices for health care (Anheier and Hammack, 2013, p. 140). These endeavors, along with the global health initiative of the Gates Foundation, still provoke the age-old criticism that philanthropy serves pet projects more than the public good and does little to change long-term trends in global health outcomes. Even so, the self-governance of the nonprofit sector and the industry-wide coordination of large foundations have helped Gates and others avoid the congressional vilification once aimed at the Rockefellers and the widespread cynicism that greeted the tax tricks of Henry and Edsel Ford.

[*See also* **Higher Education and Science; Medicine; National Institutes of Health; Public Health;** *and* **Research and Development (R&D).**]

BIBLIOGRAPHY

Abir-Am, Pnina. "The Disclosure of Physical Power and Biological Knowledge in the 1930s: A Reappraisal of the Rockefeller Foundation's 'Policy' in Molecular Biology." *Social Studies in Science* 12 (1982): 341–381, and replies, ibid., 14 (1984): 225–263. The original article and the responses give both sides of the debate over how to apply poststructuralist approaches to the history of foundations.

Anheier, Helmut K., and David C. Hammack, eds. *American Foundations: Roles and Contributions.* Washington, D.C.: Brookings Institution Press, 2010. A collection of essays written by many of the leading scholars in philanthropic studies.

Anheier, Helmut K., and David C. Hammack. *A Versatile American Institution: The Changing Ideals and Realities of Philanthropic Foundations*. Washington, D.C.: Brookings Institution Press, 2013. Anheier and Hammack follow their edited collection of essays with a single-volume treatment of how the broader universe of foundations, and not just the wealthiest, developed as institutions. Their argument synthesizes recent contributions to the history of foundations and the appendices include the most concise, up-to-date review of the definitions, data, and scholarly literature on foundations.

Apple, Rima D. "Patenting University Research: Harry Steenbock and the Wisconsin Alumni Research Foundation." *Isis* 80, no. 3 (1989): 374–394.

Arnove, Robert F., ed. *Philanthropy and Cultural Imperialism: The Foundations at Home and Abroad*. Boston: G. K. Hall, 1980. A collection of essays that summarizes the Marxist critiques of foundations as a hegemonic force ascendant in the early 1980s.

Berliner, Howard S. *A System of Scientific Medicine: Philanthropic Foundations in the Flexner Era*. New York: Tavistock, 1985.

Bill and Melinda Gates Foundation. http://www.gatesfoundation.org/.

Bonner, Thomas Neville. *Iconoclast: Abraham Flexner and a Life in Learning*. Baltimore: Johns Hopkins University Press, 2002. An excellent recent biography on the life and work of Flexner.

Boris, Elizabeth T., and Council on Foundations. *Philanthropic Foundations in the United States: An Introduction*. Washington, D.C.: Council on Foundations, 2000. This industry publication is a good guide to how grant making functions and how foundations define their own work.

Brown, E. Richard. *Rockefeller Medicine Men: Medicine and Capitalism in America*. Berkeley: University of California Press, 1979. Emphasizes the economic self-interest of the Rockefeller Foundation.

Carnegie Corporation of New York. http://carnegie.org/. Includes links to the various other Carnegie charities including the Carnegie Institution for Science and the Carnegie Foundation for the Advancement of Teaching.

Carter, Susan B., Scott Sigmund Gartner, Michael R. Haines, Alan L. Olmstead, Richard Sutch, and Gavin Wright, eds. *Historical Statistics of the United States: Earliest Times to the Present*, millennial ed. New York: Cambridge University Press, 2006. http://hsus.cambridge.org. Described in the editors' preface as "the standard source for quantitative indicators in American history," this work includes compilations of available data on foundations, health outcomes, and government spending over the course of centuries.

Cueto, Marcos, ed. *Missionaries of Science: The Rockefeller Foundation and Latin America*. Bloomington: Indiana University Press, 1994.

Curti, Merle, and Roderick Nash. *Philanthropy in the Shaping of American Higher Education*. New Brunswick, N.J.: Rutgers University Press, 1965. One of the earliest systematic studies by professional historians on the role of foundations in education.

Dowie, Mark. *American Foundations: An Investigative History*. Cambridge, Mass.: MIT Press, 2001. Dowie argues that modern foundations need to reform to become more responsible stewards of democracy. Although his aim is policy recommendation, he gives a fair-minded historical summary and a lucid overview of the contours of contemporary debate.

Ettling, John. *The Germ of Laziness: Rockefeller Philanthropy and Public Health in the New South*. Cambridge, Mass.: Harvard University Press, 1981. The most comprehensive volume on the effort to fight hookworm in the South.

Ford Foundation. http://www.fordfoundation.org/.

Fosdick, Raymond B. *The Story of the Rockefeller Foundation*. Introduction by Stephen C. Wheatley. [1952] New Brunswick, N.J.: Transaction Publishers, 1989. Fosdick served as president of the Rockefeller Foundation but gives the most complete account of its inner workings prior to 1950. Wheatley's 1989 introduction adds valuable context and insight.

"Foundation Directory Online." The Foundation Center. http://fconline.foundationcenter.org/. The Foundation Center's comprehensive online database for information on specific foundations as well as for corporate donors and other public charities in the nonprofit sector.

Foundation Center. "Key Facts on U.S. Foundations." New York: The Foundation Center, 2013. http://foundationcenter.org/gainknowledge/research/keyfacts2013/. An annual report summarizing the data available online at Foundation Stats.

"Foundation Stats: Guide to the Foundation Center's Research." The Foundation Center. http://data.foundationcenter.org/. Whereas the online *Directory* allows searches for individual foundations,

Foundation Stats compiles aggregate data on the number of foundations and on grant giving broken down by various categories.

Fox, Daniel M. "Abraham Flexner's Unpublished Report: Foundations and Medical Education, 1909–1928." *Bulletin of the History of Medicine* 54 (1980): 475–496.

Fox, Daniel M. *The Convergence of Science and Governance: Research, Health Policy, and American States.* Berkeley: University of California Press, 2010. Although Fox's book focuses on health policy and the state, his analysis of governance, broadly construed, also encompasses the evolving role that foundations played in public health over the twentieth century.

Friedman, Lawrence Jacob, and Mark D. McGarvie, eds. *Charity, Philanthropy, and Civility in American History.* New York: Cambridge University Press, 2003. This collection of essays serves as an excellent overview to the methodologies and historiographical issues in the history of foundations.

John D. and Catherine T. MacArthur Foundation. http://www.macfound.org/.

Karl, Barry D., and Stanley N. Katz. "The American Private Philanthropic Foundation and the Public Sphere, 1890–1930." *Minerva* 19 (1981): 236–270. The work of Katz and Karl spurred interest about philanthropy among historians, earning them lasting influence within the historiography of foundations.

Kevles, Daniel J. "Foundations, Universities, and Trends in Support for the Physical and Biological Sciences, 1900–1992." *Daedalus* 121, no. 4 (1992): 195–235. The best article-length introduction to the history of university/foundation relationships throughout the twentieth century.

Kohler, Robert E. *Partners in Science: Foundations and Natural Scientists, 1900–1945.* Chicago: University of Chicago Press, 1991. Based on extensive archival research in the files of the large foundations, Kohler focuses on the natural sciences to analyze the organizational influence of philanthropy.

Lagemann, Ellen Condliffe, ed. *Philanthropic Foundations: New Scholarship, New Possibilities.* Philanthropic Studies. Bloomington: Indiana University Press, 1999. The contributors lay the groundwork and set the agenda for the then-emerging field of philanthropy history.

Macdonald, Dwight. *The Ford Foundation: The Men and the Millions.* New York: Reynal, 1956. A compilation of articles published in the *New Yorker*, Macdonald's book mixes basic facts about Ford Foundation operations and history with social commentary.

Magat, Richard, ed. *Philanthropic Giving: Studies in Varieties and Goals.* Yale Studies on Nonprofit Organizations. New York: Oxford University Press, 1989. As a former foundation employee, Magat aims this collection at philanthropic stakeholders looking to understand and improve foundation policies.

Mowery, David C. *Ivory Tower and Industrial Innovation: University-industry Technology Transfer before and after the Bayh-Dole Act in the United States.* Stanford, Ca.: Stanford University Press, 2004.

National Science Foundation, National Center for Science and Engineering Statistics. *National Patterns of R&D Resources: 2011–12 Data Update.* Detailed Statistical Tables NSF 14-304. Arlington, Va.: Author, 2013. http://www.nsf.gov/statistics/nsf14304/. Includes data on the money spent for research and development across all sectors of the U.S. economy and all sources of funding dating back to 1953.

Nielsen, Waldemar A. *The Big Foundations.* New York: Columbia University Press, 1972. Nielsen's largely critical assessment of the 33 largest foundations includes detailed portraits of each institution.

The NIH Almanac. The National Institutes of Health. http://www.nih.gov/about/almanac/.

Reeves, Thomas C. *Freedom and the Foundation; the Fund for the Republic in the Era of McCarthyism.* New York: Alfred A. Knopf, 1969. The Fund for the Republic, a creation of the Ford Foundation, published reports investigating the origins of McCarthyism. Reeves tells the story of the political controversy aimed at foundations as a result.

Robert Wood Johnson Foundation. http://www.rwjf.org/.

Rockefeller Foundation. http://www.rockefellerfoundation.org/.

Sealander, Judith. *Private Wealth & Public Life: Foundation Philanthropy and the Reshaping of American Social Policy from the Progressive Era to the New Deal.* Baltimore: Johns Hopkins University Press, 1997. Sealander pioneered the argument that, contrary to the heated controversies surrounding them, foundations most served to maintain the political status quo.

Solovey, Mark. *Shaky Foundations: The Politics-Patronage-Social Science Nexus in Cold War America*. New Brunswick, N.J.: Rutgers University Press, 2013. Solovey follows in the footsteps of a growing body of literature on the patronage system in science by extending the analysis to knowledge creation in the social sciences.

Wheatley, Steven Charles. *The Politics of Philanthropy: Abraham Flexner and Medical Education*. History of American Thought and Culture. Madison: University of Wisconsin Press, 1988. A compelling political history of the implementation of the Flexner report recommendations.

Zunz, Olivier. *Philanthropy in America: A History*. Princeton, N.J.: Princeton University Press, 2012. The most thorough and accessible single-volume introduction to the history of philanthropy. Zunz covers all forms of charity but foundations play a major role in his story.

Kevin A. Walters

4-H CLUB MOVEMENT

In 1914, the Cooperative Extension Service of the Department of Agriculture established a Rural Youth Division by bringing together a number of independent clubs for farm boys and girls under the direction of country agricultural agents. The term "4-H" (head, heart, hands, and health), popularized by Extension Agent Gertrude Warren, became the clubs' official name in 1919. The 4-H clubs, which proved highly popular, served as a useful vehicle for training young people in advanced farming and home economics techniques often resisted by their parents.

In the 1920s a Chicago-based private organization, the National 4-H Service Committee, solicited contributions from corporate donors to underwrite 4-H prizes and established rules for participation in 4-H projects and county-fair competitions emphasizing animal husbandry, crop production, and home economics. In 1948 the Cooperative Extension Service established a second private organization, the National 4-H Foundation, in Washington, D.C., primarily to underwrite international farm youth exchanges, an

activity the 4-H Service Committee proved reluctant to sponsor. The two organizations sustained an uneasy working relationship until 1976, when they merged into a single National 4-H Council.

In the 1960s, as the farm population continued its long decline, 4-H added to its list of sponsored projects a number of hobby activities designed to appeal to small-town and urban youth. Although 4-H continued to promote traditional farm-related programs, it placed considerable emphasis on leadership training and community-development projects. As the twenty-first century began, 4-H continued to rank among the nation's largest youth organizations and the only one that was federally sponsored.

[*See also* Agricultural Education and Extension; Agricultural Technology; Agriculture, U.S. Department of; *and* Home Economics Movement.]

BIBLIOGRAPHY

Wessel, Thomas, and Marilyn Wessel. *4-H, An American Idea, 1900–1980: A History of 4-H*. Chevy Chase, Md.: National 4-H Council, 1982.

Thomas Wessel and Marilyn Wessel

FRANKLIN, BENJAMIN

Although the common view of Benjamin Franklin's contributions to science and technology regards them as practical and commonsensical—as with the invention of lightning rods and bifocals—his engagement with science was in fact almost always theoretical and experimental because he wanted to make contributions to the knowledge of physical nature that resembled those of European men of science. He succeeded to an astonishing degree. His theorization of electricity and of population dynamics in North America had significant impact in his time and they survive as intellectual components of physics and of biology today. For this reason, Franklin is probably one of the most important scientific theorists who stood between

Isaac Newton (whom he emulated) and Charles Darwin (whom he influenced).

Born in 1706, Franklin was an excellent example of how science had become, by the early eighteenth century, part of an educated person's cultural inheritance. The findings of an earlier generation of experimental men of science, especially Robert Boyle and Isaac Newton, were widely publicized and the reading public was encouraged to keep up with newer discoveries. These assumptions had spread to urbanized parts of Europe's overseas empires, including Franklin's native Boston, where he trained as a printer and read widely. When he continued his apprenticeship in London, he aspired to meet Newton and did manage to meet Hans Sloane, later president of the Royal Society of London, one of the oldest scientific societies in the world.

From being a consumer of scientific culture, Franklin became a disseminator. After settling in Philadelphia from 1726 onward, he published information about science and medicine in his newspaper, the *Pennsylvania Gazette*, and his almanac, *Poor Richard*. More substantively, he invited his almanac readers to use telescopes and microscopes to participate in science themselves. Franklin helped to foster a public culture that was intellectually engaged and scientifically literate in several other ways, as a founder of the Library Company (1731), the American Philosophical Society (1743), and the Philadelphia Academy (1749; now the University of Pennsylvania).

Had Franklin been less ambitious, he might have accepted a role as someone who provided information about America to European men of science. That was a common path for colonists with scientific interests, who would dutifully compile data, as for astronomical phenomena that could not be observed in Europe, or else gather specimens or give reports on natural history. Franklin made his higher goals clear in an unusual pamphlet, *An Account of the New-Invented Pennsylvanian Fire-Places* (1744). This self-published piece described the earliest version of what would become the free-standing "Franklin" stove, but it also reported a series of experiments that Franklin had performed with heat, with notes on the specialist literature on heat (and experimentation)

that he had been reading. This connection between the practical and the theoretical would become his hallmark.

Franklin next wanted recognition from the scientific experts in Europe. Beginning in the 1740s, he made multiple attempts to publish something under his name in the *Philosophical Transactions* of the Royal Society of London, finally succeeding in their publication for 1751 to 1752. In parallel, he welcomed a London benefactor's gift to the American Philosophical Society: an apparatus that generated an electrical charge, which made possible experiments with electricity. Although Franklin performed those experiments with three collaborators, he was the official reporter who wrote letters to contacts in London describing the procedures, and he gathered these pieces together into his *Experiments and Observations on Electricity*, which appeared in London in 1751.

That work was hugely influential in its definition of electricity as a single manifestation that took two forms, positive or negative. (Earlier theories of electricity had posited that it occurred in many different forms, depending on the circumstances of its generation.) This simple and elegant theory was eventually universally persuasive. Continued use of positive and negative to describe electrical charges, as well as adoption of several other of Franklin's terms (including "battery"), show the lasting influence of his theory, not only for electricity, but also for magnetism. His experiments also led him to develop the lightning rod, a metal projection that, when grounded by extending into the earth, could convey a lightning strike harmlessly away from a building. With this invention, as with his stove, Franklin had made good on a promise that modern science, even in its seemingly abstract experimental forms, would have practical application.

Franklin was widely lauded as a brilliant savant. He became a fellow of many of Europe's learned societies and his *Experiments and Observations* would run through multiple editions. Those essays would gather up his maturing thoughts on electricity, as well as on weather systems. His reputation would be invaluable to the American patriots who, after 1776, needed a recognizable and influential spokesman to broker French support for their War of Independence against Great Britain. Although the "Electrical American" was thus ideologically charged, it was another of Franklin's theories that made an openly political claim. In 1751, Franklin had composed some "Observations on the Increase of Mankind." In this, he hypothesized that, because North America offered plentiful land to its settler inhabitants, they could marry young and produce families larger than those found in Europe. In subsequent iterations of the essay, Franklin settled on the estimate that the colonial population was doubling in size every 20 years. (The first U.S. census, of 1790, would bear him out.) The implication was obvious and widely quoted: the population balance in the British Empire would shift from one side of the Atlantic Ocean to the other and make British rule of America ridiculous.

This early contribution to the science of demography would have an unintended but powerful impact on the two Victorian theorists of evolution. Both Alfred Russel Wallace and Charles Darwin read Thomas Malthus, *An Essay on the Principle of Population* (1798), which had taken up Franklin's doubling-every-20-year prediction. Each of the evolutionists used that idea to imagine the competition for material resources as a limitation on reproduction and therefore the impetus for changes in natural species. Just as his electrical theories had made a permanent contribution to physics, so Franklin's theory about population dynamics had made a permanent contribution to modern biology.

[*See also* **American Philosophical Society; Biological Sciences; Electricity and Electrification; Evolution, Theory of; Physics; Popularization of Science;** *and* **Science.**]

BIBLIOGRAPHY

Chaplin, Joyce E. *Benjamin Franklin's Political Arithmetic: A Materialist View of Humanity.* Washington, D.C.: Smithsonian Institution, 2009. Traces Franklin's influence on Darwin.

Chaplin, Joyce E. *The First Scientific American: Benjamin Franklin and the Pursuit of Genius.* New York: Perseus Books, 2006. An analysis of Franklin's life in science.

Cohen, I. Bernard. *Franklin and Newton: An Inquiry into Speculative Newtonian Experimental Science and Franklin's Work in Electricity as an Example Thereof.* Philadelphia: American Philosophical Society, 1956. The major work on Franklin's contributions to physics.

Stearns, Raymond Phineas. *Science in the British Colonies of America.* Urbana: University of Illinois Press, 1970. A survey of scientific activities in British America, pre-1776.

Joyce E. Chaplin

FULTON, ROBERT

(1765–1815), builder of America's first commercially successful steamboat, was born in Lancaster County, Pennsylvania. Displaying precocious artistic and mechanical talent, he studied art in Philadelphia and in 1786 traveled to England, where he studied at London's Royal Academy of Art with the American-born artist Benjamin West, in whose home he lived for a time. His interests soon turned to marine engineering, however. He secured several patents and in 1796 published *A Treatise on the Improvement of Canal Navigation.* In 1800, having moved to France and gained support from the government of Napoleon I, he built a workable submarine, the *Nautilus.* He also demonstrated the military effectiveness of torpedoes, a term he coined in its modern usage.

Meanwhile, the Scotsman James Watt in 1776 had invented a stationary steam engine that marked a significant improvement on earlier models beginning with Thomas Newcomen's steam-powered pump to drain water from coal mines (1712). Inventors on both sides of the Atlantic soon adapted steam engines to marine navigation. As early as 1787 John Fitch demonstrated a steam-powered boat on the Delaware River. Others, including William Symington in Scotland and John Stevens in New Jersey, experimented with steamboats as well. A breakthrough came in 1803 when the American Oliver Evans developed a more efficient high-pressure, double-action piston steam engine. Amid this inventive ferment, Robert Fulton in 1802 contracted with the U.S.

ambassador to France, Robert R. Livingston, a prominent New York political leader, to build a steamboat capable of transporting passengers and goods. Fulton conducted his initial tests in 1803 on the Seine River in Paris. Returning to the United States, he successfully tested a steamboat, the *Clermont,* on the Hudson River in August 1807. In 1808 the New York legislature granted Fulton and Livingston a monopoly of steam navigation on New York State waters, and the partners soon established a highly profitable commercial steamboat service between Albany and New York City. (In 1806 Fulton married Robert Livingston's niece, Harriet Livingston; they had four children.) The 1803 Louisiana Purchase opened the vast North American interior to settlement and commercial development, increasing the demand for improved transportation. Turning to the West, Fulton in 1811 built in Pittsburgh the *New Orleans,* a steamboat that plied the Mississippi River using an engine based on Oliver Evans' more efficient design. He also developed a steam-powered warship to defend New York harbor in the War of 1812 and served on the Erie Canal Commission from 1811 until his early death four years later.

Robert Fulton's mechanical genius and his demonstration of the steamboat's commercial and military viability won him lasting fame, including a statue in the U.S. Capitol's Statuary Hall and representation on several commemorative postage stamps. Although he was not the inventor of the steamboat, he nevertheless ranks among those mechanically gifted tinkerers-innovators of the early national era who, before the advent of railroads, advanced the efficient water-based transport of people and goods, facilitating a transportation revolution that proved crucial to the nation's early social, commercial, and industrial development.

[*See also* **Canals and Waterways; Maritime Transport; Military, Science and Technology and the; Shipbuilding;** *and* **Steam Power.**]

BIBLIOGRAPHY

Hunter, Louis C. *Steamboats on the Western Rivers: An Economic and Technological History.* Cambridge, Mass.: Harvard University Press, 1949.

Philip, Cynthia O. *Robert Fulton: A Biography.* New York: Franklin Watts, 1985.

Sale, Kirkpatrick. *The Fire of His Genius: Robert Fulton and the American Dream.* New York: Free Press, 2001.

Paul S. Boyer

G

GATES, WILLIAM H., III

See Computers, Mainframe, Mini, and Micro; Software.

GELL-MANN, MURRAY

(1929–), physicist. From the end of World War II until the U.S. Congress canceled funding for the Superconducting Super Collider (SSC) in 1993, high-energy particle physics—the effort to identify the fundamental building blocks of matter— was the most dramatic area of physical research and it captured much of the attention of both physicists and members of the general public. Within this 48-year stretch, from about 1955 to 1970, Murray Gell-Mann dominated high-energy particle physics.

Gell-Mann was born in New York City. From his childhood, Gell-Mann's practice of perusing encyclopedias coupled with his near fail-safe memory endowed him with an air of authority on many subjects. His interest in physics, however, did not develop until he entered Yale University as an undergraduate; even then, he was reluctant to major in physics. He went to graduate school at the Massachusetts Institute of Technology and did his dissertation with Victor F. Weisskopf. However, it was not until he went to the University of Chicago that Gell-Mann's creativity in physics began to blossom.

In 1953, Gell-Mann made his first major contribution to particle physics. Starting in 1947, "strange" particles were discovered that defied understanding. Gell-Mann invented a new quantum number and conservation law he called the conservation of strangeness. The long-standing problem was solved and his conservation law became a permanent part of physics.

In 1955, Gell-Mann decided to go to the California Institute of Technology (Caltech) because Richard Feynman was there. At Caltech he collaborated with Feynman on one important paper concerning the weak interaction. Gell-Mann remained at Caltech until his retirement, where he made many major contributions to particle physics. Two contributions are noteworthy because, as he did with "strangeness," he added new terms to physics.

By 1960 the number of particles that interact very strongly had proliferated dramatically and there was no organizing principle to bring order. Gell-Mann created a system he called the Eightfold Way (independently created by Yuval Ne'eman) from which he generalized to a scheme that allowed him to predict the Ω^- particle (also predicted by Ne'eman). Another outgrowth of his Eightfold Way was the most important contribution he made to particle physics: in 1964 he proposed that the proton and neutron consisted of three particles he called quarks. The quark idea was proposed independently by George Zweig. Because quarks had to be fractionally charged—a gigantic departure from orthodox physics—Gell-Mann fell short of actually predicting them. In 1969, Gell-Mann won the Nobel Prize in Physics for his contributions and discoveries concerning the classification of elementary particles and their interactions.

After retirement in 1993, Gell-Mann moved from simplicity to complexity, from elementary particles to complex systems. He cofounded the Santa Fe Institute, which opened its doors in 1987. Gell-Mann became chairman of the board of the institute and resides in Santa Fe today.

[*See also* **Feynman, Richard; Nobel Prize in Biomedical Research; Physics;** *and* **Quantum Theory.**]

BIBLIOGRAPHY

Gell-Mann, Murray. *The Quark and the Jaguar: Adventures in the Simple and the Complex.* New York: W. H. Freeman, 1994.
Johnson, George. *Strange Beauty: Murray Gell-Mann and the Revolution in Twentieth-Century Physics.* New York: Alfred A. Knopf, 1999.

John S. Rigden

GENDER AND SCIENCE

Eighteenth-century Enlightenment natural philosophy underwrote the U.S. Constitution and also came to justify the exclusion of women, along with slaves and non-property-owning men, from the rights of citizenship, higher education, the professions, science, and public life more generally. Exclusion rested on demonstrating a lesser natural endowment in reason. This way of thinking has been so strong that the Harvard University president Lawrence H. Summers used women's lack of "intrinsic aptitude" in math and science as a justification for their underrepresentation at elite universities still in 2005.

The U.S. history of women and gender in science has been fiercely interdisciplinary. Historians have studied the lives of men and women scientists within the context of institutions that for centuries held women at arm's length. Sociologists have analyzed the mechanisms of subtle gender bias in scientific institutions. Biologists have examined how science has studied intelligence, sex, and race. Cultural historians have explored normative understandings of femininities and masculinities. Philosophers and historians of science have examined how gender has influenced the content and methods of the sciences. This essay treats the history, sociology, and philosophy of gender in American science, primarily over the past 40 years, by tracing three themes: the history of women's participation in science, the history of gender in scientific institutions, and the history of gender in knowledge. Although it is useful to distinguish these themes for analytical purposes, they are closely tied to one another. Emerging evidence suggests that women will *not* become equal participants in science until institutions are restructured and gender analysis is integrated into research.

The History of Women in Science. The first strand of literature focuses on the participation of women as historical actors. Although the study of women in science has historical roots, it first became an important field of U.S. scholarship in the 1980s. Encyclopedias, such as *Hypatia's Heritage*

(1986), *Women of Science* (1990), and *Out of the Shadows* (2006), documented exceptional women and their contributions. Important intellectual biographies appeared on the Nobel–prize winning cytogeneticist Barbara McClintock (Keller, 1983), astronomer Maria Mitchell (Bergland, 2008), physicist Lise Meitner (Sime, 1997), and computer scientist Grace Hopper (Beyer, 2009), among many others. Autobiographies such as that of physicist Fay Ajzenberg-Selove added insight into women's daily experiences (*A Matter of Choices*, 1994). Biographies led to a greater recognition of women: Marie Curie (1867–1934), the first person—man or woman—to win two Nobel prizes, was honored by the French government in 1995 by having her ashes enshrined in the Panthéon, the resting place of French heroes, such as Voltaire, Rousseau, and Victor Hugo.

Much of this early scholarship fits the "history of great men" model with women measured against male norms. Margaret Rossiter's *Women Scientists in America* (Vol. 1, 1982; Vol. 2, 1995; Vol. 3, 2012) broke this mold by shifting the focus from the exceptional woman to the more usual patterns of women working in science. In 2014, her three-volume work remained unsurpassed for the history of women in American science. As Rossiter documented, women embarked on modern careers in science after the women's movements of the 1870s and 1880s propelled them into universities. As women gradually gained admittance to graduate schools—by the twentieth century a prerequisite for serious work in science—they began flooding into PhD programs in all fields. By the 1920s, their numbers were at a historic high in the United States. Between 1930 and 1960, however, the proportion of women PhDs plunged as a result of the rise of fascism in Europe and the Cold War and McCarthyism in the United States. Women did not regain their 1920s levels of participation in academic science until the 1970s.

Women made some gains in faculty ranks during World War II. The number of women in teaching positions rose from 12 percent in 1942 to 40 percent in 1946 (Rossiter, 1995, p. 10). After the war, however, in a process that Rossiter has called the remasculinization of science, the "old girls" were moved aside. Universities seeking to increase their prestige raised salaries, reduced teaching loads, hired more PhDs, and restored faculty positions to men. One university president is quoted as saying, "we do not want to bring in more [women] if we can get men" (Rossiter, 1995, p. 36).

Women's lot was further worsened in the postwar period by the GI Bill, which provided qualified veterans with university tuition and a living allowance. Of the nearly 8 million veterans flooding American universities in the postwar period, only 400,000 were women (Rossiter, 1995, p. 31). Women for the most part missed out on the great age of postwar American science, during which record growth occurred in terms of monies spent, persons trained, and jobs created.

This early history of women in U.S. science dispels the myth of inevitable progress: women's participation cannot be characterized as a march of steady progress but as cycles of advancement and retrenchment. Women's situation has changed along with the fortunes of war and peace, politics and economies, and climates of opinion.

Data Collection. The Soviet Union's *Sputnik* launch in 1957 unleashed a frenzy of recruitment into science, fueled by the sense that the United States required more scientists to retain its competitive edge. In this atmosphere, women and minorities figured as valuable national resources. Women took important steps forward with the 1963 Equal Pay Act and the 1964 Civil Rights Act that outlawed discrimination on the basis of race, sex, or national origin in education and employment.

This national legislation, coupled with the renewed women's movements of the 1960s and 1970s, propelled women into science. Women scientists began forming key action groups, such as the American Association for Women Scientists. The 1972 Equal Employment Opportunity Act allowed disgruntled women to sue, and sue they did. By 1978 the federal government had forced four large public universities—Ohio State, Purdue, Michigan, and Wisconsin—to compensate women for discrimination. Importantly, the threat of large lawsuits led universities to comply with the law (Rossiter, 2012).

In more positive developments, the National Science Foundation (NSF) established a number of programs, including a 1980 Task Force on Programs for Women, to examine barriers to women's full participation. Over the next several decades, the NSF focused support for women's careers by increasing women's research funding, teaching women how to negotiate, setting up mentoring networks, and the like—or, more generally, making women more competitive in a man's world (Rosser, 2008).

In the 1980s, national governments began collecting sex-disaggregated data to monitor trends in women's careers in science. In 1982, the NSF published its first congressionally mandated biennial report, *Women and Minorities in Science and Engineering* (persons with disabilities were added later). These data revealed both "vertical" and "horizontal" segregation with respect to women's and minorities' positions in science. Vertical segregation describes disparities between men and women by level of seniority in a particular field. Data consistently showed that women's representation has been lowest at the highest institutional levels. This was true in the 1980s and, although improved, remained true in 2014. One of the many consequences of vertical segregation is that women and minorities have been substantially underrepresented as academic "gatekeepers," including grant evaluators, hiring and promotion committee members, and journal editors.

Horizontal segregation describes how men and women cluster in different subfields of science. Women, for example, have held a majority of PhDs in the life sciences, whereas men have held a majority in the physical sciences (NSF, 2011). Looking more closely at subfields within the life sciences, women at the doctoral level have reached parity in molecular biology, microbiology, genetics, neuroscience, and ecology, but not in biophysics or bioinformation. In the physical sciences, women have been close to parity in ocean and marine science, but not in chemistry, astronomy, computer science, or physics. One of the many consequences of horizontal segregation is a lack of mentors in fields where women are poorly represented.

NSF data, however, do not report on fields of science originally created by women that were not in the early twenty-first century considered "science." These included midwifery, nursing, and home economics. Home economics programs were established at many U.S. universities, including the University of California, Berkeley, Cornell, and Iowa State, in the early twentieth century (Rossiter, 1982). These programs gained prestige during World Wars I and II with their foundational research, for example, on E vitamins and government ration programs. After 1945, home economics programs rapidly collapsed and, by 1970, many had been abolished, renamed, or absorbed into traditional science departments.

It is interesting to note that men fared better in traditionally female fields, such as nursing, than women do in traditionally male fields, such as physics or engineering. Women earned the overwhelming majority of PhDs in nursing (91 percent in 1991 and 90 percent in 2011). Men, however, have held between 4 percent and 5 percent of professorships in nursing across this period and over 4 percent of nursing school deanships (American Association of Colleges of Nursing, 2012, p. 4). In the early twenty-first century, it would be uncommon for women scientists to earn less than 10 percent of the PhDs in a particular field, but hold 4 percent of full professorships.

As women PhDs increasingly moved into universities but failed to reach the top, questions arose about women's lack of success. In 1979, the National Research Council blamed the low numbers of women in science on the "sex discrimination practiced for many years in some graduate science departments" (National Research Council, 1979, p. 19). Jonathan Cole countered this line of argument in his book *Fair Science* (1979), contending that the failure of women to rise to the top resulted from their lesser contribution to scientific knowledge, their lower productivity, lesser citation rates, and so forth. According to Cole, science is "fair"; women are to blame for their poor showing.

Cole's claims spawned a minor industry among sociologists and university administrators aimed at producing precise measures of scientific productivity, the coin of the realm garnering resources and rewards. As the decades progressed, analysts came to realize that men outproduce women because they tend to hold higher ranks than women at more prestigious research universities. Men as a group have outproduced women statistically because a few well-placed men turn out large numbers of papers. These men benefited from what sociologists call "cumulative advantage"— that is, those who do well professionally amass the resources to do even better in the future. Men have been more likely to be among the academic elite with endowed chairs, generous funding, spacious and modern labs, collaborators worldwide, membership in national and foreign academies, and multiple prestigious prizes.

Debates began to subside in 1992 when J. Scott Long astonished many with his finding that, although men publish more, the average paper by a woman is cited 1.5 times more often than the average paper by a man. Sociologists came to agree that gender differences in productivity have declined significantly over the past decades. Yu Xie and Kimberlee Shauman's *Women in Science: Career Processes and Outcomes* (2003) capped off these studies. They argued that differences in productivity between women and men can be traced to gender inequalities in access to resources within the academy and, outside the academy, to gender inequalities in family roles. Xie and Shauman concluded that change is needed in both areas.

Whereas women's careers in academic science have been well studied, little scholarship treats women in U.S. industrial science. Laurel Smith-Doerr's *Women's Work: Gender Equality vs. Hierarchy in the Life Sciences* (2004) documented that women PhD scientists working in small biotech companies have a higher probability of leading a research team than do women PhDs working in hierarchically organized institutions, such as academia or large pharmaceutical companies. Smith-Doerr's key finding was that women's careers flourish in small firms that run on teamwork and interfirm networks. Patenting was also an issue. Sue V. Rosser found that across all fields of science, the percentage of women obtaining patents is less than that of men relative to their representation in a particular area (Rosser, 2011).

Data collection culminated perhaps in 1999 with the famous Massachusetts Institute of Technology report. Hard data—in this instance, measuring lab space in feet and inches—showed that the 17 tenured women at the Massachusetts Institute of Technology commanded less lab space than their men colleagues. The report caused a national sensation—and led to numerous reforms.

Much of the analysis of women's place in science also applied to minorities. Importantly, women minorities often encountered the "double bind" of sexism and racism. Vivienne Malone Mayes's experience at the University of Texas is a case in point. In 1962, Mayes became the third African American woman ever in the United States to earn a PhD in mathematics (the first two were awarded in 1949). She found that her race made her ineligible for teaching and banned her from some classrooms. Her race also barred her from the cafe where her adviser and classmates met for informal discussions. Only after winning the fight to desegregate the cafe did she discover that women, whatever their race, were not welcome. Reflecting on her experiences some years later, she wrote, "I was the only black and the only woman…my isolation was absolute and complete" (Mayes, 1975).

By the 1870s, U.S. women had gained admission to universities, and by the 1920s, they had gained admittance to PhD programs. Yet despite these changes, women have not advanced in scientific professions as fast or as far as men, even after discrimination became illegal. A second strand of literature—the history of gender in scientific institutions—has looked beyond women's careers to the history of institutional reform that has supported women's efforts to achieve equality in science.

The History of Gender in Scientific Institutions. Does science have a gender? Many have argued that it should. In the seventeenth century, Sir Francis Bacon, a founding father of modern science, called upon the Royal Society of London to raise a "masculine" philosophy (as the new science was called). Even the great English feminist Mary Wollstonecraft, in her efforts to create equality between the sexes, called for women to become "more masculine and respectable." In 1985 U.S. physicist and philosopher Evelyn Fox Keller declared that Western science is "masculine," not only in its practitioners but also in its ethos and substance. Discussions about gender ignited in the 1980s and turned attention away from women—their triumphs, trials, and tribulations—toward the history of gender in the cultures and structures of scientific institutions.

A culture is more than institutions, legal regulations governing a profession, or a series of degrees or certifications. It consists in the unspoken assumptions and values of its members. Despite claims to value neutrality, the sciences have identifiable cultures whose customs and folkways have developed over time. Many of these customs developed historically in the absence of women and also in opposition to their participation.

As decades of scholarship have demonstrated, gender defines powerful fault lines in American culture. Gender in the cultures and institutions of science is significant because women's long legal prohibition from scientific institutions was buttressed by elaborate gender ideologies that coded behaviors and activities as appropriately masculine or feminine. Unearthing assumptions surrounding gender in science has helped unearth unspoken notions about who can become a scientist and what science is all about.

Masculinity and femininity do not map directly onto sex (nor should they). In the same way that the sciences have many cultures and subcultures, so women and men come from various classes, geographic locations, and ethnic backgrounds. Gendered characteristics are not innate, but neither are they arbitrary: They are formed by historical circumstances. Two key developments in Western science and society—the professionalization of science and the privatization of the family—were crucial to structuring the historic clash between the cultures of science and of femininity. The eighteenth-century rise of modern democracies prescribed separate spheres of competence for men and women and structured gender norms, ideologies, and stereotypes (Schiebinger, 1989). Many of the problems women have faced in science, such as balancing work and family, date from this period and are common across professions.

Whereas historians have documented the rise of separate spheres and their legacies, sociologists have detailed how masculinities and femininities continue to play out in the day-to-day workings of contemporary U.S. science. For example, an important study published in the *Proceedings of the National Academy of Sciences* (2012) looked at gender bias in hiring and promotion decisions. A carefully devised experiment showed that both male and female biology, chemistry, and physics professors have been significantly more likely to hire a man applicant versus a woman applicant with the same academic record. Scientists in the study gave "John" a competence score of 4 of 7 points and "Jennifer" a score of 3.3; they also offered John a higher starting salary although the application materials were exactly the same. The only thing that differed was the applicants' name.

Different behaviors and manners have often been expected of men and women. Whereas men have shouldered the burden of exuding confidence, women have been tasked with appearing modest. The norm of feminine modesty has penalized women in a number of ways when they enter the professional world. For example, women might have been penalized more than men when negotiating for salary and resources. These norms and expectations may, in turn, influence behavior: A 2007 study revealed that 7 percent of women candidates for new jobs negotiated compared with 57 percent of men (Bowles et al., 2007, p. 85). Much the same is true in academic leadership. Characteristics associated with leadership have been viewed as incongruent with women's gender roles. Leadership in U.S. science and society has required a certain display of assertiveness and competence. Women, who are properly modest and thus not assertive, have tended to be considered incompetent. However, women who display assertiveness could be considered unpleasant. Reflecting upon the 2008 U.S. elections, this catch-22 might be titled the "Hillary Clinton" syndrome.

Interestingly, mothers have been viewed as less competent than women who were not mothers. A 2007 experiment evaluated application materials of paired women job candidates who differed only with respect to parental status (the application materials were the same throughout the study). Evaluators scored mothers as less competent and committed than nonmothers and awarded them lower starting salaries.

Overall, mothers have earned less than women who were not mothers; this holds true for whites and African Americans (Correll and Benard, 2007). Although this study found that men were not penalized for, and sometimes benefited from, being fathers, other studies have emphasized differences between fathers who actively fathered and those who did not. What has been identified as a "daddy penalty" revealed that managers with professional wives have earned some 25 percent less than male managers with stay-at-home wives.

Gender asymmetries can also influence student evaluations. An experiment in which professional actors delivered lectures to a physics class found that students judged men actors to have a better grasp of the material than women actors—although the actors delivered identical scripted lectures. The belief that the men actors were more knowledgeable was most prominent among men students, but existed among women students as well (Bug, 2010). Similarly, a 2007 study of 886 recommendations for job applicants in chemistry and biochemistry found that letters for men and women were more alike than different. However, letters written for men tended to have more standout descriptors, such as "the most gifted," "best qualified," or "rising star," than those written for women. An earlier study showed that letters for men emphasize research whereas those for women emphasized teaching (Schmader et al., 2007, pp. 509, 513).

Gender stereotypes have held to some degree across ethnic groups, but not always. One Asian American woman, discussing the perplexities raised by stereotyping, remarked, "I've found the generalization 'girls can't do math' balanced nicely by the adage 'Asians are all math brains.' My presumed poor verbal skills as the child of immigrants was countered by the assumption that women are geniuses with words" (Knecht, 1993).

Transforming Institutions. A large restructuring of U.S. universities occurred during the late twentieth and early twenty-first century surrounding dual-career academic hiring. Since the eighteenth century, what North Americans commonly call "individuals" have, in fact, been heads of households—typically, male professionals with mobile family units consisting of a stay-at-home wife and children. With many a professional woman, by contrast, comes a professional man, very often in a field of study close to her own. *Dual-Career Academic Couples: What Universities Need to Know* (2008), a study of 30,000 faculty at 13 top research universities across the United States, found that 83 percent of women scientists in academic couples are partnered with another scientist compared with 54 percent of men scientists. With women entering the scientific labor force in large numbers since the 1970s, universities increasingly developed dual-career hiring policies. This has been one important component in recruiting and retaining women scientists.

Many institutional reforms in the early twenty-first century were supported by the NSF ADVANCE program, launched in 2001. Unlike earlier NSF programs that focused on individual women's careers, ADVANCE assisted institutional transformation aimed at improving women and underrepresented minorities' success in science and engineering. This program has made the United States a global leader in institutional transformation.

A particularly successful ADVANCE program was the University of Michigan's Strategies and Tactics for Recruiting to Improve Diversity and Excellence (STRIDE) program aimed at removing gender and ethnic bias from hiring practices. Michigan increased its hires in science and engineering of women, averaging 14 percent pre-STRIDE to around 30 percent post-STRIDE. In this program, distinguished senior science and engineering faculty (five men and four women) studied the social science and historical literature on bias in hiring, evaluation, and promotion. This STRIDE committee (whose members were compensated by the university for their time) then prepared a presentation for departmental hiring committees based on this literature. Not only did hiring practices improve, but also, by virtue of the fact that these newly trained "gender experts" were permanent and respected members of science and engineering faculties, the climate of opinion surrounding gender issues improved dramatically (Lavaque-Manty and Stewart, 2008).

As of 2014, much remained to be done to restructure research and educational institutions to achieve gender and ethnic equality. The goal has been to create institutions where both men's and women's careers can flourish. This second approach to gender equality, however, often has sought to reform institutions while assuming that the knowledge issuing from those institutions is gender neutral. Restructuring institutions is important but must be coupled with efforts to eliminate gender bias from basic and applied research. Change is needed also at a third level: the knowledge level.

The History of Gender in Knowledge. Many people have been willing to concede that women have not been given a fair shake, that social attitudes and scientific institutions need to be reformed. They may have also been willing to concede that women have been excluded in subtle ways. They have stopped short, however, from analyzing how gendered practices have structured knowledge. Is the question of gender in science merely one of institutional organization and opportunities for women, or does it impact science itself? How has the exclusion of women and minorities from science shaped human knowledge?

The power of Western science—its methods, techniques, and epistemologies—has been celebrated for producing objective and universal knowledge, transcending cultural restraints. With respect to gender, race, ethnicity, and much else, however, science is not value neutral.

Body politics. Western science arose as part of the great eighteenth-century democratic revolutions in the United States, France, and Haiti. Women, along with slaves and non-property-owning men, were excluded from the rights of citizenship, higher education, and the professions—including science.

"Nature" and its laws played a pivotal role in the rise of modern democracies. Natural law philosophers—in both Europe and the United States—attempted to ground newly emerging democracies in nature. Within this framework, an appeal to natural rights could be countered only by proof of natural inequalities (Schiebinger, 1989).

Science, as an exacting way to know and interpret nature, took on a special role in these debates. The rise of a belief in meritocracy based on individual ability inaugurated a complementary belief

that social inequalities resulted not from systematic discrimination (discussed earlier) but from intrinsic inabilities within certain groups. The claim of science to objectivity was the linchpin holding together a system that rendered women's and minorities' exclusion from science invisible and made this exclusion appear fair and just (Schiebinger, 1993). The brand of scientific sexism and scientific racism that first emerged toward the end of the eighteenth century has continued to structure arguments surrounding women's participation in science in the early twenty-first century.

Feminist biologists—notably Ruth Bleier, Ruth Hubbard, and Richard Lewontin—attacked this type of biological determinism in the 1970s. These scholars importantly demonstrated that biology is not the static bedrock of organic life that biological determinists would like it to be. They showed instead how cultural factors interact with and shape biology (at the same time that biology interacts with and shapes culture). This effort to fight science with science culminated in Anne Fausto-Sterling's magnificent *Myths of Gender* (1985, 1992), which debunked the (pseudo) science involved in IQ debates.

In the early twenty-first century, fights against scientific racism and scientific sexism diverged in important ways. *Revisiting Race in a Genomic Age* (2008), for example, showed that race is not genetic in the way sex is. Most genetic markers, for instance, do not differ sufficiently by race to be useful in medical research. Biologists Marcus Feldman and Richard Lewontin traced differences among humans to divergent ancestral geographic regions rather than genetics. This volume documented how debates about race are part of larger political and moral struggles about what it means to be human. In this same period, studies of sex expanded to encompass queer theory, transsexuality, and transgenderism. In 2000, Fausto-Sterling published her *Sexing the Body* that challenged the strict dualism of sex. Fausto-Sterling argued for five sexes—males, merms, herms, ferms, and females, but ultimately argued that what is meant by sex should encompass a full continuum of human sexuality.

More recently, Anne Fausto-Sterling published her "Bare Bones of Sex" (2005) and "Bare Bones of Race" (2008) to emphasize how biology is not purely physical but interacts with environment and culture to "shape the very bones that support us." Her point is that a complex disease, such as osteoporosis, emerges over the life course in response to how biology and culture interact in "specific lived lives." Fausto-Sterling sought to do away with dualisms—of nature/nurture and sex/gender—and introduced instead a dynamic systems approach to health and disease.

U.S. scholars of race and gender have also turned a critical eye to the role of body politics in colonialism and postcolonialism. Schiebinger (1993, 2004) analyzed Western anatomies of race that directed bioprospecting in the eighteenth-century Atlantic world. Sandra Harding expanded on her earlier feminist and postcolonial work in important ways by asking, *Is Science Multicultural?* (1998).

Retheorizing science. In the 1980s, major feminist epistemologists, such as Evelyn Fox Keller, Donna Haraway, and Sandra Harding, began challenging the idea that scientific knowledge is objective. Keller (1985) applied object/relations theory to reveal how science is "masculinist"; her goal was to reclaim science as a human as opposed to a "masculine" project and to rethink Western divisions of intellectual and emotional labor. Hilary Rose (1994) extended this argument in her *Love, Power, and Knowledge*. Donna Haraway (1988) famously identified traditional notions of objectivity as "a view from nowhere."

Science theorists called for adding a greater understanding of cultural contexts to scientific research (Haraway's "situated knowledge" [1988] and Harding's "strong objectivity" [1991]). The idea is that what natural scientists have called "objectivity" is often weak objectivity because scientists have failed to analyze the politics of knowledge; that is, they have failed to understand how research priorities are set and who benefits (and who does not) from a particular line of research.

Standpoint theory of the 1990s was central to feminist epistemology. Importantly, standpoint theory argued for a new starting point for scientific research: marginal lives—those of women, people of color, gays and lesbians, and other groups lacking social and economic privilege. Standpoint theory interacted strongly with postmodernism to call into question the singular category "woman," employed in much mainstream feminism, that ignored the powerful intersections of gender, race, and class (Harding, 2004).

The 1980s and 1990s also saw important debates about creating a feminist science. There were many visions of what might constitute feminist science—these often included aspects of feminist methodologies and values, such as reflexivity, anti-reductionism, taking seriously "a female point of view," and inclusive and cooperative ethics (Fedigan, 1997). Ultimately, however, feminists feared that an idealized feminist science (if successful) would become isolated from mainstream science. Creating a feminist science has largely been abandoned in favor of "doing science as a feminist" (Longino, 1987; Wylie, 2007), recognizing that this approach will be as diverse as the variants of feminism. Sue Rosser has cogently distinguished at least eight feminist approaches to science—from liberal to socialist, African American, and existentialist (Rosser, 2011).

Gender bias in science. In the late 1980s and 1990s, scholars began documenting how gender inequalities, built into the institutions of science, have influenced the knowledge issuing from those institutions. Specific examples (discussed later) highlighted how science has been shaped by gender, ethnicity, socioeconomic disparities, war, imperialism, and cultural ideologies and how these factors, in turn, have been molded by science. Scholars also documented how gender bias has limited scientific creativity, excellence, and benefits to society.

Major critiques treated primatology. Haraway's *Primate Visions* (1989) presented a thorough political, social, intellectual, and economic account of the history of primatology from the 1890s to the late twentieth century. This work was part of the remaking of primatology over the past several decades—a remaking so foundational that primatologist Linda Fedigan (1997) argued that primatology is a "feminist science." In her work with Shirley Strum (Strum and Fedigan, 2000), Fedigan analyzed, for example, the "baboonization" of primate life—that is, how baboons became the model animal for primatology research. From the 1950s to the 1970s, baboons were widely studied to the exclusion of other species, such as bonobos, that are genetically more similar to humans. The study of baboons—and the implications drawn for primal human nature and society—provided ready explanations for human warfare, violence, and aggression that meshed well with Cold War politics. In this instance, the choice of subject interjected a potent antifeminist element, highlighting and reinforcing notions of male dominance.

Other works documented gender bias across the sciences—and the consequences for both science and society. Emily Martin (1991) showed how the notions of the "active sperm" and "passive egg" influenced biological understandings of human conception. Schiebinger (1993) analyzed how gender shaped eighteenth-century natural history, showing, for example, how Linnaean nomenclature—the coining of the term *Mammalia*, for instance—helped legitimate women's place in the restructuring of Western societies. This study emphasized, in particular, the contingencies of scientific knowledge and especially what is foregone in the choice of one pathway over another. Nelly Oudshoorn (1994) famously demonstrated how estrogens and androgens came to be sexed female and male, respectively, in the 1920s and 1930s, although researchers recognized the presence in both sexes of both types of hormones. More recently, Sarah Richardson (2008) has analyzed the history of genetics to show how the Y chromosome came to be seen as the sex-determining factor.

Issues surrounding biodiversity, environmentalism, and conservation also have important gender dimensions. Patricia Howard (2003) told how gendered divisions of labor specific to particular cultures have led men and women in those cultures to cultivate knowledge of specific environments and plant species and their uses and ecologies. For example, women have been responsible for food production and medical care in much of the developing world and, consequently, have possessed unique intellectual resources (such as knowledge about the medicinal properties of plants) as well as material resources (such as seeds for specific strains of crops).

A rich complex of studies in gender in science has surrounded archaeology, paleoanthropology, and evolutionary theory. Margaret Conkey (2007), among others, has challenged gender roles—calling into question origin stories about "man the hunter" and "woman the gatherer" that both build on and reinforce Western gendered-based divisions of labor. In archeology, reevaluating "tools" to include objects used for nutting, leatherworking, grain harvesting, and woodworking led theorists to better understandings of human evolution (Gero and Conkey, 1991). In paleoanthropology, analyzing sex in fossil remains, such as the famed australopithecine "Lucy," called into question the convention of sexing relatively small skeletal remains female and, as a consequence, identifying the places they are found as home sites (Hager, 1997).

Running through critiques of science are analyses of gendered language. Word choices, analogies, metaphors, and narratives have functioned to construct as well as describe—they had both hypothesis-creating and proof-making functions. Words and models have had the power to influence the direction of scientific practice, the questions asked, the results obtained, and the interpretations made. "Sharing a language means sharing a conceptual universe" within which assumptions, judgments, and interpretations of data can be said to "make sense" (Keller, 1992). Zoologists, for example, often have referred to herds of animals, such as horses, antelope, and elephant seals, as "harems" (Lancaster, 1975). Embedded in the word harem are robust assumptions about social organization such that researchers often fail to "see" what lies outside the logic of the metaphor. Researchers, for example, who questioned the notion of a harem found that female mustangs range from band to band, often mating with a stallion of their choice (Brown, 1995).

Gendered innovations in science.

A new international project, launched in 2009, moved beyond identifying gender bias to employing gender analysis as a *resource* to create new knowledge. Funded by Stanford University, the European Commission, and the NSF *Gendered Innovations in Science, Medicine, Engineering, and Environment* is built on 40 years of gender in science scholarship (Schiebinger et al., 2011–2013). A robust collaboration among historians, gender scholars, scientists, and engineers has developed practical methods of gender analysis for basic and applied research in science and engineering. Sex and gender analysis acted as yet further controls—one set among many—providing critical rigor in scientific research, policy, and practice. Sex and gender, and their interactions, have been analyzed in each step throughout the research process: from setting priorities to making funding decisions, establishing project objectives, developing methodologies, gathering and analyzing data, evaluating results, securing patents, transferring ideas to markets, and drafting policies.

Gendered Innovations has also developed some 20 case studies as concrete illustrations of how sex and gender analysis lead to new knowledge. The case study on stem-cell research offers one example of how analyzing sex can lead to important breakthroughs. Analyzing sex encouraged researchers to identify and report the sex of cell lines; prospectively design experiments for meaningful analysis of sex differences of results (not all sex differences will be significant); and record data in ways that allow for meta-analysis. In animal studies, for example, the regenerative potentials of pluripotent stem cells differed by sex, cell type, and the tissue being repaired.

The case study on genetics of sex determination built on insights from Emily Martin, Anne Fausto-Sterling, and Sarah Richardson. For decades the study of sex determination was focused on research on testis determination in the male. As a result, the ovary-producing female pathway had—until the early twenty-first century—gone largely unstudied. Since about 2010, new models of sex determination have focused research on the ovarian developmental pathway and, in the process, refined our understanding of testes determination. New empirical data show that both ovarian and testis development occurred along active, genetically regulated pathways.

Examples also come from the developing world. Reliable, efficient access to water serves as a good example. Analyzing gendered divisions of labor has helped researchers understand who in a community holds the knowledge required for a particular project. In much of sub-Saharan Africa, water procurement has been women's work. Consequently, many women have had detailed knowledge of soils and their water yields. Engaging these women via participatory research has enhanced the success of community-managed water services.

Historians and the Future of American Science. Historians often study the past, but historians and the understandings of complex historical change they create may also change the future. U.S. history of science, as an academic field, has changed dramatically since the 1960s. Women entered the field in the 1970s and 1980s as graduate students and professors. Methodological revolutions have produced social history and its correlates—women's history, the history of sexuality, and—everyday life in addition to new brands of intellectual and gender histories—all of which opened doors to reevaluating knowledge making. Moreover, history (including history of science) took the "anthropological turn," further complicated by the "cultural" and "literary" turns that required new and sophisticated analysis of representations (of both image and word). All of these developed hand in hand with academic sex and gender analysis (as documented earlier).

The promise of history of science, when founded in the 1920s, has not materialized (that is, history of science has not been broadly embraced by scientists as an equal partner in deliberations about future directions in science or vice versa). Despite this larger failing, the field of women and gender in science is one area where scholars have worked across disciplines to change not only the history of science but also science itself. Much of the push and pull in this area continues to be mediated by NSF funding (a theme traced throughout this essay).

What is the state of play in the early twenty-first century? As documented earlier, the NSF has, since the 1980s, supported individual women's careers in science. The NSF has, since 2001, spent some $138 million to combat subtle gender bias and to transform research institutions so that both women and men can succeed in science. The challenge for the NSF in the twenty-first century is to move to the knowledge level—to fix the knowledge. Forty years of scholarship has demonstrated how gender bias in science makes American science partial (rather than impartial). Gender bias in science can be wasteful, dangerous, and harmful: for example, between 1997 and 2000, 10 drugs were withdrawn from the U.S. market because of life-threatening health effects; eight of these posed greater health risks for women than for men. More importantly, perhaps, gender bias in science can hamper creativity. What is the price of doing nothing? What kinds of discoveries have been missed because bias keeps us from seeing clearly?

Although the United States in the early twenty-first century was the global leader in policy in institutional transformation (overcoming gender bias in scientific institutions), the European Union's Directorate-General for Research & Innovation was the global leader in policy supporting the integration of gender analysis into science. Since 2003, the European Commission's calls for proposals have encouraged researchers to specify "whether, and in what sense, sex and gender are relevant in the objectives and the methodology of the project" (European Commission, 2003, 2010). These policies were confirmed and expanded in Horizon 2020, the European Commission's current funding framework (European Commission, 2014).

Historians, policy makers, and scientists have been increasingly working to integrate sex and gender analysis into science. This paradigm shift has been reinforced by academic gatekeepers: funding agencies have encouraged grantees to incorporate sex and gender analysis into research. Hiring and promotion committees have evaluated researchers and educators on their success in implementing gender analysis. Editors of peer-reviewed journals have required sophisticated use of sex and gender methodology when selecting papers for publication (for a list of such policies, see Schiebinger et al., 2011–2014, Policy portal). Importantly, university curricula have been revamped to include the results of sex and gender analysis across basic and applied science courses.

Historians often study how human knowledge—what we know, what we value, what we consider important—changes in specific cultural moments. Historians may be part of this cultural moment by helping to incorporate methods of sex and gender analysis into the research objectives and methodologies. This process may also facilitate women gaining greater equality in science and science leadership. As four decades of scholarship suggest, gender analysis sparks creativity by asking new questions and opening new areas to research. Can we afford to ignore such opportunities?

[*See also* **Engineering; Gender and Technology; Medicine; Science;** *and* **Technology.**]

BIBLIOGRAPHY

American Association of Colleges of Nursing. *Annual Report.* Washington, D.C.: American Association of Colleges of Nursing, 2012.
Bergland, Renée. *Maria Mitchell and the Sexing of Science.* Boston: Beacon Press, 2008.
Beyer, Kurt. *Grace Hopper and the Invention of the Information Age.* Cambridge, Mass.: MIT Press, 2009.
Bowles, Hannah, Linda Babcock, and Lei Lai. "Social Incentives for Gender Differences in the Propensity to Initiate Negotiations: Sometimes It Does Hurt to Ask." *Organizational Behavior and Human Decision Processes* 103 (2007): 84–103.
Brown, Nancy Marie. "The Wild Mares of Assateague." *Research/Penn State* 16, no. 4 (1995).
Bug, Amy. "Swimming against the Unseen Tide." *Physics World Archive* (August 2010): 16–17.
Conkey, Margaret. "Questioning Theory: Is There a Gender of Theory in Archeology?" *Journal of Archaeological Method and Theory,* 14 (2007): 285–310. Most authors in this thematic special issue are archaeologists who analyze issues surrounding "doing archaeologist as a feminist."
Correll, Shelley, and Stephen Benard. "Getting a Job: Is There a Motherhood Penalty?" *American Journal of Sociology,* 112 (2007): 1297–1338.
European Commission. *Stocktaking 10 Years of "Women in Science" Policy by the European Commission, 1999–2009.* Luxembourg: Publications Office of the European Union, 2010.
European Commission. *Vademecum: Gender Mainstreaming in the Sixth Framework Programme—Reference Guide for Scientific Officers and Project Officers.* Brussels: Directorate-General for Research, 2003.

European Commission. *Vademecum on Gender Equality in Horizon 2020*. Brussels: RTD-B7 Science with and for Society, 2014.

Fausto-Sterling, Anne. "The Bare Bones of Race." *Social Studies of Science*, 38 (2008): 657–694.

Fausto-Sterling, Anne. "Bare Bones of Sex: Part I, Sex & Gender," *Signs: Journal of Women in Culture and Society* 30, no. 2 (2005): 1491–1528.

Fausto-Sterling, Anne. *Sexing the Body: Gender Politics and the Construction of Sexuality*. New York: Basic Books, 2000.

Fedigan, Linda. "Is Primatology a Feminist Science?" In *Women in Human Evolution*, edited by Lori Hager, pp. 56–75. New York: Routledge, 1997.

Gero, Joan, and Margaret Conkey, eds. *Engendering Archaeology: Women and Prehistory*. Oxford: Blackwell, 1991. This early volume contributed to bringing questions about women and gender into archaeology.

Hager, Lori, ed. *Women in Human Evolution*. New York: Routledge, 1997. This important volume includes many practitioners in the field and analyzes how taking women into consideration requires rethinking aspects of evolutionary theory.

Haraway, Donna. "Situated Knowledges: The Science Question in Feminism and the Privilege of Partial Perspectives." *Feminist Studies* 14 (1988): 575–599. This key essay—in large part a response to Sandra Harding's early work—has been anthologized several times.

Harding, Sandra. *The Feminist Standpoint Theory Reader: Intellectual and Political Controversies*. New York: Routledge, 2004. This collection of essays from major theorists reevaluates standpoint after a decade of debate.

Harding, Sandra. *Is Science Multicultural? Postcolonialisms, Feminisms, and Epistemologies*. Bloomington: Indiana University Press, 1998.

Harding, Sandra. *Whose Science? Whose Knowledge? Thinking from Women's Lives*. Ithaca, N.Y.: Cornell University Press, 1991. This collection of essays follows Harding's *Science Questions in Feminism*. Here she expands her arguments to include race and global perspectives.

Howard, Patricia. *Women & Plants: Gender Relations in Biodiversity Management & Conservation*. London: Zed Books, 2003.

Keller, Evelyn Fox. *A Feeling for the Organism: The Life and Work of Barbara McClintock*. New York: W. H. Freeman, 1983.

Keller, Evelyn Fox. *Reflections on Gender and Science*. New Haven, Conn.: Yale University Press, 1985. Keller's was one of the first important books to launch current debates about gender in science.

Keller, Evelyn Fox. *Secrets of Life, Secrets of Death: Essays on Language, Gender and Science*. New York: Routledge, 1992.

Knecht, Kathryn. "Letter to the Editor." *Science* 261 (1993): 409 (slightly modified).

Lancaster, Jane. *Primate Behavior and the Emergence of Human Culture*. New York: Holt, Rinehart, and Winston, 1975.

Lavaque-Manty, Danielle, and Abigail Stewart. "A Very Scholarly Intervention: Recruiting Women Faculty in Science and Engineering." In *Gendered Innovations in Science and Engineering*, edited by Londa Schiebinger, pp. 165–181. Stanford, Calif.: Stanford University Press, 2008.

Longino, Helen. "Can There Be Feminist Science? *Hypatia* 3 (1987): 51–64.

Martin, Emily. "The Egg and the Sperm: How Science Has Constructed a Romance Based on Stereotypical Male-Female Roles." *Signs: Journal of Women in Culture and Society* 16 (1991): 485–501.

Mayes, Vivienne Malone. "Black and Female." *Association for Women in Mathematics Newsletter* 5 (1975): 4–6.

Moss-Racusin, Corinne, John Dovidio, Victoria Brescoll, Mark Graham, and Jo Handelsman. "Science Faculty's Subtle Gender Biases Favor Male Students." *Proceedings of the National Academy of Sciences*, 109 (2012): 15474–16479.

National Research Council, Committee on the Education and Employment of Women in Science and Engineering. *Climbing the Academic Ladder*. Washington, D.C.: National Academy of Sciences, 1979.

National Science Foundation. *Women, Minorities, and Persons with Disabilities in Science and Engineering: 2011*. Washington, D.C.: Division of Science Resource Statistics, 2011.

Oudshoorn, Nelly. *Beyond the Natural Body: An Archaeology of Sex Hormones*. London: Routledge, 1994. Oudshoorn provides an excellent co-constructionist analysis of how biologists identified and defined hormones in female and male bodies.

Richardson, Sarah. "When Gender Criticism Becomes Standard Scientific Practice: The Case of Sex Determination Genetics." In *Gendered Innovations in Science and Engineering*, edited by Londa

Schiebinger, pp. 22–42. Stanford, Calif.: Stanford University Press, 2008.

Rosser, Sue V. *Breaking into the Lab: Engineering Progress for Women in Science.* New York: New York University Press, 2012. In this collection of essays, Rosser reviews 30 years of work on women in science. In addition to analyzing new areas, such as women's relative representation in patenting, Rosser draws from her experience as a scientist, NSF program officer, and high-level university administrator to provide unique insights.

Rosser, Sue V. "Building Two-Way Streets to Implement Policies That Work for Gender in Science." In *Gendered Innovations in Science and Engineering,* edited by Londa Schiebinger, pp. 182–197. Stanford, Calif.: Stanford University Press, 2008.

Rossiter, Margaret. *Women Scientists in America: Before Affirmative Action, 1940–1972.* Baltimore: Johns Hopkins University Press, 1995.

Rossiter, Margaret. *Women Scientists in America: Forging a New World since 1972.* Baltimore: Johns Hopkins University Press, 2012.

Rossiter, Margaret. *Women Scientists in America: Struggles and Strategies to 1940.* Baltimore: Johns Hopkins University Press, 1982. These three excellent volumes represent a foundational history of women careers in American science.

Schiebinger, Londa. *Has Feminism Changed Science?* Cambridge, Mass.: Harvard University Press, 1999. Written for scientists, this book brings together issues concerning women's participation in science, gender in cultures of science, and gender in the substance or results of science.

Schiebinger, Londa. *The Mind Has No Sex? Women in the Origins of Modern Science.* Cambridge, Mass.: Harvard University Press, 1989. Schiebinger shows how women were trained and willing to take their place in the scientific revolution of the seventeenth and eighteenth centuries and how the emerging new sciences were used to exclude women.

Schiebinger, Londa. *Nature's Body: Gender in the Making of Modern Science.* Boston: Beacon Press, 1993. This book examines examples of gender bias in early modern science.

Schiebinger, Londa. *Plants and Empire: Bioprospecting in the Atlantic World.* Cambridge, Mass.: Harvard University Press, 2004. Schiebinger inserts gender into the global expansion of science in the eighteenth century.

Schiebinger, L., I. Klinge, I. Sanchez de Madariaga, and M. Schraudner, eds. *Gendered Innovations in Science, Health & Medicine, Engineering, and Environment,* 2011–2014. http://genderedinnovations.stanford.edu.

Schmader, Toni, Jessica Whitehead, and Vicki Wysocki. "A Linguistic Comparison of Letters of Recommendation for Male and Female Chemistry and Biochemistry Job Applicants." *Sex Roles* 57 (2007): 509–514.

Sime, Ruth. *Lise Meitner: A Life in Physics.* Berkeley: University of California Press, 1997.

Strum, Shirley, and Linda Marie Fedigan. *Primate Encounters: Models of Science, Gender, and Society.* Chicago: Chicago University Press, 2000. This important volume by primatologists evaluates the foundational changes that gender studies brought to the field of primate studies.

Wylie, Alison. "Doing Archaeology as a Feminist: Introduction." *Journal of Archaeological Method and Theory* 14 (2007): 209–216.

Londa Schiebinger

GENDER AND TECHNOLOGY

The centuries since Europeans began settling the New World have been characterized by both enormous technological change and significant transformation of gender roles. These two phenomena have also shaped each other in a myriad of ways, large and small. Over time, Americans built ideas about gender into the material fabric of their lives. In turn, the interaction of people and things has often reinforced (or sometimes subverted) norms of gendered behavior. Thus gender has been a pervasive element of American technological history and, at the same time, technology has functioned to express, enforce, and sometimes redefine categories of gender.

Despite the ubiquity of gender–technology interactions, historians have chosen to give most of their attention to exploring a few large and not so large themes. Histories of reproductive technologies, housework, sexual division of paid labor, and consumption have well-developed literatures (Lerman et al., 2003). Despite a flurry of interest in early American technology during the 1980s, most scholars have also tended to concentrate their research on more recent time periods (McGaw, 1994). As a consequence, historians of technology interested in this earlier period have often relied on scholarship from other historical subfields, particularly material culture studies and American social history. A strong feminist tradition among science and technology studies scholars has also produced a very large body of ethnographies and sociological studies, some of which document the role of gender in the recent history of technologies such as the Internet (Hopkins, 1999, Wajcman, 2004).

Gender Analysis and the History of Technology.

The relationship between social ideas of maleness and femaleness and "ways of making and doing things," to use historian Melvin Kranzberg's definition of technology (1959), is likely as old as human culture. But the constructs "gender" and "technology" are much newer. The term technology can be traced back to nineteenth-century Europe, where it initially described the study of useful arts. Usage only expanded to include material things in the early twentieth century (Oldenziel, 1999). The term gender is even newer and was not coined by historians. Beginning in the mid-1980s, they adopted it from the social sciences.

Clinical psychologists first imported the word gender from linguistics in the early 1960s as they struggled to explain transsexualism: the phenomenon of individuals who had the biological attributes of one sex but a psychological identification with the other (Meyerowitz, 2002). Gender, in this context, was understood to be a form of *identity*, psychological and social, signaled most immediately through the symbolic use of clothing and gender-appropriate personal grooming. At this early date, psychologists also recognized that gender identity could be *performative*, meaning that maleness or femaleness is communicated through behaviors ranging from subtle inflections of voice or gesture to choice of profession or leisure activities. Although they would not have put it this way at the time, establishment of a believable gender identity requires transgendered people to deftly deploy a wide range of everyday technologies.

Psychologists' efforts to distinguish between biological sex and gender coincided with a growing women's movement intent on questioning claims about women's supposedly inherent roles and limitations. By the 1970s, feminist scholars had built up a body of empirical research demonstrating that many "sex roles" and "sex types" were social constructs, often used to further patriarchal or capitalist ends. Historically grounded studies of women's paid and unpaid labor growing out of Marxist feminism proved particularly influential, not only on feminist discourse of the time, but also on a later generation of scholars interested in the history of technology. At this early juncture, feminist scholars struggled not only to understand how patriarchy had shaped what was then termed sex roles, but also to find an appropriate language with which to express their ideas.

Following in the path of psychologists, anthropologists and some sociologists began to adopt the term gender, while broadening its meanings to suit the concerns of their disciplines. They recognized that gender could be used not only as noun, but also as a verb. Thus gendering can be analyzed as a social process through which qualities of maleness and femaleness are assigned to people, things, or processes. Gender, these social scientists pointed out, is often *relational*, meaning that maleness and femaleness define each other. In fact, all human societies seem to make distinctions between men's activities and women's activities. Physical objects are not only designated as belonging to a particular gender, but also sometimes considered to have gender in and of themselves (Nelson, 1997). For instance, in Western Europe, ships were traditionally gendered female but assigned to the management and care of a male-gendered profession: sailors. Meanwhile, structural sociologists added yet another analytical dimension to the use of gender analysis. They observed that gender is often literally and figuratively built into social institutions such as schools and professions. These institutions serve a powerful function in reinforcing gender norms. But they can also be used as a tool for social restructuring.

Many historians of technology first encountered gender analysis in the 1980s when Joan Scott presented it to the historical profession as a "useful category of historical analysis." A pioneering generation of feminist scholars had already called attention to the challenges of writing women's stories into the master narrative of technological history. Their guiding analytical insight was that patriarchy had historically functioned to exclude most women from the technological activities considered particularly valuable by their cultures (and historians of technology) such as engineering, while devaluing activities designated as feminine. Much of this early work fell into the category of "recovery history," consisting of efforts to identify female engineers and inventors. Ruth Cowan's landmark history of housework, *More Work for Mother*, suggested a different way forward. Scholars might look for technology in women's domains, rather than looking for women where male-gendered technological activities were already taking place. Cowan's book appeared in 1983, just before the language and methods of gender analysis began to really take hold in feminist scholarship.

For the generation of scholars who followed, gender analysis, coupled with Cowan's insight that historians should be looking beyond the machine shop and engineering firm, offered a powerful means for reimagining history of technology as a field that went far beyond a critique of patriarchy or recovery history. Instead, widening the scope of historically significant technologies, activities, and actors could open the way for a broadening of the overall breadth of the history of technology. Coupled with the concept of "social construction," coming into use in the sociology of science and technology, gender analysis suggested a variety of powerful historical questions. If, for instance, technology is indeed socially constructed, how are cultural ideas about masculinity and femininity literally designed into the size, shape, color, and function of various objects and technological systems? How does gender figure into the creation, transmission, and valuation of technological knowledge or "skill"? And most importantly, how does gender affect the historical process of "mutual shaping" between technology and society?

Up until this point, the close association of gender analysis with feminism had meant that, in practice, gendered history was often synonymous with women's history. Attending to the ways men had historically claimed technological activities such as engineering to be inherently masculine allowed a smaller group of scholars to profitably reimagine some of the most traditional parts of the field. Others stretched conventional definitions of technology and technological activities to make visible a much wider range of actors. Most importantly, they peopled what Ruth Schwartz Cowan described as the "consumption junction" (1987) with a rich array of consumers, users, and mediators, while showing how each of those roles was powerfully gendered.

The growing visibility of gender analysis has not been without controversy, even among scholars who might seem like natural advocates. Some feminist historians, who worked hard to document struggles of female pioneers in male technical fields and the more pervasive effect of patriarchy in shaping technology, have worried that gender analysis, with its inclusion of both masculinity and femininity, might become yet another means for excluding women from technology's history. Others have pointed out that gender is indeed a category of historical analysis, but far from the being the only one. Attention to race, class, and the inside of technology's "black box" also have a place. Overeager gender scholars have also lent some credence to the saying "when all you have is a hammer, everything looks like a nail." More often, gender remains underutilized as a category of analysis. Although feminist and gender-based critiques of the scope and focus of the field helped open the door to a wider range of topics, much remains to be done.

Gender and Technology in American History.

In early America, the word "technology" did not exist. Instead European colonists (and their Native American neighbors) thought about their efforts to measure and manipulate their physical environment in terms of a set of smaller categories such as tools, industries, and useful knowledge. With no self-consciousness, they attached gendered meanings to nearly all the material things and activities in their daily lives. In their passage across the Atlantic, Europeans brought with them not only habits and assumptions about gender, but also ideologies. They viewed distinctive roles for men and women as natural, immutable, and as befits a profoundly religious society, sanctioned by God.

Looking backward, historians have often described gender relations in early America as patriarchal. In the simplest sense, this meant that men had more social and political power than women. Patriarchy had many negative consequences. Some men abused their power. Many women lived in a chronic state of uncertainty because their society allowed them little say over their own property and labor and even over their own bodies. However, our ancestors were more likely to view the gendered divisions of their lives not in negative terms, but rather as representing a kind of bargain in which men and women both had responsibilities toward each other and, in turn, benefitted what was owed them. In practice, early America's gender system included a degree of flexibility and negotiation. Particularly under duress, gendered conventions gave way to necessity: men cooked and sewed; women plowed and fought. It is therefore perhaps not surprising that both men and women worked hard to construct and police boundaries surrounding the gendering of both technological knowledge and material objects. This process began in childhood and continued until death.

In our own world, the deployment of technology to ensure appropriate gender socialization begins soon after birth, as parents, family members, and friends choose clothing and toys deemed culturally appropriate for a particular baby's biological sex. Early Americans made almost no effort to visually signal whether infants were male or female. Even toddlers of both sexes wore dresses and sported long curls—a practice that continued through the Victorian era. The serious business of teaching a child to be male or female began when he or she became self-aware and was deemed old enough to have some capacity to reason. In this patriarchal society, this transition was signaled by "breeching" boys, that is, putting them into long pants (Calvert, 1992).

Small children learned appropriate social behavior and the rudiments of adult skills by observation and informal imitation. Those lessons included nearly constant reinforcement of gender norms. Toys and books—the didactic technologies of a later age—were seldom in evidence before 1750 and still absent in many early nineteenth-century households. A growing number of children did enjoy some form of schooling, particularly if they were free, white, and lived in the Northeastern states. However, even in the New England colonies where free schooling was most available, a striking gendered difference is apparent in levels of literacy and numeracy. Girls often received instruction only in reading. Boys were more likely to be taught to both read and write—skills that not only prepared them for the world of commerce, but also empowered them to share information with a wide range of people beyond their immediate community. In the nineteenth century, the literacy gap between men and women gradually closed, but at the same time numeracy—the language of commerce and technological innovation—increasingly came to be gendered male.

By adolescence, the divide between masculine and feminine technological domains was firmly established for both early American children and their descendants. Enslaved girls might be sent out into the fields to work alongside their brothers. But free white girls of every social class remained in the same kinds of spaces where they had spent their earliest years, gradually learning the techniques of running a household. Boys followed their fathers out into the fields or, in urban settings, the streets and workshops. Some parents bound their sons (or very rarely, daughters) out as apprentices, usually at the age of 12 or 13. The practice of apprenticeship not only functioned as a means of passing on specialized technological skills from generation to generation, but also allowed parents to choose an appropriate male role model for their adolescent sons.

As children became adults, they gradually acquired an increasingly complex and subtle understanding of the gendered conventions of their society as expressed in technical skill and material culture. Journeymen learned to impress prospective employers with appropriate verbal expressions of manly competence to open the door to employment. Newly married husbands and wives negotiated the overlapping spaces between male and female spheres, often following the example of family and community in deciding which invisible boundaries mattered and which did not. Young men who found themselves in all-male environments such as ships and prisons (or even workshops) also learned more complicated lessons. Cooking, sewing, and laundering—technological skills firmly gendered female in domestic settings—became the responsibility of men. But mastery of feminine tasks did not necessarily correlate with feminization. Instead, the youngest and least skilled found themselves treated as symbolically, and sometimes sexually, female.

Until the late eighteenth century, the knowledge and techniques associated with reproduction remained the most sacrosanct area of female technical expertise. Midwives played a particularly important role, not only assisting in births, but also providing remedies and advice about everything from unwanted pregnancies to infected breasts. However, in the latter part of the eighteenth century, the availability of forceps to ease the birth process as well as the growing authority of the medical profession opened the door to regendering expertise around reproduction. Male physicians rather than female midwives attended an increasing number of births. We now know that their invasive techniques spread infection, resulting in higher mortality rates for both mothers and infants.

The intimate struggle between midwives and male doctors over birthing techniques and technologies signaled a much broader destabilization of early America's gender system. Until the late eighteenth century, both gender roles and the tools and material contexts for everyday life changed very slowly. Ideologies born of the Enlightenment and the American Revolution, an emerging market economy, and the social and technological dislocations of industrialization all worked to undermine the status quo. Although a few enlightened voices argued for a radical rethinking of gender roles, more often, capitalists and patriarchs repurposed existing gender conventions to justify new arrangements to explain, for example, why women should be paid less for their wage labor than men. Middle-class women's

unpaid labor in the household also underwrote the growth of a male managerial class (Boydston, 1990).

New ideologies, first *republican motherhood* and then *separate spheres*, reified a somewhat different patriarchal gender bargain. Both recast women as inherently virtuous as long as they kept to their proper roles as wives, mothers, and keepers of the home. At the same time, a spirit of critical reasoning born of the Enlightenment opened the door for critics of existing gender relations and experiments in alternatives. Ideologies often do not describe historical realities, but the idea of separate spheres did reflect a growing nineteenth-century divide between the home and outside world as sites of gendered technological activity.

The Industrialization of Everything. Although industrialization is often identified with the invention and spread of the factory system, it was, in fact, a far more pervasive economic, social, and technological process that transformed nearly every aspect of everyday life. Gender informed what processes and places were industrialized first, who had control over the process, how men and women participated, and even the size and shape of machines and the nature of goods produced. Many historians have observed that the spatial, ideological, and economic separation of domestic work from wage labor constitutes a distinctive and particularly important characteristic of industrialization.

Throughout the nineteenth century, the growing middle class prescribed the creation of domestic spaces as refuges from an increasingly impersonal, immoral, and ruthlessly capitalistic outside world. Aided by a growing number of labor-saving devices as well as domestic servants, women could focus on overseeing and maintaining a haven for children and male family members. The reality proved more complicated. New domestic technologies and manufactured products, such as cookstoves and milled grain, often saved labor previously done by men as unpaid labor for their families. Meanwhile, rising standards of cleanliness and consumption resulted in an overall pattern of "more work for mother." For middle- and upper-class women, more work often involved cooking, sewing, cleaning, or supervising servants carrying out those tasks.

Working-class women of this period confronted an even harsher dilemma: should they work for wages and, if so, what tasks were appropriate for their skills and domestic responsibilities? Family economies often depended upon wages from all family members capable of work. But gender conventions determined that women would be responsible for cooking, cleaning, and child care. In a further irony, labor-saving services such as steam laundries, lauded for saving middle-class women's energies for more noble pursuits, largely employed working-class women who might spend 10 hours a day over a tub or mangle before going home to their own families. At the end of the nineteenth century, urban reformers occasionally remarked on the sad squeak of a clothes-line pulley as women who had worked all day to ease another woman's labor hung out their own laundry.

The earliest factories let the market decide, employing entire families. But this strategy, used by Samuel Slater and others in the textile mills of New England, resonated poorly with idealists worried that a childhood spent piecing threads would not prepare citizens of a republic for their civic responsibilities. The Lowell Mills was America's most famous response to the challenge of creating wealth while preserving democracy. In the heyday of the experiment, young women were engaged to toil at loom and spindle. The experience would prepare them for republican motherhood, advocates rationalized, by teaching them the habits of industry. Their wages could subsidize fathers and brothers who could thereby remain on the land as yeomen farmers. The Lowell Associates made a virtue of necessity—justifying the employment of low-paid female labor in ideological terms.

Where they could, factory owners utilized technological innovation to replace skilled, difficult-to-control white men with machines and inexpensive female workers. They justified what scholars would later call the "sex-typing" of labor with a series of gendered stereotypes: women were more tolerant of tedious, repetitive tasks but had no aptitude for the complexities of building

or repairing machinery. Machinery was designed and constructed to reinforce the gendering of specific tasks, typically by scaling machines and tools to fit larger male bodies.

However, the story of masculinity, industrialization, and technological change is more complicated than just a steady litany of deskilling. Men were quick to claim that understanding and operation of the most prestigious technologies of the age—the railroad, the steamboat, and the machine shop—were beyond the capabilities of women. Competition for well-paying jobs associated with these technologies came mostly from other men, not women, because replacing a man with a woman at the wheel of a ship or the throttle of a locomotive was unthinkable for even the greediest of capitalists. Instead, male workers delineated very different forms of masculinity in association with different kinds of technological work. At the bottom of the hierarchy was the "rough" masculinity of canal diggers, track layers, and lumberjacks—men whose principle skills involved hard physical labor and indifference to constant risk. Craftsmen, operators of complicated machines such as locomotives, and a vaguer category of "skilled labor" occupied a vast middle ground (Horowitz, 2001).

Increasingly, a third category of men called "engineers" earned their livelihood in occupations that required them to know about complex, modern technologies without necessarily having the full skillset necessary to actually do or make something using that technology. In the nineteenth century, engineering in all its various forms emerged as the quintessential masculine technical profession. Although apprenticeship on the shop floor remained a pathway to the profession into the twentieth century, it was the command of abstract knowledge, particularly mathematics, that set engineers apart from other men (Oldenziel, 1999).

As the century wore on, various kinds of institutions and organizations played an expanding role in gendering technological knowledge. Gender ideologies that emphasized women's prescribed social role as caregivers and mothers rationalized their exclusion from scientific, medical, and technological education. Pseudo-scientific arguments that postulated a biological difference between male and female brains or claimed that too much thinking would render women unfit to bear children provided further justification. Particularly determined middle-class women fought back by creating all-female centers of higher education or by disguising scientific and technological education in the language of domesticity. Beginning in the late 1880s, the social movement of progressivism gave this first cohort of college-trained women an opportunity to apply newly acquired skills. Some turned their attention to the social and environmental ills caused by unfettered industrialization. Female chemists and engineers created the fields of home economics. Female architects designed spaces for collective living that would free their middle-class peers from domestic drudgery.

Cultures of Consumption and Other Negotiations.

For many people living in late nineteenth-century America, the physical presence of new technologies in their lives—streetcars, gaslight, electricity—made the rapid pace of technological change obvious. Gender ideologies and gender conflicts were also more likely to be expressed in material form—whether different kinds of bicycles for men and women or the introduction of public restrooms clearly marked and segregated by gender. As the next century began to take shape, a new element became increasingly important in the continuing discourse over the gendering of technology and technological knowledge. For the first time in human history, large numbers of people could directly access a range of sophisticated "personal" technologies through the marketplace (Corn, 2011). Consumer devices ranging from telephones and automobiles to elective surgery gave individuals the unprecedented ability to challenge and manipulate existing gender norms. Most people did not consciously seize the opportunity. Rather, they were gradually (and sometimes reluctantly) swept along by the market, the prescriptions of experts, and the trend-setting example of a few individuals. Adults, settled in their existing lives, were most likely to cling to existing gender norms. Instead, outside forces forced open an already-widening

rupture in the process whereby children and adolescents learned how to be male or female primarily by observing and modeling themselves on parents and other adults with whom they had day-to-day interaction.

Mass communications technologies provided an important means through which both children and adults could obtain information, both prescriptive and descriptive, about the gendering of technology and the use of technology to construct gender. The extraordinary expansion of print culture in the late nineteenth century created a first wave, followed in the 1910s and 1920s by cinema and radio. Although it has proved extraordinarily difficult to document exactly how audiences received and acted upon the messages embedded in advertisements, popular novels, radio soap operas, and films, it is clear that they are part of a larger cultural phenomenon that washed away the last vestiges of isolated communities where people learned everything they needed to know about the relationship between gender and technology from their immediate communities and surroundings.

In the first half of the twentieth century, automobility became a particularly conspicuous site for negotiation, experimentation, and public demonstration of the relationship between technology and gender. Wealthy men were the first to realize the new technology's possibilities as a means of public displays of courage and technological competence. They rapidly abandoned fast-driving horses, the preferred means through which the previous generation exhibited masculine skill and bravado, for the motor car. Almost as quickly, the automobile press began to assert (contrary to a growing body of data) that women were inherently dangerous and incompetent drivers. Various groups of feminists also seized on the automobile as a symbol of the "new woman"—liberated by technology and her own enlightenment from dependence on men.

Whereas these public contests over technology and gender attracted much publicity, in everyday life a more complicated scenario emerged. A wide range of people from farmers to suburbanites adopted the automobile as a tool for facilitating everyday tasks. Many of those uses were gendered along the lines set down in earlier eras such as child care, maintenance of the household, and planting and harvesting cash crops. Necessity and convenience overruled essentialist arguments about why women should not be allowed behind the wheel. Driving rapidly became an accepted part of being a farmwife or a housewife, although most women quickly learned to slide over into the passenger seat if an adult male was in the car.

Bent on reaching as wide a market as possible, American car manufacturers created vehicles that were gender-neutral in terms of accommodating male and female bodies. Nearly everyone appreciated an automobile that was easy to drive, even if innovations such as electrical starters were sometimes portrayed as particularly beneficial for women. At the same time, automobile manufacturers constructed advertising and marketing strategies involving a variety of gender stereotypes: "Motor Cars that match Milady's mode—yes, her every mood!" the Paige Motor Car Company told readers of *The Ladies' Home Journal* in 1927.

In the nineteenth century, competence in operating, maintaining, and repairing machinery was identified with working-class masculinity. Combined with formal education and abstract knowledge, mechanical skills provided a path to mobility into the middle class for the engineering profession, but no one expected even the most manly of middle-class men to know much about machines, let alone get his hands dirty (Gelber, 1997). The proliferation of consumer machines, beginning with the automobile, shifted this expectation. This new definition of middle-class masculinity required male consumers to at least know how to kick the tires and talk knowledgeably about carburetors, transmissions, and the differences between Fords and Cadillacs even if they could barely find the dipstick to measure the oil. Manufacturers added "features" to a wide array of consumer technologies to give male consumers something to talk about (Horowitz and Mohun, 1998).

At the same time, however, technological innovation over the twentieth century functioned to destabilize gendered job categories involving paid labor. The proliferation of novel technologies in twentieth-century workplaces forced continual decision making about whether these devices should be assigned to men or women. Was a typewriter a man or a woman? What about a computer? Sometimes economics drove the process as capitalists replaced well-paid, "skilled" male labor with poorly paid, "unskilled" female labor. In other industries, such as steel manufacturing and railroads, economics and culture coincided in decisions to continue relying on more expensive male labor.

The century's two world wars also played an important role in undermining assumptions and rationalizations about which kinds of work naturally belonged to American men or women. The all-male environments of armies and navies had long offered men an exemption from stigma of engaging in female-gendered work. But the total mobilization characteristic of industrialized warfare also required the lifting of at least some gendered barriers that had kept women out of male-gendered jobs. Because policy makers and most of the public found the idea of female soldiers anathema, employing women civilians in jobs previously gendered male offered the only alternative. Consequently, during World War II, women gained access to technologies and technological knowledge previously barred to them. They also found out that the skills needed to rivet a ship or drive a truck were actually not that difficult to learn. Demobilization after both wars was consequently characterized by a sudden, often wrenching, and ultimately incomplete return to previous gender norms. Eventually, equal employment opportunity legislation was required to permanently wedge open the door that had been unlocked in World War II.

The later-twentieth-century relationship between technology and gender was also characterized by the ways in which technology has been deployed to modify the appearance and function of gendered bodies. Historically, many cultures have employed technology to make individual bodies conform to gendered ideals of beauty, ranging from the use of lead to whiten women's complexions in Ancient Greece to the insertion of a lip plate among the Suri people of East Africa. Twentieth-century America differs from these examples in the extent and commodification of bodily modification as well as the use of science-based technologies. Hair removal by X-ray, surgical breast enhancements, and the use of steroids and electrical stimulation by male body-builders to achieve a hypermasculine body type are only a few examples. However, medical interventions in fertility and childbearing have arguably had a much more profound effect on gendered social roles, especially for women (Tone, 2001). As noted at the beginning of this essay, it was the growing success of sex-reassignment surgery in the 1960s that prompted clinical psychologists to find a way to describe the difference between biological sex and socially or psychologically based ideas of maleness and femaleness.

Technology Shapes a Gender-free Future?

In 1991, in a deliberately provocative essay entitled "A Cyborg Manifesto: Science, Technology, and Socialist-Feminism in the Late Twentieth Century," historian of science Donna Haraway suggested that at century's end, inhabitants of postindustrial nations had all become "cyborgs"—"hybrids of machine and organism." Rather than critique and push away technological innovation as a tool of patriarchy, Haraway suggested that feminists should embrace becoming cyborgs to usher in a "post-gender world" (Hopkins, 1999). This utopian vision confirms how powerful the relationship between gender and technology has been in the United States and, in fact, every society. But the historical evidence overwhelmingly suggests that technological change does not result in the erasure of gender. Rather, it becomes part of the redefinition of gender. Nowhere is this more apparent than in the proliferation of online worlds where video-game characters and second-life avatars often manifest exaggerated biological, behavioral, and sartorial markers of gender difference. People who enter these worlds may engage in a bit of technological cross-dressing, but they rarely choose an ungendered identity. Moreover, as social psychologist Sherry Turkle has shown, digital technologies are just as likely as sailing ships and butter churns to be socially constructed in gendered ways (Turkle, 2005).

[*See also* **Anthropology; Engineering; Ethics and Professionalism in Engineering; Gender and Science; Lowell Textile Mills; Military,** **Science and Technology and the; Motor Vehicles; Printing and Publishing; Radio; Slater, Samuel; Social Sciences;** *and* **Technology.**]

BIBLIOGRAPHY

Boydston, Jeanne. *Home and Work: Housework, Wages, and the Ideology of Labor in the Early Republic.* New York: Oxford University Press, 1990. Landmark study of the influence of domestic ideologies on the market value of women's work.

Calvert, Karin. *Children in the House: The Material Culture of Early Childhood, 1600–1900.* Boston: Northeastern University Press, 1992. Definitive work on the material culture of childhood, including clothing, toys, and books. Places material culture in the context of social and women's history.

Corn, Joseph. *User Unfriendly: Consumer Struggles with Personal Technologies, from Clocks and Sewing Machines to Cars and Computers.* Baltimore: Johns Hopkins University Press, 2011.

Cowan, Ruth. "The Consumption Junction." In *The Social Construction of Technological Systems: New Directions in the Sociology and History of Technology,* edited by Wiebe Bijker. Cambridge, Mass.: MIT Press, 1987. Important essay arguing for the importance of putting consumers and users of technology in the center of analyses of the social construction of technology.

Cowan, Ruth. *More Work for Mother: The Ironies of Household Technology from the Open Hearth to the Microwave.* New York: Basic Books, 1983. Pioneering work on domestic technology with a strong central thesis that still provokes discussion.

Gelber, Steven. "Do-It-Yourself: Constructing, Repairing and Maintaining Domestic Masculinity." *American Quarterly* 49, no. 1 (1997): 66. A rare example of historical scholarship examining masculinity in the domestic sphere.

Hopkins, Patrick. *Sex/Machine: Readings in Culture, Gender, and Technology.* Bloomington: Indiana University Press, 1999. Offers a wide selection of important articles including early selections from feminist historians of technology and Haraway's "Cyborg Manifesto."

Horowitz, Roger. *Boys and Their Toys: Masculinity, Technology, and Class in America.* New York: Routledge, 2001. Includes a useful introduction and several important articles on masculinity and paid labor.

Horowitz, Roger, and Arwen Mohun. *His and Hers: Gender, Consumption, and Technology.* Charlottesville: University Press of Virginia, 1998. A good starting place for understanding the wide range of topics relating gender, technology, and consumption.

Kranzberg, Melvin. "At the Beginning." *Technology and Culture* 1 (1959): 1.

Lerman, Nina, Arwen Mohun, and Ruth Oldenziel. *Gender & Technology: A Reader.* Baltimore: Johns Hopkins University Press, 2003. An anthology of important articles on gender and technology, mostly republished from the journal *Technology and Culture.* Includes a detailed bibliographic essay.

McGaw, Judith A. *Early American Technology: Making and Doing Things from the Colonial Era to 1850.* Chapel Hill: University of North Carolina Press, 1994. Especially useful for the introductory essay and Susan Klepp's article on abortificients.

McShane, Clay. *Down the Asphalt Path: The Automobile and the American City.* New York: Columbia University Press, 1994. Includes a chapter on the gendering of automobility.

Meyerowitz, Joanne. *How Sex Changed: A History of Transsexuality in the United States.* Cambridge, Mass.: Harvard University Press, 2002. Explains the history of the adoption of gender analysis in psychology.

Nelson, Sarah. *Gender in Archaeology: Analyzing Power and Prestige.* Walnut Creek, Calif.: AltaMira Press, 1997.

Oldenziel, Ruth. *Making Technology Masculine: Men, Women, and Modern Machines in America, 1870–1945.* Amsterdam: Amsterdam University Press, 1999. Explains how engineering and the term "technology" came to be gendered as male.

Scott, Joan Wallach. "Gender as a Useful Category of Historical Analysis." *American Historical Review* 91(December 1986): 1053–1075. Landmark essay that introduced the concept of gender analysis to the historical profession.

Tone, Andrea. *Devices and Desires: A History of Contraceptives in America.* New York: Hill and Wang, 2001.

Turkle, Sherry. *The Second Self: Computers and the Human Spirit.* Twentieth anniversary ed. Cambridge, Mass.: MIT Press, 2005. Important early study by a psychologist of gendered interactions with computers.

Wajcman, Judy. *TechnoFeminism.* Cambridge, U.K.: Polity, 2004.

Arwen P. Mohun

GENETICS AND GENETIC ENGINEERING

Genetics is the scientific study of heredity and variation in living organisms. Genetic engineering, which is also referred to as gene manipulation, gene cloning, recombinant DNA technology, and genetic modification, is the application of genetic principles to create products that do not exist naturally. A complete picture of the history of genetics and genetic engineering in the United States requires knowledge of not only the intellectual pursuits and disciplinary infrastructure but also the societal views of the applications of genetic principles and genetic engineering in biology, medicine, and agriculture.

The Origin of Genetics. The originator of these sciences is usually identified as Gregor Mendel, who was working in a monastery in Brünn when he performed extensive experiments with pea plants and published his findings in 1866. Mendel established nomenclature and laws that influenced early experiments of transmission genetics. Transmission, or classical, genetics examines how traits pass from one generation to the next. Dominant factors, according to Mendel, occurred more often and were transmitted in their entirety; recessive traits stemmed from factors occurring less frequently and were latent. In 1900, three botanists, Hugo de Vries in the Netherlands, Carl Correns in Germany, and Erich von Tschermak-Seysenegg in Austria, independently verified aspects of Mendel's work, prompting a number of American researchers to perform investigations that confirmed and extended Mendel's laws.

Mendelian genetics drew the interest of embryologists, morphologists, agriculturalists, breeders, and other biologists. Some recognized its potential to provide information vital to evolutionary theory and many realized the benefits of genetic principles for improving society. The inability to explain inheritance, specifically how information passes from parent to progeny, was one inadequacy of the Darwinian theory of evolution. Although knowledge of genetics made little headway during the first decade of the twentieth century, the discipline began taking shape. Genetics, genes, and several other words still used today were coined in the 10 years following the rediscovery of Mendel's work.

Eugenics and Genetics.

Before the rediscovery and expansion of Mendel's work, eugenics gained traction, first in Britain and soon thereafter in the United States. Eugenics, a word coined by Francis Galton in 1883, translates to "good in birth" or "noble in heredity." The eugenics movement sought to improve the physical and mental capabilities of human beings through selective breeding. Positive eugenics is the promotion of breeding among physically and mentally superior human beings and negative eugenics prevents procreation among those exhibiting traits considered undesirable.

In the early twentieth century eugenics had two meanings—the study of human genetics and the application of genetic laws to guide human reproduction. As decades passed, the definition of eugenics changed. A negative connotation developed as the result of policies passed, including sterilization laws at the state level and immigration laws at the national level in the United States. The Holocaust provided a horrific example of what eugenic policies can lead to. Many policies were enacted prematurely and based on either incomplete scientific knowledge or a poor understanding of the science. Human genetics came to denote the scientific study of heredity and variation in human beings, whereas eugenics came to be defined as the practical application of genetic principles to improving the human population.

Eugenics gained monetary support from private foundations and philanthropists in large part because it fit well with long-held notions about inequalities among human races and supported the motivations of public-health and urban reform movements of the late nineteenth and early twentieth centuries. The Eugenics Record Office at Cold Spring Harbor, for example, was funded by the Carnegie Institute of Washington and a wealthy widow, Mary Harriman. Workers based at the Eugenics Record Office helped shape eugenic research and laws during its 30 year existence.

Academic Support for Genetics.

To be academically successful, a discipline needs not only intellectual contributions but also infrastructure in the forms of journals, societies, and university programs that serve the collective needs of its members. Geneticists started two journals, *Journal of Heredity* in 1910 and *Genetics* in 1916, and founded the Genetics Society of America in 1932, all of which were important to the discipline's growth and centralization. Geneticists comprised a small community during the first third of the twentieth century and most American contributions originated from one of a few academic centers. Thomas Hunt Morgan and E. B. Wilson oversaw genetics education at Columbia University. William E. Castle and E. M. East directed the Bussey Institution at Harvard University. H. S. Jennings and Raymond Pearl were at Johns Hopkins University. From these three institutions emerged the first generation of biologists trained as geneticists.

At Columbia University after 1910, T. H. Morgan and his students established the fundamentals of so-called "classical" transmission genetics and greatly expanded genetic knowledge by performing experiments with *Drosophila*. Their contributions laid foundations for the modern evolutionary synthesis, which merged genetics and Darwinian evolution. In the early years, researchers sought abnormal traits that could be tracked over several generations. After breeding many generations of fruit flies, Morgan found one fly with white rather than red eyes. Tracking white-eyed flies over several generations, he found that certain traits occur together routinely. Morgan concluded that genes were physical entities residing on chromosomes like beads on string; when a chromosome underwent an aberration such as crossover, the genes located close to one another remained on the same chromosome. Morgan's conclusion led two members of his laboratory, A. H. Sturtevant and C. B. Bridges, to begin mapping the specific locations of more genes on chromosomes. Understanding the importance of abnormalities to genetic studies, H. J. Muller, also Morgan's student, used X-rays to induce mutations in organisms and demonstrated that the frequency of gene mutations was proportional to the organism's exposure. Morgan and Muller received the Nobel Prize in Physiology and Medicine in 1933 and 1946, respectively. Morgan and his students took a reductionist approach, believing that with a better understanding of the basics, genetics would eventually elucidate issues involving evolution and development.

The reductionist approach was attacked most notably by German émigré Richard Goldschmidt, who was at the University of California from 1936 to his retirement in 1947. German practitioners typically concentrated on developmental genetics, thinking about the whole organism and the role of genes in the evolution of species and the development of individuals. This approach stemmed naturally from the German educational system that discouraged specialization and produced scholars with a wide breadth of knowledge.

International Perspectives. Political events contributed to the momentum of genetics in the United States, and again, Goldschmidt's career was representative. He had been director of the Kaiser Wilhelm Institute for Biology for 15 years before losing his job because of Nazi laws, and in 1936 he moved permanently to the United States. German émigré geneticists Curt Stern and Ernst Caspari also relocated permanently to the United States, in 1932 and 1938, respectively. Important to the growth and development of genetics was the small size and close ties between members of this international community. In addition to a number of foreign geneticists permanently relocating to the United States, many participated in exchange programs, spending extended periods in foreign laboratories, and attended conferences such as the International Congress on Genetics held about every five years.

Russia's unique approach to genetics also proved pivotal to contributions made in the United States. Theodosius Dobzhansky trained in Kiev and brought an appreciation for population genetics with him when he moved permanently to the United States in 1927. Spending more than 10 years in Morgan's laboratory, Dobzhansky performed field and laboratory studies of *Drosophila* and in his 1937 book, *Genetics and the Origin of Species*, explained how the principles of genetics,

when applied to populations, allowed one to study mechanisms of evolution. Geneticists' discussions shifted from the roles of individuals to populations, and evolutionary change became understood as the result of mutations existing within the gene pool. Dobzhansky is considered one of the chief architects of the modern synthesis that unified plant, animal, and human evolution; his work fused field, laboratory, and mathematical approaches and merged knowledge of evolution on micro and macro levels. Biologists began applying the principles of genetics to evolution during the 1930s, and by the end of World War II a new understanding of evolution had been achieved with genetics at its center.

Research and Development. The biological sciences established a collective identity during midcentury. The modern evolutionary synthesis intellectually unified many biological research programs and World War II provided a reason to capitalize on the burgeoning unification. In the United States during the first half of the twentieth century the biological sciences were largely segregated, with many academic centers operating separate botany and zoology departments. Funding for the biological sciences favored research with applications to improve society, which ranged from agriculture and livestock to eugenics and medicine. Geneticists had long enjoyed monetary support not only from private foundations and philanthropists but also from federal agencies like the Public Health Service, National Institutes of Health, and Department of Agriculture. When the U.S. government was deliberating postwar funding for scientific research, biologists testified that they deserved comparable federal support for basic research as was enjoyed by physicists, chemists, and mathematicians. Scientists' congressional testimonies in 1945–1946 about the establishment and structure of a governmental agency to support scientific research, what is today the National Science Foundation, resulted in making biological research an entity separate from medicine and on par with physics, chemistry, and mathematics.

Cold Spring Harbor, after closing the Eugenics Record Office in 1939, continued to be a major site for genetic research and a place where scientists made foundational contributions to molecular biology. Phage research undertaken by Yugoslav-born Milislav Demerec and cytogenetics pursued by American-born Barbara McClintock contributed to the Cold Spring Harbor Laboratory's influence in molecular genetics. Demerec studied mutations in genes in the 1930s and, after being named the Laboratory's director in 1941, shifted his research to mutations in bacteria. In the 1930s McClintock provided proof of crossover, a concept previously suggested by T. H. Morgan, and in the 1950s showed that genes can jump locations. She won a Nobel Prize in 1983 for her work. The Laboratory's annual symposium covering a range of topics in biophysics and genetics incited original research and the dispersion of genetic principles across an array of research areas. Closely associated with Cold Spring Harbor Laboratories and its annual courses were pioneers of bacteriophage genetics, namely German-born Max Delbrück, Italian-born Salvador Luria, and American-born Alfred Hershey. Working independently and together, they enhanced the understanding of the genetic structure and replication mechanism of viruses using bacteriophage, a virus that infects and replicates in bacteria, for which they shared a Nobel Prize in 1969.

Genetics and Molecular Biology. Investigations of the chemical properties of hereditary substances were another growing area of genetic research beginning in the 1930s that fell under the umbrella of molecular biology. Before researchers turned their attention to deoxyribonucleic acid (DNA), proteins were considered the components of chromosomes that provided the key to inheritance. Chemist Linus Pauling of the California Institute of Technology (Caltech) seemed likely to discover the structure of DNA because of his intimate knowledge of the structural chemistry of proteins and their building blocks, amino acids. Pauling and three colleagues showed in 1949 that individuals suffering from sickle-cell anemia, an inherited disease of the blood, have hemoglobin with an abnormal structure. Using electrophoresis to analyze blood samples, they determined that individuals with full-blown disease

were homozygous recessive and those with sickle-cell trait were heterozygous. A few months earlier, James V. Neel of the University of Michigan had come to the same conclusion by comparing blood samples of children with sickle-cell anemia with those of their parents. Tracing traits through a family's lineage dominated studies of human heredity before the mid-twentieth century for lack of other tactics and was a common approach to eugenic investigations. A significant feature of Pauling and his colleagues' accomplishment was that their technique required an individual rather than two or more familial generations. Pauling called sickle-cell anemia a "molecular disease" to connect form and function: a molecule with an abnormal structure can impair a person's health.

A major turning point came in 1953, when American James D. Watson and Englishman Francis Crick announced that the structure of DNA was a double-helix formed from four nucleotides (adenine, thymine, guanine, and cytosine) and speculated that DNA's structure led to an understanding of its replication. In 1957, Matthew Meselson and Franklin W. Stahl of Caltech confirmed Watson and Crick's hypothesis that one strand of DNA produces a complementary strand. Like the rediscovery of Mendel's laws, Watson and Crick's findings opened up several rewarding research pathways, including techniques that would eventually be used to decipher the genetic codes of many organisms and the human genome. Biochemical research of protein synthesis led investigators to figure out the genetic code during the 1960s. Shortly after the structure of DNA was announced, physicist George Gamow suggested that a string of three nucleotides in DNA determines which amino acid is produced. Crick responded with what would turn out to be an accurate hypothesis that genes code proteins. Figuring out the steps of this process was the result of investigations conducted in several countries, including England and the United States.

The German Johann Heinrich Matthaei, while at the National Institutes of Health and with some assistance from Marshall Nirenberg, made an important breakthrough in 1961. Matthaei synthesized a long chain of RNA composed entirely of uracil, which he called "poly U." Like DNA, RNA has adenine, cytosine, and guanine, but instead of thymine it has uracil. Matthaei's work proved that poly U made a protein, confirming that many triplets of nucleotides synthesize a string of amino acids that then fold into configurations to make proteins. In additional experiments, Matthaei demonstrated that poly U made only one amino acid, phenylalanine, and not any of the other 19 amino acids.

The poly U experiment convinced Crick and others that a biochemical approach led to answers. Working with Sydney Brenner, Crick soon verified that a triplet of uracil, UUU, codes for phenylalanine. By late 1966 it was known which nucleotide triplet codes for which amino acid and which triplets mark the end of a chain. The genetic code had been deciphered.

Genetic Engineering.

Genetic engineering developed in the 1970s and 1980s with the advent of recombinant DNA technology. Paul Berg of Stanford University accomplished the first step in 1971 with a gene-splicing experiment that linked DNA fragments of two different organisms. In 1973, Stanley Cohen of Stanford University and Herbert Boyer of University of California at San Francisco spliced together DNA fragments and cloned them inside a bacterium host. The Cohen–Boyer experiment helped initiate the biotechnology industry because cloning opened up new possibilities in gene therapies for diseases and disorders. Boyer became a major player in the corporate side, founding Genentech in 1976.

Technological developments and investment were crucial to genetic engineering. In the 1970s Walter Gilbert and Allan Maxam at Harvard University and Fred Sanger at Cambridge University developed procedures for sequencing nucleic acids, and in the early 1980s Marvin Carruthers of the University of Colorado devised a technique for adding new bases one by one and producing a predetermined DNA sequence. Carruthers and Leroy Hood of Caltech invented new technologies that automated DNA sequencing, and soon it was possible to conceive of deciphering an organism's entire genome. The Human Genome Project, which lasted from 1990 to 2003, aimed to decode the twenty thousand to twenty-five

thousand human genes and the 3 billion bases in human DNA. It was an international effort to which the United States made a large financial commitment. From the rediscovery of Mendel's laws to sequencing the human genome, the twentieth century witnessed radical transformations in biology, medicine, and industry as a result of genetics and genetic engineering.

Ethics and Policy in Genetics.

The applications of genetic principles and genetic engineering have raised ethical and policy issues. Before genetics was a rigorous science, eugenics gained followers, as discussed earlier. More recently, genetic screening and counseling have raised concerns that couples will "play God" and design their children. In addition to controversies about genetically modified humans, there are also concerns about genetically engineered plants and animals. Attempts to regulate genetically engineered seeds have stemmed from the potentiality for an environmental disaster if the modified genes that make crops heartier were to be picked up by weeds and concerns that large corporations are overpowering small-scale farming.

A change witnessed over the course of the history of genetics and genetic engineering is the advent of preemptive discussions on the research, ethical, legal, social, and policy issues that the novel scientific and technological findings generate. After the invention of recombinant DNA technology, scientists worried about potential hazards of genetic engineering and pushed for the development of research guidelines. A moratorium on using recombinant DNA technology was issued until after a conference to draft guidelines was held. Many scientists whose contributions had shaped this field, including Paul Berg, Herbert Boyer, and James D. Watson, authored the request that appeared in *Science* on 26 July 1974. About one hundred molecular biologists met in 1975 at the Asilomar Conference Grounds to deliberate on how research could go forward safely. Two aims were to have scientists devise the guidelines (and not governmental officials) and to focus solely on the responsible conduct of research (and not the larger social and ethical issues of recombinant DNA technology). The Human Genome Project instituted an Ethical, Legal, and Social Implications (ELSI) research program that addressed the complex issues associated with the projected scientific achievement. A goal was to address the implications while simultaneously performing the scientific research so that solutions to controversial issues could be devised before the science and technology became available. The history of genetics and genetic engineering has repeatedly proven to be a contentious topic in the public sphere, so the initiative shown by geneticists and molecular biologists to consider the implications of their research findings is commendable.

[*See also* **Agricultural Technology; Agriculture, U.S. Department of; Biological Sciences; Biotechnology; Botany; Chemistry; Cloning; Delbrück, Max; Disease; DNA Sequencing; Dobzhansky, Theodosius; Environmentalism; Ethics and Medicine; Eugenics; Evolution, Theory of; Germ Theory of Disease; Higher Education and Science; Human Genome Project; Hybrid Seeds; Journals in Science, Medicine, and Engineering; McClintock, Barbara; Medicine; Molecular Biology; Morgan, Thomas Hunt; National Institutes of Health; National Science Foundation; Nobel Prize in Biomedical Research; Pauling, Linus; Public Health; Public Health Service, U.S.; Research and Development (R&D); Science; Sickle-Cell Disease; Societies and Associations, Science; Technology; Watson, James D.;** *and* **Zoology.**]

BIBLIOGRAPHY

Appel, Tobey. *Shaping Biology: The National Science Foundation and American Biological Research, 1945–1975.* Baltimore: Johns Hopkins University Press, 2000.

Berry, Roberta M. *The Ethics of Genetic Engineering.* New York: Routledge, 2007.

Bud, Robert. *The Uses of Life: A History of Biotechnology.* Cambridge, U.K.: Cambridge University Press, 1993.

Dunn, L.C. *A Short History of Genetics: The Development of Some of the Main Lines of Thought: 1864–1939.* Ames: Iowa State University Press, 1965, 1991.

Hughes, Sally Smith. *Genentech: The Beginnings of Biotech.* Chicago: University of Chicago Press, 2011.

Kay, Lily E. *The Molecular Vision of Life: Caltech, the Rockefeller Foundation, and the Rise of the New Biology.* New York: Oxford University Press, 1993.

Kevles, Daniel J., and Leroy Hood, eds. *The Code of Codes: Scientific and Social Issues in the Human Genome Project.* Cambridge, Mass.: Harvard University Press, 1992.

Krementsov, Nikolai. *International Science between the World Wars: The Case of Genetics.* New York: Routledge, 2005.

Larson, Edward J. *Evolution: The Remarkable History of a Scientific Theory.* New York: Modern Library, 2004.

Morange, Michel. *A History of Molecular Biology.* Translated by Matthew Cobb. Cambridge, Mass.: Harvard University Press, 2000.

Paul, Diane B. *Controlling Human Heredity: 1865 to Present.* New York: Humanities Books, 1995.

Melinda Gormley

GEOGRAPHY

Geography is generally defined as the study and description of the earth's surface, although over the course of the nation's history the field has encompassed much more. Through the nineteenth and twentieth centuries, geography has taken on multiple roles, many of which advanced the interests of the state and shaped national identity.

Intellectual Origins. The creation of the United States brought calls to extend that independence into the realm of knowledge as well as politics. Just as Noah Webster created grammars and spellers to teach "American" English to young citizens, Jedidiah Morse insisted that an American geography must be authored by one of its countrymen. Morse's *Geography Made Easy* (1784) and *American Universal Geography* (1789) dominated the market for geography textbooks and shaped the outlook of young Americans in the early nineteenth century. Geography was a popular study for both boys and girls in the early nineteenth century, conceived as a path to literacy and civic nationalism and as a subject in itself. Morse's texts were succeeded by those of Emma Willard and William Woodbridge, which cemented geography's place at the center of the American curriculum.

These indigenous sources of American geographical knowledge were shaped by Europeans, most importantly Alexander von Humboldt and Karl Ritter. Humboldt and Ritter reconceptualized the meaning of geography in the early nineteenth century. Ritter began to think about geography as the relationship among vegetation, animal life, and physical geography. Humboldt worked in a similar fashion to expand the vision of contemporary scientists from description and classification to the study of distribution, relationships, and laws of operation. Both men influenced not only the study of geography in schools, but also the agenda of American scientists in the antebellum era. Their ideas were extended by Arnold Guyot, who immigrated to the United States in 1848 and became one of the nation's first university geographers in 1854, at the College of New Jersey (later Princeton University). Guyot also contributed extensively to Joseph Henry's ongoing efforts at the Smithsonian to develop a national system of weather observation.

Another robust dimension of nineteenth-century geographical knowledge was medical geography, a field sparked by the recurrence of epidemics. Medical geography sought the environmental sources of disease, and in an era before germ theory, it offered tremendous promise and hope to discern the causes of yellow fever, cholera, and other mortal diseases. The best known practitioner of the field was physician Daniel Drake, who authored several studies of the Mississippi Valley geography and its relationship to these diseases. The interest in medical geography was strongest in the antebellum era, when expansion into the west incorporated entirely new territory into the nation, and before the advent of germ theory weakened its explanatory power.

Institutional Origins. The field of geography was largely undefined in institutional terms prior to the rise of universities in the latter half of the nineteenth century. In 1851, the American

Geographical and Statistical Society (later the American Geographical Society) was founded by a group of men in New York who saw geography and statistics as important sources of power for a nation expanding westward. The Society attracted a broad range of men from business and politics, as well as cultural leaders such as the historian George Bancroft, its first president. The founding of the National Geographic Society (1888) created another forum for exchanging ideas regarding exploration, science, and the gradually expanding responsibilities of federal science. Both organizations were committed to advancing science and to extending the nation's reach in the west (the American Geographical and Statistical Society) and abroad (the National Geographic Society).

Despite this relationship to state power, geography was somewhat handicapped in the institutional setting of the new universities after the Civil War. G. Stanley Hall, the influential educator, characterized geography as "the sickest of all sick topics of the curriculum," one which had "all the unity of a sausage." This sense that geography was meant for those "who crave to know something, but not too much, of everything" persisted in the twentieth century. In part it resulted from geographers' claims that their subject was "the mother of all sciences," one that encompassed other fields such as geology. Yet the new university system was one predicated not on fields of breadth, but rather of depth, and specialization quickly became the watchword of the new university disciplinary system (Hall, 1911, p. 556).

The earliest generation of professional, university-trained geographers was keenly aware of this problem of identity. William Morris Davis, commonly regarded as the father of American university geography, was actually trained in geology because no geography departments existed until the first was established at the University of California in 1898. Davis and his mentor, Nathanial Southgate Shaler, developed a course of training at Harvard in physical geography, which emphasized the surface features of the earth. Davis used Darwinian principles of evolution to conceptualize the landscape as dynamic and evolving through processes such as soil erosion. Davis also worked to create a separate disciplinary and professional identity for geographers by establishing the Association of American Geographers in 1904. By 1914 over fourteen universities in the United States offered at least six courses in geography. Most important among these were Harvard University, Yale University, the University of Pennsylvania, and the University of Chicago (Martin, 2005, pp. 339–340).

Twentieth-Century Developments. Davis trained many geographers at the turn of the century, but this younger generation became primarily interested in studying the relationship between humans and physical geography. This idea at the core of the field also enabled geography to maintain its position between the social and natural sciences and to claim "disciplinary" status. Yet it was just this claim of intellectual breadth that neighboring sciences began to question. Despite this slight suspicion, geography grew in several intellectual directions in the twentieth century. Leading the study of the human–environment relationship after the turn of the century were Ellen Semple, Ellsworth Huntington, and Isaiah Bowman. Evolutionary ideas—both of Charles Darwin and of Jean-Baptiste de Monet de Lamarck—were used by Semple and Huntington to explain the influence of the natural environment on human development and especially the advent of civilizations around the world. At precisely the same time, historian Frederick Jackson Turner used geographical approaches to develop his theory of American history as influenced by the concepts of the frontier and geographical sections. Each of these scholars conceived of geography as linking nature and culture. Indeed, geography gradually moved in the early twentieth century from the earth toward the social sciences.

World War I had both an intellectual and a practical effect upon American geography. The war broke down the imperial structures that had existed around the world, but it also advanced the careers of individual geographers, most notably Isaiah Bowman. As director of the American Geographical Society, Bowman was asked to join the "Inquiry" committee, the group of scholars gathered by President Wilson's administration to prepare for the postwar peace at Paris by study-

ing the state of political, social, and economic geography around the world, especially in Europe. Bowman was also typical of his fellow geographers when, after the war, he retreated from the concept of geography as the influence of the physical environment over human behavior and instead began to consider the reverse. After the war, geographers engaged in an intense examination of the meaning and definition of their field, which produced important disciplinary statements of Richard Hartshorne and Carl Sauer.

After World War II, Harvard University closed its geography department, which had been the origin of the discipline under Davis at the turn of the century. Several other institutions did the same, although overall the number of geography programs rose after the war as a result of the general expansion of higher education. The advent of computer technology in subsequent decades ignited geographers' interest in quantitative approaches to their work—specifically the adoption of statistical analysis—and gave them a new tool to search for spatial patterns. In a parallel development, the core identity of geography as a regional study declined since midcentury, whereas new thematic approaches gained popularity, such as urban, cultural, feminist, historical, and economic geography. The specialty groups within the Association of American Geographers proliferated, the most popular being Geographic Information Systems, urban geography, and remote sensing (Martin, 2005, pp. 425–426).

Another development in recent decades has been the return of a political and reformist sensibility to geography. Perhaps this interest has been advanced by the democratic thrust of the Internet and Geographic Information Systems because such tools enabled geographers to think widely about structures of power, the politics of space, and the role of mapping. This reformist strain within geography also reflected an ongoing self-examination within the field, with geographers drawing on social theory to consider the political meaning and effects of their work. The diversity in geography today reflects a theme that has persisted within the discipline for over a century, with practitioners contesting the very nature of their enterprise.

[*See also* Cartography; Cholera; Disease; Geology; Henry, Joseph; Mathematics and Statistics; Smithsonian Institution; *and* Yellow Fever.]

BIBLIOGRAPHY

Daniels, Stephen, Dydia DeLyser, J. Nicholas Entrikin, and Doug Richardson, eds. *Envisioning Landscapes, Making Worlds: Geography and the Humanities.* New York: Routledge, 2011.

Hall, G. Stanley. *Educational Problems,* p. 556. New York: Appleton & Co., 1911.

Hall, G. Stanley. "The Ideal School." In *Proceedings of the NEA.* Buffalo, N.Y.: National Education Association, 1901.

James, P. E., and C. F. Jones, eds. *American Geography: Inventory and Prospect.* Syracuse, N.Y.: Syracuse University Press, 1954.

Martin, Geoffrey J. *All Possible Worlds: A History of Geographical Ideas.* New York: Oxford University Press, 2005.

Morin, Karin. *Civic Discipline: Geography in America, 1860–1890.* Surrey, England: Ashgate, 2011.

Schulten, Susan. *The Geographical Imagination in America, 1880–1950.* Chicago: University of Chicago Press, 2001.

Schulten, Susan. *Mapping the Nation: History and Geography in Nineteenth-Century America.* Chicago: University of Chicago Press, 2012.

Susan Schulten

GEOLOGICAL SURVEYS

Geological surveys are among the earliest examples of government-supported science in the United States, with state surveys predating the federal geological survey by nearly half a century. The North Carolina Geological Survey, begun in 1823, generally ranks as the first, although North Carolina supported only one individual (Denison Olmsted) part-time and did not fund an ongoing institution. In other respects, however, North Carolina's survey typified state surveys to come. First, it was temporary and ended, with a report, in 1825. Nearly all early state surveys worked in fits and starts, flourishing in good times and disappearing during hard times. The longest continuously operating survey, in New York, was founded in 1836. Second, the North Carolina survey, although not formally a part of a university, was conducted by a university professor (Olmstead taught at the University of North Carolina at Chapel Hill). The association between state surveys and universities became common and, in many cases, endured. Finally, the North Carolina survey sought practical results. Emerging from a program of internal improvements within the state, it was designed primarily to produce geologic information for engineering applications, such as building roads and bridges. Geological surveys in other states focused on minerals and mining or the relationship between soils and agriculture.

By 1850, 20 states had established geological surveys. Others added surveys in the 1850s and 1860s, particularly in the burgeoning West. Several western surveys, such as Nevada's and California's, focused on mining and minerals. Between 1850 and 1900, thirteen western states created geological surveys, and three more territories—Arizona, New Mexico, and Oklahoma—started surveys soon after achieving statehood in the early twentieth century. Many of these surveys, in both the West and the East, pursued two simultaneous, sometimes conflicting missions: generating practical results for utilitarian application while producing basic scientific knowledge, such as geologic maps or paleontologic reports.

Although the federal government did not establish a national survey until 1879, geologists were often associated with pre–Civil War land and railroad surveys of the Old Northwest and the trans-Mississippi West, frequently under the Army Corps of Topographical Engineers. In the postwar years Washington funded a number of western geological and geographical surveys: Clarence King's army-sponsored Geological and Geographical Exploration of the Fortieth Parallel (begun in 1867); George M. Wheeler's Geographical Surveys West of the 100th Meridian (1869), also undertaken by the army; John Wesley Powell's Geographical and Topographical Survey of the Colorado River (1870), funded by the Department of the Interior; and Ferdinand V. Hayden's Geological and Geographical Survey of the Territories (1873).

King, Powell, and Hayden were all experienced geologists, and Wheeler, an army lieutenant, recruited the geologist Grove Karl Gilbert to assist him. This proliferation of western surveys prompted Congress in 1879 to create a centralized U.S. Geological Survey (USGS), first under the direction of King and later under Powell, who made it the nation's premier scientific institution of the late nineteenth century.

By 1900 the state surveys were also becoming more permanent. The word "survey," at least as applied to state surveys, came to be used as a noun (referring to a survey as an organization), implying an enduring existence. Relations between the state surveys and the USGS were not always smooth, although over time, the various state surveys and the USGS worked out a relative division of labor. The USGS generally concentrated on issues that cut across state lines, such as topographic mapping and surface-water resources. The USGS became known for a stream-gauging network that measured streamflow, for national water-quality studies, and for mapping on public lands and much of the west.

During the first half of the twentieth century, having completed the reconnaissance phase of their work and having gained a good idea of their state's general geology, many state surveys now undertook more detailed investigations of smaller areas. County studies were common, as were studies of minerals, especially petroleum in states such as Texas and metallic minerals in Nevada, Washington, and Wyoming.

After 1950, many state geological surveys found their role changing. Some, such as Michigan's, were given increasing regulatory responsibility. Others, such as Utah's, were incorporated into state departments of natural resources, making them less research oriented and more an arm of state government. Some surveys, such as Kentucky's and Illinois', were a university division. Even those surveys that were removed from the university environment usually maintained an academic connection.

During the 1960s and 1970s, concern about environmental issues—such as the relationship between subsurface geology and groundwater contamination, groundwater depletion, radioactive-waste disposal, and a host of other problems—gave

geological surveys a renewed relevance. Individual surveys carved out scientific niches. The Kansas Geological Survey developed computer software and pioneered the use of computerized cartography and the application of quantitative methods to geologic problems. State surveys in Utah and Colorado focused on geologic hazards. The Texas Bureau of Economic Geology did extensive petroleum research often funded by external grants and contracts. In the 1980s and 1990s, state surveys redoubled their efforts in geologic mapping, working with the USGS to develop funding for the National Cooperative Geologic Mapping Act and collaborating on geographic information systems and other new cartographic methods. In 1981, Genevieve Atwood was appointed the state geologist of Utah, the first woman to head a geological survey in the United States.

The 1990s and the early twenty-first century were a time of budgetary difficulty for both the USGS and many of the state geological surveys. In 1994, the USGS was targeted for elimination by the Republican Contract with America. Although the USGS survived and today has a budget approaching $1 billion annually, it was forced to defend itself in a difficult political battle. In some respects, this was the beginning of a time of change for the USGS. In 1996, the National Biological Service was merged into the USGS. In 2009, Marcia McNutt was named the first female director of the USGS and in 2010 she reorganized the agency according to mission areas, including climate and land-use change, ecosystems, energy and minerals, hazards, and water.

Budgetary issues were equally challenging on the state side. State budgets saw shortfalls in the early part of the twenty-first century, and several state geological surveys saw dramatic declines in state funding, to the point that their existence was threatened. The state surveys identified the preservation of geologic data as another major priority for a cooperative program with the USGS. The level of support for data preservation never achieved anything close to that for geologic mapping. Like all scientific organizations, state and federal surveys had to adapt to electronic methods of making their reports and data available. At the same time, important new advances in technology, such as three-dimensional seismic imaging, computer models, and enhanced remote sensing techniques, provided better understanding of the earth's surface and subsurface than ever.

Still, the budgets for state surveys across the nation totaled approximately $245 million in 2011, and the ongoing challenges of energy, water, and the environment provided funding opportunities; many of the state surveys began extensive work in geologic sequestration of carbon dioxide and geothermal energy, while continuing more traditional programs in water and minerals and geologic hazards (Allis, 2011, p. 1). State funding decreased, but grant and contract funding from the federal government and other sources increased. Total staffing in state surveys was 1,980 in 2011, down from a peak of 2,900 in the 1980s (Allis, 2011, p. 2). The average state survey had a staff of 40 and a total annual budget of just over $4 million, although they ranged in size from Texas (with an annual budget of $31 million) to several state surveys with annual budgets of less than $1 million per year (Allis, 2011, p. 4). Studies showed that a large percentage of staff members within surveys were in their fifties, a demographic similar to that found throughout the geosciences by the American Geological Institute and the American Association of Petroleum Geologists, creating ongoing workforce challenges for the near future.

Although both the state and the federal survey faced budgetary challenges and questions about an aging workforce, they continued to be as relevant as the problems they addressed in energy, water, climate change, and other environmental issues. Some sought to reinvent themselves, some were threatened with extinction, and others flourished, continuing to be a permanent but constantly changing feature of America's scientific landscape.

[*See also* **Army Corps of Engineers, U.S.; Cartography; Environmentalism; Geography; Geology; Mining Technology; Paleontology; Powell, John Wesley;** *and* **Science.**]

BIBLIOGRAPHY

Aldrich, Michele L. "American State Geological Surveys, 1820–1845." In *Two Hundred Years of Geology in America*, edited by Cecil J. Schneer. Hanover, N.H.: University Press of New England, 1979.

Cobb, James C. *Association of American State Geologists Centennial History: 1908–2008*. Alexandria, Va.: Association of American State Geologists, 2008.

Goetzmann, William H. *Exploration and Empire: The Explorer and the Scientist in the Winning of the American West*. New York: Alfred A. Knopf, 1971.

Gundersen, Linda, et al. *Geology for a Changing World 2010–2020: Implementing the U.S. Geological Survey Science Strategy*. Reston, Va.: U.S. Geological Survey, 2011.

Manning, Thomas G. *Government in Science: The U.S. Geological Survey, 1867–1894*. Lexington: University of Kentucky Press, 1967.

Merrill, George. *Contributions to a History of American State Geological and Natural History Surveys*. Washington, D.C.: U.S. Government Printing Office, 1920.

Millbrooke, Anne Marie. *State Geological Surveys of the Nineteenth Century*. PhD diss. University of Pennsylvania, 1981.

Socolow, Arthur A., ed. *The State Geological Surveys: A History*. Grand Forks, N.D.: Association of American State Geologists, 1988.

Worster, Donald. *A River Running West: The Life of John Wesley Powell*. New York: Oxford University Press, 2001.

Rex C. Buchanan

GEOLOGY

A familiar figure with which to begin a history of American geology is William Maclure (1763–1840), a wealthy Scots immigrant who in 1809 published a widely circulated map and report on the geology of the United States and thereafter gained the appellation "the father of American geology" (Merrill, 1924). Maclure, like many geologists in the first three decades of the nineteenth century, interpreted American rocks using European theories, and, accordingly, he followed the taxonomy and classificatory system proposed by Abraham Werner (1749–1817), professor of mineralogy at the state mining academy in Freiburg,

Germany. Werner identified layers of rock (called formations) by their lithological composition (e.g., granite, limestone) and grouped them together into four classes—primary, transition, secondary, and alluvial—each of which corresponded to a particular moment in the earth's history. Werner thought the earth was originally covered by a universal ocean out of which the primary rocks, which included granite, precipitated first and were found on mountaintops, such as the Appalachians. Transition rocks were deposited on top of these and were generally found tilted along mountainsides. Next were the flat-lying, sedimentary rocks such as sandstones and limestones, which made up the secondary class, and finally alluvial formations, the youngest rocks, including basalt, which capped off the entire stack of strata.

Maclure modified Werner's "Neptunist" theory with the competing "Vulcanist" explanation of the earth's origin proposed by the Scots philosopher James Hutton (1726–1797), who had theorized that rocks, especially granite and basalt, were of subterranean or volcanic origin, not aqueous depositions. Maclure's willingness to combine and adapt Neptunist and Vulcanist theories proved to be characteristically American. Benjamin Silliman (1779–1864), professor of natural history and chemistry at Yale College and the foremost teacher of geology in the first decades of the nineteenth century, for example, took a very similar theoretical stance. Silliman rejected Hutton's cyclical theory of the earth, an ahistorical process whereby rocks are continually eroded, sediments consolidated, and land uplifted, for Werner's directionalist theory in which older rocks and hence older time periods were different and distinct from younger rocks and younger time periods. Silliman and most antebellum American geologists held to a linear history of the earth, which conformed closely to their Christian religious views. In regard to method, Americans likewise followed Werner's practices rather than Hutton's philosophizing. They went into the field to identify, describe, name, classify, and correlate rock strata, which was known as stratigraphy, the dominant activity of most geologists throughout the nineteenth century.

By the standards of the time, Maclure's map was crude, and one of the first Americans to attempt a more detailed stratigraphy was Amos Eaton (1776–1842), a former student of Silliman and an itinerant lecturer in science. Under the patronage of the wealthy patron Stephen van Rensselaer III (1764–1839), Eaton surveyed upstate New York and the region bordering the 363-mile-long Erie Canal. The scale of the phenomena (a recurrent theme in American geology) convinced Eaton that American geology needed an American nomenclature and classificatory system. Between 1818 and 1830, Eaton proposed no fewer than nine different taxonomies, but in all of them he continued to identify and describe rock formations lithologically and then group them within Werner's four classes. The limitations of this system were made known by a younger cohort of American geologists, some of whom (e.g., James Hall, 1811–1898) had been Eaton's students at the Rensselaer School (est. 1824). By the 1830s, the dominant method for identifying, describing, and classifying sedimentary rocks was by fossil content, not composition.

In the first half of the nineteenth century, geology was the most popular science in America. It was enlightening, morally uplifting, and exciting. Geology encompassed the exploration of unmapped lands and the discovery of exotic rocks, minerals, and fossils and, to many Americans, it meant the revelation of God's plan in nature. Simultaneously, geology's popularity was founded on its unorthodox and unsettling findings. Extinct animals, for example, had caught Americans' attention ever since the painter and natural historian Charles Wilson Peale (1741–1827) had unearthed and reconstructed a mastodon in 1800, which he then put on display in his Philadelphia museum. Public lectures and lyceums along with new books and college courses publicized the strange animals and plants of a deep and distant past.

Such geological revelations seemed to challenge the Biblical account of the earth, and although most antebellum geologists were able to reconcile Genesis and geology by adopting a nonliteral interpretation of Scripture, ordinary Americans were not so easily assured. Reconciling Noah's Flood, on the other hand, was easier for early nineteenth-century geologists and their audiences. But by midcentury, the explanation for erratic boulders, striated rocks, and long, linear deposits of gravel turned to ice rather than to water. The Swiss naturalist Louis Agassiz (1807–1873), who came to the United States in the 1840s, pictured thick sheets of ice covering huge parts of North America. By the 1870s, Thomas Chrowder Chamberlain (1843–1928), chief geologist of the Wisconsin geological survey and later (1887–1892) president of the University of Wisconsin, had identified and named numerous episodes of advancing and retreating continental glaciers. American geological thought during most of the nineteenth century was predisposed toward catastrophist explanations; consequently, the British geologist Charles Lyell (1797–1875) and his neo-Huttonian uniformitarianism had limited purchase among Americans.

Another source of geology's popularity was its promise of mineral wealth. Ever since the sixteenth century, explorers of the Americas had been looking for gold, and undoubtedly such explorations spurred the natural historians of the seventeenth and eighteenth centuries to collect and classify mineral and rock specimens for their cabinets of curiosities. By the nineteenth century coal was coming to rival gold as the most sought-after mineral resource. Thus, although Eaton had found no coal during his 1820s reconnaissance, the possibility that it existed elsewhere in New York prompted the state legislature to sponsor a full state survey in 1836 (Aldrich, 2000). Surveys are synonymous with nineteenth-century geology in the United States and in Europe, and antebellum state surveys provide one of the best examples of the ways and means by which public (i.e., government) policy indirectly stimulated private economic interests and influenced the development of a scientific discipline. American surveys were enacted by state politicians, but it was often the geologists themselves who drafted the actual legislation directing them to discover, describe, and distribute information about the location and value of natural resources. On occasion, state legislators asked for more than maps and reports on the rocks and minerals; they wanted catalogs of plants, animals, minerals, waters, and, on very rare occasions, also fossils. In these instances, surveys (such as that of New York) encompassed all of natural history and thereby contributed to the sciences of botany, zoology, mineralogy, chemistry, and paleontology as well as to geology.

Whether natural historical or geological, surveys were justified in public pronouncements and in political debates by their usefulness. Not only would valuable resources (gold, coal, building stones, mineral fertilizers, artesian springs, birds, fish, hardwoods) be made known, but also such information would forestall the ignorant from reckless investment in ill-founded projects or even outright swindles (Lucier, 2008). In addition, all surveys were intended to enlighten and uplift ordinary citizens. As vehicles of internal improvement, surveys fit with other such public works as the building of schools, roads, canals, bridges, and ports. All these projects were funded by states, not the federal government, which possessed neither the political will nor the constitutional right to appropriate monies to specific improvements. But if state patronage brought financial wherewithal and public recognition to geology, legislatures could also suspend such support. Geologists needed to respond to the demands of politicians and their constituents, and as a result antebellum American geology was far more democratic than its European counterparts. Historians have sometimes misinterpreted this accountability for practicality, which has led to a misrepresentation of American geology as more Baconian than theoretical. Although survey geologists emphasized economic results for political reasons, they did not neglect theory (Aldrich, 2000).

Before the Civil War, 29 of the 34 states of the Union began surveys, all of which were designed to improve agriculture by studying soils and fertilizers; transportation by locating routes for roads, canals, and railroads; mining by finding gold, coal, and iron ore; and construction by identifying building stone. North Carolina commissioned the first survey in 1823 and South Carolina followed the next year. The impetus for these surveys, as well as for those of Georgia (1836) and Virginia (1836–1841), was agriculture, but a stronger push proved to be the gold rush to the southern Appalachians. By contrast, in the northern Appalachian states of Maryland (1833–1841) and Ohio (1836–1841), as well as in Connecticut (1835–1842), Massa-chusetts (1830–1841), and Rhode Island (1839–1840), coal was the mineral objective, although no appreciable amounts were ever found in New England. In Pennsylvania (1836–1842), finding soft (bituminous) coal in the western part of the state was not such a pressing concern, but lawmakers wanted information regarding the suitability of eastern hard (anthracite) coal as fuel for heating, steam engines, and making iron. In New York (1836–1842), not finding coal presented problems that required clever politicking to preserve funding for discovering other resources. In general, antebellum geologists were quite successful at dealing with the political pressures for practical results. The exception was the California survey (1860–1874), whose director, Josiah Dwight Whitney (1819–1896), chose to focus on geological mapping instead of studies of the diminishing gold supplies or the potential oilfields. After one too many lectures on the difference between a scientific survey and a prospecting party, the California legislators pulled the funding (White, 1968; Lucier, 2008).

Despite being temporary and task-specific agencies at the mercy of politics, antebellum surveys made significant long-term contributions to science. At a time when there were few colleges and no mining schools for the education of geologists, surveys became the principal institution for training in fieldwork and for publishing research. And although not all surveys produced final reports, most did issue yearly accounts which, taken together, display the major scientific result of ante-bellum research: the geological mapping of most of the North American continent east of the Mississippi river. In this regard, the New York system—a taxonomy developed by the state geologists for the identification, ordering, and naming of the oldest fossiliferous rocks in America—became the standard of reference for future stratigraphic correlations with western parts of the continent and with the Paleozoic in Europe (Aldrich, 2000). An equally important innovation was the contour map, a graphic device developed by the Pennsylvania survey to unravel coal basins; it became the methodological tool for understanding Appalachian mountain–type folding and

subsequently the conceptual foundation for the subfield of structural geology. Coal and petroleum geology were also American specialties, and the detailed stratigraphy of American coal fields served as the rationale for the introduction of two other systems: the Pennsylvanian and the Mississippian, which are the only terms of American origin in the modern geological time scale (Lucier, 2008).

The challenges antebellum geologists faced in trying to correlate rocks from one state to the next spurred them to organize on a national level. In 1840, members of various surveys formed the Association of American Geologists and Naturalists to work out rules for naming rocks, and in 1848 this group enlarged its membership to include other disciplines and became the American Association for the Advancement of Science (AAAS), an organization often identified by modern scholars with the professionalization of American science. However, it would be more accurate to identify survey geologists with a particular type of professional practice more in tune with the nineteenth century, namely consulting. Making a living from geology (or any science) was difficult and forced nineteenth-century geologists to cobble together several survey and teaching positions, and often they accepted commissions from companies and capitalists to do private surveys for a fee. These consulting engagements were among the first forms of professional science involving fees for expertise (Lucier, 2008).

The summation of antebellum survey geology found its clearest expression in the *Manual of Geology* (1862) by James Dwight Dana (1813–1895), professor of geology at Yale. Dana argued that America, by virtue of its size and structure, exemplified the science of geology (meaning a complete record of the earth's history) better than the rocks found anywhere else on the planet. Such nationalistic pride echoed Eaton's earlier admonishments and highlighted the parity, or even superiority, felt by many American geologists toward their European colleagues. Dana, along with Hall, was also responsible for one of the major theoretical frameworks of nineteenth-century geology, the geosyncline, an explanation of how mountains arose and continents grew. According to Dana and

Hall, a seaward trough bordering continents had filled with thick sediments, the weight of which warped the crust downward and thus forced adjacent landward rocks to warp upward into mountains, such as the Appalachians (Dott, 1979).

In the aftermath of the Civil War, geologists focused their attention on the trans-Mississippi region, and the federal government took over from the states as the leading patron of the science. The most expensive and widely publicized projects were the federal surveys of the West, which, unlike the U.S. government expeditions undertaken before the war, were not primarily military in purpose. In 1867 the U.S. Congress authorized two surveys: the Geological Exploration of the Fortieth Parallel under Clarence King (1842–1901) and the Geological and Geographical Survey of the Territories under Ferdinand V. Hayden (1829–1887). King's party followed the route of the first transcontinental railroad (completed in 1869), whereas Hayden's covered Nebraska, Wyoming, and Colorado. In 1871 Congress commissioned two more parties: the Geographical Surveys West of the One Hundredth Meridian under Lieutenant George M. Wheeler (1842–1905), Corps of Engineers, and the Geological and Geographical Survey of the Rocky Mountain Region under John Wesley Powell (1834–1902). Of all the western surveys, Powell's was the most sensational, with a headline-grabbing boat trip down the Colorado River through the Grand Canyon, but King's was the more scientific. King, for example, introduced microscopic petrography, a method for the identification, naming, and classification of igneous and metamorphic rocks, to the United States (Moore, 2006). All four surveys began topographical and geological maps of the West, and it was this overlap and competition that led Congress to consolidate them under the U.S. Geological Survey (USGS) in 1879.

The establishment of the USGS marks a watershed in the history of American geology by creating a permanent government agency to support it (Rabbitt, 1979). King became the first director and set a policy that focused research on western gold and silver mining. The USGS conducted detailed surveys of particular mining districts, and among the best examples are George F. Becker's

(1847–1919) 1882 monograph on the Comstock Lode of Nevada and Samuel F. Emmons's (1841–1911) massive 1886 monograph on Leadville, Colorado. When King stepped down after two years, Powell took over and redirected the USGS back toward mapping the entire country. The new stratigraphic maps resulted in a proliferation of new formations, and in 1888 the Geological Society of America (GSA) was formed in response to demands (both national and international) to coordinate taxonomy and nomenclature. Under Powell, the USGS also began studies of irrigation, land use, and landforms. The topography of the arid, treeless West provided vivid evidence of the interaction of surface waters (e.g., rivers and lakes) and vertical movements of the earth's crust. In 1889, Clarence Dutton (1841–1912), a geologist working for the USGS mainly in the Colorado Plateau region, coined the term isostasy, meaning "equal standing," to describe a general equilibrium within the earth's crust whereby lighter-weight (less dense) mountains were balanced by adjacent greater-weight (more dense) lowlands. In 1890, Grove Karl Gilbert (1843–1918), a close friend of Dutton and Powell who rose to the position of senior geologist in the USGS, published his monograph on the drainage of the ancient Lake Bonneville, and later Gilbert introduced the concept of the graded river, an explanation of whether rivers erode or deposit sediments (Pyne, 1980). William Morris Davis (1850–1934) of Harvard University synthesized these ideas about how rivers create landforms into a theory of landscape evolution or geomorphology, and by the early twentieth century, American geological thought had swung away from catastrophist explanations and toward uniformitarian and multiple-working hypotheses. Adherence to the concepts of isostasy and the geosyncline, for example, explain in large part American's rejection of Alfred Wegener's (1880–1930) theory of continental drift (Oreskes, 1999).

In 1894 Powell ran afoul of a cost-cutting Congress and had to resign. Under the third director, the paleontologist Charles Walcott (1850–1927), the USGS returned to King's priority of mining geology, which became the largest, most innovative division within the agency under the direction of S. F. Emmons. In addition, Walcott broadened the

USGS's purview to include research on water resources and other economic minerals, particularly petroleum. This big-tent approach made the USGS the dominant force in American geology by the turn of the twentieth century, but it also raised fears of scientific monopoly. In 1905, Frederick Ransome (1868–1935), a member of Emmons's prestigious mining geology division, founded *Economic Geology* as an independent journal, and by World War II *Economic Geology* had the largest circulation of any geological publication in the world. In 1920 Ransome helped establish the Society of Economic Geologists (SEG) as an affiliate of the Geological Society of America. Petroleum geologists, however, did not want affiliation with the Society, and in 1917 they established their own independent organization, the American Association of Petroleum Geologists (AAPG), which after World War II became the largest geological society in the world. Petroleum geologists' separate professional identity had much to do with their employment; unlike economic geologists, most worked for oil companies, not the USGS. By the 1920s, oil companies were leaders in geological research and innovators of such specialties as economic paleontology, microlithology, sedimentology, and, especially, exploration geophysics, a set of theories and practices developed out of World War I artillery range–finding techniques for mapping subsurface structures like salt domes along the Gulf Coast (Owen, 1975).

Similarly, economic geology made important contributions to both theory and practice. Understanding the origin and occurrence of metal ores required new studies of metamorphic and igneous rocks (the usual locations of metals) and subterranean heat, pressure, and solutions, in other words, the chemistry and physics of rocks, or, as these interdisciplinary studies became known in the early twentieth century, geochemistry and geophysics. For example, while working on the gold-mining camps of the West, four USGS geologists, C. Whitman Cross (1854–1949), Joseph P. Iddings (1857–1920), Louis V. Pirsson (1860–1919), and Henry S. Washington (1867–1934), devised a quantitative chemical–mineralogical classification and nomenclature for igneous rocks, the CIPW system. But by the early twentieth century,

the leading research in economic geology was no longer being conducted under the auspices of the USGS. In 1907 an undistinguished scientist, George O. Smith (1871–1944), became the fourth director, and the USGS became increasingly practical in orientation. In 1912, Waldemar Lindgren (1860–1939), one of the founders of *Economic Geology*, resigned as chief geologist of the USGS and became head of the Department of Geology at the Massachusetts Institute of Technology. Likewise, George Becker, Whitman Cross, and Henry Washington helped to establish the Geophysical Laboratory of the Carnegie Institution of Washington (CIW, founded in 1905) as a world leader in experimental petrology, the investigation, using high pressure and high temperatures, of the chemical composition of rocks. In the 1920s at the CIW, for example, Norman L. Bowen (1887–1956) developed a model for the cooling of magmas, in which he explained the evolution of igneous rocks as a regular order of deposition of crystals from a melt, the eponymous "Bowen's reaction series" (Servos, 1990).

In the early 1920s the CIW also established a Seismological Laboratory in Pasadena, California, to study earthquakes. Since the late nineteenth century, geophysicists had been recording the pathways and travel times of earthquake waves (primary, or pressure, and secondary, or shear) in an attempt to understand the earth's interior. Hugo Benioff (1899–1968) joined the laboratory in 1924, where he designed some of the world's most sensitive seismographs to detect and locate the source of earthquakes. Using these instruments, the seismologist Charles Richter (1900–1985) and the German American Beno Gutenberg (1889–1960), professor of geophysics at the California Institute of Technology, devised a standard measure of the relative magnitudes of earthquakes, the Richter scale. By the 1940s, seismology had provided some of the best evidence of the earth's internal structure—solid core, liquid outer core, mantle, asthenosphere (plastic layer in the upper mantle), and crust (Oldroyd, 1996).

American seismology also provided crucial evidence in support of plate tectonics. In the 1950s, Benioff's investigations of earthquakes in the Pacific showed a correlation between depth and distance from an oceanic trench; earthquakes occurred along a fault plane or "surface" moving down into the mantle. Other investigations of the sea floor conducted by the Scripps Institution of Oceanography (est. 1903), Woods Hole Oceanographic Institution (est. 1930), and Lamont–Doherty Earth Observatory (est. 1949) revealed a worldwide network of mid-oceanic ridges. In 1960 Harry Hess (1906–1969), a former U.S. Navy officer and professor of geology at Princeton University, proposed that oceanic ridges were created by the upwelling of material from the mantle, which then spread across the sea floor. The new oceanic basins created by the spreading sea floor sink back into the mantle, thereby forming oceanic trenches and causing very deep earthquakes. Further evidence of sea floor spreading came from geomagnetic studies across mid-oceanic ridges that found alternating strips of normally and reversely polarized material (in other words, as the lava cooled it was magnetized in the direction of the earth's field) running parallel to and on each side of the ridge. In 1968 W. Jason Morgan (b. 1935), professor of geophysics at Princeton University, published a theory of plate tectonics that described the earth's outer layer (the lithosphere) as composed of 10 or so rigid plates that move with convection currents in the mantle. New material from the mantle rises along mid-oceanic ridges, and the old edges of plates are subducted back into the mantle, creating oceanic trenches. Since the 1970s plate tectonics has served as the paradigmatic explanation of the earth's history, and geophysics has replaced stratigraphy as the dominant method for studying the earth.

Beside methodology and the plate tectonics paradigm, the other major change in American geology during the late twentieth and early twenty-first centuries was scope. Beginning with the lunar explorations of the 1960s, the earth became increasingly viewed as much more than a collection of old rocks and fossils; it was seen as a dynamic planet, and understanding its interconnected systems involves the integration and coordination of numerous sciences, including oceanography, meteorology, and astronomy. The scale of phenomena is so large, in fact, as to require a new name—earth sciences.

[*See also* **Agassiz, Louis; Agricultural Technology; American Association for the Advancement of Science; Army Corps of Engineers, U.S.; Astronomy and Astrophysics; Biological Sciences; Botany; Canals and Waterways; Chemistry; Dana, James Dwight; Geography; Geological Surveys; Geophysics; Meteorology and Climatology; Mining Technology; Oceanography; Paleontology; Petroleum and Petrochemicals; Plate Tectonics, Theory of; Powell, John Wesley; Railroads; Religion and Science; Scripps Institution of Oceanography; Silliman, Benjamin;** *and* **Zoology.**]

BIBLIOGRAPHY

Aldrich, Michele L. *New York State Natural History Survey, 1836–1842: A Chapter in the History of American Science.* Ithaca, N.Y.: Paleontological Research Institute, 2000. The best study of an antebellum state survey. It addresses the scientific, social, and political contexts of the largest and most influential survey of the time.

Barrow, Mark V., Jr. *Nature's Ghosts: Confronting Extinction from the Age of Jefferson to the Age of Ecology.* Chicago: University of Chicago Press, 2009. The best study of the meanings of animal extinction in America.

Dott, Robert H., Jr. "The Geosyncline—First Major Geological Concept 'Made in America.'" In *Two Hundred Years of Geology in America*, edited by Cecil J. Schneer, pp. 239–264. Hannover, N.H.: University Press of New England, 1979.

Lucier, Paul. *Scientists and Swindlers: Consulting on Coal and Oil in America, 1820–1890.* Baltimore: Johns Hopkins University Press, 2008. The best work on the relations of science and capitalism in the nineteenth century. It shows how the emergence of a new profession—scientific consulting—was key to the development of the coal, kerosene, and petroleum industries as well as to the rise of the American scientific specialties of structural geology and petroleum geology.

Merrill, George P. *The First One Hundred Years of American Geology.* New Haven, Conn.: Yale University Press, 1924. The standard work on nineteenth-century American geology. It established the major contours of the dominant narrative and identified the principal characters in the American story.

Moore, James Gregory. *King of the 40th Parallel: Discovery in the American West*. Stanford, Calif.: Stanford University Press, 2006. Best recent work on King's geology. It focuses mainly on King's survey across the West and ends with King's tenure as the first director of the U.S. Geological Survey.

Oldroyd, David. *Thinking about the Earth: A History of Ideas in Geology*. London: Athlone Press, 1996. The most comprehensive overview of the history of the earth sciences. It traces the development of ideas about the earth from antiquity up to the present. Although it is largely an intellectual history, its best chapters address the field practices of nineteenth-century geology and the instruments of early twentieth-century geophysics.

Oreskes, Naomi. *The Rejection of Continental Drift: Theory and Method in American Earth Science*. New York: Oxford University Press, 1999. The best study of the history of plate tectonics in the American context.

Owen, Edgar Wesley. *Trek of the Oil Finders: A History of Exploration for Petroleum*. Tulsa, Okla.: American Association of Petroleum Geologists, 1975. The most thorough study of any subdiscipline of the earth sciences. It follows the growth of petroleum geology from the first oil well in western Pennsylvania in 1859 across the globe and up to the 1970s with the rise of multinational corporations.

Pyne, Stephen J. *Grove Karl Gilbert: A Great Engine of Research*. Austin: University of Texas Press, 1980. One of the very few studies of a late-nineteenth-century American geologist.

Rabbitt, Mary C. *Minerals, Lands, and Geology for the Common Defense and General Welfare*. Vol. 1, *Before 1879*; Vol. 2, *1879–1904*; Vol. 3, *1904–1939*. Washington, D.C.: U.S. Government Printing Office, 1979, 1980, 1986. The first volume is a detailed history of the most important scientists, surveys, and theories of nineteenth-century American geology. The second and third volumes trace the development of American geology through the work of the U.S. Geological Survey.

Servos, John W. *Physical Chemistry from Ostwald to Pauling: The Making of a Science in America*. Princeton, N.J.: Princeton University Press, 1990. One of the best histories of a scientific discipline. It discusses the intellectual and institutional development of physical chemistry, including the related discipline of geochemistry.

White, Gerald T. *Scientists in Conflict: The Beginnings of the Oil Industry in California*. San Marino, Calif.: Huntington Library, 1968. Detailed account of the bitter personal rivalry between the Yale chemist Benjamin Silliman Jr. and the California geologist Josiah Dwight Whitney over prospects for petroleum production in southern California in the 1860s.

Worster, Donald. *A River Running West: The Life of John Wesley Powell*. New York: Oxford University Press, 2001. The most thorough biography of the second director of the U.S. Geological Survey, particularly of the politics of science in late nineteenth-century Washington, D.C.

Paul Lucier

GEOPHYSICS

Few practitioners of geophysics—the physics of the earth—worked in the United States throughout the nineteenth century. But the field grew rapidly in the twentieth century, propelled by private foundations (which saw its potential to address major unsolved questions, such as the nature of terrestrial magnetism), the petroleum industry (which valued geophysical methods for revealing difficult-to-detect oil and natural gas reserves), and the federal government (whose long-standing support increased greatly during World War II and the Cold War, when the earth sciences became crucial for national security). By the early twenty-first century, geophysicists (among them climate scientists) were active players in efforts to understand the behavior of physical systems (such as the circulation of ocean waters) on our home planet and throughout the solar system.

Limited Beginnings. Geophysics was not a well-represented research field in the United States in its first century after independence, despite several notable practitioners. Benjamin Franklin did fundamental investigations of the electrical nature of lightning, described the Gulf Stream as an ocean current, and advanced ideas on the role of moist, heated air as a cause for storms and equator-to-pole atmospheric circulation. The self-taught Joseph Henry, who rose from humble beginnings to become the first secretary of the Smithsonian Institution, pursued pioneering research on electricity and magnetism. In the 1830s, William C. Redfield, James Pollard Espy, and Robert Hare advanced competing theoretical interpretations of the nature and causes of storms, and 20 years later Matthew F. Maury, director of the U.S. Naval Observatory and Hydrographic Office, developed theories for the circulation of ocean currents while mapping the deep sea bed of the North Atlantic Ocean.

Yet these were isolated peaks: much U.S. physical sciences research at the time was scattered, limited, and of modest significance. Indeed, throughout much of the nineteenth century, geophysics was a handmaiden of the then broadly encompassing realm of geography, which encompassed such subjects as meteorology, oceanography, terrestrial magnetism, and gravity. Many investigations had utilitarian rather than theoretical ambitions, including careful determinations of latitude, longitude, and variations between local landmarks and magnetic north (crucial for accurate surveying on land and for improved navigation at sea). Closely related concerns led to the founding of the U.S. Coast Survey in 1807, whose first director, the Swiss-born Ferdinand Hassler, undertook geodetic triangulations to accurately map the character of the eastern shore. Although a few late nineteenth-century geophysicists such as George Ferdinand Becker (U.S. Geological Survey) argued that geophysics was ideally suited to elucidate the past history and present conditions of the earth, including the origins of petroleum and ore deposits, past climatic conditions, the globe's elasticity, and its initial formation (cosmogony), relatively few U.S. earth scientists pursued studies of this kind, largely because many, in contrast to their European counterparts, were deficient in mathematics.

Not until the turn of the twentieth century did geophysics gain a substantial institutional toehold in the United States. In part this came about through private philanthropy. The steel magnate Andrew Carnegie endowed the Carnegie Institution of Washington, whose leaders sought to "take possession of the vacant ground between geology and physics and geology and chemistry" by founding a Division of Terrestrial Magnetism, funding studies of solar–terrestrial relations at its Mt. Wilson Observatory, and sending the nonmagnetic ship *Carnegie* on worldwide cruises to explore magnetic variations (University of Wisconsin, 1903). At the same time, Jesuit scholars—who had begun recording seismic disturbances by 1909 to gain acclaim in a culture that celebrated science—began making fundamental contributions to seismology. James B. Macelwane, appointed the first professor of geophysics at St. Louis University, founded the Jesuit Seismological Association, making St. Louis a major center for seismological research. His efforts paralleled new research programs in seismology established in California after the 1906 San Francisco earthquake, including at the University of California, Berkeley, and the California Institute of Technology.

Geophysical research expanded rapidly through the first third of the twentieth century. In 1919, in the aftermath of World War I, the first U.S. professional society for the geophysical sciences (the American Geophysical Union) was established as an affiliate of the National Academy of Sciences—the same year that the first international body (the International Union of Geodesy and Geophysics) was organized. Soon thereafter, program managers at the Rockefeller Foundation's General Education Board, the most influential patron for U.S. science in the interwar years, began supporting research in this field, including a major grant for geophysical research at Harvard University that supported the high-pressure research of later Nobel Laureate Percy W. Bridgman. A separate $3 million Rockefeller Foundation grant in 1930 established the Woods Hole Oceanographic Institution, devoted to physical oceanography. Foundation leaders did so because they felt convinced that interdisciplinary fields offered unusual promise for scientific advances.

A distinct branch of geophysics called applied geophysics—the application of earth sciences instruments and methods (including seismic, gravimetric, and magnetic techniques) to explore for oil and natural gas deposits—emerged in the 1920s. These approaches soon proved fruitful: the Oklahoma-based Geophysical Research Corporation, a division of the Amerada Petroleum Corporation, identified 11 oil-rich geological formations called salt domes along the Gulf Coast in just nine months in 1925, breaking previous records. Universities including the Colorado School of Mines began establishing training programs in exploration geophysics, and new firms began mass-producing specialized geophysical equipment for seismic prospecting, among them Geophysical Service Incorporated (established in 1930), the progenitor of Texas Instruments. Later, geophysicists became involved in efforts to estimate natural resource abundances within the United States (including petroleum, natural gas, and rare minerals), and resource-extraction firms vigorously expanded their use of geophysical methods into the twenty-first century.

Expanding Earth Sciences: World War II, the Cold War, and Beyond.

Although geophysics was burgeoning in the early twentieth century, even as late as 1945, as geophysicists H. W. Straley and M. King Hubbert noted, no North American university offered "an advanced geophysical curriculum embracing the eight geophysical branches recognized by the American Geophysical Union" (*AIME Transactions*, 1945, p. 398). Yet advances in geophysics during World War II soon changed this situation. Physical oceanographers had played a crucial role in aiding the D-Day landing of the Allies in Normandy in 1944 (by accurately predicting surf and swell conditions), and the U.S. geophysicist W. Maurice Ewing, by discovering the "sound channel" in the oceans, enabled the military to use this layer to communicate over great distances. Recognizing the importance of geophysics for emerging Cold War defense needs, Pentagon officials argued that all facets of the earth sciences were vital for U.S. national security: atmospheric conditions and minute variations in Earth's gravity affected the paths of guided missiles, submarine and anti-submarine warfare depended on understanding oceanographic phenomena, and the prospect of altering local environments—weather control— seemed a potentially potent tool. This caused an unprecedented expansion of the earth sciences in the United States, supported by the Office of Naval Research, the newly established Air Force Cambridge Research Laboratory, and other government agencies.

Another result was the creation of broadly conceived new institutes of geophysics connected to U.S. universities, designed to address the new demands for knowledge by various patrons, particularly the U.S. military. New geophysical institutes were established at the University of California, Los Angeles, the University of Alaska at Fairbanks, and Columbia University by 1949; other universities soon followed suit. These institutes were interdisciplinary, trained students to work broadly across component fields of the earth sciences, and addressed the critical shortage of workers in geophysics available to the U.S. government by the early 1950s. Research funding surged further by the early 1960s, as advances in Soviet submarine technology helped convince government leaders to invest more in oceanography.

Heightened state interest in the earth sciences also led them to support efforts to understand worldwide geophysical phenomena (including global atmospheric circulation, which meteorologists regarded as essential for improving the accuracy of weather forecasting). Building on the International Polar Years of 1882–1883 and 1932–1933, American geophysicists helped launch the International Geophysical Year of 1957–1958, the largest international scientific undertaking of the twentieth century, involving thousands of scientists from 67 nations.

President Dwight D. Eisenhower understood the IGY's importance to national security—its planned satellite launches ultimately established the principle of satellite overflight—but the IGY, partly supported by the new U.S. National Science Foundation, also reflected a strong national commitment to inventory Earth's physical phenomena. One result was that American scientists provided crucial data and analysis that led to the theory of plate tectonics, the most significant twentieth-century conceptual advance in the earth sciences. As expeditions by the National Aeronautics and Space Administration (NASA) began reaching the moon and other bodies in the solar system by the 1960s, geophysicists also began placing the "earth sciences" in larger planetary contexts.

The importance of understanding geophysical phenomena for U.S. national security continued well beyond the end of the Cold War, even as definitions of what mattered most shifted over time. Whereas in 1947 Pentagon officials had urged earth scientists to assess polar warming (crucial for assessing Soviet agricultural yields and ice in northern harbors) and the White House recruited geophysicists to assist in negotiating what became the Limited Nuclear Test Ban Treaty (because their expertise was vital for verification), in the post–Cold War years physical environmental scientists became vital players in assessing stratospheric ozone depletion, human contributions to climate change, and related global concerns. U.S. geophysicists remained active in both research and policy issues at the start of the twenty-first century: by 2010 the American Geophysical Union included some 60,000 members, making it the largest professional earth sciences association on the planet.

[*See also* **Bridgman, Percy; Chemistry; Electricity and Electrification; Environmentalism; Franklin, Benjamin; Geography; Geological Surveys; Geology; Global Warming; Henry, Joseph; Instruments of Science; International Geophysical Year; Mathematics and Statistics; Maury, Matthew Fontaine; Meteorology and Climatology; Military, Science and Technology and the; Mining Technology; Missiles and Rockets; National Academy of Sciences; National Aeronautics and Space Administration; National Science Foundation; Nobel Prize in Biomedical Research; Nuclear Weapons; Oceanography; Petroleum and Petrochemicals; Physics; Plate Tectonics, Theory of; Rockefeller Institute, The; Satellites, Communications; Science;** *and* **Societies and Associations, Science.**

BIBLIOGRAPHY

Barth, Kai-Henrik. "The Politics of Seismology: Nuclear Testing, Arms Control, and Transformation of a Discipline." *Social Studies of Science* 33 (2003): 743–781. On the role seismologists played in the 1960s as the Limited Nuclear Test Ban Treaty was negotiated.

Bowler, Peter J. *The Earth Encompassed: A History of the Environmental Sciences.* New York: W. W. Norton, 1992. One of the best general histories of the environmental sciences, although not primarily focused on the United States.

Doel, Ronald E. "Constituting the Postwar Earth Sciences: The Military's Influence on the Environmental Sciences in the USA after 1945." *Social Studies of Science* 33, no. 5 (2003): 635–666. An analysis of how national security concerns influenced the development of the physical environmental sciences during the Cold War.

Doel, Ronald E. "The Earth Sciences and Geophysics." In *Science in the Twentieth Century,* edited by John Krige and Dominique Pestre, pp. 361–388. London: Harwood Academic Publishers, 1997. An overview of the intellectual and institutional growth of the earth sciences, including in North America.

Dupree, A. Hunter. *Science in the Federal Government: A History of Policies and Activities.* Baltimore: Johns Hopkins University Press, 1986. Paperback edition of classic work that explores the role of science, including the earth sciences, within the American state.

Fleming, James Rodger. *Meteorology in America, 1800–1870.* Baltimore: Johns Hopkins University Press, 1990. A broad overview of the practice of early American meteorology.

Geschwind, Carl-Henry. *California Earthquakes: Science, Risk, and the Politics of Hazard Mitigation.* Baltimore: Johns Hopkins University Press, 2001. How local natural phenomena propelled the rise of a key branch of geophysics.

Good, Gregory A., ed. *Sciences of the Earth: An Encyclopedia of Events, People, and Phenomena.* New York: Garland Publishing, 1998. An ambitious and comprehensive source for a vast range of geophysical topics, with extensive coverage of North America.

Harper, Kristine C. *Weather by the Numbers: The Genesis of Modern Meteorology.* Cambridge, Mass.: MIT Press, 2008. Examines the practice of modern American meteorology and the development of numerical weather prediction.

Livingstone, David. *The Geographical Tradition: Episodes in the History of a Contested Enterprise.* Cambridge, Mass.: Wiley–Blackwell, 1993. A wide-ranging and insightful account of the emergence of geography as a field, including areas of research now considered branches of the earth sciences.

Oreskes, Naomi. *The Rejection of Continental Drift: Theory and Method in American Earth Science.* New York: Oxford University Press, 1999. Explores how commitment to particular computational approaches in geodesy helped persuade North American earth scientists to reject the possibility of drift through the mid-twentieth century.

Oreskes, Naomi, and Ronald E. Doel. "Physics and Chemistry of the Earth." In *The Cambridge History of Science,* Vol. 5, *Modern Physical and Mathematical Sciences,* edited by Mary Jo Nye, pp. 538–552. New York: Cambridge University Press, 2002. Addresses factors and ideas that influenced the emergence of the modern earth sciences.

Reingold, Nathan. *Science in Nineteenth Century America: A Documentary History.* Chicago: University of Chicago Press, 1964. Includes lengthy introductory sections that in part analyze the geophysical tradition in America.

Straley, H. W., and M. K. Hubbert et al. "The Professional Training of Geophysicists: Report of the Geophysical Educational Committee, Mineral Industry Education Division." *AIME Transactions* 164 (1945): 397–420.

Van Hise, Charles R. "Report of the Adviser for Geophysics to the Executive Committee of the Carnegie Institution of Washington." Box 1, General Files, C. K. Leith Collection, Division of Archives. Madison: University of Wisconsin–Madison, n.d. (circa October 1903).

Ronald E. Doel

GERM THEORY OF DISEASE

The phrase "germ theory of disease" first came into widespread use among English-speaking scientists and physicians in the 1870s and 1880s to designate two related propositions: first, that living micro-organisms were the cause of infectious diseases, and second, that these micro-organisms always arose from a previous case of the disease. First advanced as a speculative theory in the late 1860s, it was vigorously debated among European and American scientists during the 1870s and 1880s. Largely because of increasingly strong laboratory proofs offered in its defense, American physicians accepted the germ theory as scientific truth by the 1890s. Bacteriology, the new scientific discipline it inspired, became a powerful force in American public health by the turn of the century.

In early modern Europe, writers such as the Italian Girolamo Fracastoro in the fifteenth century and the German Athanasius Kircher in the seventeenth century speculated that living beings too small to see with the naked eye might be the source of epidemic disease, but as of the mid-1800s, this so-called animacular hypothesis had no scientific credibility. Most medical authorities believed complex interactions of air, water, and soil conditions produced chemical "ferments" of disease that then spread from person to person. When in the late 1850s the French chemist Louis Pasteur discovered that the ferments responsible for sour beer and wine were living organisms, he suggested that the same might be true of human and animal diseases. Intrigued by his findings, a broad circle of scientists began to investigate the role of micro-organisms, variously referred to as microbes, microzymes, cryptograms, and germs. By the 1870s, the phrase germ theory of disease had come into general use to refer to the work of not only Louis Pasteur but also the Scottish surgeon Joseph Lister, the German physician Jacob Henle and his student Robert Koch, and the English physicist John Tyndall. On a lecture tour to the United States in 1870, John Tyndall played an especially important role in popularizing the germ theory to an American audience.

In both the United States and Europe, many careful observers of human disease found both the germ theory and the experimental evidence offered to support it unconvincing. Dissenters found the emphasis on specific micro-organisms far too simplistic to account for the complexity of disease patterns. Medical and scientific journals in the 1870s and 1880s carried heated debates about the new theory. The growing sophistication of experimental methods slowly built up acceptance of the germ theory of disease. The leader in this regard was Robert Koch, who developed superior methods for culturing and isolating bacterial disease agents, staining and fixing them for microscopic analysis, and using animal subjects to verify their causal role in producing disease. Using these methods, Koch isolated the cause of anthrax in 1876, wound infection in 1877, tuberculosis in 1882, and cholera in 1884. Using similar methods, other researchers isolated the causes of gonorrhea in 1879, typhoid in 1880, and syphilis in 1905.

Although the American medical press followed the European developments, Americans had little direct role in the early proofs of the germ theory. But starting in the 1870s, young researchers began to go to Germany, France, and England to be trained in the new methods and then returned to teach them in the United States. Koch's methods, learned directly from him or by reading accounts of his work, were especially influential. The essence of the new bacteriology was reflected in Koch's postulates, which laid out a series of steps necessary to prove a specific micro-organism was responsible for a specific disease: first, researchers had to find the micro-organism in a sick animal but not a well one; isolate and purify a culture of that micro-organism; produce the same symptoms by injecting the culture into a well animal; and then show that the second generation of micro-organism was identical to the first. After 1900, Americans played a more visible role in microbiological research, particularly in establishing the importance of insect vectors such as flies and mosquitoes, as well as the existence of healthy human carriers such as the famous Typhoid Mary Mallon, who was first arrested and quarantined in 1907.

By the late 1890s, scientists already suspected the existence of micro-organisms too small to be seen with even the best microscopes. Referred to as "filterable viruses" because superfine filters could capture them, they only became visible with the invention of the electron microscope in the late 1930s. But using Koch's postulates, researchers were able to isolate the viral agents responsible for yellow fever (1901), rabies (1903), smallpox (1906), polio (1908), and influenza (1933). Reflecting the growing recognition of viruses and other nonbacterial disease agents, by the late 1930s, the term microbiology gradually replaced bacteriology as the discipline's preferred designation.

From the 1890s to the 1930s, the germ theory remained a better foundation for diagnosis and prevention than it did for treatment. With the sole exception of Paul Ehrlich's salvarsan to treat syphilis, researchers repeatedly failed in early efforts to develop effective antimicrobial drugs until the discovery of the sulfa drugs in the late 1930s, followed by penicillin in the 1940s and broad-spectrum antibiotics in the 1950s. Antiviral drugs remain limited in their capacity to cure diseases in the early twenty-first century. Researchers were more successful in early efforts to develop vaccines that conferred immunity against infection. Paradoxically, inoculation and vaccination for smallpox predated acceptance of the germ theory. By the early 1900s, there was also a diphtheria antitoxin and a typhoid vaccine.

The germ theory probably had its greatest impact through the new scientific legitimacy it conferred on late nineteenth-century public-health departments, which were struggling to expand their powers. Despite their concern that it was too simplistic, the older generation of sanitarians realized that the germ theory could be used to reinforce their efforts to improve municipal sanitation. The theory's advocates also conceded that exposure alone did not determine infection and that lifestyle and heredity could make some individuals and groups more susceptible to disease than others. In cities struggling with unprecedented rates of poverty and immigration, the new bacteriology provided a template for urban cleanliness. Even before American laboratories began to contribute original research, Americans gained notice for their practical uses of the new germ theory, for example, in the work of the German-trained Hermann Biggs in New York City at the turn of the century. Through Progressive-Era public-health campaigns and school programs, the importance of exacting personal, household, and neighborhood cleanliness became an important facet of American culture. Coinciding as it did with the early years of American empire, the acceptance of germ theory and bacteriology also shaped American policy in Cuba, Puerto Rico, and the Philippines.

In many ways, the fear of germs helped deepen existing patterns of prejudice against the poor, immigrants, and people of color. In the United States, the system of racial segregation known as

Jim Crow built upon heightened germ fears. At the same time, the argument that "germs know no color or class line" inspired reformers' efforts to improve living and working conditions for the poor. The Progressive-Era anti-tuberculosis movement played a critical role in raising awareness not only of the "white plague" but also of the social inequalities that produced it. Last but not least, the commercial invocations of germs and their dangers through product advertising played a critical role in spreading new standards of personal and public hygiene.

By the late 1920s, public-health measures had succeeded in significant reductions in the death rates from infectious diseases, so that heart disease and cancer had replaced tuberculosis as the leading causes of death in the United States. Infectious diseases by no means lost their capacity to terrify, as evident in the World War I influenza epidemic, in which an estimated 675,000 Americans died, as well as the smaller but still frightening outbreaks of infantile paralysis, also known as polio. With the introduction of antibiotics and polio vaccines after 1950, many Americans felt that the problem of infectious diseases had been solved, but that complacency would be destroyed by the AIDS epidemic in the 1980s.

[*See also* **Disease; Medicine: From the 1870s to 1945; Medicine: Since 1945;** *and* **Public Health.**]

BIBLIOGRAPHY

Anderson, Warwick. *Colonial Pathologies: American Tropical Medicine, Race, and Hygiene in the Philippines.* Durham, N.C.: Duke University Press, 2006. Groundbreaking reconsideration of American public health in general and bacteriology in particular from the standpoint of U.S. imperial history.

Leavitt, Judith Walzer. *Typhoid Mary: Captive to the Public's Health.* Boston: Beacon Press, 1997. Masterful treatment that explores both ethnic and gendered dimension of the famous case of Mary Mallon.

Oshinsky, David M. *Polio: An American Story.* New York: Oxford University Press, 2006. Broad history of one of the most influential twentieth-century epidemic diseases.

Roberts, Samuel Kelton, Jr. *Infectious Fear: Politics, Disease, and the Health Effects of Segregation.* Chapel Hill: University of North Carolina Press, 2009. Ambitious, insightful account of the links between public-health science and racial segregation.

Tomes, Nancy. "American Attitudes toward the Germ Theory of Disease: Phyllis Allen Richmond Revisited." *Journal of the History of Medicine* 52, no. 1 (January 1997): 17–50. A focused account of early medical debates over the germ theory in the United States compared to Europe.

Tomes, Nancy. *The Gospel of Germs: Men, Women, and the Microbe in American Life.* Cambridge, Mass.: Harvard University Press, 1998. A broadly framed account of the scientific, social, and cultural reception of the germ theory of disease in the United States.

Nancy Tomes

GERONTOLOGY

In 1903, in his book *The Nature of Man: Studies in Optimistic Philosophy,* Elie Metchnikoff stressed the need for a new field that he termed "gerontology." Its focus, he asserted, would be the scientific study of old age. Before Metchnikoff's declaration, others throughout the nineteenth century had worked on defining the stage of life through a variety of scientific approaches: individuals such as Adophe Quetelet, Francis Galton, and the American physician George Beard had employed statistics to designate the start of senescence; the Paris school of medicine, and especially Marie François Xavier Bichat, had examined changes in the tissue to define the aging process. Metchnikoff's declaration was then followed in 1909 by Austrian physician I. L. Nascher, who called for a medical specialty—"geriatrics"—that would be devoted to the diseases of old age.

Throughout the early twentieth century, however, the negative connotations linking aging to death and the difficulties of separating normative aging from pathological processes limited the response to these appeals. In 1922, the American physician G. Stanley Hall added to the calls for a scientific study of old age; his work *Senescence* underscored the relationship between old age and degeneration.

Beginning in the 1930s, gerontology started to develop roots within the American scientific community. In 1937, E. V. Cowdry invited a group of 20 highly regarded experts to a conference at Woods Hole, Massachusetts, to examine the current state of knowledge about old age. The resulting 1939 volume, Cowdry's *Problems of Ageing*, presented aging through a multidisciplinary approach, although as John Dewey's introduction made clear, the goal was to create a field linked with a common language. The book explored the relationship between aging and degeneration; it attempted to define the natural process of aging; it examined the question of aging as a "problem" as well as reached beyond the biological constructs of old age to consider its social, psychological, and cultural dimensions. Yet, by its focus, and as reflected in the title, it continued to view aging as a problem defined by debility and poverty and best solved through medical and professional intervention.

This conception of aging reflected both the social and the medical realities of the era. In the early twentieth century, as infectious diseases were eliminated and public-health measures affected morbidity, the leading nineteenth-century causes of death—pneumonia and flu, tuberculosis, and intestinal disease—decreased and were replaced by heart disease, cancer, and stroke. The life expectancy of the Americans began to rise significantly. In 1900 life expectancy at birth was 49.2 years; in 1940 it was 63.6; the number of aged in the population almost tripled from 3.1 million to 9 million (Shrestha, 2006, p. 3). At the same time, government studies, such as the 1934 publication *The Need for Economic Security in the United States*, seemed to prove that poverty among the old was a widespread and rapidly growing problem. Social advocates joined medical experts and government officials in their call for Social Security and other measures for support of the aged.

The growing postwar awareness and concern for the study of aging was then formalized through the first professional organization, the Gerontological Society of America. Founded in 1945, the society's mission was to "promote the scientific study of aging." Reflecting the multiplicity of interests, in 1952 it divided into four sections: biological sciences; psychological and social sciences; social work and administration; and health societies. In 1949, the International Association of Gerontology was established and the U.S. Public Health Service's Section on Aging, which had been created in 1941, became the National Institute on Aging in 1975.

This developing institutional interest in gerontology was tied to the rapidly expanding numbers of the elderly in the United States and awareness of future demographic trends. By 2000, average life expectancy had risen to 77.5 years, with the added years coming at the end of the life cycle; the number of aged people had increased to nearly 35 million. Individuals over 85 years of age, the "old-old," became the fastest growing segment of the population (Scommegna, 2004).

These growing numbers and the diversity of the elderly population had an impact upon the perspective and definition of gerontology. The early classification of aging as a problem to be solved and the postwar debate over what constituted normative aging have generally been replaced by a life-course and cross-cultural perspective. Aging, for gerontologists, is a process rather than an end; it is an integral part of life-span development. Although this approach may mean that gerontology does not constitute the distinct and separate "science" envisioned by Metchnikoff, it has evolved over the century into a broad, multidisciplinary field dedicated to research, service, and education. As the National Institute of Aging stated in 1986, gerontology can perhaps best be defined as the "study of aging from the broadest perspective."

[*See also* **Alzheimer's Disease and Dementia; Cancer; Cardiology; Death and Dying; Disease; Hall, G. Stanley; Influenza; Life Expectancy; Medicare and Medicaid; Medicine; National Institutes of Health; Public Health; Public Health Service, U.S.; Science; Social Sciences;** *and* **Tuberculosis.**]

BIBLIOGRAPHY

Achenbaum, W. Andrew. *Crossing Frontiers: Gerontology Emerges as a Science.* New York: Cambridge

University Press, 1995. Examines the establishment of gerontology in the United States in the twentieth century. The author finds that gerontology did not become the science envisioned by its early advocates; rather, its multidisciplinary approach challenged traditional specialization and boundaries.

Achenbaum, W. Andrew, and Levin, Jeffrey S. "What Does Gerontology Mean?" *The Gerontologist* 29, no. 3 (1989): 393–400. Argues that although gerontology has continued to grow over the half decade, given its multiple agendas and approaches, little consensus has emerged on its definition or scope.

Ackerknecht, Edwin H. *Medicine at the Paris Hospital, 1794–1848.* Baltimore: Johns Hopkins University Press, 1967. Focusing on the key figures in France in the early nineteenth century, this study examines the development of modern medicine through a new pathological approach.

Cowdry, E. V., ed. *Problems of Ageing: Biological and Medical Aspects.* Baltimore: Williams and Wilkins Co., 1939. This classic gerontological study introduces an interdisciplinary approach to the study of old age. This work brings together the leaders in the field in the 1930s.

Gerontological Society of America. "History." https://www.geron.org/about-us/history (accessed 16 October 2012).

Haber, Carole. *Beyond Sixty-Five: The Dilemma of Old Age in America's Past.* New York: Cambridge University Press, 1983. Looks at the establishment of gerontological medicine, old age, homes, and pensions in the nineteenth and early twentieth centuries based on social and economic changes that affected the old.

Hall, G. Stanley. *Senescence: The Last Half of Life.* New York: D. Appleton & Co., 1922. After writing the key work on adolescence, Hall turns to examine old age and calls for experts to devote attention to the last stage of life. The work, however, looks at senescence largely in terms of decay and decline.

Metchnikoff, Elie. *The Nature of Man: Studies in Optimistic Philosophy.* New York: G. P. Putnam's Sons, 1903. Looks at disease, old age, and death and argues that the infirmities of old age can be overcome through "phagoctyes" that destroy debilitating microbes in the body.

Nascher, I. L. "Geriatrics." *New York Medical Journal* 90, no. 8 (August 1909): 428–429. Nascher names the field of geriatrics and calls for physicians to concentrate on the disease of the old.

Scommegna, Paola. "U.S. Growing Bigger, Older, and More Diverse." *Population Reference Bureau,* 2004, n.p. http://www.prb.org/Articles/2004/USGrowingBiggerOlderandMoreDiverse.aspx (accessed 16 October 2012). Examines changes in the U.S. population since 1950, especially in terms of the elderly.

Shrestha, Laura B. "Life Expectancy in the United States." CRS Report for Congress. 2006. http://passitonwell.com/wordpress/wp-content/uploads/downloads/2012/11/expectancyus.pdf (accessed 16 October 2012). Traces life expectancy in the United States and provides useful charts on the changes.

Carole Haber

GIBBS, JOSIAH WILLARD

(1839–1903), foremost mathematical physicist in nineteenth-century America. Born in New Haven, Connecticut, the son and namesake of a professor of philology at Yale College, Gibbs was the first of a new breed of physicists in America who were also highly trained in advanced mathematics. He received the first engineering PhD awarded by an American institution, the Sheffield Scientific School at Yale, in 1863. Gibbs worked first as an unpaid tutor at Yale for three years and then took a three-year intellectual tour of Europe. He studied physics and mathematics in Paris, Berlin, and Heidelberg. Gibbs returned to his beloved New Haven in the summer of 1869, where he resumed his position at Yale. He became professor of mathematical physics at Yale in 1871, remaining in that position until his death.

Equally at home parsing Latin and doing mathematics, Gibbs became interested in many different areas of physics and mathematics. He conducted research in engineering and mechanics and spent the years from 1873 to 1876 studying thermodynamics. Gibbs also studied vector analysis and statistical mechanics, as well as the electromagnetic theory of light. He is attributed with laying much of the theoretical foundation for thermodynamics and the field of physical chemistry. The "Gibbs phase rule"—specifying the number of solids, liquids, and gases present in complex chemical processes at equilibrium—established the significance of thermodynamics in determining the direction of chemical reactions and in uncovering chemical equilibrium. Gibbs published his main contribution to physics in the obscure pages of the *Transactions of the Connecticut Academy of Arts and Sciences:* "On the Equilibrium of Heterogeneous Substances" (1876–1878), a two-part work that became the cornerstone of the field of physical chemistry.

James Clerk Maxwell, the great Scottish physicist, early recognized Gibbs's brilliance and alerted chemists to the implications of his research. Gibb's work became well known in Europe, where he was elected to and received honors from major physical and mathematical societies. His impact on the scientific community in the United States at first was smaller; American scientists were less aware of the importance of physical theory. A full appreciation of his achievement came only posthumously. It took a new generation of industrial and pharmaceutical chemists wielding the Gibbs phase rule to rediscover and champion his contributions. Because he made no assumptions about the ultimate nature of matter and based his system on a strictly phenomenological treatment of substances, his work avoided later controversies about the interpretation of thermodynamic functions and the statistical nature of quantum theory.

[*See also* **Chemistry; Mathematics and Statistics; Physics; Quantum Theory;** *and* **Science.**]

BIBLIOGRAPHY

Garber, Elizabeth A. "Gibbs, Josiah Willard (1839–1903)." In *The History of Science in the United States: An Encyclopedia,* edited by Marc Rothenberg, pp. 242–243. New York and London: Garland Publishing, 2001.

Seeger, Raymond J. *Men of Physics: J. Willard Gibbs: American Mathematical Physicist Par Excellence.* Oxford and New York: Pergamon Press, 1974.

Wheeler, Lynde Phelps. *Josiah Willard Gibbs: The History of a Great Mind.* New Haven, Conn.: Yale University Press, 1952.

David A. Tomlin;
updated by Elspeth Knewstubb and
Hugh Richard Slotten

GLOBAL WARMING

The late twentieth century saw global warming emerge as an imminent, deleterious environmental phenomenon. However, one hundred years earlier, Swedish physical chemist Svante Arrhenius had argued that carbon dioxide released during fossil fuel combustion might warm the atmosphere by absorbing long-wave radiation from the earth and reradiating it back—moderating harsh winters and producing a longer growing season for his fellow Scandinavians. Others, including Smithsonian Institution secretary Charles Greeley Abbot, thought atmospheric water vapor was a more likely source of global warming, but a lack of data and instruments precluded extensive scientific investigations into climate change.

Early twentieth-century climatologists—working under geography's disciplinary umbrella—were concerned with classifying Earth's perceived stable climate regimes. Meteorologists were trying to put their discipline on a firmer physics-based footing to produce valid short-term weather forecasts. The climate was static for them as well. Then, in 1938, British steam engineer and amateur meteorologist Guy Stewart Callendar stepped into this static climate milieu and argued that the thousands of tons of carbon dioxide being pumped into the atmosphere every minute by the burning of fossil fuels were substantially changing the chemical composition of Earth's atmosphere, were responsible for 60 percent of the $0.9°F$ ($0.5°C$) global temperature increase in the preceding one hundred years, and would be responsible for future temperature increases. And warming was seen as good: reducing winter heating bills, preventing glacial growth, and lengthening the planting season in higher latitudes, whereas the extra carbon dioxide would help plants grow larger.

Meteorologists were not impressed, disputing the accuracy of nineteenth-century carbon dioxide measurements and accusing Callendar of neglecting the global circulation's dispersion of pollutants. But Callendar refused to back down, and by 1941 he convinced meteorologists and climatologists that carbon dioxide's heat absorption properties were far more important than previously thought, outstripping the influence of water vapor on global climate.

By the late 1940s, meteorologists were still not very concerned about global warming, but the U.S. military was. The apparent warming on both sides of the Atlantic and glacial retreat on Iceland gained the attention of Pentagon staffers concerned that Arctic ice might break up, allowing Soviet vessels easier access to North America—a national security concern.

The International Geophysical Year (IGY, 1957–1958) provided an opportunity to systematically measure carbon dioxide levels. An infrared gas analyzer was installed at the Mauna Loa Observatory on the "Big Island" of Hawai'i—a clear-air site far removed from industrial pollution. Charles Keeling, a young atmospheric chemist from the Scripps Institution of Oceanography in San Diego, California, collected and analyzed data from both Mauna Loa and Antarctica, finding that carbon dioxide concentrations increased during the 18-month research window (and exhibited a significant 3% seasonal change between spring and fall as vegetation leafed out, lived, and died). Having confirmed Callendar's earlier results, the data-gathering effort continued post-IGY and resulted in the now-familiar saw-tooth Keeling curve showing the steady rise of carbon dioxide levels in the atmosphere, which spurred additional climate change research.

The World War II–assisted increase in observation stations, coupled with the availability of electronic digital computers, allowed climate researchers to analyze global temperature data more efficiently starting in the 1960s. They found that temperature fluctuations, both up and down, had been occurring since at least the 1940s, but focused on the overall upward trend. Global weather forecast models were still in their infancy, and climate models would need to await both greater computing power and better knowledge of atmospheric processes.

Evidence of regional warming, particularly in high latitudes, started to mount in the early 1980s. Although studies pointed to warming, there was no agreement on how much and how fast the temperature would rise, nor was there agreement on societal impacts of global warming. Unable to predict the future, paleoclimatologists looked back

in time using mud and ice cores as climate proxies, finding that carbon dioxide levels had been as low as 200 ppm during the ice age and 275 ppm before the Industrial Revolution. (By the early 1980s the level was 331 ppm, and it reached 396 ppm in 2012.) The proxy temperature data could then be used to test the effectiveness of climate models to predict the future.

By the late 1980s the controversy surrounding the effect of carbon dioxide concentrations on global warming was less about the scientific mechanism than about warming's extent and possible amelioration. The arguments often boiled down to the accuracy of climate models that rarely produced the same answer and would only verify long after their creators and users had died. Their inconsistency provoked disputes, particularly when policy decisions based on model output would affect present-day economics and societal norms. These concerns triggered international political interest in global warming and led the World Meteorological Organization and the United Nations Environmental Programme to establish the Intergovernmental Panel on Climate Change (IPCC), which was tasked with assessing "the scientific, technical, and socioeconomic information relevant for the understanding of the risk of human-induced climate change." After analyzing extant published research, IPCC scientists concluded in 1990 that the global temperature was increasing, possibly because of natural processes, and that it might take another 10 years to determine the influence of human activity. By 1994, they concluded that greenhouse gases were playing the most important role in climate change, and in 1995—the year of a record-setting global temperature—they reported that human activity was a definite influence.

The Fourth IPCC Assessment (2007) reported that the "warming of the climate system is unequivocal" and "most of the observed increase in global average temperatures since the mid-20th century is very likely due to the observed increase in anthropogenic greenhouse gas concentrations." By then, the global temperature had been broken once again, in 2005 (2010 tied with 2005). The Fifth Assessment Report, due to be completed in 2014, was expected to update the knowledge of climate change and the concomitant implications for sustainable development. In the meantime, arguments over global warming and how to address it have continued to focus on political, economic, social, and ethical considerations that affect all of Earth's inhabitants.

[*See also* **Computers, Mainframe, Mini, and Micro; Environmentalism; Environmental Protection Agency; Geography; Geophysics; International Geophysical Year; Meteorology and Climatology;** *and* **Scripps Institution of Oceanography.**]

BIBLIOGRAPHY

Edwards, Paul N. *A Vast Machine: Computer Models, Climate Data, and the Politics of Global Warming.* Cambridge, Mass.: MIT Press, 2010. How scientists use models to understand atmospheric processes and global warming.

Fleming, James Rodger. *Historical Perspectives on Climate Change.* New York and Oxford: Oxford University Press, 1998. An examination of how perceptions of climate changed from the late nineteenth century through the mid-twentieth century and an analysis of the proposed causes of climate change and their acceptance or rejection by the scientific community.

Intergovernmental Panel on Climate Change. http://www.ipcc.ch (accessed 18 October 2012). Includes copies of all IPCC documents and reports and links to sites with more information on climate change.

National Climatic Data Center. http://www.ncdc.noaa.gov/oa/ncdc.html (accessed 18 October 2012). Current and archived climate data from all over the world, as well as information on weather and climate events.

Schneider, Stephen H. *Global Warming: Are We Entering the Greenhouse Century?* San Francisco: Sierra Club Books, 1989. An informative, nontechnical book by the late climate scientist explaining the global warming controversy.

Singer, S. Fred. *Hot Talk, Cold Science: Global Warming's Unfinished Debate.* Oakland, Calif.: Independent Institute, 1997. A contrarian view of global warming by meteorologist Singer, a global warming skeptic.

Weart, Spencer R. *The Discovery of Global Warming.* 2d ed. Cambridge, Mass.: Harvard University Press, 2008. Background and current status of the global warming debate; the companion website contains additional, up-to-date information.

Weart, Spencer R. *The Discovery of Global Warming.* Center for History of Physics, American Institute of Physics. http://www.aip.org/history/climate/index.htm (accessed 18 October 2012). The continuously updated, fully sourced hypertext version of Weart's book, which contains many scholarly resources for researchers.

Kristine C. Harper

GODDARD, ROBERT H.

(1926–1945), physicist and rocket enthusiast. In 1926 he launched what was probably the first liquid-propellant rocket to fly successfully. Earlier, he had written a paper entitled "A Method of Reaching Extreme Altitudes" that the Smithsonian Institution published in 1920. It was later to have great influence on those interested in developing rockets for space travel. As a result of this paper, his 1926 launch, and his later highly inventive work on rockets, he is widely and correctly regarded as one of the three principal pioneers of rocketry and spaceflight.

Goddard virtually invented cooling for the combustion chamber, controlling the rocket through movable vanes in the exhaust, pumps that were light in weight to force the propellants into the combustion chamber, igniters for the propellants, and injectors to atomize and mix the propellants for proper ignition, among other technologies. But because he did not want his discoveries to be used by others, he published few details of his work during his lifetime.

In 1916 he had applied to the Smithsonian for funding of his research, predicting that his rockets would reach 100 to 200 miles (161 to 322 kilometers) in altitude for scientific research. Between then and 1941, he received over $200,000 in funding from a variety of civilian sources, but the highest altitude any of his rockets ever reached was an estimated 8,000 to 9,000 feet (2,438 to 2,743 meters), well short of his goal. Part of the reason was his working with a small number of technicians in secrecy rather than with other rocket engineers.

Equally importantly, he failed to test his innovative designs systematically. Educated as a physicist, he ignored the standard practices of engineers. By the time details of his innovations were published in 1948, other rocket developers had surpassed his achievements. Although his research was significant, it influenced rocket development less than it could have. However, the National Aeronautics and Space Administration's Goddard Space Flight Center was named after him and his 214 patents made his widow a wealthy woman.

[*See also* **Engineering; Missiles and Rockets; National Aeronautics and Space Administration; Physics; Smithsonian Institution; Space Program;** *and* **Space Science.**]

BIBLIOGRAPHY

Clary, David A. *Rocket Man: Robert H. Goddard and the Birth of the Space Age.* New York: Hyperion, 2003.

Hunley, J. D. "The Enigma of Robert H. Goddard." *Technology and Culture: The International Quarterly of the Society for the History of Technology* 36, no. 2 (April 1995): 327–350.

J. D. Hunley

GOODYEAR, CHARLES

(1800–1860), the American inventor of vulcanization, a process rendering rubber impervious to acids, bases, and heat and controlling the level of elasticity. His innovation made possible the first widely available thermoset plastic.

Goodyear became a professional inventor after his family's hardware store went bankrupt. He enrolled patents for items including improved molasses gates, before becoming fascinated with rubber around 1834. The American rubber industry, centered in New England and New York, began around 1800. It was a high-profit, boom-and-bust industry. Many manufacturers hired inventors to solve rubber's well-known problems.

Goodyear experimented with chemical and mechanical methods, including a partially effective

acid–gas process. He purchased Nathaniel Hayward's patent for metallic rubber (a surface cure using sulfur and sunlight) and probably discovered in 1838–1839 that heating this compound in a stove cured the rubber thoroughly. The exact details of when, what, how, and by whom this discovery was made fueled legal battles throughout Goodyear's life. Goodyear spent five years learning to control the process (heating a mixture of rubber, sulfur, and white lead) before receiving a patent in 1844. The patent was reissued twice, in 1849 and 1860. Both times the reissue broadened the definition of the invention, encompassing developments made by others. Horace Day, already in dispute with Goodyear over machinery for making elasticated textiles, owned one of the patents rendered invalid by the 1849 reissue.

A shifting coalition of manufacturers and investors supporting Goodyear vied with a similar coalition supporting Day in the courts throughout the remainder of Goodyear's life. In the notorious "Great India Rubber Case," in 1852, Daniel Webster spoke for Goodyear. The courts found in Goodyear's favor in this and most of the more than one hundred related lawsuits. The validity of the patent itself was never directly tested. Goodyear's coalition gained a patent extension in 1858 on the grounds of Goodyear's poverty (which was debatable). The second reissue of the patent in 1860, on the inventor's death, broadly claimed vulcanization as a whole, including its effect on gutta percha, a substance unknown to Euro-America in 1839.

[*See also* **Chemistry; Machinery and Manufacturing;** *and* **Plastics.**]

BIBLIOGRAPHY

Goodyear, Charles, and Thomas Hancock. *A Centennial Volume of the Writings of Charles Goodyear and Thomas Hancock.* Boston: American Chemical Society, 1939.

Guise-Richardson, Cai. "Redefining Vulcanization: Charles Goodyear, Patents, and Industrial Control, 1834-1865." *Technology & Culture* 51 (2010): 357–387.

Cai Guise-Richardson

GOULD, STEPHEN JAY

(1941–2002), evolutionary biologist, historian of science, and science popularizer. Gould's interests were both broad and deep. A polymath and polyglot, he ranged across numerous theaters of human knowledge and culture. Especially because his works in the history of the human and biological sciences, including *The Mismeasure of Man*, attracted huge academic and lay audiences, he could be considered one of the most influential historians of science (the first being Thomas Kuhn). His talents as a writer and lecturer for a lay audience earned him the sort of intellectual celebrity generally reserved for literary figures. In addition to 25 books, Gould wrote three hundred essays over a 25 year period for a monthly popular science column in *Natural History Magazine*. He logged more than five hundred scholarly articles, with his most signal contributions to evolutionary science concerning stochastic models, morphology, and speciation, and he became especially known for the theory of punctuated equilibria. He was also a champion of underdog theories, most dramatically via his early support for the theory that an asteroid caused the Cretaceous-Tertiary (K-T) mass extinction.

Gould was born and died in New York City, but spent his entire professional career in Cambridge, Massachusetts. In 1963, he graduated from Antioch College in Yellow Springs, Ohio, whose integrated work-study curriculum placed him as an undergraduate with subsequent graduate advisor Norman Newell at Columbia University and the American Museum of Natural History. Hired upon completion of his PhD in 1967, Gould quickly became one of Harvard University's youngest full professors, earning the status in 1973.

In 1972, with Niles Eldredge, Gould originated punctuated equilibria, a theory on the origin of species. Initially scorned, it soon took root and became widely held, without significant opposition. In essence, the theory holds that stasis is the backdrop against which evolution takes place, with significant morphological change occurring rarely, rapidly, and in small, local populations, which then grow and migrate to larger home ranges. His theory thus predicted a fossil record that accurately tracks the history of species in any given area—a pattern of equilibrium punctuated by an abrupt transition to a daughter species that quickly achieves its own equilibrium.

This macroevolutionary theory draws on Ernst Mayr's model of peripatric speciation. According to this model, the larger the population, the more subpopulations form at the periphery of its home range. Most subpopulations go extinct or are reabsorbed by the parent population. However, on rare occasions, a geographically isolated daughter population remains genetically isolated long enough to speciate. With detailed analyses from paleontology, developmental biology (e.g., heterochrony), European-style morphological theory (e.g., allometry), and population genetic models informed by the then-recent discoveries of massive hidden variation in the genomes of morphologically stable species (e.g., pioneering work on electrophoresis by Richard Lewontin and Jack Hubby), Eldredge and Gould concluded that such speciation would happen rapidly because of changes in selective regime concomitant to the new environment the subpopulation was occupying and, especially, because of intense genetic drift (including founder effect, bottlenecks, and straight generational drift). As the new species adapted to its niche and began to expand its territory, the effect of drift would become less dramatic and the species would settle into morphological stasis. The newly evolved species would migrate into areas suiting its adaptations. The geographic area in which the actual speciation event took place became, then, only a tiny fraction of the species' eventual full range, and the brief geological moment of its birth was dwarfed by the species' subsequent history. Finding the specific small region and brief instant of a species' birth would be exceedingly rare, whereas finding areas in which the geological record recorded the sudden in-migration of the new species to a previously unoccupied area would be quite common. This evolutionary model contrasts with phyletic gradualism, which expects slow, gradual transformation of large populations over long stretches of geologic time.

Gould's intellectual orientation drove him to confront what he considered dire fallacies associated with evolution in both the broader culture and the scientific work of his colleagues and predecessors, namely notions of biological progress, determinism, gradualism, and adaptationism. As a result, much of his most influential work engages directly with these misconceptions to wit, punctuated equilibria counters gradualism; the "Spandrels of San Marco" article (co-authored with Lewontin, with some four thousand scholarly citations) argues against adapationism; his book *Wonderful Life* contradicts the narrative of progress over the course of evolutionary history; and his widely read *The Mismeasure of Man* is a stinging rebuke of biological determinism. This last book also resonates most directly with his left-leaning political views, which were motivated especially by his thoroughgoing antiracism, itself a heritage from his Marxist father as well as a deeply held personal philosophy. As a student at Antioch, he participated in direct civil rights activism. He later adopted a more scholarly approach to social justice, using his facility with the tools of science (especially statistics and evolutionary biology) to demonstrate the vacuity of purportedly scientific claims used to justify cultural views, social practices, and policy strictures that functioned to thwart individual achievement because of race, class, and IQ.

Gould also took on a prominent role defending evolutionary biology from attacks by creationists—perhaps most dramatically as an expert witness, in 1981, on the part of the plaintiffs in the creationism trial *McLean v. Arkansas Board of Education*. Gould considered himself an agnostic and advocated mutual respect between science and religion. His analysis concluded that these two fields of human endeavor were best regarded as "nonoverlapping magisteria," with each asking different questions and supplying fundamentally different kinds of explanations—science examining the laws of nature and religion exploring the moral realm.

[*See also* Biological Sciences; Creationism; Evolution, Theory of; Paleontology; Popularization of Science; Psychological and Intelligence Testing; Race Theories, Scientific; Religion and Science; Science: Since 1945; Sociobiology and Evolutionary Psychology; *and* Zoology.]

BIBLIOGRAPHY

Ruse, Michael. "Gould, Stephen Jay." *eLS*, April 2014. New York: John Wiley & Sons.
York, Richard, and Brett Clark. *The Science and Humanism of Stephen Jay Gould.* New York: Monthly Review Press, 2011.

Patricia Princehouse

GRAHAM, SYLVESTER

(1794–1851), health reformer. Born in West Suffield, Connecticut, as the 17th child in the family of a Presbyterian minister, young Graham suffered through years of repeated illnesses, including a severe nervous collapse at age 29. By his early thirties he had recovered sufficiently to enter the Presbyterian ministry in New Jersey, where he acquired a reputation as a powerful and successful evangelist, especially when speaking on his favorite subject of temperance. Before long Graham began adding the blessings of a meatless diet to his material on temperance. By the spring of 1831 he was lecturing independently in Philadelphia at the Franklin Institute on "the Science of Human Life," a broad spectrum of topics ranging from proper diet to control of the natural passions. Within a year he moved to New York City, where he lectured for an entire year.

The cholera epidemic of 1832 vaulted Graham and his program of health reform into the national spotlight. Several months before the disease reached the shores of North America, he revealed to a New York audience, estimated at two thousand, an almost sure way to ward off an attack: by abstaining "from flesh-meat and flesh soups, and from all alcoholic and narcotic liquors and substances, and from every kind of purely stimulating substances, and [by observing] a correct general regimen in regard to sleeping, bathing, clothing, exercise, the indulgence of the natural passions, appetites, etc." After the epidemic had subsided, he happily reported "that of all who followed my pre-scribed regimen uniformly and consistently, not one fell a victim to that fearful disease, and very few had the slightest symptoms of an attack." During the 1830s Graham visited most of the major East-ern cities and won a widespread following among those Americans who had lost faith in the more traditional methods of preserving health. In 1839 he wrote out his oft-repeated *Lectures on the Science of Human Life*, borrowing liberally but without acknowledgment from the French pathologist François J. V. Broussais. The best way to stay healthy, advised Graham in his *Lectures,* was to avoid all stimulating and unnatural foods and to subsist "entirely on the products of the vegetable kingdom and pure water"—"the only drink that man can ever use in perfect accordance with the vital properties and laws of his nature." An ideal food, and one that came to be associated with Gra-ham's name, was bread made from unbolted wheat flour and allowed to sit for 24 hours (which was nothing like the present-day Graham cracker). If, after following his abstemious regimen, a person did succumb to illness, the cardinal rule to re-member was that "ALL MEDICINE, AS SUCH, IS ITSELF AN EVIL."

The public outcry against Graham's strange re-forms was more than matched by its outrage at his views on sex, which first appeared in *A Lecture to Young Men on Chastity* (1834). In fact, one of his fellow reformers was convinced that "while the public odium was ostensibly directed against his anti–fine flour and anti–flesh eating doctrines, it was his anti-sexual indulgence doctrines, in reality, which excited the public hatred and rendered his name a by-word and a reproach." As Stephen Nis-senbaum has pointed out, this work broke with the older moralistic literature on the subject in two ways: it was based largely on scientific rather than biblical arguments, and it focused not on the sins of adultery and fornication but on the previ-ously neglected problems of masturbation and marital excess, which Graham defined for most people as intercourse more than once a month.

[*See also* **Food and Diet; Health and Fitness;** *and* **Medicine.**]

BIBLIOGRAPHY

Nissenbaum, Stephen. *Sex, Diet, and Debility in Jacksonian America: Sylvester Graham and Health Reform.* Westport, Conn.: Greenwood Press, 1980.

Ronald L. Numbers

GRAY, ASA

(1810–1888), botanist. A native of upstate New York, Gray earned an MD degree from the Fair-field (New York) Medical School in 1831, at a time when medical schools required little prelimi-nary education and awarded degrees after only eight months of coursework. Gray briefly prac-ticed medicine but increasingly turned his atten-tion to botany. In 1834 he moved to New York City to collaborate with botanist John Torrey. The two worked on a number of projects including the *Flora of North America* (1838–1843). In 1842, Gray accepted Harvard College's offer of a profes-sorship of natural history with special responsi-bility for botany. Although Gray quit teaching in 1873, he remained on the Harvard faculty until his death in 1888. Gray, together with Torrey, introduced the new natural system of botanical classification in use in Europe, replacing the Lin-naean system. His various textbooks brought him national fame, whereas his scientific work won him international acclaim as America's leading taxonomic botanist. He synthesized the findings of previous generations of botanists, notably John Bartram and Cadwallader and Jane Colden, to produce the country's first comprehensive flora. An especially important study, published in the *Memoirs of the American Academy of Arts and Sci-ences* in 1859, focused on the similarities between the floras of America and Japan. In this, Gray argued that species common to both regions came from common ancestors that had moved south during the glacial period. Gray's ideas in this work were influenced by his correspondence with Charles Darwin.

Gray is best remembered for the prominent role he played in the American debates over Dar-win's *Origin of Species* (1859). Gray, who had met Darwin during a visit to England in 1851 and had subsequently corresponded with him, sought to ensure that his English friend received "fair play" in North America. An orthodox, if not particularly pious, Congregationalist Christian, Gray also strove to present Darwin's theory in a theistic light. He proposed that the unexplained organic variations on which natural selection acted be attributed to divine providence, a suggestion that Darwin rejected. Gray not only urged a "special origination" in connection with the appearance of human beings, but also expressed skepticism about the ability of natural selection "to account for the formation of organs, the making of eyes, &c." Although he described himself as "one who is scientifically, and in his own fashion, a Darwinian," he confessed to a friend that his theistic interpretation of evolution was "*very anti-Darwin*." Gray presented his views in two influential books: *Darwiniana* (1876), a collection of previously anonymous essays and reviews, and *Natural Science and Religion* (1880), based on a series of lectures at Yale Divinity School.

[*See also* **Agassiz, Louis; Bartram, John and William; Biological Sciences; Botany; Colden, Cadwallader and Jane; Evolution, Theory of; Religion and Science;** *and* **Science.**]

BIBLIOGRAPHY

Dupree, A. Hunter. *Asa Gray, 1810–1888*. Cambridge, Mass.: Belknap Press of Harvard University Press, 1959.

Keeney, Elizabeth. "Gray, Asa (1810–1888)." In *The History of Science in the United States: An Encyclopedia,* edited by Marc Rothenberg, pp. 244–245. New York and London: Garland Publishing, 2001.

Numbers, Ronald L. *Darwinism Comes to America.* Cambridge, Mass.: Harvard University Press, 1998. An important work addressing the history of Darwinism in America that refutes many misconceptions about debates about Darwinism, the impact on scientists' religious ideas, the Scopes trial of 1925, and the regional and denominational distribution of pro- or antievolutionary sentiment.

Ronald L. Numbers; updated by Elspeth Knewstubb and Hugh Richard Slotten

GROUP PRACTICE

The history of group practice in American medicine presents a challenge because of the number of practice formats that are included in the rubric. For the purposes of this discussion, a group practice is an organization that serves as the primary employer of a number of health-care workers who collaborate, using the resources of the organization, for the principle goal of providing patient care. Donald Madison described three institutions, which this definition excludes, as antecedent to group practice: academic clinics, industrial clinics, and dispensaries. There are still three basic approaches to group practice that must be considered: the multispecialty or clinic group, designed to provide a comprehensive medical encounter; the specialty group designed to ensure comprehensive, ongoing coverage of a special area of medical practice; and the organization, physical or virtual, designed as a means of providing prepaid or partially paid health-care coverage, the insurance group. Although the histories of the three types of group practice overlap and all three continue to flourish, their periods of prominence are sequential.

Group practice in all its formats has been made practical by legislative creation of factious legal entities—partnerships, corporations, and more modern professional associations. Group practices grew out of earlier forms of associations of physicians, especially the partnership, formal and informal. Well-established practitioners have always had more demands on their time than they could meet and many have employed or otherwise become associated with younger practitioners, who can share the burden while continuing to learn their profession and develop their own practices. The most common form of such intergeneration association has been familial. As the idea of emergent surgical care evolved in the late nineteenth century, a new association of peers, frequently siblings or classmates, became increasingly common. These two traditions combined in the practice of William W. Mayo and his sons in late nineteenth-century Minnesota.

The Mayo Clinic, the paradigm of the clinic group, evolved out of the family business in two stages—first, the 1892 decision to keep the family or general practice of the senior Dr. Mayo as he retired led to the recruitment of associates who were not surgeons, although surgical associates were eventually recruited as well; then, after 1900, experts in nonsurgical areas of medicine moved the Mayo Clinic into significant medical consultation. The final step in the development of the Mayo Clinic model was the 1919 dissolution of the Mayo partnership and creation (by the state legislature at the request of the Mayo brothers) of the Mayo Property Association (after 1964, the Mayo Foundation) as the owner of the clinical practice, removing all the physician consultants from a direct fiduciary role with the patients.

Other physicians, particularly surgeons, formed partnerships, and as the twentieth century progressed, more and more partnerships and associations of various sorts evolved. The development of suburbs without work centers and hospitals resulted in an acceleration of patients coming to the doctor's office or clinic on trolleys and trains because doctors were tied to a place, particularly surgeons and obstetricians, who needed to be near hospitals, and those practitioners dependent on laboratory analysis, such as consulting pediatricians and internists. Many physicians did not form legal partnerships but simply shared the expenses of office and staff because most state law forbade the practice of medicine by corporations, and partnerships enhanced rather than reduced liability. The work of Neil Shumsky et al. (1986) on the location of medical practice in San Francisco is highly suggestive and needs to be extended by careful study of the roles of hospitals and physicians' and surgeons' buildings, as well as transportation and other nonmedical factors in various cities. As patients came to doctor's offices and as technology (especially X-ray and the laboratory) impacted practice, the need for collaborations to meet the standard of care became more common; Shumsky et al. suggest that many of these arrangements were impacted by the physical geography of the emerging city; if so, then there may be differences among urban centers based on nonmedical factors. By 1932, the American Medical Association (AMA) estimated there were almost three hundred group practices in the United States with

an average of five physicians; most were specialty-related collaborations, but specialization was still evolving (see Stevens [1976] for these and subsequent statistics on the emergence of group practice).

Whereas specialty groups were growing with cities, the clinic groups developed in smaller towns, particularly in the Midwest. Railroad transportation, along with other cultural factors, led to the creation of a large number of surgery-based "clinics." Many of these, like the Mayo Clinic, grew into multispecialty clinics, particularly after physicians experienced the collaboration of internal medicine, clinical pathology, and psychiatry in World War I. George Crile, who returned from the war to found the Cleveland Clinic, captured the intellectual importance of multispecialty work in his phrase "to function as a unit." These multispecialty, clinic group practices formed the core of a management conference, which met for the first time in 1926 with 14 clinics represented. In 1946, the Medical Group Management Association drew 67 clinics to its annual meeting. There were always some clinics that were not multispecialty, such as the Menninger Clinic (psychiatry) and the Joslin Clinic (endocrinology).

Specialization, based in the American Board system of the 1930s, grew rapidly after World War II and the legal creation of the Professional Association or Professional Corporation in many states in the 1960s facilitated the rapid grow of single-specialty group practices. Therapeutic and diagnostic progress encouraged the growth and subspecialization of internal medicine, and by 1975 there would be almost eight thousand group practices in the United States, the vast majority being specialty groups.

Specialization and urbanization both contributed to a growing access problem in American medicine, and prepaid group practices evolving from the Depression were a partial answer. The AMA fought prepaid groups but the new innovation enjoyed popularity with patients. With the Health Maintenance Organization Act of 1972, prepaid groups received a privileged place in American health policy. Although HMOs were originally conceived as a different, heavily preventive approach to practice, they quickly became another insurance scheme and, to meet the terms of the act, hospitals and insurance companies began to form virtual groups or panels that shared little beyond accounting services—these insurance groups, under a variety of names, were the fastest growing groups of the late twentieth century and continued to evolve in the early twenty-first century.

[*See also* American Medical Association; Health Insurance; Health Maintenance Organizations; Hospitals; Mayo Clinic; Medical Specialization; Medicine; Medicine and Technology; Public Health; Railroads; *and* Technology.]

BIBLIOGRAPHY

Madison, D. L. "Notes on the History of Group Practice: The Tradition of the Dispensary." *Medical Group Management Journal* 37, no. 5 (September–October 1990): 52–54, 56–60, 86–93. Madison advocates the nineteenth-century charitable dispensary as an important source of physician collaboration and so the emergence of group practice, an idea that may merit more exploration.

Shumsky, N. L., Bohland, J., and Knox, P. "Separating Doctors' Homes and Doctors' Offices: San Francisco, 1881–1941." *Social Science and Medicine* 23, Vol. 10 (1986): 1051–1057. In these related papers, Shumsky et al. began to explore the impact of urban geography on medical practice, an idea that demands more work on other cities.

Smith, D. C. "Modern Surgery and the Development of Group Practice in the Midwest." *Caduceus* 2 (1988): 1–34. The authors' earlier elaboration of many of the ideas in this article.

Stevens, E. B. *The History of the Medical Group Management Association.* Inglewood, Calif.: Medical Group Management Association, 1976. A valuable source of statistics and guide to the archival records of the community of business managers associated with group practice.

Dale Smith

GUYOT, ARNOLD HENRY

(1807–1884), physical geographer and geologist. Guyot was born into a Swiss family of French Protestant background. Four years after earning a PhD from the University of Berlin in 1835 for a dissertation on "The Natural Classification of Lakes," he accepted the professorship of history and physical geography in the academy of Neuchâtel, Switzerland, where he was associated with Louis Agassiz, an acquaintance since youth. In 1848 he followed Agassiz to the United States. Six years later, after winning a reputation as the foremost geographer in America, the College of New Jersey (Princeton) installed him in a specially created chair in physical geology, a position he occupied until his death.

A devout Presbyterian who as a youth had briefly studied for the ministry, he devoted considerable effort to harmonizing the Bible and science, especially the nebular hypothesis of the origin of the solar system and the recently discovered geological evidence of the antiquity of life on earth. His innovative harmonizing scheme, which interpreted the "days" of Genesis 1 as immense cosmic epochs, first appeared in print in 1852; it quickly garnered influential supporters in both science and theology. Although he assigned most of the work of creation to divine ordained laws of nature, he resisted accepting the transmutation of one species into another. After Agassiz's death in 1873 he became the most prominent American anti-evolutionist in the scientific community—according to one hyperbolic report, the *only* "working naturalist of repute in the United States…that is not an evolutionist." A long-time sufferer of severe writer's block, he delayed until the year of his death the publication of his book *Creation; or, The Biblical Cosmogony in the Light of Modern Science* (1884), in which he insisted on only the special creation of matter, life, and humans. "Evolution from one of these orders into the other—from matter into life, from animal life into the spiritual life of man—is impossible." His long-time friend James Dwight Dana suspected that Guyot in his later years "was led to accept, though with some reservation, the doctrine of evolution through natural causes."

[*See also* Agassiz, Louis; Creationism; Dana, James Dwight; Evolution, Theory of; Geography; Geology; Paleontology; Religion and Science; *and* Science.]

BIBLIOGRAPHY

Numbers, Ronald L. *Creation by Natural Law: Laplace's Nebular Hypothesis in American Thought.* Seattle: University of Washington Press, 1977.

Numbers, Ronald L. *Darwinism Comes to America.* Cambridge, Mass.: Harvard University Press, 1998.

Wilson, Philip K. "Influences of Alexander von Humboldt's *Kosmos* in the Swiss-American Geohistorian and Creationist Arnold Guyot's *Earth and Man* (1849)." *Omega: Indian Journal of Science and Religion* 4 (2005): 33–51.

Ronald L. Numbers

HALE, GEORGE ELLERY

(1868–1938), astronomer and creator of scientific institutions. The son of a wealthy Chicago businessman, Hale became fascinated with science at an early age. His father supported his son's interest by presenting him with a small scientific laboratory and telescope. Hale earned a bachelor's degree in physics from the Massachusetts Institute of Technology (MIT) in 1890. He was especially interested in applying developments in physics, especially spectroscopy, to problems in astronomy. His senior thesis discussed the design for a spectroheliograph, an instrument for photographing the sun at different wavelengths that would allow astronomers to study different layers of the sun's atmosphere.

In 1892, after an extended visit to Europe to meet with major scientists and attend lectures, Hale accepted a position on the faculty of the newly created University of Chicago. The president, William Rainey Harper, supported his efforts to build a major research observatory. With funding from his father, Hale had already built an observatory at his family home in the Kenwood section of Chicago (known as the Kenwood Observatory). A wealthy Chicago businessman, Charles T. Yerkes, donated the necessary funds to construct a major observatory for the university with a 40-inch refractor, the largest in the world at that time.

Hale believed large telescopes were necessary for progress in astronomical research, especially in stellar astrophysics. He played an important role in convincing the Carnegie Institution of Washington, D.C., to support the establishment, in 1904, of the Mount Wilson Observatory in Southern California. The observatory included not only solar telescopes but also a 60-inch and, after 1918, a one-hundred-inch reflecting telescope.

Hale became a leader of the astronomy community in the United States. He was instrumental in establishing astrophysics as a separate discipline. In 1891, he cofounded the journal *Astronomy and*

Astrophysics. Renamed the *Astrophysical Journal* in 1895, it became a leading research journal in the field of astronomy.

Hale's most important research dealt with solar and stellar astrophysics. Using spectroscopic observations, he investigated the structure, circulation, and composition of the sun and other stars. Hale was particularly interested in the phenomenon of sunspots. Using instruments he helped design, he discovered the role of magnetic fields in sunspot structure.

Hale was not only important for American astronomy but also played a leading role in the establishment of general institutions supporting science in the United States. After World War I, he worked to make the National Academy of Sciences more significant nationally through the construction of a permanent building in Washington, D.C. In an effort to link the federal government to the scientific expertise of the country, Hale also was a driving force behind the creation of the National Research Council in 1916. His institution-building activities extended internationally as well. He helped organize the specialized International Union for Cooperation in Solar Research (crucial in the founding of the International Astronomical Union after World War I) and the more general International Council of Scientific Unions.

Poor health forced Hale to resign as the director of the Mount Wilson Observatory in 1923, but he continued to work to gain support for large telescopes. Most important, in 1928 he succeeded in convincing the International Education Board of the Rockefeller Foundation to fund a two-hundred-inch reflecting telescope. Construction of the telescope was not completed until 1949, 11 years after Hale's death. The Hale two-hundred-inch telescope at the Palomar Observatory in Southern California was operated jointly by the California Institute of Technology, which Hale had also played an important role in establishing, and the Carnegie Institution of Washington, the owner of the Mount Wilson Observatory.

[*See also* **Astronomy and Astrophysics; Hubble, Edwin Powell; Hubble Space Telescope; Instruments of Science; Journals in Science, Medicine, and Engineering; National Academy of Sciences;** **Physics; Research and Development (R&D); Rockefeller Institute, The; Science;** *and* **Societies and Associations, Science.**]

BIBLIOGRAPHY

Adams, Walter S. "George Ellery Hale." *Biographical Memoirs of the National Academy of Sciences,* 21 (1939): 181–241.

Wright, Helen. *Explorer of the Universe: A Biography of George Ellery Hale.* New York: Dutton, 1966.

Wright, Helen, J. N. Warnow, and C. Weiner, eds. *Legacy of George Ellery Hale: Evolution of Astronomy and Scientific Institutions, in Pictures and Documents.* Cambridge, Mass.: MIT Press, 1972.

Hugh Richard Slotten

HALL, G. STANLEY

(1844–1924), psychologist and educator. Born and raised in rural western Massachusetts, Granville Stanley Hall attended several rural one-room schools. He then taught at rural schools himself for a year, but decided to pursue higher education. He attended Williston Academy in Easthampton and then entered Williams College in 1863, where he developed an interest in the theory of evolution and in philosophy. He received his BA in 1867. After a brief stint at Union Theological Seminary, he studied philosophy and physics at Bonn, Berlin, and Heidelberg, in Germany. In 1876 he began graduate work at Harvard where, under William James's direction, he earned the first American doctorate in psychology. Returning to Germany, he studied the new experimental, physiological psychology under Wilhelm Wundt at Leipzig.

When Hall returned to America in 1880, he found the field of education the most receptive to the new psychology. His lectures on pedagogy interested Daniel Coit Gilman, president of the recently founded Johns Hopkins University in Baltimore, who in 1883 appointed him America's first full-time professor of psychology and pedagogy. There Hall established the first U.S. psychological laboratory, attracting students such as

John Dewey and James McKeen Cattell, who would make lasting contributions to the profession, and founded the *American Journal of Psychology* (1887). He remained the journal's editor for many years. In 1888, Hall was named president of Clark University in Worcester, Massachusetts, a graduate research institution modeled on Johns Hopkins. While carrying out his administrative duties, he launched other psychological journals, helped organize the American Psychological Association (1892), and played a role in introducing psychoanalysis to the United States by bringing Sigmund Freud and Carl Jung to Clark in 1909.

Hall's most influential work focused on the mental development of children and adolescents. He used a modified and enlarged version of the questionnaire method developed by the German philosopher Moritz Lazarus to establish what children think and know. This study of children inspired Hall to work on educational methods. He took part in developing psychological and intelligence testing for children, which led to improved teaching methods. Hall's most influential publication, which came out of his study of child development and education, was the two-volume *Adolescence: Its Psychology and Its Relations to Physiology, Anthropology, Sociology, Sex, Crime, Religion and Education* (1904). This pioneering contribution to child study and educational psychology reflected changing perceptions of human development in American culture.

[*See also* Cattell, James McKeen; Evolution, Theory of; Higher Education and Science; Journals in Science, Medicine, and Engineering; Psychological and Intelligence Testing; Psychology; *and* Social Sciences.]

BIBLIOGRAPHY

Ross, Dorothy. *G. Stanley Hall: The Psychologist as Prophet*. Chicago: University of Chicago Press, 1972.

Schultz, Duane P., and Sydney Ellen Schultz. *A History of Modern Psychology*, 10th ed. Belmont, Calif.: Wadsworth, Cengage Learning, 2011.

Shore, Elizabeth Noble. "Hall, Granville Stanley." In *Complete Dictionary of Scientific Biography*, Vol. 6, pp. 52–53. Detroit: Charles Scribner's Sons, 2008.

Kerry W. Buckley;
updated by Elspeth Knewstubb

HALSTED, WILLIAM

(1852–1922), surgeon. After graduation from medical school at the College of Physicians and Surgeons in New York City in 1877, William Stewart Halsted served as house physician at New York Hospital. He then spent two years studying at the Germanic clinics of Central Europe. Returning to New York in 1880, he practiced surgery at several hospitals and maintained a private practice while conducting a series of research studies. Experiments with the recently discovered local anesthetic effects of cocaine resulted in Halsted's becoming addicted to that drug in 1885. He spent 1886 in a sanatorium in Providence, Rhode Island, in an attempt to conquer this addiction. He was able to stop using cocaine, but continued as a user of morphine. In 1888, Halsted was invited to do laboratory research at the Johns Hopkins Hospital in Baltimore under the sponsorship and oversight of William H. Welch, the professor of pathology, where he established a reputation for excellence in surgery and surgical experimentation.

Halsted was an early Listerian, but he preferred the idea of aseptic, rather than antiseptic surgical techniques, eventually introducing sterilized clothing, rubber gloves, and masks into the operating theaters at Johns Hopkins. He perfected several new surgical techniques, including the intestinal suture, and developed a number of instruments and procedures for maintaining healthy tissue, which he believed would best assist healing and fight infection. Halsted was appointed surgeon in chief to the Johns Hopkins Hospital and professor of surgery at the medical school, which opened in 1893.

During Halsted's 30-year tenure in these positions, he founded a distinctive school of surgery, characterized by emphasis on the physiological basis of disease and therapy, laboratory and clinic research, and meticulous operative technique, all of which were marked advances over previous practices. A number of the men who completed Halsted's rigorous training established departments of their own at other universities, thus spreading his teachings throughout the academic community. He also devised the residency framework upon which modern surgical training is based and made important contributions to the operative treatment of breast cancer, thyroid disease, and inguinal hernia. By the time of his death, the methodology that is still called Halstedian had become the predominant style in American surgical teaching and clinical work. It is for this reason that William Halsted has been called "the Father of American Surgery."

[*See also* Hospitals; Medical Education; Medicine; Pharmacology and Drug Therapy; Surgery; *and* Welch, William H.]

BIBLIOGRAPHY

Bynum, Helen. "Halsted, William." In *Dictionary of Medical Biography*, edited by W. F. Bynum and Helen Bynum, Vol. 3, pp. 605–606. Westport, Conn. and London: Greenwood Press, 2007.
Crowe, Samuel J. *Halsted of Johns Hopkins: The Man and His Men.* Springfield. Ill.: Thomas, 1957.
Nuland, Sherwin B. "Medical Science Comes to America: William Stewart Halsted of Johns Hopkins." In *Doctors: The Biography of Medicine*, pp. 386–421. New York: Alfred A. Knopf, 1988.

Sherwin B. Nuland;
updated by Elspeth Knewstubb

HEALTH AND FITNESS

The pursuit of health and fitness in America has long connoted more than a desire for the absence of sickness, illness, or disease; it implied a search for the strength, energy, and vitality that were believed to occur naturally in human beings in a right relationship with nature or by way of divine intervention and to result in a profound sense of well-being and spiritual harmony. Whereas medication and the ministrations of a physician suggested the need for some external and "unnatural" agent, advocates of health and fitness typically embraced self-reliance mediated through spiritual law in their search for an idealized state in which mind, body, and spirit united in perfect harmony. Health, therefore, became not only the prerequisite but also the justification for an abundant life on earth or in some transcendent realm.

Colonial Era and the Nineteenth Century. The quest for health in early America unfolded within a worldview that intertwined the natural and supernatural worlds. This conjunction, described in such works as Cotton Mather's *Angel of Bethesda* (1724) and *Primitive Physick* (1747) by the British Methodist leader John Wesley, manifested itself most obviously through a belief in a direct connection between sin and sickness, morality and health. Health came not only through a correct balance of bodily fluid, or humor, and proper attention to hygiene, but also by maintaining a proper relationship to God and ascertaining the divine purposes for one's life.

The Second Great Awakening, beginning in the late eighteenth century, witnessed waves of religious fervor that by mid-nineteenth century found revivalism and reform linked in a kaleidoscope of social movements struggling against not only poverty, slavery, and war but also meat, alcohol, and unhealthful dress. Fed by millennial enthusiasm, armed with "science for the common man," and appealing to a widespread desire for improvement through education and self-control, reformers set about to perfect the United States by saving bodies as well as souls.

Food and dietary reform loomed large in this effort. Antebellum Americans drank vast quantities of alcohol and consumed appallingly rich, fatty meals of meats and heavy desserts washed down with coffee. Fruits and green vegetables were a rarity. Corn and pork formed the staples of the rural diet, whereas urban populations relied more on bread and beef. Potatoes, turnips, cabbage, and later tomatoes completed the basic American diet.

Dismayed by these food-and-drink habits, reformers launched a temperance movement and sought to change eating patterns. In the 1830s the Presbyterian preacher and health reformer Sylvester Graham (1794–1851) attacked the American diet, warning that intoxicating drink, stimulating spices, coffee, tea, and meat should be avoided because they debilitated the mind and body and overstimulated the gross and sensual side of human nature. Both as remedy for and as prevention of the sickness and immorality of society, Graham advocated coarse whole-wheat bread, pure water, and a vegetarian diet. Three midcentury technological innovations—the ice cutter, refrigerated rail cars, and canning—made such dietary reforms increasingly feasible by improving the distribution and year-round availability of perishable and seasonal foods.

Beginning in the 1840s, Grahamite health reformers forged links with water-cure enthusiasts like Mary Gove Nichols (1810–1884) who sought to cure bodily ills with various water treatments and who believed they were following nature's way of maintaining health. During the middle decades of the century, more than two hundred water-cure establishments catered to a wide public, including many women in search of better health, social exchange, and cures for a wide range of "female complaints." Hydropaths believed that the sick body's self-curative powers could be vitalized by copious amounts of water, taken internally and externally, a nonstimulating diet, sunlight, exercise, and relaxation. In the words of hydropathic reformer Russell T. Trall (1812–1877), who built a "hygienic system" on such principles, "All healing or remedial power is inherent in the living system" and "health is found only in obedience" to the divinely ordained laws of nature (*The Hygienic System*, 1872).

The spiritual impulse underlying the nineteenth-century crusade for health became explicit among the adherents of Mormonism and Seventh-day Adventism, who not only aspired to a purity of spirit that would enable good to triumph over evil, but also believed that much sickness could be prevented by righteous living and self-discipline. Advocating natural remedies and eschewing alcohol, coffee, tea, tobacco, and all

stimulants, which polluted the body and corrupted the spirit, Adventists and Mormons spiritualized their rules of healthful living. Later studies demonstrated the health benefits of these practices, documenting a correlation between the Mormon and Adventist lifestyles and a lower incidence of cancer, heart disease, and other life-shortening diseases.

Numerous nineteenth-century reformers popularized the many dimensions of health reform. Dioclesian Lewis (1823–1886) preached the power of physical exercise and gymnastics; Catharine Beecher (1800–1878) warned of the dangers of corsets and tight lacing; Horace Fletcher (1849–1919) attributed his prodigious physical strength to the thorough chewing of food; and Adventist John Harvey Kellogg (1852–1943) invented flaked breakfast cereals, founded a medical school, and established numerous sanitariums devoted to the gospel of healthful living, most notably at Battle Creek, Michigan.

Other nineteenth-century health reformers, however, cared less about achieving a perfect harmony of mind, body, and spirit through discipline and right practice than about sublimating or ignoring the body. Such reformers and mystical teachers included mesmerists, hypnotists, Christian Scientists, members of the Emmanuel Movement, spiritualists, and numerous mind healers who summoned mental gymnastics, mesmeric fluids, or the disembodied spirits of the dead to relax nervous tensions, relieve spiritual ennui, or build spiritual health and fitness. Mary Baker Eddy (1821–1910), founder of Christian Science, radically denied the existence of matter, sin, sickness, and death and encouraged the denial of the body and the affirmation of spirit while rejecting all physical means to achieve health. Less radical metaphysical health reformers, including Warren Felt Evans (1817–1889), Emma Curtis Hopkins (1849–1925), and Ernest Holmes (1887–1960), formed the so-called New Thought movement of the late nineteenth and twentieth centuries, whose followers believed in a mind–body relationship that allowed the mind to heal mental and physical maladies and maintain physical health.

The Twentieth and Twenty-First Centuries.

The health and fitness impulses of nineteenth-century Americans reached a zenith with the turn-of-the-century progressive movement. Allying scientific management, education, public health, and religion with their middle- and upper-class sensibilities, progressives worked to reform corrupt city and state governments, improve the infrastructure of urban social services, reduce venereal disease by combating prostitution, enact a constitutional amendment imposing nationwide prohibition, and save the immigrant masses from their sordid ways. Their often paternalistic efforts to instruct working-class Americans, especially recent immigrants, in matters of personal hygiene, principles of scientific nutrition, and the dangers of intemperance demonstrated their commitment to matters of health.

The "muscular Christianity" advocated by popular evangelists such as Dwight L. Moody (1837–1899) and organizations such as the Young Men's Christian Association (YMCA) continued the colonial and nineteenth-century Protestant insistence on developing mind, body, and spirit. But other ideological currents of the Gilded Age and beyond revealed a changing understanding of "fitness" as no longer simply evidence of divine favor and spiritual well-being. Health and fitness advocates drew Americans to regimens that altered physical appearances in ways that conformed to new cultural ideals of strength and beauty and that served as socioeconomic markers of an individual's ability to survive the competitive Darwinian struggle for existence in urban–industrial America.

The post–World War II years witnessed an epidemiological shift from infectious diseases to ailments related to lifestyle and old age. This shift resulted in part from the impact of antibiotics on infectious diseases, but the increasing number of Americans holding white-collar jobs and leading sedentary lives contributed to obesity and its associated debilities. Moreover, scientific investigators began to produce compelling evidence of the health benefits of exercise, a proper diet, and the avoidance of tobacco.

In response to these developments President Dwight D. Eisenhower in 1956 established the President's Council on Youth Fitness (after 1968, the President's Council on Physical Fitness and Sports) to combat the lack of physical fitness among children and young people. In the 1960s the federal government began to warn citizens of the health dangers of cigarettes, and the Child Nutrition Act of 1966 refocused the national school lunch program from farm subsidy to the nutritional needs of children. But these efforts lacked the religious overtones that had characterized many earlier health-reform movements. Science and secular morality had replaced religion and virtue as the primary motivators for national health reform; and physical health, greater longevity, and economic productivity had replaced moral perfection as the primary purpose of healthful living. However, the temporary increase in physical-fitness activities and the decline in smoking were not uniform; white, affluent Americans with a postsecondary education embraced those behaviors more often than did the poor and blue-collar workers, the less educated, and members of ethnic minority groups.

Despite these government efforts to encourage physical exercise and promote nutritional diets, a 2001 *Call to Action* by U.S. Surgeon General David Satcher declared obesity an American health crisis of "epidemic proportions" and called for diet and exercise initiatives to counter its expensive and deadly costs. Federal and state efforts targeted fatty and sugar-laden foods, overeating, and sedentary lifestyles. Some health reformers criticized the fast-food industry and bemoaned the decline in physical education programs in schools, whereas others, unwilling to identify obesity as a disease or a social evil, argued that Americans just needed more self-discipline and stronger moral character.

Much of the explosive interest in health, diet, and fitness that characterized the late twentieth and early twenty-first centuries was concentrated among aging members of the so-called baby-boom generation. The continuing popularity of the vogue for jogging that began in the 1970s, the health and fitness clubs that proliferated in the 1980s, and the mountain biking enthusiasm of the 1990s all illustrated the faddish side of the late twentieth-century exercise and fitness movement and revealed the increasingly commercial and technological aspect of the quest for health and fitness.

The spiritual impulse for health and fitness persisted, however, in the growth of the so-called New Age movement that began during the 1970s and 1980s. An echo of the New Thought movement of the previous centuries, this movement envisioned a new age of personal enlightenment, health, and well-being achieved by following an eclectic mixture of esoteric Western and Eastern wisdom. A broad range of educated, affluent Americans meditated daily and adhered to popularized versions of Hindu, Buddhist, and Yogic diet, physical exercise, and sexual practice. They also embraced a holistic understanding of the relationship between humans and nature that intersected with the modern environmentalist movement. These sensibilities contributed to the growth in popularity of the holistic, integrative, and "natural" eating habits of vegetarianism, slow food, and other alternative food movements.

Despite these aging baby-boomer efforts to ensure eternal youth, cognitive decline resulting from aging, including Alzheimer's disease and other forms of dementia, grew to epidemic proportions by the early twenty-first century. Of unknown cause, this increase in age-related dementia stimulated an immense brain fitness industry. Health reformers advocated diets, vitamins, herbs, brain-teaser exercises, spiritual healing, and treatments of dubious efficacy in an effort to halt or slow its onset and to ameliorate its symptoms. The retention of mental acuity and attainment of increased longevity, so deeply rooted in America's religious and cultural history, had become badges of status and the means to a better life either here or in the hereafter.

[See also **Alcohol and Alcohol Abuse; Alzheimer's Disease and Dementia; Disease; Food and Diet; Food Processing; Graham, Sylvester; Household Technology; Hygiene, Personal; Life Expectancy; Medical Education; Medicine; Obesity; Penicillin; Public Health; Race and Medicine; Refrigeration and Air Conditioning; Religion and Science; Science; Sexually Transmitted Diseases;** *and* **Technology.**]

BIBLIOGRAPHY

Brandt, Allan M., and Paul Rozin, eds. *Morality and Health.* New York: Routledge, 1997.

Cayleff, Susan E. *Wash and Be Healed: The Water-Cure Movement and Women's Health.* Philadelphia: Temple University Press, 1987.

Goldberg, Philip. *American Veda: From Emerson and the Beatles to Yoga and Meditation How Indian Spirituality Changed the West.* New York: Random House Digital, 2010.

Goldstein, Michael S. *The Health Movement: Promoting Fitness in America.* New York: Twayne, 1992.

Green, Harvey. *Fit for America: Health, Fitness, Sport, and American Society.* Baltimore: Johns Hopkins University Press, 1988.

Griffith, R. Marie. *Born again Bodies: Flesh and Spirit in American Christianity.* Berkeley and Los Angeles: University of California Press, 2004.

Harrington, Anne. *The Cure Within: A History of Mind–Body Medicine.* New York: W. W. Norton & Company, 2008.

Kluger, Richard. *Ashes to Ashes: America's Hundred-Year Cigarette War, the Public Health, and the Unabashed Triumph of Philip Morris.* New York: Random House Digital, 1996.

Levenstein, Harvey A. *Revolution at the Table: The Transformation of the American Diet.* Berkeley and Los Angeles: University of California Press, 1988.

Numbers, Ronald L., and Darrel W. Amundsen. *Caring and Curing: Health and Medicine in the Western Religious Traditions.* New York: Macmillan, 1986.

Verbrugge, Martha H. *Able-Bodied Womanhood.* New York and Oxford: Oxford University Press, 1988.

Whorton, James C. *Crusaders for Fitness: The History of American Health Reformers.* Princeton, N.J.: Princeton University Press, 1982.

Rennie B. Schoepflin

HEALTH INSURANCE

The history of health insurance in America is largely a twentieth-century story. Well beyond 1900, paying for medical care, with few exceptions, remained a private activity between patients and their physicians and hospitals. Widespread interest in sickness insurance—as it was originally called—did not develop in the United States until the 1910s, and then attention focused on compulsory, not voluntary, insurance. Inspired by the rapid spread of government-sponsored sickness-insurance plans in Europe, the progressive American Association for Labor Legislation in 1912 set up a committee to prepare a model bill for introduction in state legislatures (on the assumption that the U.S. Constitution prohibited a federal plan). The model bill required the enrollment of most manual laborers earning $100 a month or less and provided for both income protection and medical care. Although many physicians greeted this plan enthusiastically, sentiment turned against the measure during World War I, and by 1920, when the American Medical Association (AMA) formally declared its opposition, the campaign for compulsory health insurance—or socialized medicine, as it was sometimes called—was dead.

The Great Depression, beginning in 1929, again brought health insurance to the fore. As hospital receipts and physician income plummeted, interest in voluntary health insurance grew. In December 1929, the Baylor University Hospital in Dallas, Texas, announced a plan to sell hospitalization policies to the city's school teachers for 50 cents a month. Other hospitals around the country adopted similar plans, and within a few years groups of hospitals were banding together to offer what came to be called Blue Cross insurance. The AMA, which urged Americans "to save for sickness" rather than purchase insurance, initially opposed this development "as being economically unsound, unethical and inimical to the public interests." In 1937, however, in the face of a renewed push for compulsory health insurance, the AMA finally approved group hospitalization plans—as long as they left the payment of physicians out of the scheme. By this time the AMA was working on a physician-controlled plan to provide medical (as opposed to hospital) insurance. The resulting medical-society plans, which started in the Pacific Northwest, took the name Blue Shield. By 1952 over half of all Americans owned some health insurance, and although insurance benefits paid only 15 percent of private expenditures for health care, prepayment plans were being hailed

by a presidential commission as "the medical success story of the past fifteen years."

With voluntary plans failing to protect so many Americans, the perennial debate over compulsory health insurance flared again. Pro-insurance reformers had been bitterly disappointed when President Franklin Delano Roosevelt failed to include health insurance in the 1935 Social Security Act. To remedy this omission, the Social Security Board in 1943 drafted a bill to provide health insurance to all persons paying Social Security taxes, as well as to their families. Despite the strong backing of President Harry S. Truman, neither this nor subsequent versions were enacted by Congress. The election of a Republican administration in 1952 temporarily ended the debate over compulsory health insurance.

The 1960 election of John F. Kennedy, a Democrat, revived discussion of the government's responsibility to provide adequate health care for its citizens. Despite strong opposition from organized medicine, Kennedy's successor, Lyndon B. Johnson, persuaded Congress in 1965 to include health insurance as a Social Security benefit (Medicare) and to provide for the indigent through grants to the states (Medicaid). Ironically, after years of warning that government health insurance would ruin the medical profession financially, physicians found that Medicare and Medicaid—by bringing in more patients, raising fees, and facilitating bill collecting—greatly increased their income.

The last third of the twentieth century witnessed numerous attempts to solve the twin problems of access to health care and its ever-increasing cost. In the private sector the most notable development was the rapid growth of health maintenance organizations after 1973, when Congress passed the bipartisan Health Maintenance Organization Act. By 1990, the California-based Kaiser Permanente, which had pioneered in developing prepaid group-practice arrangements, had enrolled more than 6.5 million members. Proposals offered by different advocacy groups ranged from a national health service on the left, in which medical workers would become salaried employees, to income-tax credits for the purchase of commercial health insurance on the right. None of these efforts succeeded, including President Bill Clinton's ill-fated health security plan in 1993–1994, which would have covered all Americans through large health-insurance purchasing cooperatives. Instead, the United States continued its piecemeal response to health-care coverage, with considerable experimentation occurring at the state level. By the end of the century, nearly 84 percent of Americans enjoyed health-insurance coverage, but an estimated 44 million remained uninsured, some by choice (over 8 percent of the uninsured earned more than $75,000 annually), but most by necessity.

A significant, although partial, solution to the absence of universal coverage came in 2010 with the passage of the Patient Protection and Affordable Care Act, known popularly as Obamacare because of its promotion by President Barack Obama. Featuring a provision known as an individual mandate, earlier adopted by the state of Massachusetts, it required individuals not covered by employer or government plans to purchase health insurance or pay a fine. It also required states to establish health-insurance exchanges for such persons (as well as for small businesses), although a subsequent Supreme Court decision allowed states to opt out of this provision. Projections estimated that the new law would cover some 30 million previously uninsured Americans, leaving 25 to 28 million still lacking health insurance. Despite the promises of some politicians, there was little reason to expect the new law to reduce health-care costs in the United States.

[*See also* **Group Practice; Health Maintenance Organizations; Medicare and Medicaid; Medicine: From the 1870s to 1945; Medicine: Since 1945;** *and* **Public Health.**]

BIBLIOGRAPHY

Cunningham, Robert, III, and Robert M. Cunningham Jr. *The Blues: A History of the Blue Cross and Blue Shield System.* DeKalb: Northern Illinois University Press, 1997.

Derickson, Alan. *Health Security for All: Dreams of Universal Health Care in America.* Baltimore: Johns Hopkins University Press, 2005.

Gordon, Colin. *Dead on Arrival: The Politics of Health Care in Twentieth-Century America.* Princeton, N.J.: Princeton University Press, 2003.

Hirshfield, Daniel S. *The Lost Reform: The Campaign for Compulsory Health Insurance in the United States from 1932 to 1943.* Cambridge, Mass.: Harvard University Press, 1970.

Hoffman, Beatrix. *Health Care for Some: Rights and Rationing in the United States since 1930.* Chicago: University of Chicago Press, 2012.

Johnson, Hayes, and David S. Broder. *The System: The American Way of Politics at the Breaking Point.* Boston: Little, Brown and Co., 1996. An analysis of the Clintons' efforts to enact universal health care.

Numbers, Ronald L. *Almost Persuaded: American Physicians and Compulsory Health Insurance, 1912–1920.* Baltimore: Johns Hopkins University Press, 1978.

Numbers, Ronald L., ed. *Compulsory Health Insurance: The Continuing American Debate.* Westport, Conn.: Greenwood Press, 1982.

Poen, Monte M. *Harry S. Truman versus the Medical Lobby: The Genesis of Medicare.* Columbia: University of Missouri Press, 1979.

Starr, Paul. *Remedy and Reaction: The Peculiar American Struggle over Health Care Reform.* New Haven, Conn.: Yale University Press, 2011.

Starr, Paul. *The Social Transformation of American Medicine.* New York: Basic Books, 1982.

Ronald L. Numbers

HEALTH MAINTENANCE ORGANIZATIONS

Dr. Paul Ellwood Jr. and his colleagues in the Nixon administration coined the phrase "health maintenance organization" (HMO) in 1970 to describe a service that would "combine insurance and health care in a single organization," like the Kaiser Permanente prepaid group practices. Those began when industrialist Henry J. Kaiser hired physicians to care for his employees who were building the Grand Coulee Dam in 1938 and who worked in his steel mills during World War II. By the 1960s Kaiser Permanente had evolved into the nation's largest prepaid health plan.

Local consumer organizations and labor unions in various cities joined forces to establish and operate prepaid health plans. These included the Group Health Association, Inc., of Washington, D.C., in 1937; the Group Health Cooperative of Puget Sound in Seattle, Washington; and the Health Insurance Plan of Greater New York in the 1940s. Notable additions in the 1970s were the Group Health Cooperative of Eau Claire (Wisconsin) and Group Health Cooperative of South Central Wisconsin.

Kaiser Permanente and consumer cooperatives served as models for HMOs. Both groups aimed to control costs while improving quality. They required subscribing patients to have a primary-care physician who would coordinate their care and medications; promote routine preventative measures and early intervention; and make referrals to specialists as needed. Limited availability of hospital beds forced doctors to avoid unnecessary and needlessly prolonged hospitalizations. Savings and quality were further enhanced by well-trained nurses and staff who educated patients and performed other specialized tasks.

The U.S. Social Security Act of 1965 established Medicare and Medicaid to cover millions of previously uninsured elderly, infirm, and poor Americans. That legislation, as well as rapid postwar growth in private insurance, increased the demand for health-care services and caused dire shortages of medical staff, facilities, and funding.

The first HMOs developed in response to this crisis. Some, such as Marshfield Clinic's Greater Marshfield Community Health Plan in central Wisconsin, involved medical group practices. Academic physicians at such institutions as Harvard, Johns Hopkins, and Yale started others. These HMOs partnered with nonprofit Blue Cross and Blue Shield plans and other insurance carriers to offer comprehensive care for a prepaid fee. Some HMOs owned hospitals and employed doctors; others contracted with independent physicians and hospitals. Unlike traditional private insurance companies, HMOs modified payment incentive structures and tried to influence medical decision making to ensure that high-quality, cost-efficient care was available to more people.

For their part, congressional lawmakers introduced dozens of bills, many involving prepaid care. Both Republicans and Democrats feared that charging HMO subscribers a set monthly fee, regardless of their health status, would lead to rationing of essential services. Their concerns resulted in the HMO Act of 1973, a bipartisan compromise with few financial incentives for potential HMO sponsors and many restrictive mandates. HMO development was further hobbled by the Nixon administration, which sought to fund as few prepaid plans as possible. Consequently, three times as many prepaid plans were started during the four years before the Act's passage as during the four years following its implementation.

The HMO movement gathered momentum when employers sought to reduce their costs for employee insurance during the economic recession in the early 1980s. In an effort to promote private investment, President Ronald Reagan's administration discontinued federal funding of HMOs in 1981. Many Blue Cross and Blue Shield plans shed their nonprofit status and, along with dozens of other for-profit insurance companies, jumped into the market with prepaid insurance products that borrowed some organizational attributes of HMOs. Competition among these insurance companies facilitated the transformation of health care into a corporate industry.

The term "managed care" entered the American lexicon as newly trained lay insurance administrators employed business management tools to monitor and control physician decisions about patient care to decrease costs and maximize profits.

Economic pressures and rapidly escalating health-care costs eventually forced many large companies to bypass insurance carriers entirely by developing self-funded plans. Some contracted directly with providers, recalling the early Kaiser Permanente Health Plan.

Medicaid and Medicare encountered problems integrating with prepaid care plans in the 1980s, but by 2011 managed care covered 71.4 percent of Medicaid recipients and had greatly reduced per-patient costs. To promote greater HMO participation in Medicare's prepaid plan, "Medicare Advantage," federal reimbursements were higher than for traditional care. Nevertheless, that plan covered only 24 percent of program recipients in 2011 (The Henry J. Kaiser Foundation State Health Facts at www.statehealthfacts.org, accessed January 2012).

Enrollment in nonprofit HMOs and other types of prepaid managed care organizations peaked in 1999 with 104.6 million members in 902 plans. By July 2010, enrollment dropped to 67.2 million members in 452 plans and accounted for only 22 percent of the health-insurance market. Market penetration varied greatly from state to state and showed no regional patterns. The HMO movement did not fulfill its promise, but its influence revolutionized the financing of U.S. health care and offered valuable tools for promoting quality and containing costs (Sanofi-Aventis Managed Care Digest Series, Vol. 14, 2010–2011, p. 8; and The Henry J. Kaiser Foundation State Health Facts at www.statehealthfacts.org, accessed January 2012).

[See also Foundations and Health; Health and Fitness; Health Insurance; Hospitals; Medical Specialization; Medicare and Medicaid; Nursing; and Public Health.]

BIBLIOGRAPHY

Coombs, Jan Gregoire. The Rise and Fall of HMOs: An American Health Care Revolution. Madison: University of Wisconsin Press, 2005.
Hendricks, Rickey. A Model for National Health Care: The History of Kaiser Permanente. New Brunswick, N.J.: Rutgers University Press, 1993.
Luft, Harold S. Health Maintenance Organizations, Dimensions of Performance. 2d ed. New Brunswick, N.J.: Transaction Books, 1987.
Starr, Paul. The Social Transformation of American Medicine: The Rise of a Sovereign Profession and the Making of a Vast Industry. New York: Basic Books, 1982.
                                        Jan Gregoire Coombs

HEATING TECHNOLOGY

Homes and public buildings in early America depended on a central hearth or open fireplace for heating. The hearth provided a gathering place for the family and served for cooking as well as heating. Many families heated only one room during the winter; even so, fireplace heat often proved inadequate. Cast-iron stoves did not come into general use until the second quarter of the nineteenth century. Even Benjamin Franklin's famous Franklin stove of the 1740s, although efficient, had not been widely adopted. By the 1830s, however, cast-iron stoves became popular as the scarcity of firewood in urban areas converged with improvements in stoves themselves and the increasing accessibility of stoves and coal.

Centrally installed coal-fired furnaces appeared in the homes of wealthy Americans during the latter half of the nineteenth century. Built-in steam or forced-air heating systems did not become the norm until well into the twentieth century, however, when furnaces powered by other fuels, such as natural gas, electricity, and heating oil, became available. As urban areas and the middle class grew and utility companies expanded their distribution networks in the twentieth century, especially after World War II, heat consumption in homes and public buildings kept pace.

The steady increase in heating consumption ended in the early 1970s. With the energy crisis and a slowing of improvements in heating technologies, consumers became increasingly concerned with fuel and energy efficiency. "Alternative" heating methods, such as passive solar heating, became popular, although they represented only a small percentage of heating sources. By the early 1990s, natural gas led the heating fuels, followed by electricity, fuel oil, and all other sources.

Throughout the history of heating, older technologies have overlapped with new ones. Consumers in abundantly wooded rural areas continued to use fireplaces even as urban dwellers adopted stoves. Coal furnaces coexisted with cast-iron stoves, and gas and electric furnaces overlapped with coal. As late as the 1950s, coal and wood were still widely used heating fuels. Moreover, in a nation as vast and diverse as the United States, wide regional differences have existed in the choice of heating devices and fuels. In warm climate zones, heating has been less of a concern than keeping the indoors cool.

Shaped by markets and institutions, heating technologies have in turn influenced cultural standards. For example, eighteenth-century Americans largely rejected stoves because of cost, perceived dangers, and the abundance of wood fuel for fireplaces. In the twentieth century, by keeping rates low and advertising to targeted groups, utility companies encouraged Americans to consume more heat. Whereas eighteenth- and even nineteenth-century Americans endured smoky and unevenly heated homes and public buildings, by the post–World War II era Americans had become accustomed to automatic heating (and, increasingly, cooling) systems and were taking a well-regulated and comfortable indoor environment for granted.

[*See also* Electricity and Electrification; Environmentalism; Food and Diet; Franklin, Benjamin; Home Economics Movement; Household Technology; *and* Refrigeration and Air Conditioning.]

BIBLIOGRAPHY

Rose, Mark H. *Cities of Light and Heat: Domesticating Gas and Electricity in Urban America.* University Park: Pennsylvania State University Press, 1995.
Strasser, Susan. *Never Done: A History of American Housework,* pp. 50–67. New York: Pantheon, 1982.

Libbie J. Freed

HENRY, JOSEPH

(1797–1878), physicist and first director of the Smithsonian Institution. Born in Albany, New York, Joseph Henry attended the Albany Academy between 1818 and 1822. He was professor of mathematics and natural philosophy at the Albany Academy from 1826 to 1832 and professor of natural philosophy at the College of New Jersey, now Princeton University, from 1832 to 1846. He was chosen as the first secretary (director) of the Smithsonian in 1846. He served as chairman of the National Academy of Sciences from 1868 until his death in 1878 and as president of the American Association for the Advancement of Science from 1846 to 1850. He was also a member of the Light House Board from 1852 to 1871 and chair from 1871 to 1878.

Henry was the first American since Benjamin Franklin to gain an international reputation in experimental physics, especially electromagnetic induction. His pioneering work on the nature of electricity, magnetism, and the interaction between the two was valued both for its own sake and because it demonstrated how understanding nature could lead to technological advancement. His research proved essential for the development of the telegraph. Among his other discoveries, while a professor at Princeton, were the concept of the transformer and the oscillatory character of the discharge of a capacitor. He invented new ways to protect buildings from lightning. He also conducted research in electromagnetic screening, examined molecular cohesion using soap bubbles, and made the first measurements of temperature differences between sunspots and other regions on the surface of the sun. In recognition of his discoveries in electromagnetic induction, the unit of inductance, the "henry," was named after him.

As director of the Smithsonian, he set aside his personal research program to become an administrator and spokesperson for the importance of basic scientific research. At the Smithsonian, Henry established a program of direct support of research, scholarly publication, and international exchange. He supported research not only in the natural sciences, but also in anthropology and ethnography.

Often compared with Franklin, Henry became a larger-than-life symbol of American accomplishment in science and the acclaimed father of modern electrical technology, including the electric motor and the telephone. The premier American physicist of the mid-nineteenth century and an able institution builder, Henry, as a leader of American science, sought to establish rigorous standards for the scientific community and to increase the level of public support for research. His early accomplishments in the laboratory and his important discoveries accorded him the prestige and respect necessary for his success as a science administrator and spokesperson.

[*See also* **American Association for the Advancement of Science; Electricity and Electrification; Franklin, Benjamin; Mathematics and Statistics; National Academy of Sciences; Physics; Science; Smithsonian Institution; Technology; Telegraph;** *and* **Telephone.**]

BIBLIOGRAPHY

Bruce, Robert V. *The Launching of Modern American Science 1846–1876.* New York: Alfred A. Knopf, 1987.
Moyer, Albert E. *Joseph Henry: The Rise of an American Scientist.* Washington, D.C.: Smithsonian Institution Press, 1997.
Rothenberg, Marc. "Henry, Joseph (1797–1878)." In *The History of Science in the United States: An Encyclopedia,* edited by Marc Rothenberg, pp. 261–262. New York and London: Garland Publishing, 2001.

Marc Rothenberg;
updated by Elspeth Knewstubb

HEWLETT, WILLIAM

See **Computers, Mainframe, Mini, and Micro.**

HIGHER EDUCATION AND SCIENCE

Since the early eighteenth century, science and higher education have had a reciprocal relationship, but the terms of that relationship have varied with historical circumstance. Initially, influence flowed from science, across the Atlantic, and into the halls of American colleges, where it was fortunate to find a partial and somewhat precarious home. Over time, colleges and universities nurtured scientists who, in turn, were able to contribute to the advancement of science. Science nonetheless largely constituted an external and independent factor, and the timing and accuracy of its incorporation were critical variables in the evolution of higher education. Only in the twentieth century does the balance gradually shift so that the number, expertise, and resources of university scientists become a significant factor in the advancement of science. This relationship is thus central to the history of science and the history of American higher education.

Colonial Colleges.   The growth of science in the colonial college was slow and contested, yet dramatic. It was during the pre-Revolutionary era that the first real break with the fixed, scholastic curriculum was made. This break with scholasticism allowed science the room to flourish as an academic subject. It was prompted by three factors: first, the growth of Newtonian physics forced pedagogical innovations; second, the American Enlightenment fostered new intellectual footholds for science in the colleges; and finally, the creation of medical schools offered an alternative venue for biology and chemistry to develop.

The development of Newtonian physics spurred major changes to traditional instruction. First, colleges needed to hire new faculty capable of describing the universe in Newtonian terms. Newtonian physics demanded the elevation of mathematics instruction beyond the arithmetic included in the classical course. These innovations were sometimes supported by external donations to the colleges, as was the case at Harvard with the founding of the Hollis Professorship of Mathematics and Natural Philosophy in 1727. This founding was accompanied by the rapid acquisition of a large collection of philosophical apparatus (scientific instruments). Once Harvard assumed early leadership in the sciences, other colonial colleges felt pressure to adopt similar innovations. John Winthrop was appointed to the Hollis chair in 1739, and during his distinguished 40-year career he was the nearest colonial exemplar of a professional scientist. His course in "experimental philosophy" explicated basic Newtonian principles, and in 1761 he led an expedition to Newfoundland to observe the transit of Venus. He was an outspoken defender of scientific reasoning against the superstitions of the age. In 1762 he was named a member of the Royal Society.

The introduction of Newtonian science altered the pedagogy of colonial colleges by introducing demonstration lectures. Philosophical apparatus were required to deliver these lectures. These consisted of assorted balances, pulleys, and inclined planes to demonstrate mechanical principles; air pumps and vessels for creating vacuums; optical instruments to show Newton's theories of light and color; and microscopes, thermometers, and barometers. Winthrop demonstrated experiments with electricity in the 1740s, before Benjamin Franklin popularized that subject. For the foremost Newtonian science, astronomy, Harvard acquired a collection of telescopes, and other colleges followed when able. Yale president Thomas Clap (1738–1766), although a rigid Puritan, was devoted to astronomy. The most enlightened college leaders, provost William Smith of the College of Philadelphia (1754–1778, now the University of Pennsylvania) and president John Witherspoon of the College of New Jersey (1768–1795, now Princeton University), were strong supporters of science. Smith taught the subject himself, was secretary of the American Philosophical Society, and teamed with David Rittenhouse (a respected scientist and instrument maker) to measure the 1769 transit of Venus. Witherspoon made appointing a professor of natural philosophy his highest priority and outfoxed Smith to purchase the famed Rittenhouse orrery (a mechanical model of the solar system). Both these Scotsmen helped to embed Scottish moral philosophy in the American college, which rationalized a scientific approach to the natural world. Beyond this vague warrant, the colonial colleges did little to institutionalize or advance science. Even Winthrop's accomplishments produced no progeny. The situation was somewhat different in medicine.

Interest in establishing medical schools arose among physicians in New York and Philadelphia in the 1760s. In both cases, impetus came from recent medical graduates of Edinburgh University. The medical school in Edinburgh had only been organized in 1726, but it quickly became renowned for its excellent instruction and the favored locus for American students. William Shippen Jr. and John Morgan, both sons of wealthy Philadelphia families, took MDs at Edinburgh in 1762 and 1763, respectively. In 1765 both were appointed professors in what became the first school of medicine. Other appointments followed of physicians with similar backgrounds, including Benjamin Rush, who became the most authoritative figure in early American medicine. By 1769 a full medical faculty of five was offering courses—all Edinburgh graduates. Nor was the medical school a financial burden to the college; the professors' only compensation came from student fees.

Physicians in the eighteenth century were a highly stratified profession in which standing was determined primarily by the amount and nature of education they received; however, patients "rarely benefited…in proportion to the amount of training of the physician" (Rothstein, 1972, p. 38). In fact, medical science made little progress during the age of Enlightenment. Physicians, whether practitioners or professors, made diagnoses on the basis of superficial symptoms, like fever, and were hopelessly muddled regarding causation and cures. European-trained doctors nonetheless had great confidence in academic medicine. Their desire to introduce the "unequalled Lustre" of European medical education made the school a center for learning, unfortunately much of it erroneous.

The Early Republic. The American Revolution and the difficult conditions that followed disrupted the intellectual momentum of the major colonial colleges. Massachusetts enshrined Harvard as the "University at Cambridge" in the state constitution, and aspirations were high that "republican universities" would extend Enlightenment rationalism to "useful" and philosophical subjects. But these hopes were disappointed. The end of the eighteenth century marked the low point for American colleges and American science as well. The failure of republican universities left the colleges beset by poverty, stagnation, and rebellious youths; and the French Revolution and European wars isolated American science. In 1800 perhaps 20 positions existed for scientists in America. Before 1820, American science was sustained by a handful of individuals. This low state was epitomized by the neglect of astronomy—the favored science of the colonial colleges. In the early nineteenth century only elementary textbooks were used, and those were hopelessly out of date. Not until after 1830 would the importation of European texts begin the rehabilitation, and soon the revitalization, of that subject.

The flame of science was largely kept alive in American colleges through the cultivation of chemistry. The new "French chemistry" inspired by Antoine Lavoissier was expounded by a few pioneers, but was difficult to incorporate into the college course. Columbia appointed Samuel Mitchill as the first professor with that title in 1792, but he resigned in 1801 as the college offerings were reduced to the classical course. John MacLean, a Scotsman with European training, settled in Princeton and was soon made professor of chemistry (1798–1812), but he complained of little opportunity to teach his subject. Chemistry was a core subject of medical schools (where Mitchill also taught), which accounted for most early appointments. In 1802 Yale's pious president, Timothy Dwight, appointed Benjamin Silliman as professor of chemistry and natural history. He was chosen for his character and piety, but had no knowledge of the subject. Silliman went to the medical school of the University of Pennsylvania to learn his trade under James Woodhouse. The medical school was the largest and most prestigious in the country; in fact, for the first quarter century it was the country's largest institution of higher education. Medical schools did not necessarily appoint scientists to their chemistry chairs because they preferred MDs. However, Woodhouse's chair was eventually filled by Robert Hare, a brilliant innovator who had been Silliman's fellow student. The succeeding generation of American chemists were nearly all trained at the Pennsylvania or Harvard medical schools or by Benjamin Silliman at Yale.

Silliman completed his chemical education in England in 1806 and began 50 years of teaching and lecturing at Yale. He was famed for the spectacular effects produced in demonstration lectures, delivered to students, the public, and the Yale medical school. An active scientist, Silliman had his greatest impact on the development of academic chemistry. In 1818 he founded the *American Journal of Science (and Arts)*, which became the premier American scientific journal. He also founded a scientific dynasty at Yale. His son, Benjamin Silliman Jr., was drawn to applied chemistry and played an important role in launching the scientific school at Yale. His son-in-law, James Dwight Dana, replaced Silliman upon his retirement in 1850 and greatly contributed to the development of geology and zoology. And Dana's son Edward also became a Yale professor of chemistry and editor, like his father and uncle, of the *American Journal of Science*. Perhaps the elder Silliman's greatest contribution was to steer American chemistry toward mineralogy and geology, fields that would spearhead the intellectual and institutional advancement of academic science.

Science in the Antebellum Colleges.
Science in American colleges grew larger, denser, and more up to date in the decades of the 1820s and 1830s, but this growth was quite uneven. The proliferation of colleges after 1820 produced many small, precarious institutions where science teaching would long be rudimentary at best. Among established colleges in the East, growth in enrollments and endowments allowed the expansion of faculties. Despite ubiquitous acceptance of the fixed, classical course, consecrated by the Yale Reports of 1828, a disproportionate number of new positions were in science. Professors of chemistry and natural history were appointed, and those positions were then split. Mathematics was similarly separated from natural philosophy, which soon gave rise to physics. By the 1840s a college faculty of nine would be likely to contain four professors in science. Whether they would be filled with scientists was another matter. Men with scientific training were exceedingly scarce, as Thomas Jefferson noted when trying to staff the University of Virginia in 1824. Colleges were largely staffed with their own graduates. Whether some of them acquired advanced training or were capable of contributing to their fields could be largely happenstance.

Robert McCaughey (1974) has documented the gradual professional orientation of the Harvard faculty on the basis of specialized training prior to appointment, scholarly publications, external professional ties, and above all dedication to one's field. The 1821 faculty consisted of heavily inbred Harvardian pedagogues with the exception of George Ticknor, an independent Boston Brahmin, and William Peck, an amateur naturalist who was appointed to a chair in natural history endowed by local enthusiasts. In the 1845 faculty, however, five of the nine professors fulfilled professional roles, two of them scientists. Benjamin Peirce was a local product and a brilliant mathematical astronomer; and Asa Gray was an outsider whose talents as a naturalist had been recognized by the university. The academic ambitions of the Latin professor, Charles Beck (a German political refugee with a PhD), foreshadowed the important role that philology would soon play in academic professionalization.

The severe depression of 1837–1843 caused a hiatus in academic development in the United States, but the two decades that followed saw marked change in science and higher education. In 1840, the first professional scientific organization was founded, the Association of American Geologists and Naturalists (AAGN). Geology had gained a foothold in the colleges beyond Yale and was further stimulated by state-sponsored surveys to discover mineral wealth, particularly coal. Unlike previous learned societies, the AAGN sought only members "devoted to Geological research with scientific views and objects" (Kohlstedt, 1976, p. 67). In 1847 the AAGN resolved to expand into the American Association for the Advancement of Science (AAAS) to popularize science and expand its teaching in the colleges. The AAAS was organized as a far more inclusive organization. However, its leadership came heavily from scientists associated with colleges. Meaningful scientific activity was still quite concentrated. Among college-educated leaders, one-third graduated from Harvard or Yale. Moreover, Harvard dominated in physical science and natural history, whereas Yale ruled chemistry and geology. Another 8 percent were products of the U.S. Military Academy, the only institution teaching engineering and French mathematics prior to 1840. Active scientists were also concentrated in New England and the Mid-Atlantic states. Although a good number of natural scientists also possessed MDs, the proprietary medical schools of the Jacksonian era taught little up-to-date science.

The leaders of American science had become acutely sensitive to the need for advanced scientific training beyond the classical AB degree. A few "resident graduates" continued their studies at Yale and Harvard, but after 1850 a larger number of aspiring scientists were going to Europe to study, especially at German universities—a nagging reminder of the dependence and inferiority of American science. How advanced study might be organized, let alone financed, was a conundrum. The fixed classical curriculum had no room for additional or advanced courses, and schemes for scientific education all assumed that it would be left intact. In 1851 the leading figures in the AAAS developed a scheme to establish a German-style university in Albany that would teach advanced students strictly in science. Their proposal was seriously considered in the New York State Legislature, but when no funding was forthcoming the plan collapsed. That same year, Henry Tappan published *University Education* (1851), the fruit of his observations of Prussian universities and his experience at New York University. Tappan too envisioned a university dedicated to science and scholarship, for which the colleges would offer mere preparation. He hypothesized that only New York City offered the concentration of wealth and talent to realize such an enterprise. Instead, he was named president in 1852 of the reorganized University of Michigan. Tappan achieved what was possible before being fired in a trustee coup in 1863: he established a true Bachelor's of Science course and later added engineering; he raised funds for an observatory and hired a German PhD as astronomer; and he attempted to found graduate education for at least the Master's degree.

Schools of Science. The problem of advancing teaching and research in basic, theoretical science in the United States was confounded by a prevailing belief in "useful knowledge." This notion, which stemmed from the Enlightenment, held that basic scientific discoveries could be translated directly into practical applications for common workers. It was a powerful stimulus of support for science, but the bane of actual scientific institutions. American universities in fact had an implicit model for expanding into fields beyond the scope of a liberal education, namely, professional schools. Departments or schools for law and divinity had evolved earlier in the century by appointing a professor, gathering students, raising funds for more appointments, and conferring degrees. These units were self-supporting, were not open to college undergraduates, and did not require a college degree for admission. This pattern was duplicated in different ways at Yale and Harvard to produce separate "schools of science."

At the request of the Sillimans, Yale in 1846 authorized separate scientific classes in agricultural and applied chemistry. When the Greek professor also sought formal classes for his advanced students, Yale responded by creating the Department of Philosophy and the Arts (1847), an echo of a European faculty of philosophy, but purposely designed as a parallel professional school. That year it offered eight unrelated courses, six by regular faculty and two by the unpaid science professors. In 1852 a professor of civil engineering was added, and the Yale Scientific School began awarding Bachelors of Philosophy degrees after two years of courses and an examination. The school came of age after 1860, when the gifts of Joseph Sheffield finally allowed it to pay faculty salaries. The course was extended to three years and broadened to several science and engineering fields, and an entrance exam was established to provide some control over student qualifications. Now called the Sheffield Scientific School, it offered the first true undergraduate course in science. It also awarded the first American PhDs in 1861. Yale's evolutionary approach produced a course of study that future Harvard president Charles W. Eliot would soon recognize as the most successful exemplar of the "New Education" (*Atlantic Monthly*, February and March, 1869).

Harvard's Lawrence Scientific School had greater resources from the outset, thanks largely to textile magnate Abbott Lawrence, but less of an educational focus. Lawrence supported the hiring of the renowned Swiss naturalist, Louis Agassiz, who brought prestige but little interest in students. Few classes were offered, so that students largely studied individually with professors. Admissions standards were lenient and most students enrolled for a year or less. The Lawrence School embellished Harvard's scientific resources but contributed less than it might have to science education. Ironically, the independent success of the Sheffield School would be an impediment to its integration with Yale College for 60 years, but the lack of definition of the Lawrence School eased its incorporation into Harvard College under President Eliot (1869–1909).

Schools of science nonetheless became temporary solutions for incorporating science alongside the classical college in the third quarter of the nineteenth century. They were established with less success at Dartmouth, Princeton, and numerous other colleges, where they were typically plagued by the inferior preparations of their students and an invidious separation from the classical course. On the eve of the Civil War, a basic undergraduate education in science might be had at Yale, Harvard, and Michigan; and in engineering at Yale, Union College, West Point, and Rensselaer Polytechnic Institute.

Land-Grant Universities. The landscape of science in higher education changed when Justin Morrill achieved the passage of the Land Grant Act of 1862. Although ostensibly designed to promote "agriculture and mechanic arts"—"without excluding other scientific and classical studies and including military tactics"—the initial legacy of the Morrill Act was the creation of new institutions and not the modernization of science. Daniel Coit Gilman of the Sheffield Scientific School became the foremost spokesman for the idea that the new land-grant universities should be our "national schools of science." However, aside from his own institution, which received the Connecticut land-grant designation, the new land-grant colleges tended to be claimed by agricultural interests with quite practical expectations. As these expectations were disappointed, their hostility to science and liberal arts grew more strident, compromising the early development of the colleges in many states. This situation only improved after the 1887 Hatch Act funded agriculture experiment stations and the Second Morrill Act of 1890 provided these colleges with an invaluable infusion of funds. Agriculture then led the institutionalization of scientific research, particularly at those land-grant institutions that were not state flagship universities.

Engineering too made little progress at land-grant universities before these developments. The two exceptions were Cornell University and the Massachusetts Institute of Technology (MIT). MIT received only one-third of the state's land-grant funds, but they were decisive in realizing the vision of William Barton Rogers of a school to teach modern engineering. MIT's engineering programs combined solid scientific preparation with specialized practical instruction in several branches of engineering, and its graduates soon dispersed to build America's railroads and cities. In New York, the gift of Ezra Cornell allowed president Andrew Dickson White (1868–1885) to erect a true university with advanced scientific instruction.

The Academic Revolution. The last three decades of the nineteenth century experienced the most decisive transformation in the history of American higher education. Described by Laurence Veysey as the *Emergence of the American University* (1965), it consisted of the erection of universities dedicated to graduate education and the advancement of knowledge on the base of undergraduate colleges largely oriented toward student socialization and collegiate culture. This dichotomy reflected the dual mission of American universities. The academic revolution nevertheless transformed three complementary organizational spheres. First, the coalescence and formal organization of academic disciplines provided the intellectual foundation. Second, the reformation of existing universities, led by Charles W. Eliot, president of Harvard (1869–1909), achieved the assimilation of disciplinary knowledge. This process was accelerated by the founding of new, endowed universities, led by Johns Hopkins University (f. 1876), explicitly

dedicated to research and graduate education. Third, the transformation throughout higher education to a curriculum based on the disciplines made the American university the fount of academic knowledge.

Charles W. Eliot was already an expert on the new scientific education when he was elected president, and his inaugural speech left no doubt about his dedication to an unfettered quest for knowledge. He gradually dismantled the traditional classical course until undergraduate study was nearly entirely elective by the 1880s, thus allowing professors to offer advanced courses in their fields of expertise. With recitations and rote learning banished, Harvard sought distinguished scholars for all its teaching positions. The Lawrence Scientific School was phased into the College faculty, and, in 1890, the Harvard Faculty of Arts and Sciences was formed to include all nonprofessional faculty, who also taught in the Graduate School. This was the basic structure assumed by the American university, which allowed a large body of undergraduates to support a learned faculty that also taught small numbers of graduate students and performed research.

Harvard arrived at this formula through trial and error, guided by Eliot's strategic vision. The new universities tried various schemes before evolving toward the same result. Cornell, which opened in 1868, was the first university to be intentionally structured in a modern way. Andrew Dickson White had been inspired by Henry Tappan's ambitious reforms at the University of Michigan. He instituted a system of major subjects that allowed the teaching and learning of advanced subjects. He was more successful initially in establishing excellent science departments than in fulfilling the land-grant mandate in agriculture and engineering. But in the 1880s Cornell pioneered the advancement of those fields as well.

Daniel Coit Gilman left Yale after being passed over for the presidency and accepted the challenge of making the University of California a "national school of science" (1872–1874). However, he became frustrated by the agricultural interests that opposed academic science and attempted to claim the land-grant funds. He welcomed the offer to become president of the newly chartered

Johns Hopkins University. There he abandoned the land-grant mission of applied science in favor of emulating the pure research focus of German Universities—now recognized as the world leaders in science and humanistic scholarship. He assembled the best scholars and scientists he could hire and aimed, above all, to train PhDs. Until 1890, Hopkins had more graduate than undergraduate students and produced more PhDs than Harvard and Yale combined. Hopkins professors were leaders in establishing several of the disciplinary associations, and the university helped them found academic journals. In the 1880s, the decade in which the largest number of Americans earned German PhDs, Hopkins set the standard for research training in the United States and a challenge to other universities.

Hopkins psychology professor, G. Stanley Hall, sought to improve on this model when he became president of the new Clark University (1889) by creating an all-graduate institution. He was able to recruit a faculty of distinguished scientists and appeared to make a brilliant start. However, the imperious Hall quickly alienated his faculty and the donor, Jonas Clark, who expected an undergraduate college to soon follow. In 1891, most scientists departed, 15 of them recruited to the new University of Chicago. By 1890 research and graduate education were widely touted as urgent goals for American higher education, but graduate universities proved too narrow a base to support the large and diversified faculty required for a disciplinary curriculum, even with gilded-age philanthropy.

The University of Chicago, which opened in 1891, was the most consequential of the new philanthropic universities. Its organizing force was William Rainey Harper, a Yale professor and brilliant scholar of Near-Eastern Languages, the most erudite Baptist in the country, if not the world. Fellow Baptist John D. Rockefeller entrusted him with responsibility for fashioning a major graduate university. Harper crafted an eclectic institution of great distinction, consciously intended to be a capstone of sorts for Chicago and the Midwest. He recruited top scientists and scholars for graduate education, but also included a large undergraduate college. He established a press to publish scholarly works and a summer school to provide wider opportunities for advanced studies and encouraged junior colleges to become feeders for the university. Harper cajoled increasing donations from the ever-wealthier Rockefeller, but he also appealed to community leaders in Chicago with the university's multiple offerings.

The new universities and the academic disciplines grew symbiotically. Between 1882 and 1905 the national associations of all the major disciplines were organized, as well as most of their initial journals. The forces shaping each discipline differed somewhat, but all sought to place the guidance of intellectual standards in the hands of the most competent professionals. With academic reputations established through publication and recognition, the emerging research universities were able to evaluate the relative contributions of scientists. Or rather, academic departments were entrusted with evaluating scholars in their respective fields. Concomitant with the professionalization of the disciplines was the rise of autonomous academic departments, charged with defining and upholding academic standards. The research universities not only led this process, but also, until well into the twentieth century, they were the only American institutions that could cultivate faculties capable of research. A survey in 1906 of the "Leading American Men of Science" found 60 percent of known scientists in 15 research universities, half of those at Harvard, Columbia, Chicago, and Cornell (Geiger, 2004, p. 39). Still, only the wealthiest of this group were able to provide adequate support for research in science.

In 1901 Andrew Carnegie resolved to address America's inferiority in research with a major gift for the advancement of knowledge. Despite pleas to support universities, he instead endowed the Carnegie Institution of Washington (CIW, 1902). Indeed, the CIW regarded professors as overburdened with teaching and colleges too penurious to support research. Rockefeller too bypassed universities when he sought to further medical knowledge by endowing The Rockefeller Institute for Medical Research (1901). The devastating survey of medical education by Abraham Flexner (Carnegie Foundation, *Report on Medical Education in the United States and* Canada, 1910) confirmed that only the medical schools at Harvard and Johns Hopkins were capable of undertaking research. But the philanthropists before World War I underestimated the academic scientific community. Its chief asset was size—the large and growing number of scientists, now well trained, holding faculty positions that allowed them to participate to some degree in scientific investigation. And an increasing number of American scientists had achieved distinction. This was apparent in physics, perhaps the most competitive international discipline.

A comparison of national physics communities, circa 1900, revealed the United States to have the largest number of academic physicists and a higher rate of growth than any European country, although in intellectual productivity it was far behind world-leading Germany (Forman et al., 1975). American physicists Henry Rowland of Johns Hopkins and Albert Michelson of the University of Chicago had achieved international reputations. In chemistry, Harvard's Theodore Richards was eminent enough to be offered a professorial chair at Göttingen. At MIT, Arthur Noyes established the Research Laboratory in Physical Chemistry (1903), half supported with his own funds, which became renowned for its precise measurements of physical properties and soon for training young scientists who later would carry their expertise to Berkeley and Caltech. Thomas Hunt Morgan, a Hopkins PhD (1891), took the study of genetics to Columbia in 1904 and created a modern collaborative research group that pioneered the new study of chromosomes. These and numerous less-honored scientists added to American achievements in geology and astronomy. Still, the resources needed for the vigorous growth of university science were largely lacking, and the prowess of academic scientists was not fully recognized until World War I.

A Privately Funded System of University Research.

During America's brief engagement in the war, university scientists were largely assigned to government installations, where they contributed research on submarine detection, ballistics, communications, and wartime production. More significant, this experience changed perceptions of the value of university science. Following the war, this newfound appreciation encouraged support for academic research by technological industries and, especially important, the large, general-purpose foundations.

A key role in all these developments was fulfilled by George Ellery Hale. Possessing only a BA from MIT (1890), he became the country's first astrophysicist through his own discoveries in solar astronomy. Appointed to the faculty of the new University of Chicago (1892), he convinced traction magnate Charles Yerkes to fund an observatory, which he directed. In 1904 he induced the CIW to build the Mount Wilson Observatory above Pasadena, where he relocated. Elected to the National Academy of Sciences in 1902, Hale was frustrated by the torpor of this honorific body. Sensing the need to mobilize American science for the war effort, he campaigned to establish the National Research Council (NRC) in 1916. As chairman of the NRC, Hale oversaw the wartime research committees that allocated scientific tasks. He also sought to continue this role of providing strategic direction for American science after the Armistice.

Encouraging the collaboration of academic science and industrial research became a fundamental goal of the NRC, which included leaders of the principle industrial research laboratories—AT&T, General Electric, Du Pont, and Eastman Kodak. The scientific leadership was pervaded by an ideology of science that held that the nation could benefit greatly from the support and coordination of basic scientific research. Legislation was considered to provide funds for states to found engineering research stations, after the agricultural model, but failed to be enacted. The private interests represented in the NRC, in fact, feared that federal involvement would bring political interference. Instead, they favored a privately supported system of university research, orchestrated by industry, foundations, and the NRC. Specifically, they wished to support "best-science" practices for directing research, support from private capital, and oversight from a kind of interlocking directorate of industry and nonprofit leaders in the governing boards of the relevant institutions, including private universities.

Foundations embraced the role of advancing university research tentatively at first, but massively by the end of the 1920s. Hale was instrumental in inducing the Carnegie Corporation to donate $5 million for a building and endowment for the NRC (1919), which allowed the organization to serve as both broker and facilitator of science policy. The Rockefeller Foundation gave $500,000 for postdoctoral fellowships in the sciences (1919, later expanded), to be awarded on merit by NRC committees. This pattern of utilizing an intermediary body to serve and coordinate academic fields was soon duplicated by the Social Science Research Council, the American Council on Education, and the American Council of Learned Societies. Foundation fellowships made a particularly crucial contribution, as the Guggenheim Foundation and the International Education Board supplemented the Rockefeller awards. More than 130 young scientists received postdoctoral fellowships in physics alone during the 1920s, and 50 of them studied in leading European centers. In addition, awards to Europeans brought scientists to the United States, where they found far better career opportunities in American universities. Foundation support brought this new generation of academic scientists to the frontiers of research and thus played a crucial role in the rise of American science.

Industry played several roles in the advancement of academic science. Leaders of technology corporations served in the highest councils of the NRC, and numerous industrialists, like Kodak founder George Eastman, supported universities through their philanthropy. Direct research connections tended to be confined to chemistry and engineering. The universities of Michigan and Illinois, for example, established engineering departments to perform contract research. This approach was pursued most aggressively by MIT throughout the 1920s. However, by the end of the decade it was largely ostracized by the scientific establishment for being too committed to applied research for industry. In 1930 Princeton physicist Karl Compton was named president with a clear mandate to rejuvenate basic research at the institute. Scientific leaders like George Hale nonetheless believed that industry had an obligation to support academic research because it was the ultimate beneficiary. In 1925 he led the organization of a National Research Fund intended to raise $20 million in contributions from major corporations. Despite elite backing, pledges were difficult to attract, and the commitments that were made were abandoned at the onset of the Depression.

Hale was more successful in connecting with industry through the California Institute of Technology. When he became director of the Mount Wilson Observatory, Hale joined the board of Throop College of Technology in Pasadena and resolved to raise it to a major scientific institution. He persuaded his wartime NRC colleagues, Arthur Noyes (1919) and Robert Millikan (1920), to join him, and Caltech became a new center of American science. Millikan, the country's most renowned scientist, was awarded the Nobel Prize in 1923 for measuring the charge of an electron. The resources to make Caltech into a scientific powerhouse came first from Southern California boosters on the board, who believed that science could spur the development of the region, and then from Hale's foundation connections. Caltech met the expectations of all these backers by focusing basic research on areas of scientific challenge and practical usefulness, such as aviation, high-voltage transmission, and petroleum engineering.

The privately funded system of university research received an enormous stimulus from Rockefeller trusts in the latter 1920s. From 1923 to 1929, Beardsley Ruml directed grants in the social sciences, and Wickliffe Rose did the same for the natural sciences. Both made commitments of increasing magnitude until their funds were incorporated into the reorganized Rockefeller Foundation at the end of the decade. Rose famously declared his intention to "make the peaks higher," a strategy that approximated most foundations' tacit approach. Although sounding rather crass today, this policy was highly effective in aiding American science finally to overcome its provincial status.

In 1910 the science journalist Edwin Slosson portrayed 14 *Great American Universities*: the venerable universities of Harvard, Yale, Princeton, Columbia, and Pennsylvania; the philanthropic foundings, Cornell, Johns Hopkins, Stanford, and Chicago; and the state universities of California, Illinois, Michigan, Minnesota, and Wisconsin. Along with MIT and Caltech, these institutions constituted the research universities of the interwar years (Geiger, 2004). To varying degrees, they sought to encourage scientific research, graduate education, and the kind of academic quality on which these activities depend. Not until the late 1930s were additional institutions able to make serious efforts to emulate this model. Some of these interwar research universities were the "peaks" that foundation philanthropy sought to raise, but all required internal efforts to become or remain competitive in research.

Harvard remained the preeminent American university during the interwar years. President Abbott Lawrence Lowell (1909–1933) emphasized undergraduate education, and his rather inbred faculty may have been slipping somewhat. However, Harvard's inherent wealth and resources guaranteed scientific leadership and the patronage of foundations. When chemist James Conant replaced Lowell in 1933, Harvard once again became committed to scientific leadership. The University of Chicago was the largest magnet for foundation support. Ruml was a Chicago PhD (1917), and he accurately judged the university to be furthest advanced in his goal of bringing scientific rigor to the study of society. Rose made the largest single foundation commitment to Chicago's Oriental Institute. Foundation largesse undoubtedly helped to sustain Chicago's rigorous commitment to science during the quixotic presidency of Robert Maynard Hutchins (1929–1951). Although interwar Princeton has been described as a finishing school for wealthy undergraduates, a Rockefeller grant virtually forced it to become a research university. A matching Rockefeller grant greatly enhanced Princeton's science capacity with endowed professorships, soon filled with eminent European scientists. Princeton established a small but elite graduate program, which was reinforced when the Institute for Advanced Study located there (1932).

Foundations rarely made grants to public universities before the 1930s, and then in rather small amounts. Only a few of these universities were able to derive scientific funding from other sources. At the University of Wisconsin, the legislature forbade acceptance of foundation money, which it considered tainted. However, in 1925 university scientist Harry Steenbock donated his discovery for vitamin D enhancement to the university, which established the Wisconsin Alumni Research Foundation (WARF). By the 1930s, income to WARF was available to support scientific research. At the University of Michigan, a large gift from Horace Rackham established the graduate school and an endowment for research. Such windfalls strengthened research at both schools, but state universities depended more heavily during the 1920s on growth—in students, faculty, and state appropriations. They chiefly hired new PhDs or NRC fellows and nurtured their maturation as scientists. No university was more successful with this approach than the University of California. The departments of chemistry and physics, in particular, made determined efforts to advance by hiring the best available young scientists—and not their own graduates, as was common elsewhere. E. O. Lawrence was hired away from Yale in 1928, largely on his promise, which was amply fulfilled. In 1929 J. Robert Oppenheimer joined the faculty after extensive study among European physicists. Oppenheimer was in the forefront of theoretical physics, and Lawrence received a Nobel Prize in 1939 for developing the cyclotron. Physics at California succeeded in rivaling the top-rated private universities.

The 1920s constituted a watershed for American science and American universities. During that decade the universities became definitively the "Home of Science." As late as mid-decade, an official inquiry still found "obstacles to university research": difficulty in attracting top scientific talent, the burden of excessive teaching, etc. However, these problems were largely transcended at the research universities mentioned earlier. Scientific research had become a major commitment, supported by external and internal funds. By the end of the decade, the United States had achieved scientific parity with the leading European nations. When Albert Einstein joined the Institute for

Advanced Study in 1933, even Europeans considered America the new center of the natural sciences. The European scientists who followed Einstein in fleeing Nazi persecution merely enriched the university-based scientific profession that had emerged in the previous decade.

The decade of the 1930s presented challenges of a different sort. Smaller awards replaced large foundation grants for specific projects. Scientific leaders felt that this now-thriving enterprise was malnourished for resources. Many now looked to the federal government to provide a New Deal for academic research. They presented a compelling case in a report, *Research—A National Resource* (1938), but to no avail. A major federal presence in the university research system had to await the exigencies of World War II.

The Federal Government and Postwar University Research.

The mobilization of university science for World War II began in 1940 with the National Defense Research Committee (NDRC) under Vannevar Bush. Although most wartime research was carried out at universities, most scientists relocated to special laboratories that focused expertise. Work on the largest effort, the Manhattan Project to build the atomic bomb, migrated from Chicago and Berkeley to the isolated weapons laboratory at Los Alamos, New Mexico. The next largest effort was devoted to radar and located at MIT. Other huge laboratories were established at Johns Hopkins for proximity fuses and Caltech for rocket and jet propulsion. The NDRC was responsible for assigning other research projects to teams of scientists at other locations. Medical research was least disruptive, with smaller projects awarded to scientists largely in their own laboratories. The most significant aspect of wartime research for future science was that it was destined to continue. Hostilities or not, the country was compelled to keep extending understanding of atomic energy, radar, jet propulsion, fuses, and myriad other technologies. However, a strong sentiment at the end of the war suggested that the achievements of American science rested on the strength of the prewar scientific establishment and that federal patronage would be necessary to sustain future advances.

Bush cogently argued this case in *Science—The Endless Frontier* (1945), calling for the establishment of a federally supported National Research Foundation.

Bush's insistence on independence from political influence doomed Congressional passage of the foundation in 1946. The National Science Foundation, when it was finally created in 1950, was a shell of its original conception. Instead, federal postwar research largely remained in its wartime channels. The Atomic Energy Commission (AEC) inherited the charge of the Manhattan Project and responsibility for all atomic research. Wartime medical research was incorporated into the Public Health Service and the National Institutes of Health. Among the armed services, the navy was most eager to exploit postwar science, and it created the Office of Naval Research (ONR) as a special unit for this purpose. Concerned above all with establishing working relationships with academic scientists, the ONR concluded open contracts with research universities to which investigator-initiated projects could be appended as task orders. For the remainder of the 1940s it served as an ersatz national science foundation, the chief federal patron of basic science.

The postwar research system hugely enlarged university research but, with the exception of medicine funding, came entirely from the defense establishment. The large wartime laboratories at MIT, Johns Hopkins, Caltech, and elsewhere continued to operate as quasi-government installations. Basic research was supported to varying degrees by ONR and by the AEC in nuclear physics. Universities possessed differing capacities and eagerness to accommodate these new conditions. MIT embraced a continuation of its wartime role wholeheartedly. Its original radar lab expanded in size and scope, and the Institute soon extended its expertise to applications and weapons development. Harvard took the opposite position, eliminating its wartime labs in sonar and radar countermeasures and eschewing (for a time) all but the most basic research. No institution benefited more than the University of California. The AEC gave Lawrence and his epigone a virtual blank check for nuclear research, and Berkeley emerged as the world leader in this most prestigious and visible field. Under the subsequent leadership of the chancellor, Clark Kerr, the campus built upon this distinction to become the nation's most highly rated university. At nearby Stanford, the engineering dean Frederic Terman used federal contracts for defense electronics to leverage a modest research institution into the top ranks of the American universities. Universities that were slow to adapt to the postwar research system soon realized that they had to participate or become scientific backwaters.

Federal largesse could make research into an autonomous university activity. This was not an entirely novel situation. University museums and observatories had always been separately administered with their own sources of funding, and this had also been true of the occasional prewar institute. Now this pattern was greatly expanded with special units that blended federal purposes with academic spin-off benefits. Universities thus created centers and institutes—at the University of California they were called "organized research units." However, the proliferation of such units meant that the amount of research performed was no longer related to the amount of teaching or the number of teachers.

Academic research grew steadily in the 1950s. Congress regularly increased appropriations to NIH for medical research, and postwar foundation patronage of the social sciences was magnified by the entry of the giant Ford Foundation. However, science leaders were dismayed by the continual reliance on defense agencies, which in turn became focused more narrowly on weapons development. The debate over appropriate funding for academic science changed abruptly in October 1957, with the Soviet launch of *Sputnik I*. Science and education now became part of Cold War competition—and the United States appeared to be losing. Federal policy quickly endorsed an expansion of basic research. A report chaired by Berkeley physicist Glenn Seaborg (1960) emphasized research as a crucial investment; academic research and graduate education were essential to the national welfare and hence a responsibility of the federal government. Moreover, it recommended doubling the number of

first-rate centers of academic science from 15 to 20 universities to 30 to 40.

From 1960 to 1968 spending for academic research quadrupled; the national effort as a portion of the gross domestic product (GDP) doubled. The creation of the National Aeronautics and Space Administration (NASA) as the civilian space agency provided an additional channel of federal patronage. The National Science Foundation was finally given the resources required for its original mission. And Congressional appropriations for NIH were each year more bounteous. The unique feature of this expansion was the support given to developing the scientific base. Specific science development programs were intended to build faculty and infrastructure at emerging research universities. Doctoral students receiving federal support quadrupled to 60,000 from 1961 to 1966. The number of PhD graduates, which averaged 9,000 in the mid-1950s, exceeded 30,000 in the early 1970s. These graduates not only staffed the country's burgeoning colleges and universities, but also populated a larger number of research universities. The projection of the Seaborg Report was certainly exceeded. Universities that had been only marginally involved in organized research in the 1950s became full participants in the academic research economy by 1970. Established research universities were perhaps even greater beneficiaries, expanding graduate programs, scientists, and sponsored research. Although critics would allege that undergraduate education suffered as a result, there can be no doubt of the powerful impact on the advancement of science. As in the 1920s, a fairly extravagant national investment in universities produced a striking relative and absolute advancement of American science.

Like all golden ages, the post-*Sputnik* boom could not last. Federal support for research leveled off after 1968. The overall decline in research dollars was modest, in part because of continued ample funding for medical research. The sharpest cuts occurred in programs supporting science development. By the early 1970s, the country no longer needed additional research universities or more PhDs. Retrenchment became the watchword for much of the decade as both private and public universities faced fiscal constraints, exacerbated by rising inflation. However, academic research tended to consolidate rather than decline. Moreover, there were two striking areas of progress in this otherwise dismal decade. Late-developing universities in the so-called Sun Belt significantly advanced their academic standings. For many, enrollments and appropriations continued to grow, and now there was an oversupply of academic talent from which to recruit. Second, and of greater long-term significance, basic discoveries in the manipulation of genetic materials laid the foundation for a revolution in the scientific understanding of the most fundamental biological processes—a scientific revolution equivalent to the breakthroughs in atomic physics in the first half of the century.

The genomic revolution in biotechnology occurred with the development of techniques for manipulating DNA molecules to join them together in new combinations and to clone them in bacteria. These techniques allowed genes to be isolated, replicated, and inserted into different organisms. In the 40 years since the breakthroughs of the early 1970s, recombinant DNA technology has transformed understanding of fundamental life processes and provided tools for affecting them. Thus, the biotechnology revolution has revolutionized not only theory and practice in the life sciences, but also applications to medicine and agriculture. This revolution had its origins in university laboratories, but unleashed forces that transformed academic research as well. The early pioneers never considered patenting the tools and techniques that drove these discoveries, but in 1974 Stanley Cohen of Stanford and Herbert Boyer of U.C. San Francisco applied for a patent claiming ownership over recombinant DNA (issued in 1980). Boyer subsequently teamed with a venture capitalist to found Genentech, which created and marketed human insulin as the first bioengineered pharmaceutical product. By 1980, the gold rush was on to patent and commercialize the fruits of biotechnology. Although the commercialization of the life sciences that ensued has many critics, the inherent value of the cornucopia of innovations issuing from biotechnology made commercialization inevitable.

Science and Universities in the Current Era, 1980–2010.

The revolution in biotechnology converged with deep concern for the competitiveness of the American economy to spur a reorientation of the research economy in the 1980s. This new direction was signaled by the Bayh–Dole Act (University and Small Business Patent Procedures Act, 1980), giving universities title to inventions resulting from federally sponsored research and enjoining them to commercialize them. The act was explicitly intended to utilize the results of university research for economic development. Before Bayh–Dole, 25 universities had internal intellectual property offices for patenting and licensing; by 15 years later, every major university had one. Bayh–Dole itself was merely the most prominent of a series of enactments by federal and state governments intended to mobilize academic research to develop and transfer technology to industry. This effort, however, derived credibility and urgency from the revolution in biotechnology. Few university patents generated income, but some biotechnology licenses produced large windfalls for universities.

Patenting and licensing were the most visible outward manifestations of a reorientation of university research. Biotechnology was the most prominent of the "science-based technologies"—areas of pure science with clear commercial potential. A far larger movement supported research in collaboration with industry or in areas deemed ripe with future economic potential. The National Science Foundation led the effort to force-feed the development of emerging science-based technologies, particularly nanotechnology, and state governments also joined this effort hoping to stimulate academic research that would contribute to local economies. The portion of university research funded by industry roughly doubled in the 1980s, from 3 to 6 percent, but did not rise beyond that level. However, the great rapprochement of universities and industry was a marriage encouraged and sometimes arranged by government policies.

The growing economic relevance of academic research has been a boon to the country's research universities. American society has provided relatively abundant resources in the expectations of furthering innovation and, ultimately, the competitiveness of American industry. Not that these resources have been targeted only on research with commercial potential; rather, the federal science agencies, leading universities, and large corporate funders have generally realized the necessity of maintaining a healthy, balanced academic research system focused primarily on basic science. The result has been one of the most remarkable features of the current era: from 1968 to 1982 academic research grew from roughly $8 to $12 billion constant dollars. Since then it has risen on average by more than $1 billion each year. University research's share of the GDP increased by 50 percent in the in the early twenty-first century, and the proportion of basic research remained stable during this expansion.

Although there has been a secular trend toward the extension of research to greater numbers of universities, the bulk of research funding still flows to the laboratories of the same universities. Over the current era, this has produced a pronounced intensification of research as the volume of research has increased far more than enrollments or faculty. More than ever, research has become an autonomous mission, only loosely linked (if at all) with undergraduate education. Research universities have become critical centers of the knowledge economy, advancing the frontiers of knowledge while providing multiple services as repositories and disseminators. They have set the standard for what are now dubbed, with undisguised envy, world-class universities.

[*See also* **Agassiz, Louis; Agricultural Experiment Stations; American Association for the Advancement of Science; American Philosophical Society; Astronomy and Astrophysics; Atomic Energy Commission; Biological Sciences; Biotechnology; Bush, Vannevar; Chemistry; Compton, Karl Taylor; Dana, James Dwight; DNA Sequencing; Eastman, George; Einstein, Albert; Electricity and Electrification; Flexner Report; Genetics and Genetic Engineering; Geology; Gray, Asa; Hale, George Ellery; Hall, G. Stanley; High Schools, Science Education in; Human Genome Project; Jefferson, Thomas; Journals in Science, Medi-**cine, and Engineering; Lawrence, Ernest O.; Manhattan Project; Mathematics and Statistics; Medical Education; Medicine; Michelson, Albert Abraham; Millikan, Robert A.; Morgan, Thomas Hunt; Morrill Land Grant Act; National Academy of Sciences; National Aeronautics and Space Administration; National Institutes of Health; National Science Foundation; Nobel Prize in Biomedical Research; Noyes, Arthur Amos; Oppenheimer, J. Robert; Physics; Popularization of Science; Public Health Service, U.S.; Research and Development (R&D); Rittenhouse, David; Rockefeller Institute, The; Rogers, William Barton; Rowland, Henry A.; Rush, Benjamin; Science; Seaborg, Glenn T.; Silliman, Benjamin; Social Science Research Council; Space Program; Space Science; *and* Zoology.]**

BIBLIOGRAPHY

Benson, Keith R. "From Museum Research to Laboratory Research: The Transformation of Natural History into Academic Biology." In *The American Development of Biology*, edited by Ronald Rainger, Keith R. Benson, and Jane Maienschein. Philadelphia: University of Pennsylvania Press, 1988.

Bruce, Robert V. *The Launching of Modern American Science, 1846–1876*. Ithaca, N.Y.: Cornell University Press, 1987.

Caullery, Maurice. *Universities and Scientific Life in the United States*. Cambridge, Mass.: Harvard University Press, 1922.

Daniels, George H. *American Science in the Age of Jackson*. New York: Columbia University Press, 1968.

Elliot, Clark A., and Margaret W. Rossiter. *Science at Harvard University: Historical Perspectives*. Bethlehem, Penn.: Lehigh University Press, 1990.

Forman, Paul, J. L. Heilbron, and Spencer R. Weart. *Physics Circa 1900: Personnel, Funding, and Productivity of the Academic Establishments*, Vol. 5, Historical Studies in the Physical Sciences. Princeton, N.J.: Princeton University Press, 1975.

Geiger, Roger L. *Knowledge and Money: Research Universities and the Paradox of the Marketplace*. Stanford, Calif.: Stanford University Press, 2004.

Geiger, Roger L. *Research and Relevant Knowledge: American Research Universities since World War II*. New Brunswick, N.J.: Transaction Publishers, 2004.

Geiger, Roger L. "The Rise and Fall of Useful Knowledge: Higher Education for Science, Agriculture,
and the Mechanic Arts, 1850–1875." In *The
American College in the Nineteenth Century*, edited
by Roger L. Geiger. Nashville, Tenn.: Vanderbilt
University Press, 2000.

Geiger, Roger L. "Science, Universities and
National Defense, 1945–1970." *Osiris* 7 (1992):
94–116.

Geiger, Roger L. *To Advance Knowledge: The Growth
of American Research Universities, 1900–1940*.
New Brunswick, N.J.: Transaction Publishers,
2004.

Geiger, R. L., and Creso M. Sá. *Tapping the Riches of
Science: Universities and the Promise of Economic
Growth*. Cambridge, Mass.: Harvard University
Press, 2008.

Greene, John C. *American Science in the Age of
Jefferson*. Ames: Iowa State University Press, 1984.

Guralnick, Stanley M. *Science and the Ante-Bellum
American College*. Philadelphia: American Philosophical Society, 1975.

Heilbron, John L., and Robert W. Seidel. *Lawrence
and His Laboratory: A History of the Lawrence
Berkeley Laboratory*. Berkeley: University of California Press, 1989.

Hindle, Brooke. *The Pursuit of Science in Revolutionary America, 1735–1789*. Chapel Hill: University of North Carolina Press, 1956.

Kevles, Daniel J. *The Physicists: The History of a Scientific Community in Modern America*. New York:
Random House, 1995.

Kohler, Robert. *Partners in Science: Foundations and
the Natural Scientists, 1900–1945*. Chicago: University of Chicago Press, 1991.

Kohlstedt, Sally Gregory. *The Formation of the American Scientific Community: The American Association for the Advancement of Science, 1848–1860*.
Urbana: University of Illinois Press, 1976.

Leslie, Stuart W. *The Cold War and American Science:
The Military–Industrial–Academic Complex at
MIT and Stanford*. New York: Columbia Univer-
sity Press, 1993.

Lowen, Rebecca S. *Creating the Cold War University:
the Transformation of Stanford*. Berkeley: University of California Press, 1997.

McCaughey, Robert. "The Transformation of American Academic Life: Harvard University 1821–1892."
Perspectives in American History VIII (1974):
239–332.

Oleson, Alexandra, and John Voss, eds. *The Organization of Knowledge in Modern America, 1860–1920*.

Baltimore: Johns Hopkins University Press,
1979.

Reynolds, Terry S. "The Education of Engineers in
America before the Morrill Act of 1862." *History
of Education Quarterly* 32, no. 4 (Winter 1992):
459–482.

Rossiter, Margaret W. *The Emergence of Agricultural
Science: Justus Liebig and the Americans, 1840–
1880*. New Haven, Conn.: Yale University Press,
1975.

Rossiter, Margaret W. *Women Scientists in America:
Before Affirmative Action, 1940–1972*. Baltimore:
Johns Hopkins University Press, 1995.

Rossiter, Margaret W. *Women Scientists in America:
Forging a New World since 1972*. Vol. 3. Baltimore: Johns Hopkins University Press, 2012.

Rothstein, William G. *American Physicians in the
Nineteenth Century: From Sects to Science*. Baltimore: Johns Hopkins University Press, 1972.

Sapolsky, Harvey M. *Science and the Navy: The History of the Office of Naval Research*. Princeton,
N.J.: Princeton University Press, 1990.

Servos, John W. *Physical Chemistry from Ostwald to
Pauling*. Princeton, N.J.: Princeton University
Press, 1990.

Storr, Richard J. *The Beginnings of Graduate Education in America*. Chicago: University of Chicago
Press, 1953.

Stratton, Julius, and Loretta H. Mannix. *Mind and
Hand: The Birth of MIT*. Cambridge, Mass.: MIT
Press, 2005.

Veysey, Laurence. *The Emergence of the American
University*. Chicago: University of Chicago Press,
1965.

Roger L. Geiger

HIGH SCHOOLS, SCIENCE EDUCATION IN

The history of education in American high schools
is a story of continuous reform. Over the years,
practitioners, researchers, and policy makers have
advanced a variety of pedagogical and curricular
innovations to remedy the deficiencies of those
that were themselves once hailed as enlightened
and innovative (Tyack and Cuban, 1995). The
history of science education—since the appearance of science in high schools in the early

1800s—is no exception. Although the initial argument for the inclusion of science in the curriculum was rooted in the practical value it held for students in a young and expanding nation, other claims have been made as well over the past two hundred years for its inclusion in schools. Justifications for teaching science, the form it assumed, and the audiences it sought to reach have varied considerably from one era to the next, and these variations can be traced to everything from changes in the scientific profession itself to the shifting demographics of public schooling to, even more fundamentally, the range of social, cultural, and political forces that have shaped the interrelationships among science, schools, and the public.

Two notable themes are apparent when considering the historical place of science in American high schools. First is a continuing tension in the perceived value of science as a subject of study—between the immediate, utilitarian value inherent in knowing how natural processes operate in the world (that is, being able to *do* things to meet human needs) and what some have referred to as the disciplinary value of science, the idea that study of science promotes more abstract goals related to morality, virtue, analytical thinking, or aesthetics. From the beginning of high school science education, these perceptions have vacillated between the extremes of the practical-abstract continuum. At times, the view of science as "practical" has dominated public policy and educational conversations, whereas at other times, the ability of science study to ensure public virtue and intellectual discipline has held sway among science education advocates.

A second theme, however, relates to a discernable historical shift, with World War II marking the watershed moment. Prior to the 1940s, the form of science education in high schools was predicated on the idea that science had something of value to offer students and the general public, that science provided tools (in either its content or its methods) that could be used in other venues or for purposes outside of institutional science. With the unprecedented dramatic increase in federal investment in scientific research and development for national security and economic development following the war, science edu-

cation increasingly aimed at sustaining the professional science community itself, rather than drawing from science to meet the needs of the general public. There was a shift, in other words, from teaching science to make the everyday lives of students and citizens better to teaching science to ensure the success of the scientific enterprise first and foremost. The drawing of broad themes from American history on any topic is, of course, an exercise in simplification, and such is the case with science teaching in American high schools. Some valuable insights might be had based on this account nevertheless.

Early Science Teaching and the High School. Historians typically point to the founding of Boston's English Classical School in 1821 as marking the arrival of the first public high school in the United States. Yet that event by no means heralded the widespread adoption of a new educational institution. The development and dissemination of what many would recognize as the typical American high school happened more gradually throughout the nineteenth century. From the early 1800s through the Civil War, a wide variety of schools provided formal instruction—some tax supported, but most tuition based—to students who had completed their common school studies and were seeking "higher" schooling for a variety of reasons. These types of schools included seminaries, academies, and collegiate institutes. Distinctions among schools at this level and institutions of even higher learning—the nation's colleges and universities—were difficult to make throughout this period (Reese, 1995; Tolley, 2001). What was clear from the outset, however, was the place of science instruction in the courses of study these schools offered. In the industrializing, market economy of nineteenth-century America, science in its various manifestations was viewed by its patrons and citizens more generally as a subject of great utility, one that had, in addition, a natural moral and disciplinary foundation (Slotten, 1991).

The prevalence of science teaching in early secondary schooling runs counter to a commonly held perception that high schools and academies focused primarily on the study of Greek and Latin, the classical subjects that prepared students for college or university matriculation. Although schools for this purpose certainly existed, the fraction of the population that attended college during this period was far too small to support the comparatively larger number of secondary schools in the country. These high schools, academies, and seminaries survived only by concentrating their efforts on studies that had more immediate value—practical subjects that promised to contribute to the advancement of trade, commerce, and the mechanical arts (Krug, 1969; Reese, 1995). Geography was the most common early science course offered, suited as it was to these practical goals (Schulten, 2001). As the United States expanded westward across the North American continent during the early 1800s, knowledge of geography seemed particularly useful and contributed to feelings of national pride. Other subjects with similar practical virtues were gradually introduced as well, including natural philosophy (an early variant of physics), astronomy, chemistry, and botany. All of these were commonly taught in the years prior to the Civil War (Keeney, 1992).

Although offered in most secondary schools, these early science courses were not uniformly attended by all pupils. Enrollments varied by geographic area and by gender in particular. Although in the early twenty-first century the sciences typically have been viewed as boys' subjects (at least since the second half of the twentieth century), female academies and seminaries, especially in the South during the middle decades of the 1800s, enrolled more students in the sciences than comparable male-only institutions. This is explained partly by the fact that boys from families of higher socioeconomic status were more likely to be engaged in preparatory work for college, which, given the entrance requirements of the time, entailed mastery of the classical curriculum. With colleges generally not open to women before the Civil War, pursuing classical study made little sense. Girls were encouraged instead to pursue more practical subjects and those that possessed disciplinary and moral value. This resulted in higher numbers of girls in science courses up through the 1890s when colleges began to matriculate women (Tolley, 2002).

The perceived moral and disciplinary attributes of science were important factors in the subject's incorporation into the secondary school curriculum during the nineteenth century for boys and girls alike (Hollinger, 1984; Slotten, 1991; Guralnik, 1975). If narrow utilitarian advantage were all the sciences could offer, these subjects would likely have been limited to trade, vocational, and specialized engineering schools. Advocates of chemistry, botany, and zoology, however, insisted that their subjects had the power to reveal the work of God in nature. The study of science, they argued, would bring students closer to the divine, which was no small aim in the era of religious revivalism and natural theology of the early and middle 1800s (Ahlstrom, 1972). Complementing this spiritual benefit was the belief that science had the ability to develop the intellectual faculties of the mind in the same way that study of the classical languages did. Psychological theories viewed the mind as an organ that, much like a muscle, could be developed through exercise. Grappling with the precise, rigorous structure of scientific knowledge, many believed, produced mental discipline of the highest level, at least on par with that derived from classical study.

As the forces of industrialization extended across the United States, the sciences began to occupy a central place in the high school curriculum. This trend reflected the growing American faith in science and technology as a means of economic and social advancement. The spiritual and intellectual virtues were also recognized and promoted, but it was the practical payoff in the end that secured the place of science in the curriculum. In a collection of widely circulated essays published in 1861, the British political theorist and polymath Herbert Spencer famously asked "What Knowledge Is of Most Worth?" His answer, reflecting the spirit of the time, was "science," knowledge of which by his reckoning could be fruitfully applied to nearly all of life's affairs. In 1862 Congress passed the Morrill Land Grant Act, which led to the establishment of state universities charged with disseminating useful knowledge in agriculture and engineering nationwide (Geiger, 1998). These operated alongside private schools devoted to the advancement of the mechanical arts and industry, such as the Massachusetts Institute of Technology and the Stevens Institute in New Jersey, which grew up in the decade after the Civil War.

This rising tide of interest in the practical led a number of states in the 1870s to institute natural science requirements in their elementary teacher certification exams. This, in turn, generated even greater emphasis on science in the high schools, which were the institutions (along with normal schools) that were primarily responsible for the preparation of school teachers in the lower grades (Ogren, 2005). By the middle 1880s, the public high schools had eclipsed private academies and seminaries as the primary institution for secondary education in the United States, and the sciences were firmly established as central components of their curriculum (Krug, 1969; Reese, 1995).

From Textbooks to Laboratories.

The relationship between forms of subject-matter knowledge and its social value is worth examining carefully in the case of science. The perceived utility of the sciences in the middle decades of the nineteenth century derived, for the most part, from the factual knowledge about the world that scientists sought to accumulate. Accordingly, some educators classified subjects such as chemistry, botany, and physics as "information-giving" subjects, whereas the value of other subjects, such as mathematics and the languages (whether modern or classical), derived from their ability to "discipline" young minds. This division was seen as particularly appropriate in the common schools and high schools where practical application was regarded more highly than intellectual exercise (although science was believed capable of both). Seeing the contributions of science this way—a view common in the 1860s and 1870s—led textbook authors and teachers to a pedagogy of transmission. Teaching science, in other words, entailed the careful study of the systematized knowledge of the subject in question, typically through the canonical textbooks of the time. Instruction most often involved student rote memorization followed by recitation of the facts learned (Reese, 1995). More innovative teachers might have in-

terspersed their lessons with a scientific display or demonstration. Alternatively, they may have organized their lesson around an object brought into the classroom following the Oswego method that came out of upstate New York in the 1860s, although such practices were far from common.

During the 1880s, some university science faculty and high school science teachers pushed to move beyond sterile textbook approaches to science instruction. Invoking Harvard zoologist Louis Agassiz's oft-repeated maxim to "study nature, not books," educators in the second half of the nineteenth century pushed for students to confront nature directly, either in the field or in the laboratory (Owens, 1985; Kohlstedt, 2010). Agassiz himself led Boston-area teachers into the field to learn from nature firsthand at his island summer school off the coast of Massachusetts. In this same spirit, laboratory work had been introduced earlier at specialized technical schools such as Rensselaer Polytechnic and the Massachusetts Institute of Technology, founded in 1824 and 1864, respectively. Such instruction, however, was aimed primarily at preparing students seeking careers in the industrial and technical fields opening up in the United States at the time (Angulo, 2008; Kremer, 2011).

Laboratory instruction as a means of general education, or liberal study, however, came only after the incorporation of teaching laboratories into the larger universities and liberal arts colleges of the United States. In the 1850s and 1860s, Harvard University set up such laboratory space in chemistry and physics with the encouragement of its president, Charles Eliot. As the land-grant universities got their footing following passage of the Morrill Act, they too instituted laboratory instruction in many of their introductory science classes (Geiger, 1998; Keeney, 1992). Johns Hopkins University, founded in Baltimore in 1876, provided perhaps the epitome of a university committed to the laboratory teaching ideal, serving as both a model of instructional approach and an incubator for science faculty dedicated to teaching via laboratory methods (Owens, 1985).

The impetus for the shift from textbook study to laboratory teaching can be traced in part to the professionalization of science in the United States during this period (Higham, 1979). This professionalizing impulse reflected a growing enthusiasm for the German research ideal that young scientists brought back with them from their studies abroad during the 1870s and 1880s. German universities in the second half of the nineteenth century embraced a research and teaching ideology grounded in the ideals of *Lehrfreiheit*, the freedom to teach as one saw fit, and *Wissenschaft*, the pursuit of broad-based inquiry or research. American scientists studying in places like Göttingen, Berlin, and Munich returned to their own institutions in the states with an enthusiasm for a higher education dedicated to pure research and learning. Following their German academic role models, they were less concerned about the immediate utility or practical application of their research than they were with exploring the fundamental questions in their field. The American version of the German ideal was heavily focused on the natural sciences and laboratory work, which was believed to be central to the advancement of science both in research and in teaching. Teaching laboratories were, in fact, well developed already in the United States. But the growing community of American scientists increasingly viewed teaching laboratories as an essential element of the modern research university (Veysey, 1970; Geiger, 1998; Roberts and Turner, 2000).

Laboratories in the High Schools. Given the blurred boundary between colleges and universities and the secondary schools of the time, it is not surprising that high school science teachers rapidly adopted laboratory teaching. Seeking to promote this new instructional approach as widely as possible, university faculty and high school science teachers wrote new textbooks designed for use in laboratory settings. In addition, the growing number of graduate programs at American universities produced more and more new science PhDs trained in the new methods and committed to the research ideal. These were students who, upon seeking gainful employment, often took up positions in the new high schools being built in towns and communities as part of the growing expansion of public schooling at the time

(Olesko, 1995; Kohler, 1996; Rudolph, 2005, "Epistemology for the Masses"). The porous boundary between university and high school thus allowed the flow of not only the new professional vision of science with its particular, German-inspired methods, but also personnel—the teachers and textbook writers—who refashioned emerging high school science teaching practices in ways that mirrored those in the university (Hoffman, 2011; Turner, 2011). As much as these various agents of change moved school science practice toward the new ideal, college admissions requirements were perhaps the more powerful lever for reform; in the mid-1880s, revised college admission requirements had a profound effect on the spread of laboratory instruction in the sciences.

Many scholars, as well as science educators who lived through the transition, point to the introduction of a laboratory option in the entrance requirements for physics introduced at Harvard University in 1886 as the tipping point in the widespread adoption of laboratory methods (Rudolph, 2005, "Epistemology for the Masses"). Aspiring Harvard students electing the laboratory option were required to demonstrate their manipulative skills in a laboratory practical exam on campus, in addition to submitting a laboratory notebook documenting their completion of 40 specific physics exercises, often completed during the student's senior year of high school. The physicist Edwin Hall, who had been recruited by the Harvard president, Eliot, in part for his experimental talents, was charged with overseeing the development and implementation of the laboratory examinations. In hiring Hall, Eliot tapped into the growing network of scientists who were committed to laboratory teaching, people like the Johns Hopkins chemist Ira Remsen and the physicist Henry Rowland (also at Hopkins), under whom Hall had earned his PhD Hall was a powerful advocate of both laboratory teaching and Harvard's new examination option. He widely advertised the revised admissions policy in scientific and educational circles, fully expecting area high schools to adjust their course offerings to meet the new expectations (Rosen, 1954; Moyer, 1976; Rudolph, 2005, "Epistemology for the Masses").

The Harvard admissions requirement in physics was, of course, not solely responsible for the adoption of laboratory methods in American high schools. Some scholars have argued that Hall's role in the transformation of science teaching during this period has been overstated and that the new laboratory-based physics curriculum was, in reality, an amalgam of ideas and materials that were in the air at the time (Turner, 2011). It would be a mistake to argue otherwise; there were without question numerous factors, both in and outside of Cambridge, that contributed to what many have described as the national "craze" for laboratory teaching that took hold at the end of the nineteenth century. What should not be minimized in all this, however, is both the real and the symbolic role that Hall and Harvard assumed in the perceptions of this particular educational reform movement (Mann, 1909).

Although not as well chronicled as the move to laboratory teaching in the physical sciences, the biological sciences experienced a similar transformation of teaching practices in the high school. The push for reform came on two fronts, one from the plains state of Nebraska and the other from overseas in the United Kingdom. The eminent botanist Charles Bessey was the domestic agitator for reform. Teaching and writing initially at Iowa State College in Ames, Bessey became a strong voice for laboratory teaching following his move to the University of Nebraska in Lincoln in 1884. He worked closely with teachers in the region, providing guidance on how to set up and run a school laboratory, lending specimens to local classrooms, and teaching summer school courses that modeled appropriate pedagogical practice. His popular high school botany textbooks had far greater reach, spreading the laboratory approach to all corners of the country. From across the Atlantic, the influential zoologist Thomas Henry Huxley busily shaped the study of the biological sciences as well. Huxley had a strong interest in promoting greater public understanding of science, and his 1876 textbook, *A Course of Practical Instruction in Elementary Biology*, set a new standard for a laboratory-based approach to life science instruction, quickly sparking the publication of a number of American textbooks

following Huxley's template (Tobey, 1981; Keeney, 1992).

Although the adoption of hands-on teaching methods in high school subjects like botany, zoology, and physiology lagged somewhat behind the changes in physics and chemistry, by the end of the century, students in the life sciences spent considerable time engaged in laboratory and field work. They gathered and dissected specimens, built herbaria, learned the intricacies of microscopic observation, and assembled museum-like classroom displays among their many activities (Benson, 1988; Conn, 1998; Rudolph, 2012). For many educators, such work continued to be justified by the remnants of natural theology—a worldview that held that the study of nature was in and of itself virtuous, revealing as it did the beauty and wisdom of God's design in nature. The content of high school courses in zoology and botany, however, was increasingly organized around seeing the morphological features of plants and animals as environmental adaptations and understanding the evolutionary relationships among organisms and their place in broader systems of classification (Pauly, 2000; Rudolph, 2012).

The move to laboratory instruction across all subjects in American high schools was solidified in 1893 with the publication of the *Report of the Committee of Ten on Secondary School Studies*. Under the leadership of the Harvard president Charles Eliot, this report became the de facto educational standards document of the time. The final version included recommendations from subcommittees covering all the primary high school subjects. In the sciences the list spanned physics, astronomy, chemistry, botany, zoology, physiology, physical geography, geology, and meteorology. As Eliot summarized, "All the Conferences on scientific subjects dwell on laboratory work by the pupils as the best means of instruction," an emphasis he heartily endorsed (Committee of Ten, 1894, p. 18). From the publication of the Committee of Ten report through the first decade of the twentieth century, laboratory methods spread throughout the country. By the early 1900s, the most common science subjects in high schools were botany, zoology, chemistry, physics, and physiography (or geography), all taught with an emphasis on laboratory study. Laboratory teaching, advocates insisted, not only captured the essence of scientific work, but also provided an effective path to mental discipline and moral rectitude, much as the study of classical languages had years earlier. Such thinking marked a clear shift from earlier justifications for science teaching derived from the practical utility of factual knowledge about the natural world.

The Progressive Vision of Science Education.

Early in the twentieth century, a new vision of science education emerged to challenge the course of study and pedagogical approach endorsed by the Committee of Ten. Informing this vision was the rapid expansion of the educational enterprise that began at the university level during the 1890s and extended with amazing speed to high schools over the subsequent decades. This expansion resulted in a push for a radically different kind of science education, designed to address student everyday needs and interests rather than following the abstractions at the heart of the scientific research establishment.

The explosion in high school enrollments during this period was nothing short of astounding. From 1885 to the end of the century, the total number of secondary school students rose from approximately 132,000 to nearly 650,000, a fivefold increase in a period of only 15 years. During the first two decades of the twentieth century, school enrollments further quadrupled, reaching 2,757,000 by 1920 (Rudolph 2005, "Epistemology for the Masses," pp. 354–356). The flood of students stimulated a school building boom that, over the two decades spanning the turn of the twentieth century, saw an average of one new high school built every day to accommodate the growing numbers of students (Reese, 2011, p. 181). As enrollments increased, a consensus emerged among educators that the masses of students now populating the schools had decidedly different educational needs than their predecessors. The percentage of students engaged in high school study as preparation for college (although always small) shrank even further to single digits. Even more troubling to them was the fact that overall student enrollments in the sciences were declining at an even greater rate proportional to enrollments in other traditionally academic school subjects like Latin (Rudolph, 2005, "Turning Science to Account," pp. 359–362). The majority of students now in attendance increasingly were viewed as needing more practical and personally relevant instruction in contrast with the formal disciplinary studies that had been common to that point (Krug, 1969; Kliebard, 2004).

As the student population grew, so too did the number of high school teachers, who were prepared in the growing number of graduate schools, universities, and colleges across the country. These individuals found common cause teaching their new student clientele and began to rethink the mission of the schools. In an era of professionalization, they formed associations, organized regional and national meetings, and launched journals dedicated to articulating an alternative vision of what a high school education should accomplish. In the sciences, this professional energy was channeled into the Central Association of Science and Mathematics Teachers, founded in 1901, the most prominent and far-reaching organization of its kind. Science teacher leaders from this group drew on ideas from the young fields of educational psychology and child study promoted by individuals such as G. Stanley Hall of Clark University (who wrote the book *Adolescence* in 1904) and John Dewey, then at the University of Chicago, to inform their thinking about what school science education should look like in the new century. The needs and interests of students, they insisted, should chart the course of the science curriculum. The logical organization of the disciplines, although important, would take a backseat to the everyday experiences of children that would provide the personal motivation and social justification for learning science (Ross, 1992; Rudolph, 2005, "Turning Science to Account"; Smuts, 2006).

The push for change began with what was called the "new movement among physics teachers." Beginning in 1905, leaders of this reform group—mostly teachers and officials of the Central Association of Science and Mathematics Teachers—called for a radically different approach to physics teaching that would break free of the standard 40 descriptive laboratory exercises put forward by Edwin Hall and Harvard University. Drawing on the work of the new psychologists, they argued that physics teaching needed to be relevant for students. The dry, quantitative laboratory exercises of the past should be replaced by more qualitative exercises or even replaced entirely by teacher demonstrations or illustrated lectures that made real connections between physics and the world in which students lived (Moyer, 1976; Olesko, 1995; Rudolph, 2005, "Epistemology for the Masses"). High school chemistry experienced similar pressures to shift away from the sterile academic presentation of the discipline toward more practical and applied treatments that were likely to appeal to the non-college-bound students of the subject (DeBoer, 1991; Cotter, 2008).

The Appearance of Biology and General Science.
During this period of curricular unrest two entirely new school science subjects made their debut—biology and general science. Prior to the turn of the century, the life sciences appeared in the schools primarily in the form of zoology and botany and, where necessary, physiology (a subject mandated in many states as a means of checking the spread of alcohol consumption) (Pauly, 1991; Zimmerman, 1999). Although educators debated which of these subjects provided the best introduction to the life sciences for high school students (most often botany with its easily accessible laboratory material and more direct connection to students' lives won out), schools increasingly offered the two subjects in consecutive semesters to provide a general, year-long survey of living organisms.

Whether studying plants or animals, students were instructed in true natural history fashion to assemble collections of insects and plants, dissect specimens, and make careful drawings to record their observations. The focus of learning was typically on having students appreciate the marvels of adaptation of organisms to environment as well as to understand the broader relationships among organisms—that is, how they fit into a natural order, a system of classification. Such schoolwork

was initially pursued within the framework of natural theology (as testimony to the wisdom of the Creator) but shifted over time—with little change in actual classroom activity—to embrace (and demonstrate) evolutionary relationships as evolutionary perspectives increasingly took hold in the sciences and in American culture more generally (Pauly, 2000; Numbers, 1998; Rudolph, 2012).

By 1905, the fused botany-zoology curriculum had transformed into the more familiar school subject, biology. This new subject first appeared in the urban high schools of the northeast, where large numbers of immigrants streamed into the growing public school systems. As part of the reform movement aimed at student interests and social relevance, the new biology textbooks covered many of the traditional topics common to earlier botany and zoology courses while expanding their scope to topics such as hygiene, personal nutrition, ventilation, urban sanitation, and sex and reproduction (although this last topic was taught indirectly without reference to humans). By the mid 1920s, biology had become a staple of the high school curriculum, most often required at the sophomore level and designed to teach students, following progressivist thinking, how to live better. The conditions of the time created a pressing need, biology educators asserted, for adolescents and those new to America to themselves adapt to their increasingly urban industrial environment (Pauly, 1991).

Joining biology in the high schools during this period was general science. Offered most frequently as an introduction to science for high school freshmen, the development of this course was a response to what many in the new professional education establishment saw as the overspecialization of the discipline-focused courses of physics and chemistry. Borrowing heavily from the early psychological work on student interest and John Dewey's writing on the nature of rational thought, proponents of the general science course sought to engage students in a survey of science topics common in everyday experience. This approach dispensed with disciplinary structure altogether, foregrounding instead a universal method of problem solving modeled on scientific thinking.

This approach, proponents insisted, not only captured the essence of science as an enterprise, but also provided something of real value and utility for the majority of students attending high school. Here was a skill, they maintained, that had practical payoff in students' lives.

From the emergence of general science as a distinct course in the 1910s through the 1930s, the problem-solving approach described above quickly transformed into what came to be known as project-based teaching. This pedagogical approach—still following the trend toward the practical—had students study, among other things, the best methods for ventilating their classroom, the proper operation of a home furnace, or methods of drinking-water filtration. General science pursued by means of the project method ended up, in many ways, as a course in civic engineering, at least in the urban areas where it first took hold (Heffron, 1995; Rudolph, 2005, "Epistemology for the Masses," 2005, "Turning Science to Account").

The changing curricular emphasis of these subjects on the everyday and utilitarian, in contrast to the disciplinary and academic, aligned with broader curricular shifts occurring in education. The most well-known marker of this shift was the publication in 1918 of the National Education Association's *Committee on the Reorganization of Secondary Education,* commonly referred to as the Cardinal Principles Report. This was a national statement of educational policy that turned sharply from the recommendations of the Committee of Ten, which had set the pattern for high school instruction 30 years earlier. In place of calls for the focused study of disciplinary knowledge and praise of the virtues of rigorous laboratory work, the Cardinal Principles Report called for teaching that would promote, among other things, health, worthy home membership, citizenship, ethical character, and worthy use of leisure time. The only goal mentioned in the report having an academic cast was the "command of fundamental processes," which covered the skills of reading, writing, and mathematical calculation. The subcommittee report on science, which was published two years after the main report, echoed these goals, placing heavy emphasis on general

science as one powerful way of accomplishing them (DeBoer, 1991; Kliebard, 2004; Rudolph, 2005, "Turning Science to Account").

The new subjects of biology and general science enjoyed tremendous success in the early decades of the twentieth century. From 1910 to the mid-twentieth century, general science textbooks and courses proliferated. Hundreds of new textbooks appeared on the market, and every state in the union could point to school districts offering the new course. In 1941, nearly three-quarters of all science courses offered at the freshman level were general science. The spread of biology as a high school subject was equally impressive, with the subject making up 79 percent of all science offerings at the sophomore level. These two school subjects, developed as they were for the general education of the average citizen, represented a faith and optimism in the role science could play in the progressive amelioration of the difficult conditions of the new industrial age. Chemistry and physics, although serving a smaller number of students at the junior and senior levels of high school, similarly sought to appeal to the everyday interests of students but at the same time introduced more formal disciplinary structures suitable for those in the college-preparatory track. Through all these subjects, educators believed that an understanding of scientific thinking and its application to the material and social problems of everyday life could do much to improve the human condition.

The Scopes Trial and Its Fallout. The progress science education made during this period by extending its secular vision of society through the dissemination of mechanistic accounts of natural processes along with a universal method of thinking was not welcomed by all. There were social and cultural groups in the United States that resisted the worldview associated with this progressivist manifestation of science. None objected more strongly, perhaps, than the conservative Christian groups that coalesced into the religious fundamentalist movement during the early 1920s. Leaders of this movement viewed modernist thinking of the time with its evolutionary assumptions as a threat to traditional moral values and as a support for militarism and even Communism in one form or another. This cultural clash came to a spectacular head with the trial of John Scopes in Dayton, Tennessee, in 1925.

The Scopes trial pitted famed courtroom litigator and agnostic Clarence Darrow against the fading political giant and anti-evolution crusader William Jennings Bryan in a debate over the "truth" of scripture versus the "truth" of science. This courtroom contest was one of the most significant media events of the first half of the twentieth century. The proceedings of those hot July days in Tennessee transfixed people across the United States and beyond. On the one side was Darrow, not only advancing the modern, scientific account of life on earth and the historical path of its evolution, but also representing more broadly the power of science as a means of social improvement—a view that permeated progressivist thinking. On the other side was the three-time Democratic presidential candidate and gifted orator Bryan, who even in his declining years commanded national attention. Bryan, who readily accepted the antiquity of life on earth and interpreted the "days" of Genesis as vast geological epochs, focused his objections on human evolution, which, he believed, undermined Christian ethics and provided cover for the social and economic elite through its "survival of the fittest" mantra. Even more troubling to Bryan was the damage he believed such thinking would do to the country's religious faith; and teaching evolution to the nation's impressionable school children was a practice likely to actively undermine that faith.

Although there was much give and take over where the truth really lay (the Bible or science), both in the streets of Dayton and in the national press (just what the town boosters had hoped for in orchestrating the trial), the legal point at issue was more mundane, having to do with whether the state could regulate the content of the curriculum. Scopes, having admitted to teaching from a biology textbook that included evolution in violation of state law, was convicted (although it was set aside on a technicality) (Larson, 1997; Marsden, 2006; Shapiro, 2008).

The impact of the Scopes trial on high school biology teaching has been much debated. It had been widely held that the event marked a significant retreat from the teaching of evolution in the schools. Terms like "evolution," "natural selection," and "Darwin" nearly disappeared from the tables of contents, indexes, and glossaries of most high school biology textbooks after 1925 as the big publishing companies of the time scrubbed from their books easy-to-spot references to the controversial topic in an effort to maintain sales. Scholars have offered different accounts of this curricular pullback from evolution. Although some have indeed viewed the elimination of evolution and related vocabulary as an abdication of sustained engagement with the topic, others have argued that these changes were merely superficial and did little to overturn the broad scientific and cultural commitment to the progressive evolutionary ideas the biology textbook authors were most invested in. The content of biology as it appeared in textbooks throughout the first half of the twentieth century, in other words, was the story of the evolutionary progress of life on earth, humans included, whether the specific words appeared or not (Skoog, 1979; Pauly, 1991; Larson, 1997; Ladouceur, 2008).

From the 1930s to the 1950s, science education followed the general trend toward the personal and the practical as advocated by the Cardinal Principles Report of 1918, but all in the context of the country's democratic political ideology. With universal high school education increasingly accepted as a social norm, school districts continued to cater to student interests and the wide range of student ability in the nation's classrooms. All the standard high school science subjects from general science at the freshman level through biology, chemistry, and physics in the upper grades included heavy doses of applied science and technology; textbooks were filled with examples of objects and appliances students were likely to encounter in their daily lives. One could read about everything from the biological principles of proper nutrition and personal hygiene to the physics of home refrigerators and industrial steam shovels. The focus in science education on personal and social problems, projects, and activities reflected the dominant progressive view of education in these decades leading up to World War II.

With respect to the process of science, students were regularly fed a diet of the "scientific method," which typically consisted of five steps (beginning with the identification of a problem and ending with its resolution) often taught by rote that could be algorithmically applied to nearly any life situation or natural phenomenon. Skill in rational thinking was deemed essential for public engagement in democratic processes, and the safeguarding of democracy was seen as a high priority as totalitarian regimes spread across the globe in the late 1930s and 1940s (DeBoer, 1991; Rudolph, 2005, "Epistemology for the Masses"; Hollinger, 1990; Thurs, 2011).

Sputnik and the Cold War.

During World War II the business of schooling centered on wartime mobilization. Few innovations were introduced that had any lasting effect. The years following the war, however, present a different picture altogether as American society experienced dramatic economic, social, and cultural changes. The federal government began to heavily invest in scientific research and development; the baby boom that started in the 1950s drew increasing attention to national educational policy and infrastructure; the launch of the Soviet satellite Sputnik ratcheted up national security concerns; and the popular counterculture movement that emerged later in the 1960s forced citizens and policy makers to rethink the role of science in society.

The American high school was naturally swept up in these changes, and science education, in particular, was a popular target of reform. Through the newly established National Science Foundation (NSF), which was founded in 1950, the federal government took on unprecedented financial and bureaucratic roles in the development and dissemination of new curricular materials in the sciences. Spurred by a commitment to academic excellence and growing concerns about Soviet scientific advances, federal administration officials and legislators saw improving science education for the baby-boom generation as a national priority (Atkin and Black, 2003; Reese, 2011).

Science education was far from the only area undergoing profound changes during the 1950s—federal education and scientific research policy were transformed as well. Of special relevance, a fundamental shift occurred in the realm of education policy, as educational issues that had traditionally been handled locally were moved onto the national stage. The Supreme Court's 1954 decision in Brown v. Board of Education of Topeka, Kansas, which held that segregated schools for African Americans and whites were inherently unequal, positioned the federal government as a force to implement change at the local level (Patterson, 2001). In addition, federal legislation, such as the National Defense Education Act (NDEA), passed by Congress during the Eisenhower administration in response to Sputnik, and the Elementary and Secondary Education Act in 1965, as well as the establishment of institutions like the NSF and the upgrading of the Office of Education into a cabinet-level body in 1953, placed education, science education, and science policy squarely on the federal agenda (Kaestle and Smith, 1982; Kleinmann, 1995; Kevles, 2001; Urban, 2010). Taken together, these various agencies and laws transformed the relationship among scientists, educators, schools, curricula, and students. The legacy of these changes, especially the active involvement of scientists recruited to develop new science curriculum materials and the role of the federal government in science and science education, were still important in the early twenty-first century.

Science education scholars have tended to mark the launch of Sputnik in 1957 and the subsequent passage of the NDEA as the turning point in federal involvement in science education (Clowse, 1981; Kaestle, 2001; Urban, 2010). The impetus for reform, however, occurred several years earlier. Concerns about the adequacy of high school physics, for example, were voiced as early as 1951 in meetings of the government's Science Advisory Committee in the White House Office of Defense Mobilization. These conversations led the Massachusetts Institute of Technology physicist Jerrold Zacharias to assemble the Physical Science Study Committee (PSSC) in 1956.

Riding the wave of popular support for scientists after their military successes during World War II, Zacharias used this group to push for an updated high school physics curriculum. Using PSSC as the vehicle for reform, Zacharias, with the support of the NSF, led a group of nationally recognized physicists in the development of a new, cutting-edge high school physics course. In addition to upgrading the quality of high school teaching in this subject, PSSC offered a concrete way for the scientists to assert their social capital, promote their view of disciplinary knowledge, and defend the utility of science to society. In this case, what they deemed useful to society was basic, or "pure," science rather than applied science or engineering. Drawing from the military models and organizational structures they experienced during the war, these scientists hoped that the new science education they were constructing would, in the long term, ensure continued federal financial support for basic scientific research, a goal they believed was essential to the survival of the United States in the dangerous nuclear age (Rudolph, 2002, "From World War to Woods Hole").

Their pedagogical positions were buttressed by popular books such as Arthur Bestor's scathing attack on the progressive curriculum of the prewar years in *Education Wastelands* (1953) and Jerome Bruner's book *The Process of Education* (1960). The latter provided a psychological framework that was used to justify the new curricular focus on disciplinary knowledge in preference to the past emphasis on what was increasingly viewed by the public as the "soft" personal and applied aspects of school subjects (Rudolph, 2002, *Scientists in the Classroom*).

The Golden Age of Science Education. The PSSC was not the only scientist-led group assembled to reform the high school science curriculum. Spurred by the crisis atmosphere generated in the wake of *Sputnik*, the Massachusetts Institute of Technology–based project was soon followed by other curriculum development projects directed by scientists from top research universities across the country. In biology, the Florida zoologist Arnold Grobman and the Johns Hopkins

geneticist Bentley Glass launched the Biological Sciences Curriculum Study at the University of Colorado in Boulder in 1958. In chemistry two projects were initiated, the Chemical Bond Approach, under Arthur Scott of Reed College, and the more popular Chemical Education Materials Study, led by the Berkeley Nobel laureate Glenn Seaborg. Other new curricula followed in the 1960s, including the Earth Science Curriculum Project and Introductory Physical Science, among others.

The approach taken by what many refer to as the alphabet curricula was grounded on the belief that as long as these courses accurately reflected the discipline under study and teachers were properly trained to use the new material, then student achievement would surely follow. A common theme among these projects was an emphasis on what came to be referred to as "scientific inquiry"—the process through which scientists arrived at their knowledge about the world. Scientific inquiry was often taught through direct participation in laboratory activities, although text-based exercises in data analysis and the careful reading of original scientific work were used as well. The idea underpinning the emphasis on inquiry—that the process of science could only be properly understood and appreciated from within the disciplinary structures that informed it—stood in sharp contrast to the views advanced by the Progressive-Era educators who believed that the methods of scientific thinking could be applied to any project or problem independent of the discipline from which they originated. According to the scientists involved in these new projects, the idea of a universal "scientific method"—one that could be applied in some blind, algorithmic way independent of context—was a myth. Science was an enterprise that involved a measure of craft and creativity, best practiced by scientists themselves and properly funded by the general public.

Many have described this period, from the late 1950s through the 1960s, as the "golden age" of high school science education in the United States. And it was golden from a resource perspective to be sure. Federal money to improve science teaching flowed at all levels, from curriculum development to teacher training to classroom materials. The NSF, with the full backing of Congress, took the lead on the curriculum and teacher-quality front, funding the scientist-led textbook writing projects as well as scores of summer institutes for teachers across the country (Krighbaum and Rawson, 1969). The U.S. Office of Education provided resources directly to local school districts through the NDEA. Among the NDEA's provisions was Title III, which provided hundreds of millions of dollars for schools to purchase science equipment and apparatus and to modernize their laboratories and teaching facilities (Urban, 2010; Rudolph, 2012).

The Humanistic Turn. Despite the warm welcome they received in some quarters, the discipline-centered curricula had its detractors. By the mid-1960s some educators and scientists alike began to complain that curricula like PSSC reached only an elite group of students—mainly upper-middle-class, Caucasian males, leaving women, minorities, and those without any particular aptitude for science behind. They faulted the strict disciplinary notion of science presented in the textbooks, which, they believed, precluded more nuanced and humanistic elements of science that might appeal to students not already on the science track. Partly in response to critiques like these, the NSF began supporting a second wave of curriculum efforts in the mid-1960s. The most prominent of these was Harvard Project Physics (HPP), a high school physics curriculum led by the Harvard physicist and historian of science Gerald Holton along with F. James Rutherford and the Harvard education professor Fletcher Watson. HPP sought to provide an alternative to the technical physics of PSSC by presenting the subject through a more humanistic lens. Its developers drew specifically from the relatively new field of history of science to provide a richer social and cultural context of the subject's development over time.

Other second-wave projects included initiatives to extend reforms to the lower grades through curricula such as Science–A Process Approach and the Elementary Science Study, as well as to move beyond the natural sciences to the social

sciences, as was done with the project Man: A Course of Study (MACOS), directed by the Harvard psychologist Jerome Bruner. Proponents of these science curricula claimed that by softening the image of science, they could increase student enrollments and interest in science courses. Furthermore, according to many educators, courses like PSSC were ignorant of teachers' real needs in terms of materials and instructional flexibility. These curricular alternatives were presented as a way to give teachers choices that would allow them to respond to the needs of their students (DeBoer, 1991; Dow, 1991).

The transition in the mid-1960s from the narrow focus on the technical practices of disciplinary research to a more open and humanistic science education was aligned with the broader cultural and social movements of the period that questioned the value and practices of the professional science establishment. This happened at both ends of the political spectrum. The Biological Sciences Curriculum Study's brash reinsertion of evolution in biology textbooks sparked considerable reaction among a revived religious community on the right that rose up to battle the scourge of Darwin much as it did in Dayton, Tennessee, 40 years earlier. And on the left people across the country were staging sit-ins against the Vietnam War, demanding that universities cut their research ties to military contractors and private corporations and founding organizations devoted to directing scientific knowledge and resources toward the public good. People also took to the streets, demanding equal rights for all. Science was not immune from such critiques. Not only did everyday citizens begin to question the power and authority of science, but professional scientists joined the chorus, challenging science's premise of a better world as well (Vettel, 2006; Moore, 2008). By the 1970s the reforms of the 1960s had failed, ultimately collapsing under the weight of high expectations and the difficulties of changing long-standing school structures and classroom practices. Despite their failure in the short term, they did fundamentally alter, in more subtle ways, the perceived role of science teaching in American culture.

STS and Scientific Literacy. If the 1960s were the heyday of science education reform—an era of lavishly funded curriculum projects directed by high-status, nationally renowned scientific researchers with the help of officials at the highest levels of government—the 1970s were nearly the opposite. The enthusiasm for fixing high school science teaching through large-scale curriculum reform and new teacher-training programs waned in the face of entrenched school practices, social unrest, and a new skepticism of the social value of science.

Funding for new projects was a particular challenge during this time. Money for curricular development from governmental agencies like the NSF began drying up in the early 1970s and was finally cut off in 1975 after public objections to MACOS, which, critics argued, undermined family values and promoted cultural relativism. Federal funding of educational materials for use in local schools had been a sensitive issue from the beginning. The controversy over MACOS alongside a faltering national economy and public disillusion with scientists in the mid-1970s made pulling back federal involvement in curriculum work an easy decision. Scientists slowly migrated away from the world of high school science education, leaving it the sole domain of educators once more. With the country moving into a period of détente with the Soviet Union and facing new challenges on economic and political fronts, the sense of urgency that had attended science education in the post-*Sputnik* era largely disappeared. Promoting student excellence and achievement in science receded in importance, and educators in many ways were freed from federal and political expectations (Dow, 1991; Milam, 2013).

Following in the spirit of those who had advanced a student-centered, socially relevant science curriculum from the 1920s through the 1950s, educators in the 1970s were once again free to assert their vision of what high school science might accomplish. They argued for a curriculum that would engage students with scientific and technological issues that were personally meaningful. This meant infusing science teaching with the dominant social and political issues of the day. Issues of racial, social, and economic inequity

moved from the political world to the classroom, as did concerns over the environment. This reversed the disciplinary approach to science teaching advocated by scientists in the 1960s, which educators believed had touched only a narrow slice of privileged, college-bound students and had failed to meet the educational needs of the majority (DeBoer, 1991; Moore, 2008).

Immersing students in pressing technological and societal issues, such as world hunger, population growth, water resources, and energy shortages, was not a novel approach to science teaching. Dubbed "Science Technology and Society" (STS), it had a long tradition that, although overshadowed by the disciplinary turn taken by scientists in the 1960s, remained of interest to science educators throughout the twentieth century. In 1958, the Stanford education professor Paul DeHart Hurd, a leading advocate of the STS approach, argued for a type of science education that would promote understanding science in its social and political context, a goal that came to be widely known as "scientific literacy." The idea of literacy, then and now, is highly ambiguous, and it meant different things to the various science and educational interest groups with a stake in science education. Like an empty vessel, everyone poured their understanding of what science education should be into their understanding of scientific literacy. For most, in fact, the term continued to carry with it connotations of specialized knowledge (DeBoer, 1991).

The phrase scientific literacy had gone largely unnoticed in the pages of education journals for most of the 1960s. In the 1970s, however, it gained notoriety as a powerful slogan that aimed to capture Hurd's notion of an understanding of science in which one's personal interest and the real-world functionality of an individual's scientific knowledge were of primary importance. In 1971, the National Science Teachers Association (NSTA), the leading professional organization of high school science teachers (established in 1944), identified scientific literacy as the number one goal of science education. Their focus on relating science to everyday life was, in part, a reaction against scientists' insistence on building student understanding of disciplinary knowledge and scientific process, which was the predominant learning goal of the curriculum projects brought to market a decade earlier. The new focus on scientific literacy harkened back to science education's progressive roots. Through a variety of curricular projects and concerted public advocacy, classroom science teachers and university education professors succeeded in shifting the focus of science education away from scientists and the discipline-centered, large-scale reforms of the 1960s toward science education for personal and social needs.

The most explicit efforts to reformulate science education in the 1970s could be seen in the multiple attempts to develop STS content for science classrooms. The teaching units developed during this period were smaller and less ambitious in some ways than the richly funded material from the NSF era. And although advocates of STS education tended to agree on the overarching goals of science education, their approaches varied. In general, they agreed that science education should take a "humanistic approach," in which the material taught would enable students to make connections between science and the wide range of human endeavor. Similarly, the method of science teaching changed as well, from immersing students in cutting-edge laboratory activities toward instruction that promoted student choice and decision making with the ultimate goal being social action (often related to solving environmental problems). Some educators argued for a more explicit incorporation of personal and social values into science education. In those classrooms, students would be presented with a dilemma and asked to work through a solution on their own. This approach lent itself to argumentation, disagreement, and discussion, and the goal was a type of engagement that would promote social consciousness and a willingness to make changes in the world (Aikenhead, 2003).

By the close of the 1970s, one could find a mix of approaches to science education—at times, science teaching was structured around social issues, and at other times, the primacy of the disciplines was maintained, with an emphasis on the mastery of science content first and concern with social application deferred. Yet despite these differences,

most science educators agreed that science needed to be presented within its social and cultural context. Educators on both sides of the debate were keen to emphasize student needs by framing subject matter through pressing social issues. They would soon encounter, however, the political realities of the 1980s when a new group of policy makers urged that science education should be viewed, once again, as a way to advance national interests—this time, however, interest centered on the American economy rather than national security.

Inquiry and the Standards Movement. A renewed wave of attention to science education rippled through the country following the 1983 release of the Reagan administration's Commission on Educational Excellence report, *A Nation at Risk*. Echoing the public clamor for science education reform following the launch of *Sputnik*, the report insisted on greater attention to academic subject matter to compete with the Japanese economic juggernaut. This report set off a new round of reform initiatives and ushered in the educational standards era in the United States. The report complained that the U.S. educational system had spent the previous decade on the sidelines and had "squandered the gains in student achievement made in the wake of the *Sputnik* challenge." Yet unlike during the height of the Cold War, the stakes this time revolved around economic concerns. "Our once unchallenged preeminence in commerce, industry, science, and technological innovation," the report opened, "is being overtaken by competitors throughout the world." The Commission argued that this was a systemic failing and blamed politicians and educators alike who had "lost sight of the basic purposes of schooling, and of the high expectations and disciplined effort needed to attain them" (National Commission on Excellence in Education, 1983).

The response from educators and scientists alike was to draft several policy documents that would define the scope of a basic science education for all citizens. In the late 1980s, the American Association for the Advancement of Science (AAAS), the leading professional science society in the country, initiated Project 2061, an operation dedicated to generating a vision and blueprint for science education reform in the United States. Recognizing the difficult challenge before them, they set the year 2061 (when Halley's comet would return to the earth's sky) as their target date. In 1989 they released a manifesto describing what the general public should know about science, *Science for All Americans*. Three years later, the NSTA put out its own set of standards, *Scope, Sequence, and Coordination of Secondary School Science*, which was itself followed by the more detailed, age-graded version of the AAAS standards, *Benchmarks for Science Literacy* (1993). The presence of competing standards documents was a point of concern among policy makers and science education advocates, which prompted the National Research Council (NRC) (the working arm of the National Academy of Sciences, a long-standing quasi-governmental and powerful science advisory agency) to step in and lay out the *National Science Education Standards* in 1996. In this work, the NRC largely endorsed the work of the AAAS, effectively consigning the NSTA curricular guidelines to a work of small influence (Collins, 1998).

Both the NRC and the AAAS documents overlooked much of the science education work of the 1970s. Instead, they drew heavily from the disciplinary approach pursued in the 1960s. (The primary architect of AAAS's Project 2061, in fact, was F. James Rutherford, one of the key figures in the development of Harvard Project Physics years earlier.) Two central themes explicitly framed the science curriculum put forth within these texts: inquiry and the nature of science. The inclusion of inquiry reprised the focus on inquiry central to the post-*Sputnik* curriculum reform projects. The meaning of inquiry in the new standards documents, however, had shifted—the term now referred primarily to a method of instruction involving student hands-on engagement with science projects or activities. Left behind were the more fundamental learning goals related to student understanding of scientific inquiry as it operated within disciplinary contexts.

Complementing this emphasis on inquiry in these standards documents was a focus on student understanding of the nature of science, a key learning goal that had emerged during the early days of the standards era (the early 1980s), partly in response to the conservative push for teaching creation science alongside evolution in the schools. (This was yet another skirmish in the ongoing debates over evolution in the schools) (Numbers, 2006). Many believed that understanding the nature of science—elements of which included knowing that science depends on empirical evidence, is subject to change as new evidence comes to light, and so on—would provide students with the skills to distinguish science from nonscience and ensure that topics like creationism would be clearly seen as "unscientific." Inquiry and the nature of science, many hoped, would work in concert in the curriculum, with students engaging and participating in science through inquiry and inquiry activities undergirding and providing lessons on the nature of science and the process of knowledge production.

These documents, the *Benchmarks for Scientific Literacy* in particular, did much to shape the content and sequence of high school science programs during the 1990s and 2000s. Local school districts and state education departments worked to match their curricula and classroom instruction to the prescriptions laid out in these texts. They arguably had even greater impact on the content of science textbooks. Not long after the ink was dry on the standards, publishers leapt to adjust their books and touted their alignment with the national science standards as a key marketing strategy.

Standards, Testing, and Global Competition. By the mid-1990s, science education was increasingly driven by concerns over quality and accountability, viewed through the lens of student performance on standardized tests. Mediocre student performance on a series of international science and math assessments in the 1980s, highlighted in the *Nation at Risk* report, led to a government push for far-reaching accountability systems. In the late 1980s, the White House and the National Governor's Association approved the Goals 2000 initiative, which called for U.S. students to be first in the world in science and mathematics achievement by the year 2000. During this same period, the National Science Board, an independent, presidentially appointed advisory body representing the U.S. science and engineering community, began tracking the low state of public understanding of science using a fact-based metric in its biannual *National Science and Engineering Indicators* reports.

These surveys consistently found appalling low levels of public understanding of basic science content and scientific process. Continued poor performance by U.S. students on tests at a variety of levels, from the international *Trends in International Mathematics and Science Study* and the *Programme for International Student Assessment* as well as domestic assessments such as those done as part of the *National Science and Engineering Indicators* throughout the 1990s and 2000s, provided ammunition to critics who argued for greater emphasis on science and mathematics in the name of global economic competition. The result was, not surprisingly, a greater focus on science content knowledge in schools and more testing to ensure its mastery (Fuhrman, 2003; Toch, 1991; Vinovskis, 2008).

By the turn of the twenty-first century, educational accountability was a national issue. Passage of the No Child Left Behind Act in 2001 required that all school districts demonstrate annual yearly progress in student standardized test scores across multiple student groupings in a common set of academic subjects. The first years of the law required achievement gains only in mathematics and reading, reducing subjects like science and social studies to the margins. (In 2013, it remained too early to tell how more recent testing in science—started in 2007, but not included in calculations of annual yearly progress—would affect the amount of time devoted to teaching that subject.) Classroom experiences of children have reflected this new emphasis on testing and accountability. Many educators and parents have bemoaned the rote learning taking place in classrooms, whereas others have applauded the ability of these tests to transform pedagogy. Much of the concern over curricular narrowing—the result of teachers focusing instruction on material likely to appear on the exams—has been limited to the pre–high school grades where concerted efforts

have been made to improve student test performance to avoid the legislative sanctions that result from failure to make annual yearly progress.

At the high school level, the emphasis on accountability has been combined with the neoliberal emphasis on education as a private good to push greater adoption of Advanced Placement (AP) course offerings in academic subjects, particularly the sciences. Although poor districts have been measured by the results of No Child Left Behind, wealthy suburban school districts have increasingly measured their success—and sold the strength of their high schools—by how many AP courses are available to students and how well their students perform on the corresponding AP exams, with high scores enabling students to secure college credit for the subject tested. The AP course curricula and tests, in their aim to mirror introductory, college-level courses, focus heavily on mastery of disciplinary content rather than on scientific process or epistemology. The increased attention on test performance at all levels and across all schools—from elementary through high school and from poor to wealthy districts—has pushed students toward fact-based, content-focused learning outcomes. Although this emphasis has aligned somewhat with the science concepts and knowledge included in the various national standards documents, it has also marginalized efforts—equally present in the standards—to have students learn about the relationship between science and cultural issues, learning goals that were more prominent in the 1970s (Sadler et al., 2010).

The urgency and pace of reform has picked up notably since the turn of the twenty-first century with a renewed emphasis on scientific and technical workforce issues. Pulitzer prize–winning journalist Thomas Friedman's popular book, *The World Is Flat*, published in 2005, reignited national concerns over global competition in science and technology, which led to another National Academy of Sciences report two years later, *Rising above the Gathering Storm: Energizing and Employing America for a Brighter Economic Future* (2007). This report coincided with the passage of the America Competes Act that same year, legislation that many have described as a modern-day equivalent of the 1958 NDEA. Yet another call for more attention to science and mathematics education was made by the Carnegie Institute for Advanced Study Commission 2009 policy statement, *The Opportunity Equation: Mobilizing for Excellence and Equity in Mathematics and Science Education.* This report, as well as others, called for a smaller set of common science standards that could be used as a guiding framework by all states and that would apply across the relevant science, technology, engineering, and mathematics disciplines. The first step toward realizing this goal was the NRC publication of *A Framework for K-12 Science Concepts and Core Ideas: Practices, Crosscutting Concepts, and Core Ideas* in 2011.

This flurry of reports and government action since 2005 set off a new sense of crisis similar to that of the 1980s, but this time with countries like India and China taking Japan's place as the economic threat to American prosperity. An overriding concern with scientific capacity and technological innovation clearly took hold as the predominant motivation for national investment in science education, especially in the high schools. In many ways, this focus on the vocational and utilitarian aims of science teaching echoed the original utilitarian justifications for teaching science in schools made in the middle 1800s. The main difference, however, lay in the shift in emphasis of the public justification for science teaching in high schools. In the era prior to World War II, the study of science in schools was viewed as something with a tangible benefit to students and citizens, be it moral, cultural, or practical. Arguments in 2012 cast the value of science education in terms of national security and global competitiveness, and state and private interests increasingly developed the social and administrative machinery to turn science classrooms toward such ends.

[*See also* **Engineering; Higher Education and Science; Medical Education; Military, Science and Technology and the; Museums of Science and Natural History; Popularization of Science; Religion and Science; Science;** *and* **Technology.**]

BIBLIOGRAPHY

Ahlstrom, S. E. *A Religious History of the American People.* New Haven, Conn.: Yale University Press, 1972.

Aikenhead, G. "STS Education. A Rose by Any Other Name." In *A Vision for Science Education: Responding to the Work of Peter Fensham,* edited by R. Cross, pp. 49–58. London: Routledge-Falmer, 2003.

American Association for the Advancement of Science. *Benchmarks for Scientific Literacy.* New York: Oxford University Press, 1993.

American Association for the Advancement of Science. *Science for All Americans.* Washington, D.C.: Author, 1989.

Angulo, A. J. *William Barton Rogers and the Idea of MIT.* Baltimore: Johns Hopkins University Press, 2008.

Atkin, J. M., and Black, P. *A History of Curricular and Policy Change.* New York: Teachers College Press, 2003.

Benson, K. R. "From Museum Research to Laboratory Research: The Transformation of Natural History into Academic Biology." In *The American Development of Biology,* edited by Ronald Rainger, Keith R. Benson, and Jane Maienschein, pp. 49–83. Philadelphia: University of Pennsylvania Press, 1988.

Carnegie Corporation of New York and Institute for Advanced Study Commission on Mathematics and Science Education. *The Opportunity Equation: Mobilizing for Excellence and Equity in Mathematics and Science Education for Citizenship and the Global Economy.* New York: Carnegie Corporation of New York, 2009.

Clowse, B. *Brainpower for the Cold War: The Sputnik Crisis and National Defense Education Act of 1958.* Westport, Conn.: Greenwood Press, 1981.

Collins, A. "National Science Education Standards: A Political Document." *Journal of Research in Science Teaching* 35 (1998): 711–727.

Committee of Ten. *Report of the Committee of Ten on Secondary School Studies.* Washington, D.C.: National Education Association of the United States, 1894.

Committee on Science, Engineering, and Public Policy. *Rising above the Gathering Storm: Energizing and Employing America for a Brighter Economic Future.* Washington, D.C.: National Academies Press, 2006.

Conn, S. *Museums and American Intellectual Life, 1876–1926.* Chicago: University of Chicago Press, 1998.

Cotter, D. "A Disciplinary Immigrant. Alexander Smith at the University of Chicago, 1894–1911." *Annals of Science* 65, no. 2 (2008): 221–256.

DeBoer, G. *A History of Ideas in Science Education: Implications for Practice.* New York: Teachers College Press, 1991.

Dow, P. *Schoolhouse Politics: Lessons from the Sputnik Era.* Cambridge, Mass.: Harvard University Press, 1991.

Fuhrman, S. "Riding Waves, Trading Horses. The Twenty-Year Effort to Reform Education." In *A Nation Reformed? American Education 20 Years after A Nation at Risk,* edited by D. T. Gordon, pp. 7–22. Cambridge, Mass.: Harvard Education Press, 2003.

Geiger, R. "The Rise and Fall of Useful Knowledge: Higher Education for Science, Agriculture, and the Mechanic Arts." *History of Higher Education Annual* 18 (1998): 47–65.

Geiger, R. L. *To Advance Knowledge: The Growth of American Research Universities, 1900–1940.* New York: Oxford University Press, 1986.

Guralnik, S. M. *Science and the Ante-bellum American College.* Philadelphia: American Philosophical Society, 1975.

Heffron, J. M. "The Knowledge Most Worth Having: Otis W. Caldwell (1869–1947) and the Rise of the General Science Course." *Science & Education* 4, no. 3 (1995): 227–252.

Higham, J. "The Matrix of Specialization." In *Organization of Knowledge in Modern America, 1860–1920,* edited by Alexandra Oleson and John Voss, pp. 3–18. Baltimore: Johns Hopkins Press, 1979.

Hoffman, M. "Learning in the Laboratory: The Introduction of 'Practical' Science Teaching in Ontario's High Schools in the 1880s." In *Learning by Doing: Experiments and Instruments in the History of Science Teaching,* edited by P. Heering and R. Wittje, pp. 177–205. Stuttgart: Franz Steiner Verlag, 2011.

Hollinger, D. A. "Free Enterprise and Free Inquiry: The Emergence of Laissez-Faire Communitarianism in the Ideology of Science in the United States." *New Literary History* 21 (1990): 897–919.

Hollinger, D. A. *Inquiry and Uplift: Late Nineteenth-Century American Academics and the Moral Efficacy of Scientific Practice.* Bloomington: Indiana University Press, 1984.

Kaestle, C. "Federal Aid to Education since World War II: Purposes and Politics." In *The Future of the Federal Role in Elementary and Secondary Education,* edited by Jack Jennings, pp. 13–39. Washington, D.C.: Center for Education Policy, 2001.

Kaestle, C. F., and M. S. Smith. "The Federal Role in Elementary and Secondary Education, 1940–1980." *Harvard Education Review* 52, no. 4 (1982): 384–408.

Keeney, E. *The Botanizers: Amateur Scientists in Nineteenth-Century America.* Chapel Hill: University of North Carolina Press, 1992.

Kevles, D. J. "Principles, Property Rights, and Profits: Historical Reflections on University/Industry Tensions." *Accountability in Research* 8 (2001): 12–26.

Kleinmann, D. J. *Politics on the Endless Frontier: Postwar Research Policy in the United States.* Durham, N.C.: Duke University Press, 1995.

Kliebard, H. M. *The Struggle for the American Curriculum, 1893–1958,* 3d ed. New York: Routledge-Falmer, 2004.

Klopfenstein, K., and Thomas, M. K. "Advanced Placement Participation: Evaluating the Policies of States and Colleges." In *AP: A Critical Examination of the Advanced Placement Program,* edited by Philip M. Sadler, Gerhard Sonnert, Robert H. Tai, and Kristin Klopfenstein, pp. 167–188. Cambridge, Mass.: Harvard Education Press, 2010.

Kohler, R. E. "The Ph.D. Machine: Building on the Collegiate Base." In *The Scientific Enterprise in America: Readings from Isis,* edited by Ronald L. Numbers and Charles E. Rosenberg, pp. 98–122. Chicago: University of Chicago Press, 1996.

Kohlstedt, S. G. *Teaching Children Science: Hands-On Nature Study in North America, 1890–1930.* Chicago: University of Chicago Press, 2010.

Kremer, R. L. "Reforming American Physics Pedagogy in the 1880s: Introducing 'Learning by Doing' via Student Laboratory Exercises." In *Learning by Doing: Experiments and Instruments in the History of Science Teaching,* edited by P. Heering and R. Wittje, pp. 243–280. Stuttgart: Franz Steiner Verlag, 2011.

Krighbaum, H., and Rawson, H. *An Investment in Knowledge: The First Dozen Years of the National Foundation's Summer Institutes Programs to Improve Secondary School Science and Mathematics Teaching 1954–1965.* New York: New York University Press, 1969.

Krug, E. A. *The Shaping of the American High School, 1880–1920.* Madison: University of Wisconsin Press, 1969.

Labaree, D. F. *The Making of an American High School: The Credentials Market and the Central*

High School of Philadelphia, 1838–1939. New Haven, Conn.: Yale University Press, 1988.

Lacy, T. "Examining AP: Access, Rigor, and Revenue in the History of the Advanced Placement Program." In *AP: A Critical Examination of the Advanced Placement Program* edited by Philip M. Sadler, Gerhard Sonnert, Robert H. Tai, and Kristin Klopfenstein, pp. 17–48. Cambridge, Mass.: Harvard Education Press, 2010.

Ladouceur, R. P. "Ella Thea Smith and the Lost History of American High School Biology Textbooks." *Journal of the History of Biology* 41, no. 3 (2008): 435–471.

Larson, E. L. *Summer for the Gods: The Scopes Trial and America's Continuing Debate over Science and Religion.* New York: Basic Books, 1997.

Larson, E. J. *Trial and Error: The American Controversy over Creation and Evolution,* 3d ed. New York: Oxford University Press, 2003.

Mann, C. R. "Physics Teaching in the Secondary Schools of America." *Science* n. s. 30, no. 779 (1909), 789–798.

Marsden, G. M. *Fundamentalism and American Culture.* New York: Oxford University Press, 2006.

Milam, E. L. "Public Science of the Savage Mind: Contesting Cultural Anthropology in the Cold War Classroom." *Journal for the History of the Behavioral Sciences,* 49, no. 3 (2013): 306–330.

Moore, K. *Disrupting Science: Social Movements, American Scientists, and the Politics of the Military, 1945–1975.* Princeton, N.J.: Princeton University Press, 2008.

Moyer, A. E. "Edwin Hall and the Emergence of the Laboratory in Teaching Physics." *Physics Teacher* 14, no. 2 (1976): 96–103.

National Commission on Excellence in Education. *A Nation at Risk: The Imperative for Educational Reform.* Washington, D.C.: Government Printing Office, 1983.

National Education Association. *Cardinal Principles of Secondary Education.* Washington, D.C.: U.S. Government Printing Office, 1918.

National Education Association. *Reorganization of Science in Secondary Schools: A Report of the Commission on the Reorganization of Secondary Education.* Washington, D.C.: U.S. Government Printing Office, 1920.

National Research Council. *A Framework for K-12 Science Education: Practices, Crosscutting Concepts, and Core Ideas.* Washington, D.C.: National Academies Press, 2011.

Numbers, R. L. *The Creationists: From Scientific Creationism to Intelligent Design,* expanded ed. Cambridge, Mass.: Harvard University Press, 2006.

Numbers, R. L. *Darwinism Comes to America.* Cambridge, Mass.: Harvard University Press, 1998.

Ogren, C. A. *The American State Normal School: An Instrument of Great Good.* New York: Palgrave Macmillan, 2005.

Ogren, C. A. "Where Coeds Were Coeducated: Normal Schools in Wisconsin, 1870–1920." *History of Education Quarterly* 35, no. 1 (1995): 1–26.

Olesko, K. M. "German Models, American Ways: The 'New Movement' among American Physics Teachers, 1905–1909." In *German Influences on Education in the United States to 1917* edited by Henry Geitz, Jurgen Heideking, and Jurgen Herbst, pp. 129–153. New York: Cambridge University Press, 1995.

Owens, L. "Pure and Sound Government: Laboratories, Playing Fields, and Gymnasia in the Nineteenth-Century Search for Order." *Isis* 76, no. 2 (1985): 182–194.

Patterson, J. T. *Brown v. Board of Education: A Civil Rights Milestone and Its Troubled Legacy.* New York: Oxford University Press, 2001.

Pauly, P. J. *Biologists and the Promise of American Life: From Meriwether Lewis to Alfred Kinsey.* Princeton, N.J.: Princeton University Press, 2000.

Pauly, P. J. "Development of High School Biology: New York City, 1900–1925." *Isis* 82, no. 4 (1991): 662–688.

Purcell, E. A. *The Crisis of Democratic Theory: Scientific Naturalism and the Problem of Value.* Lexington: University Press of Kentucky, 1973.

Ravitch, D., and Vinovskis, M. *Learning from the Past: What History Teaches Us about School Reform.* Baltimore: Johns Hopkins Press, 1995.

Reese, W. J. *America's Public Schools: From the Common School to "No Child Left Behind.* Baltimore: Johns Hopkins Press, 2011.

Reese, W. J. *The Origins of the American High School.* New Haven, Conn.: Yale University Press, 1995.

Rosen, S. "A History of the Physics Laboratory in the American Public High School (to 1910)." *American Journal of Physics* 22 (1954): 200.

Ross, D. *The Origins of American Social Science.* New York: Cambridge University Press, 1992.

Ruben, J. A. *The Making of the Modern University: Intellectual Transformation and the Marginalization of Morality.* Chicago: University of Chicago Press, 1996.

Roberts, J. H., and Turner, J. *The Sacred and the Secular University*. Princeton, N.J.: Princeton University Press, 2000.

Rudolph, J. L. "Epistemology for the Masses: The Origins of 'The Scientific Method' in American Schools." *History of Education Quarterly* 45, no. 3 (2005): 341–376.

Rudolph, J. L. "From World War to Woods Hole: The Use of Wartime Research Models for Curriculum Reform." *Teachers College Record* 104 (2002): 212–241.

Rudolph, J. L. "Inquiry, Instrumentalism, and the Public Understanding of Science." *Science Education* 89, no. 5 (2005): 803–821.

Rudolph, J. L. *Scientists in the Classroom: The Cold War Reconstruction of American Science Education*. New York: Palgrave Macmillan, 2002.

Rudolph, J. L. "Teaching Materials and the Fate of Dynamic Biology in American Classrooms after *Sputnik*." *Technology & Culture* 53 (2012): 1–36.

Rudolph, J. L. "Turning Science to Account: Chicago and the General Science Movement in Secondary Education, 1905–1920." *Isis* 96, no. 3 (2005): 353–389.

Sadler, P. M., G. Sonnert, R. J. Tai, and K. Klopfenstein, eds. *AP: A Critical Examination of the Advanced Placement Program*. Cambridge, Mass.: Harvard Education Press, 2010.

Schulten, S. *The Geographical Imagination in America, 1880–1950*. Chicago: University of Chicago Press, 2001.

Shapiro, A. R. "Civic Biology and the Origin of the School Antievolution Movement." *Journal of the History of Biology* 41, no. 3 (2008): 409–433.

Shapiro, A. R. *Trying Biology: The Scopes Trial, Textbooks, and the Antievolution Movement in American Schools*. Chicago: University of Chicago Press, 2013.

Skoog, G. "Topic of Evolution in Secondary School Biology Textbooks, 1900–1977." *Science Education* 63, no. 5 (1979): 621–640.

Slotten, H. R. "Science, Education, and Antebellum Reform: The Case of Alexander Dallas Bache." *History of Education Quarterly* 31, no. 3 (1991): 323–342.

Smuts, A. B. *Science in the Service of Children, 1893–1935*. New Haven, Conn.: Yale University Press, 2006.

Thurs, D. P. "Scientific Methods." In *Wrestling with Nature: From Omens to Science*, edited by P. Harrison, R. L. Numbers, and M. H. Shank, pp. 307–336. Chicago: University of Chicago Press, 2011.

Tobey, R. C. *Saving the Prairies: The Life Cycle of the Founding School of American Plant Ecology, 1895–1955*. Berkeley: University of California Press, 1981.

Toch, T. *In the Name of Excellence: The Struggle to Reform the Nation's Schools, Why It's Failing, and What Should Be Done*. New York: Oxford University Press, 1991.

Tolley, K. "The Rise of the Academies: Continuity or Change?" *History of Education Quarterly* 41, no. 2 (2001): 225–239.

Tolley, K. *The Science Education of American Girls: A Historical Perspective*. New York: RoutledgeFalmer, 2002.

Turner, S. C. "Changing Images of the Inclined Plane: A Case Study of a Revolution in American Science Education." In *Learning by Doing: Experiments and Instruments in the History of Science Teaching*, edited by P. Heering and R. Wittje, pp. 207–242. Stuttgart: Franz Steiner Verlag, 2011.

Tyack, D. B., and Cuban, L. *Tinkering toward Utopia: A Century of Public School Reform*. Cambridge, Mass.: Harvard University Press, 1995.

Tyack, D. B., & Hansot, E. *Managers of Virtue: Public School Leadership in America 1820–1980*. New York: Basic Books, 1982.

Urban, W. J. *More Than Science and Sputnik: The National Defense Education Act of 1958*. Tuscaloosa: University of Alabama Press, 2010.

Vettel, E. J. *Biotech: The Countercultural Origins of an Industry*. Philadelphia: University of Pennsylvania Press, 2006.

Veysey, L. R. *The Emergence of the American University*. Chicago: University of Chicago Press, 1970.

Vinovskis, M. *From a Nation at Risk to No Child Left Behind: National Education Goals and the Creation of Federal Education Policy*. New York: Teachers College Press, 2008.

Zimmerman, J. *Distilling Democracy: Alcohol Education in America's Public Schools, 1880–1925*. Lawrence: University Press of Kansas, 1999.

John L. Rudolph
and David Meshoulam

HIGHWAY SYSTEM

The American highway system is a twentieth-century creation. Before 1916, roads remained a local responsibility. Bicyclists and railroad executives lobbied for better roads in the 1890s, and by 1910 the automotive industry echoed those demands. Reflecting many elements of the Progressive-Era reform movement (especially central administration by technical experts), the Federal-Aid Highway Act of 1916 created the first national road system, funded by $50 million over five years for construction of rural roads for mail deliveries. Consistent with federalism, the bill mandated cooperation between state highway departments and the Bureau of Public Roads (BPR, later the Federal Highway Administration). Thus states designed, built, and maintained the roads, whereas federal engineers inspected and approved plans, specifications, and construction. Initially costs were shared equally, although BPR engineers always exercised greater influence than this 50–50 balance would suggest, owing to their superior technical expertise.

In 1921, Congress shifted federal emphasis from rural mail delivery to a national network of primary and secondary roads between cities. As first mapped in 1923, this federal-aid highway system totaled 169,000 miles—7 percent of the nation's highway mileage. Under a numbering scheme adopted in 1925, odd-numbered roads ran north to south starting at the Atlantic coast, whereas even-numbered highways ran east to west beginning at the Canadian border. Federal appropriations averaged about $75 million annually during the 1920s, increased during the Depression, and jumped significantly after 1945. Gasoline taxes, first introduced in Oregon in 1919, rapidly became the primary source of state funds. System additions included extensions into urban areas (1938), additional secondary roads (1940), and the interstate system authorized in 1944. The states and BPR first designated interstate routes in 1947, although special funding began only in 1954. Massive appropriations for the 42,500 miles of National System of Interstate and Defense Highways came in 1956, with the federal government assuming 90 percent of the original estimated cost of $25 billion. When the last section opened in 1991, the total cost had climbed to $114 billion (adjusted for inflation) and the system had expanded to 46,720 miles.

This highway network altered many facets of American life, facilitating the development of a mobile culture and the rise of standardized national fast-food and motel franchises. Long-distance family summer vacations spent in campgrounds or motels became common, and long-haul trucking replaced railroads as primary freight haulers. The decline of central-city business districts, the "malling" of America, rapid postwar suburbanization, and new patterns of land use all resulted from easy access to express highways after 1950.

At the same time, the scale of interstate highway construction inside cities engendered significant resistance from neighborhood groups and environmentalists. After 1960, a "Freeway Revolt" made headlines and its partisans won numerous court cases, halting interstate construction projects in such cities as San Francisco, Boston, New Orleans, and Philadelphia. Although several disputed freeways were eventually completed in the late 1980s, the momentum for road construction was lost. Even so, the federal-aid system in 2006 encompassed 985,128 miles of the national total of more than 4 million miles of roads and highways. Annual federal appropriations for this system alone increased from $5 million in 1916 to $585 million in 1954 and more than $38 billion in 2006, when total highway spending surpassed $161 billion (U.S. Department of Transport, 2006).

[See also Bicycles and Bicycling; Engineering; Environmentalism; Environmental Protection Agency; Food and Diet; Ford, Henry; Motor Vehicles; Postal Service, U.S.; Railroads; Roads and Turnpikes, Early; Subways; and Urban Mass Transit.]

BIBLIOGRAPHY

Rose, Mark H. Interstate: Express Highway Politics, 1939–1989. Revised ed. Knoxville: University of Tennessee Press, 1990.

Seely, Bruce E. *Building the American Highway System: Engineers as Policy Makers*. Philadelphia: Temple University Press, 1987.

U.S, Department of Commerce, Bureau of Public Roads. *Highway Statistics*, p. 130. Washington, D.C.: U.S. Department of Commerce, 1954.

U.S. Department of Transport, Federal Highway Administration. *America's Highways, 1776–1976: A History of the Federal-aid Program*. Washington, D.C.: U.S. Government Printing Office, 1976.

U.S. Department of Transportation, Federal Highway Administration. *Dwight D. Eisenhower National System of Interstate and Defense Highways: Interstate Funding*. http://www.fhwa.dot.gov/programadmin/interstate.cfm#interstate_funding (accessed 26 January 2012).

U.S. Department of Transportation, Federal Highway Administration. *Expenditure of Federal Funds Administered by the Federal Highway Administration during Fiscal Year 2009*. http://www.fhwa.dot.gov/policyinformation/statistics/2009/fa3.cfm (accessed 26 January 2012).

U.S. Department of Transportation, Federal Highway Administration. *Highway Statistics*. http://purl.access.gpo.gov/GPO/LPS4717 (accessed 26 January 2012).

U.S. Department of Transportation, Federal Highway Administration. *Highway Statistics, 2006, Public Road Length—2006 1/ Miles by Ownership and Federal-Aid Highways—National Summary*. http://www.fhwa.dot.gov/policy/ohim/hs06/pdf/hm16.pdf (accessed 26 January 2012).

U.S. Department of Transportation, Federal Highway Administration. *Highway Statistics 2009, Public Road and Street Length 1980–2009—Miles by Functional System—National Summary 1*. http://www.fhwa.dot.gov/policyinformation/statistics/2009/hm220.cfm (accessed 26 January 2012).

U.S. Department of Transportation, Federal Highway Administration. *Highway Statistics 2009, Public Road Mileage—VMT—Lane Miles, 1920–2009*. http://www.fhwa.dot.gov/policyinformation/statistics/2009/vmt421.cfm (accessed 26 January 2012).

U.S. Department of Transportation, Federal Highway Administration. *Total Disbursements for Highways, by Function*. http://www.fhwa.dot.gov/policyinformation/statistics/2009/disb.cfm (accessed 26 January 2012).

Bruce E. Seely

HISTORY OF SCIENCE SOCIETY

Although the History of Science Society (HSS) began as an international society (and remains one), it is rooted in American soil. Established in January 1924 in Boston and incorporated in the District of Columbia in January 1925, the HSS's origins differ from that of many academic societies in the United States in two important ways, both of which have to do with *Isis*, the society's major journal. First, the vestiges of the HSS lay in Belgium, not the United States, with George Sarton (1884–1956), a mathematician with a romantic turn of mind. Sarton saw in World War I the urgent need for a civilizing education, a "New Humanism" in which the beauty of science and the beauty of the humanities become linked in common course (Pyenson, 2006, p. 320). Writing decades before C. P. Snow's *The Two Cultures*, Sarton had begun pushing his new humanism in 1912 with the founding of *Isis*, declaring on its cover that it was an "International Review Devoted to the History of Science and Civilization." The first issue appeared in Belgium in 1913 and reflected the decidedly European nature of *Isis*, with articles written in French, Italian, and German. And so the second difference for the HSS is that its major journal precedes the founding of the society by more than a decade.

The American founding of the HSS can be traced to 1916 to Lawrence J. Henderson, a chemist at Harvard, who was then working as an associate of the Carnegie Institution. In that year he invited Sarton to Harvard for a two-year lectureship, which would lead to Sarton settling permanently at Harvard in 1920, where, working steadily in Harvard's Widener Library, he continued to edit *Isis* for the next 32 years. Vital support from the Carnegie Institution supplemented Sarton's meager income as he struggled to maintain the journal out of his own funds. *Isis*'s unsteady future led a group of supporters to assemble at the American Academy of Arts and Sciences in 1924 and form an international society, the HSS, to support *Isis*, making the HSS, in effect, a subscription society (Cohen, p. S29, 1999).

The influence of America on the new society manifested itself from the beginning. Of the 505 founding members, some 75 percent resided in America, with the lion's share of the remaining minority coming from France, England, and Germany (Brasch, 1925, pp. 158–164; Numbers, 2009, p. 103). The American roots of the HSS became firmly established shortly after the society's incorporation in 1925: it affiliated with the American Association for the Advancement of Science that same year and the American Historical Association soon after that and became a constituent member of the American Council of Learned Societies in 1926. But even with these affiliations, the HSS did not become an American society.

Like the American Council of Learned Societies, which was established in 1919 to represent the United States in the Union Académique Internationale, the HSS remains fixed in its international outlook. The society has taken a leading role in the U.S. National Committee for the International Union for the History and Philosophy of Science—a UNESCO organization. The HSS continues its emphasis on the universal nature of science, with the banner of *Isis* still declaring it to be an "International Review," and its international membership has increased since its founding days, with nearly a third of its members living outside of the United States. And although few historians of science now agree with Sarton's declaration that science is the only human activity that is cumulative and progressive, Sarton's vision of science informed by the humanities continues. This remains an international outlook and even though the HSS is American in the sense that its officers, its editors, and its members have mostly been American, they have been Americans with a global turn of mind.

[*See also* **American Association for the History of Medicine;** *and* **Society for the History of Technology.**]

BIBLIOGRAPHY

Brasch, Frederick E. "List of the Foundation Members of the History of Science Society." *Isis* 7, no. 3 (1925): 371–393.

Cohen, I. Bernard, "The *Isis* Crises and the Coming of Age of the History of Science Society." *Isis* 90, supplement (1999): S28-S42.

History of Science Society. http://www.hssonline.org.

Numbers, Ronald L. "The *American* History of Science Society or the *International* History of Science Society? The Fate of Cosmopolitanism since George Sarton." *Isis* 100, no. 1 (March 2009): 103–107.

Pyenson, Lewis. *The Passion of George Sarton: A Modern Marriage and Its Discipline.* Philadelphia: American Philosophical Society, 2006.

Robert J. Malone

HITCHCOCK, EDWARD

(1793–1864), geologist, pastor, educator, author, college president, was born in Deerfield, Massachusetts, and died in Amherst, Massachusetts. Hitchcock founded and served as the first president of the Association of American Geologists (1840), which expanded, in 1848, to become the American Association for the Advancement of Science. He was also a member of the American Philosophical Society, the American Academy of Arts and Sciences, and the National Academy of Sciences (inaugural class, 1863).

From 1818 to 1821 and 1825 to 1826, Hitchcock studied theology (with Nathaniel Taylor) and natural sciences (with Benjamin Silliman) at Yale College. In 1826 Hitchcock joined the faculty of Amherst College (est. 1821), where he remained for life, lecturing in natural history, chemistry, botany, natural theology, Christian discipleship, and, most notably, geology, paleontology, and mineralogy. From 1845 to 1854 he also served as Amherst College president.

Hitchcock completed the earliest state-funded geological and mineralogical surveys (Massachusetts, Rhode Island, Vermont). He also analyzed fossilized "giant bird" tracks (later understood as dinosaurian) in the Connecticut River Valley (*Fossil Footmarks of the United States*, 1848; *Ichnology of New England*, 1858). He authored the first geology textbook specifically tailored for the college classroom (*Elementary Geology*, 1840).

Nurtured in catastrophism and diluvialism, Hitchcock developed a hesitant respect for Charles Lyell's geological analysis by interpretation of "causes now in 'operation'"—nevertheless, Hitchcock never abandoned the idea that some geological processes of the past simply do not remain currently in operation. Seeing relevant markings in Europe and comparing them with markings in North America, Hitchcock accepted the general outline of Louis Agassiz's glacial theory.

Hitchcock maintained a deep commitment to a "unity of truth" epistemology. Religious faith, rightly held, will never permanently conflict with natural science, rightly understood—any "conflicts" would prove only apparent and temporary (*Religion of Geology*, 1851). For religious Americans, Hitchcock seemed to open a way for embracing a scientific view of nature without sacrificing faith. In particular, Hitchcock suggested a biblically credible and scientifically compatible framework for embracing the great age of the earth and universe, as well as for understanding the Genesis Flood as historically real yet geologically imperceptible.

Although he saw Darwin's "Development Hypothesis" as philosophically questionable ("renders doubtful and unnecessary the existence of a deity") and scientifically weak (lack of intermediate forms, etc.), Hitchcock nevertheless insisted, in a signature statement, that "the real question is, not whether these hypotheses accord with our religious views, but whether they are true" ("The Law of Nature's Constancy Subordinate to the Higher Law of Change," *Bibliotheca Sacra* (1863) 20:670:489–561).

Hitchcock married Orra White (1796–1863) in 1821. They raised four daughters and two sons, the latter (Edward and Charles) undertaking significant careers in science. Orra White Hitchcock developed into an accomplished artist whose drawings illustrated many of Hitchcock's works.

[*See also* Agassiz, Louis; American Association for the Advancement of Science; American Philosophical Society; Botany; Chemistry; Dinosaurs; Evolution, Theory of; Geological Surveys; Geology; National Academy of Sciences; Paleontology; Religion and Science; Science; *and* Silliman, Benjamin.]

BIBLIOGRAPHY

Guralnick, Stanley M. "Geology and Religion before Darwin: The Case of Edward Hitchcock, Theologian and Geologist (1793–1864)." *Isis* 63 (1972): 529–543.

Hitchcock, Edward. *Reminiscences of Amherst College: Historical, Scientific, Biographical, and Autobiographical.* Northampton, Mass.: Bridgman & Childs, 1863. Hitchcock's publications are listed on pp. 378–393.

Hitchcock, Mary Lewis Judson [Mrs. Edward Hitchcock, 1831–1918], and Dwight Whitney Marsh. *The Genealogy of the Hitchcock Family*, pp. 443–447. Amherst, Mass.: Press of Carpenter and Morehouse, 1894.

Rodney L. Stiling

HIV/AIDS

June 5, 1981, marks the official start of the human immunodeficiency virus (HIV) epidemic. On that date, the Centers for Disease Control and Prevention (CDC) published what has become a landmark communication in its *Morbidity and Mortality Weekly Report* (*MMWR*). Written by Michael Gottlieb, a young infectious disease doctor, and colleagues, it alerted the public-health community that between October 1980 and May 1981 five young and previously healthy homosexual men had been treated in Los Angeles hospitals for biopsy-confirmed *Pneumocystis carinii* pneumonia (PCP). From this information, the CDC suggested a possible link between PCP and homosexual sex or "lifestyle." As if to reinforce that point, Gottlieb's paper was closely followed by another from New York City and San Francisco; it reported that in the 30 months preceding July 1981, Kaposi's sarcoma (KS) had been diagnosed in 26 gay males, 26 to 51 years of age. KS was known as a rare cancer that occurred primarily in elderly males and in immunosuppressed organ recipients. Its appearance in a relatively large number of young men was startling, as was that of PCP in individuals without a clinically based cause for immunodeficiency. As they read of the reports, doctors on both coasts recognized similar patients who had passed through their emergency rooms and services since the late 1970s. Others remembered young patients who had died of infections that were difficult to diagnose and devastating in their course.

Over the next 18 months, the news of the burgeoning epidemic became darker, heralded by the headlines of the *MMWR*. On July 9, 1982, the CDC announced that KS and opportunistic infections had been diagnosed in Haitians. A week later the *MMWR* reported PCP in hemophiliacs. On December 10, it alerted its readers to the possible transmission of the new disease, now called acquired immune deficiency syndrome (AIDS), through blood transfusions. On December 17, 1982, and January 7, 1983, respectively, the agency reported unexplained immunodeficiency in infants and in female sex partners of men with AIDS. What emerged was a profile of the epidemic as the burgeoning of a new sexually transmitted and blood-borne disease. In 1982, the extent to which AIDS had spread was unknown. That would remain so until the responsible viral agent would be discovered and characterized and a blood test developed and used in serosurveys.

Investigating an Unknown Disease.
That research began in mid-1981, when the CDC initiated a special task force charged with surveillance on KS and opportunistic infections. Its purpose was to confirm that the observed disorder was new and that all new cases were verified. To determine whether KS had occurred before 1980 in young individuals, the task force queried epidemiologists at state and local tumor registries. Because the CDC was the sole supplier of pentamidine, a drug used to treat PCP, its own files could reveal whether the infection had been seen previously in adults without an underlying illness. By August 1981, the CDC requested that all state health departments report all suspected cases of KS and PCP.

What was the cause of this new disorder? What relationship did it bear to sex? What induced immunosuppression in previously healthy young gay men? As a start, the CDC performed a brief survey in San Francisco, New York, and Atlanta of 420 men attending clinics for sexually transmitted diseases. The 35 cases of KS or PCP culled from this unrepresentative sample were interviewed in depth with the hope of developing scientific leads. Researchers found that these men, all homosexuals, had had many sexual partners the previous year and had frequently used recreational drugs like marijuana, cocaine, and amyl or butyl nitrite. The rate of nitrite use was closely associated with the number of partners, suggesting a relationship between the two. It was also possible that nitrite use was a confounding factor, appearing to be linked to PCP and KS because it was part of the men's sexual activity (a median number of 87 partners in the past year). But that 86 percent of the gay and bisexual men in the CDC's survey had used amyl nitrite in the previous five years, compared to only 15 percent of heterosexual males, was a striking finding, and amyl nitrite became one of the first putative determinants to be investigated. Amyl nitrite seemed worth examining, particularly because it appeared to be a component of the "gay lifestyle" hypothesis that was riveting the epidemiologic researchers at the time. From 1981 until the middle of the decade, researchers pursued the association, positing that amyl nitrite might predispose homosexual men to immune deficiency.

An alternative hypothesis, also investigated by the CDC, was the possibility that the new syndrome was caused by cytomegalovirus (CMV), a microbe suspected of both being sexually transmitted and a cause of KS. A small clinical study published in 1981 by Michael Gottlieb and colleagues found a high rate of CMV in homosexual men with KS or PCP; the latter group also suffered from a low count of T4 lymphocytes (also known as T4 helper cells). Although it was possible that CMV infection might result from T4-cell deficiency and the reactivation of a dormant infection, the authors of this study preferred to suspect the virus, based on earlier research that found much higher rates of CMV infection in gay compared to heterosexual men.

A third hypothesis focused on multiple factors that, in concert, overloaded the immune system and led to its dysfunction. An editorial in the *New England Journal of Medicine* posited that the joint effects of persistent, sexually transmitted viral infection (possibly CMV) and a recreational drug like amyl nitrite precipitated immunosuppression in genetically predisposed males. In the *Journal of the American Medical Association*, Dr. Joseph Sonnabend, who treated many of the early AIDS cases in Greenwich Village, proposed a model in which repeated sexual contact with many partners exposed a subgroup of homosexual men to CMV and allogeneic sperm, over time leading to a damaged and suppressed immune system. In his article, as in the gay press, Sonnabend indicted the "unprecedented level of promiscuity" over the past decade in urban enclaves like the Village. However, he could not explain why the same disease should be seen in Haitians and hemophiliacs. Instead, he looked to possible alternative factors, suggesting a list of variables that, ironically, presaged the arguments of future HIV denialists like Peter Duesberg, a prominent Berkeley professor of molecular and cell biology—namely, malnutrition, recreational drugs, and acute viral infections.

As epidemiologic evidence accumulated, the lifestyle hypothesis—despite its initial appeal—became increasingly untenable. The occurrence of PCP in hemophiliacs who had no other underlying disease but were dependent on factor VIII therapy raised the possibility that blood was a vehicle for a transmissible agent. That theory was strengthened by a CDC report of the appearance of immunodeficiency and opportunistic infection in a 20-month-old child who had previously received multiple transfusions from a donor later found to have AIDS.

On March 4, 1983, a Public Health Service (PHS) interagency report in the *MMWR* formalized a major shift in the conceptualization of the epidemic. The weight of the evidence, including cases of immunodeficiency and opportunistic infections in the female sex partners of bisexuals and intravenous drug–using men, and of their children, pointed to the existence of an infectious agent. Although the microbe was unknown, it appeared from the case distribution to be analogous to hepatitis B, a virus transmitted sexually, parenterally, and through blood and blood products. In this reconfiguration of the known variables, life-

style did not drop out, but instead became an indirect cause of AIDS.

The hepatitis B model suggested a direction for public-health intervention. The PHS recommended that actions known to limit the spread of the hepatitis B virus be applied to the new epidemic. In particular, the PHS strongly advised against sexual contact with persons suspected of having AIDS. In addition, it asked groups at higher risk of the disease not to donate blood or plasma and encouraged doctors to recommend to surgical patients that they donate their own blood for transfusions purposes. Finally, the PHS called for the development of blood-screening procedures.

In the same *MMWR*, the CDC made reference to "high-risk groups" whose members carried a greater probability of infection and of causing infection, carrying a microbe that could be transmitted through the vehicle of bodily fluids. Although the CDC stressed that "each group contains many persons who probably have little risk of acquiring AIDS," in reality no such distinction could be drawn. In the absence of a screening or diagnostic test, risk group designation was, in effect, synonymous with carrier status for its members.

One of the results of creating high-risk groups was to reinforce the linkage between AIDS and socially "marginal" members of society. Conceptually, although each group represented a threat to the rest of the community, public-health strictures were to delimit the contamination. The hope was that the epidemic could be cordoned at the margins of the social body, protecting against a residue who were "different" from the majority. A risk group designation, however inadvertently, created a source of blame and a target for discrimination. Gays, in particular, were targeted. The conservative commentator and former advisor to presidents Nixon and Reagan, Patrick Buchanan, for example, taunted the group: "The poor homosexuals," he wrote, "they have declared war upon Nature, and now Nature is exacting an awful retribution" (quoted in Brandt, 1988, p. 155).

Stigmatization and fear of HIV was not limited to IV drug users and gays, especially when the public felt vulnerable. Thirteen year-old Ryan White, who developed AIDS as a result of treatment for hemophilia, was refused admission to an Indiana public school in 1985. White's plight was contested in his school system and in others. In New York City, parents boycotted public schools in two districts attended by infected children. In the late 1980s, White became a poster child for the fight against AIDS discrimination. By 1990, as panic subsided, Ryan White's death was followed by congressional legislation commemorating his courage and perseverance, the Ryan White Comprehensive AIDS Resources Emergency Act. Renewed in 1996, 2000, 2006, and 2009, its purpose has been to fund services, including pharmaceuticals, for those HIV-infected people lacking the financial resources or health coverage they require.

From Subjects to Participants: Redefining Biomedical Research.

The unique constellation of forces unleashed by AIDS—sociopolitical as well as clinical—had a profound impact on the conduct of research designed to address the scientific challenges posed by the epidemic. In 1978, the National Commission for the Protection of Human Subjects of Biomedical and Behavioral Research issued its Belmont Report; this codified a set of ethical principles that ought to inform the work of researchers. Those norms provided the foundations for regulations subsequently enacted by the U.S. Department of Health and Human Services and the Food and Drug Administration. At the core of those guidelines was the radical distinction between research designed to produce socially necessary, generalizable knowledge and therapy designed to benefit individuals. Against the former, individuals—but especially those who were socially vulnerable—needed protection against conscription.

During the 1980s, AIDS forced a reconsideration of this formulation. The role of the randomized controlled trial, the importance of placebo controls, the centrality of academic research institutions, the dominance of scientists over subjects, the sharp distinction between research and therapy, and the protectionist ethos of the Belmont Report were all brought into question.

Although scholars concerned with the methodological standards of sound research and ethicists committed to the protection of research subjects played a crucial role in the ensuing discussions, both as defenders of the received wisdom and as critics, the debate was driven by the articulate demands of those most threatened by AIDS. Most prominent were groups such as the People with AIDS Coalition and the AIDS Coalition to Unleash Power (ACTUP), organizations made up primarily of white, gay men. But advocates of women's, children's, and prisoners' rights also made their voices heard. What was so stunning—disconcerting to some and exciting to others—was the rhythm of challenge and response. Rather than the careful exchange of academic arguments, there was the mobilization of disruptive and effective political protest.

The threat of death hovered over the process. As Carol Levine noted in her 1988 essay "Has AIDS Changed the Ethics of Human Subjects Research?" "the shortage of proven therapeutic alternatives for AIDS and the belief that trials are, in and of themselves, beneficial have led to the claim that people have a *right to be* research subjects. This is the exact opposite of the tradition starting with Nuremberg—that people have a *right not to be* research subjects" (p. 172). That striking reversal resulted in a rejection of the model of research conducted at academic centers, with restrictive (protective) standards of access and strict adherence to the "gold standard" of the randomized controlled trial. Blurring the distinction between research and treatment—"A Drug Trial Is Health Care Too"—those insistent on radical reform sought to open wide the points of entry to new "therapeutic" agents both within and outside of clinical trials; they demanded that the paternalistic, ethical warrant for the protection of the vulnerable from research be replaced by an ethical regime informed by respect for the autonomous choice of potential subjects who could weigh for themselves the possible risks and benefits of new treatments for HIV infection. Moreover, the revisionists demanded a basic reconceptualization of the relationship between researchers and subjects. In place of protocols imposed from above, they proposed a more egalitarian and democratic model in which negotiation would replace scientific authority.

The number of newly diagnosed cases of AIDS rose dramatically in the first decade of the epidemic, peaking in 1991. So too did the number of deaths, which reached their apogee in 1995. Indeed, one prominent physician treating HIV-infected patients observed that "death came to define the AIDS epidemic." Although it was clear the biology of HIV placed everyone at potential risk, the social distribution of the epidemic told a very different story. As had been true from the early 1980s, those affected most were gay and bisexual men, intravenous drug users, and their sexual partners who bore the burden of morbidity and mortality. Increasingly, the epidemic affected disproportionate numbers of African Americans. Although many middle-class men had been diagnosed with AIDS, the disease was increasingly linked, like many infectious disorders, to those in the lower socioeconomic strata.

The first years of the epidemic, so characterized by therapeutic limits and a unique social distribution of disease, had been marked by a public-health "exceptionalism" designed to maximize the cooperation of those most at risk of the disease. In fact, when the exceptionalist perspective first took hold, many of its proponents viewed it as an opportunity to address the traditional authoritarian dimensions of public health; they asserted that modern public health need not involve limitations on the rights of individuals. Indeed, they argued that a public health that respected rights would be both more effective and more ethical. But, by the late 1980s, the hold of this perspective had already been subject to erosion as the era of therapeutic impotence yielded to an enhanced capacity to manage opportunistic infections, even if the picture surrounding antiretroviral therapy remained bleak.

The Search for Antiviral Therapy. Beginning in February 1986, 282 individuals with AIDS or serious symptoms associated with HIV infection were enrolled in a randomized, double-blind controlled trial of the drug AZT under the auspices of the National Institute of Allergy and Infectious Disease. Six months later the study was halted by its data-monitoring board on the ethical grounds that those on AZT were clearly benefit-

ting compared to others only receiving a placebo. Of 137 participants, 19 of the latter had died, compared to 1 of the 145 on the new agent. What was so striking was that many of those on AZT underwent a stunning revitalization, gaining weight and energy: opportunistic infections began to clear away as patients' T-cell counts rose. By 1988, however, it had become clear that the positive effects of AZT were short-lived, with patients whose conditions had improved returning to a downward, fatal course as viral resistance increased. For doctors and patients who had become enthusiastic, the limitations of AZT were devastating. Although some physicians reversed course and discounted the drug entirely, research continued to show its power to slow AIDS progression. In 1989, another clinical trial—this time to determine the effect of AZT on asymptomatic HIV-infected patients—was halted early; again a data-monitoring board found there was sufficient positive evidence to make the trial's continuation ethically unsupportable.

Enthusiasm was challenged in 1993, when the Concorde trial, a European study of AZT that mirrored that of the United States, reported its results. Unlike its counterpart, the Concorde trial had continued to run its full course of three years. Its findings contradicted the American results and stood as a sharp rebuke to those who had claimed that life could be prolonged if AZT treatment was started early in asymptomatic patients. In the years between 1989 and 1993, however, many who treated AIDS had begun to view AZT as a drug of limited utility, given viral resistance. Concorde seemed to confirm their worst fears. Others, in examining the study, found it methodologically flawed. But as a consequence of its negative results, it was no longer clear when to initiate AZT therapy or whether to use it at all.

The Waning of the Epidemic? In the early 1990s, more than a decade after the onset of the epidemic, there was a pervasive sense, reinforced by the Concorde results, that the effort to meet the clinical challenge of AIDS had stalled. Although there had been progress in the management of HIV-related opportunistic infections like PCP and tuberculosis, attempts to identify powerful antiretroviral agents had achieved little that could add dramatically to the life expectancy of those with AIDS. But in 1994, there was new reason for limited hope. In February of that year, the Data Safety and Monitoring Board of the National Institute of Allergy and Infectious Diseases recommended the interruption of clinical trial 076, designed to determine whether the administration of AZT to pregnant women infected with HIV could reduce the rate of maternal-fetal viral transmission. Women who had received AZT in the trial had a transmission rate of 8.3 percent; among those who had received a placebo, 25.5 percent of the newborns were infected.

Despite its remarkable success in reducing the risk of transmitting HIV infection from pregnant women to their babies, AZT had by 1994 come to symbolize the lag in therapeutic progress. Almost four years had elapsed between its approval by the FDA and the licensing of ddI, a second drug in the same class as AZT. However, within the next two and a half years, two additional antiretroviral drugs were licensed, ddC and d4T. At the end of 1995, 3TC was approved. It was the availability of this range of drugs that made possible a breakthrough, a conceptual shift in the treatment of AIDS. Single drugs would inevitably fail as viral resistance developed. But a combination of agents, used judiciously, could forestall that inevitability.

It was not, however, the combination of drugs in the same class as AZT that was to signal a radical transformation in the therapeutic world of those confronting HIV. Rather, it was an entirely new class of drugs known as protease inhibitors that promised to reduce the blood level of HIV to undetectable levels. In January 1996, at a national conference in Washington, that news of the protease inhibitors was first publicly presented. Six months later, its therapeutic effects were presented to an enthusiastic audience of thousands of physicians, researchers, and AIDS activists attending the XI International AIDS Conference in Vancouver, Canada. There researchers presented results demonstrating that the protease inhibitors could radically reduce viral load and provide patients with periods during which they were symptom free. On one hand, as experience with the new drugs deepened, doctors and patients became acutely aware of burdensome or toxic side effects that sometimes made the protease inhibitors difficult to tolerate. Among the effects of the new drugs were high blood sugar levels and diabetes, altered body fat distribution (lipodystrophy), higher blood lipid levels and greater possibility of coronary heart disease, and stomach or liver toxicity. On the other hand, doctors learned to manage these side effects and the underlying HIV disease itself; HIV/AIDS, treated with "cocktails" of the new drugs, became a chronic disease.

The degree to which the new drug therapies prevented HIV infection in newborns or extended the survival in those infected tended to strengthen the growing sense that HIV/AIDS, so long a vector of dread, was on the verge of being subdued in the United States. Increasingly, physicians considered the possibility that AIDS was about to become "normalized": to lose its extraordinary status as an exceptional clinical and public-health problem. With new antiretrovirals, AIDS would join the class of potentially fatal but well-managed chronic disorders. And physicians who had dedicated themselves to treat AIDS, to become "AIDS doctors" when others avoided the disease and its sufferers, would be reabsorbed into the hierarchies and conventions of the medical mainstream.

In 1998, three years after the introduction of the protease inhibitor drug cocktails, the transformation was captured in a year-end editorial in the *New England Journal of Medicine*:

> The good news continues in the battle against AIDS. In the United States, the age-adjusted death rate among people with human immunodeficiency virus (HIV) in 1997 was less than 40 percent of what it was in 1995. The 16,685 deaths in 1997 represent the lowest annual total in nearly a decade.... Not only has mortality from AIDS decreased, but so has the incidence of AIDS among those who have HIV infection.... The dramatic declines in morbidity and mortality due to AIDS in Western nations are the result of

the widespread use of potent combinations of antiretroviral drugs.

—Steinbrook, p. 126

Between 1993 and 1998, AIDS diagnoses fell by 45 percent, meaning far fewer infected individuals progressed to the last stages of the disease; from 1995 to 1998, deaths from HIV infection plummeted 63 percent. Thereafter, diagnoses and mortality remained unchanged, annually averaging 38,279 and 17,489 individuals, respectively. Between 1996 and 2008, the prevalence of AIDS in Americans over the age of 13 doubled, as infected people successfully lived with the HIV virus.

—CDC, 2011, p. 689

To public-health officials concerned about the incidence of infection, the failure to reduce new cases represented both a failure and a challenge. Those behavioral approaches to interrupting the spread of HIV that had come to characterize efforts in the United States were clearly inadequate to the problem. Since it was clear that those who were unaware of their HIV status were more likely to transmit infection than those who were diagnosed, the CDC moved in 2006 to recommend that HIV screening of adults and adolescents be routinized and that the standard of consent be shifted from the opt-in model to an opt-out approach. But significant potential breakthroughs had to await new scientific research results.

It had long been known that individuals with undetectable viral loads were less likely to transmit HIV to their sexual partners. A study in 2011 demonstrated that in HIV-discordant couples the provision of antiretroviral therapy to infected individuals, regardless of serum CD4 count (a mark of immune status), radically reduced the risk of infection. This opened up the possibility of a chemotherapeutic intervention with broad public-health implications. Concomitantly, researchers found that providing antiretroviral drugs to uninfected individuals could significantly lower the risk of acquiring HIV infection. Both sets of findings provided new hope for prevention efforts that had remained stalled for years.

In the mid-1990s, fundamental transformations occurred as therapeutic prospects like the use of AZT to dramatically reduce maternal-fetal transmission and the emergence of antiretroviral cocktails radically transformed every dimension of the HIV epidemic. Under these altered circumstances, the "exceptionalist" perspective would be subject to further challenge. Nevertheless, even as critical elements of public-health policy would assume more traditional dimensions, the legacy of exceptionalism would remain clear in the extent to which the claims of individual rights and the importance of consultations with stakeholders had become defining features of contemporary public health theory and practice.

The therapeutic advances of the mid-1990s brought to an end so much that came to define AIDS in its first years, at least in countries such as the United States. A disease that seemed to reveal the limits of medicine would become a complex, sometimes difficult to manage long-term disease. And with that change the narratives of death and suffering that had defined AIDS would, in time, become something increasingly difficult to recall. But as the toll exacted by HIV in the advanced industrialized nations took on more "normal" dimensions, attention would shift to the extraordinary and stark picture of the epidemic's course elsewhere, especially in sub-Saharan Africa and parts of Asia. In these resource-poor areas, the therapeutic breakthroughs that had transformed HIV/AIDS in other parts of the world remained out of reach of all but a few, focusing global attention on issues of life and death arising from inequality and inequity. In fact, the AIDS epidemic, which had sensitized Europeans and North Americans to the global threat of invasive infectious diseases, also catalyzed a global health movement in which the wealthier countries, however hesitantly, began to recognize that they could not remain indifferent in the face of global disparities in health resources and the consequent flood of HIV mortality. During the early twenty-first century, Western foundations, universities, nongovernment organizations, and nations working through the World Health Organization and, in the case of the United States, the President's Emergency Plan for AIDS, began to export trained personnel, education, pharmaceuticals like anti-retrovirals, and other resources to poor countries. It is in these economically challenged countries that the next chapter in the history of AIDS will be written.

[*See also* **Centers for Disease Control and Prevention; Disease; Ethics and Medicine; Medicine: Since 1945; National Institutes of Health; Pharmacology and Drug Therapy; Public Health; Public Health Service, U.S.; Sexually Transmitted Diseases;** *and* **World Health Organization.**]

BIBLIOGRAPHY

Astor, Gerald. *The Disease Detectives.* New York: New American Library, 1984.

Bayer, Ronald. *Private Acts, Social Consequences: AIDS and the Politics of Public Health,* p. 24. New York: Free Press, 1989.

Bayer, Ronald. "Public Health Policy and the AIDS Epidemic: An End to AIDS Exceptionalism?" *New England Journal of Medicine* 324 (1991): 1500–1504.

Bayer, Ronald, and Gerald M. Oppenheimer. *AIDS Doctors: Voices from the Epidemic.* New York: Oxford University Press, 2000.

Berridge, Virginia. *AIDS in the UK: The Making of Policy, 1981–1994.* Oxford: Oxford University Press, 1996.

Brandt, Allan. "AIDS: From Social History to Social Policy." In *AIDS and the Burdens of History,* edited by Elizabeth Fee and Daniel M. Fox, pp. 147–171. Berkeley: University of California Press, 1988. *New York Post,* May 24, 1983, as quoted by Allan Brandt.

Centers for Disease Control and Prevention. "HIV Surveillance—United States, 1981–2008." *Morbidity and Mortality Weekly Report* 60 (2011): 689–693.

Centers for Disease Control and Prevention. "Kaposi's Sarcoma and *Pneumocystis* Pneumonia among Homosexual Men—New York City and California." *Morbidity and Mortality Weekly Report* 30 (1981): 305–308.

Centers for Disease Control and Prevention. "*Pneumocystsis* Pneumonia—Los Angeles." *Morbidity and Mortality Weekly Report* 30 (1981): 250–252.

Centers for Disease Control and Prevention. "Prevention of Acquired Immune Deficiency Syndrome (AIDS): Report of Inter-agency Recommendations." *Morbidity and Mortality Weekly Report* 32 (1983): 101–104.

Chambré, Susan M. *Fighting for Our Lives: New York's AIDS Community and the Politics of Disease.* New Brunswick, N.J.: Rutgers University Press, 2006.

Cohen, Myran S. "Prevention of HIV-1 Infection with Early Antiretroviral Therapy." *New England Journal of Medicine* 365 (2011): 493–505.

Concorde Coordinating Committee. "Concorde: MRC/ANS Randomized Double-Blind Controlled Trial of Immediate and Deferred Zidovudine in Symptom-free HIV Infection." *Lancet* 343 (1994): 871–881.

Connor, Edward M., et al. "Reduction of Maternal-Infant Transmission of Human Immunodeficiency Virus Type I with Zidovudine Treatment." *New England Journal of Medicine* 331 (1994): 1173–1180.

Epstein, Steven. *Impure Science: AIDS, Activism, and the Politics of Knowledge.* Berkeley: University of California Press, 1996.

Gostin, Lawrence O. *The AIDS Pandemic: Complacency, Injustice, and Unfulfilled Expectations.* Chapel Hill: University of North Carolina Press, 2004.

Gottlieb. Michael S., et al. "Pneumocystis carinii Pneumonia and Mucosal Candidiasis in Previously Healthy Homosexual Men: Evidence of a New Acquired Cellular Immunodeficiency." *New England Journal of Medicine* 305 (1981): 1425–1431.

Harden, Victoria A. *AIDS at 30: A History.* Washington, D.C.: Potomac Books, 2012.

Iliffe, John. *The African AIDS Epidemic: A History.* Oxford: James Currey Ltd., 2006.

Levine, Carol. "Has AIDS Changed the Ethics of Human Subjects Research?" *Law, Medicine, and Health Care* 16 (1988): 167–173.

Oppenheimer, Gerald M. "In the Eye of the Storm: The Epidemiological Construction of AIDS." In *AIDS: The Burdens of History,* edited by Elizabeth Fee and Daniel M. Fox, pp. 267–300. Berkeley: University of California Press, 1988.

Oppenheimer, Gerald M., and Ronald Bayer. *Shattered Dreams? An Oral History of the South African AIDS Epidemic.* New York: Oxford University Press, 2007.

Shilts, Randy. *And the Band Played On: Politics, People, and the AIDS Epidemic.* New York: St. Martin's Press, 1987.

Sonnabend, Joseph, et al. "Acquired Immunodeficiency Syndrome, Opportunistic Infections, and Malignancies in Male Homosexuals: A Hypothesis of Etiologic Factors in Pathogenesis." *Journal of the American Medical Association* 249 (1983): 2370–2374.

Steinbrook, Robert. "Caring for People with Human Immunodeficiency Virus Infection." *New England Journal of Medicine* 339 (1998): 126–128.

Verghese, Abraham. *My Own Country.* New York: Simon & Schuster, 1994.

Volberding, Paul, et al. "Zidovudine in Asymptomatic Human Immunodeficiency Virus Infection." *New England Journal of Medicine* 322 (1990): 941–949.

Gerald M. Oppenheimer and Ronald Bayer

HMOs

See Health Maintenance Organizations.

HOME ECONOMICS MOVEMENT

A social movement, discipline, and profession, crystallized in the writings of Catharine Beecher, a leading mid-nineteenth-century educator. Beecher stressed that women needed scientific knowledge to fulfill their socially sanctioned roles of wife, mother, and homemaker. In the late nineteenth and early twentieth centuries, universities and land-grant colleges, as well as elementary and secondary schools, institutionalized these principles as domestic science, domestic economy, or household science. The most significant force behind home economics in this period was Ellen Swallow Richards, a graduate of Vassar College and the Massachusetts Institute of Technology. At ten conferences Richards convened at Lake Placid, New York, between 1899 and 1908, educators and social reformers debated the role of the growing movement. The last conference established the movement's professional organization, the American Home Economics Association (later renamed the American Association of Family and Consumer Sciences).

Richards and her followers believed that the application of science to domestic problems could save society from the social disintegration they saw at the turn of the century. This program of science education for women had benefits and limitations. On the positive side, women could pursue science degrees in higher education. Moreover, the movement's social-reform impetus encouraged women to apply this knowledge in the wider arena of public life. For example, nutrition was not only the healthful feeding of one's family; it was also educating other women, constructing healthful dietaries for schools and institutions, generating the knowledge base underlying these efforts, and more. Thus, professional avenues such as dietician, nutritionist, and extension specialist opened for women. On the negative side, the existence of home economics departments enabled schools to direct women interested in science into a sex-segregated educational track and sex-segregated occupations, with relatively low prestige and limited resources.

Through the twentieth century, home economics was closely associated with public school education and cooperative extension service. These affiliations and federal legislation, namely the Smith–Lever Act (1914), which created home economics extension, and the Smith–Hughes Act (1917), which funded home economics education in the elementary and secondary schools, had similarly contradictory effects on the path of home economics. Although they brought in additional resources, they also deflected attention from scientific research, further separating home economics from more prestigious science departments.

Throughout the twentieth century, home economists were employed in academic, educational, governmental, and business settings. Greater occupational opportunities expanded the field, but also fragmented it. At the end of the twentieth century and the beginning of the twenty-first, home economics, also labeled family and consumer sciences and human ecology, was struggling to construct a clear identity for itself.

[*See also* **Agricultural Education and Extension; Food and Diet; Gender and Science; Gender and Technology; Higher Education and Science; High Schools, Science Education in; Household Technology; Hygiene, Personal; Morrill Land Grant Act; Richards, Ellen Swallow;** *and* **Social Sciences.**]

BIBLIOGRAPHY

Elias, Megan J. *Stir It Up: Home Economics in American Culture.* Philadelphia: University of Pennsylvania Press, 2008.

Rossiter, Margaret. *Women Scientists in America: Struggles and Strategies to 1940.* Baltimore: Johns Hopkins University Press, 1982.

Stage, Sarah, and Virginia B. Vincenti, eds. *Rethinking Home Economics: Women and the History of a Profession.* Ithaca, N.Y.: Cornell University Press, 1997.

Rima D. Apple

HOOVER DAM

On the Colorado River southeast of Las Vegas, Hoover Dam, ranked as one of America's "seven modern civil engineering wonders" by the American Society of Civil Engineers, was dedicated 30 September 1935 by President Franklin D. Roosevelt. The dam's ideological origins lie in the Progressive Era's interest in utilizing America's natural resources more intelligently, including managing rivers for irrigation, hydroelectric power, and flood control. In 1928 Congress authorized construction of a dam on the Colorado. Seven abutting states supported the project. California senator Hiram Johnson and William Mulholland, head of the Los Angeles Water and Power Department, helped shepherd the bill through Congress. Originally planned for Boulder Canyon, the site was shifted to Black Canyon. Preliminary work began in 1931 and actual dam construction in 1933.

Under Roosevelt, the project became part of the New Deal's dam-construction program, including those of the Tennessee Valley Authority in the Southeast and the Grand Coulee and Bonneville on the Columbia River, the Glen Canyon on the Colorado, and many others in the West. (Political differences sparked a controversy over the name: generally described as "Boulder Canyon Dam" in the planning stage, it was called Hoover Dam—for President Herbert Hoover—in 1930–1933 and Boulder Dam during the New Deal. Congress made "Hoover Dam" official in 1947.)

The Interior Department's Bureau of Reclamation contracted the monumental project to a private consortium known as the Six Companies. This consortium employed a total of 21,000 men and built a new town, Boulder City, to house them. A 33-mile railroad spur hauled materials and equipment from Las Vegas. A 222-mile transmission line brought power from San Bernardino. Four tunnels blasted through the canyon walls diverted the Colorado during construction. Workers called "high scalers" were lowered down the canyon walls to dynamite unstable rock. Low-paid "muckers" shoveled up the debris.

The dam itself comprises a series of vertical columns of over two hundred massive rectangular concrete blocks, eventually rising to a height of 726 feet. The large dump buckets of concrete were lowered into the forms by a complex network of cables. Overall, 5 million barrels of Portland cement and 4.5 million cubic yards of aggregate (hauled from nearby alluvial deposits and processed on site) went into making the blocks. Interlocked by horizontal and vertical grooves and sealed by grouting, the blocks created a solid mass holding back the river's pressure. In keeping with New Deal aesthetics, artists ornamented the dam's accessible upper area with plaques, bas reliefs, and two thirty-foot winged figures.

Supervised by the brilliant, hard-driving Quebec-born civil engineer Frank T. Crowe, the project raced forward. The contractors' security operatives crushed a strike in 1931–1932. The work was extremely dangerous, and safety standards were lax. The official fatality toll of 96 killed on site does not include injured workers who died later, those felled by heat stroke in the desert sun, or those who succumbed to deadly exhaust fumes in the diversion tunnels.

Hoover Dam water and power promoted agriculture in arid regions and spurred the postwar growth of Las Vegas, Phoenix, Los Angeles, and other cities. Lake Mead, formed by the dam, became a popular recreational site. With rising environmental awareness, the West's network of dams came to be viewed more skeptically for its destruction of wetlands, animal habitats, and spawning streams. But iconic Hoover Dam, a much-visited tourist site, remained a source of pride.

In 2010, highway officials opened the two-thousand-foot Colorado River Bridge, soaring nine hundred feet above the river and offering a spectacular view of the dam, eliminating a bottleneck on Interstate 93. Some compared this five-year, $240-million project with Hoover Dam itself as reassuring evidence of the nation's continuing engineering know-how.

[*See also* **Agricultural Technology; Dams and Hydraulic Engineering; Electricity and Electrification; Engineering; Ethics and Professionalism in Engineering; Hydroelectric Power; Rivers as Technological Systems;** *and* **Tennessee Valley Authority.**]

BIBLIOGRAPHY

Hiltzik, Michael A. *Colossus: Hoover Dam and the Making of the American Century.* New York: Free Press, 2010.
U.S. Department of the Interior, Bureau of Reclamation, Lower Colorado Region, Hoover Dam. http://www.usbr.gov/lc/hooverdam/ (accessed 29 February 2012).

Paul S. Boyer

HOSPITALS

For over one hundred years the American general-care hospital occupied a central role in the American health-care delivery system; it represented and promised to deliver all that was possible in modern medicine. Almost all late-twentieth-century and early-twenty-first-century Americans had at least one experience with a hospital: by the mid-twentieth century over half of the population was born in one, and at the end of the century almost all would be. But the prominence of the general hospital for so long obscures the fact that there have been many other kinds and models of hospitals and, moreover, that the earliest general hospitals were marginal to the medical experience of most Americans. Hospitalization was a last resort for most Americans requiring medical care or treatment in the eighteenth and much of the nineteenth century. Fears about hospitals as death houses and even places where physicians routinely experimented on patients abounded and were not entirely unjustified. Until the last quarter of the nineteenth century, medical care in an institutional setting was overwhelmingly an experience for people with no other choice.

Hospitals of all kinds—chronic-care institutions, general hospitals, specialized hospitals for specific diseases, and slave hospitals—increased in the early nineteenth century. Hospitals continued to proliferate after the Civil War and were organized by local governments, philanthropic and religious groups, physicians, and workers. General hospitals in particular grew in number. The federal government was briefly involved with the organization of hospitals through the Freedmen's Bureau and more long-term with the development of facilities for veterans that included medical care, but the federal government was not involved in local hospital organization until the twentieth century. Still, medical treatment in a hospital continued to remain outside the medical experience of Americans who could afford care and treatment elsewhere in the late nineteenth century. By the 1920s, that was no longer the case. The population of general hospitals expanded to include patients of all social classes who sought care for treatments and procedures only available in newly organized hospitals boasting of the best that modern medicine had to offer. This was not an egalitarian development: hospital care in the United States in the modern era was always characterized by distinctions of class, race, and region.

In addition, institutions providing chronic care retained the legacy of early hospitals, usually overcrowded and grim and very much in the shadows of the American health-care system.

Early American Hospitals.

The earliest American hospitals were established in Spanish and French colonies. Spanish new world hospitals were part of a colonial apparatus dealing with matters of medicine, public health, and religion. By the end of the seventeenth century Quebec City in New France had several hospitals organized and run by religious communities of women. The same was true of Charity Hospital in New Orleans (1736), the first hospital in what would become the United States.

Early colonists in British North America built few hospitals. Some were temporary and erected for emergency situations like epidemics and wars. The British Army and later the Continental Army organized military hospitals. Local governments organized quarantine hospitals in ports and during epidemics (often known as pest hospitals), and physicians offered inoculation for smallpox in what were also often referred to as hospitals.

The first permanent institution in the British North American colonies founded exclusively for medical care was the Pennsylvania Hospital in Philadelphia, organized in 1751 by Benjamin Franklin (1706–1790) and the Philadelphia physician Thomas Bond (1712–1784.) Like other colonial physicians with formal training, Bond had studied abroad. One of his motives for founding the hospital was to give American doctors the opportunity to gain wider clinical experience with patients. (In 1765, the University of Pennsylvania organized the first medical school in the mainland colonies attached to an institute of higher learning. It was located just a few blocks away from Pennsylvania Hospital, and students observed patients in the hospital as part of their training.) The Pennsylvania Hospital was a private charity hospital, as was New York Hospital, which was granted a charter in 1771. Colonial almshouses normally also served as hospitals by providing minimal medical care for the destitute, often sickly, who were forced to seek shelter in them, and this contributed to the public perception of hospitalization as a last resort.

In the immediate decades following the American Revolution, the role of hospitals in American medical practice did not change significantly. Port cities maintained maritime hospitals to quarantine travelers and mariners who showed symptoms of diseases considered contagious (not always successfully), and almshouses remained the only refuge for the sick poor. Epidemics created public-health crises, which swelled almshouse wards and occasionally prompted the temporary quartering of patients in hospitals. During the yellow fever epidemic of 1793, the New York almshouse organized fever wards; in Philadelphia, where the death toll reached five thousand, the city's Guardians of the Poor opened a temporary hospital on a property known as Bush Hill.

Antebellum Hospital Development.

Hospitalization continued to be marginal to American medical care through the Civil War; most people with any choice in the matter, as in the past, were nursed at home by family members and received medical treatment from visiting midwives, physicians, and surgeons. Hospitals remained outside the medical world of most practitioners too. Most physicians did not have hospital training; hospital observation and practice were highly coveted and the domain of a few elite practitioners. However, the number of hospitals increased in the early nineteenth century as social developments—geographic mobility, immigration, and urbanization—created new circumstances that strained traditional methods of caretaking. An increasing number of Americans, especially immigrants, no longer lived near family members and the local communities where they were born. Moreover, in the optimistic reform spirit of the 1830s and 1840s that had its origins in the early-nineteenth-century religious revival known as the Second Great Awakening, many Americans came to view institutional care in a positive light given the alternatives and for some conditions, particularly mental ones, as a therapeutic environment. The convergence of these factors motivated reform-minded Americans of the era to engage in a flurry of institution building. Alternately called asylums and hospitals, they were organized through private benevolent initiatives as well as through local, mostly municipal, governments that opened medical facilities separate from the old almshouses. After first separating the sick from other inmates, the Philadelphia General Hospital was organized as an institution separate from the almshouse in 1825; similarly, after a cholera epidemic, Bellevue Hospital in New York was organized in 1849.

Care of the Mentally Ill and Hospital Development.

Prominent among the new institutions were hospitals for people referred to at the time as feeble-minded. (The first public hospital exclusively for people who in modern language would be referred to as mentally ill was established in Williamsburg, Virginia, in the 1770s by the colonial government, known later as the Virginia Eastern Asylum. In Philadelphia, the charter for Pennsylvania Hospital included provision for the insane, but neither institution admitted many such patients during their early years.) New antebellum institutions organized under private auspices included the Friend's Asylum in Philadelphia (1813), McLean Asylum in Boston (founded in 1818 but acquired its new name in 1826), and the Hartford retreat in Connecticut (1824) but these too never cared for many patients (Grob, 1994, pp. 20, 31). The mentally ill who were not in the care of family were more often committed to an almshouse or public hospital where they were held in cells, restrained, and minimally clothed. After observing the appalling conditions of mentally ill people confined to a Boston jail, reformer Dorothea Dix initiated a national campaign in the 1840s to change the way the mentally ill were cared for. Subsequently, state governments assumed an active role; reform efforts were aided by contemporary fears that insanity was on the rise and was no longer simply an individual problem; it was a social problem that required a governmental response.

Although the first state mental institutions were established before the Civil War, construction increased markedly during the late nineteenth century. Called either hospitals or asylums, they were very large and located in relatively remote regions. People suffering from mental disease continued to live in these institutions well into the next century, long after the methods and optimism associated with their foundation were discredited and abandoned.

In the United States, Dr. Thomas Story Kirkbride, who had worked at the Friend's Asylum, would develop and popularize a European theory of a regulated life involving hard work in institutions specifically for the care of the insane. Kirkbride's treatment plan depended on a specific architectural design and an institutional setting outside of the city in park-like surroundings. His model, which was replicated throughout the country, elaborated on all aspects of the asylum, from patient schedules, administrative details, and ward organization to architectural design, construction materials, and plumbing, and was accompanied by the emergence of the treatment of mental conditions as a subspecialty within medicine. Asylum proponents and superintendents were primarily responsible for the direction of the treatment of the mentally ill in the mid-nineteenth century, which combined custodial and therapeutic approaches. The term psychiatry would not come into use until the twentieth century; the late-nineteenth-century practitioners were called alienists (Noll, 2011, p. 181).

Supported as they were by public funds, these institutions were intended to be self-sustaining. They were located on sites of several hundred acres with working farms. Patients were housed in wards differentiated by sex and symptoms, and they were required, as part of their treatment, to work. Women inmates sewed and did housekeeping chores, whereas men worked on the farm and grounds and did maintenance and repair work. Recreation was also part of the routine; outside speakers were invited to entertain and educate. The institutions encouraged patients to pursue creative endeavors; most institutions produced a newspaper or journal. Although Kirkbride's methods fell out of favor during the twentieth century, the massive and imposing buildings this era produced remained state hospitals for the insane until the 1980s when psychiatrists advocated deinstitutionalization.

Slave Hospitals.

Slave hospitals (also called Negro hospitals) proliferated in the antebellum era as slave owners (particularly after the closing of the international slave trade in 1808) saw financial incentive in some issues of health among enslaved people, and medical colleges and physicians recognized a lucrative avenue of income and clinical experience and experimentation. (Facilities that provided medical care and treatment for enslaved people originated in the eighteenth century; the earliest varied from small shacks to larger quarantine facilities, especially in malarial areas, on plantations.) Proslavery advocates often used health care and general health in their arguments, comparing the well-being of enslaved people with that of northern wage laborers, but the reality was that medical practice by white physicians on enslaved people was about maintaining valuable property. Fears among enslaved people about what could and might happen in slave hospitals, and in treatment administered by white doctors, were strong and for good reason. Located in cities, small towns, and railroad junctions along the slave-trading routes, enslaved people were held and readied for sale in hospitals and used as bodies for doctors to explore. A well-documented example is the Alabama surgeon J. Marion Sims, who experimented with numerous operations (without anesthetic) on enslaved women (and later immigrant women) to develop a technique to correct the condition known as vesicovaginal fistula, a tearing in the vaginal wall that can follow a difficult delivery and results in a painful and debilitating condition. Until recently, Sims was lauded as the founder of modern gynecology without any acknowledgment of the circumstances in which he achieved his reputation (McGregor, 1998, pp. 214–218).

Charity Hospitals. Interest in the plight of pregnant women and abandoned infants, referred to as foundlings, led to the establishment of lying-in, maternity and foundling homes, and hospitals. Like the mentally ill, these were especially vulnerable populations who often ended up in almshouses because of no other alternative. Many abandoned infants found and deposited at almshouses did not survive, and maternity patients risked contracting the very contagious puerperal fever, a postpartum infection that frequently circulated through the lying-in wards in public general hospitals and was a major cause of death among childbearing women.

Reformers organized maternity hospitals as charities for women who might have to seek institutional care when they gave birth. Most institutions made clear in their charters that they were for the exclusive care of women who were not responsible for the circumstances in which they found themselves: patients were required to conform to admission requirements that categorized them as deserving of charity care; wet nurses, for example, were usually acceptable patients. Women who sought medical care in these hospitals often did so in exchange for clinical observation by physicians and medical students, and risked their lives in doing so, because of the prevalence of puerperal fever in maternity wards where infection spread rapidly.

Lay Protestant women and Roman Catholic women belonging to religious communities organized and managed most foundling homes and maternity hospitals. The Roman Catholic Sisters of Charity established the first foundling hospital in the United States, St. Mary's Asylum for Widows, Foundlings, and Infants, in Buffalo, New York, in 1852. Other cities set up similar institutions during the following decades. Women who organized and managed these and other kinds of hospitals, both lay and religious, gained unique access to a wide range of civic responsibilities. Granted state charters for their institutions, they were members of a corporation, could buy and lease land, and could manage all aspects of their institutions, including the financial needs.

As gynecology became a medical specialty in the mid-nineteenth century, women's hospitals were organized for specifically female medical conditions, although early discussion did not reference the term gynecology (McGregor, 1998, p. 133). Gunning S. Bedford established the first gynecological clinic in the United States in 1841 in New York City; Marion Sims founded the New York Women's Hospital, also in New York City, in 1855 (Morantz-Sanchez, 1999, p. 91).

The rising number of female physicians in the United States after 1850 contributed to a concurrent increase in women's hospitals in the latter decades of the nineteenth century. Many female physicians chose to specialize in medical care of women and children; at the same time, women's hospitals (and dispensaries) were part of a movement to provide medical education for women and access to postgraduate clinical training. Women's hospitals also provided female physicians with a supportive professional network of likeminded female physicians (Morantz-Sanchez, 1999, pp. 166–168).

Compared to Europe, American foundling hospitals had a relatively brief history. In New York, for example, three of the four foundling hospitals established in the nineteenth century would close in the early twentieth century (Miller, 2008, p. 224). Serious concerns about infection and changing ideas about unwed mothers and children made them unpopular with reformers and, in later years, with professional social workers.

However, institutions for the care of children of all ages increased in both number and size after the Civil War. Although adoption and foster care became the preferred method of care for dependent children in the early twentieth century, orphanages remained a dominant form of care (Hacsi, 2009, p. 227). Most children who would spend time in an American orphanage were not by strict definition orphaned; they more frequently had one parent. By the early twentieth century many orphanages were quite large, often located out of town and, like the asylums for the care of the mentally ill, functioned in a self-contained world economically and physically, with their own working farms and medical facilities.

General hospitals, institutions that cared for patients of all ages and of both sexes who suffered from a wide variety of illnesses and conditions, were also among the private benevolent hospital initiatives in antebellum period. For the most part they offered chronic care and hospital stays could be lengthy, lasting weeks or even months. In some cases the institutions operated as de facto old-age homes. Many were organized by religious and national groups to provide an alternative to other hospitals, both religious and municipal, where treatment included heavy doses of proselytizing by Protestant clergy and lay visitors. Jews and Roman Catholics both organized their own hospitals to provide care in a different religious environment. Although at the time this was considered a therapeutic advantage, most of these hospitals took pains to note that the physical medical practice of the institution was mainstream. Mullanphy Hospital (1828) was the earliest Roman Catholic Hospital in the United States; the first Jewish hospital was the Jewish Hospital in Cincinnati (1850). In addition to the care they offered, Catholic and Jewish hospitals very self-consciously delivered a message that the religious groups were taking care of their own so they would not be a burden on general private and public charity.

On the eve of the Civil War, the different types of hospitals all operated as part of the charity services of most cities. They differed from almshouses because physical and mental conditions were requirements for admission but, like almshouses, they usually remained a last option for medical care and treatment.

Hospitals and Medical Care in the Civil War. During the Civil War, the Confederate and Union armies organized hospitals on an unprecedented level in numbers and size: approximately 150 by the Confederate army and over 200 by the Union army (Cunningham, 1993, pp. 26–27; Rosenberg, 1987, p. 98). Each army entered the conflict with a small medical organization, but by the end of the war both sides had developed very organized (and similar) procedures and facilities to deal with the staggering number of battlefield casualties. Except for the use of ether as an anesthetic, which allowed for a longer surgical procedure but also probably increased the possibility of infection, the therapeutics of medicine remained very much what it had been during previous military conflicts. However, there were several noteworthy aspects of Civil War hospital care that changed markedly. Military hospitals were organized according to the relatively new idea Florence Nightingale had so successfully publicized following the Crimean War (1853–1856) that order, cleanliness, and ventilation were critical components in creating a relatively infection-free hospital environment, and the Civil War experience revealed that hospitals could be safe and efficient providers of medical care. Moreover, for the first time in the history of the United States, men from all social classes experienced hospital care. Still, the Civil War did not immediately usher in a medical or popular movement to expand hospital care to the general population. In the words of the historian Charles Rosenberg, the war "exerted a substantial but elusive influence on the development of American hospitals" (Rosenberg, 1987, p. 99). The number of hospitals in the United States during the following decades continued to increase, but although hospitals grew in size, they remained primarily charity institutions through the 1870s and into the 1880s even as, especially in smaller religious hospitals, patients often paid for some portion of their care.

Rise of the General Hospital.

As a result of continued and rapid industrialization and urbanization, the number of hospitals in the second half of the nineteenth century grew at a greater rate than in the antebellum era. A study conducted by the American Medical Association in 1873 counted 178 hospitals of various kinds; by 1914 there were approximately 5,000 hospitals in the United States (*Historical Statistics of the United States Millennial Online Edition*, "Hospitals and Beds, by Type of Hospital 1909–1953"). As in the earlier period, they were founded under public and private auspices and served various populations.

Even as the number of hospitals increased, most Americans who received medical care never saw the inside of a hospital. Far more Americans who did not have the option of care by a private physician received treatment in dispensaries, which were clinics that provided medicines and vaccinations and performed minor surgeries, including pulling teeth. Like hospitals, dispensaries were organized and run under both public and private benevolent auspices, and some were organized for particular purposes and patients. Many physicians received practical training in dispensaries. Although the earliest dispensaries were freestanding (founded by private charities and public authorities), later in the nineteenth century hospitals frequently also opened dispensaries.

Late-century hospitals included hospitals for miners and railroad workers. Railroad hospitals grew out of earlier medical services provided by companies for accident victims (passengers and workers), using contracted physicians and surgeons. They flourished in hub cities (smaller clinics proliferated along rail lines), and workers often participated in prepayment fee plans, (Aldrich, 2001, p. 255). Western miners' unions organized hospitals in response to the very meager and poor care provided by mining companies to workers. Miners were forced to pay for the limited care the companies provided; mandatory deductions were taken from pay. In contrast, miners ran the hospitals they built; hospital directors were elected from the membership (Derickson, 1988, pp. 125–127).

By the end of the nineteenth century, new understandings of tuberculosis (identification of the tuberculosis bacillus by Robert Koch in 1882) and the development of new regiments of care contributed to the organization of specialized institutions for tuberculosis patients. Tuberculosis hospitals would be called sanitariums and, later, sanatoriums, with the change made to suggest a curative rather than simply caretaking environment (Bates, 1992, p. 185; Caldwell, 1988, pp. 70–71). Dr. Edward L. Trudeau opened the first in the United States in 1885 in Saranac Lake, New York. Their numbers increased in the early twentieth century as tuberculosis was identified as a major public-health threat; they were organized by local governments and private groups, often as charity hospitals. Rest in the outdoor air and a regimented schedule and carefully controlled diet were characteristic of treatment that also might include surgery for collapsed lungs. Stays were often lengthy. The death rate from tuberculosis was declining in the early twentieth century and was significantly lower in 1945 than it had been in 1904 (from 188.1 to 40.1 per 100,000), but cure remained elusive until the introduction of effective drug therapies (streptomycin, para-aminosalicylic acid) made these institutions obsolete, mostly by the 1960s (Caldwell, 1988, pp. 12–14).

At the end of the nineteenth century general hospitals were fast becoming more prominent among American hospitals; by the early 1920s general hospitals accounted for over 65 percent of the over six thousand hospitals in the United

States (*Historical Statistics of the United States Millennial Online Edition,* "Hospitals and Beds, by Type of Hospital 1909–1953"). Economics played a part in this transformation: the economic depression at the end of the century strained the benevolent base that had supported a variety of hospitals in earlier decades; in light of public health and welfare concerns, general hospitals (including privately run hospitals) were more likely to receive public financing. At the same time, the general hospital changed internally: hospital stays shortened and patients with chronic conditions, notably tuberculosis, were no longer admitted, but general hospitals also took patients suffering from infectious diseases, which had not generally been the case in the earlier period. In historical language, the general hospital became modern in the period between approximately 1880 and 1920; it assumed a form and pivotal place in American health care that continued into the twenty-first century.

There were three different categories of twentieth-century general hospitals that created a specifically American system of hospitals: public hospitals financed and run by public authorities, usually by cities but in some cases by counties; proprietary, for-profit hospitals, usually organized and run by physicians; and voluntary hospitals, run by private groups. Voluntary hospitals continued to rely on benevolence but many also received public funds. Increasingly, most patients would be required to pay for the cost of their care (as some had in the earlier period).

Trained Nurses and the Modern Hospital. One of the first changes that contributed to the remaking of the general hospital was the introduction of a new kind of nurse who became the backbone of the hospitals' labor force. Traditionally, hospital nurses had often been patients or untrained employees considered domestic workers. Although the overwhelming majority of the nursing volunteers in the Civil War, both male and female, nursed without any formal training, the war introduced a new standard for nursing in the United States. Roman Catholic sisters who had nursed in Catholic hospitals prior to the war were an exception, but sisters who had not nursed

also went to war and nursed. Dorothea Dix, superintendent of the U.S. Army Nurses, sought to impose requirements on volunteer nurses but experience in nursing was not one of them (Sarnecky, 1999, pp. 12–13).

Reform-minded women who had been involved in Civil War relief efforts took from their war experiences a belief that female nurses had a role to play in improving the increasing numbers of American hospitals after the war. At the same time, they sought to create a new occupational niche for the increasing numbers of young women who were flocking to cities in search of work. Inspired by the nursing reform movement in England and the model proposed by Florence Nightingale, their efforts resulted in the opening of hospital-oriented nurse-training programs in 1873 in Boston, New Haven, and New York City. In these schools, as well as others that followed, student nurses lived together in hospital housing where they were closely monitored.

By the early decades of the twentieth century, most general hospitals had established nursing schools and, through the 1930s, relied on the students to provide most nursing services in the hospital. Although a small group of graduate nurses remained in staff positions in hospitals, most student nurses intended to nurse patients who would pay them privately. The number of nurses training in hospital schools increased almost 70 percent in the decade after 1915 (Department of the Interior, *Statistics of Nurse Training Schools, 1926–1927,* p. 1). The overwhelming majority of pupils (and graduates) were female.

Nursing reformers also had strong opinions about the proper character required for nursing. Requirements for admission often included a reference letter from a clergyman, and life in the nurses' residence and on the wards was carefully monitored. Nursing students were expected to be modest, deferential, and above all obedient; nursing historians have noted a military-like respect for rank and hierarchy in training school protocol (Melosh, 1982, pp. 49–50; Reverby, 1987, pp. 65–66). In these efforts as well as through attempts to impose educational requirements and state licensing of graduate nurses, nursing leaders sought to legitimize the trained nurse and distance nursing from an earlier tradition of nursing, which they viewed as thoroughly disreputable. Nursing leaders did not always speak for the rank and file of nurses who resented and often opposed what was called professionalization because of its implications for the work they did. However, all nurses would struggle together in efforts to carve out their own sphere of work and autonomy in medical practice both inside and outside of hospitals.

Surgery and Technology in the Emergent General Hospital. Technical and scientific innovations were only partially responsible for other important developments leading to the modern hospital, including the shift from chronic to acute care, the increased role of physicians, and the expansion of the patient population. Treatment and services provided in the hospital changed as a result of new understandings and therapeutics, but there was nothing inherent in these developments that made the new procedures necessarily or exclusively hospital based. Rather, change occurred because practitioners and patients put their faith in what could be accomplished with new science and technology in a hospital setting.

The first change in the treatment of patients was an increase in the number and kinds of surgeries performed in hospitals. With the new understanding of the germ theory of disease and the widespread adoption of antiseptic and asepsis techniques and procedures by the 1880s, surgery became less dangerous because of the lessened risk of postsurgical infection. (Joseph Lister's antiseptic techniques reached American medicine in the early 1870s and were widely accepted by the end of the decade.) The adoption of these procedures did not require surgical procedures to take place in a hospital setting; doctors' offices could also be sterile. By the end of the century, however, many surgeons were limiting their surgeries to hospitals. That trend continued, and by the early twentieth century surgical cases accounted for half of admission in most general hospitals (Rosenberg, 1987, pp. 247, 381). By the 1920s, surgery was firmly entrenched in hospital services and care, and surgical (and obstetrical) patients continued to account for large percentages of

hospital admissions (Stevens, 1989, p. 106). The change began with some specific operations. Appendectomies, for example, became relatively routine and usually successful by the end of the nineteenth century.

Innovations in medical procedures also influenced general hospital development although developments moved more slowly. With the introduction of diphtheria antitoxin in the late 1890s, medical practitioners could successfully treat that dreaded disease in a hospital setting, if they diagnosed it properly. Moreover, new diagnostic tools in the laboratory as well as X-rays (introduced after 1896) transformed not only the routine practice of medical care but also the public perceptions about the importance of scientific and technical developments in medicine. The result was a new focus on the hospital as a scientific institution.

As this occurred, hospital stays shortened and chronic patients were no longer accepted. Some hospitals opened convalescent homes, nursing homes, and specialty hospitals (particularly for tuberculosis patients) to accommodate those patients who had previously filled wards. Horse-drawn ambulances were introduced in many urban hospitals by the 1890s. Other technologies were also instrumental in remaking the institution: adding machines made accounting and cost calculations more efficient, punch card technology encouraged more elaborate record keeping, and washing machines and gas stoves facilitated large-scale housekeeping procedures.

Physicians, Hospitals, and Medical Education. These innovations created a new role for physicians in hospitals and more closely linked hospitals to medical education. And as physicians gained status in hospitals, they also actively encouraged the new innovations. In early hospitals nurses played the most important role in the treatment of patients; doctors were marginal to the everyday goings-on. Only a few doctors were allowed to visit the wards or admit patients; hospital privileges were the purview of a few very well-connected urban physicians. Because early hospitals primarily operated as benevolent institutions, matrons and trustees usually admitted patients on the basis of medical and financial need as well as the worthiness of a patient's character. In the modern hospital doctors assumed a much more prominent and activist role, not only admitting patients and delivering care but also making decisions about the direction and future of the hospital they were associated with.

This change in the role of physicians in hospitals began in the last quarter of the nineteenth century concurrent with the development by American physicians of medical specialties. Hospitals participated in this process by creating new specialized departments beginning with anesthesiology, clinical pathology, and radiology. As this occurred, physicians increasingly sought admitting privileges at hospitals, which, in turn, were particularly interested in obtaining affiliations with doctors who could bring in paying patients (Rosenberg, 1987, pp. 252–256).

African American and female physicians continued to have very limited access to most hospitals as students, medical faculty, or practitioners (Gamble, 1995, pp. 28–32; More, 1999, pp. 108–109; Morantz-Sanchez, 1985, pp. 157, 165, 333). Beginning in the 1890s, African American doctors, like female doctors, opened their own hospitals to offer professional opportunities denied elsewhere, but their hospitals were also organized to provide up-to-date medical care for African American patients who also faced discrimination at other hospitals. Early African American hospitals included Provident Hospital in Chicago (1891) and Douglass Hospital in Philadelphia (1895).

As the number and size of hospitals increased during the late nineteenth century, so too did the number of physicians with hospital affiliations. More patients meant hospitals needed more hospital staff, including doctors. Moreover, the patient population changed as hospitals increasingly served surgical patients who required closer postoperative observation by doctors. At the same time, the growing number of patients paying for their own medical treatment expected closer observation by physicians and the benefit of technical improvements in hospitals such as gas and electric lighting, which also made 24-hour care and monitoring possible. By the turn of the twentieth century most American hospitals depended on the services provided by medical and surgical physicians in various specialties who supervised and trained junior house officers hired by the hospitals.

Medical students also joined the hospital routine in the early twentieth century. Some hospitals had always had some involvement with medical education; lectures and demonstrations were conducted in amphitheaters, and attending doctors brought students to observe their treatment of patients but these practices were not regularized. Efforts to make practical ward experience an educational requirement for physicians gained momentum in the last quarter of the nineteenth century among reformers interested in changing the existing system of medical training. They advocated a requirement that medical schools affiliate with universities and other reputable educational institutions (also called colleges) run independently by groups of doctors.

Reformers were concerned not only about the enormous increase in the number of American physicians at the end of the century, which they worried might lead to a surplus, but also about the potential for such abuses as unnecessary surgeries or unqualified surgeons with hospital access. Reformers looked to hospital training with physician control as a way to standardize and upgrade medical education in the United States and to give doctors more power in hospitals. Not surprisingly, the reformers, notably the American Medical Association (1847), viewed the increase in the number of hospitals as an opportunity to advocate for medical school control of hospitals. They favored the model adopted by the relatively new Johns Hopkins School of Medicine (1893), which had its own hospital and controlled all the staff appointments.

Most hospitals, however, were not eager to comply with this system. Trustees anticipated encroachment on both their authority and their patients' rights. They also feared that doctors would prioritize their wards for teaching. Still, hospital trustees and managers began to see the benefits of participating in medical education; the promise of science that went with linking education and research proved a powerful motivator. Larger hospitals decided to affiliate with medical schools first; smaller voluntary and municipal hospitals adopted this model over the next few decades. Training in a hospital became an integral and required component of medical education. Contributing to this trend, the Carnegie Foundation published an influential study of medical education in 1910 (known as the Flexner Report after the author Abraham Flexner) that advocated the reform model of hospital (and laboratory) training, and the number of students training in hospitals as interns increased in the following decade. In 1914, one-half of medical school graduates could avail themselves of internships in hospitals; within 10 years internships were available for all graduates, although internships at larger teaching hospitals were more competitive (Ludmerer, 1985, pp. 82, 93).

In the following decades, state-licensing boards mandated hospital training in the curriculum, and schools unable to comply found it hard to continue and subsequently closed. A number of those schools had served specific populations who would be denied access to other medical schools. The numbers of women and African Americans attending medical school declined significantly, and the numbers of black and female practicing physicians subsequently also declined (Gamble, 1995, pp. 29-32; More, 1999, pp. 98–99).

Following World War I, optimism in the principles of scientific management found a receptive audience in hospital managers (Stevens, 1989, p. 103). Standardization in almost every aspect of hospital organization and management became a goal as much as a means to an end. A lively hospital literature developed featuring new and efficient methods of record keeping, dietary and laundry service, and, very critically, billing. As the modern hospital shed its charity roots, it increasingly relied on paying patients. Hospital superintendents established a professional organization in 1899. After reorganization during the next decade this institution became the American Hospital Association. Catholic hospitals formed their own Catholic Hospital Association in 1915, and African American hospitals founded the National Hospital Association eight years later.

Standardization and Hospital Growth.

World War I formalized many of the changes that had taken place in American hospitals. During the war, some American hospitals very literally went to Europe with the American Expeditionary Force. The American military organized personnel and materials into portable hospitals complete with motorized ambulances and portable X-ray machines. (Individuals and communities organized and sent hospitals too.) The American medical system during the war consisted of a series of hospitals where patients would be assessed and treated; unlike earlier wars, emergency treatment was administered early in the process. A medic initially identified wounded soldiers on the scene and was responsible for ordering them transported by stretcher to a rear dressing station. An ambulance then transported them to an evacuation hospital, often located near a railroad; from there, the wounded were relocated to a base hospital further removed from the battle.

The Red Cross spearheaded efforts to recruit nurses into the Army Nurse Corps, which had been established in 1901 and had first sent nurses to war in Mexico in 1916. Technological change accounted for some improvements in treatment. The Carrel-Dakin technique was an effective new way to clean wounds, and X-rays used for the first time in battle allowed for better identification of the location of bullets and shrapnel. However, technology also created the new condition of shellshock, a term coined by soldiers to describe the psychological reaction to constant bombardment by artillery.

During the flu epidemic of 1918, hospitals were inundated with flu patients; the event intensified the broadened use of the hospital among all social classes and set a new precedent for the admission of patients with infectious diseases. Moreover, the epidemic revealed that there was little community-wide organization of health services. During the next decade, medical leaders emphasized a need for efficiency and coordination among hospitals and other health-care providers. Specific details about how this might be accomplished were ambiguous, but postwar reformers were successful in continuing standardization in hospital care and also in bringing new services and treatments into the hospital (Bristow, 2012, pp. 47–48; Stevens, 1989, p. 102).

The emergent American modern hospital was presented and generally understood to be an inevitable result of convergent developments in science and medical practice; however, the modern hospital that assumed such a pivotal role in American medicine in the next decades was the result of deliberate efforts by influential physicians who had served in the war. Drawing on their experiences with wartime hospitals, many spearheaded efforts after the war to continue the development of the hospital as a standardized acute-care facility rather than an institution primarily emphasizing prevention.

Proponents of a greater emphasis on prevention had favored more outreach in hospital dispensaries and in public-health efforts in both the home and the workplace. A few hospitals organized efforts to follow up on patients after discharge to monitor and advise; however, outpatient services were for the most part marginalized in hospital organizations because they generally carried a stigma of being identified with the charity model of medical care from an earlier era.

At in-patient facilities during the 1920s many more patients paid for their care than ever before. The institutions increasingly functioned as community-minded organizations serving a consumer population. To meet that end the hospital changed internally; rooms became smaller and were designed for two to four patients. Wards that were the centerpiece of hospitals in the earlier period began to disappear although larger urban (especially municipal) hospitals did not follow this pattern because they did not attract as many middle- and upper-class patients. A patient's bed was no longer the only focus of medical treatment; critical action in the modern hospital also took place in operating theaters, X-ray facilities, and laboratories.

Obstetric services grew dramatically during the 1920s as childbirth was increasingly viewed as a potential medical problem requiring hospital care. Twilight sleep, a drug regimen for women in labor, became the standard procedure for deliveries, largely through the efforts of women who agitated for it after its introduction in the United States in 1914. Twilight sleep required close supervision; the hospital was marketed and accepted as the best possible place to administer it safely. As a result, an increasing number of women from all social classes chose to deliver their babies in hospitals.

As the American general hospital developed a standard organization and its numbers grew, hospitals organized for special populations and by specific groups confronted a dilemma of identity. Religious hospitals struggled to maintain their traditional mission; the new emphasis on scientific and standardized care complicated their efforts to maintain identification as offering superior care because of religious and cultural factors. By the 1930s urban Americans who had choices in hospital care usually chose a hospital because of its location or a physician's affiliation, not because of its cultural or religious origins or management.

A major change took place in the nursing staff of American hospitals during the 1930s as nurses increasingly worked in hospitals after graduation, rather than in private duty or public health as they had in the previous decades. These full-time graduate nurses increasingly replaced nurses in training as the main providers of general-duty nursing care. By 1940 graduate nurses were fast replacing private-duty and student nurses on the floor in general hospitals (Reverby, 1987, p. 175). Although their numbers had decreased, hospital training schools continued to educate most nurses. College-based nurses' education grew after the war: the first associate degree program, intended to be a middle ground between a baccalaureate and a hospital-trained nurse, was introduced in 1952. The number of nurses with bachelor's degrees increased rapidly in the 1960s (Lynaugh, 2008, p. 21; Melosh, 1982, pp. 207).

As hospital care became central to the delivery of acute medical care in the United States, the cost of hospital care became part of a debate about compulsory national health insurance (Lubove, 1986, pp. 70–71). Many hospital leaders initially supported a compulsory plan (for workers, not the entire population) because it appeared to have the potential to attract paying patients. After World War I, however, their support waned, and opponents successfully attacked, in the midst of the Red Scare, any plan for government or mandatory involvement in health insurance as evidence of creeping socialism. The first hospital insurance plan successfully enacted was voluntary and private. Organized by Baylor University Hospital in 1929 for school teachers in Dallas, the hospital insurance program eventually became Blue Cross.

By the eve of World War II, American hospitals oversaw the birth and death of an increasing number of Americans, but strong regional differences remained in medical services, including the availability of hospital care. Hospitals were much fewer and further apart in rural areas and in the southern United States than in other parts of the country. Southern hospitals often did not admit African American patients, and northern hospitals were usually segregated.

Medical innovations developed during the interwar period and adopted by military doctors during World War II included new surgical techniques, the use of blood plasma, sulfanilamide drugs and, after 1943, penicillin. World War II prompted the building of new hospitals with federal funds. As migrants following war work moved into areas lacking medical infrastructure, Congress responded with the Lanham Act of 1941, which committed funds to creating services, including hospitals, in regions of the country involved in the defense industry. Over eight hundred hospitals, both voluntary institutions and institutions operated by local governments, were constructed during the war. Although there were concerns that this was unprecedented federal involvement in hospital development, the Lanham Act established an important precedent for federal funding of private nonprofit hospitals (Stevens, 1989, p. 209). Another wartime federal initiative, the Emergency Maternity and Infant Care program of 1943, which covered childbirth costs, was influential in the continuing growth in the number of women who delivered their babies in hospitals. By 1955, 95 percent of all births in the United States occurred in hospitals (Leavitt, 1986, p. 260).

American Hospitals Following World War II.
Unlike World War I when neither civilian nor military hospitals adjusted or expanded to meet the needs of returning veterans, World War II led to an expansion of veteran's services, including access to hospitals. In the Hospital Survey and Construction Act of 1946 (referred to as the Hill-Burton Act after its sponsors) Congress provided funds through the individual states to analyze hospital needs and construct hospitals. This legislation had enormous consequences for the American health-care delivery system; it cemented the central role of hospitals in that system and the continued emphasis on the hospital as the bastion of the most up-to-date medical science and practice.

Hill-Burton initiatives expanded hospital services to previously unrepresented areas, particularly in the South and in smaller communities. Continuing postwar trends, hospital admissions increased, as did the size of hospitals, and the medical care provided was overwhelmingly acute care and short term. Even as the federal government set guidelines for hospital construction, establishing, for example, the number of hospital beds required based on the population of a community, federal policies supported the traditional American system of hospital development based on local and private initiative; the majority of American hospitals remained voluntary nonprofit institutions, although in some regions for-profit were more common (Stevens, 1989, p. 232).

The decade and a half following World War II witnessed the flowering of the modern American hospital as a scientific and community-based institution. Responding to new technology and the increase in the number of Americans participating in third-party payment plans (in large part from labor negotiations that included health benefits in worker contracts), hospitals expanded physically and introduced new services, including the intensive care ward and specialized coronary units. But as in the earlier period, factors influencing hospital reorganization involved more than the introduction of new technology or intellectual understandings of disease. The earliest formally organized intensive-care units in the 1960s, for example, which would come to be characterized by high levels of technology, were not initially technologically driven; they were introduced into American hospitals as a remedy for a nursing shortage (Bulander, 2010, pp. 630–631). Critically ill patients were separated from others to receive very close and labor-intensive monitoring by nurses who were given a much greater degree of autonomy and authority than in other situations. Further development came about when more patients could pay for this critical care, as it was enhanced by technological advancement and became a specialty among physicians. Intensive-care unit development was not without critics: the placement of patients so closely together challenged traditional gender and racial separations, raised issues of privacy, and was thought by some to perpetuate a ward atmosphere harkening back to the days when hospital patients were mostly charity patients.

As hospital emergency services expanded, emergency medicine developed as a specialty and ambulance service medicalized. (As late as the 1960s ambulance service usually simply meant transport by untrained volunteers. Morticians also commonly provided ambulance service because their hearses could accommodate a stretcher.) National concern about highway safety facilitated the development of federal requirements in hospital transport through the Highway Safety Act (1966) and the Emergency Medical Services Systems Act (1973), which specifically established standard treatment and training for emergency medical services (Simpson, 2013, pp. 168, 196).

Postwar hospitals, both old and new, continued racial segregation and exclusion in hiring practices; Hill-Burton funding did not require any accommodation to racial equality. Race continued to be a factor delineating the delivery of medical care in the United States. In one notorious example of medical racism, African American men involved in a syphilis study begun in the 1920s and continued through the 1970s were deliberately denied treatment when it became available. Henrietta Lacks, a cancer patient at Johns Hopkins Hospital who died from her disease in the 1950s, had no idea doctors would use tissue samples taken for research without her permission for decades to follow without any notification or compensation to her or her family (Jones, 1993, p. 200; Skloot, 2010, pp. 3–4). Most American hospitals remained segregated until the Civil Rights Act of 1964 and the Medicare legislation of 1965.

Efforts to organize hospital workers were met with tremendous opposition by hospital administrators armed with federal legislation. Under the National Labor Relations Act of 1935 (Wagner Act), which supported union efforts in many other industries, hospitals were defined as charities and hospital workers' right to unionize was not protected. By the late 1950s hospitals were already very large employers, and the number of hospital workers would increase even more rapidly in the next decades (Stevens, 1989, pp. 180, 237, 282). Moreover, many hospital employees were black, Hispanic, and female and ongoing successful union efforts were linked to the growing civil rights movement (Fink and Greenberg, 1989, p. 78; Sacks, 1988, p. 38). Amendments in 1974 to the Taft-Hartley Act (1947), which had exempted voluntary hospitals from collective bargaining requirements, facilitated union efforts (Stevens, 1989, p. 303). Prominent among health-care workers' efforts to organize was the National Union of Health and Hospital Workers (1199), which began as a pharmacist's union in New York City in the 1930s (Fink and Greenberg, 1989, p. 20).

The Impact of Medicare and Medicaid. Optimism that the new federal programs would equalize the playing field in American health care was short-lived. By the 1960s there was a growing consensus among all segments of the population that the health-care system, and specifically hospitals, needed reforming. Growth and expansion brought duplication of services in areas with several hospitals. Moreover, the elderly and poor, nonparticipants in third-party plans for payment, were unable to pay the rising costs of hospital care.

In this context the issue of government involvement in providing payment for health-care services once again became important in the political landscape. Beginning in 1950, Congress established procedures whereby states received federal funds to distribute for the health care (including hospitalization) of persons receiving public assistance. However, the program was very small and it did not address the needs of people who could not afford health care and were not on public assistance. Although several legislative proposals in the 1950s to address the problem of health costs were unsuccessful, they did reorient the discussion of government health insurance from an older compulsory model of government health insurance to a government-funded model targeted at the elderly and financed through Social Security. Several bills came before Congress in the 1960s that ultimately resulted in the Medicare and Medicaid legislation of 1965. Congress drew a sharp distinction between two groups of Americans: the elderly, who by right of age were all entitled to the program's benefits, and other Americans who could receive federal assistance for medical care based on a financial means test. (In 1972 Medicare benefits were extended to the disabled and to kidney dialysis and transplant patients.)

Medicare and Medicaid recognized the central role of hospitals in the health-care delivery system, and the infusion of federal funds led to a further expansion of hospital services, particularly for the elderly. Federal expenditures to hospitals after 1965 quickly and vastly exceeded estimates. At the same time, hospital costs rose as treatments changed. New techniques, especially in surgery and recovery, radiation, and chemotherapy, meant more expensive hospital stays. As hospitals sought to reduce costs by shortening patient stays, the modern hospital increasingly functioned as an acute-care facility, a development that echoed the reorganization of hospitals that had occurred earlier in the century.

The optimism surrounding the implementation of Medicare and Medicaid as the means for addressing woeful inadequacies in the delivery of hospital services to all Americans was relatively short lived. By the 1970s, critics began attacking hospitals for unacceptable levels of spending. Hospital reform efforts mainly stressed cost containment, and during the remainder of the twentieth century, hospital administrators primarily functioned as chief operating officers of their institutions.

Urban hospitals serving populations that did not fit into the federal programs and who did not have their own insurance were particularly handicapped by financial problems. In some cities voluntary hospitals closed or moved, often in the face of tremendous opposition from community leaders and staff. Critics severely attacked municipal hospitals for providing glaring substandard services. (Several even lost accreditation.) At the same time, the number of beds in for-profit hospitals increased as hospital admissions, particularly for surgical procedures, increased dramatically relative to the population between 1965 and 1980. For-profit hospitals were especially successful in the South and the West. Most of these institutions were part of centrally run systems, and their organization heralded an overall movement among hospitals to consolidate and form large chains (Stevens, 1989, pp. 296–298).

The federal government initiated a new system of reimbursement for Medicare payments in 1983. This decision led to a fundamental change in the role of government in hospital care. Under the original system, Medicare had reimbursed hospitals for incurred costs; under the new system specific reimbursements were attached to specific diagnoses arranged in what were called diagnosis-related groups. In some ways, this system can be seen as an extension of the scientific management approach introduced earlier in the century by Progressive-Era reformers. But it also expanded federal involvement in hospital care and introduced a nationally regulated system of hospital care.

This marked a profound change. One mistaken assumption about American hospitals is that for most of their history they were privately or voluntarily supported and operated. Most voluntary hospitals had received public funding (although not federal funding) for most of their history, but with the institution of diagnosis-related group rules, the government (specifically the federal government) became more than a supporter of hospitals; it set rules for hospitals to follow when treating patients.

American Hospitals at the Close of the Twentieth Century.

Hospitals continued to experience financial problems through the 1990s. They responded by shortening hospital stays and limiting the use of procedures that had first been introduced as ideally suited to the modern hospital. Surgery was routinely performed on an ambulatory basis, and testing and treatment took place in doctor's offices as well as in new centers, notably radiation clinics and dialysis and rehabilitation centers. One major problem noted by the American Hospital Association at the century's end was staff shortages, especially in nursing.

The consolidation among American hospitals continued as a survival strategy; the previous century's hospital founders likely would have marveled at some of the combinations of merged institutions that existed at the end of the twentieth century and the beginning of the twenty-first, which were created by different groups to maintain distinctions. They would certainly have been astounded by the number of hospitals, totaling over six thousand in 1997 even with closings and mergers (*Historical Statistics of the United States Millennial Online Edition*, "Hospitals and Beds, by Ownership or Control, 1946–1997") and their size, technology, and available treatments, as well as the scale of the crises many confronted. But they likely would not have been surprised to see that hospitals remained contested territory and the focus of controversy. Understanding the reasons for hospitals and the issues important for developing and understanding quality hospital care still depended on a range of concerns that was more complex than therapeutics narrowly defined.

[*See also* **American Medical Association; Animal and Human Experimentation; Biochemistry; Cardiology; Death and Dying; Disease; Ethics and Medicine; Flexner Report; Forensic Pathology and Death Investigation; Foundations and Health; Group Practice; Health Insurance; Health Maintenance Organizations; Higher Education and Science; Indian Health Service; Mayo Clinic; Medical Education; Medical Malpractice; Medical Specialization; Medicare and Medicaid; Medicine; Medicine and Technology; Mental Health Institutions; Mental Illness; Midwifery; Molecular Biology; National Institutes of Health; National Medical Association; Native American Healers; Nobel Prize in Biomedical Research; Nursing; Organ Transplantation; Pharmacology and Drug Therapy; Public Health; Public Health Service, U.S.; Race and Medicine; Research and Development (R&D); Surgery; Tuberculosis; Tuskegee Syphilis Study;** *and* **War and Medicine.**]

BIBLIOGRAPHY

Aldrich, Mark. "Train Wrecks to Typhoid Fever: The Development of Railroad Medicine Organizations, 1850–World War I." *Bulletin of the History of Medicine* 75, no. 2 (Summer 2001): 254–289.

Bates, Barbara. *Bargaining for Life: A Social History of Tuberculosis, 1876–1938.* Philadelphia: University of Pennsylvania Press, 1992.

Borst, Charlotte B. *Catching Babies: The Professionalization of Childbirth, 1870–1920.* Cambridge, Mass.: Harvard University Press, 1995.

Bristow, Nancy. *American Pandemic: The Lost Worlds of the 1918 Influenza Epidemic.* New York: Oxford University Press, 2012.

Brown, Thomas. J. *Dorothea Dix: New England Reformer.* Cambridge, Mass.: Harvard University Press, 1998.

Buhler-Wilkerson, Karen. *No Place Like Home: A History of Nursing and Home Care In the United States.* Baltimore: Johns Hopkins University Press, 2001.

Bulander, Robert E. "'The Most Important Problem in the Hospital': Nursing in the Development of the Intensive Care Unit, 1950–1965. *Social History of Medicine* 23, no. 3 (December 2010): 621–638.

Caldwell, Mark. *The Last Crusade: The War on Consumption, 1862–1954.* New York: Atheneum, 1988.

Crosby, Alfred W. *America's Forgotten Pandemic: The Influenza of 1918.* Cambridge, U.K.: Cambridge University Press, 1989.

Cunningham, H. H. *Doctors in Gray: The Confederate Medical Service.* Baton Rouge: Louisiana State University Press, 1958, 1960. Reprint, 1993.

Department of the Interior, Bureau of Education. *Statistics of Nurse Training Schools, 1926–1927.* Washington, D.C.: U.S. Government Printing Office, 1928.

Derickson, Alan. *Worker's Health Workers' Democracy: The Western Miners' Struggle, 1891–1925.* Ithaca, N.Y.: Cornell University Press, 1988.

Dowling, Harry F. *City Hospitals: The Undercare of the Underprivileged.* Cambridge, Mass.: Harvard University Press, 1982.

Downs, Jim. *Sick from Freedom: African-American Illness and Suffering during the Civil War and Reconstruction.* New York: Oxford University Press, 2012.

Ettinger, Laura E. *Nurse Midwifery: The Birth of a New Profession.* Columbus: Ohio State University Press, 2006.

Fink, Leon, and Greenberg, Brian. *Upheaval in the Quiet Zone: A History of Hospital Workers; Union, Local 1199.* Urbana: University of Illinois Press, 1989.

Gamble, Vanessa Northington. *Making a Place for Ourselves: The Black Hospital Movement, 1920–1945.* New York: Oxford University Press, 1995.

Grob, Gerald N. *The Mad among Us: A History of the Care of America's Mentally Ill.* New York: Free Press, 1994.

Hacsi, Timothy A. "Orphanages as a National Institution: History and Its Lessons." In *Home Away from Home: The Forgotten History of Orphanages,* edited by Richard B. McKenzie, pp. 227–248, 284. New York: Encounter Books, 2009.

Hine, Darlene Clark. *Black Women in White: Racial Conflict and Cooperation in American Nursing.* Bloomington: Indiana University Press, 1989.

Hirshfield, Daniel S. *The Lost Reform: The Campaign for Compulsory Health Insurance in the United States from 1932 to 1943.* Cambridge, Mass.: Harvard University Press, 1970.

Historical Statistics of the United States Millennial Online Edition. http://hsus.cambridge.org.

Howell, Joel D. *Technology in the Hospital: Transforming Patient Care in the Early Twentieth Century.* Baltimore: Johns Hopkins University Press, 1995.

Jones, James H. *Bad Blood: The Tuskegee Syphilis Experiment.* New York: Free Press, 1981, 1993.

Katz, Michael. *In the Shadow of the Poorhouse: A Social History of Welfare in America.* New York: Basic Books, 1986.

Kauffman, Christopher J. *Ministry and Meaning: A Religious History of Catholic Health Care in the United States.* New York: Crossroad, 1995.

Kenny, Stephen. "'A Dictate of Both Interest and Mercy?': Slave Hospitals in the Antebellum South." *Journal of the History of Medicine and Allied Sciences* 65, no.1 (January 2010): 2–47.

Kraut, Alan M., and Deborah A. Kraut. *Covenant of Care: Newark Beth Israel and the Jewish Hospital in America.* New Brunswick, N.J.: Rutgers University Press, 2007.

Leavitt, Judith Walzer. *Brought to Bed: Child-bearing in America, 1750 to 1950.* New York: Oxford University Press, 1986.

Litoff, Judy Barrett. *American Midwives: 1860 to the Present.* Westport, Conn.: Greenwood Press, 1978.

Long, Diana Elizabeth, and Golden, Janet, eds. *The American General Hospital: Communities and Social Contexts.* Ithaca, N.Y.: Cornell University Press, 1989.

Ludmerer, Kenneth. *Learning to Heal: The Development of American Medical Education from the Turn of the Century to the Era of Managed Care.* New York: Basic Books, 1985.

Lubove, Roy. *The Struggle for Social Security, 1900–1935.* Pittsburgh: University of Pittsburgh Press, 1986; Cambridge, Mass.: Harvard University Press, 1968.

Lynaugh, Joan E. "Nursing the Great Society: The Impact of the Nurse Training Act of 1964." *Nursing History Review* 16 (2008): 13–28.

Maynard, Aubre de L. *Surgeons to the Poor: The Harlem Hospital Story.* New York: Appleton-Century-Crofts, 1978.

McBride, David. *Integrating the City of Medicine: Blacks in Philadelphia Health Care, 1910–1965.* Philadelphia: Temple University Press, 1989.

McCauley, Bernadette. *Who Shall Take Care of Our Sick?: Roman Catholic Sisters and the Development of Catholic Hospitals in New York.* Baltimore: Johns Hopkins University Press, 2005.

McGregor, Deborah Kuhn. *From Midwives to Medicine: The Birth of American Gynecology.* New Brunswick, N.J.: Rutgers University Press, 1998.

Melosh, Barbara. *"The Physician's Hand": Work, Culture, and Conflict in American Nursing.* Philadelphia: Temple University Press, 1982.

Miller, Julie. *Abandoned: Foundlings in Nineteenth-Century New York City.* New York: New York University Press, 2008.

Morantz-Sanchez, Regina Markell. *Conduct Unbecoming a Woman: Medicine on Trial in Turn-of-the-Century Brooklyn.* New York: Oxford University Press, 1999.

Morantz-Sanchez, Regina Markell. *Sympathy and Science: Women Physicians in American Medicine.* New York: Oxford University Press, 1985.

More, Ellen. S. *Restoring the Balance: Women Physicians and the Profession of Medicine, 1850–1995.* Cambridge, Mass.: Harvard University Press, 1999.

Noll, Richard. *American Madness: The Rise and Fall of Dementia Praecox.* Cambridge, Mass.: Harvard University Press, 2011.

Numbers, Ronald. L. *Almost Persuaded: American Physicians and Compulsory Health Insurance, 1912–1920.* Baltimore: Johns Hopkins University Press, 1978.

Numbers, Ronald. L., ed. *Compulsory Health Insurance: The Continuing Debate.* Westport, Conn.: Greenwood Press, 1982.

Opdyke, Sandra. *No One Was Turned Away: The Role of Public Hospitals in New York City since 1900.* New York: Oxford University Press, 1999.

Pearson, Reggie L. "'There Are Many Sick, Feeble, and Suffering Freedmen': The Freedmen's Bureau's Health-Care Activities during Reconstruction in North Carolina, 1865–1868." *The North Carolina Historical Review* 79, no. 2 (April 2002): 141–181.

Reverby, Susan M. *Examining Tuskegee: The Infamous Syphilis Study and Its Legacy.* Chapel Hill: University of North Carolina Press, 2009.

Reverby, Susan M. *Ordered to Care: The Dilemma of American Nursing, 1850–1945.* Cambridge, U.K.: Cambridge University Press, 1987.

Rosenberg, Charles E. *The Care of Strangers: The Rise of America's Hospital System.* New York: Basic Books, 1987.

Rosner, David. *A Once Charitable Enterprise: Hospitals and Health Care in Brooklyn and New York, 1885–1915.* Cambridge, U.K.: Cambridge University Press, 1982.

Rothman, David J. *The Discovery of the Asylum: Social Order and Disorder in the New Republic.* Boston: Little, Brown and Company, 1971.

Sacks, Karen Brodkin. *Caring By the Hour: Women, Work, and Organizing at Duke Medical Center.* Urbana: University of Illinois Press, 1988.

Sarnecky, Mary T. *A History of the U.S. Army Nurse Corps.* Philadelphia: University of Philadelphia Press, 1999.

Skloot, Rebecca. *The Immortal Life of Henrietta Lacks.* New York: Crown Publishers, 2010.

Simpson, Andrew T. "Transporting Lazarus: Physicians, the State, and the Creation of the Modern Paramedic and Ambulance, 1955–73." *Journal of The History of Medicine and Allied Sciences* 68, no. 2 (April 2013): 163–197.

Smith, Susan B. *Japanese American Midwives: Culture, Community, and Health Politics, 1880–1950.* Urbana: University of Illinois Press, 2005.

Starr, Paul. *The Social Transformation of American Medicine.* New York: Basic Books, 1982.

Stevens, Rosemary. *In Sickness and in Wealth: American Hospitals in the Twentieth Century.* New York: Basic Books, 1989.

Tomes, Nancy. *The Art of Asylum-Keeping: Thomas Story Kirkbride and the Origin of American Psychiatry.* Philadelphia: University of Pennsylvania Press, 1994.

Toner, J. M. "Statistics of Regular Medical Associations and Hospitals of the United States." *American Medical Association, Transactions* 24 (1873): 314–333.

Vogel, Morris J. *The Invention of the Modern Hospital: Boston 1870–1930.* Chicago: University of Chicago Press, 1980.

Wall, Barbra Mann. *American Catholic Hospitals: A Century of Changing Markets and Missions.* New Brunswick, N.J.: Rutgers University Press, 2011.

Wall, Barbra Mann. *Unlikely Entrepreneurs: Catholic Sisters and the Hospital Marketplace, 1865–1925.* Columbus: Ohio State University Press, 2005.

Warner, John Harley. *The Therapeutic Perspective: Medical Practice, Knowledge and Identity in America, 1820–1885.* Cambridge, Mass.: Harvard University Press, 1986.

Washington, Harriet A. *Medical Apartheid: The Dark History of Medical Experimentation on Black Americans from Colonial Times to the Present.* New York: Harlem Moon, 2006.

Zink, Brian J. *Anyone, Anything, Anytime: A History of Emergency Medicine.* Philadelphia: Mosby Elsevier, 2005.

Bernadette McCauley

HOUSEHOLD TECHNOLOGY

Industrialization transformed American society, permanently altering the tools and work processes of preindustrial America. New technologies helped change how households fed, clothed, cleaned, and cared for their members, although these technologies affected men, women, and children differently. In addition, the rich and the urban tended to gain access to new technologies before the poor and the rural. Despite the unevenness and inequities of these changes, developments in household technology played an important role in elevating the American standard of living through the nineteenth and twentieth centuries.

From the Colonial Era to 1920.

In the Colonial Era, most households used simple tools to maintain their living standards, which for all but the very rich consisted of simple diets, limited wardrobes, and low standards of cleanliness. Open, wood-burning hearths provided heat and a place to cook, whereas candles gave light. Women prepared and preserved food, made medicines, and used spinning wheels, looms, and needles to turn wool and flax into clothing. Men farmed, cut and hauled wood, whittled, and sewed leather items. Metalware such as kettles, pots, axes, and knives eased food preparation, wood gathering, and agricultural labor. Servants, slaves, and children, as well as male and female heads of house, provided household labor. Occasional reliance on people outside the household who produced and repaired metalware and sold staples such as salt and lime linked relatively self-sufficient households to the developing market economy.

Through the nineteenth century, the mass production of goods by new industries removed many traditionally male tasks from homes and placed most household technology in the hands of women and servants. Beginning in the 1830s, versions of Benjamin Franklin's 1740s cast-iron stove, modified to include ovens and stove-top hot plates, began replacing open hearths and altering cooking practices. By 1850, many households were purchasing coal and commercially ground flour, eliminating traditional male tasks like gathering wood, shelling corn, pounding grain, and, increasingly, farming itself. New technology diversified women's housework as well, removing some jobs and adding others. Kerosene eliminated the job of making candles. Purchasing textiles reduced long hours spent spinning and weaving, and after the Civil War Isaac Singer's manufacture of a practical, treadle-operated sewing machine allowed women to sew family clothes without hiring seamstresses. The shift from leather and woolen clothing to cotton garments boosted standards of cleanliness, but added the arduous weekly task of hand-cleaning laundry. As culinary stan-

dards advanced, women invested more time preparing more varied meals

Between 1880 and 1920, private industries began providing even more of the goods and services that households had traditionally produced. As increasing numbers of Americans moved from rural to urban areas, for example, many families began purchasing goods and services they had previously provided themselves: foodstuffs from grocery stores, health care from physicians and hospitals, and ready-made clothing from department stores. As municipalities developed water systems, many homes acquired running water, water heaters, sanitary fixtures, and indoor bathrooms. Following Thomas Edison's invention of electric lights in 1879 and the first electric power station in 1882, many urban families gradually switched from kerosene lamps to electric light bulbs. By 1910, after the Westinghouse Corporation introduced alternating-current motors, industries began to mass-produce electric fans, sewing machines, washing machines, and vacuum cleaners for a national market, boosting the total value of electrical household appliances and supplies to $16 million (*Historical Statistics of the United States* [HSUS], Vol. II, p. 700). By 1920, when 35 percent of American homes had electricity and devices using resistance-coil heaters like irons and toasters had become widely available, the total value of electrical household appliances more than quadrupled to $83 million (HSUS, Vol. II, p. 700). A small but growing number of families also owned telephones and automobiles.

Household Technology after 1920.

Household technology became big business in the early 1920s when General Electric (GE) began to shift its focus from making large electrical equipment for use by big industries to making household technologies marketed directly to consumers. To consolidate the shift, the company launched a coordinated national advertising campaign to raise brand awareness among consumers and foster a national "electrical consciousness." Other electric appliance makers responded with their own campaigns, inundating consumers with images of electric appliances functioning as staples of "modern" home life. When GE began the assembly-line production of home refrigerators at the end of 1926, it enjoyed rapid success, controlling nearly one-third of the market by 1929 (Marchand, 1989, pp. 188–191). Over the course of the 1920s, the total value of electrical household appliances and supplies more than doubled from $83 million to $177 million (HSUS, Vol. II, p. 700).

As household technologies spread during the interwar decades, the national standard of living rose significantly, especially in urban areas. New electric refrigerators, for example, enhanced the ability to preserve food in the home and had become widely enough disseminated by the 1930s that the Birdseye Corporation was able to make frozen food widely available. Owners of electric dishwashers and other new kitchen appliances found cooking and cleaning easier, if still time-consuming. Radio, the phonograph, and later television brought free entertainment into homes around the nation. The Great Depression of the 1930s initially undercut sales of electrical household appliances, which plummeted from $160 million in 1930 to just $82 million in 1933, but sales began to rebound thereafter, reaching new records of $218 million in 1935 and $333 million in 1936 (HSUS, Vol. II, p. 700). Wholesale sales of all electrical goods and appliances followed a similar pattern, falling from $847 million in 1929 to $276 million in 1933 and then rising steadily to $788 million in 1939 (HSUS, Vol. II, p. 851). During the same period, the spread of labor-saving technologies such as automatic washing machines made it easier and more acceptable for housewives—still the primary operators of household technologies—to perform domestic work without hired help.

After World War II, sales of electronics and electrical household appliances exploded, jumping from $788 million in 1939 to $4.3 billion in 1948, $13.6 billion in 1967, $68.6 billion in 1997, and $84.6 billion in 2008 (HSUS, Vol. II, p. 851; *Statistical Abstract of the United States: 2003*, p. 661; *Statistical Abstract of the United States: 2010*, p. 645). New products, such as microwave ovens and compact-disk (CD) players—and then personal computers, digital cameras, and mobile phones—continued to appear at the end of the twentieth century and into the beginning of the twenty-first century. Personal computers in particular, whose retail sales reached $22.9 billion in 2008, made significant inroads across all regions and social levels, with Nielsen, the media research company, estimating that 73 percent of all households in the United States owned some sort of personal computer in 2008 (Jackson et al., 2011, p. 20). In combination with Internet access, personal computers connect households to a variety of networks—personal, professional, social, and commercial—that create new, very different opportunities for their users, the implications of which were not yet clear in the early twenty-first century.

[*See also* **Agricultural Technology; Computers, Mainframe, Mini, and Micro; Edison, Thomas; Electricity and Electrification; Electronic Communication Devices, Mobile; Food and Diet; Franklin, Benjamin; Gender and Technology; Health and Fitness; Home Economics Movement; Internet and World Wide Web; Motor Vehicles; Radio; Refrigeration and Air Conditioning; Technology;** *and* **Television.**]

BIBLIOGRAPHY

Bushman, Richard L. *The Refinement of America: Persons, Houses, Cities.* New York: Alfred A. Knopf, 1992.

Cowan, Ruth Schwartz. *More Work for Mother: The Ironies of Household Technology from the Open Hearth to the Microwave.* New York: Basic Books, 1983.

Historical Statistics of the United States, Colonial Times to 1970. 2 vols. Washington, D.C.: U.S. Dept. of Commerce, Bureau of the Census, 1975.

Jackson, Barcus, Caroline Howard, and Phillip Laplante. "Use of Technology in the Household: An Exploratory Study." *International Journal of Strategic Information Technology and Applications* 2 (October–December 2011): 20–29.

Larkin, Jack. *The Reshaping of Everyday Life, 1790–1840.* New York: Harper & Row, 1988.

Marchand, Roland. "The Inward Thrust of Institutional Advertising: General Electric and General

Motors in the 1920s." *Business and Economic History* 18 (1989): 188–196.

May, Elaine Tyler. "The Commodity Gap: Consumerism and the Modern Home." In *Consumer Society in American History: A Reader*, edited by Lawrence B Glickman, pp. 298–313. Ithaca, N.Y.: Cornell University Press, 1999.

McGaw, Judith, ed. *Early American Technology: Making and Doing Things from the Colonial Era to 1850*. Chapel Hill: Published for the Institute of Early American History and Culture, Williamsburg, Virginia, by the University of North Carolina Press, 1994.

Statistical Abstract of the United States. Washington, D.C.: U.S. Census Bureau, 1878–2011.

Strasser, Susan. *Never Done: A History of American Housework*. New York: Pantheon Books, 1982.

Williams, James C. "Getting Housewives the Electric Message: Gender and Energy Marketing in the Early Twentieth Century." In *His and Hers: Gender, Consumption, and Technology*, edited by Roger Horowitz and Arwen Mohun, pp. 95–113. Charlottesville: University Press of Virginia, 1998.

Christopher W. Wells

HUBBLE, EDWIN POWELL

(1889–1953), generally regarded as the leading observational cosmologist of the twentieth century. Hubble was born in Marshfield, Missouri, although he was raised largely in Illinois. His father worked in the insurance business and his mother was a homemaker.

As an undergraduate at the University of Chicago, Hubble won a Rhodes Scholarship to Oxford and studied jurisprudence. He returned to science in 1914 when he became a graduate student in astronomy at the University of Chicago's Yerkes Observatory. He completed his dissertation in 1917 on the classification of spiral nebulae, or what we would now call galaxies (a topic to which he would return later in his career). In 1919, he joined the staff of the superbly equipped Mount Wilson Observatory in California.

With the aid of the Observatory's one hundred-inch telescope (the most powerful in the world), Hubble made a series of observations of a class of stars known as Cepheid variables in one of the spiral nebulae. The changing brightness of these stars enabled him to calculate the distance to the spiral. He found that it lay well beyond our own galaxy, thereby establishing that there are indeed galaxies of stars beyond our own Milky Way star system.

Later in the 1920s, along with his collaborator Milton Humason, he embarked on an important program of research in which he measured the distances to galaxies and Humason measured the redshifts in their light. These researches led to Hubble being widely regarded as the discoverer of the expanding universe (although if any one person can be said to deserve the credit, it is in fact the Belgian astronomer and mathematician Georges Lemaître).

Hubble reckoned it was possible to separate observations from theories in a straightforward manner and very much saw himself as an observer. He was sceptical of theory in general and in particular of the expanding universe because, unlike the great majority of astronomers, he was never persuaded that the redshifts indicate a galaxy's speed away from the earth. The researches of his most productive period, 1923–1936, were summarized in his 1936 book *The Realm of the Nebulae*.

[*See also* **Astronomy and Astrophysics; Hubble Space Telescope; National Aeronautics and Space Administration; Science; Space Program; Space Science;** *and* **Technology.**]

BIBLIOGRAPHY

Christianson, Gale. *Edwin Hubble: Mariner of the Nebulae*. Chicago: University of Chicago Press, 1995.

Smith, Robert W. *The Expanding Universe: Astronomy's "Great Debate" 1900–1931*. New York: Cambridge University Press, 1982. Paperback ed., 2010.

Robert W. Smith

HUBBLE SPACE TELESCOPE

The Hubble Space Telescope (HST) is a large, automated space observatory. A number of spaceflight visionaries had envisaged the possibilities of telescopes in space even before World War II. Serious planning, however, began for what would become the HST in the late 1960s. The project received strong backing from the White House, but its coalition of supporters was put to the test during a battle for congressional approval between 1974 and 1977. In cost-saving moves, Hubble's design became less ambitious and a partner was added, the European Space Agency.

The result of the efforts of two space agencies (the National Aeronautics and Space Administration [NASA] and the European Space Agency), many thousands of people, hundreds of companies, the Space Telescope Science Institute, and universities and government labs across the United States and Europe, the HST cost around $2 billion (1990 dollars) to build. Constructed to work in the optical, ultraviolet, and near-infrared wavelengths of the electromagnetic spectrum, at Hubble's heart is a reflecting telescope with a ninety-four and one-half inch–diameter primary mirror. Although this size of mirror was relatively small by the standards of the biggest ground-based telescopes of the 1990s, astronomers had pressed for the building of such a space telescope because they believed that its position above the earth's atmosphere would more than make up for its relatively modest light-gathering capabilities. The telescope was also explicitly planned to be serviced and maintained by the Space Shuttle. A number of its key elements, including its scientific instruments, were therefore designed so they could be readily exchanged in orbit by space-suited astronauts.

The HST's journey into space was delayed by the grounding of the Shuttle fleet after the destruction of *Challenger* in 1986. When it was finally launched in 1990, astronomers and engineers were shocked to find that the telescope's primary mirror had been ground to the wrong shape. But after repairs by visiting Shuttle astronauts in 1993, the HST went on to perform at least as well as expected and it became a pathbreaking tool for many areas of astronomical research, a symbol of scientific excellence, and the most famous telescope ever built. At the time of writing it is still in operation.

[*See also* Astronomy and Astrophysics; *Challenger* Disaster; Hubble, Edwin Powell; National Aeronautics and Space Administration; Space Program; *and* Space Science.]

BIBLIOGRAPHY

DeVorkin, David, and Robert W. Smith. *Hubble: Imaging Space and Time.* Washington, D.C.: Smithsonian National Air and Space Museum, in association with National Geographic, 2008.

Smith, Robert W. *The Space Telescope: A Study of NASA, Science, Technology and Politics.* New York: Cambridge University Press, 1989, and expanded paperback ed., 1993.

Robert W. Smith

HUMAN GENOME PROJECT

Launched as a small "initiative" at the U.S. Department of Energy (DOE) in 1985, the Human Genome Project (HGP) was a political response to the needs of the American biotechnology industry, an effort to find a post–Cold War mission for the DOE's system of national laboratories, and a dramatic manifestation of molecular biology's rising scientific power. After early debates about the feasibility and desirability of a "crash program" to map the human genome—all genetic material in the 24 human chromosomes—Congress in 1986 approved public funding through both the DOE and the National Institutes of Health. Japan, Germany, France, China, the United Kingdom, and other countries provided additional support for the project. Led by the United States, the International Human Genome Sequencing Consortium completed and published the full sequence of the human genome in April 2003. The database serves as a medical and scientific resource analogous to the nineteenth-century geological maps of the American West. Researchers can use the final product to investigate the inheritance of specific traits over several generations, including the inheritance of genetic diseases.

The HGP proved extremely controversial. Its scientific strategies, funding, relationship to the biotechnology industry, impact on young scientists, patenting issues, and long-term social and ethical implications all generated significant public debate. Members of Congress worried about the possible misuse of genetic information produced by the project, and critics characterized the project as a new form of colonial exploitation of resources—that is, of Third World blood and DNA.

The potential ethical ramifications led to congressional support for a novel funding arrangement: beginning in 1989, the HGP appropriations include funding set aside of 5 percent for scholarly studies of the ethical and social implications of the scientific work itself. Although memorable for its scientific impact, the project was thus also noteworthy as an experiment, however flawed, in socially responsible science. The HGP produced many new genetic discoveries and remarkable insights into the complexities and contrarieties of genomic structure. Despite the safeguards, however, it continued to provoke widespread fears of a "new eugenics" facilitated by high-tech gene mapping and predictive diagnostics.

At a White House conference in June 2000, officials of the National Human Genome Research Institute and of the Celera Genomics Corporation, a private firm also working on mapping the human genome, made a simultaneous announcement that a "working draft" of the DNA sequence of the human genome had been completed. In February 2001, the International Consortium published the first draft of the human genome in the journal *Nature*. A surprising result of this first draft was the discovery that researchers had significantly overestimated the total number of human genes.

Another especially important aspect of the project was the major effort by the federal government to transfer technology to private industry through various means, including licensing technologies developed by the project to private companies. The HGP thus not only had important implications for basic scientific research but also has played a significant role in stimulating the development of the biotechnology industry in the United States.

[*See also* Biological Sciences; Biotechnology; DNA Sequencing; Ethics and Medicine; Eugenics; Genetics and Genetic Engineering; Medicine; Molecular Biology; National Institutes of Health; Research and Development (R&D); Science; *and* Technology.]

BIBLIOGRAPHY

Cook-Deegan, Robert M. *The Gene Wars: Science, Politics, and the Human Genome Project*. New York: W. W. Norton & Co., 1994.
Human Genome Project Information, Genomic Science Program, Department of Energy Microbial Genomics. http://www.ornl.gov/sci/techresources/Human_Genome/home.shtml (accessed 26 October 2012).
Kevles, Daniel J., and Leroy Hood, eds. *The Code of Codes: Scientific and Social Issues in the Human Genome Project*. Cambridge, Mass.: Harvard University Press, 1992.
McElheny, Victor K. *Drawing the Map of Life: Inside the Human Genome Project*. New York: Basic Books, 2010.

Susan Lindee;
updated by Hugh Richard Slotten

HUTCHINSON, G. EVELYN

(1903–1991), ecologist. George Evelyn Hutchinson was born and spent his childhood in Cambridge, England. After Gresham's School in Norfolk, he studied zoology at Cambridge University, receiving double firsts in 1925. He then went to the Naples Zoological Station with a Rockefeller research fellowship and then to a lectureship at the University of Witwatersrand in South Africa. Together with Grace Pickford, he studied the physical and biological aspects of the South African shallow lakes; limnology thus became his special field, later incorporated into ecology. In 1928 he received an instructorship at Yale University and remained at Yale as professor and researcher until 1991, when he returned to England.

Hutchinson is widely considered the father of modern ecology. Together with his graduate and postdoctoral students, he initiated several fields within ecology and was particularly important in providing ecology with a new theoretical basis. His special expertise was in water bugs (Hemiptera). His research on these and other aquatic organisms formed a basis for much of his well-known theoretical work, including his theory of the ecological niche as a multidimensional hypervolume (Hutchinson, 1957).

He was the first to use radioisotopes as tracers in lake ecosystems, initiating the large field of radiation ecology. Together with postdoctoral student Raymond Lindeman, he began the field of ecosystem ecology. He moved the field of biogeochemistry, in which he did a great deal of research, into an important place within ecology. Using the earlier work of Gause, Lotka, and Volterra, he and his students, including Robert MacArthur, Lawrence Slobodkin, and Egbert Leigh, developed population and mathematical ecology. His four-volume *A Treatise of Limnology* remains an important resource in the field to which he contributed many new ideas and research. Hutchinson won many environmental awards, largely for his basic research. He also had important roles in relation to practical environmental and biodiversity issues (Slack, 2011). Hutchinson was noted for his unusual (and encouraging) teaching methods; many of his more than 50 graduate and postdoctoral students became leading ecologists.

Hutchinson was also one of the twentieth-century's best writers of science. In 1932 he was the lead biologist on the Yale Expedition to North India (now Ladakh). He published an account of his journey and his research on the very high altitude lakes as *The Clear Mirror*, a literary success. For many years he wrote essays (Marginalia) on a wide variety of scientific topics for the *American Scientist*. Many were collected into his books of essays (e.g., "The Ecological Theater and the Evolutionary Play" [1965]). In addition to all the ecology students of whom he called himself the proud "father," his legacy includes 328 books and papers.

[*See also* **Ecology;** *and* **Zoology.**]

BIBLIOGRAPHY

Edmondson, Yvette H. *Limnology and Oceanography* 16, no. 2 (1971). Dedicated to G. Evelyn Hutchinson. Reprinted in 1991 as Vol. 36, no. 3.

Hutchinson, G. Evelyn. *The Clear Mirror: A Pattern of Life in Goa and in Indian Tibet.* Cambridge, U.K.: Cambridge University Press, 1936.

Hutchinson, G. Evelyn. "Concluding Remarks." *Cold Spring Harbor Symposium Quantitative Biology* 22 (1957): 415–427.

Hutchinson, G. Evelyn. *The Ecological Theater and the Evolutionary Play.* London and New Haven, Conn.: Yale University Press, 1965.

Hutchinson, G. Evelyn. "Homage to Santa Rosalia or Why Are There So Many Kinds of Animals?" *American Naturalist* 93 (1959): 145–159.

Hutchinson, G. Evelyn. *A Treatise on Limnology,* Vol. 4. *The Zoobenthos.* New York: John Wiley & Sons, 1993. Volumes 1–3 were published in 1957, 1967, and 1975; this volume was published posthumously.

Kingsland, Sharon. *Modeling Nature. Episodes in the History of Population Biology.* Chicago: University of Chicago Press, 1985.

Levin, Simon. *Complexity and the Commons. Fragile Dominion: Complexity and the Commons.* Cambridge, Mass.: Perseus Books, 1999.

Lewin, Roger. "Santa Rosalia Was a Goat." *Science* 22 (1983): 636–639.

Lovejoy, Thomas E. "George Evelyn Hutchinson." *Biographical Memoirs Fellows of the Royal Society* 57 (2011): 167–177.

Skelly, David K., David M. Post, and Melinda D. Smith, eds. *The Art of Ecology: Writings of G. Evelyn Hutchinson.* New Haven, Conn.: Yale University Press, 2010.

Slack, Nancy G. "Are Research Schools Necessary? Contrasting Models of Twentieth-Century Research at Yale led by Ross G. Harrison, Grace E. Pickford, and G. Evelyn Hutchinson." *Journal of the History of Biology* 36 (2003): 501–529.

Slack, Nancy G. *G. Evelyn Hutchinson and the Invention of Modern Ecology.* New Haven, Conn.: Yale University Press, 2010.

Nancy G. Slack

HYBRID SEEDS

Plant breeding has been practiced for thousands of years, but prior to the early twentieth century, farmers still planted their corn crop with seed saved from the previous year's harvest. They chose the best looking corn, believing that it was the strongest. That began to change in 1906 when a geneticist in the United States named G. H. Shull began experimenting with the inheritance traits of corn. Over the next 20 years processes for in-breeding and seed selection led to the introduction of a commercially viable hybrid seed in 1924. Henry Wallace, future U.S. secretary of agriculture (1933–1940) and vice president of the United States (1941–1945), began selling a hybrid corn seed called "Copper Cross" in 1924. Two years later he cofounded the Hi-Bred Corn Company (later Pioneer Hi-Breed), the first company created to develop and sell hybrid seed.

Hybrid corn is, at its simplest, created by crossing two different strands of corn to create a new strand. The earliest hybrids were developed by removing the male flowers from the corn so that the fertilization of the female flowers could be controlled. The early success of cross-breeding led to single crosses, produced by crossing two inbred lines, and then to double crosses, which are produced by crossing two different single crosses (and having the traits of all four original parents). Repeated over several generations of seed, hybrid corn proved to be more resistant to pests, weather, and disease and disproved the theory that the best looking corn was the strongest corn.

Early efforts to sell hybrid corn were relatively unsuccessful, but productive corn crops planted by Pioneer in the 1930s during the Great Depression, at a time when naturally pollinated crops failed at a high rate, began to change the minds of farmers. By the 1950s, nearly all of the U.S. corn crop was planted with hybrid seed.

In Mexico in the 1940s, scientist Normal Borlaug and his colleagues developed a drought-hardy, rust-resistant wheat strain and crossed it with a dwarf Japanese strain to produce a hybrid short enough to survive the wind. Borlaug's dwarf wheat sparked what is now known as the "green revolution," so dramatically increasing yields in Mexico that over the next several decades new varieties were introduced in Pakistan, India, and other areas of the world. Pakistan's wheat crop grew from a pre–green revolution total of 4.6 million tons to 8.4 million tons just four years later ("Norman Borlaug—Nobel Lecture"). In 1970, Borlaug was awarded the Noble Peace Prize for his work.

Genetically Modified Organisms. In the 1980s, transgenic technology pushed hybridization by artificially injecting genes into a plant to create transgenic crops, better known as genetically modified organisms, or GMOs. Genetically engineered crops were first commercialized in the United States in 1996, and according to the U.S. Department of Agriculture, more than 90 percent of planted corn, soybeans, and upland cotton in the United States in 2011 was genetically modified ("Adoption of Genetically Engineered Crops in the US: Extent of Adoption"). A number of variations have since been introduced, including stacked corn, a hybrid that includes various levels of genes that resist corn borers, rootworm, and pesticides.

Globally, the development of genetically modified crops has been a hotly contested issue. Advocates, including the World Bank and the Food and Agriculture Organization of the United Nations, point to increased yields and pest, weed, and weather resistance and consider global adoption of genetically modified crops a primary path to feed a world whose population was expected to grow from 6 billion people in 2011 to 9 billion people by 2050. Opponents argue that genetically modified crops have unknown environmental impacts, that they promote the evolution of "super weeds," and that unknown human hazards are imminent.

In 2011, many countries were still deciding whether genetically modified crops should be grown and whether food made from genetically modified crops could be sold within their boundaries. In 1998, the European Union banned the introduction of new genetically modified crops. In 2004, the ban was lifted for the introduction of a genetically modified sweet corn called Bt-11, but the ban still existed for most other crops seven years later. In poverty-stricken countries like those on the African continent, debate continued not only over the ethics of enhanced seed, but also over the escalating costs that put the technology beyond the reach of most farmers.

In 2010, it was estimated that more than 148 million hectares globally was planted with genetically modified crops, up from 1.7 million hectares in 1996. Twenty-nine countries were growing genetically modified crops, led by the United States (66.8 million hectares), Brazil (25.4), Argentina (22.9), India (9.4), Canada (8.8), China (3.5), Paraguay (2.6), Pakistan (2.4), South Africa (2.2), and Uruguay, with 1.1 million hectares ("Global Status of Commercialized Biotech/GM Crops: 2010").

[*See also* **Agricultural Technology; Biological Sciences; Food and Diet; Genetics and Genetic Engineering;** *and* **Nobel Prize in Biomedical Research.**]

BIBLIOGRAPHY

"Adoption of Genetically Engineered Crops in the US: Extent of Adoption." U.S. Department of Agriculture. http://www.ers.usda.gov/data-products/adoption-of-genetically-engineered-crops-in-the-us.aspx (accessed 10 April 2012).

Culver, John C., and John Hyde. *American Dreamer: A Life of Henry A. Wallace.* New York: W. W. Norton & Company, 2000.

Fitzgerald, Deborah Kay. *The Business of Breeding: Hybrid Corn in Illinois, 1890–1940.* Ithaca, N.Y.: Cornell University Press, 1990.

"Improving Corn." U.S. Department of Agriculture. http://www.ars.usda.gov/is/timeline/corn.htm (accessed 10 April 2012).

International Service for the Acquisition of Agri-Biotech Applications. "Global Status of Commercialized Biotech/GM Crops: 2010." http://www.isaaa.org/resources/publications/briefs/42/executivesummary/default.asp (accessed 10 April 2012).

"Norman Borlaug—Nobel Lecture." Nobelprize.org. http://www.nobelprize.org/nobel_prizes/peace/laureates/1970/borlaug-lecture.html/ (accessed 10 April 2012).

"The Technology Challenge." Food and Agriculture Organization. http://www.fao.org/fileadmin/templates/wsfs/docs/Issues_papers/HLEF2050_Technology.pdf (accessed 10 April 2012).

Neil Dahlstrom

HYDROELECTRIC POWER

Electricity generated through the use of waterwheels or hydraulic turbines is known as hydroelectric power. In the early 1880s, small water-powered mills were utilized to produce direct-current (DC) electricity. However, the full potential of hydroelectric power was not realized until the proliferation of alternating-current (AC) power systems in the 1890s. In contrast to DC (which could not be transmitted efficiently more than about 10 miles), polyphase AC systems proved capable of transmitting power hundreds of miles. As a result, AC systems allowed the development of large waterpower sites in remote locations far removed from urban markets.

America's first polyphase AC hydroelectric power system came online in 1893 near San Bernardino, California. California subsequently led the nation in long-distance hydroelectric power development; Fresno received power over a 35-mile-long transmission line in 1896, and by 1901 San Francisco was connected to generating plants in the Sierra Nevada mountains, more than 140 miles away. In the eastern United States, Niagara Falls became the focus of hydroelectric power development; in 1896, AC power was first transmitted over a 22-mile-long line connecting Niagara Falls to Buffalo, New York.

In the early twentieth century, the conservationist Gifford Pinchot championed a movement advocating government regulation of hydroelectric power systems built by privately owned utilities. In the 1920s, the struggle between public and private interests over control of the electric power industry focused on the Muscle Shoals (or Wilson) Dam on the Tennessee River in northern Alabama. The government had started the Muscle Shoals project during World War I to manufacture nitrates used in explosives. After the war, Henry Ford proposed buying the dam for general industrial purposes. However, public-power supporters in Congress blocked the transfer of control into private hands. In 1933, President Franklin Delano Roosevelt successfully incorporated Wilson Dam—and several other proposed dams—into the newly created and publicly administered Tennessee Valley Authority (TVA).

In the West, hydroelectric power constituted a key component of the Boulder Canyon Project. Authorized in 1928 by President Calvin Coolidge, this project included federal financing of the Hoover Dam across the Colorado River near Las Vegas, Nevada. During Roosevelt's New Deal, many other large-scale hydroelectric power dams were built in the West by the federal government. These included Grand Coulee Dam in Washington State, Shasta Dam in California, and Marshall Ford Dam in Texas.

After World War II, still more hydropower dams were built (including Glen Canyon Dam in northern Arizona), but the economic importance of hydroelectricity waned as fossil-fuel and nuclear-generating plants grew in size and number. By the end of the twentieth century, hydroelectric power accounted for about 10 percent of electricity used in the United States; concurrently, public concern over environmental costs associated with reservoir construction prompted a movement to remove hydroelectric dams to restore river valleys to a more natural state.

[*See also* **Dams and Hydraulic Engineering; Electricity and Electrification; Environmentalism; Ford, Henry; Hoover Dam; Nuclear Power; Rivers as Technological Systems;** *and* **Tennessee Valley Authority.**]

BIBLIOGRAPHY

Hubbard, Preston J. *Origins of the TVA: The Muscle Shoals Controversy, 1920–1932.* Nashville, Tenn.: Vanderbilt University Press, 1961.
Hughes, Thomas P. *Networks of Power: Electrification in Western Society, 1880–1930.* Baltimore: Johns Hopkins University Press, 1983.

Donald C. Jackson

HYGIENE, PERSONAL

Early European visitors to the United States frequently commented on the absence of reliable supplies of soap and water and the prevalence of mud and manure, flies and insects, and disgusting tobacco stains (from both spitting and chewing). More than four of five Antebellum-Era Americans lived in hygienically primitive situations on small farms or in country villages. Even in the few big cities, where cholera and typhoid fever epidemics signaled the need for water and sewer systems and stimulated massive public-works construction, changes in personal and domestic cleanliness practices came slowly.

American housewives tried to keep their homes tidy and their families clean. But much of what would later be considered essential, such as bathing and frequent washing of clothes, was not thought important. Because indoor plumbing was rare, basic hygiene was extremely laborious. To wash a load of clothes, housewives and hired girls (slaves in the South) had to carry full buckets of water some distance, cut wood to heat it, and then lift and hang heavy, wet laundry to dry. Soap was typically homemade from ashes, lye, and rendered animal fat.

Improvements in hygiene habits were rooted in technological innovations and changed attitudes. From the early nineteenth through the mid-twentieth centuries, the gradual spread of municipal water and sewage systems and the growing availability of indoor plumbing, hot-water heaters, washing machines, and commercially manufactured soaps—developments that extended outward from urban centers to small towns and rural regions and down the social scale from the elite to the poor—marked a major transition in hygiene, making possible higher standards of personal cleanliness, less exhausting laundry procedures, and the sanitary disposal of human wastes. Americans also heeded the advice and warnings of early nineteenth-century reformers and sanitarians such as Catharine Beecher and the New England educator William Alcott (1798–1859). The work of the U.S. Sanitary Commission and women volunteers during the Civil War proved especially important in stimulating a national campaign for better personal hygiene. Because disease killed more soldiers than guns or cannons, sanitarians effectively demonstrated that sanitation was the war's crucial weapon. By wrapping cleanliness in the mantle of victory and patriotism, they taught army doctors, soldiers, and loved ones on the home front that poor personal hygiene, which led to rampant camp diseases, was a fearsome enemy.

Nevertheless, hygienic problems continued to threaten vast numbers of Americans and immigrants who flocked to industrializing cities after the war. In congested working-class neighborhoods, the health of these newcomers was clearly at risk as outbreaks of cholera, typhoid, and yellow fever took a heavy toll. Through the combined efforts of public-health reformers, city officials, settlement houses, public-works engineers, teachers, and employers, immigrants and native-born migrants from the countryside learned the significance of personal hygiene—first to their health and then to their opportunities for upward mobility.

By the 1920s, reformers concerned with immigrant assimilation successfully made cleanliness a hallmark of being "American." But it was the producers of hygiene products and household cleaning appliances, for whom cleanliness meant profits, who persuaded American consumers to accept nothing less than perfection, to look for "the cleanest clean possible." Incessant advertising appeals in magazines, on radio, and later on television created a culture of cleanliness that by the 1950s set Americans apart. With houses cleaner and bodies better groomed than ever before, cleanliness became an obsession. Dependence on daily showers, sensitivity to body odors, desire for immaculately clean houses, and preoccupation with teeth that gleamed distinguished Americans as a people.

During the mid-1960s, however, as more married women (traditionally the quintessential agents of cleanliness) entered the workforce, they spent less time at housecleaning, and most husbands chose not to pick up the slack. Environmentalists, feminists, and members of the counterculture for reasons of their own also turned their backs on what had become an obsession. Thus, by the early twenty-first century, Americans were less likely to be swayed by the old rationales for chasing dirt, yet they still took delight in their daily showers, "natural" soaps, and luxurious bathrooms.

[*See also* Cholera; Disease; Electricity and Electrification; Health and Fitness; Heating Technology; Household Technology; Military, Science and Technology and the; Public Health; Radio; Refrigeration and Air Conditioning; Television; Typhoid Fever; War and Medicine; *and* Yellow Fever.]

BIBLIOGRAPHY

Cowan, Ruth Schwartz. *More Work for Mother*. New York: Basic Books, 1983.

Duffy, John. *The Sanitarians*. Urbana: University of Illinois Press, 1990.

Hoy, Suellen. *Chasing Dirt*. New York: Oxford University Press, 1995.

Palmer, Phyllis. *Domesticity and Dirt*. Philadelphia: Temple University Press, 1989.

Vinikas, Vincent. *Soft Soap, Hard Sell*. Ames: Iowa State University Press, 1992.

Williams, Marilyn Thornton. *Washing "The Great Unwashed."* Columbus: Ohio State University Press, 1991.

Suellen Hoy

ILLUMINATION

All societies have developed methods of illumination beginning with fire until electric lighting emerged in the middle of the nineteenth century. In general, this has been a history of expansion from domestic lighting and street illumination to commercial and industrial uses.

Domestic Lighting. While lighting might appear a necessity, the perception of needs has changed radically since the pre-industrial times when expense restricted illumination almost exclusively to indoor use, and the night was sharply demarcated from the day. Light and fire seemed inseparable, indeed identical. People relied on tallow candles, floating tapers that burned assorted greases, and lamps that burned fuels such as lard and turpentine. Whale oil, cleaner but more expensive, was in such demand by the early nineteenth century that whaling ships circled the globe chasing their prey. Oil and natural gas were used in some parts of ancient China for light. Kerosene came into use in Persia during the ninth century and was also known in the Caspian Sea area and the Arab world. Distillation techniques spread to Europe, but oil was not important for lighting until oil production began in the middle nineteenth century in Poland, Canada, and the United States. Kerosene lamps were widely used until the 1920s (longer in rural areas), but were gradually replaced by incandescent electric lighting based on the Edison system. Available first in New York after 1882, it spread only gradually. In ca. 1910, 15 percent of all American homes had electricity. By 1930 the level topped 90 percent in cities and reached 50 percent in irrigated farm regions, but was only 10 percent in the countryside. New Deal programs completed rural electrification after 1935. In densely populated nations, such as The Netherlands, and in large cities such as Berlin and London, electrification took place at the same pace as in the United States.

The shift to electricity had implications for home design. Gas lighting consumed oxygen and produced heat, rooms had to be aired out, and doors were advisable so that rooms could be aired out separately and minimize drafts that could blow out flames and cause houses to fill with explosive gas. Burning manufactured gas produced soot, and its use coincided with dark wallpapers and furnishings. In contrast, electric lights were unaffected by the wind and gave off no soot or gases. Their adoption facilitated open floor plans and a lighter color scheme.

Street Lighting. In preindustrial times, people who ventured out at night had to carry torches or lanterns. In 1682 Amsterdam was one of the first cities to erect streetlamps and hire lamplighters, an innovation that soon spread to Dresden, Berlin, and other German cities, but did not reach Paris for a century. On Pall Mall, in London, gas lighting first appeared in 1807 and proved so popular that by 1820 the city had almost three hundred miles of gas mains. In the United States, Benjamin Franklin organized the first public lighting in 1751, placing lanterns on Philadelphia's streets. Gaslight came in 1816, when Baltimore, like London, began to make gas from coal. By 1828 New York's Broadway was brilliantly illuminated with gas flares, and, as in London, a new phenomenon emerged, nightlife, which intensified with the spread of lighting. Commercial gas works spread to every industrial country, and by 1860 the United States alone had 183 urban gas lighting companies.

Natural gas, cleaner than coal gas, was known in ancient China. It was used in several buildings in Fredonia, New York, in 1821, but it was long ignored as a light source because of difficulties in getting it to consumers. Early oil drillers intentionally discharged gas into the air or burned it off. Only in the 1870s did natural gas begin to overtake manufactured gas in some areas; but before pipelines were developed electrification seized much of the market. Charles Brush's powerful arc lights, first displayed in 1878 in Cleveland, were common in cities and at expositions for a generation, whereas incandescent lighting was little used outdoors before ca. 1900. The arc light made possible illumination comparable with a full moon by erecting lights on towers much higher than gas streetlights. Some cities adopted such systems in the 1880s and 1890s, but they were more suited to a smaller walking city than to one with tall buildings or heavy traffic. Gradually, as incandescent lighting became more powerful, it displaced both gas and arc lighting. As it spread, Americans debated whether it should be a public or a private utility. Boston, New York, and Chicago opted for private power, but others, notably Cleveland and Los Angeles, did not.

Industrial and Commercial Lighting. Artificial illumination expanded the workday. Early factories closed at dusk, but as lighting improved and became less costly investors realized that longer hours or even several shifts were possible. Gas light, dangerous for use in cotton mills, was tried in newspaper offices and other venues. Californian gold mines adopted arc lighting in the 1870s, and incandescent light was safe for cotton and flour mills and soon spread wherever visual acuity mattered. After ca. 1900 artificial light entered banks, offices, libraries, and schools, and the ancient distinction between day and night eroded, a process that accelerated during 1940s war production. Yet World War II also saw full blackouts of all major cities from Tokyo to London, in a generally unsuccessful attempt to hide from enemy bombing. Children growing up during that war were the last to experience darkness on a regular basis. Those born since then find illumination natural, and night is almost unknown.

During the same time that electric lighting spread to industries, it was also adopted in commercial districts, lengthening the day for shoppers. Where utilities were privately owned, they sold street lighting systems to merchants directly or more often to associations of merchants from particular streets or districts. This led to competition for public attention, including the use of flashing advertising signs and spectacular lighting of buildings. The Singer Tower in New York illustrates this trend. The world's tallest building in 1908, it was brilliantly illuminated and stood out above the New York skyline. Such dramatic lighting allowed one corporation to gain maximum public attention, and other skyscraper owners soon followed suit. The competitive use of lighting developed rapidly in the United States but was confined in Britain to fewer districts and largely prohibited in some places, notably Paris, which instead bathed the city in white light. A few cities, like Venice, scarcely intensified their lighting at all.

Coinciding with the rise of electrification itself, spectacular lighting was one of the most striking features of world's fairs between the late 1870s and 1939. Each exposition sought to outdo the previous event and usually succeeded, because the technology of lighting improved rapidly. The arc lights of the 1870s and 1880s were eclipsed by incandescent lights in the 1890s, and they in turn were outdone by colored lights in impressionist hues during the 1900s, only to be overpowered by spot lights, search lights, and electrical fireworks after 1905. These new technologies were immediately adopted in amusement parks and theater districts, and some were transferred to the lighting of skyscrapers and public monuments. By the 1920s cities had electrified skylines that were widely reproduced in photographs. Their nighttime aspect often was more alluring than by day because the lighting emphasized impressive buildings, monuments, and bridges, whereas areas of blight and poverty became unimportant blanks.

This dynamic landscape found its fullest expression in the United States, where it constantly changed in accord with whatever most attracted and stimulated the public. In the new night space ca. 1915 there were strong contrasts, few shadows, and little sense of depth. The rows of streetlights provided some sense of perspective, but the lighted billboards and flashing advertising signs were not designed to a common scale and did not always permit a clear sense of the relations between front and back. The sheer power of the lighting arrays, combined with air pollution, blotted out most of the stars and outshone the moon, which had become unimportant as a source of light. Illuminated landmarks provided visual orientation most effectively to those who already knew the city by day. To a tourist, this new night space was a jumble with an uncertain sense of scale, and it was both exhilarating and disorienting. Europeans generally controlled private advertising displays more tightly, with zoning laws that restricted the size and location of electrical signs

to shops and a few enclaves of neon. In the early twenty-first century, the lighting during an evening walk along the Seine still emphasized churches, museums, and public buildings.

Systematic Problems.   To supply large populations with service, electrical utilities connected themselves into grids. Their ingenious engineering shunts power around to balance the load at each local utility. But the interconnections that make the grid possible also enable cascading system failures, spreading blackouts from one city and state to another. Some of the most spectacular examples occurred in the United States in 1965, 1977, and 2003, when blackouts spread at almost the speed of light and forced millions into temporary self-sufficiency. Although most blackouts last only a few hours or days, in 1998 Auckland, New Zealand, suffered a 10-week blackout, the longest in recent times.

The arguably excessive use of urban lighting has contributed to global warming and hastened the rapid depletion of fuel supplies. In the early twenty-first century, more efficient light-emitting diode (LED) lights and better design helped to solve these problems. Lower-intensity lighting is more sustainable and will recover the lost experience of the night sky.

[*See also* Edison, Thomas; Electricity and Electrification; Franklin, Benjamin; Household Technology; Rural Electrification Administration; *and* Skyscrapers.]

BIBLIOGRAPHY

Nye, David E. *Electrifying America: Social Meanings of a New Technology, 1880–1940.* Cambridge, Mass.: MIT Press, 1990.

Nye, David E. *When the Lights Went Out: A History of Blackouts in America.* Cambridge, Mass.: MIT Press, 2010.

Schivelbusch, Wolfgang. *Disenchanted Night: The Industrialization of Light in the Nineteenth Century.* Berkeley: University of California Press, 1988.

Schlereth, Thomas J. "Conduits and Conduct: Home Utilities in Victorian America." In *American Home Life, 1880–1930,* edited by Jessica H.

Foy and Thomas J. Schlereth. Knoxville: University of Tennessee Press, 1992.

David E. Nye

INDIAN HEALTH SERVICE

Medical encounters between Europeans and American Indians began in the earliest years of colonization. Edward Winslow treated Massasoit for a digestive complaint in 1623. Boston colonists nursed many Massachusetts during the 1633 smallpox epidemic. Although colonial officials praised such acts, they never adopted formal policies about Indian health. Organized efforts started after Independence. Thomas Jefferson vaccinated a Miami delegation in 1801; he tried to send vaccine with Lewis and Clark in 1803. Physicians at army posts provided sporadic care to nearby tribes. Although they aimed to protect their soldiers, Indians presumably benefitted. The Office of Indian Affairs (OIA), established by Congress in 1824, supported missionaries who provided nursing and medical care. The 1832 Indian Vaccination Act allocated $12,000 and led to the vaccination of roughly 50,000 Indians. That year Congress ratified the first treaty that promised medical care. Two dozen similar treaties were signed by 1871. Subsequent Indian health programs have been funded as "gratuity appropriations."

As it imposed the reservation system in the 1870s, the federal government increased its health programs. By 1900 the OIA employed over 80 physicians (roughly one per reservation) and had built several hospitals. Despite this expansion, rampant epidemics and malnutrition persisted. Indian health services were compromised by inadequate staffing and supplies. When health surveys revealed abysmal conditions in the 1910s, Congress increased its appropriations. The 1921 Snyder Act established a formal basis for the Indian medical programs. By 1922 the OIA operated seventy-three hospitals with positions for two hundred physicians and three hundred nurses, administrators, and field matrons. In 1924 it created a Medical Division and hired professional staff from the Public Health Service (PHS).

Conditions, however, did not improve. The 1928 Meriam Report found high rates of tuberculosis and other diseases. Many programs remained focused on protecting whites from the diseases of Indians and not on improving Indian health.

Reformers sought to transfer the Medical Division from the OIA to the PHS, but the PHS and many Indian leaders opposed this move. Instead, invigorated by the New Deal, the OIA built more hospitals, improved sanitary conditions, hired more physicians, and organized campaigns against infant mortality, tuberculosis, and trachoma. World War II, however, diverted staff and funding. Postwar investigations again found appalling conditions. Congress responded with more funding and, despite continuing opposition from the PHS and Oklahoma tribes, transferred responsibility for Indian health to the PHS, creating the Indian Health Service (IHS) in 1955.

The IHS succeeded where past efforts had failed. With better hospitals, a professional medical staff, and new ancillary services (e.g., community health workers, telemedicine), the IHS orchestrated dramatic reductions in tuberculosis (down 96 percent by 1990), infant mortality (down 92 percent), and many other leading causes of death. It has also granted tribes more authority over their health services. With a budget of over $4 billion in the early 2000s, it provided health services to over 2 million Indians in 35 states. Despite progress, difficult problems remained. Life expectancy among Indians was five years lower than the national average. Although absolute rates had decreased, marked mortality disparities persisted for tuberculosis, diabetes, substance abuse, injuries, suicide, and many others. Dramatic disparities existed among different Indian populations as well. The IHS struggled to provide care not only to Indians living on large, rural reservations, but also to growing numbers of urban Indians. Dire poverty undermined health and health care for both groups. To make matters worse, the IHS received much less funding per enrollee than did the other federal health programs (Medicare, Medicaid, veterans, or active military). In the early twenty-first century, more still needed to be done to eradicate once and for all the health inequalities that have afflicted American Indians for centuries.

[*See also* Diabetes; Disease; Health and Fitness; Hospitals; Jefferson, Thomas; Lewis and Clark Expedition; Life Expectancy; Medicare and Medicaid; Medicine; Native American Healers; Public Health; Public Health Service, U.S.; Smallpox; *and* Tuberculosis.]

BIBLIOGRAPHY

Jones, David S. *Rationalizing Epidemics: Meanings and Uses of American Indian Mortality since 1600.* Cambridge, Mass.: Harvard University Press, 2004.

Kunitz, Stephen J. *Disease Change and the Role of Medicine: The Navajo Experience.* Berkeley: University of California Press, 1989.

Rife, James. *Caring & Curing: A History of the Indian Health Service.* Landover, Md.: PHS Commissioned Officers Foundation for the Advancement of Public Health, 2009.

Trannert, Robert A. *White Man's Medicine: Government Doctors and the Navajo, 1863–1955.* Albuquerque: University of New Mexico Press, 1998.

David S. Jones

INFLUENZA

Influenza is a highly contagious disease of humans, swine, horses, other mammals, and species of domestic and wild birds. In humans, symptoms include fever, cough, runny nose, sore throat, headache, muscular aches, fatigue, and in severe cases bronchitis and pneumonia. Scientists are increasingly viewing influenza as a group of viruses that continually mutate and evolve within an ecosystem of multiple hosts and environments. Influenza occurs in both endemic and epidemic forms and is thus marked by common, annual cycles as well as serious epidemic events. Endemic influenza appears seasonally in countries around the globe, circling from the northern to the southern hemisphere. According to the U.S. Centers for Disease Control, from 1976 to 2006 deaths from influenza in the United States ranged from three thousand to forty-nine thousand annually.

The origin of influenza in humans is difficult to determine because flu symptoms are similar to those of other respiratory diseases. Influenza's association with Old World domestic animals such as pigs and chickens and the vulnerability of indigenous peoples in the Americas and Australasia suggest that it was confined to Europe until the late fifteenth century, when it became one of the diseases that devastated native populations in the Caribbean and the Americas.

Pandemic of 1918. Although at times deadly, influenza paled as a threat compared with other lethal diseases that preyed on humans such as smallpox, yellow fever, cholera, and tuberculosis. But in 1918, an extremely virulent influenza appeared, causing one of the deadliest pandemics in history. Intensifying the suffering of the World War I (1914–1918), influenza sickened at least one quarter of the world's population, killing 2 to 4 percent of them. The origin of the 1918 influenza remained a mystery for nearly eight decades; the responsible virus was not identified until 1997.

The prevailing theory is that influenza first appeared in the American Midwest in March 1918 and spread to soldiers in training camps across the country, where medical officers reported high influenza rates but few deaths. The virus then traveled to Europe, probably aboard troopships, and in May and June sickened thousands of soldiers along the Western Front. This virulent strain exploded in August 1918 in several Atlantic ports. Contemporaries called it "Spanish influenza" because Spain was one of the few European countries not at war and therefore did not censor reports of illness within its borders.

Given the short incubation period, historians have tracked the spread of influenza with precision. As ships or trains carrying sick passengers arrived in various communities, officials reported a virtual explosion of feverish, coughing patients, many requiring hospitalization. Symptoms included a quick onset of illness, high fever, severe headaches, torpor, nosebleeds, a blood-producing cough, and cyanosis—a bluish cast to the skin caused by lack of oxygen in the blood. In the United States influenza arrived first in Boston, striking army trainees at nearby Camp Devens,

Massachusetts, the week of 7 September 1918. From there it swept south and west, following wartime transportation routes. It hit Kansas on 21 September and northern California and Texas on 27 September, and by the week of 16 October the epidemic had spread nationwide. Sickness rates averaged 25 percent, but varied from 15 percent of residents in Louisville, Kentucky, to a staggering 53 percent in San Antonio, Texas.

The arrival of influenza often caused social paralysis, rendering thousands of people pale and helpless, flooding hospitals with patients, and crowding morgues. Officials closed schools, government offices, theaters, and churches in an effort to prevent the spread of disease, and in many countries, so many nurses and physicians were in military service that communities had to recruit medical workers out of retirement or training schools. Influenza also depleted the labor force so that some war industries suspended operations. The U.S. Army canceled the October 1918 draft call, but given the wartime emergency, the Wilson administration declined to significantly reduce crowding on troopships and continued the national campaign of public rallies to sell war bonds.

Medical professionals did their best to save their patients, but in an era before antibiotics, they lacked effective tools. Treatment included bed rest, a light diet, aspirin for fever and pain, and keeping the patient warm in hopes of preventing pneumonia. Both military and civilian caregivers also tried whiskey, patent medicines, a range of home remedies, and injections of various serums. Preventive measures included wearing masks and the use of throat and nasal sprays and myriad vaccines, but short of a complete, prolonged quarantine, nothing worked. Patients usually recovered after a few days unless pneumonia developed. Some victims suffered permanent lung damage or mental symptoms such as prolonged fatigue, depression, sleeplessness, and anxiety.

The impact of the pandemic on World War I is still being weighed. Some historians believe that if influenza had not incapacitated entire divisions of the German army in the summer of 1918, the German government may have succeeded in forcing France and Britain to negotiate peace terms before the United States could come to their aid. Once the American Expeditionary Force did launch a massive campaign (26 September to 11 November 1918) at the Meuse-Argonne, influenza dramatically reduced the number of soldiers who could fight, overwhelmed medical services, and increased casualty rates. More American soldiers died of disease than in combat during the war. By mid-November the flu had subsided in Europe, but reappeared in January and February 1919 and sickened participants at the Paris Peace Conference. This third wave, less powerful but still deadly, again swept the globe, but by mid-1919 influenza had probably infected almost all susceptible hosts and thus either evolved to a more benign form or burned out.

Recent research suggests that the lethality of the 1918 virus resulted from its ability to penetrate deep into the lung tissue. This caused some victims to succumb quickly to viral pneumonia. The majority, however, died after a longer period from secondary pneumonia caused by a variety of bacteria, including Pfeiffer's bacillus. Whereas normal influenza usually kills the youngest and oldest people, creating a "U-shape" mortality curve, the 1918 virus was especially deadly for people ages 20 to 40, producing was a "terrible W" curve of high mortality for the young and old at the extremes of the demographic spectrum, with an unusual peak at its center. One theory for this is that the viral attack stimulated such a strong immune response that people with especially robust immune systems—like young adults—released so many antitoxins or cytokines to fight it that the fluids flooded the lungs, producing pneumonia.

Scholars are still refining estimates for the global mortality. With one quarter of the world falling ill, even a low case mortality rate of 2 to 4 percent generated shocking figures. Britain, France, and Germany each lost 250,000 to influenza, and in the midst of civil war, Russia lost some 450,000. During the roughly two years of the epidemic, an estimated 25 million Americans became ill and 675,000 died, causing the average life expectancy to drop by 12 years in 1918.

Vaccine Development and Global Surveillance. Scientists have struggled to identify the origin and cause of the devastating 1918 epidemic and have experimented with various treatments and vaccines. A review of this research by Edwin Jordan at the University of Chicago (1926) noted no consensus on these matters. In 1933, however, three British researchers identified the influenza virus A, and in 1940 the American Thomas Francis Jr. discovered the influenza virus B. Remembering the 1918 catastrophe as Europe again descended into war, the U.S. Army established an epidemiological board to investigate and control influenza and other diseases. Its Commission on Influenza headed by Thomas Francis of the University of Michigan began a crash program to develop a vaccine with a consortium of university researchers, pharmaceutical companies, and military personnel. In 1943 the Commission conducted a large-scale vaccine trial involving 15,000 students in army training programs at nine universities across the country. Researchers injected half of the students with vaccine and the other half with control material. When influenza made the rounds of the universities in November and December 1943, the vaccinated individuals had a 2.22 percent sickness rate compared with 7.11 percent in controls, an estimated efficacy rate of almost 70 percent. Based on these results, army officials hoped they had found a preventive vaccine for influenza similar to that already available for diseases such as smallpox, typhoid, and tetanus. The War Department vaccinated the entire army in October and November 1945.

In 1947, however, when the vaccine failed to protect the army and other vaccinated populations from an influenza outbreak, scientists began once again to debate the nature of the virus. Scientists dominated by members of the American Influenza Commission argued that viral variation was finite and that the various strains needed only to be identified and captured in a vaccine to provide long-term protection, conclusions rejected by some non-American researchers.

Given the prospect of an effective vaccine and the changing nature of the influenza virus, surveillance became an increasingly important strategy in tracking viral strains. Although the system generally was able to track flu strains for vaccines against seasonal influenza, it proved to be an imperfect system for the early detection of pandemics. By 1953 the World Influenza Center, established by the World Health Organization (WHO) in 1947, had 54 cooperating laboratories in 42 countries throughout the Americas and Europe, but in April 1957, the system failed to detect a serious outbreak in China. When Maurice R. Hilleman, a scientist at the Walter Reed Army Research Center, read about the outbreak in the newspaper, he obtained samples of the virus and determined that it had evolved from an H1N1 strain to H2N2, with both hemagglutinin and neuraminidase having changed. Fearing that a strain to which humans had no immunity could kill millions, Hilleman notified public-health authorities around the world and developed a vaccine.

Before the vaccine could provide protection, what became known as the Asian flu killed 1.5 to 2 million people worldwide. In 1968 influenza surveillance failed again to detect an incipient pandemic. By the time researchers had developed a vaccine against a virus that first appeared in Hong Kong, it had swept through much of the world. Although milder than the 1957 strain, the Hong Kong flu claimed a million deaths worldwide. In early 1976, when a young soldier died from influenza–pneumonia at Fort Dix, New Jersey, memories of 1918 and subsequent pandemics motivated public-health officials to launch a national vaccination campaign. When no epidemic materialized and some vaccinated individuals developed Guillain–Barré syndrome, a paralytic and sometimes fatal disease, public support for annual flu vaccination plummeted.

The appearance in the 1980s of another deadly disease, HIV/AIDS, as well as the emergence of new diseases and the development of drug resistance in others, renewed interest in epidemics. Global terrorism and destructive natural disasters also raised awareness about how catastrophic events could traumatize societies, but few people worried about influenza. In 1997 scientists under the leadership of Jeffrey Taubenberger at the U.S. Armed Forces Institute of Pathology identified the virus responsible for the 1918 pandemic as the H1N1. Despite this discovery and decades of global surveillance, scientists did not foresee the influenza H1N1 outbreak of 2009. The appearance of a virus similar to the killer of 1918 set off dire warnings from the WHO and stimulated a number of national vaccine campaigns. Although the dreaded pandemic did not develop, it provided a reminder that influenza continues to be a threat.

[*See also* **Centers for Disease Control and Prevention; Disease; HIV/AIDS; Medicine; Nursing; Public Health; Public Health Service, U.S.; War and Medicine;** *and* **World Health Organization.**]

BIBLIOGRAPHY

Bristow, Nancy. *American Pandemic: The Lost Worlds of the 1918 Influenza Epidemic.* New York: Oxford University Press, 2012. An excellent social history of the pandemic in the United States, focusing on varied experiences of men and women, the rich and poor, and various racial and ethnic groups in the country.

Byerly, Carol R. *The Fever of War: The Influenza Epidemic in the U. S. during World War I.* New York: New York University Press, 2005. Focuses on the experience of army medical personnel and the impact of influenza on the American conduct of the war.

Crosby, Alfred. *America's Forgotten Pandemic: The Influenza of 1918.* Cambridge, U.K.: Cambridge University Press, 1989. First published in 1976 as *Epidemic and Peace,* 1918, the first and most enduring comprehensive analysis of the pandemic in the United States.

Jordan, Edwin. *Epidemic Influenza: A Survey.* Chicago: American Medical Association, 1927. The first comprehensive overviews of the 1918 pandemic.

Stern, Alexandra Minna, Martin S. Cetron, and Howard Markel. "The 1918–1919 Influenza Pandemic in the United States: Lessons Learned and Challenges Exposed." *Public Health Reports* 125, suppl. 3 (April 2010): 1–144. Articles providing an in-depth analysis of the pandemic in the United States.

Woodward, Theodore E. *Armed Forces Epidemiological Board: The Histories of the Commissions.* Washington, D.C.: Walter Reed Army Medical Center Borden Institute, 1994. Provides a history of the Army Commission on Influenza with excerpts from many original documents.

Carol R. Byerly

INSTRUMENTS OF SCIENCE

As with science in general, Americans have used scientific instruments for a broad range of philosophical, practical, and pedagogical purposes, and although many of these instruments were imported from abroad, this essay focuses on the most important instruments made in the United States. Several trends are clear: over time there has been a steady increase in the number and sorts of instruments made and used, the phenomena they could capture, and the precision they could attain, as well as a steady decrease in the instruments' intellectual and visual transparency. Cost and size are more difficult to characterize: some instruments could only be acquired with public funds, whereas others were within the reach of individual scientists, both professional and amateur; some were incredibly large, whereas others, because of miniaturization of components, have decreased in size. Most routine scientific work has been done with discrete instruments that were commercially available, and most experiments have relied on apparatus assembled and modified for the problem at hand.

Surveying. Because American lands were vast and had to be laid out quickly, most surveyors used chains to measure distances and surveyors' magnetic compasses to measure angles; these instruments originated in Europe but were not sufficiently precise for most European demands. Rowland Houghton's "new theodolate" received a patent from the Massachusetts Bay Colony in 1735, thus becoming the first patented American surveying instrument. The vernier compass developed by mathematicians and instrument makers in and around Philadelphia late in the century allowed surveyors to compensate for magnetic deviation, trace lines relative to true north, and retrace old lines if the original deviation were known. Around the same time, in places where brass was scarce and accuracy was not so important, some Americans made do with locally made wooden compasses.

The surveyors' transit, introduced around 1830, proved useful on railroad surveys and other engineering projects. The solar compass, a compass with an attachment that allowed surveyors to determine north by reference to the sun, was devised by William A. Burt, a U.S. Deputy Surveyor working in Michigan in the 1830s.

Because surveyors often worked far from urban centers, American-made instruments tended to be robust and able to withstand rough usage. The American precision instrument industry flourished in the nineteenth century, thanks in large part to skilled immigrants, and it floundered in the mid-twentieth century when instruments became increasingly electronic.

Astronomers had long understood that light could be used to measure the distance from one point to another if its time of passage could be measured, but the development of electronics during and after World War II made the implementation of the idea practicable. The first electronic distance measuring instruments (EDMs) were produced in Sweden and South Africa, but found ready acceptance in the United States, originally in connection with such federal projects as positioning the tracking cameras at launch sites for missiles. American firms were producing EDMs by the 1970s. These were substantially more expensive than chains and tapes, but also substantially more accurate, and they quickly displaced traditional instruments. Development of the Global Positional System, which vastly extended the EDM principle, began in the late 1960s by scientists and engineers working under the aegis of the U.S. Navy and the U.S. Air Force.

Geodesy. Charles Mason and Jeremiah Dixon, who landed in Philadelphia in 1763 to establish the long-disputed boundary between Pennsylvania and Maryland, brought the best geodetic instruments then to be had: a zenith sector (for determining latitude) and a transit and equal-altitude instrument (for determining longitude and running the line). Left behind after their departure, these English instruments were used and copied by such American geodesists and instrument makers as David Rittenhouse and Andrew Ellicott. The U.S. Coast Survey initially purchased its precision instruments from the leading London shops. That situation began to change in the 1830s when the Survey hired William Würdemann, the first of a long line of German mechanicians who settled in the United States.

Geophysics. Alexander Dallas Bache, who became director of the Coast Survey in 1843, asked Joseph Saxton, a scientific mechanician who joined the Survey at this time, to develop self-recording tide gauges of the sort that had recently come into use in Britain. When Saxton gauges installed on the Pacific Coast captured evidence of the tsunami caused by a Japanese earthquake in December 1854, Bache understood that they could be used for seismology as well as for routine charting purposes.

Inspired by William Thomson in Scotland, who had found a mechanical way to analyze the periodic motion of tides as the sum of a series of periodic motions, William Ferrel of the U.S. Coast and Geodetic Survey (as the Coast Survey was known after 1878) designed a harmonic analyzer that served as a tide predictor. An improved version, designed by Rollin Harris, a Survey mathematician, and Ernst Fisher, the Survey's chief instrument maker, remained in use for some 50 years. Albert Michelson and Samuel Stratton, both of the University of Chicago, developed a harmonic analyzer for use with light waves.

Working at the Carnegie Institution's Seismological Laboratory in southern California in the 1920s, Harry Wood and John Anderson designed a torsion seismometer that was widely used for local earthquake observations; and Hugo Benioff designed a vertical seismograph and a strain instrument to record the stretching of the earth's surface. The Geophysical Research Corporation was organized in 1925 by Everette DeGolyer, a petroleum prospector who had become convinced of the practical advantages of seismic techniques and who wanted an American firm that would design and produce suitable instruments; a spin-off firm, Geophysical Service, Inc., would later become Texas Instruments. In the late 1950s, convinced that seismographs could detect underground nuclear bomb tests, the Department of Defense provided funds for a worldwide network of standard seismograph stations; these were equipped with instruments made by the Sprengnether Instrument Company in St. Louis and the Geotechnical Corporation in Dallas.

Because compass surveying was so important, Americans became interested in terrestrial magnetism at an early date, making observations with European dip circles and magnetometers. The Department of Terrestrial Magnetism, organized by the Carnegie Institution of Washington in 1904, designed and produced a universal magnetometer and deployed examples of the instrument around the world.

America's first significant gravity program, conducted under the aegis of the Coast Survey, began in the 1870s when Charles S. Peirce bought a German gravity pendulum and worked out protocols for its use. This program expanded greatly in the 1890s when, under the leadership of T. C. Mendenhall, the Survey developed a more portable pendulum apparatus and used it around the country. Learning that gravitational anomalies could indicate petroleum deposits, American prospectors began using European torsion balances in the 1920s; American instrument makers began designing gravimeters for this purpose in the 1930s. Realizing the importance of the geoid, the equipotential surface of the Earth's gravity field that best fits the global mean sea level (in a least-square sense), for controlling missiles and artificial satellites, the American military establishment bought many gravimeters during the Cold War and sent gravimetric teams everywhere except behind the Iron Curtain.

Meteorology. Immigrants of Italian descent whose families had been making glass instruments for several generations produced thermometers, barometers, and hydrometers used to test the purity of alcoholic and other liquids. The Army Signal Corps and its successor, the U.S. Weather Bureau, bought many instruments from Julien P. Friez, an Alsatian immigrant who established a machine shop in Baltimore. Working at the National Bureau of Standards in the 1930s, Harry Diamond

and his associates developed a radiosonde that met the needs of the U.S. Navy and that would become a standard Weather Bureau instrument.

Navigation. Americans made some compasses and backstaffs in the eighteenth century and some octants and sextants in the nineteenth century. The development of bubble sextants and other instruments for aerial navigation, which began around the time of World War I, was an international enterprise, but much was done by men affiliated with the U.S. Navy. In situations such as in an airplane, the natural horizon appears far below the navigator. To get around this problem, the navigator uses a bubble sextant and brings the image of a celestial body to the edge of the bubble, rather than to the horizon.

Astronomy. David Rittenhouse, a clockmaker in Norriton, Pennsylvania, built a transit instrument for observing the transit of Venus of 1769. Henry Fitz, a mechanic in New York, began making telescopes in the 1840s, selling them to wealthy colleges and individuals. Alvan Clark, an artist in Massachusetts, began grinding lenses around 1850. Five times during the next half century, Alvan Clark & Sons would produce the optics of what were then the largest refractors in the world. Warner & Swasey, a machine tool firm in Cleveland, made the mounts for many telescopes and transit instruments. Henry Draper's reflecting telescope with a primary mirror made of glass and coated with silver was made and publicized in the 1860s. Notable reflecting telescopes from the early twentieth century include the sixty-inch and one-hundred-inch instruments on Mount Wilson and the two-hundred-inch on Mount Palomar. In the late twentieth century, greater light-gathering power was gained by such sophisticated instruments as the Multiple Mirror Telescope on Mount Hopkins in Arizona and the Keck Segmented Mirror Telescope on Mauna Kea in Hawaii. The Hubble Space Telescope, built by the National Aeronautics and Space Administration (NASA) with help from the European Space Agency and operated by the Space Telescope Science Institute, was placed in a low earth orbit in 1990; it was built with a primary mirror of

2.4-meter aperture, as well as cameras, spectrographs, and photometers.

Americans have been involved with astronomical photography since the 1840s and with astronomical spectroscopy since the 1860s. Notable here were the diffraction gratings produced on engines designed and constructed by Lewis Rutherfurd of New York and Henry Rowland of Baltimore. Rutherfurd also designed micrometers for measuring astronomical photographs. The charge-coupled device that was invented by Willard Boyle and George Smith at Bell Laboratories in 1969 allowed the digitization of optical images and found countless scientific, military, and industrial applications.

Americans also worked to expand the range of light that could be captured and studied. Samuel Langley, a solar physicist at the Allegheny Observatory, invented the bolometer for determining the heat in the solar corona and initiated a major bolometric program at the Smithsonian Astrophysical Observatory. Charles Greeley Abbot, Langley's assistant and successor, developed a silver disk pyrheliometer for this purpose. William Coblentz at the National Bureau of Standards designed a thermopile and measured the infrared radiation from stars. Seth Barnes Nicholson and Edison Pettit, at the Mount Wilson Solar Observatory in the early 1920s, used a vacuum thermocouple to measure the heat from the moon and other celestial bodies. The Mauna Kea Observatory was established in 1967 and soon became a leading site for ground-based infrared astronomy. The Infrared Astronomical Satellite was launched in 1983.

In 1933, while using a powerful antenna that he built at the Bell Laboratories in New Jersey, Karl Jansky detected radio signals from various directions in space and discovered cosmic radio noise. Grote Reber, an engineer in Illinois, built an antenna in 1937 and made a survey of celestial radio signals. Public funds became available for radio astronomy after World War II. The National Radio Astronomy Observatory at Green Bank, West Virginia, was home to the world's largest fully steerable radio telescope. The Arecibo Observatory in Puerto Rico boasted the world's largest and most sensitive radio telescope.

Very-long-baseline interferometry became practical in 1967 when teams of scientists in the United States and Canada realized that tape recorders and precise oscillators eliminated the need for a cable connection between the two receivers. Irwin Shapiro, a physicist at the Massachusetts Institute of Technology (MIT), realized that by reversing the process he could use interferometric observations to measure the distance between two widely spaced receiving antennas, thereby solving a geodetic problem of importance to scientists and the military. The very long baseline array, 10 large antennas in New Mexico able to be moved from one place to another on the site, became operational in 1993.

Study of the other end of the spectrum began in the 1930s when scientists at the Naval Research Laboratory hypothesized that solar X-rays may cause disturbances in radio communication. Practical investigation that developed after World War II, when rockets became available for scientific research, expanded dramatically after the launch of *Sputnik* in 1957, the establishment of NASA in 1958, and the outpouring of government support for science. Uhuru, the first X-ray astronomy satellite, was launched in 1970. The first High Energy Astronomy Observatory was launched in 1977. Because the atmosphere prevents gamma rays from reaching the earth, this radiation must be studied from balloon and space-based platforms. Notable here is the Compton Gamma Ray Observatory that NASA launched in 1991.

In the 1960s, Joseph Weber, at the University of Maryland, built a device that he believed detected evidence of gravity waves produced in nuclear reactions in the sun. The Laser Interferometer Gravitational Wave Observatory, an exceptionally large and ambitious project that was begun in the early 1990s with support from the National Science Foundation, aimed to capture gravitational waves of cosmic origin.

Physics. Benjamin Franklin became fascinated with electricity in the 1740s and replicated basic European experiments. He invented the lightning conductor—the first American research instrument—and showed that it could be used to capture atmospheric electricity as well as to protect structures from damage by lightning. He also encouraged Americans to produce glass tubes and electrostatic generators much more cheaply than they could be had from abroad. Joseph Henry's powerful electromagnets, developed in the 1830s, were used for science and industry—making possible, in particular, Morse's electromagnetic telegraph. Robert Hare, professor of chemistry at the University of Pennsylvania, developed a powerful electrochemical battery that he termed a calorimotor. Edward S. Ritchie, a commercial instrument maker in Boston, developed powerful induction coils in the late 1850s. Working with a graduate student at the University of Chicago in the early 1900s, Robert Millikan built a device that measured the charge of an electron; the Millikan-Palmer oil drop apparatus, designed by Frederic Palmer Jr. of Haverford College and produced by CENCO, a Chicago firm that catered to the educational market, allowed students across the country to repeat this Nobel Prize–winning experiment.

In the 1870s, after men in Thomas Edison's laboratory modified a European Sprengel pump, they could pull a sufficient vacuum to create incandescent light bulbs that would not burn out so quickly. By improving the diffusion pump, Irving Langmuir, at the General Electric Research Laboratory, was able to produce high-vacuum tubes.

After becoming professor of physics at the newly established Johns Hopkins University, Henry Rowland built an apparatus with which he obtained a new determination of the mechanical equivalent of heat. In 2012, Rowland's instrument was in the National Museum of American History, as were several other instruments mentioned in this essay.

Albert Michelson, the first American to win a Nobel Prize in science, developed an interferometer to measure very small differences in the speed of light; in 1887, using an improved version of this instrument, Michelson and Edward Morley showed that the speed of light was the same in all directions and was not affected by the supposed ether. Michelson later developed echelon spectroscopes that surpassed the resolving power of the best diffraction gratings.

Ernest Fox Nichols, a physicist at Dartmouth College, built a radiometer in 1901 and used it to demonstrate that light exerts pressure. Working at the MIT Radiation Laboratory during World War II, Robert Dicke developed a microwave radiometer with lock-in detection that could measure the difference between the noise temperature of the source to which it was directed and the noise temperature of the instrument itself.

The mass spectrometer—the term covers a wide range of instruments used in numerous fields of science and industry—originated in England around the time of World War I, but inspired work around the world. Alfred Nier, in the United States, designed the mass spectrometers used for uranium separations during the Manhattan Project of World War II and those on the Viking Landers that NASA sent to sample the atmosphere of Mars.

Charles Townes and his students at Columbia University assembled the first operating masers and lasers in the 1950s and stimulated other physicists at academic, industrial, and military laboratories to further work in this field.

The case of nuclear magnetic resonance, the predecessor of magnetic resonance imaging, shows how difficult it can be to assign authorship or nationality to complex instrumentation. Much of the theoretical groundwork was done by I. I. Rabi at Columbia University in the 1930s and by Felix Bloch and Edward Purcell at the MIT Radiation Laboratory during and immediately after World II. Raymond Damadian, a New York physician, showed that the technique could be used to distinguish healthy cells from cancerous ones. The first practical instruments, however, were developed by EMI, a British electronics firm.

While working at the Los Alamos National Laboratory, Clyde Cowan and Frederick Reines imagined that the elusive neutrino hypothesized by Wolfgang Pauli would be created in a nuclear reactor and may be captured by a suitable detector; they succeeded at the Savannah River National Laboratory in 1956. In the early 1970s, building on his experience studying neutrinos produced at the Brookhaven National Laboratory, Raymond Davis Jr. installed a detector in a gold mine in South Dakota and captured evidence of solar neutrinos.

Eric Cornell and Carl Wieman and their students at the University of Colorado used such newly developed techniques as laser cooling and magnetic trapping to bring a gas of two thousand rubidium atoms to a temperature less than 100 billionths of a degree above absolute zero. This caused the atoms to lose their individual identities for a full 10 seconds, thus producing a quantum-mechanical state that Albert Einstein, building on the work of Satyendra Nath Bose, had predicted in the mid-1920s and that became known as the Bose-Einstein condensation. Cornell and Wieman (along with Wolfgang Ketterle, an MIT physicist who extended their work) shared the 2001 Nobel Prize in Physics for this accomplishment.

Chemistry. Henry Troemner, a German machinist who came to the United States in 1832 and settled in Philadelphia, produced balances for scientific, commercial, and governmental purposes. William Ainsworth began making balances in Denver in 1880, at the height of the Colorado gold rush. Both firms remained in business in the early twenty-first century.

The photoelectric spectrophotometer developed by Arthur Hardy, a physicist at MIT, and produced by General Electric, was highly favored by those who could afford it from the 1930s through the 1970s. The DU spectrophotometer that Arnold Beckman introduced in 1941 was much less expensive and found much more widespread use. The same was true with Beckman's pH meters.

Microscopes and Other Optical Instruments. Charles A. Spencer, a self-taught optician in Canastota, New York, in the mid-nineteenth century showed that Americans could produce microscope objectives comparable to the best European ones. German immigrants, such as Joseph Zentmayer in Philadelphia and Bausch & Lomb in Rochester, New York, also contributed to the American microscope industry. Under the leadership of Vladimir Zworykin, a Russian electrical engineer who had immigrated to the United States and kept up with European scientific literature, the Radio Corporation of America produced its first commercial electron microscope in 1940.

Following the development of long-range artillery in the late nineteenth century, military establishments around the world began purchasing such optical instruments of fire control as telescopes, binoculars, telescopic sights, range finders, and periscopes. In World War I, when German instruments could no longer be had, the American military commissioned these instruments from American firms.

Industry. The Constitution gave Congress the power to fix the standard weights and measures needed to ensure fair commerce. A small federal program, begun in 1807 under the aegis of the nascent Coast Survey, expanded in the 1830s when Congress recognized the need for standard weights and measures used in the several customhouses and in the several mints to regulate the coinage; this program eventually morphed into the National Bureau of Standards. Standards were also kept in the several states. Others were available for commercial purposes.

Although the use of instruments for quality and process control began in antiquity, the practice expanded markedly in the nineteenth century as industrialists came to rely on chemists and their instruments. Frederick Bates, a sugar scientist at the National Bureau of Standards, designed the first American saccharimeter, or polarimetric instrument suitable for sugar analysis, in the early years of the twentieth century; the actual instruments were manufactured in Prague. George Saybolt, organizer of the Inspection Laboratory of the Standard Oil Co. of New Jersey, designed the first important American viscosimeter (similar, it must be said, to European designs) in the 1880s. He later designed a colorimeter suitable for use with refined petroleum.

By the 1830s, Americans were manufacturing a host of rulers and other devices needed for practical tasks ranging from carpentry to commerce. As electricity gained industrial and commercial importance in the late nineteenth century, high-quality ammeters, voltmeters, light meters, and the like were produced by the Weston Electrical Instrument Company in New Jersey and the meter division of General Electric under the leadership of Elihu Thomson. American ingenuity can also be seen in the water meters, current meters, and other devices that monitor so many aspects of modern life.

Biology. *Drosophila melanogaster*, the common fruit fly, became a standard experimental organism around 1900, largely through the work of William Castle at Harvard University and Thomas Hunt Morgan at Columbia University. The Jackson Laboratory in Bar Harbor, Maine, bred most of the white mice used for biological research.

Polymerase chain reaction is a method of identifying and reproducing a gene or segment of DNA. The work was done by hand until 1983 when Kary Mullis began developing a way to automate the process. The resultant thermal cycler facilitated gene sequencing and analysis and brought Mullis a Nobel Prize. John Sanford and colleagues at Cornell University developed a gene gun that could deliver foreign DNA into living plants and formed a firm to market the device.

Psychology, Physiology, and Anatomy. At Columbia University in the 1890s, James McKeen Cattell developed tests to measure such mental traits as short-term memory and sensory perceptions. Henry Goddard worked with French intelligence tests and revised them for use in America. Lewis Terman and his colleagues at Stanford University issued the first edition of the "Stanford Revision and Extension of the Binet-Simon Intelligence Scale" in 1916. Robert Yerkes of the University of Minnesota produced the Army Alpha test that was widely used on recruits in World War I.

The Drunkometer described in 1931 by Rolla Harger of the University of Indiana was based on the discovery that the amount of alcohol in a sample of exhaled breath is proportional to the amount in a blood sample. The Breathalyzer designed in 1954 by Robert Borkenstein, also of Indiana, allowed fairly precise analysis and became widely used in the United States and elsewhere.

German colony counters, used by biologists to determine the number of bacteria colonies in a given sample, came to American attention at an early date. A. H. Stewart of Philadelphia developed a counter with electric illumination that became commercially available in the early twentieth century. The Quebec colony counter, although designed in Canada, was manufactured, distributed, and improved by the American Optical Company in Buffalo, New York. The Coulter counter, developed by Wallace Coulter in Chicago in 1948, counts the number of cells in a sample.

The AutoAnalyzer, the first successful commercial continuous-flow colorimeter, was invented by Leonard T. Skeggs, developed by Technicon, and introduced to the market in 1957. Although expensive, it was widely used in clinical laboratories to measure the amount of chemicals in blood, urine, or other body fluids.

The respiration calorimeter, a device that measured the energy provided by different foods and led to a growing awareness of food calories, was developed in the 1890s by Wilbur Atwater, a chemist at Wesleyan University in Connecticut who received support from the U.S. Department of Agriculture.

Body-scanning devices clearly exemplify the collaborative and international nature of many scientific instruments. In 1961, William Oldendorf, an American physician who was able to apply techniques from one field to another, applied for a patent for what would later become positron emission tomography (PET scan). Allan Cormack had the idea for what would become X-ray computed tomography (CAT scan) while working as a physicist in South Africa in the 1950s, but published his papers on the subject while working in the United States in the 1960s. He would later share the 1979 Nobel Prize in Physiology or Medicine with the British electrical engineer Godfrey Hounsfield.

Computers. The first American computers were developed during and shortly after World War II and used to solve problems of interest to the military establishment. The Harvard Mark I, built under the leadership of Howard Aiken, was funded by IBM and the U.S. Navy. ENIAC, the first fully functional electronic calculator, was built at the Moore School of Electrical Engineering at the University of Pennsylvania and used for complex calculations associated with the design of the H-bomb. The IAS machine was a computer built under the leadership of John von Neumann, a mathematics professor at Princeton University and the Institute for Advanced Study;

funds were provided by the Institute, the Atomic Energy Commission, and several military agencies. The development of the transistor in 1947 by three men at Bell Laboratories eventually made possible computers that were substantially more compact and more power efficient.

Particle Accelerators, Particle Detectors, and Atomic Clocks.
The high-voltage electrostatic generator invented by Robert Van de Graaff was widely used for nuclear physics research in the 1930s. Ernest Lawrence, at the University of California at Berkeley, developed the cyclotron and won a Nobel Prize in physics in 1939. A host of increasingly powerful accelerators of various types would follow in the postwar period.

In the 1920s, with funding from the Carnegie Corporation, Robert Millikan and Arthur Compton acquired portable electrometers and recorded the geographical variation of cosmic ray intensity. Donald Glaser built the first bubble chamber at the University of Michigan in 1952 and used it to detect cosmic rays. John Wood and other physicists working with Luis Alvarez at Berkeley followed this with a series of increasingly large and powerful chambers and filled with an assortment of liquids.

These extremely precise and accurate instruments use an atom or molecule experiencing a transition between two well-defined energy states to develop a constant-period electromagnetic oscillation. The ammonia absorption clock, announced by the National Bureau of Standards in 1949, was the first successful device of this sort. The Atomichron, the first commercial atomic clock, was conceived and promoted by Jerrold Zacharias at MIT, developed at the National Co. in suburban Boston, and unveiled in 1956.

Pedagogy.
David Rittenhouse built two remarkable orreries, a mechanical model of the solar system named after the Earl of Orrery, in the late eighteenth century; one of these elaborate astronomical clocks went to the College of New Jersey and one to the University of Pennsylvania. American firms began making globes, both terrestrial and celestial, in the 1830s. By midcentury, Ameri-

can educators could use American apparatus to demonstrate many mechanical, electrical, electromagnetic, and pneumatic phenomena. This market mushroomed in the latter years of the century as the population expanded, high school attendance increased even faster, and the laboratory method came into fashion. William Gartner, an instrument maker from Germany who established a shop near the University of Chicago, made a host of precision instruments for lecture rooms and laboratories. Other firms offered especially rugged instruments suitable for student use.

Many American research scientists became involved with educational matters in the aftermath of *Sputnik*. The Physical Sciences Study Committee, a project begun at MIT, aimed to motivate students to interrogate nature on their own, often with equipment they made themselves. With a ripple tank, the signature device of the committee, students could observe wave phenomena under various conditions.

[*See also* **Biological Sciences; Chemistry; Geophysics; Medicine; Medicine and Technology; Meteorology and Climatology; Physics; Science;** *and* **Technology.**]

BIBLIOGRAPHY

Robert Bud and Deborah Warner, eds. *Instruments of Science. An Historical Encyclopedia.* New York and London: Garland Publishing, 1998.

Deborah J. Warner

INTEGRATED CIRCUIT

See **Solid-State Electronics.**

INTELLIGENCE, CONCEPTS OF

In 1923, the American psychologist Edwin G. Boring (1886–1968), faced with the problem of defining intelligence, famously explained that intelligence is what intelligence tests test. This seemingly circular characterization was not an off-the-cuff remark, but a serious attempt to deal with one of the more vexing issues confronting early twentieth-century psychologists. From the development of the modern intelligence test by the French psychologist Alfred Binet in 1905, the practice of measuring intelligence had grown rapidly, especially in the Anglo-American world. However, little consensus had emerged concerning the actual capability that the tests sought to measure. Some argued that intelligence referred to an individual's potential for learning or adapting to new situations, others that it denoted ability to solve problems or generate abstract ideas. And all questioned whether it was one thing or many, produced by heredity or environment, and shared with animals or uniquely human.

Interest in some mental attribute characterizing an overall ability to think or reason can be traced, in the West at least, to Aristotle's definition of human beings as creatures who reason. Intelligence became a sustained topic of scientific inquiry during the late eighteenth century, when various European naturalists began to compare human beings systematically with other animals as part of their grand taxonomic projects. Such comparisons often focused on differences in intelligence and were linked by many to the notion of a "great chain of being" stretching from the simplest organisms through human beings and up to God. During the late eighteenth and early nineteenth centuries, various physical features were used to rank intelligence, with cranial capacity and brain weight predominating, especially through the labors of the American Samuel G. Morton (1799–1851). By measures such as these, a number of nineteenth-century scientists—including the so-called American school of anthropology, led by Morton—sought to demonstrate the inferiority of Africans by suggesting that their intelligence was closest of all human groups to that of the apes. By the century's end, however, this research program had largely been abandoned because variations within groups proved to be much more significant than differences among them.

Apart from comparisons of races or groups and medical assessments of profound mental de-

ficiency, intelligence as a personal and differential characteristic elicited little concern until the later nineteenth century. Before then, American mental philosophers spoke of the intellect largely in terms of the universal attributes of human reason and conceived of the mind as possessing a wide variety of faculties or powers, a certain subset of which could be grouped under the general term "intelligence." The advent of evolutionism, however, heightened the importance of an organism's overall mental power because it was considered a central factor influencing progressive adaptation to the environment. At the same time, the expansion of primary education rendered differences in individual intellect more visible, and the development of the new so-called "scientific" psychology in the late nineteenth century placed a premium on analyzing the mind in terms of quantitative methods and laboratory techniques.

The first successful psychological technique for quantifying differences in individual mental ability came early in the twentieth century. Asked in 1904 to participate in a French commission on children falling behind in the classroom, Binet and his colleague, Théodore Simon, created a scale of tests to identify those lagging in intellectual development. Focused on the higher mental processes and implicitly viewing intelligence itself as singular and quantifiable, the 1905 Binet–Simon Intelligence Scale was constructed as a sequence of age-related tasks that quantified the test-taker's intellectual level vis-à-vis his or her chronological peers. Using the Binet–Simon scale as a model, the American psychologists Henry H. Goddard and Lewis M. Terman began in the 1910s to refine the test further and link it more closely to arguments for the biological and inheritable nature of intelligence. Terman's 1916 Stanford–Binet test, which rapidly became the dominant instrument for assessing mental ability, reported scores in terms of a chronologically invariant measure, the intelligence quotient (IQ), a ratio of mental age to chronological age developed by the German psychologist William Stern. Frequently revised, the Stanford–Binet test remained a leading instrument for the individual measurement of intelligence at the end of the twentieth century, although after the 1930s it was somewhat eclipsed by two scales created by David Wechsler, the Wechsler Intelligence Scale for Children (WISC) and the Wechsler Adult Intelligence Scale (WAIS).

The outbreak of World War I gave American psychologists the opportunity to promote their services and products to a broader public. Using assembly line–like methods, which included the multiple-choice question and group testing, psychologists administered mental tests to approximately 1.75 million army recruits and cited this experience in the postwar period to convince schools and industry that intelligence testing offered an efficient means of assessing students and staff. Although critics such as Walter Lippmann and William C. Bagley challenged the tests as flawed and inimical to democracy and many business leaders shifted to measures of personality, by the end of the 1920s the concepts of intelligence and "intelligence assessment" were well entrenched. New instruments continued to be developed, and by the 1940s one test, the Scholastic Aptitude Test (SAT), had become a gatekeeper for admission to many colleges and universities. In addition, organizations such as the Educational Testing Service (ETS) of Princeton, New Jersey, which administers the SAT, flourished. Founded in 1947 under the leadership of Henry Chauncey, ETS quickly became the nation's leading testing agency. In 1998, with 2,300 employees and revenues of nearly $500 million, the nonprofit ETS administered the SAT and other tests to some 9 million persons, signifying the continued power of the concept of intelligence.

Discussions of intelligence have historically focused on two main issues. First, is intelligence one thing or many? Beginning in 1904, the British psychologist Charles Spearman used factor analysis to argue that intelligence consists of a single mental trait he called "general intelligence" (g). Edward L. Thorndike of Columbia University, by contrast, contended that intelligence is thoroughly heterogenous, whereas L. L. Thurstone of the University of Chicago argued that intelligence consists of a small number of relatively independent abilities. Later, Philip E. Vernon offered a hierarchical conception of intelligence in an attempt to achieve consensus on this issue. Nonetheless, alternative theories have abounded: in the 1960s and beyond, Joy P. Guilford argued that intelligence is composed of 150 independent factors; Howard Gardner discerned seven discrete types of intelligence; and Robert J. Sternberg contended that intelligence is triarchically organized. All the while, Spearman's (g) has continued to attract many proponents.

The second continuing issue focuses on the problem of nature versus nurture. The nineteenth-century interest in craniometry and the ranking of species and races had assumed that intelligence was both biological and inheritable. In the early twentieth century, advocates of eugenics fortified this conviction and enlisted it to justify a variety of social programs, from ability tracking in high schools to sterilization of "the unfit." Studies by the anthropologist Franz Boas in the 1910s, however, suggested that the environment plays a significant role in shaping intelligence, a position strengthened by subsequent work in the 1920s to 1940s, especially at the Iowa Child Welfare Research Station, and by a number of African American psychologists and sociologists. In the post–World War II era, both heredity and environment have received much attention. Research on identical twins has been interpreted by some psychologists as demonstrating the existence of a close connection between intelligence and heredity. At the same time, data on the worldwide increase in IQ scores (the Flynn effect) and the demonstrated influence of nutrition and home conditions on intelligence suggested the equally strong role of environmental factors. Although some interpreted the positive correlation of IQ scores with socioeconomic status as evidence of the meritocratic nature of Western societies, others argued that it underscored the culture-bound character of all intelligence-measurement instruments, especially when applied to the issue of IQ and race. Whether critical or supportive of intelligence testing, however, few have challenged the notion that intelligence itself is a characteristic of relevance to negotiating the contemporary world.

[*See also* **Anthropology; Boas, Franz; Eugenics; Evolution, Theory of; Military, Science** and Technology and the; Psychological and Intelligence Testing; Psychology; Race Theories, Scientific; *and* Social Sciences.]

BIBLIOGRAPHY

Carson, John. *The Measure of Merit: Talents, Intelligence, and Inequality in the French and American Republics, 1750–1940.* Princeton, N.J.: Princeton University Press, 2007.

Danziger, Kurt. *Naming the Mind: How Psychology Found Its Language.* London: Sage, 1997.

Degler, Carl N. *In Search of Human Nature: The Decline and Revival of Darwinism in American Social Thought.* New York: Oxford University Press, 1991.

Fancher, Raymond E. *The Intelligence Men: Makers of the IQ Controversy.* New York: W. W. Norton, 1985.

Gould, Stephen Jay. *The Mismeasure of Man.* New York: W. W. Norton, 1981.

Jacoby, Russell, and Naomi Glauberman, eds. *The Bell Curve: History, Documents, Opinions.* New York: Times Books, 1995.

Sokal, Michael M., ed. *Psychological Testing and American Society, 1890–1930.* New Brunswick, N.J.: Rutgers University Press, 1987.

Zenderland, Leila. *Measuring Minds: Henry Herbert Goddard and the Origins of American Intelligence Testing.* Cambridge, U.K.: Cambridge University Press, 1998.

John Carson

INTERNAL COMBUSTION ENGINE

Internal combustion engines detonate a fuel–air mixture in a combustion chamber and direct the force of the resulting explosion to move pistons or the blades of a turbine.

Although internal combustion engines have found their greatest use in motor vehicles—a broad category that includes road vehicles, aircraft, and motor boats—the earliest internal combustion engines provided stationary power for farmers and small manufacturers.

Patents and experimental prototypes date to the late eighteenth and early nineteenth centuries, but Etienne Lenoir, a Belgian engineer living in Paris, introduced the first commercially successful internal combustion engine in 1860. Lenoir's two-stroke engine resembled a horizontal double-acting steam engine that replaced steam with a mixture of coal gas and air ignited by a spark.

The German manufacturer Nicolaus Otto built a working four-cycle engine in 1876, which drew in the fuel–air mixture on the first stroke, compressed it on the second, combusted it on the third, and emptied the chamber of exhaust on the fourth. Compression significantly increased the engine's power. Because Alphonse Beau de Rochas had described the four-stroke engine in theoretical terms in 1862, courts ruled against Otto's attempts at patent enforcement, opening development to other manufacturers.

From the mid-1880s onward, inventors adapted internal combustion engines to road vehicles, launching that automobile age. In contrast to steam and electric motors, internal combustion engine designs advanced rapidly before World War I. Improved electric ignition systems, carburetors, cooling systems, pump-powered lubrication methods, and mechanically operated inlet valves boosted performance. These years also saw the creation of the diesel engine (in 1893), which achieves superior thermal efficiencies from lower-quality fuel using compression rather than a spark to ignite its fuel–air mixture.

After tetraethyl lead went into commercial production in 1926 as a gasoline additive, automakers experimented with higher compression ratios, steadily boosting the average horsepower of American automobiles from roughly twenty in the 1920s to roughly one hundred by 1950. Other interwar changes included greater use of six- and eight-cylinder engines, mechanical fuel pumps, downdraft carburetors, and automatic chokes. The same period also saw substantial advancements in gas turbine engines and jet propulsion for use in aircraft.

Since the end of World War II, major developments have included the turbo-diesel in Germany, the Wankel rotary piston engine and CVCC stratified-charge engine in Japan, and the first sophisticated electronic engine controls in the United States. After 1970, environmental regulations prompted greater attention to reducing emissions of harmful chemicals, spurring development of crankcase ventilation systems and catalytic converters. The oil shock of 1973 and the Corporate Average Fuel Economy (CAFE) standards of 1975 prompted increased attention to fuel efficiency.

[*See also* **Agricultural Technology; Airplanes and Air Transport; Motor Vehicles; Petroleum and Petrochemicals;** *and* **Steam Power.**]

BIBLIOGRAPHY

Flink, James J. *The Automobile Age.* Cambridge, Mass.: MIT Press, 1988.
Newcomb, T. P., and R. T. Spurr. *A Technical History of the Motor Car.* New York: A. Hilger, 1989.
Rae, John B. *The American Automobile: A Brief History.* Chicago: University of Chicago Press, 1965.

Christopher W. Wells

INTERNATIONAL GEOPHYSICAL YEAR

The International Geophysical Year (IGY) spanned 18 months from July 1957 through December 1958. During that time, 67 nations participated in an array of coordinated earth science observations and field experiments. The nongovernmental International Council of Scientific Unions and the United Nation's intergovernmental World Meteorological Organization coordinated research, which was executed by a complex network of public and private research institutions, with logistical support provided at times by militaries.

The origins of the IGY are as complex as the interests of the science diplomat Lloyd Berkner. Berkner first proposed the idea based on the International Polar Year (IPY) model (Needell, 2000). Since 1882 the IPY had encouraged Arctic and Antarctic researchers to coordinate explorations and share observations in the fields of meteorology, oceanography, glaciology, geomagnetism,

aurora and airglow, and latitude/longitude determination. In 1952, the International Council of Scientific Unions issued an invitation for participation in a third IPY to be conducted for the cyclical 11-year peak of solar activity.

Because of limited interest, a second invitation was sent, this time suggesting a more comprehensive study of the earth. The newly named IGY expanded its scope of exploring the earth's fluid envelope beyond the poles to eventually ranging into fields of interest including solar activity, cosmic ray physics, ionospheric phenomena, geomagnetism, seismology, and nuclear radiation (using "atomic tracers" to understand oceans and weather as well as measuring pollution from weapons and power plants). Researchers employed technologies iconic of mid-twentieth-century science—rockets, satellites, atomic tracers, submarines, and radio communications hardware—to collect environmental data that would (among many other things) eventually refine the design and use of these same instruments.

The IGY legacy might best be characterized by its unprecedented collection and archiving of scientific data. The United States and the Soviet Union offered to set up World Data Center repositories from which researchers could collect, analyze, and publish observations. In light of this, much ado was made of the virtues of scientific cooperation and the possibility of a Cold War détente. Yet contrary to what the popular press often implied, IGY scientists represented not their national governments but their respective scientific institutions, nor did the exchange of scientific data necessitate consolidation of national scientific programs.

Indeed, despite its scientific significance, the IGY has often functioned as little more than a parenthetical event precipitating the launch of *Sputnik* and the ensuing space race. In this context, scholars have emphasized how participation in the IGY benefited the national security state. First, it established the legality of satellite overflight. Second, the contribution of "scientific earth satellites" to the IGY allowed the United States and the Soviet Union to showcase their satellite networks as symbols of technological preeminence and scientific internationalism (McDougall, 1985).

Recently historians have begun to revisit the IGY, exploring its diverse scientific activities, the micropolitics of daily operations, its driving personalities, and its legacies. In this context, themes of nationality and the state remain important factors for analyzing contributions of long-established participants in international science as well as the decolonizing world (Launius et al., 2010).

[*See also* **Berkner, Lloyd; Meteorology and Climatology; Missiles and Rockets; Oceanography; Radio; Satellites, Communications; Science; Space Program;** *and* **Space Science.**]

BIBLIOGRAPHY

Launius, Roger, James Rodger Fleming, and David DeVorkin, eds. *Globalizing Polar Science: Reconsidering the International Polar and Geophysical Years.* New York: Palgrave Macmillan, 2010.

McDougall, Walter. *The Heavens and the Earth: A Political History of the Space Age.* New York: Basic Books, 1985.

Needell, Alan. *Science, Cold War, and the American State: Lloyd V. Berkner and the Balance of Professional Ideals.* Washington, D.C.: Smithsonian Institution, 2000.

Angelina Callahan

INTERNET AND WORLD WIDE WEB

The Internet operating in 2012 evolved over more than four decades. Its core protocols were developed during the late 1970s and early 1980s, but until the early 1990s public and commercial use of the network was restricted. In the 1990s the World Wide Web, a new network application, catalyzed the spread of the Internet into homes and businesses, and it rapidly subsumed or eclipsed other networks. By the 2010s Internet connections were being built not only into computers but also into cellphones, televisions, cars, DVD players, and aircraft.

Defining the Internet. The Internet has its origins in the concept of "internetworking." It is not a single network but the interconnection of many different networks operated all over the world by telecommunications companies, corporations, governments, universities, and other organizations. These constituent networks exchange data using the Transmission Control Protocol (TCP)/Internet Protocol (IP) protocol suite to deliver packets of data from one computer to another, in whichever networks they are part of. TCP/IP largely displaced other protocols widely used within local networks during the 1980s and 1990s.

The Internet's success is a result of its remarkable simplicity and flexibility. TCP/IP can transmit messages over many different kinds of physical connections, including satellite links, fiber-optic cables, modems, Ethernet cables, cellular data links, and local Wi-Fi networks. Many of these were unknown in the 1970s when the protocol suite was being designed. Likewise, many different applications exchange data over the Internet using TCP/IP. These applications include email, Web browsing, streaming video, file downloads, and video games. Because TCP/IP transmits data packages without reference to their content, the invention and spread of new applications has not required fundamental redesigns of the Internet.

Packet Switching. One common myth holds that the Internet was built for military use and was intended to survive a nuclear war. Neither of these assertions is literally true, although both reflect in distorted form elements of the actual history.

Communication across a network involves connecting between two or more points on the network. In traditional telephone networks this meant moving wires to create an electrical circuit between two telephones for the duration of the call. The Internet relies instead on the "packet-switching" approach. Messages are broken down into small packets of data, and each one is addressed to its recipient and dispatched separately, to be copied from one node to another on the way to their final destination. Their full route is not necessarily mapped out in advance, and if one link in the path stops working, the other computers will realize that packets have been delayed and try to resend them by an alternative and perhaps less direct path.

This approach to networking was first described by RAND Corporation analyst Paul Baran in 1961, under an Air Force contract to explore communications systems able to function during wartime disruption. However, the first packet-switched networks to be built were for civilian research use and no attempt was made to fortify them against attack.

APRPANET Origins. The first packet-switched network was the ARPANET, the direct predecessor of the Internet. It was named after its sponsor, the Advanced Research Projects Agency (ARPA), part of the U.S. Department of Defense. During the 1960s ARPA was the most important source of federal funds for academic research in the emerging discipline of computer science. Its resources were focused on researchers in a small number of institutions such as the Massachusetts Institute of Technology (MIT), the University of California at Berkeley, and the University of Michigan. ARPA directors were drawn from the same elite community and had the flexibility to fund projects of personal interest without a formal process of peer review. J. C. R. Licklider, an ARPA program director in the early 1960s, had formulated a novel vision of "man–computer symbiosis," the use of computers as tools to support creative work. This required the development of new, interactive computer systems quite different from those then in widespread use. Licklider also articulated a vision for what he had jokingly called an "intergalactic computer network," which, under the direction of his successors Ivan Sutherland and Bob Taylor, led to the planning and construction of ARPANET.

ARPA justified the network as a way to make the research groups it supported more efficient. Researchers had used ARPA funds to obtain a variety of different computer systems, running different operating systems and hosting a variety of different programs and information resources. A packet-switched national network would allow researchers to share resources, improve communication, and cut back on duplicate hardware and the effort wasted on redundant software. A terminal connected to a computer in Los Angeles could dispatch keystrokes to a computer at MIT and receive back its responses.

By the end of 1969, four sites had been connected to the network: the University of California at Los Angeles, the Stanford Research Institute, the University of California at Santa Barbara, and the University of Utah. By 1971 around two dozen sites were connected, many of which had finished writing and debugging the programs needed to make the connection useful.

A Network of Computers. Compared with other networks of the 1960s, the ARPANET was unusual in connecting computers together, rather than just linking a single computer to a number of terminals. It was also unusual in being designed to transmit data sequences of any kind, rather than being tightly engineered to support a particular application. Finally, it was unusual in supporting many different kinds of computers and so could not achieve compatibility by having each group use the same hardware and software. Instead, its designers created functional specifications for the exchange of data and left it up to each group to produce its own software to implement them.

One part of the network configuration at each site was standardized: the interface message processor (IMP). The IMP mediated between each network connection and up to four local computers. It was a specially adapted minicomputer built by Bolt, Beranek, and Newman (BBN) under a 1968 contract to implement and operate the network. Applications running on the main computers were largely shielded from the underlying complexity of the network.

Email: The Key Application. The leading ARPANET application was, unexpectedly, electronic mail (later usually shortened to "email"). The history of ARAPNET email gives us a good sense of the network's distinctive features. Email was already widely used to send messages between different user accounts on a single computer. "Network mail" service began in 1971 when Ray Tomlinson of BBN modified an existing electronic mail program to copy mail messages across the network and into the mailbox files of users on other computers. He built the new mail-delivery mechanism around an established protocol for file transfer. The Internet philosophy has always been to get something working quickly and let it evolve during use, rather than to spend years of committee work trying to produce a design to handle every possible need.

Subsequent email innovation took place on two levels: programs and protocols. Because the ARPANET relied on published protocols for compatibility, a particular user could shift at any time to improved mail programs with new features and still swap messages with colleagues. Mail programs evolved rapidly, for example, by integrating the sending and reading of messages within the same program and adding options such as reply, forward, or save messages.

The parallel evolution of email transmission protocols and formatting standards is preserved in a series of documents known as requests for comment (RFCs). These were the centerpiece of an exceptionally informal approach to standards within the ARPANET community, used not only for email but also for all aspects of the network. RFCs fulfilled many roles and were archived on the network itself. Some were humorous; others presented new ideas, discussion, proposed specifications, or agreed standards. The RFC process was open to all, although in this era the community interested in the network was small, technically minded, and fairly homogenous.

In some cases, such as the standardization of message headers such as "From" and "Date," the email-related RFC documents existed primarily to codify existing practice. Others defined new practices. The most important of these, issued in 1982 by Jon Postel, described what remains the standard Internet email transmission method: the Simple Mail Transfer Protocol.

Internetworking and TCP/IP. During the late 1970s and early 1980s the ARPANET became the foundation for the Internet. ARPA-funded researchers investigated new protocols capable of interconnecting networks based on communication media with very different characteristics such as radio links, fast local networks, and long-distance data lines. This required the reliable transmission of messages over unreliable links.

A group led by Vinton Cerf and Robert Kahn developed a more flexible replacement for the ARPANET's Network Control Protocol, used for communication between computers and the network IMPs. The two new protocols were TCP, run on computers connected to the Internet to control the sending and receiving of messages, and IP, to handle the routing of individual data packets over the network. After experiments during the 1970s, version 4 of the TCP/IP protocol suite was approved in 1980 and in 2012 still accounted for most Internet traffic.

Domain Names. IP introduced the basic system of numerical addresses still in use in the early twentieth century. An address such as 129.89.43.3 identifies both the computer to which a data packet should be sent and the network on which it could be found. However, ARPANET users were accustomed to specifying computers and associated email addresses by name, for example, "ucbvax" or "fred@princeton," rather than by numerical addresses. In 1983 this capability was adapted for the Internet through a new system of hierarchical domains. A new Domain Name System, and associated protocols for name resolution, replaced the centralized list of ARPANET-connected computers formerly maintained by a Network Information Center. "Top-level domains" had either national codes, such as .uk, or functional codes, such as .mil, .gov, and .edu. Organizations received domain names such as ibm.com or mit.edu to cover their assigned blocks of numerical addresses, with delegated responsibility to allocate those addresses to individual computers and operate the electronic directory that would translate names within that domain to numbers.

This system did still rely on centralized control to create new domains and allocate addresses to them. For many years Jon Postel ran what came to be known as the Internet Assigned Numbers Authority. As the Internet commercialized during the 1990s, the process was formalized, with a greater governmental role in control of the top-level domains and companies competing to offer domain registration services. Commercialization of the Internet sparked many legal disputes over ownership of valuable domain names such as "sex.com" or the registration of company names as domains by opportunistic "cybersquatters." Domain name resolution remains the most centralized aspect of the Internet, and failures of the "root nameservers" (some caused by hackers) have shut down much of the network on several occasions.

ARPANET to Internet. During the 1970s and early 1980s, the Department of Defense increasingly used the ARPANET for its own purposes and was eager to interconnect its current and planned data networks. This drove the rapid conversion of the ARPANET over to TCP/IP, driven by a 1982 deadline. TCP/IP's separation of different network functions into independent layers made it simpler than existing approaches, as well as more flexible with respect to network scale and underlying communications hardware and software. ARPANET users still struggled initially to implement the new protocols, in part because their "host" computers now had to deal with routing and transmission reliability issues from which the IMPs had previously shielded them.

By the mid-1980s the Internet consisted of two main networks: the research-oriented ARPANET and the more secure military MILNET. TCP/IP was gaining popularity rapidly, in part because high-quality implementations had been added to popular operating systems such as Unix. The National Science Foundation's NSFnet began operation in 1986 and was rapidly expanded and upgraded. With high-speed connections and international reach, it quickly replaced ARPANET as the hub of the Internet, expanding the three hundred networks connected to its "backbone" in mid-1988 to more than five thousand by early 1992. In 1990 the ARPANET was officially shut down.

Many of the most widely used Internet applications of the late 1980s and early 1990s were direct descendants of those pioneered on the ARPANET. These included electronic mail, file transfer (via the File Transfer Protocol), and remote login (via the Telnet Protocol). These established applications were joined by newer ones. Wide-Area Information Servers (WAIS) were used to search text databases held on different computers. Internet Relay Chat (IRC) allowed a small group of users to type messages simultaneously on a shared screen. News was a decentralized system of discussion groups begun on USENET, a separate network not based on Internet technologies. Its assimilation into the Internet exemplifies the latter's growing importance as a way of transmitting information between different kinds of networks via ad hoc "gateways."

Academic to Commercial. Beginning as early as the 1960s, politicians and computer experts began discussing the eventual creation of a universal data network open to all kinds of users. However, the Internet only emerged as a likely candidate for this role in the early 1990s. NSFnet's expansion involved the creation of the world's most advanced-speed long-distance data lines. The underlying technology was developed in collaboration among universities, the federal government, and companies such as IBM and MCI.

In 1992 Congress approved the interconnection of NSFnet with commercial networks, propelling the development of Internet service providers offering network access to businesses and individuals. By 1993 NSF decided to end its control of the Internet backbone, instead encouraging commercial networks to interconnect with each other directly. The NSFnet backbone was shut down in 1995.

Legacies of the Internet's Academic Roots. The speed with which the Internet displaced networks designed for commercial use is surprising. It inherited from its academic roots a characteristic set of strengths and weaknesses that together go a long way to explain not only the rapid success of the commercial Internet but also its persistent problems and weaknesses.

1. Unlike networks designed for commercial use, Internet technologies provided no way to charge users according to the network resources they consumed or compensate the providers of

network services. Thus, for example, Internet users were not billed by their network provider for use of intercontinental connections or for reading premium content online.

2. The Internet transmitted data packets without regard to their content, having been designed to serve a research community rather than to perform a specific task (often called the "end-to-end principle").

3. The Internet relied more on social mechanisms than on technical ones to provide security and eliminate troublemakers, having been designed for a homogeneous population of highly educated users given access through their employers or universities. This is why issues such as spam and attacks from hackers proved so hard to deal with after commercialization.

4. The Internet was designed to support many different machine types, achieving compatibility through shared protocol rather than shared code.

5. Any computer connected to the Internet could send data as well as receive it. This "peer-to-peer" operation meant that a computer could turn into a file server, an email server, or (later) a Web server simply by running a new program. Users were expected to publish online information and provide services to each other, rather than to rely on a single central collection of resources.

6. The Internet integrated many different communications media. The abstraction of TCP/IP from media-specific aspects of communication has made it possible to extend the Internet over many new kinds of links such as cellular telephones, Wi-Fi, and fiber-optic cables.

Invention of the Web. The ambitiously named World Wide Web, including prototype browser and server software, was created by Tim Berners-Lee, a British computer specialist with a PhD in physics working at the European particle physics lab CERN. The first website went live on 6 August 1991. The project never received an official budget, eventually scraping together a total of around 20 person-years of work. Not all of it was authorized and interns conducted much of the work.

The lack of resources forced Berners-Lee to rely on proven technologies and systems. Three crucial standards defined the Web, each of which survived essentially intact from his 1991 prototype into the browsers used by millions five years later.

1. The HyperText Transfer Protocol (HTTP), used by Web browsers to request pages and by Web servers to transmit them over the Internet. Like earlier application protocols, this ran over TCP/IP.

2. The HyperText Markup Language (HTML). The browser decodes the HTML instructions to display Web pages on the screen. HTML was based on the Standard Generalized Markup Language (SGML), intended as a universal and extendable way of encoding information on document structure within text files.

3. The Uniform Resource Locator (URL; originally Universal Resource Identifier). This built on the existing Domain Name System, adding prefixes such as nttp:// (newsgroups), ftp:// (file transfer), and telnet:// (remote logins) to identify specific services. The prefix http:// was used to identify Web pages.

As a hypertext system the Web was quite crude and ignored some of the questions then preoccupying researchers, such as what to do when a page that has been linked to is updated or changed. Berners-Lee's greatest contribution was to produce a simple method for an author to link to a page on the other side of the world as easily as to another part of the same document. Users clicking the link could follow it anywhere on the Internet.

The Web functioned initially as a catalog for the disparate resources scattered across the Internet. Browser programs could manipulate many of these directly and for the first few years were used largely to access existing resource types such as text files, gopher sites, and newsgroups.

Adoption of the Web. As browsers became more popular, the incentive to create Web pages grew. The process was catalyzed by the creation, in 1993, of the Mosaic browser by Marc Andreessen and Eric Bina at the University of Illinois Urbana–Champaign's National Center for Supercomputing Applications. This was set up as part of the National Science Foundation's push to make supercomputing resources available over the Internet. Unlike Berners-Lee's prototype, Mosaic was able to run on a variety of popular computer platforms. Internet software was uniquely easy to distribute because anyone with a network connection could download it.

For organizations already connected to the Internet it was easy to set up a website: just install a simple Web server program on an existing computer. The peer-to-peer architecture of the Internet made this much easier than in existing electronic publishing systems based on centralized and tightly controlled servers.

Commercialization of the Web.

A new browser, Netscape Navigator, was released in 1994 by a startup company and spread with enormous speed. Netscape's initial public offering of stock the next year was accompanied by a surge in its price, initiating an era of massive enthusiasm for, and investment in, almost anything that could be called an "Internet company."

The resulting flood of money drove rapid advances in Internet software, Internet retail, the use of the Internet to exchange information between businesses, the speed and reach of Internet backbone networks, and adoption by organizations of the Web and Internet email for internal and external communication. Microsoft made development of its own Web software a key priority, building Internet capabilities into its dominant Windows operating system.

The Clinton administration seized on the Internet, promoting policies to address a "digital divide" within society between those with access to it and those without. Bipartisan faith in the transformative power of the Internet made this a politically palatable proposal at a time when initiatives explicitly justified as subsidies for minorities or the poor were no longer acceptable.

By the time the bubble in Internet stocks collapsed in 2001, the Internet's position in the daily life of tens of millions of Americans was established.

Developments since 2000.

It is hard to do justice here to the history of the Internet in the early twenty-first century. Fundamentally, the Internet has been so thoroughly assimilated into so many different aspects of daily life that we generally think of it as "the Internet" only when something has broken and we are suddenly reminded of the material infrastructure underpinning online services. In 2012, specific Internet-enabled services were seen as having different modes of operation and cultural meanings: Kindle, Netflix, blogs, YouTube newspapers, eBay, electronic banking, Twitter, online dating, Google, online education, filesharing, instant messaging, Skype, Facebook, Expedia, or email. As the Internet became part of the foundation of everything, the history of the Internet became the history of everything.

However, one can at least point to a few underlying trends that supported these new services. One is the growing maturity of Web technologies as a platform on which computer applications of all kinds can be built. Early Web browsers loaded a screen of information, perhaps with some controls to select choices or enter information. Taking an action triggered the loading of another page. By 2005, applications such as Google Maps began to allow users to interact as smoothly and rapidly with Web-based applications as with traditional desktop applications. This reflects a gradual evolution of browsers, bandwidth, and standards rather than a sudden leap of the kind attempted in the late 1990s when programs written in the Java programming language were expected to revolutionize user interaction with websites.

Meanwhile, by 2012 personal computers were increasingly being replaced, or at least supplemented, for Internet access by tablet computers such as Apple's spectacularly popular iPad. This meant Web services were increasingly accessed through custom "app" programs optimized for computers of this type, rather than through a general-purpose Web browser.

Another underlying shift was the proliferation of Internet access methods and devices. In the mid-1990s someone in need of home Internet service would require a new and expensive computer hooked up to a slow modem. Over the next 15 years, high-speed Internet was delivered widely across the country to homes and small businesses, public spaces started to provide Internet access via Wi-Fi, and "smartphones" acquired powerful Internet capabilities. Cheap laptops, dubbed "netbooks," provided enough computer power for most Internet applications. Devices from digital picture frames to cars were connected to the Internet to exchange data, in what some predicted would soon become an "Internet of things" in which the objects around users were constantly interacting with humans and with other objects.

The most striking technical feature of the Internet remained its openness to new applications. A new application such as the Twitter message service, launched in 2006, could spread rapidly without any need for its creators to make deals with Internet service providers. The principle that these companies must treat data packets from all sources equally has been called "network neutrality" and the proper role of the Federal Communication Commission in enforcing it became the subject of considerable debate in recent years. New services delivered over the Internet competed with the telephone and cable television businesses, and so commercial interest of companies such as Time Warner Cable in limiting or controlling threats to their business is clear. This represents a fundamental challenge to the Internet's tradition as a universal communications infrastructure on which innovative applications and services can be built.

[*See also* **Airplanes and Air Transport; Computer Science; Computers, Mainframe, Mini, and Micro; Defense Advanced Research Projects Agency; Electronic Communication Devices, Mobile; Military, Science and Technology and the; Motor Vehicles; National Science Foundation; Satellites, Communications; Science; Software; Technology; Telegraph; Telephone;** *and* **Television.**]

BIBLIOGRAPHY

Abbate, Janet. *Inventing the Internet.* Cambridge, Mass.: MIT Press, 2000. Concise, reliable, and analytical scholarly history of the ARPANET and early Internet. Little coverage after 1990.

Abbate, Janet. "Privatizing the Internet: Competing Visions and Chaotic Events, 1987–1995." *IEEE Annals of the History of Computing* 32, no. 1 (2010): 10–22. Describes the creation of the NSFNET backbone and its replacement by commercially managed services, the key event in the reinvention of the Internet as a largely commercial network.

Aspray, William, and Paul Ceruzzi, eds. *The Internet and American Business.* Cambridge, Mass.: MIT Press, 2008. Various authors explore aspects of the 1990s adoption of the Internet by business, including my chapters on Web technologies, email, and the Internet search business.

Baran, Paul. "On Distributed Communications Networks." *RAND Corporation papers, document P-2626.* 1962. http://www.rand.org/pubs/papers/P2626.html (accessed 17 December 2012). One of a series of papers outlining the concepts behind packet-switched networks.

Berners-Lee, Tim, and Mark Fischetti. *Weaving the Web: The Original Design and Ultimate Destiny of the World Wide Web by Its Inventor.* San Francisco: Harper Audio, 1999. Bland, but valuable for context and detail.

Frana, Philip L. "Before the Web There Was Gopher." *IEEE Annals of the History of Computing* 26, no. 1 (2004): 20–41. Gopher flourished briefly just before the rise of the Web, providing some similar capabilities.

Gillies, James, and Robert Cailliau. *How the Web Was Born: The Story of the World Wide Web.* Oxford: Oxford University Press, 2000. Best history of the early Web (to 1995), focused on events at CERN and the first browsers with digressions into the broader history of the Internet and personal computing.

Hafner, Katie. *Where Wizards Stay Up Late.* New York: Simon and Schuster, 1996. Journalistic history of the ARPANET and early Internet. Readable, accurate, and less frothy than the title suggests.

Kita, Chigusa Ishikawa. "JCR Licklider's Vision for the IPTO." *IEEE Annals of the History of Computing* 25, no. 3 (2003): 62–77.

Licklider, J. C. R. "Man–Computer Symbiosis." *IRE Transactions on Human Factors in Electronics* HFE-1 (1960): 4–11. Seminal statement of the vision behind ARPA's early computing work.

Norberg, Arthur, and Judy E. O'Neill. *Transforming Computer Technology: Information Processing for the Pentagon, 1962–1982.* Baltimore: Johns Hopkins University Press, 1996. Covers ARPA's support for computing, with a chapter devoted to ARPANET.

Russell, Andrew L. "'Rough Consensus and Running Code' and the Internet-OSI Standards War." *IEEE Annals of the History of Computing* 28, no. 3 (2006): 48–61. Describes the distinctive, informal standard setting process of the early Internet, captured in the title, and contrasts it with the intergovernmental approach taken by supporters of the main set of rival standards.

Schewick, Barbara van. *Internet Architecture and Innovation*. Cambridge, Mass.: MIT Press, 2010. A painstakingly detailed examination of the development of the Internet as a communications architecture and its relationship to policy debates on network neutrality.

Waldrop, Mitch. *The Dream Machine: JCR Licklider and the Revolution That Made Computing Personal*. New York: Viking Press, 2001. Sprawling journalistic account, covering not only Licklider but also most of the key technologies and developments behind interactive, networked computing.

Zittrain, Jonathan. *The Future of the Internet—And How to Stop It*. New Haven, Conn.: Yale University Press, 2008. Argues for the "generative" nature of personal computers and the early Internet and against trends toward the imposition of increased commercial control. Insightful and includes fragments of recent history.

Thomas Haigh

IRON AND STEEL PRODUCTION AND PRODUCTS

Iron and steel are closely related substances. Iron is an element. Add carbon to it and it can become the compound known as steel. In the nineteenth century, steel was defined as having between 0.2 and 1.0 percent carbon. Anything with a carbon content of 2 percent or more was considered cast iron. Any pig iron processed down to contain a lower carbon content than steel was considered wrought iron.

Without readily available chemical testing, trained iron puddlers made the decision whether the metal had reached the stage from which they could turn it into their intended final product. Because it took an enormous amount of work to reach any required carbon range and an enormous

amount of skill to recognize when that range had been reached, steel was relatively rare around the world until the latter part of the nineteenth century. As iron and steel became more important to industrialization in both the United States and Great Britain, iron workers became labor's aristocracy.

Such skilled workers turned molten iron and steel into a wide range of metal products. As the name implies, cast iron was poured into molds shaped like the desired final product. This was done at a foundry. The most common use for cast iron was cook stoves. Wrought iron was generally produced in blacksmith shops by heating, hammering, and reheating metal on a forge. Horseshoes were a common product produced from wrought iron. Because of its rarity, steel was used mostly for small specialty products like cutlery, swords, or razors. In the era of mass production, separate industries arose to create steel end products like nails or tin cans.

After 1850, a series of changes in technology developed around the world made steel easier to produce and therefore cheaper. For example, iron makers began to heat the air blast before blowing it into the furnace where the metal was because this saved time. They also began to switch the fuel they used from charcoal (derived from wood) to coal. Although these innovations cut costs, the amount of pig iron that any puddler could work depended upon his strength. Therefore, the only way to expand operations was to build more furnaces and hire more expensive skilled labor. This led to a considerable amount of research about how to mechanize as much of the metal-making process as possible.

Bessemer Process and the Open-Hearth Furnace.
The invention of the Bessemer process not only worsened the position of labor in this industry, but also aided a rapid transition from iron to steel in many end-product categories. The Englishman Henry Bessemer is generally credited with the idea of making steel by blowing air through the molten metal to manipulate the amount of carbon it contained. Whereas Bessemer first did this in 1858, the American William Kelly obtained a patent for a similar process that he developed earlier, independent of Bessemer. Unfortunately,

he went bankrupt before doing anything with his invention. The first commercially successful production facilities with Bessemer converters appeared in the mid-1860s and early 1870s. Designed primarily by the American engineer Alexander Holley, they were arranged with the blast furnace very close to the Bessemer converter to make the process as continuous as possible, thereby increasing efficiency.

Holley designed his mills with the specific needs of rail production in mind. Thanks to the Bessemer process, steel quickly replaced iron as the primary metal for rails because it was cheaper than iron, and steel rails lasted longer once steel-making technology was perfected. As late as 1872, most steel rails used for America's railroads were imported. The spread of the Bessemer process in America changed that, thereby fueling a huge railway-building spree during the last decades of the nineteenth century. Steel also served as an essential component for building locomotive engines and cars.

The German engineer Charles Siemens first proposed the creation of a regenerative open-hearth furnace to make steel in 1861. The heat for melting the ore could come from outside the furnace, and that heat could be recycled, thereby making the process more efficient. The future mayor of New York City, Abram Hewitt, introduced the process to the United States in 1868. It caught on quickly because it allowed producers to use cheaper ore and get what they perceived to be a better quality product. Americans, particularly Andrew Carnegie, invested heavily in equipment that used this process because it allowed them to further increase output and lower prices.

The most important product coming from these new open-hearth furnaces was structural steel destined for skyscrapers and other parts of the modern infrastructure. As demand for steel from railroads slackened, the demand for steel to build in urban areas exploded. Using structural steel as a skeleton, construction firms could erect taller buildings than ever before in record time. Another important product made in open-hearth furnaces was armor plate for battleships. Not only did it protect American ships enough so that its navy could switch to an offensive strategy for the first time in that nation's history, but also American steel companies exported it.

International Competition and Technological Development. In America, iron was mostly produced by small firms scattered across many regions of the country. As steel gradually displaced iron, economies of scale tended to favor technologically adept firms like the one owned by Andrew Carnegie. If Carnegie could not convince a competitor to join forces with them, he often used his technological advantages to undersell that competitor and drive them out of business. J. P. Morgan organized the massive U.S. Steel Company in 1901 in large part so that his steel-related interests did not have to compete with those of Carnegie. By the early twenty-first century, international competition destroyed this once-powerful oligopoly.

The other two important steel-producing countries after the United States around the turn of the twentieth century were Great Britain and Germany. Although the British industry was in decline around 1900, it still produced armor plate for the navy and a lot of tin plate in Wales, used for cans and roof shingles. In 1904, the largest German steelmakers organized a syndicate. Going into World War I, no section of the industry in that country controlled the entire steel production process from beginning to end like in the United States. Despite overcapacity and older technology, Germany still played the role of America's greatest competitor in world markets during this era.

After World War I, the next great technical innovation in steel production was the basic oxygen furnace (BOF), developed by the Swiss engineer Robert Durrer in 1948 and commercialized by two Austrian companies in the early 1950s. This device was similar to the Bessemer converter except it used oxygen rather than air and the gas entered the furnace from the top rather than the bottom. Because oxygen contained no impurities, the time it took to make high-quality steel in a BOF dropped considerably compared with earlier methods of steel production. This technology spread quickly, especially in countries with growing steel industries like Japan. Because of the success of BOFs around the world, the Bessemer process almost entirely disappeared.

Technologies that improved steel production helped make more countries competitive with the United States, as did other nontechnological factors. Continual labor strife in the American steel industry also contributed to even American buyers turning to foreign steel during the 1960s and 1970s. By the mid-1970s, hundred-year-old American mills began to close, their product unable to compete with cheap imported steel. U.S. Steel's decision to buy Marathon Oil in 1982 served as a potent symbol that the American steel industry had no future.

In fact, the American steel industry did have a future. That future was based upon electric furnaces. Electric furnaces were not new. They first appeared during the 1920s in the United States, replacing the old crucible steel production process that had been used primarily for expensive specialty products. They are particularly good for melting scrap metal. Therefore, when scrap prices dropped in the wake of deindustrialization, electric furnaces spread quickly, especially during the 1990s. In the early twentieth century, most steel mills in the United States no longer made steel at all. Even older mills tended to use scrap because it was so much cheaper than making new steel from scratch.

In 2012, the most dynamic parts of the American steel industry were the minimills—small, nonunion operations that use electric furnaces to melt scrap steel and reform it in small batches into a variety of products. The best known of these firms was Nucor, based in North Carolina. Countries like China and Brazil were important steel producers in the early twenty-first century. The 2004 decision to sell a Cleveland rolling mill to a Chinese firm for disassembly and rebuilding in that country suggests that steel production changed in some ways in the early twenty-first century, but in other ways had not changed at all.

[*See also* **Building Technology; Electricity and Electrification; Empire State Building; Military, Science and Technology and the; Railroads; Shipbuilding; Skyscrapers;** *and* **Technology.**]

BIBLIOGRAPHY

Gordon, Robert B. *American Iron 1607–1900*. Baltimore: Johns Hopkins University Press, 1996.

Hoerr, John P. *And the Wolf Finally Came*. Pittsburgh, Pa.: University of Pittsburgh Press, 1988.

Misa, Thomas J. *A Nation of Steel: The Making of Modern America 1865–1925*. Baltimore: Johns Hopkins University Press, 1995.

Rees, Jonathan. *Managing the Mills: Labor Policy in the American Steel Industry during the Nonunion Era*. Lanham, Md.: University Press of America, 2004.

Rogers, Robert P. *An Economic History of the American Steel Industry*. New York: Routledge, 2009.

Tiffany, Paul A. *The Decline of American Steel: How Management, Labor and Government Went Wrong*. New York: Oxford University Press, 1988.

Jonathan Rees

J

JACOBI, MARY PUTNAM

(1842–1906), physician, feminist. One of the most distinguished woman physicians of the nineteenth century, Mary Putnam Jacobi is remembered for both her professional achievements in medicine and her activism on behalf of women. Born Mary Corinna Putnam in 1842, she was the first child of the publisher George Palmer Putnam. Raised in and around New York City, she grew up immersed in her father's world of literature. But as a young woman, she decided to pursue medicine and studied at three institutions, earning three degrees. She first attended the New York College of Pharmacy (1863) and then the Female (later Woman's) Medical College of Pennsylvania (1864). She worked briefly as an intern at the New England Hospital for Women and Children in Boston. She traveled to Paris in 1866 and became the first woman admitted to the École de Médecine, graduating with high honors and a second MD (1871).

She returned to New York City, where she practiced medicine for over 30 years. After establishing a private medical office, she turned her attention to teaching. She was a lecturer and professor at the Woman's Medical College of the New York Infirmary, her home base from 1871 to 1889, and a lecturer at the New York Post-Graduate Medical School from 1882 to 1885. She worked as an attending physician at the New York Infirmary for Women and Children, helped establish the Pediatric Clinic at Mt. Sinai, and served as a visiting physician at St. Mark's Hospital. Simultaneously, she carried out research, favoring scientific models of medicine and advocating experimentation, including vivisection. She was also a prolific writer, publishing nine books and over 120 articles on diverse subjects, including physiology, neurology, pathology, obstetrics, and gynecology, as well as nonmedical topics.

Jacobi's research focused heavily on women's health issues, including anemia, hysteria, uterine and ovarian diseases, and menstruation. Her famous essay, *The Question of Rest for Women during*

Menstruation (1877), won the Harvard Boylston Prize for 1876. Her medical work merged with political activism when she advocated for women's higher education and medical coeducation, labor reform, and, in the 1890s, woman suffrage. She also studied children's health, at times in collaboration with her husband, Abraham Jacobi, known as the "father of pediatrics." Together, they revised *Infant Diet* (1874), a manual of childhood nutrition. Childhood illness was both a professional cause and a personal tragedy for the Jacobis. Their only son, Ernst, died of diphtheria after Abraham Jacobi had spent years trying to understand the disease and stop its spread in New York City.

Mary Putnam Jacobi led the charge for women in the medical profession and was highly respected both within and beyond the women's medical community. She gained the support of many medical men, who admitted her into some of the profession's most important medical societies. Notably, she was the first woman admitted to the New York Academy of Medicine and chaired its section on neurology. She represented women's ability to practice medicine and conduct scientific research, equal to men, and believed science was an ally of women's rights. In 1906, she died as the result of a brain tumor.

[*See also* **Animal and Human Experimentation; Gender and Science; Medical Education; Pediatrics;** *and* **Physiology.**]

BIBLIOGRAPHY

Bittel, Carla. *Mary Putnam Jacobi and the Politics of Medicine in Nineteenth-Century America.* Chapel Hill: University of North Carolina Press, 2009.

Harvey, Joy. "Clanging Eagles: The Marriage and Collaboration between Two Nineteenth-Century Physicians, Mary Putnam Jacobi and Abraham Jacobi." In *Creative Couples in the Sciences.* Edited by Helena M. Pycior, Nancy G. Slack, and Pnina G. Abir-Am, pp. 185–195. New Brunswick, N.J.: Rutgers University Press, 1996.

Morantz-Sanchez, Regina. *Sympathy and Science: Women Physicians in American Medicine.* 2d ed. Chapel Hill: University of North Carolina Press, 2000.

Wells, Susan. *Out of the Dead House: Nineteenth-Century Women Physicians and the Writing of Medicine.* Madison: University of Wisconsin Press, 2001.

Carla Bittel

JEFFERSON, THOMAS

(1743–1826), lawyer, politician, and naturalist. A slave-holding planter from Virginia, Jefferson served as the principal author of the Declaration of Independence, as governor of Virginia (1779–1781), as the first U.S. secretary of state (1790–1793), as vice president under President John Adams (1797–1801), as the third president of the United States (1801–1809), and as the founder of the University of Virginia (1819). Throughout his life he avidly cultivated natural history, both personally and as a patron.

Although the deistic Jefferson, damned in some circles as a "howling atheist" for his free-thinking ways, found geology too speculative for his taste, he took a special interest in "the huge fossil Bones" being dug up both in Europe and in North America. Unlike some of his devout Christian friends, he believed they were the remnants of giant prehistoric elephants or mammoths, not antediluvian humans. In the 1780s Jefferson wrote his landmark natural history of Virginia, *Notes on the State of Virginia* (1787). In the section on anthropology, he discussed the "aborigines" of North America, especially the mystery of their origin. "Great question has arisen from whence came those aboriginal inhabitants of America?" he observed. He tried to solve the puzzle himself, but eventually gave up in frustration, concluding that the answer must be "Ignoro."

In 1796 Jefferson, who always professed a greater love for science than for politics, acceded to the presidency of the American Philosophical Society, calling his election "the most flattering incident of my life." The society, he said magnanimously, comprehended "whatever the American World has of distinction in Philosophy & Science in general." Despite his enthusiasm for science, during his two terms as president of the United States

Jefferson was limited by constitutional contraints from doing as much as he would have liked to promote science. As one disappointed surveyor noted at the time, "The President of the United States is both a lover of Science, and a man of science himself; but he has no power by his Constitution to aid any branch of philosophy, mechanics, or literature unless it be done at his cost" (Bedini, 1990, p. 323).

Jefferson did, however, find one way around the perceived constitutional ban against federal support for science. Shortly after becoming president in 1801—two years before purchasing the immense Louisiana territory from the French—he began making arrangements to send a "Corps of Discovery" to the Pacific Ocean. In negotiating with the Spanish and French for permission to pass through their lands, he emphasized the *scientific* nature of the expedition. In confidentially requesting congressional funding, however, he avoided the politically charged subject of federally funded science and stressed instead that information about the West was needed for *commercial* purposes, which the U.S. Constitution condoned.

To lead the corps, he chose his personal secretary, the 29-year-old Meriwether Lewis, a captain in the U.S. Infantry. Lewis, in turn, selected an old army friend, William Clark, a Kentuckian four years his senior, to serve as co-commander. "The object of your mission," Jefferson told Lewis, "is to explore the Missouri river, & such principal stream [sic] of it, as, by it's [sic] course & communication with the water of the Pacific ocean may offer the most direct & practicable water communication across this continent, for the purposes of commerce." As Lewis subsequently informed Clark, another primary goal was to make a "friendly and intimate acquaintance" with the Indians of the West in hopes of cultivating trade with them. "The other objects of this mission," he continued, "are scientific … to collect the best possible information relative to whatever the country may afford as a tribute to general science." During their heroic adventure, which lasted from 1804 to 1806, Clark focused on map-making and Lewis on natural history. He not only identified dozens of new species of plants and animals but also sketched elegant illustrations of them. Lewis and Clark returned to Washington as national heroes.

Shortly before completing his second term as president, Jefferson expressed his immense relief at leaving public office. "Nature intended me for the tranquil pursuits of science, by rendering them my supreme delight," he confided to a friend. "But the enormities of the times in which I have lived, have forced me to take part in resisting them, and to commit myself on the boisterous ocean of political passions" (Bedini, 1990, pp. 396–397). Back home in Virginia, Jefferson resumed his experiments in scientific agriculture and meteorology and began designing the buildings and grounds of the University of Virginia, regarded by many as the greatest architectual achievement in America. When the university opened in 1825, four of the six founding faculty members taught mathematics or science, including the professor of anatomy and medicine, who occupied the first full-time appointment in medicine in the country.

[*See also* American Museum of Natural History; American Philosophical Society; Higher Education and Science; Lewis and Clark Expedition; Meteorology and Climatology; Museums of Science and Natural History; *and* Science.]

BIBLIOGRAPHY

Bedini, Silvio A. *Thomas Jefferson: Statesman of Science.* New York: Macmillan, 1990.
Greene, John C. *American Science in the Age of Jefferson.* Ames: Iowa State University Press, 1984.
Jefferson, Thomas. *Notes on the State of Virginia,* edited by William Peden. Chapel Hill: University of North Carolina Press, 1955.
Thomson, Keith. *Jefferson's Shadow: The Story of His Science.* New Haven, Conn.: Yale University Press, 2012.

Ronald L. Numbers

JOBS, STEVE

See Computers, Mainframe, Mini, and Micro; Software.

JORDAN, DAVID STARR

(1851–1931), educator, naturalist, peace activist, and eugenicist. Jordan was born in a farmhouse near Gainesville, New York, in 1851. He graduated from Cornell University in 1872 with a Masters of Science in botany. The preponderance of his later scientific work, however, focused on ichthyology. Indiana University hired Jordan as professor of natural history in 1879 and six years later appointed him as university president. In 1891 Leland Stanford hired Jordan as the first president of Leland Stanford Jr. University. Jordan served as president of Stanford until he was appointed as the university's chancellor from 1913 to 1916. While at Stanford, Jordan established a highly influential school of ichthyologists, known for their large collection of specimens and prodigious rate of publication.

Through his classes, lectures, and writings on bionomics, heredity, eugenics, education, and the importance of peace, Jordan established himself as one of the leading intellectuals of the United States. Jordan's evolutionary theories stressed the importance of geographic distribution in speciation. He formulated "Jordan's Law," which states that the nearest related life form is not found in the same or a faraway region, but on the other side of a barrier. Jordan contributed to a growing American eugenics movement with his writings on the degeneration of isolated human populations, the eugenic dangers of war, and the development of Anglo-Saxon superiority. Jordan served as president of the Eugenics Committee of the American Breeders Association, director of the National Education Association, director of the World Peace Foundation, president of the Sierra Club, and president of the California Academy of Sciences. Jordan's theories on race, evolution, and dynamic individualism proved fertile ground for American naturalist authors such as Jack London, who referred to Jordan as a modern hero in his novel *Sea Wolf*. Jordan died in 1931 of unspecified causes at his home on the Stanford University campus.

[*See also* Biological Sciences; Botany; Eugenics; Evolution, Theory of; Geography; Higher Education and Science; Race Theories, Scientific; Science; Sierra Club; *and* Zoology.]

BIBLIOGRAPHY

Barton, Warren Evermann. "David Starr Jordan." *Science* 74, no. 1918. New Series (October 2, 1931): 327–329. An obituary published by one of Jordan's students.

Burns, Edward McNall. *David Starr Jordan: Prophet of Freedom*. Stanford, Calif.: Stanford University Press, 1953. One of the few biographies of Jordan currently out of print.

Jordan, David Starr. *The Days of a Man: Being Memories of a Naturalist, Teacher, and Minor Prophet of Democracy*. Yonkers-on-Hudson, N.Y.: World Book Co., 1922. Jordan's copious two-volume autobiography.

London, Jack. *The Sea Wolf*. New York: Penguin Books, 1965. First published in 1904 by the Macmillan Company.

Thurtle, Phillip. *The Emergence of Genetic Rationality*. Seattle and London: University of Washington Press, 2007. Jordan plays a key role as Thurtle investigates the changes in American society that led to genetic understanding.

Phillip S. Thurtle

JOURNALS IN SCIENCE, MEDICINE, AND ENGINEERING

Journals are the physical representation of a complex and essential process in the scientific enterprise. They are the point of contact among researchers, publishers, editors, readers, and repositories. As such, they represent the mechanism through which personal information is transformed into public knowledge. This transformation has several components including the formal presentation and reification of data; establishment of priority of discovery; certification of correctness and relevance; and dissemination and preservation for posterity. How all these functions are accomplished depends on such factors as the size and structure of the scientific community; the kind of research being done; and the economics and technology of publishing. Consequently, there are distinct periods in the history of periodicals; even the perception of what constitutes a scientific, technical, or medical journal is period specific.

Pre-1880. Henry Carrington Bolton's exhaustive 1897 *Catalogue of Scientific and Technical Periodicals 1665–1895* lists 8,556 unique titles, of which 1,140 were published in the United States. But these numbers and Bolton's list must be used advisedly as a characterization of nineteenth-century scientific publishing. First, only a fraction of these periodicals would be considered scientific journals in the modern meaning of the term. Bolton included agricultural and gardening magazines as well as botanical journals, magazines devoted to the gaslight industry as well as geology and engineering journals, and general-interest magazines as well as medical journals. He did so because all these varied publications contained technical and research articles. Additionally, most of the titles in his *Catalogue* were very short lived. The *Catalogue* reflects the fluidity of the nineteenth-century scientifically active population and its publishing efforts—energy more than strength. Additionally, until late in the nineteenth century, the proceedings and transactions of learned societies were the most abundant and important specialized scientific publications and Bolton usually excluded them from his *Catalogue*.

Publications of Learned Societies. Most scientific research of this time was pursued by independent investigators or by government employees; it was not college or university based. Although many scientists were also academics, they looked to their local societies rather than their employers for research support and to provide a publishing outlet. Society publications were of varying quality because they generally published only the discoveries of their members (and rarely communications from correspondents elsewhere). Consequently, they accepted whatever was offered to them to fill their pages and keep members happy. Society membership might be limited by race, class, gender, social, ethnic, or economic standing, but rarely scientific accomplishment or area of in-

terest. These periodicals were local in their character, reflecting a highly regionalized structure to the early to mid-nineteenth-century scientific community. The Boston Society of Natural History published work by Bostonians; similarly, the Academy of Natural Sciences of Philadelphia did the same for Philadelphians.

The parochial nature of much of American science during this period was not the only spur to local publication; researchers often preferred publishing their work in local magazines or society publications because it afforded them close control over the final appearance of their work. Oversight of illustrators and engravers was especially important for naturalists, where the illustration was as important as the text. This authorial control was so important to the renowned botanist and St. Louis resident George Engelmann that he published almost exclusively in the *Transactions* of the Academy of Science of St. Louis once that series began in 1860.

Specialized Journals. Although general magazines and local society publications dominated scientific publishing for much of the nineteenth century, there were also several attempts to establish independent, commercially viable journals. Specialized journals appeared earlier and with greater success in medicine, although these retained a regional flavor. Samuel Lathan Mitchill's *Medical Repository*, published in New York from 1797 to 1824, was the first commercial medical journal in the United States. Over the next two decades, it was joined by Benjamin Barton's *Philadelphia Medical and Physical Journal* (1804–1809), David Hosack's and John W. Francis's *American Medical and Philosophical Register* (New York, 1804–1814), *The New England Journal of Medicine and Surgery* (Boston, 1812–) founded by John Gorham, James Jackson, and John C. Warren, and *The Western Quarterly Reporter of Medical, Surgical and Natural Science* (Cincinnati, 1822–1823) founded by Daniel Drake. Each title published material reflective of the medical philosophy and theories of its editors.

George Baron's *Mathematical Correspondent* (New York, 1804–1806) and Robert Adrain's *The Analyst, or Mathematical Companion* (Philadelphia,

1814) were the earliest of several attempts at commercial, specialized science journals. Archibald Bruce founded the first nonmathematical specialized scientific journal in the United States, the *American Mineralogical Journal* (New York, 1814). All ventures of this type were short-lived and generally less successful than their medical counterparts. These early attempts at founding specialized scientific journals were evidence of an energetic and optimistic scientifically active population, but one that was not sufficiently developed to sustain ongoing publication of specialized titles. The country could provide neither the financial resources nor sufficient copy. Editors encountered fatal shortages of articles as well as of subscribers.

In engineering, the situation was somewhat different. Most engineering periodicals were company or industry organs whose intention was to publish authoritative work that simultaneously supported the goals of their sponsors. The principal exception was the *Journal of the Franklin Institute*, which advertised itself toward the end of the nineteenth century as "the only technological journal published without pecuniary interest."

The American Journal of Science and Arts (Silliman's Journal).

Yale geologist Benjamin Silliman took a new approach to the scientific journal that reflected a more accurate assessment of the capabilities of the scientific community of the day when he founded *The American Journal of Science and Arts* (New Haven, 1818 to the present). Silliman explicitly sought to overcome the regional bias in publishing and actively refused to ally his journal with any of the urban centers of scientific activity. During its early years Silliman's journal, as it was often known, published more geology and mineralogy than anything else, but also solicited and accepted articles in all fields, including engineering and practical arts. His approach seemed to be the best route to a successful, independent scientific publication at the time, but it was still a struggle. Silliman had to actively solicit quality articles and struggled to increase the number of subscribers; but at the same time, he had to try to limit the costs of publication,

which entailed reducing the number of pages as well as controlling the number of copies printed. Additionally, he had to discourage others who were considering founding rival journals—there was simply neither sufficient research to fill nor adequate demand to support more than one such journal. This situation held true until midcentury when an emerging national scientific community found its voice in the journal *Science*.

1880 to World War II: The Rise of National Professional Science Journals.

The closing decades of the nineteenth century saw some fundamental changes in American scientific activity, which resulted in turn in new models of scientific publishing. The emergence of modern research universities in the United States, beginning with Johns Hopkins University, combined with a new emphasis on laboratory science to create an exclusive scientific community composed of specialists who established national, professional associations. This new academic environment (combined with improved communication via rail) increased the need for specialized research journals and made possible the economic structures that made them viable.

Journals of this new type were more narrowly focused than their predecessors; they were written by and for specialists in their discipline; they were national, not regional in orientation and generally became the embodiment of American research in the discipline both within the United States and abroad. Initially, they were edited by one or two individuals who were able to solicit, evaluate, and select articles for publication. Those editors therefore possessed extraordinary authority as arbiters both of the content of a field and of the way in which research was presented. In this period it was typical for editors to serve for terms of a decade or more, during which their journals, and hence the disciplines they served, bore that individual's stamp. The prestige of a title and of the articles it contained depended on the reputation of the editor.

Specialization implied a narrow subscription base. Despite their national orientation, there were simply not enough subscribers to enable most journals of this type to survive as independent commercial publications. Instead, some of these journals, like the *American Journal of Mathematics* and *The American Chemical Journal*, began university-sponsored publications. Others, such as the *Journal of the American Chemical Society* (1879), were sponsored by professional associations and supported in part by membership dues. There were financial advantages to both models and many journals eventually combined them, becoming a society journal published by an academic press. Alternatively, some societies, including the American Chemical Society, chose to remain their own publisher. It often took years for these arrangements to sort themselves out. The *Physical Review*, for example, founded in 1893, passed through a variety of ownership and publishing models until it settled wholly under the auspices of the American Physical Society in 1913.

Post–World War II. The decades after World War II witnessed changes in the extent, function, and economics of science journals. Beginning after the war, and especially following the 1957 launch of *Sputnik*, the number and size of American research universities grew at an unprecedented rate. With the support of federal funds, this was the start of the large-scale collaborative research programs collectively known as "Big Science." Discipline after discipline began both to subdivide into ever more distinct specializations and to spin off entirely new fields of research. The combination of federal research dollars with a publish-or-perish requirement for secure employment led to an explosion in the number of research articles written and in the number of highly specialized journals being published.

The relationship among journals, authors, and readers also changed. Broadly speaking, it was becoming more important for researchers (and their careers) to publish articles in reputable journals than it was to receive and read those journals. With greater specialization, scientists became less interested in following (and less able to follow) developments outside their research focus. Within specific research areas, scientists developed more targeted and rapid means of disseminating relevant research results so that they would not have to suffer the sometimes years-long delay between the completion of research and the publication of an article. Preprints, circulating copies of accepted article drafts prior to their publication, are perhaps the best example of alternative means of dissemination of research (and credit for that research) among a specialist community. Journals were becoming more important as publications of record and as a means of disseminating knowledge outside the research community than they were for communication among practitioners within that community. Among scientists, then, the value of a journal did not lie solely, or even primarily, in its connection to readers. Perhaps for these reasons, it was becoming less important to personally subscribe to and receive relevant journals. For university-based scholars it was often enough for the journals to be in the library.

The management of journals and consequently the nature of their authority also changed during this period. Gradually, and at different times in different disciplines, journal editors found that they lacked the expertise and the time to evaluate all the submissions they received and they modified or expanded their existing editorial apparatus accordingly. The most consequential innovation was the introduction of peer review into the article evaluation process. In general, journals adopted peer review as a work management practice as a sort of extension of editorial expertise. But it soon took on an unanticipated role as the publishing practice that defined a journal as scholarly. Rather than simply being an aid to making an informed editorial decision, peer review came to serve as the means by which the originality, accuracy, and significance of an article were established. Over time, the reputation of a journal was predicated less on the judgment of its editor and more on the rigor of its peer review process.

Economics of Journal Publishing. This period also saw the general adoption of two transformative practices in the economics of publishing, page charges and institutional pricing. The American Physical Society first began charging authors to publish their articles beginning in 1930. The practice spread, especially after the war, and even more after 1961 when federal agencies explicitly permitted the use of federal research funds to pay them. By 1976, the practice was widespread among society-sponsored journals, but was less used by independent journals published by university presses. Page charges frequently paid for around half the cost of publication, allowing a journal to keep its individual subscription rate relatively low and consequently the number of subscribers relatively high.

The practice of charging libraries and other institutions a higher subscription rate seems to have come about gradually and to have been adopted at different times by different journals. In some cases, the original idea was that library copies should cost more because they were read by multiple users. In others it appears to have been an extension of the idea that nonmembers paid a higher subscription rate for a society publication than members.

In a different economic climate, university presses beginning in the early 1970s were increasingly required to be more financially responsible, if not self-supporting. At the same time, the scholarly output of American scientists continued to increase. This disparity between funding for research and publication led to the introduction into the American scientific, technical, and medical publishing of the practice of escalating subscription prices that had been pioneered by commercial science journal publishers in Germany in the 1930s. For a number of reasons, libraries were more resistant than individuals to canceling subscriptions in response to higher prices. Combined with page fees, high institutional subscriptions meant that for the first time in American history, the publication of scientific periodicals could be profitable on a large scale. (This did not mean that individual titles were guaranteed to succeed.) Whether they were American or European, the for-profit publishers of American scientific, technical, and medical journals sent institutional prices for journals through the roof, provoking a budgetary upheaval in research libraries that became known as the serials crisis. Some authorities assert that the potential for profit has created a perverse quantity-over-quality incentive, leading to the publication of an increasing proportion of articles of little or no research value.

The Twenty-first Century. In the early twenty-first century there was a growing recognition that the existing model of scientific, medical, and technical journal publishing was unsustainable. It was being critiqued on several fronts. Research (and several noteworthy cases of fraud) had called into question the efficacy of peer review as a process to identify and certify the most important research. With the dominance of the vast undertakings of Big Science, the very concept of authorship became a topic of debate. When dozens or even hundreds of authors are listed, as they frequently were, for a brief article, what does authorship mean? Who deserves the credit for the article, who is responsible for its accuracy, and who takes the blame if fraud is alleged? It was initially hoped that electronic publishing would lower the subscription cost, but that was based on the assumption, since proved incorrect, that there was a correlation between publishing costs and subscription costs. Some research also suggested that electronic presentation makes articles so easily discoverable that it paradoxically narrows the breadth and scope of research in the published literature. Furthermore, efficiencies of scale in publishing have led to the dominance of a relative handful of international megapublishers.

In the early twenty-first century, many alternatives were explored, including self-publishing as well as several variants on open-access publishing. In general, open-access journals transfer publishing costs to entities other than the reader or a subscribing institution. As a result, the journal is made freely available to all potential readers. The concept of the journal or journal issue as discretely identifiable objects was also being called into question. Some publishers promoted the idea that the impact and authority of an article depend more on the publisher than on which of the publisher's journals contained the article. Alternatively, some repositories were developed with the idea that the authors' institutional affiliations matter more than the publishing journal or press.

[*See also* **Internet and World Wide Web; Popularization of Science; Printing and Publishing; Research and Development (R&D); Science Journalism; Silliman, Benjamin;** *and* **Societies and Associations, Science.**]

BIBLIOGRAPHY

Abel, Richard E., and Lyman W. Newlin, eds. *Scholarly Publishing: Books, Journals, Publishers, and Librarians in the Twentieth Century.* New York: John Wiley & Sons, 2002. Particularly useful because it offers interpretations of a moment of change from the diverse perspectives of participants.

Ad Hoc Committee on Economics of Publication, D. H. Michael Bowen, chairman, ed. *Economics of Scientific Journals.* Bethesda, Md.: Council of Biology Editors, 1982.

Baatz, Simon. "Squinting at Silliman: Scientific Periodicals in the Early American Republic." *Isis* 82 (1991): 223–244.

Biagioli, Mario, and Peter Galison, eds. *Scientific Authorship: Credit and Intellectual Property in Science.* New York and London: Routledge, 2003. Essays by the editors were especially useful for this article.

Bolton, Henry Carrington. *A Catalogue of Scientific and Technical Periodicals. 1665–1895. Together with Chronological Tables and a Library Check-list.* Washington, D.C.: Smithsonian Institution, 1897. Available online at http://www.hathitrust .org/ (accessed 10 April 2012).

Burnham, John C. "The Evolution of Editorial Peer Review." *JAMA* 263 (1990): 1323–1329. Seminal research on this topic, this article is part of a special issue comprising papers presented at the First International Congress on Peer Review in Biomedical Publication.

Elliott, Clark A. *History of Science in the United States: A Chronology and Research Guide.* New York and London: Garland Publishing, 1996. The essential timeline and introduction to research in the field. For the nineteenth century, specifically, see also his website: *History of Science in the United States: Research Aids for the Nineteenth Century.* http:// historyofscienceintheunitedstates-19thcentury. net/ (accessed 10 April 2012).

Evans, James A. "Electronic Publication and the Narrowing of Science and Scholarship." *Science* 321, no. 5887 (2008): 395–399. A fascinating and suggestive study of how behavioral responses to electronic publishing have limited the scope of literature searches.

Greco, Albert N., Robert M. Wharton, Hooman Estelami, and Robert F. Jones. "The State of Scholarly Journal Publishing: 1981–2000." *Journal of Scholarly Publishing* 37 (2006): 155–214.

Hartman, Paul. *A Memoir on the Physical Review: A History of the First Hundred Years.* New York: American Institute of Physics, 1994.

Jagodzinski, Cecile M. "The University Press in North America: A Brief History." *Journal of Scholarly Publishing* 40 (2008): 1–20.

Science 209, no. 4452 (2008). This is the journal's centennial issue, consisting of articles on its history.

Daniel Goldstein

JUST, ERNEST EVERETT

(1883–1941) cell biologist and early African American pioneer in science. Ernest Everett Just was a pioneering American scientist whose discoveries in biology paved the way for modern understandings of cell behavior. Just is also recognized as an early African American trailblazer in science. Born in Charleston, South Carolina, in 1883, Ernest Just's father died young and his mother, Mary, struggled to support the family. A leader in Charleston's black community, she profoundly influenced Ernest. Because of the dearth of opportunities in the South, Just attended Kimball Union Academy in New England and graduated early. In 1903 Just enrolled at Dartmouth College in New Hampshire, where he pursued his interest in biology. Just graduated first in his class in 1907 with a degree in zoology and was awarded honors in botany, history, and sociology.

After Dartmouth, Just took a position at Howard University, a historically black university in Washington, D.C., and by 1908 became the head of its biology department, a position he held until his death in 1941. Because of racial discrimination, blacks could only obtain faculty positions at historically black colleges and universities. However, by 1909 Just began to spend summers conducting research at the renowned Woods Hole in Massachusetts, where he thrived as a result of being part of this research community. At the suggestion of his mentor at Woods Hole, Frank R. Lillie, Just eventually pursued doctoral work at the University of Chicago. After taking a year's leave from Howard, Just received his PhD in 1916. A prolific scientist, Just's work received wide acclaim. By 1920, Just had been elected to several scientific societies including the American Society of Zoologists and the American Association for the Advancement of Science. Yet, Just's success did not distract him from the importance of working to expand opportunities for other African Americans in the academy: he helped found Omega Psi Phi, the first predominantly African American fraternity to be founded at a historically black university.

Despite international recognition for his groundbreaking work, Just encountered severely limited opportunities as an African American scientist in the United States. He attempted to secure faculty positions at U.S. universities but was denied. After research stints in Germany, France, and Italy, Just eventually immigrated to Europe. In 1939, Just published *The Biology of the Cell Surface*. This book was published a year before Germany invaded France, where Just lived. Just spent a short time in an internment camp before the U.S. State Department freed him. Just returned to the United States in 1940 in declining health and died of cancer in 1941.

[*See also* **American Association for the Advancement of Science; Biological Sciences; Botany; Higher Education and Science; Race and Medicine; Science;** *and* **Zoology.**]

BIBLIOGRAPHY

Kessler, James. H. *Distinguished African American Scientists of the 20th Century.* Westport, Conn.: Greenwood Publishing Group, 1986.

Manning, Kenneth. R. *Black Apollo of Science: The Life of Ernest Everett Just.* New York: Oxford University Press, 1983.

Mark Robinson

K

KÁRMÁN, THEODORE VON

(1881–1963), aeronautical engineer and applied mathematician. Born in Budapest, Hungary, and educated there and in Germany, he was the son of Maurice von Kármán, a professor of philosophy and educational reformer, and Helen Kohn. He received his BS in mechanical engineering in 1902. He taught hydraulics at the Palatine Joseph Polytechnic in Budapest until 1906, when he went to the University of Göttingen. There he studied under Felix Klein, Ludwig Prandtl, and David Hilbert. He obtained a PhD in engineering in 1908, with a thesis on the strength of columns in large structures.

Throughout his career, von Kármán combined industrial consulting, university teaching, and a practical approach to the application of mathematical theory to engineering problems. After witnessing an airplane trial in 1908, he became fascinated by aerodynamics. He became professor at the Technical University in Aachen, Germany,

in 1913 and, with an interruption caused by World War I, continued to direct the Aachen Aerodynamics Institute until 1930. He investigated laminar fluid flow and discovered Kármán vortices.

After World War I, von Kármán encouraged a renewal of international scientific exchange by sponsoring the first International Congress on Aerodynamics and Hydrodynamics in Austria in 1922, Holland in 1924, and many more in later years. At the 1924 meeting, he met Robert Andrews Millikan, president of California Institute of Technology in Pasadena (Caltech), who invited him in 1926 to consult on the design of a wind tunnel for Caltech. After two years of splitting his time between Caltech and Aachen, in 1930 von Kármán accepted a position in Pasadena as director of the Daniel Guggenheim Aeronautical Laboratory. He remained based in the United States for most of the rest of his career and he became a U.S. citizen in 1936.

During the 1930s and 1940s von Kármán helped design commercial airplanes for Douglas, including the DC-3. He and his students designed and built jet engines and rockets. He worked with the Air Corps Jet Propulsion Research Project early in World War II and helped found Aerojet Engineering Corporation. In 1944, Caltech's Jet Propulsion Laboratory was established with von Kármán as director. During the war and into the 1950s von Kármán also consulted with the U.S. government and with the North Atlantic Treaty Organization (NATO) on aircraft, rockets, and the future of warfare.

Von Kármán never married, but he maintained close family ties. When he moved to the United States, in part to escape growing pro-Nazi and anti-Semitic feeling in Germany, he brought his sister and mother with him. His sister, Dr. Josephine de Kármán, helped him entertain and she also organized his affairs, thus freeing him to concentrate on his work.

Theodore von Kármán died in Aachen, where he had traveled for a rest cure at a spa.

[*See also* **Airplanes and Air Transport; Engineering; Mathematics and Statistics; Millikan, Robert A.;** *and* **Missiles and Rockets.**]

BIBLIOGRAPHY

Dorn, Michael H. *The Universal Man: Theodore von Kármán's Life in Aeronautics.* Washington, D.C.: Smithsonian Institution Press, 1992.
Dryden, Hugh L. "Theodore von Kármán." *Biographical Memoirs* 38 (1965): 344–384.

Gregory A. Good

KELLER, HELEN

(1880–1968) was born in Tuscumbia, Alabama. Keller lost her vision and hearing at the age of 19 months. As she approached school age, unable to communicate well and increasingly unruly, her parents contacted Alexander Graham Bell for advice. This ultimately led to their engaging Anne Sullivan, a graduate of the Perkins School for the Blind, as Helen's teacher.

Sullivan began by spelling words on Keller's palm. At first these had no meaning, but in a sudden realization Keller connected the feeling of water with the letters w-a-t-e-r and proceeded quickly to learn language. In 1888, after a year together, teacher and pupil moved to Boston to be near the Perkins School. Perkins Director Michael Agnanos publicized Keller's accomplishments and the teenager developed friendships with Mark Twain and other notables of the time.

Keller learned to read raised print but came to advocate the braille code of embossed dots. Influenced by Bell, she abjured American Sign Language and instead took speech lessons. She matriculated at Radcliffe in 1900, graduating *cum laude* in 1904. Throughout her schooling Keller depended on Sullivan to interpret printed matter and lectures.

Keller continued living with Sullivan after Sullivan's 1905 marriage to the socialist literary critic John Macy. Her own socialist beliefs were important to Keller; she joined the Socialist Party in 1909 and the Industrial Workers of the World in 1912. Briefly engaged to a comrade in 1916, the relationship was ended by her mother. Keller later expressed regret that she had never married.

With Macy's assistance, she published *The Story of My Life* in 1903. Subsequent writings dealt not only with her life and disabilities but also with her political and religious beliefs. In the early 1920s, after their income from Keller's writing had diminished and Sullivan's marriage had deteriorated, Keller and Sullivan developed a vaudeville act. On the Orpheum circuit Keller met Sophie Tucker and other entertainers.

In 1924 Keller joined the staff of the recently formed American Foundation for the Blind and for much of the remainder of her life she toured on its behalf. After Sullivan's death in 1936, Polly Thomson, who had been part of their household since 1914, became Keller's assistant. Between 1946 and 1957 Keller and Thomson made seven overseas trips. Thomson died in 1960, Keller suffered a stroke in 1961, and a nurse-companion lived with Keller through the end of her life. She died in Easton, Connecticut.

[*See also* **Bell, Alexander Graham; Blindness; Assistive Technologies and; Deafness; Ophthalmology;** *and* **Optometry.**]

BIBLIOGRAPHY

Brooks, Van Wyck. *Helen Keller: Sketch for a Portrait.* New York: Dutton, 1956.

Nielsen, Kim E. *The Radical Lives of Helen Keller.* New York: New York University Press, 2004.

Edward Morman

KILBY, JACK

See Solid-State Electronics.

KILLIAN, JAMES RHYNE, Jr.

(1904–1988), science and education administrator, was best known for his service as president of the Massachusetts Institute of Technology (MIT) from 1948 to 1959 and as the first special assistant to the president for science and technology, or science adviser, under Dwight D. Eisenhower while on leave from MIT from 1957 to 1959. In both positions he helped shape American science, technology, and educational policy during the Cold War despite his lack of advanced training in science or technology.

Born in Blacksburg, South Carolina, Killian went to MIT and received a bachelor's degree in management in 1926. He started his career at the MIT magazine *Technology Review* after graduation, becoming editor in 1930. In 1939 he became executive assistant to the institute's physicist president Karl Compton, worked as the liaison between the institute and the radar-making Rad Lab during World War II, and eventually succeeded Compton in 1948, a remarkable recognition of his management skills.

In 1953–1954, Killian played his first major national policy role when he chaired the Technological Capabilities Panel of the Science Advisory Committee of the Office of Defense Mobilization (ODM-SAC) in the Executive Office of the President. With Eisenhower's approval, the panel's recommendations resulted in a decisive acceleration of the U.S. missile programs and the launching of reconnaissance satellites and the U-2 spy planes. This performance also led Eisenhower to tap Killian to head the influential President's Board of Consultants on Foreign Intelligence Activities.

Killian was thrust onto the national stage in 1957 when Eisenhower appointed him as the White House's first full-time science adviser to help coordinate American science, technology, space, and defense policy in the aftermath of the Soviet launching of the satellite *Sputnik*. Killian was also elected chairman of the newly reconstituted President's Science Advisory Committee (PSAC) on the basis of the old ODM-SAC. He soon wore a third hat as chairman of the new Federal Council for Science and Technology that consisted of representatives from major federal agencies. With vital support from the PSAC, Killian played a key role in the establishment of the National Aeronautics and Space Administration, the reorganization and centralization of the Department of Defense, and the launching of the negotiations that would eventually lead to the Limited-Test-Ban treaty in 1963. A political moderate and an adroit interlocutor at the interface between science and government, Killian gained Eisenhower's trust and helped open the golden age of presidential science advising.

Killian returned to MIT in 1959 as chairman of its board but remained active in Washington as a member (and later consultant) of the PSAC and as chairman of the President's Foreign Intelligence Advisory Board under John F. Kennedy. In the 1960s and 1970s, Killian, as chairman of the Carnegie Commission on Educational Television and then of the Corporation for Public Broadcasting, devoted much of his energy to the establishment of public broadcasting in the United States. In 1973, after President Richard Nixon abolished the PSAC over policy disagreements, Killian chaired a "blue ribbon" committee of the National Academy of Sciences, whose report was in part responsible for the establishment of the White House Office of Science and Technology Policy in 1976.

[*See also* **Compton, Karl Taylor; Higher Education and Science; Journals in Science, Medicine, and Engineering; Military, Science and Technology and the; Missiles and Rockets;**

National Academy of Sciences; National Aeronautics and Space Administration; Nuclear Weapons; Office of Science and Technology Policy; Physics; President's Science Advisory Committee; Satellites, Communications; Science; Space Program; Space Science; Technology; *and* Television.]

BIBLIOGRAPHY

Killian, James R., Jr. *The Education of a College President: A Memoir.* Cambridge, Mass.: MIT Press, 1985.

Killian, James R., Jr. *Sputnik, Scientists, and Eisenhower: A Memoir of the First Special Assistant to the President for Science and Technology.* Cambridge, Mass.: MIT Press, 1977.

Wang, Zuoyue. *In Sputnik's Shadow: The President's Science Advisory Committee and Cold War America.* New Brunswick, N.J.: Rutgers University Press, 2008.

Zuoyue Wang

KING, CLARENCE RIVERS

(1842–1901), geologist, explorer, writer, and the first director of the U.S. Geological Survey (USGS). A magnetic personality with wide-ranging intellectual interests, he impressed contemporaries as one of the foremost talents of his day. King was born in Newport, Rhode Island, to parents with deep New England roots. His father, who hailed from a merchant family that thrived in the China trade, died when King was six. At 18, King enrolled in Yale's Sheffield Scientific School. In 1863, he joined the California Geological Survey headed by Josiah Dwight Whitney, an experience that formed the basis of his classic book *Mountaineering in the Sierra Nevada* (1872).

While still in his twenties, King persuaded Secretary of War Edwin Stanton to back his proposal for a survey of the Great Basin, the arid region between the Sierra Nevada and Rocky Mountains. In 1867, Congress authorized funding for the King-led Fortieth Parallel Survey under the auspices of the War Department. King made scien-

tific research a top priority, selecting an expedition team that included geologists, a zoologist, and a botanist, among others. Fieldwork for the Fortieth Parallel Survey resulted in the publication of highly regarded studies on geology, mining, paleontology, ornithology, and the region's flora, and advanced the science of microscopic petrography. King's influential *Systematic Geology* (1878) offered an account of the formation of the North American Cordillera. Challenging the prevailing uniformitarianism of his era, King described the forces—some "catastrophic," others more gradual—that forged Western landscapes from the Precambrian to the Quaternary period.

The Fortieth Parallel Survey set new standards for expeditionary science and topographical mapping of the U.S. West, while making pioneering use of photography. King and colleagues earned further accolades for uncovering a diamond hoax that nearly cost investors millions of dollars. Most importantly, the survey marked a key transition from military to civilian control of Western exploration. In 1879, Congress passed a measure creating the USGS to consolidate Western surveys in a single, civilian office. President Rutherford B. Hayes nominated King to lead the agency. As USGS director, King emphasized "economic geology" in the service of mining, while continuing to support basic research. He resigned in 1881 to pursue private mining interests, allowing explorer John Wesley Powell to take charge of the USGS.

A legendary raconteur, King was a fixture in the social worlds of New York, Washington, and London. However, his mines and other business ventures faltered, leaving him on the edge of insolvency. In the late 1880s, he began a clandestine relationship with Ada Copeland, an African American nursemaid in New York City. The couple had five children. Fearing a scandal, King took pains to conceal the relationship and did not reveal his true name to Ada until just before his death from tuberculosis in 1901.

Historian Henry Adams had predicted that his friend Clarence King "would die at eighty the richest and most many-sided genius of his day." Despite precocious successes in science, literature, and government, King spent his final decades in pursuit of a fortune that eluded him.

[*See also* Botany; Cartography; Geological Surveys; Geology; Mining Technology; Paleontology; Photography; Powell, John Wesley; Science; *and* Zoology.]

BIBLIOGRAPHY

Adams, Henry. *The Education of Henry Adams*. Oxford: Oxford University Press, 1999.

Goetzmann, William H. *Exploration and Empire: The Explorer and the Scientist in the Winning of the American West*. New York: Alfred A. Knopf, 1966.

Nelson, Clifford M., and Mary C. Rabbitt. "The Role of Clarence King in the Advancement of Geology in the Public Service, 1867–1881." In *Frontiers of Geological Exploration of Western North America*, edited by Allen E. Leviton, Peter U. Rodda, Ellis Yochelson, and Michele L. Aldrich, pp. 19–35. San Francisco: Pacific Division of the American Association for the Advancement of Science, 1982.

Sachs, Aaron. *The Humboldt Current: Nineteenth-Century Exploration and the Roots of American Environmentalism*. New York: Viking, 2006.

Sandweiss, Martha A. *Passing Strange: A Gilded Age Tale of Love and Deception across the Color Line*. New York: Penguin Books, 2009. Sandweiss explores King's "double life," describing how the sandy-haired, blue-eyed geologist fabricated an identity as a "black" railway porter named James Todd in his relationship with Ada Copeland, while continuing to circulate among political, artistic, and scientific elites as Clarence King.

Wilkins, Thurman. *Clarence King: A Biography*. Revised and enlarged ed. Albuquerque: University of New Mexico Press, 1988.

John Suval

KINSEY, ALFRED

(1894–1956), zoologist, sex researcher, reformer. Born in Hoboken, New Jersey, Kinsey was a sickly child. At his parents' insistence he spent much time attending the Methodist church, thoroughly absorbing the tenets of evangelical Protestantism. After two fruitless years during which he followed his father's wishes to study engineering at Stevens Institute of Technology, he broke with his domineering father and transferred to Bowdoin College in Maine. Majoring in biology, he graduated magna cum laude in 1916. Three years later he received a doctorate in zoology from Harvard. In 1920 he joined the faculty of Indiana University. There he abandoned religion, raised a family, and won a reputation as a respected teacher and preeminent authority on gall wasps.

Kinsey's work on gall wasps consumed the first two decades of his academic life and was innovative in that he collected thousands of specimens, rather than only a few intended to be representative. From the perspective of evolutionary biology, it was important to study entire populations of organisms and the variations among them. Kinsey collected and measured thousands of specimens. He published several papers and two books on his findings. He is, however, better remembered for his study of human sexuality.

Kinsey's study of sexuality began in 1938, when he began to teach a course on marriage to undergraduates at Indiana. He found little existing scientific study of human sexual behavior and aimed to fill this gap. Privately, Kinsey experienced sadomasochistic and homoerotic urges deeply at odds with conventional morality. Like many closeted homosexuals of his day, he lived a double life, pursuing same-sex liaisons at every opportunity. Rejecting society's judgment that homosexuality was abnormal, he studied human sexuality, using essentially the same taxonomic methodology he had developed in his gall-wasp research. Attracting grants from the National Research Council and The Rockefeller Foundation, he founded the Institute for Sex Research at Indiana University in 1947. Defended against critics by the university president, Kinsey and his staff interviewed thousands of subjects nationwide. The value of the massive data they compiled remains contested, however, because of methodological and sampling flaws. In the studies that made him famous, *Sexual Behavior in the Human Male* (1948) and *Sexual Behavior in the Human Female* (1953), Kinsey and his coworkers shattered the conspiracy of silence surrounding sexuality. Showing that millions of American men and women routinely violated middle-class morality, these books sparked sustained debates about sexual mores and practices.

Along with the library and archive of his institute, Kinsey left three important legacies, summed up in his beliefs that human sexuality can be studied scientifically, that social and legal policies relating to sex should be informed by scientific knowledge, and that society should cultivate tolerance in the face of such diversity in sexual behavior.

[*See also* Biological Sciences; Evolution, Theory of; Religion and Science; Rockefeller Institute, The; Sex and Sexuality; Sex Education; Sexually Transmitted Diseases; Social Sciences; *and* Zoology.]

BIBLIOGRAPHY

Bancroft, John. "Kinsey, Alfred Charles." *Complete Dictionary of Scientific Biography*, Vol. 22, pp. 123–130. Detroit: Charles Scribner's Sons, 2008.

Gathorne-Hardy, Jonathan. *Sex the Measure of All Things: A Life of Alfred C. Kinsey*. Bloomington: Indiana University Press, 2000.

Jones, James H. *Alfred C. Kinsey: A Public/Private Life*. New York: W. W. Norton, 1997.

Ruemann, Miriam G. *American Sexual Character: Sex, Gender, and National Identity in the Kinsey Reports*. Berkeley: University of California Press, 2005.

James H. Jones;
updated by Elspeth Knewstubb

L

LADD-FRANKLIN, CHRISTINE

(1847–1930), mathematician, psychologist, and logician. She was born Christine Ladd in Connecticut to parents Eliphalet and Augusta (Niles) Ladd. In 1865 Ladd-Franklin graduated as valedictorian from the Wesleyan Academy, a coeducational school in Wilbraham, Massachusetts. In 1866 Ladd-Franklin enrolled in Vassar College and majored in mathematics. After graduating in 1869, she taught secondary science and mathematics in Pennsylvania and began publishing in the mathematics journals *Educational Times of London* and *The Analyst: A Journal of Pure and Applied Mathematics*.

In 1878 Ladd-Franklin applied to Johns Hopkins University as C. Ladd and was accepted with a fellowship. When the trustees realized she was a female, they tried to reverse their decision. The renowned British mathematics professor James Joseph Sylvester, a distinguished faculty member, supported her attendance as his pupil.

Thus, she retained her funding but not the title of fellow. Ladd-Franklin completed her degree requirements in 1882; however, she did not receive her actual Doctor of Philosophy degree until 44 years later because the school thought it would hurt their reputation to award an advanced degree to a woman. Her dissertation "The Algebra of Logic" was published in the 1883 *Studies in Logic by Members of the Johns Hopkins University*. It demonstrated her still-significant work in logic, including the reduction of all syllogisms to one formula or antilogism, as she named it.

After leaving graduate school, Ladd-Franklin expanded her research into color vision theory and psychology. The Ladd-Franklin color-sensation theory is considered a fundamental principle. In 1903 she became the first female faculty member in the arts and sciences at Johns Hopkins; however, she taught only one class a semester and on a year-to-year contract—an arrangement that would be repeated at Columbia University where she taught from 1915 until her death in 1930. She was recognized as one of 50 leading psychologists in the first edition of the *American Men of Science* (1906). Throughout her life, she published in leading scientific journals, wrote and lectured widely in both logic and psychology, made significant and lasting contributions in all of her chosen fields, and remained a strong advocate for women in higher education. Ladd-Franklin married Fabian Franklin in 1882 and they had one daughter, Margaret Ladd-Franklin, who lived to adulthood.

[*See also* **Gender and Science; Higher Education and Science; Journals in Science, Medicine, and Engineering; Mathematics and Statistics;** *and* **Psychology.**]

BIBLIOGRAPHY

Cadwallader, Thomas C., and Joyce V. Cadwallader. "Christine Ladd-Franklin (1847–1930)." In *Women in Psychology: A Bio-bibliographic Sourcebook,* edited by Agnes N. O'Connell and Nancy Felipe Russo. New York: Greenwood Press, 1990.
"Christine Ladd-Franklin." *Vassar Encyclopedia.* http://vcencyclopedia.vassar.edu/alumni/christine-ladd-franklin.html (accessed 19 March 2012).
Scarborough, Elizabeth, and Laurel Furumoto. *Untold Lives: First Generation of American Women Psychologists.* New York: Columbia University Press, 1987.

Leslie N. Sharp

LANGMUIR, IRVING

(1881–1957), industrial chemist, atmospheric scientist, Nobel laureate. The son of Charles Langmuir and Sadie Langmuir (née Comings), Langmuir was born on 31 January 1881 in Brooklyn, New York. He was educated in the public schools of Philadelphia, New York, and Paris. After earning his BS degree in metallurgical engineering from the Columbia University School of Mines (1903), he received his PhD degree from the University of Göttingen under Nobel laureate Walther Nernst (1906).

In 1906 Langmuir became instructor of chemistry at Stevens Institute of Technology in Hoboken, New Jersey. In 1909 he joined the General Electric (GE) Research Laboratory in Schenectady, New York, where he became associate director in 1929, retired in 1950, and remained as a consultant until his death.

In 1919 Langmuir outlined his "concentric theory of atomic structure," building on Gilbert Newton Lewis' cubical atom theory and embroiling him in a priority dispute. Although Langmuir, who here devised the term "covalent," was largely responsible for popularizing the Lewis–Langmuir theory, credit for the theory of valence shells and the octet rule belongs primarily to Lewis.

At GE Langmuir studied low-pressure chemical reactions and emission of electrons by hot filaments in a vacuum and introduced the concept of a monolayer. He received 63 patents, invented the high-vacuum tube and gas-filled incandescent lamp, and discovered atomic hydrogen and used it to invent the plasma welding process. In 1989 he was inducted into the Inventors Hall of Fame.

The holder of numerous awards, 15 honorary doctorates, and president of the American Chemical Society (1929) and the American Association for the Advancement of Science (1943), Langmuir received the 1932 Nobel Prize in Chemistry "for his discoveries and investigations in surface chemistry," the first industrial chemist to win this award.

In 1938 Langmuir turned to atmospheric science and meteorology. He discovered a wind-driven surface circulation in the sea (Langmuir circulation). He devised cloud seeding to induce rain, a still controversial technique. Much in demand as a semipopular lecturer, in 1953 he coined the term "pathological science" to denote research tainted by unconscious bias.

During World War II Langmuir participated in national defense scientific research programs, including radar, naval sonar to detect submarines, a smoke-maker to screen troops and ships, and de-icing methods for aircraft wings. After a short illness, he died of a heart attack on 16 August 1957 in Woods Hole, Massachusetts. His name is enshrined in, among others, Langmuir

College, Langmuir Laboratory for Atmospheric Research, Alaska's Mount Langmuir, and the ACS's Irving Langmuir Award in Chemical Physics (founded in 1931) and *Langmuir*, a journal established in 1985 for surface and colloid chemistry.

[*See also* American Association for the Advancement of Science; Chemistry; Engineering; Meteorology and Climatology; Military, Science and Technology and the; Nobel Prize in Biomedical Research; *and* Societies and Associations, Science.]

BIBLIOGRAPHY

Bacon, Egbert K. "Irving Langmuir 1881–1957." In *American Chemists and Chemical Engineers*, edited by Wyndham D. Miles, pp. 288–289. Washington, D.C.: American Chemical Society, 1976.

Langmuir, Irving. *The Collected Works of Irving Langmuir*, 12 vols, edited by C. Guy Suits. Oxford: Pergamon Press, 1960–1962. Volume 12, by Albert Rosenfield, includes a complete biography based on Langmuir's personal diaries, letters, and reminiscences by his family and colleagues, 134 references to biographical source material, and many of his popular and semipopular articles and addresses.

Nobel Foundation. *Nobel Lectures Including Presentation Speeches and Laureates' Biographies, Chemistry 1922–1941*, pp. 281–327. Amsterdam, London, and New York: Elsevier Publishing Co., 1966. Langmuir's Nobel lecture, "Surface Chemistry," appears on pp. 287–325.

George B. Kauffman

LATROBE, BENJAMIN

(1764–1820), one of the earliest professional architects and engineers in the United States. Born in Britain, he attended schools of the Moravian faith, first in Britain and then in Germany, and obtained some training under a Prussian engineer. Settling in London, Latrobe extended his architectural and engineering skills, apparently working briefly under the engineer John Smeaton and obtaining a substantial apprenticeship with the architect Samuel Pepys Cockerill. Striking out on his own, he became London's Surveyor of the Police Offices and obtained commissions for two country houses. The death of his wife in 1793 and an inheritance of Pennsylvania land led him to immigrate to the United States, where after his arrival in 1796 Latrobe readily found architectural commissions, creating or redesigning numerous public buildings and about 60 private residences. The Baltimore Cathedral (finished in 1821), with its double-shell dome, is the most outstanding surviving example of his architectural work, although his engineering achievements exist only in memory. The most important of the latter was undoubtedly the Philadelphia Waterworks (completed in 1801), which not only was the first substantial American urban water system, but also was one of the earliest successful uses of steam power in the United States. Subsequently, Latrobe promoted steam power for industrial uses and (in collaboration with Robert Fulton) was an early builder of steamboats in Pittsburgh. He died in New Orleans while completing the first steam-powered waterworks for that city.

Latrobe's engineering and architectural skills were buttressed by his knowledge of acoustics, hydrology, lighting, and materials, drawing particularly on French sources. As a result, he strove for high-level designs and stronger materials that sometimes put him at odds with business patrons who wanted quick and cheap solutions.

Displaying interests ranging from geology to botany to animal behavior, Latrobe was a talented naturalist whose careful descriptions and drawings filled his notebooks. He was often in the company of other Americans of a scientific bent, as well as European savants traveling in the United States, and was elected to and active in the American Philosophical Society.

[*See also* American Philosophical Society; Botany; Building Technology; Engineering; Fulton, Robert; Geology; Science; Steam Power; *and* Zoology.]

BIBLIOGRAPHY

Carter, Edward C., II, ed. *The Papers of Benjamin Henry Latrobe*. 9 vols. New Haven, Conn.: Yale University Press, 1977–1994.

Fazio, Michael W., and Patrick A. Snadon. *The Domestic Architecture of Benjamin Henry Latrobe*. Baltimore: Johns Hopkins University Press, 2006.

Darwin H. Stapleton

LAW AND MEDICINE

See Medical Malpractice.

LAW AND SCIENCE

We tend to think of modern science and law in dichotomies. Science deals with nature, law with society; science is objective, directing our knowledge; law is normative, directing our actions; science is an open-ended search for truth, law is a close-ended search for justice. Yet, it is equally true that nature and society, knowledge and action, truth and justice are all mutually constitutive. Thus, although science and law are indeed fundamentally distinctive cultures, they are at the same time also deeply connected institutions, heavily invested in each other (Jasanoff, 1995). Scientific advances have shaped the legal terrain, from regulation and tort to evidence and patents. Similarly, the law has had no less of an impact on science by clarifying the character of legitimate scientific practices and adjusting the institutions and social relations required for its successful application (Golan, 2004b).

Historians have unearthed plenty of records of premodern legal requests for expert advice on topics such as causes of diseases, nature of wounds, meaning of celestial signs or Latin phrases, construction and navigation of ships, identification marks of witches, and much more (Learned Hand, 1901). Twenty-first century courts of law may no longer consider many of these advices to be scientific, but those interested in history should keep in mind that well into the eighteenth century the meaning of "science" was far more inclusive, designating all fields of systematic knowledge, including the law, which was considered among the highest branches of science, second only to theology.

Both science and law have changed much since the eighteenth century. The ancient meaning of science still excretes some influence in modern culture. A case in point is the primary method of American modern legal pedagogy, the "case method," which considers the law an empirical science that mines its data from appellate cases to derive from them general principles of justice (Schweber, 1999). But overall the modern connotation of science has been narrowed to the experimental and quantifiable investigation of natural phenomena and transmuted in the process from a spiritual calling to a mainstay of modern material culture. Meanwhile, the law has changed no less. Early in the eighteenth century, American colonial judges served at the pleasure of the British Crown, which abolished trial by jury in the colonies. By the end of the century the United States won its independence, reduced its judges to umpires, and adopted with great zeal the English adversarial system and its institution of the jury as a mainstay of liberty and self-government.

During the nineteenth century, both science and law increasingly developed the sort of authority they came to enjoy in modern society. The scientists solidified their control over the representation of nature and turned the scientific method into the yardstick of demonstrable truth. Meanwhile, the lawyers solidified their control over the representation of society and turned the legal process into a benchmark of demonstrable justice. To be sure, these were not unrelated developments. Science provided the law with evidence and facts, whereas the law provided science with a stage and authority; together they supplanted religion as the main purveyors of truth, justice, and power in modern society.

The Growing Role of Science in the Courts. Early in the nineteenth century, it was still hard to find signs of scientific life in the

American courtroom, other than traditional figures such as the nautical and medical experts. This changed quickly during the latter part of the nineteenth century, as the United States was going through a rapid industrial development, driven by science-based innovations such as electricity, new chemicals and alloys, the internal combustion engine, and new communication technologies such as the telegraph and radio. The growing application of scientific principles to the business of everyday life inevitably increased the number of legal cases that involved technological and scientific argumentation and required the courts to seek specialized advice. By the end of the nineteenth century, scientific and technological experts had become prevalent in the courtroom, where they untangled for court and jury the complexities of the technoscientific evidence central to the mounting litigious activities in modern matters such as energy (first gas and then electricity), environment (pollution and contamination), public health (food and drug adulteration, water supply, sewage treatment), communication, transportation, agriculture, mining, industry, malpractice, and insurance (Scheiber, 1987; Okun, 1986; De Ville, 1990).

The forensic sciences prospered as well. At the beginning of the nineteenth century, they only employed basic microscopy and toxicology. By the end of the century they not only pushed microscopy and toxicology to a superior level but also incorporated physics, mineralogy, zoology, botany, and what used to be called anthropometry. Late nineteenth-century forensic scientists detected stains and forgeries using infrared and ultraviolet light; traced minute quantities of arsenic by their organic chemical prints and inorganic substances by their line spectra; differentiated between humans and other species by the shape and size of their red-blood corpuscles; reconstructed important characteristics of a corpse from partial clusters of bones; and photographed the insides of things with the mysterious X-rays (Golan, 2004a).

The late nineteenth-century legal system provided important patronage to the fledgling American scientific community. In an era when scientific expertise provided only a limited means of livelihood, legal functions such as the production of

evidence, expert testimony, arbitration, and counseling constituted lucrative employment opportunities for a growing number of men of science. Equally important, the steady legal demand for additional and better scientific evidence spurred important technical and theoretical developments in various scientific fields, including medicine, microscopy, chemistry, geology, and electricity, and focused scientific attention on the key problems of standardization, accuracy, and reliability (Mohr, 1993, Golan, 2004b).

The role of science in the courtroom continued to increase during the twentieth century. Early in the century scientific experts learned to identify people by their fingerprints, firearms by their ballistics prints, and blood relations by blood groups (Cole, 2001; Rudavsky, 1999). By midcentury, the police forces had embraced science and technology as key to their modernization; the U.S. Supreme Court restricted the acquisition of evidence via the traditional violent interrogation techniques; and the federally sponsored crime laboratories flooded the courts with innovative scientific techniques such as electronic microscopes, truth tests and sera, voice prints, neutron activation analysis, and much more. Late in the twentieth century X-ray technology was joined by a host of innovative imaging technologies such as computerized tomography (CT), positron emission tomography (PET), and magnetic resonance imaging (MRI and functional MRI). Molecular biologists learned to identify people and determine their genetic relationships by their DNA fingerprints (Dumit, 1999; Lynch, 2008).

The social sciences did not lag far behind. As the American legal profession slowly warmed up to the idea that the law was not a deductive science from first principles but an organic part of greater society, it increasingly came to accept the advice of those sciences that studied society and its business and mores (Moynihan, 1979). At the start of the twentieth century no social science was yet allowed into the courtroom. By the end of the century, the courts had incorporated a multitude of social science expertise on a wide variety of problems: economists testified in matters such as antitrust litigation and work discrimination, anthropologists and sociologists on social practices and cultural norms, social workers on kids' best interest, and psychologists on the causes and prospects of violence, the validity of eyewitness testimony or repressed memories, and much more.

The Malaise of Science in the Courts.

The growing role of science in the legal process reflected the growing role of science and technology in American society at large, but at the same time it has also perpetuated the marginalization of science within the legal process. Summoned from their private laboratories to the public courtroom, nineteenth-century men of science hoped to represent there the authoritative laws of nature. Instead, they found themselves isolated in the witness box, away from the decision-making process, browbeaten and set against each other by the lawyers. It quickly became apparent that what thrives in the temperate climate of the laboratory could not survive the heat of the adversarial courtroom. The results too often were an embarrassing display of scientific definitions in disarray, inconsistent experimental results, and a parade of scientific experts zealously opposing each other from the witness stand (De Ville, 1990; Mohr, 1993; Golan, 2004b).

The late nineteenth-century American scientific community toiled hard on its professional status and on promoting the scientific method as the yardstick of truth and men of science as the impartial keepers of this truth. Ironically, the success of these efforts only served to cast serious doubts on the integrity of the scientific experts appearing in court. If these are legitimate representatives of science, the public asked, how can they disagree so frequently and sharply? And because these experts were paid lavishly for their opinion, their partisanship was interpreted as a sign of moral corruption, of prostituting their science to the highest bidder. By 1870, the first systematic study of scientific expert testimony in American courts reported "[an] unmistakable tendency on the part of eminent judges and jurists to attach less and less importance to testimony of this nature" and explained it by "the surprising facility with which scientific gentlemen will swear to the most opposite opinions upon matters falling within their domain" (Anonymous, 1870). By

the century's end, a fashionable legal witticism spoke of three kinds of liars: the common liar, the damned liar, and the scientific expert (Foster, 1897–1898).

The growing mistrust of science, and even more so of the integrity of the men of science, in such an important cultural domain as the legal system deeply troubled the young American scientific community. Most scientific commentators agreed that the opposing views of the partisan experts in court reflected no real scientific disagreement and blamed the opposition in court on the adversarial legal procedures. Most legal commentators rejected the allegations. Scientific disagreements in court, one commentator noted, were usually not about facts but of opinions, on which men of science, like everyone else, could have legitimate differences. But regardless of whether the experts themselves or the legal mode of their deployment should be blamed, almost all commentators concurred that the frequent disagreement among the scientific witnesses was detrimental to both justice and science and that if partisanship and charlatanism could be somehow be swept aside, scientific disagreement would subside significantly (Golan, 2004b).

The reform of expert testimony in court became a central topic in the meetings of the various scientific, medical, and legal associations that mushroomed in late nineteenth-century America. Many bills were drafted to remedy the evils of scientific expert testimony. For the selection of experts, reformers suggested that they be chosen by the court, either reserving or denying the right of the parties to call additional witnesses; that the selection of the courts be unassisted or made from an official list chosen in some other manner; and that the official list be either permanent or special for each case. In regard to the examination of witnesses, reformers recommended that the court conduct the examination, with or without the right of the parties to cross-examine, or there be no examination at all and that the expert would submit a report. In regard to decisions where experts disagree, it was recommended that a jury of experts be selected or that an expert sit with the judge during the trial and advise him (for a long list of reform proposals, see Dooley, 1942).

Unfortunately, the proposed reforms have all gone against one or more of the postulates of the American legal system. Getting rid of the jury contradicted the constitutional right to a trial by a jury of one's peers; and allowing the court to call in assessors or witnesses independent of the parties ran against another two equally fundamental postulates—the right of the parties to control the evidence and the passivity of the court in factual matters. Hence, even those in the legal profession who empathized with the frustration of the scientific community considered the suggested reforms remedies worse than the disease itself. Consequently, most reform bills did not pass the legislative stage, and the few that did were promptly held unconstitutional by the courts (Kidd, 1914–1915).

The judiciary was well aware of the need to protect the lay jury from scientific charlatanism. But it was able to do little about it. The institution of the jury was highly cherished in colonial America, and the enduring political philosophy of self-government kept this enthusiasm alive throughout the nineteenth century. Fears of undue judicial influence on the jury were met by constitutional restrictions on the power of the trial judges. By the end of the nineteenth century, in 21 of the 49 states and territories comprising the United States, judges were expressly forbidden by constitutional provisions to charge the jury on questions of facts. And in about half of the remaining 28 states and territories, the courts had voluntarily adopted the same restriction. Only in federal courts and a minority of state courts were judges allowed to comment on the weight of the evidence in their charge to the jury (Anonymous, 1964; Horowitz, 1991).

The courts were also unable to lay down a precise rule for determining who was and who was not a competent expert. Scientific titles and diplomas carried little meaning during the nineteenth century, and in most cases it was hard for the trial judge to satisfy himself as to the qualifications of the proffered expert. Unable to discriminate with any reasonable degree of accuracy between experts and charlatans, the actual practice of the courts came to be to admit almost anyone presented as an expert and count on cross-examination to expose quackery and for the jury to be the judge of the ensuing battles between the scientists and the lawyers. No one trusted the lay jury to be able to do this job well. The scientific community complained bitterly about the absence of a judicial hand that could guide the jury in its difficult task of assessing the scientific evidence ("Science in the Courts," 1872). But the legal profession considered it a fair price to pay for an open market of expert opinion that was considered vital for good and wise governance and the best protection from the abuse of political power (Anonymous, 1870).

Nineteenth-Century Attempts to Control the Performance of Science in Court. Nineteenth-century American courts tried to improve the performance of science in the adversarial courtroom. Unable to check the selection of the experts or to guide the jury's assessment of their evidence, the nineteenth-century American courts' strategy was based on the application of the "rules of evidence" to regulate the processes through which the experts communicated their information in court.

One major evidentiary doctrine sought to protect the credulous jury from being uncritically influenced by the expert by preventing the expert from giving his opinion upon the "ultimate issue" (i.e., the specific factual issue before the jury). Alas, the application of this doctrine bred much confusion and led to absurd consequences. In theory, it made irrelevancy a ground for admission and relevancy grounds for exclusion. In practice, the ultimate issue was often exactly what the expert testimony was all about. In an insanity defense, the ultimate issue was the mental state of the defendant; in a malpractice suit the ultimate issue was whether the patient was treated properly; in forgery cases, the ultimate question was often the genuineness of a certain document, and so on. In these cases and others, the doctrine seemed to exclude expert evidence exactly where it was most needed. Consequently, the courts developed various ad hoc options to allow the witnesses to give their opinion on the ultimate issue (McCormick, 1941).

One popular approach differentiated between general and specific causation and allowed the expert to opine only on the first (i.e., whether the alleged cause could potentially produce the alleged result) and leave it to the jury to decide whether causation was actually produced in the specific case under their consideration. To enable this, a secondary evidentiary doctrine came into play. Under the "hypothetical question" doctrine, the expert's testimony was given in the form of answers to hypothetically framed questions. These questions specified a set of factual premises, already submitted in evidence, and the expert was asked to draw his conclusions from them, assuming that they were true. This cumbersome technique was justified on the following grounds: as a means of enabling the expert to apply his general knowledge to facts that were not within his personal knowledge; allowing the jury to recognize the factual premises upon which the scientific opinion was based; and allowing the expert to give his opinion on the ultimate issue without "invading" the province of the jury. The jury was then instructed to credit the opinion given only if it believed these premises (Ladd, 1952).

Sound in theory, the technique broke down in practice. If counsel was required to recite all the relevant facts, the question became intolerably lengthy. If allowed to select a partial set of the facts, it prompted one-sided hypotheses. Designed and controlled by the interested parties, the hypothetical question was used more as a means to manipulate the facts of the case than to clarify them for the jury. Forced to assume as true any cleverly defined transcript of the facts of the case, the expert was frequently manipulated to give an answer against his true conviction.

Even the old and powerful hearsay doctrine that attempted to limit the testimony of ordinary witnesses to information based solely on their personal observations turned out to be problematic in the context of scientific testimony. The caution of the courts in admitting hearsay evidence, and the fear of misleading the jurors with scientific statements they were hardly competent to assess, had led many early nineteenth-century American courts to reject what many considered the most natural resource of scientific information—standard textbooks, reports, and the like. The use of such documents in court was excluded by the

hearsay doctrine on the premise that they were statements not made under oath or that the author was not available for cross-examination. As with other doctrines, the courts had slowly devised ways to work around this. Some courts permitted the use of scientific treatises, but only to discredit an expert. Others allowed the expert to "refresh his memory" by reading from standard works. Other courts yet allowed publications of exact science, assuming their statements to be well established, and excluded other treatises, especially medical works. Confusion and inconsistency, again, were rampant (Wigmore, 1892).

By the end of the nineteenth century, the debate over the causes and solutions to the malaise of science in the courts had been picking up steam for almost half a century, and the remedies suggested were as numerous as prescriptions for the cure of rheumatism and generally about as useful. With no resolution in sight, the two camps grew belligerent. What had seemed before the Civil War to be a vital civil function of science had become by the end of the century a source of deep discontent. Instead of bringing the legal and the scientific communities closer together, forensic science was drawing them further apart. The courts were growing increasingly weary of the scientific partisanship displayed in the courtroom, whereas the scientific community remained frustrated by the awkward position it came to occupy in the adversarial courtroom (Mohr, 1993). Still, the deadlock did not hold back the tide of science-rich litigation and inevitably the growing deployment of science in the courtroom. The pressure for a solution was therefore mounting. Something eventually had to give way in the sacred triangulation of the adversarial system: the party's right to control the evidence, the jury's right to decide all factual debates, or the passive position of the court.

Twentieth-Century Admissibility Standards of Scientific Evidence. In 1905, Michigan passed a statute that embodied the mildest version of the most popular reform suggestion—allowing the court to nominate its own experts. The statute did not preclude the parties from using their own witnesses but provided in criminal cases for the additional appointment by the court of no more than three disinterested experts to investigate and testify to their findings at the trial. Nevertheless, the Michigan Supreme Court considered it no part of the duties of the court to select witnesses and held the statute to be in violation of two state constitution provisions: the separation of powers and due process (*People v. Dickerson*, 1910).

The decision dealt a serious blow to those advocating the reform of expert testimony by means of statutory enactment. Accepting that the experts should remain party-chosen and the jury the final trier of the facts, early twentieth-century legal scholars started to ponder the possibility of involving the scientific community in the effort to control the performance of science in the courts. Their renewed hopes of succeeding where their predecessors had so miserably failed hinged on a clear change in the market of scientific expertise, created by the rising professional culture in America.

By the second decade of the twentieth century a wide spectrum of expertise—from chemists, physicists, and engineers to architects, surveyors, actuaries, realtors, insurers, and accountants—came to be dominated by professional associations that developed codes of ethics, standards of education, training, and practice, and defined minimum qualifications of certification either through their own examinations or through those of the various state boards of examiners (Haskell, 1984). The legal profession took notice and began to explore ways in which the courts could take advantage of this newly standardized market of expertise to check the quality of the scientific experts.

No special rule for the admissibility of scientific evidence existed in the early decades of the twentieth century. Like all other evidence, scientific evidence was evaluated according to traditional evidentiary criteria: the qualifications of the witness, the relevancy of the evidence, and its helpfulness to the trier of fact. In 1923, the Court of Appeals of the District of Columbia was the first to come up with a ruling dedicated to the admissibility of scientific evidence. In what later came to be known as the Frye ruling, the appellate federal court affirmed the decision of the lower

court, in a highly charged murder case, to exclude from evidence the result of a newly invented lie-detector test. The Court of Appeals of the District of Columbia put forward the following analysis (Frye, 1923):

> Just when a scientific principle or discovery crosses the line between the experimental and demonstrable stages is difficult to define. Somewhere in this twilight zone the evidential force of the principle must be recognized, and while courts will go a long way in admitting expert testimony deduced from a well recognized scientific principle or discovery, the thing from which the deduction is made must be sufficiently established to have gained general acceptance in the particular field in which it belongs. We think that the systolic blood pressure deception test has not yet gained such standing scientific recognition among physiological and psychological authorities as would justify the courts in admitting expert testimony deduced from the discovery, development, and experiments thus far made.

Proposing to look outside the courtroom for a general acceptance in the particular field to which the proffered expertise belonged, the Frye opinion offered a potent departure from the deadlock of scientific expert testimony. The jury would still be the final trier of facts, and the experts would still be party-chosen; but the judicial ability to control the play of science in the courtroom would be enhanced by a new admissibility test that would take advantage of the newly standardized market of expertise to evaluate the proffered expertise against the standards of the relevant community.

Originating in the extreme case of the lie-detector and containing no precedential citations, the Frye opinion long remained an isolated solution to a particular problem. For the next three decades, American courts applied the general acceptance test suggested by Frye only to exorcize from criminal trials evidence derived from various lie-detection and truth serum schemes. That began to change after World War II, when trial judges found the general acceptance test a convenient tool to decide the admissibility of the new species of scientific evidence offered by the up-and-coming crime laboratories. By the early 1950s, the general acceptance test, nicknamed *Frye*, was already proclaimed in the legal literature as the main criterion for the admissibility of novel types of scientific evidence. By the 1970s, it had become the *sine qua non* in practically all of the criminal courts that considered the question of the admissibility of new scientific evidence. By the 1980s, the courts expanded the use of Frye from criminal to civil proceedings, thereby completing its transformation from a special judicial device designed to check controversial new technologies to a general judicial device for pretrial screening of scientific evidence (Giannelli, 1980).

The expanding judicial control over the play of science in their courts met with increased criticism. The earliest attacks considered judicial screening of the scientific evidence an unnecessary procedure that deprived the jurors of their right to decide for themselves what facts are valuable. Frye's proponents argued that it finally provided the courts with a uniform method for ensuring the reliability of the scientific evidence. However, "the thing from which the deduction is made" has meant different things to different courts at different times. The ambiguities inherent in determining the particular field to which a new scientific development belongs, and in deciding how to measure its general acceptance, left ample room for judicial discretion. Consequently, Frye ended up having not one but many "general acceptance" criteria, which the courts seemed to apply in a selective manner, according to their views about the reliability of the particular scientific technique before them (Imwinkelried, 1981–1982).

Meanwhile, a new twist entered the plot. In 1975, after much agitation, the rules of evidence that federal judges follow were finally codified. Completely disregarding Frye, the newly enacted *Federal Rules of Evidence* prescribed no special test to ensure the reliability of scientific evidence, new or old. Instead, casting the widest net possible, the enacted *Federal Rules of Evidence* provided that:

> If scientific, technical, or other specialized knowledge will assist the trier of fact to understand the evidence or to determine a fact in issue, a witness qualified as an expert by knowledge, skill, experience, training, or education, may testify thereto in the form of opinion or otherwise.
> —*Federal Rules of Evidence* 1975, Rule 702

Having left open the question of how one defines "scientific, technical, or other specialized knowledge," the *Federal Rules of Evidence* was generally regarded as prescribing a flexible judicial consideration of the proffered scientific evidence. At the same time, because the enacted *Federal Rules of Evidence* did not state an explicit intent to abandon the Frye rule, some federal and almost all state courts remained committed to the general acceptance criterion for the admissibility of scientific evidence.

The debate concerning the proper judicial standard for the admissibility of scientific evidence intensified during the late 1980s and early 1990s, energized by growing fears of tort litigation explosion. Since the 1970s, dangerous drugs, industrial defects, environmental pollutants, and other technological breakdowns all became the subject of prolonged litigation with ever-escalating financial stakes. In the great majority of these cases, the central legal questions were of risk and causation, which invariably turned upon scientific evidence and revealed again the all-too-familiar sight of scientific experts producing in court conflicting data and contradictory conclusions.

The Changing Notions of Risk and Causation. Modern science has offered lawmakers two distinct modes of deciding causation and calculating risks: an experimental reductive science, built on the strength of the laboratory, and an observational statistical science built on the power of big numbers. The first option dominated until the middle of the twentieth century. Public and medical attention was focused on infectious diseases—each caused by a specific microbiological agent. Fighting infectious diseases was a job for the laboratory—to isolate the specific causal organism, study it, and devise the best means to fight back. Observational science served in this campaign merely by informing experts of geographical and social patterns of the disease.

By the middle of the twentieth century the balance began to shift. The battle against infectious diseases seemed to have been won, at least in the developed world. Public and medical attention increasingly turned to a new pattern of diseases—noninfectious, chronic, with long latency, and poorly understood etiology—diseases such as blood pressure, cancer, or heart problems. Previously considered inevitable failures of the aging organism, they now began to top the medical charts. Experimental science was frustrated with these diseases. Their mechanisms kept eluding the researchers. They seemed to involve multiple causes and effects, and their long latency made experimentation difficult. Statistical science, on the other hand, proved much more flexible. Biostatisticians adapted their computational strategies to a distributed, multivariate model of causation that seemed to better fit the nature of these new diseases, where a cause could have many effects and an effect many causes (Golan, 2012).

During the late 1950s and early 1960s, a cluster of British and American epidemiological studies first implicated cholesterol and smoking as significant causal factors for heart disease and, in the case of smoking, also for lung cancer. Running ahead of experimental research, these studies introduced a new lexicon that abstained from causal claims and appealed only to what came to be known as "risk factors"—environmental, social, and other patterns that are statistically correlated with higher incidence of disease. This may not be the best science, the epidemiologists argued, but in a growing number of cases it is the best science could offer decision makers in the late-modern era of latent and irreducible causes and chronic diseases.

The pragmatic program of epidemiology was warmly embraced by the expanding regulatory regime of the late twentieth century. Legislators, judges, administrators, and public health officers have found epidemiology, with its quantified, population-based logic, perfectly placed to provide them with estimates of the prevalence of otherwise irreducible health problems, investigate their probable sources, identify those groups with

elevated risks, and target them with preventive measures. The parallel growth of medical registries and computer technology allowed for the deployment of increasingly sophisticated statistical techniques in the search for increasingly smaller risks in increasingly larger populations. The epidemiologists traded up their mechanical rulers for software programs and got comfortable with the new tools of multivariate correlation and regression and exotic tests of statistical significance. By the end of the twentieth century, the reduction of causes to a distributed network of risk factors had become prevalent and increasingly informed medical research as well as regulatory and legal action (Susser, 1985, 1996).

As regulation was taking center stage in American polity, the capacity of science to provide reasons good enough to legitimize regulatory action was closely scrutinized and frequently contested. The laboratory-based science of the regulatory agencies included *in vitro* studies, which examine the effects of chemical agents on various organic materials, from cells and bacteria to DNA and proteins, in an attempt to understand the biochemical mechanisms involved. It is a long way from molecules to humans, however, and other researchers have taken a shortcut by performing *in vivo* (animal) studies. This reduces some difficulties but introduces new ones. Unable to experiment directly with humans, the experimentalists run their studies on other mammals, which are fed larger-than-life doses, to shorten the experiment and augment its effects. This strategy requires the extrapolation from short and intense exposures of one species to chronic, low-level exposures of another species, *Homo sapiens*. The setting of exposure standards required the scientists to also add arbitrary safety factors to protect the more susceptible subpopulations. Epidemiological studies were also conducted but were expensive, insensitive, and prone to a host of methodological problems from selection biases to confounding variables (McClellan, 1999).

During the 1970s, as the young regulatory agencies began to churn out their safety and exposure standards, both industry and civil action groups challenged the science behind the standards—industry in an attempt to moderate the standards, activists to step them up. The ensuing legal battles revealed the fragility of the science involved. The notorious nonlinearity of physiological systems was mobilized to undermine the extrapolations from high to low doses and from short to long exposures; and the poorly understood interspecies and intrahuman variations were called upon to show that the justification of the standards went beyond scientific and technical competence (Tesh, 2000).

Eager to protect the regulatory regime, the legal system responded by adopting various versions of the precautionary doctrine, which admitted the fragility of the science involved but justified the right of the authorities to act upon it, based on the ever-pressing need to regulate potential risks before they turn into actual harms. The legitimacy of such a regulatory regime, the courts prescribed, resided in the deployment of the best scientific tools available; and by the end of the twentieth century these tools included statistical techniques, built on the power of big numbers.

Tort, Junk Science, and the Development of Evidence Law. In the absence of unequivocal science, American courts have been willing to put forward the precautionary doctrine to legitimize regulatory action. The judges required a relatively low level of scientific proof to justify regulative action. Anything more, the judges recognized, would leave but few environmental regulations standing. But they were reluctant to extend the same leniency to the private sphere of tort litigation. The claims made in this sphere were about actual harm and were treated differently. Tort law required the plaintiff to offer a persuasive proof of a concrete and actual harm caused by the defendant. Anything less, the courts held, would be unfair to the defendant, who should not be forced to pay for injuries he or she did not cause.

Tort, a branch of private law that deals with personal injury claims, had prided itself on its tradition of individualized approach. Its clients were rightful citizens whose property could not be arbitrarily appropriated, without the careful exercise of human judgment on a case-by-case basis (Holmes, 1881). To that end, nineteenth-century tort law cultivated a theory of causality as reductive as that of the science of infectious disease. To exist, a legal cause had to be reduced to a causal agent. The plaintiff's burden of proof, like that of the scientific experimentalist, was to single out the causal agent and demonstrate the chain of events that linked the agent's actions to the plaintiff's injury. If a specific causal agent could not be uniquely determined; if the plaintiff could show only that the defendant's action might have caused the harm; or if another indistinguishable potential cause existed, the courts dismiss the claim for the failure to prove specific causation.

As the twentieth century progressed, tort law became less private and more public, and by the end of the century the "statistical victim" became tort's biggest client (Jasanoff, 2002). With the new client came new practices: individual care gave way to economy of scale, and eyewitness testimony gave way to statistical evidence. These were uneasy changes for tort law and they presented the legal mind with a host of difficult problems regarding the differences between statistical correlation and legal causation, the circumstances in which we could pass from one to the other, and how and by whom these should be decided (Gold, 1986).

The reductive model of causation has worked quite well in traditional tort cases such as accidents or assaults, where a single causal agent could be verified via eyewitness accounts and other demonstrable evidence. This was not the case, however, in a growing range of environmental, work-safety, and product liability cases that came to be known by the end of the 1970s as "toxic tort" cases. These cases involved injuries of the kind that have frustrated experimental science, chronic, with long latency and poorly understood etiology, injuries that could not be comfortably reduced to a single cause.

In the absence of direct or experimental proof of cause and effect, the courts increasingly turned in toxic tort cases to statistical evidence. This was especially true for the new phenomenon of mass toxic tort litigation that began to emerge in the late 1970s and clustered together large crowds with various case histories, all claiming to be harmed by the same exposure or by the same standardized mass-marketed product. Here, legislators, lawyers, and judges again found statistics' quantified and population-based analysis conducive to their needs. Until the 1970s statistical evidence could hardly be found in tort cases. But by the 1980s it was already announced as "the best (if not the sole) available evidence in mass exposure cases." And by the start of the 1990s, judges were dismissing tort cases for not relying on solid epidemiological evidence (*In re: Agent Orange*, 1984; *Daubert*, 1993).

The mutually constitutive rise of mass tort litigation and statistical science shaped much of the relation between law and science in the late twentieth-century American courtroom. The complexities of statistical evidence provided a new target for the traditional fears regarding the jury's capacity to handle complex evidence. The complexities of mass tort litigation legitimated a new judicial role, less arbitral and more managerial in kind, with an eye toward both economic consequences and broad social and political implications. And the unprecedented financial stakes in mass tort litigation induced powerful economic players to put their weight behind the campaign to reform the legal procedures for handling science (Schuck, 1987).

By the 1990s the alarm was sounded that America's courts were being swamped by junk science, produced by an unholy alliance between opportunistic attorneys and unscrupulous experts, aiming to milk the deep pockets of the corporations. Accordingly, the judges were urged to raise the bar and rely on the conservative general acceptance test to protect the credulous jury from pseudoscientific experts and the deep-pocketed corporations from greedy lawyers (Huber, 1991).

Having never before addressed the legal processing of scientific evidence, the U.S. Supreme Court visited the topic on three separate occasions during the 1990s, all of them product liability cases (*Daubert*, 1993; *Joiner*, 1997; *Kumho*, 1999). Known as the *Daubert Trilogy*, these three Supreme Court opinions announced the arrival of a new era in the relations between law and science. The federal judges were directed to follow

the *Federal Rules of Evidence*, but the Supreme Court rejected the flexible let-it-all-in interpretation of these rules. Instead, the Supreme Court read the *Federal Rules of Evidence* as delegating to the trial judge the responsibility to ensure that any scientific evidence admitted into the courtroom be reliable. To help the trial judge in the difficult task of differentiating between good and bad science, the Supreme Court offered a flexible recipe of four nonexclusive factors that could be used by the trial judge in determining the quality of the scientific evidence proposed:

1. Testability: whether the suggested theory or technique had been tested.

2. Peer review: whether the suggested theory or technique had been subjected to peer review.

3. Standardization/error rate: whether there are standards controlling the technique's operation and established estimates of its error rate.

4. General acceptance (the Frye rule): the degree to which the theory or technique has been generally accepted in the scientific community.

And so, by the end of the twentieth century, the growing judicial scrutiny of science in the courtroom had reached its peak. The trial judge, who had long remained passive in the play of science in the adversarial courtroom, became an active gatekeeper, responsible for preventing junk science from reaching the jury and bamboozling them. Ironically, the growing judicial scrutiny of scientific evidence has not been driven by the deteriorating judicial faith in the powers of science to reveal truth. Indeed, despite the persistent malaise of deploying science in the adversarial courtroom, the legal profession has never wavered in its trust in science.

The steadfast belief in the powers of science to reveal the truth induced the judiciary to interpret conflicting scientific evidence not as a legitimate debate, but as a sign of moral decadence. During the twentieth century, as science became influential and contested, it pushed the judiciary into an increasingly active role in the effort to differentiate good from bad science and exclude the latter from the courtroom. Consequently, at the beginning of the twenty-first century, American lay judges found themselves deeper than ever before in the scientific territories of biostatistics, error rates, and experimental protocols, charged with the ungraceful task of weighing the merit of highly specialized scientific claims. How well could the lay judges meet these challenges? Would their new gatekeeping role lead to a better science? Would a better science lead to better law? These are some of the major concerns that have occupied the early twenty-first century discussion about the relations between science and law.

[*See also* **Animal and Human Experimentation; Cancer; Disease; Electricity and Electrification; Electronic Communication Devices, Mobile; Environmentalism; Environmental Protection Agency; Ethics and Medicine; Forensic Pathology and Death Investigation; Genetics and Genetic Engineering; Health Insurance; Mathematics and Statistics; Medical Malpractice; Motor Vehicles; Public Health;** *and* **Social Sciences.**]

BIBLIOGRAPHY

Legal Cases

Daubert v. Merrell Dow Pharmaceuticals, Inc. (1993) U.S. Reports 509: 579.

Diamond v. Chakrabarty. U.S. Reports (1980) 447: 303.

Frye v. United States. Federal Reports (1923) 293: 1013–1014.

General Electric Co v. Joiner. U.S. Reports (1997) 522: 136.

Kumho Tire Company, Ltd. v. Carmichael. U.S. Reports (1999) 526: 137.

People v. Dickerson. Northwestern Reports (1910) 129: 198.

Primary Works

Alder, Ken. *The Lie Detectors: The History of an American Obsession.* New York: Free Press, 2007.

Anonymous. "The Changing Role of the Jury in the Nineteenth Century." *Yale Law Journal* 74 (1964): 170–197.

Anonymous. "Expert Opinion on Ultimate Facts." In "Notes on Legislation." *Iowa Law Review* 26 (1941): 825–840.

Anonymous. "Expert Testimony." *American Law Review* 5 (1870): 227–246, 428–442.

Bayh–Dole Act, or Patent and Trademark Law Amendments Act. Title 35 of the U.S. Code, 1980.

Brodeur, Paul. Outrageous *Misconduct: The Asbestos Industry on Trial.* New York; Pantheon, 1985.

Caudill, David S., and Lewis H. LaRue No Magic Wand: The Idealization of Science in Law. Lanham, Md.: Rowman & Littlefield, 2006.

Choate, J. H. "Trial by Jury: Annual Address before the American Bar Association." *American Law Review* 33 (1898): 285–314.

Cole, Simon A. *Suspect Identities: A History of Fingerprinting and Criminal Identification.* Cambridge, Mass.: Harvard University Press, 2001.

De Ville, Kenneth Allen. *Medical Malpractice in Nineteenth-Century America: Origins and Legacy.* New York: New York University Press, 1990.

Dooley, D., ed. *Index to State Bar Association Reports and Proceedings,* 176–177. New York: Voorhis, 1942.

Dumit, Joseph. "Objective Brains, Prejudicial Images." *Science in Context* 12 (1999): 173–201.

Faigman, David L. *Laboratory of Justice: The Supreme Court's 200-Year Struggle to Integrate Science and the Law.* New York: Times Books, 2004.

Faigman, David L. *Legal Alchemy: The Use and Misuse of Science in the Law.* New York: W. H. Freeman and Co., 1999.

Federal Rules of Evidence. New York: Federal Judicial Center, 1975.

Foster, William L. "Expert Testimony—Prevalent Complaints and Proposed Remedies." *Harvard Law Review* 11 (1897–98): 169–186.

Giannelli, Paul. "The Admissibility of Novel Scientific Evidence: *Frye v. United States,* A Half-Century Later." *Columbia Law Review* 80 (1980): 1197–1250.

Golan, Tal. "The Emergence of the Silent Witness: The Legal and Medical Reception of X-rays in the USA." *Social Studies of Science* 34 (2004a): 469–499.

Golan, Tal. "Epidemiology, Tort, and the History of Causation in the Twentieth-Century American Courtroom." In *Worldly Science: Instruments, Practices, and the Law,* edited by Mario Biagioli and Jessica Riskin. New York: Palgrave MacMillan, 2012.

Golan, Tal. *Laws of Man and Laws of Nature: A History of Scientific Expert Testimony.* Cambridge, Mass.: Harvard University Press, 2004b.

Golan, Tal, ed. *Science and Law.* Special volume of *Science in Context,* 243 pp. Cambridge, U.K.: Cambridge University Press, 1999.

w

Gold, Steve. "Causation in Toxic Torts: Burdens of Proof, Standards of Persuasion, and Statistical Evidence." *Yale Law Journal* 96 (1986): 376–402.

Goldberg, Steven. *Culture Clash: Law and Science in America.* New York: New York University Press, 1994.

Green, Michael D. *Bendectin and Birth Defects: The Challenges of Mass Toxic Substances Litigation.* Philadelphia: University of Pennsylvania Press, 1996.

Gross, Hans. *Criminal Investigation: A Practical Handbook for Magistrates, Police Officers, and Lawyers.* Madras, India: Krishnamachari, 1906.

Haney, C. "Criminal Justice and the Nineteenth-Century Paradigm." *Law and Human Behavior* 6 (1982): 191–235.

Haskell, T. L. *The Authority of Experts: Studies in History and Theory.* Bloomington: Indiana University Press, 1984.

Holmes, Oliver Wendell. *The Common Law.* Boston: Little, Brown, 1881.

Horowitz, A. "Changing Views of Jury Power: The Nullification Debate, 1787–1988." *Law and Human Behavior* 15 (1991): 165–182.

Huber, Peter. *Galileo's Revenge: Junk Science in the Courtroom.* New York: Basic Books, 1991.

Imwinkelried, Edward. "A New Era in the Revolution of Scientific Evidence—A Primer on Evaluating the Weight of Scientific Evidence." *William and Mary Law Review* 23 (1981–1982): 261–290.

"In re: Agent Orange Product Liability Litigation." *Federal Supplements* 597 (1984): 749.

Jasanoff, Sheila. *Designs on Nature: Science and Democracy in Europe and the United States.* Princeton, N.J.: Princeton University Press, 2005.

Jasanoff, Sheila. *Science at the Bar: Law, Science, and Technology in America.* Cambridge, Mass.: Harvard University Press, 1995.

Jasanoff, Sheila. "Science and the Statistical Victim: Modernizing Knowledge in Breast Implant Litigation." *Social Studies of Science* 32 (2002): 37–69.

Johnson-McGrath, Julie. "Speaking for the Dead: Forensic Pathologists and Criminal Justice in the United States." *Science, Technology, and Human Values* 20 (1995): 438–459.

Kidd, A. "The Proposed Expert Evidence Bill." *California Law Review* 3 (1914–1915): 216–226.

Ladd, M. "Expert Testimony." *Vanderbilt Law Review* 5 (1952): 414–431.

Learned Hand. "Historical and Practical Considerations Regarding Expert Testimony." *Harvard Law Review* 15 (1901): 40–58.

Lucier, Paul. "Court and Controversy: Patenting Science in the Nineteenth Century." *The British Journal for the History of Science* 29 (1996): 139–154.

Lynch, Michael, Simon A. Cole, Ruth McNally, and Kathleen Jordan. *Truth Machine: The Contentious History of DNA Fingerprinting.* Chicago: University of Chicago Press, 2008.

McClellan, Roger O. "Human Health Risk Assessment: A Historical Overview and Alternative Paths Forward." *Inhalation Toxicology* 11, no. 14 (1999): 477–518.

McCormick, Charles T. "Expert Testimony as an 'Invasion of the Province of the Jury.'" *Iowa Law Review* 26 (1941): 819–840.

McCormick, Charles T. "Some Observations upon the Opinion Rule and Expert Testimony." *Texas Law Review* 23 (1954): 128–130.

Mnookin, Jennifer. "The Image of Truth: Photographic Evidence and the Power of Analogy." *Yale Journal of Law and the Humanities* 10, no. 1 (Winter 1998): 1–74.

Mnookin, Jennifer. "Science and Law." In *The Oxford Companion to American Law,* edited by Kermit L. Hall. New York: Oxford University Press, 2002.

Mohr, James C. *Doctors and the Law: Medical Jurisprudence in Nineteenth-Century America.* Oxford: Oxford University Press, 1993.

Moynihan, Daniel P. "Social Science and the Courts." *National Affairs* 54 (1979): 12–31.

Okun, Mitchell. *Fair Play in the Marketplace: The First Battle for Pure Food and Drugs.* Dekalb: Northern Illinois University Press, 1986.

Rudavsky, Shari. "Separating Spheres: Legal Ideology v. Paternity Testing in Divorce Cases." *Science in Context* 12 (1999): 123–138.

Sanders, Joseph. *Bendectin on Trial: A Study of Mass Tort Litigation.* Ann Arbor: University of Michigan Press, 1998.

Scheiber, Harry N. "The Impact of Technology on American Legal Development, 1790–1985." In *Technology, the Economy, and Society: The American Experience,* edited by J. Colton and S. Bruchey, pp. 83–124. New York: Columbia University Press, 1987.

Schuck, Peter H. *Agent Orange on Trial: Mass Toxic Disasters in the Courts.* Cambridge, Mass.: Belknap Press of Harvard University Press, 1987.

Schweber, Howard. "Law and the Natural Sciences in Nineteenth-Century American Universities." *Science in Context* 12 (1999): 101–121.

"Science in the Courts." *Scientific American* (1872): 167.

Susser, M. "Choosing a Future for Epidemiology: I. Eras and Paradigms." *American Journal of Public Health* 86 (1996): 668–677.

Susser, M. "Epidemiology in the United States after World War II: The Evolution of Technique." *Epidemiological Reviews* 7 (1985): 147–177.

Tesh, Sylvia. *Uncertain Hazards*. Ithaca, N.Y.: Cornell University Press, 2000.

Tighe, Janet A. *A Question of Responsibility: The Development of American Forensic Psychiatry, 1838–1930*. PhD thesis. University of Pennsylvania, 1983.

Wigmore, J. H. "Scientific Books in Evidence." *American Law Review* 26 (1892): 390–403.

Tal Golan

LAWRENCE, ERNEST O.

(1901–1958), physicist, Nobel laureate. Ernest Orlando Lawrence invented the cyclotron and created a new style of physics in his University of California Radiation Laboratory (UCRL). Later he helped develop the atomic bomb and promoted Edward Teller's development of the hydrogen bomb.

Lawrence was born in Canton, South Dakota, and attended the universities of South Dakota, Minnesota, and Chicago before earning his Ph.D. (1925) at Yale University, where he became an instructor. In 1928, he moved to the University of California, Berkeley, where he planned and developed the cyclotron, a magnetic resonance particle accelerator. Supported by the university, federal funds, and private philanthropy, he created the UCRL, which by the outbreak of World War II had produced three cyclotrons of increasing size and power. In 1939, he won the Nobel Prize in Physics for his invention of the cyclotron. During World War II Lawrence turned the laboratory's capability to the separation of uranium isotopes to provide the critical material for the first atomic bomb and played central roles in creating the Los Alamos Laboratory, where his colleague, J. Robert Oppenheimer, led the effort to design the

bomb, and the MIT Radiation Laboratory, where microwave radar was developed.

After World War II, Lawrence built a much larger laboratory, with new accelerators such as Louis Alvarez's proton linear accelerator and Edwin M. McMillan's synchrotron. Between 1949 and 1955, with McMillan and William Brobeck, he oversaw the construction of a 6 billion–electron-volt proton accelerator, the Bevatron. This machine permitted the discovery of many subatomic particles in the 1950s and 1960s. At the same time, Lawrence built massive production accelerators for enriching depleted uranium and, at the insistence of Edward Teller, a second nuclear weapons laboratory at Livermore, the Ernest O. Lawrence Livermore National Laboratory. In 2012, UCRL became the Ernest O. Lawrence Berkeley National Laboratory.

More than any other scientist, Lawrence created modern big physics, with its large laboratories; collaborative, interdisciplinary teams of scientists and engineers; and massive particle accelerators and detectors. His abilities and enthusiasm made him one of the most influential scientists in the United States during the World War II and Cold War eras. He died shortly after returning from the 1958 Geneva Conference on nuclear arms control.

[*See also* **Atomic Energy Commission; Manhattan Project; Military, Science and Technology and the; National Laboratories; Nobel Prize in Biomedical Research; Nuclear Regulatory Commission; Nuclear Weapons; Oppenheimer, J. Robert; Physics; Science;** *and* **Teller, Edward.**]

BIBLIOGRAPHY

Heilbron, John L., and Robert W. Seidel. *Lawrence and His Laboratory: A History of the Lawrence Berkeley Laboratory*. Vol. 1. Berkeley: University of California Press, 1989.

Herken, Greg. *Brotherhood of the Bomb: The Tangled Lives and Loyalties of Robert Oppenheimer, Ernest Lawrence, and Edward Teller*. New York: Henry Holt and Co., 2002.

Robert W. Seidel

LEDERBERG, JOSHUA

(1925–2008), Nobel laureate geneticist and microbiologist. Joshua Lederberg was born 23 May 1925, in Montclair, New Jersey, one of three sons of Zvi and Esther Lederberg. A 1941 graduate of Stuyvesant High School in New York City, he entered Columbia University at age 16 and joined the accelerated wartime Navy V-12 program. In 1944 he received a BA in zoology and started medical school at Columbia.

In response to the seminal publication by Avery, McCarty, and McLeod (1944) on DNA-medicated transformation in pneumococcus, Lederberg and his mentor, Francis J. Ryan, who was a prominent biochemist, attempted to demonstrate this phenomenon in *Neurospora*. Lederberg soon turned his attention to sexual phases of bacteria and isolated mutants of *Escherichia coli* to study possible recombination events. Edward Tatum had a collection of mutants of *E. coli* and Lederberg proposed a collaboration to study possible genetic recombination in that organism. Within six weeks of joining Tatum at Yale in early 1946, Lederberg had obtained evidence of recombination in Tatum's *E. coli* K12 strain. This discovery resulted largely from his lucky choice of one of the rare, recombination proficient strains of *E. coli*.

By the fall of 1947 Lederberg was awarded a PhD from Yale University and received an offer of a faculty position in the Department of Genetics at the University of Wisconsin. Lederberg and his wife, Esther Zimmer Lederberg, also a geneticist, spent the next 11 years in Madison, where they continued to explore the genetics of bacteria and bacteriophages. He never returned to his medical studies.

Lederberg received the Nobel Prize in Physiology or Medicine in 1958 for his work in bacterial genetics. He also made significant contributions in computer science, artificial intelligence, and exobiology. Lederberg's broad interests, incisive mind, and openness to new ideas coupled with a strong social conscience and unusual organizational skills all combined to make him a forceful spokesperson for science as well as a leading public intellectual.

In 1959 Lederberg moved to Stanford University as the chair of genetics and expanded his interests into the application of computers to biological problems including artificial intelligence, structural chemistry, and information management. He developed a "dendritic algorithm," dubbed DENDRAL, which is regarded as a significant contribution in computer science. His interests encompassed the challenges of space biology and he coined the term "exobiology" for this new field.

Lederberg assumed the presidency of the Rockefeller University in 1978, a position that allowed him a platform as a spokesperson for science and public policy. He retired in 1990, but continued as one of the nation's most visible and respected public intellectuals in science.

[*See also* **Artificial Intelligence; Biological Sciences; Computer Science; Genetics and Genetic Engineering; Germ Theory of Disease; Medicine; Nobel Prize in Biomedical Research; Space Science;** *and* **Zoology.**]

BIBLIOGRAPHY

Bradley, S. Gaylen. "Joshua Lederberg, 1925–2008." In *Biographical Memoirs*. Washington, D.C.: National Academy of Sciences, 2009.

Strick, James E. "Creating a Cosmic Discipline: The Crystallization and Consolidation of Exobiology, 1957–1973." *Journal of the History of Biology* 37 (Spring 2004): 131–180.

Wolfe, Audra J. "Germs in Space: Joshua Lederberg, Exobiology, and the Public Imagination, 1958–1964." *Isis* 93 (June 2002): 183–205.

William C. Summers

LEE, TSUNG-DAO

(Li Zhengdao in pinyin, 1926–), a leading theoretical physicist, a prominent Chinese American, and one of the most influential scientists in U.S. and Chinese science and educational policy.

Lee was born in Shanghai, China, to father Li Junkang, a fertilizer factory manager, and mother Zhang Mingzhang. He received primary education at home but was forced into exile inland during the Japanese invasion in the 1940s. He studied physics at Zhejiang University for one year before enrolling in the Southwest Associated University in Kunming in 1944. He did so well at Southwest that even before graduation his professors selected him as part of a Chinese Nationalist government mission to the United States in 1946 to learn how to make an atomic bomb.

Because secrecy requirements made that mission impossible, Lee enrolled instead at the University of Chicago, pursuing graduate studies in physics with the eminent nuclear physicist Enrico Fermi. After receiving his PhD, in late 1949, with a thesis on white dwarf stars, Lee spent several months at Yerkes Observatory in southeastern Wisconsin before taking up an assistant professorship at the University of California, Berkeley, where he married Jeanette Chin. A year later, he moved to the Institute for Advanced Study at Princeton, New Jersey, before settling down two years later at Columbia University; he started as an assistant professor of physics in 1953 but was quickly promoted to full professor in 1956.

That same year Lee collaborated with Chen Ning Yang, a fellow Chinese student at Chicago who was then a physicist at the Princeton institute, to formulate a theory that would mark a turning point in modern physics. Studying the behaviors of the so-called "strange particles," they proposed that the long-accepted left–right parity broke down in a nuclear process called "weak interactions." The idea was met with strong resistance initially, but soon was confirmed by an experiment conducted by Chien Shiung Wu, Lee's Chinese American colleague at Columbia, in collaboration with scientists at the Bureau of Standards in Washington, D.C. Lee and Yang won the Nobel Prize for Physics in 1957.

Lee and Yang continued their fruitful partnership on many important topics in physics, facilitated by Lee's spending two years from 1960 to 1962 at the Princeton institute. But the partnership ended in 1962 when personal friction, in part over credit for their joint scientific discoveries, reached a breaking point. Lee returned to Columbia, where he remains a professor of physics, conducting research on a wide range of areas

from particle physics to high-temperature superconductivity to dark energy. He also played an active role in the development of the Relativistic Heavy-Ion Collider at the Brookhaven National Laboratory in Long Island that has produced important discoveries since its completion in 2000.

The reopening of U.S.–China relations in the early 1970s allowed Lee to return to China in 1972 for his first visit back. He has since devoted much time and energy to U.S.–China scientific collaboration, including the creation of the China–U.S. Physics Examination and Application program that brought approximately one thousand Chinese physics graduate students ("Lee scholars") to the United States. Lee also played an active role in Chinese science and education policy, especially in the introduction of the postdoctoral system, the establishment of the Chinese National Natural Science Foundation, and the building of the Beijing Electron–Positron Collider.

[*See also* **Fermi, Enrico; Lee, Tsung-Dao; Nobel Prize in Biomedical Research; Nuclear Weapons; Physics;** *and* **Yang, Chen Ning.**]

BIBLIOGRAPHY

Bernstein, Jeremy. *A Comprehensible World*. New York: Random House, 1967. Contains a profile of Tsung Dao Lee and Chen Ning Yang, "A Question of Parity," first published in *The New Yorker*, 12 May 1962, pp. 49–103.

Lee, T. D. *T. D. Lee: Selected Papers*, edited by G. Feinberg. 3 vols. Boston: Birkhäuser, 1986.

Novick, Robert, ed. *Thirty Years since Parity Nonconservation: A Symposium for T. D. Lee*. Boston: Birkhäuser, 1986.

Zuoyue Wang

LEIDY, JOSEPH

(1823–1891), naturalist and polymath. Leidy was born, raised, and educated in Philadelphia, where he spent his entire professional career. Born into a fifth-generation German American family, his interest in nature from an early age persisted so that he became a natural historian of astonishing breadth, documented in over six hundred publications and books, some beautifully illustrated by him. In 1844 he received an MD degree from the University of Pennsylvania, and after a brief, unhappy period in clinical practice, he joined Professor William Horner, professor of Anatomy at Penn, to begin his career as a lecturer and investigator. In 1846 he became a member of the Academy of Natural Sciences of Philadelphia, serving as curator and president, and in 1853 succeeded Horner as professor, a position he retained for the rest of his life.

As well as being the foremost American comparative anatomist of his time with a popular textbook of human anatomy to his credit, he was one of the earliest productive American microscopists. He has been credited with the discovery of the larvae of *Trichinella* in pork, the source of human infection, thus completing the life cycle of this scourge of public health; he advocated the heating of pork to eliminate infection. Leidy was the foremost of early American microscopists and protozoologists, a discipline he founded. He also studied numerous parasites and, extending his range of interest, he was the major American authority on prehistoric fossils—vertebrate paleontology. He has been considered the "father" of all three disparate areas of research. Being a retiring man, Leidy left paleontology to avoid the vitriolic controversy between his student Edward Cope and Othniel Marsh. As a paleontologist he had shown that the horse existed in prehistoric North America and described for the first time hundreds of extinct animals that roamed the plains (camel, rhinoceros, lion, saber tooth tiger, titanotherium, oreodonts, dinosaur). Leidy was the first to describe a dinosaur in America, *Hadrosaurus foulkii*, assigning to it its modern configuration. His expertise extended to entomology, botany, and precious stones.

In 1871, he founded the department of Natural History at Swarthmore College, where he lectured. In addition to all his other responsibilities, he founded the department of biology at the University of Pennsylvania and was the head of the

Wagner Free Institute of Science of Philadelphia. Not only was he an academic research biologist, but also he was heavily involved in teaching, the administration of research institutions, and solving practical problems such as combating insect infestations of public parks. The achievements of this unheralded man of protean erudition are astonishing. He was awarded many honors and prizes and was a charter member of the National Academy of Sciences (U.S.).

[*See also* **Academy of Natural Sciences of Philadelphia; Anatomy and Human Dissection; Biological Sciences; Botany; Cope, Edward Drinker; Dinosaurs; Entomology; Marsh, Othniel Charles; National Academy of Sciences; Paleontology; Public Health; Science;** *and* **Zoology.**]

BIBLIOGRAPHY

Warren, Leonard. *Joseph Leidy, the Last Man Who Knew Everything.* New Haven, Conn.: Yale University Press, 1998.

Leonard Warren

LEOPOLD, ALDO

(1887–1948), conservation scientist, writer, and philosopher. Following graduation from Yale University's Forest School in 1909, Leopold joined the U.S. Forest Service, where he became a leading innovator in soil conservation, range management, recreational planning, game management, and wilderness protection. His fieldwork in these years provided the foundations for understanding landscape-scale ecosystem processes such as fire and soil erosion, leading in turn to new approaches to land management. Concerned by the accelerating fragmentation of the nation's wild lands, he led efforts that in 1924 resulted in the designation of the nation's first wilderness area within the Gila National Forest in New Mexico. After 1928, Leopold devoted himself to the development of wildlife ecology and management as a distinct field, first as an independent researcher (1928–1933) and then as professor at the University of Wisconsin (1933–1948). His fundamental contribution in these years was to apply concepts from the science of ecology to the conservation of wildlife populations and habitats. His text *Game Management* (1933) was the first in the field.

Through his many nontechnical writings, including policy statements, editorials, and nature essays, Leopold defined a new approach to conservation, one that sought to blend elements of older utilitarian and preservationist traditions within a broader context of contemporary ecological understanding. He argued that successful conservation involved more than the simple economic goal of perpetual yields of discrete resources and products; rather, conservation ought to promote "the capacity for self-renewal" in "soils, waters, plants, and animals, or collectively, the land." Concerned by the accelerated pace of technological change and its impact on biotic diversity and ecological processes, Leopold in the postwar years focused his writing on the ethical aspects of human–nature relationships.

In the final years of his life, Leopold compiled many of his essays into a collection published posthumously as *A Sand County Almanac* (1949). *Sand County* became, along with Rachel Carson's *Silent Spring* (1962), a basic text for the modern environmental movement. Especially influential was its capstone essay, *The Land Ethic*, in which Leopold argued for an expansion of the sphere of human ethical concern to include the natural world. Leopold's writings have remained influential, providing important foundations for such emerging interdisciplinary fields as environmental history, ecological economics, environmental ethics, restoration ecology, and conservation biology.

[*See also* **Carson, Rachel; Conservation Movement; Ecology; Environmentalism; Fish and Wildlife Service, U.S.; Fisheries and Fishing; Forestry Technology and Lumbering; Forest Service, U.S.; Muir, John; Science;** *and* **Sierra Club.**]

BIBLIOGRAPHY

Flader, Susan. *Thinking Like a Mountain: Aldo Leopold and the Evolution of an Ecological Attitude toward Deer, Wolves, and Forest.* Madison: University of Wisconsin Press, 1994.
Meine, Curt. *Aldo Leopold: His Life and Work.* Madison: University of Wisconsin Press, 2010.
Newton, Julianne Lutz. *Aldo Leopold's Odyssey.* Washington, D.C.: Island Press/Shearwater Books, 2006.

Curt Meine

LEWIS AND CLARK EXPEDITION

In a January 1803 message to Congress, President Thomas Jefferson called for an expedition up the Missouri River and west to the Pacific. With the Louisiana Purchase later that year, the project took on even greater significance. Jefferson chose the army captain Meriwether Lewis (1774–1809) to lead the expedition. Lewis selected as his partner a fellow officer, William Clark (1770–1838). More than 40 men, including York, Clark's slave, composed the Corps of Discovery as it started up the Missouri in a keelboat and three canoes on 14 May 1804. By late October the expedition had reached present-day central North Dakota, where the members established their winter quarters, Fort Mandan.

In April 1805, Lewis and Clark sent the keelboat downriver before resuming their journey west, accompanied by a young Shoshone woman, Sacagawea (1786–1812), and her French Canadian husband. They reached the source of the Missouri and advanced up a tributary, the Jefferson, before having to abandon their boats. Using horses obtained from Sacagawea's Shoshones, the expedition crossed the Continental Divide at Lemhi Pass and surmounted the Bitterroot Mountains via the Lolo Trail. At the Clearwater River they entrusted their horses to the Nez Percés, built canoes, and floated down the Clearwater, Snake, and Columbia rivers to the Pacific, which they reached on 18 November 1805. They named their winter quarters Fort Clatsop.

In late March 1806, the corps started home. At the mouth of Lolo Creek the expedition split, Clark's contingent returning as they had come and Lewis's group advancing directly east to the falls of the Missouri, where the units were reunited. On 23 September 1806, after an absence of twenty-eight months, the Corps of Discovery arrived at St. Louis.

The Lewis and Clark Expedition, which produced extensive published records and journals, was one of the most successful in the annals of world exploration. It destroyed the concept of an all-water route to the Pacific and helped fix in the public mind the vast extent of the Louisiana Purchase territory and the Pacific Northwest. The expedition made important geographical discoveries based on detailed readings of longitude and latitude. Lewis was mainly responsible for the crucial astronomical readings; Clark drafted the maps and charts. They also recorded plants and animals, noting their relationship to the physical environment. The expedition identified a number of species new to Western science. And Lewis and Clark made important ethnological and linguistic observations of American Indians, including some of the first observations of Indian groups in the Upper Missouri River Valley, the nearby Rocky Mountain region, and the Northwest Coast.

[*See also* Jefferson, Thomas; *and* Science.]

BIBLIOGRAPHY

Ambrose, Stephen E. *Undaunted Courage: Meriwether Lewis, Thomas Jefferson, and the Opening of the American West*, New York: Simon and Schuster, 1996.

Cox, Robert S., ed. *The Shortest and Most Convenient Route: Lewis and Clark in Context*. Philadelphia: American Philosophical Society, 2004.

Cutright, Paul Russell. *Lewis and Clark: Pioneering Naturalists*. Urbana: University of Illinois Press, 1969.

Moulton, Gary E., ed. *The Journals of Lewis and Clark Expedition*.13 vols. Lincoln: University of Nebraska Press, 1983–2001.

Seefeldt, Douglas, Jeffrey L. Hantman, and Peter S. Onuf, eds. *Across the Continent: Jefferson, Lewis and Clark, and the Making of America*. Charlottesville: University of Virginia Press, 2005.

Richard A. Bartlett;
updated by Hugh Richard Slotten

LIFE EXPECTANCY

One of the greatest achievements of the contemporary era has been the reduction of death rates and the prolongation of human life. One way of summarizing this mortality transition is life expectancy, which expresses the average number of years of life remaining to a person at some age, often at birth. This measure is derived from life tables and can be calculated from data at a point in time for persons of different ages (period life expectancy) or by following the same groups of people over time as they age (cohort life expectancy). The data used usually come from census counts by age and sex, as well as from vital statistics of deaths, also by age and sex. But other data, such as genealogies and family reconstitutions, can be used, as can other methods.

Life expectancy in the United States has evolved through several stages. During the early colonial period, life expectancy was relatively short, death rates were high and variable, and epidemics of infectious disease were common. Life expectancy at birth generally ranged from 20 to 30 years. By the late seventeenth century, conditions had begun to improve, and by the late eighteenth century, mortality conditions were quite favorable by world standards. Thomas R. Malthus commented in 1798 that mortality conditions were quite benign in the new United States and had been for a while. Mortality was lowest in New England, with life expectancy at birth, abbreviated as "e(0)," at 35 to 60 years; was more severe in the middle colonies, where the e(0) was 30 to 45 years; and was highest in the South, where the e(0) was 25 to 35 years. Gradually, epidemic diseases such as measles and smallpox became endemic and joined malaria, dysentery, pneumonia, and bronchitis and tuberculosis as major causes of endemic, baseline mortality. Infectious and parasitic diseases accounted for most deaths and continued to do so until the twentieth century, when degenerative diseases such as cancer, cardiovascular diseases, and diabetes became dominant.

The Mortality Transition. The sustained mortality transition did not commence in the United States until about the 1870s. Life expectancy likely reached a high point in the late eighteenth century and then declined until the later nineteenth century. For example, genealogical data yield an expectation of life at age 10 (abbreviated "e(10)") of almost 57 years for white males in 1790–1794, but that expectation had declined to 48 years by 1855–1859. These results are supported by data on human stature, another indicator of physical well-being. Heights of Civil War military recruits, West Point cadets, college students, and others (mostly males) declined from those born in the 1830s to those born in the 1870s, consistent with a deteriorating disease environment. Information on specific cities with adequate vital statistics (New York, Boston, Philadelphia, Baltimore, New Orleans) reveals constant or rising mortality prior to the Civil War, with substantial mortality peaks as a result of cholera—which first appeared in the United States in 1832—typhoid fever, and yellow fever.

During the nineteenth century, sources of data improved. The U.S. Census, a federal mandate, was taken decennially from 1790. Questions about mortality in the year prior to the Census were asked in the censuses of 1850–1900. But collection of vital statistics was left to state and local governments and consequently was uneven. In 1842, Massachusetts became the first state to commence comprehensive registration of births, deaths, and marriages. Quality was good by about 1855. Several states followed suit, but in 1900 the Death Registration Area was formed with only 10 states and the District of Columbia. The entire United States was not covered until 1933.

By the middle of the nineteenth century there was enough information to make reasonable national estimates of life expectancy. By 1850, e(0) for the white population was about 38 years,

e(10) was about 47 years, and the infant mortality rate (deaths below age one per one thousand live births) stood at 217. The sustained mortality transition for the nation as a whole began only in the 1870s. The e(0) for whites overall changed little between 1850 and 1880 and then began to rise from about 40 years in 1880 to about 52 years in 1900, about 69 years in 1950, and about 78 years in 2006. Much of the change in the first half of the twentieth century was caused by the continued decline in infectious and parasitic diseases, from 43 percent of total deaths in 1900/1902 to just 7.5 percent in 1949/1951. A considerable portion of the increase in expectation of life in the later twentieth century was the result of improvements in longevity at older ages. For example, expectation of life at age 60 improved from 17 years in 1949/1951 to 22.4 years in 2006.

Variations between Groups. The black population suffered a substantial mortality disadvantage, although blacks were protected somewhat by their more rural residence earlier in the twentieth century. (About 80 percent of the black population lived in rural areas in 1900, in contrast to 58 percent of the white population.) In 1900 the e(0) for blacks was about 20 percent lower than that for whites, and their infant mortality rate was about 54 percent higher. The situation had been even worse around 1850, when blacks, mostly slaves, had an estimated e(0) of 23 to 40 percent lower than that for whites—and an estimated infant mortality rate of about 350—a full 61 percent higher than that for whites. Although between 1850 and 1900 the absolute differences in the infant mortality rate between blacks and whites had narrowed to about 8 infant deaths per 1,000 live births, the relative difference had grown. The black infant mortality rate, 13.4, was twice as high as that for whites, 5.6. In terms of e(0), by 2006 the difference between whites and blacks had narrowed, although the e(0) for the black population, 73 years, was still five years below that of the white population.

Overall, in the early twenty-first century, the United States did not stand well in the world in terms of its ranking of expectation of life and the infant mortality rate. According to World Bank data for 2006, the United States ranked 44th in the world for e(0) and 46th for the infant mortality rate. Even if the whole American population had the same infant mortality rate as the white population (5.6), the rank would improve only to 38th. The reasons are multiple, but a lack both of widely distributed, basic medical care and of public-health interventions, especially prenatal care, has played a role. Nonetheless, the absolute differences between the United States and the rest of the world have not been large.

In terms of other mortality differentials, women have tended to live longer than men. In 1850, women had an e(0) that was 6 percent higher than that of men, a gap that had narrowed to only about 2 percent by 1900. But women outlived men by almost 7 percent, or five years, by 2006.

Differences between rural and urban people were historically also large. In the nineteenth century, cities were distinctly less healthy places to live. Around 1830, the e(10) was 51 years in forty-four New England towns, 42 percent higher than the average for Boston, New York City, and Philadelphia (at 35.9 years). By 1900, the probability of a child surviving to age five was 22 percent worse in urban than in rural areas. This urban penalty disappeared by 1920, when the improved public-health programs and investments in urban America overcame the disadvantages of crowding, problems with water supplies and sewerage disposal, food contamination, lack of rubbish removal, and the large influx of immigrants. Among the foreign born, life expectancy historically had usually been lower than that of native-born whites, partly because the foreign born tended to have a lower socioeconomic status and partly because they tended to concentrate in urban areas.

Finally, regional variations in mortality were substantial at the first point for which they can be observed for the nation as a whole—around 1900. The lowest mortality areas were in the Midwest, and the highest mortality was found in the South and New England. These regional differences converged during the twentieth century thanks to the spread of public-health programs and medical care.

[*See also* **Biological Sciences; Cancer; Cardiology; Cholera; Diabetes; Disease; Gender and Science; Malaria; Public Health; Public Health Service, U.S.; Race and Medicine; Smallpox; Tuberculosis; Typhoid Fever;** *and* **Yellow Fever.**]

BIBLIOGRAPHY

Bell, Felicitie C., Alice H. Wade, and Stephen C. Goss. *Life Tables for the United States Social Security Area 1900–2080*. Actuarial Study No. 107. Baltimore: U.S. Department of Health and Human Services, Social Security Administration, Office of the Actuary, 1992.

Easterlin, Richard A. "The Nature and Causes of the Mortality Revolution." In *Growth Triumphant: The Twenty-First Century in Historical Perspective*, pp. 69–82. Ann Arbor: University of Michigan Press, 1996.

Haines, Michael R. "The American Population, 1790–1920." In *The Cambridge Economic History of the United States*, Vol. 2, edited by Stanley L. Engerman and Robert E. Gallman. Cambridge, U.K.: Cambridge University Press, 1998.

Haines, Michael R., ed. "Vital Statistics." In *Historical Statistics of the United States*, edited by Susan Carter, Scott Sigmund Gartner, Michael R. Haines, Alan L. Olmstead, Richard Sutch, and Gavin Wright, Vol. 1, chapter Ab. Millennial ed. Cambridge, U.K.: Cambridge University Press, 2006.

Kunitz, Stephen J. "Mortality Change in America, 1620–1920." *Human Biology* 56, no. 3 (September 1984): 559–582.

Pope, Clayne L. "Adult Mortality in America before 1900: A View from Family Histories." In *Strategic Factors in Nineteenth Century American Economic History: A Volume to Honor Robert W. Fogel*, edited by Claudia Goldin and Hugh Rockoff, pp. 267–296. Chicago: University of Chicago Press, 1992.

Preston, Samuel H., and Michael R. Haines. *Fatal Years: Child Mortality in Late Nineteenth-Century America*. Princeton, N.J.: Princeton University Press, 1991.

Michael R. Haines

LINDBERGH, CHARLES

(1902–1974), aviator. Lindbergh burst upon the world stage on 20–21 May 1927 when he piloted his single-engine Ryan monoplane, *The Spirit of St. Louis,* solo across the Atlantic. Although this was the signature achievement of his life, Lindbergh's impact went well beyond his epic flight. Reared on a farm in Little Falls, Minnesota, the son of a farm-bloc congressman, Lindbergh in 1920 enrolled as an engineering student at the University of Wisconsin. He dropped out after two years, learned to fly, and spent the summer of 1923 barnstorming through the West. Seeking more experience and training, Lindbergh enlisted as a U.S. Army flying cadet; trained in San Antonio, Texas; and graduated first in his class in 1925. After the military, he found employment as an air-mail pilot, the most demanding and dangerous type of flying in this period.

Learning of a $25,000 prize to fly from New York to Paris, Lindbergh saw an unmatched aviator's challenge. But he also sought to further the cause of aviation and to demonstrate the capabilities and reliability of the airplane. His successful thirty-three and a half hour, nonstop flight from Roosevelt Field on Long Island, New York, to Le Bourget Field just outside Paris not only vaulted Lindbergh to instant fame but also contributed to renewed enthusiasm and investment in the nascent U.S. aviation industry and commercial air transport. Other crucial building blocks in the 1920s contributed to later American preeminence in aerospace, such as the Air Commerce Act of 1926 and the Guggenheim Fund for the Promotion of Aeronautics, but Lindbergh's achievement provided an important catalyst.

Lindbergh's impact on aeronautics continued in the 1930s as he made several pioneering transoceanic flights with his wife, Anne Morrow Lindbergh, for Pan American Airways and other emerging airlines. He was an early and ardent supporter of the research of the rocket pioneer Robert H. Goddard. During World War II, Lindbergh served as a consultant to several aircraft manufacturers and as a civilian adviser to the U.S. military. In 1944, as a civilian, he flew 50 combat missions in the Pacific theater and shot down one Japanese fighter. After the war he served as a special consultant for research and development to the U.S. Air Force.

The 1932 kidnapping and murder of the Lindberghs' infant son, and the ensuing trial and execution of Bruno Hauptmann in 1935, stirred nearly as much public and media attention as Lindbergh's Atlantic crossing. On the eve of World War II, as a leading spokesman for the noninterventionist pressure group the America First Committee, Lindbergh drew widespread criticism for highly publicized statements regarding the superiority of German airpower, for his consistent failure to denounce German atrocities, and for anti-Semitic utterances made in a speech delivered in Des Moines, Iowa, on 11 September 1941. Lindbergh later restored his reputation somewhat as an activist in the environmentalist movement. A talented and prolific writer, he wrote seven books and numerous articles.

[*See also* **Airplanes and Air Transport; Environmentalism; Goddard, Robert H.;** *and* **Military, Science and Technology and the.**]

BIBLIOGRAPHY

Berg, A. Scott. *Lindbergh.* New York: G. P. Putnam's Sons, 1998.

Crouch, Tom D., ed. *Charles A. Lindbergh: An American Life.* Washington, D.C.: Smithsonian Institution Press, 1977.

Davis, Kenneth S. *The Hero: Charles A. Lindbergh and the American Dream.* Garden City, N.Y.: Doubleday, 1959.

Lindbergh, Charles A. *Autobiography of Values.* New York: Harcourt Brace Jovanovich, 1976.

Peter L. Jakab

LINGUISTICS

Linguistics as practiced in America implies three overlapping subject categories: any and all work in the field done by Americans (defined here in terms of post-European settlement), distinctive contributions to the field made by Americans, and research done on indigenous American languages.

Amerindian research began with the work of European missionaries in the colonial period, such as the English Puritan John Eliot's 1666 grammar of the Massachusetts language, prepared as an aid to Bible translation. In the nineteenth century interest turned to ethnology: the American Philosophical Society of Philadelphia sponsored work on the classification of Amerindian languages (hence peoples) via grammatical analysis, whereas the Smithsonian Institution and the U.S. Bureau of Ethnology focused on compiling native vocabularies, a "salvage" method pursued in light of the advancing reduction of Indian populations.

By the start of the twentieth century, American linguists tended, as had long been done in Europe, to model their field on natural-scientific methodologies—originally those of the life and earth sciences and ultimately those of mathematics and physics. The aim was to increase the field's autonomy, and hence prestige, by separating linguistics from humanistic scholarship and restricting investigation to the phenomena of "language itself." A forerunner in this regard was the Yale linguist William Dwight Whitney (1827–1894), who stressed the "uniformitarian" notion (borrowed from geology) of continuity in the kinds of forces that have shaped language throughout history. This thesis became a basic underpinning of later "synchronic" approaches to language study, as opposed to the then dominance of historical linguistics.

The early twentieth century saw the emergence of linguistic structuralism, which viewed languages synchonically, as internally organized systems; it also gave priority to form over meaning. The Swiss linguist Ferdinand de Saussure's posthumous *Cours de linguistique générale* (1916) launched this perspective internationally, yet a beginning had already been made in the United States in the investigation of Amerindian languages developed by the German-born anthropologist Franz Boas (1858–1942). Continued in field studies conducted by Boas's students, this approach sought to analyze living languages according to their built-in systems of organization.

The preeminent American linguists of the first half of the twentieth century were Edward Sapir (1884–1939) and Leonard Bloomfield (1887–1949), who converged on the idea that each language was composed of a distinctive set of meaningful sound units (*phonemes*), the analysis of which yielded a profile of a language's characteristic structure. In addition, Sapir's psychologistic outlook inspired the so-called "Sapir–Whorf hypothesis," the controversial notion that different languages differently structure a speech community's mental perception of reality.

Bloomfield took a different tack by treating mental activity as well as social context as irrelevant to linguistic investigation, a delimiting of the object of study that shaped the research agenda for the next generation. Bloomfield's successors, including Zellig Harris (1909–1992) and Charles Hocket (1916–2000), among others, produced distributive analyses of the grammatical elements in various languages based on an exhaustive review of the linguistic environments in which each element occurred. The presentation of this work was characterized by the use of algebraic formulae, tabular displays, and statistics.

Structuralist methods spread globally in this period through the work of the Summer Institute of Linguistics (SIL), established in the 1930s to train researchers in the study of unrecorded languages as a prerequisite for Bible translation. Kenneth L. Pike (1912–2000) contributed investigations of numerous aboriginal languages and, in the 1950s, developed for this purpose his *tagmemic* method of analysis.

Transformational-generative grammar, the logicodeductive approach associated with Noam Chomsky (b. 1928) and his followers, attained mainstream status in the 1960s. In a significant break with the past, Chomsky focused on language's mental wellsprings, in contrast with what he considered the structuralists' concern with mere surface patterns. Yet although he aimed to discover the innate mental structures that govern language universally, he also regarded those structures as determining even particular grammatical forms—this opposed to the Bloomfieldians' assumption that each language's distinctive features resulted from social–behavioral learning. Working in a more empirical mode, Joseph Greenberg (1915–2001) sought linguistic universals through the comparison of large groups of related languages, chiefly in Africa.

Rejecting what they regarded as a pervasive focus on abstract language systems (including the Chomskyan method), William Labov (b. 1927) and his associates at Columbia University launched in the late 1960s the field of sociolinguistics, a bid to study actual speech behavior, including subgroup vernaculars. By the beginning of the twenty-first century, sociolinguistics had spawned multiple subfields interacting with both diachronic and structuralist perspectives, all sharing roughly equal status with generativist linguistics. Problems of regional and local variation, the role of social contact, and random occurance versus regularity remain contested within sociolinguistics, pushing the larger question of the extent to which linguistic analysis can filter out the influence of arbitrary volition. Meanwhile, the Chomskyan notion that the autonomous structure of language provides the key to the structure of the mind helped inspire "cognitive science," an umbrella field drawing together work in linguistics, psychology, computer science, and artificial intelligence.

[*See also* **American Philosophical Society; Anthropology; Artificial Intelligence; Boas, Franz; Computer Science; Mathematics and Statistics; Physics; Psychology; Smithsonian Institution;** *and* **Social Sciences.**]

BIBLIOGRAPHY

Alter, Stephen G. *William Dwight Whitney and the Science of Language*. Baltimore: Johns Hopkins University Press, 2005.

Andresen, Julie Telel. *Linguistics in America, 1769–1924: A Critical History*. London and New York: Routledge, 1990.

Hymes, Dell, and John Fought. *American Structuralism*. New York: Mouton, 1975. The most complete treatment available of this mid-twentieth-century movement.

Joseph, John E. *From Whitney to Chomsky: Essays in the History of American Linguistics*. Amsterdam and Philadelphia: John Benjamins, 2002. A collection of the author's meticulous work, covering the crucial period since the late nineteenth century.

Koerner, E. F. K. *Toward a History of American Linguistics*. London and New York: Routledge, 2002.

Koerner, E. F. K., and R. E. Asher, eds. *Concise History of the Language Sciences: From the Sumerians to the Cognitivists*. Tarrytown, N.Y.: Elsevier Science, 1995. Valuable chapters on American structuralism, twentieth-century linguistics, and the Chomskyan revolution.

Stephen G. Alter

LITERATURE AND SCIENCE

On the morning of 26 June 2000, President William Jefferson Clinton held a press conference to announce the historic completion of a draft of the map of the human genome. He delivered his announcement from the East Room, the largest room in the White House, often used for press conferences. The president made much of the location for this event, summoning another achievement that he likened to the mapping of the genome. "Nearly two centuries ago," he declaimed, "in this room, on this floor, Thomas Jefferson and a trusted aide spread out a magnificent map, a map Jefferson had long prayed he would get to see in his lifetime." During his presidency, Jefferson had partitioned the room to serve as the bedroom and office of the aide, Meriwether Lewis, and there, Clinton suggested, Jefferson and Lewis viewed the map that chronicled the "expedition across the American frontier all the way to the Pacific" and "defined the contours and forever expanded the frontiers of our continent and our imagination" (Clinton, 2000).

The comparison between the map of the genome and the map of the expedition seems serendipitous, a rhetorical gesture resulting from the location of the press conference and the use of the term "map." Whereas Lewis and Clark's map represents their sense of what they saw as they crossed the continent from Saint Louis to the Pacific Ocean, the map of the human genome depicts the relative location of genes on a chromosome.

As cognitive tools, however, both maps manifest ways of knowing the world, and, like all maps, they share the implication of discovery, obscuring the acts of interpretation through which the mapmakers produce and make sense of information, territorial and biological. Lewis and Clark's map helped to produce "America"; Clinton marvels that the genome map is "of even greater significance....Without a doubt,...the most important, most wondrous map ever produced by human kind" in its depiction of "the language in which God created life" (Clinton, 2000). If the first map manifested the destiny of a nation, the second map charts nothing less than the blueprint of humanity. The maps are important, he suggests, for what they make imaginable.

For him both maps represent a paradigm shift defined, in Thomas Kuhn's words, as new "ways of seeing the world and of practicing science in it" (Kuhn, 1996, p. 4). Such radical changes, however, are never sudden. Rather, they mark a saturation point following gradual but widespread shifts in assumptions about the world: social, geopolitical, cultural, scientific. Changing textual and visual vocabularies and storylines promote those transformations until they become conventional. With their particular attention to language, image, and storyline, literary works amplify the changes, facilitating introspection as they imaginatively engage with the mutating assumptions of a paradigm shift. They thereby offer insight into the literariness of life: the way the human imagination makes sense of the world and negotiates the complexities of lived experience.

The nation's earliest political figures understood the political power of the imagination. Noah Webster's call for an "America...as independent in *literature* as she is in *politics*, as famous for *arts* as for *arms*" was characteristic of the widespread summons for the citizens of the new nation to help articulate "America" into existence through the distinction of its achievements and the uniqueness of its cultural production. The new nation was, in their view, a territory and an approach, a new way of seeing and being in the world. The sciences were not exempt; as the historian Perry Miller notes, a cultural observer "could still happily use 'science' and 'letters' as interchangeable terms" in early America. In "Science—Theory and Approach," the final section of his incomplete *The Life of the Mind in America, from the Revolution to the Civil War*, Miller describes the dilemma of the scientist as both world- and nation-maker through the words of the nineteenth-century man of letters William Barton, who, eulogizing his uncle, the astronomer, inventor, and surveyor David Rittenhouse, called the natural scientist "a citizen of the world" who nonetheless embraces the "'spirit of patriotism which ever stimulates a good man to contribute his primary and most important services to his own country'" (Barton, quoted in Miller, 1965, p. 270). For the most fervent nationalists, that patriotism found expression in the achievement of "Americans" in this universal field as well as in its distinctive landscape, from geological formations to its unique flora and fauna. With the pithy phrase "nature's nation," Miller marked the terms through which the settler colony forged a national identity. And the relationship to a terrain that held such ideological significance was the subject of key works in the field of American literary and cultural studies from Miller's contemporary Henry Nash Smith's *The Virgin Land* (1950) through Leo Marx's *The Machine in the Garden* (1964) to Richard Slotkin's *Regeneration through Violence* (1973).

This essay chronicles paradigm shifts in scientific ideas about nature and the human through attention to literary works that engage them. Early works on the relationship of science and literature, such as Ronald E. Martin's *American Literature and the Universe of Force* (1983) and Lisa Steinman's *Made in America: Science, Technology, and American Modernist Poets* (1987), focused on how literary authors consciously engaged scientific theories and technological innovations. In more recent scholarship, such as N. Katherine Hayles's numerous studies—including *How We Became Posthuman: Virtual Bodies in Cybernetics* (1999), *My Mother Was a Computer: Digital Subjects and Literary Texts* (2010), and *How We Think: Digital Media and Contemporary Technogenesis* (2012)—and Mark Seltzer's *Bodies and Machines* (1992), literary works offer insight into, and even contribute to, the processes through which paradigms shift. With their attention to language, image, and storyline, literary works register how ideas circulate among media, fields, and disciplines to shape a collective imagination. This chapter brings the concerns of the earlier work in American studies together with the work in science and literature to chronicle the shifting stories about the world as they unfold in both science and literature from the emergence of the nation through its changing place in a world and, subsequently, a planetary system.

National History, Naturally. In the 1780s, the educator Jedidiah Morse, lamenting his students' ignorance of the geography of their nation, authored a textbook, *American Geography* (1789), designed to countermand the "imperfect…accounts of America" that had been published by "Europeans" who "have too often suffered fancy to supply the place of facts, and thus have led their readers into errors, while they professed to aim at removing their ignorance." Viewing knowledge of "this country" as a privilege of its inhabitants, who can replace Europeans' "fancy" with "facts," he believed America's role in the world made that emendation a national responsibility (Morse, 1796, p. iii). *American Geography* argued that geographic knowledge did not stop with a description of a territory. Rather, it established the *place* of America as it retold its story, ranging from an understanding of the cosmos and the planet (astronomy and natural history) to historical accounts of the "discovery of America" by Christopher Columbus and the "unaccountable caprice of mankind" that has "perpetuated the error" attributing that distinction to the "adventurer" Americus Vespucius and assigning his name to the country, leaving "mankind…to regret an act of injustice, which, having been sanctioned by time, they can never redress" (Morse, 1796, p. 73).

As Clinton recognized, Lewis and Clark's expedition had a significant role in the story of America. Jefferson was both an avid natural historian and a consummate politician, and he understood that the resources of the land were as fascinating scientifically as they were valuable commercially. Following the Louisiana Purchase, he secured Congressional funds for the expedition and commissioned Lewis, an army captain, to lead the team. Their charge was to find a transcontinental passage across which goods and people could travel as well as to catalog the unique flora, fauna, geology, and native inhabitants of this wilderness to prepare for its cultivation. In the process, they would claim the uncharted land both in the name and as the manifestation of the nascent nation.

The story of America was told not only by geographers, explorers, and naturalists, but also across the spectrum of American letters, as writers responded to calls such as Webster's to construct a national imaginary. A new nation needs a history, so it is not surprising that historical fiction would proliferate in the early years of the Republic, with narrative poets and novelists from James Wallis Eastburn and Robert Sands to James Fenimore Cooper, Washington Irving, Lydia Maria Child, and Catharine Maria Sedgwick responding to the nationalist fervor that followed the War of 1812. The early years of British settlement proved especially rich for these novelists, who infused the historical past with the character and needs of their own moment. The historian Joyce Chaplin has shown how the mandates of British settler colonialism in North America shaped contemporary views of nature, race, and science. The need to establish themselves as "natives," for example, prompted colonists to develop increasingly hierarchical ideas about heritable racial differences that rooted British superiority to native bodies and cultures in "nature." By the early nineteenth century, the need for that entitlement gave way to the more pressing concern of distinguishing American citizens from British subjects. Identification with the (romanticized) natives facilitated the distinction. The historical romances superimposed those views, manifesting a historical ambivalence toward natives and nature, as they turned natural history into national history.

Although a cultivated landscape displayed the national power to subdue the forces of nature, many indigenous peoples resisted their coterminous "cultivation," so, in these fictional histories, Indians had both to disappear and to bestow their blessing on the nascent nation. Ostensible evidence of such bestowals took the form of names, customs, or a particular kind of knowledge of the land to which settlers and their descendants laid claim. These works of fiction told a story of America that featured the cultivation of the land and the incorporation of native culture as a legacy of an obligingly disappearing, even mythical, people.

The first published work of Lydia Maria Frances (not yet Child), *Hobomok* (1824), exemplifies this mythical construction. The novella celebrates the transformation of the landscape that displays national achievement. The (male) author persona "glow[s]" with "national pride" contemplating the "long train of associations… connected with [New England's] picturesque rivers, as they repose in their peaceful loveliness, the broad and sparkling mirror of the heavens,— and with the cultivated environs of her busy cities, which seem everywhere blushing into a perfect Eden of fruit and flowers" (Child, 1986 [1824], p. 5). "Nature's nation" manifests (and legitimates) itself through a metamorphosis that makes cultivated nature somehow more natural than the untouched land—it is the environs of the cities that turn into "a perfect Eden." Literature is central to the process, as the "scenes [are] rendered classic by literary associations" (Child, 1986 [1824], p. 5). Those transformational associations mark the difference between the settlers and the "unlettered Indians," whom the author imagines as arriving "at no other conclusion than that the English were the favorite children of the Great Spirit," and the failure of "the various tribes" to "rise in their savage majesty, and crush the daring few who had intruded upon their possessions, is indeed a wonderful exemplification of the superiority of intellect over mere brutal force" (Child, 1986 [1824], p. 29).

The land and its inhabitants turn *Hobomok's* protagonist, reluctant British emigrant Mary Conant, into an American. Believing her British royalist fiancé has died and finding her Puritan father rigid in his piety, she marries the Wampanoag Hobomok and goes to live with him in the woods, where she bears him a son. When his wife's fiancé miraculously reappears, Hobomok obligingly relinquishes her to her countryman, whom he knows has retained her heart all along. Announcing that "'Hobomok will go far

off among some of the red men in the west…[so] Mary may sing the marriage song in the wigwam of the Englishman" (Child, 1986 [1824], p. 139), he departs "with a bursting heart…murmur[ing] his farewell and blessing, and forever passe[s] away from New England" (Child, 1986 [1824], p. 141). Mary returns from her liminal sojourn with Hobomok, having absorbed the Americanizing values of the land, to a father softened by her figurative death and prepared to accept her marriage to Charles.

As in Chaplin's formulations, race surfaces as an insurmountable difference, superseding the religious differences that had once set Mr. Conant against his daughter's fiancé. Although Charles wants Mary to accompany him back to England, however, the "mixed race" son of Hobomok has Americanized his mother, who explains that she "'cannot go to England'" because her son "'would disgrace [her],'" and she would never leave him "'for love to him is the only way that [she] can now repay [her] debt of gratitude.'" The child assimilates the family, and the memory of Hobomok, too, is assimilated as a blessing for a proto-nation depicted in organic terms (nature's nation): his "devoted, romantic love…was never forgotten by its object; and his faithful services to the 'Yengees' are still remembered with gratitude; though the tender slip which he protected, has since become a mighty tree and the nations of the earth seek refuge beneath its branches" (Child, 1986 [1824], p. 150).

The passage of time and increasing "removal" of the native presence in the United States augmented the romanticizing of their legacy. On one hand, the nation was baptized with the blood of defeated "savages"; on the other hand, the native legacy survived in a privileged relationship to nature that distinguished Americans from their effete European forebears. No one painted more repugnant pictures of native savagery than James Fenimore Cooper, and no one more fully romanticized indigenous peoples through portraits of an innate nobility and poetry as well as understanding of the natural world. In a footnote appended to the 1831 edition of *The Last of the Mohicans* (originally 1826), Cooper distinguishes between native and European names for natural places, with the former registering appreciation for the land. Whereas European names reflected political hierarchies, "nearly all of [the native] appellations were descriptive of the object. Thus, a literal translation of the name of this beautiful sheet of water, used by the tribe that dwelt on its banks, would be 'The Tail of the Lake.' Lake George, as it is vulgarly, and now indeed legally called, forms a sort of tail to Lake Champlain, when viewed on the map. Hence the name" (Cooper, 1937 [1831], p. 2).

By the early nineteenth century, the ambivalence that constructed native knowledge of the land as ambiguously ignorant and privileged had resolved into an enduring convention of nostalgia for indigenous knowledge and the conviction of European superiority marked by settlers' having subdued nature and natives. The irony was not lost on native commentators, who denounced the hypocrisy of settler culture, including science and religion. In "An Indian's Looking-Glass for the White Man" (1833), the Pequot Methodist minister William Apess expressed a characteristic critique of Anglo hypocrisy when he wondered why settlers refused a "common education" to indigenous peoples: "Perhaps some unholy, unprincipled men would cry out, the skin was not good enough; but.... I would ask if there cannot be as good feelings and principles under a red skin as...under a white? And...is it not on the account of a bad principle, that we who are red children have had to suffer so much as we have?" (Apess, 2007 [1833], p. 76). Imagining a congress of nations, Apess notes the paucity of white skins and magnitude of white crimes—against natives and "humanity" generally, against Indians' "lawful rights, that nature and God require them to have" (Apess, 2007 [1833], p. 77). Such charges increasingly forced white apologists to work harder to justify the unjustifiable, while Indian Removal allowed white America to assume the guise of the inhabitants they displaced. Cooper's settlers conquer land and enemies by "emulating the patience and self-denial of the practiced native warriors" (Cooper, 1937 [1826], p. 11).

Nature's Secrets. Conquered land and peoples nonetheless represented a continuing threat

of eruption, frequently finding expression in the literary imagination as cautionary tales about the danger of scientific "conquest" and the human hubris that motivates it. Mary Shelley famously depicted the dangerous dabbling of a scientist obsessed with attaining the power to create life— and conquer death. When Dr. Frankenstein's sensitive creature awakens into a world in which, lacking social context, he meets only with fear and loathing, he turns "monstrous," vengefully destroying everyone his creator loves.

The danger of prying into nature's secrets was not lost on writers on the other side of the Atlantic. In a nation that celebrated itself as an example of the magnitude of human achievement, the human encounter with nature was a reminder of its limits. Against the backdrop of the emerging institutionalization and prestige of science in mid-nineteenth-century America, the scientists of Nathaniel Hawthorne's "The Birthmark" (1846) and "Rappaccini's Daughter" (1844) display the single-mindedness that leads them to value scientific knowledge over their closest relationships and to destroy what they most love. Hawthorne sets "The Birthmark" at the end of the previous century when "the comparatively recent discovery of electricity, and other kindred mysteries of nature, seemed to open paths into the region of miracle, [and] it was not unusual for the love of science to rival the love of woman, in its depth and absorbing energy" (Hawthorne, 1982 [1846], "The Birthmark," p. 764). The protagonist's obsession with his beautiful wife's barely visible birthmark manifests his belief in the perfectibility of humankind and in the possibilities of science, and he commits the Faustian error of seeking knowledge beyond what he can manage. He removes the birthmark at the cost of her life. Even more disturbing is the misanthropic Dr. Rappaccini's deliberate sacrifice of his daughter when he turns her and his beautiful garden monstrous in his obsessive quest for knowledge. Rappaccini's "spiritual...love of science" (Hawthorne, 1982 [1844], "Rappaccini's Daughter," p. 982) expresses anxiety about a nation's seemingly soulless drive for progress.

For Hawthorne's contemporary, Charles Darwin, "monsters" bore witness not to human hubris, but to nature's anomalies that may be harbingers of the future and catalysts of change. The 1839 publication of the journal documenting Darwin's geological and zoological observations while he was a naturalist on the HMS *Beagle* offered the earliest insights that would eventually lead him to write *The Origin of Species* (1859). Widely read on both sides of the Atlantic, Darwin's writings challenged conventional ideas about specialization and classification, particularly about the nature and place of the human, offering a new creation story for the modern world. Within the longue durée of natural history, national and even human history was but a moment of time.

Fiction writers were quick to understand the dramatic potential of Darwin's theories as they explored their impact on the concept of the human. Herman Melville's ocean voyages, which furnished content for most of his fiction, had brought him into contact with many of the locations about which Darwin wrote, and "The Encantadas, or Enchanted Isles" (1854), a series of vignettes based on his journey around the Galapagos Islands, begins with a wry response to the natural historian. Whereas Darwin's scientific gaze saw the desolate volcanic islands as the source of a wealth of information about natural history, Melville's literary eye focuses on the insight they can afford into humankind in its social rather than biological aspects. Enchanted these islands must be, muses Melville, because to his naked eye they are mountains of cinder amid "the vacant lot of the sea" (Melville, 1984 [1854], p. 764). Whereas Darwin saw the endless unfolding of evolution in this landscape, Melville describes a place to which "change never comes; neither the change of seasons nor of sorrows," and, devoid of most zoological life, it hosts mostly reptiles, making "the chief sound of life [t]here … a hiss" (Melville, 1984 [1854], p. 765). The description leads the narrator to quip that "in no world but a fallen one could such lands exist" (Melville, 1984 [1854], p. 766).

But the serious implications of Darwin's theories pierce the wryness of Melville's tone when he describes how the islands convey the ephemerality of all things, the reality of death in the illusion of life: "Nothing can better suggest [than the volcanic islands] the aspect of once living things malignly crumbled from ruddiness into ashes. Apples of Sodom, after touching, seem these isles" (Melville, 1984 [1854], p. 767). And nothing makes Melville's point better than his use of the Galapagos tortoises. Whereas the readily recognizable differences among tortoises on each of the islands was central to Darwin's emerging theories, for Melville the monsters are emblematic of human insignificance. The experience of the Galapagos has left the narrator of the sketches distrustful of human knowledge—gained through his senses or through collective memory—and he muses that he may be "the occasional victim of optical delusion concerning the Gallipagos" when "in scenes of social merriment, and especially at revels held by candle-light in old-fashioned mansions," he is haunted by the hallucination of a "gigantic tortoise, with 'Memento * * * *' burning in live letters upon his back" (Melville, 1984 [1854], p. 768). The centrality of the tortoise to Darwin's theories makes that reptile a fitting emblem of human evolution and thereby human vanity. Compared to the longevity of the tortoise, human life is fleeting.

In Melville's "Two Sides to a Tortoise," when crewmembers bring three tortoises onto his ship, the narrator goes rhetorically overboard in establishing their connection with the prehistoric past. They are "antediluvian-looking tortoises" that "seemed hardly of the seed of earth," ambassadors from the past: "mystic creatures suddenly translated by night from unutterable solitudes to our peopled deck.… They seemed newly crawled forth from beneath the foundations of the world.… The great feeling inspired by these creatures was that of age:—dateless, indefinite endurance" (Melville, 1984 [1854], pp. 770–771). They are almost synonymous with the earth itself, as the narrator, seemingly "an antiquary of a geologist," peruses their shells, "citadel[s] wherein to resist the assaults of Time," for traces of creatures long extinct (Melville, 1984 [1854], p. 771). The narrator pictures "these three straightforward monsters, century after century, writhing through the shades, grim as blacksmiths; crawling so slowly and ponderously, that not only did toadstools and all fungous things grow beneath their feet, but a sooty moss sprouted upon their backs" (Melville, 1984 [1854], p. 772). But his tortoise-inspired reveries are disrupted when, the next day, he sits down to "a merry repast from tortoise steaks and tortoise stews" and helps to fashion their shells into "soup-tureens" and "salvers" (Melville, 1984 [1854], p. 772). Humankind triumphs over the agents of memory, but the insights nonetheless permeate the sketches, as they do all of Melville's fiction. He is keenly aware of the ephemerality of the life of both individual and species and, like Hawthorne, of the limitations of human knowledge of the world, social as well as natural.

Human beings may well be insignificant, but their unique ability to imagine, and hence to perceive beyond the constraints of vision, distinguished them for the writers associated with the New England–based movement known as Transcendentalism, including Ralph Waldo Emerson, Margaret Fuller, Amos Bronson Alcott, Orestes Brownson, and Henry David Thoreau. Science could be an important instrument in the quest for transcendence, but it required an awareness of its limitations. Science and religion converge in Emerson's early, anonymously published, although influential 1836 essay "Nature," when he prescribes a journey into the solitude of the woods, where "we return to reason and faith" and learn to perceive with the soul rather than the eyes. In what may be his most cited passage, Emerson articulates the transcendental ideal: "Standing on the bare ground,—my head bathed by the blithe air, and uplifted into infinite space,—all mean egotism vanishes. I become a transparent eye-ball; I am nothing; I see all; the currents of the Universal Being circulate through me; I am part or particle of God" (Emerson, 1983 [1836], p. 10).

The writer ostensibly casts off subjectivity as he becomes the perfect observer, estranged from social ties—"the name of the nearest friend sounds then foreign and accidental: to be brothers, to be acquaintances,—master or servant, is then a trifle and a disturbance"—and discovering instead (pace Darwin) "an occult relation between man and the vegetable" (Emerson, 1983 [1836], pp. 10, 11). This ideal is in accord with the mid-nineteenth-century emergence of a new emphasis on the objective observer in the sciences that Lorraine Daston and Peter Galison document, but for Emerson such observation leads to the perception of "the reverential withdrawing of nature before its God" (Emerson, 1983 [1836], p. 33). Nature yields secrets to those who abandon themselves to, rather than seek to harness, them: "when a faithful thinker, resolute to detach every object from personal relations, and see it in the light of thought, shall, at the same time, kindle science with the fire of the holiest affections, then will God go forth anew into the creation" (Emerson, 1983 [1836], p. 47).

The eponymous figure of his 1844 essay "The Poet" was just such a thinker. The poet "alone knows astronomy, chemistry, vegetation, and animation, for he does not stop at these facts, but employs them as signs. He knows why the plain, or meadow of space, was strown with these flowers we call suns, and moons, and stars; why the deep is adorned with animals, with men, and gods; for, in every word he speaks he rides on them as the horses of thought" (Emerson, 1983 [1836], p. 456). Science, for an older Emerson, pierced convention. "Science corrects the old creeds," he asserted in "Progress of Culture," a Phi Beta Kappa address delivered in 1867, "sweeps away, with every new perception, our infantile catechisms; and necessitates a faith commensurate with the grander orbits and universal laws which it discloses" (Emerson, 2010 [1867], p. 120). But science must be only the beginning of that process. The limitation of science, he would later write, is that it "does not know its debt to imagination" (Emerson, 2010 [1876], p. 10). Isolating objects to observe and explain them was like "hunting for life in graveyards," and Emerson abjured science for being "false by being unpoetical.... Reptile or mollusk or man or angel only exists in system, in relation." By contrast, "the metaphysician, the poet, only sees each animal form as an inevitable step in the path of the creating mind" (Emerson, 2010 [1876], p. 10).

Thoreau similarly understood science as the beginning of a process that artists would extend. In an 1842 essay entitled "The Natural History of Massachusetts," he called science "brave, for to know is to know good.... What the coward overlooks in his hurry, she calmly scrutinizes," but "she" does so to "break[] ground like a pioneer for the array of arts that follow" (Thoreau, 1990 [1842], p. 148). The arts restored an intuition that contemporary science was in danger of losing. In *A Week on the Concord and Merrimack Rivers* (1849), Thoreau lamented, "the most prominent scientific men of our country, and perhaps of this age, are either serving the arts and not pure science or are performing faithful but quite subordinate labors in particular departments.... There is wanting constant and accurate observation with enough of theory to direct and discipline it. But, above all, there is wanting genius. Our books of science, as they improve in accuracy, are in danger of losing the freshness and vigor and readiness to appreciate the real laws of Nature, which is a marked merit in the ofttimes false theories of the ancients" (Thoreau, 1985 [1849], p. 296). Science, he believed, should lead to a sensual imagination "as well fitted to penetrate the spaces of the real, the substantial, the eternal, as...outward [senses] are to penetrate the material universe" (Thoreau, 1985 [1849], p. 313).

The poetry of Emily Dickinson similarly celebrates that imagination, and no one better embodies the eponymous poet of Emerson's essay. An eager student of science, she nonetheless insisted in her poetry on the debt science owed to imagination. With her typical playfulness, she fashions a contest between them in a poem about the very kind of classification in which she delighted as a collector and cataloger of flora:

> *"Arcturus" is his other name—*
> *I'd rather call him "Star."*
> *It's very mean of Science*
> *To go and interfere!....*
> *I pull a flower from the woods—*
> *A monster with a glass*
> *Computes the stamens in a breath—*
> *And has her in a "class"!*
> *(Dickinson, 1999 [1859], p. 61)*

Dickinson's poetry works, as Fred D. White observes, "to counteract scientific reductionism, which tempts us into thinking that science can

present reality whole and undistorted" (White, 1992, p. 123). When Dickinson invokes "a certain slant of light," for example, she describes more than an observation. The word "certain" evokes an experience of light that she assumes as a common association:

> *There's a certain Slant of light,*
> *Winter Afternoons—*
> *That oppresses, like the Heft*
> *Of Cathedral Tunes—*
> *(Dickinson, 1999 [1862], p. 142)*

As it assumes a shared response to that certain slant of light and to cathedral tunes, the poem ties observation to imagination; it affirms a collective experience of and investment in the world that is produced through the imagination and captured by poetry.

The quest for truth for Dickinson, as for the Transcendentalists, always led back to the soul, which was ultimately beyond the reach of scientific understanding. A late poem typifies Dickinson's poetics as well as her philosophy, comparing the solitude of space, sea, and. even Death to a "profounder site": "That polar privacy/A soul admitted to itself—" (Dickinson, 1999 [undated], p. 610). Dickinson scholars disagree about whether the poem ends on that tantalizing line or with the recursive oxymoron "Finite infinity," but either ending conveys the inadequacy of reason as a means fully to access experience.

Dickinson and the Transcendentalists asserted the limitations of science at a moment that witnessed the emergence of a new science with a distinctly political agenda: the ethnological study of racial difference in the context of American slavery. Emerson could celebrate the transcendental ideal in which distinction between master and servant is "a trifle and a disturbance," but for an enslaved person, or any descendent of Africans in the mid-nineteenth-century United States, daily experience belied such an assertion. The idea of polygenesis—that races represented different species with unique origins—circulated widely in the work of mid-nineteenth-century ethnologists, including Samuel George Morton and George R. Gliddon. Scientific assertions of the biological inferiority of Africans justified the oppressive hierarchies of the peculiar institution.

Assertions of the humanity of enslaved and oppressed persons, and the inconsistency of white Americans on this issue, punctuate the work of African American authors, including Frederick Douglass, Harriet Wilson, Harriet Jacobs, Hannah Crafts, and William Wells Brown. "Must I undertake to prove that the slave is a man?" asks Douglass in an 1852 speech. "That point is conceded already.... The slaveholders themselves acknowledge it in the enactment of laws for their government.... It is admitted in the fact that southern statute books are covered with enactments forbidding, under severe fines and penalties, the teaching of the slave to read or write. When you can point to any such laws, in reference to the beasts of the field, then I may consent to argue the manhood of the slave" (Douglass, 1982 [1852], p. 369). Stylistically and thematically, these works explore the shaping of fundamental assumptions and the blindness of even the best-intentioned white Americans to the terms of black oppression. In Wilson's *Our Nig* (1859), for example, Frado, the young black female protagonist, listens as her white benefactors "discuss the prevalent opinion of the public, that people of color are really inferior; incapable of cultivation and refinement. They would glance at Nig, which promised so much if rightly directed" (Wilson, 1983 [1859], p. 73). Their inscription in those prevailing racist narratives is evident in their condescension. In the United States, debates about the nature of the human were inevitably inflected by racial slavery and its aftermath as the nation struggled to make sense of an institution that so fundamentally contradicted its defining precept: "all men are created equal."

Humanity's Borders. The scientific study of social inequities took a new form as unprecedented domestic and global migrations put human beings in increasing contact. The founding editor of the *American Journal of Sociology*, Albion Small, introduced the first issue in 1895 with the observation, "*In our age the fact of human association is more obtrusive and relatively more influential than in any previous epoch*. Men are more definitely

and variously aware of each other than ever before. They are also more promiscuously perplexed by each other's presence.... Whatever modern men's theory of the social bond, no men have ever had more conclusive evidence that the bond exists" (Small, 1895, p. 1). Sociology, also known as "the science of society," was one among several fields of inquiry—social sciences—that emerged to study the nature and impact of that bond even as a national frame of reference strained against increasingly global networks. The mutual inflection of literature and the social sciences in this period is evident in the literary fascination with eponymous protagonists' movement through expanding social environments—in works such as Henry James's *Daisy Miller* (1878) and *Portrait of a Lady* (1881), William Dean Howells's *The Rise of Silas Lapham* (1885), Frances Watkins Harper's *Iola Leroy* (1892), Stephen Crane's *Maggie: A Girl of the Streets* (1893), Theodore Dreiser's *Sister Carrie* (1900), Jack London's *Martin Eden* (1909), James Weldon Johnson's *Autobiography of an Ex-Colored Man* (1912), Sui Sin Far (Edith Maud Eaton)'s *Leaves From the Mental Portfolio of an Eurasian* (1909), Abraham Cahan's *Yekl* (1896) and *The Rise of David Levinsky* (1917), and numerous others—and in social scientists' use of literary works to understand human motivations and actions as well as the environments they create. Fiction writers at the turn of the twentieth century influenced as they worked through theories of human nature. The characters in the fiction of self-styled realists such as James, Howells, and Edith Wharton interact within social environments of their own making, whereas the fate of characters in the fiction of writers influenced by the ideas of naturalism—including Crane, London, Frank Norris, and Hamlin Garland—is determined more by biological and environmental forces beyond their control.

The turn of the twentieth century witnessed the proliferation of fiction by authors who experienced social marginalization because of the impact of human migrations on such factors as race and ethnicity, gender and sexuality, class, and religion. Their protagonists' experience of life on these margins offered valuable insight for social scientists who believed that the mechanisms of social control were most accessible to study in social interstices. The influential University of Chicago sociologist W. I. Thomas consulted the fiction of the Lithuanian Jewish immigrant Cahan for its insights into ghetto life, and Thomas's description of the danger of "the girl coming from the country to the city" in search of employment reads like a plot summary of Theodore Dreiser's *Sister Carrie* (Thomas, 1906, p. 42). Thomas's colleague Robert Park drew on literary works such as immigrant autobiographies to exemplify the lived experience—and anguished struggle—of cultural hybridity experienced by the eponymous figure of his influential essay, "The Marginal Man." In the character "types" and predictable patterns of social behavior explored in the literary works, social scientists recognized a kind of novelization of life writ large.

Whereas sociologists and anthropologists studied the formation of collective beliefs and assumptions, psychologists typically focused on the individual mechanisms of transmission. For the philosopher William James, who taught Harvard University's first psychology classes and wrote a textbook for the nascent discipline, habits of attention best explained how beliefs circulated. "Men have no eyes but for those aspects of things which they have already been taught to discern," he wrote in *Principles of Psychology*. "*The only things which we commonly see are those which we preperceive*, and the only things which we preperceive are those which have been labeled for us, and the labels stamped in our mind" (James, 1983 [1890], p. 420).

James wrote these words at the end of a century that had witnessed a proliferation of visual technologies and widespread cultural fascination with the mechanics of vision, evident in the popularity of optical illusions—parlor games, magicians—in the middle of the century and the veritable obsession with the earliest moving pictures later in the century. Visual technologies made the quotidian appear strange and magical, and the insights they produced catalyzed the formal and stylistic experimentation in turn-of-the-century visual and literary arts. The French painter Paul Cézanne, for example, revolutionized painting in his exploration of vision and perspective, encouraging the consideration of sight itself by depicting the overlapping perspectives that ocular physiology produces but the human brain corrects when it fuses the views of each eye into a single (binocular) perspective.

One of James's best-known students, the sociologist W. E. B. Du Bois, turned these insights into a literary meditation on the nature of racism. In his 1897 essay, "Strivings of the Negro People," which became the first chapter of his 1903 *The Souls of Black Folk*, Du Bois described his "double-consciousness" as a black American, as "this sense of always looking at one's self through the eyes of others, of measuring one's soul by the tape of a world that looks on in amused contempt and pity" (Du Bois, 1986 [1903], p. 364). Du Bois's readers often miss his emphasis on the *sense* of seeing oneself through another's eyes—the awareness of what he calls "second-sight"—which he describes as a "gift." Since everyone pre-perceives, the experience of selfhood always reflects the internalization of others' perceptions, but, he contends, the demeaning misperceptions to which black Americans are subject make that process more visible for them than for their white counterparts. A changing world system made such discrepancies increasingly visible and troubling to those living, as Du Bois put it, "within the Veil" (Du Bois, 1986 [1903], p. 359). "The problem of the twentieth century," he predicted, would be "the problem of the color-line" as people of color would come to recognize the global nature of the networks that oppressed them (Du Bois, 1986 [1903], p. 372). Increasingly, he maintained, they would turn technologies of oppression into tools of liberation.

Visual technologies elucidate the role of pre-perception in the experience of selfhood in such works as James Weldon Johnson's *Autobiography of an Ex-Colored Man* (1912) and Zora Neale Hurston's *Their Eyes Were Watching God* (1937) in which a mirror and a photograph, respectively, manifest their protagonists' discovery of their "blackness." The features of Johnson's fictional protagonist change before his eyes when, after being told that he is not white, he examines himself in a mirror for evidence of his racialized traits. Hurston's Janie initially fails to recognize herself in a photograph of herself with her white playmates. Knowing herself only through her position in relation to them, she, like Johnson's anonymous protagonist, discovers the social construction of race in America.

Writers of the period often turned to literary experimentation to foster readers' self-consciousness about their deepest biases and assumptions. Another well-known student of James, Gertrude Stein, studied habits of attention in her classwork with him and in laboratory work with his associate, Hugo Münsterberg. After a brief foray in medical school, she turned her psychosocial insights into literary work famous for its apparent incomprehensibility. The behaviorist B. F. Skinner, believing he had discovered the key to her work in her earliest writings (about her lab experiments), announced that Stein's literary output was nothing more than automatic writing. In fact, her prose was conscious and deliberate, but designed to break the habits of attention she had studied and to force readers to confront both their desire for meaning and the cultural determinants through which they produce it: what they see and what they fail to see as a result of their tutored "pre-perceptions."

Although Stein often insisted that her writing was realist in its attention to the mechanics of human attention, perception, and consciousness—its emphasis on how and what we *really* see—her belief that human beings could envision and transcend their perceptual limitations through art made her a central figure in the movement known as Modernism. With other Modernists, from Ezra Pound and Ernest Hemingway to Filippo Marinetti and Mina Loy, she was keenly aware of the scientific and technological innovations and geopolitical developments that were making the world strange and fascinating, and her formal and stylistic experiments registered the influence of technological innovations. Looking back on her early writing, she observed, "I was doing what the cinema was doing....I of course did not think of it in terms of the cinema, in fact I doubt whether at that time I had ever seen a cinema, but, and I cannot repeat this too often, any one is of one's period and this our period was undoubtedly the period of the cinema and series

production. And each of us in our own way are bound to express what the world in which we are living is doing" (Stein, 1935, p. 177). Sensory overload, for her, was to be celebrated, not lamented, as generating creativity and, ultimately, attention to one's own habits of attention.

Among her closest friends and associates, painters such as Pablo Picasso and Georges Braque were similarly experimenting with attention and perception, and she sought to translate their visual experiments into words. The first entry in Stein's volume entitled *Tender Buttons* (1914), "A Carafe, That Is a Blind Glass," reads like a literary still life:

> A kind in glass and a cousin, a spectacle and nothing strange a single hurt color and an arrangement in a system to pointing. All this and not ordinary, not unordered in not resembling. The difference is spreading.
> —Stein, 1997 [1914], p. 3

The piece works through association, juxtaposition, and composition. It is not nonsense, as Skinner may assert, but a challenge to the conventions of sense making. Stein's literary experiments encourage readers to experience their yearning for meaning and to see themselves straining to make sense; in that way, she invites them into the meaning-making process. Meanings proliferate: "Cousin" draws "kin" out of "kind." Meaning is, like kinship, a system of social relations, constantly modified with each new piece of information. It is an "arrangement," both a configuration and an agreement. This process works according to a "system"; it is not arbitrary or automatic. With the verbal forms that end each sentence, Stein uses grammar—the ultimate conventional system of meaning making—to suggest tension between the proliferation that interests her and the writerly convention signaled by the period, which marks an endpoint at the end of the sentence. The proliferation of meanings is not anarchic. A still life is still (yet) life. The life is in the artistry: the composition that leads to more active contemplation. This introductory "button" shows readers how to read associatively, attending to their own roles as viewers in creating the work of art and in generating the response. It places responsibility for creating art, making meaning, and seeing life differently onto the spectator—responsibility for seeing themselves in the act of seeing, which includes seeing their own inattention, or blindness.

As the twentieth century wore on, Stein and her cohort watched the world explode into war, and they witnessed the acceleration of the technological and scientific developments that war demands. They expressed dismay as innovations optimistically hailed as records of human achievement metamorphosed into instruments of human devastation. Literary works registered the nature of individual and collective trauma in figures such as F. Scott Fitzgerald's Nick Carraway, who comes back from the war wanting "the world to be in uniform and at a certain moral attention forever," and Ernest Hemingway's Nick Adams, whose breakdown and recovery are chronicled in *In Our Time* (Fitzgerald, 2004 [1925], p. 2). In the process, literary works manifested proliferating questions about the nature of the human that ranged from the metaphysical to the biological. The concept of the unconscious articulated by Sigmund Freud gained particular traction because it helped to explain the strange phenomenon that medical professionals dubbed "shell shock." The unconscious, as Freud explained it, suggested a strikingly literary "self," compulsively reenacting a role in a script long since written and knowable only through careful interpretation of language and actions that betrayed buried meanings. Freud's writings, most notably *The Interpretation of Dreams* (1899), helped to fashion enduring reading habits and analytic practices.

The Future of the Human.

If World War I raised the specter of human annihilation, that distinct possibility hovered ominously in the atomic wake of the next global conflict. Nuclear war certainly posed the threat of human extinction in the decades following the war, but it also crystallized anxieties about human finitude attendant upon new scientific theories and radical geopolitical transformations. The postwar world was increasingly not only a global, but also a planetary one, and new conceptions of nature and the human accompanied the changing temporal and geographical scales. Whereas nuclear devastation underscored humanity's embeddedness in the intricate, far-reaching connections of the web of life, the "evolutionary synthesis" manifested the inevitability of human metamorphosis. Conjoining the insights of natural historians and geneticists, the synthesis offered a causal mechanism for natural selection, showing how genetic mutations could ultimately lead to changes in a population, even a new species. Evolutionary changes, of course, required a long imagination, since they would be measured in millennia, but imminent threats made human extinction a more immediate concern. Cultural observers who wondered whether humanity could restrain its self-destructive impulses long enough to witness even the turn of the next century were soon joined by environmentalists who heralded threats to the fragile web of life from the indiscriminate use of chemical toxins and the exhaustion of natural resources. The biologist Rachel Carson was far from the first person to sound the alarm about the widespread danger of chemical pesticides and herbicides, but her best-selling 1962 *Silent Spring*, originally published in the *New Yorker*, effectively dramatized the immediacy of the threat to humanity and to the planet. By the end of the decade, the United Nations Economic and Social Council had issued a report warning that the continuation of current rates of pollution and resource use threatened "the future of life on earth" ("Crisis of Human Environment").

Literary and cinematic works during this time took advantage of the drama of end-of-the-world scenarios, ranging from nuclear war and environmental devastation to comets, viruses, and alien invasions. But these scenarios typically crystallized more implicit challenges to the conception of the human. Both wars accelerated innovations in science and technology, including cybernetics, robotics, genetics, and neuroscience, that offered new insights into the hardwiring of human beings. At the same time, those global conflagrations fueled geopolitical transformations, as noted by the political scientist Harold Isaacs, founder of the Association for Third World Studies, who observed that the many "new states carved out of the old empires since 1945 [and] made up of

nonwhite peoples newly out from under the political, economic and psychological domination of
white rulers" had left people "stumbling blindly
around trying to discern the new images, the
new shapes and perspectives these changes have
brought, to adjust to the painful rearrangement of
identities and relationships which the new circumstances compel" (Isaacs, 1969, p. 235), the
problem, as Du Bois had foretold, of the color line.
These scientific, social, and geopolitical changes
inspired calls for new definitions of the human
and accompanying narratives of humanity that
would ensure more effective recognition of intrinsic human equality, as well as a more just politics of life, by political philosophers as diverse as
the Martinican psychoanalyst Frantz Fanon and
the German Jewish philosopher Hannah Arendt.

The emergence of science fiction as a distinct
mass genre following World War II constituted a
response to those calls. Stories of planetary destruction and exploration, of alien encounter and
human evolution, of time travel and mind control,
registered the fantasies and anxieties of a rapidly
transforming world. The mythic nature of some
of the defining works of the genre, such as the
first three works of the Russian immigrant Isaac
Asimov's saga of galactic empire, the Foundation
series (1951–1953), and the English Arthur C.
Clarke's story of human evolution, *Childhood's
End* (1953), helped to set the tone for the genre.
As a whole, science fiction explored the parameters of the definition of the human as it grappled
with the biological and political contingency that
was fundamental both to the human condition
and to planetary existence.

The idea of the human, in science fiction, often
came into focus when it was threatened. The theme
of alien possession, which the writer and editor
John Campbell introduced in his 1938 novella
Who Goes There?, proliferated in the 1950s in such
works as Robert Heinlein's 1951 *The Puppet Masters*, the 1953 film *Invasion from Mars*, Philip K.
Dick's 1954 "The Father Thing," and, most notably,
Jack Finney's *The Body Snatchers*, a 1954 *Collier's*
serial that appeared in book form in 1955 and
spawned numerous cinematic and fictional in-
carnations. Featuring a small-town physician's
dawning recognition that aliens are turning his

friends and neighbors into emotionless pod
people, Finney's novel powerfully captured the
implications of contemporary challenges to the
concept of the human, such as had been articulated by the mathematician Norbert Wiener—
"we are not stuff that abides, but patterns that
perpetuate themselves"—against the backdrop
of small-town nostalgia and Cold War paranoia
(Wiener, 1954 [1950], p. 96). The many retellings
of this story in the 1950s and into the twenty-first
century attest to its widespread resonance. This
broad appeal lies at least partly in the affirmation
of humanity as something intangible and beyond
definition, evident in the ability to experience feelings, something worth fighting for and ultimately,
the novel suggests, inalienable. Against the increasing biologization of the human and the social
conformity of the moment, *The Body Snatchers*
reassured readers that the spark of humanity could
not be extinguished.

The numbing effect of conformity and its
danger was a prominent subject of both fiction
and social science in the aftermath of the war.
Anxieties about this loss found their most literal
expression in a fear of brainwashing following
reports of its use on U.S. prisoners during the
Korean War. Resulting from research yielding
new insight into the workings of the human brain,
this form of mind control was especially disturbing in its depiction of an extreme form of the
normal processes of socialization, as in the work of
the psychiatrist Joost Meerloo, the social psychologist Erik Erikson, and the Hollywood publicist
turned author Richard Condon. In his 1959 novel,
The Manchurian Candidate, Condon dramatized
the psychological insights by chronicling the
transformation of a desperately unhappy man into
an unknowing assassin. The metamorphosis is
facilitated by the psychological makeup of the
victim of this plot, which is contrived by Communist scientists from China and Russia in conjunction with his psychologically damaged mother
(a victim of father-daughter incest).

Condon was hardly alone in putting a perverse
family drama at the heart of accounts of deaden-
ing—or deadly—conformity. Repressed family
secrets nearly destroy families and communities
in best-selling novels of the mid-1950s, such as

Sloan Wilson's *The Man in the Grey Flannel Suit* (1955) and Grace Metalious's *Peyton Place* (1956). And Freudian theories of repression take on deadly proportions of historical significance for Arthur Miller, as for Condon, in Miller's rewriting of the Salem witch trials in *The Crucible* (1953), the play that gave an era its most potent metaphor. Although the social commentary evinced in these works has been well documented, their reflection of and on competing scientific theories of human automatism has typically been overlooked.

Critical discussion of the formal and stylistic experimentation of the mid-twentieth century has similarly privileged the social concerns of artistic works over their engagement with contemporary scientific theories. Yet, the "starving, hysterical, naked" minds careening wildly through Allen Ginsberg's "Howl" (1956), the "mad ones" manically on the road in Jack Kerouac's *On the Road* (written 1951, published 1957), and the junkies crawling through the muck of William Burroughs's *Naked Lunch* (1959) are all searching as much for a language—and manifestation—of human agency against the increasingly mechanistic formulation of the human as for human expression and connection from within the confines of a stifling corporate conformity. The writers of the Beat movement sought liberation in the excesses of their lives as in their prose, believing that to stand still was to risk mental capture. For Burroughs, that meant cutting and pasting his manuscripts to avoid the structuring mechanisms of his own unconscious, a technique he explored in the early 1960s in his science fictional Nova trilogy. Even the language experiments of Gertrude Stein risked reproducing the social forms that shaped the unconscious for the writer who lived, Burroughs confessed, sounding like the protagonist of *The Body Snatchers*, "with the constant threat of possession, and a constant need to escape from possession, from Control" (Burroughs, 1985, p. xxii).

An avid reader of science, Burroughs was aware of a central premise of mind control that grew out of one of the key insights of his historical moment. Emerging from such areas of research as cybernetics, genetics, and virology was the idea that

human beings were elaborate information systems composed of circulating messages that were subject to external influence. That was the acknowledged premise of the burgeoning public relations industry, a chief architect of which was Freud's nephew, Edward Bernays. The insight was also central to the literary experiments of postmodernism, which were characterized by a radical self-consciousness about the process of meaning making, and of communication generally, as Burroughs's work exemplified.

The accelerating pace of scientific and technological innovation as the century progressed further underscored the inadequacy of any definition of the human. Machines that could learn raised new questions about human programmability, whereas the creation of immortal cell lines—and especially the effort to patent them—sparked debates about the "ownership" of human body parts. As the "code book" of DNA became more readable, genetic information revealed "the human" to be a matter of interpretation. But scientists were not only interpreters, they were also increasingly creators; the successful transplantation of genetic material across species (recombinant DNA) in the 1970s brought extraordinary insights into life itself while it summoned the specters of Doctors Frankenstein and Moreau and the possibility of creatures born in the laboratories of the human imagination—beings without precedent or name. "What traits ultimately define a human? Where is the borderline of 'humanness'?" one prominent cultural commentator ominously asked in 1987, and "if we do not know how to define 'human,' what about 'human rights'?" (Toffler, 1987, p. 20).

Although experimental modernists such as Stein were important precursors for postmodern writers, the challenge to the human registered in their experimental work as a play of ultimately unmoored signifiers, with the self in particular rendered as a mise en abyme. No longer a tool in the control of human beings, language for them produces its own unstable meanings—dangerously, as in Samuel Delany's *Babel-17* (1966), and maddeningly, as Oedipa Maas discovers in Thomas Pynchon's *The Crying of Lot 49* (1966). As humankind seemed on the verge of unlocking nature's remaining secrets, postmodern literary works registered a profound skepticism about what human beings could really know. These works are often darkly humorous, as surfaces turn recursively to mirrors and plots tauntingly lead toward an ever-elusive resolution. Pynchon's protagonist wanders through coincidences and conceptual landscapes as she attempts to discern whether she has discovered a plot, is the victim of an elaborate ruse, or has lost her mind. At one point a group of children in Golden Gate Park tell her they are dreaming their gathering; frustrated that they ignore her questions, "Oedipa, to retaliate, stop[s] believing in them" (Pynchon, 1966, p. 96). In John Hersey's 1946 journalistic account of the aftermath of the Hiroshima bombing, one of the survivors recalls having to remind himself that "the slimy living bodies" he was helping were "human beings" (Hersey, 1985 [1946], p. 45); in the fiction of postmodernists such as Burroughs and Pynchon, that literal unrecognizability gave way to metaphysical uncertainty.

Octavia Butler brilliantly dramatized the nature of the anxieties surrounding the question of the human in the age of biotechnology in her speculative fiction of the late 1980s, the Xenogenesis trilogy. Staging an encounter between the human survivors of nuclear war and the aliens who rescued them and restored the planet, she explores the clash of worldviews between radical humanistic and ecological perspectives. The Oankali are peace loving, respectful of all living things, and medically sophisticated, and their most honorific title is "treasured stranger." They offer the human survivors health and longevity, as well as full integration into their world, but they insist on the one thing the human beings cannot accept: they must integrate biologically as well as socially with their rescuers, whom they come to see as captors. Whereas the Oankali have genetic memories and can access the stages of their evolution, which they have expedited through constant cross-breeding with the alien species they encounter, the human beings consider the untainted reproduction of their species into an indefinite future essential to their sense of humanity. The evolution proposed by the Oankali represents extinction to them.

Butler's trilogy cannily manifests the social and geopolitical contexts for biotechnological innovation and anticipates the nature of the debates

surrounding research produced under the auspices of the Human Genome Project, funded the year after the trilogy appeared. The Oankali embody the claims of geneticists, such as Luigi Luca Cavalli-Sforza, that the human genome was a story written in deep time—like the "Memento" inscribed on the back of Melville's ghostly tortoise—and that it could revise the history of the evolution of the species and its biological and ecological connection to all living things. Genetic information as well as the possibilities offered by genetic manipulation thereby challenged the conventional definitions of the human, biological and social. Central to such challenges was the question of race, which surfaces implicitly but powerfully in Butler's trilogy as, for example, when the human beings articulate their anxieties about their loss of humanity as an exclusion from history—from memories of the past as well as from the future. Orlando Patterson (1982) called such an exclusion "social death" in reference to the strategies by which white America sought to exclude enslaved persons from full humanity. For Butler's humans, historical memory of racial slavery implicitly inflects their understanding of the challenge to humanity posed by the Oankali. The humans are not in any allegorical way enslaved; rather, the trilogy offers an analysis of how history intrinsically structures the concept of "humanity."

Clinton's 2000 address announcing the mapping of the genome, with which this essay began, attests to Butler's prescience. His cautionary words about the potential consequences of the new information similarly summon the implicit history of racial slavery: "increasing knowledge of the human genome must never change the basic belief on which our ethics, our government, our society are founded," he warns. "All of us are created equal, entitled to equal treatment under the law. After all,... one of the great truths to emerge from this triumphant expedition inside the human genome is that in genetic terms, all human beings, regardless of race, are more than 99.9 percent the same." Butler's trilogy offers insight into why a new understanding of the human would raise concerns about the stability of political institutions founded on a tacit assumption about "human" entitlements and why those concerns would find expression in the threat of racism. The human beings in the trilogy manifest a tenacious attraction for the very divisions—racial, ethnic, national—that would have destroyed them but for the intervention of the Oankali.

The tenacity is evident in Clinton's address as well, despite its claims to worldliness. With a satellite image of British Prime Minister Tony Blair in the background, Clinton offers a global vision, advocating the need to "ensure that new genome science and its benefits will be directed toward making life better for all citizens of the world, never just a privileged few." But the tenacity of the national, as manifested by Butler's humans, is evident in the staging of the event and, ultimately, in the rhetoric of the speech. A presidential address to the nation is highly ritualized. "Hail to the Chief" announces the president and accompanies him to the podium, where the presidential seal serves as a visual reminder of the national context. Yet beyond the ritual, the tenacity of the national surfaces in the slippage of the first-person plural pronoun from the *human* "we" who must ensure the universal benefits to the *national* (American) subject whose ethics, government, and society are founded on the "proposition," as Abraham Lincoln had famously called it, that "all of us are created equal."

"The universe is made of stories," observed the poet Muriel Rukeyser, "not atoms." Of course, it is made of both (Rukeyser, [1968], p. 465). And both evolve. Bodies change, and ideas about bodies—about the human, about life—change as well. Humanity is not a fixed concept; "the human" cannot be defined—scientifically or metaphysically—for all time and in all places. Whereas life evolves slowly, however, stories clash and can therefore evolve rapidly. The stories written in the human genome are subject to interpretation and thereby to change. Clinton's story is ultimately a story about the triumph of American science, reminiscent of the story Perry Miller identified in early America. Butler shows one consequence of that story and also the possibility of telling others. If genomics offers insight into the long history of human existence, literature shows us the role of stories in making sense of that existence; it shows us the power of the human imagination, which is to say the power to change the stories and, with them, the way we inhabit the world.

[*See also* Environmentalism; Evolution, Theory of; Geography; Human Genome Project; Jefferson, Thomas; Lewis and Clark Expedition; Medicine; Popularization of Science; Psychology; Race Theories, Scientific; Religion and Science; Science; Science Fiction; Social Sciences; Technological Enthusiasm; *and* Technology.]

BIBLIOGRAPHY

Apess, William. "An Indian's Looking-Glass for the White Man" (1833). In *American Indian Nonfiction: An Anthology of Writings, 1760s–1930s*, edited by Bernd C. Peyer, pp. 75–80. Norman: University of Oklahoma Press, 2007.

Burroughs, William S. *Queer* [written 1951-3]. New York: Viking Press, 1985.

Butler, Octavia. *Adulthood Rites*. New York: Warner Books, 1988.

Butler, Octavia. *Dawn*. New York: Warner Books, 1987.

Butler, Octavia. *Imago*. New York: Warner Books, 1989.

Carson, Rachel. *Silent Spring*. New York: Houghton Mifflin, 1962.

Chaplin, Joyce E. *Subject Matter: Technology, the Body, and Science on the Anglo-American Frontier, 1500–1676*. Cambridge, Mass.: Harvard University Press, 2001.

Child, Lydia Maria. *Hobomok and Other Writings on Indians* [1824]. Edited by Carolyn L. Karcher. New Brunswick, N.J.: Rutgers University Press, 1986.

Clinton, William Jefferson. Press conference, 26 June 2000. A recording of the press conference can be accessed at http://www.ornl.gov/sci/techresources/Human_Genome/project/clinton1.shtml, and a copy of the transcript can be found at http://clinton3.nara.gov/WH/New/html/genome-20000626.html. The quotations are from that transcript. Excerpted transcripts were published widely. See, for example, "Reading the Book of Life: White House Remarks on Decoding of Genome." *New York Times* (27 June 2000), F8.

Condon, Richard. *The Manchurian Candidate* [1959]. New York: Four Walls Eight Windows, 2003.

Cooper, James Fenimore. *The Last of the Mohicans: A Narrative of 1757* [1826]. New York: Charles Scribner's Sons, 1937.

"Crisis of Human Environment." United Nations Economic and Social Council, *Report of the*

Secretary General on Problems of the Human Environment, pp. 4–6, 47th Session, Agenda Item 10, 26 May 1969.

Daston, Lorraine J., and Peter Galison. *Objectivity.* Brooklyn, N.Y.: Zone Books, 2007.

Dickinson, Emily. "'Arcturus' is his other name" [117]. In *The Poems of Emily Dickinson*, p. 610, edited by R. W. Franklin. Cambridge, Mass.: Belknap Press of Harvard University Press, 1999.

Dickinson, Emily. "There is a solitude of space" [1696]. In *The Poems of Emily Dickinson*, p. 610, edited by R. W. Franklin. Cambridge, Mass.: Belknap Press of Harvard University Press, 1999.

Dickinson, Emily. "There's a certain Slant of light" [320]. In *The Poems of Emily Dickinson*, pp. 142–143, edited by R. W. Franklin. Cambridge, Mass.: Belknap Press of Harvard University Press, 1999.

Douglass, Frederick. "What to the Slave Is the Fourth of July?" [1852] In *Frederick Douglass Papers.* Vol. 2. Edited by John Blassingame, pp. 359–388. New Haven, Conn.: Yale University Press, 1982.

Du Bois, W. E. B. *The Souls of Black Folk* [1903]. In *W. E. B. Du Bois: Writings*, pp. 357–547. New York: Library of America, 1986.

Emerson, Ralph Waldo. "Nature" [1836]. In *Ralph Waldo Emerson*, pp. 5–49. New York: Library of America, 1983.

Emerson, Ralph Waldo. "The Poet" [1844]. In *Essays: Second Series. Ralph Waldo Emerson*, pp. 611–761. New York: Library of America, 1983.

Emerson, Ralph Waldo. "Poetry and Imagination" [1876]. In *The Collected Works of Ralph Waldo Emerson, Vol. VIII: Letters and Social Aims*, edited by Ronald A. Bosco, Glen M. Johnson, and Joel Myerson. Cambridge, Mass.: Belknap Press of Harvard University Press, 2010.

Emerson, Ralph Waldo. "The Progress of Culture" [1867]. In *The Collected Works of Ralph Waldo Emerson, Vol. VIII: Letters and Social Aims*, pp. 108–123, edited by Ronald A. Bosco, Glen M. Johnson, and Joel Myerson. Cambridge, Mass.: Belknap Press of Harvard University Press, 2010.

Finney, Jack. *The Body Snatchers.* New York: Dell, 1955.

Fitzgerald, F. Scott. *The Great Gatsby* [1925]. New York: Charles Scribner's Sons, 2004.

Hawthorne, Nathaniel. "The Birthmark" [1846]. In *Nathaniel Hawthorne: Tales and Sketches*, pp. 764–780. New York: Library of America, 1982.

Hawthorne, Nathaniel. "Rappaccini's Daughter" [1844]. In *Nathaniel Hawthorne: Tales and Sketches*, pp. 975–1005. New York: Library of America, 1982.

Hersey, John. *Hiroshima* [1946]. New York: Vintage Books (Random House), 1985.

Indigenous Peoples Council on Biocolonialism. http://www.ipcb.org.

Isaacs, Harold. "Color in World Affairs." *Foreign Affairs* 47, no. 2 (January 1969): 235–250.

James, William. *The Principles of Psychology* [1890]. Cambridge, Mass.: Harvard University Press, 1983.

Kuhn, Thomas. *The Structure of Scientific Revolutions* [1962]. 3d ed. Chicago: University of Chicago Press, 1996.

Melville, Herman. "The Encantadas, or Enchanted Isles" [1854]. In *Herman Melville*, pp. 764–818. New York: Library of America, 1984.

Miller, Perry. *The Life of the Mind in America. Book III: Science—Theoretical and Applied*, pp. 267–326. Massachusetts Historical Society: Harcourt, Brace and World, 1965.

Morse, Jedidiah. *The American University Geography, or, a View of the Present State of All the Kingdoms, States, and Republics in the Known World and of the United States of America in Particular. In Two Parts.* [1789]. Boston: Isaiah Thomas and Ebenezer T. Andrews, 1796.

Park, Robert. "Human Migration and the Marginal Man." *American Journal of Sociology* 33, no. 6 (May 1928): 881–893.

Patterson, Orlando. *Slavery and Social Death: A Comparative Study.* Cambridge, Mass.: Harvard University Press, 1982.

Pynchon, Thomas. *The Crying of Lot 49.* New York: HarperPerennial, 1966.

Rukeyser, Muriel. "The Speed of Darkness" [1968], *Collected Poems.* Edited by Janet E. Kaufman and Anne F. Herzog with Jan Heller Levi. Pittsburgh, Pa.: University of Pittsburgh Press 2005: 465–468.

Small, Albion. "The Era of Sociology." *American Journal of Sociology* 1, no. 1 (July 1895): 1–15.

Stein, Gertrude. "Portraits and Repetition." In *Lectures in America*, pp. 163–206. New York: Random House, 1935.

Stein, Gertrude. *Tender Buttons: Objects, Food, Rooms* [1914]. Mineola, N.Y.: Dover Publications, 1997.

Thomas, William I. "The Adventitious Character of Woman." *American Journal of Sociology* 12, no. 1 (July 1906): 32–44.

Thoreau, Henry David. "The Natural History of Massachusetts" [1842]. In *The Essays of Henry David Thoreau*, pp. 145–163, edited by Richard Dillman. Albany, N.Y.: NCUP, Inc., 1990.

Thoreau, Henry David. *A Week on the Concord and Merrimack Rivers* [1849]. In *Henry David Thoreau*, pp. 1–319. New York: Library of America, 1985.

Toffler, Alvin. "What Is Human Now?" *Christian Science Monitor.* 4 June 1987, pp. 20, 27.

Webster, Noah. "Letter to John Canfield" [1783]. In *Letters of Noah Webster*, edited by Harry R. Warfel. New York: Library Publications, 1953.

White, Fred D. "'Sweet Skepticism of the Heart': Science in the Poetry of Emily Dickinson." *College Literature* 19, no. 1 (February 1992): 121–128.

Wiener, Norbert. *The Human Uses of Human Beings: Cybernetics and Society* [1950]. Reprint, Garden City, N.Y.: Doubleday Anchor Books, 1954.

Wilson, Harriet E. *Our Nig; or Sketches from the Life of a Free Black* [1859]. New York: Vintage Books, 1983.

Priscilla Wald

LOWELL TEXTILE MILLS

The cotton textile mills of Lowell, Massachusetts, were the most famous factories in the United States in the first half of the nineteenth century. From them emanated innovations in technology, the organization of work, and business practices that made signal contributions to industrial capitalism in the United States.

In the early 1820s, Boston capitalists, organized as the Boston Manufacturing Company of Waltham, sought a site for expansion. They purchased land, a transportation canal, and water-power rights at the Pawucket Falls of the Merrimack River in East Chelmsford. There they began manufacturing printed cotton cloth in 1823.

Implementing their grand vision, the mill owners incorporated the town of Lowell in 1826, naming it for the late Francis Cabot Lowell, a founder of the Waltham venture. High profits led to rapid expansion, and by 1850 the Lowell mills, employing more than 10,000 workers, were the nation's leading textile-manufacturing center. With a population of 33,000, Lowell was the second largest city in Massachusetts.

Mill towns patterned after Lowell arose across New England and collectively came to constitute the Waltham–Lowell system. Large, red brick, water-powered mills housed all the machinery needed to manufacture cotton cloth from raw cotton. Employing a work force consisting of native-born single daughters of Yankee farmers, the mills erected boardinghouses for their workers. Combining corporate paternalism with monthly cash wages, the owners of the Lowell mills sought to industrialize without replicating the social ills associated with English factory towns in this era. Later, immigrant workers replaced native-born young women.

The Lowell mills offered the first major source of wage work for women in the nation. After the Civil War, Lowell occupied a less important place in the textile industry and the industrial economy. Employment and production in Lowell grew until World War I but declined thereafter, as textile production shifted to the South. By 1980 only scattered, minor textile production continued in Lowell, the dominant center of the early American industrial revolution. In 1978, Congress created the Lowell National Historical Park on the site of a restored mill and associated buildings.

[*See also* **Gender and Technology; Machinery and Manufacturing;** *and* **Technology.**]

BIBLIOGRAPHY

Dalzell, Robert F., Jr. *Enterprising Elite: The Boston Associates and the World They Made.* Cambridge, Mass.: Harvard University Press, 1987.

Dublin, Thomas. *Women at Work: The Transformation of Work and Community in Lowell, Massachusetts, 1826–1860.* Revised ed. New York: Columbia University Press, 1979, 1994.

Thomas Dublin

LUMBERING

See **Forestry Technology and Lumbering.**